Wiley Encyclopedia of
Electrical and
Electronics
Engineering

Volume 12

Wiley Encyclopedia of
Electrical and
Electronics
Engineering

Volume 12

John G. Webster, Editor

Department of Electrical and Computer Engineering
University of Wisconsin–Madison

A Wiley-Interscience Publication

John Wiley & Sons, Inc.

New York · Chichester · Weinheim · Brisbane · Singapore · Toronto

Editorial Staff

Editor: John G. Webster

Publisher: Kim H. Kelly

Sponsoring Editor: George J. Telecki

Managing Editor: Kimi J. Sugeno

Logo designed by Gottlieb John Marmet.

This book is printed on acid-free paper. ⊗

Copyright © 1999 by John Wiley & Sons, Inc.   All rights reserved.

Published simultaneously in Canada.

For ordering and customer service, call 1-800-CALL-WILEY.

Library of Congress Cataloging-in-Publication Data:
Wiley Encyclopedia of electrical and electronics engineering / John G. Webster, editor.
     p. cm.
   Includes index.
   ISBN 0-471-13946-7 (set : cloth : alk.paper)
   1. Electric engineering–Encyclopedias. 2.
Electronics–Encyclopedias. I. Webster, John G., 1932-
   TK9 .E53 1999
   621.3'03–dc21

                                                          98-44761
                                                          CIP

Printed in the United States of America

10 9 8 7 6 5 4 3 2 1

MAGNETIC MEDIA, IMAGING

Magnetic imaging is used to examine small-scale magnetic features within materials, often on the submicrometer scale. There are several techniques available that give complementary information about the magnetization within and at the surface of a material and the magnetic field outside the material. These techniques can be used for a range of magnetic materials, but this article emphasizes applications to magnetic storage media. Although many of the examples cited refer to hard disk media, the techniques are equally applicable to flexible or magnetooptical media. Magnetic imaging provides insight into the magnetic structure of patterns of data written onto media and the mechanisms for magnetization reversal in media.

We will discuss the capabilities, limitations, and resolution of various magnetic imaging techniques based on different physical principles. Lorentz transmission electron microscopy (LTEM) and electron holography (EH) are transmission electron microscope techniques sensitive to the magnetic field experienced by a beam of electrons passing through a sample. Scanning electron microscopy with polarization analysis (SEMPA) and the magnetooptical Kerr effect (MOKE) are sensitive to the magnetization state of the material near the surface of a sample. Bitter patterns, magnetic force microscopy (MFM), and EH are used to determine the magnetic field outside the medium. For each technique, we give a brief description of the principle on which it is based, discuss the information that it provides, and describe its advantages and limitations. We compare these methods and assess future developments in magnetic imaging.

BITTER PATTERNS

Early domain images were obtained using the Bitter method (1,2). Bitter patterns are formed by applying a colloidal suspension of fine ferromagnetic particles to the surface of the ferromagnetic material of interest. The particle pattern delineates the magnetic field lines at the surface and is observed in an optical microscope. The suspension, traditionally a precipitate of fine Fe_3O_4 particles with a dispersant, can be placed directly on the magnetic surface or applied using an applicator pen. Both the optical microscope and the particle size limit submicron analysis. Thus, low-frequency bit patterns on a hard disk can be observed for location purposes, but micromagnetic details cannot be analyzed. To achieve higher-resolution imaging, submicron ferromagnetic particles can be dispersed on the surface and observed by scanning or transmission electron microscopy. This allows magnification of up to 20,000× and resolution below 0.1 μm, but further advances remain limited by formation, dispersion, and alignment of the ultrafine magnetic particles (3). However, 80 nm resolution has recently been reported using the Bitter technique (4) by forming 20 nm magnetic particles by sputtering and depositing them onto written media in a vacuum chamber and, then observing the sample by scanning electron microscopy.

MAGNETOOPTICAL KERR EFFECT

The magnetooptical Kerr effect is commonly used to study magnetic structure (5–10). In this technique, plane-polarized light is reflected from the surface of a magnetic material. The plane of polarization is rotated by the Kerr angle, typically of order 1°, on reflection from the surface, with the sense of the rotation dependent on the direction of magnetization. By passing the reflected beam though a polarizer, magnetic contrast is observed such that domains or domain walls show as different gray levels. This is performed using an optical microscope with magnification of up to about 1000×. By varying the geometry of the optical system, the contrast can be made sensitive to domain walls or to the magnetization of the magnetic domains themselves, and measurements may also be performed in transmission mode for transparent materials. In transmission mode, the rotation of the plane of polarization of light is known as the Faraday effect. MOKE, like the optical Bitter method, is limited by the wavelength of light, but has been particularly useful for imaging domain patterns in samples such as permalloy pole pieces in recording heads. Advantages of the MOKE technique include the ability to do high-frequency dynamic imaging and to image through thick transparent materials in a nondestructive way without need for special sample preparation.

Near field optical microscopy (NFOM) or scanning near-field optical microscopy (SNOM) combined with Kerr magnetometry has recently been shown to offer dramatic improvement over standard MOKE resolution (7–10). Figure 1 shows a NFOM image of bits written onto a Co/Pt multilayer film (8). The best-case resolution in NFOM is below 50 nm, which is approaching the useful range for imaging details of recording media having bit sizes that are currently on the order of 200 nm × 2000 nm. Resolution improvement beyond this may be difficult owing to the combination of rapid resolution loss with increasing probe– or lens–media spacing, finite lens size, and media surface roughness. Continuing improvement of the NFOM technique may be anticipated as it is developed for other applications such as optical data storage technology. MOKE will continue to be used for analysis of larger-scale domain patterns in recording heads and in particular for dynamic measurements.

LORENTZ TRANSMISSION ELECTRON MICROSCOPY

Lorentz transmission electron microscopy (11–14) is based on a transmission electron microscope (TEM), in which an electron source emits electrons, which are accelerated by an accurate electric field and focused by a series of condenser lenses, generating suitable illumination at a specimen. The electron beam is transmitted through the thin specimen and focused by the objective lens to produce both a diffraction pattern at the back focal plane of the lens and an image at the image plane of the lens. By use of appropriate apertures and imaging conditions, a variety of data can be obtained including bright-field, dark-field, and high-resolution images and diffraction patterns. In LTEM, the imaging conditions are se-

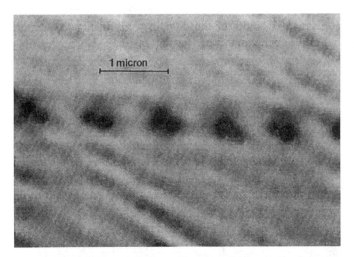

Figure 1. Magnetooptical Kerr effect near-field optical microscopy image of five individual 0.5 μm diameter magnetic domains in a Co/Pt multilayer film with out-of-plane magnetization (8). Reprinted with permission of the authors and the American Institute of Physics.

lected to display magnetic contrast in the specimen. This is based on deflection of the electrons by the Lorentz force. The Lorentz force is given by the vector product $-e(\boldsymbol{v} \times \boldsymbol{B})$, where e is the electron charge, $\boldsymbol{v}$ the electron velocity, and $\boldsymbol{B}$ the magnetic flux density. This deflects the electrons in a direction perpendicular to both $\boldsymbol{B}$ and $\boldsymbol{v}$. The total deflection angle is proportional to the in-plane component (i.e., the component perpendicular to the electron trajectory) of $\boldsymbol{B}$ integrated along the electron trajectory. If we neglect the effect of the field outside the specimen and assume that the magnetization is constant through the thickness of the specimen then the angular deflection is proportional to the in-plane component of the magnetization and to the sample thickness.

The deflection angle of the electrons can be detected in several ways.

1. *Low-angle diffraction mode.* The angular deflection of electrons results in a shift of the electron beam and can be measured directly from the displacement of the transmission spot (the center spot in the diffraction pattern) at the back focal plane of the imaging lens.

2. *Foucault mode.* The image is observed with an aperture placed in the diffraction pattern at the back focal plane. This aperture is off-centered, allowing the passage only of beams deflected in a certain direction. Domains with magnetizations that deflect electrons in that direction will appear bright while others appear dark. By manipulating the aperture position, one can identify the in-plane component of the magnetization in different domains of the specimen.

3. *Fresnel mode.* The image is observed in an out-of-focus condition. The Lorentz force effectively deflects the electrons as they are transmitted through a magnetic domain. This deflection is not visible in the in-focus image, but as the image is defocused, the domain image is shifted normal to its magnetization direction, which causes different domain images to overlap or move apart, giving rise to magnetic contrast wherever the magnetization component parallel to a domain bound-

ary changes magnitude and/or direction across the boundary.

4. *Differential phase contrast mode.* This method is implemented in a scanning transmission electron microscope, in which incident electrons are focused to a fine probe and scanned across the specimen. The angular deflection of the beam is measured using a quadrant detector. These signals quantitatively measure the in-plane magnetization component in the small analyzed area. By scanning the probe across the specimen, a map of the in-plane magnetization component can be determined.

Foucault and Fresnel modes are simple to implement and are generally used for qualitative measurement. The former detects magnetization inside domains while the latter is sensitive to variation of magnetization due to, for example, domain walls. The differential phase contrast mode can provide quantitative information about the magnetization distribution in the specimen. The wave nature of electrons also allows the use of interference methods to detect magnetic information. Such methods produce interferograms between a reference beam and a sample-modulated beam, from which both the phase and amplitude of the electron wave exiting the specimen can be extracted. The major interference method is electron holography, which is discussed separately in this article. Coherent Foucault imaging, in which the opaque aperture used in the Foucault mode is replaced with a phase-shifting aperture, is also possible.

In a conventional TEM the specimen is placed as close as possible to the objective lens to achieve high resolution (about 0.2 nm or better resolution) and high magnification. However, the magnetic field from the lens distorts or erases the magnetization pattern in a magnetic sample, so during LTEM the objective lens is turned off and the intermediate lens serves as the imaging lens. Ideally, an additional lens is installed farther from the specimen. The lens resolution is, however, reduced to about 2 nm to 3 nm. A field emission electron source is also desirable to obtain optimum magnetic contrast. Sample preparation consists of cutting a 3 mm diameter sample and thinning it until it is transparent to electrons. Since the angular deflection of the electrons is proportional to both the magnetization and the magnetic film thickness, the ideal LTEM specimen will have uniform thickness but needs to be thin enough (<50 nm to 100 nm) to be transparent to electrons. Most TEM specimen preparation methods produce wedge-shaped specimens, but a combined mechanical thinning and chemical etching technique has been developed for Co-alloy/Cr hard-disk media that produces a large area of uniform film suitable for LTEM (15).

Many applications of these methods have been demonstrated. For instance, magnetization reversal processes in TbCo-biased spin valves (16) and magnetization vortices in CoCrTa hard-disk media (15) have been imaged by the Fresnel mode. Domain walls and magnetization processes in NiFe have been imaged at high resolution by the Foucault mode (17,18). Stray magnetic fields outside write heads have been imaged by both differential phase contrast (DPC) and Foucault modes (19). Tomographic reconstruction of the three-dimensional magnetic field was performed by analyzing a set of DPC images taken in different directions. However, the need for sophisticated TEM facilities has limited the use of LTEM

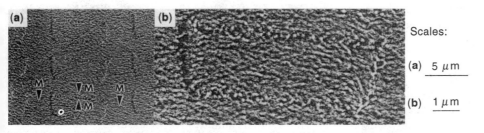

Scales:

(a) 5 µm

(b) 1 µm

Figure 2. Fresnel LTEM image of low-density data tracks written in an initially dc-erased CoCrTa/Cr longitudinal hard disk. Parts (a) and (b) represent different magnifications. Arrows show the magnetization direction. Tracks run from top to bottom of the figure (15). Reprinted with permission of the authors and the Institute of Electrical and Electronic Engineers.

to relatively few laboratories. Figure 2 shows a Fresnel mode image of magnetic bits written in a magnetic hard disk (15). Boundaries between the magnetic bits can clearly be seen. Within the bits and in the intertrack regions, ripple patterns indicate local fluctuations in magnetization direction. Details on a scale of 50 nm to 100 nm may be resolved.

SCANNING ELECTRON MICROSCOPE–BASED TECHNIQUES

Magnetic imaging due to the Lorentz force can also be carried out using a scanning electron microscope (SEM). In a conventional SEM, the sample is scanned by a beam of electrons of energy between about 1 kV and 20 kV. The yield of secondary (low-energy) electrons emitted from the surface gives an image of surface topography, while the yield of primary (high-energy backscattered) electrons is sensitive to atomic number and hence composition. Deflection of electrons by magnetic fields in or near the sample additionally provides information on the magnetization distribution in the sample (20). In type I imaging, deflection of secondary electrons by the magnetic field above the sample surface is observed using an asymmetric detector. This has been applied to detect, for instance, stray fields outside magnetic recording heads (21). In type II imaging, the deflection of obliquely incident electrons by internal magnetic fields can be observed. The electrons are deflected towards or away from the surface depending on the direction of magnetization, and this alters the yield of secondary electrons. The different electron yield from domains of different magnetization leads to contrast between domains. Contrast arising from domain walls can also be obtained from electrons incident normally or obliquely. These methods can be used to probe the depth dependence of magnetic structure (22). Imaging is done under ultrahigh-vacuum conditions with sample preparation limited to removal of surface layers or contaminants.

Spin-polarized SEM (SEMPA) is a SEM-based technique that can provide information on the three-dimensional magnetization vector of the sample (23–29). It is based on the observation that secondary electrons emitted from a ferromagnet retain their original spin orientation as they leave the sample. A vector map of magnetization can be measured by analyzing the three components of polarization using a finely focused electron beam. SEMPA is surface sensitive due to the small escape depth of secondary electrons of a few nanometers. SEMPA can probe the surface magnetization vector with a best-case resolution of about 20 nm, but poor efficiency makes for slow data aquisition and scans take 30 min to a few hours. Resolution is affected only slightly by surface topography or fringing fields outside the sample.

The SEMPA instrument consists of an electron source and optical column with 10 keV to 50 keV accelerating voltage, a secondary-electron collector, and a set of three orthogonal spin detectors made from gold targets. Spin detectors have been designed to measure both high-energy electrons (20 keV to 100 keV), which are insensitive to the cleanliness of the gold target surface, or low-energy electrons (around 100 eV), for which the detectors are less bulky but require extremely clean gold surfaces. Figure 3 shows a SEMPA image of a magnetic hard disk written with bits at a density of 100 and 240 kfci (kiloflux changes per inch) (39×10^3 and 94×10^3 flux changes cm^{-1}) (28). This image shows the component of magnetization parallel to the data track. A further development of SEMPA is represented by spin-polarized low-energy electron microscopy (SPLEEM). The physics and resolution of SPLEEM are similar to SEMPA (30,31) but SPLEEM offers parallel detection, hence higher data rates, athough there is greater environment sensitivity owing to the use of electrons with energies below 10 eV.

ELECTRON HOLOGRAPHY

Electron holography (EH) is a method based on TEM, in which both the amplitude and the phase of the electron beam are recorded as it passes through a sample. Conventional TEM records only the amplitude of the transmitted beam. By detecting phase shifts of the transmitted beam caused by electric or magnetic fields within the sample, high-resolution images of the electric or magnetic field distribution may be made. Electron holography was proposed in the 1940s (32,33)

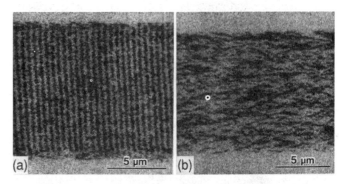

Figure 3. SEMPA image of data tracks written in a hard disk at (a) 100 kfci and (b) 240 kfci after removal of the carbon overcoat. The tracks run left to right. The component of magnetization parallel to the track is shown. In (b), significant percolation is visible and bits cannot be resolved (28). Reprinted with permission of the authors and Oxford University Press.

but has become more widely used as high-intensity field-emission electron sources have been developed. A range of holographic techniques and applications has been described (34–36).

EH can be used for quantitative measurements of magnetic field distribution within or near a sample at high resolution. The technique uses a TEM with an electrostatic biprism (37), which splits the incident electron beam. Part of the beam passes through the sample while a reference beam passes through a hole in the sample (the absolute mode) or through an adjacent region of the sample (the differential mode) (38). The beams are recombined to form a hologram consisting of interference fringes. Phase changes to the electron beam passing through the specimen are detected as shifts in the fringes. For a sample of uniform thickness and composition, shifts in the fringes correlate directly with the magnetic field within the plane of the sample averaged through the sample thickness, so, for example, magnetic domain walls can be imaged as shifts in the fringes. The hologram needs to be reconstructed optically or by computer simulation to yield the phase differences that contributed to it. By measuring phase changes caused by a reference sample, for instance, a nickel film of known thickness, the system can be calibrated so that the magnetic field within the sample can be measured quantitatively.

This method has been used to image magnetization within films, for instance, to show Néel walls within a Ni sample and magnetization in the Co layers of a Co/Pd multilayer (38). It can also be used to image magnetic fields outside samples such as magnetic force microscopy tips (39) and magnetic heads (40). The spatial resolution of EH can be of the order of 1 nm (34). Although this technique requires specialized equipment and analysis, it is valuable in providing quantitative data for comparison with other techniques such as Lorentz microscopy and MFM.

MAGNETIC FORCE MICROSCOPY

Magnetic force microscopy (MFM) has become the most important tool for imaging magnetization patterns in a large number of technologically and scientifically important applications. This includes imaging of magnetic recording media (41), recording heads (42), biomagnetic structures (43), and numerous other materials (44). In many cases, postprocessing of MFM data provides insights into the micromagnetic details of magnetization processes, for instance, quantification of track-edge percolation phenomena in magnetic thin film media (45). Significantly, MFM can be used to study magnetization reversal in individual particles by imaging in a varying externally applied field (46). Examples of MFM images of recorded data tracks are given in Fig. 4 (45).

MFM was developed as an extension of atomic force microscopy (AFM) (47). MFM relies largely on the same instrumentation techniques as AFM but uses a magnetic tip to sense stray magnetic fields above the sample surface. The essential elements of the instrument are a magnetic tip, which is mounted on a cantilever, piezoelectric motors for raster scanning and for control of vertical tip motion, and laser optics for detection of vertical tip response (48). Tips are commonly silicon pyramids coated with a CoCr film (49). Other tip geometries have been developed to improve resolution (50,51). Furthermore, in order to image soft magnetic materi-

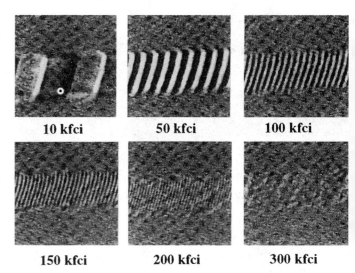

Figure 4. MFM image of data tracks written on a longitudinal hard disk at a range of densities (1 kfci = 390 bits cm^{-1}) (45). Reprinted with permission of the authors and the American Institute of Physics.

als, tips have also been coated with soft or superparamagnetic materials (52). Imaging is performed in ambient atmosphere with minimal sample preparation.

Modern implementations of MFM typically interleave a line scan in contact mode (53) for acquisition of the surface topography with a scan where the tip is servoed at a controlled lift height (a few tens of nanometers) above the specimen surface to measure long-range tip–sample interactions (54). The interleaving technique allows for effective separation of topographic information from the magnetic signal. In many MFM designs, detection sensitivity is increased by resonance detection techniques. The tip is excited at the free-resonance frequency of the cantilever and changes of the vibration amplitude, frequency, or phase due to the tip–sample interaction are measured (55,56). Lateral resolution is typically about 40 nm, though resolution as good as 10 nm has been reported (57).

The magnetic interaction between the tip and the specimen is due to a coupling with energy density $-\boldsymbol{H} \cdot \boldsymbol{M}$ where $\boldsymbol{H}$ is the magnetic field above the specimen and $\boldsymbol{M}$ is the tip magnetization. The force on the tip is given by the negative gradient of the energy density integrated over the tip volume. This force causes a deflection of the cantilever or, in the case of resonance detection, leads to a detuning of the free-resonance conditions. The minimum detectable force gradient for a cantilever with spring constant 1 N/m, resonance frequency 100 kHz, vibration amplitude 10 nm, and quality factor 200 is estimated to be better than 10^{-4} N/m (54). If the phase shift is measured, the MFM signal measures the gradient of the force on the tip, which is proportional to the second derivative of the magnetic field above the sample surface, from which the magnetization pattern in the medium must be inferred. The magnetization within the medium cannot be deduced uniquely from the MFM image, so in practice a magnetization pattern is assumed and the calculated field derivatives are compared with the MFM image until agreement is reached (44). It is also possible to image the magnetic field strength directly by using a feedback loop to control an externally applied field (58).

CONCLUSIONS

With bit lengths in hard disk media now below 200 nm, magnetization features of interest in magnetic recording are generally not observable by conventional optical techniques. High-resolution techniques including NFOM, SEM, SEMPA, SPLEEM, EH, Lorentz TEM, and MFM are being pushed to their limits in the competition to provide the most convenient, highest-resolution magnetic images. Instrument design and observation techniques are advancing rapidly, but at the same time, the requirements (higher-resolution images obtained from thinner, lower-moment films having higher coercivity, smaller structural features, and smaller magnetized bits) become more stringent. Thus, imaging needs continue to be at the very limit of our capabilities. SEMPA, SPLEEM, and NFOM instruments continue to be in prototype mode, making comparisons and tool use difficult. Some interesting applications of these techniques have been shown, but for now it appears that LTEM and MFM will continue to be used in most studies of magnetic media. For imaging needs related to magnetic recording heads, such as examination of domain patterns in inductive pole pieces, MOKE is the dominant technique due to its simplicity and applicability to dynamic measurements.

Considering high-density recording media, LTEM continues to offer the highest-resolution two-dimensional map of thin-film magnetization patterns, with resolution in the best cases as good as 10 nm for optimized samples and imaging conditions. However, samples must be carefully selected and are often specially made to allow useful analysis. Imaging can be impeded by grain morphology, substrate topography, substrate material compatibility with sample preparation, and magnetic layer thinness. As the layer thicknesses of real production hard-disk media decrease, the magnetization–thickness product does not provide sufficient electron deflection for the highest-resolution LTEM.

MFM, which measures surface magnetic field independent of film thickness, has become the workhorse of the industry. Its major advantages are that sample preparation is minimal, topographic information is recorded simultaneously, and resolution is good. Approximately 40 nm resolution images are now almost routine on real media, and there is promise of higher resolution with design of sharper probe tips. As a surface technique, MFM resolution is not affected by decreasing magnetic layer thickness, but because MFM measures field gradients above the surface, the results are not necessarily interpretable as bulk magnetic structure. MFM is nondestructive to the media, allowing direct comparisons between micromagnetic properties and recording performance, and the effects of an external applied field can be measured. TEM and MFM continue to have a place in magnetic media research and development. TEM and other electron microscopy techniques will continue to be used in larger research laboratories, while MFM is increasingly important in applications related to analysis of recorded bit patterns.

BIBLIOGRAPHY

1. F. Bitter, On inhomogeneities in the magnetization of ferromagnetic materials, *Phys. Rev.,* **38**: 1903–1905, 1931.

2. F. Bitter, Experiments on the nature of magnetism, *Phys. Rev.,* **41**: 507–515, 1932.

3. K. Goto, M. Ito, and T. Sakurai, Studies on magnetic domains of small particles of barium ferrite by colloid-SEM method, *Jpn. J. Appl. Phys.,* **19**: 1339–1346, 1980.

4. O. Kitakami, T. Sakurai, and Y. Shimada, High density recorded patterns observed by high resolution Bitter scanning electron microscope method, *J. Appl. Phys.,* **79**: 6074–6076, 1996.

5. W. Rave, R. Schafer, and A. Hubert, Quantitative observation of magnetic domains with the magneto-optical Kerr effect, *J. Magn. Magn. Mater.,* **65**: 7–14, 1987.

6. W. W. Clegg et al., Development of a scanning laser microscope for magneto-optic studies of thin magnetic films, *J. Magn. Magn. Mater.,* **95**: 49–57, 1991.

7. E. Betzig et al., Near-field magneto-optics and high density data storage, *Appl. Phys. Lett.,* **61**: 142–144, 1992.

8. T. J. Silva, S. Schultz, and D. Weller, Scanning near-field optical microscope for the imaging of magnetic domains in optically opaque materials, *Appl. Phys. Lett.,* **65**: 658–660, 1994.

9. M. W. J. Prins et al., Near-field magneto-optical imaging in scanning tunneling microscopy, *Appl. Phys. Lett.,* **66**: 1141–1143, 1995.

10. C. Durken, I. V. Shvets, and J. C. Lodder, Observation of magnetic domains using a reflection-mode scanning near-field optical microscope, *Appl. Phys. Lett.,* **70**: 1323–1325, 1997.

11. H. W. Fuller and M. E. Hale, Determination of magnetization distribution in thin films using electron microscopy, *J. Appl. Phys.,* **31**: 238–248, 1960.

12. R. H. Wade, Transmission electron microscope observations of ferromagnetic grain structures, *J. Phys. (Paris) Colloque,* **C2**: 95–109, 1968.

13. J. P. Jakubovics, Lorentz microscopy and applications (TEM and SEM), in U. Valdre and E. Ruedl (eds.), *Electron Microscopy in Materials Science,* Luxembourg: Commission of the European Communities, 1975, Vol. 4, pp. 1303–1403.

14. J. N. Chapman, The investigation of magnetic domain structures in thin foils by electron microscopy, *J. Phys. D.,* **17**: 623–647, 1984.

15. K. Tang et al., Lorentz transmission electron microscopy study of micromagnetic structures in real computer hard disks, *IEEE Trans. Magn.,* **32**: 4130–4132, 1996.

16. J. N. Chapman, M. F. Gillies, and P. P. Freitas, Magnetization reversal process in TbCo-biased spin valves, *J. Appl. Phys.,* **79**: 6452–6454, 1996.

17. B. Y. Wong and D. E. Laughlin, Direct observation of domain walls in NiFe films using high resolution Lorentz microscopy, *J. Appl. Phys.,* **79**: 6455–6457, 1996.

18. S. J. Hefferman, J. N. Chapman, and S. McVitie, In-situ magnetizing experiments on small regularly-shaped permalloy particles, *J. Magn. Magn. Mater.,* **95**: 76–84, 1991.

19. I. Petri et al., Investigations on the stray fields of magnetic read-write heads and their structural reasons, *IEEE Trans. Magn.,* **32**: 4141–4143, 1996.

20. K. Tsuno, Magnetic domain observation by mean of Lorentz electron microscopy with scanning techniques, *Rev. Solid State Sci.,* **2**: 623–658, 1988.

21. J. B. Elsbrock and L. J. Balk, Profiling of micromagnetic stray fields in front of magnetic recording media and heads by means of a SEM, *IEEE Trans. Magn.,* **20**: 866–868, 1984.

22. I. R. McFadyen, Magnetic-domain imaging techniques, *Proc. 51st Annu. Meet. Micr. Soc. Amer.,* **51**: San Francisco, CA: San Francisco Press, 1993, pp. 1022–1023.

23. K. Koike and H. Hayakawa, Observation of magnetic domains with spin-polarized secondary electrons, *J. Appl. Phys.,* **45**: 585–586, 1984.

24. J. Unguris et al., High resolution magnetic microstructure imaging using secondary electron spin polarization analysis in a SEM, *J. Microsc.,* **139**: RP1–2, 1985.

25. J. Unguris et al., Investigations of magnetic microstructures using scanning electron microscopy with spin polarization anlysis, *J. Magn. Magn. Mater.,* **54–57**: 1629–1630, 1986.

26. D. R. Penn, Electron mean-free-path calculations using a model dielectric function, *Phys. Rev. B,* **35**: 482–486, 1987.

27. H. P. Oepen and J. Kirschner, Magnetization distribution of 180° domain walls at Fe(100) single crystal surfaces, *Phys. Rev. Lett.,* **62**: 819–822, 1989.

28. H. Matsuyama and K. Koike, Twenty-nm resolution spin polarized scanning electron microscope, *J. Electron. Microsc.,* **43**: 157–163, 1994.

29. M. R. Scheinfein et al., Scanning electron microscopy with polarization analysis (SEMPA), *Rev. Sci. Instrum.,* **61**: 2501–2506, 1990.

30. H. Pinkvos et al., Spin-polarized low-energy electron microscopy study of the magnetic microstructure of ultra-thin epitaxial cobalt films on W (110), *Ultramicrosc.,* **47**: 339–345, 1992.

31. M. S. Altman et al., Spin polarized low energy electron microscopy of surface magnetic structure, *Mater. Res. Soc. Symp. Proc.,* **232**: 125–132, 1991.

32. D. Gabor, A new microscopic principle, *Nature,* **161**: 777–778, 1948.

33. D. Gabor, Microscopy by reconstructed wave fronts, *Proc. R. Soc. London A,* **197**: 454–487, 1949.

34. A. Tonomura, Applications of electron holography, *Rev. Mod. Phys.,* **59**: 639–669, 1987.

35. J. M. Cowley, Twenty forms of electron holography, *Ultramicrosc.,* **41**: 335–348 1992.

36. A. Tonomura, *Electron Holography,* Berlin: Springer-Verlag, 1993.

37. G. Mollenstedt and H. Duker, Beobachten und Messungen an Biprisma-Interferenzen mit Elektronwellen, *Z. Phys.,* **145**: 377–397, 1956.

38. M. Mankos, M. R. Scheinfein, and J. M. Cowley, Quantitative micromagnetics: electron holography of magnetic thin films and multilayers, *IEEE Trans. Magn.,* **32**: 4150–4155, 1996.

39. D. G. Streblechenko et al., Quantitative magnetometry using electron holography: Field profiles near magnetic force microscope tips, *IEEE Trans. Magn.,* **32**: 4124–4129, 1996.

40. Y. Takahashi et al., Observation of magnetic induction distribution by scanning interference electron microscopy, *Jpn. J. Appl. Phys.,* **33**: L1352–1354, 1994.

41. D. Rugar et al., Magnetic force microscopy: general principles and application to longitudinal recording media, *J. Appl. Phys.,* **68**: 1169–1183, 1990.

42. S. H. Liou et al., High resolution imaging of thin film recording heads by superparamagnetic magnetic force microscopy tips, *Appl. Phys. Lett.,* **70**: 135–137, 1997.

43. R. B. Proksch et al., Magnetic force microscopy of the submicron magnetic assembly in a magnetotactive bacterium, *Appl. Phys. Lett.,* **66**: 2582–2584, 1995.

44. E. D. Dahlberg and J. G. Zhu, Micromagnetic microscopy and modelling, *Phys. Today,* **48** (April): 34–40, 1995.

45. M. E. Schabes et al., Magnetic force microscopy study of track edge effects in longitudinal media, *J. Appl. Phys.,* **81**: 3940–3942, 1997.

46. R. O'Barr et al., Preparation and quantitative magnetic studies of single-domain nickel cylinders, *J. Appl. Phys.,* **79**: 5303–5305, 1996.

47. D. Sarid, *Scanning Force Microscopy,* Oxford: Oxford University Press, 1994.

48. K. Babcock et al., Magnetic force microscopy: recent advances and applications, *Mater. Res. Soc. Symp. Proc.,* **355**: 311–322, 1995.

49. K. Babcock et al., Optimization of thin film tips for magnetic force microscopy, *IEEE Trans. Mag.,* **30**: 4503–4505, 1994.

50. M. Ruehrig et al., Electron beam fabrication and characterization of high resolution magnetic force microscope tips, *J. Appl. Phys.,* **79**: 2913–2919, 1996.

51. G. D. Skidmore and E. D. Dahlberg, Improved spatial resolution in magnetic force microscopy, *Appl. Phys. Lett.,* **71**: 3293–3295, 1997.

52. P. F. Hopkins et al., Superparamagnetic magnetic force microscopy tips, *J. Appl. Phys.,* **79**: 6448–6650, 1996.

53. Q. Zhong et al., Fractured polymer/silica fiber surface studied by tapping mode atomic force microscopy, *Surf. Sci.,* **290**: L688–692, 1993.

54. P. Gruetter, H. J. Mamin, and D. Rugar, Magnetic force microscopy, in R. Wiesendanger and H.-J. Guentherodt (eds.), *Scanning Tunneling Microscopy II,* Springer Series in Surface Sciences, Berlin: Springer-Verlag, 1992, vol. 28, pp. 151–207.

55. Y. Martin, C. C. Williams, and H. K. Wickramasinghe, Atomic force microscope-force mapping and profiling on a sub-100 Å scale, *J. Appl. Phys.,* **61**: 4723–4729, 1987.

56. T. R. Albrecht et al., Frequency modulation detection using high-Q cantilevers for enhanced force microscopy sensitivity, *J. Appl. Phys.,* **69**: 668–673, 1991.

57. P. Gruetter et al., 10-nm resolution by magnetic force microscopy on FeNdB, *J. Appl. Phys.,* **67**: 1437–1441, 1990.

58. R. Proksch et al., Quantitative magnetic field measurements with the magnetic force microscope, *Appl. Phys. Lett.,* **69**: 2599–2601, 1996.

C. A. Ross
Massachusetts Institute of Technology

M. E. Schabes
IBM Almaden Research Center

T. Nolan
Komag Inc.

K. Tang
IBM Research

R. Ranjan
Seagate Magnetics

R. Sinclair
Stanford University

MAGNETIC MEDIA, MAGNETIZATION REVERSAL

This article describes the magnetization reversal (magnetization switching) processes and the time-dependent reversal phenomena that occur when data are stored in magnetic recording media. These reversal processes determine the response of the medium to the applied magnetic field during the writing of data and the stability of written data patterns over the lifetime of the medium. Because magnetization reversal at low fields is a thermally activated process, the time scale over which the writing field is applied affects the switching field or coercivity of the medium, and thus the field that the write head needs to apply in order to write the data. Additionally, time-dependent magnetization reversal can lead to "superparamagnetic" behavior in which the written patterns are thermally unstable and read-back signals decay with time. These considerations apply to all magnetic media, but are most prominent in hard-disk media, and most of the examples described in this article refer to hard-disk media. Superparamagnetic and time-dependent effects are becoming increasingly important with the continuing increase in storage density and data rates in recording systems and are expected to limit the ultimate storage density achievable in hard-disk media.

MECHANISM OF MAGNETIZATION REVERSAL IN MAGNETIC MEDIA

Information is written onto hard-disk magnetic recording media when a magnetic write head passes near the surface of the recording medium. The head is magnetized by a varying current whose polarity changes according to a digitized stream of data. The resulting head field saturates the medium to form domains of opposing magnetization directions. The precise location and definition of these magnetization transitions is of great importance for the achievable linear density of data. Any fluctuation in the mean position of the transitions introduces noise, while the definition of the lateral bit boundaries limits the achievable track density.

The switching mechanism of the recording medium, that is, the reversible and irreversible processes that occur as the magnetization changes in response to an applied field, determines the recording performance of the system. The switching behavior depends upon the structure of the medium— whether, for instance, it is made from a continuous thin film with in plane magnetization directions (as in a hard disk), perpendicular magnetization (as in perpendicular hard-disk media or magneto-optical media), or from magnetic particles embedded in a matrix (as in flexible media). In continuous thin-film media, the deposition conditions are usually chosen to produce a film consisting of small magnetic grains separated by nonmagnetic grain boundaries. Therefore both thin-film and particulate media may be treated as assemblies of small, single-domain particles subject to magnetic interactions. The switching mechanism of the medium depends on the switching behavior of the individual particles and also, importantly, on interactions between the particles via magnetostatic (long-range) or exchange (short-range) forces.

The switching mechanism of an isolated single-domain particle depends on the particle anisotropy, size, and magnetization. The simplest description of magnetization reversal is that the magnetization vector rotates uniformly (coherently) as the external field is varied. This coherent rotation model was described by Stoner and Wohlfarth (1) and was extended (2,3) to include thermally activated switching over a single energy barrier. However, magnetic particles often show incoherent switching in which the magnetization within the particle is not uniform (4), and the energy barrier may be complex. Recent magnetic force microscopy (MFM) investigations of magnetization reversal have helped elucidate the behavior of isolated magnetic particles (5), which is necessary for understanding of the switching behavior of assemblies of interacting particles.

Even if the switching behavior of individual particles is understood, the interactions between particles leads to complex collective behavior of the medium. For example, if the

data track is initially prepared in a dc-erased state, so that the magnetization direction is uniform and parallel to the data track, new data will be written onto the track either in or against the initial magnetization direction. These are known as "easy" or "hard" transitions, respectively, and their physical location with respect to the recording head differs because of the superposition of the field from the previously erased track and the field from the head. Similarly, the effect of preexisting magnetization patterns leads to nonlinear bit shift in newly recorded data (6). It can therefore be difficult to calculate the magnetic response of a recording medium.

The behavior of magnetic media has been investigated using micromagnetic calculations by treating the medium as an assembly of interacting Stoner–Wohlfarth single-domain particles (7–9). This assumption has been shown to be valid for grains with sizes comparable to the magnetic exchange length or smaller (10), which is the case for thin-film media. For pure cobalt and permalloy (NiFe) the exchange length is 7 nm and 20 nm, respectively. The reversal dynamics can then appropriately be described by solving coupled Landau-Lifshitz equations (11). It has been shown that the solutions of this nonlinear system of ordinary differential equations have a rich structure that captures the collective magnetization processes. The magnetization direction of adjacent grains is found to be correlated, and the size of these correlated regions governs the noise properties of the medium. Exchange interactions enlarge magnetic correlations and therefore increase noise and large-scale percolation structures, in which the jaggedness of the bit boundary is large compared to the bit width (12). As an example, Fig. 1 (13) shows how the presence of exchange coupling between grains affects the magnetization patterns at a written transition. Theory has also shown that magnetostatic interactions lead to the development of magnetization vortices in which the magnetization locally follows a circular pattern. These vortices have been imaged experimentally (14).

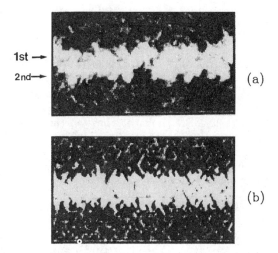

1st →
2nd →

(a)

(b)

Figure 1. Micromagnetic simulation of two transitions (bits) written into a hard disk, (a) with and (b) without exchange coupling. The track direction is top to bottom. The gray scale represents the magnetization direction in the film plane parallel to the track direction, with white parallel and black antiparallel (13). Reprinted with permission of the authors and the Institute of Electrical and Electronic Engineers.

Application of this modeling has yielded a detailed understanding of the magnetization processes occurring during the writing process. However, with the need for high-frequency writing and with the continued decrease of grain size it is apparent that the deterministic Landau–Lifshitz time evolution now needs to be augmented by more sophisticated dynamical models that also take into account thermal fluctuations (15).

TIME DEPENDENCE OF MAGNETIZATION PROCESS

There has been considerable study of time-dependent magnetic phenomena (16–19,19a). Recently, changes in coercivity with field sweep rate have become important in the design of high-density recording media (20–32). Analysis of time-dependent phenomena treats magnetic media as consisting of an assembly of single-domain particles having magnetic anisotropy energy K and switching volume V. K represents the tendency of magnetization to lie along an "easy axis," and V is the volume of the particle that needs to switch its magnetization to initiate reversal of the magnetization direction in the particle. K includes contributions from magnetocrystalline anisotropy K_u as well as from the shape anisotropy of the particle and any magnetostrictive effects. V is equal to the physical particle size if the magnetization rotates coherently during reversal. V may be smaller than the physical particle volume if switching is incoherent, or may exceed the particle volume if interactions cause particles to switch cooperatively. The ease of switching the magnetization depends on the energy KV needed to rotate the magnetization so that it is perpendicular to the easy axis. The susceptibility of media to thermal effects (i.e., spontaneous reversal of the magnetization) therefore increases as the parameter KV/kT decreases, where k is Boltzmann's constant and T the temperature. KV is measured either from the sweep-rate dependence of coercivity or from the time dependence of magnetization in an applied reverse field.

Switching volumes can be measured from the time dependence of magnetization $M(t)$ in an initially saturated sample placed in a reverse magnetic field H_{rev} (16,18,35,36). The magnetization often follows a logarithmic relation such that

$$M_r(t, H_{rev}) = M_0(t, H_{rev}) + S(H_{rev})\ln(t) \quad (1)$$

where M_0 is the initial remanent magnetization and $M_r(t)$ is the remanent magnetization at time t at H_{rev}, and S is the magnetic viscosity. The functional form of this relation results from a superposition of exponential decays for different populations of magnetic particles. Figure 2 shows a linear dependence of magnetization on $\ln(t)$ (24). The activation volume V^* is then related to S by

$$V^*(H) = \chi_{irr} kT/M_s S(H) = kT/M_s H_f \quad (2)$$

where χ_{irr} is the irreversible susceptibility. This is the rate of change of irreversible magnetization with applied field that can be found by differentiation of the remanence curve for the material, measured using a vibrating sample magnetometer (VSM) or alternating gradient force magnetometer (AGFM). H_f is known as the fluctuation field, a fictitious field whose effect on the magnetization is equal to the effect of thermal fluctuations.

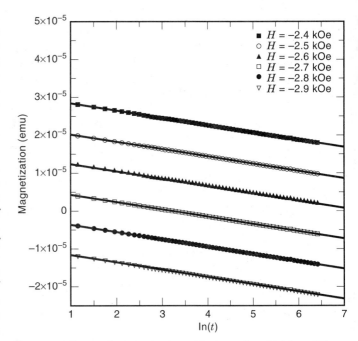

Figure 2. Decay of magnetization with time t for a Fe/Pt multilayer sample initially saturated at 10 kOe, then held in a reverse field of between 2.4 kOe and 2.9 kOe (24). Reprinted with permission of the authors and the American Institute of Physics.

The logarithmic relation is followed for systems that have a wide distribution of energy barriers for reversal of the magnetic particles (19a). A narrow distribution of barrier heights leads to a nonlinear dependence on $\ln(t)$ but the switching volume can still be calculated from a series of time-dependent curves separated by small increments in applied field (27). These measurements are commonly done in an AGFM. It should be noted that V^* has a different physical interpretation from V defined in Eq. (4), below, since V^* has an explicit field dependence. V can be identified with $V^*(H_c)$ (20).

Switching volumes determined using this method have been reported for a range of materials, including barium ferrites (37), iron oxides (38), α-iron particles (38,39), and CoCrTa (27) and CoCrPt (41) hard-disk media. This analysis can also be applied to continuous magnetic films which switch by domain-wall motion, for instance rare-earth/transition metal amorphous films used in magneto-optical media (40,40a,40b,40c). In this case, V^* represents the volume of part of a domain wall pinned at a pinning center.

Thermally activated switching leads to a time-dependent coercivity. The variation of coercivity with the scan rate of the external applied field is commonly analyzed using a method described by Sharrock (20). This assumes an assembly of noninteracting single-domain particles whose magnetic switching is thermally assisted. The particles have uniaxial magnetic anisotropy of magnitude K and saturation magnetization M_s. The coercivity H_c is given by

$$H_c(t) = H_0\{1 - [(kT/KV)\ln(At)]^n\} \quad (3)$$

where half of the particles reverse their magnetization in time t when the applied field is $H_c(t)$. A is a constant related to the attempt frequency of switching, of the order of 10^9 s^{-1}. The geometrical factor n takes a value between 0.5 and 0.7

(1,33) depending on the angle between the easy axis of the particles and the applied field. H_0 is given by $2xK/M_s$, where x is a geometrical constant related to n and has a value between 0.5 and 1. K can be found from H_0 or from other measurements such as torque magnetometry.

The model was developed originally for particulate media such as iron oxide or ferrite particles, but has more recently (24,26,28,30) been applied to thin-film media, for which the assumption of noninteracting particles is even less justifiable. However, Monte Carlo simulations of switching in thin-film media (15,23,34) indicate that an equation of the form of Eq. (3) still describes the variation of measured coercivity with sweep rate, but with a higher exponent of $n = 1$ (23) or $n = 0.735$ (15) due to interactions between the magnetic grains. An exponent of $n = 1$ reduces Eq. (3) to the form

$$H_c(t) = C + (kT/M_sV)\ln(t') \qquad (4)$$

which is linear in $\ln(t')$, where t' is the rate of change of the field and C is a constant. This linear form has also been used to analyze time dependence of coercivity in media (24). It may be difficult to distinguish between Eq. (3) and Eq. (4) unless the coercivity measurements cover several decades of scan time. Using Eq. (3) or (4), V can be found from the scan-rate dependence of H_c using an AGFM or a Kerr-effect looper that measures the hysteresis loop of the sample. Comparison between V and the physical particle size (determined by microscopy) yields information about the switching mechanism and magnetic interactions between particles.

Until recently, there had been relatively little application of this analysis to thin films, partly because its validity has not been well established for interacting systems. Figure 3 shows an application of Eq. (3) to data from a CoCrTa/Cr hard disk (30). There are many other examples in the literature. For instance, in the case of noninteracting, aligned particles of ferrite particulate media with $n = 0.5$, the switching

volume was found to be smaller than the physical volume, implying incoherent rotation of the magnetization (25). Analysis using an exponent of $n = 1$ in CoCrTaPt media (26) and in Fe/Pt multilayers (24) showed switching volumes smaller than the grain volume. However, in CoSm deposited on a Cr underlayer (29), the switching volume was similar to the Cr grain size and considerably larger than the CoSm crystal size, and in CoCrPt/Cr the switching volume was also found to be considerably larger than grain size (27). A comparison of V with $V^*(H_c)$ in Fe/Pt multilayers showed reasonable agreement (24).

LONG-TERM STABILITY OF MAGNETIZATION PATTERNS

The stability of written magnetization patterns is related to the time-dependent magnetic effects and magnetic viscosity described above. If the product KV/kT in the media is sufficiently small, spontaneous magnetic switching of the particles can occur and the amplitude of the read-back signal decays. This phenomenon is of practical concern if there is significant decay of the signal during the lifetime of the media.

Long-term stability and time-dependent coercivity of media have been estimated by performing Monte Carlo simulations of the behavior of arrays of magnetic grains (15,23,34). These simulations show the difference between the effective coercivity of the medium for long-term data storage, relevant to signal readback, compared to the effective coercivity during writing, where the applied field varies at high frequencies. Using parameters suitable for 10 Gbit/in.² longitudinal recording, the reading and writing coercivities differ significantly if KV/kT is less than about 100. The decay of written bits in longitudinal media was shown to depend strongly on KV/kT, on the recording density, and to a lesser extent on the grain aspect ratio and on interactions between grains given by M_s/H_k. H_k is the anisotropy field, which is the field needed to rotate the magnetization away from the easy axis. Choosing the stability criterion that the magnetization midway between two transitions at 400 kfci (400 thousand flux changes per inch) must exceed 0.4 memu/cm² after 6 months, the suitability of various values of K and M_s can be plotted on a map, Fig. 4 (15), based on a grain size of 10 nm. Only region 5 is suitable for stable recording; in region 1 M_s (and the signal) is too small, in region 2 transitions overlap, magnetostatically reducing the signal, in region 3 the signal decays thermally, and in region 4 the head field cannot saturate the medium. For the 10 Gbit/in.² example cited, optimum signal was obtained for reading coercivities of about 3500 Oe and writing coercivities of about 5000 Oe.

Thermally assisted switching of the magnetic domains occurs rapidly at head-on transitions in longitudinal media because there are strong demagnetizing fields tending to promote magnetization reversal. In perpendicular media, the demagnetizing fields at the transition are less strong and the media are believed to have better thermal stability. Stability is also enhanced by the greater film thickness (and grain volume) in perpendicular media (42). In longitudinal media, simulations show that signal loss is likely to be significant for values of KV/kT less than about 60. This is not an issue for existing media, which have ratios of order 2000, but as grain sizes and film thicknesses continue to decrease, signal decay will become an increasing problem.

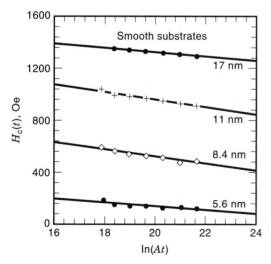

Figure 3. The dependence of coercivity H_c on the rate of change of the magnetic field for CoCrTa hard-disk media. The field scan rate is proportional to $1/t$, and A is a constant. As the field scan rate increases, the measured coercivity increases. Data are shown for four different thicknesses of CoCrTa films (30). Reprinted with permission of the authors and the Institute of Electrical and Electronic Engineers.

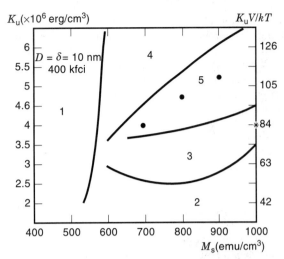

Figure 4. Calculated regimes of different behavior for a hard disk medium with 10 nm grain size D, 10 nm film thickness δ, and a range of values of magnetocrystalline anisotropy K_u and saturation magnetization M_s. The bullets indicate the maximum readback signal for different M_s values. In region 5 the film would be suitable for 10 Gbit/in.^{2} longitudinal recording (15). Reprinted with permission of the authors and the Institute of Electrical and Electronic Engineers.

Approaches to making higher-density media are centered on increasing K to improve thermal stability and increasing M_s by composition modification to increase signal strength, while maintaining small grain size in order to minimize noise. K is controlled by modifying the film or particle composition, for instance, by adding Pt to the CoCr-based alloys used in hard-disk media, or by cobalt doping of iron oxide particles or by increasing particle aspect ratios for flexible media. Since K is affected by the processing conditions of the material, its value for thin films and for bulk materials may be different. K and M_s cannot be increased arbitrarily without making the medium more difficult to write, so the choice of medium parameters is related to the design of the recording head, particularly the saturation magnetization of the pole pieces (34).

CONCLUSIONS

Although considerations of switching behavior and thermal stability apply to all types of magnetic recording media, a great deal of attention is now being focused on thermal stability in hard-disk media, because hard-disk systems have the highest storage density. In the quest to produce higher density hard-disk media, there has been a continuing trend to increase coercivity and anisotropy and to reduce magnetic film thickness, in order to reduce the transition width and increase the linear density. Grain size has simultaneously decreased in order to reduce the transition noise. As the superparamagnetic limit is approached (i.e., the thermal decay of the signal amplitude becomes significant with reducing grain size) it will be necessary to continue to increase the anisotropy of the materials used. This makes the medium more difficult to write, so that recording heads are required to produce a higher write field. High anisotropy materials, including barium ferrite (43) and Co—rare earth compounds such as CoSm (44), are being considered for advanced recording media. At present there are significant fabrication problems.

Ferrite films, for instance, require high-temperature processing that is not compatible with the use of aluminum hard-disk substrates. However, there is intense study of the fabrication and properties of these materials. Simultaneously there are developments in high saturation materials for recording heads, such as iron nitrides, in order to write the media (45).

Existing thin-film hard-disk longitudinal media materials are predicted to be capable of supporting recording densities of at least 40 Gbit/in.^{2} (34,46). This limit will likely be higher in practice as the aspect ratio of the bits is reduced by reducing the track spacing faster than the linear bit spacing. For higher densities this evolutionary development will ultimately be limited, and new data storage schemes may be introduced. There is considerable research on perpendicular hard-disk media, which are believed to have superior high-density noise properties and thermal stability (42), but so far perpendicular hard-disk media have had few commercial applications. On a longer developmental time scale, one possible scheme is the use of patterned media, in which data are stored in discrete patterned magnetic domains (46,47). These patterned media may be writtten and read by ultrasmall recording heads or by probe tips based on MFM technology. Since the boundaries of these bits are physically defined, there is no transition noise, and the constraints on the coercivity and grain size of the medium are reduced. The superparamagnetic limit then refers to the size of the bit rather than to the size of the grains of which it is composed, allowing a very high recording density to be achieved.

BIBLIOGRAPHY

1. E. C. Stoner and E. P. Wohlfarth, A mechanism of magnetic hysteresis in heterogeneous alloys, *Philos. Trans. Roy. Soc. London, Ser. A,* **240**: 599–642, 1948.

2. M. L. Néel, Theorie du trainage magnetique des ferromagnetiques en grains fins avec applications aux terres cuites, *Ann. Geophys.,* **5**: 99–136, 1949.

3. W. F. Brown, Thermal fluctuations of a single-domain particle, *Phys. Rev.,* **130**: 1677–1686, 1963.

4. A. Aharoni and J. P. Jakubovics, Field induced magnetization structure in small isotropic spheres, *IEEE Trans. Magn.,* **32**: 4463–4468, 1996.

5. R. O'Barr et al., Preparation and quantitative magnetic studies of single-domain nickel cylinders, *J. Appl. Phys.,* **79**: 5303–5305, 1996.

6. H. N. Bertram, *Theory of Magnetic Recording,* Cambridge, UK: Cambridge Univ. Press, 1994.

7. G. F. Hughes, Magnetization reversal in cobalt-phosphorus films, *J. Appl. Phys.,* **54**: 5306–5313, 1983.

8. J. G. Zhu and H. N. Bertram, Recording and transition noise simulatons in thin-film media, *IEEE Trans. Magn.,* **24**: 2706–2708, 1988.

9. J. J. Miles and B. K. Middleton, A hierarchical micromagnetic model of longitudinal thin film recording media, *J. Magn. Magn. Mater.,* **95**: 99–108, 1991.

10. M. Schabes, Micromagnetic theory of non-uniform magnetization processes in magnetic recording particles, *J. Magn. Magn. Mater.,* **95**: 249–288, 1991.

11. T. C. Arnouldussen, in L. L. Nunnelley (ed.), Noise in Digital Magnetic Recording, Singapore: World Scientific, 1992.

12. H. N. Bertram and J. G. Zhu, Fundamental magnetization processes in thin film recording media, in H. Ehrenreich and D. Turnbull (eds.), *Solid State Physics Review*, New York: Academic Press, 1992, vol. 46, p. 271.

13. J. G. Zhu, Noise of interacting transitions in thin film recording media, *IEEE Trans. Magn.*, **27**: 5040–5042, 1991.

14. K. Tang et al., Lorentz transmission electron microscopy study of micromagnetic structures in real computer hard disks, *IEEE Trans. Magn.*, **32**: 4130–4132, 1996.

15. P. L. Lu and S. H. Charap, High density magnetic media design and identification: susceptibility to thermal decay, *IEEE Trans. Magn.*, **32**: 4130–4132, 1996.

16. R. Street and J. C. Wooley, A study of magnetic viscosity, *Proc. Phys. Soc. London, Sec. A*, **62**: 562–572, 1949.

17. D. K. Lottis, R. M. White, and E. D. Dahlberg, Model systems for slow dynamics, *Phys. Rev. Lett.*, **67**: 362–365, 1991.

18. R. Street and S. D. Brown, Magnetic viscosity, fluctuation fields and activation energies, *J. Appl. Phys.*, **76**: 6386–6390, 1994.

19. E. D. Dahlberg et al., Ubiquitous non-exponential decay: the effect of long-range couplings?, *J. Appl. Phys.*, **76**: 6396–6401, 1994.

19a. R. Street and J. C. Wooley, Magnetic viscosity under discontinuously and continuously variable field conditions, *Proc. Phys. Soc.*, **B65**: 679–696, 1952.

20. M. P. Sharrock, Time-dependence of switching fields in magnetic recording media, *J. Appl. Phys.*, **76**: 6413–6418, 1994.

21. W. D. Doyle, L. He, and P. J. Flanders, Measurement of the switching speed limit in high coercivity magnetic media, *IEEE Trans. Magn.*, **29**: 3634–3636, 1993.

22. L. He, W. D. Doyle, and H. Fujiwara, High speed switching in magnetic recording media, *J. Magn. Magn. Mater.*, **155**: 6–12, 1996.

23. P.-L. Lu and S. H. Charap, Magnetic viscosity in high density recording, *J. Appl. Phys.*, **75**: 5768–5770, 1994.

24. C. P. Luo, Z. S. Shan, and D. J. Sellmyer, Magnetic viscosity and switching volumes of annealed Fe/Pt multilayers, *J. Appl. Phys.*, **79**: 4899–4901, 1996.

25. Y. J. Chen et al., Thermal activation and switching in c-axis aligned barium ferrite thin film media, *IEEE Trans. Magn.*, **32**: 3608–3610, 1996.

26. C. Gao et al., Correlation of switching volume with magnetic properties, microstructure and media noise in CoCr(Pt)Ta thin films, *J. Appl. Phys.*, **81**: 3928–3930, 1997.

27. P. Dova et al., Magnetization reversal in bicrystal media, *J. Appl. Phys.*, **81**: 3949–3951, 1997.

28. T. Pan et al., Temperature dependence of coercivity in Co-based longitudinal thin film recording media, *J. Appl. Phys.*, **81**: 3952–3954, 1997.

29. E. W. Singleton et al., Magnetic switching volumes of CoSm thin films for high density longitudinal recording, *IEEE Trans. Magn.*, **31**: 2743–2745, 1995.

30. C. A. Ross et al., Microstructural evolution and thermal stability of thin CoCrTa/Cr films for longitudinal magnetic recording media, *IEEE Trans. Magn.*, **34**: 282–292, 1998.

31. R. W. Chantrell, G. N. Coverdale, and K. O'Grady, Time dependence and rate dependence of the coercivity of particulate recording media, *J. Phys. D*, **21**: 1469–1471, 1988.

32. M. El-Hilo et al., The sweep-rate dependence of coercivity in particulate recording media, *J. Magn. Magn. Mater.*, **117**: L307–L310, 1992.

33. H. Pfeiffer, Determination of the anisotropy field distribution in particle assemblies taking into account thermal fluctuations, *Phys. Status Solidi A*, **118**: 295–306, 1990.

34. S. H. Charap, P. L. Lu, and Y. He, Thermal stability of recorded information at high densities, *IEEE Trans. Magn.*, **33**: 978–983, 1997.

35. R. W. Chantrell et al., Models of slow relaxation in particulate and thin film materials, *J. Appl. Phys.*, **76**: 6407–6412, 1994.

36. R. W. Chantrell, Magnetic viscosity of recording media, *J. Magn. Magn. Mater.*, **95**: 375–378, 1991.

37. J. M. Gonzalez et al., Magnetic viscosity and microstructure: particle size dependence of the activation volume, *J. Appl. Phys.*, **79**: 5955–5957, 1996.

38. G. Bottoni, D. Candolfo, and A. Ceccetti, Interaction effects on the time dependence of the magnetization in recording particles, *J. Appl. Phys.*, **81**: 3809–3811, 1997.

39. F. Li and R. M. Metzger, Activation volume of *a*-Fe particles in alumite films, *J. Appl. Phys.*, **81**: 3806–3808, 1997.

40. T. Thomson and K. O'Grady, Temperature dependence of activation volumes in Tb-Fe-Co magneto-optic thin films, *IEEE Trans. Magn.*, **33**: 795–798, 1997.

40a. F. D. Stacey, Thermally activated ferromagnetic wall motion, *Aust. J. Phys.*, **13**: 599–601, 1960.

40b. T. G. Pokhil, B. S. Vredensky, and E. N. Nikolaev, Slow motion of domain walls in amorphous TbFe films, *Proc. SPIE*, **1274**: 305–315, 1990.

40c. T. Pokhil and E. N. Nikolaev, Domain wall motion in RE-TM films with different thickness, *IEEE Trans. Magn.*, **29**: 2536–2538, 1993.

41. D. H. Han et al., Time decay of magnetization in longitudinal CoCrTa/Cr high density thin film media, *IEEE Trans. Magn.*, **33**: 3025–3027, 1997.

42. Y. Hirayama et al., Development of high resolution and low noise single-layered perpendicular recording media for high density recording, *IEEE Trans. Magn.*, **33**: 996–1001, 1997.

43. A. Morisako, M. Matsumoto, and M. Naoe, Sputtered hexagonal barium ferrite films for high density magnetic recording media, *J. Appl. Phys.*, **79**: 4881–4883, 1996.

44. Y. Liu, Z. S. Shan, and D. J. Sellmyer, Lorentz microscopy observation of magnetic grains in CoSm/Cr films, *IEEE Trans. Magn.*, **32**: 3614–3616, 1996.

45. M. K. Minor, B. Viala, and J. A. Barnard, Magnetostriction and thin film stress in high magnetization magnetically soft FeTaN thin films, *J. Appl. Phys.*, **79**: 5005–5008, 1996.

46. R. White, R. H. New, and R. F. W. Pease, Patterned media: A viable route to 50 Gb/in² and above for magnetic recording?, *IEEE Trans. Magn.*, **33**: 990–995, 1997.

47. S. E. Lambert et al., Recording characteristics of submicron discrete magnetic tracks, *IEEE Trans. Magn.*, **23**: 3690–3692, 1987.

C. A. Ross
Massachusetts Institute of Technology

M. E. Schabes
IBM Almaden Research Center

T. Nolan
Komag, Inc.

K. Tang
IBM Research Laboratories

R. Ranjan
Seagate Magnetics

R. Sinclair
Stanford University

MAGNETIC METHODS OF NONDESTRUCTIVE EVALUATION

In nondestructive evaluation (NDE), measurements are made in such a way that after the measurements are completed, the specimen is not physically altered as a result of the measurement. In the case of magnetic measurements this means that the specimen, if magnetized, can always be demagnetized and restored to its original state. A nondestructive measurement is therefore just what the label says—nondestructive.

Magnetic methods of nondestructive evaluation are used to find three major types of information: (1) detection and characterization of macroscopic flaws in a specimen such as cracks, corrosion pits, or inclusions; (2) characterization of microstructural features such as creep damage, plastic deformation, grain size, and compositional features; (3) characterization of residual stress and residual stress distribution in a specimen. In most cases, a magnetic technique is used typically for only one of these uses; in some cases, it may be used for several.

Our discussion in this review will categorize magnetic methods primarily according to the purpose for which the methods are used. Thus, the article will be divided into three major sections according to the type of information sought. Not included in our discussion will be NDE measurement techniques involving eddy currents, as such techniques are used mostly for nonmagnetic materials. A discussion of promising future magnetic NDE techniques will be included in a final major section.

Review articles on various aspects of magnetic NDE that might also be consulted are found in Refs. 1–20.

MAGNETIC METHODS FOR CRACKS, CORROSION PITS, OR INCLUSIONS

Magnetic Particle Inspection (MPI)

Magnetic particle inspection (MPI) was developed in the 1930s by Magnaflux Corporation (21). The method was based on the chance discovery that iron filings tended to collect close to flaws in steels during the grinding process. Magnaflux Corporation turned this observation into a successful commercial method of locating flaws in steels (1,21).

Magnetic particle inspection (MPI) depends on the leakage of magnetic flux at the surface of a ferromagnetic material at locations of surface-breaking or near-surface flaws (see Fig. 1). In order for the method to be effective, the magnetic material has to be magnetized in the vicinity of the flaw. This is

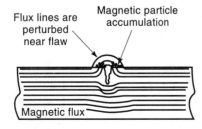

Figure 1. Magnetic particle accumulation in the leakage flux produced by a flaw (after Ref. 10).

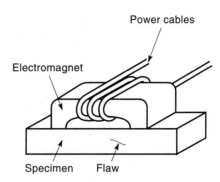

Figure 2. The yoke method for magnetizing a specimen (after Ref. 10).

accomplished usually in one of two ways: (1) via a magnetic field injected into a specimen by a magnetic "yoke" wound with current-carrying coils (see Fig. 2); (2) by using contact electrodes called "prods," which inject currents directly into the specimen, in turn magnetizing the specimen. In the case of the "yoke," the magnetic field circulates into the specimen parallel to the surface, running from one yoke pole piece to the other. Flaws in the material break up this parallel field pattern in the vicinity of the flaw to produce a field gradient and magnetic force, which holds iron filings near the flaw. With "prods," the injected current generates a circulating magnetic field, as given by the "right-hand rule" (22). Flaws distort the field pattern, again producing field gradients and forces that hold filings near the flaws. In both cases, the best indication is given when the field is perpendicular to the largest flaw dimension, either the crack length or the most prolately shaped side of the flaw in the case of corrosion pits or inclusions.

The MPI method is reliable, when used correctly, for finding surface and near-surface flaws of sufficient macroscopic size and gives an indication of the location and length of the flaw. The field must be strong enough to hold the particles applied. Very shallow cracks can be missed, as can subsurface cracks, if the leakage fields are weak. The magnetic particles best used are ones that are fine enough and have a high enough permeability to be held (21). The component being tested can be almost any size or shape, although care is needed with complex geometries.

The method has limitations. For best results, the magnetic field must lie perpendicular to the flaw direction. Flaws can be overlooked by misorientation of the field or by using a field that is not strong enough to hold the particles. Finally, while the length of the flaw is obtainable, depth of flaw can only be guessed (unsatisfactorily) by the amount of powder accumulated.

Various enhancements have been added (1,10). These include wet techniques, such as water-borne suspensions known as "magnetic inks." Also, fluorescent magnetic powders often give clearer indication of smaller flaws when viewed under ultraviolet light (23). Another method is a magnetic tape, which is placed over the area to be inspected (24). The tape is magnetized by the strong surface field, the gradients of which leave imprints of flux changes at defect locations. A quantitative flux leakage reading is obtained by inspecting the tape with a Hall probe or fluxgate magnetometer.

The tape is particularly useful in places hard to inspect by MPI.

Magnetic Flux Leakage (MFL)

As with the MPI method, the magnetic flux leakage (MFL) method depends on the perturbation of magnetic flux caused by surface or near-surface flaws. The MFL method differs from MPI in that it utilizes a flux-detecting device to detect the perturbations associated with the flaw. Another name for the MFL method is the "magnetic perturbation" (MagPert) method (7).

The MFL method offers extra information because the flux density components in three directions, parallel and perpendicular to the flaw direction and normal to the surface, can be measured. Usually, however, only components parallel to the surface are actually measured.

The method gained acceptance after a practical flux-leakage measuring system was developed (25), which was capable of detecting surface and subsurface flaws on the inner surface of steel tubes, a location unsuitable for the MPI method. The MFL technique is now even more developed in that it can be used for both detection and characterization of flaws (7, 26–28).

The leakage flux probe is usually an induction coil or a Hall probe. The probe is accompanied by a magnet that magnetizes the specimen in its vicinity. As the probe is scanned across the specimen surface, detected flux density anomalies indicate flaw location. Figure 3 shows the use of such a probe both (a) to detect a crack and (b) to detect a region of low

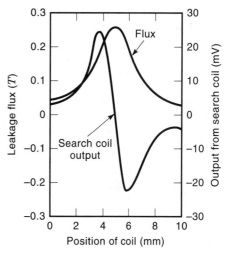

Figure 4. Leakage flux variation and search coil output with distance across a crack (after Ref. 27).

permeability. Figure 4 shows variation of leakage flux density as the probe moves across a crack and the voltage in a search coil moving at constant speed through this changing flux density (27).

The use of the search coil sensor is based on Faraday's law of induction, which states that the voltage induced in the coil is proportional to the number of turns in the coil multiplied by the time rate of change of the flux threading through the coil (7). To produce a voltage, either the coil must be in motion or the flux density must be changing as a function of time. For MFL, the moving coil is used to sense spatial changes in leakage flux. If the coil is oriented to sense flux changes parallel to the specimen surface in the direction x, then as the coil moves through the spatially perturbed flux above a flaw, the induced emf is given by (7)

$$V = N\frac{d\Phi}{dt} = NA\frac{dB_p}{dx}\frac{dx}{dt} \qquad (1)$$

where A is coil cross-sectional area, N is number of turns, B_p is flux density component parallel to the surface, dx/dt is constant coil velocity, and $d\Phi/dt$ is rate of change of magnetic flux $\Phi = B_pA$ that is threading the coil. From Eq. (1), the coil voltage V is proportional to the flux density gradient along the direction of coil motion times the coil velocity.

The Hall sensor does not detect the flux gradient, but measures directly the component of flux itself in a direction perpendicular to the sensitive area of the device (7,27). Because the Hall sensor response is not dependent on probe motion, a variable scanning speed can be used. In air, the Hall sensor is often used to measure magnetic field $H_p = B_p/\mu_0$, where μ_0 is the permeability of free space, and where the Hall sensor is oriented to measure field and flux density components parallel to the specimen surface. The Hall sensor is used to measure H_p because it has a small sensitive area that can be placed very close to the specimen surface. As H_p is continuous across the surface boundary, the H_p measured is equivalent to H_p in the specimen. The Hall sensor is more difficult to fabricate and more delicate than induction coil sensors.

The disadvantage of the MFL method, compared with MPI, is that scanning a leakage flux detection probe across the sur-

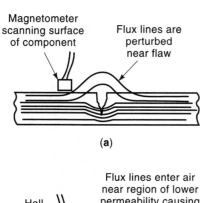

(a)

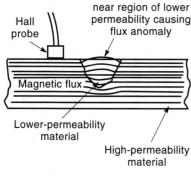

(b)

Figure 3. Using magnetic flux leakage (a) to detect flaws and (b) to detect regions of different permeability (after Ref. 10).

face of a specimen can be quite time-consuming. The MPI method, on the other hand, can check large areas of a specimen quite quickly.

The MFL method is quite useful if the location of the flaw is known with a fairly high probability because the MFL method can be used to characterize flaws as to size and depth (26–28). Also, if a matrix of scanners can cover the entire surface of the specimen in one pass, the MFL method offers advantages in that it can be done systematically. This is the case when a circular ring of leakage flux scanners is placed on a "pig" inside a pipeline and is sent through the pipeline to detect corrosion and other flaws on the inside of the pipeline (29,30). For pipeline, MFL is preferred because the inside of a pipe is hard to inspect visually.

Interpretation of MFL in terms of flaw size and shape was made possible by quantitative leakage field modeling. In early papers, Shcherbinin and Zatsepin (31,32) approximated surface defects by linear magnetic dipoles and by calculating the dipole magnetic fields. In this way, expressions were obtained for both the normal and tangential components of the leakage flux density. Numerical computations fit experimental data surprisingly well. In the case of leakage fields due to inclusions and voids, a model by Sablik and Beissner (28) was developed, which approximated voids and inclusions as either prolate or oblate spheroids. Expressions for the three components of flux density were used to study the effects of oblateness and size and depth of defect. Another analysis was that of Edwards and Palmer (33), who approximated a crack as a semielliptic slot and computed the leakage flux density components and the forces on magnetic particles, so that quantitative analysis of MPI might also be tenable.

Significant progress in leakage field computation was made by Hwang, Lord, and others (34–37), who used finite element modeling methods. The leakage field profiles obtained for a simple rectangular slot agreed excellently with observation (34). Other finite element calculations (35) showed how different defect shapes and geometries affected the leakage field signals. A similar technique, finite difference modeling, was also used for computing leakage fields (38).

Reviews of leakage field calculations and interpretation of measurements have been given by Dohman (39) and Holler and Dobmann (40). These authors have discussesd both detection and sizing. Owston (41) reported on differences in leakage flux signals between fatigue cracks and artificial flaws such as saw slots, and also on leakage flux as a function of liftoff (i.e., distance of detector from specimen surface). Forster (42) has discussed the correlations of observed magnetic leakage field measurements with expectations based on finite element modeling and shown notable discrepancies.

One complication in the leakage field computations is residual stress around the defect. The residual stress produces a distribution of permeability changes about the defect, in effect creating a new magnetic geometry. This is why the leakage field of a fatigue crack differs from that of a slot. Recent papers have attempted both theoretically (43) and experimentally (44,45) to deal with the effect of residual stress on leakage field signals from corrosion pits.

A leakage field detection technique, called the electric current perturbation (ECP) technique (46,47), has also been used for nonmagnetic materials. It involves either injecting (via electrodes) or inducing (via a coil) an electric current in the vicinity of a flaw. The current passes around the defect, creating a characteristic leakage field signal in the flux density produced by the flow of current. Clearly the technique has its origins in MFL.

The MFL is used primarily by the oil and gas industry for inspections inside tubulars, such as gas pipelines, downhole casing and other steel piping (7,29,30). The cylindrical geometry aids in characterizing defects. The MFL has also been used for irregularly shaped parts, such as helicopter rotor blade D-spars (48), and for bearings and bearing races (49).

Magnetostrictive Sensors (MsS)

A relatively new technique for nondestructively locating flaws well away from the position of the sensor is called "magnetostrictive sensing" (MsS) (50,51). Current in an ac coil, axisymmetrically surrounding a steel pipe in the presence of a static axial magnetic field, generates an electromagnetic field in the pipe wall, which magnetostrictively generates oscillating strains and hence elastic waves. These elastic waves travel down the pipe wall in both directions from the coil location and reflect off defects such as corrosion pits or deep cracks. The returning reflected elastic waves magnetostrictively produce changing magnetization underneath a detector coil, also in a static axial magnetic field, and the changing magnetization induces an emf in the detector coil. The time lag between generation and detection gives information about where the defect is located, and the shape of the signal gives clues about the nature of the defect (52,53).

The technique not only offers the possibility of inspecting long sections of pipe or tubing, as much as 500 m away, but also works when there are bends and elbows in the pipe (53). The wave generation coils, sensing coils, and field coils all are suspended axisymmetrically on plastic forms around the pipe, so that the coupling of coils to the pipe does not involve direct contact (51). In addition, the inspection is volumetric and senses defects and flaws well inside the pipe wall and not just near the surface (53). Bridge cable strands (54) and concrete reinforcement bars (55) also can be inspected with the technique.

Figure 5 shows the transmitter and receiver coils on a steel pipe specimen and a block diagram showing the basic electronic circuitry. The bias magnets, operating together with transmitter and receiver coils, can be large current dc coils surrounding the ac coils. (In Fig. 5 there is one bias coil enveloping both receiver and transmitter.) The bias magnets can also be arrays of magnetic circuit modules spaced equally around the pipe (56). Figure 6 shows this latter configuration. Figure 7 shows reflected waves from welds in a pipe and from the far end of the pipe. Because second reflections of weld reflections are seen, careful interpretation is required.

Russian investigators began to consider the possibility of using magnetostrictive wave generation for sensing reflected waves from rod and pipe defects and developed theoretical models for the magnetostrictive wave generation in the long wavelength approximation (57,58). More recently, Sablik and Rubin (59) have produced a model, from which numerical results can be extracted for any frequency, without restriction. The more recent model reproduces the dispersion spectrum measured in pipes (59,60). The issue of wave amplitude and signal-in to signal-out still needs more work for a complete match with the experiment (61).

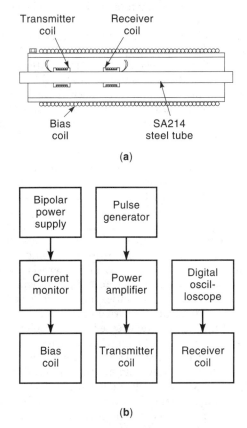

(a)

(b)

Figure 5. Magnetostrictive wave setup: (a) schematic diagram showing a steel tube surrounded by both transmitter and receiver in the axial bias field of a large dc coil; (b) block circuit diagram for bias coil, transmitter coil, and receiver coil (after Ref. 61).

Figure 6. Photograph shows an array of MsS magnetic bias field modules placed around a 406.6 mm outside diameter steel pipe. Installing an encircling ac coil on a continuous pipe is accomplished with a ribbon coil that can be strapped onto the pipe (after Ref. 56).

The MsS technique is promising. More research is needed to understand the effect of nonlinear, hysteretic magnetization and magnetostriction on the efficiency of the generation and sensing process. In addition, both defect identification and characterization need to be better addressed.

The MsS technique is similar to that of EMATS, which are used to generate elastic waves from electromagnetic waves in nonferrous media. EMATS rely on the Lorentz force to couple electromagnetic waves to the nonferrous metal; whereas the MsS approach relies on magnetostrictive coupling, which in ferromagnets is larger than Lorentz coupling. In addition, EMATS are typically meander coils, whereas the MsS geometry is cylindrical.

MAGNETIC METHODS FOR MICROSTRUCTURAL FEATURES

Microstructural NDE via Hysteresis Loop Parameters

All ferromagnetic materials exhibit hysteresis in the variation of flux density B with magnetic field H. Hysteresis means that as the field in a specimen is increased from $-H_\mathrm{m}$ to $+H_\mathrm{m}$, the $B(H)$ at each value of H is different from the value that exists when the field is decreased from $+H_\mathrm{m}$ to $-H_\mathrm{m}$. In other words, the flux density depends on the history of the H variation as well as on the value of H itself. Figure 8 shows this history dependence for a field varying between $-H_\mathrm{m}$ and

$+H_\mathrm{m}$. The B vs H plot is known as the magnetic hysteresis loop.

In Fig. 8 are indicated various parameters associated with the hysteresis. The remanence B_r is defined as the nonzero flux density still remaining in the material when the field in the material is brought from its maximum value H_m back to zero. The coercivity H_c is the additional amount of field in the

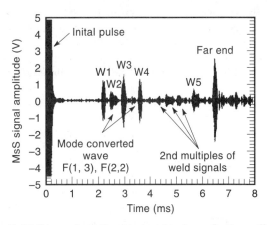

Figure 7. MsS trace from the detector showing reflections off welds, 2nd multiples of the weld signals, and the reflection from the far end of the pipe (after Ref. 53).

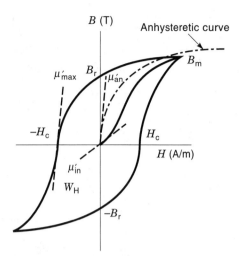

Figure 8. Hysteresis plot showing various hysteresis parameters–coercivity (H_c), remanence (B_r), maximum flux density (B_m), maximum differential permeability (μ'_{max}), initial differential permeability (μ'_{in}), the anhysteretic curve, and the initial differential anhysteretic permeability (μ'_{an}). The area inscribed by the hysteresis curve is the hysteresis loss (W_H) (after Ref. 6).

opposite direction that has to be applied before the remaining flux density in the material is finally brought back to zero. (Technically speaking, the terms remanent flux density and coercive field should be used for unsaturated loops, whose maximum flux density B_m is less than the saturation value B_s.) Note that the slope of the $B–H$ curve, known as the differential permeability, is typically a maximum at the coercive field H_c, and so the maximum differential permeability μ'_{max} is another characteristic of the hysteresis loop. The path taken on the $B–H$ plot when an unmagnetized specimen is brought to maximum field H_m is known as the initial magnetization curve, and the slope of the initial magnetization as the field begins to increase from $H = 0$ is known as the initial differential permeability μ'_{in}.

The area enclosed by the hysteresis loop in Fig. 8 is the hysteresis loss W_H. It has that name because the loop area is the magnetic energy that must be inputted if the material is to be completely cycled around the loop. This energy loss is associated with irreversible motion of magnetic domain walls inside the material and appears in the material as heat. The hysteresis is thus due to irreversible thermodynamic changes that develop as a result of magnetization.

Another feature, depicted in Fig. 8, is the anhysteretic curve. If a large amplitude ac field is superimposed on a constant dc bias field H and if the ac amplitude is gradually decreased, the material's flux density will tend to a value on the anhysteretic curve. By changing the bias field to H' and repeating the procedure, another point on the anhysteretic curve is obtained. At saturation, both the anhysteretic and the hysteresis curve have the same end point value. Note that the anhysteretic curve is single-valued. The slope of the anhysteretic curve at $H = 0$ is another hysteresis parameter and is called the initial differential anhysteretic permeability μ'_{an}. A paper by Sablik and Langman (62) discusses the experimental attainment of the anhysteretic curve in both the absence and the presence of mechanical stress.

All the hysteresis parameters are known to be sensitive to such factors as stress, plastic strain, grain size, heat treatment, and the presence of precipitates of a second phase, such as iron carbide in steels. Excepting stress, all of these factors refer to microstructural conditions in the material. In addition, microstructural changes can be produced by the application of stress at high temperatures; such changes, referred to as creep damage, generally involve degradation of the material so that it is more susceptible to mechanical failure such as cracking and rupture. The presence of creep damage can be sensed by characteristic magnetic property changes—that is, changes in the hysteresis parameters. Similarly, cyclic application of stress eventually results in microstructural changes that will eventually lead to mechanical failure. These microstructural changes due to cyclic stress application are known as fatigue damage, and they too are associated with characteristic magnetic property changes. Because hysteresis parameters are sensitive to microstructure and microstructural changes, the measurement of hysteresis loops for a given specimen becomes an NDE technique for characterizing microstructure. Proper systematics must be followed, so as to sort one microstructural effect from another.

One effect that must be addressed is the problem of demagnetizing effects (63) due to finite geometries and magnetic pole formation at both ends of the specimen, leading to a reduction in effective local field in the material by $-D_m M$, where D_m is known as the demagnetization factor and is dependent on sample shape. If this effect, due to finiteness and shape of the sample, is not properly addressed, apparent changes in sample magnetic properties may in fact be caused by geometrical instead of microstructural effects.

With proper care and systematics, NDE via hysteresis has had great success in evaluation of the condition of steel components. Mikheev (64) has used magnetic parameters to determine the quality of heat treatment of steels and to evaluate hardnesses of steels (65). In most cases, Mikheev used a coercivity measurement to characterize the material, and correlations were made between coercivity and chemical composition, microstructure, hardness, and heat treatment. Figure 9 is an example showing one such relationship—that H_c increases linearly with carbon content when carbon precipitates are lamellar in shape but nonlinearly when the precipitates are spheroidal (64). The effect of grain size and hardness on hysteresis parameters was studied by Kwun and Burkhardt (66), who looked at these effects in alloy steels.

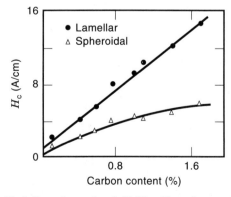

Figure 9. Variation of coercive field H_c with carbon content for lamellar and spheroidal precipitates (after Ref. 65).

Ranjan et al. (67,68) have also looked at grain size effects as well as carbon content effects in decarburized steels.

Another possible use of hysteresis parameters is in detection of creep damage (19). It is known that creep damage causes a general reduction in the values of hysteresis parameters—that is, remanence, coercivity, maximum differential permeability, and hysteresis loss (69). The reason is twofold. During creep, voids move out of the grains to the grain boundaries where they coalesce and form cavities; the cavities become magnetically polarized, creating a demagnetizing field that results in a decreased local field and decreased overall flux density and remanence (69). Also, during creep, dislocations, which normally act as domain wall pinning centers, move out of the grains to the grain boundaries, producing a reduction in coercivity (69). These effects have been modeled (19,69), using a modification of the Jiles–Atherton model of hysteresis (70). In addition, in the case where the creep damage is distributed nonuniformly, the model can be incorporated into a finite element formalism (71). This is important in the case of seam welds in steam piping because in such welds there is a greater weld width on the inside and outside of the pipe wall, but a smaller weld width in the wall interior. The stress in the pipe due to steam loading gets concentrated at the weld V inside the wall, and that is where creep damage begins. The finite element simulation shows that one can detect this creep damage, even when it is interior to the wall, by recording the induced secondary emf of a magnetic C-core detector (71).

Yet another potential use of hysteresis parameters is in NDE of fatigue damage caused by cyclic stress. In this case, the coercivity follows a very specific pattern. During the early stages of the fatigue process, the coercivity increases gradually, following a logarithmic dependence on the number N of stress cycles as: $H_c - H_{c_\eta} = b \ln N$. After a long period, a point is reached where the end failure process begins, after which the coercivity increases very rapidly (72,73). Figure 10 shows these effects. When the coercivity reaches the stage where it starts increasing rapidly, it is time to remove the test specimen from service before it fails. Remanence shows the opposite effect and decreases linearly with $\ln N$ (73).

Another microstructural effect is plastic deformation, where slipping and movement of dislocations under large stress results in dimensional changes in a specimen after the stress is removed. The stress at which plastic deformation starts is called the yield point. Swartzendruber et al. (74) have shown that in low carbon steels, the coercivity increases as the square root of plastic strain, where strain is defined as change in length divided by length.

It is anticipated that the monitoring of hysteresis parameters will be one of the preferred future methods in monitoring creep damage, fatigue damage, and plastic deformation.

Microstructural NDE via Barkhausen Effect and Magnetoacoustic Emission

The Barkhausen effect (BE) and magnetoacoustic emission (MAE) are related effects. The BE results from irreversible step-like changes in magnetization, produced mainly by sudden movement of 180° domain walls. The discontinuous change of magnetization generates a noise-like BE voltage proportional to the time derivative of the magnetic flux into a pick-up coil placed near the material being magnetized. The

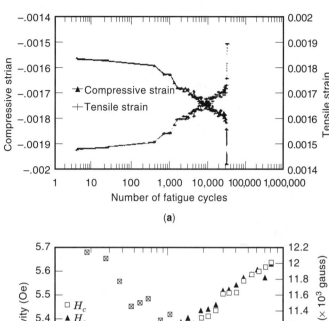

(a)

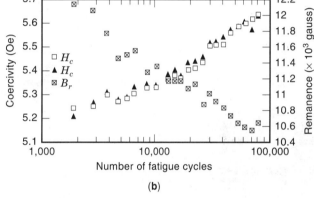

(b)

Figure 10. For A533B structural steel is depicted: (a) tensile and compressive strain as a function of number of stress cycles at fixed stress amplitude (272 MPa); (b) coercivity and remanence as a function of number of stress cycles (after Ref. 73).

MAE is caused by microscopic changes in local strain induced magnetoelastically during the discontinuous motion of non-180° walls. The acoustic waves so generated can be detected by a piezoelectric transducer bonded to the test material.

The BE is one of the most important magnetic NDE methods for investigating intrinsic properties of magnetic materials. Since its discovery in 1919 (75), the BE has been the subject of numerous investigations. The literature prior to 1975 is reviewed by Stierstadt (76) and by McClure and Schröder (77), who treat primarily the physical basis of the BE and its detection techniques. Reviews of Barkhausen applications in NDE include Refs. 2–6, 13, 78–81.

The as-received BE signal is influenced by BE electromagnetic wave signal propagation conditions in the material (82) as well by transducer properties (77,80,83). A typical BE voltage signal (U_e) induced in a pickup coil (wound on a low carbon steel bar) is shown in Fig. 11 (84). The signal was recorded during a half cycle of magnetic field H sweep, during which H increased at a constant time rate. The high frequency component (U_s) of the U_e signal, transformed to a dc-like envelope voltage signal U_b, depicts the BE intensity envelope. Its maximum usually occurs at or near coercivity field strength H_c. The MAE intensity envelope reveals mostly two maxima in the "knee" region of hysteresis. A BE and MAE measuring setup for NDE is shown schematically in Fig. 12. A biasing C-shaped magnet is cycled at low frequency (gener-

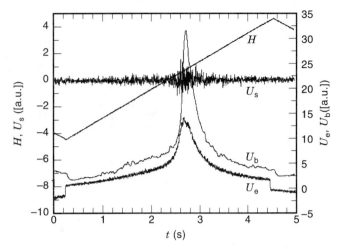

Figure 11. Schematic diagram showing time variation of the applied field H, the pickup coil voltage U_e, the BE component voltage U_s, and its envelope U_b (after Ref. 84).

ally less than 100 Hz). The BE voltage detected inductively is analyzed by a signal processor in the frequency range 1–300 kHz. The MAE probe (piezoceramic transducer, mostly resonant) provides the strain signal, which is analyzed in a range from 20 kHz to 1 MHz. It was established (85) that mostly large movement of non-180° DW or creation and annihilation of DW generates the MAE signals (86–89) while the BE signals are due to all kinds of DW movement, though mainly 180° in steel (85,90).

The various methods of BE and MAE signal processing and analysis can be separated into three main groups: (1) power spectrum analysis (77,91–93); (2) individual pulse analysis; and (3) integrated pulse analysis. Individual pulse analysis leads to pulse amplitude distribution (94), pulse amplitude and pulse duration distribution (95), and autocorrelation functions (96). The rms-like voltage (97) envelope of rectified pulses (98) and pulse count rate (67,99) signatures can be recorded as a function of applied field strength as part of individual pulse analysis. Single parameter evaluations of BE intensity such as mean value of rms (100), mean pulse amplitude (2,78), total number of pulses (101), and rectified single envelope maximum level are also used. Due to progress

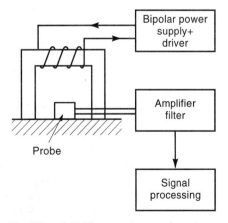

Figure 12. The BE and MAE transmitter and receiver system (after Ref. 4).

in microelectronics, BE measurement sets have been transformed from laboratory-like sets (102) to portable sets (78,81,103) and small compact (101) sets.

The BE intensity was also modeled in various ways. The rms-like parameter level was evaluated by Sakamoto et al. (104) and the power spectrum function by Alessandro et al. (92,93). Kim et al. have used a wall potential energy model for pulse amplitude (94).

The BE and MAE are dependent on the density and nature of pinning sites within the material (78). Precipitation of solute carbon as carbide is easily detected by BE analysis (67,78). Increase of particle size increases the stress field around the particle and the associated pinning effect causes a rise in BE intensity. The BE intensity maximum appears when the particle size is comparable with DW width (105). For larger particle size than DW width, new closure domains appear on the precipitate surface and magnetic structure becomes more detailed, leading to a decrease of BE pulse amplitude and BE intensity (78,97). The MAE intensity increases with increase of precipitate concentration located at grain boundaries (105).

Grain size affects the magnetic properties in two ways: first, by change of domain structure due to generation of closure domains (reverse spikes) at the grain boundaries and, second, by change of DW motion conditions because grain boundaries present obstacles to movement of DW (67). The DW moves further between pinning sites in increased size grains. An increase of grain size leads thus to an increase of rms peak level and total number of counts of the BE (67), pulse amplitude (106), pulse duration (107), and mean rms level (108). The role of grain size on BE power spectrum was discussed by Bertotti et al. in a theory in which BE is connected to the statistical properties of random local coercive fields experienced by a moving DW (109). The MAE signal intensity follows the same trend with grain size as does the BE signal (67,107).

Texture direction can be evaluated using BE due to the strong orientation effect of BE intensity (78). Komatsubara and Porteseil (110) found that the integral of BE power losses increases as a function of misorientation of grains against magnetization direction. Tiitto (108) found that BE level patterns vary systematically over a wide range of tested textures. Krause et al. (111) argued that the angular dependence of BE intensity in Si-Fe oriented steel was modulated by anisotropic internal fields that moderate 180° DW motion.

Plastic deformation changes considerably the BE intensity (78). The zero-stress BE intensity decreases within small tensile plastic strain (5,78,91,112,113) and increases for compressive plastic strain (5,13) indicating "compressive-like" and "tensile-like" residual stress, respectively, due to plastic strain. The BE and MAE intensities are reduced during annealing of plastically deformed steel due to dislocation density decrease (112).

The BE intensity is correlated with hardness level (4,5). A decrease of hardness of hardened parts is accompanied by an increase of BE intensity (5). Two frequency bands of BE signal filtering were used in order to evaluate surface hardening depth (4,114).

New approaches to NDE of microstructure are possible using MAE, in which Barkhausen pulses due to non- 180° DW jumps are detected during sample loading within the elastic range of stress (6,78,79,89,95,99,115). A distribution function of internal stress at microstructural defects is obtainable directly from strain dependence of the MAE intensity (99,116).

The BE and MAE are used to analyze microstructure change due to various thermal treatments (78,97,113,117). A decrease of dislocation density with tempering time was found to be correlated with different BE and MAE dependences on time (117). Buttle et al. tested the result of heat treatment of strained iron and have proposed simultaneous measurement of BE and MAE for NDE characterization of materials microstructure (112).

Structural degradation of industrial materials due to fatigue or creep is an important NDE application of the Barkhausen effect (2,11). Sundstrom and Torronen report preliminary results indicating that BE can be used for in-service inspection of high-temperature pipelines of ferritic materials used in power stations (2). The as-observed decrease of the BE intensity in the overheated areas of tested tubes was well correlated with reduction of hardness level. Lamontanara et al. have tested the influence of cycling load and plastic deformation on BE properties in boiler tubes correlating change of BE parameters with fatigue damage (11). Monotonic decrease of BE intensity was observed for power station tubes as a function of their exploitation time (118). Similar decrease in Barkhausen signal due to fatigue was found by Chen et al. (119).

Other applications of BE include grinding, shot-peening, and crack propagation. Grinding operations provide microstructure changes which can be detected by means of BE inspection (5,120). Tiitto looked at increase of BE intensity as an indication of grinding burns on a camshaft valve lobe (5) and ball bearing surface (108). Shot peening as a surface treatment for extending fatigue life can be controlled by means of BE inspection (108,121). McClure et al. (122) and Battacharya and Schröder (123) used both BE and MAE to detect discontinuous changes in magnetization of ferromagnets caused by fatigue crack propagation.

The BE and MAE methods have been clearly established as viable NDE techniques for NDE microstructural changes evaluation. The physical mechanisms for microstructural influences on the Barkhausen effect need to be further elucidated. Also, measurement conditions and signal processing should be delineated carefully so as to establish NDE procedure standards.

MAGNETIC NDE OF RESIDUAL STRESS

Inhomogeneous heat treatment due to welding and inhomogeneous plastic deformation during fabrication can leave strong residual stresses inside steel components (12). These stresses affect component service life because they can add to applied loads causing fatigue and failure. The residual stresses might also be beneficial. For example, railroad wheels have compressive residual stress built into the wheel rims to inhibit crack formation. Through braking and general use, the compressive stress in the wheel rim can change to tensile stress, which can cause cracks in the rim to widen (16,124). For these and many reasons, an NDE method for measuring stress is sought.

Of the NDE methods for measuring residual stress (125), none currently gives a complete map of stress field inside each component. In this section, we discuss magnetic properties that correlate with stress and note how they are used for NDE of residual stress.

Using Hysteresis Parameters for Residual Stress NDE

The effect of stress on magnetization has been studied for many years (126–137). The vast majority of work has been one-dimensional (i.e., stress axis, applied magnetic field, and magnetization all collinear along the same axis). Recently, noncoaxial stress, field, and magnetization has been investigated (137,138). In addition, there have been magnetic studies where the stresses are biaxial, that is, two stresses act independently along two perpendicular axes (18,139–146).

Two major types of magnetic processes have been studied—σH processes and $H\sigma$ processes. In σH processes, stress (σ) is first applied and then H is varied while stress is kept constant. A typical σH process is a magnetic hysteresis loop taken at constant stress. This would be the way NDE hysteresis measurements would be conducted. In the $H\sigma$ process, the field is first set at a constant nonzero value, and then magnetization varies as applied stress is varied (128,135). Discussion here is restricted to σH processes, as they apply to NDE measurements.

Three major model types account for hysteresis in materials undergoing the σH process. First, a macroscopic model has been developed by Sablik and Jiles (and others) for polycrystalline ferromagnets (17,132,136,137,147). Second, a micromagnetic model, taking into account domain wall types, has been developed for crystals and polycrystals by Schneider et al. (135), Schneider and Richardson (140), and Schneider and Charlesworth (148). Third, another micromagnetic model, using an energy formulation and a statistical formulation for the domains, is due to Hauser and Fulmek (134) and Hauser (149,150), who applied it mostly to crystalline and grain-oriented Fe(Si) alloy steel. In addition, Garshelis and Fiegel have proposed a simple nonhysteretic model for stress effects on magnetic properties (151). Early models were also given by Brown (152,153), and Smith and Birchak (154). A recent model relating stress to magnetic properties in thin magnetic films has been developed by Callegaro and Puppin (155).

The effects of stress on ferromagnetic materials is complicated, as several factors must be considered. For instance, it must be known whether the stress is within the elastic range of the material or whether it is plastically deforming the material. Also, one must know something about the nature of the magnetostriction—whether, for example, it is positive or negative.

Magnetostriction refers to the change in dimensions of ferromagnetic materials as they are magnetized. The relative change in dimensions is quite small ($\sim 10^{-5}$ or 10^{-6}) for most ferromagnetic materials and depends on the strength and orientation of the applied field. A material with positive magnetostriction increases in length along the magnetization direction. Conversely, a material with negative magnetostriction decreases in length along the magnetization direction.

A tensile stress, applied to a material with positive magnetostriction, will generally increase the magnetic induction B. The stress produces an effective magnetic field that acts in conjunction with the applied magnetic field and in effect adds to it. A compressive stress, applied to a material with positive magnetostriction, generally decreases magnetic induction B.

In this case, the stress produces an effective magnetic field that opposes the applied magnetic field.

Many ferrous alloys have a mixed type of magnetostriction depending on the applied magnetic field and stress. Iron, under zero stress, has a positive magnetostriction up to about 250 Oe (20 kA/m); above this, it has a negative magnetostriction (13). In alloys, the field that produces a change in sign of the magnetostriction will be different depending on stress and material composition. Nickel has a negative magnetostriction (126); this is true also of some ferrites (156).

In low magnetic fields and under tensile stress in the elastic range, ferrous alloys with positive magnetostriction show increased magnetic induction with increased applied stress. The effect is nearly linear until a magnitude of stress is reached at which the induction reaches a maximum, after which for higher stresses, induction is smaller. This appearance of this maximum under tensile stress is called the Villari effect (137,157). An explanation often advanced for this effect is a change in sign of the magnetostriction with increased stress (13). More exactly, it would be a change in sign of the derivative of the bulk magnetostriction with respect to the bulk magnetization ($d\lambda/dM$), because the stress effective field H_σ is predicted to be proportional to $d\lambda/dM$ (17,136,137,158). Other explanations have been also given for the Villari effect (135). One of these is that as tensile stress is increased, the domains antiparallel to the magnetization tend to shrink and disappear, with the result that the magnetostriction tends not to change as much under tensile stress, leading to a shrinking $d\lambda/dM$ which tends to produce a maximum in the magnetic induction (147). Under compression, a stress-caused demagnetization term $-D_\sigma M$ comes about because compression produces spatial divergence of magnetization near grain boundaries (viz., $|\nabla \cdot M| \neq 0$), which in turn results in magnetic poles at the grain boundaries, producing a demagnetization field that subtracts from the magnetic induction (147). The effect of the demagnetization term is so strong for applied compressive stresses that it counteracts the effect of reduction in $d\lambda/dM$. For this reason, the Villari extremum is not seen under compressive stress. The behavior is thus asymmetric (147). The fact that a maximum appears in the magnetic induction under tensile stress and coaxial field H complicates NDE measurement of stress.

Even more complications come about when stress and field are noncoaxial (i.e., stress axis and field are at an angle θ with respect to each other). Generally, in an isotropic or polycrystalline ferromagnetic material, an angle θ can be found for which stress causes no change in magnetic properties when stress and field are noncoaxial (137,138). If the field is *perpendicular* to the stress axis, then for σH processes, increased stress under tension produces *decreased* magnetic induction for each value of H, and vice versa for compression (130,131,137,138). Furthermore, the Villari extremum shifts to negative (i.e., compressive) stresses (137). Because of all these extra complications, an NDE measurement of stress must be carefully designed.

The process is simplified if the direction of the stress axis is known (for, in that case, the magnetic field can be applied parallel to the stress axis) and a low enough field can be applied that the magnetic properties vary canonically with stress. In such a situation, for positive magnetostriction, the remanence B_r should be increased linearly with increased tensile stress and decreased linearly with increased compressive

stress; similarly, for B_{max} and M_{max} (136). The coercivity, on the other hand, is decreased with increased compressive stress (137).

In the small field–small stress range, one can obtain the stress from the hysteresis parameters, provided one calibrates the parameters for the unstressed material and knows the stress axis direction. This works well if there is little variation in properties from sample to sample (as often happens for commercial steels). If the stress axis direction is unknown, the angular variation of the magnetic properties can be used to determine the stress axis direction (137).

Measuring stress using hysteresis parameters is a *bulk* stress measurement, usually with an error that is ± 5–10% of yield point stress. The bulk measurement is unlike X-ray and Barkhausen noise measurements, which yield only *surface* stress (2,5).

Generally speaking, hysteresis loops, true to the specimen, are measured when a cylindrical specimen is wrapped inside an excitation coil, with a secondary coil also wrapped around the specimen in the center of excitation coil, and with a Hall probe positioned next to the specimen surface to measure H. This approach, called the permeameter approach, most assuredly measures the intrinsic magnetic properties of the specimen. Alternating current I is applied to the excitation coil and the resulting alternating magnetic induction B in the specimen induces a voltage in the secondary coil, the signal from which is then phase-adjusted to be in synchronization with the Hall voltage detected, so that a hysteresis loop can be generated. Quasi-dc properties can be determined by using low frequencies of the order 0.5 Hz to 2 Hz.

An NDE field probe, in most cases, must be small and portable, and so, out in the field, the probe is usually a C-core, that is, an electromagnet, in the shape of either a circular or squared-off C, for which the pole pieces are designed to be flush with the sample. The hysteresis loop, measured with a secondary coil wrapped around one of the pole pieces close to the specimen, is not the true hysteresis loop of the specimen; however, that does not matter if stress is being measured because the loop that is measured is influenced by the stress acting on the specimen and will vary proportionally to the true stress-influenced hysteresis loop of the specimen. Hence, the variations of the hysteresis loop obtained with a C-core can still be calibrated with the stress, provided the C-core is always flush against the specimen. If there is a variable air space (liftoff) between the C-core and the specimen, there will be flux leakage and the method will not be reliable. Thus, NDE of stress (usually with a C-core) must also contend with this liftoff issue (159).

There are many papers that discuss NDE of stress by using one or more of the hysteresis parameters. The earliest seems to be that of Ershov and Shel (160), who used magnetic permeability to measure tensile stress in steel with the magnetic field both perpendicular and parallel to the stress axis. Abiku and Cullity (161), and later Abiku (162), used permeability to measure stress in steel and nickel. Musikhin et al. (163) used the coercive field as an NDE indicator of stress. Devine (164) described the detection of stress in railroad steels via many magnetic property measurements, using remanence, coercivity, maximum differential permeability, and hysteresis loss. For the case of biaxial stress, a method has been suggested by Sablik (18) for measuring the difference in the biaxial stresses using hysteresis parameter measurements for cases where the parameters are measured first with the magnetic field parallel to one of the stress axes and then perpendicular to it. For an absolute measurement of the stresses along both axes, a technique with the magnetic field perpendicular to the biaxial stress plane is discussed (146).

Nonlinear Harmonic Method (NLH)

The magnetic induction B (magnetic flux density) of a ferromagnetic material, when subjected to a sinusoidally varying field H, is not sinusoidal but distorted. This is due to nonlinear, hysteretic variation of the magnetic induction with field H. Figure 13 illustrates how a sinusoidal H produces a nonsinusoidal induction B because of magnetic hysteresis (9).

The distorted waveform of B contains odd harmonics. The reason that only odd harmonics of B are present is because B must satisfy (165)

$$B(t \pm (T/2)) = -B(t) \qquad (2)$$

because, as seen in Fig. 13, the waveform of B repeats itself over the second half of the cycle but is negative. In Eq. (2), T is the period of the waveform and is equal to $T = 1/f = 2\pi/\omega$, where ω is angular frequency and f is frequency. Each harmonic has a period $\tau_n = T/n$. Thus, when $n = 2$, $B(t \pm (T/2)) = B(t \pm \tau_n) = B(t)$ is not Eq. (2); when $n = 3$, $B(t \pm (T/2)) = B(t \pm (3\tau_n)/2)) = -B(t)$. It is seen that only odd n can satisfy Eq. (2).

Because stresses influence the magnetic hysteresis, it follows that the harmonic content of the magnetic induction is also sensitive to the stress. With NLH, these harmonic frequencies are detected, and their amplitudes are related to the state of stress in the material (166,167).

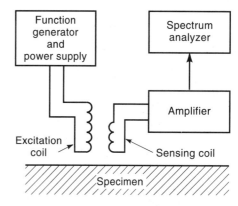

Figure 14. Block diagram illustrating nonlinear harmonics instrumentation (after Ref. 9).

The NLH instrumentation is shown schematically in Fig. 14 (9). The magnetic field H is applied to the specimen with an excitation coil and the resulting magnetic induction measured with a sensing coil. A C-core setup or wrapped coil about a cylindrical specimen can be used. The sinusoidal excitation current is supplied by a function generator (or oscillator) and power amplifier. The induced voltage in the sensing coil is amplified and, via the use of filters, the harmonic signals are separated from each other, amplified, and analyzed. Typically, the third, fifth and possibly seventh harmonic signal can be displayed. However, only the third harmonic signal is used for stress determination, because that usually has the largest amplitude.

The harmonic amplitudes depend not only on stress, but also on relative orientation of stress and field. When positive magnetostriction applies, the harmonic amplitude increases with tension when stress and field axes are parallel and decreases when they are perpendicular.

The NHL technique senses stresses with sensing depth near the ac skin depth. Because skin depth is a function of frequency, the sensing depth can be varied with the frequency. By using quasi dc frequencies, a near bulk measurement is also possible.

The NLH measurements are sensitive to factors unrelated to stress, such as microstructure, heat treatment, and material variables. If a C-core is used, then the possibility of an air space between probe and sample can cause problems, particularly on a curved surface such as a pipe. All of these other factors must be considered when doing the NLH measurement.

This technique is usually effective to a range of stress of up to about 50% of the yield stress, and the accuracy of the technique is about $\pm$ 35 MPa ($\pm$ 5 kpsi). At stress levels of higher than 50% of yield stress, the NHL response tends to saturate. With this technique, stress can be measured while scanning at high speed [~10 m/s, (or approximately 30 ft/s)] (167). This technique thus has a potential for rapidly surveying stress states in pipelines or continuously welded rail (167,168). A simple model for simulating NLH analysis of stress may be found in Ref. 158.

Stress Induced Magnetic Anisotropy (SMA)

In the absence of stress, a polycrystalline ferromagnetic material without texture will have isotropic magnetic properties

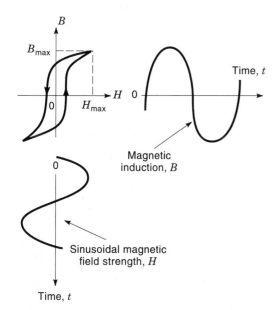

Figure 13. Distortion of the magnetic induction B caused by nonlinearity in hysteresis. The curve for B consists of a fundamental and higher order odd harmonics (after Ref. 9).

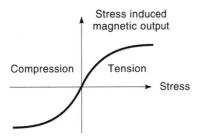

Figure 15. B_n/B_p ratio against stress, showing continuous change from compressive to tensile loads (after Ref. 20).

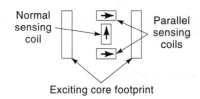

Figure 16. Concept diagram for Langman's SMA measurement (after Ref. 20).

independent of the direction of measurement. In the presence of stress, this is no longer true, and the material becomes magnetically anisotropic, an effect known as stress-induced magnetic anisotropy (SMA) (141). For mild steel, the peak magnetic flux density ratio between directions parallel and perpendicular to the stress axis can be as high as 5 (130). As mentioned earlier, the physics for understanding the difference in magnetic response in different directions has been developed by Sablik et al. (137). Although any magnetic method for measuring stress could be considered an SMA technique, the term is usually reserved for techniques that simultaneously measure magnetic properties in perpendicular directions.

In a series of papers (16,130,139,169,170) Langman describes an SMA technique based on measuring the angle between magnetic field intensity H and magnetic flux density B. Magnetic permeability μ, a scalar in an isotropic ferromagnetic material but dependent on the magnitude of H, becomes a tensor in the presence of stress and $\boldsymbol{H}$ and $\boldsymbol{B}$ vectors no longer are parallel in the general case. $\boldsymbol{B}$ will be canted with respect to $\boldsymbol{H}$. Langman's SMA technique detects the induced flux normal to the applied field, which in a magnetically isotropic material would be zero. For analysis purposes, Langman considers the ratio between the flux density component (B_n) normal to the applied field and the flux density component (B_p) parallel to it (viz., B_n/B_p). In Fig. 15, it is shown how the ratio B_n/B_p changes continuously as the stress changes from compression (negative stress) to tension (positive stress). For biaxial stresses, this ratio is proportional to the algebraic difference of the two biaxial stresses (viz., $B_n/B_p = f(\sigma_1 - \sigma_2)$, where σ_1 is the stress along one axis and σ_2 is the stress along the other). A plot similar to that of Fig. 15 is found, where now the abscissa is $\sigma_1 - \sigma_2$. Note that the behavior seen in Fig. 15 is linear at low values of stress, becoming nonlinear at about 1/3 of the yield strength, after which the response shows a tendency to saturation. A similar behavior was encountered also for NLH in the last section. An error margin reported for this type of measurement is 20 MPa in the difference between principal stresses and roughly 5° in their direction, when the stresses are biaxial (20).

Figure 16 shows a concept diagram for Langman's SMA probe. The pole pieces of the core of the magnetizing coil induce a strong field in one direction. A modulation frequency of between 30 and 80 Hz is used, which is equivalent to an inspection depth of about 0.5 mm in mild steel. Two air-cored pickup coils parallel to and close to specimen's surface H are placed on either side of a third air-cored pickup coil, which is perpendicular to exciting field H. The outputs of these coils are translated into the ratio B_n/B_p. The whole rig is rotated

obtaining a sinusoidal output at twice the rotation rate with extremes in value at 45° to the principal stress axes, in the case of biaxial stress. By pinpointing the directions in which these extremes occur, one can locate the stress axes. Typical excitation fields are of the order of several hundred A/m (139). Difficulties are encountered when the inspection surface is not flat.

Another type of SMA probe, used in Japan, is known as a magnetic anisotropy sensor (MAS) (171–173). It differs from Langman's probe in that it consists of two perpendicularly positioned magnetic cores instead of one magnetic core and three air cores. Figure 17 shows the basic construction. In the case of the Kashiwaya MAS probe (171), the detector core has an air space (liftoff) between its pole pieces and the specimen. The finite air space makes the detector less sensitive to variations in liftoff. In the case of the Wakiwaka et al. (172) and Kishimoto et al. (173) probe, neither exciter core nor detector core has any built-in liftoff. These authors provide an analysis, which considers the reluctances of flux paths and analyzes the result from the point of view of an equivalent magnetic circuit. Again, the MAS output voltage is largest at 45° with respect to the principal stress axes, having a cloverleaf representation on a polar plot (see Fig. 18). Note that the cloverleaf increases in size as frequency is increased. Figure 19 shows a plot of the MAS output voltage vs stress, for relatively low stress values, for which the response is fairly linear.

Most applications of SMA (or MAS) have been in mild steel for the railroad industry. Measurements of railroad rail longitudinal stresses were performed in Japan (171). Stress differ-

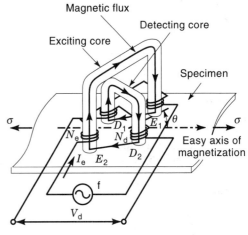

Figure 17. Basic construction of the MAS sensor (after Ref. 173).

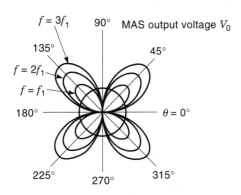

Figure 18. Calculated frequency dependence of magnetic anisotropy signal patterns (after Ref. 173).

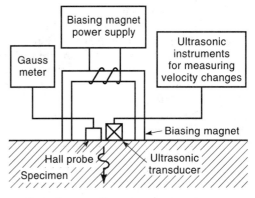

Figure 20. Block diagram of instrumentation for measuring MIVC (after Ref. 9).

ences due to day-night heating and cooling of railroad rail were measured. A three-point probe (174) for curved surfaces was used for biaxial stress measurement in a railroad car axle. Langman used his SMA technique to do field studies of stresses in railway wheels (16).

Magnetically Induced Velocity Changes (MIVC) for Ultrasonic Waves

In MIVC, the dependence of the elastic moduli on the magnetization is exploited as an NDE technique. One utilizes this dependence by passing ultrasonic waves through the magnetized material and measuring change in transit time between when the material is magnetized and when it is not. The elastic moduli are affected not only by a magnetic field but also by stress (which changes the magnetization). Thus, stress changes the MIVC. Indeed, the MIVC for ultrasonic waves is not only dependent on stress but also on the angle between the stress direction and the direction of the applied magnetic field (175,176). The characteristic stress dependence of the MIVC is used for stress determination (9,14,175–178).

Figure 20 shows a diagram of the instrumentation for measuring MIVC. An electromagnet supplies a biasing magnetic field H to the specimen. The applied field H is measured with a Hall probe. A transducer transmits ultrasonic waves to the specimen and detects signals reflected from the back of the specimen. For surface waves, separate transmitting and receiving transducers are used. The shift in arrival time of the received ultrasonic wave, caused by velocity change due to

field and stress, is measured using the phase comparison technique (175).

Typical plots are seen in Fig. 21 for longitudinal MIVC = $\Delta v/v_0$ against applied field H, where v_0 is the velocity in the absence of H, and $\Delta v = v - v_0$ is the change in velocity due

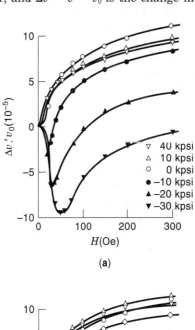

(a)

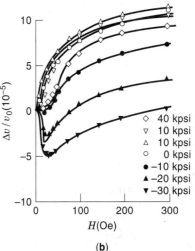

(b)

Figure 21. Longitudinal wave velocity vs magnetic field H for various stress levels in A-514C steel with H (a) parallel and (b) perpendicular to the stress axis (after Ref. 177).

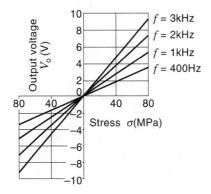

Figure 19. Calculations of MAS output voltage V_0 vs stress σ for low-carbon steel (after Ref. 173).

to the presence of H. Figure 21(a) applies when H is parallel to the stress axis for positive magnetostriction materials. It is noted in Fig. 21(a) that under uniaxial tension ($\sigma > 0$), the MIVC is decreased from its $\sigma = 0$ value but stays positive and gets larger with increased H, ultimately saturating. Compression ($\sigma < 0$) results in a reduction of the MIVC from its $\sigma = 0$ value, but with the MIVC starting out negative, reaching a minimum, and then increasing, finally reaching positive values at large H. The more negative the compression, the deeper the minimum. For $\boldsymbol{H}$ perpendicular to the stress axis [Fig. 21(b)], the MIVC behavior under compression is similar but, under tension, the MIVC is larger at small tensions, and smaller at large tensions than the case for $\sigma = 0$.

The detailed dependence of the MIVC on stress varies depending on the sign of stress (tensile or compressive, i.e., positive or negative), the stress type (uniaxial or biaxial), the angle between the stress axis and the applied magnetic field, the wave mode used (shear, longitudinal, or surface), and material grades. Generally, NDE studies (14,177,178) have shown that an unknown stress in the material can be characterized (magnitude, direction and sign) utilizing the known stress dependences of the MIVC. The MIVC has been used to measure residual welding stresses (177), residual hoop stresses in railroad wheels (178), and through-wall detection of biaxial stresses in operating pipelines (14). In the case of biaxial stress, it is found that the MIVC works better for compression than for tension (145). Thus, MIVC would complement other measurements of biaxial stress that work between under tension than compression (146).

The MIVC technique can be used to measure bulk or surface stresses by applying bulk (shear or longitudinal) or surface ultrasonic waves. A measurement can be made in a few seconds. However, because MIVC depends on material, reference calibration curves need to be established for the material. The technique has the advantage of being insensitive to variations in texture and composition of nominally the same material. The accuracy in stress measurement is similar to that of other methods ($\pm$ 35 MPa or $\pm$ 5 kpsi). One disadvantage of the technique is that a relatively large electromagnet is needed to magnetize the part under investigation, which may be cumbersome in practical application. Also, because of the difficulty in magnetizing parts of complex geometry, the application of the technique is limited to relatively simple geometries.

Barkhausen Effect and Magnetoacoustic Emission for Residual Stress NDE

The Barkhausen effect (BE) and magnetoacoustic emission (MAE) are sensitive to stress, making them important, truly nondestructive, portable and fast alternative NDE tools for residual stress measurements. The origin of stress dependence lies in the interaction between strain and local magnetization. Under uniaxial stress, the results consistently show that tension increases, while compression decreases, the BE intensity for positive magnetostriction materials (77,78). Shibata and Ono (179) and Burkhardt et al. (180), in early works on MAE, revealed that MAE intensity decreases under *both* tensile and compressive stress. Rautioho et al. tested the impact of microstructure on the stress dependence of BE (181). Examples of studies of uniaxial load on BE and MAE intensity are reported in Refs. 182–185.

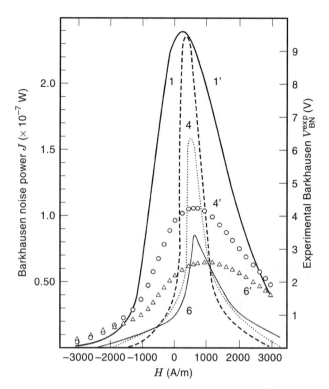

Figure 22. Measured BE intensity (1,4,6) and computed BE power (1′,4′,6′) as a function of H with stress (in MPa) acting as: (1) + 114, (4) −37, (6) − 162 (after Ref. 187).

The impact of uniaxial stress on BE intensity was modeled by Tiitto (5,108), utilizing statistical consideration of the domain magnetization vector distribution under load. Sablik (186) applied the magnetomechanical hysteresis model of Sablik and Jiles (136) to compute the normalized BE signal from the derivative of the irreversible component of magnetization, utilizing the BE power spectrum model of Allesandro et al. (92). Sablik's model was recently applied to fit results of BE measurements (187). Figure 22 shows the result of comparison of experimental and computed BE envelopes for uniaxial load of low carbon steel.

The BE intensity depends on the angle between uniaxial load and magnetization direction. When stress and field directions are parallel, tension causes an increase in BE intensity while compression causes a decrease; when the field is perpendicular to the stress axis, the effects of stress are reversed (78). The angular dependence of BE was evaluated by Kwun (138) and by Sablik (17). Stress dependence of BE intensity has been applied to load sensor design (81,188).

With biaxial load, the transverse tensile stress mostly decreases and compressive stress increases the BE intensity (78). In practice, biaxial calibration of BE intensity as a function of applied strains utilizes cross-shaped samples and four bending point modes of load (108,189).

Evaluation of stress due to welding is an example of BE and MAE industrial application (101,108,190–192). The BE intensity measurements are performed at a given point in two directions: along and across the weld seam, assuming that one of these directions is parallel to the main stress axis. Enhancement of stress resolution was achieved using the number of BE pulses as the BE intensity parameter (193). Figure

23 presents a residual stress distribution in a direction perpendicular to weld seam line, evaluated via the number of BE pulses. The BE results for weld stress analysis have been confirmed by NDE X-ray analysis as well by the hole drilling method (108,193). Stress dependence of BE signal in pipeline steels were tested by Jagadish et al., using rms signal, pulse height distribution and power spectra (194). Special application of BE stress evaluation to roll surface inspection is reported in Ref. 78.

Since the discovery that the Barkhausen effect can be used for NDE of stress (195), the Barkhausen effect has become one of the usual techniques for NDE of residual stress (100,101) and commercial apparatus sets are available for this usage. The technique has a drawback in that the measured stress distributions are near the surface. The high-frequency electromagnetic signals generated by domain wall motion in the interior of the specimen are quickly attenuated by eddy currents before they reach the surface. The effective depth for stress detection is about 0.5 mm. When bulk stress evaluation is needed, another NDE method might be better.

The MAE can in principle be used for bulk residual stress measurement because the acoustic waves generated by domain wall motion do not attenuate as rapidly as electromagnetic BE waves. However, there are problems. First, the MAE transducer used for detection requires a couplant to the specimen, whereas the BE technique is noncontacting. There are also signals coming from surface-reflected waves that must be separated from signals coming directly from moving domain walls. Also, MAE intensity is not a monotonic function of strain or stress in the near stress region [In fact, it peaks at zero stress (179,180)]. More research is needed on MAE.

Other Magnetic Methods for Residual Stress NDE

Another NDE method for stress evaluation is the incremental permeability technique (4). This refers to a time varying change of the magnetic field superimposed on bias field H. In effect, the flux density is varied along a minor loop originating

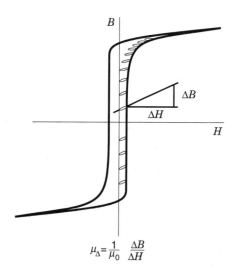

$$\mu_\Delta = \frac{1}{\mu_0} \frac{\Delta B}{\Delta H}$$

Figure 24. Plot showing how the incremental permeability is obtained from the average slope of a minor loop (after Ref. 4).

at a field point H on a major hysteresis loop. Figure 24 shows this type of variation, starting at many different points on the major loop (4). Each of the different minor loops has an average slope $\Delta B/\Delta H$, which depends on the stress that is acting. The slope is called the incremental permeability. Division by μ_0 produces the incremental "relative" permeability. Incremental permeability is dependent not only on stress but also on microstructure, and has been used for magnetic NDE assessment of hardness (196). The incremental permeability is called "reversible" permeability (197) because variation of the field along the minor loop is small enough that the change in magnetization is due to domain wall bowing and bending, which is a reversible process. Imposition of a radio frequency time-varying signal on the bias field H yields essentially the same technique, but when rf is used, it is called *magabsorption* (8,197). In that case, the time variation of the B–H slope (permeability) is monitored on an oscilloscope, and the maximum slope (instead of the average slope) is correlated with stress. A magnetomechanical hysteresis model has been given for the magabsorption (197).

Yet another variant involving the use of permeability for stress measurement is the differential effective permeability (DEP) technique (15). In this case, the initial permeability is effectively used to measure stress. A small-amplitude time-varying $H(t)$ and $H = 0$ produces what is known as a Rayleigh loop. The Rayleigh loop effectively corresponds to a minor loop in the incremental permeability technique, except that it is centered about $H = 0$. The slope of the loop depends on stress. The DEP technique has been used for biaxial stress management (15).

Another approach, which has not yet been fully implemented, is to exploit the dependence of the magnetostriction on stress. Although the dependence of magnetostriction on stress tends to be more nonlinear than many of the other properties, magnetostriction offers the possibility of additional NDE characterization in a multiparameter investigation (102).

PROMISING NEW MAGNETIC TECHNIQUES

One potentially new NDE technique is magnetic force microscropy (MFM) (198). The MFM involves sensitively mapping

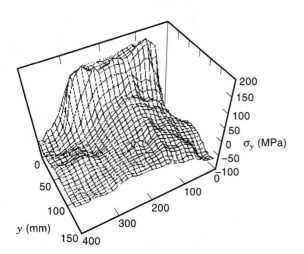

Figure 23. A 3D presentation of the residual stress over the welded plate in a direction perpendicular to the weld seam, as evaluated from the BE measurements (after Ref. 189).

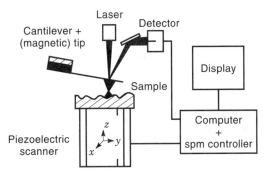

Figure 25. Schematic for a scanning probe magnetic force microscope (after Ref. 198).

surface magnetic fields (and thus surface topography) over a 20 μm size area (for example). The scanning probe has a sharp point and is mounted on a weak cantilever. (See Fig. 25.) Magnetic interactions between the scanning tip and the specimen cause the cantilever to deflect. This deflection is detected by reflecting a laser beam off the back of the cantilever into a position-sensitive photodiode. A map of the cantilever deflection obtained from the photo output gives an image of the surface topography. Because the instrumentation can be transportable, MFM is promising as an NDE tool for determining surface microstructure. A less sensitive form of MFM should help in measuring magnetic leakage fields and in locating cracks, corrosion pits, and other microscopic flaws. Two drawbacks remain: (1) frequent replacement of the tip, which wears easily; and (2) the necessity for calibration of the tip at the start and end of each daily usage. Another issue concerns the need for surface preparation. Also, MFM and related techniques at present are very costly, and interpretation is still currently difficult.

A second technique that shows promise for NDE involves use of commercial high-T_c SQUIDs. The high-T_c superconducting quantum interference device (SQUID) needs liquid nitrogen as a coolant (rather than liquid helium). This means that it could be used as a portable field device as liquid nitrogen is fairly cheap. Because a SQUID can measure magnetic flux densities very precisely (199), it could be used with nonferrous metals as a magentic leakage field detector in the case of the electric current perturbation technique (199). Another use would be detection of fatigue damage in nonferromagnetic stainless steels (200), where fatigue causes formation of ferritic steel regions, which are ferromagnetic and enhance the overall magnetic field detected for the stainless steel.

BIBLIOGRAPHY

1. F. W. Dunn, Magnetic particle inspection fundamentals, *Mater. Eval.,* **35**: 42–47, Dec. 1977.

2. O. Sundstrom and K. Torronen, The use of Barkhausen noise analysis in nondestructive testing, *Mater. Eval.,* **37**: 51–56, Feb. 1979.

3. J. F. Bussiere, On-line measurement of the microstructure and mechanical properties of steel, *Mater. Eval.,* **44**: 560–567, Apr. 1986.

4. P. Holler, Nondestructive analysis of structure and stresses by ultrasonic and micromagnetic methods. In J. F. Bussiere, J. P. Monchalin, C. O. Ruud and R. E. Green, Jr., (eds.), *Nondestruc-*

tive Characterization of Materials II, New York: Plenum, 1987, pp. 211–225.

5. K. Tiitto, Use of Barkhausen effect in testing for residual stresses and defects. In W. B. Young (ed.), *Residual Stress in Design, Process, and Materials Selection,* Metals Park OH: ASM Int'l 1987, pp. 27–36.

6. D. C. Jiles, Review of magnetic methods for nondestructive evaluation, *NDT International,* **21**: 311–319, 1988.

7. R. E. Beissner, Magnetic field testing. In S. R. Lampman and T. B. Zorc (eds.), *Metals Handbook,* Vol. 17, Metals Park, OH: ASM Int'l, 1989, pp. 129–135.

8. W. L. Rollwitz, Magaborption NDE. In S. R. Lampman and T. B. Zorc (eds.), *Metals Handbook,* Vol. 17, Metals Park, OH: ASM Int'l 1989, pp. 144–158.

9. H. Kwun and G. L. Burkhardt, Electromagnetic techniques for residual stress measurements. In S. R. Lampman and T. B. Zorc (eds.), *Metals Handbook,* Vol. 17, Metals Park, OH: ASM Int'l, 1989, pp. 159–163.

10. D. C. Jiles, Review of magnetic methods of nondestructive evaluation (Part 2), *NDT International,* **23**: 83–92, 1990.

11. J. Lamontanara et al., Monitoring fatigue damage in industrial steel by Barkhausen noise, *Nondestr. Test. Eval.,* **8–9**: 603–614, 1992.

12. D. J. Buttle and T. M. Hutchings, Residual stress measurement at NNDTC, *Brit. J. NDT (now Insight)* **34**: 175–182, 1992.

13. M. K. Devine, The magnetic detection of material properties, *J. Metals,* 24–30, Oct. 1992.

14. H. Kwun, Application of magnetically induced velocity changes of ultrasonic waves for NDE of material properties, *Nondestr. Test. Eval.,* **10**: 127–136, 1992.

15. C. B. Scruby et al., Development of non-invasive methods for measurement of stress in welded steel structures, *Eur. J. NDT,* **3** (2): 46–54, 1993.

16. R. A. Langman and P. J. Mutton, Estimation of residual stresses in railway wheels by means of stress-induced magnetic anisotropy, *NDT&E International,* **26**: 195–205, 1993.

17. M. J. Sablik, Hysteresis modeling of the effects of stress on magnetic properties and its application to Barkhausen NDE. In *Current Topics in Magnetics Research,* Vol. 1, Trivandrum, India: Research Trends, 1994, pp. 45–57.

18. M. J. Sablik, Modeling the effects of biaxial stress on magnetic properties of steels with application to biaxial stress NDE, *Nondestr. Test. Eval.,* **12**: 87–102, 1995.

19. M. J. Sablik and D. C. Jiles, Magnetic measurement of creep damage: modeling and experiment. In M. Prager and R. E. Tilley (eds.), *Nondestructive Evaluation of Utilities and Pipelines,* Vol. 2947, SPIE Proc., Bellingham, WA: SPIE, 1996, pp. 166–176.

20. J. A. Alcoz, S. Nair, and M. J. Sablik, Electromagnetic methods for stress measurement, *Nondestructive Testing Handbook,* Vol. 9, R. K. Stanley and P. O. Moore (eds.), Columbus, OH: ANST, 1996, pp. 421–430.

21. C. E. Betz, *Principles of Magnetic Particle Testing,* Chicago: Magnaflux Corp., 1967.

22. P. A. Tipler, *Physics,* New York: Worth Publishers, 1976, p. 858.

23. Y. F. Cheu, Automatic crack detection with computer vision and pattern recognition of magnetic particle indicators, *Mater. Eval.,* **42**: 1506–1511, 1984.

24. F. Forster, Developments in magnetography of tubes and tube welds, *Nondestructive Testing,* **8**: 304–308, 1975.

25. C. H. Hastings, A new type of flaw detector, *ASTM Proc.,* **47**: 651–664, 1947.

26. K. F. Bainton, Characterizing defects by determining leakage fields, *NDT International,* **10**: 253–257, 1977.

27. R. E. Beissner, G. A. Matzkanin, and C. M. Teller, NDE application of magnetic leakage field methods. *SwRI Report NTIAC-80-1*, NTIAC, Southwest Research Institute, San Antonio, TX, 1980.

28. M. J. Sablik and R. E. Beissner, Theory of magnetic leakage fields from prolate and oblate spheroidal inclusions, *J. Appl. Phys.*, **53**: 8437–8450, 1982.

29. T. A. Bubenik et al., Magnetic flux leakage (MFL) technology for natural gas pipeline inspection, *Gas Research Institute Report 91-0367*, GRI, Chicago, IL, 1992.

30. P. A. Khalileev and P. A. Grigorev, Methods of testing the condition of underground pipes in main pipelines, *Sov. J. NDT*, **10**: 438–459, 1974.

31. N. N. Zatsepin and V. E. Shcherbinin, Calculation of the magnetic field of surface defects, I. Field topography of defect models, *Sov. J. NDT*, **2**: 385–393, 1966.

32. V. E. Shcherbinin and N. N. Zatsepin, Calculation of the magnetic field of surface defects, II. Experimental verification of the principal theoretical relationships, *Sov. J. NDT*, **2**: 394–399, 1966.

33. C. Edwards and S. B. Palmer, The magnetic field of surface breaking cracks, *J. Phys. D*, **19**: 657–673, 1986.

34. J. H. Hwang and W. Lord, Finite element modeling of magnetic field-defect interactions, *J. Test. Eval.*, **3**:21–25, 1975.

35. W. Lord et al., Residual and active leakage fields around defects in ferromagnetic materials, *Mater. Eval.*, **36**: 47–54, July 1978.

36. D. L. Atherton and W. Czura, Finite element calculations on the effect of permeability variation on magnetic flux leakage signals, *NDT Int.*, **20**: 239–241, 1987.

37. D. L. Atherton, Finite element calculations and computer measurements of magnetic flux leakage patterns for pits, *Brit. J. NDT*, **30**: 159–162, 1988.

38. B. Brudar, Magnetic leakage fields calculated by the method of finite differences, *NDT Int.*, **18**: 353–357, 1985.

39. G. Dobmann, Magnetic leakage flux techniques in NDT: a state of the art survey of the capabilities for defect detection and sizing. In W. Lord (ed.), *Electromagnetic Methods of NDT*, New York: Gordon and Breach, 1985, pp. 71–95.

40. P. Holler and G. Dobmann, Physical analysis methods of magnetic flux leakage. In R. S. Sharpe (ed.), *Res. Techniques NDT*, Vol. IV, New York: Academic Press, 1980, pp. 39–69.

41. C. N. Owston, The magnetic flux leakage technique of nondestructive testing, *Brit. J. NDT*, **16**: 162–168, 1974.

42. F. Forster, New findings in the fields of nondestructive magnetic field leakage inspection, *NDT Int.*, **19**: 3–14, 1986.

43. R. E. Beissner et al., Analysis of mechanical damage detection in gas pipeline inspection, *Proc. Conf. Prop. Applic. Magnetic Materials,* Illinois Institute of Technology, Chicago, IL, May 1996.

44. T. W. Krause et al., Variation of the stress dependent magnetic flux leakage signal with defection depth and flux density, *N.D.T.&E. Int.*, **29**: 79–86, 1996.

45. T. W. Krause et al., Effect of stress concentration on magnetic flux leakage signals from blind hole defects in stressed pipeline steel, *Res. Nondestr. Eval.*, **8**: 83–100, 1996.

46. R. E. Beissner et al., Detection and analysis of electric current perturbation caused by defects. In G. Birnbaum and G. Free (eds.), *Eddy Current Characterization of Materials and Structures*, ASTM ATP 722, Philadelphia: ASTM, 1981, pp. 428–446.

47. R. E. Beissner, M. J. Sablik, and C. M. Teller, Electric current perturbation calculations for half-penny cracks. In D. O. Thompson and D. E. Chimenti (eds.), *Rev. Progr. In Quant. NDE,* Vol. 2B, New York: Plenum, 1983, pp. 1237–1254.

48. J. A. Birdwell, F. N. Kusenberger, and J. R. Barton, Development of magnetic perturbation inspection system (A02G5005-1) for CH-46 rotor blades, *P.A. No. CA375118,* Technical Summary Report for Vertol Division, The Boeing Company, 1968.

49. J. R. Barton, J. Lankford, and P. L. Hampton, Advanced nondestructive testing methods for bearing inspection, *SAE Trans.*, **81**: 681–696, 1972.

50. H. Kwun and C. M. Teller, Nondestructive evaluation of pipes and tubes using magnetostrictive sensors, U.S. Patent No. 5,581,037, December 1996.

51. H. Kwun, J. J. Hanley, and C. M. Teller, Performance of a noncontact magnetostrictive AE sensor on steel rod, *J. Acoust. Emission*, **11**: 27–31, 1993.

52. H. Kwun and A. E. Holt, Feasibility of underlagging corrosion detection in steel pipe using the magnetostrictive sensor technique, *NDT&E Int.*, **28**: 211–214, 1995.

53. H. Kwun and J. J. Hanley, Long-range, volumetric inspection of tubing using the magnetostrictive sensor technique, *Proc. 4th EPRI Balance-of-Plant Heat Exchanger NDE Symposium,* Jackson Hole, Wyoming, 1996.

54. H. Kwun and C. M. Teller, Detection of fractured wires in steel cables using magnetostrictive sensors, *Mater. Eval.*, **52**: 503–507, 1994.

55. K. A. Bartels, H. Kwun, and J. J. Hanley, Magnetostrictive sensors for the characterization of corrosion in rebars and prestressing strands. In *Nondestructive Evaluation of Bridges and Highways,* SPIE Conf. Proc. 2946, SPIE, Bellingham, WA, 1996, pp. 40–50.

56. H. Kwun, Back in style: magnetostrictive sensor, *Technology Today,* Southwest Research Institute, San Antonio, TX, Mar. 1995, pp. 2–7.

57. V. G. Kuleev, P. S. Kononov, and I. A. Telegina, Electromagnetoacoustic excitation of elastic longitudinal cylindrical waves in ferromagnetic bars, *Sov. J. NDT*, **19**: 690–698, 1983.

58. V. D. Boltachev et al., Electromagnetic-acoustic excitation in ferromagnetic pipes with a circular cross-section, *Sov. J. NDT,* **25**: 434–439, 1989.

59. M. J. Sablik and S. W. Rubin, Modeling magnetostrictive generation of elastic waves in steel pipes. I. Theory, *Int. J. Appl. Electromagnetics and Mechanics,* submitted 1998.

60. H. Kwun and K. A. Bartels, Experimental observation of elastic waves dispersion in bounded solids of various configurations. *J. Acoust. Soc. Am.*, **99**: 962–968, 1996.

61. M. J. Sablik, Y. Lu, and G. L. Burkhardt, Modeling magnetostrictive generation of elastic waves in steel pipes. II. Comparison to experiment, *Int. J. Appl. Electromagn. Mech.* submitted 1998.

62. M. J. Sablik and R. A. Langman, Approach to the anhysteretic surface, *J. Appl. Phys.*, **79**: 6134–6136, 1996.

63. S. Chikazumi and S. H. Charap, *Physics of Magnetism,* Malabar, FL: R. E. Krieger Publ. Co., 1984, pp. 19–24.

64. M. N. Mikheev, Magnetic structure analysis, *Sov. J. NDT*, **19**: 1–7, 1983.

65. M. N. Mikheev et al., Interrelation of the magnetic and mechanical properties with the structural state of hardened and tempered products, *Sov. J. NDT*, **18**: 725–732, 1983.

66. H. Kwun and G. L. Burkhardt, Effects of grain size, hardness and stress on the magnetic hysteresis loops of ferromagnetic steels. *J. Appl. Phys.*, **61**: 1576–1579, 1987.

67. R. Ranjan, D. C. Jiles, and P. K. Rastogi, Magnetoacoustic emission, magnetization and Barkhausen effect in decarburized steel, *IEEE Trans. Magn.*, **22**: 511–513, 1986.

68. R. Ranjan, D. C. Jiles, and P. K. Rastogi, Magnetic properties of decarburized steels: an investigation of the effects of grain

size and carbon content, *IEEE Trans. Magn.*, **23**: 1869–1876, 1987.

69. Z. J. Chen et al., Assessment of creep damage of ferromagnetic material using magnetic inspection, *IEEE Trans. Magn.*, **30**: 4596–4598, 1994.

70. D. C. Jiles and D. L. Atherton, Theory of ferromagnetic hysteresis. *J. Magn. Magn. Mater.*, **6**: 48–61, 1986.

71. M. J. Sablik et al., Finite element simulation of magnetic detection of creep damage at seam welds, *IEEE Trans. Magn.*, **32**: 4290–4292, 1996.

72. Z. J. Chen, D. C. Jiles, and J. Kameda, Estimate of fatigue exposure from magnetic coercivity, *J. Appl. Phys.*, **75**: 6975–6977, 1994.

73. Z. Gao et al., Variation of coercivity of ferromagnetic material during cyclic stressing, *IEEE Trans. Magn.*, **30**: 4593–4595, 1994.

74. L. J. Swartzendruber et al., Effect of plastic strain on magnetic and mechanical properties of ultraslow carbon sheet steel, *J. Appl. Phys.*, **81**: 4263–4265, 1997.

75. H. Barkhausen, Two phenomena revealed with help of new amplifiers. *Pzysikalischte Zeitschrift*, **20**: 401–403, 1919.

76. K. Stierstadt, The magnetic Barkhausen effect. In *Springer Tracts in Modern Physics*, **40**: 2–106, 1966 (in German).

77. J. C. McClure and K. Schröder, The magnetic Barkhausen effect, *CRC Critical Reviews in Solid State Sciences*, **6**: 45–83, 1976.

78. G. A. Matzkanin, R. E. Beissner, and C. M. Teller, The Barkhausen effect and its applications to nondestructive evaluation, *SWRI Report No NTIAC-79-2*, 1979.

79. S. Segalini, M. Mayos, and M. Putignani, Application of electromagnetic methods to steel microstructure control. *Memoires et Etudes Scientifique, Revue de Metallurgie*, October 1985, pp. 569–575 (in French).

80. W. L. Vengrinovich, Magnetic noise spectroscopy, In *Minsk-Science*, Minsk, 1991, 284 pp (in Russian).

81. T. Piech, Technical application of Barkhausen effect, *PNPS 475*, ISSN 0208-7979, Technical University of Szczecin, Szczecin, 1992, 160 pp (in German).

82. A. Zentkova and M. Datko, Propagation of the electrodynamic disturbance following a Barkhausen jump in metallic ferromagnetic samples. I Infinite medium, *Czech, J. Phys.*, **B24**: 310–321, 1974.

83. V. M. Vasiliev et al., Some computation and design problems of induction transducers for the detection of Barkhausen jumps, *Defektscopiya*, **2**: 73–83, 1986.

84. B. Augustyniak, Magnetomechanical effects, *Rapport TEMPRA*, GEMPPM, INSA de Lyon, 1995, 90 pp (in French).

85. J. Mackersie, R. Hill, and A. Cowking, Models for acoustic and electromagnetic Barkhausen emission. In J. Boogaard and G. M. van Dijk (eds), *Non-Destr. Test. Proc. 12th World Conf.*, Amsterdam: Elsevier Science Publ., 1989, pp. 1515–1518.

86. M. M. Kwan, K. Ono, and M. Shibata, Magnetomechanical acoustic emission of ferromagnetic materials at low magnetization levels (type I behavior), *J. Acoustic Emission*, **3**: 144–156, 1984.

87. M. M. Kwan, K. Ono, and M. Tibet, Magnetomechanical acoustic emission of ferromagnetic materials at low magnetization levels (type II behavior), *J. Acoustic Emission*, **3**: 199–210, 1984.

88. M. Guyot, T. Merceron, and C. Cagan, Acoustic emission along the hysteresis loops of various ferro- and ferrimagnets, *J. Appl. Phys.*, **63**: 3955–3957, 1988.

89. B. Augustyniak, Magnetomechanical emission. *Acoustic Emission*, J. Malecki, J. Ranachowski (eds.), IPPT-PAN Warsaw, 1994, pp. 417–445 (in Polish).

90. A. D. Beale et al., Micromagnetic processes in steels, *Mat. Res. Soc. Symp. Proc.*, Materials Research Society, 1991, pp 313–318.

91. D. G. Hwang and H. C. Kim, The influence of plastic deformation on Barkhausen effects and magnetic properties in mild steel, *J. Phys. D.*, **21**: 1807–1813, 1988.

92. B. Alessandro et al., Domain-wall dynamics and Barkhausen effect in metallic ferromagnetic materials, I. Theory, *J. Appl. Phys.*, **68**: 2901–2907, 1990.

93. B. Alessandro et al., Domain-wall dynamics and Barkhausen effect in metallic ferromagnetic materials. II. Experiments, *J. Appl. Phys.*, **68**: 2908–2915, 1990.

94. H. C. Kim, D. G. Hwang, and B. K. Choi, Barkhausen noise in 5% Mo-75.5% Ni permalloy with rolling texture, *J. Phys. D.*, **21**: 168–174, 1988.

95. B. Augustyniak, Magnetomechanical effects research for their application in nondestructive evaluation of ferromagetic materials, *Rapport ATP de France*, Nr 717, Technical University of Gdansk, 1996 (in French).

96. L. Basano and P. Ottonello, Use of time-day correlators and wave-shaping techniques in the statistical analysis of Barkhausen pulses, *J. Magn. Magn. Mater.*, **43**: 274–282, 1994.

97. C. Gatelier-Rothea et al., Role of microstructural states on the level of Barkhausen noise inpure iron and low carbon iron binary alloys: *Nondestructive Test. Eval.*, **8–9**: 591–602, 1992.

98. D. J. Buttle et al., Magnetoacoustic and Barkhausen emission in ferromagnetic materials, *Philos. Trans. R. Soc. London*, **A320**: 363–378, 1986.

99. B. Augustyniak and J. Degauque, New approach to hysteresis process investigation using mechanical and magnetic Barkhausen effects, *J. Magn. Magn. Mater.*, **140–144**: 1837–1838, 1995.

100. American Stress Technologies, Inc. Stresscan 500 C operating instructions, Pittsburgh Pennsylvania, 1988.

101. B. Augustyniak, M. Chmielewski, and W. Kielczynski, New method of residual stress evaluation in weld seams by means of Barkhausen effect, *Proc. XXIV National Conf. NDE*, PTBN i DT, Poznan-Kiekrz, 1995, pp. 9–17 (in Polish).

102. D. C. Jiles, Integrated on-line instrumentation for simultaneous automated measurements of magnetic field, induction, Barkhausen effect, magnetoacoustic emission, and magnetostriction, *J. Appl. Phys.*, **63**: 3946–3949, 1988.

103. A. Parakka and D. C. Jiles, Magnetoprobe: a portable system for non-destructive testing of ferromagnetic materials, *J. Magn. Magn. Mater*, **140–144**: 1841–1842, 1995.

104. H. Sakamoto, M. Okada, and M. Homma, Theoretical analysis of Barkhausen noise in carbon steels, *IEEE Trans. Magn.*, **23**: 2236–2238, 1987.

105. D. J. Buttle et al., Magneto-acoustic and Barkhausen emission from domain-wall interaction with precipitates in incolay 904, *Philos. Mag. A*, **55**: 735–756, 1987.

106. R. Rautioaho, P. Karjalainen, and M. Moilanen, Coercivity and power spectrum of Barkhausen noise in structural steels, *J. Magn. Magn. Mater.*, **61**: 183–192, 1986.

107. R. Ranjan et al., Grain size measurement using magnetic and acoustic Barkhausen noise, *J. Appl. Phys.*, **61**: 3199–3201, 1987.

108. S. Tiitto, Magnetoelastic Barkhausen noise method for testing of residual stresses. American Stress Technologies, Inc., Pittsburgh, PA, 1989.

109. G. Bertotti, F. Fiorillo, and A. Montorsi, The role of grain size in the magnetization process of soft magnetic materials, *J. Appl. Phys.*, **67**: 5574–5576, 1990.

110. M. Komatsubara and J. L. Porteseil, Barkhausen noise behavior in grain oriented 3% SiFe and the effect of local strain, *IEEE Trans. Magn.*, **MAG-22**: 496–498, 1986.

111. T. W. Krause et al., Correlation of magnetic Barkhausen noise with core loss in oriented 3% Si-Fe steel laminates, *J. Appl. Phys.*, **79**: 3156–3167, 1996.

112. D. J. Buttle et al., Magneto-acoustic and Barkhausen emission: their dependence on dislocations in iron, *Philos. Mag. A,* **55**: 717–734, 1987.

113. A. J. Birkett et al., Influence of plastic deformation on Barkhausen power spectra in steels, *J. Phys. D,* **22**: 1240–1242, 1989.

114. C. Bach, K. Goebbels, and W. Theiner, Characterization of hardening depth by Barkhausen noise measurements, *Mater. Eval.,* **46**: 1576–1580, 1988.

115. L. Malkinski, Z. Kaczkowski, and B. Augustyniak, Application of Barkhausen effect measurements in piezomagnetic study of metallic glasses, *J. Magn. Magn. Mater.,* **112**: 323–324, 1992.

116. B. Augustyniak and J. Degauque, Microstructure inspection by means of mechanical Barkhausen effect analysis, *J. de Physique,* IV, **C8**: 527–530, 1996.

117. P. Deimel et al., Bloch wall arrangement and Barkhausen noise in steels 22 NiMoCr 3 7 and 15 MnMoNiV 5 3, *J. Magn. Mater.,* **36**: 277–289, 1983.

118. B. Augustyniak, Results of recent progress in new NDT methods of ferromagnetic materials, *IFTR Reports IPPT PAN Warsaw 1996,* 1/1996 (in Polish).

119. Z. J. Chen, A. Strom, and D. C. Jiles, Micromagnetic surface measurements for evaluation of surface modifications due to cyclic stress, *IEEE Trans. Magn.,* **29**: 3031–3033, 1993.

120. P. Gondi et al., Structural characteristics at surface and Barkhausen noise in AISI 4340 steel after grinding, *Nondestr. Test. Eval.,* **10**: 255–267, 1993.

121. W. A. Theiner and V. Hauk, Nondestructive characterization of shot peened surface states by the magnetic Barkhausen noise method. In J. Boogaard and G. M. van Dijk (eds.), *Non-Destr. Test., Proc. 12th World Conf.,* Amsterdam: Elsevier, 1989, pp. 583–587.

122. J. C. McClure Jr., S. Bhattacharya, and K. Schröder, Correlation of Barkhausen effect type measurements with acoustic emission in fatigue crack growth studies, *IEEE Trans. Magn.,* **MAG-10**: 913–915, 1974.

123. S. Battacharya and K. Schröder, A new method of detecting fatigue crack propagation in ferromagnetic specimens, *J. Test. Eval.,* **3**: 289–291, 1975.

124. S. Nishimura and K. Tokimasa, Study on the residual stresses in railroad solid wheels and their effect on wheel fracture, *Bull. JSME,* **19**: 459–468, 1976.

125. J. F. Shackelford and B. D. Brown, A critical review of residual stress technology, *Intl. Adv. NDT,* **15**: 195–215, 1990.

126. R. M. Bozorth, *Ferromagnetism,* Chap. 13, NJ: AT&T, 1978 (reprinted from 1951), pp. 595–712.

127. R. M. Bozorth and H. J. Williams, Effect of small stresses on magnetic properties, *Rev. Mod. Phys.,* **17**: 72–80, 1945.

128. D. J. Craik and M. J. Wood, Magnetization changes induced by stress in a constant applied field, *J. Phys. D,* **3**: 1009–1016, 1970.

129. A. J. Moses, Effect of stress on d.c. magnetization properties of permendur, *Proc. IEE,* **122**: 761–762, 1975.

130. R. Langman, Measurement of the mechanical stress in mild steel by means of rotation of magnetic field strength, *NDT Int.,* **14**: 255–262, 1981.

131. R. Langman, The effect of stress on the magnetization of mild steel at moderate field strengths, *IEEE Trans. Magn.,* **21**: 1314–1320, 1985.

132. M. J. Sablik et al., Model for the effect of tensile and compressive stress on ferromagnetic hysteresis, *J. Appl. Phys.,* **61**: 3799–3801, 1987.

133. I. J. Garshelis, Magnetic and magnetoelastic properties of nickel maraging steels, *IEEE Trans. Magn.,* **26**: 1981–1983, 1990.

134. H. Hauser and P. Fulmek, The effect of mechanical stress on the magnetization curves of Ni and FeSi single crystals at strong fields, *IEEE Trans. Magn.,* **28**: 1815–1825, 1992.

135. C. S. Schneider, P. Y. Cannell, and K. T. Watts, Magnetoelasticity for large stresses, *IEEE Trans. Magn.,* **28**: 2626–2631, 1992.

136. M. J. Sablik and D. C. Jiles, Coupled magnetoelastic theory of magnetic and magnetostrictive hysteresis, *IEEE Trans. Magn.,* **29**: 2113–2123, 1993.

137. M. J. Sablik et al., A model for hysteretic magnetic properties under the application of noncoaxial stress and field, *J. Appl. Phys.,* **74**: 480–488, 1993.

138. H. Kwun, Investigation of the dependence of Barkhausen noise on stress and the angle between the stress and magnetization direction, *J. Magn. Magn. Mater.,* **49**: 235–240, 1985.

139. R. Langman, Measurement of the mechanical stress in mild steel by means of rotation of magnetic field strength—Part 2: Biaxial stress, *NDT Int.,* **15**: 91–97, Apr. 1982.

140. C. S. Schneider and J. M. Richardson, Biaxial magnetoelasticity in steels, *J. Appl. Phys.,* **53**: 8136–8138, 1982.

141. D. J. Buttle et al., Comparison of three magnetic techniques for biaxial stress measurement. In D. O. Thompson and D. E. Chimenti (eds.), *Rev. Progr. In Quant. NDE,* Vol. 9, New York: Plenum, 1990, pp. 1879–1885.

142. R. Langman, Magnetic properties of mild steel under conditions of biaxial stress, *IEEE Trans. Magn.,* **26**: 1246–1251, 1990.

143. K. Kashiwaya, Fundamentals of nondestructive measurement of biaxial stress in steel utilizing magnetoelastic effect under low magnetic field, *Jap. J. Appl. Phys.,* **30** (11A): 2932–2942, 1991.

144. M. J. Sablik et al., Micromagnetic model for biaxial stress effects on magnetic properties, *J. Magn. Magn. Mater.,* **132**: 131–148, 1994.

145. M. J. Sablik, H. Kwun, and G. L. Burkhardt, Biaxial stress effects on hysteresis and MIVC, *J. Magn. Magn. Mater.,* **140–144**: 1871–1872, 1995.

146. M. J. Sablik, R. A. Langman, and A. Belle, Nondestructive magnetic measurement of biaxial stress using magnetic fields parallel and perpendicular to the stress plane. In D. O. Thompson and D. E. Chimenti (eds.), *Rev. Progr. In Quant. NDE,* Vol. 16B, New York: Plenum, 1997, pp. 1655–1662.

147. M. J. Sablik, A model for asymmetry in magnetic property behavior under tensile and compressive stress in steel, *IEEE Trans. Magn.* **33**: 3958–3960, 1997.

148. C. S. Schneider and M. Charlesworth, Magnetoelastic processes in steel, *J. Appl. Phys.,* **57**: 4196–4198, 1985.

149. H. Hauser, Energetic model of ferromagnetic hysteresis, *J. Appl. Phys.,* **75**: 2584–2597, 1994.

150. H, Hauser, Energetic model of ferromagnetic hysteresis. 2. Magnetization calculations of (110)[001] FeSi sheets by statistic domain behavior, *J. Appl. Phys.,* **77**: 2625–2633, 1995.

151. I. J. Garshelis and W. S. Fiegel, Recovery of magnetostriction values from the stress dependence of Young's modulus, *IEEE Trans. Magn.,* **22**: 436–438, 1986.

152. W. F. Brown, Jr., Domain theory of ferromagnetics under stress, part I, *Phys. Rev.,* **52**: 325–334, 1937.

153. W. F. Brown, Jr., Domain theory of ferromagnetics under stress, part II, *Phys. Rev.,* **53**: 482–489, 1938.

154. G. W. Smith and J. R. Birchak, Internal stress distribution theory of magnetomechanical hysteresis—an extension to include effects of magnetic field and applied stress, *J. Appl. Phys.,* **40**: 5174–5178, 1969.

155. L. Callegaro and E. Puppin, Rotational hysteresis model for stressed ferromagnetic films, *IEEE Trans. Magn.,* **33**: 1007–1011, 1997.

156. A. Bienkowski and J. Kulikowski, The dependence of the Villari effect in ferrites on their magnetocrystalline properties and magnetostriction, *J. Magn. Magn. Mater.,* **26**: 292–294, 1982.

157. E. Villari, Change of magnetization by tension and by electric current, *Ann. Phys. Chem.,* **126**: 87–122, 1865.

158. M. J. Sablik et al., A model for the effect of stress on the low frequency harmonic content of the magnetic induction in ferromagnetic materials, *J. Appl. Phys.,* **63**: 3930–3932, 1988.

159. Z. J. Chen et al., Improvement of magnetic interface coupling through a magnetic coupling gel, *IEEE Trans. Magn.,* **31**: 4029–4031, 1995.

160. R. E. Ershov and M. M. Shel, On stress measurement by means of the magnetoelastic method, *Industrial Laboratory,* **31**: 1008–1011, 1965.

161. S. Abiku and B. D. Cullity, A magnetic method for the determination of residual stress, *Experimental Mech.,* **11**: 217–223, 1971.

162. S. Abiku, Magnetic studies of residual stress in iron and steel induced by uniaxial deformation, *Jap. J. Appl. Phys.,* **16**: 1161–1170, 1977.

163. S. A. Musikhin, V. F. Novikov, and V. N. Borsenko, Use of coercive force as an indicator parameter in nondestructive measurement of mechanical stresses, *Sov. J. NDT,* **23**: 633–635, 1988.

164. M. K. Devine, Detection of stress in railroad steels via magnetic property measurements, *Nondestr. Test. Eval.,* **11**: 215–234, 1994.

165. H. Kwun and G. L. Burkhardt, Effects of stress on the harmonic content of magnetic induction in ferromagnetic material, *Proc. 2nd Nat'l Seminar NDE Ferromagnetic Materials,* Dresser-Atlas, Houston, TX, 1986.

166. H. Kwun and G. L. Burkhardt, Nondestructive measurement of stress in ferromagnetic steels using harmonic analysis of induced voltage, *NDT Int.,* **20**: 167–171, 1987.

167. G. L. Burkhardt and H. Kwun, Application of the nonlinear harmonics method to continuous measurement of stress in railroad rail. In D. O. Thompson and D. E. Chimenti (eds.), *Rev. Progr. Quant. NDE,* Vol. 7B, New York: Plenum, 1988, pp. 1413–1420.

168. H. Kwun, G. L. Burkhardt, and M. E. Smith, Measurement of the longitudinal stress in railroad rail under field conditions using nonlinear harmonics. In D. O. Thompson and D. E. Chimenti, (eds.), *Rev. Progr. Quant. NDE,* Vol. 9, New York: Plenum, 1990, pp. 1895–1903.

169. R. Langman, Measurements of the mechanical stress in mild steel by means of rotation of magnetic field strength—part 3. Practical applications, *NDT Int.,* **16**: 59–65, 1983.

170. R. Langman, Some comparisons between the measurement of stress in mild steel by means of Barkhausen noise and by rotation of magnetization, *NDT Int.,* **20**: 93–99, 1987.

171. K. Kashiwaya, H. Sakamoto, and Y. Inoue, Nondestructive measurement of residual stress using magnetic sensors, *Proc. VI Intl. Congress Experimental Mech.,* Society for Experimental Mechanics, Bethel, CT, 1977, Vol. I, pp. 30–35.

172. H. Wakiwaka, M. Kobayashi, and H. Yamada, Stress measurement using a magnetic anisotropy sensor utilizing ac magnetization, *IEEE Transl. J. Magn. in Japan,* **6**: 396–401, 1991.

173. S. Kishimoto et al., Conversion theory of magnetic anisotropy sensor, *IEEE Trans. J. Magn. in Japan,* **7**: 269–273, 1992.

174. K. Kashiwaya, Y. Inoue, and H. Sakamoto, Development of magnetic anisotropy sensor for stress measurement of curved surface. In J. Boogaard and G. M. Van Dijk (eds.), *Proc. 12th World Conf. On Non-Destructive Testing,* Amsterdam: Elsevier, 1989, pp. 601–606.

175. H. Kwun and C. M. Teller, Stress dependence of magnetically induced ultrasonic shear wave velocity change in polycrystalline A-36 steel, *J. Appl. Phys.,* **54**: 4856–4863, 1983.

176. H. Kwun, Effects of stress on magnetically induced velocity changes for ultrasonic longitudinal waves in steel, *J. Appl. Phys.,* **57**: 1555–1561, 1985.

177. H. Kwun, A nondestructive measurement of residual bulk stresses in welded steel specimens by use of magnetically induced velocity changes for ultrasonic waves, *Mater. Eval.,* **44**: 1560–1566, 1986.

178. M. Namkung and D. Utrata, Nondestructive residual stress measurements in railroad wheels using the low-field magnetoacoustic test method. In D. O. Thompson and D. E. Chimenti (eds.), *Rev. Progr. Quant. NDE,* Vol. 7B, New York: Plenum, 1988, pp. 1429–1438.

179. M. Shibata and K. Ono, Magnetomechanical acoustic emission—a new method for non-destructive stress measurements, *NDT Int.,* **14**: 227–234, 1981.

180. G. L. Burkhardt et al., Acoustic methods for obtaining Barkhausen noise stress measurements, *Mater. Eval.,* **40**: 669–675, 1982.

181. R. Rautioaho, P. Karjalajnen, and M. Moilajnen, The statistical contribution of magnetic parameters to stress measurements by Barkhausen noise. In H. Fujiwara, T. Abe, and K. Tanaka (eds.), *Residual Stresses-III, Science and Technology,* Vol. 2, London: Elsevier, 1989, pp. 1087–1092.

182. D. J. Buttle et al., The measurement of stress in steels of varying microstructure by magnetoacoustic and Barkhausen emission, *Proc. R. Soc. London,* Ser. A, **414**: 469–496, 1987.

183. M. Nankung et al., Uniaxial stress effects on magnetoacoustic emission. In B. R. McAvoy (ed.), *Proc. IEEE 1989 Ultrasonics Symposium,* Montreal 1989, Vol. 2, [IEEE 1089], pp. 1167–1170

184. C. Jagadish, L. Clapham, and D. L. Atherton, Influence of uniaxial elastic stress on power spectrum and pulse height distribution of surface Barkhausen noise in pipeline steel, *IEEE Trans. Magn.,* **26**: 1160–1163, 1990.

185. M. G. Maylin and P. T. Squire, The effects of stress on induction, differential permeability and Barkhausen count in a ferromagnet, *IEEE Trans. Magn.,* **26**: 3499–3501, 1993.

186. M. J. Sablik, A model for the Barkhausen noise power as a function of applied magnetic field and stress, *J. Appl. Phys.,* **74**: 5898–5900, 1993.

187. M. J. Sablik and B. Augustyniak, The effect of mechanical stress on a Barkhausen noise signal integrated across a cycle of ramped magnetic field, *J. Appl. Phys.,* **79**: 963–972, 1996.

188. T. Piech, Application of the Barkhausen effect to mechanical stress measurements in ferromagnetics. In E. Czoboly (ed.), *Proc. 9th Congress of Materials Testing,* Budapest 1986, Vol. 2, 1986, pp. 495–496.

189. S. Tiitto, Magnetoelastic testing of biaxial stresses, *Experimental Techniques,* pp. 17–22, July/August 1991.

190. W. A. Theiner and P. Deimel, Non-destructive testing of welds with the 3MA-analyzer, *Nucl. Eng. Design,* **102**: 257–264, 1987.

191. K. Tiitto et al., Evaluation of the stress distribution in welded steel by measurement on the Barkhausen noise level, *Proc. Conf. Practical Applic. Residual Stress Technology,* Indianapolis 1991, pp. 55–59.

192. B. Augustyniak, New approach in Barkhausen effect application to residual stress evaluation, *Nondestr. Testing,* Polish Society for NDT, **5**: 17, 1996 (in Polish).

193. B. Augustyniak and W. Kielczynski, Comparison of non-destructive methods of residual stress evaluation in weld seams, *Proc.*

25th Nat'l Conf. on NDT, Szczyrk 1996, PTBN&DT SIMP, Warsaw 1996, Zeszyty Problemowe, 1, 235, 1996 (in Polish).

194. C. Jagadish, L. Clapham, and D. L. Atherton, Effect of bias field and stress on Barkhausen noise in pipeline steels, *NDT Int.,* **22**: 297–301, 1989.

195. R. L. Pasley, Barkhausen effect—an indication of stress, *Mater. Eval.,* **28**: 157–161, 1970.

196. W. A. Theiner and H. H. Willems, Determination of microstructural parameters by magnetic and ultrasonic quantitative NDE. In C. O. Ruud and R. E. Green, Jr., (eds.), *Nondestr. Methods for Mater. Property Determination,* New York: Plenum, 1984, pp. 249–258.

197. M. J. Sablik, W. L. Rollwitz, and D. C. Jiles, A model for mag-absorption as an NDE tool for stress measurement, *Proc. 17th Symp. on NDE,* San Antonio, TX, NTIAC, Southwest Research Institute, San Antonio, TX, 1989, pp. 212–223.

198. K. Babcock et al., Magnetic force microscopy: recent advances and applications. In D. G. Demczyk, E. Garfunkel, B. M. Clemens, E. D. Williams, and J. J. Cuomo (eds.), *Evol. of Thin Film and Surf. Struct. and Morphology,* MRS Proceedings, Vol. 335, Pittsburgh: Materials Research Society, 1995, pp. 311–321.

199. A. C. Bruno, C. H. Barbarosa, and L. F. Scavarda, Electric current injection NDE using a SQUID magnetometer, *Res. Nondestr. Eval.,* **8**: 165–175, 1996.

200. M. Lang et al., Characterization of the fatigue behavior of austenitic steel using HTSC-SQUID, *QNDE Conference,* Univ. San Diego, San Diego, CA, July 1997.

M. J. Sablik
Southwest Research Institute

B. Augustyniak
Technical University of Gdansk

MAGNETIC MICROWAVE DEVICES

In this article we discuss the following topics on magnetic microwave devices: yttrium–iron–garnet (YIG) film devices, magnetostatic waves, magnetooptic devices, absorbing layers, antireflection layers, and nonlinear responses. The article is organized as follows: The section entitled "Theoretical Background" presents a general theoretical background underlying the physics for the operation of ferrite components in microwave devices. Discussions include the derivation of the Polder permeability tensor, the effective fields associated with electron spin motion, the general dispersion spectrum for electromagnetic waves propagating in a bulk magnetic medium, magnetostatic waves admitted by the geometry of a YIG film, and the nonlinear instabilities for spin waves occurring at high power. The section entitled "Magnetostatic Waves and YIG Film Devices" describes YIG film devices based on the operation of magnetostatic waves (MSWs), including delay lines, filters, directional couplers, and resonators. The section entitled "Magnetic Microwave Nonlinear Devices" depicts nonlinear magnetic devices of frequency selective power limiters, signal-to-noise enhancers, amplitude correctors, and ferrimagnetic echoing devices. The section entitled "Magnetooptic Devices" discusses magnetooptic Kerr and Faraday effects and describes the operation of magneto-optic Bragg diffraction devices. Finally, design of microwave absorbing layers and antireflection layers is briefly mentioned in the section entitled "Antireflection Layers and Absorbing Layers."

THEORETICAL BACKGROUND

This section provides a theoretical background underlying the physics that allows for the operation of a magnetic microwave device. A magnetic microwave device generally requires the use of an insulating magnetic ferrite material so that magnetization or spin motion is coupled to Maxwell equations without inducing much eddy-current loss at high frequencies. Also, in order to eliminate domain wall motion, single-domain operation is preferred at radio frequencies, and the ferrite material needs to be magnetized to saturation using an external direct current (dc) magnetic field. Alternatively, effective fields arising from either the crystalline or shape anisotropy of the ferrite material may be used to fulfull the bias requirement of the magnetic device. Thus, under small-signal approximations the electromagnetic property of the ferrite is described by a tensor permeability whose off-diagonal elements result in novel applications of nonreciprocal devices, for example. Most important, the permeability tensor can be varied by adjusting the bias field strength, resulting in tunability of the microwave device over frequencies.

A magnetic microwave device is normally operating in the frequency range from 0.1 GHz to 40 GHz or higher, and its performance can be interpreted in terms of the spin/magnetization motion of the ferrite material where coupling to optical/photon modes, elastic/phonon modes, or exchange/magnon modes may be utilized. Depending on the regime of applications, a microwave magnetic device may be distinguished as either a magnetodynamic device or a magnetostatic-wave device. The first class of devices includes circulators, isolators, filters, phase shifters, patch antennas, and so on, whose operation must be analyzed using the full set of Maxwell equations. The second class consists of mainly high-quality single-crystal yttrium–iron–garnet film devices where the propagation of magnetization waves involves a wavelength comparable to the film thickness. As such, the displacement currents can be omitted in Maxwell equations. This renders the so-called magnetostatic approximation, which implies that the resultant radiofrequency RF-magnetic field can be derived from a scalar potential. Important magnetostatic devices include delay lines, filters, resonators, echo lines, and other nonlinear devices, whose operation complements their low-frequency counterparts below 2 GHz involving surface acoustic wave (SAW) devices. Magnetodynamic devices are not discussed in this article.

In the following subsections we first derive the coupling between the magnetization field and the other electromagnetic fields, giving rise to a Polder permeability tensor for the ferrite material under small-signal approximation. Effective fields are then introduced in the equation of motion, allowing for the coupling of the magnetization field with the other physical fields required for transducer applications. Based on the frequency versus wave-number dispersion diagram, the propagation of magnetization waves can be divided into three zones upon which magnetic microwave devices are conventionally defined at several regimes. Magnetostatic waves are then discussed, whose dispersion diagrams are described in terms of the bias-field configuration relative to the YIG-film device geometry. Finally, spin-wave instability is briefly mentioned, delineating the high-power threshold that a ferrite device can operate before a cascading energy transfer occurs between the input rf power and the parametric excitation of spin waves.

Polder Permeability Tensor

In a source-free medium, Maxwell equations take the form

$$\nabla \times \boldsymbol{h} = j\omega\epsilon\boldsymbol{e}, \quad \nabla \times \boldsymbol{e} = -j\omega\boldsymbol{b}$$
$$\nabla \cdot \boldsymbol{b} = 0, \quad \nabla \cdot \boldsymbol{e} = 0 \tag{1}$$

where $\boldsymbol{e}$ and $\boldsymbol{h}$ are the RF electric and magnetic fields and $\boldsymbol{b}$ is the RF magnetic induction field. In Eq. (1), ϵ denotes the permittivity and the time dependence of the RF quantities is assumed to be $\exp(j\omega t)$. For a linear isotropic medium one may define a constant μ, the permeability, so that $\boldsymbol{b}$ and $\boldsymbol{h}$ are linearly proportional to each other:

$$\boldsymbol{b} = \mu\boldsymbol{h} \tag{2}$$

Equation (2) holds true if the medium is diamagnetic ($\mu < \mu_0$) or paramagnetic ($\mu > \mu_0$). Here μ_0 denotes the permeability of vacuum. For a ferromagnetic or a ferrimagnetic medium the relationship between $\boldsymbol{b}$ and $\boldsymbol{h}$ is neither linear nor isotropic. However, under small-signal approximation the linear relationship between $\boldsymbol{b}$ and $\boldsymbol{h}$ may be assumed, but the scalar permeability needs to be replaced by a tensor. That is, Eq. (2) becomes

$$\boldsymbol{b} = \underline{\underline{\mu}}\boldsymbol{h} = \mu_0(\boldsymbol{m} + \boldsymbol{h}) \tag{3}$$

Here, $\boldsymbol{m}$ denotes the RF magnetization field, and $\underline{\underline{\mu}}$ is called the permeability tensor (1).

In a magnetic substance the net magnetic dipole moment per volume, or the magnetization, denoted as $\boldsymbol{M}$, is nonzero due to spontaneous magnetization of the material. Denote the angular momentum per volume of the medium be $\boldsymbol{J}$. The time rate change of angular momentum can be equated with the applied torque, and this implies

$$\frac{\partial \boldsymbol{J}}{\partial t} = \mu_0\boldsymbol{M} \times \boldsymbol{H} \tag{4}$$

Here, $\boldsymbol{H}$ denotes the internal magnetic field within the volume. From both classical mechanics and quantum mechanics, the relationship between $\boldsymbol{J}$ and $\boldsymbol{M}$ is linear, which can be expressed as

$$\boldsymbol{M} = \gamma\boldsymbol{J} \tag{5}$$

In Eq. (5) γ is called the *gyromagnetic ratio,* which can be written as

$$\gamma = -\frac{g|e|}{2m_e} \tag{6}$$

where g is the Landé g factor, and e and m_e are charge and mass of an electron, respectively. Classically, $g = 1$, for orbital angular momentum, and $g = 2$, for spin angular momentum. Quantum mechanically, g can take a noninteger value between 1 and 2 due to the interaction between the spin and the orbital motion of the electron (2). However, for magnetic transition metal ions, Fe, Co, and Ni, the orbital motion of $3d$ electrons is usually quenched and hence $g \approx 2$. This im-

plies $\gamma = -1.76 \times 10^{11}$ C/kg. Combining Eqs. (4) and (5), we derive, therefore, the following constitutive equation for a magnetic medium:

$$\frac{\partial \boldsymbol{M}}{\partial t} = \gamma\mu_0\boldsymbol{M} \times \boldsymbol{H} \tag{7}$$

We now assume that the magnetic medium is magnetized to saturation either by an externally applied magnetic field or by an internal anisotropy field, or both. Let the saturation magnetization be denoted as M_{S}. We separate the dc and the rf components of $\boldsymbol{M}$ and $\boldsymbol{H}$ as follows:

$$\boldsymbol{M} = \boldsymbol{M}_0 + \boldsymbol{m}, \quad \boldsymbol{H} = \boldsymbol{H}_0 + \boldsymbol{h} \tag{8}$$

Here, capital letters denote dc quantities, and lowercase denote RF quantities. Under the small-signal assumption, $|\boldsymbol{m}| \ll |\boldsymbol{M}_0| \approx M_s$, $|\boldsymbol{h}| \ll |\boldsymbol{H}_0|$, Eq. (7) can be linearized to yield

$$\frac{\partial \boldsymbol{m}}{\partial t} = \gamma\mu_0 M_S \boldsymbol{i}_z \times \left(\boldsymbol{h} - \frac{H_0}{M_S}\boldsymbol{m}\right) \tag{9}$$

In Eq. (9) we have assumed $\boldsymbol{H}_0$, and hence $\boldsymbol{M}_0$, to be along the z axis whose unit vector is denoted as $\boldsymbol{i}_z$. From Eqs. (3) and (9), we derive, assuming again the $\exp(j\omega t)$ time dependence, following the Polder tensor permeability

$$\underline{\underline{\mu}} = \mu_0\begin{pmatrix} \mu & -j\kappa & 0 \\ j\kappa & \mu & 0 \\ 0 & 0 & 1 \end{pmatrix} \tag{10}$$

The Polder tensor elements μ and κ are given as

$$\mu = 1 + \frac{\omega_z\omega_m}{\omega_z^2 - \omega^2} \tag{11}$$

$$\kappa = \frac{\omega\omega_m}{\omega_z^2 - \omega^2} \tag{12}$$

and ω_z and ω_m are defined as

$$\omega_z = |\gamma|\mu_0 M_{\mathrm{S}} \tag{13}$$

$$\omega_m = |\gamma|\mu_0 H_0 \tag{14}$$

Equations (1), (3), and (10) sufficiently describe the general behavior of a linear microwave magnetic device.

Effective Fields

In Eq. (7) the magnetic field $\boldsymbol{H}$ is the internal field effectively experienced by the spins in the magnetic medium. In other words, an effective field is defined if there exists a coupling between the magnetization motion of the medium and the other physical field quantities. The coupling energy density is denoted as $w(\boldsymbol{M}, \partial\boldsymbol{M}/\partial x_i)$, which may show dependence on the magnetization $\boldsymbol{M}$, the magnetic strains $\partial\boldsymbol{M}/\partial x_j$, or both. For example, the externally applied magnetic field, $\boldsymbol{H}_a$, can be associated with the Zeeman energy density, $w = \mu_0\boldsymbol{H}_a\cdot\boldsymbol{M}$. For other couplings the resultant effective fields can be derived from the following Lagrangian equations (3,4):

$$(\boldsymbol{H}_{\text{eff}})_i = \frac{1}{\mu_0}\left[-\frac{\partial w}{\partial M_i} + \sum_{j=1}^{3}\frac{\partial}{\partial x_j}\frac{\partial w}{\partial(\partial M_i/\partial x_j)}\right], \quad i = 1, 2, 3$$
(15)

The associated energy flux is

$$s_i = \sum_{j=1}^{3}\frac{-\partial w}{\partial(\partial M_j/\partial x_i)}\frac{\partial M_j}{\partial t}, \quad i = 1, 2, 3$$
(16)

so that

$$\frac{\partial \boldsymbol{w}}{\partial t} + \nabla \cdot \boldsymbol{s} = 0$$
(17)

In general, $\boldsymbol{H}$ in Eq. (7) consists of the following components:

$$\boldsymbol{H} = \boldsymbol{H}_a + \boldsymbol{H}_D + \boldsymbol{H}_A + \boldsymbol{H}_E + \boldsymbol{H}_S + \boldsymbol{h}_{\text{RF}} + \boldsymbol{h}_d + \boldsymbol{h}_G$$
(18)

where

$\boldsymbol{H}_a$ = externally applied magnetic field (parallel to the z axis)
(19)

$\boldsymbol{H}_D$ = dc demagnetizing field
(20)

$\boldsymbol{H}_A = (2K/\mu_0 M_S)\boldsymbol{i}_z$ = uniaxial anisotropy field (along the z axis)
(21)

$\boldsymbol{H}_E = (2A/\mu_0 M_S^2)\nabla^2\boldsymbol{M}$ = magnetic exchange field
(22)

$\boldsymbol{H}_S$ = magnetoelastic field
(23)

$\boldsymbol{h}_{\text{RF}}$ = externally applied RF driving field
(24)

$\boldsymbol{h}_d$ = RF dipolar field
(25)

$\boldsymbol{h}_G = (-\lambda/\gamma\mu_0 M_S)\partial\boldsymbol{M}/\partial t$ = Gilbert damping field
(26)

K, A, and λ are, respectively, the (uniaxial) anisotropy constant, exchange stiffness, and Gilbert damping constant. We note that although $\boldsymbol{H}_A$, $\boldsymbol{H}_E$, and $\boldsymbol{H}_S$ are written in capital letters, they may contain both dc and RF components.

The dc demagnetizing field, $\boldsymbol{H}_D$, which results from the shape anisotropy, can be solved analytically only for an ellipsoidally shaped body. In this case a demagnetizing-factor tensor $\underline{\underline{\boldsymbol{N}}}_D$ can be calculated so that (1)

$$\boldsymbol{H}_D = -\underline{\underline{\boldsymbol{N}}}_D\boldsymbol{M}_0$$
(27)

where $\boldsymbol{M}_0$ denotes the dc component of the magnetization expressed in Eq. (8). For the limiting case of a thin flat ferrite slab lying on the x–y plane, Eq. (27) becomes

$$\boldsymbol{H}_D = -\boldsymbol{M}\cdot\boldsymbol{e}_z\boldsymbol{e}_z$$
(28)

Equation (21) denotes the effective field associated with a uniaxial anisotropy. For other anisotropy fields, $\boldsymbol{H}_A$ can be derived from Eq. (15) using the appropriate energy density of the anisotropy. For example, the corresponding energy density for a cubic anisotropy is

$$w_A = K_1(\alpha_1^2\alpha_2^2 + \alpha_2^2\alpha_3^2 + \alpha_3^2\alpha_1^2) + K_2\alpha_1^2\alpha_1^2\alpha_1^2 + \cdots$$
(29)

where α_i, $i = 1, 2, 3$, is the directional cosine of the magnetization, $\boldsymbol{M}$, with respect to the ith cubic axis, and K_1 and K_2 are the associated anisotropy constants (5).

By using Eq. (15) the magnetoelastic field can be derived from the following magnetoelastic energy density

$$w_S = b_1(\alpha_1^2\epsilon_{11} + \alpha_2^2\epsilon_{22} + \alpha_3^2\epsilon_{33} \\ + 2b_2(\alpha_1\alpha_2\epsilon_{12} + \alpha_2\alpha_3\epsilon_{23} + \alpha_3\alpha_1\epsilon_{31})$$
(30)

where α_i, $i = 1, 2, 3$, is the directional cosine of the magnetization, $\boldsymbol{M}$; ϵ_{ij}, $i, j = 1, 2, 3$, is the strain field, and b_1 and b_2 are the magnetoelastic coupling constants (3). The magnetoelastic coupling measures the response that a strain signal or an acoustic signal is converted into a magnetic signal in a magnetoelastic transducer device, or vice versa.

The dipolar field $\boldsymbol{h}_d$ denotes the RF field associated with the RF magnetization field $\boldsymbol{m}$ in Eq. (8), which needs to be solved from Maxwell equation, Eq. (1). When expressed in the spectral domain $\boldsymbol{h}_d$ relates to $\boldsymbol{m}$ as follows:

$$\boldsymbol{h}_d = (\underline{\underline{\boldsymbol{\mu}}}/\mu_0 - \underline{\underline{\boldsymbol{!}}})^{-1}\boldsymbol{m}$$
(31)

where $\underline{\underline{\boldsymbol{\mu}}}$ is the Polder tensor derived earlier in this section and $\underline{\underline{\boldsymbol{!}}}$ denotes the identity tensor. Under magnetostatic approximation, $\boldsymbol{h}_d$ satisfies the following magnetostatic equations

$$\nabla \cdot \boldsymbol{h}_d = -\nabla \cdot \boldsymbol{m}$$
(32)

$$\nabla \times \boldsymbol{h}_d = 0$$
(33)

subject to suitable boundary conditions. Thus, $\boldsymbol{h}_d$ is solved from $\boldsymbol{m}$ in almost the same way that the dc demagnetizing field $\boldsymbol{H}_D$ is solved from $\boldsymbol{M}_0$. In the literature $\boldsymbol{h}_d$ is sometimes called the RF demagnetizing field.

The Gilbert damping field can be effectively accounted for if one replaces H_0 in Eq. (8) by $H_0 + (j\lambda/\gamma\mu_0)\omega$, or, equivalently (1),

$$H_0 \rightarrow H_0 + \frac{j\Delta H}{2}$$
(34)

where ΔH denotes the ferromagnetic resonance (FMR) linewidth. The Gilbert damping term is identical to the Landau–Lifshitz form in first order, which is introduced into the equation of motion, Eq. (7), phenomenologically to account for the damping torque experienced by the spins undergoing precessing motion. Equation (34) describes the measured magnetic loss of a magnetic microwave device very well, provided that the applied RF frequency is not too different from the frequency that ΔH was measured. However, the physical meaning of the Gilbert damping, as related to the relaxation processes in the medium, is lacking, in contrast with the other damping forms—for example, the Bloch–Bloembergen damping (1).

Finally, we have to specify the boundary conditions on $\boldsymbol{m}$ in the presence of an exchange field given by Eq. (22). The (direct) exchange coupling is associated with the overlapping integral which relates the spin–spin interaction for two electron spins at neighboring atomic sites. The exchange constant A is larger than 0 for ferromagnetic coupling, and A is smaller than 0 for ferrimagnetic and for antiferromagnetic couplings. The spatial boundary condition on $\boldsymbol{m}$ can be derived from the equation of motion, Eq. (7), which requires that the quantity

$$\mu_0 A M \times \frac{\partial \boldsymbol{M}}{\partial n}$$
(35)

needs to be continuous across the ferrite boundaries. Here $\boldsymbol{n}$ denotes the direction normal to the outward surface of the boundary. The time boundary condition is derived from the energy conservation law. That is, from Eq. (16) we require the outward energy flux

$$-\mu_0 A \frac{\partial \boldsymbol{M}}{\partial n} \cdot \frac{\partial \boldsymbol{M}}{\partial t}$$
(36)

to be continuous across the material boundaries. However, instead of Eq. (36), it is popular in the literature to use the following spin-pinning condition at the material boundaries:

$$\frac{\partial \boldsymbol{M}}{\partial t} = 0$$
(37)

Since A is a microscopic quantity and at the boundary layers the environment there is quite different from that of the bulk, the pinning condition, Eq. (37), might be more realistic than the one representing the macroscopic average, Eq. (36).

Dispersion Curves for Bulk Modes

Equations (1), (3), and (10) can now be solved for a bulk ferrite medium for plane wave solutions. The resultant dispersion relation, ω versus k ($= 2\pi/\lambda$), is shown in Fig. 1. Here, k is the wave number and λ is the wavelength. In Fig. 1 the k-space is conventionally divided into three zones. For the small-k region, $k < k_1$ (≈ 0.1 cm^{-1}), electron spin motion is strongly coupled with the RF electromagnetic fields so that the full set of Maxwell equations is required to solve the dispersion relations. This region is called the *retarded zone* or *magnetodynamic zone*, and most magnetic microwave devices making use of bulk ferrite materials are operational in this region—for example, circulators, isolators, phase shifters, resonators, and so on. The next region comprises of intermediate k values, k_1 (≈ 0.1 cm^{-1}) $< k < k_2$ ($\approx 10^6$ cm^{-1}), known as the *magnetostatic wave zone*. In this region $\omega \ll k(\epsilon\mu_0)^{-1/2}$ and hence the displacement current, $\omega\epsilon\boldsymbol{e}$, can be ignored in Maxwell equations. That is

$$\nabla \times \boldsymbol{h} \approx 0$$
(38)

Equation (38) is called the *magnetostatic approximation*. A magnetostatic wave device usually require the use of a high-quality single-crystal magnetic film such as YIG whose thickness determines the wavelength of the resultant magnetostatic waves prevailing in the device structure. The last region is for $k > k_2$ ($\approx 10^6$ cm^{-1}), which is called the *spin wave zone*. In this region the dispersion curves grow proportional to k^2, as dictated by the effective exchange field, $\boldsymbol{H}_E$, in Eq. (22). Although not many practical microwave devices are designed in this region, the spectrum of spin waves is important in the sense that the normal spin precessing motion will break up into spin waves at the onset of instability when a magnetic microwave device is driven beyond a high power threshold.

For a given wave propagation direction, k, Eqs. (1), (3), and (10) imply two plane wave solutions. Similar to the plane wave solutions in an isotropic medium, the three vectors, $\boldsymbol{e}$, $\boldsymbol{b}$, and $\boldsymbol{k}$, for each mode in an anisotropic magnetic medium are still mutually perpendicular to each other. However, unlike the isotropic case, the two modes in the magnetic medium are nondegenerate, possessing different effective permeabilities and polarizations. Due to the wrong sense in polarization, one mode is weakly coupled to the photon waves, and hence its dispersion curve represents little departure from that of the (uncoupled) photon modes. This dispersion curve is shown in Fig. 1 as a straight (short-) dashed line in the retarded zone. The other mode couples strongly to the photon waves, giving rise to distortion of the dispersion curves in the retarded zone.

For the strongly coupled mode, two branches show up, depending on whether the bias magnetic field is applied above or below FMR. These two branches are shown in Fig. 1 as lower and upper curves, respectively. The upper branch, curve (3) in Fig. 1, lies entirely in the retarded zone and shows very little variation with respect to the wave propagation directions. That is, the propagation of strongly coupled electromagnetic waves biased below FMR is nearly isotropic in the magnetic medium. However, the lower branch, curves (1) and (2), depends strongly on the wave propagation directions. When k is parallel to the z axis, the dispersion curve is shown as curve (1) in Fig. 1; and when k is perpendicular to the z axis, the dispersion curve is shown as curve (2). For other propagation directions, the dispersion curves are distributed between these two curves, and for this reason the region bounded by curves (1) and (2) in Fig. 1 is usually referred to as the *spin-wave manifold*. In the literature, curve

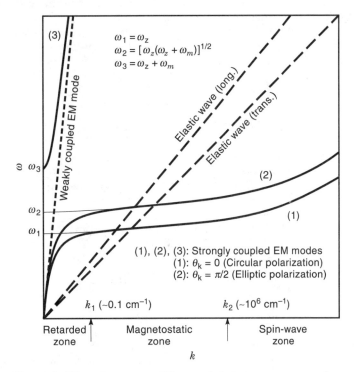

$$\omega_1 = \omega_z$$
$$\omega_2 = [\omega_z(\omega_z + \omega_m)]^{1/2}$$
$$\omega_3 = \omega_z + \omega_m$$

(1), (2), (3): Strongly coupled EM modes
(1): $\theta_k = 0$ (Circular polarization)
(2): $\theta_k = \pi/2$ (Elliptic polarization)

k_1 (~0.1 cm^{-1}) k_2 (~10^6 cm^{-1})

Retarded zone   Magnetostatic zone   Spin-wave zone

Figure 1. Dispersion curves of the coupled photon–magnon modes. The k space has been divided in three zones for retarded, magnetostatic, and spin wave modes. Phonon dispersion curves are also shown in the figure.

(1) is known as the *Kittel mode* and curve (2) is known as the *Voigt mode*. A Kittel mode possesses right-hand circular polarization, whereas a Voigt mode is associated with elliptic polarization. In Fig. 1, θ_k denotes the angle between k and the z axis, which is designated as the applied field direction, ω_1, ω_2, and ω_3 are given as

$$\omega_1 = \omega_z \tag{39}$$

$$\omega_2 = [\omega_z(\omega_z + \omega_m)]^{1/2} \tag{40}$$

$$\omega_3 = \omega_z + \omega_m \tag{41}$$

and ω_1 and ω_2 are the limiting values of the magnetostatic modes in the retarded zone.

The elastic modes are also shown in Fig. 1 as straight (long-) dashed lines. There are two kinds of phonon modes, longitudinal phonons and transverse phonons (6). In the presence of magnetoelastic coupling, b_1 and b_2 are nonzero in Eq. (30), and the phonon or acoustic modes will couple to the spin-wave, or magnon, modes. For the coupled case the dispersion curves of the phonons and the magnons will avoid running across each other in the same fashion that the photon modes and the magnon modes detour each other in the retarded zone region as shown in Fig. 1 (7). (If one views the uncoupled dispersion lines of two modes as two intersecting straight lines, the coupled dispersion lines resemble the two branches of a hyperbola using the two original lines as asymptotes.) In Fig. 1 the uncoupled magnon modes in the retarded zone are shown as dotted lines, extending curves (1) and (2) smoothly from the magnetostatic wave zone, intersecting the photon line, and ending at ω_1 and ω_2 of the ω axis.

Magnetostatic Waves in a Magnetic Layer

Wave propagation and dispersion in a magnetic layer can be derived in a manner similar to that in Fig. 1 except that boundary conditions need to be considered at the layer/air interfaces. For most device applications the excited waves have wavelengths on an order comparable to the layer thickness. As such, the magnetostatic approximation, Eq. (38), applies, which implies that the RF-magnetic field can be derived from a scalar potential, and hence the dispersion calculations are greatly simplified. Figure 2 shows such a dispersion diagram. When compared with Fig. 1, we see that the retarded zone has been pushed away into the $k = 0$ line in Fig. 2, and the exchange coupling showing k^2 dependence in the large-k region has been neglected. However, the magnetostatic dispersion is not a horizontal line, as depicted in the magnetostatic-wave zone of Fig. 1. The finite curvature of the dispersion curves shown in Fig. 2 is due to the finite thickness of the magnetic layer, d, which is distributed roughly in the region bounded by the two vertical lines $k = 0$ and $k = 2\pi/d$.

Magnetostatic waves can be volume waves and surface waves. For a volume wave the RF magnetization varies sinusoidally along the thickness direction, whereas for a surface wave it varies exponentially in this direction. Thus, a volume wave penetrates the whole thickness of the magnetic layer, whereas a surface wave is concentrated near the surface and the film-substrate interface. For a forward wave the dispersion increases monotonically with k so that the group velocity, $d\omega/dk$, is positive. This is in contrast with a backward wave where $d\omega/dk$ is negative. Thus, for a forward wave the

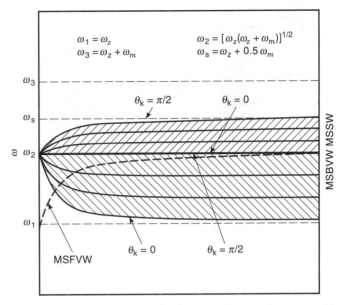

Figure 2. Dispersion curves of magnetostatic waves in a magnetic layer. The MSFVW is shown as a heavy dotted line, rising from ω_1 to ω_2 as k increases from 0 to ∞. The MSBVW and MSSW are shown hatched depending on the propagation direction of the magnetostatic waves. θ_k denotes the angle between the wave propagation direction and the applied field direction.

transmitted power is along the same direction as wave propagation, whereas the power transmitted by a backward wave is opposite to the wave propagation direction.

When the external field is applied normal to the layer plane, magnetostatic waves will propagate isotropically with the wave propagation direction lying on the layer plane. This branch of waves is called *magnetostatic forward volume waves* (MSFVWs), whose dispersion is shown in Fig. 2 as a heavy dotted line. The MSFVW mode undergoes uniform precessing motion at the Kittel frequency ω_1 for $k = 0$, and the frequency increases thereafter, approaching the bulk limit of the Voigt frequency ω_2 as k goes to infinity. Dispersion of this kind can be readily understood based on the spin motion occurring in the layer.

However, when the external bias field is applied in the layer plane, anisotropy results in general except at $k = 0$ where the uniform processing motion occurs at the Voigt frequency ω_2. When k increases further, two kinds of modes are possible: magnetostatic backward volume waves (MSBVWs) and magnetostatic surface waves (MSSWs). Depending on the propagation angle θ_k, MSBVW dispersion will decrease as k increases, approaching the respective bulk-mode limit as k goes to infinity. Therefore, unlike MSFVW, MSBVW occupies a finite area in the dispersion diagram, which is shown hatched between ω_1 and ω_2 in Fig. 2. For device applications, MSBVW is usually generated at $\theta_k = 0$, i.e., the excited MSBVW is collinear with the applied field direction, since it provides the widest frequency bandwidth among all the MSBVW propagation directions.

For MSSW the dispersion falls within the forbidden area of the bulk modes extending from ω_2 to ω_3 (see Figs. 1 and 2). MSSW frequencies increase as k increases, reaching limiting values at large k. The propagation of MSSW is anisotropic; and the largest frequency occurs at $\theta_k = \pi/2$, which gives rise to a limiting frequency $\omega_s = \omega_z + 0.5\,\omega_m$, known as the *Damon and Eshbach frequency*. The most popularly used MSSW is for $\theta_k = \pi/2$, which requires the MSSW to propagate transverse to the applied field direction and results in the widest frequency band for MSSW device applications.

In Fig. 2, MSFVW and MSBVW are shown only for the lowest-order volume modes. Higher-order volume waves are also possible. A high-order volume wave, which assumes additional nodal points along the thickness of the magnetic layer, will converge to the same frequency as the lower-order waves at large k, except that the curvature of the dispersion curve is reduced. In general, magnetostatic waves are generated in single-crystal YIG films epitaxially grown on gadolinium–gallium–garnet (GGG) substrates. Very often, a dielectric superstrate, for example, alumina, is covered on top of the YIG film to facilitate the excitation of magnetostatic waves. For these applications the dispersion diagram of Fig. 2 remains qualitatively unchanged. However, when a metal ground plane is placed on top of the superstrate shown in Fig. 3 the dispersion of MSSW, but not MSFVW and MSBVW, will change. The influence of a metal plane is that the MSSW dispersion curve will grow in a convex manner, increasing initially from ω_2 at $k = 0$ to a maximum value less than ω_3 followed by decreasing to ω_s as k goes to infinity. When a second metal ground plane is added to the bottom side of the GGG substrate shown in Fig. 3 the propagation of MSSW becomes even nonreciprocal. That is, the dispersion curve is different depending on whether the wave propagation is along the $+k$ or the $-k$ direction. In Fig. 3, magnetostatic waves may be excited by using a microstrip line, consisting of only the top ground plane, or a stripline, consisting of both the top and the bottom ground planes. For a multilayered system containing alternating magnetic and dielectric layers, the bulk and the surface modes form a band structure in almost the same way that atomic energy levels are crowded into energy bands when atoms are brought together to form a periodic lattice (8,9).

Finally, let us discuss the propagation loss of a delay line. When an observer is traveling with the wave down the delay line for a delay time $t = \tau_d$, the electric field is

$$\boldsymbol{e} = \boldsymbol{e}_0 \exp[2\pi j(f + j\Delta f)\tau_g] \tag{42}$$

where $\boldsymbol{e}_0$ denotes the initial amplitude at $t = 0$. This implies that the propagation loss in decibels is

$$\alpha = 40\pi\,(\log_{10} e)\Delta f \tau_g \tag{43}$$

In Eq. (42), f and Δf denote the real and the imaginary part of frequency. Δf can be related to the linewidth of the device to be used as a resonator cavity weakly coupled by a feeder line circuit. Thus, we have

$$\Delta f = \Delta f_\mathrm{m} + \Delta f_\mathrm{d} + \Delta f_\mathrm{c} \tag{44}$$

where Δf_m, Δf_d, and Δf_c denote, respectively, contributions from magnetic loss, dielectric loss, and conductor loss. As discussed in deriving Eq. (34), Δf_m may be identified as half the FMR linewidth multiplied by a volume filling factor F_m denoting the volume ratio of the ferrite material relative to the total volume enclosing the resonating cavity:

$$\Delta f_\mathrm{m} = F_m |\gamma| \mu_0 \Delta H / 2 \tag{45}$$

The other two linewidths Δf_d and Δf_c can be estimated in a normal way dealing with a dielectric loss cavity (see, for example, Ref. 10). If we assume that magnetic loss dominates and approximate $F_m \approx 1$, Eqs. (43) and (45) imply

$$\alpha \approx 76.4\tau_\mathrm{d}\Delta H \tag{46}$$

where τ_d is in microseconds and ΔH is in oersteds. Equation (46) was originally derived in Ref. 11 for an MSW delay line. However, since delay time is measured as group delay, τ_d expressed in Eqs. (42), (43), and (46) shall be multiplied by a factor v_g/v_k, where v_g denotes the group velocity ($= d\omega/dk$) and v_k denotes the phase velocity ($= \omega/k$) for wave propagation.

Spin-Wave Instability

Finally, let us discuss briefly the spin-wave instability occurring at the high power threshold of a magnetic microwave device (12). The spectrum of spin waves shown in Fig. 1 has proved to play a dominant role in the relaxation processes. When an RF field is applied, it drives the spins into precessing motion, which in turn couples with spin-wave propagation, dumping energy into lattice vibration via spin–spin and spin–lattice relaxation processes. The coupling to spin waves must originate from the nonlinear terms in the equation of motion, Eq. (7). The second-order terms come from $\boldsymbol{h} \times \boldsymbol{m}$ and the third-order terms come from either $\boldsymbol{h}\boldsymbol{m} \cdot \boldsymbol{m}$ or $\boldsymbol{m}\boldsymbol{m} \cdot \boldsymbol{m}$. For a spin-wave component, $\boldsymbol{m}_k$, it generates two accompanying $\boldsymbol{h}$ fields, the RF exchange field, denoted as $\boldsymbol{h}_\mathrm{E}$, from Eq. (22), and the dipolar field, $\boldsymbol{h}_\mathrm{d}$, from Eqs. (32) and (33). These two $\boldsymbol{h}$ fields then feed back and couple nonlinearly with the original $\boldsymbol{m}_k$ field to generate instability if a threshold power is reached. The $\boldsymbol{m} \cdot \boldsymbol{m}$ term arises from the constraint that the magnitude of $\boldsymbol{M}$ needs to be a constant (equal to $\boldsymbol{M}_\mathrm{S}$). That is, in Eq. (8) the longitudinal component $\boldsymbol{M}_0$ is replaced by

$$M_0 \approx M_\mathrm{S}(1 - \boldsymbol{m} \cdot \boldsymbol{m}/2M_s^2) \tag{47}$$

As a consequence, the dc demagnetizing field, if not zero, will add cubic nonlinearity to the equation of motion, giving rise to adverse effects in influencing the spin-wave instability.

Under FMR measurements, the experiments showed a subsidiary absorption at high excitation power which occurs at a dc field less than that required by the resonance condition. Also, as the input power increases, the resonance absorption peak broadened accordingly, rendering premature

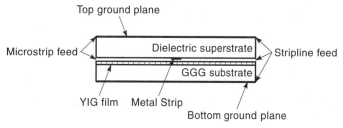

Figure 3. Magnetostatic wave excitation configuration. The YIG film is deposited on top of the GGG substrate. A superstrate can be used to provide microstrip excitation configuration. A bottom ground plane can also be deposited on the GGG substrate to provide stripline excitation configuration.

saturation of the main resonance. Suhl (12) showed that the subsidiary peak arises from a spontaneous transfer of energy from the uniform precessing motion of spins to spin waves of half the resonance frequency, $\omega_k = \omega/2$. This first-order instability is caused by the second-order interaction between the dipolar field, $\boldsymbol{h}_d$, and the spin wave, $\boldsymbol{m}_k$. The second-order instability responsible for the broadened and declined saturation of the main resonance peak comes from a catastrophic energy transfer from the uniform precessing motion of spins to spin waves of the same resonance frequency, $\omega_k = \omega$. This instability is brought about by the third-order interaction between the exchange field $\boldsymbol{h}_E$ and the spin waves, $\boldsymbol{m}_k$ and $\boldsymbol{m}_k$. The resultant input field thresholds for these two instabilities are, respectively,

$$h_d^{th} = \frac{2\Delta H_k}{\omega_m}\sqrt{(\omega - \omega_z)^2 + \omega_{\Delta H}^2} \qquad (48)$$

$$h_{Ex}^{th} = h_d^{th}\sqrt{\frac{\omega_m}{2\omega_{\Delta H}}} \qquad (49)$$

In Eqs. (48) and (49)

$$\omega_{\Delta H} = |\gamma|\mu_0 \Delta H \qquad (50)$$

and ΔH_k is the linewidth of the spin wave which is introduced phenomenologically to parameterize the energy transfer from the spin wave $\boldsymbol{m}_k$ to lattice vibration. Since $\omega_m \gg \omega_{\Delta H}$, it implies

$$h_{Ex}^{th} \gg h_d^{th} \qquad (51)$$

as we expect.

MAGNETOSTATIC WAVES AND YIG FILM DEVICES

There exists continuing demand for signal processing devices which can be used for radar detection, electronics communication, and instrumentation applications. At UHF frequencies, surface acoustic wave (SAW) devices have been widely used, providing phase shifting, time delaying, and other analog signal processing functions. However, at higher frequencies above 2 GHz, SAW devices are inefficient due to device fabrication and increased insertion loss. At microwave or even millimeter wave frequencies, signal processing devices have been largely achieved utilizing the newly developed magnetostatic wave (MSW) technology providing similar functional performance as SAW devices. Additional advantages include low insertion loss, large bandwidth up to 1 GHz, ease of fabrication, frequency tuning, dispersion shaping, and nonlinear operation.

In contrast to SAWs, MSWs are very dispersive and they can be controlled easily by means of an external magnetic field. In principle, three basic types of MSWs are distinguished: forward volume MSWs (MSFVWs), backward volume MSWs (MSBVWs), and surface MSWs (MSSWs). An MSFVW is excited in the ferrite material, usually a YIG film, magnetized perpendicularly to the film plane. An MSBVW is excited in a YIG film and the magnetization direction is in the film plane along the wave propagation direction. An MSSW is also associated with a transverse magnetization, but the wave propagation is perpendicular to the direction of magnetiza-

tion. All of these three types of MSWs can be effectively used in fabricating microwave devices, such as delay lines, tunable filters, phase shifters, resonators, noise suppressors, amplitude correctors, and so on.

In order to reduce wave propagation loss, high-quality single-crystal yttrium–iron–garnet (YIG) films are usually used for the fabrication of MSW devices. When YIG single-crystal films are epitaxially grown on gadolinium–gallium–garnet (GGG) substrates, the ferromagnetic resonance (FMR) linewidth, ΔH, can be as narrow as 0.3 Oe at 9 GHz, and 0.6 Oe at 20 GHz. Using Eq. (46) the propagation loss in an MSW delay line is 23 and 46 dB per microsecond at 9 and 20 GHz, respectively. Therefore, for a typical delay line application, the propagation loss requiring 200 ns delay will be, respectively, 4.6 and 9.2 dB at 9 and 20 GHz. This compares very favorably with other kinds of delay lines such as a coaxial cable; at 9 GHz a 200 ns coaxial cable would require a length of 50 m, resulting in 30 dB loss in propagation.

MSWs are excited within YIG films using either the microstrip or the stripline transducer circuits. Due to the coupling between the guided electromagnetic modes of the transducers and the spin waves in the YIG films, magnetostatic waves are generated, traveling down from the input to the output transducer to perform signal processing functions, for example. In addition to straight microstrip/stripline transducers, meander lines, gratings, and interdigital and unidirectional transducers can also be effectively used to couple in and out the MSW signals. Short-circuited and open-circuited microstrip transducers are commonly used for wideband MSW device applications. The entire MSSW, MSFVW, or MSBVW frequency band can be excited by using narrow (10 μm) microstrip transducers. For narrowband devices, meander lines and gratings can be used. These transducers can be designed with 50 to 75 Ω input impedance over wide frequency bands, and matching circuits can be used to reduce mismatch losses.

In order to reduce spurious reflection of MSWs from the YIG film edges, MSW terminations or absorbers need to be utilized. These terminations can appear in the form of ferrite powders, or iron/permalloy rods, or even recording tapes, or GaAs thin films may be placed at the YIG edges. However, the simplest way to avoid MSW interference is to cut the YIG edges into angles other than 90° such that the reflected beam is directed away from the active area of the MSW device (13). Among many MSW applications we will briefly discuss in this section the operation of MSW delay lines, filters, directional couplers, and resonators. Other nonlinear and magnetooptic MSW devices will be discussed in the sections entitled "Magnetic Microwave Nonlinear Devices" and "Magnetooptic Devices," respectively.

MSW Delay Lines

Figure 4 shows the commonly used flip-over configuration for MSW delay lines. Microstrip lines serving as the transmitter and receiver for MSWs are fabricated using a photolithographic technique on top of a dielectric superstrate such as alumina, sapphire, fused silica, or duroid material. The high-quality crystal YIG film epitaxially grown on GGG substrate is brought in contact with the transducers via a spacer layer, and the overall dielectric/spacer/YIG/GGG assembly is placed between the poles of the biasing magnet. The biasing mag-

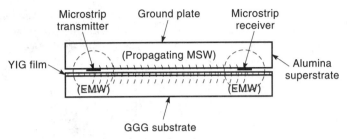

Figure 4. Schematic showing the MSW delay line configuration. Microstrip transducers are used to couple in and out the MSWs. Depending on the direction of the bias magnetic field, all types of MSWs can be excited and detected using the present circuit configuration.

netic field can be directed on the YIG film plane either parallel to or perpendicular to the microstrip lines, or perpendicular to the YIG film plane, to provide MSBVW, MSSW, or MSFVW delay line operation, respectively. Therefore, due to the coupling between the electromagnetic waves (EMWs) induced by the microstrip lines and the MSWs, the microstrip transmitter will excite MSWs which travel down the delay line structure to be picked up by the microstrip receiver. Let the distance between the transmitter and the receiver be D, and thus the time delay for this device is

$$\tau_d = D/v_g \tag{52}$$

where $v_g = d\omega/dk$ denotes the group delay velocity of MSWs.

Delay line elements exhibiting a linear dependence of the delay time, τ_d, on frequency are key components in pulse compression radar, microscan receiver, and Fourier transform systems. In general, τ_d is highly dispersive, depending nonlinearly on frequency. Various methods exist that allow τ_d to show linear dependence on frequency over wide-frequency bands (14–17). For example, at X-band it is possible to show 1 GHz bandwidth for a linearly dispersion delay line by using a thin YIG film with thickness 20 μm covered with a thin dielectric superstrate of the same thickness (14). Alternatively, the linear dependence of τ_d can be obtained by varying the separation distance between the ground plane and the YIG film (15), using the bias field gradients (16), or deploying multiple YIG films (17).

Nondispersive wideband delay lines are potential devices replacing phase shifters at microwave frequencies, providing electronic tuning capability for phased array antenna and other signal processing component applications. Therefore, we require the time delay, τ_d, to be independent of the bias field strength and, hence, the frequency over a wide frequency bandwidth. A possible solution to this requirement is to cascade two wideband linearly dispersive delay lines with opposite propagation characteristics. That is, the first device operates for forward volume MSWs and the second device operates for backward MSWs such that they compensate for each other to provide nondispersive dependence on frequencies. Other methods make use of nonuniform bias field (16) and multilayer structure (17), as discussed for the construction of a linearly dispersive delay line device.

MSW Filters

Filtering of electronic signals is performed as a frequency-selective process realized in the frequency domain. In princi-

ple, any delay line configuration can be viewed as a frequency filter structure provided that the following features are emphasized: low insertion loss occurring at the passband and high attenuation occurring at the stop bands. The filter characteristics can be feasibly obtained by controlling the transducer dimensions and the YIG/ground plane separation. While short-circuited straight microstrip lines are generally used as transducers for a wideband filter, multielement grating transducers, such as those shown in Fig. 5, are used to synthesize narrowband filters. As such, narrowband filters of bandwidth 30 MHz tunable from 3 to 7 GHz have been successfully demonstrated (18). Similarly, by carefully adjusting the width of the short-circuited microstrip transducer and the YIG film thickness, one can obtain a wideband filter tunable from 0.3 to 12 GHz with stopband rejection better than 45 dB (19). The advantage of using an MSW filter is that the passband can be tuned by varying the strength of the bias magnetic field.

MSW Directional Couplers

Figure 6 shows the schematic of an MSW directional coupler. In Fig. 6, two YIG films are deployed face to face, separated by a dielectric spacer. Ground planes are deposited on the outer sides of the GGG substrates and two multistrip lines are used as transducers, coupling in and out microwave power though exchange of MSWs. By careful design of the microstrip line-spacing as well as the dimension of the dielectric spacer, the characteristic of a directional coupler can be obtained. Operation at full power coupling is possible (20), and hence the directional coupler can be equivalently used as a bandpass filter. Also, by varying the bias field strength the power transferring coefficient of the directional coupler can be consequently adjusted.

MSW Resonators

Although the MSW bandpass filters discussed above are useful for many applications, there are occasions which require

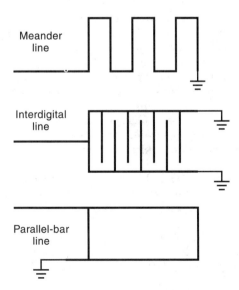

Figure 5. Microstrip circuit showing the geometry for multielement grating transducers used for excitation and receiving MSWs over a narrow frequency band for tunable filter applications.

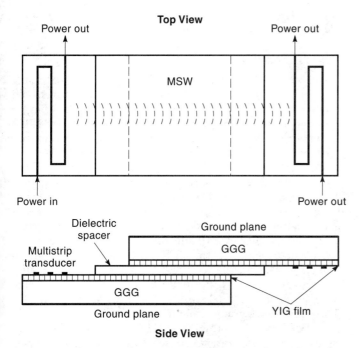

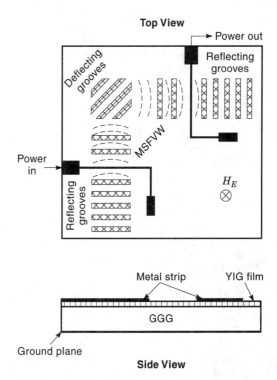

Figure 6. Schematic showing the MSW directional coupler configuration. Microstrip multistrip transducers are used to couple in and out the MSWs. All types of MSWs are possible depending on the bias field direction. The coupling coefficient between the input and output ports can be varied by changing the strength of the bias magnetic field.

Figure 8. Schematic showing the configuration of an MSFVW resonator. Two cavities are placed at a 90° angle, which are coupled via MSFVWs passing through an obliquely oriented groove grating array. The bias field is applied perpendicular to the YIG film plane.

considerable signal selectivity over as narrow as possible a passband width. For example, tunable MSW resonators can be used as the frequency-selective elements in tunable oscillator circuits in the microwave frequency bands. MSSW resonators can be constructed by placing reflective metal gratings at the edges of the resonating cavity. Alternatively, grooves may be cut on the YIG film surface using wet etching or ion bombardment to form an MSSW cavity. This is shown in Fig. 7, in which two arrays of straight grooves are etched parallel to

but on the two sides of the microstrip transducer lines; and MSSWs, once generated, bounce back and forth indefinitely between these two groove arrays at resonance. As reported in Ref. 21, MSSW resonators fabricated in this manner have shown a tuning capability between 2 GHz and 5 GHz, exhibiting a loaded Q of 500 and an off-resonance rejection level of 15 dB.

For MSFVWs the bias magnetic field is perpendicular to the YIG film plane, and hence the propagation of MSFVWs is isotropic in all the directions in the film plane. As such, the MSSW resonator shown in Fig. 7 is not suitable for MSFVW applications; it will result in poor off-resonance isolation, since at off-resonance the propagation of MSFVWs at slightly tilted angles may still satisfy the resonance condition. To avoid this drawback, a new configuration which involves two cavities coupled by MSFVWs was proposed in Ref. 22, as shown in Fig. 8. Each cavity consists of two etched-groove gratings and a single microstrip transducer. The two cavities are placed at a 90° angle and are coupled by a 45° obliquely oriented grating capable of deflecting the incident MSW beam by a 90° angle. The resonators reported in Ref. 22 exhibited insertion losses between 20 and 32 dB and a loaded Q value of 290 to 1570 over a tuning range of 2 GHz to 11 GHz.

The resonator structures discussed so far involve reflection surfaces or mirrors, which is complex in the sense that they require groove or metal-strip arrays to be fabricated on the YIG film surface. To avoid this complexity, it is also possible to directly use the straight edges of the YIG film to form an MSW cavity resonator. Figure 9 shows such a straight-edge resonator (SER) device, where the YIG/GGG resonator, which is of a rectangular shape cut by a dicing saw, is placed on top

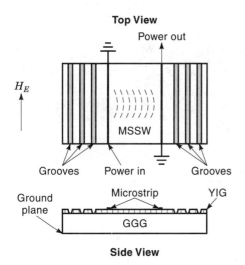

Figure 7. Schematic showing the configuration of an MSSW resonator. Reflecting arrays of grooves are cut on the YIG film surface to form an MSSW cavity resonator. The bias field is applied parallel to the microstrip transducer direction.

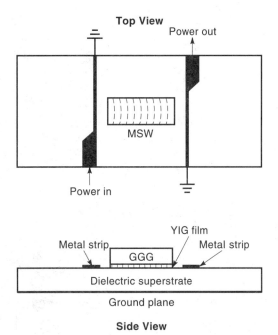

Figure 9. Schematic showing the configuration of a two-port MSW straight-edge resonator. The bias field can be applied perpendicular to the microstrip transducers in the YIG film plane or perpendicular to the YIG film plane for MSFVW or MSSW operation, respectively.

of the dielectric superstrate coupled in and out by the short-circuited microstrip transducers on both sides of the resonator. Depending on the bias field direction, both MSSW and MSFVW SERs can be constructed. Thus, MSSWs propagate along the surface of the YIG film and are reflected back onto the surface at the straight edges. A standing wave pattern results and a high-Q resonance is obtained. In this manner a MSSW SER was reported in Ref. 23, exhibiting an insertion loss of 3.1 dB and a sideband suppression level better than 20 dB tunable from 1 GHz to 22 GHz.

Finally, we consider the MSW structure of Fig. 10, which depicts a one-port resonator circuit. The circuit of Fig. 10 can be fabricated using photolithographic techniques, and, hence, cutting of reflective grooves on the YIG film is avoided. In Fig. 10, MSWs are excited within the metal window coupled in and out via the microstrip transducer. In order to form total reflection of the MSWs at the window boundary, and hence to achieve a high Q value, the geometry of the resonator needs to be carefully designed. The circuit of Fig. 10 allows for operation for all types of MSWs, and simple MSW resonators can be constructed.

MAGNETIC MICROWAVE NONLINEAR DEVICES

As discussed in the section entitled "Theoretical Background," nonlinear interaction terms arise from the equation of motion for the magnetization vector for which the small-signal assumptions no longer hold true. Under these conditions, the magnetization field will couple to itself, resulting in spin-wave interactions between different wave numbers and frequencies. In this section we discuss how this nonlinear phenomenon can be utilized for device applications. Among many important nonlinear devices, we will discuss frequency-selec-

tive limiters, signal-to-noise enhancers, amplitude correctors, and ferrimagnetic echoing devices.

Frequency-Selective Power Limiter

Frequency-selective limiters have been demonstrated using YIG spheres and ferrite slabs in the waveguide, coax, and stripline configurations (24). We will discuss here the limiter operation using a single-crystal yttrium–iron–garnet (YIG) film arranged in the configuration shown in Fig. 3, except that a meander line is normally used instead of a straight microstrip line. Also, a thick YIG film is preferred, because it implies an abundant spin-wave spectrum allowing the guided electromagnetic waves to react sufficiently with spin waves to induce instability at high power. All types of ferrite frequency-selective limiters operate analogously: When a transmission line is loaded with a ferrite element, the transmitted power cannot exceed a threshold value beyond which catastrophic energy transfer occurs between the guided RF electromagnetic fields and the spin precessing motion. This phenomenon has been discussed in the section entitled "Theoretical Background," and it is generally known as *Suhl's spin-wave instability* (12). Figure 11 depicts a simple picture explaining the operation of a frequency-selective power limiter.

We note that Suhl's spin-wave instability occurs when the RF field is perpendicular to the dc bias field. However, in some limiter applications the instability can also be induced by an RF field component parallel to the dc field. Spin-wave instability of this kind is known as parallel pumping, which, similar to Suhl's first-order instability, occurs at a magnon frequency one-half the frequency of the RF-pumping signal, $\omega_k = \omega/2$. The threshold for parallel pumping is

$$h_{\mathrm{P}}^{\mathrm{th}} = 2\Delta H_k \frac{\omega}{\omega_m} \tag{53}$$

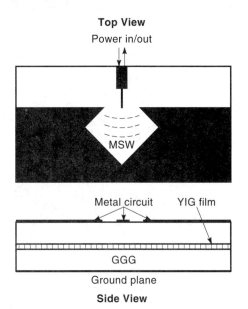

Figure 10. Schematic showing the configuration of a one-port MSW groove-free resonator. The bias field can be applied either perpendicular to the microstrip transducer in the YIG film plane, or perpendicular to the YIG film plane for MSFVW or MSSW operation, respectively.

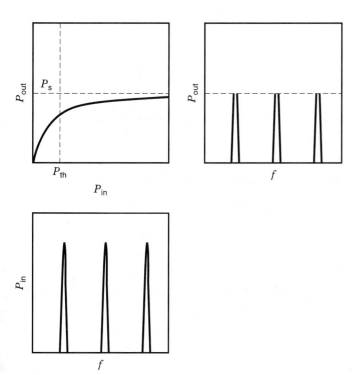

Figure 11. Power transmitted through a nonlinear frequency selective limiter. Input power spectrum is shown at the bottom left, and the output power spectrum is shown at the top right. The relationship between the input and the output powers is shown at the top left.

which compares closely in magnitude to h_d^{th}, the threshold for Suhl's first-order instability shown in Eq. (48).

As shown at the top left of Fig. 11, the relationship between the input power, P_{in}, and the output power, P_{out}, is roughly linear only when P_{in} is smaller than P_{th}, the threshold power. When P_{in} is larger than P_{th}, P_{out} becomes gradually saturated at P_s, and the excess power, $P_{in} - P_{out}$, couples into spin-wave motion to be ultimately converted into lattice vibration and hence dissipated as heat. Thus, when P_{in} is given as a function of frequency, as shown at the bottom left of Fig. 11, the excess power of $P_{in} - P_{th}$ will be removed by the limiter; the resultant output characteristic is shown at the top right of Fig. 11. We note that Fig. 11 only shows an idealized operation. In reality the P_s level shown in the P_{out} versus P_{in} diagram is not a constant; it increases slightly with P_{in} when P_{in} exceeds P_{th}. As such, instead of showing chopped-head peaks, the output will consist of rounded-head transmission peaks in the P_{out} versus f diagram shown in Fig. 11.

As an example, in Ref. 25 a 57 μm thick YIG film was coupled to a 25 μm wide microstrip meander line of characteristic impedance 50 Ω fabricated on a high dielectric constant substrate. This limiter, which operated in the 2 GHz to 4 GHz range, showed a limiting range of 25 dB, a threshold input power level of 0 dBm, and a small signal loss of 7 dB. The upper frequency limits were not sharply defined, since the limiting power decreased with increasing frequency, but operation up to 8 GHz is possible with the present device.

The reason that the limiter device reported in Ref. 25 failed to operate at high frequencies can be explained in terms of the onset condition giving rise to Suhl's spin-wave instabilities. As discussed in the section entitled "Theoretical Background," Suhl's first-order instability occurs for spin waves

possessing a frequency one-half the applied RF frequency. Thus, above 8 GHz, instability will occur in the spin waves of frequencies larger than 4 GHz, which are located above the spin-wave manifold extending the frequency range from 2 to 4 GHz. The second-order spin-wave instability, which involves spin waves of the same frequency as the applied RF signals, has a higher threshold and is not used in frequency-selective power limiters. Thus, a power limiter is operational only when half the applied frequency falls within the spin-wave manifold region bounded by the two frequencies ω_1 and ω_2 shown in Fig. 1. Since ω_1 and ω_2 can be tuned by varying the applied field strength, the limiter is therefore termed a frequency-selective device.

Signal-to-Noise Enhancers

Epitaxially grown single-crystal YIG films have been used to fabricate frequency-selective limiters and signal-to-noise enhancers. Although the construction of these two devices is very similar, as shown in Fig. 3, they perform opposite signal processing functions. The limiter presents low attenuation to low-intensity signals and high attenuation to high-intensity signals while the signal-to-noise enhancer attenuates weak signals more severely than strong signals. However, the major difference results from the origin of nonlinearities admitting the operation of these two devices. For a limiter the nonlinear coupling is related to the onset of spin-wave instabilities, whereas the nonlinear behavior of a signal-to-noise enhancer comes from the generation of magnetostatic waves (MSWs). The occurrence of the latter is at a much lower power level than that of the former.

Although a power limiter and a signal-to-noise enhancer are constructed using similar configuration as shown in Fig. 3, they are operating under different physical principles. While the former device requires the insertion loss to increase with input power, the latter requires the insertion loss to decrease with input power. A power limiter circuit will couple most efficiently with spin waves, whereas a signal-to-noise enhancer will avoid this by operating at a frequency ω so that $\omega/2$ is located well beyond the spin-wave manifold region. On the contrary a signal-to-noise enhancer will couple tightly with magnetostatic waves, for example, MSSW, while for a power limiter it is generally not the case. As a result, meander lines are therefore commonly used in power limiter circuits.

Let us examine the configuration of Fig. 3. Here we assume that the biasing magnetic field is applied parallel to the microstrip direction such that MSSWs are excited propagating perpendicular to the strip. Since the generation of MSSW in this configuration is very efficient, the input RF power is almost entirely consumed for the generation of MSSWs at low input-power levels. However, as the input power increases, the amplitudes of MSSWs increase accordingly until reaching saturation beyond which no more conversion into MSSWs is appreciable. Figure 12 shows that P_{MSSW} increases with the input power, P_{in}, linearly in the initial region but saturates at large P_{in}. The functional dependence of P_{MSSW} is similar to that of the output power from a limiter device shown in Fig. 11. The output power from the signal-to-noise enhancer is $P_{in} - P_{MSSW}$, which is shown as the heavy line in Fig. 12. Thus, from Fig. 12 we conclude that weak signals will be damped more by the generation of MSSWs than strong signals.

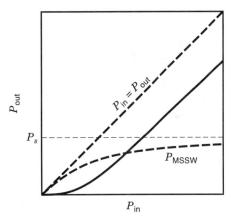

Figure 12. The relationship between the input power, P_{in}, and the output power, P_{out}, of a signal-to-noise enhancer. The converted MSSW power, P_{MSSW}, is also shown, and $P_{out} = P_{in} - P_{MSSW}$.

In Fig. 13 the top diagram shows the power spectrum output from a signal-to-noise enhancer device based on the input spectrum shown at the bottom of the figure. It is seen in Fig. 12 that noises, appearing at low power levels, are damped out, leaving alone the high level signals with improved signal-to-noise ratio. Again, Fig. 12 shows an idealized situation that P_s does not depend on the input power level, P_{in}. In reality, P_s will increase slightly with P_{in} when saturation is approached. As an example, Ref. 26 shows a signal-to-noise enhancer device centered at 3.3 GHz with a bandwidth of 800 MHz which exhibited 16 dB less attenuation when the input power is increased from −6 to +10 dBm.

Amplitude Correctors

For signal processing at microwave frequencies, broad-band amplifiers are needed whose characteristics are desired to show linear dependence on the amplitude of the input signals. However, for most power amplifiers including microwave

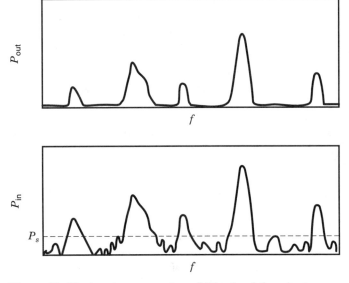

Figure 13. The input power spectrum (bottom) and the output power spectrum (top) applied to and transmitted from a signal-to-noise enhancer, respectively.

traveling wave tubes (TWTs) the amplification deteriorates at high power, resulting in reduced power amplification for high-power input signals. One possible way to resolve this problem is to compensate the input power with a corrector which attenuates more input power at low input-power levels. This is exactly the same characteristics that we have discussed for a signal-to-noise enhancer, and hence it can be equally used as an amplitude corrector device. It is shown in Ref. 27 that an amplitude corrector operated at 3 GHz with low and high signal suppression level of 4.2 and 1 dB, respectively, and the threshold power level was 100 mW. This device was equipped with an O-type TWT operating from 2.5 GHz to 3.6 GHz, and correct output characteristics have thus been obtained.

Nonlinear Ferrimagnetic Echoing Devices

Echo phenomena are characterized by the reradiation of the input signals stored in a nonlinear system through the agitation of a consequently applied pump pulse. Observation of ferrimagnetic echoes was first reported in 1965 by Kaplan in polycrystalline yttrium–iron–garnet (YIG) samples (28). Amplified echoes were thereafter reported in cylinders and truncated spheres of YIG crystals (29) and in single-crystal YIG films (30). Echo experiments offer a possibility of a novel approach to performing important signal processing functions, such as nondispersive time delay and pulse correlation in the frequency range below 10 GHz. With the demand of electronic technology advances there is now a renewed interest in the use of ferrimagnetic echoing devices.

Among many nonlinear systems capable of producing an echoed signal (e.g., cyclotron echo, plasma echo, molecular echo, phonon echo, and spin echo), only ferrimagnetic echo can show amplification. This feature renders the ferrimagnetic echo phenomena in a very unique position for device applications. Ferrimagnetic echo is mainly concerned with the reservoir of the electron spin system provided by a YIG crystal which is nearly the perfect medium for signal storage. That is, in view of the extremely narrow FMR linewidth (0.33 Oe at X band) the damping action accompanying the spin motion in the YIG crystal is very small; and once the spins are put in motion, they will continue the motion indefinitely in time exhibiting very little attenuation. The lifetime of the magnons in a YIG crystal is very long, usually exceeding 1 μs. We imagine that at the time instant $t = -\tau$ a signal pulse is applied to the YIG crystal setting the spins in precessing motion. Before the spin motion damped out, a pump pulse is applied at the time instant $t = 0$ which is so intense that nonlinear interaction is aroused in the spin system. Due to the (odd) cubic nature of the interaction, the process of time conjugation is recalled, which reverses the time scale for the stored signal such that the spins begin to precess in the opposite direction. As a consequence, at time $t = \tau$ (≤ 1 μs) the original signal pulse recovers, which appears as the image of the original signal echoed back by the pump pulse.

However, we must emphasize that if the cubic interaction is local in nature, no amplification is possible for the echoed signals, as occurs in the other nonlinear systems involving only isolated echoing sites. For ferrimagnetic echo the nonlinear interaction is brought about by the dipolar field interacting with the nonlocal spin waves showing a long-range dependence [this nonlocal interaction can be described in terms of a Green's function whose kernel shows a $1/r$ dependence (31,32)]. As such, the echoed signal can get amplification, absorbing power from the pump pulse not only to restore the shape of the original signal, but also to amplify the signal to high intensity. The amplification gain can be as large as 100, as measured experimentally (28) and calculated theoretically.

Ferrimagnetic echo experiments are carried out using crystal YIG bulk or thin film materials in the presence of a high magnetic-field gradient ($\approx$1 kOe/cm). In order to obtain high amplification gain the YIG material needs to be clamped between two polycrystalline poles to effectively suppress the demagnetizing field. Theoretically it has been demonstrated that the dc demagnetizing field has adverse effect in reducing the echo gain to zero (31,32). This result is consistent with Suhl's finding (12) that the dc demagnetizing field is apt to enhance spin-wave instability, resulting in subsidiary absorption during microwave resonance measurements.

Finally, we want to point out the similarities between the ferrimagnetic echo signals and the intermodulation noises observed in a ferrite circulator junction (33). Two RF signals at adjacent frequencies f_1 and f_2 will couple each other to form intermodulation noises at frequencies $2f_1 - f_2$ and $2f_2 - f_1$. Intermodulation signals grow rapidly with power, which may be identified as clicking noises in a telephone line. Actually, both the echoing signals and the intermodulation noises are generated through the same cubic interaction terms in the equation of motion. The only difference is that echo is a phenomenon in the time domain, whereas intermodulation is manifested in the frequency domain. While the dc demagnetizing field has been shown to have adverse effects in influencing the echo gain in the time domain, it has also been demonstrated that the same demagnetizing field will enhance the intermodulation level in a circulator junction (33). Therefore, the dc demagnetizing field needs to be suppressed or minimized in nonlinear studies concerning ferrite materials dealing with either the spin-wave instability, the echoing gain, or the intermodulation noises.

MAGNETOOPTIC DEVICES

At optical frequencies the Polder permeability tensor of a ferrite specimen is nearly isotropic with the diagonal element $\mu \approx 1$ and off-diagonal element $\kappa \approx 10^{-5}$, as can be calculated from Eqs. (11) and (12). Though small, the resultant magnetic anisotropy or gyrotropy can be measured using a laser optical beam. Upon incidence, the reflected and the transmitted beams will carry the magnetization information of the specimen, resulting not only in a rotation in polarization, but also in a change in reflectivity and transmission. This is called magnetooptic Kerr and Faraday effects for reflection and transmission measurements, respectively. Kerr and Faraday effects have been used in observing the dynamic processes of domain-wall motion in a ferromagnetic metal or a ferrimagnetic insulator sample, respectively (34).

The most important devices utilizing magnetooptic coupling for electronic signal processing applications concerns the scattering process between photons and magnons. This process is called Bragg diffraction, which in a ferrimagnetic medium a photon of momentum $\boldsymbol{P}$ is scattered by a magnon of momentum $\boldsymbol{k}$ to result in a photon of momentum $\boldsymbol{P} + \boldsymbol{k}$. This interaction is second order in nature and is described by

the term $\gamma\mu_0\, \boldsymbol{m}_k \times \boldsymbol{h}_P$ in Eq. (7). Here $\boldsymbol{m}_k$ denotes the magnetization field of the magnon and $\boldsymbol{h}_P$ is the magnetic field of the photon. As such, the photon wave is said to be modulated by the magnetization wave, carrying along with it the electronic information after scattering. Due to the nature of a second-order interaction, Bragg diffraction between photons and magnons, or MSWs in a magnetic film, is not very prominent, and only about 4% light diffraction was observed experimentally with a 7 mm interaction length for MSSW excitation approaching saturation (35).

Optical techniques are being increasingly utilized to meet the ever-growing data rate requirements of signal processing and communication applications. A key element to such applications has been acoustooptic modulators based on Bragg diffraction between photons and phonons. A large time-bandwidth product (TBW; i.e., time delay of acoustic signal in traversing the optical beam-times-signal bandwidth) is usually desirable. However, acoustic waves cannot be efficiently excited at frequencies above 2 GHz. Instead, magnetooptic devices offer the potential of large TBW modulation directly at microwave frequencies. The diffraction of guided optical waves by MSW is analogous to optical diffraction by a SAW and has the potential to enhance a wide variety of integrated optical applications such as spectrum analyzer, optical filters, deflectors, switches, and convolvers.

The basic theories of the MSW–optical interaction, including the development of expressions for optical diffraction efficiency and coupling factor as a function of the MSW power and other relevant parameters, has been derived in Ref. 35. The theory applies to the collinear configuration, shown in Fig. 14, with the MSW traveling in a direction parallel or antiparallel to the optical beam, as well as to the transverse configuration, shown in Fig. 15, where the MSW travels at an angle to the optical beam. In Figs. 14 and 15 prisms are used to guide optical beams onto the YIG layers which are transparent to light wave propagation. In Figs. 14 and 15 the MSW configuration may be replaced by the superstrate structure shown in Fig. 3. To achieve good optical properties the YIG film is usually highly doped, mainly by bismuth. As such, Bragg diffraction of guided optical waves by MSSWs and MSFVWs has been demonstrated in the configurations shown in Figs. 14 and 15, and the conversions between TE and TM

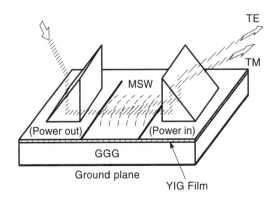

Figure 14. Collinear configuration with codirectional MSW and optical beams. Due to Bragg diffraction between the optical beam and the MSWs the original TM guided optical beam is scattered into a TE beam with deflected angle of propagation.

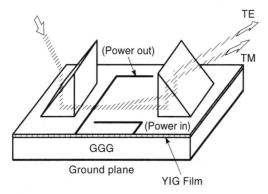

Figure 15. Transverse configuration for MSW and optical beams. Bragg diffraction of the guided optical wave by an MSW induces conversion between orthogonally polarized optical modes.

modes have been experimentally observed from 1 to 6.5 GHz (35).

These MSW-optical devices are currently in an early stage of development, with their basic feasibility having been demonstrated. However, they are expected to lead to a variety of high-performance integrated optical signal processing devices.

ANTIREFLECTION LAYERS AND ABSORBING LAYERS

To date, microwave or millimeter-wave antireflection layers and absorbing layers are almost exclusively used for radome design applications. Due to the highly classified nature of this topic, not much data have been published in the literature. In this section we discuss only the concepts that lead to the construction of microwave or millimeter-wave antireflection and absorbing layers. No explicit design parameters are given here. By definition, an antireflection/absorbing layer is to be placed on top of a substrate such that an incident microwave or millimeter wave beam will be totally transmitted/absorbed when passing across the layer without causing reflection. The layer shall be functional over a frequency range as wide as possible to be independent of the incident angle and polarization of the incident beam.

To realize the design of an antireflection layer, we consider first the case of normal incidence. According to the transmission line theory one concludes immediately that the first order solution would require the layer to behave as a quarter-wave transformer. This implies that the layer shall possess a thickness equal to one-quarter the wavelength with a characteristic impedance

$$Z_L = (Z_0 Z_S)^{1/2} \qquad (54)$$

where

$$Z_0 = (\mu_0/\epsilon_0)^{1/2} \qquad (55)$$

$$Z_S = (\mu_S/\epsilon_S)^{1/2} \qquad (56)$$

Z_0 (Z_S) is the characteristic impedance of air (substrate), and ϵ_0 (ϵ_S) and μ_0 (μ_S) are the permittivity and permeability of air (substrate), respectively. When wider bandwidth is desired, higher-order solutions are required, and this results in a

multilayer system with progressively changing electromagnetic parameters matching the impedance difference between air and substrate.

For an absorbing layer design the air impedance can also be matched by a multilayer system by progressively increasing damping parameters such that the series of the layers satisfies the transformer matching requirement. That is, we use the same impedance transformer theory for the design of absorbing layers, except that the impedances of the layers are now complex numbers, since the permittivity and permeability of the layers are complex numbers. As such, the incident wave damps out when it passes through the layer system before arriving at the substrate.

When oblique-angle incidence is considered, the impedance transformer theory can be generalized using the transfer matrix technique (36). The present problem becomes one of determining an optimal layer system design allowing for polarization-independent operation over as wide a frequency band and angle range on beam incidence as possible. Also, the design task is subject to a very important constraint requiring the thickness of the layers to be minimal, since at microwave and/or millimeter frequencies a practical layer system shall be thin enough compared with the wavelength in air. The design is in general not a trivial problem, and efficient computer algorithms are needed.

Once an optimal multilayer system is determined, the remaining task is to synthesize it using real materials. Unfortunately, nature does not provide general materials covering the whole range of electromagnetic parameters. Instead, artificial materials need to be developed. The first kind of artificial materials includes particle composites that contain dielectric particles (37), metal-shelled particles (38), and ferrite particles (39,40), and graphite powders are embedded in a matrix epoxy such that the effective permittivity, permeability, and/or conductivity of the composite can be controlled over the desired frequency range by adjusting the mixing fraction of the particles.

The second class of artificial materials is quite new; it involves periodic patterns of metal strips or grooves (surface relief gratings) to be fabricated on top of a layer surface. In fact when electromagnetic waves interact with periodic structures much finer than the wavelength, they do not diffract, but instead reflect and transmit as if they are encountering a nonstructured medium. Effective field theory describes the interaction between electromagnetic waves and such subwavelength structures by representing a region of subwavelength heterogeneity in terms of a homogeneous material possessing a single set of effective electromagnetic parameters: permittivity, permeability, and conductivity (41). Actually, the antireflection structures consisting of surface relief gratings can be found on the cornea of certain night-flying moths, and the first scientists investigating antireflection structured surfaces for application in the visible or near infrared portion of the spectrum worked to replicate moth eye surfaces (42).

BIBLIOGRAPHY

1. B. Lax and K. J. Button, *Microwave Ferrites and Ferrimagnetics*, New York: McGraw-Hill, 1962.

2. J. C. Slater, *Quantum Theory of Matter,* 2nd ed., New York: McGraw-Hill, 1972.

3. H. How, R. C. O'Handley, and F. R. Morgenthaler, Soliton theory for realistic domain wall dynamics, *Phys. Rev. B,* **40** (7): 4809, 1989.

4. C. Vittoria, *Microwave Properties of Magnetic Films,* World Scientific, Singapore, 1993.

5. S. Chikazumi and S. H. Charap, *Physics of Magnetism,* New York: Wiley, 1964.

6. C. Kittel, *Introduction to Solid State Physics,* 4th ed., New York: Wiley, 1971.

7. F. R. Morgenthaler, Dynamic magnetoelastic coupling in ferromagnets and antiferromagnets, *IEEE Trans. Magn.,* **MAG-8**: 130 (1972).

8. H. How and C. Vittoria, Surface retarded modes in multilayered structures: parallel magnetization, *Phys. Rev. B,* **39** (10): 6823, 1989.

9. H. How and C. Vittoria, Bulk and surface retarded modes in multilayered structures: antiparallel magnetization, *Phys. Rev. B,* **39** (10): 6831, 1989.

10. R. E. Collin, *Foundations for Microwave Engineering,* New York: McGraw-Hill, 1966.

11. C. Vittoria and N. D. Wilsey, Magnetostatic wave propagation losses in an anisotropic insulator, *J. Appl. Phys.,* **45**: 414, 1974.

12. H. Suhl, The theory of ferromagnetic resonance at high signal powers, *J. Phys. Chem. Solids,* **1**: 209, 1957.

13. V. L. Taylor, J. C. Sethares, and C. V. Smith, Jr., MSW terminations, *Proc. IEEE Ultrasonic Symp.,* 562, 1980.

14. M. R. Daniel, J. D. Adam, and T. W. O'Keeffe, Linearly dispersive delay lines at microwave frequencies using magnetostatic waves, *Proc. IEEE Ultrasonic Symp.,* 806, 1979.

15. K. W. Chang, J. M. Owens, and R. L. Carter, Linearly dispersive delay control of magnetostatic surface wave by variable ground-plane spacing, *Electron. Lett.,* **19**: 546, 1983.

16. F. R. Morgenthaler, Field gradient control of magnetostatic waves for microwave signal processing applications, *Proc. RADC Microwave Magn. Workshop,* 133, 1981.

17. L. R. Adkins and H. L. Glass, Dispersion control in magnetostatic delay lines by means of multiple magnetic layer structures, *Proc. IEEE Ultrasonic Symp.,* 526, 1980.

18. W. S. Ishak and K. W. Chang, Magnetostatic wave devices for microwave signal processing, *Hewlett-Packard J.,* 10, 1985.

19. J. D. Adam, An MSW tunable bandpass filter, *Proc. IEEE Ultrasonic Symp.,* 157, 1985.

20. J. P. Castera and P. Hartemann, Adjustable magnetostatic surface wave multistrip directional coupler, *Electron. Lett.,* **16**: 195, 1980.

21. J. P. Castera, New configurations for magnetostatic wave devices, *Proc. IEEE Ultrasonic Symp.,* 514, 1980.

22. J. P. Castera and P. Hartemann, A multipole magnetostatic wave resonator filter, *IEEE Trans. Magn.,* **MAG-18**: 1601, 1982.

23. K. W. Chang and W. S. Ishak, Magnetostatic forward volume waves straight-edge resonators, *Proc. IEEE Ultrasonic Symp.,* 473, 1986.

24. G. S. Uebele, Characteristics of ferrite microwave limiters, *IEEE Trans. Microw. Theory Tech.,* **7**: 18, 1959.

25. S. N. Stizer and H. Goldie, A multi-octave frequency selective limiter, *IEEE MTT-S Dig.,* 326, 1983.

26. J. D. Adam and S. N. Stizer, A magnetostatic signal to noise enhancer, *Appl. Phys. Lett.,* **36**: 485, 1980.

27. W. S. Ishak, E. Reese, and E. Huijer, Magnetostatic wave devices for UHF band applications, *Circuits Syst. Signal Process.,* **4**: 285, 1985.

28. D. E. Kaplan, Magnetostatic mode echo in ferromagnetic resonance, *Phys. Rev. Lett.,* **14** (8): 254, 1965.

29. D. E. Kaplan, R. M. Hill, and G. F. Herrmann, Amplified ferrimagnetic echoes, *J. Appl. Phys.,* **40** (3): 1164, 1969.

30. F. Bucholtz, D. C. Webb, and C. W. Young, Jr., Ferrimagnetic echoes of magnetostatic surface wave modes in ferrite films, *J. Appl. Phys.,* **56** (6): 1859, 1984.

31. H. How and C. Vittoria, Theory on amplified ferrimagnetic echoes, *Phys. Rev. Lett.,* **66** (12): 1626, 1991.

32. H. How and C. Vittoria, Amplification factor of echo signals in ferrimagnetic materials, *IEEE Trans. Microw. Theory Tech.,* **MTT-39**: 1828, 1991.

33. H. How, T.-M. Fang, C. Vittoria, and R. Schmidt, Nonlinear intermodulation coupling in ferrite circulator junctions, *IEEE Trans. Microw. Theory Tech.,* **MTT-45**: 245, 1997.

34. J. M. Florczak and E. D. Dahlberg, Detecting two magnetization components by the magneto-optical Kerr effect, *J. Appl. Phys.,* **67**: 7520, 1990.

35. A. D. Fisher, Optical signal processing with magnetostatic waves, *Circuits Systems Signal Process.,* **4**: 265, 1985.

36. H. How, W. Hu, and C. Vittoria, Planar circuits in stratified dielectric/magnetic layered structures—An application to a slot line fabricated on a ferrite substrate, *IEEE Trans. Microw. Theory Tech.,* 1997, to be published.

37. L. Rayleigh, On the influence of obstacles arranged in rectangular order upon the properties of a medium, *Philos. Mag.,* **34**: 481, 1982.

38. H. How, W. A. Spurgeon, and C. Vittoria, The Microwave Properties of Conducting Spherical Shells, Technical Report, U.S. Army Materials Technology Laboratory, MTL TR 90-7, February 1990.

39. H. How and C. Vittoria, Demagnetizing energy and magnetic permeability tensor of spheroidal magnetic particles dispersed in cubic lattices, *Phys. Rev. B,* **43** (10): 8094, 1991.

40. H. How and C. Vittoria, The polder tensor of spherical magnetic particles in cubic lattices: An exact solution in multipole expansion, *Phys. Rev. B,* **44** (17): 9362, 1991.

41. D. H. Raguin and G. M. Morris, Antireflection Structured Surfaces for the Infrared Spectral region, *Appl. Opt.,* **32**: 1154, 1993.

42. P. B. Clapham and M. C. Hutley, Reduction of lens reflection by the 'moth eye' principle, *Nature (London),* **244**: 281, 1973.

HOTON HOW
ElectroMagnetic Applications

MAGNETIC MODELING

THE CLASSICAL PREISACH FRAMEWORK

The first hysteresis model that recognized that in order to describe minor loop behavior, the model must have memory, was described by the Hungarian born physicist, Franz (Ferenc) Preisach, in his landmark paper almost 60 years ago (1). However, the novelty of Preisach's approach was not recognized and applied to describe magnetic hysteresis until the late 1950s (2). It is noted that an approach identical to Preisach's was independently discovered and developed for many years to describe adsorption hysteresis (3), without recognizing the similarities with Preisach's original approach. The unified formal mathematical treatment of hysteresis in general has been defined recently (4).

In the Preisach method, each elementary particle has a rectangular hysteresis loop and, as an isolated particle has

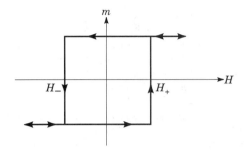

Figure 1. Classical Preisach representation of an isolated particle.

"up" and "down" switching fields, H_+ and H_- are equal in magnitude, as illustrated in Fig. 1. Since energy is dissipated during hysteresis, the hysteresis loop is always traversed in the counter clockwise direction; therefore, $H_+ \geq H_-$ for all particles in the medium. From a physical point of view the "up" and "down" saturated states are identical; therefore, the magnitude of the magnetization, m, in these two states is the same.

The Preisach method is a hysteresis model to describe the behavior of a collection of interacting particles with different up and down switching fields. In an assembly of particles, the field that an individual particle "sees" is not the external field but the sum of the external field and the interaction H_i due to other particles in the medium. In the classical Preisach representation, the effect of the interaction field H_i is to shift the hysteresis of the particle by H_i, as illustrated in Fig. 2.

The classical Preisach technique uses a statistical approach to describe the hysteretic many-body problem. The method assumes that the magnetic medium is composed of a continuum of "elementary particles," each of which characterizes the average behavior of an ensemble of particles. The *Preisach calculation plane* (also called the Preisach plane) takes the up and down switching fields as the coordinate axes. A point on the Preisach plane with coordinates (H_+, H_-) corresponds to the particle whose elementary hysteresis loop switches up and down at H_+ and H_-, respectively. Magnetic materials exhibit saturation-type hysteresis; therefore, it is reasonable to assume that all elementary particles switch to their up state above a positive saturating field H_S and switch down below a negative saturating field, $-H_S$. The choice of the value of H_S depends on the material that is being characterized. With these considerations, the physical region of the Preisach plane and representative elementary particles

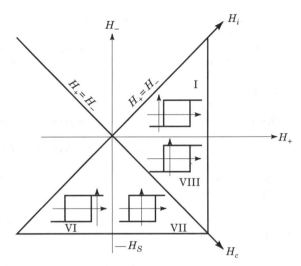

Figure 3. Representative elementary-particle hysteresis loop in each octant of the physically realizable region of the classical Preisach plane.

in it are shown in Fig. 3. We will call regions I, VI, VII, and VIII the first, second, third, and fourth quadrants of the Preisach plane, respectively.

The *normalized Preisach density function,* or Preisach function for short, $p(H_+, H_-)$ is a three-dimensional density function defined over the Preisach plane, as illustrated in Fig. 4. The value of $p(H_+, H_-)$ is the fraction of the magnetization contribution by the point (H_+, H_-) to the total normalized magnetization m. The density function is zero outside the Preisach plane and, from magnetic symmetry, $p(H_+, H_-) = p(-H_-, -H_+)$. Thus,

$$\iint\limits_{H_+ \geq H_-} p(H_+, H_-)\, dH_+ dH_- = 1 \qquad (1)$$

In most cases one simply illustrates the Preisach function by its contour plot or by a representative curve of the contour plot.

The normalized Preisach function can also be defined as $p(H_i, H_c)$, where H_i is the interaction field and H_c is the critical

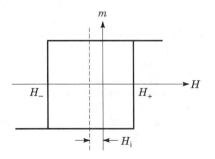

Figure 2. Classical Preisach representation of a particle in the presence of an interaction field H_i.

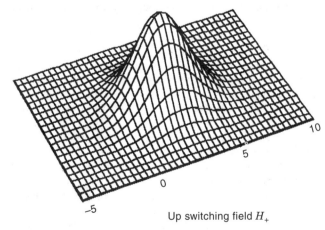

Figure 4. Illustration of the normalized Preisach function in the (H_+, H_-) coordinate system.

field, and normalization requires that

$$\int_{H_c > 0} \int_{-\infty}^{\infty} p(H_i, H_c)\, dH_i dH_c = 1 \qquad (2)$$

In magnetic media, the critical field distribution and the interaction field distribution are independent of each other, that is,

$$p(H_i, H_c) = p_i(H_i) p_c(H_c) \qquad (3)$$

The (H_i, H_c) and (H_+, H_-) coordinate systems are related by

$$H_i = \frac{H_+ + H_-}{2}, \qquad H_c = \frac{H_+ - H_-}{2} \qquad (4)$$

and

$$H_+ = H_i + H_c, \qquad H_- = H_i - H_c \qquad (5)$$

It is often convenient to use one or the other of these two classical Preisach coordinate systems.

From a control point of view, the classical Preisach model can also be viewed as a parallel connection of rectangular two-state hysterons (4) with a distribution of up and down switching fields, as illustrated in Fig. 5.

CALCULATION OF THE MAGNETIZATION USING THE CLASSICAL PREISACH MODEL

We will now show how the classical Preisach model is used to compute the magnetization due to the applied field sequence illustrated in Fig. 6(a). The magnetization state of a hysteretic system depends on the magnetization history; therefore, the initial magnetization state of the system has to be known. For this example, we will assume that the initial state is negative saturation; other types of initial states will be discussed later. Using the classical Preisach representation, negative saturation means that every elementary particle on the Preisach plane is in its down state.

Let us first compute the magnetization at $H = H_*$, illustrated in Fig. 6(b). As discussed previously, the location of each elementary particle on the classical Preisach plane is determined by its up switching field H_+ and its down switching field H_-. Thus, elementary particles to the left of the vertical line that intersects the H_+ axis at H_* will switch to their

up state, indicated by the + sign, since they all have up switching fields less than the applied field H_*. Elementary particles to the right of the vertical line have up switching fields greater than the applied field; therefore, they will remain in their previous magnetization state. In this example, since the initial state was negative saturation, particles in this region will remain in their negative state, as indicated by the − sign.

The case when the field reaches $H = H_1$ is illustrated in Fig. 6(c). The calculation is similar to that described previously, but since the applied field is larger, the positively magnetized region, indicated by the + sign, has grown further at the expense of the negatively magnetized region, indicated by the − sign.

As the field is reduced from H_1, let us consider the case when $H = 0$, illustrated in Fig. 6(d). At this field, all elementary particles above the horizontal line corresponding to the applied field $H = 0$, which in this case coincides with the H_+ axis, will switch to their down state, since their switching field is greater than the applied field. All particles below the line will remain in their previous magnetization state. Now, since the "medium" has been exposed to a magnetization history, the positively and negatively magnetized regions are separated by a staircase line.

The case when the field reaches $H = H_2$ is illustrated in Fig. 6(e). The calculation is similar to that described previously, but since the applied field was further reduced, the negatively magnetized region, has grown at the expense of the positively magnetized region.

The classical Preisach plane corresponding to the final state of the magnetizing process at $H = H_3$ is illustrated in Fig. 6(f). It is seen that the positively and negatively magnetized regions are separated by a staircase line. The coordinates of the two vertices of the staircase line are (H_1, H_2) and (H_3, H_2).

Thus, the normalized magnetization m, through use of the classical Preisach model, is computed by

$$nm = \int\!\!\int_{R_+} p(H_+, H_-)\, dH_+ dH_- - \int\!\!\int_{R_-} p(H_+, H_-)\, dH_+ dH_- \qquad (6)$$

where R_+ and R_- denote the positively and negatively magnetized regions of the Preisach plane, respectively.

One may also write the preceding Preisach integral as

$$nm = \int\!\!\int_{H_+ \geq H_-} Q(H_+, H_-) p(H_+, H_-)\, dH_+ dH_- \qquad (7)$$

where the state variable, $Q(H_+, H_-)$, takes on the value $+1$ and -1 in the positively and negatively magnetized regions, respectively.

As shown in Fig. 1, the magnetization of an elementary Preisach particle switches discontinuously at the particle's up and down switching field. However, since the Preisach plane is comprised of a distribution of such elementary particles, the magnetization computed by the classical Preisach model is in general a smooth function of the applied field. The smoothness of the computed hysteresis curve depends on the number of field points at which the magnetization is computed and the resolution with which the Preisach density

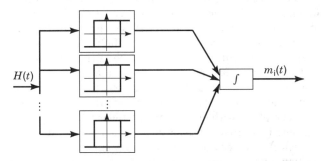

Figure 5. Classical Preisach model as a collection of parallely connected rectangular hysterons with a distribution of up and down switching fields.

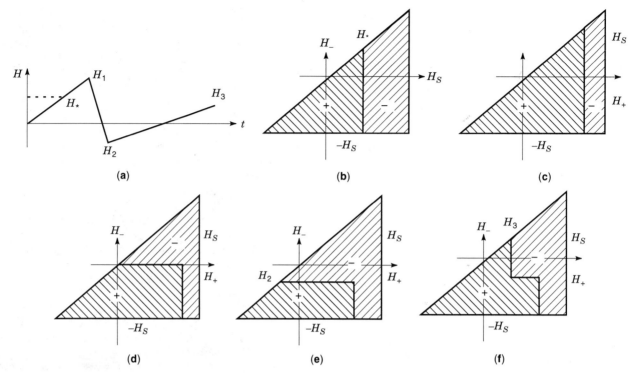

Figure 6. Applied-field sequence to illustrate magnetization calculation with the classical Preisach model, starting from (a) an initial magnetization state of negative saturation. Division of Preisach plane at (b) $H = H_*$, (c) $H = H_1$, (d) $H = 0$, (e) $H = H_2$, and (f) $H = H_3$.

function is discretized. Computational issues of the model, which are beyond the scope of this article, are discussed in Ref. 5.

The hysteresis loop computed by the classical Preisach corresponding to the applied field sequence of Fig. 6 is illustrated in Fig. 7.

PROPERTIES OF THE CLASSICAL PREISACH MODEL

Irreversible, Locally Reversible, and Apparent Reversible Magnetization

With use of Fig. 3, it is seen that the model assumes that the elementary particles comprising the medium have perfectly

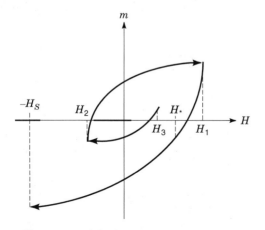

Figure 7. Illustration of the hysteresis curve corresponding to the magnetizing process of Fig. 6 using the classical Preisach model.

square hysteresis loops. This means that the magnetization of each particle is always in its saturated state at $\pm m_S$ and jumps discontinuously from one state to the other at the particle's switching fields. In other words, since the magnetization of each elementary particle is constant until its switching field is reached, the magnetization is purely irreversible.

The locally reversible magnetization, which is defined using the stored energy in the system, is discussed in detail in Ref. 6. However, at this point it needs to be pointed out that during a locally reversible process, no energy is dissipated, that is, the locally reversible magnetization component stores the energy supplied to it by the field and returns it when the field is reversed. Furthermore, due to energy considerations, the reversible magnetization has to be zero when the applied field is zero. Since the magnetization of the elementary particles of the classical Preisach model is either constant or switches hysteretically, it does not describe locally reversible magnetizing processes.

Let us again consider the magnetizing process illustrated in Fig. 6(a), the corresponding intermediate Preisach states shown in Fig. 6(b)–(f) and hysteresis curve given by Fig. 7. We will first focus on the portion of the process where $0 < H < H_1$. A representative Preisach state of this process is illustrated in Fig. 8. It is seen that, as the applied field is reduced from H_1 towards zero, only the cross-hatched region in the first quadrant is switched, since this is the only region of the Preisach plane with elementary particles whose down switching field is positive. Thus, although each elementary Preisach particle has a rectangular hysteresis curve, the appropriate portion of the computed hysteresis curve of the collection of such particles will be descending, as illustrated in Fig. 7. Similarly, for the portion of the magnetizing process where

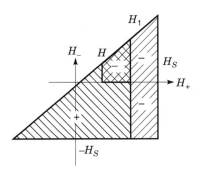

Figure 8. Preisach state of the magnetizing process in Fig. 6(a) for the case $0 < H < H_1$.

$H_2 < H < 0$, the computed hysteresis curve will be ascending due to the contribution of elementary Preisach particles in the third quadrant, as illustrated in Fig. 7.

Let us consider the two magnetizing processes illustrated by the dashed line and the solid line in Fig. 9. The final magnetization state corresponding to the solid lines was given in Fig. 6(f). Using the method presented previously to calculate the magnetization, it is easy to show that the final magnetization corresponding to dashed lines is identical to that of the solid lines. However, it is noted that at any intermediate field value, such as $H = H_*$, the magnetization corresponding to the two processes will be different. Furthermore, it also directly follows from the Preisach method of calculating the magnetization that the magnetization computed by the two processes will be identical at the local extrema H_1, H_2, and H_3. This property, referred to as rate independence, means that the calculated magnetization depends only on the local extrema of the applied field sequence.

Let us now consider the applied field sequence illustrated in Fig. 10(a). The positively and negatively magnetized regions of the Preisach plane corresponding to the local maximum H_1 is shown in Fig. 10(b). The Preisach plane corresponding to the local minimum H_2 is shown in Fig. 10(c). The Preisach plane corresponding to the local maximum H_3, whose magnitude is greater than H_1, is shown in Fig. 10(d). It is seen that this field has deleted the effect of the previous smaller local maximum. In general, it directly follows that each local maximum in the applied field deletes the vertices of the staircase line separating the positively and negatively magnetized regions of the Preisach plane, whose corresponding H_+ coordinates are below this local maximum. Similarly, each local minimum in the applied field deletes the vertices of the staircase line separating the positively and negatively magnetized regions of the Preisach plane, whose corresponding H_- coordinates are above this local minimum. This prop-

erty of the classical Preisach model, which will be referred to as the deletion property, means that a larger applied field completely erases the effect of previous smaller fields. In other words, this property means that a recording system modeled by the classical Preisach model has perfect overwrite.

In general, it is now seen that at any field value H, the classical Preisach plane consists of a negatively and a positively magnetized region that are separated by a staircase line, as illustrated in Fig. 11. The vertices of this staircase line have H_+ coordinates that correspond to previously undeleted local maxima and H_- coordinates that correspond to previously undeleted local minima. The region below the staircase line is magnetized positively; the region above the staircase line is magnetized negatively. The magnetization of the region indicated by "u" will be unaffected by the magnetizing process; therefore, the magnetization of this region is determined by the *initial magnetization state* of the system. In other words, the classical Preisach model stores the magnetization history in the staircase line whose shape depends on the applied field history. Thus, we will call the staircase line corresponding to a particular value of the applied field and a given initial state the *classical Preisach state of the system*. In some cases it may be convenient to represent the classical Preisach state of the system in terms of the state variable $Q(H_+, H_-)$; however, there is obviously a one-to-one correspondence between this and the staircase line.

Let us consider the magnetizing process illustrated by the solid lines in Fig. 12(a). Starting at negative saturation, the field is increased to H_1, where it is reversed. The field is then reduced to H_A, where it is again reversed. Finally, after reaching H_B, the field is cycled between H_A and H_B, traversing a minor loop. The Preisach plane corresponding to this process is illustrated in Fig. 12(b). It is seen that, when traversing this minor loop, only the double shaded triangular region enclosed by H_A, H_B, and the H_i axis of the Preisach plane is switched. It is seen that in the classical Preisach model, a minor loop becomes stable after the first reversal at H_B. Upon subsequent cycling between H_A and H_B, the same minor loop is traversed. It is also seen that

$$m_B - m_A = 2 \int\!\!\int_R p(H_+, H_-) \, dH_+ dH_- \qquad (8)$$

where m_A and m_B denote the magnetization at H_A and H_B, respectively, and the region of integration, R, is the triangular region enclosed by H_A, H_B, and the H_i axis.

Let us now consider the magnetizing process illustrated by the dashed lines in Fig. 12(a). Starting at negative saturation, the field is increased to H_2, where it is reversed and is cycled between H_A and H_B, traversing a minor loop, as discussed previously. The Preisach plane corresponding to this process is illustrated in Fig. 12(c). Similarly to the previous case, only the triangular region enclosed by H_A, H_B, and the H_i axis of the Preisach plane is switched during the traversal of this minor loop. In fact, this region is identical to the corresponding region in Fig. 12(b). Furthermore, by comparing the corresponding regions of Fig. 12(b) and (c), it is seen that the identically shaded negatively magnetized regions are also identical. The positively magnetized region of Fig. 12(c) is greater than that of Fig. 12(b) and the negatively magnetized

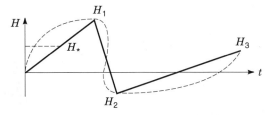

Figure 9. Illustration of rate independence of magnetization calculated by the classical Preisach model.

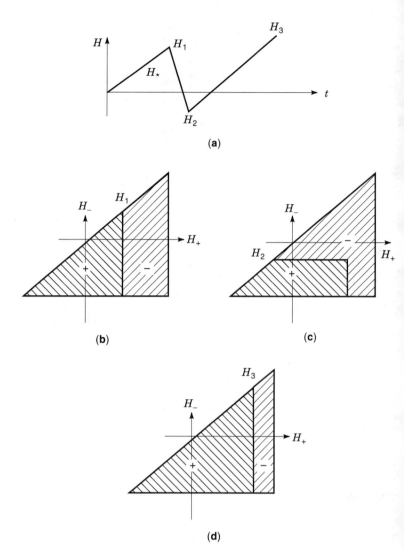

Figure 10. Applied-field sequence illustrating (a) the deletion property of the classical Preisach model. (b) Preisach plane at the local maximum H_1. (c) Preisach plane at the local minimum H_2. (d) Preisach plane at the local maximum H_3, which has deleted the effect of H_1.

region is less than that of Fig. 12(b). Thus, it is seen that the magnetization at *any* field H after the initial reversal point H_1, corresponding to the process illustrated by the solid lines will have a magnetization that is a constant less than the magnetization at the same field H, corresponding to the mag-

netizing process illustrated by the dashed lines. This constant difference in magnetization is given by

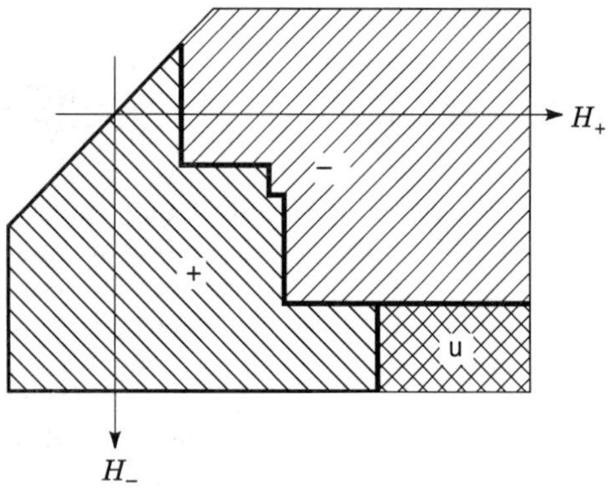

$$\Delta M = M(H_2) - M(H_1) \tag{9}$$

where $M(H_2)$ and $M(H_1)$ denote that magnetization at the reversal points H_2 and H_1, respectively.

In general, it is now seen that minor loops computed by the classical Preisach model close after the first traversal of the minor loop. In other words, minor loops computed by the classical Preisach model *do not accommodate*. We can also see that minor loops obtained by cycling between the same of pair of field extrema but originating at different initial reversal points, such as those illustrated in Fig. 12(a), will have identical shapes, in other words, minor loops computed by the classical Preisach model are *congruent* to each other.

In the previous paragraphs, we have shown that the classical Preisach model possesses the deletion property of applied-field extrema and that minor loops are congruent. It has been mathematically proven that these two properties form *the necessary and the sufficient conditions* for a process to be representable by the classical Preisach model (3).

Figure 11. Illustration of positively and negatively magnetized regions and the staircase line separating these regions. The region whose magnetization is unaffected is indicated by "u."

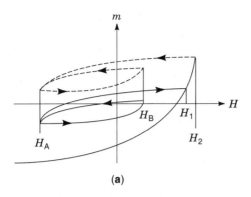

(a)

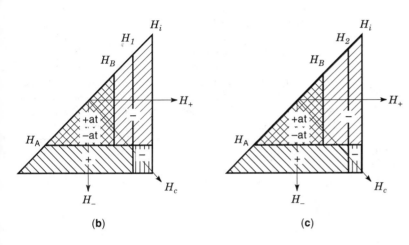

(b)   **(c)**

Figure 12. (a) Magnetizing process illustrating congruent minor loops computed by the classical Preisach model. (b) Preisach plane corresponding to reversal at H_1 and cycling between H_A and H_B. (c) Preisach plane corresponding to reversal at H_2 and cycling between H_A and H_B.

IMPROVED PREISACH-BASED MODELS

Since most magnetic materials do not exhibit some of the properties of the classical Preisach model, in order to develop a physically derivable model is necessary to extend it. We will only discuss physically derived extensions of the model computing static magnetic hysteresis. Physically derived dynamic extensions of the model have been developed by G. Bertotti et al. A detailed description of mathematically motivated extensions to the model are discussed in detail in Ref. 7.

A fundamental assumption of the classical Preisach model is that a constant Preisach function exists for the material to be modeled. However, since this function is a statistical description of the interaction field distribution and the critical field distribution, and the interaction field is a function of the magnetization state of the particles, the question arises whether a "stable" Preisach function exists. An analytical study of interacting Stoner–Wohlfarth particles showed that, if in addition to the up and down switching fields, the Preisach function is also a function of the magnetization, then the physically derived Preisach function is always statistically stable. Furthermore, it was shown that the standard deviation of this Preisach function is constant and its expected value is linearly proportional to the magnetization. This model, called the moving model (8) is a physically derived Preisach-based hysteresis model. As shown in Fig. 13, the moving model is a magnetization-dependent Preisach model with positive feedback. The feedback constant is the material-dependent moving constant whose value depends on the shape, orientation, and reversal mode of the particles that make up the medium.

Another fundamental limitation of the Preisach model is that since the elementary hysterons that make up the model are rectangular, the model only computes the irreversible component of the magnetization. For a detailed description of the irreversible, reversible, and locally reversible magnetization, and the energy relations in Preisach-based hysteresis models, which are beyond the scope of this article, see Ref. 6. In the complete moving hysteresis (CMH) model (9), each point on the Preisach plane is represented by a more realistic, nonrectangular hysteresis loop, as illustrated in Fig. 14. This realistic hysteron can be broken up into a rectangular classical Preisach-like hysteron computing the irreversible magnetization and to a single-valued nonlinear function computing the locally reversible component of the magnetization. In order to give the model a physical foundation, the CMH model includes the material-dependent moving parameter, as discussed previously.

SUMMARY

Hysteresis models based on the classical Preisach model have recently evolved to the point where they can be utilized to

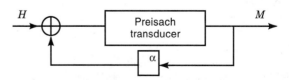

Figure 13. Functional block diagram of the moving Preisach model. The feedback is the moving parameter α, and the box denoted *Preisach transducer* is shown in detail in Fig. 5.

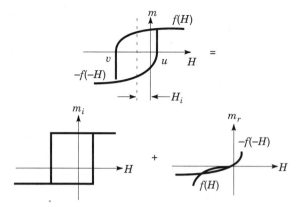

Figure 14. Elementary Preisach particle in the CMH model. Decomposition of an interacting hysteron into the sum of a purely irreversible and a purely locally reversible component.

model, predict, and explain experimental phenomena that cannot be described using any other modeling technique. This article was limited to describing the fundamentals of the model. For a detailed description of hysteresis models, including vector hysteresis modeling, modeling of accommodation and aftereffect and robust identification methods to measure the model parameters, see Ref. 10.

BIBLIOGRAPHY

1. F. Preisach, Uber die magnetische Nachwirkung, *Z. Phys.,* **94**: 277–302, 1935.

2. G. Schwantke, The magnetic tape recording process in terms of the Preisach representation, *Frequenz,* **12**: 383–394, 1958. English translation in *J. Audio Eng. Soc.,* **9**: 37–47, 1961.

3. I. D. Mayergoyz, Mathematical models of hysteresis, *Phys. Rev. Lett.,* **56**: 1518–1521, 1986.

4. M. A. Krasnoselskii, *Systems with Hysteresis,* (in Russian), Moscow: Nauka, 1983. English translation: Berlin: Springer, 1989.

5. F. Vajda and E. Della Torre, Efficient numerical implementation of complete-moving-hysteresis models, *IEEE Trans. Magn.,* **29**: 1532–1537, 1993.

6. E. Della Torre, Energy relations in hysteresis models, *IEEE Trans. Magn.,* **28**: 2608–2610, 1992.

7. I. D. Mayergoyz, *Mathematical Models of Hysteresis,* Berlin: Springer, 1991, p. 44.

8. E. Della Torre, Effect of interaction on the magnetization of single domain particles, *IEEE Trans. Audio Electroacoust.,* **14**: 86–93, 1965.

9. F. Vajda, E. Della Torre, and M. Pardavi-Horvath, Analysis of reversible magnetization-dependent Preisach models for recording media, *J. Mag. Magn. Mater.,* **115**: 187–189, 1992.

10. E. Della Torre, *Magnetic Hysteresis,* to be published.

Reading List

G. Bertotti and V. Basso, Considerations on the physical interpretation of the Preisach model for ferromagnetic hysteresis, *J. Appl. Phys.,* **73**: 5827–5829, 1993.

D. H. Everett, A general approach to hysteresis, Part 4: An alternative formulation of the domain model, *Trans. Faraday Soc.,* **51**: 1551–1557, 1955.

Ferenc Vajda
ADE Technologies

MAGNETIC NOISE, BARKHAUSEN EFFECT

The change of the magnetic state of a ferro- or ferrimagnetic body under the effect of a slowly changing externally applied magnetic field is a complex process, involving reversible and irreversible changes of the magnetization. Due to irreversible magnetization processes, any change of the magnetic state is accompanied by losses, manifested in the presence of the magnetic hysteresis. The reason for the losses is that any real material has a defect structure, responsible for the details of the hysteresis loop. The magnetization changes discontinuously, by jumps from defect to defect, giving rise to magnetic noise, which can be made audible through a headphone. This process and the corresponding noise is named after its discoverer, H. Barkhausen (1919) as *Barkhausen jumps* and *Barkhausen noise*.

The magnetization process (M) of a ferromagnetic (or ferrimagnetic) body is represented by a nonlinear and multivalued function of the applied magnetic field (H), the magnetic hysteresis loop. Figure 1 shows the magnetic hysteresis loop of a magnetic composite of nanometer size iron particles embedded in a nonmagnetic ZnO matrix. At any given temperature the maximum attainable magnetization is called the saturation, or spontaneous magnetization, M_s. Reducing the magnetic field from saturation to $H = 0$ brings the magnetic body to the remanent state with the remanent magnetization (or remanence) of M_r. Further reducing the field in the negative direction down to the coercive field (coercivity), at $-H = -H_c$, the magnetization will be reduced to $M = 0$.

The equilibrium state of the magnetization corresponds to the minimum of the free energy of the magnetic material. The most important contributions to the free energy are due to (1) the quantum mechanical exchange energy, responsible for the collective ordering of the individual spin magnetic moments of the electrons; (2) the magnetocrystalline anisotropy energy, favoring some crystalline directions for the magnetization direction with respect to others, giving rise to the easy and hard directions of magnetization; (3) magnetostrictive stresses, that is, the length change of a magnetic material in the presence of mechanical stress; (4) the Zeeman energy of the interaction with the externally applied magnetic field(s); (5) the magnetostatic energy due to the creation of internal and external demagnetizing fields, depending on the shape and size of the magnetic body; and (6) the energy of interaction of the magnetization with defects in the magnetic body. Due to (6)

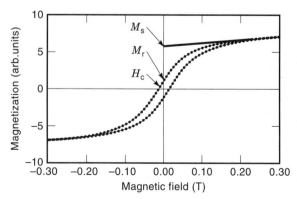

Figure 1. Magnetic hysteresis loop of a nanocomposite of iron in ZnO. M_s saturation magnetization, M_r remanent magnetization, H_c coercivity (M. Pardavi-Horvath, unpublished data).

Figure 2. Magnetic domain structure in $H = 0$ applied magnetic field of a defect-free 5 μm thin single crystalline yttrium iron garnet film. The magneto-optic contrast is due to the opposite direction of the magnetization in black-and-white domains with respect to the surface normal. The periodicity of the domain structure in 10 μm (Microphotograph in polarized light).

the details of the magnetization curve of two apparently similar pieces of the same magnetic material might differ substantially. In the absence of any applied magnetic field, the equilibrium state of a magnetic body would be that of zero magnetization. The existence of the remanent magnetization, the basis of the operation of any permanent magnet or magnetic recording device, is the result of the defect structure in the material, and/or shape and size distribution of the particles or grains. These features are controlled by the technology of manufacturing the material. However, any magnetic material can be brought to the $M = 0$ state in $H = 0$, by demagnetizing it. The most frequent process of demagnetization is to apply an ac magnetic field of an amplitude large enough to saturate the material, and slowly and gradually reduce the amplitude to zero.

In the demagnetized state the distribution of the magnetization inside of the magnetic body is nonuniform. The magnetic structure consists of domains, small regions of different magnetization direction (1). The vectorial sum of all of the domain magnetizations is zero in the absence of any applied magnetic field. The geometry and the size of the domain structure is governed by the interactions, delineated previously. The domain structure can be observed by a number of techniques, including magnetic force microscopy, Lorentz electron microscopy, or in polarized light, utilizing the magneto-optic Faraday or Kerr effects. In some cases the domain structure can be very simple, consisting of only two sets of domains, magnetized "up" and "down," as in the case of a 5 μm thin, single crystalline ferrimagnetic garnet sample, a microphotograph of which, taken in polarized light via the magneto-optic Faraday effect, is shown in Fig. 2. The areas of dark and light contrast correspond to the directions of the magnetization perpendicular to the sample surface, directed "downward" and "upward," respectively. In $H = 0$ the two regions have equal volumes, such that the total magnetization $M = 0$. The boundary line between the domains is the domain wall (DW), where the direction of the magnetization is changing gradually between the two domains whose magnetization directions differ by 180°. The size and shape of the domains and the position of the DWs correspond to the minimum free energy configuration. In a flawless material the domain structure is very regular by crystal symmetry, as seen

in Fig. 2 for a single crystal grown along the [111] crystalline direction. The threefold symmetry of that axis is evident.

Upon applying an external magnetic field to a previously demagnetized magnetic body, its magnetization follows the initial (virgin) magnetization curve. Assuming a perfectly homogeneous, flawless material without any preferred axis, the process of magnetization change might proceed reversibly, with no losses involved (1,2). In such a case, the field will exert a force on the domain magnetizations and increase the magnetizations of the domains lying near to the field direction by increasing their volume. This magnetization change happens via the motion of the DWs to new positions. This wall motion is reversible up to a certain magnetic field value and if the field is reduced to zero, the walls move back to their original position and M returns to zero ($M_r = 0$) in $H = 0$. In an ideally perfect material this reversible wall motion would dominate the magnetization process up to magnetic field values near to saturation. Then the magnetization vectors of the remaining domains would start turning into the direction of the field, completing the magnetization process by rotation. This magnetization process is illustrated in Fig. 3. For a thin slab of a magnetic material, having an easy axis of the magnetization along the side of slab, the magnetic field is applied along this easy axis.

In any real material there are always microstructural defects: inclusions of foreign phases, voids, dislocations, inhomogeneous stresses, irregular surfaces and interfaces, grain boundaries, cracks, and so on, such that the free energy of the material is not uniform over the body, and the conditions for magnetization change will vary locally. The free energy might have local minima (or maxima) at some defects, where the DWs will be pinned. They can only move and increase (decrease) the magnetization when the magnetic field changes enough to provide a force sufficient to overcome the energy barriers between the defects. The domain wall will then discontinuously jump from one pinned position to another. The irreversible processes of discontinuous DW displacement are

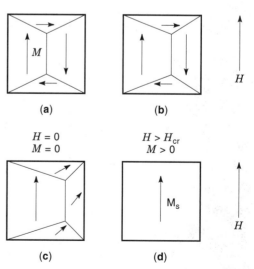

Figure 3. Magnetization processes by wall motion and rotation. (a) Demagnetized state: $\boldsymbol{H} = \boldsymbol{0}$, $\boldsymbol{M} = \boldsymbol{0}$; (b) Irreversible wall motion: $\boldsymbol{H} > \boldsymbol{0}$, domains having favorable magnetization direction with respect to the applied magnetic field are growing; (c) The magnetization in domains with unfavorable direction is rotating toward the magnetic field direction; (d) Saturation: domains disappeared, magnetization aligned with the applied field.

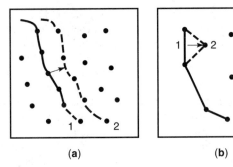

(a)   **(b)**

Figure 4. Magnetization change by irreversible wall motion: Barkhausen jumps of domain walls (lines) pinned at defects (dots). (a) Wall pinned at many defects, simultaneous breakaway with a large magnetization jump. (b) Domain wall motion by individual jumps from defect to defect. 1 is the original DW position in $H = 0$; 2 is the DW position in $H > H_{cr}$.

called Barkhausen jumps. In a macroscopic body, the defects are distributed more or less randomly in space and energy. When the magnetic field reaches the critical value for the irreversible DW motion of the weakest defect, $H = H_{cr}$, the wall will move until it reaches the next defect, where it will be pinned again. Upon increasing the field there might be another DW pinned at another defect, having a critical field equal to the applied field, and now that wall will jump to the next available energy minimum, and it will be pinned at the next defect. Depending on the density of the defects and the strength of the interaction between defects and DWs, it might happen that a wall will be pinned by many similar defects and it will move, breaking away from many defects simultaneously, giving rise to a large magnetization jump. Figure 4 shows these two different scenarios for Barkhausen jumps. Figure 5 shows strong DW pinning at inclusions in a garnet crystal, near magnetic saturation. The applied field is high enough to reduce the black domains to a very small area, but it is not enough to overcome H_{cr} of the inclusions.

A finite size ferromagnetic material is usually broken into many magnetic domains. Any material, including the demag-

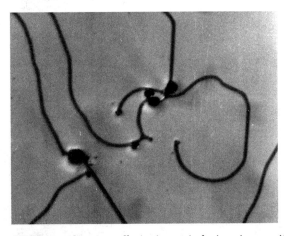

Figure 5. Strong domain wall pinning at inclusions in an epitaxial garnet crystal in applied magnetic field near to saturation, where the magnetization $M = M_{up} - M_{down} \leq M_s$. The black stripe domains correspond to M_{down}. Upon applying a magnetic field $H \gg H_c$, most of the walls disappear, only those pinned by the strongest defects (black dots) are left.

netized state, is not, in general, a unique state, as there is a very large number of possible domain arrangements summing to a particular M value. Consequently, the sequence of the Barkhausen jumps is probabilistic. Moreover, the critical field for a jump depends not only on the strength of the actual defect-DW interaction, but on the magnetic state of the neighboring domains too. These domains, like small magnets, create their own magnetic fields at the site of their neighbors (magnetostatic, dipole fields). Depending on the direction of the magnetization of a given domain, it can help or prevent the externally applied magnetic field in supplying the energy for the jump. This energy, needed to overcome the barriers caused by the defects, results in hysteresis losses.

The discrete and irregular nature of the magnetization process causes hysteresis loss and noise in magnetic devices. To reduce these losses, the material should be uniform. The losses can be reduced, even in the presence of a large defect density, if the defects are uniformly distributed and/or they interact very weakly with the moving DWs. In this case the DWs are moving smoothly across the material, with no sudden jumps in the magnetization. The other limiting case is when the material has a very narrow distribution of defects yielding a rectangular hysteresis loop. In contrast, for a good permanent magnet very active and strong pinning centers have to be created to keep the magnetization from changing, that is, avoid spontaneous demagnetization.

EQUILIBRIUM MAGNETIZATION DISTRIBUTION

Domains and Domain Walls

The equilibrium state of the macroscopic magnetization of any magnetic body is the state with minimum total free energy (1–5). The term in the free energy responsible for the existence of ordered magnetic states is the exchange interaction between the electrons' magnetic moments. The magnitude of the atomic magnetic moments is constant at a given temperature. The atomic magnetic moments are rendered parallel (antiparallel) to each other due to the exchange interaction, giving rise to ferromagnetic (antiferromagnetic, ferrimagnetic) order. The thermal energy tends to randomize the direction of the atomic magnetic moments, and at the Curie temperature, T_c, it overcomes the exchange energy, and there is no magnetic order above T_c. The magnitude of T_c is proportional to A, the exchange constant, characterizing the strength of the exchange interaction. For iron, $A = 1.49 \times 10^{-11}$ J/m.

The origin of magnetocrystalline anisotropy energy is the interaction between crystalline electric fields and atomic magnetic moments (spin-orbit coupling), rendering these moments parallel to certain, easy crystallographic directions. The simplest form of the magnetocrystalline anisotropy is the uniaxial anisotropy, with one easy axis. The anisotropy energy density can be characterized by a single anisotropy constant K_u. For the hexagonal, uniaxial cobalt at room temperature $K_u = 4.1 \times 10^5$ J/m^3. For cubic crystals the anisotropy is weaker and characterized by two constants, K_1 and K_2, where usually $K_2 \ll K_1$. For iron $K_1 = 4.8 \times 10^4$ J/m^3.

In the absence of an applied magnetic field, the equilibrium state of the distribution of the atomic moments can be found from the minimalization of the total energy, that is, it is determined by the competition of the anisotropy and exchange energies, giving rise to the magnetic domain structure. The change of the magnetization direction between adjacent domains is not abrupt, because it would require a large investment of energy against the exchange energy. However, if the transition is very broad then many magnetic moments would be along unfavorable directions from the point of view of anisotropy energy. The result is the minimum energy situation with a formation of a DW of width:

$$\delta_{\mathrm{w}} = \pi \sqrt{A/K}$$

To create a DW, a new, magnetic surface has to be formed. This surface energy of the DW energy is given by:

$$\gamma_{\mathrm{w}} = 4\sqrt{AK}$$

There are other contributions to the free energy of a magnetized body, which have to be taken into account when calculating the distribution of the magnetization. These are the magnetostrictive stresses, the Zeeman energy of the interaction with applied magnetic field, and the magnetostatic energy due to production of demagnetizing fields, depending on the shape and size of the magnetic body.

The total magnetization of the body is the vector sum of the domain magnetizations, $\boldsymbol{M} = \sum \boldsymbol{m}_i v_i$, where $\boldsymbol{m}_i$ is the magnetization vector in each domain directed along ϕ_i with respect to field direction, and v_i is the volume of the ith domain. A magnetic material is saturated when all the atomic magnetization vectors are parallel to the applied magnetic field ($e_i = \cos\phi_i = 0$), that is, all the domain magnetizations are turned into the field direction $\boldsymbol{M} = \boldsymbol{M}_r$. (See Fig. 3.) Before saturation any change of the total magnetization can be described by:

$$\delta\boldsymbol{M} = \delta\sum \boldsymbol{m}_i v_i = \sum \boldsymbol{e}_i \delta|m_i| + \sum \boldsymbol{e}_i \delta v_i + \sum v_i |m_i| \delta \boldsymbol{e}$$

The first term is zero because the magnitude of the magnetic moments is constant; the second term describes the change of the volume of the domains via DW motion, and the third term is the contribution of the rotation of the magnetization angle ϕ_i inside the domains to the change the magnetization.

DOMAIN WALL—DEFECT INTERACTION

Domain Wall Pinning Coercivity

The equilibrium magnetization distribution in any $\boldsymbol{H}$ field is expected to be determined by the minimum energy configuration, when all the contributing energies are taken into account. In a perfectly uniform material the DW energy is the same everywhere and the position and the distance between the DWs is uniform. However, real materials are not uniform. Material properties differ from point to point, and as a consequence, the material constants A and K are not constants, but they are functions of the position $\boldsymbol{r}$: $A(\boldsymbol{r})$ and $K(\boldsymbol{r})$. The result is that at some sites the DW energy is locally lowered (or increased) and extra energy is needed to move the wall from that position, that is, to change the magnetization. In a one-dimensional case, the energy needed to move a DW of volume v ($v = hL\delta_{\mathrm{w}}$, height h, length L, width δ_{w}, specific DW energy γ_{w}) by dx has to be compensated by the change of the total

DW energy (5,6):

$$2HMdx = d(v\gamma_{\mathrm{w}})$$
$$= \gamma_{\mathrm{w}}\,dv + v\,d(\gamma_{\mathrm{w}}) \simeq \gamma_{\mathrm{w}} d(hL\delta_{\mathrm{w}}) + v\,d\sqrt{A(x)K(x)}$$

The critical field for the start of the motion of a given DW, that is, the first Barkhausen jump, is:

$$H_{\mathrm{cr}} = \frac{1}{2M_{\mathrm{s}}} \frac{d(v\gamma_{\mathrm{w}})}{dx}$$

For the whole material, the average critical field, that is, the coercive force of the start of the DW motion, for the case when the DW volume remains constant, is given by the rms value of the critical fields of all defects:

$$H_{\mathrm{c}} = \overline{H_{\mathrm{cr}}^2}^{1/2} = \frac{1}{2hLM_{\mathrm{s}}} \overline{(d\gamma_{\mathrm{w}}/dx)^2}^{1/2}$$

Thus the necessary condition for the occurrence of Barkhausen jumps is the presence of a gradient of the DW energy. Figure 6 shows a one-dimensional sketch of the energy landscape of a magnetic body, due to the gradient of the DW energy. The DW is located in an energy minimum, pinned by the defect. Upon reaching the critical field of interaction with a defect, the DW breaks away, jumping to a new metastable equilibrium position. The DW can freely move and the magnetization changes continuously if the defects along its path are weaker than the previous maximum in the gradient. The variation of the DW volume across the surface might have a significant contribution to hysteresis losses due to surface roughness effects, as in the case of soft magnetic thin films in recording heads (6).

The change of the DW energy due to local variations of anisotrophy and exchange energy causes DW pinning at localized defects. These defects are responsible for Barkhausen jumps and hysteresis losses. During the magnetization process the energy from the applied magnetic field has to over-

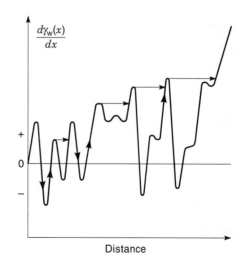

Figure 6. One-dimensional variations of the DW energy landscape of a magnetic body due to the position dependence of anisotropy and exchange energies, caused by localized crystalline defects. Arrows denote the jump of the domain walls to the next position of equal energy.

come the energy of interaction of the moving domain wall with defects in the magnetic body. Due to the statistical nature of the defect distribution, the magnetization curve of two apparently similar pieces of the same magnetic material might differ substantially, depending on the technology of production. Figure 7 is a measured hysteresis loop of 25 quasi-identical magnetic garnet particles. Each particle is seen to be magnetized at different fields, illustrating the distribution of critical fields (7). The mean value of the critical fields for DW motion is scaled with the coercivity of the material. However, for the macroscopic critical field, the coercivity is a statistical parameter, and it has a certain distribution and standard deviation. The standard deviation of the critical field distribution characterizes the spectrum of the defects based on the strength of DW-defect interaction. More accurate, statistical models, given in Ref. 5, take into account that the process of pinning/depinning takes place only over a distance of one domain width w_d, so there will be a $\ln(w_d/\delta_w)$ factor in all formulas for DW pinning. If the defect size is very small (10^{-7} m) compared to the DW width, then the moving DW averages out the defect potential, and there is no irreversible magnetization and hysteresis loss associated with a very fine microstructure. On the other hand, when the defects are very large compared to δ_w, then the energy associated with the inclusions will be reduced due to the presence of secondary domain structure and the associated discontinuous DW motion will be negligibly small and the irreversible magnetization losses will be reduced.

Inhomogeneous Stress Fields

One of the main contributors to Barkhausen noise is related to the stress sensitivity of magnetic materials. The magnetoelastic behavior is characterized by the magnetostrictive strain $\lambda = dl/l$, the relative shape and size change of a magnetized body upon change of the magnetization, or vice versa the change of magnetization upon deformation (1, 5). For single crystals the strain might be different in different crystallographic {hkl} directions, with different λ_{hkl}. For polycrystalline materials the average value of λ is used. For most of the magnetic materials $\lambda \cong 10^{-5}$ to 10^{-6}. Elastic stresses σ couple to magnetostriction causing an effective stress-induced aniso-

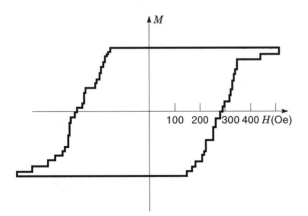

Figure 7. Measured hysteresis loop of 25 quasi-identical magnetic garnet particles, illustrating the sequence of Barkhausen jumps corresponding to the distribution of critical fields of individual particles. (From Ref. 8.)

tropy of the material:

$$K_\sigma = -\frac{3}{2}\lambda\sigma\cos^2\phi$$

where ϕ is the direction of the stress relative to the magnetization. The total anisotropy of the magnetic body is the sum of the magneto-crystalline and stress-induced parts: $K = K_u + K_\sigma$. Localized fluctuations of the internal stresses $\sigma(r)$ cause fluctuations in the DW energy through $K(r)$ and contribute to the coercivity, that is, to the critical fields for Barkhausen jumps.

If there are long range one-dimensional stress fluctuations $\sigma(x)$ in the material, with an average spatial periodicity (wavelength) Λ, large compared to the DW width, then the DW energy will fluctuate as

$$\gamma_w = 4\sqrt{A[K_u + (3/2)\lambda\sigma(x)]}$$

causing a critical field for the DW motion of:

$$H_{cr}^\sigma = \frac{3\lambda\sigma_w(x)}{4M_s}\frac{d\sigma(x)}{dx}$$

This is the basic equation of the stress theory of Barkhausen jumps and coercivity. If $\sigma(x)$ is known then H_{cr} can be calculated.

Assuming a quasi-sinusoidal stress field with an amplitude of ($\sigma_0 \pm \Delta\sigma/2$) in the form of

$$\sigma(x) = \sigma_0 + \frac{\Delta\sigma}{2}\sin(2\pi x/\Lambda)$$

the critical field for the start of the DW motion is given by

$$H_{cr}^\sigma = p\frac{\lambda\overline{|\sigma_i|}}{2M_s}$$

where p depends on the ratio of σ_w/Λ. Although the usual effect of the stress fields of the defects is that the anisotropy of the material is locally reduced, thus creating an effective energy well for the DW, it was demonstrated that it is possible to have anisotropy barriers with locally higher DW energy, preventing the change of the magnetization very efficiently (8). Figure 8 shows such a case, where the stress field from an inclusion in the nonmagnetic substrate penetrates a magnetic crystal and the DWs are repulsed by both stress fields, preventing saturation of the magnetization at fields much higher than the theoretical saturation field.

Domain Wall Interaction with Dislocations

A special case of the stress effects on magnetization processes is due to the presence of dislocations (9,10,11,12,13,14). Plastic deformation produces defects, such as dislocations, stacking faults, and point defects, each associated with an internal stress field, which will affect the Barkhausen noise. For low deformation the number of dislocations increases linearly with shear stress τ, and depends on the material's shear modulus G and the magnitude of the Burgers vector **b** characterizing the dislocation. The moving DW interacts with the stress field of the dislocation, depending on the direction and the distance between them. The dislocation exerts a force **f**

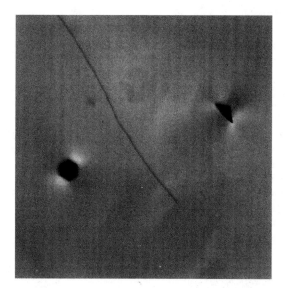

Figure 8. Repulsion of domain walls in a magnetic garnet material by the stress fields of triangular and hexagonal Ir inclusions, originating in the nonmagnetic substrate. The inclusion's stress field penetrates the magnetic layer and interacts with the domain walls. The sample is near saturation, only the thin black domain remains due to the very strong interaction with the stress field.

upon the DW according to its line element dl and the magnetostrictive stress tensor σ of the DW: $f = -dl \times \sigma\mathbf{b}$. The calculation of this interaction for a general case is not trivial.

For a simple case of a 180° DW, lying in the $x-y$ plane and a dislocation running parallel to the plane of the DW, the geometry shown in Fig. 9, the critical field is calculated in Ref. 10. For the simple case of the change of the magnetization across the DW $\phi = \pi/\delta_w$, $\theta = 0$. The edge and screw components of the dislocations are $b\sin\alpha$ and $b\cos\alpha$. For an edge dislocation $\alpha = \omega = 90°$, for a screw dislocation $\alpha = 0$. For a dislocation running parallel to the x axis $\epsilon = 0$. The angle between $\mathbf{b}$ and the z axis is ω. The component of the force f_z is given as

$$f_z = (3/2)Gb\lambda(\sin 2\phi\cos\alpha + 2\sin^2\phi\sin\omega\sin\alpha)$$

The DW interacts with the dislocation only as long as the dislocation is inside of the wall, because $f_z = 0$ for $\phi = 0$, consequently, $f_{max} = (3/2)Gb\lambda$. To move the DW by dz, $dW = f\,dz$ work should be performed. The energy necessary

to get the whole DW over the single dislocation barrier is $W = (3/2)Gb\sigma_w\lambda$. The critical field for the DW motion from a single straight edge dislocation is:

$$H_{cr}^d = (3/2M_s)Gb\lambda$$

For a dislocation density of N m^{-2} H_{cr} will be proportional to $N^{1/2}$, that is, the linear defect density.

In some cases the measured H_{cr} is much higher than the expected from the above equation. This is due to the fact that, depending on material parameters, the DW might be elastically deformed, that is, stretched between two dislocations before breaking free (1,5,15). Assuming that the amplitude of DW bulging is one-quarter of the distance between dislocations, the critical field for the DW jump is:

$$H_{cr}^d = \sqrt{2N}\gamma_w/M_s$$

Domain Wall Pinning at Inclusions and Voids

In developing new magnetic materials the control of the microstructure is of primary importance. Nonmagnetic grain boundaries are essential for magnetic recording media in decreasing the exchange coupling between grains; the properties of permanent magnets are defined through inclusions of phases with magnetic properties different from that of the matrix, while soft magnetic materials should be as homogenous as possible. In case of nonmagnetic inclusions or voids, the DW is pinned because the local DW energy is reduced by the missing volume when the DW includes the inclusion or void. At the same time, depending on the size of the defects, there will be free magnetic poles on the surface of the inclusions and voids, producing demagnetizing (stray) fields with a significant energy contribution (10,16,17,18,19).

Small Inclusions, $d \ll \delta_w$. When a spherical inclusion of size d, volume $v = \pi d^3/6$, is located outside of the DW it behaves like a dipole, having a dipole energy of $W_D \approx (1/3)M_s^2 v$. If it is inside of the DW, then it behaves as a quadrupole, and $W_Q = W_D/2$. If the interaction depends only on the distance between the inclusion (or void) and DW, then $d\gamma_w/dx \approx d^3\gamma_w/\delta_w^2$, and the critical field for the start of the DW motion, due to the volume effect for $d \ll \delta_w$ is given by (10):

$$H_{cr}^i = \frac{2.8\gamma_w}{M_s w}\left[\frac{d}{\delta_w}\right]^{3/2}v^{1/2}\left[\ln\frac{2w}{\delta_w}\right]^{1/2}$$

The contribution due to the demagnetizing field effects:

$$H_{cr}^D = \frac{2.8M_s d^{7/2}}{w\delta_w^{5/2}}v^{1/2}[\ln(2w/\delta_w)]^{1/2}$$

The critical fields H_{cr}^i and H_{cr}^D are comparable when $d \approx \delta_w/4$.

Large Inclusions, $d \gg \delta_w$

For large inclusions there would be a very large demagnetizing field contribution due to the strong dependence on the size of the inclusions, which is disadvantageous from the point of view of energy minimalization. The consequence is that a wedge-shaped secondary domain structure is developed around large inclusions to dilute the density of free poles over

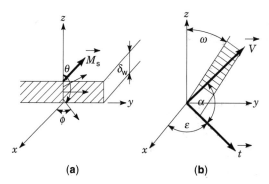

(a)   **(b)**

Figure 9. Geometry of the DW dislocation interaction. M_s saturation magnetization, δ_w DW width, V Burger's vector of the dislocation.

a larger surface of the new DWs. The irreversible motion of the DW starts before it would be completely torn off the inclusion, at a field:

$$H_{cr}^{I} = \gamma_w v^{2/3} / (2M_s d)$$

The theory of inclusions was originally developed for spherical inclusions. The case of elliptical inclusions was treated in Refs. 18 and 19, showing that the strongest demagnetizing field effects are due to flat inclusions lying perpendicular to the plane of the DW. The critical field is for the case of extended planar inclusions, width d, including magnetic inclusions, H_{cr} depends on the relative magnitude of the anisotropy and exchange constants of the two materials (A and A', and K and K'), and it is given in Ref. 20:

$$H_{cr}^{i} = \frac{2Kd}{3^{3/2}M_s\delta_w}[A/A' - K'/K], \quad for \; d \ll \delta_w$$

$$H_{cr}^{i} = \frac{KA'}{2M_sA}[A/A' - K'/K], \quad for \; d \gg \delta_w$$

The calculation of critical fields can be solved analytically for simple cases only (21,22,23). Numerical methods of micromagnetism are very powerful in calculating the critical fields for DW pinning for any geometry and combination of defects. The most cumbersome part of the calculation, the magnetostatic fields due to the nonzero divergence of magnetization at the defects, can be handled easily by numerical methods. Details of the pinning process can be revealed by these methods, like the repulsion of the approaching DW by the dipole field of small inclusions in high magnetization materials (24,25). Figure 5 shows the domain structure of a garnet single crystal in $H \gg H_c$, near to saturation, where the strongest defects keep the last DWs pinned (26,27,28).

IRREVERSIBLE MAGNETIZATION PROCESSES

In a ferromagnet with a domain structure irreversible magnetization takes place by DW motion, due to the sudden changes of the magnetization as the applied magnetic field reaches the critical field of depinning a DW from a defect, as given in the previous section. On reducing the size of the magnetic body down to the typical size of a single domain, no DW can be formed and, consequently, there is no DW motion, and no irreversible losses associated with it. However, there are still irreversible magnetization changes due to the sudden rotation of the magnetic moment of the single-domain particles at the switching field of the particle. The sequence of individual rotation events of particle magnetizations gives rise to the hysteresis loops of particulate materials, usually described by Preisach models (1,7,29). The energy barrier for magnetization switching is related to the effective anisotropy field of the particles: $H_{cr}^{rot} = 2 K_{eff}/M_s$, a fairly large value. These irreversible wall motion or rotational magnetization processes are the cause for the hysteresis losses. The measure of the energy loss associated with the hysteresis can be determined from the area under the hysteresis loop:

$$W_H = \oint_{loop \; area} HdM$$

The loss associated with an individual Barkhausen jump dW_B is determined in a similar way, by the area of the minor loop of one jump. Integrating over all individual jumps gives the total loss, W_H.

The shape of the hysteresis loop is expected to be determined by the sequential events of depinning the DWs from the defects, or the sequence of the individual switching events of particles by rotation. The sequence would depend only on the strength of the interaction between a defect and a DW. In this case the shape of the hysteresis loop would be reproducible and no random noise would be associated with the magnetization process. At the absolute zero of temperature $T = 0$ K, with no interaction between the particles, this could be a valid assumption. However, the shape of the hysteresis loop and the associated losses are influenced by several other irreversible processes. First of all, at *finite temperatures* temperature fluctuations add a random noise to the magnetization process, randomly raising or decreasing the energy barriers what the moving DW should overcome in order to change the magnetization (30,31). Moreover, the height of the barrier, that is, the strength of the interaction between the DW and the defect, might depend on the temperature itself.

At any given temperature if one waits long enough there will be a finite probability that a temperature fluctuation appears, large enough to push the DW over the next barriers. This characteristic time delay between an instantaneous change in the applied field and the subsequent change of the magnetization is the magnetic aftereffect (4,30). The higher the temperature the more effective is the assistance of temperature fluctuations. The slower the change of the field, the more chance the DW has to jump over the energy barrier, waiting for the moment of the temperature fluctuation assisted decrease in the energy barrier. The critical field of the start of the DW motion at one given defect depends on the temperature and time: $H_{cr} = H_{cr}(T, t)$. As a result, the faster the rate of change of the field, the larger the field needed to change the magnetization, that is, the coercivity H_c is increasing with frequency. The defect-DW interaction is influenced by the aftereffect, leading to an apparent frequency dependence.

Another effect that can't be neglected is the shape and size dependence of the magnetization curve. If a magnetic body of finite size is magnetized, free magnetic poles occur on the surface, producing demagnetizing field, H_d, which opposes the direction of the magnetization. The demagnetizing field is proportional to the magnetization, $\boldsymbol{H_d} = -N\boldsymbol{M}$, where N is the demagnetizing factor (1,2,32). When the magnetization changes in a small region of the body, the demagnetizing field changes too, and the slope of the hysteresis loops changes, depending on the demagnetizing factor and the permeability $\mu = d\boldsymbol{B}/d\boldsymbol{H} = d(\mu_0 \boldsymbol{H} + \boldsymbol{M})/d\boldsymbol{H}$. This change causes an apparent change in the magnetization at a given applied field. A more serious problem is that the magnetic body is not homogeneously magnetized. The direction and the magnitude of the domain magnetizations vary, and the permeability is not a well-defined quantity. The effect of the sample's shape and size makes the comparison of experimental data very difficult.

Eddy current losses can't be neglected in the case of ac magnetization processes of metallic magnets. When the magnetization is changed in a conductor, eddy currents and, associated with it, a time-dependent magnetic induction $B(t)$ is produced. Integrating $B(t)$ over the sample surface, the

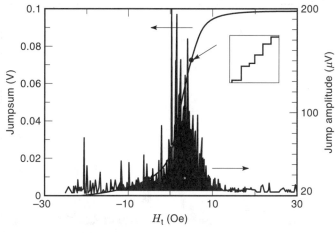

Figure 10. Barkhausen jump statistics of a low carbon steel specimen, and the integrated jumpsum signal. The inset is the measured signal at higher magnification. (Courtesy of L. J. Swartzendruber, NIST, unpublished data).

change of the flux in time, that is, the voltage induced by the eddy current is obtained. Due to interference from this eddy-current induced signal, the measurement of the Barkhausen noise in metallic magnets is not a simple task (33).

BARKHAUSEN NOISE

Any discontinuous, sudden change of the magnetization in a slowly changing magnetic field, due to the irreversible break-away of a DW from a pinning center or an irreversible act of rotation, induces a voltage spike in a coil surrounding the specimen. The sequence of these pulses is the Barkhausen noise. The pulses are characterized by their number in the applied field interval, magnitude of the induced voltage, width of the pulse, and the integrated voltage, the *jumpsum*. The change of the magnetization during one jump is related to the integral under the area of the resultant pulse. The steepest part of the hysteresis loop, around the coercivity H_c contains the largest number and highest amplitude pulses. The defect structure influences both mechanical and magnetic properties of the materials. The distribution of jump size, shape, duration, and the power spectrum of the Barkhausen noise is directly related to the microstructure of the magnetic material, and carries very important information for nondestructive testing. Figure 10 shows the applied magnetic field dependence of the Barkhausen noise data for a steel specimen in terms of pulse heights of the induced voltage, and the integrated voltage (jumpsum), which is related to the magnetization (33).

Stress Effects

The defect DW interaction is very sensitive to the stress due to stress-induced anisotropy in materials with finite magnetostriction. Upon an application of stress the DW energy landscape and the critical field for individual DW jumps change and, as a result, the shape of the hysteresis loop and the permeability change with stress as well. After proper calibration, and over a wide range of stress, the Barkhausen noise can often be used as a measure of the strain state of the material (34,35).

In materials with strong uniaxial anisotropy, the equilibrium domain structure is relatively simple. In a stressed long wire or ribbon, domain magnetizations will be aligned along the long axis and the magnetization process is dominated by one huge Barkhausen jump. The hysteresis loop of such a material is nearly square. For a wire with large magnetostriction, $\lambda > 0$, under tensile stress σ along the axis, there is a stress-induced uniaxial anisotropy: $K_s = (3/2)\sigma\lambda$. The large change in magnetic induction upon changing the magnetic field will induce a large, narrow pulse of voltage in a coil wound around the sample, used in many sensors and devices. Stress can be induced in amorphous magnetostrictive wires during heat treatment, and can be used as inductors. Figures 11 and 12 show the effect of magnetostriction and mechanical stress on Barkhausen noise statistics upon magnetization of a highly magnetostrictive amorphous metallic ribbon in a free state and when the same ribbon is slightly stretched along its axis (36). The change of the Barkhausen jump spectrum from

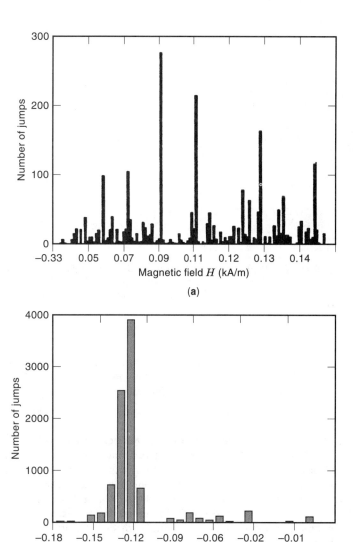

(a)

(b)

Figure 11. Effect of magnetostriction and mechanical stress on Barkhausen noise statistics upon magnetization of a highly magnetostrictive amorphous metallic ribbon (a) Free ribbon; (b) Ribbon under weak tensile stress (M. Pardavi-Horvath, unpublished data).

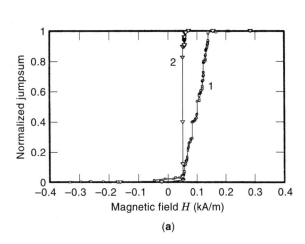

(a)

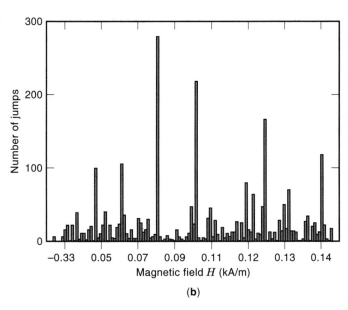

(b)

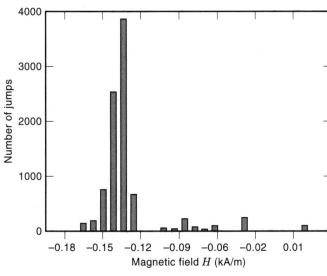

(c)

Figure 12. (a) Integrated signal of individual jumps for the sample in Fig. 12. 1 is the free ribbon; 2 is the ribbon under tensile stress. (b) Number of jumps for 1. (c) Number of jumps for 2. (M. Pardavi-Horvath, unpublished data.)

numerous small jumps to a very few huge jumps is evident. The integral of the jumps over the magnetic field gives a curve similar to the magnetization curve (37).

Temperature Dependence

The Barkhausen effect has a strong temperature dependence. The DW energy itself depends on the magnetic anisotropy, a strong function of temperature, leading to the increase of the DW energy and the coercivity on lowering the temperature. The energy of temperature fluctuations play a lesser role at low temperatures, and the characteristic time constant of the magnetic aftereffect is increasing with decreasing temperature. As the probability for a jump decreases, the DW has to wait longer for a kick from the thermal fluctuations, and the number of jumps in a given field and time interval decreases (33). On the contrary, upon increasing the temperature the coercivity decreases, and due to strong thermal activation processes, it's easier for a DW to make a jump. At the same time the thermal energy becomes larger than some of the

DW-defect interaction energies, many of the low energy defects are "turned off"" and the width of the defect-DW interaction spectrum decreases too. The magnetic material becomes magnetically softer.

Statistics of Barkhausen Jumps

The statistics of Barkhausen jumps has been the topic of wide ranging investigations for many years. The static interaction of *one* defect with *one* domain wall is just the beginning of the theoretical treatment of the magnetization process of a magnetic material. In a real material, a very large number of statistically distributed defects interact with a large number of DWs, resulting in a complex Barkhausen noise behavior. The process depends on temperature, time, sample size, and shape (38,39). For the interpretation of the Barkhausen signal the functional form of the energy landscape is needed. All these unknown parameters can be taken into account as fluctuations in the DW energy, and described by an ensemble of stochastic Langevin functions. The model generates both hysteresis loops and Barkhausen jump distributions, showing a power-law behavior for small jumps and a rapid cut-off at large jump sizes, in agreement with the experimental data (40,37,41).

Unfortunately, along the hysteresis loop different types of magnetization processes take place. Therefore, the treatment is usually restricted to the constant permeability region around the coercivity. The assumption used in predicting the noise power spectrum (42) is that both dB/dH and dH/dt are constant. This approach was extended to the whole hysteresis loop assuming that the Barkhausen activity in terms of the jumpsum is proportional to the differential susceptibility, which can be determined from hysteresis models (43).

An important question is whether the individual jumps are correlated or not. For statistically independent jumps the theory predicts that the power density is constant at low frequencies and decays as $1/f^2$ at high frequencies. Some experimental data agree with this prediction, however, in some cases there is a maximum at low frequencies. The analysis of the Barkhausen noise signal characteristics, as autocorrelation, jump amplitude correlation, jump time and amplitude correlation, and power density show no evident deviation from random noise behavior. Evidently, a different mechanism is responsible for the low frequency losses (44).

The progress in nonlinear dynamics and chaos, relatively easily applicable to the behavior of domains and domain walls, leads to simple nonlinear models for DW dynamics, taking into account the viscous damping of the DW moving under the effect of a harmonic external force in the field of defects. It was shown that Barkhausen noise can be characterized by a low fractal dimension (45,46). Barkhausen jumps show the attributes of self-organized criticality, with the distribution of lifetimes and areas of discrete Barkhausen jumps following a power-law behavior (47,48).

Experimental Techniques

Modern Barkhausen spectrometers are based on the same principle of Faraday's law of induced electromotive force (emf), as used by Barkhausen in 1919. According to Faraday's law, whenever there is a magnetic flux change dB/dt there will be an induced voltage V, proportional to dB/dt and the number of turns of a measuring pickup coil. The difference is that the original amplifier is now substituted by a digital computer controlled data acquisition system and digital storage oscilloscope (33,44,49). Usually the size and the duration of Barkhausen pulses is measured and statistically registered. The results are frequency distributions. The entire sequence of Barkhausen pulses can be considered as a stochastic process and then the spectral density of the noise energy, related to the total irreversible magnetization reversal is measured (33,39). Figure 13 shows the block diagram of a typical apparatus. The equipment consists of a magnetizing system used to sweep the sample through the full hysteresis loop from negative saturation to positive saturation, at a frequency of typically 0.05 Hz; a Hall probe to measure the magnetic field; a surface coil to measure the induced voltage and/or another pick-up coil, surrounding the sample, for power spectrum measurements. The measurement is controlled and the data are analyzed by a digital computer. The bandwidth of the system is from 100 Hz up to 100 kHz. A discriminator is used to reject signals below a certain noise level. At each

instant of time, t_n, the value of the magnetic field, H_n and the induced voltage is V_n measured. So the amplitude of jumps versus field, the integrated signal, the jumpsum, and the rate of the jumpsum can be determined. The jumpsum rate is qualitatively similar to the rms noise. Power density spectra and autocorrelation functions can also be obtained from the data. For a stationary random process, the power density spectrum is the Fourier transform of the autocorrelation function, $\Phi(t) = \langle V(t) \times V(t + \tau) \rangle$, so that statistical properties of the Barkhausen noise can be easily investigated by this technique (49).

A widely used method in studying the Barkhausen effect is to measure the power as a function of the frequency of cycling the field. This technique yields information about the intervals between jumps (33). The frequency of the Barkhausen jumps is related to the average DW velocity, and it contains information about the average time it takes for a moving DW to reach the next barrier through the average distance between defects.

Utilizing the optical activity of magnetic materials it is possible to identify individual defects, pinning mechanisms, and measure actual DW velocities. In transparent magnetic materials individual DW-defect interactions can be investigated via the magneto-optical Faraday-effect (26,27,28), as illustrated in Figs. 2, 6, 7, and 9. In metals, the Kerr effect offers a possibility to visualize the defect-DW interaction, as it was done for several metallic film magneto-optical recording media, in order to study the cause of media noise in readout (50). It was shown that the distance between pinning centers is about 0.4 μm and the DW "waits" up to several seconds before the next jump. The pinning time decreases exponentially with increasing driving field.

Nondestructive Testing

When applicable the measurement of magnetic Barkhausen noise is a fast, reliable, and simple technique for nondestruc-

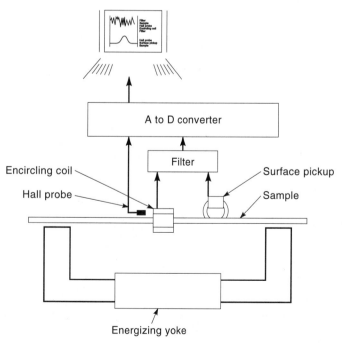

Figure 13. Block diagram of Barkhausen noise measurement. (Courtesy of L. J. Swartzendruber, NIST.)

tive material evaluation, as compared to x-ray diffraction, ultrasound, and other more sophisticated measurements (49,51). The Barkhausen effect is especially well-suited for the study of steel, one of the most important structural materials. Commercial instruments to characterize materials based on Barkhausen noise are now available. The Barkhausen spectrum can be, for example, related to the grain size of the material, so it can be used for grain size measurement after cold-rolling (52). The stress sensitivity of the DW motion makes it very convenient to study residual stresses, using the Barkhausen emission under stress. Depending on the bandwidth of the measurement, either the surface or near subsurface layers can be tested (34). The Barkhausen emission test is used in the grinding industry to check the residual stress due to thermal damage and microhardness (53). Plastic deformation increases the dislocation density, an effect easily detected by the DW-dislocation interaction induced critical field increase. There is a possibility to use deformed thin wires of iron and iron–nickel alloys as strain gages in the 10^{-3} range with a sensitivity of $\pm 5\%$ (35).

Barkhausen noise plays an important role in magnetic recording. The head-to-medium velocities in hard drives may range from 1 to 50 m/s, with data frequencies from 100 kHz to 50 MHz; for tape systems the velocity might be as low as 25 mm/s at 1 Hz frequency. These parameters are very near to the range of Barkhausen noise. The main source of the medium and head noise is the Barkhausen noise. The nonuniformity of media leads to a localized, regular noise, superimposed on the random noise from magnetization switching processes. For particulate materials the noise is related to the shape, size, and orientation dispersion of the particles; for multigrain thin film media the signal-to-noise ratio (SNR) depends on the microstructure, texture, and grain cluster size. In soft magnetic write-read head materials the DW pinning and the related hysteresis losses are the most important source of noise (50,54).

FLUX PINNING AND LOSSES IN SUPERCONDUCTORS

The basic parameters defining the transition from the superconducting to the normal state are the critical temperature T_c, critical magnetic field H_c, and critical current density J_c. Type II superconductors possess the highest possible critical parameters. Destruction of their bulk superconductivity occurs at the upper critical field H_{c2}. If no applied magnetic field is present, then J_c up to 10^{11} Am^{-2} can be reached. High current density and small energy dissipation is the main reason for industrial applications of superconductors in magnets, electrical machinery, and power transmission lines. Ideal superconductors are lossless, however, using Type II superconductors for practical applications dissipative processes are of great importance. In Type I superconductors dissipation is negligible up to very high frequencies, where the electromagnetic field destroys the Cooper pairs, responsible for superconductivity. Losses in ideal Type II superconductors are associated with viscous motion of vortex lattices. Losses in the most important class of superconductors, the Type II nonideal superconductors, are due to magnetic hysteresis, similar to the case of ferromagnetic materials. In superconductors the irreversible motion of magnetic flux lines through pinning centers causes the hysteresis. The problem is more complicated in

technical materials: superconducting cables are multifilament, multicore, twisted, stabilized composites where the coupling losses to the normal matrix play a significant role in dissipation (55).

Vortex Structure

Superconductors (SC) are characterized by the penetration depth of the weak magnetic field λ, the coherence length ξ, and the Ginzburg-Landau parameter $\kappa = \lambda/\xi$. Superconductors can be classified according to the magnitude of κ. Type I SCs have $\kappa < 1/\sqrt{2}$, and the external field is totally screened by diamagnetic supercurrents flowing in the penetration depth distance from the surface (Meissner effect). For Type II SCs $\kappa > 1/\sqrt{2}$. The N/S boundary is in equilibrium only in high applied magnetic field. A type II SC below the first critical field H_{c1} behaves as a Type I material. In a field of $H > H_{c1}$ it is energetically more favorable to break into the mixed state of alternating N and SC phases, consisting of thin, normal core vortex lines, parallel to the external field, with circulating paramagnetic supercurrents. Each vortex carries a quantum of flux $\phi_0 = hc/2e$, where h is the Planck constant, c is the light velocity and e is the electron charge. The size of a vortex is typically a few hundred nanometer. The vortex lattice is in some way analogous to the magnetic domain structure of ferromagnetic materials, as discussed previously. Upon increasing the applied magnetic field H, the number of vortices increases as the vortex lattice period becomes smaller. At $H = H_{c2} = \sqrt{2}\,\kappa H_c$ the vortex cores overlap and the material becomes a normal metal. H_{c2} of SC for practical applications is on the order of 10^2 T.

Vortex Motion and Flux Jumps

Many technical applications of SC require high current densities and low losses, which can be accomplished by immobilizing the vortices by pinning them at defects. The vortices have a complex magnetic field around them, decaying exponentially within the distance λ. If the SC is placed in an applied field H, parallel to its surface, then supercurrents J_m will be generated by a vortex located at x_0:

$$J_m = \lambda^{-1} H \exp(-x_0/\lambda)$$

and a repulsive Lorentz force will act on the vortex, $F_m = J_m \phi_0$. This force changes with the distance, thus changing the free energy of the vortex system. If there is an energy gradient, that is, a local change in the potential for the vortex motion, then the vortex can be pinned at the defect lowering the free energy of the system. There is an energy barrier to the penetration of vortices in and out of the SC at the boundary, and vortex motion into the SC can be prevented by surface treatment. Ideal Type II SCs are thermodynamically reversible if there are no defects in the material to act as pinning centers. But even in the absence of bulk pinning, there is always the surface, itself an irregularity, so there is no ideal SC. In the absence of pinning centers, the vortex lattice's motion is like a hydrodynamic flow in a viscous medium. Dissipation is due to the viscous damping process. The power necessary to move the vortices is $W_{SC} = \rho_n J_s^2$, where ρ_n is the specific resistance. Because there are always thermal fluctuations, and J_c is decreasing with increasing T, any slight increase in the temperature decreases the critical current and by preserving the total current, the current distribution has to change, and as a consequence, the magnetic field distribution generated by the current distribution is changes too. According to Maxwell's law whenever there is a flux change there will be an electric field, and this field does the work to move the vortices, that is, a flux jump is observed. Flux jump instabilities are very dangerous because they can transform the SC into the N state completely. Thermally activated flux flow is one of the main reasons for the low current density in new high-T_c SCs.

Another case of flux jumping occurs due to magnetic field fluctuations. As the applied field increases, critical shielding currents are induced to prevent the penetration of flux into the SC until a certain field H^* (or magnetic flux density associated with that field B^*) value , where flux jumps occur. For thin filaments the flux completely penetrates the material, and there is no flux jumps even for applied fields much greater than H^*. The constraint on the thickness d of the SC material is that:

$$\mu_0 J_c d < B^* = [3\mu_0 S(T_c - T)]^{1/2}$$

where S is the volume heat capacity of the material (56).The distribution of the induction is usually described by the Bean-London model of critical state, where the magnitude of the critical current is directly related to the gradient of the local induction $J_c = dB/dx$. A change in the penetration occurs only when the change in the external field exceeds the surface shielding fields created by shielding currents, according to Maxwell's equations. This picture of SC hysteresis losses is valid up to about 10^3 Hz. In the Bean–London model the contribution of the vortex structure is not included. A complete analytical solution for the distribution of current and induction in hard SCs is still missing.

Nonideal Type II SCs

Real materials always have spatial inhomogeneities acting as pinning centers that prevent the penetration of flux into the SC upon increasing the applied field; or preventing the exclusion of flux upon decreasing field, giving rise to hysteresis and associated irreversible losses. There is a remanence flux frozen in the SC even in the absence of external field. In analogy with hard magnets, these nonideal Type II SCs are called *hard SCs*.

The pinning centers are similar to those in magnetic materials, preventing the motion of domain walls, such as: point defects of inclusions, voids, precipitates of second phase, line defects like dislocations, grain boundaries, twin boundaries, and so on. The bigger the difference between the properties of the defect and the SC the larger is the pinning effect, again similar to the case of the strong DW pinning for high DW energy gradient in ferromagnets.

There is no complete, exact theory of pinning in SCs, and the methods used to describe the vortex-defect interactions are very similar to those dealing with DW pinning. Strong pinning and high current density is achieved mostly by microstructural developments through technology of preparation, compositional modifications, and different treatments (for example, irradiation). In intermetallic SCs (Nb_3Sn, Nb_3Ge) the pinning centers are related to the fine grained structure on the order of a few nm, and precipitation of oxides or carbides

at grain boundaries. In the high-T_c cuprate SCs vortex pinning is linked to the layered structure of the material, the vortices are interacting strongly within the layers (pancake vortices) and weakly between layers. As a result the electrical properties are highly anisotropic. The pancake vortices are very mobile. Increasing the interlayer coupling would make a more effective pinning with higher (and isotropic) current densities (57).

In an ideal SC the vortex structure is a regular two-dimenisonal network. Due to the very inhomogeneous structure of the hard SCs, the vortex lattice is distorted, following the microstructural features, that is, the energy landscape of the material. For high defect densities and high current densities the vortex lattice becomes "amorphous", the lattice is "melted".

BIBLIOGRAPHY

1. Soshin Chikazumi and Stanley H. Charap, *Physics of Magnetism,* New York: Wiley, 1964.
2. Richard M. Bozorth, *Ferromagnetism,* New York: IEEE Press, 1993.
3. Ami E. Berkowitz and Eckart Kneller (ed.), *Magnetism and Metallurgy,* New York: Academic Press, 1969.
4. H. Kronmüller, *Magnetisierunskurve der Ferromagnetika* I. In Alfred Seeger (ed.), *Moderne Probleme der Metallphysik,* Vol. 2, Ch. 8, Berlin: Springer Verlag, 1966.
5. H. Träuble, *Magnetisierungskurve der Ferromagnetika* II. In Alfred Seeger (ed.), *Moderne Probleme der Metallphysik,* Vol. 2, Ch. 9, Berlin: Springer Verlag, 1966.
6. M. Pardavi-Horvath and Hyunkyu Kim, Surface roughness effects on the coercivity of thin film heads. *J. Korean Magn. Soc.,* **5**: 663–666, 1995.
7. M. Pardavi-Horvath, A simple experimental Preisach model system. In G. Hadjipanayis (ed.) *Magnetic Hysteresis in Novel Magnetic Materials,* Amsterdam: Kluwer Publ., 1997.
8. J. A. Jatau, M. Pardavi-Horvath, and E. Della-Torre, Enhanced coercivity due to local anisotropy increase. *J. Appl. Phys.,* **75**: 6106–08, 1994.
9. H. Träuble, in Ref. 3, pp. 622–685.
10. H. Träuble, in Ref. 5, pp. 257–279.
11. Martin Kersten, Über die Bedeutung der Versetzungsdichte für die Theorie der Koerzivkraft rekristallisierter Werkstoffe, *Z. Angew Phys.,* **8**: 496–502, 1956.
12. K.-H. Pfeffer, Zur Theorie der Koerzitivfeldstärke und Anfangssusceptibilität, *Phys. Stat. Sol.,* **21**: 857–872, 1967.
13. K.-H. Pfeffer, Wechselwirkung zwischen Versetzungen und ebenen Blochwänden mit starrem Magnetisierungsverlauf, *Phys. Stat. Sol.,* **19**: 735–750, 1967.
14. K.-H. Pfeffer, Mikromagnetische Behandlung der Wechselwirkung zwischen versetzungen und ebenen Blochwänden, *Phys. Stat. Sol.,* **21**: 837–856, 1967.
15. Horst-Dietrich Dietze, Theorie der Blochwandwölbung mit Streufeldeinfluss. *Z. Phys.,* **149**: 276–298, 1957.
16. L. J. Dijkstra and C. Wert, Effect of inclusions on coercive force of iron, *Phys. Rev.,* **79**: 979–985, 1950.
17. R. S. Tebble, The Barkhausen effect. *Proc. Phys. Soc., London,* **B86**: 1017–1032, 1995.
18. E. Schwabe, Theoretische Betrachtungen über die Beeinflussung der ferromagnetischen Koerzivkraft durch Einschlüsse mit rotationellelliptischer Form, für den Fall, dass deren Abmessungen klein gegen die Dicke Blochwand sind, *Ann. Physik,* **11**: 99–112, 1952.

19. K. Schröder, Magnetisierung in der Umgebung unmagnetischer Einschlüsse in Ferromagnetika, *Phys. Stat. Sol.,* **33**: 819–830, 1969.

20. D. I. Paul, Extended theory of the coercive force due to domain wall pinning, *J. Appl. Phys.,* **53**: 2362–2364, 1982.

21. P. Gaunt, Ferromagnetic domain wall pinning by a random array of inhomogeneities, *Phil. Mag. B,* **48**: 261–276, 1983.

22. Xinhe Chien and P. Gaunt, The pinning force between a Bloch wall and a planar pinning site in MnAlC, *J. Appl. Phys.,* **67**: 2540–2542, 1990.

23. Wolfgang Prause, Energy and coercive field of a porous ferromagnetic sample with Bloch walls. *J. Magn. Magn. Mater.,* **10**: 94–96, 1979.

24. M. A. Golbazi et al., A study of coercivity in Ca-Ge substituted epitaxial garnets, *IEEE Trans. Magn.,* **MAG-23**: 1945, 1987.

25. E. Della Torre, C. M. Perlov, and M. Pardavi-Horvath, Comparison of coercivity calculations of anisotropy and exchange wells in magnetooptic media, *J. Magn. Magn. Mater.,* **104–107**: 303–304, 1992.

26. M. Pardavi-Horvath, Defects and their avoidance in LPE of garnets, *Progress in Crystal Growth and Characterization,* **5**: 175–220, 1982.

27. M. Pardavi-Horvath, Coercivity of epitaxial magnetic garnet crystals, *IEEE Trans. Magn.,* **MAG-21**: 1694, 1985.

28. M. Pardavi-Horvath and P. E. Wigen, Defect and impurity related effects in substituted epitaxial YIG crystals, Advances in Magneto-Optics, *J. Magn. Soc. Jpn.,* **11**: S1, 161, 1987.

29. Franz Preisach, Untersuchungen über den Barkhauseneffekt, *Ann. Physik,* **5**: 737–799, 1929.

30. Isaak D. Mayergoyz and Can E. Korman, Preisach model with stochastic input as a model for magnetic viscosity, *J. Appl. Phys.,* **69**: 2128–2134, 1991.

31. I. D. Mayergoyz and G. Friedman, The Preisach model and hysteretic energy losses, *J. Appl. Phys.,* **61**: 3910–3912, 1987.

32. Xiaohua Huang and M. Pardavi-Horvath, Local demagnetizing tensor calculation for rectangular and cylindrical shapes, *IEEE Trans. Magn.,* **32**: 4180–4182, 1996.

33. John C. McClure, Jr. and Klaus Schröder, The magnetic Barkhausen effect, *CRC Critical Rev. Solid State Sci.,* **6**: 45–83, 1976.

34. D. C. Jiles, P. Garikepati, and D. D. Palmer, Evaluation of residual stress in 300M steels using magnetization, Barkhausen effect and X-ray diffraction techniques. In Donald O. Thompson and Dale E. Chimenti (Eds.) *Rev. Progr. Quantitative Nondestructive Evaluation,* **8B**: 2081–2087, Plenum Press: New York, 1989.

35. A. H. Wafik, Effect of deformation on Barkhausen jumps of fine wires of iron, nickel and iron-nickel alloy, *J. Magn. Magn. Mater.,* **42**: 23–28, 1984.

36. M. Pardavi-Horvath, unpublished data.

37. R. D. McMichael, L. J. Swartzendruber, and L. H. Bennett, Langevin approach to hysteresis and Barkhausen jump modeling in steel, *J. Appl. Phys.,* **73**: 5848–5850, 1993.

38. Richard M. Bozorth, Barkhausen effect in iron, nickel and permalloy. I. Measurement of discontinuous change in magnetization, *Phys. Rev.,* **34**: 772–784, 1929.

39. U. Lieneweg and W. Grosse-Nobis, Distribution of size and duration of Barkhausen pulses and energy spectrum of Barkhausen noise investigated on 81% nickel-iron after heat treatment, *Int. J. Magn.,* **3**: 11–16, 1972.

40. B. Alessandro et al., Domain-wall dynamics and Barkhausen effect in metallic ferromagnetic materials, *I. Theory, J. Appl. Phys.,* **68**: 2901–2907, 1990.

41. F. Y. Hunt and R. D. McMichael, Analytical expression for Barkhausen jump size distributions, *IEEE Trans. Magn.,* **30**: 4356–4358, 1994.

42. G. Bertotti, G. Durin, and A. Magni, Scaling aspects of domain wall dynamics and Barkhausen effect in ferromagnetic materials, *J. Appl. Phys.,* **75**: 5490–5492, 1994.

43. D. C. Jiles, L. B. Sipahi, and G. Williams, Modeling of micromagnetic Barkhausen activity using a stochastic process extension to the theory of hysteresis, *J. Appl. Phys.,* **73**: 5830–5832, 1993.

44. L. J. Swartzendruber et al., Barkhausen jump correlations in thin foils of Fe and Ni, *J. Appl. Phys.,* **67**: 5469–5471, 1990.

45. B. Alessandro, G. Bertotti, and A. Montorsi, Phenomenology of Barkhausen effect in soft ferromagnetic materials, *J. Physique,* **49** (C8): 1907–1908.

46. H. Yamazaki, Y. Iwamoto, and H. Maruyama, Fractal dimension analysis of the Barkhausen noise in Fe-Si and permalloy, *J. Physique,* **49** (C8): 1929–1930.

47. P. J. Cote and L. V. Meisel, Self-organized criticality and the Barkhausen effect. *Phys. Rev. Lett.,* **67**: 1334–1337, 1991.

48. J. S. Urbach, R. C. Madison, and J. T. Markert, Reproducibility of magnetic avalanches in an Fe-Ni-Co magnet, *Phys. Rev. Lett.,* **75**: 4964–4967, 1995.

49. L. J. Swartzendruber and G. E. Hicho, Effect of sensor configuration on magnetic Barkhausen observations, *Res. Nondestr. Eval.,* **5**: 41–50, 1993.

50. S. Gadetsky and M. Mansuripur, Barkhausen jumps during domain wall motion in thin magneto-optical films, *J. Appl. Phys.,* **79**: 5667–5669, 1996.

51. G. V. Lomaev, V. S. Malyshev, and A. P. Degterev, Review of the application of the Barkhausen effect in nondestructive inspection, *Sov. J. Nondestructive Testing,* **20**: 189–203, 1984.

52. S. Titto, M. Otala, and S. Säynäjäkangas, Non-destructive magnetic measurement of steel grain size. *Non-Destructive Testing,* **9**: 117–120, 1976.

53. H. Gupta, M. Zhang, and A. P. Parakka, Barkhausen effect in ground steel, *Acta Mater.,* **45**: 1917–1921, 1997.

54. C. Denis Mee and Eric D. Daniel (eds.), *Magnetic Recording,* New York: McGraw-Hill, 1987.

55. V. Kovachev, *Energy Dissipation in Superconducting Materials,* Clarendon Press: Oxford, 1991.

56. Lawrence Dresner, *Stability of Superconductors,* Plenum Press: New York, 1995.

57. George W. Crabtree and David R. Nelson, Vortex physics in high-temperature superconductors, *Physics Today,* **38–45**: April 1997.

Reading List

A rich source of up-to-date information on Barkhausen noise, magnetic domain wall pinning, and hysteresis losses are the issues of the *J. Appl. Phys.,* and *IEEE Trans. Magn.,* publishing the material of the annual conferences on magnetism. The *IEEE Trans. Appl. Supercond.* is suggested as a source of current information on flux pinning in superconductors.

MARTHA PARDAVI-HORVATH
The George Washington University

MAGNETIC NONDESTRUCTIVE TESTING.   See MAGNETIC METHODS OF NONDESTRUCTIVE EVALUATION.

MAGNETIC PARTICLES

Ever since the development of the magnetic compass revolutionized navigation in the second century C.E., applications of magnetic materials have become essential in most branches of engineering. In many of these applications the magnetic materials are utilized in the form of microscopic magnetic particles. Some modern representative examples from electrical and electronics engineering are nonvolatile storage of information on magnetic tapes and disks, magnetic inks, refrigerators that make use of the magnetocaloric effect, ferrofluid vacuum seals, and the microscopic machines known as microelectromechanical systems (MEMS). It is likely that these current and emergent engineering applications of microscopic and nanoscopic magnetic particles will be joined by many more in the next few decades.

Even though magnetic particles have been used for a very long time, our understanding of their physical behaviors is relatively new and still limited. In addition, the current understanding of the physical properties of magnetic particles, and the engineering applications that are made possible by these properties, cannot easily be described without sophisticated mathematics. Some of the reasons for this are as follows:

1. The magnetism of magnetic particles is fundamentally quantum-mechanical in origin. Hence, understanding the properties of magnetic particles requires the use of quantum mechanics, either directly or by the inclusion of quantum-mechanical effects into phenomenological models.

2. Magnetism involves vector quantities, which have a direction as well as a magnitude. The magnetic field H (given in units of amperes per meter, A/m), the magnetic moment per unit volume of a magnetic substance M (given in units of A/m), and the magnetic induction or magnetic flux density B (given in units of webers per square meter, Wb/m^2, or tesla, T) are all vector quantities, and their interrelationships must be expressed in terms of vector and tensor equations.

3. For ferromagnetic particles, which are the ones used in most engineering applications, the relationships between the three vector fields H, M, and B depend on the history of the particle—how the field and magnetization have varied in the past, and possibly how the particle was manufactured.

4. Engineering applications of magnetic particles are determined by physical properties that depend on energy considerations that originate from a number of different physical mechanisms. For example, in spherical particles the coercive field (defined later) is due primarily to the energy associated with crystalline anisotropy, whereas in elongated needlelike single-domain particles the dipole-dipole interactions (the magnetostatic energy) are the most important in determining the coercive field.

Commonly these complications are what make magnetic particles so useful in engineering applications. For example, the dependence of M on the detailed history of the particle is what makes magnetic recording possible.

Most engineering applications utilize large numbers of magnetic particles, which may differ in size, orientation, composition, and so on. However, a detailed understanding of the behavior of such an assortment of particles requires knowledge of the properties of individual particles. The properties of systems consisting of large numbers of particles can be obtained by an appropriate averaging over the size, orientation, shape, type, and location of individual particles.

This article will focus on single magnetic particles. It is only in very recent years that it has become possible to study the behavior of individual particles experimentally. This is due to the development of better methods to produce well-characterized magnetic particles and to the development of magnetic measurement techniques with resolutions at the nanometer scale. These ultrahigh resolution techniques include magnetic force microscopy (MFM) (1), micro-SQUID devices (2), Lorentz transmission electron microscopy (3), and giant magnetoresistive (GMR) measurements (4). These experimental advances are complemented by the emergence of computational science and engineering techniques, such as micromagnetic and Monte Carlo simulations, which allow detailed comparisons between the predictions of models for magnetic particles and experimental data. In this article both high-resolution experimental data for single particles and numerical results from model simulations will be used in the figures to illustrate the physical phenomena described.

TYPES OF MAGNETIC MATERIALS

At the atomic level, a magnetic material is an arrangement of local magnetic moments. Fundamentally, each such magnetic moment is a quantum-mechanical quantity arising from either the intrinsic spin or the orbital motions of the electrons of an atom or molecule. Each of these localized magnetic moments, often simply called "spins" for brevity, may be thought of as a small bar magnet. The microscopic structures of the most widely studied types of magnetic materials are shown schematically in Fig. 1. This figure shows the lowest-energy arrangement (the ground state) of the local magnetic moments or spins in (a) ferromagnetic, (b) antiferromagnetic, and (c) ferrimagnetic materials. The total magnetizations due to the spin arrangements shown in Fig. 1 can be found by adding the vectors together. In a ferromagnet the spins add up to a large vector. For an antiferromagnet they add up to zero. For a ferrimagnet there is some cancellation, but the spins still add up to a nonzero value.

The alignment of the spins in Figs. 1(a) to 1(c) corresponds to a temperature of absolute zero. At nonzero temperatures, the alignment of the spins is somewhat random due to thermal fluctuations. At temperatures above a critical temperature, T_c (which is different for different materials), the thermal fluctuations are so large that the spin arrangement becomes essentially random, the total magnetization is zero, and the material becomes paramagnetic [as shown in Fig. 1(d)]. For ferromagnets in zero applied magnetic field the magnetization vanishes at T_c and remains zero above T_c. Since most engineering applications of magnetic materials use ferromagnets, the remainder of this article will focus on ferromagnetic particles.

RELATIONSHIPS BETWEEN THE VECTOR FIELDS

The fundamental relationship between the vector fields in SI units is given by

$$B = \mu_0(H + M) \tag{1}$$

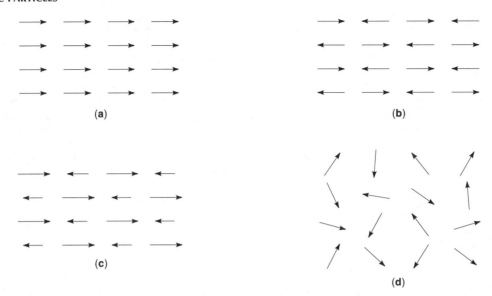

(a)

(b)

(c)

(d)

Figure 1. Schematic representation of the arrangement of local magnetic moments or spins in different types of magnetic materials. In (a) all the spins are aligned, and the material is a ferromagnet. The total magnetization, given by the sum of all the spin vectors, is large. In (b) the nearest-neighbor spins are all antialigned, so the total magnetization is zero. This arrangement is that of an antiferromagnetic material. In (c) the nearest-neighbor spins are also antialigned, but here the spins on each sublattice have different lengths. Hence the total magnetic moment is nonzero in this ferrimagnetic material. In (d) the directions of the spins are randomly distributed due to thermal fluctuations so that the total magnetic moment is zero in zero field. This represents a paramagnet. Above a critical temperature, T_c (which is different for different materials), all magnetic materials become paramagnetic.

where $\mu_0 = 4\pi \times 10^{-7}$ H/m (henry per meter) is the permeability of free space. Equation (1) is true for all materials, even for nonlinear ones.

For a linear, homogeneous and isotropic material, the relationship between $\boldsymbol{M}$ and $\boldsymbol{H}$ is linear and given by

$$\boldsymbol{M} = \chi_m \boldsymbol{H} \qquad (2)$$

where χ_m is a material-dependent quantity called the magnetic susceptibility. In this case it is possible to write

$$\boldsymbol{B} = \mu_0[1 + \chi_m]\boldsymbol{H} = \mu_0 \mu_r \boldsymbol{H} = \mu \boldsymbol{H} \qquad (3)$$

where μ is the permeability of the medium and the parameter μ_r is its relative permeability.

In ferromagnetic materials, however, the relationship between the three vector fields is generally nonlinear and history dependent. Thus simple relationships such as those of Eq. (2) and Eq. (3) are not justified. In this case it is necessary to talk about a hysteresis loop (5,6). Figure 2 shows a hysteresis loop, with identification of the intrinsic coercive field H_c, the remanent (spontaneous) magnetization M_r, and the saturation magnetization M_s (the maximum magnetization of a magnetic particle in a strong field). The area of the hysteresis loop corresponds to the work that must be done in taking the magnetic material through one cycle of the applied field. This work is converted to heat and represents a major source of energy loss in devices such as transformers and motors. It was first studied systematically by Ewing, Warburg, and Steinmetz (6) over a century ago. The hysteresis loop shown in Fig. 2 is a nonidealized loop for a small magnetic particle at finite temperature, obtained from a computer simulation of a model magnetic system. Thermal fluctuations in experimental hysteresis loops of nanoscale magnetic particles have been seen in many studies, and they necessitate a probabilistic interpretation of quantities such as the intrinsic coercive field (1,2). Such hysteresis loops indicate that magnetic particles may find applications as nonlinear network elements with intrinsic noise.

One useful classification of magnetic materials for engineering purposes is into soft and hard magnets. Soft magnetic materials are usually used for cores of transformers, generators, and motors, and the heads in magnetic tape and disk devices; applications that require a low coercive field, H_c, small hysteresis loop area to minimize heat generation, or high permeability. Hard magnetic materials are used for electric sensors, loudspeakers, electric meters, magnetic recording media, and other uses that require a high coercive field, a high remanence, or a large hysteresis loss.

Not all hysteresis loops have the shape shown in Fig. 2. In fact, for single-crystal ferromagnets the shape of the hysteresis loop is usually dependent on the orientation of the applied field with respect to the crystalline axes. For materials composed of many different magnetic particles or grains, the nonlinear effects of each particle must be added to obtain a composite hysteresis loop. For such composite materials, or for bulk materials with impurities, the hysteresis loop is not smooth but contains small jumps, called Barkhausen jumps, that correspond to successive switching of small regions of the material (7). These loops can sometimes be parameterized using a model called the Preisach model (8).

MAGNETIC ENERGIES

The energies that are relevant to the properties of magnetic materials arise from a variety of physical effects. Which ones are most important for a particular engineering application depend on the composition of the particular piece of material, its mesoscopic structure (grain size, local stress, etc.), its surface properties, and its size and shape.

Exchange Energy

The exchange energy comes from the quantum-mechanical overlap and hybridization of the exchange integrals between atoms and molecules. Since the interaction constant J is due to the overlap of orbitals, it is a short-range interaction that often does not extend beyond nearest-neighbor pairs of lattice sites. For ferromagnetic materials, Fig. 1(a), $J > 0$, while for antiferromagnetic or ferrimagnetic materials, Fig. 1b(c), $J < 0$. For most materials J is somewhat temperature and stress dependent, due to changes in interatomic distances. It is also dependent on the local environment of a particular magnetic atom. Consequently exchange interactions for atoms at a surface, near a grain boundary, or near an impurity atom may be different from the exchange interaction of the same kind of atom in the bulk of a cryst●l. Often the Hamiltonian for a magnetic material is written as a sum of Heisenberg terms, such as $\mathscr{H} = J\boldsymbol{S}_i \cdot \boldsymbol{S}_j$, where the three-dimensional spins $\boldsymbol{S}_i$ and $\boldsymbol{S}_j$ are located at nearest-neighbor sites i and j of the crystal lattice.

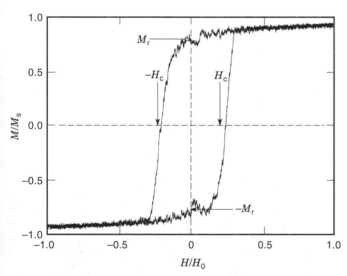

Figure 2. A hysteresis loop for a model of a single-domain uniaxial magnetic particle, the Ising model. An external magnetic field $H_0 \sin(\omega t)$ is applied, and the magnetization is recorded as a function of time. Shown on the loop is the location of the average intrinsic coercive field, H_c. This is the field that must be applied to make the magnetization equal to zero. Also shown is the remanent, or spontaneous, magnetization, M_r. This is the magnetization when the applied field is equal to zero. In this particular model the value of M_r is known exactly. The saturation magnetization, M_s, corresponds to all the spins aligned, as in Fig. 1(a). The fluctuations on the hysteresis loop are due to random thermal noise, which is typically seen in experimental hysteresis loops for small particles at nonzero temperatures as well. This loop is for a temperature of $0.97T_c$. The loop would become smoother at lower temperatures and for larger particles than the 64×64 lattice used here. The loop area can be a complicated function of H_0, ω, and temperature (5). The data used to generate this figure are from a Monte Carlo simulation, courtesy of Dr. Scott W. Sides.

Magnetocrystalline Anisotropy Energy

When magnetic atoms are arranged on a crystal lattice, their spins have a lower energy if they are aligned along certain directions. These are called the easy axes. The directions that require the highest energy for the orientation of the magnetic spins are the hard directions or hard axes. Consequently, certain magnetization directions are preferred. One consequence of magnetocrystalline anisotropy is that hysteresis loops, such as the one shown in Fig. 2, depend on the angles between the applied magnetic field and the crystal axes. In magnetic materials that are not perfect crystals, the orientation of the easy axes in different grains will most likely be different.

For example, iron is a cubic crystal, and the three cube edges are the easy axes. The anisotropy energy of iron is expressed as

$$U_K = K_1(\alpha_1^2\alpha_2^2 + \alpha_2^2\alpha_3^2 + \alpha_3^2\alpha_1^2) + K_2\alpha_1^2\alpha_2^2\alpha_3^2 \quad (4)$$

where the α's are the direction cosines for the angles between the cube edges and the direction of the magnetization. The anisotropy coefficients K_1 and K_2 depend on temperature and are zero above T_c.

Dipole–Dipole Energy

The classical interaction energy between two magnetic dipoles $\boldsymbol{m}_1$ and $\boldsymbol{m}_2$ (which are vector quantities) is given by $U(\boldsymbol{m}_1, \boldsymbol{m}_2) = (\mu_0/4\pi r^3)\,[\boldsymbol{m}_1 \cdot \boldsymbol{m}_2 - 3(\hat{\boldsymbol{r}} \cdot \boldsymbol{m}_1)(\hat{\boldsymbol{r}} \cdot \boldsymbol{m}_2)]$, where $\boldsymbol{r} = r\hat{\boldsymbol{r}}$ is the vector from $\boldsymbol{m}_1$ to $\boldsymbol{m}_2$. This gives a long-range interaction between the dipoles. Since each spin in Fig. 1 represents a magnetic dipole, this dipole-dipole interaction must be considered in dealing with magnetic particles. In many engineering applications where there are no time-varying external fields, the microscopic motion of each individual spin (due to precession and random thermal fluctuations) need not be taken into account, and the familiar equations of magnetostatics are recovered. Consequently, the dipole-dipole interaction is sometimes called the magnetostatic energy. In particular, from magnetostatics the energy stored in the fields is given by

$$U_M = -\frac{1}{2} \int \boldsymbol{M} \cdot \boldsymbol{H}\, dV \quad (5)$$

where the volume integral is over all ferromagnetic bodies. This may be understood as the integral of the interaction of each dipole with the field $\boldsymbol{H}$ created by all the other magnetic dipoles.

In general, the field inside a uniformly magnetized ferromagnetic body is not uniform, except in very special cases, such as ellipsoids. For an ellipsoid that is uniformly magnetized, the demagnetizing field inside the ellipsoid can be written as $\boldsymbol{H}_d = -\boldsymbol{N}\boldsymbol{M}$ where $\boldsymbol{N}$ is a tensor. The demagnetizing field is due to the magnetic field caused by all the other dipoles in the particle. When $\boldsymbol{M}$ is parallel to one of the principal axes of the ellipsoid, the tensor is diagonal, and its three diagonal elements are called the demagnetizing factors. Then the magnetostatic self-energy of a uniformly magnetized ellipsoid of volume V is given by

$$U_M = \frac{V}{2}(N_x M_x^2 + N_y M_y^2 + N_z M_z^2) \quad (6)$$

since the tensor is diagonal. The demagnetizing factors may not be equal, so some directions will be preferred. This gives a shape anisotropy term in the energy of a ferromagnet. Note that the preferred directions from shape anisotropy can be different from the preferred directions for magnetocrystalline anisotropy. For more complicated geometries, a numerical method of solving for $\boldsymbol{H}_d$ and evaluating the integral in Eq. (5) is recommended.

The dipole-dipole interaction is responsible for the formation of magnetic domains. These domains are equilibrium regions where the magnetization is predominantly oriented in a single direction. They are separated by domain walls. Domain walls cost some energy due to the exchange and magnetocrystalline energy terms. However, this energy cost is balanced by the reduction in the magnetostatic energy resulting from the reduction in the total magnetization. Hence large magnetic particles break up into domains magnetized in different directions and separated by domain walls. Domain walls are classified according to the orientation of the spins in the domain-wall region.

For small enough particles the energy cost of a domain wall is always higher than the gain resulting from the reduction in the magnetostatic energy. As a result, sufficiently small particles consist of a single magnetic domain. Nanometer-sized cobalt particles created by electron beam lithography may have only one or a few domains. Figure 3 shows pictures of an array of such particles. Figure 3(a) shows the particles themselves, while Fig. 3(b) shows the magnetic structure of the same particles. Schematic illustrations of the domain structures seen are shown in Figs. 3(c) and 3(d). Even though all of the particles have similar sizes, small differences in their manufacturing histories nevertheless cause their domain structures to be significantly different. This illustrates both the history dependence of magnetic domains and the sensitivity of the magnetization to small changes in the geometry, environment, and composition of a magnetic particle.

Substrate and Surface Energies

The surface of a magnetic particle can influence its physical behavior as much or even more than the bulk. This is because the energies associated with the exchange interactions and the magnetocrystalline anisotropy are extremely sensitive to the local environment. The environment of a magnetic atom is different at the surface and in the bulk. Surface effects become increasingly important as the particle size is decreased, due to the increased surface-to-volume ratio.

Surface and interface effects can have important engineering applications. For example, GMR materials can be grown by coupling a ferromagnetic film or particle to an antiferromagnetic bulk, leading to an exchange interaction between the interfacial spins of the two types of materials. This gives an exchange bias that leads to an asymmetric hysteresis loop. This is highly desirable because it enhances the signal-to-noise ratio of GMR read heads used in magnetic recording applications.

Surfaces can also affect the response of a magnetic particle to changes in its environment. For example, if the direction of the external field is reversed, reversal of the magnetization of the particle may be initiated at the surface. This can make

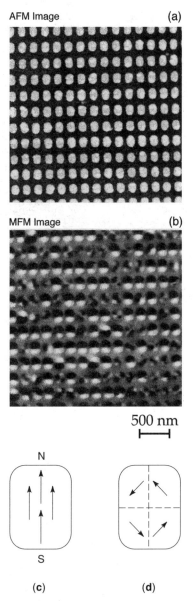

Figure 3. Nanometer-sized magnetic particles of cobalt (Co). These were grown as a polycrystalline film on a GaAs substrate, and electron beam (e-beam) lithography was used to make isolated particles. The substrate in this case should contribute minimally to the magnetization of the particles. The size of these particles is roughly 170 nm, which is about as small as can currently be manufactured using e-beam lithography. Although the particles look uniform, as seen in the atomic force microscope (AFM) image (a), due to the history dependence of magnetic materials different magnetizations are seen in the magnetic force microscope (MFM) image (b). This image measures the magnetic field outside the magnetic particles caused by their individual magnetizations. The dark region corresponds to the north pole and the light region corresponds to the south pole. About two-thirds of the particles are seen to be single domain, while the others have more complicated magnetic arrangements. The domain patterns shown are for the particles as manufactured, before any additional magnetic field was applied. They are schematically illustrated in the sketches (c) and (d). Thanks to Prof. Andrew Kent of the New York University Physics Department for the unpublished AFM and MFM images.

the coercive field of small particles significantly smaller than it would be in the absence of surfaces.

Magnetostriction and Magnetomechanical Effects

When a magnetic particle is exposed to a magnetic field, its physical dimensions change. This effect, which was discovered in 1842 by Joule, is called magnetostriction. It is related to magnetocrystalline anisotropy and both effects are mainly due to quantum-mechanical spin-orbit coupling.

The application of stress to a magnetic material results in a strain response. The sensitivity of the local magnetization to the local environment means that strain in the lattice will change the local energies (the exchange energy, the magnetocrystalline anisotropy energy, and the dipole–dipole energy), which can alter the behavior of the magnetization locally. This is the inverse magnetostriction effect, commonly called the magnetomechanical effect.

The application of an oscillating magnetic field causes an oscillatory change in the linear dimensions of a magnetic material, which can produce sound waves in surrounding materials. One application of this effect is in magnetostrictive transducers, with uses in sonar, medical technology, and ultrasonic cleaning. The humming sound from electrical transformers is also mainly generated by magnetostriction in the magnetic particles that make up the core.

Temperature

The energy scales associated with the effects discussed previously must be compared with the thermal energy $k_B T$, which causes the randomness of the spins seen in Fig. 1(d). Here k_B is Boltzmann's constant, and the temperature T is given in kelvins. If the thermal energy is comparable to the other energies, it can affect engineering applications. For instance, at nonzero temperature there is some probability that the orientation of the magnetization of a particle will change spontaneously. This can lead to loss of data integrity in magnetic recording media (9).

DYNAMICS OF MAGNETIC PARTICLES

The magnetization dynamics of magnetic particles are important for engineering applications. For example, consider a bit of information to be stored on one of the single-domain particles shown in Fig. 3. The coding may be that a north–south orientation corresponds to Boolean 0 and a south-north orientation to Boolean 1. Engineering questions involving the magnetization dynamics include the following: (1) How strong a field must be applied to change a 0 to a 1 or vice versa? (2) How long can a particle encoded with a bit be exposed to a stray field without losing data integrity, and how strong a stray field can be tolerated? (3) If a bit is written today, how long can it be trusted to remain as it was written?

Metastability

Arguably the most important concept needed to understand the dynamics of magnetic particles is metastability. One familiar example of this ubiquitous natural phenomenon is the supercooling of water. That metastability should be relevant to magnetism is suggested by the fact that the magnetization M of a magnetic particle is history dependent.

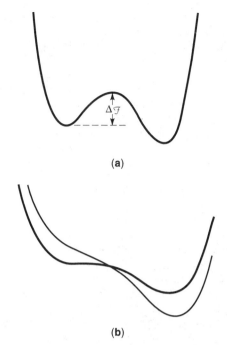

Figure 4. A metastable well is illustrated in (a). Even though the lowest-energy state is on the right, an energy barrier $\Delta\mathscr{F}$ must be overcome before the metastable state on the left can decay to the equilibrium state. As the external magnetic field is increased, the barrier decreases, and it becomes zero at a particular field called the nucleation field, H_{nucl}, as illustrated by the heavy curve in (b). There is no metastable state for higher fields, as illustrated by the light curve in (b).

Figure 4(a) shows the standard picture of metastability in a bistable system, such as a uniaxial single-domain magnetic particle. There are two free-energy minima, one of which is only a local minimum (the metastable state) and one of which is the global minimum (the equilibrium state). If the system is in the metastable state, a free-energy barrier, $\Delta\mathscr{F}$, must be overcome before the system can relax to the equilibrium state. The average lifetime of the metastable state, τ, is given by

$$\tau = \tau_0 \exp\left(\frac{\Delta\mathscr{F}}{k_B T}\right) \qquad (7)$$

where τ_0 is an inverse attempt frequency. In magnetic particles τ_0 is typically taken to be on the order of an inverse phonon frequency, usually $\tau_0 \approx 10^{-10}$ s. More detailed analysis shows that τ_0 depends on the curvature of the free energy at both the metastable minimum and the saddle point (the maximum that separates the metastable and equilibrium states).

Once the system has been prepared in the metastable state, how can it get out? There are two possible answers to this question. An analogy is a pitcher half filled with water, which is held above a sink. The water is the system, the pitcher is the metastable state, and the sink is the equilibrium state. One way to get the water into the sink is to shake the pitcher to make the water splash out. The more vigorously the pitcher is shaken, the faster the water will splash out. This method of escape from a metastable state is analogous to a magnetic material at a nonzero temperature. As seen from Eq. (7), the lifetime increases as the height of the barrier increases, whereas it decreases as the temperature increases. Another way of getting the water into the sink is to tip the pitcher. As the pitcher continues to tip, at some point the water starts to flow out. This corresponds to a way of escaping from the metastable state of a magnetic particle, which is valid even at zero temperature. In particular, the barrier $\Delta\mathscr{F}$ depends on the applied field H_{appl}, and changing H_{appl} corresponds to tipping the pitcher. At a particular value of the field $\Delta\mathscr{F}$ equals zero, and the metastable state disappears. This situation is illustrated in Fig. 4(b). The field that must be applied for $\Delta\mathscr{F}$ to vanish is often called the nucleation field, H_{nucl}.

This pitcher analogy also illustrates an important consideration concerning the escape from a multidimensional metastable well at zero temperature. In particular, how far the pitcher must be tipped before the water starts to spill out depends on the direction in which it is tipped. The smallest angle is needed if the pitcher is tipped to make the water spill from the spout—the lowest part of the rim. Similarly, the nucleation field, H_{nucl}, often depends on the direction of the applied field. However, for thermally driven escape the decay always proceeds across the saddle point (analogous to the spout) as the temperature (analogous to the amplitude of the shaking) is increased, or as the waiting time is increased at fixed temperature.

The waiting time is related to another physical quantity of interest, the probability that a magnetic particle starting in the metastable state at $t = 0$ has never left it at time t. This probability is often called $P_{not}(t)$. It is only in the last decade that it has become possible to measure $P_{not}(t)$ for individual single-domain magnetic particles (1,2). For thermal escape over a single barrier one has

$$P_{not}(t) = \exp(-t/\tau) \tag{8}$$

where τ is given by Eq. (7).

Equation (7) is the fundamental equation needed to understand the dynamics of a magnetic particle. All one requires is a knowledge of $\Delta\mathscr{F}$. This, however, requires a knowledge of the energy of the spin configuration at the saddle point. Since $\Delta\mathscr{F}$ enters in an exponential, the value of τ is extremely sensitive to small changes in $\Delta\mathscr{F}$.

Coherent Rotation

Consider a spherical single-domain uniaxial magnetic particle with magnetocrystalline anisotropy coefficient K and a uniform magnetization $\boldsymbol{M}$ in an applied magnetic field along the easy axis. If all the spins in the particle point in the same direction, the total energy is $E = KV \sin^2(\theta) - M_s VH \cos(\theta)$. Here the saturation magnetization M_s makes an angle θ with the easy axis, and the volume of the particle is V. In this case it is possible to obtain the zero-temperature free-energy barrier exactly—namely,

$$\Delta\mathscr{F} = KV\left[1 + \left(\frac{HM_s}{2K}\right)^2\right]$$

In this model all the spins rotate coherently and act like one large spin. Consequently, particles that behave this way are called superparamagnetic particles. Note that the zero-temperature energy barrier has been assumed to be valid at finite temperatures in this model of a superparamagnet. Since superparamagnetism implies escape over a single barrier, $P_{not}(t)$ is given by Eq. (8).

The particle volume V enters the metastable lifetime in an exponential, so a small change in particle size can lead to extremely large changes in τ. For example, for an iron sphere with a radius of 115 Å the lifetime $\tau \approx 0.07$ s. If the radius is 140 Å then $\tau \approx 1.5 \times 10^5$ s [42 hours] (10).

The coherent rotation mode described here is often called the Néel–Brown reversal mode (10) since Néel derived Eq. (7) with $\Delta\mathscr{F} = KV/k_BT$ and Brown wrote a differential equation for a random walk in the metastable well to obtain a nonconstant prefactor for Eq. (7). It is also possible to obtain a zero-temperature hysteresis curve using the same assumptions—namely, a uniform magnetization and that only the external field and a uniaxial anisotropy are important. This is called the Stoner–Wohlfarth model (10). This model gives an upper bound for the intrinsic coercive field as $H_c \leq 2K/M_s$.

An equivalent analysis for the case in which shape anisotropy is important can also be performed. Again, with the assumption that all the spins always point in the same direction, the analysis is the same except that K now arises from shape anisotropy. The assumption that the spin configuration at the saddle point has all spins pointing in the same direction is only sometimes valid. There are other zero-temperature reversal modes where the zero-temperature saddle point has spins in other configurations. Examples include modes descriptively named buckling, curling, and fanning. The dominant mode depends on the geometry and size of the magnetic particle.

Nucleation and Growth

As the volume V increases, the average rate for magnetization reversal via coherent rotation quickly becomes too small to be practically important. This was illustrated by the example with iron particles discussed previously. Other reversal modes with lower free-energy barriers can then come into play, especially for highly anisotropic materials, in which domain walls are relatively thin.

At nonzero temperatures, thermal fluctuations continually create and destroy small "droplets" of spins aligned with the applied field. The free energy of such a droplet consists of two competing parts: a positive part due to the interface between the droplet and the metastable background, and a negative part due to the alignment of the spins in the droplet with the field. For a droplet of radius R, these terms are proportional to R^{d-1} and $-|H|R^d$, respectively. Here d is the spatial dimension of the particle, and these relations hold both for three-dimensional particles ($d = 3$) and for particles made from a thin film ($d = 2$). Small droplets most likely shrink. However, if a droplet manages to become larger than a critical radius, $R_c \propto 1/|H|$, it will most likely continue to grow and eventually bring the whole particle into the stable magnetization state. The free-energy barrier associated with such a critical droplet is proportional to $1/|H|^{d-1}$, which is *independent* of the particle size! Since the droplets can only grow at a finite speed, in sufficiently large particles or for sufficiently strong fields, new critical droplets may nucleate at different positions while the first droplets formed are still growing.

This droplet switching mechanism gives rise to three regimes of field strengths and particle sizes:

1. For sufficiently weak fields and/or small particles, a critical droplet would be larger than the entire particle. As a result, the saddle point configuration consists of an interface that cuts across the particle, so that $\Delta\mathscr{F} \propto V^{(d-1)/d}$, independent of $|H|$ to lowest order. The behavior in this regime is effectively superparamagnetic, even though the dependence of $\Delta\mathscr{F}$ on the particle size is somewhat weaker than predicted for uniform rotation. In this regime both Eqs. (7) and (8) are valid.

2. For stronger fields and/or larger particles, the magnetization reverses by the action of a single droplet of the stable phase. The free-energy barrier is independent of the particle volume, but because the droplet can nucleate anywhere in the particle, the average lifetime is inversely proportional to V. We call this decay regime the single-droplet regime. A series of snapshots of a computer simulation of single-droplet decay is shown in Fig. 5. In this regime Eq. (8) is still valid.

3. For yet stronger fields and/or larger particles, the decay occurs via a large number of nucleating and growing droplets. In this regime, which we call the multidroplet regime, the average lifetime is independent of V. A series of snapshots of a computer simulation of multidroplet decay is shown in Fig. 6. In this regime Eq. (8) is no longer valid, and $P_{\text{not}}(t)$ takes the form of an error function (11,12).

Switching Fields

Two rather similar quantities that are often measured for magnetic particles are the switching field, H_{sw}, and the intrinsic coercive field, H_{c}. The former is defined as the magnitude of the field for which a particle switches with probability 1/2 within a given waiting time after field reversal. The latter is the value of the field at which the magnetization crosses the field axis, as shown in the hysteresis loop in Fig. 2. Both depend weakly on the time scale of the experiment (waiting time or field cycle time), but they are qualitatively similar over a wide range of time scales. A collection of experimentally measured coercive fields for various materials are shown as functions of particle size in Fig. 7(a) (11). The increase in H_{c} with particle size for small sizes is due to the superparamagnetic behavior of the particles, whereas the decrease for larger sizes is due to the dipole-dipole interaction, which causes large particles to break up into multiple domains. For particles in the nanometer range, which are single-domain in equilibrium, the crossovers between the three nucleation driven magnetization reversal mechanisms described previously give rise to very similar size dependences (12,13), as seen from the computer simulation data in Fig. 7(b).

Domain Boundary Movement

If the magnetic particle is multidomain in zero field, the application of a magnetic field will cause the domain wall(s) to move. In this case the dynamics of the magnetization are dominated by the domain-wall movement. Typically there will be pinning sites due to impurities, grains, and surfaces that the domain wall must overcome before it can move. These obstacles can either be overcome by applying a sufficiently large field or by waiting for random thermal fluctuations to move the domain wall past the pinning centers. This is analo-

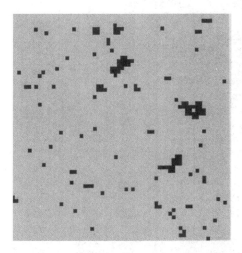
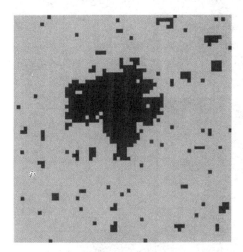
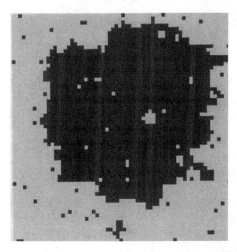

Figure 5. Three snapshots from a computer simulation of the single-droplet switching mechanism for a model of a particle made from a highly anisotropic, uniaxial ultrathin magnetic film. Time increases from top to bottom in the figure. Data courtesy of Dr. György Korniss.

gous to the zero-temperature and finite-temperature reversal mechanisms in single-domain particles. If the domain wall moves due to random thermal fluctuations, the magnetization of the particle will change slowly with time, a phenomenon called magnetization creep (13).

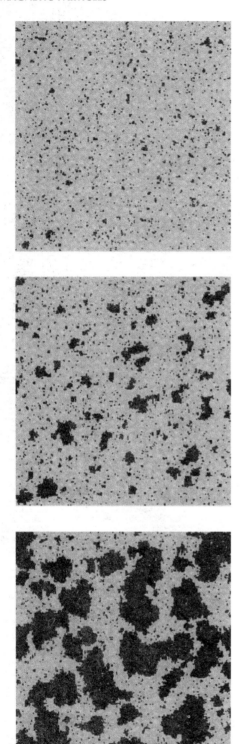

Figure 6. Three snapshots from a computer simulation of the multi-droplet switching mechanism for a model of a particle made from a highly anisotropic, uniaxial ultrathin magnetic film. Time increases from top to bottom in the figure. Data courtesy of Dr. György Korniss.

Magnetic Viscosity

Consider a large number of identical noninteracting particles. If a strong field is applied and then quickly removed, the remanent magnetization will decay with time as $M_r(t) = M_r(0)\exp(-t/\tau)$ as particles cross the barrier separating the two equilibrium states. However, in the typical case the particles are not identical, and there is a distribution of lifetimes, $\mathscr{P}(\tau)$. In this case the time decay is

$$M_r(t) = M_r(0)\int_0^\infty \mathscr{P}(\tau)\exp(-t/\tau)\,d\tau \tag{9}$$

Under some circumstances and for certain specific distribu-

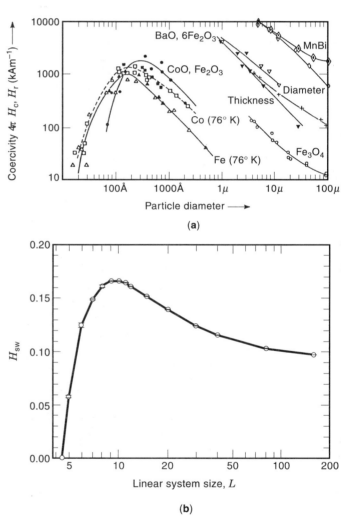

Figure 7. Plots of the intrinsic coercive and switching fields versus particle size. As explained in the text, these fields can be considered to be roughly the same. (a) The intrinsic coercive field, H_c, for magnetic particles of various materials, shown versus the particle diameter. Reproduced from (11) with permission from John Wiley & Sons, Ltd. (b) For a simple model for a uniaxial single-domain particle, the $L \times L$ two-dimensional Ising model, simulations show a maximum in H_{sw} (in units of J) versus L, even when there are no dipole-dipole interactions so that particles of all sizes remain single domain. This effect is due to different nucleation decay mechanisms for particles of different size (12,13), as described in the text.

tions $\mathscr{P}(\tau)$, Eq. (9) can be approximated by $M_r(t) \approx C - S \ln(t/\bar{t}_0)$, where C, S, and $\bar{t}_0$ are constants. This logarithmic decay of the magnetization is called magnetic viscosity. It must be emphasized that this logarithmic equation is a valid approximation *only* under certain specific circumstances and even then *only* for a limited range of t (10).

MAGNETOCALORIC EFFECTS

Magnetocaloric effects of magnetic particles have engineering applications principally in refrigeration. Current applications are mostly in ultralow temperature refrigeration below a few kelvins, but near-term applications close to room temperature seem promising. Given that at constant volume the $P \, dV$ work term is zero, the first law of thermodynamics becomes

$$\text{đ}Q = dU - \text{đ}w = dU - \mu_0 \boldsymbol{H} \cdot \text{d}\boldsymbol{M} \tag{10}$$

where U is the internal energy of the magnetic particle, and Q and w represent heat and magnetic work, respectively. The symbol đ denotes an infinitesimal change rather than a true differential. This must be considered because of the history dependence of $\boldsymbol{M}$. For an adiabatic change, $\text{đ}Q = TdS = 0$, where S is the entropy. Then the relation

$$\Delta T = -\frac{T}{C_H} \left(\frac{\partial M}{\partial T}\right)_H \Delta H \tag{11}$$

can be obtained (14), where C_H is the heat capacity at constant field. Thus an increase in H produces a rise in temperature, and vice versa, which is the magnetocaloric effect. This can be utilized as a cooling mechanism by placing the magnetic material in a strong applied field and then turning off the applied field. This will lead to a decrease in temperature due to adiabatic demagnetization according to Eq. (11).

ACKNOWLEDGMENTS

This work was supported by the US National Science Foundation Grants No. DMR-9520325 and DMR-9871455, and by Florida State University through the Center for Materials Research and Technology, the Supercomputer Computations Research Institute (US Department of Energy Contract No. DE-FC05-85ER25000), and the Department of Physics. Supercomputer time at the National Energy Research Supercomputer Center was provided by the US Department of Energy.

BIBLIOGRAPHY

1. M. Lederman, S. Schultz, and M. Ozaki, Measurement of the dynamics of the magnetization reversal in individual single-domain ferromagnetic particles, *Phys. Rev. Lett.,* **73**: 1986–1989, 1994.

2. W. Wernsdorfer et al., Experimental evidence of the Néel-Brown model of magnetization reversal, *Phys. Rev. Lett.,* **78**: 1791–1794, 1997.

3. C. Salling et al., Measuring the coercivity of individual sub-micron ferromagnetic particles by Lorentz microscopy, *IEEE Trans. Magn.,* **27**: 5184–5186, 1991; C. Salling et al., Investigation of the magnetization reversal mode for individual ellipsoidal single-domain particles of γ-Fe$_2$O$_3$, *J. Appl. Phys.,* **75**: 7989–7992, 1994.

4. V. Cros et al., Detection of the magnetization reversal in submicron Co particles by GMR measurements, *J. Magn. Magn. Mater.,* **165**: 512–515, 1997.

5. S. W. Sides, P. A. Rikvold, and M. A. Novotny, *J. Appl. Phys.,* **83**: 6494–6496, 1998.

6. J. A. Ewing and C. P. Steinmetz, On effects of retentiveness in the magnetization of iron and steel, *Proc. Roy. Soc. (London),* **34**: 39–45, 1882; E. Warburg, Magnetische Untersuchungen, *Ann. Phys. Chem. (Neue Folge),* **13**: 141–164, 1882; C. P. Steinmetz, On the law of hysteresis, *Trans. Am. Inst. Electr. Eng.,* **9**: 3–51, 1892.

7. O. Perkovic, K. Dahmen, and J. P. Sethna, Avalanches, Barkhausen noise, and plain old criticality, *Phys. Rev. Lett.,* **75**: 4528–4531, 1995.

8. I. D. Mayergoyz, *Mathematical Models of Hysteresis,* New York: Springer-Verlag, 1991; C. E. Korman and P. Rugkwamsook, Identification of magnetic aftereffect model parameters: comparison of experiment and simulations, *IEEE Trans. Magn.,* **33**: 4176–4178, 1997.

9. T. Yogi and T. A. Nguyen, Ultra high density media: gigabit and beyond, *IEEE Trans. Magn.,* **29**: 307–316, 1993; P. L. Lu and S. Charap, Thermal instability at 10 Gbit/in^2 magnetic recording, *IEEE Trans. Magn.,* **30**: 4230–4232, 1994; I. Klik, Y. D. Yao, and C. R. Chang, Chain formation, thermal relaxation and noise in magnetic recording, *IEEE Trans. Magn.,* **34**: 358–362, 1998; J. J. Lu and H. L. Huang, Switching behavior and noise in obliquely oriented recording media, *IEEE Trans. Magn.,* **34**: 384–386, 1998; E. N. Abarra and T. Suzuki, Thermal stability of narrow track bits in 5 Gbit/in^2 medium, *IEEE Trans. Magn.,* **33**: 2995–2997, 1997; Y. Hosoe et al., Experimental study of thermal decay in high-density magnetic recording media, *IEEE Trans. Magn.,* **33**: 3028–3030, 1997.

10. A. Aharoni, *Introduction to the Theory of Ferromagnetism,* Chap. 5. Oxford: Oxford Univ. Press, 1996.

11. R. S. Tebble and D. J. Craik, *Magnetic Materials,* London: Wiley, 1969.

12. H. L. Richards et al., Magnetization switching in nanoscale ferromagnetic grains: description by a kinetic Ising model, *J. Magn. Magn. Mater.,* **150**: 37–50, 1995.

13. P. A. Rikvold et al., Nucleation theory of magnetization reversal in nanoscale ferromagnets, in A. T. Skjeltorp and D. Sherrington (eds.), *Dynamical Properties of Unconventional Magnetic Systems,* NATO Science Series E: Applied Sciences, Vol. 349, Dordrecht: Kluwer, 1998, pp. 307–316.

14. D. Craik, *Magnetism, Principles and Applications,* , Chap. 1. New York: Wiley, 1995.

Reading List

A. Aharoni, *Introduction to the Theory of Ferromagnetism,* Oxford: Oxford Univ. Press, 1996.

H. N. Bertram, *Theory of Magnetic Recording,* Cambridge, UK: Cambridge Univ. Press, 1994.

S. Chikazumi, *Physics of Magnetism,* Malabar, FL: Krieger, 1978.

D. Craik, *Magnetism, Principles and Applications,* New York: Wiley, 1995.

J. C. Mallinson, *The Foundations of Magnetic Recording,* 2nd ed., Boston: Academic Press, 1993.

For a review of recent experimental studies of magnetization switching in magnetic nanoparticles, see the introduction of H. L. Richards et al., Effects of boundary conditions on magnetization switching in kinetic Ising models of nanoscale ferromagnets, *Phys. Rev. B,* **55**: 11521–11540, 1997.

M. A. Novotny
P. A. Rikvold
Florida State University

MAGNETIC RECORDING.   See MAGNETIC TAPE RE-
CORDING.

MAGNETIC RECORDING HEADS

A crucial component of any magnetic storage system is the
magnetic recording head. It writes, reads, and erases the information on the magnetic medium. Magnetic recording
heads are used for both digital applications, as in computer
hard disk drives, and for analog applications, as in audio or
video recorders. Many of the recent advances in data storage
are a direct result of improvements in materials and fabrication technology of recording heads. These advances have, in
turn, spawned many of the remarkable new developments in
software, entertainment, and communication applications
available today. For a comprehensive review of recording and
recording heads see Refs. 1–7.

BASIC OPERATION

Geometry of Storage Drives

The most common classes of magnetic recording systems currently available are tape and disk drives (Fig. 1). Recording
heads used for tape applications are designed for either linear
or helical formats. Linear tape heads have a rounded surface
over which the tape is guided. The tape medium, generally a
thin polyester ribbon on which a thin magnetic film is deposited, is in contact with the head and either the head or tape
is moved back and forth to access different tracks. This configuration permits mechanisms utilizing parallel operation of
multiple read, write, and erase heads either along or across
the track. Tape heads vary from low performance audio
heads, used in cassette recording, to high performance digital
heads, used for information storage.

Helical scan drives are used in applications where high frequency signals are recorded, such as video cassettes. Multiple
inductive heads are mounted on a rotating drum, and the
head sweeps across the tape at a high velocity. The tape

moves at a speed such that for each rotation of the drum, the
tape has moved incrementally to the next data track. There
are at least two heads on the drum with the write gaps at an
angle relative to each other. Each successive track that is
written partially overwrites the preceding track, and because
adjacent tracks are written at different angles there is little
interference between tracks. This is called azimuthal recording, used in high data rate applications. Typical track
widths for video heads are 125 μm to 250 μm and head velocities relative to the medium are 25 m/s.

In rigid disk drives the recording medium is composed of
ferromagnetic material deposited on a thin metal or glass
disk. The recording head is mounted on a slider that effectively "flies" above the surface of the disk on an air bearing.
The slider moves radially over the disk, to access data and
servo tracks configured in concentric circles on the disk. Unlike tape heads, there is no intentional contact between the
head and the medium. The slider has an air bearing surface
(ABS) that rests against the disk when it is stationary (Fig.
2). As the disk comes up to speed, air is forced between the
ABS and the medium, raising the slider above the disk surface. The flying distance between the head and the medium
has decreased as storage densities have increased, since the
recording system's areal density capability is determined in
large part by this distance. At areal densities in disk drives
today, 0.5 Gbit/in.^{2} to 2 Gbit/in.^{2}, the distance is in the range
of 10 nm to 50 nm and is determined by the disk rotational
rate (typically 5400 to 10,000 rpm) and the geometry of the
rails on the slider.

Floppy disk recording heads share features of both disk
and tape drives. The ferromagnetic recording medium is deposited on a flexible plastic disk. As in rigid disk drives, the
head moves across a rotating disk to access the tracks. However, the head makes contact with the flexible medium (no air
bearing), similar to a tape drive. The relatively low storage
density of the standardized floppy format, coupled with the
low cost of production of ferrite recording heads, has discour-
aged floppy drive manufacturers from advancing to thin film
inductive technology.

Figure 2 shows the configuration of a disk drive. The recording element is located at the rear of the slider. A stainless

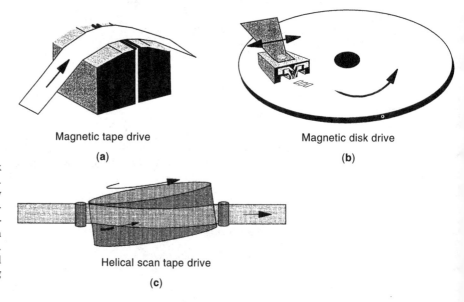

Magnetic tape drive

(a)

Magnetic disk drive

(b)

Helical scan tape drive

(c)

Figure 1. Schematic of magnetic tape and disk
drives: Arrows indicate direction of movement.
(a) The tape head in a linear drive is stationary
as the ribbon medium moves across it, in contact with the head. (b) The tape head in a helical scan drive is mounted on a rotating drum
which sweeps the head across the moving tape.
(c) The disk recording head moves in a radial
direction, flying above the surface of a rotating
disk.

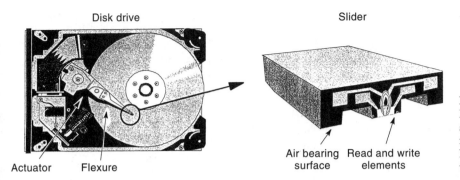

Disk drive

Slider

Actuator   Flexure

Air bearing   Read and write
surface   elements

Figure 2. Schematic of a slider in a disk drive: Read and write elements are deposited on the slider, which has a machined air bearing surface. The slider is mounted on a flexure and the head is positioned over the medium by the actuator.

steel flexure connects the head to an actuator that moves the head to access the recorded tracks on the medium. A gimbal is used between the head and the flexure to allow a small amount of motion of the head (pitch and yaw) so as to follow the surface of the medium. Typical recorded trackwidths are 2 μm to 5 μm and the head velocity relative to the medium is 10 m/s to 25 m/s.

Writing Function

The basic mechanism of "writing" on magnetic media has remained unchanged since first developed in 1898 by Valdemar Poulsen (8). His invention, the telegraphone, recorded audio signals by switching the magnetization in a steel wire by using the magnetic field generated by a simple electromagnet. Contemporary write heads use the same principle. The head consists of wire coils wrapped around a ring-shaped magnetic core with a gap (Fig. 3). When the core is magnetized by passing a current through the windings (Ampere's Law), a magnetic fringing field extends beyond the gap. This fringing field is used to switch the magnetization of a small region on the ferromagnetic medium. In digital recording, data are written (or encoded) as regions of alternating magnetization in the ferromagnetic medium. In analog recording, the information is written using both the direction and the magnitude of the magnetization in the medium.

The geometry of the ring and the gap region determines the size of the ferromagnetic region switched in the medium,

and hence determines how closely the switched regions can be placed together (the areal density). The coercivity of the medium itself will also determine how closely the regions can be placed, the higher the coercivity, the closer the possible spacing. Improvements in storage density have resulted from reducing the pole tip size (producing smaller bits), and generating larger fields (allowing higher coercivity media).

Reading Function

"Reading" information stored in the medium is achieved by sensing the transitions of the magnetic domains in the medium. These transitions produce fringing fields that extend above the surface of the medium. Reading is accomplished by using one of two principles: magnetic induction or magnetoresistance.

Inductive technology was the first to be developed. Inductive heads sense the time rate of change of the flux from the media, often using the same head that is used for writing [Fig. 4(a)].

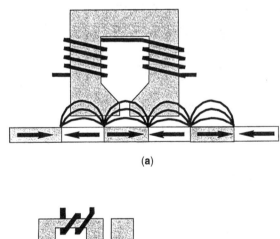

(a)

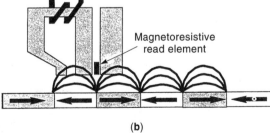

Magnetoresistive
read element

(b)

Figure 4. Read process. (a) Inductive heads sense the time rate of change of the magnetic flux emanating from the medium. Head components are as in Figure 3. (b) Magnetoresistive heads use a sensor to detect the field directly. A separate gap is needed for an inductive write head.

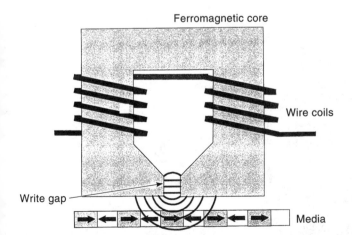

Ferromagnetic core

Wire coils

Write gap

Media

Figure 3. Write process: Current flowing in coils surrounding a soft ferromagnetic core induces a field within the core. Fringing fields emanating from the gap in the core magnetize the ferromagnetic medium, "writing" information on the disk.

When the gap of the ring head is moved across the fringe fields emanating from the medium, a voltage is induced in the coils. This voltage is proportional to the time rate of change of the magnetic flux that circulates past the coils (Faraday's Law). The variation in the voltage is decoded to obtain the original information, either the data itself, or the servo information, which gives the position of the head relative to the data track.

Since the signal from inductive heads is proportional to the time rate of change of the magnetic flux, a relatively large flux moving at a high velocity with respect to the head is required. Industry trends toward smaller drives, along with the relentless quest for higher areal density, have rendered the output from inductive heads insufficient. Smaller disks translate to lower disk velocity ($v = \omega \times r$, where r is the radius), and higher density results in smaller fringing flux from magnetic transitions in the medium.

In response to this challenge, head designers and manufacturers have incorporated magnetoresistive (MR) read elements into recording heads [Fig. 4(b)]. The signal output of an MR head is directly proportional to the flux from the medium's fringing fields, and is velocity independent. Consequently, increased sensitivity is obtained by the choice of materials in the MR element and not by the increased velocity of the medium. The trend toward smaller, high density media has spurred development of new magnetoresistive materials, and MR heads with much higher outputs than inductive heads have been developed and marketed.

TYPES OF RECORDING HEADS

Inductive Heads

Inductive heads are found in all types of magnetic storage systems. The head geometry is different in each application, depending on choices of either tape or disk media, high or low storage density, and analog or digital recording. However, there are two main categories based on the method of fabrication: either the heads are machined from bulk ferrite material and are called *ferrite* heads, or they are made from thin film processes and are called *thin film inductive* (TFI) heads.

Ferrite heads historically came first, followed by thin film technology. Ferrite heads are machined directly from bulk ferrite and the wires are wound manually. Their primary advantage is that the heads can be made very inexpensively. However, it is difficult to machine narrow trackwidths. In addition, the large volume of the core material results in high inductance. This limits the storage density and data rates achievable by the heads.

A major step forward in recording-head technology was the development of thin film processing techniques for making both the core and the windings. Thin film inductive heads were first introduced in 1979 in IBM's 3370 disk drive (9). These heads consist of a ferromagnetic bottom pole, a coil and a ferromagnetic top pole, all defined with photolithography (Fig. 5). Thin film processing results in a number of improvements. The trackwidth, defined by the width of the pole tips, is now controlled by lithography rather than by a machining process. Also, the small size of the poles reduces the inductance, enabling use of the heads at higher frequencies. While the process of making the heads is quite complicated, the dimensional control achieved with lithography is much more reproducible than possible with machining. In addition, many heads are made simultaneously on each wafer, lowering manufacturing costs. Increasing the number of coil turns remains a processing challenge, and today's heads have three or four layers of stacked coils.

Anisotropic Magnetoresistive Heads

To keep pace with the increases in areal density driven by the hard disk market, alternative head technologies have been developed. A major improvement has been the replacement of inductive readback with magnetoresistive readback (10). MR heads have separate read and write elements (Fig. 6). An inductive thin film head is still used for writing, but a separate gap with a magnetoresistive sensor is used for reading the magnetic information. The MR sensor detects the flux from the medium directly. The resistance of the sensor varies with applied field and a sense current is used for detection. Figure 7 shows a view of an MR head from the air bearing surface.

The resolution of the head along the track is determined by the distance between two magnetic shields placed on either

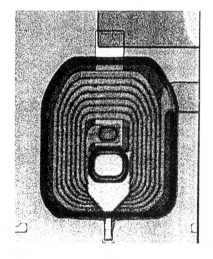

Figure 5. A thin film inductive head. (a) Plane view during wafer fabrication (b) Cross-sectional schematic showing coils surrounded by the magnetic poles. (a) courtesy Data Storage Systems Center, Carnegie Mellon University.)

(a)   (b)

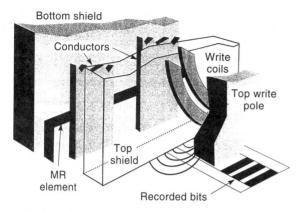

Bottom shield
Conductors
Write coils
Top write pole
Top shield
MR element
Recorded bits

Figure 6. Anatomy of a magnetoresistive (MR) recording head: Elements are depicted in order of deposition: bottom shield, magnetoresistive device elements, conducting leads, top shield (also functions as bottom write pole), write coils, and top write pole. Elements are separated by a dielectric, typically Al_2O_3.

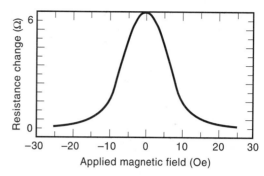

Figure 8. Anisotropic magnetoresistance response to an applied magnetic field: Data are taken from a patterned film, no biasing technique is applied. Current flows parallel to the zero field magnetization, and the field is applied transverse to this direction.

side of the element. A dielectric is used to separate the element from the shields. The total distance between the shields determines the achievable linear storage density. A common practice is to have one of the shields serve as the bottom pole of the inductive head reducing the total number of magnetic layers fabricated in the head. The resolution across the track, which determines the track density, is controlled either by the distance between the two conductors used to sense the resistance, or by the width of the magnetically active region of the magnetoresistive element.

One major advantage of MR heads is that the readback sensitivity is determined by the choice of the magnetoresistive material. Heads currently in production use materials based on the anisotropic magnetoresistance (AMR) effect. The resistance either increases or decreases depending on the relative orientation of the magnetization of the material to the current direction. The field response of a patterned AMR film

is shown in Fig. 8. The 10 nm to 20 nm thick AMR films commonly used have magnetoresistive effects on the order of 2.5% at low applied fields.

The physical origin of the magnetoresistance effect lies in spin orbit coupling. As the direction of the magnetization rotates, the electron cloud about each nucleus deforms slightly. This deformation changes the amount of scattering undergone by the conduction electrons when traversing the lattice (Fig. 9). A heuristic explanation is that the magnetization direction rotates the closed orbit orientation with respect to the current direction. If the orbits are in the plane of the current, there is a small cross-section for scattering, giving a low resistance state. If the orbits are oriented perpendicular to the current direction, the cross-section for scattering is increased, giving a high resistance state.

The response of a magnetoresistive element to an applied field is nonlinear. Thus, it is necessary to bias the MR material into a region that is linear (that is, offset the effect from zero field) such that the response is single valued with respect to the applied field, since both positive and negative fields are sensed from the medium. There are two pertinent methods currently used to bias the MR response for use in magnetic read heads: the soft adjacent layer (SAL) and dual stripe (DS) configurations.

Soft Adjacent Layer Heads. SAL heads use a thin film of soft magnetic material placed next to the MR element to bias the device into a linear region (11). The magnetization of this film is constrained to be transverse to the current direction, effectively applying a dc field locally to the MR element [Fig. 10(a)]. The field from the sense current is sufficient to rotate the magnetization of the SAL film to the transverse direction. When the SAL layer is magnetized, it produces a fringing field (H_F) that rotates the magnetization of the MR material with respect to the current. Since the change of resistance in an MR head goes as $\cos^2\theta$, where θ is the angle between the magnetization direction and the current, the optimum bias configuration is for the MR material's magnetization to be rotated near 45 degrees with respect to the current direction. This condition puts the zero field resistivity at a value midway between the maximum and minimum of the resistance. Thus the resistance will either increase or decrease when sensing positive or negative fields.

A difficulty with this biasing technique is that it leads to cross-track asymmetry. Since the magnetization of the sense

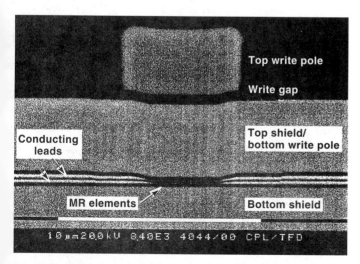

Top write pole
Write gap
Conducting leads
Top shield/ bottom write pole
MR elements
Bottom shield
10μm200kU 840E3 4044/00 CPL/TFD

Figure 7. SEM photograph of a dual stripe MR head at the air bearing surface: The two MR elements are discernible in the bottom gap. The active area of the device is defined by the conducting leads.

Figure 9. Schematic demonstrating the physical origins of anisotropic magnetoresistance: Current flows along the long direction of the bar. Dark gray ovals represent the scattering cross-sections of the bound electronic orbitals. For fields parallel to the current flow, electrons see a greater scattering cross-section than for fields perpendicular to the current direction.

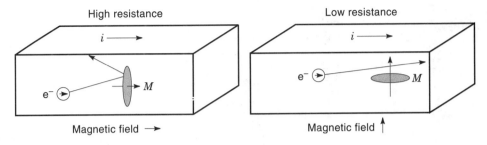

film is rotated away from the long axis of the MR element, and since the flux flow in the element is perpendicular to the direction of magnetization, there will be an asymmetry in response of the head as it moves across the track. This asymmetry either needs to be taken into account when designing the tracking function of the head or it can be removed by more complicated head geometries.

Dual Stripe Heads. A dual stripe head uses two MR sense elements that are electrically isolated from each other (12). The fields from the sense currents again rotate the magnetization, but in this case both elements are rotated equally and in opposite directions [Fig. 10(b)]. An advantage of the dual stripe design is that the response of the head is differential between the responses of the individual MR elements. The field from a transition in the medium rotates the magnetization of both head elements in the same direction, resulting in the resistance of one element increasing and the resistance of the other element decreasing. Because the nonlinearity of the two elements is essentially second order, the subtraction of the two parabolic responses linearizes the signal. The two elements are detected differentially, giving a greater response than a single element. Differential detection also removes any common mode signals originating from stray pickup or thermal noise. In addition, the cross-track asymmetry present in SAL heads is not present in the dual stripe design, since the asymmetry of one element cancels that of the other.

The greater readback signal of MR heads has enabled designs for higher storage density drives. However, design challenges have hampered the incorporation of MR elements into disk drives, evidenced by the time needed for the technology to advance to large scale production. Only in 1997, 26 years after the first MR head was demonstrated (10), have the majority of manufactured disk heads used MR technology. Problems that arise relate to the difficulties in obtaining the correctly biased state, in stabilizing the magnetic elements, and in obtaining high yield manufacturing processes. These topics will be discussed later.

Giant Magnetoresistive Heads

Giant Magnetoresistive Phenomenon. Giant magnetoresistance (GMR) was discovered in 1988 by Baibich et al. (13), in antiferromagnetically coupled multilayers of Fe/Cr. In this structure, thin layers of magnetic material are separated by layers of nonmagnetic material. The magnetic layers are coupled through the nonmagnetic layers either ferromagnetically or antiferromagnetically, depending on the thickness of the nonmagnetic layers. Magnetoresistance effects of up to ~50% were observed at low temperatures. This effect was subsequently found to occur in a number of multilayer magnetic film systems.

The GMR effect requires a method to change the relative orientations of the magnetization in adjacent magnetic layers, and requires that the thickness of the films must be less than the mean free path of the electrons. Figure 11 shows a schematic of the multiplayer structures. The GMR effect can be qualitatively understood on the basis of a two fluid picture of the conduction process in a magnetic metal.

The conduction electrons are divided into two classes: those with spin parallel to the local magnetization and those with spin antiparallel. The resistance of the material is determined by the scattering processes to which the electrons are subject. Strong scattering processes produce a short mean free path and large resistance, weak processes produce long mean free paths and lower resistance. GMR effects are produced when the scattering processes for one spin orientation of the conduction electrons is more effective than for the other spin orientation. In the two fluid picture, electrons with spin oriented parallel to the magnetization of the metal have a lower resistance than those with spin are oriented antiparallel. The high resistance state of the GMR materials occurs when the magnetic layers are antiferromagnetically aligned, so that electrons experience strong scattering where the magnetization of the material is opposite to the spin orientation. The low resistance state is obtained when a magnetic field strong enough to overcome the antiferromagnetic coupling is

Figure 10. Biasing methods of three different MR heads. (a) The SAL MR head is biased by an adjacent magnetic film. (b) A dual-stripe head uses two sense elements. The current in each element biases the other element. (c) The spin valve head is self biased when the sense film is orthogonal to the pinned film.

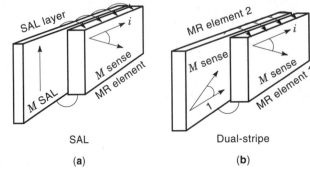

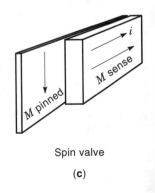

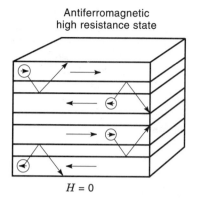
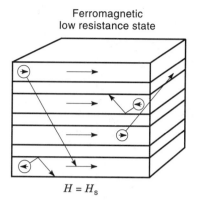

Antiferromagnetic
high resistance state

Ferromagnetic
low resistance state

$H = 0$

$H = H_s$

Figure 11. Schematic of the high and low resistance states of a GMR multilayer system: The magnetic layers, with arrows indicating the direction of the magnetization, are separated by nonmagnetic conducting layers. In zero applied field, the magnetic layers couple antiferromagnetically. Beyond the saturation field, the magnetization of the layers line up ferromagnetically. (See text for explanation.)

applied, and rotates the magnetization of the layers to a ferromagnetic configuration. When the magnetic layers are ferromagnetically aligned, only half of the conduction electrons experience strong scattering processes, while the other half experience weak scattering processes, with the net effect of reducing the overall resistance of the material.

For a multilayer to be attractive as an MR head sensor, it must have not only a large $\Delta\rho/\rho$, but also must have a large sensitivity to a magnetic field. The original Fe/Cr system requires extremely large fields (20 kG) to rotate the magnetization to the ferromagnetic configuration, and is therefore unattractive as a recording head device. Schemes have been developed, however, where *uncoupled* magnetic films can be switched from the antiparallel to parallel configuration. These devices have been termed *spin valve structures.*

Spin Valve Heads. Spin valves were developed to provide more control over the magnetics of the GMR multilayers (14). These structures contain two thin magnetic films separated by a nonmagnetic, conducting spacer layer. An antiferromagnetic "pinning" layer is exchange coupled to one of the magnetic films (pinned layer) in order to hold the orientation of the magnetization constant. The other magnetic layer (sense layer) is free to switch back and forth in the presence of a magnetic field. The principle for the magnetoresistance is similar to the GMR multilayer; that is, spin dependent scattering gives a low resistance state when the magnetic layers are ferromagnetically aligned, while a high resistance state is obtained in the antiferromagnetic configuration (Fig. 12).

The resistance in the spin valve structure depends on the angle between the magnetizations in the two magnetic layers, and is independent of the current direction, unlike the situation for AMR materials. The resistance varies as $\cos(\theta)$ where

θ is the angle between the magnetization of the two layers. A typical value for the magnetoresistance in a spin valve material suitable for a recording head is 10%. To bias the structure in a recording head, the magnetization of the pinned film is oriented 90 degrees to the magnetization of the sense film [Fig. 10(c) and Fig. 13]. Optimizing the spin valve heads such that the many magnetic fields present in the structures (demagnetizing fields, exchange coupling fields, fields due to currents, anisotropy fields, stabilizing fields) are balanced requires extensive development and only recently have prototypes been demonstrated (15).

HEAD MATERIALS AND PROCESSING

The design of a magnetic recording head, and the selection of its constituent materials, are largely dictated by two key requirements of the recording system: areal density and frequency response (data rate). Head design (for example, choice of inductive or magnetoresistive technology) narrows the list of prospective materials, but even for a particular design the list of candidate materials can be quite large. Materials selection for recording heads plays an important role in the performance of the device, since critical parameters such as magnetic permeability, saturation magnetization, magnetostriction, and coercivity vary widely among ferromagnetic materials. The following section provides a description of recording head materials, as well as fabrication processes used in head construction.

Ferrite Heads

In current low end applications, a simple "ring" head often satisfies the density and frequency response criteria. Conven-

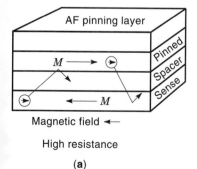

AF pinning layer

$M \longrightarrow$

$\longleftarrow M$

Magnetic field ◄—

High resistance

(a)

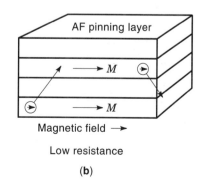

AF pinning layer

$\longrightarrow M$

$\longrightarrow M$

Magnetic field —►

Low resistance

(b)

Figure 12. Principle of operation of a spin valve system. Two magnetic layers are separated by a nonmagnetic conducting spacer. The magnetization of the top layer is pinned by exchange coupling to an antiferromagnet layer, while the magnetization of the bottom magnetic layer (sense layer) is free to rotate in response to a magnetic field. (Not shown in the self biased state.)

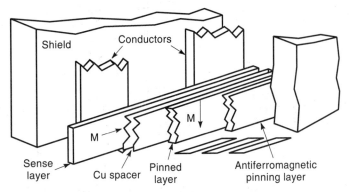

Figure 13. Anatomy of a spin valve read head: The magnetoresistive device consists of a magnetic sense layer, a Cu spacer, a magnetic pinned layer, and an antiferromagnetic pinning layer. The trackwidth is defined by the separation of the conductors. The device is deposited between magnetic shields.

tional ring heads are made of either iron alloy or ferrite cores. The cores are machined in two parts and bonded together to form the gap. The gap distance is typically 0.2–2.0 μm, and determines the resolution of the head along the length of the track. Fine Cu wire is wound around the core to complete this simple structure. A schematic of a ring head is shown in Fig. 3. One problem with ring heads made from iron alloy cores is the degradation of head performance at high frequencies due to eddy currents in the conductive magnetic cores. Eddy currents are reduced by lamination of iron alloy cores, or by use of high resistivity ($\rho > 0.1$ Ω cm) ferrite cores. Materials requirements of the core include magnetic properties, such as high permeability, high saturation magnetization, and low coercivity; as well as mechanical properties, such as the ability to be easily machined and high hardness, which correlates to the wear resistance of the core material.

Ferrites are alloys of the general form $MO(Fe_2O_3)$, where M is a divalent metal such as Mn, Zn, Ni, Co, or Fe. If M = Fe, then the chemical formula becomes Fe_3O_4, known as magnetite, or lodestone in ancient times. However, magnetite does not have sufficiently high permeability for use in most sensors, so Fe is replaced with Mn, Zn, or Ni, or some combination of these three elements. These mixed ferrites are man-

ufactured by sintering particles of Fe_2O_3 and MO at high temperature and pressure. The magnetic properties of commonly used mixed ferrites are compared with lamination core materials in Table 1. One shortcoming of ferrites is low saturation magnetization, limiting the magnitude of the write field the head can produce. Manufacturers of ferrite heads have addressed this problem by depositing a layer of high moment material on either side of the gap. Such heads are called metal-in-gap (MIG) heads, and combine the best attributes of ferrite and high moment lamination materials. The marriage of ferrite and high moment materials in the form of MIG heads has extended the life of wire-wound core heads. However, another more fundamental deficiency of machined ring heads remains: the difficulty in manufacturing heads capable of high density recording using mechanical machining. Recording heads in modern disk drives read and write information at track densities of nearly 350 tracks/mm, corresponding to write pole widths of less than 3 μm.

Thin Film Inductive Heads

In 1979 researchers at IBM applied advanced semiconductor processing techniques to the fabrication of magnetic recording heads, and the thin film inductive (TFI) head was born. With this new technology hundreds, if not thousands, of heads are processed simultaneously on a single substrate using thin film deposition and photolithographic pattern definition. The feasibility of this approach required identifying a ferromagnetic material that is both amenable to thin film process techniques and is able to satisfy the stringent requirements of the recording transducer. Permalloy, an alloy of $Ni_{81}Fe_{19}$ (atomic percent), is particularly well suited for the thin film core since it can be deposited by electroplating, and has very attractive magnetic properties, as shown in Table 1.

The process of fabricating a TFI head is understood by referring to Fig. 5(b). First, the bottom half of the core is defined on the substrate by electroplating permalloy into a photoresist stencil, followed by sputter deposition of a dielectric gap (typically Al_2O_3 or SiO_2). Next, a Cu spiral coil is plated on top of the bottom core and gap, and this coil is completely encapsulated by an insulator, such as photoresist. Several layers of coils are added to generate sufficient inductive sig-

Table 1. Magnetic Materials Used in Magnetic Cores of Inductive Recording Heads: Initial Permeability (μ_0), Coercive Field (H_c), Saturation Magnetization (M_S), Resistivity (ρ), and Vickers Hardness

Material	μ_0	H_c (Oe)	M_S (kG)	ρ ($\Omega \cdot$ cm)	Vickers Hardness
Mo permalloy[1]	20,000	0.025	8	100×10^{-6}	120
Alfenol[2]	8,000	0.038	8	150×10^{-6}	290
Sendust[3]	10,000	0.025	10	85×10^{-6}	480
NiZn ferrite[4]	300–1500	0.15–0.35	4–4.6	10^5	900
MnZn ferrite[4]	3,000–10,000	0.15–0.20	4–6	5	700
MnZn ferrite[5]	400–1000	0.05	3–5	>0.5	
Plated $Ni_{45}Fe_{55}$	1700	0.4	16	48×10^{-6}	
Plated $Ni_{81}Fe_{19}$	4000	0.3	10	24×10^{-6}	

[1]4% Mo 17% Fe 79% Ni
[2]16% Al 85% Fe
[3]5.4% Al 9.6% Si 85% Fe
[4]Hot Pressed
[5]Single crystal
Source: Ref. 2, p 6.24 and Ref. 16.

nals for read head applications. Finally, the top half of the permalloy core is plated using another photolithographically defined stencil. A clear advantage of this approach over traditional machining process is that the density capability of the head is determined by distances defined by either film thickness (the gap) or photolithography (the plated cores). The dimensional control afforded by these processes is far superior to that of machined parts, enabling production of heads with core widths on the order of 1 μm. Another advantage of thin film processing is the relatively small inductance, and corresponding superior high frequency performance, of the small NiFe core as compared to that of machined blocks of ferrite or laminated iron alloy cores.

Permalloy has served as the industry standard pole material in thin film heads for more than ten years, but steadily advancing recording densities and data rates have begun to push NiFe to its limit. On the density front, coercivities of recording media in excess of 2500 Oe, a necessity for high density storage, are placing increased demands on the amount of flux needed from the head. In order to write effectively, the head must produce a field roughly twice as large as the coercivity of the medium. Permalloy has a saturation magnetization M_S of 10 kG, but losses in the head, particularly that due to the spacing between the head and the medium, reduce the field at the medium substantially. The need for larger write fields in plated thin films heads prompted the development of heads incorporating $Ni_{45}Fe_{55}$, the composition of peak magnetic moment (M_S = 16 kG) in the Ni–Fe system (16). This NiFe alloy is well-suited as the writing element in conjunction with magnetoresistive readout (later in this article), but its higher anisotropy and magnetostriction limit its use in TFI heads. High M_S alloys, such as Fe–N (17), FeTaN (18), FeAlN (19), each with saturation magnetization of about 20 kG, are being actively studied for writing on future generations of high coercivity media. Further improvement of the frequency response of NiFe is in jeopardy due to eddy current damping. Solutions to this problem are being addressed on two fronts: (1) laminating the pole material with dielectric spacer layers, and (2) investigating more resistive high moment alloys [such as amorphous CoZrTa (20)].

Magnetoresistive Heads

The addition of a magnetoresistive read element creates a more complicated device than the standard TFI head, since an inductive element is still required for writing. Referring to Fig. 6, standard construction of such a dual element head begins with the deposition of a ferromagnetic shield, followed by the bottom half of the read gap dielectric. The MR element and contacts are then defined, after which the top half of the read gap and top shield are deposited. In this merged pole structure, the top shield of the read head also serves as the lower core, or pole, of the inductive write element. Finally, the remainder of the TFI write head is fabricated on top of the top shield. Unlike the TFI head, where the thick (2 μm to 4 μm) ferromagnetic core and Cu coil are plated, the critical sensor materials in an MR head are deposited by sputtering (21).

Biasing techniques incorporated in MR heads were described previously, and now some of the associated materials considerations will be discussed. The soft adjacent layer, or SAL design, introduces additional materials into the read head; whereas the dual stripe design simply replicates an MR element. Both designs use permalloy for the MR element(s). Key attributes of the SAL film are: low $\Delta R/R$, high M_S, high permeability, low magnetostriction, and high resistivity. Ternary alloys NiFeX (22) (examples of X are Nb, Rh, or Zr) and amorphous Co-based alloys such as CoZrMo (23) are used for the SAL film. For proper biasing, the MR element and SAL film must be magnetically decoupled by interposing a thin layer of high resistivity metal (for instance, Ta) between them.

For the dual stripe biasing technique, one key materials challenge is maintaining electrical isolation between the two MR elements that are separated by a dielectric film of 50 nm or less. Fabrication of dual stripe heads also taxes the photolithographic alignment process, since overlaid elements are defined in separate patterning steps, and the track-widths must be lined up precisely.

In addition to its relatively large magnetoresistance ($\Delta\rho/\rho \sim 2\%$) and high permeability (μ = 2000), permalloy has another characteristic in its favor: zero magnetostriction. Magnetostriction is the dimensional change, or strain, a material undergoes when exposed to a magnetic field. The magnetostriction of permalloy is typically less than 1×10^{-6}, whereas that of many other ferromagnetic alloys can be in the 10^{-5} range. Conversely, if a magnetostrictive material is stressed, it develops magnetoelastic anisotropy energy density (E_a) according to the equation:

$$E_a = -3\lambda\sigma \cos^2 \theta$$

where σ is the stress in dynes/cm^2, λ is the magnetostriction in units cm/cm, and θ is the angle of the magnetization relative to the axis of applied stress. If both σ and λ are of the same sign, then the magnetoelastic energy will tend to create an easy axis parallel to the applied stress, whereas, if the two are of opposite sign, the magnetostriction will contribute an anisotropy orthogonal to the applied stress. Since the magnetic anisotropy of NiFe is already quite low, additional magnetoelastic anisotropy can either destabilize or reduce the sensitivity of the MR element. Control of magnetostriction, primarily through permalloy composition (24), is very important for producing a stable MR transducer.

Unlike an inductive head, a magnetoresistive sensor operates as a parametric amplifier: the voltage output is proportional to the sense current. The maximum tolerated current is determined by the power dissipation in the head, which in turn dictates the operating temperature. Current densities of about 2×10^7 A/cm^2 are commonly used, resulting in a temperature rise of several tens of degrees above ambient. Smaller dimensions permit higher current densities, but introduce concerns over electromigration. Fortunately, NiFe is not very susceptible to electromigration, although the contact lead metallurgy may be. Common contact lead metals are W and Ta/Au/Ta.

Another aspect considered in the design and fabrication of MR heads is the propensity of ferromagnetic materials to form magnetic domains. Any free magnetic charges present near the edges of the device, or in defects generated in manufacturing, will produce demagnetizing fields which create magnetic domains in the sense element. These domains are a source of noise (Barkhausen noise) and must be eliminated in the active area of the sensor. Various techniques, used individually or in combination, are used to stabilize the sense ele-

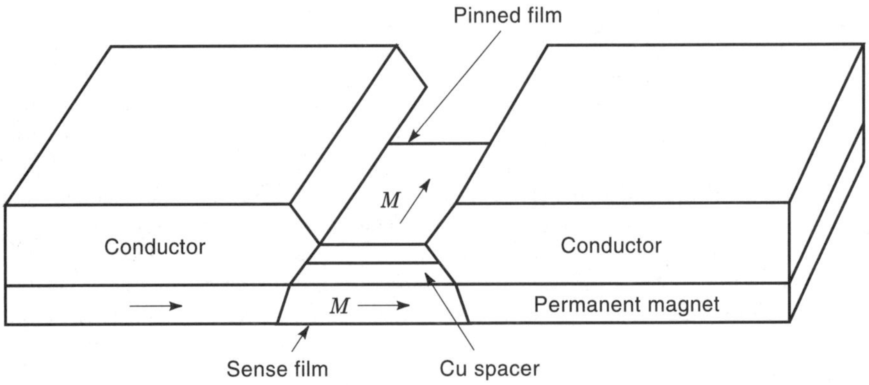

Figure 14. Permanent magnet stabilization: Permanent magnet thin films are deposited adjacent to the spin valve sense layer. The field from the permanent magnets creates a single domain sense element, greatly reducing noise in the response of the device.

ment. The material anisotropy is increased to overcome the destabilizing fields caused by magnetostatic effects at the ends of the MR element. The element height is reduced, producing a demagnetizing field constraining the magnetization direction along the long axis of the sense element. Antiferromagnetic (AF) exchange coupling pins the sense element magnetization at the track-edge, reducing the probability of multiple domains at the track-edge (25). Finally, thin permanent magnet films are fabricated at the track-edge (Fig. 14), providing a longitudinal stabilization field (26).

The two most common domain stabilization techniques are (1) AF exchange coupling the sensor to an antiferromagnetic outside the active region, and (2) fabricating permanent magnetic films on both sides of the MR element patterned to exactly the desired trackwidth. Properties of MnFe, MnNi, NiO, and IrMn antiferromagnets used for stabilization are listed in Table 2, and will be described in more detail later. The permanent magnetic stabilization method uses CoPt-based alloys similar to those used in high coercivity magnetic disks. By controlling the product of magnetization and thickness of the permanent magnet film, the magnitude of the longitudinal stabilizing field is tailored according to sensor requirements.

One difficulty with MR heads is the possibility of shorting the element(s). Since a sense voltage is applied across the MR head, momentary contact between the head and the medium may short the element. Asperities may scratch the air bearing surface, possibly shorting the element to the shield. Asperities also produce thermal spikes in the element, causing a resistance variation in the head. These problems have been mitigated by overcoating the air bearing surface of the slider with a thin (<15 nm) protective film of amorphous carbon. Such overcoats also have dramatically improved the tribological characteristics of the interface between the head and the disk.

Table 2. Properties of Antiferromagnetic (AF) Materials in NiFe (25 nm)/AF Exchange Couples: Exchange Field (H_X), Blocking Temperature (T_B), Resistivity (ρ), and Corrosion Resistance

Material	H_X (Oe)	T_B (°C)	ρ ($\mu\Omega \cdot$ cm)	Corrosion Resistance
MnFe	50	150	130	poor
MnNi	120	>400	190	moderate
NiO	30	220	>10^{10}	good
IrMn	100	250	200	moderate

Magnetoresistive materials for use in recording heads have been the object of research since the early 1970s, and while scientists have searched for materials superior to NiFe, no obvious successor has been found. Alternate materials are elusive primarily because of the numerous properties they must possess in addition to large magnetoresistance: low coercivity, low magnetic anisotropy, low magnetostriction, and high permeability. However, in 1988 the field of magnetoresistive research was thrown wide open with the discovery of giant magnetoresistance (GMR) in multilayer films.

Giant Magnetoresistive Materials

Giant magnetoresistive materials were discovered only recently, but are already in the product plans of most head manufacturers. A common denominator of all GMR films is the interaction of at least two magnetic layers with magnetization vectors that can be rotated with respect to each other. As discussed previously, if the two regions are close enough (<10 nm), electrons can traverse the nonmagnetic spacer layer separating the ferromagnetic (FM) layers without loss of spin orientation. The electrons of one spin state will be scattered preferentially over electrons of the other spin state depending upon the relative angle between the magnetization vectors. A low resistance condition corresponds to parallel magnetization, and a high resistance to antiparallel magnetization, independent of the direction of current.

The GMR structure best suited for recording applications is the spin valve (14). Both the sense and pinned films are typically NiFe, Co, or bilayers of NiFe and Co, and in contrast to the SAL and DS sensors, the spacer material must be a low resistivity metal, such as Cu.

The magnetization in the pinned film is fixed by exchange coupling to an AF film. One of the primary challenges for the spin valve design is producing sufficiently large pinning field to ensure that the pinned film magnetization remains perpendicular to the sense layer magnetization, particularly in the operating environment of elevated temperature and alternating magnetic fields. A key attribute of AF exchange coupling is it produces a unidirectional anisotropy, meaning the pinned layer has a unique easy magnetization direction. In such a system, the magnetization direction of the pinned FM layer in zero field is independent of field history, an advantage for fabricating sensors with a predictable magnetization state. Two of the most important parameters of the AF/FM exchange couple are exchange field strength (H_X) and blocking temperature (T_B, the temperature at which the exchange field goes to zero). The exchange field should be as large as possible. The blocking temperature must be high enough to prevent loss of pinning at the operating temperature, but low enough to set the exchange field orientation without causing diffusion at the interfaces of the spin valve layers. Properties of some of the most widely used antiferromagnetic alloys are presented in Table 2.

Spin valves have been fabricated using each of the materials listed in Table 2 as pinning layers. Those with MnFe (27) suffer from sensitivity to corrosion and low T_B, which limits the operating temperature (i.e., current density) of the device. NiO-based spin values (28) have barely adequate exchange field strength, but otherwise have ideal attributes for use in a spin valve. An insulating antiferromagnetic is particularly advantageous since there is no shunt current loss. The chief disadvantage of MnNi is that high temperature processing ($T > 250°C$) is required to establish the AF phase (29), which can lead to degradation in magnetoresistance. IrMn is a relative newcomer to the list of exchange alloys in spin valve heads (30), but its properties appear to satisfy the requirements of a spin valve sensor.

An MR or spin valve head contains many materials other than the MR elements themselves that affect device performance and the manufacturing yield. The substrate, ferromagnetic shields, gap dielectric, magnetic stabilization and lead all play important roles in MR head performance. For instance, the insulator that electrically isolates the sensor from the shields is typically 50 to 100 nm of Al_2O_3. However, a dielectric with higher thermal conductivity would permit higher sense currents, improving the output of the head. The substrate material, forming the body of the slider, is generally a hard ceramic material such as $CaTiO_3$, SiC, or Al_2O_3–TiC, and is chosen based on its thermal, mechanical, and tribological properties. Lead metallurgy must be thoughtfully chosen to ensure that electromigration is not a problem and that device resistance is kept as low as possible.

THEORY OF OPERATION

The basic head requirement is to write and read the magnetization states in the magnetic medium. In many high density applications, the write and read elements are separate, and each is individually optimized for peak performance. Both head and media materials have comparable magnetic moments, but their coercive fields differ substantially. The head material is "soft," returning to zero-magnetization state after removal of the applied field; whereas the medium material is hard, once pushed to a saturated state of magnetization it retains its magnetization state upon removal of the field. The medium thus has hysteretic characteristics, represented by such parameters as the saturation magnetization M_S, the coercive squareness of hysteresis loop S^*, and the coercivity H_c.

Write Modeling

Although the writing process is intuitively simple, analyzing it is quite complex. As the magnetic field from the head magnetizes the medium, the induced magnetization opposes the head field by generating a demagnetizing field. The head di-

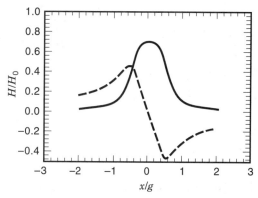

Figure 15. Calculated fields produced in the medium from an inductive head: Solid and dashed lines represent the x and y components, respectively, of the field induced in the medium. Results are derived from the Karlquist expressions, Equations (1) and (2).

mensions, and the separation between the ABS and the medium, are crucial parameters in writing. A simple but informative analytical method is first presented, followed by the incorporation of more complex behavior, and finally, a more realistic numerical approach is discussed.

In the simple analysis, assume the medium is very thin and only responds to the longitudinal component of the head field. The assumption of a very thin medium implies no perpendicular component of magnetization in the medium. Also assume that the head pole pieces are of infinite thickness. With these assumptions, the head field in the medium is given by the Karlqvist expression (31)

$$H_x(x,y) = \frac{1}{\pi}H_o\left[\arctan\left(\frac{\frac{g}{2}+x}{y}\right) + \arctan\left(\frac{\frac{g}{2}-x}{y}\right)\right] \quad (1)$$

$$H_y(x,y) = \frac{1}{2\pi}H_o\log\left[\frac{\left(\frac{g}{2}-x\right)^2+y^2}{\left(\frac{g}{2}+x\right)^2+y^2}\right] \quad (2)$$

where H_x and H_y are the longitudinal and perpendicular components of the field seen by the medium, H_o is the magnetic field at the ABS of the head, and g is the gap distance. Figure 15 shows a plot of the longitudinal and perpendicular components of the head field at a distance of $g/4$ from the ABS of the head. If the medium was all negatively magnetized, and a large positive head current is suddenly applied to the head coil, the medium's magnetization will become positive. Once the head moves away, or the head current is reduced to zero, the magnetization will relax to a point on the hysteresis loop vertical axis. Figure 16 shows a pictorial presentation of a head field contour and the hysteresis loop. As the head field goes to zero, different points in the medium return to different values of remanent magnetization, as depicted by different arrows in the hysteresis loop. The medium's magnetization thus goes through a transition region. If the magnetization change is assumed to be linear, then the slope of magnetization, dM/dx, can be written as M_r/a_L, where a_L is called a transition parameter. A first estimate of the transition parameter is obtained from

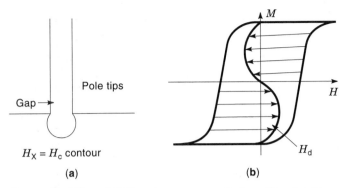

(a)   **(b)**

Figure 16. Effect of field contour on magnetic transition. (a) Shape of the fringe field contour below the gap of the write poles. (b) The medium transition will have a distribution of remanent magnetization, depending on the relative position of the point on the medium to the head contour field.

$$\frac{dM}{dx} = \left(\frac{dM}{dH}\right)\left(\frac{dH}{dx}\right) \tag{3}$$

and

$$a_L = M_r \Bigg/ \left(\frac{dM}{dH}\right)\left(\frac{dH}{dx}\right) \tag{4}$$

A small transition parameter, that is, a sharp change in magnetization in the medium, requires a low remanent magnetization, a square hysteresis loop (dM/dH), and a large head field gradient (dH/dx). Although Eq. 4 approximates the recording process, in reality, transitions are not linear and often have long tail sections. An arctangent or hyperbolic tangent are often better representations.

As the medium is magnetized, the regions of magnetization variation accumulate magnetic charge which produces another component of magnetic field called the demagnetizing field (H_d). The demagnetizing field is obtained from the Maxwell equation

$$\nabla \cdot H_d = \nabla \cdot M$$

When the written track-width is large compared to the dimension in the longitudinal direction, and the medium's magnetization is arctangent, then the previous expression can be solved for the demagnetizing field. If the transition is represented by an arctangent transition, then the longitudinal and the perpendicular components of the demagnetizing field are given by (32). Figure 17 shows the shape of the arctangent transition and the associated demagnetizing field.

$$H_d^x(x,y) = -\frac{M_r}{\pi}$$

$$\times \left[\arctan\left(\frac{a+y+\frac{\delta}{2}}{x-x_o}\right) + \arctan\left(\frac{a-y+\frac{\delta}{2}}{x-x_o}\right) \right.$$

$$\left. -2\arctan\left(\frac{a}{x-x_o}\right) \right] \tag{5}$$

$$H_d^y(x,y) = \frac{M_r}{2\pi} \log \frac{\left[a+\frac{\delta}{2}-y\right]^2 + (x-x_o)^2}{\left[a+\frac{\delta}{2}+y\right]^2 + (x-x_o)^2} \tag{6}$$

Once the head field and the demagnetizing field are understood, the transition parameter can be determined. In Eq.(3) the total field can be written as

$$H = H_h + H_d$$

where H_h is the applied head field, given by Eqs. (1) and (2), and H_d is the demagnetizing field given by Eqs. (5) and (6). Referring to Eq. (3), the slope of the hysteresis loop at the coercive field can be written as

$$\left.\frac{dM}{dH}\right|_{X_o} = \frac{M_r}{H_c(1-S^*)}$$

where S^* represents the squareness of the loop at the coercivity point. At this point, there is enough information to solve Eq. (3) for any head field, spacing and loop shape. In 1971, Williams and Comstock published a now widely used paper on an analytical expression of the transition width (33). Eq. (4) implies a large head field gradient is required to obtain a sharp transition. The Williams–Comstock analysis assumes the transition occurs at the point of maximum head field gradient. That is, for every head-medium spacing the head current is adjusted so that it reaches a value slightly larger than the coercivity (called the remanent coercivity, H_r) at exactly the same spatial point where the head field gradient is also maximum. With this assumption, the following expression is obtained for the transition parameter.

$$a = \frac{(1-S^*)\left(d+\frac{\delta}{2}\right)}{\pi Q}$$

$$+ \sqrt{\left[\frac{(1-S^*)\left(d+\frac{\delta}{2}\right)}{\pi Q}\right]^2 + \frac{M_r \delta \left(d+\frac{\delta}{2}\right)}{\pi Q H_c}}$$

In the previous expression Q is a function of head field gradient and its value varies very slightly around 0.75. For a high moment, thin, longitudinal recording medium, the last term in this expression dominates and the transition parameter re-

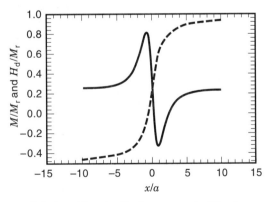

Figure 17. Calculated longitudinal component of the demagnetizing field and the resulting transition shape: Variation of magnetization in the transition region (dashed) and longitudinal component of the demagnetizing field (solid) normalized to the remanent magnetization (M_r) is plotted as a function of position normalized to the transition width a.

duces to

$$a \cong \sqrt{M_r \delta \left(d + \frac{\delta}{2} \right) \bigg/ (\pi Q H_c)}$$

or equivalently,

$$\frac{a}{\delta} = \sqrt{\left(\frac{M_r}{H_c} \right) \left(\frac{d}{\delta} + 0.5 \right) \bigg/ (\pi Q)}$$

The transition parameter is thus a strong function of M_r/H_c and d/δ. For a state of the art longitudinal recording, M_r/H_c is close to 5 and d/δ is close to 2.5. For these values a/δ is about 2.5 and for typical thin film longitudinal media the transition parameter is only 50 nm.

Although we have outlined the basic principles through simple analysis, actual operation is much more complex. The hysteresis behavior of the medium itself is still the subject of active research. One method of simulating the hysteretic behavior of the medium has been through a phenomenological model called the Preisach model, in which the medium is assumed to comprise an assembly of idealized particles. With a proper switching density distribution of these particles and careful consideration of the switching history of the particles, the major and minor loops can be modeled quite effectively. Also, the arctangent magnetization assumption breaks down for most cases, and one must perform a numerical analysis that involves finding a self-consistent solution. In this iterative method, the magnetization and field are carefully computed until the solutions converge. A two-dimensional application has been reported by Bhattacharyya, Gill, and Simmons (34). A more elaborate and careful three-dimensional work has been done by Davidson (35). In this work, Davidson uses the vector Preisach model for the medium, a three-dimensional head field through finite element analysis, and also an exhaustive self-consistent method to calculate the medium's magnetization. This work clearly shows that in thin film disks, there are substantial regions where the magnetization is not longitudinal and these regions can introduce distortions in cross-track characteristics which may limit recording density. Other methods for calculating the magnetization of the medium produced by the head fields are based on micromagnetics, and both two-dimensional (36) and three-dimensional (37) models have been developed.

Inductive Head Readback

When a magnetic medium with spatially varying magnetization moves under the inductive head, the time variation of flux through the windings of the coil generates a voltage according to Faraday's law of induction. Regions of alternating magnetization produce alternating, nearly Lorentzian, pulses in the head. The voltage produced in the coils can be written as

$$V(x) = -NWv \frac{d\phi}{dx}$$

where N, W, v, and ϕ represent the number of turns of the readback element, trackwidth of the head, velocity of the magnetic medium (head is conventionally stationary), and the flux linked by the windings of the head, respectively. In mod-

ern high density recording, the value of W has been reduced to increase the tracks per unit length and M_r has been reduced to lower the transition parameter and thus increase linear density of recording. Both of these factors reduce the readback voltage. The industry has tried to compensate by increasing the number of coils and the velocity of the medium.

To determine the flux linking the coils of the inductive head, a very useful formulation called reciprocity is used. The principle of reciprocity states that the flux linking the coil can be found through a correlation integral of the head sensitivity function (magnetic field produced in the recording medium for a total current of unity flowing through the coils of the head) and the medium's magnetization. For the head sensitivity function, the Karlqvist expression is used. The reciprocity relationship can be used to determine the flux per unit trackwidth

$$\phi(x, y) = NE\mu_0 \int_{-\infty}^{\infty} dx' \int_{-\delta/2}^{\delta/2} h(x' + x, y') M(x', t') \, dy$$

where $h(x, y)$ represents the head sensitivity function, which is the head field for unit current, and E represents the efficiency of the head. After integration the readback voltage can be written as

$$V(x) = \frac{2}{\pi g} NWE v \mu_0 M_r \delta \left\{ \arctan \left[\left(\left(\frac{g}{2} + x \right) \bigg/ (d + a) \right) \right. \right.$$
$$\left. \left. + \arctan \left(\left(\frac{g}{2} - x \right) \bigg/ (d + a) \right) \right] \right\}$$

The peak voltage is found by putting $x = 0$ in the above expression and the pulsewidth at half maximum (PW_{50}), an important parameter for determining the channel performance, is

$$\text{PW}_{50} = \sqrt{g^2 + 4(d + a)(d + a + \delta)} \qquad (7)$$

where d is the head-medium separation and δ is the medium thickness. Note that in the PW_{50} expression, g^2 and the second term should be comparable in magnitude, and thus contribute almost equally to the pulsewidth. To reduce pulsewidth, one has to reduce the read head gap, as well as the flying height, medium thickness, and the transition parameter. For a 0.2 μm read gap, 50 nm transition parameter and 20 nm medium thickness, one achieves a pulsewidth of 297 nm.

The Karlqvist head field, though very useful, does not truly apply for thin film heads with narrow pole thicknesses. Thin film heads show distinct undershoots, which are not predicted by the above analysis. While the write analysis is still correct (the medium is not affected by the small change of field at pole edges), the readback is very sensitive to these effects. Lindholm has developed formulae for the head field applicable to thin film heads with finite pole tip thickness (38). This field, when correlated with the medium's magnetization, clearly shows the undershoots observed experimentally in thin film heads. Lindholm's expression is also used to investigate the writing at the edges of the track, where the magnetization often has a substantial transverse component. When the control of the head geometry is important finite element methods (FEM) are useful and many commercial packages are now available.

Magnetoresistive Head Readback

Analytic and Numerical Methods. The resistance of MR heads changes when a magnetic field rotates the magnetization of the sensor relative to the current direction. The magnetization rotation, measured from the easy axis of the film, can be written as

$$\sin\theta = \left(\frac{H_y}{H_k}\right)$$

where H_y is the perpendicular component of the field at the head and H_k is the crystalline anisotropy field of the MR element. The resistivity of the MR film is $\rho = \rho_o - \Delta\rho\sin^2\theta$, where ρ_0 is the zero field resistivity, giving the resistance variation with applied magnetic field as

$$\rho = \rho_0 - \Delta\rho\left(\frac{H_y}{H_k}\right)^2$$

For NiFe, the most widely used magnetoresistive material, $\Delta\rho/\rho_0$ is approximately 2% and H_k is 4 to 5 Oe.

For most practical applications, the magnetic read heads are in a shielded structure. This complicates the analysis, since the additional magnetic layers also conduct magnetic flux. In 1974 Potter (39) showed that a shielded MR head can be thought of as two Karlqvist heads connected back to back. Consequently, many of the expressions derived for inductive heads can be used with minor modifications for MR heads as well. An MR head with a shield to shield distance $2g$ has a PW_{50} identical to an inductive head with a distance between the top and bottom poles (write gap) of g.

The transmission line method correctly accounts for the decay of medium flux in the MR film. An application of the transmission line to SAL and DS MR films has been proposed by Bhattacharyya and Simmons (40). Although not completely analytical in nature, this paper derives a closed-form expression for the flux flow in the MR film for any shielded MR head. For an MR film of thickness t, located symmetrically between the shields of separation $2g$, the decay length of flux is $(\mu g t)^{1/2}$. This expression is in units of length, and is considered in determining the height used in high performance MR heads. Taller elements have material that is not influenced by the magnetic field of the medium, which only serves to shunt the current, reducing $\Delta R/R$. Small MR heights are desirable for high output voltages, but have difficulties in proper biasing and in reproducible manufacturing.

Micromagnetic Methods. While the transmission line method is useful in some respects, it does not include the crystalline anisotropy, ferromagnetic exchange, magnetostatic coupling between films, or magnetic saturation effects. Micromagnetic analysis accounts for these factors. In steady state micromagnetic analysis, one uses the torque equation: $\boldsymbol{M} \times \boldsymbol{H} = 0$. As applied to magnetic read heads, this equation becomes:

$$\hat{\boldsymbol{m}}(\boldsymbol{r}) \times \left[\boldsymbol{H}_{\mathrm{a}}(\boldsymbol{r}) + \boldsymbol{H}_{\mathrm{D}}(\boldsymbol{r}) + H_k m_z(\boldsymbol{r})\hat{\boldsymbol{z}} + \frac{2A}{M_{\mathrm{s}}}\nabla^2\hat{\boldsymbol{m}}\right] = 0$$

The first term within brackets includes the applied field due to current, the medium, and any other field such as perma-

nent magnet or exchange stabilization. The second term is the demagnetizing field, the third, the crystalline anisotropy field, and the fourth, the exchange energy field. The medium flux is computed separately. One way to compute the medium flux is to find a suitable Green's function for the geometry, and then integrate the normal derivative of Green's function with the medium charges. A good reference for such work is a paper by Yuan, Bertram, and Bhattacharyya (41). Here the authors used the micromagnetic formulation to study the off-track asymmetry of shielded SAL MR heads, and showed the importance of proper biasing and MR film heights. The micromagnetic method has also been extended to analyze the mechanisms for domain nucleation in MR heads (42).

Micromagnetic analysis has been applied to spin valve devices (43). The magnetic configuration of a spin valve head is shown in Figs. 10(c) and 13. The pinned film's magnetization is held perpendicular to the ABS, whereas the sense film's magnetization is along the long axis of the head at zero field. The medium's field rotates the sense layer's magnetization in a transverse direction. Spin valves generate signals 2 to 3 times larger than SAL or DS MR heads. Spin valve design is more complicated than MR heads since the coupling field (between the pinned and free layer), the demagnetizing field, and the sense current field must sum to a net zero field in the sense layer. Figure 18 shows the magnetization patterns of the pinned and free films obtained from a three-dimensional micromagnetic analysis similar to that reported by Yuan et al. (41).

RECORDING HEAD TESTING

Recording heads are tested at various stages in the fabrication process. At wafer level, films used in the fabrication of the head are tested to determine if they have the required magnetic and mechanical characteristics. After devices are patterned, the devices or test structures can be measured to monitor their properties. After fabrication into heads, their output signals are analyzed using equipment similar to actual disk drives. In research and development of recording heads many advanced measurement tools are employed. These are often very labor intensive, and not done on a regular basis, but provide valuable insight into the functioning of prototype recording heads.

Wafer Level Testing

Sheet films of the head structure are tested using a variety of methods. The magnetization of the pole materials is measured either using an inductive $B–H$ loop tester or a vibrating sample magnetometer. Parameters of importance are: saturation magnetization, related to the total field output of the head; coercivity, related to noise during readback; magnetic anisotropy, related to the efficiency of a write head and the sensitivity of a read head. Permeability is measured as a function of frequency to assess the high frequency capability of the pole materials. Other common measurements include resistivity, magnetostriction, stress, hardness, and film thickness.

For fabrication of an MR head, the magnetoresistance of the film is monitored by varying the drive field and monitoring the resistance of a long bar test structure. This shape gives a uniform direction of current flow in the material, im-

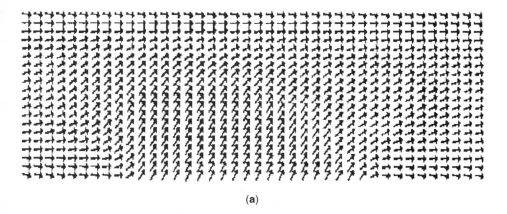

(a)

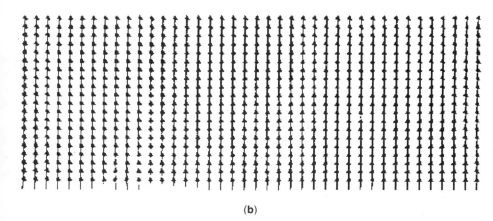

(b)

Figure 18. Result of a micromagnetic analysis: Micromagnetic models are used to calculate the distribution of magnetic flux in the magnetic layers of a read head. The arrows indicate the local direction of the magnetization in these films. (a) The magnetization of the sense film is rotated by the field emanating from the medium. (b) The magnetization of the pinned film is held perpendicular to the ABS.

portant since the magnetoresistance in an AMR head depends on the angle of the magnetization with respect to the current. A magnetic field often is applied in two orthogonal directions to fully characterize the response of the element. The hard axis curve is obtained for a field applied transverse to the direction of the material anisotropy (defined by device shape or induced anisotropy), and an easy axis curve is obtained for a field applied parallel to the direction of anisotropy. An example of the magnetoresistive response to an applied field for a spin valve film is shown in Fig. 19. In addition to magnetoresistance, such an R-H loop provides the coercivities of both the pinned and sense layers, as well as the exchange pinning field and magnetic coupling field between pinned and sense films.

After the films have been patterned into devices, the device response to an applied field is an important tool for determining final head performance. The amount of noise in the device, and the linearity of the response is evaluated. A field response curve is shown in Figure 20 for an unshielded spin valve device. Care must be taken in relating this response to that of a recording head sensing the medium, since the field from a transition excites the MR element in a manner that is different from a uniformly applied field. Nevertheless, this tool is useful to analyze the magnetic characteristics and ascertain whether the MR element is behaving as expected from modeling and if it is magnetically stable.

Fabricated Head Testing

A spin stand test bed is used to evaluate the performance of heads and media in a simulated disk drive operating environ-

ment. The heads are in the final assembled state, with the air bearing machined on the slider, and the slider mounted on the head gimbal assembly. The assembly is affixed to the flexure and the head is flown over the disk. Parameters tested include signal amplitude, resolution, signal to noise ratio, and so on. Measured parameters are discussed in detail below. Spin stand testing is used for verification of theory, testing new technologies, evaluating new head designs, assessing the quality of heads or media, and as a failure analysis tool.

A typical spin stand test bed is diagrammed in Fig. 21. The channel electronics include a preamplifier, amplifier and filter, and read write control electronics. With the increasing pressure for higher density and faster data rate, the channel electronics are also required to improve, resulting in stringent specifications for wide bandwidth and low noise. The migration to MR technology in disk drives precipitated the introduction of a new family of preamps with bandwidths beyond 100 MHz and input noise as low as 0.45 nV/Hz$^{1/2}$. The read channel electronics on the test stand can also include a filter, equalizer, and circuitry to provide signals for time domain measurements. Spin stand testers typically are able to do both parametric tests and bit error rates (BER). An example of an output waveform from a spin stand tester is shown in Fig. 22 for an isolated transition in the medium. The readback waveform for a series of isolated transitions is shown in Fig. 23.

Several types of tests were designed to determine the performance of either the disk or the head, although frequently the measurement results are the combination of both head

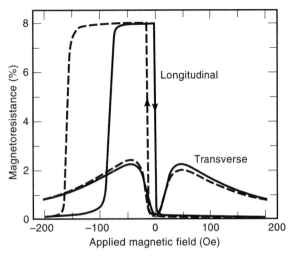

Figure 19. Uniform field response of a spin valve film: The percentage change in resistance is plotted vs. field, applied both parallel (longitudinal) and perpendicular (transverse) to the pinned layer's magnetization. Refer first to the longitudinal response. For negative fields greater than the pinning exchange field (here, $H_X \sim -125$ Oe), the magnetization of both layers are parallel to the negative field direction, and the resistance is low. As the field drops below H_X, the pinned layer's magnetization (M_{PL}) switches back, while the sense layer's magnetization (M_{SL}) still points in the negative field direction. M_{PL} and M_{SL} are antiparallel and the resistance is high. As the field becomes positive, the M_{SL} also switches. M_{PL} and M_{SL} are again parallel, resulting in the low resistance state. For the transverse curve, at low applied fields, M_{SL} rotates perpendicular to M_{PL}. The magnetoresistance increases slower since M_{SL} is rotated perpendicular to the magnetic anisotropy direction; its magnitude is lower since the maximum angle between M_{SL} and M_{PL} is 90 degrees. The resistance decreases with higher fields as M_{PL} is also rotated, and the magnetizations become parallel again.

and disk properties. An example is the measurement of the PW_{50} of an isolated pulse. The PW_{50} is the width of the pulse at half amplitude [Eq. (7)] and depends on the read gap of the head, the transition parameter a, thickness δ of the medium, and also head and disk separation d. Separating the roles of the different parameters in determining the PW_{50} is a challenge for the experimentalist.

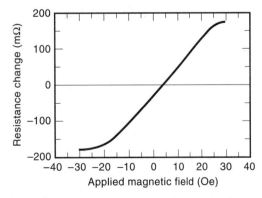

Figure 20. Transfer curve of a spin valve device: The change in resistance (in mΩ) of a spin valve head is plotted. The device is self biased: it has a linear response about zero field, and is single valued with respect to field orientation. The response saturates at low fields, ~25 Oe.

Typical Test Parameters

- *Isolated pulse.* A simple measurement useful in determining the quality of the head/ medium system. The amplitude and PW_{50} are obtained from this measurement (Fig. 22). Pulse shape and the peak curvature are used to predict how the system will behave at higher density.

- *Track average amplitude (TAA).* An average of the readback signal strength from the head and related to parameters such as the head efficiency and the amount and distribution of flux from the recording medium.

- *Resolution.* Defined as the ratio of the 2f TAA to the 1f TAA. This test relates to the linear density along the track of a head/medium system. 1f and 2f refer to the frequencies used to write the data.

- *Overwrite (OW).* Measures the ability of the head/medium system to erase previous data by writing new data over it. Expressed in dB, the overwrite value is the ratio of the 1f signal before and after being overwritten with a 2f signal. This parameter is important because any remaining data will become coherent noise and will degrade the readback signal.

- *Roll off curve.* Multiple TAA measurements made over a range of linear density (Fig. 24) at sufficient write current to ensure saturation of the medium. The amplitude of the TAA decreases at higher transition densities because of interactions between adjacent pulses during both write and read processes.

- *Signal to noise ratio (SNR).* Relates to the medium noise in the disk. Figure 25 is an example of SNR measurement of a thin film disk. It shows the effect of noise increasing and signal decreasing as density increases.

- *Amplitude modulation.* The amount of low frequency modulation around the track, expressed as a percentage of TAA. It relates to the mechanical and magnetic circumferential uniformity of the disk.

- *Track profile.* The TAA measurement as the head is displaced across a written track. This test is to determine the track density ability of the head.

With the emergence of MR head technology, more extensive tests are required to ensure quality and performance. Problems specific to MR heads are: (1) baseline shift, where the reference voltage of the head shifts between measuring positive and negative pulses; (2) amplitude and baseline popping, where Barkhausen noise produces voltage spikes in the output; and (3) pulse height asymmetry of positive and negative signals, due to the nonlinear nature of the transfer curve.

The MR head is also sensitive to the field at the trackedge, and the response in this region is characterized by measuring the cross-track profile. The response of the head at the edge of the track is important for determining the position of the head during tracking, and a linear response as a function of cross-track position is desired. In addition, a more careful look at the cross-track profile, called the microtrack profile, gives information on the response uniformity of the sensor in the active area. These results indicate whether the MR element is in a single or multidomain state. For more details on these and other testing parameters, the reader is referred to standards published by the *International Disk Drive Equipment and Materials Association* (IDEMA) (44).

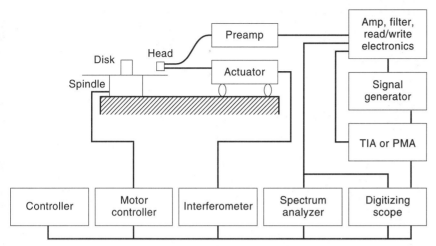

Figure 21. Schematic of a spin stand tester: The actuator positions the head over the disk, which is rotated by the spindle. The rotation speed is dictated by the motor controller, the head position is monitored by the interferometer. The head output is run through a preamplifier, amplifier, and filter. The signal generator is used to write information on the disk. The time interval analyzer (TIA) and phase margin analyzer (PMA) are used to assess the window margin of the bit error rate. The digitizing scope plots the output of the head, and the spectrum analyzer determines the features of the output pulses.

Advanced Measurement Methods

Magnetic properties and the signal readback are only two of the aspects monitored in characterizing recording heads. Comprehensive performance evaluation of a recording head also includes characterizing its flying behavior over the disk. The flying height, the spacing between the head and the medium, is measured using a method based on optical interferometry. Other optical techniques provide pictures of the curvature of the entire surface. Atomic force microscopy is used to provide detailed maps of the air bearing region. This gives important information on the amount of pole tip recession (how far the element is recessed relative to the encapsulating dielectric at the air bearing surface) that occurs during polishing of the head.

Characterization of magnetic domains in the write heads and the sense elements of MR heads is an area of continuing improvements in resolution and frequency response. Bitter fluid methods were initially used to decorate domain walls but this technique suffers from a lack of resolution, and is not dynamic. Kerr microscopy (45), where the polarization of reflected light is rotated by the magnetization of the domains, has long been a standard tool for imaging domains and for obtaining magnetization versus field information. However, this method is limited by the resolution of the light. This limitation has inspired novel measurement techniques based on electron microscopes to probe magnetic phenomena at smaller and smaller dimensions; for example, scanning electron microscopy with polarization analysis (SEMPA) (46), coherent Foucault imaging (47), and Lorentz microscopy (48). Methods utilizing synchrotron X rays are also being developed (49), providing not only high resolution but also element specific information.

New methods also have been developed to study the field distribution surrounding the write poles. By placing a thin magnetic film just above the ABS of a write head, Kerr microscopy can be used to sense the resulting magnetization pattern (50) when the write coil is energized. This method has been refined for use at high frequencies with pulsed lasers, and high resolution with solid immersion lenses. Another method uses microloops to sense the field inductively (51).

FUTURE DIRECTIONS

The growth in the areal density of storage products necessitates continuous improvements in recording heads. Scaling down the critical dimensions of the recording system will provide some improvement. Reduction in size will not be easy, as the dimensional tolerances become increasingly hard to maintain in manufacturing. For instance, a ten percent tolerance in read width, while easy to specify with trackwidths of ten

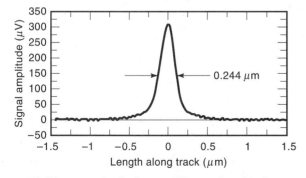

Figure 22. Response of a dual stripe MR recording head to a single transition in the medium: Signal amplitude (μV) is plotted vs. the length along the track (μm). The signal is averaged many times to remove medium and electronic noise. The pulse shape is nearly Lorenzian, with a PW$_{50}$ of 0.224 μm.

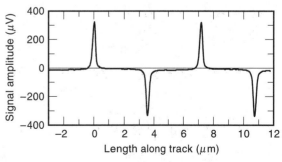

Figure 23. Response of a dual stripe MR recording head to multiple transitions in the medium: Signal amplitude (μV) is plotted vs. the length along the track (μm). Pulses for positive and negative field transitions are very symmetrical.

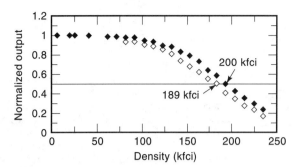

Figure 24. Roll off curve: Normalized signal output of the same head is plotted as a function of density (kilo-flux changes per inch) for two disks with differing coercivity and medium thickness. Heads operate in the region of 50% to 100% of maximum output. The density cutoff of a head depends on the media characteristics.

micrometers, becomes much more difficult at submicron dimensions. Advances in photolithography will be necessary to fabricate smaller and smaller heads.

New materials, such as the GMR materials, will provide some of the increase in head output required to read smaller bits. Major efforts are underway to invent new MR devices, such as ones based on colossal MR materials or on spin tunneling junctions (52). Detailed understanding of the magnetic properties of small structures is essential since the amount of magnetic material in either the sensor or in a recorded bit becomes minute. Averaging of properties is no longer sufficient to produce a clean response. Other active areas of research are on high moment and high frequency soft materials for write heads, and materials with high thermal conductivity and thermal stability for use at high current densities in read heads.

Reduction in the flying height of the recording head over the medium is reaching physical limitations. This eliminates the possibility of increasing areal density by reducing the bit length; as a result, increases in areal density will soon come from reducing the track width. This will provide challenges to servo techniques (methods used to stay on track), and new approaches will be needed. Some possibilities are based on micromachining servo tracks on the disk, or on merging technologies such as optical servoing with a conventional head.

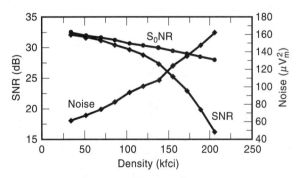

Figure 25. Effect of recording density on signal to noise ratio (SNR): SNR and noise are plotted vs. recording density. The signal amplitude decreases (see Fig. 24) as noise increases with density. The combined effect gives a precipitous drop in the SNR with increasing density. A parameter that is insensitive to the effect of the gap length and flying height is S_0NR and is the SNR as referenced to the maximum signal amplitude at low recording densities.

Magnetic media may evolve into new areas, resulting in redesigned recording heads. For instance, media with magnetization oriented perpendicular to the substrate plane are attractive at high densities. Alternatively, a patterned medium is attractive to overcome the interaction between bits. These changes have the potential to push recording densities to 100 Gbit/in.[2] and beyond, necessitating novel approaches to positioning the head and reading and writing the information. The evolution of recording heads during the past several decades has been impressive. In the future, the demands of the information storage industry ensure that recording heads will continue to evolve into even more sophisticated devices.

BIBLIOGRAPHY

1. C. D. Mee and E. D. Daniel (eds.), *Magnetic Recording,* Volume I, II, and III, New York: McGraw-Hill, 1988.
2. C. D. Mee and E. D. Daniel (eds.), *Magnetic Recording Technology,* 2nd edition, New York: McGraw-Hill, 1996.
3. R. M. White (ed.), *Introduction to Magnetic Recording,* New York: IEEE Press, 1984.
4. J. C. Mallinson, *The Foundations of Magnetic Recording,* San Diego: Academic Press, Inc., 1987.
5. F. Jorgensen, *The Complete Handbook of Magnetic Recording,* Blue Ridge Summit, PA: Tab Books, 1980.
6. P. Ciureanu and H. Gavrila, *Magnetic Heads for Digital Recording,* New York, Elsevier, 1990.
7. J. C. Mallinson, *Magneto-Resistive Heads,* San Diego: Academic Press, Inc., 1996.
8. For a brief review of the early history of recording see Reference 4, pp. 9–15.
9. Ref. 1, *Magnetic Recording,* Vol. 11, p. 21.
10. R. Hunt, A magnetoresistive readout transducer, *IEEE Trans. Magn.,* **Mag-7**: 150–154, 1971.
11. F. Jeffers, J. Freeman, R. Toussaint, N. Smith, D. Wachenschwanz, S. Shtrikman, and W. Doyle, Soft-adjacent-layer self-biased magnetoresistive heads in high-density recording, *IEEE Trans. Magn.,* **Mag-21**: 1563–1565, 1982.
12. T. C. Anthony, S. L. Naberhuis, J. A. Brug, M. K. Bhattacharyya, L. T. Tran, V. K. Hesterman, and G. G. Lopatin, Dual-stripe magnetoresistive heads for high density recording, *IEEE Trans. Magn.,* **Mag-30**: 303–308, 1994.
13. M. Baibich, J. Broto, A. Fert, F. Nguyen Van Dau, F. Petroff, P. Eitenne, G. Creuzet, A. Friederich, and J. Chazelas, Giant magnetoresistance of (001)Fe/(001)Cr magnetic superlattices, *Phys. Rev. Lett.,* **61**: p. 2472–2475, 1988.
14. B. Dieny, V. S. Speriosu, S. Metin, S. S. P. Parkin, B. A. Gurney, P. Baumgart, and D. R. Wilhoit, Magnetic properties of magnetically soft spin-valve structures, *J. Appl. Phys.,* **69**: 4774–4779, 1991.
15. C. Tsang, R. E. Fontana, T. Linn, D. E. Heim, V. S. Speriosu, B. A. Gurney, and M. L. Williams, Design, fabrication and testing of spin-valve read heads for high density Recording, *IEEE Trans. Magn.,* **Mag-30**: 3801–3806, 1994.
16. N. Robertson, B. Hu, and C. Tsang, High performance write head using NiFe 45/55, 1997 Digests of Intermag '97, paper AA02.
17. C. Chang, J. M. Sivertesen, and J. H. Judy, Structure and magnetic properties of RF reactively sputtered iron nitride thin films, *IEEE Trans. Magn.,* **23**: 3636, 1987.
18. N. Ishiwata, C. Wakabayashi, and H. Urai, Soft magnetism of high-nitrogen-concentration of FeTaN films, *J. Appl. Phys.,* **69**: 5616, 1991; G. Qui, E. Haftek, and J. A. Barnard, Crystal struc-

ture of high moment FeTaN materials for thin film recording heads, *J. Appl. Phys.*, **73**: 6573, 1993.

19. S. Wang, K. E. Obermyer, and M. K. Kryder, Improved high moment FeAlN/SiO₂ laminated materials for recording heads, *IEEE Trans. Magn.*, **27**: 4879, 1991.

20. Y. Ochiai, M. Hayakawa, K. Hayashi, and K. Aso, High frequency permeability in double-layered structure of amorphous CaTaZr films, *J. Appl. Phys.*, **63**: 5424, 1988.

21. J. L. Vossen and W. Kern, *Thin Film Processes,* Orlando, FL: Academic Press, 1978.

22. A. Okabe, et al., A narrow track MR head for high density tape recording, *IEEE Trans. Magn.*, **32**: 3404, 1996.

23. T. Maruyama, K. Yamada, T. Tatsumi, and H. Urai, Soft adjacent layer optimization for self-biased MR elements with current shunt layers, *IEEE Trans. Magn.*, **24**: 2404, 1988.

24. T. R. McGuire and R. I. Potter, Anisotropic magnetoresistance in ferromagnetic 3d Alloys, *IEEE Trans. Magn.*, **11**: 1018, 1975.

25. C. Tsang, Unshielded MR elements with patterned exchange biasing, *IEEE Trans. Magn.*, **Mag-25**: 3692–3694, 1989.

26. D. Hannon, M. Krounbi, and J. Christner, Allicat magnetoresistive head design and performance, *IEEE Trans. Magn.*, **Mag-30**: 298–300, 1990, S. H. Liao, T. Torng, T. Kobayashi, Stability and biasing characteristics of a permanent magnet biased SAL/MR device, *IEEE Trans. Magn.*, **Mag-30**: 3855–3857, 1994.

27. C. Tsang and K. Lee, Temperature dependence of unidirectional anisotropy effects in the permalloy-FeMn systems, *J. Appl. Phys.*, **53**: 2605 (1982).

28. K. Nakamoto, Y. Kawato, Y. Suzuki, Y. Hamakawa, T. Kawabe, K. Fujimoto, M. Fuyama, and Y. Sugita, Design and performance of GMR heads with NiO, *IEEE Trans. Magn.*, **Mag-32**: 3374–3379, 1996.

29. T. Lin, D. Mauri, N. Staud, C. Hwang, K. Howard, and G. Gorman, Improved exchange coupling between ferromagnetic Ni-Fe and antiferromagnetic Ni-Mn-based films, *Appl. Phys. Lett.*, **65** (1994) p. 1183.

30. H. Fuke, K. Saito, Y. Kamiguchi, H. Iwasaki, and M. Sahashi, Spin valve GMR films with antiferromagnetic IrMn layers, presented at 1996 Magnetism and Magnetic Materials Conference, Atlanta, GA, paper BD-05. (the proceedings of this conf. may be out by now in JAP)

31. O. Karlqvist, Calculation of the magnetic field in the ferromagnetic layer of a magnetic drum, *Trans. R. Inst. Technol.*, Stockholm, **86**: 3–27, 1954.

32. H. N. Bertram, *Theory of Magnetic Recording,* Cambridge: Cambridge University Press, 1994.

33. M. L. Williams and R. L. Comstock, An analytical model of the write process in digital magnetic recording, *17th Annu. AIP Conf. Proc.*, **5**: 738–42, 1971.

34. M. Bhattacharyya, H. S. Gill, and R. F. Simmons, Determination of overwrite specification in thin-film head/disk systems, *IEEE Trans. Magn.*, **Mag-25**: 4479–4489, 1989.

35. R. Davidson, R. Simmons, and S. Charap, Prediction of the limitations placed on magnetoresistive head servo systems by track edge writing for various pole tip geometries, *J. Appl. Phys.*, **79**: 5671–5673, 1996.

36. I. A. Beardsley, Modeling the record process, *IEEE Trans. Magn.*, **Mag-22**: 454–459, 1986.

37. J.-G. Zhu and H. N. Bertram, Micromagnetic studies of thin metallic films, *J. Appl. Phys.*, **63**: 3248–3253, 1988.

38. D. A. Lindholm, Magnetic fields of finite track width heads, *IEEE Trans. Magn.*, **Mag-13**: 1460, 1977.

39. R. I. Potter, Digital magnetic recording theory, *IEEE Trans. Magn.*, **Mag-10**: 502–508, 1974.

40. M. K. Bhattacharyya and R. F. Simmons, MR head manufacturing sensitivity analysis using transmission line model, *IEEE Trans. Magn.*, **Mag-30**: 291–297, 1994.

41. S. Yuan, H. N. Bertram, and M. K. Bhattacharyya, Cross track characteristics of shielded MR heads, *IEEE Trans. Magn.*, **Mag-30**: 381–387, 1994.

42. J.-G. Zhu and D. J. O'Connor, Impact of microstructure on stability of permanent magnet biased magnetoresistive heads, *IEEE Trans. Magn.*, **Mag-32**: 54–60, 1996, E. Champion and H. N. Bertram, The effect of interface dispersion on noise and hysteresis in permanent magnet stabilized MR elements, *IEEE Trans. Magn.*, **31**: 2642–2644, 1995.

43. T. R. Koehler and M. L. Williams, Micromagnetic simulation of 10 Gb/in² spin-valve heads, *IEEE Trans. Magn.*, **Mag-32**: 3446–3448, 1996.

44. International Disk Drive Equipment and Materials Association, 710 Lakeway, Suite 140, Sunnyvale, CA 94086. Also ANSI X3b7.1.

45. Z. Q. Qui and S. D. Bader, Surface magnetism and Kerr microcopy, *MRS Bulletin*, **20**: 34–37, 1995.

46. R. J. Celotta, D. T. Pierce, and J. Unguris, SEMPA studies of exchange coupling in magnetic multilayers, *MRS Bulletin*, **20**: 30–33, 1995.

47. S. McVitie and J. N. Chapman, Coherent Lorentz imaging of soft, thin film magnetic materials, *MRS Bulletin,* **20**: 55–58, 1995.

48. X. Portier, A. K. Petford-Long, R. C. Doole, T. C. Anthony, and J. A. Brug, In-situ magnetoresistance measurements on spin valve elements combined with Lorentz transmission electron microscopy, 1997 Digests of Intermag '97, paper EA-01.

49. N. V. Smith and H. A. Padmore, X-ray magnetic circular dichroism spectroscopy and microscopy, *MRS Bulletin*, **20** (10): 41–44, 1995.

50. M. R. Freeman and J. F. Smyth, Picosecond time-resolved magnetization dynamics of thin-film heads, *J. Appl. Phys.*, **79**: 5898–5900, 1996.

51. R. F. Hoyt, D. E. Heim, J. S. Best, C. T. Horng, and D. E. Horne, Direct mesurement of recording head fields using a high-resolution inductive loop, *J. Appl. Phys.*, **55**: 2241 (1984); J. A. Brug, M. K. Bhattacharyya, C. M. Perlov and H. S. Gill, Head field measurements in three dimensions, *IEEE Trans. Magn.*, **Mag-24**: 2844–2846, 1988.

52. J. A. Brug, T. C. Anthony, and J. H. Nickel, Magnetic recording head materials, *MRS Bulletin,* **21**: 23–27, 1996.

JAMES BRUG
MANOJ BHATTACHARYYA
THOMAS ANTHONY
LUNG TRAN
JANICE NICKEL
Hewlett-Packard Laboratories

MAGNETIC REFRIGERATION

The application and subsequent removal of a magnetic field causes cooling in certain materials—the magnetocaloric effect (or magnetic refrigeration). In the laboratory it is possible to pump over liquid helium and so provide considerable cooling power at ~1 K. For this reason magnetic refrigeration has usually been used to reach temperatures below 1 K. It is also the case that the high magnetic fields required are usually provided through superconducting magnets which must operate below ~5 K. There are two types of magnetic refrigerator, adiabatic demagnetization refrigerators (ADR) and nuclear demagnetization refrigerators. The ADR uses the interaction of electrons with an applied magnetic field whereas nuclear demagnetization uses the interactions of the nuclei. The temperature region of operation for an ADR depends on the magnetic material used. This can range from tens of Kelvin down to mK (1); whereas for a nuclear demagnetization refrigerator the region is 10 mK to several microkelvin. Cooling power (the amount of energy that can be absorbed) decreases strongly with decreasing temperature. For an ADR cooling powers in the region of microwatts at low temperature can be obtained whereas for nuclear demagnetization it is nanowatts or less. Such low cooling powers from nuclear demagnetization refrigerators leads to extreme measures in order to avoid unwanted heat entering the system, for example, seismic isolation. The use of nuclear demagnetization refrigerators is therefore restricted to specialized low temperature research laboratories. For ADRs with their higher cooling power such extreme measures are not required, making such refrigerators easier to use in a more typical laboratory environment, so that they are the more common of the two.

Principle of Magnetic Refrigeration

A cooling process may be regarded to be an entropy reducing process. Since entropy (or degree of disorder) of a system at constant volume or constant pressure decreases with decreased temperature, cooling can be achieved within a medium via any process which results in the decrease of entropy of that medium. For example, the liquefaction of gases is achieved by the isothermal reduction of entropy through compression of a volume V_1 at temperature T_1 to a smaller volume V_2 generating heat, which is extracted by contact with a cold reservoir, followed by adiabatic or isentropic expansion which results in cooling of the gas to below T_1. In the magnetic cooling process the disordered collection of magnetic dipoles associated with a particular ion within a medium (paramagnetic material) constitutes such a system described here. For such a material the application of a magnetic field causes alignment of the dipoles with the magnetic field and thus a reduction in entropy. The dipoles used are either electronic (electron cooling due to the electron spins) or nuclear (nuclear cooling due to the nuclear spins) depending on the required final temperature, millikelvin temperatures for electronic (ADRs) and microkelvin temperatures for nuclear demagnetization refrigerators. The principle of operation is the same the main difference being the starting temperature. For electron cooling starting temperatures of up to 20 K or more can be used whereas for nuclear cooling a starting temperature in the region of 0.01 to 0.02 K is needed in order for the magnetic interaction to dominate over the thermal energy.

Components of a Magnetic Refrigerator

An ADR is essentially composed of the following three items which need to be housed in a cryostat (2) to provide a bath temperature:

1. A paramagnetic material. This is suspended via low thermal conductivity supports within an enclosure at the bath temperature, usually pumped, liquid helium in laboratory systems. The paramagnetic material is inte-

grated with a stage or platform, upon which the experimental items under investigation may be mounted.

2. A magnet. This may either be a permanent or superconducting magnet. The latter is usually used due to their ease of operation and compactness. It is housed within the liquid helium vessel and so sufficiently cooled for superconductivity.

3. A heat switch. This is used to make and break a high thermal conductivity path between the paramagnetic material and the cold bath (e.g., liquid helium). This is used in order to cool the paramagnetic material to the starting temperature and to extract the heat of magnetization. The form of this switch may be mechanical, gaseous, or superconducting (2).

A schematic of a "classical" ADR is shown in Fig. 1. Variations on the classical form arise due to requirements to (1) increase the low temperature hold time and (2) increase the bath temperature. Increasing the low temperature hold time can be achieved by reducing the parasitic heat leak through the supports by using a second higher temperature intermediate paramagnetic material. This intercepts the heat flow from the bath to the low temperature paramagnetic material. The intermediate material comprises a higher temperature paramagnetic material which has a high heat capacity, for example, gadolinium gallium garnet. The operation of the ADR is the same as the conventional form except the intermediate paramagnetic material will demagnetize to a temperature between that of the bath and the low temperature material. Such a refrigerator is commonly called a two stage ADR (3). Increasing the bath temperature without decreasing the hold time can be accomplished by using a second ADR to cool a classical or two stage ADR. It is essentially two ADRs in series, one configured for high temperatures and the other for low temperatures. The high temperature ADR is used to cool the low temperature ADR prior to the demagnetization of that stage. This effectively simulates a lower temperature

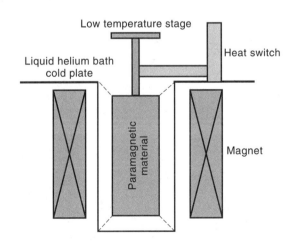

----- Low thermal conductivity supports

Figure 1. Conventional ADR schematic. A paramagnetic material is shown in a pill arrangement with a stage where samples can be attached. The pill is housed in the bore of a superconducting magnet via low thermally conducting supports. The magnet is housed in a liquid helium container and the heat switch is used to cool the pill to the temperature of the liquid helium.

bath for a conventional or two stage ADR. This kind of refrigerator has been called a double ADR (4).

A further variation of the classical ADR is the "hybrid". This consists of a conventional ADR coupled to a low temperature cryogenic stage provided by He³. The He³ stage is used to provide a temperature of 0.3 K reducing the parasitic heat load and providing a very low starting temperature for the paramagnetic material.

The composition of a nuclear demagnetization refrigerator is essentially the same as a conventional ADR except that its bath temperature must be in the region of 0.01 K. The bath is provided by either an ADR or, more typically, a He³–He⁴ dilution refrigerator (2,5).

HISTORY AND CURRENT STATUS OF MAGNETIC REFRIGERATORS

Cooling by affecting the electron spins was proposed by Debye (6) in 1926 and Giauque (7) in 1927. The first practical demonstration was by De Haas, Wiersma, and Kramers (8), Giauque and MacDougall (9) in 1933, and Kurti and Simon (10) in 1934. The nuclear demagnetization refrigerator was suggested by Gorter in 1934 and Kurti and Simon in 1935. However it was not until 1956 when Kurti et al. successfully obtained cooling from 12 mK to 20 μK via the demagnetization of nuclear spins. Since those pioneering days numerous texts on magnetic refrigeration have been published and most low temperature physics books contain a chapter describing it. For further information see Refs. 2, 5, and 11.

The advent of dilution refrigerators in the 1960s saw the demise of ADRs because of the higher cooling power and continuous operation offered by dilution refrigeration. In recent years ADRs have become more popular due to the increased use of low temperatures (0.1 to 0.01 K) for the operation of detectors for astronomy and particle physics and the development of mechanical coolers to replace the use of liquid helium (12). In astronomy the need for better signal to noise leads to low temperature detectors (10 to 100 mK) and exotic telescope locations (Antarctica, on top of high mountains, and in space). X-ray astronomy can only be conducted by space-borne instrumentation due to the absorption of X rays by the earth's atmosphere. The cost and practicalities involved require low mass systems which are gravity independent and highly reliable. The use of a consumable cryogen in space—helium—limits the duration of missions. For infrared and optical astronomy space also has enormous advantages in the form of "seeing" and sky background.

The first demonstration of an ADR in space occurred with a sounding rocket flight in 1996 (13). This was a conventional ADR comprising a liquid helium cooled magnet and an ferric amonium alum (FAA) salt pill housed in a pumped liquid helium cryostat operating at 2 K. A similar ADR is under development (14) at the Goddard Space Flight Center for the X-ray spectrometer (XRS) experiment due to fly on the Japanese ASTRO-E mission in 2000. Advances in miniaturizing ADRs and increasing their operating temperature range further are ongoing and brought about by real-use practicalities. Such development is being achieved through new materials (especially for thermal isolation), new magnet technology (low current), and high temperature paramagnetic materials. Until recently the superconducting magnet used to generate the magnetic field would have to be housed in the liquid helium bath. The advent of conduction cooled superconducting magnets, in which the magnet is cooled via conduction through the magnet housing, enables the ADR to be a unit which is simply attached to a cold plate. This enables the ADR to be connected to a mechanical cooler thereby eliminating the need for a liquid helium bath to cool the magnet and provide the bath temperature for the ADR. Cryogen-free operation could enable the ADR to become a general purpose instrument capable of autonomous computer control without the need for the user to have either cryogenic experience or special helium handling equipment.

THEORY

Certain paramagnetic materials are suitable for use as magnetic refrigerants. The magnetic ions of these materials have an interaction energy, ϵ with their crystalline environment and each other which is smaller than the average thermal energy kT. In such a situation each magnetic ion is relatively "free" resulting in a distribution of randomly oriented dipoles with $2J + 1$ degeneracy, where J is the angular momentum quantum number, that is, there are $2J + 1$ possible orientations of the ions. This gives an $R \ln(2J + 1)$ per mole contribution to the entropy of the material from the magnetic dipoles, where R is the gas constant. The entropy (S) of a paramagnetic material can be thought of comprising two components. One arises from the magnetic ion (e.g., chrome in chromium potassium alum, CPA) and is given the subscript m (S_m). The other component arises from the rest of the molecule and is refer to as the lattice ($S_{lattice}$). The total entropy is given by

$$S = S_m + S_{lattice}$$

As the temperature of the paramagnetic material is reduced the lattice contribution to the entropy of the material reduces and a point is reached where the magnetic entropy given by $R \ln(2J + 1)$ dominates. As the temperature decreases further the entropy will remain at the value given by $R \ln(2J + 1)$ until the thermal energy approaches the interaction energy ϵ at which point spontaneous ordering of the dipoles occurs, due to their own weak magnetic fields, and the entropy falls. When $\epsilon \sim k\theta$, where θ is the magnetic ordering temperature of the material (or Néel temperature), the entropy drops rapidly. At very low temperature the internal interactions between ions removes the degeneracy and the system resides in a singlet ground state of zero entropy. At a temperature greater than θ the entropy of the spin system of the magnetic ions can be reduced significantly if the interaction of the dipoles and the applied magnetic field is greater than the thermal excitation given by kT.

If the internal interaction energy is very low the magnet ions can be considered as free and the entropy (S) of a collection of magnetic ions can be given by Eq. 1. Figure 2 shows the typical form of the entropy curve as a function of temperature and applied magnetic field:

$$\frac{S(B, T)}{R} = \ln \left[\frac{\sinh(2J + 1)x/2}{\sinh(x/2)} \right] + \frac{x}{2} \coth \left(\frac{x}{2} \right) - \frac{(2J + 1)}{2} x \coth((2J + 1)x/2) \quad (1)$$

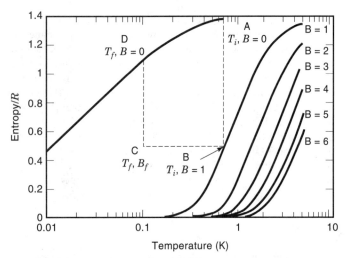

Figure 2. Typical behavior of entropy with temperature and magnetic field, with the operational sequence of an ADR. The sequence starts at point A, proceeding to point B on application of the magnetic field (heat switch engaged). Cooling, transfer to point C, occurs with the partial removal of the magnetic field (heat switch open). The temperature is held (transferring from C to D over time) by further reduction of the magnetic field at the correct rate.

where

$$x = g\beta B/kT \qquad (2)$$

and

g = spectroscopic splitting factor
β = Bohr magnetron
B = magnetic field (Gauss)
k = Boltzmann constant
T = temperature (K)

This equation is only valid if the ions are free and for high (>1 T) magnetic fields. For low magnetic fields additional terms which take into account the weak internal field have to be added. The zero magnetic field entropy curve is determined from

$$S = \int C/T \, dT$$

where

S = total entropy
C = heat capacity
T = temperature

A detailed account of paramagnetic theory as applied to magnetic refrigeration can be found in Ref. 11.

PARAMAGNETIC MATERIALS

The paramagnetic material "the refrigerant" is the core of the refrigerator. It determines the temperature to which cooling can be achieved and how much energy can be absorbed. Cooling over a wide temperature range is possible due to the varied materials available. These can be divided roughly into four temperature ranges (~10 to 40 mK, 40 to >100 mK, >0.3 K to 1 K, and >1 K). The substances widely used in these temperature ranges are detailed in Table 1. A detailed review of the first three temperature ranges can be found in Ref. 11 and for the fourth, high temperature range (15).

Refrigerant Construction (Salt Pill)

Traditionally, the refrigerant stages in an ADR have been called salt pills because in the early days all refrigerants were made from a "salt". Although other forms of paramagnetic refrigerant are now available the term salt pill has remained in common use and now, incorrectly, refers to the whole refrigerant assembly used in the ADR. The refrigerant materials commonly used and listed in Table 1 are hydrated salts (CMN, CPA, CCA, FAA, and MAS), garnets (O_{12} compounds) or perovskites (AlO_3 compounds). Salt pills comprised of the following four components, (1) a stage, to which items can be attached for cooling, (2) A thermal bus connecting the stage to the paramagnetic material, (3) the paramagnetic material, and (4) a container for the paramagnetic material. The stage and thermal bus are made from high thermally conducting material, usually copper, and the container is either a high electrical resistivity material in order to minimize eddy current heating, for example, stainless steel, or fibre glass. Garnets and perovskites, which are very large dense crystals, can be bonded via an epoxy or low temperature cement to the thermal bus. In the case of hydrated salts the thermal bus comprises many hundreds of high thermally conducting (cop-

Table 1. Magnetic Details of Some Paramagnetic Materials

Temperature Range	Material	Formation	J	g	T_n (K)
10–40 mK	Cerium magnesium nitrate (CMN)	$Ce_2Mg_3(NO_3)_{12} \cdot 24H_2O$	1/2	2	~0.01
40–>100 mK	Chromic potassium alum (CPA)	$CrK(SO_4)_2 \cdot 12H_2O$	3/2	2	~0.01
	Cesium chromic alum (CCA)	$CsCr(SO_4)_2 \cdot 12\,H_2O$	3/2	2	~0.01
	Ferric ammonium alum (FAA)	$FeNH_4(SO_4)_2 \cdot 12H_2O$	4/2	2	~0.03
>0.3 K	Manganese ammonium sulfate (MAS)	$MnSO_4(NH_4)_2SO_4 \cdot 6H_2O$	5/2	2	~0.1
>1 K	Dysprosium gallium garnet (DGG)	$Dy_3Ga_5O_{12}$	1/2	8	~0.4
	Erbium orthoaluminate (ErOA)	$ErAlO_3$	1/2	9	~0.6
	Ytterbium orthoaluminate (YbOA)	$YbAlO_3$	1/3	7	~0.8
	Gadolinium gallium garnet (GGG)	$Gd_3Ga_3O_{12}$	7/2	2	~0.8
	Dysprosium aluminum garnet (DAG)	$Dy_3Al_5O_{12}$	1/2	11	~2.5
	Dysprosium orthoaluminate (DOA)	$DyAlO_3$	1/2	14	~3.5
	Gadolinium orthoaluminate (GOA)	$GdAlO_3$	7/2	2	~3.8

per) wires attached to the stage and spreading into the salt enclosure. The hydrated salts are produced via an aqueous solution and the slow evaporation of the water. They fall into two categories, those that can be readily grown around copper wires and those that cannot. In the case where the salt will not grow around wires (CPA) small crystals have to be grown in a vessel (beaker) and then compressed along with the wires to form a solid mass which is then sealed in the pill enclosure. For the salts which will readily grow around wires (CMN, CCA, FAA, and MAS) the aqueous solution is allowed to evaporate from the enclosure which holds the wires and thermal bus. For CMN, where the orientation of the crystals to the magnetic field is important, seed crystals have to be cemented in place first with the correct orientation. The salts listed in Table 1 are all corrosive to copper and so if used this metal has to be plated with gold. A thickness of 30 to 50 μm is usually sufficient. The hydrated salts have to be hermetically sealed within the enclosure in order to ensure that water of crystallization is not lost when exposed to vacuum. Such sealing can be achieved by the use of epoxy, for example, Stycast 2850.

MAGNETIC REFRIGERATOR OPERATION

The operation of a magnetic refrigerator requires the system to be at a temperature in which the lattice entropy does not dominate the paramagnetic material in order that the applied magnetic field can reduce the entropy of the material. The process of cooling can be separated into three stages. The first is isothermal magnetization of the paramagnetic material at a temperature T_1, transferring the paramagnetic material from point A (Fig. 2) to point B. This process generates heat (magnetization energy) which has to be extracted to a heat sink at a temperature of T_1, via a heat switch. The magnetization energy (Q Joules per mole) is given by

$$Q = \int T dS \tag{3}$$

and since T is constant

$$Q = T_1[S_A - S_B] \tag{4}$$

Adiabatic demagnetization forms the second stage in which the magnetic field is reduced to a value B_f which corresponds to the desired final temperature T_f. During this stage the entropy of the paramagnetic material remains constant, resulting in cooling as given by

$$\frac{(B_i^2 + b^2)^{1/2}}{T_i} = \frac{(B_f^2 + b^2)^{1/2}}{T_f} \tag{5}$$

where

B_i = Initial magnetic field
B_f = Final magnetic field
b = Internal magnetic field associated with each magnetic ion
T_i = Initial temperature
T_f = Final temperature

The third stage can be effective in two ways, the first and most thermally efficient is isothermal demagnetization. This

provides stability at T_f by reducing the magnetic field B from B_f to zero at a rate which counteracts the thermal input from the surrounding environment. The total amount of energy the paramagnetic material can be absorbed is given by Eq. (3) and Eq. (6) next, since T is constant,

$$Q = T_2[S_C - S_D] \tag{6}$$

The duration in seconds of operation at T_f, called the hold time, is given by

$$\text{Hold Time} = \frac{n T_f[S_C - S_D]}{dQ_{th}/dt} \tag{7}$$

where

n = Number of moles of magnetic ion
dQ_{th}/dt = Total power into the paramagnetic material (e.g., parasitic heating)

Once the magnetic field reaches zero the whole process must be repeated (recycled) from stage 1. The second approach is for the magnetic field to be reduced to zero, completely demagnetizing the paramagnetic salt to a temperature of T_{min}. A constant temperature above T_{min} can be achieved by using a stage which is heated via a resistor and which has a weak thermal link to the paramagnetic material. The heater power needs to be reduced as the paramagnetic material warms under the parasitic load. Since the temperature of the paramagnetic material is not constant the energy which can be absorbed is given by Eq. (3) only. This process is less thermodynamically efficient to the isothermal process since heat is being added, however, it does not require active control of the magnetic field and therefore is simpler in some respects.

TEMPERATURE REGULATION

As previously stated the third stage of operation of an ADR, namely the holding of the final required temperature, can be achieved via demagnetizing at a rate to counteract the heat flowing into and thus warming up the salt pill. Temperature regulation is achieved via a computer controlled servo system in which the rate of reduction in magnetic field is controlled in order to maintain the temperature at the desired value. For superconducting magnets the magnetic field is proportional to current and thus it is the current that is controlled. The degree of stability is limited to the sensitivity of the thermometry and the step size in magnetic field/current. Temperature regulation of ±1 μK is possible with room temperature resistance bridge phase sensitive detection (PSD) electronics and germanium thermometers. The thermometry readout is limited by the Johnson noise associated with the resistance bridge resistors. Cooling of these resistors to approximately 10 K should make it possible to achieve ~a few hundred nanokelvin stability of an ADR when operated at a temperature of 0.1 K.

Magnetic Field Servo

Constant temperature is maintained by the reduction of electrical current (I) by a value dI every dt seconds. The value of dI/dt is determined by the servo algorithm based on how well the temperature is being maintained. With every new value of dI/dt the current is stepped down at the corresponding value of dI, (dt is kept constant) until the new value of dI/dt is calculated. Intervals of a minute or less are appropriate for most systems depending upon the application.

The current–temperature control equation (16) is:

$$dI_i/dt = (c/\Delta t)[(T_i - T_{set}) + (\Delta t/\tau) \sum (T_{i-j} - T_{set})]$$

where

i = Current value
j = previous $(i-1)$ value
dI_i/dt = ith current ramp rate
T_i = ith temperature
T_{i-j} = $i-j$th temperature
T_{set} = Servo temperature
Δt = time interval
τ = system time constant
$\sum$ = Sum over $j=0$ to $i-1$
c = system constant

SUMMARY

Millikelvin refrigeration is becoming necessary for many applications (astronomy, particle physics, material science, and biophysics). Adiabatic demagnetization refrigerators are seen as one of the most attractive ways of achieving such temperatures. Their compactness, ease of operation (turn key is a possibility since the process is purely electrical and cryogen free) and gravity independence gives many advantages over helium based apparatus. Cryogenic engineering advances now mean that an ADR can be a small bench top instrument rather than requiring a well equipped cryogenic laboratory. With this simplification and miniaturization of cryogenic instrumentation the user will no longer have to be an experienced cryogenic physicist which will open millikelvin refrigeration to a much broader community of scientists and engineers. Such an expansion process has not yet occurred with nuclear demagnetization refrigeration, however, with the constant development of lower operating temperature detectors and advances in cryogenic science and engineering it is probably only a matter of time.

BIBLIOGRAPHY

1. J. A. Barclay, *Advances in Cryogenic Engineering*, Vol. 33, New York: Plenum Press, 1988, p. 719.

2. G. K. White, *Experimental Techniques in Low-Temperature Physics*, Oxford, UK: Oxford University Press, 1989.

3. C. Hagmann and P. L. Richards, Two-stage magnetic refrigerator for astronomical applications with reservoir temperatures above 4 K, *Cryogenics, 34*: 213–226, 1994.

4. I. D. Hepburn et al., Submillimeter and Far-Infrared Space Instrumentation, *Proc. 30th ESLAB.*, ESA SP-388, 1996.

5. O. V. Lounasmaa, *Experimental Principles and Methods below 1K*, New York: Academic Press, 1974.

6. P. Debye, *Ann. Phys.*, 81: 1154, 1926.

7. W. F. Giauque, A thermodynamic treatment of certain magnetic effects. A proposed method of producing temperatures considerably below 1° absolute, *J. Am. Chem. Soc.*, 49: 1864, 1927.

8. W. J. De Haas, E. C. Wiersma, and H. A. Kramers, *Physica*, 1: 1, 1933.

9. W. F. Giauque and D. P. MacDougall, Attainment of temperatures below 1° absolute by demagnetization of $Gd_2(SO_4)_2 \cdot 8H_2O$, *Phys. Rev.*, 43: 768, 1933.

10. N. Kurti and F. E. Simon, Production of very low temperatures by the magnetic method: supraconductivity of cadmium, *Nature* (London), 133: 907, 1934.

11. R. P. Hudson, *Principles and Applications of Magnetic Cooling*, Amsterdam: North-Holland, series in Low Temperature Physics Vol. 2, 1972.

12. S. F. Kral and J. A. Barclay, *Applications of Cryogenic Technology*, vol. 10, edited by J. P. Kelly, New York: Plenum Press, 1991.

13. D. McCammon et al., A sounding rocket payload for X-ray astronomy employing high resolution microcalorimeters, *Nucl. Instrum. Meth. A, 370*: 266–268, 1996.

14. C. K. Stahle et al., Microcalorimetry arrays for high resolution soft X-ray spectroscopy, *Nucl. Instrum. Meth. A, 370*: 173–176, 1996.

15. M. D. Kuz'min and A. M. Tishin, Magnetic refrigerants for the 4.2K–20K region: garnets or perovskites, *J. Phys. D: Appl. Phys., 24*: 2039–2044, 1991.

16. G. Bernstein et al., Automated temperature regulation system for adiabatic demagnetization refrigerators, *Cryogenics, 31*: 99–101, 1991.

I. D. Hepburn
A. Smith
University College London

MAGNETIC RESONANCE

Consider a toy top on a table. If it is not rotating, it will immediately fall down because of gravity. If it is spun, however, it will rotate about the z direction keeping the angle θ constant as shown in Fig. 1. Now let a rod-shaped magnet (a nail with a small flywheel) in a magnetic field between the poles be the long axis R at an angle θ from the direction z of the field,

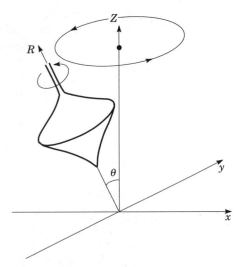

Figure 1. Precession of a top.

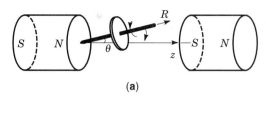

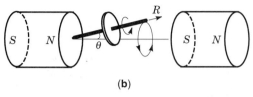

Figure 2. Small iron rod (nail) in magnetic field. To avoid the gravitational effect, the long axis is along the horizontal direction.

as shown in Fig. 2(a). It will tend to align in parallel like a magnetic compass with the line of magnetic force. If one spins it about the long axis R, however, it will rotate about z, keeping θ constant as shown in Fig. 2(b). Both of these motions are called precession and are perpetual in the absence of friction. In this case, the toy top or the rod-shaped magnet has angular momentum, and the direction of force is perpendicular to the plane defined by R and z. This force is called torque. The precession occurs also on a microscopic scale. Consider a single free electron or a single proton in a uniform field H_0 along the z axis. It has a magnetic dipole moment μ_e or μ_N, respectively, which is considered to be a tiny magnet. In this case, one does not need to spin it because it has an angular momentum a priori, and the magnetic moment is caused by the rotation of the charged particle. Because of this angular momentum, it exhibits similar motion in a magnetic field as earlier and as shown in Fig. 3. This motion is called Larmor precession. The frequency of the Larmor precession is derived from the following torque equation, which is similar to the case of a top:

$$\frac{d\mu}{dt} = \gamma[\mu \times \boldsymbol{H}] \qquad (1)$$

where γ is a coefficient called the gyromagnetic ratio and μ is either μ_e or μ_N. $\boldsymbol{H}$ is the magnetic field described as $\boldsymbol{H} = (0, 0, H_0)$. Equation (1) can be solved easily. The z component of Eq. (1) is

$$\frac{d\mu_z}{dt} = 0 \qquad (2)$$

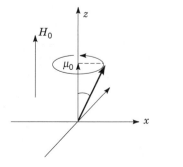

Figure 3. Precession of a magnetic moment μ.

Then one finds that μ_z, which is the z component of μ, is constant. The magnitude of μ_z depends on the initial condition and is unknown here. μ is also unknown but is given by $\cos \theta = \mu_z/|\mu|$. This is reasonable and easily understood by Fig. 3. The x and y components of Eq. (1) are described as

$$\frac{d\mu_x}{dt} = \gamma\mu_y H_0, \qquad \frac{d\mu_y}{dt} = -\gamma\mu_x H_0 \qquad (3)$$

From these equations, one can obtain

$$\mu_x + i\mu_y \propto \exp(-i\gamma H_0 t) = \cos(\gamma H_0 t) - i\,\sin(\gamma H_0 t) \qquad (4)$$

This means that the magnetic moment rotates about the z axis in the direction of a right-handed screw for $\gamma < 0$ (in the case of an electron) or the opposite way for $\gamma > 0$ (in the case of a proton) with an angular frequency

$$\omega_0 = |\gamma| H_0 \qquad (5)$$

where ω_0 is called the Larmor frequency. As will be explained later, γ is expressed as

$$\gamma = g\frac{\mu_0 e}{2m} \qquad (6)$$

where g is the so-called g-value and equal to $g_e = 2.0023$ for an electron and $g_N = 2.7896$ for a proton. e and m are the charge, which is negative for electrons and positive for protons, and the mass of the particle, respectively. μ_0 is the permeability of vacuum. The frequencies are easily calculated using $\omega - 2\pi f$ and are approximately

$$f_0 = 28.0\,\text{GHz} \quad \text{and} \quad f_0 = 42.6\,\text{MHz} \quad \text{at } B_0 = \mu_0 H_0 = 1\,\text{T} \qquad (7)$$

for moments of electrons and protons, respectively. Notice that the mass of the nucleus depends on the atom and that the g of the other nucleus is different from that of the proton.

When an ac magnetic field (electromagnetic wave at radio frequency) perpendicular to the z-axis with amplitude h_{rf} whose frequency and polarization satisfy Eqs. (4) and (5) is applied, what is its effect on the magnetic moment? First, consider a magnetic moment without an ac field in a coordinate system $x'y'z$ rotating about the z axis with ω_0 as shown in Fig. 4(a). The magnetic moment points to the fixed direction in the $x'z$ plane keeping the angle θ constant and never moves in this coordinate system. When an ac field is applied, $\boldsymbol{H}$ in Eq. (1) must be replaced by $\boldsymbol{H} = (h_{\text{rf}} \cos \omega_0 t, h_{\text{rf}} \sin \omega_0 t, H_0)$. The magnetic moment starts to rotate about the y' axis in the $x'z$ plane as shown in Fig. 4(b) in the rotating coordinate system. θ is no longer constant in this case and changes as a function of time as $\theta = 2\pi f_r t + \theta_0$, where θ_0 is the initial angle and f_r is the repetition rate given by

$$2\pi f_r = \gamma h_{\text{rf}} \qquad (8)$$

In a fixed coordinate system this means that first the precession is accelerated absorbing the electromagnetic wave power with increasing amplitude of μ_x and μ_y and with decreasing μ_z, and then with decreasing amplitude of μ_x and μ_y and with increasing μ_z toward the $-z$ direction as shown in Fig. 4(c) as

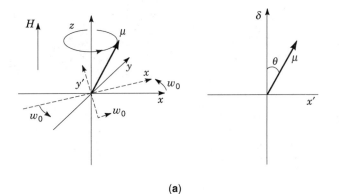

(a)

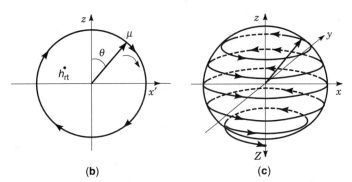

(b)   (c)

Figure 4. (a) Precession in the rotating coordinates $x'y'z$ without h_{rf}, (b) rotation of μ in the $x'z$-plane with h_{rf}, and (c) trajectory of the top of μ.

the trajectory of the top of the moment. As soon as μ completely points in the $-z$ direction, it returns until μ points in the z direction emitting electromagnetic wave power this time. Neither absorption nor emission occurs on average. Usually f_r is much smaller than f_0, so the real precession not like that in Fig. 4(c), (i.e., f_0 is 10^4 times greater than cycle of f_r). This phenomenon is called magnetic resonance in general and more directly electron spin resonance (ESR) or nuclear magnetic resonance (NMR) depending on which moment is in question. Magnetic resonance occurs when the frequency of the ac field coincides with the Larmor frequency.

As mentioned earlier, the magnetic moment of an electron is caused by the angular momentum of the electron. If one calculates μ classically, assuming that an electron is a uniformly charged sphere with radius r rotating with angular frequency ω, then μ is obtained as

$$\mu = -\frac{\mu_0 e \omega r^2}{2} = -\frac{\mu_0 e}{2m_e} L \qquad (9)$$

where e and m_e are the charge and mass of an electron, respectively, and $L = m_e \omega r^2$ is the angular momentum of an electron. We now must introduce quantum mechanics. According to this theory, L must be given by $L = s\hbar$ using Planck's constant $\hbar$, where s is called the spin angular momentum quantmum number or simply the spin and $s = 1/2$. This means an angular momentum of an electron can be only $\hbar/2$ or $-\hbar/2$. From the theory of quantum mechanical electrodynamics, however, it has been shown that g_e must be

multiplied by Eq. (9) for the magnetic moment of an electron. Then the relation between the magnetic moment of electron μ_e and spin s is

$$\mu_e = g_e \mu_B s \qquad (10)$$

where

$$\mu_B = -\frac{\mu_0 e \hbar}{2m_e} = 1.165 \times 10^{-29} [\text{Wb} \cdot \text{m}] \qquad (11)$$

is the unit of the magnetic moment of an electron and is called the Bohr magneton. The energy of the magnetic moment in a field is $-\mu_e H$, which is called Zeeman energy. Then the Hamiltonian described by the energy in quantum mechanics is written as

$$\mathsf{H} = -g_e \mu_B H s \qquad (12)$$

Because of $s = 1/2$ for an electron, only eigenstates of $\pm 1/2$ are allowed. The energy states are then shown in Fig. 5(a), and the energy difference between these two levels is

$$\Delta E = g_e \mu_B H \qquad (13)$$

In an electromagnetic wave whose frequency satisfies $\hbar \omega = \Delta E$, the electron at the ground state is excited to the upper state absorbing the electromagnetic wave energy as shown in Fig. 5(b) and then immediately comes back to the ground state emitting an electromagnetic wave as shown in Fig. 5(b) because the transition probabilities of both transitions are identical. No energy dissipation occurs in this model. This phenomenon corresponds to that explained in Fig. 4. Equation (5) is also obtained from this argument. In the case of a proton, g_e and μ_B must be replaced by $g_N = 2.7896$ and

$$\mu_N = \frac{\mu_0 e \hbar}{2m_p} \qquad (14)$$

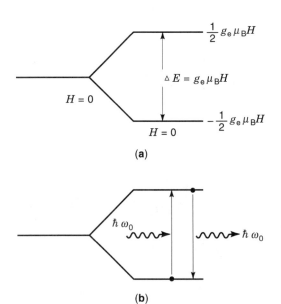

Figure 5. Schematic energy levels of a spin in magnetic field.

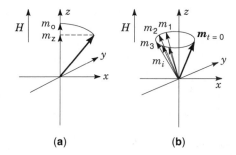

Figure 6. (a) z component of $\boldsymbol{m}$ increases up to m_0 in a time scale of T_1. (b) All moments start simultaneously at $t = 0$, but they diffuse in a time scale of T_2. $m_1, m_2, \ldots, m_i$ are each moment.

where m_p is the mass of proton. In this case, $\gamma_N = -g_N\mu_N/\hbar$ is positive because the charge of the proton is positive. In the case of an electron in an atom, it has an orbital motion around the nucleus, and it also contributes to the magnetic moment. But it is not mentioned here for simplicity.

In real materials, the magnetic resonance phenomenon is more complicated because an electron or a nucleus is located in an atom or a molecule and is no longer free. In this case, we deal with the magnetic moment $\boldsymbol{m}$, which is the average or sum of all moments on lattices in the material. Each magnetic moment in a material interacts with other moments on other lattices and with lattice vibrations (phonons). These interactions cause the relaxation phenomenon, which is another important aspect of magnetic resonance. Because of the relaxation, energy dissipation occurs in the resonance condition, and we can observe the magnetic resonance as the absorption of applied electromagnetic wave power. If a magnetic moment undergoes friction during precession, the amplitude of oscillation is supposed to damp and finally the moment becomes parallel to the z axis as shown in Fig. 6(a) instead of as shown in Fig. 3. In this case, m_z increases as a function of time t and reaches full length $m_0 = |m|$ finally. Assuming that the time rate is constant and defined as T_1, one can rewrite the equation of motion given by Eq. (2) as

$$\frac{dm_z}{dt} = -\frac{m_0 - m_z}{T_1} \tag{15}$$

and obtain $m_0 - m_z = \Delta m \exp(-t/T_1)$, where Δm is the initial difference. Because this damping comes from the interaction between spin motion and lattice vibration and then corresponds to direct dissipation of energy to the lattice, T_1 is called as "spin-lattice relaxation time" or simply pronounced "tee-one." On the other hand, the x, y components of the averaged magnetic moment m_x and m_y have a finite magnitude when each magnetic moment starts to rotate about the z axis simultaneously at $t = 0$. However, because of interactions between magnetic moments (mainly magnetic dipole interaction), the local magnetic field acting on each magnetic moment and consequently the Larmor frequency varies from site to site in the material. Then the phase of precession of each magnetic moment randomly distributes, as shown in Fig. 6(b). This effect results in decay of the x, y components of the averaged magnetic moment m_x and m_y, which finally becomes 0. The characteristic time of this decay is defined as T_2, and this is considered to be faster than T_1 because T_2 also includes the energy dissipation effect in addition to the dephasing effect of

precession. This effect modifies Eq. (3) as

$$\frac{dm_x}{dt} = \gamma m_y H_0 - \frac{m_x}{T_2}, \qquad \frac{dm_y}{dt} = \gamma m_x H_0 - \frac{m_y}{T_2} \tag{16}$$

Equations (15) and (16) are called the Bloch equations. From these equations, one can easily obtain

$$m_x + im_y \propto \exp\left(-i\omega_0 t - \frac{t}{T_2}\right)$$

$$= [\cos(\omega_0 t) - i\sin(\omega_0 t)]\exp\left(-\frac{1}{T_2}\right) \tag{17}$$

where $\omega_0 = \gamma H_0$ is used. T_2 is called the spin–spin relaxation time or simply "tee-two." This is the same as a damping oscillation. The general theory, which deals with the relaxation phenomenon more exactly, is difficult and more complicated. Because of these relaxation phenomena, the magnetic moment finally points to the z direction even if it starts to precess from some angle θ, and the magnitude of m is not kept constant during the precession because of the difference of T_1 and T_2. When an ac field $h_{rf}\,e^{i\omega t}$ is applied perpendicularly to the z direction, the averaged magnetic moment $\boldsymbol{m}$, which first points into the z direction starts to rotate at $\omega = \gamma H_0$ and θ increases. If $(\gamma h_{rf}\,T_1 T_2)^2 \ll 1$ (usually this condition is valid for ESR), not like Fig. 4(c), the Larmor precession becomes stationary, and the tilting angle θ is small. If $(\gamma h_{rf}\,T_1 T_2)^2 \gg 1$ (sometime this condition is valid for NMR), the Larmor precession is similar to Fig. 4(c). It is, however, complicated to solve the equation of motion precisely. It should be emphasized in this case that applied power of the ac field is absorbed at resonance condition by the spin system, and the absorbed power transmitted as lattice vibrations of the material via the relaxation mechanisms. This process results in temperature increase of the material. Resonance occurs not only exactly at $\omega = \gamma H_0$ but also at frequencies near $\omega = \gamma H_0$. The distribution of the resonant frequency, namely response intensity or spectrum of the characteristic oscillation as a function of ω, is obtained by the Fourier transformation of Eq. (17) as

$$1(\omega) \propto \frac{1}{(\omega - \omega_0)^2 + \left(\frac{1}{T_2}\right)^2} \tag{18}$$

and it shows a Lorentzian line shape with a half width of $\Delta\omega = 1/T_2$ (or full half width $2/T_2$) as shown in Fig. 7. By sweeping frequency of the electromagnetic wave and by observing the power dissipation in the material as a function of frequency, one can see that magnetic resonance with a line

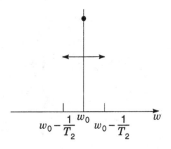

Figure 7. Spectrum of damping oscillation is Lorentzian.

Figure 8. Electrons in hydrogen molecule moving opposite in an orbit with magnetic moments caused by spins, as indicated by arrows.

shape of Fig. 7 occurs. The experimental method will be discussed later.

ELECTRON PARAMAGNETIC RESONANCE

To observe electron spin resonance, there must be isolated and independent electrons in the system. Namely, the materials we are considering must be composed of atoms or molecules that have magnetic moments. Every atom or molecule, however, does not necessarily have a magnetic moment. Electrons in an atom or in a molecule strongly couple with each other, and usually spin and orbital angular momenta are compensated by making pairs of electrons. For example, a hydrogen atom has only one electron, but it becomes a molecule when coupling with another hydrogen atom, and the angular moments of two electrons in a molecule are directly opposite as shown in Fig. 8. Then the molecule has no magnetic moment. The helium atom has two electrons whose spin and orbital angular moments are also compensated. In this manner, the net magnetic moment for most atoms and molecules disappears as a result of pairing by the strong intraatomic or intramolecular electron correlation that cancels spins and orbits. For example, in the case of iron group atoms (i.e., Sc, Ti, V, Cr, Mn, Fe, Co, Ni, and Cu in the periodic table), spin and orbital angular moments of electrons are not compensated because of Hund's rule, which was a result of the quantum mechanics. This means that these atoms have an unpaired electron and have magnetic moments. Figure 9 shows the example of Mn^{2+} and Cu^{2+}. For compounds that have these atoms as divalent or trivalent ions, the magnetic moments are isolated, and ESR can be observed. These ions are called paramagnetic ions and the ESR concerning these ions is called electron paramagnetic resonance (EPR). The atoms belonging to the palladium group, platinum group, and rare earth group also have similar properties. In these

ions, the orbital motion of electrons caused by orbital angular momentum creates a magnetic field acting on the magnetic moment caused by spin angular momentum or vise versa. So the orbital angular momentum and spin angular momentum are not completely independent. This effect is called spin-orbit coupling, and this interaction energy is of the order of 10^{-21} J. A paramagnetic ion in compounds is usually surrounded by negative ions called anions, which make a strong electric field called a crystalline field on the paramagnetic ion at the center. The energy of the crystalline field for an electron is on the order of 10^{-19} J. The electrons in the paramagnetic ion suffer electric fields from both the central nucleus and surrounding anions, and the motion of the electrons are no longer simple orbital motions. This effect gives rise to a reduction of orbital angular momentum, which is called quenching. Then the contribution of the orbital angular momentum to the magnetic moment is small and the magnetic moment μ per ion is usually expressed as

$$\mu = g\mu_B S \qquad (19)$$

where S is the total spin. This is, for example, $S = 5/2$ or $S = 1/2$ for our Mn^{2+} or Cu^{2+} ion, respectively, as easily is understood by Fig. 9. g is the so-called g-value, which reflects the effects of spin-orbit coupling and the crystalline field and is different from g_e depending on the material.

Paramagnetic ions in a crystal interact with each other by dipole interaction and exchange interaction. These interactions give rise to dephasing of Larmor precession, as mentioned previously, and result in the width of the absorption line as $1/T_2$. If the dipole interaction is dominant, the half width is approximately given by

$$\Delta\omega = 1/T_2 \cong \omega_d \qquad (20)$$

where ω_d is the sum of dipole interaction divided by $\hbar$. The sum is over all magnetic moments on the crystal lattice. If the exchange interaction is larger than the dipole interaction, the half width is approximately given by

$$\Delta\omega = 1/T_2 \cong \omega_d^2/\omega_e \qquad (21)$$

where ω_e is the nearest neighbor exchange interaction divided by $\hbar$.

The unpaired electron is also realized in organic materials as the free radical, in semiconductors as the donor or acceptor impurity, and in color centers as some special molecules like O_2 or NO. By studying EPR, one can obtain microscopic information about materials via g-value and line width.

HOW TO OBSERVE EPR (EXPERIMENTAL METHOD)

A simple method to observe EPR is described here. The equipment necessary for this experiment is an oscillator, a detector, a cavity, and a magnet. The Larmor frequency of electron spin is in the microwave region at an easily available magnetic field, namely $B = 0.1$ to 1 T, as discussed, see Eqs. (5) and (7). The most popular way is to use X-band microwaves whose wave length is around 3 cm because the size of microwave components is moderate, and the resonance field is about 0.3 T. A Gunn oscillator with detector is now commercially available and most convenient for this purpose. This

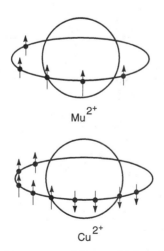

Figure 9. Electrons in 3d orbit in Mn^{2+} and Cu^{2+} ions.

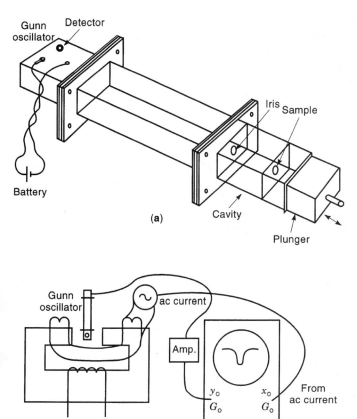

Figure 10. Experimental set-up to observe EPR.

cause of the strong exchange interaction among magnetic moments. The transition temperature T_c is called the Curie temperature. At a temperature sufficiently below T_c, the average magnetic moment m saturates and the total magnetic moment M of the specimen is given by

$$M = Ng\mu_{\mathrm{B}}S \qquad (23)$$

where N is the magnetic moment per unit volume in the specimen. The magnitude of M is comparable to the applied flux density $B_0 = \mu_0 H_0$, whereas in the case of paramagnetic materials, the averaged moment m is about 10^{-3} of the saturation moment at room temperature and at 1 T as a result of thermal fluctuation. Then the magnetic moments experience a field produced by themselves pointing in the opposite direction to the applied field. This field is called the demagnetizing field, and it must be taken into account when the equation of motion is solved. The field acting on the magnetic moment M including demagnetizing field is given for each direction as

$$H_x = -N_x M_x, \quad H_y = -N_y M_y, \quad H_z = H_0 - N_z M_z \qquad (24)$$

where Ns are the demagnetizing factor and $N_x + N_y + N_z = 1$ and H_0 is parallel to the z axis. From the equation of motion

$$\frac{d\boldsymbol{M}}{dt} = \gamma[\boldsymbol{M} \times \boldsymbol{H}] \qquad (25)$$

one can obtain the resonance conditions as

$$\frac{\omega}{\gamma} = \sqrt{\{H_0 + (N_y - N_z)M_z\}\{H_0 + (N_x - N_z)M_z\}} \qquad (26)$$

When the sample shape is spherical, $N_x = N_y = N_z = 1/3$, then

$$\frac{\omega}{\gamma} = H_0 \qquad (27)$$

In the case of a thin disk, $N_x = 1$, $N_y = N_z = 0$ or $N_x = N_y = 0$, $N_z = 1$ for an applied field parallel or perpendicular to the disk surface, respectively. In the case of a thin rod, $N_x = N_y = 1/2$, $N_z = 0$ or $N_x = 0$, $N_y = N_z = 1/2$ for an applied field parallel or perpendicular to the rod, respectively. To observe the absorption of ferromagnetic resonance, there must be energy dissipation caused by the relaxation mechanism. Different from the case of paramagnetic resonance, however, the magnetic moments are tightly bound to each other, and no dephasing effect in the xy plane is expected. So we cannot define T_2. Instead, the Landau–Lifshitz damping model is introduced for the ferromagnetic resonance. The equation of motion is described as

$$\frac{d\boldsymbol{M}}{dt} = \gamma[\boldsymbol{M} \times \boldsymbol{H}] - \lambda_{\mathrm{L}}[\boldsymbol{M} \times [\boldsymbol{M} \times \boldsymbol{H}]] \qquad (28)$$

where λ_{L} is the Landau–Lifshitz damping factor. This means that the direction of the second term is perpendicular to the direction of $\boldsymbol{M}$ in the plane made by $\boldsymbol{M}$ and $\boldsymbol{H}_0$. Then the motion of the total magnetic moment $\boldsymbol{M}$ is a damping oscillation

oscillator is commonly used for detection of speeding automobiles by policemen. Assemble the equipment as Fig. 10(a). The Gunn oscillator is connected with a short wave guide and terminated by a cavity. First, operate the Gunn oscillator and tune the cavity by moving a plunger so as to resonate at the oscillator frequency. A sample [A small amount of DPPH (α,α-diphenyl-β-pictyl hydrazyl) may be good as a test sample] is put in advance on the bottom of the cavity where the high-frequency magnetic field is strongest. Install the cavity between the poles of the magnet as shown in Fig. 10(b). A low-frequency ac (50 Hz or high) field of a few hundred microteslas must be superposed on the main dc field by modulation coils. Then sweep a magnetic field up to about 0.3 T. In this discussion, the absorption spectrum is given as a function of frequency at constant field. In the real experiment, however, changing frequency is so difficult that usually a magnetic field is swept at constant frequency. The detected signal through an amplifier is displayed on an oscilloscope as shown in Fig. 10(c). The resonance absorption line is obtained as a function of magnetic field and the half width of the resonance line must be converted by

$$\Delta H = \frac{\Delta\omega}{\gamma} \qquad (22)$$

FERROMAGNETIC RESONANCE

In the case of ferromagnetic materials, all magnetic moments point to same direction below a certain temperature T_c be-

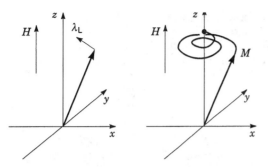

Figure 11. Direction of Landau–Lifshitz term and precession of ferromagnetic moment M.

as shown in Fig. 11. In this case, the length of M is kept constant, whereas in the case of paramagnetic resonance, the relaxation times T_1 and T_2 are independent, and the length of the averaged moment m is not constant during the motion. The magnitude of λ is related to the interaction between the motion of magnetic moments and lattice vibration and then the energy of the Larmor precession is transferred to the lattice vibration via this mechanism. The line width of FMR is expressed using this damping factor as

$$\Delta H = \frac{MH_0}{\gamma} \lambda_L \tag{29}$$

NUCLEAR MAGNETIC RESONANCE

In the case of nuclear spin, interactions with surrounding electrons, lattice vibration (phonon), other nuclear spins are weak, and a nuclear spin is considered to be almost isolated, which means that γ is almost constant and different from the

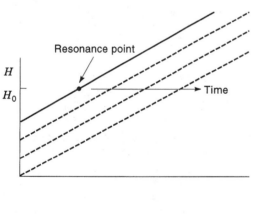

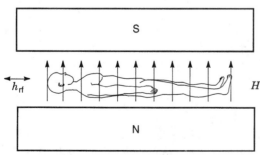

Figure 12. Schematic view of MRI.

case of EPR. The magnetic interactions with surrounding electrons can be replaced by the effective field H_{eft} which is included in Eqs. (1) and (5) as additional fields. Interactions also contribute to the change of relaxation time from that of free nucleus. Observing the shift of resonance frequency caused by this effective field and the change of relaxation time, one can obtain information about the microscopic behavior of materials. This is the reason why NMR is so useful as a probe to investigate properties of materials. Every nucleus does not necessarily have nuclear spin, and γ varies depending on the nucleus. All γ are listed in a standard table. NMR of copper nucleus is useful to investigate high T_c superconductors composed of copper oxide. Mn and Co nucleus are also important to study magnetic properties of materials composed of these atoms.

The most popular nucleus is the proton, which is the nucleus of hydrogen. All materials containing hydrogen show proton NMR. Water is the best example. To investigate molecular structures of organic molecules, polymers, proteins, and other biological materials, the proton NMR is useful and is now being used widely. In these cases, the absorption spectra of proton NMR have a complicated structure as a result interactions with neighboring atoms. By analyzing the structure of spectra, one can determine the molecular structure like neighboring atoms and distance. Because the resolution increases with increasing resonance frequency, high-field and high-frequency MNR is more useful, and now frequencies higher than 750 MHz are available in fields above 17 T by using high homogeneous superconducting magnets.

Magnetic resonance imaging (MRI) is well known as an important tool in finding tumors or other abnormal tissue in the human body. Every cell in organs contains hydrogen atoms and NMR is observable in any part of the body. But the shift or relaxation time varies depending on the organ. As shown in Fig. 12, a body is placed between the poles of a big magnet and h_{rf} is applied to it. The magnetic field has a gradient with respect to the position of the body, and the NMR is observed at only one point on the body. This gradient field is scanned, and the resonance point moves from head to foot. By analyzing the data by a computer, one can see the structure of the body. If the organ is abnormal, the density and relaxation times of NMR at the affected part are different from those at a normal part. This allows NMR to be used for diagnosis.

BIBLIOGRAPHY

1. A. Abragam, *The Principle of Nuclear Magnetism,* New York: Oxford, 1961.

2. G. E. Pake, *Paramagnetic Resonance.* New York: W. A. Benjamin, 1962.

3. C. P. Slichter, *Principle of Magnetic Resonance.* New York: Harper & Row, 1963.

MITSUHIRO MOTOKAWA
Tohoku University

MAGNETIC RESONANCE. See DIAGNOSTIC IMAGING.

MAGNETIC SEMICONDUCTORS

Magnetic semiconductors possess pronounced magnetic properties in addition to the traditional electronic and optical properties common to all semiconductors. Magnetic semiconductors contain transition-metal or rare-earth magnetic ions (such as iron, manganese, europium, or other magnetic ions) as at least one of their constituents. The presence of the magnetic ions enhances the magnetic susceptibility of the materials relative to their nonmagnetic counterparts, and in some cases these semiconductors can exhibit spontaneous magnetization in the absence of an externally applied magnetic field. The electronic and optical properties of magnetic semiconductors are usually more sensitive to magnetic fields than their nonmagnetic counterparts, displaying large magnetoresistance and magnetooptic effects in many cases. The optoelectronic properties can also show unusual physical behavior even in the absence of applied fields, such as large changes in the optical absorption or conductivity caused by changes in temperature, attributed to the effects of magnetic ordering.

Within the class of magnetic semiconductors are two principal subclasses of materials. These are the concentrated magnetic semiconductors (1) and the diluted magnetic semiconductors (2). In concentrated magnetic semiconductors the magnetic ion fills a sublattice of the crystal, as in the ferromagnetic binary compound semiconductor EuO. Diluted magnetic semiconductors can have continuously variable concentrations of magnetic ions, as in the alloy $Cd_{1-x}Mn_xTe$, for which the manganese concentration (denoted by x) can vary from the concentrated antiferromagnet MnTe to the nonmagnetic chalcogenide CdTe. A recent addition as a third subclass of magnetic semiconductors are heterogeneous magnetic semiconductors, in which magnetic precipitates are dispersed inside an otherwise nonmagnetic semiconductor host (3).

Magnetic semiconductors can exhibit ferromagnetic, antiferromagnetic, paramagnetic, or spin-glass behavior depending on their chemical constituents and concentrations and on the temperature. The diluted magnetic semiconductors may exhibit nearly all these properties within a single ternary compound, such as CdMnTe, by continuously tuning the magnetic ion concentration from the concentrated to the extremely diluted limit. For this reason, the diluted magnetic semiconductors have been the subject of extensive research.

MAGNETISM IN SEMICONDUCTORS

The properties of magnetic semiconductors are governed by the interactions of the magnetic ions with each other, called the ion–ion exchange interaction, and by the interactions of the free-carriers with the magnetic ions, called the carrier–ion exchange interaction. The ion–ion interaction governs the magnetic susceptibilities and the magnetic ordering of the magnetic semiconductors. The carrier–ion interaction governs the magnetotransport and magnetooptical properties that are unique to magnetic semiconductors.

Magnetization and Susceptibility

The strongest ferromagnetic behavior in semiconductors is observed among the concentrated magnetic semiconductors. Notable semiconductor ferromagnets are the chalcogenides of europium EuO and EuS, and the chromium chalcogenides

$CdCr_2S_4$ and $CdCr_2Se_4$. The low-field static susceptibilities of these materials exhibit Curie–Weiss behavior at temperatures above the magnetic transition temperature

$$\chi = \frac{C}{T - T_c} \tag{1}$$

where C is the Curie constant and T_c is the Curie temperature. Below the phase transition to the ferromagnetic phase, these materials exhibit a spontaneous magnetization $M(T)$ that increases with decreasing temperature.

Antiferromagnetic behavior is common among other concentrated magnetic semiconductors, such as the other chalcogenides of europium EuTe and EuSe. The concentrated manganese chalcogenides MnO, MnS, MnSe, and MnTe are all antiferromagnetic, as are the diluted magnetic semiconductors derived from these. The static low-field susceptibility at high temperature obeys a Curie-Weiss law

$$\chi = \frac{C}{T - \theta} \tag{2}$$

with a negative Curie temperature θ. The susceptibility deviates from the Curie-Weiss law as the temperature approaches the Neel temperature T_N at which the ions order antiferromagnetically.

Extremely diluted magnetic semiconductors are approximately described as a paramagnetic phase for which the magnetic moments are uncoupled and are aligned by an externally applied magnetic field. The low-field high-temperature static magnetic susceptibility in this case takes on the simple Curie form.

In many of the diluted magnetic semiconductors a cusp is observed in the static susceptibility at a characteristic temperature T_g that is a function of the composition. This cusp in the susceptibility is accompanied by irreversible thermoremanence effects in the magnetization (4). These magnetic signatures may signal a phase transition from a paramagnetic phase to a low-temperature spin-glass phase. A spin glass consists of a configuration of spins that are frozen in random orientations. The spin glass exhibits no long-range order and is often caused by frustration of antiferromagnetic order.

Superparamagnetic behavior is characterized by a high-temperature static low-field susceptibility of the Curie form, but also exhibits irreversible thermoremanence in the magnetization. Superparamagnetism is caused by clusters of aligned spins. The macroscopic spin of the cluster is aligned by an applied magnetic field, contributing to the paramagnetic susceptibility. Superparamagnetism can occur in the case of ferromagnetic precipitates in nonmagnetic semiconductors (5). Although the irreversibility shows some similarity to spin-glass behavior, it can be described classically in terms of a spin-blocking temperature in which the individual magnetic fields of the isolated clusters mutually influence the alignment of the macroscopic moments of the other clusters.

Ion–Ion Exchange Interactions

Magnetic order in the absence of an applied magnetic field arises because of exchange interactions among the magnetic ions in the semiconductor. The simplest interaction Hamiltonian can be expressed as

$$H = -\frac{1}{2} \sum_{i,j} J(\mathbf{r}_i - \mathbf{r}_j)(\mathbf{S}_i \cdot \mathbf{S}_j) \qquad (3)$$

where $\mathbf{S}_i$ is the spin of the ith magnetic ion, and $J(\mathbf{r}_i - \mathbf{r}_j)$ is the exchange integral between the spins on the ith and jth ions. For most magnetic semiconductors, the exchange occurs between pairs of magnetic ions that are too far apart for the direct exchange interaction to play a significant role. Therefore, the ions interact through the superexchange mechanism. Superexchange is an indirect exchange mechanism in which an intervening nonmagnetic ion (usually an anion) couples two magnetic ions. Superexchange most often produces antiferromagnetic coupling, but can also produce ferromagnetic ordering in some of the concentrated magnetic semiconductors.

Indirect exchange between magnetic ions, mediated through free carriers, can become important in degenerately doped semiconductors with high free-carrier concentrations. This interaction can lead to ferromagnetic ordering because the electron energy is lowered by spontaneous magnetization.

The superexchange and indirect exchange interactions lead to magnetic ordering of the magnetic ions at sufficiently low temperatures. Both ferromagnetic and antiferromagnetic order occur among the concentrated magnetic semiconductors, whereas the order in the diluted magnetic semiconductors is almost exclusively antiferromagnetic. Most transition temperatures are below room temperature, with exceptions observed for MnTe (323 K), NiO (520 K), and $CuFeS_2$ (825 K) among the antiferromagnets. The ferromagnet $CuCr_2Te_3I$ has a reported Curie temperature of 294 K (6).

Carrier–Ion Exchange Interactions

An important aspect of magnetic semiconductors is the effect of the magnetic ions on the electronic and optical properties of the free carriers. Although free carrier densities in nondegenerate semiconductors are too small for the electrons to effectively mediate indirect exchange between the magnetic ions, the free carriers themselves interact with the localized magnetic moments of the magnetic ions. The carrier exchange Hamiltonian takes on a form similar to Eq. (3):

$$H_{\text{ex}} = -\sum_{R_i} J(\mathbf{r} - R_i)(\mathbf{S}_i \cdot \boldsymbol{\sigma}) \qquad (4)$$

where R_i is the location of the ith magnetic ion, $\mathbf{r}$ is the carrier position, and σ is the carrier spin. The carrier–ion exchange is a local interaction that depends on short-range magnetic order rather than on long-range order. This fact is most dramatically illustrated by the giant red shift in the absorption edge of ferromagnetic semiconductors even when the material is in the paramagnetic phase. The electron energy is reduced by the interaction with finite ferromagnetic clusters even though the macroscopic magnetization vanishes.

The localized moments of magnetic ions arise from localized d-shell or f-shell electrons, whereas the free-carrier states arise from s-shell and p-shell valence electrons. The carrier–ion exchange interaction is therefore often referred to as the sp-d or sp-f exchange mechanism. Two common contributions to the sp-d or sp-f exchange factor $J(\mathbf{r} - R_i)$ are a

direct exchange interaction between the magnetic ion and the carrier and hybridization of the carrier wavefunction with the localized wavefunction. The direct exchange interaction is electrostatic in origin and leads to ferromagnetic interaction between the carrier and the ion. The contribution of hybridization to the exchange interaction is predominantly antiferromagnetic. It arises from the mixing of the sp wavefunctions of the anions with the localized d or f wavefunctions of the magnetic cations. Depending on the symmetries of the bands and of the localized wavefunctions for specific cases, either contribution can dominate.

Concentrated Magnetic Semiconductors

A distinction is made between those magnetic semiconductors that contain magnetic ions that fill a sublattice of the material, and those in which the magnetic sublattice is diluted by nonmagnetic ions. The former materials are called concentrated magnetic semiconductors and the latter are called diluted magnetic semiconductors.

Concentrated magnetic semiconductors share many of the optical and electronic properties in common with more traditional semiconductors, such silicon and germanium among the group IV elements, or GaAs and related III–V compounds. However, the magnetic semiconductors have significantly different crystal structure and carrier mobilities than the traditional semiconductors. Their magnetic properties and the effects of the magnetic structure on their electronic and optical structure have been studied extensively, although their development for electronic and optical applications remains at an early stage. The concentrated magnetic semiconductors are divided into two categories depending on whether they are antiferromagnets or ferromagnets at low temperature.

Antiferromagnetic Semiconductors

Many of the concentrated antiferromagnetic semiconductors have magnetic ions distributed on face-center cubic sublattices in the NaCl crystal structure. For this crystal structure, the three antiferromagnetic orderings that are most commonly observed are shown in Fig. 1. In type I antiferromagne-

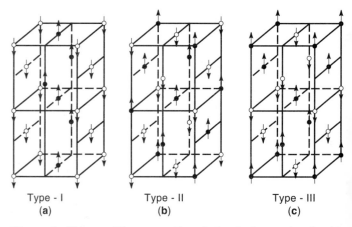

Type - I	Type - II	Type - III
(a)	(b)	(c)

Figure 1. Three antiferromagnetic ordering in face-centered cubic magnetic sublattices, showing (a) type I with tetragonal symmetry, (b) type II with trigonal symmetry, and (c) type III with tetragonal symmetry.

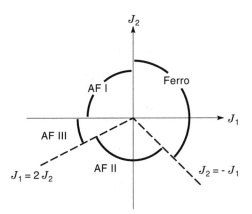

Figure 2. Magnetic phase diagram of the antiferromagnetic order depending on the nearest- and next-nearest-neighbor exchange integrals J_1 and J_2.

tism the magnetic ions form alternating planes of parallel spins perpendicular to a principal axis along $\langle 100 \rangle$, taking on tetragonal symmetry. In type II structures, which are the most common, the alternating planes of parallel spin occur orthogonally to the body diagonals along $\langle 111 \rangle$, taking on trigonal symmetry. In type III structures, the crystal structure including spin is again tetragonal with like spins distributed on a chalcopyrite sublattice. The stability of the different types of antiferromagnetic order are determined by the relative magnitudes and signs of the exchange constants J_1 and J_2 for nearest-neighbor and next-nearest-neighbor exchange interactions, respectively. The phase diagram is shown in Fig. 2 in the J_1 and J_2 plane for face-centered-cubic structures. Ferromagnetic order only occurs for positive nearest-neighbor exchange interaction and $J_1 > J_2$.

One of the most interesting and most exhaustively studied antiferromagnetic semiconductors is the rare-earth chalcogenide EuSe. In the absence of an applied magnetic field it changes from a paramagnet with decreasing temperature to an antiferromagnet at a Neel temperature of 4.6 K. However, it has a positive Curie temperature in the paramagnetic phase, indicating weak short-range ferromagnetic order. Upon the application of a magnetic field, the order changes from antiferromagnetic to ferrimagnet to ferromagnetic. This field-induced change in magnetic phase to ferromagnetic order is an example of a metamagnet. Other chalcogenides of europium, such as EuO and EuS are fully ferromagnetic. The magnetic semiconductor EuSe therefore rests on the border between antiferromagnetism and ferromagnetism. The remaining chalcogenide of europium, EuTe, behaves as an ordinary antiferromagnet with a Neel temperature of 9.58 K and a negative Curie temperature of -6 K. The exchange constants for the europium chalcogenides are given in Table 1

Table 1. Exchange Constants and Transition Temperatures for the Europium Chalcogenides

	θ (K)	$T_{c,N}$ (K)	J_1 (K)	J_2 (K)
EuO	79	69	0.606	0.119
EuS	18.7	16.6	0.228	$-0.1.2$
EuSe	8.5	4.6	0.073	-0.011
EuTe	-4.0	9.6	0.043	-0.15

From Wachter (7).

Table 2. Selected Antiferromagnetic Semiconductors

Material	T_N (K)	θ (K)
EuTe	9.58	-6
EuSe	4.6	9
Eu_3O_4	5.3	7
MnO	122	-610
α-MnS	154	-465
MnSe	173	-360
MnTe	323	-715
$MnTe_2$	80	-520
Gd_2Se_3	6	-10
NiO	520	-2600
CoO	291	-320
$LaMnO_3$	100	-500
$CuFeS_2$	825	
$CoCl_2$	25	-37
$NiCl_2$	50	-75
$ZnCr_2S_4$	18	18
$HgCr_2S_4$	60	137–142
$ZnCr_2Se_4$	20	115

From Nagev (1).

(7). The Neel and Curie temperatures for selected antiferromagnetic semiconductors are given in Table 2 (1).

Among the antiferromagnet chalcogenides are several oxides such as NiO and MnO. Many of the simple chalcogenides have NaCl crystal structures with the magnetic ions distributed on a face-centered cubic sublattice with type II antiferromagnetic order. The manganese chalcogenides are antiferromagnetic and constitute an important class of concentrated magnetic semiconductors because they represent the endpoints of the Mn-based diluted magnetic semiconductors, which are also antiferromagnetic.

Ferromagnetic Semiconductors

Ferromagnetic semiconductors have special properties not shared by the antiferromagnetic semiconductors, because the strength of the carrier–ion exchange depends on local order of the magnetic ion sublattice, which is opposite for the two cases. In ferromagnetic order, the magnetic moments align locally so that the contributions from each moment accumulates to produce large changes in the properties of the free carrier. For instance, the band-edge in ferromagnetic semiconductors experiences a pronounced red shift with decreasing temperature and shows the onset of spontaneous Faraday rotation below the Curie temperature. In addition, the sample resistivity shows a pronounced peak near the Curie temperature, and decreases significantly with increasing applied magnetic field. These effects are a consequence of the exchange interaction of the free carriers with the aligned moments of the magnetic ions.

Ferromagnetic semiconductors are not as ubiquitous as antiferromagnetic semiconductors and were discovered later. The first ferromagnetic semiconductor discovered was CrBr3 in 1960 (8), followed shortly by EuO and EuS (9). The Curie temperatures of these europium chalcogenides are 67 K and 16 K, respectively. More important for potential applications is the chromium spinel $CuCr_2Te_3I$, which has a Curie temperature near room temperature at 294 K (6). The Curie temper-

Table 3. Selected Ferromagnetic Semiconductors

Material	T_c (K)	Cell Moment (μ_B)
$CrBr_3$	37	3.85
EuO	66.8	6.8
EuS	16.3	6.87
EuB_6	8	
Eu_3P_2	25	6.8
Eu_3As_2	18	7.03
Eu_2SiO_4	5.4	
Eu_3SiO_5	9	7
Eu_3S_4	3.8	
Eu_4P_2S	24	
Eu_4P_2Se	22.5	
Eu_4As_2S	20	
$EuLiH_3$	38	
$CdCr_2S_4$	84.5–97	5.15–5.55
$CdCr_2Se_4$	130–142	5.4–6
$HgCr_2Se_4$	106–120	5.4–5.64
$CuCr_2Se_3Br$	274	5.25
$CuCr_2Te_3I$	294	4.10
$(CH_3NH_3)_2CuCl_2$	8.9	1
$KCrCl_4$	50–60	

From Nagev (1).

atures for selected ferromagnetic semiconductors are given in Table 3.

The giant red shift of the absorption edge with decreasing temperature is one of the consequences of the magnetic order in ferromagnetic semiconductors and is an effect that is unique to the magnetic semiconductors with no analog among ferromagnetic metals. This shift begins even in the paramagnetic phase in response to the formation of finite clusters of ferromagnetic spin alignment with which the free carriers interact through the sp-d or sp-f exchange mechanism. The red shift accelerates and is strongest near the Curie temperature, but continues for decreasing temperatures. The change of the energy gap as a function of temperature is shown in Fig. 3 for EuO and the spinel $HgCr_2Se_4$ (10,11). The red shift ceases when the temperature approaches $T = 0$ as the magnetization saturates. More complicated temperature dependences can also occur, including a blue shift in some cases, such as in $CdCr_2Se_4$.

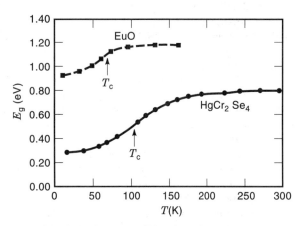

Figure 3. Red shift of the optical bandgap for EuO and $HgCr_2Se_4$ as a function of temperature (9,10). The Curie temperatures are indicated.

The resistance in nondegenerate n-type ferromagnetic semiconductors exhibits a strong peak at the Curie temperature. This effect is a consequence of the exchange-induced red shift of the conduction band-edge. For increasing temperatures below T_c, the conduction band-edge increases rapidly with respect to the Fermi level (pinned by extrinsic impurities), causing a decrease in the free-electron density. At temperatures larger than T_c, the thermal excitation of carriers from the impurity levels causes the usual increase in the carrier density that is observed for ordinary semiconductors.

Whereas nonmagnetic semiconductors usually have positive transverse-field magnetoresistance, ferromagnetic semiconductors show nearly isotropic negative magnetoresistance in which the resistance decreases under the application of a magnetic field. The drop in resistance has two origins: the reduction in magnetic fluctuations, and the change in carrier concentration with magnetic field. In the absence of a magnetic field, the local magnetization both above and below T_c fluctuates, causing carrier spin-scattering off the random orientations and magnitudes of the fluctuating moments. An applied magnetic field tends to orient the local moments, reducing the fluctuations and the scattering and increasing the conductivity of the material by increasing the carrier mobility. An applied magnetic field also shifts the conduction band-edge to lower energies closer to the impurity levels. This increases the conductivity of the material by increasing the carrier concentration. The magnitude of the magnetoresistance is largest for temperatures near T_c, where both the fluctuations and the shift of the band-edge are most sensitive to the applied field.

Faraday rotation of linearly polarized light in ferromagnetic semiconductors can attain record values under applied magnetic fields or at low temperatures. In EuS at 8 K in a field of 12 kOe the rotation can be as large as 10^6 deg/cm for some wavelengths of light (12). These values are larger than the largest values for Faraday rotation discovered in magnetic metals. Ferromagnetic semiconductors exhibit spontaneous Faraday rotation even in the absence of an applied magnetic field for temperatures below the Curie temperature, because the Faraday effect responds to the average macroscopic magnetization. The Faraday quality factor Q defines the rotation per decibel of light attenuation. In metals, because of the strong light attenuation at all wavelengths, the quality factor is typically less than 0.1 deg/dB. In the ferromagnetic semiconductors, because of the infrared transparency, the quality factor can be larger than 10^4 deg/dB, making these materials candidates for magnetooptical uses.

DILUTED MAGNETIC SEMICONDUCTORS

Diluted magnetic semiconductors are derived from the concentrated magnetic semiconductors by diluting the cation sublattice with nonmagnetic ions. This dilution process introduces alloy disorder within the sublattice, affecting the interactions by removing nearest- and farther-neighbor exchange interactions with increasing dilution. The dilution therefore changes the magnetic properties of the materials, such as magnetization and the transition temperatures. The dilution also often causes a change in crystal symmetry with important consequences for optical and electronic properties, as well as for the magnetic order.

Diluted magnetic semiconductors are important because the dilution process makes it possible to tune the magnetic properties of these materials to test theories of magnetic mechanisms in semiconductors. They are also important because they represent a means of introducing a magnetic dimension to technologically important nonmagnetic semiconductors. In the extreme dilute limit, diluted magnetic semiconductors may be viewed as being based on a nonmagnetic host to which magnetic cations are added. These semiconductors are sometimes referred to as semimagnetic semiconductors.

II–VI Based Diluted Magnetic Semiconductors

The best-studied examples of diluted magnetic semiconductors are the group II–VI semiconductors, in which a fraction x of the group II sublattice sites are replaced by magnetic ions. This class of diluted magnetic semiconductor is technologically important because the II–VI semiconductors are one of the classical semiconductor groups that are candidates for electronic and optoelectronic applications.

The structural properties of diluted magnetic semiconductors are controlled by the concentration of magnetic ions. The nonmagnetic II–VI semiconductors have tetrahedral coordination based on the sp^3 hybrid bond. They take on cubic zinc-blende or hexagonal wurzite structure depending on their composition. In the cubic phase, the cations form a face-centered cubic sublattice. The most common diluted magnetic form of the II–VI semiconductors is based on the Mn^{2+} ion that has a half-filled d-shell with a total spin $S = 5/2$. Other magnetic cations are also used, such as Cr^{2+} ($S = 4\theta2$), Fe^{2+} ($S = 4/2$) and Co^{2+} ($S = 3/2$). These diluted magnetic semiconductors have the chemical formula $C_{1-x}M_xA$, where C refers to cation, M refers to magnetic ion, and A refers to anion. Examples are $Cd_{1-x}Mn_xTe$ and $Zn_{1-x}Mn_xSe$ in which the fraction of cations replaced by magnetic cation is given by x.

Not all values of x lead to homogeneous crystal structures. For each of the original II–VI semiconductor hosts, there is a maximum value of x that is permissible for bulk growth. In some cases, the crystal structure changes from cubic to hexagonal with increasing x, up to a maximum allowable concentration before the hexagonal phase is no longer stable. This parameter space is shown for the manganese-based compound semiconductors in Fig. 4. It should be pointed out that these conditions are relaxed for nonequilibrium epitaxial growth.

In the extreme dilute limit, the diluted magnetic semiconductors are paramagnetic and the low-field and high-temperature magnetic susceptibilities exhibit Curie law dependence. The low-field susceptibility takes on the Curie form where the Curie constant is

$$C_0 = \frac{N_0(g_{\mathrm{ion}}\mu_{\mathrm{B}})^2 S(S+1)}{3k_{\mathrm{B}}} x \qquad (5)$$

where S is the spin of the magnetic ion, g_{ion} is the g factor of the magnetic ion, and N_0 is the number density of cation sites. The Curie constant depends linearly on x in the dilute limit. In the paramagnetic phase in the range of higher concentration and higher temperatures, the low-field susceptibility converts to a Curie–Weiss form where the Curie temperature is

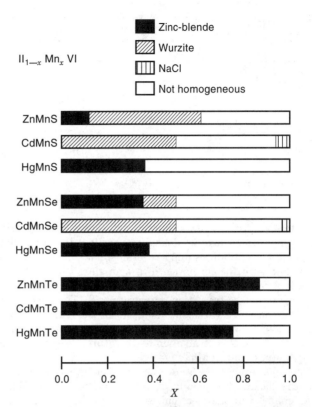

Figure 4. Composition ranges and crystal structures of the Mn-based group II–VI diluted magnetic semiconductors showing the range of homogeneous zinc-blende and wurzite structures. MnTe has a hexagonal NiAs crystal structure, whereas MnS and MnSe both have NaCl structures.

given by

$$\theta_0 = -\frac{2}{3}S(S+1)Z\frac{J}{k_{\mathrm{B}}}x \qquad (6)$$

which depends linearly on magnetic ion concentration in the dilute limit, and where J is the nearest-neighbor exchange constant, and Z is the number of nearest-neighbor cation sites ($Z = 12$ in zinc-blende and wurzite structures). The Curie temperature is negative for all the manganese-based II–VI diluted magnetic semiconductors, demonstrating that the exchange interaction is antiferromagnetic in this class of magnetic semiconductors. Measurements of the Curie temperature in principle provide a measure of the exchange interaction strength J, although other techniques give more accurate values, such as high-field magnetization, and neutron and Raman scattering. Values for the exchange constants for nearest-neighbor manganese–manganese pairs are given in Table 4 (13) for the manganese-based diluted magnetic semiconductors.

In the general case of arbitrary concentration and temperature, the magnetization can be expressed through the phenomenological expression

$$M = x_{\mathrm{eff}}N_0 g_{\mathrm{ion}}\mu_{\mathrm{B}}SB_S\frac{g_{\mathrm{ion}}\mu_{\mathrm{B}}SH}{k_{\mathrm{B}}T_{\mathrm{eff}}} \qquad (7)$$

where $B_s(y)$ is the Brillouin function and x_{eff} and T_{eff} are fittable parameters.

Table 4. Exchange Constants for Nearest-Neighbor Mn–Mn Pairs in Mn-Based Diluted Magnetic Semiconductors

Alloy	J (K)
ZnMnS	−16.1
ZnMnSe	−12.7
ZnMnTe	−9.7
CdMnS	−10.6
CdMnSe	−8.1
CdMnTe	−6.6
HgMnSe	−10.9
HgMnTe	−7.2

From Furdyna (13).

At temperatures below a characteristic value T_g that varies with composition the diluted magnetic semiconductors exhibit magnetic susceptibilities and irreversible phenomena that are consistent with the formation of a spin-glass phase. A spin glass is a random arrangement of spins that become frozen-in below the glass transition temperature T_g. The glass transition temperature varies between 0 K for the extreme dilute limit to 40 K for high concentration of magnetic ions. The variation of the glass transition temperature is shown in Fig. 5 for ZnMnTe and CdMnTe (14).

The sp-d exchange interaction between the free carriers and the localized moments of the magnetic ions produces large changes in the electronic energies of the free carriers under applied magnetic fields. The exchange interaction in Eq. (4) can be rewritten as

$$H_{ex} = \sigma_z \langle S_z \rangle x \sum_R J^{sp-d}(\mathbf{r} - R) \tag{8}$$

where S_i is replaced by the thermal average $\langle S_z \rangle$ for a field applied along the z axis. In addition, by explicitly including the magnetic ion concentration x, the sum is carried out over the full cation sublattice. With these approximations, the electron energies of the lth Landau level are

$$E_{l\uparrow} = E_g + \left(l + \frac{1}{2}\right)\hbar\omega_c + \frac{1}{2}\left(g^*\mu_B H - N_0\alpha x \langle S_z \rangle\right)$$
$$E_{l\downarrow} = E_g + \left(l + \frac{1}{2}\right)\hbar\omega_c - \frac{1}{2}\left(g^*\mu_B H - N_0\alpha x \langle S_z \rangle\right) \tag{9}$$

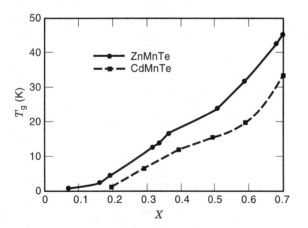

Figure 5. The spin-glass transition temperatures T_g as a function of composition for ZnMnTe and CdMnTe. Compiled in Furdyna and Samarth (14).

where E_g is the bandgap energy, ω_c is the cyclotron frequency, and g^* is the effective g factor of the conduction band electrons. The spin index refers to the electron spin orientation relative to the magnetic field. The exchange integral α for the s-like electrons of the conduction band is given by

$$\alpha = \langle S|J^{sp-d}|S \rangle \tag{10}$$

The electron energies can be expressed in terms of an effective g factor through

$$E_{l\pm} = E_g + \left(l + \frac{1}{2}\right)\hbar\omega_c \pm \frac{1}{2}g_{eff}\mu_B H, \tag{11}$$

where the effective g factor is

$$g_{eff} = g^* + \frac{\alpha M}{g_{ion}\mu_B^2 H} \tag{12}$$

which depends explicitly on the magnetization M. The effective g factor is temperature and concentration dependent, as well as field dependent, although in low-field and high-temperature situations, the ratio $M/H = \chi$ holds and is field-independent. The exchange contribution to the effective g-factor can exceed the band contribution by one to two orders of magnitude. The Zeeman splitting of the conduction band is therefore extremely large in the diluted magnetic semiconductors compared with the nonmagnetic materials.

The Zeeman splitting of the valence band is more complicated because it has Γ_8 symmetry with fourfold degeneracy. However, in this case it is possible to define an effective Luttinger parameter as

$$g_{eff} = \kappa - \frac{N_0\beta x \langle S_z \rangle}{6\mu_B H} \tag{13}$$

where β is the exchange integral

$$\beta = \langle XYZ|J^{sp-d}|XYZ \rangle \tag{14}$$

Because of the p-symmetry of the valence band states, there is strong hybridization between the anion-like valence band and the d electron states of the magnetic cation, while the s symmetry of the conduction band states prevent significant hybridization with the localized d states. Therefore the β exchange constant is larger than α in the diluted magnetic semiconductors, producing larger Zeeman effects at the valence band edge. This result is opposite to the situation in the concentrated magnetic semiconductors in which the cationlike conduction band edge interacts most strongly with the magnetic cation.

The schematic splittings of the conduction and valence band edges are shown in Fig. 6 for wide-gap diluted magnetic semiconductors in which the exchange effects dominate over Landau and band g factor effects. The transitions are shown for specific light polarizations. For light propagating parallel to the magnetic field in the Faraday geometry the two eigenstates are right and left circularly polarized light, denoted by σ_+ and σ_-, respectively. For light propagating perpendicular to the magnetic field in the Voigt geometry, the eigenmodes are linearly polarized light denoted by π. The field dependence of the exciton transitions in ZnMnTe ($x = 0.05$) is shown in

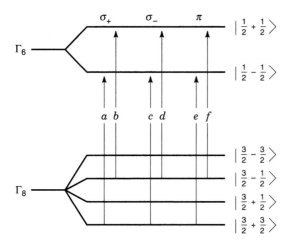

Figure 6. The symmetries and splittings of the direct interband Γ_8 to Γ_6 transitions for wide-gap diluted magnetic semiconductors. The allowed transitions for right and left circular polarization and linear polarization are shown.

Fig. 7 for circularly polarized light (15). Splittings as large as 100 meV in fields of 8 T occur at a temperature of $T = 1.4$ K in this wide-gap semiconductor. In contrast to the wide-gap diluted magnetic semiconductors, in narrow-gap or extremely dilute wide-gap diluted magnetic semiconductors the relative contributions of the exchange and band contributions to the spin splittings become comparable and the spin-state ordering changes sign.

The Faraday effect is the rotation of linearly polarized light that propagates parallel to an applied magnetic field. The rotation can be extremely large in the diluted magnetic semiconductors, and has been referred to as the Giant Faraday Effect. The rotation angle θ_F per crystal length L is given by

$$\frac{\theta_F}{L} = \frac{\sqrt{F_0}}{2\hbar c} \frac{\beta - \alpha}{g_{ion}\mu_B} M \frac{\hbar^2\omega^2}{(E_g^2 - \hbar^2\omega^2)^{3/2}} \quad (15)$$

where F_0 is a constant (16). The rotation per unit length per unit field strength is expressed through the Verdet constant.

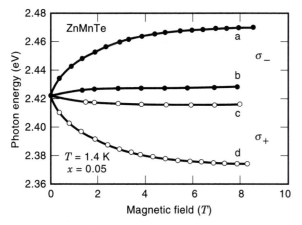

Figure 7. Energies of the a, b, c, and d transitions of Fig. 6 for ZnMnTe ($x = 0.05$) at 1.4 K in a magnetic field. From Aggarwall et al. (15).

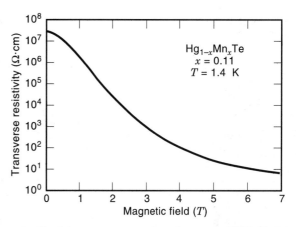

Figure 8. Giant transverse magnetoresistance in HgMnTe for $x = 0.11$ at $T = 1.4$ K. From Wojtowicz and Mycielski (17).

The Verdet constant for $Cd_{0.998}Mn_{0.002}Te$ can be as large as 3000 deg/cm·T for photon energies near the bandgap at a temperature near of 4 K (16).

The diluted magnetic semiconductors exhibit interesting magneto transport effects in addition to their magnetooptical effects. Narrow-gap diluted magnetic semiconductors, such as p-type HgMnTe, have giant negative magnetoresistance at low temperature. The transverse magnetoresistivity in $Hg_{0.89}Mn_{0.11}Te$ is shown in Fig. 8 at a temperature of $T = 1.4$ K (17). The resistivity decreases from 10^7 Ω·cm in a zero-field to 10 Ω·cm at 7 T. The giant magnetoresistance makes these materials candidates for use as recording heads for magnetic data memory. The increase in conductivity in an applied magnetic field is caused by the decrease of the acceptor binding energy combined with field-dependent changes in the defect wavefunction that produce anisotropic conductivity. Wider gap n-type CdMnSe also shows a negative magnetoresistance that has been attributed to magnetic polaron effects of the electrons.

A new direction in the optical studies of diluted magnetic semiconductors is the combination of the magnetooptic properties with photorefractive nonlinear optical effects. The interplay of the exciton-enhanced Faraday rotation with the polarization dependence of the electrooptic tensor of these materials produces interesting magnetic field dependence for photorefractive two-beam coupling (18). A more fundamental phenomenon in this respect is the connection between time-reversal symmetry and the quenching of phase-conjugate light by the application of a magnetic field (19).

MAGNETIC SEMICONDUCTOR HETEROSTRUCTURES

Semiconductors can be grown using epitaxial techniques such as by molecular beam epitaxy. This growth process opens up many possibilities for material growth and engineering. Molecular epitaxy is a layer-by-layer process that can provide layer-by-layer control to the crystal grower, making it possible to grow multiple layers of differing materials with monolayer accuracy. The diluted magnetic semiconductors can be grown by this technique, and interleaved between other magnetic layers or between nonmagnetic layers. The permutations of materials, layers, thicknesses and compositions in a single heterostructure produces a nearly limitless variety of interesting processes.

The choice of substrate for epitaxial growth is an important feature of epitaxy of the diluted magnetic semiconductors. The choice is based on several factors, including cost and availability of the substrate material, as well as on the substrate lattice constant. For this reason, many epitaxial diluted magnetic semiconductor heterostructures are grown on GaAs substrates. Gallium arsenide is not ideally lattice-matched to any of the diluted magnetic semiconductors. The closest lattice match is for the alloy ZnMnSe, but even in this case a thick buffer layer is grown first on the GaAs to allow the lattice to relax to the correct value before the heterostructure is grown. Many of the strain relaxation defects are confined to the buffer layer, allowing relatively homogeneous growth of the heterostructures. However, many threading dislocations propagate up through the heterostructures, causing potential problems for applications.

Using lattice-mismatch epitaxy, forms of the concentrated magnetic semiconductors have been grown that do not exist in nature. For instance, natural MnSe occurs in the NaCl lattice structure, while $Zn_{1-x}Mn_xSe$ occurs in a zinc-blende structure only for x up to 35%, beyond which it transforms into a hexagonal phase which is limited to values $35\% < x < 50\%$. However, when grown epitaxially on a suitable substrate, thin films of MnSe have been grown in the zinc-blende phase (20).

An interesting feature of thin epitaxial layers of concentrated and diluted magnetic semiconductors is the absence of magnetic order. Bulk materials always exhibit antiferromagnetic order below a critical temperature, which is accompanied by a cusp in the linear magnetic susceptibility. In thin films of diluted magnetic semiconductors, on the other hand, the magnetic order is missing, and the magnetic properties are found to be paramagnetic (21).

A topic of keen interest in diluted magnetic semiconductors was the relatively large bandgap of the manganese-based zinc-blende compounds. The large bandgap makes these materials potential candidates for technologically important blue–green light emitting diodes and lasers. Optically pumped stimulated emission and laser oscillations in the blue–green spectral range were first observed in ZnSe/ZnMnSe multiple quantum wells operating at low temperature (22). Currently, the most attractive II–VI materials for blue lasers are sulfur-based injection lasers operating at room temperature without magnetic constituents (23).

One of the important aspects of epitaxial growth of diluted magnetic multilayer structures is the ability to interleave magnetic layers with nonmagnetic layers. Several different phenomena can occur in this situation. One example is the production of a spin superlattice by the application of a magnetic field. In the absence of a field, the superlattice potential is small, but the field causes a large Zeeman splitting in the magnetic layer that introduces a strong superlattice potential (24). Another example is the acquisition of magnetic properties by nonmagnetic layers, such as larger than usual Zeeman splitting of the excitons confined to the nonmagnetic layer. This occurs because the wavefunction penetrates into the magnetic layers containing magnetic ions.

Continuing interest in heteroepitaxy has led to more complex multilayer structures of diluted magnetic semiconductors, such as the growth of CdMgMnTe structures which can span the full visible range by varying the magnesium concentration, including the important blue spectral range (25). Nonequilibrium epitaxial growth techniques have made it possible to grow diluted magnetic III–V semiconductors, such as (In,Mn)As (26). Growth at low substrate temperatures (200°C) produced films that formed a homogeneous alloy, while higher growth temperatures (300°C) produced materials that had inclusions of ferromagnetic MnAs clusters. More recent work has extended the diluted magnetic III–V semiconductor materials to include (Ga,Mn)As (27) and (Ga,Mn)Sb (28), as well as superlattices of magnetic and nonmagnetic layers (29).

BIBLIOGRAPHY

1. E. L. Nagev, *Physics of Magnetic Semiconductors,* Moscow: Mir Publishers, 1983.

2. J. K. Furdyna and J. Kossut (eds.), *Diluted Magnetic Semiconductors,* Boston: Academic Press, 1988.

3. J. C. P. Chang et al., Precipitation in Fe- or Ni-implanted GaAs, *Appl. Phys. Lett.,* **65**: 2801, 1994.

4. S. B. Oseroff, Magnetic susceptibility and EPR measurements in concentrated spin-glasses: CdMnTe and CdMnSe, *Phys. Rev. B,* **25**: 6584, 1982.

5. T. M. Pekarek et al., Superparamagnetic behavior of Fe3GaAs precipitates in GaAs, *J. Magn. Magn. Mat.,* **168**: 1997.

6. S. Methfessel and D. Mattis, *Magnetic Semiconductors.* Berlin: Springer-Verlag, 1968.

7. P. Wachter, Europium chalcogenides: EuO, EuS, EuSe, EuTe. In *Handbook on the Physics and Chemistry of Rare Earths,* Vol. 2, Amsterdam: North-Holland Publishing, 1979, pp. 507–574.

8. I. Tsubokawa, On the magnetic properties of a CrBr3 single crystal. *J. Phys. Soc. Jpn.,* **15**: 1664, 1960.

9. B. Matthias, R. Bosorth, and J. van Vleck, Ferromagnetic interaction in EuO, *Phys. Rev. Lett.,* **7**: 160, 1961.

10. T. Arai et al., Magnetoabsorption in single-crystal HgCr2Se4, *J. Phys. Soc. Jpn.,* **34**: 68, 1973.

11. J. Schoenes and P. Wachter, Exchange optics in Gd-duped EuO, *Phys. Rev. B,* **9**: 3097, 1974.

12. J. Schoenes, P. Wachter, and F. Rys, Magneto-optic spectroscopy of EuS, *Sol. St. Commun.,* **15**: 1891, 1974.

13. J. K. Furdyna, Diluted magnetic semiconductors, *J. Appl. Phys.,* **64**: R29–R64, 1988.

14. J. K. Furdyna and N. Samarth, Magnetic properties of diluted magnetic semiconductors: a review, *J. Appl. Phys.,* **61**: 3526, 1987.

15. R. L. Aggarwal et al., Optical determination of the antiferromagnetic exchange constant between nearest-neighbor Mn ions in ZnMnTe, *Phys. Rev. B,* **34**: 5894, 1986.

16. D. U. Bartholomew, J. K. Furdyna, and A. K. Ramdas, Interband Faraday rotation in diluted magnetic semiconductors: $Zn_{1-x}Mn_xTe$ and $Cd_{1-x}Mn_xTe$, *Phys. Rev. B,* **34**: 6943, 1986.

17. T. Wojtowicz and A. Mycielski, Magnetic field induced nonmetal–metal transition in the open-gap HgMnTe, *Physica B,* **117 & 118**: 476, 1983.

18. R. S. Rana et al., Magnetophotorefractive effects in a diluted magnetic semiconductor, *Phys. Rev. B,* **49**: 7941, 1994.

19. R. S. Rana et al., Optical phase conjugation in a magnetic photorefractive semiconductor CdMnTe., *Opt. Lett.,* **20**: 1238, 1995.

20. L. A. Kolodziejski et al., Two-dimensional metastable magnetic semiconductor structures. *Appl. Phys. Lett.,* **48**: 1482, 1986.

21. S. Venugopalan et al., Raman scattering from molecular beam epitaxially grown superlattices of diluted magnetic semiconductors, *Appl. Phys. Lett.,* **45**: 974, 1984.

22. R. B. Bylsma et al., Stimulated emission and laser oscillations in ZnSe/ZnMnSe multiple quantum wells at 435 nm, *Appl. Phys. Lett.,* **47**: 1039, 1985.

23. R. L. Gunshor and A. V. Nurmikko, eds., II-VI blue/green light emitters: device physics and epitaxial growth, in *Semiconductors Semimetals,* vol. 44, San Diego, CA: Academic Press, 1997.

24. N. Dai et al., Spin superlattice formation in ZnSe/ZnMnSe multilayers, *Phys. Rev. Lett.,* **67**: 3824, 1991.

25. A. Waag et al., Growth of MgTe and CdMgTe thin films by molecular beam epitaxy, *J. Cryst. Growth,* **131**: 607, 1993.

26. H. Munekata et al., Diluted magnetic III–V semiconductor, *Phys. Rev. Lett.,* **63**: 1849, 1989.

27. F. Matsukura et al., Growth and properties of (Ga,Mn)As: a new III–V diluted magnetic semiconductor, *Appl. Surf. Sci.,* **113–114**: 178, 1996.

28. S. Basu and T. Adhikari, Variation of band gap with Mn concentration in GaMnSb—a new III–V diluted magnetic semiconductor, *Sol. State Commun.,* **95**: 53, 1995.

29. A. Shen et al., (Ga,Mn)As/GaAs diluted magnetic semiconductor superlattice structures prepared by molecular beam epitaxy, *Jpn. J. Appl. Phys. 2, Lett.,* **36**: L73, 1997.

DAVID D. NOLTE
Purdue University

MAGNETIC SENSORS

Magnetic sensors find many applications in everyday life and in industry. They provide convenient, noncontact, simple, rugged, and reliable operations compared to many other sensors. The technology to produce magnetic sensors involves many aspects of different disciplines such as physics, metallurgy, chemistry, and electronics.

Generally, magnetic sensors are based on sensing the properties of magnetic materials, which can be done in many ways. For example, magnetization, which is the magnetic moment per volume of materials, is used in many measurement systems by sensing force, induction, field methods, and superconductivity. However, the majority of industrial sensors make use of the relationship between magnetic and electric phenomenon. A typical application of the phenomenon is the computer memory requiring the reading of the contents of a disc without making any contact between the sensor and the device. In other applications, the position of objects sensitive to magnetic fields (e.g., the metals in the ground) can be sensed magnetically. Magnetic sensors find most sensitive applications in medicine to diagnose human illnesses, as in the case of superconducting quantum interference devices (SQUID) and nuclear resonance magnetic (NMR) imaging.

The magnetic elements in sensors are used in a wide range of forms: toroids, rods, films, substrates, and coatings. Some elements are essentially free standing, whereas others are an integral part of more complex devices. In order to obtain maximum material response in magnetic sensors, the relative orientation and coupling between input measurand and magnetic properties are very important, and they are optimized at the design stages.

Many different types of magnetic sensors are available. These sensors can broadly be classified as primary or secondary. In primary sensors, also known as the magnetometers, the parameter to be measured is the external magnetic field. The primary sensors are used in biological applications and geophysical and extraterrestrial measurements. In secondary sensors, the external parameter is made from other physical variables such as force and displacement. In this article, both the primary and secondary sensors will be discussed. These sensors include inductive, eddy current, transformative, magnetoresistive, Hall-effect, metal–oxide–semiconductor (MOS) magnetic field, and magneto-optical sensors; magnetotransistor and magnetodiode sensors, magnetometers; superconductors; semiconductors; and magnetic thin films. They are offered by many manufacturers as listed in Table 1.

INDUCTIVE SENSORS

Inductive sensors make use of the principles of magnetic circuits. They can be classified as passive sensors and self-generating sensors. The passive sensors require an external power source; hence, the action of the sensor is restricted to the modulation of the excitation signal in relation to an external stimuli. On the other hand, the self-generating types generate signals by utilizing the electrical generator principle based on Faraday's Law of Induction. That is, when there is a relative motion between a conductor and a magnetic field, a voltage is induced in the conductor. Or a varying magnetic field linking a stationary conductor produces voltage in the conductor, which can be expressed as

$$e = -d\Phi/dt \text{ (V)} \tag{1}$$

where Φ is the magnetic flux.

In instrumentation applications, the magnetic field may be varying in time with some frequency, and the conductor may be moving at the same time. In many cases, the relative motion between field and conductor is supplied by changes in the measurand, usually by means of a mechanical motion.

In order to explain the operation of the basic principles of inductive sensors, a simple magnetic circuit is shown in Fig. 1. The magnetic circuit consists of a core, made from a ferromagnetic material, and a coil of n number of turns wound on it. The coil acts as a source of magnetomotive force (mmf), which drives the flux Φ through the magnetic circuit. If we assume that the air gap is zero, the equation for the magnetic circuit may be expressed as

$$\text{mmf} = \text{Flux} \times \text{Reluctance} = \Phi \times \mathscr{R} \quad \text{(A-turns)} \tag{2}$$

such that the reluctance $\mathscr{R}$ limits the flux in a magnetic circuit just as resistance limits the current in an electric circuit. By writing the magnetomotive force in terms of current, the magnetic flux may be expressed as

$$\Phi = ni/\mathscr{R} \quad \text{(Wb)} \tag{3}$$

In Fig. 1, the flux linking a single turn is expressed by Eq. (3). But the total flux linking by the entire n number of the turns of the coil is

$$\Psi = n\Phi = n^2 i/\mathscr{R} \quad \text{(Wb)} \tag{4}$$

Equation (4) leads to self-inductance L of the coil, which is described as the total flux per unit current for that particular coil. That is,

$$L = \Psi/I = n^2/\mathscr{R} \quad \text{(H)} \tag{5}$$

Table 1. List of Manufacturers

Adsen Tech, Inc. 18310 Bedford Circle La Puente, CA 91744 Fax: 818-854-2776	Motion Sensors, Inc. 786 Pitts Chapel Road Alizabeth City, NC 27909 Tel: 919-331-2080 Fax: 919-331-1666
Analog Devices, Inc. 1 Technology Way P.O. Box 9106 Norwood, MA 02062-9102 Tel: 800-262-5663 Fax: 781-326-8703	Rechner Electronics Industries, Inc. 8651 Buffalo Avenue Niagara Falls, NY 14304 Tel: 800-544-4106 Fax: 716-283-2127
Dynalco Controls 3690 N. W. 53rd Street Ft. Lauderdale, FL 33309 Tel: 305-739-4300 & 800-368-6666 Fax: 305-484-3376	Reed Switch Developments Company, Inc. P. O. Drawer 085297 Racine, WI 53408 Tel: 414-637-8848 Fax: 414-637-8861
Electro Corporation 1845 57th Street Sarasota, FL 34243 Tel: 813-355-8411 & 800-446-5762 Fax: 813-355-3120	Smith Research and Technology, Inc. 205 Sutton Lane, Dept. TR-95 Colorado Springs, CO 80907 Tel: 719-634-2259 Fax: 719-634-2601
Honeywell Dept. 722 11 West Spring Street Freeport, IL 61032 Tel: 800-537-6945 Fax: 815-235-5988	Smith Systems, Inc. 6 Mill Creek Drive Box 667 Brevard, NC 28712 Tel: 704-884-3490 Fax: 704-877-3100
Kaman Instrument Company 1500 Garden of the Gods Road Colorado Springs, CO 80907 Tel: 719-599-1132 & 800-552-6267 Fax: 719-599-1823	Standex Electronics 4538 Camberwell Road Dept. 301L Cincinnati, OH 45209 Tel: 513-871-3777 Fax: 513-871-3779
Kavlico Corporation 14501 Los Angeles Avenue Moorpark, CA 93021 Tel: 805-523-2000 Fax: 805-523-7125	Turck, Inc. 3000 Campus Drive Minneapolis, MN 55441 Tel: 612-553-7300 & 800-544-7769 Fax: 612-553-0708
Lucas 1000 Lucas Way Hampton, VA 23666 Tel: 800-745-8008 Fax: 800-745-8004	Xolox Sensor Products 6932 Gettysburg Pike Ft. Wayne, IN 46804 Tel: 800-348-0744 Fax: 219-432-0828

This indicates that the self-inductance of an inductive element can be calculated by magnetic circuit properties. Expressing $\mathcal{R}$ in terms of dimensions as

$$\mathcal{R} = l/\mu\mu_0 A \quad \text{(A-turns/Wb)} \qquad (6)$$

where l is the total length of the flux path (meters), μ is the relative permeability of the magnetic circuit material, μ_0 is the permeability of free space ($= 4\pi \times 10^{-7}$ H/m), and A is the cross-sectional area of the flux path.

If the air gap is allowed to vary, the arrangement illustrated in Fig. 1 becomes a basic inductive sensor. In this case, the ferromagnetic core is separated in two parts by the air gap. The total reluctance of the circuit now is the addition of the reluctance of the core and the reluctance of the air gap. The relative permeability of air is close to unity, and the relative permeability of the ferromagnetic material is on the order

of a few thousand, indicating that the presence of the air gap causes a large increase in circuit reluctance and a corresponding decrease in the flux. Hence, a small variation in the air gap causes a measurable change in inductance. There are many different types of inductive sensors as will be discussed next.

Linear and Rotary Variable-Reluctance Sensors

The variable-reluctance transducers are based on change in the reluctance of a magnetic flux path. These types of devices find applications particularly in acceleration measurements. However, they can be constructed to be suitable for sensing displacements as well as velocities. They are constructed in many different forms, some of which will be described in this article.

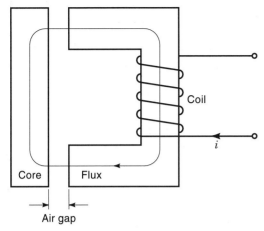

Figure 1. A basic inductive sensor consists of a magnetic circuit made up from a ferromagnetic core with a coil wound on it. The coil acts as a source of mmf, which drives the flux through the magnetic circuit and the air gap. The presence of the air gap causes a large increase in circuit reluctance and a corresponding decrease in the flux. Hence, a small variation in the air gap causes a measurable change in inductance.

Single-Coil Linear Variable-Reluctance Sensor. A typical single-coil variable-reluctance displacement sensor is illustrated in Fig. 2. The sensor consists of three elements: a ferromagnetic core in the shape of a semicircular ring, a variable air gap, and a ferromagnetic plate. The total reluctance of the magnetic circuit is the sum of the individual reluctances:

$$\mathcal{R}_T = \mathcal{R}_C + \mathcal{R}_G + \mathcal{R}_A \tag{7}$$

where $\mathcal{R}_C$, $\mathcal{R}_G$, and $\mathcal{R}_A$ are the reluctances of the core, air gap, and armature, respectively.

Each one of these reluctances can be determined by using the properties of materials involved as in Eq. (6). In this par-

ticular case, $\mathcal{R}_T$ can be approximated as

$$\mathcal{R}_T = \frac{R}{\mu_C \mu_0 r^2} + \frac{2d}{\mu_0 \pi r^2} + \frac{R}{\mu_A \mu_0 rt} \tag{8}$$

In obtaining Eq. (8), the length of the flux path in the core is taken as πR, and the cross-sectional area is assumed to be uniform with a value of πr^2. The total length of the flux path in air is $2d$, and it is assumed that there is no fringing or bending of the flux through the air gap, such that the cross-sectional area of the flux path in air will be close to that of the cross section of the core. The length of an average central flux path in the armature is $2R$. The calculation of an appropriate cross section of the armature is difficult, but it may be approximated to $2rt$, where t is the thickness of the armature.

In Eq. (8), all the parameters are fixed except the only one independent variable, the air gap. Hence, it can be simplified as

$$\mathcal{R}_T = \mathcal{R}_0 + kd \tag{9}$$

where $\mathcal{R}_0 = R/(\mu_0 r[1/(\mu_C r) + 1/(\mu_A t)]$, and $k = 2/(\mu_0 \pi r^2)$.

By using Eqs. (5) and (9), the inductance can be written as

$$L = \frac{n^2}{\mathcal{R}_0 + kd} = \frac{L_0}{1 + \alpha d} \tag{10}$$

where L_0 represents the inductance at zero air gap and $\alpha = k/\mathcal{R}_0$.

The values of L_0 and α can be determined mathematically. They depend on the core geometry, permeability, and the like, as already explained. As it can be seen from Eq. (10), the relationship between L and α is nonlinear. Despite this non-linearity, these types of single-coil sensors find applications in many areas, such as force measurements and telemetry. In force measurements, the resultant change in inductance can be made to be a measure of the magnitude of the applied force. The coil usually forms one of the components of an LC oscillator whose output frequency varies with the applied force. Hence, the coil modulates the frequency of the local oscillator.

Variable-Differential Reluctance Sensor. The problem of the nonlinearity may be overcome by modifying the single-coil system into variable-differential reluctance sensors (also known as push-pull sensors), as shown in Fig. 3. This sensor consists of an armature moving between two identical cores separated by a fixed distance of $2d$. Now, Eq. (10) can be written for both coils as

$$L_1 = \frac{L_{01}}{1 + \alpha(d - x)}, \quad L_2 = \frac{L_{02}}{1 + \alpha(d + x)} \tag{11}$$

Although the relationship between L_1 and L_2 is still nonlinear, the sensor can be incorporated into an ac bridge to give a linear output for small movements. The hysteresis error of these transducers is almost entirely limited to the mechanical components. These sensors respond to static and dynamic measurements. They have continuous resolution and high outputs, but they may give erratic performances in response to external magnetic fields. A typical sensor of this type has an input span of 1 cm, a coil inductance of 25 mH, and a coil

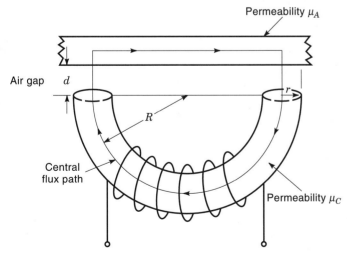

Figure 2. A typical single-coil variable-reluctance displacement sensor. The sensor consists of three elements: a ferromagnetic core in the shape of a semicircular ring, a variable air gap, and a ferromagnetic plate. The reluctance of the coil is dependent on the air gap. Air gap is the single variable, and the reluctance increases nonlinearly with the increasing gap.

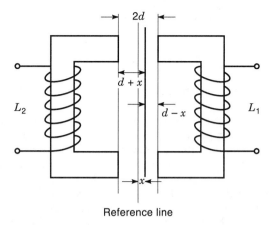

Figure 3. A variable-differential reluctance sensor consists of an armature moving between two identical cores separated by a fixed distance. The armature moves in the air gap in response to the mechanical input. This movement alters the reluctance of coils 1 and 2 thus altering their inductive properties. This arrangement overcomes the problem of nonlinearity inherent in single-coil sensors.

resistance of 75 Ω. The resistance of the coil must be carefully considered when designing oscillator circuits. The maximum nonlinearity may be limited to 0.5%.

In typical commercially available variable-differential sensors, the iron core is located half way between the two E-shaped frames. The flux generated by primary coils depends on the reluctance of the magnetic path, the main reluctance being the air gap. Any motion of the core increases the air gap on one side and decreases it on the other side. Consequently, the reluctance changes in accordance with the principles explained previously, thus inducing more voltage on one of the coils than the other. Motion in the other direction reverses the action with a 180° phase shift occurring at null. The output voltage can be modified depending on the requirements in signal processing by means of rectification, demodulation, or filtering. In these instruments, full-scale motion may be extremely small, on the order of few thousandths of a centimeter.

In general, variable-reluctance transducers have small ranges and are used in specialized applications such as pressure transducers. Magnetic forces imposed on the armature are quite large, and this limits the application severely.

Variable-Reluctance Tachogenerators. Another example of the variable-reluctance sensor is shown in Fig. 4. These sensors are based on Faraday's Law of Electromagnetic Induction; therefore, they may also be referred as electromagnetic sensors. Basically, the induced electromagnetic force (emf) in the sensor depends on the linear or angular velocity of the motion.

The variable-reluctance tachogenerator consists of a ferromagnetic toothed wheel attached to the rotating shaft, and a coil wound onto a permanent magnet, extended by a soft iron pole piece. The wheel moves in close proximity to the pole piece, causing the flux linked by the coil to change, thus inducing an emf in the coil. The reluctance of the circuit depends on the width of the air gap between the rotating wheel and the pole piece. When the tooth is close to the pole piece, the reluctance is at a minimum, and it increases as the tooth

moves away from the pole. If the wheel rotates with a velocity ω, the flux may mathematically be expressed as

$$\Psi(\theta) = \Psi_m + \Psi_f \cos m\theta \qquad (12)$$

where Ψ_m is the mean flux, Ψ_f is the amplitude of the flux variation, and m is the number of teeth.

The induced emf is given by

$$E = -\frac{d\Psi(\theta)}{dt} = -\frac{d\Psi(\theta)}{d\theta} \times \frac{d\theta}{dt} \qquad (13)$$

or

$$E = \Psi_f m\omega \sin n\omega t \qquad (14)$$

Both the amplitude and the frequency of the generated voltage at the coil are proportional to the angular velocity of the wheel. In principle, the angular velocity ω can be found from either the amplitude or the frequency of the signal. In practice, the amplitude measured may be influenced by loading effects and electrical interference. In signal processing, the frequency is the preferred option because it can be converted into digital signals easily.

The variable-reluctance tachogenerators are most suitable for measuring angular velocities. They are also used for volume flow rate measurements and the total volume flow determination of fluids.

Microsyn. Another commonly used example of variable-reluctance transducer is the microsyn, as illustrated in Fig. 5. In this arrangement, the coils are connected in such a manner that at the null position of the rotary element, the voltages induced in coils 1 and 3 are balanced by voltages induced in coils 2 and 4. The motion of the rotor in the clockwise direction increases the reluctance of coils 1 and 3 while decreasing the reluctance of coils 2 and 4, thus giving a net output voltage e_o. The movement in the counterclockwise direction causes a similar effect in coils 2 and 4 with a 180° phase shift. A direction-sensitive output can be obtained by using phase-sensitive demodulators.

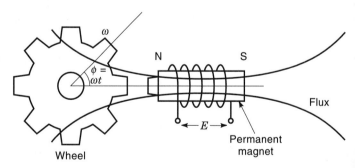

Figure 4. A variable-reluctance tachogenerator is a sensor based on Faraday's Law of Electromagnetic Induction. It consists of a ferromagnetic toothed wheel attached to the rotating shaft and a coil wound onto a permanent magnet extended by a soft iron pole piece. The wheel rotates in close proximity to the pole piece, thus causing the flux linked by the coil to change. The change in flux causes an output in the coil similar to a square waveform whose frequency depends on the speed of the rotation of the wheel and the number of teeth.

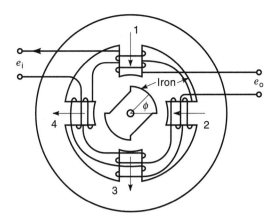

Figure 5. A microsyn is a variable-reluctance transducer that consists of a ferromagnetic rotor and a stator carrying four coils. The stator coils are connected such that at the null position, the voltages induced in coils 1 and 2 are balanced by voltages induced in coils 3 and 4. The motion of the rotor in one direction increases the reluctance of two opposite coils while decreasing the reluctance in others resulting in a net output voltage e_o. The movement in the opposite direction reverses this effect with a 180° phase shift.

Microsyn transducers are extensively used in applications involving gyroscopes. By using microsyns, very small motions can be detected giving an output signal as low as 0.01° of changes in angles. The sensitivity of the device can be made as high as 5 V per degree of rotation. The nonlinearity may vary from 0.5% to 1.0% full scale. The main advantage of these transducers is that the rotor does not have windings and slip rings. The magnetic reaction torque is also negligible.

Synchros

The term *synchro* is associated with a family of electromechanical devices. They are primarily used in angle measurements and are commonly applied in control engineering as parts of servomechanisms, machine tools, antennas, and the like.

The construction of synchros is similar to that of wound-rotor induction motors, as shown in Fig. 6. The rotation of the motor changes the mutual inductance between the rotor coil and the three stator coils. The three voltage signals from these coils define the angular position of the rotor. Synchros are used in connection with a variety of devices, such as control transformers, Scott T transformers, resolvers, phase-sensitive demodulators, and analog-to-digital (AD) converters.

In some cases, a control transformer is attached to the outputs of the stator coils such that the output of the control transformer produces a resultant mmf aligned in the same direction as that of the rotor of the synchro. In other words, the synchro rotor acts as a search coil in detecting the direction of the stator field of the control transformer. When the axis of this coil is aligned with the field, the maximum voltage is supplied to the transformer.

In other cases, ac signals from the synchros are first applied to a Scott T transformer, which produces ac voltages with amplitudes proportional to the sine and cosine of the synchro shaft angle. It is also possible to use phase-sensitive demodulations to convert the output signals to make them suitable for digital signal processing.

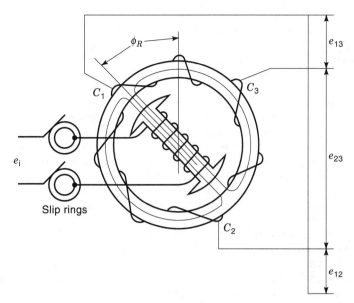

Figure 6. A synchro is similar to a wound-rotor induction motor. The rotation of the rotor changes the mutual inductance between the rotor coil and the three stator coils. The voltages from these coils define the angular position of the rotor. They are primarily used in angle measurements and are commonly applied in control engineering as parts of servomechanisms, machine tools, antennas, and the like.

Linear Variable Inductor

There is very little distinction between variable-reluctance and variable-inductance transducers. Mathematically, the principles of linear variable transducers are very similar to the variable-reluctance type of transducers. The distinction is mainly in the sensing rather than principles of operations. A typical linear variable inductor consists of a movable iron core to provide the mechanical input and the two coils forming two legs of bridge network. A typical example of such a transducer is the variable coupling transducer.

The variable-coupling transducers consist of a former holding a center-tapped coil and a ferromagnetic plunger, as shown in Fig. 7. The plunger and the two coils have the same length l. As the plunger moves, the inductances of the coils change. The two inductances are usually placed to form two arms of a bridge circuit with two equal balancing resistors. The bridge is then excited with ac of 5 V to 25 V with a frequency of 50 Hz to 5 kHz. At the selected excitation frequency, the total transducer impedance at null conditions is

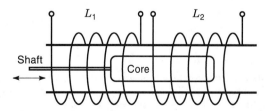

Figure 7. A typical linear variable inductor consists of a movable iron core inside a former holding a center-tapped coil. The core and both coils have the same length l. When the core is in the reference position, each coil will have equal inductances of value L. As the core moves by δl, changes in inductances $+\delta L$ and $-\delta L$ create voltage outputs from the coils.

set in the 100 Ω to 1000 Ω range. The resistors are set to have about the same value as transducer impedances. The load for the bridge output must be at least ten times the resistance R value. When the plunger is in the reference position, each coil will have equal inductances of value L. As the plunger moves by δl, changes in inductances $+\delta L$ and $-\delta L$ create a voltage output from the bridge. By constructing the bridge carefully, the output voltage may be made as a linear function displacement of the moving plunger within a rated range.

In some transducers, in order to reduce power losses resulting from the heating of resistors, center-tapped transformers may be used as a part of the bridge network. In this case, the circuit becomes more inductive, and extra care must be taken to avoid the mutual coupling between the transformer and the transducer.

It is particularly easy to construct transducers of this type, by simply winding a center-tapped coil on a suitable former. The variable-inductance transducers are commercially available in strokes from about 2 mm to 500 cm. The sensitivity ranges between 1% full scale to 0.02% in long-stroke special constructions. These devices are also known as linear displacement transducers or LDTs, and they are available in various shapes and sizes.

Apart from linear variable inductors, rotary types are also available. Their cores are specially shaped for rotational applications. Their nonlinearity can vary between 0.5% and 1% full scale over a range of 90° rotation. Their sensitivity can be up to 100 mV per degree of rotation.

Induction Potentiometer

A version of rotary-type linear inductors is the induction potentiometer, as shown in Fig. 8. Two concentrated windings are wound on stator and rotor. The rotor winding is excited with an ac, thus inducing voltage in the stator windings. The

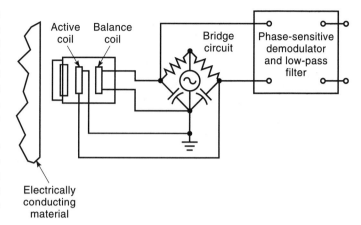

Figure 9. Eddy current transducers are inductive transducers using probes. The probes contain one active and one balance coil. The active coil responds to the presence of a conducting target, whereas the balance coil completes a bridge circuit and provides temperature compensation. When the probe is brought close to the target, the flux from the probe links with the target producing eddy currents within the target, which alter the inductance of the active coil. This change in inductance is detected by a bridge circuit.

amplitude of the output voltage is dependent on the mutual inductance between the two coils, where mutual inductance itself is dependent on the angle of rotation. For concentrated coil-type induction potentiometers, the variation of the amplitude is sinusoidal, but linearity is restricted in the region of the null position. A linear distribution over an angle of 180° may be obtained by carefully designed distributed coils.

Standard commercial induction pots operate in a 50 Hz to 400 Hz frequency range. They are small in size from 1 cm to 6 cm, and their sensitivity can be in the order of 1 V/1° of rotation. Although the ranges of induction pots are limited to less than 60° of rotation, it is possible to measure displacements in angles from 0° to full rotation by suitable arrangements of a number of induction pots. As in the case of most inductive sensors, the output of the induction pots may need phase-sensitive demodulators and suitable filters. In many inductive pots, additional dummy coils are used to improve linearity and accuracy.

EDDY CURRENT SENSORS

Inductive transducers based on eddy currents are mainly probe types containing two coils, as shown in Fig. 9. One of the coils, known as the active coil, is influenced by the presence of the conducting target. The second coil, known as the balance coil, serves to complete the bridge circuit and provides temperature compensation. The magnetic flux from the active coil passes into the conductive target by means of a probe. When the probe is brought close to the target, the flux from the probe links with the target, producing eddy currents within the target.

The eddy current density is greatest at the target surface and become negligibly small about three skin depths below the surface. The skin depth depends on the type of material used and the excitation frequency. Even though thinner targets can be used, a minimum of three skin depths may often be necessary to minimize the temperature effects. As the tar-

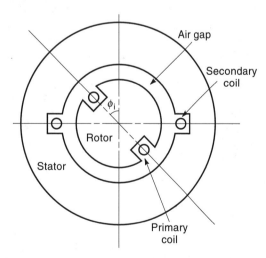

Figure 8. An induction potentiometer is a linear variable inductor with two concentrated windings wound on the stator and rotor. The rotor winding is excited with ac, inducing voltage in the stator windings. The amplitude of the output voltage is dependent on the relative positions of the coils determined by the angle of rotation. For concentrated coils, the variation of the amplitude is sinusoidal, but linearity is restricted in the region of the null position. Different types of induction potentiometers are available with distributed coils, which give linear voltages over an angle of 180° of rotation.

get comes closer to the probe, the eddy currents become stronger, causing the impedance of the active coil to change and altering the balance of the bridge in relation to the target position. This unbalance voltage of the bridge may be demodulated, filtered, and linearized to produce a dc output proportional to target displacement. The bridge oscillation may be as high as 1 MHz. High frequencies allow the use of thin targets and provide a good system frequency response.

Probes are commercially available with full-scale ranges from 0.25 mm to 30 mm with a nonlinearity of 0.5% and a maximum resolution of 0.0001 mm. Targets are usually supplied by the clients, involving noncontact measurements of machine parts. For nonconductive targets, conductive materials of sufficient thickness must be attached onto the surface by means of commercially available adhesives. Because the target material, shape, and the like influence the output, it is necessary to calibrate the system statistically for a specific target. The recommended measuring range of a given probe begins at a standoff distance equal to about 20% of the stated range of the probe. In some cases, a standoff distance of 10% of the stated range for which the system is calibrated is recommended as standard. A distance greater than 10% of the measuring range can be used as long as the calibrated measuring range is reduced by the same amount.

Flat targets must be the same diameter as the probe or larger. If the target diameter is smaller than the probe diameter, the output drops considerably, thus becoming unreliable. Curved-surface targets may behave similar to flat surfaces if the diameter exceeds about three or four diameter of the probe. In this case, the target essentially becomes an infinite plane. This also allows some cross-axis movement without affecting the system output. Target diameter comparable to the sensor could result in detrimental affects from cross-axis movements.

For curved or irregularly shaped targets, the system needs to be calibrated using an exact target that may be seen in the operation. This tends to eliminate any errors caused by the curved surfaces during the applications. However, special multiprobe systems are available for orbital motions of rotating shafts. If the curved (shaft) target is about ten times greater than the sensor diameter, it acts as an infinite plane and does not need special calibrations. Special care must be exercised to deal with electrical runout resulting from factors such as inhomogeneities in hardness, particularly valid for ferrous targets. However, nonferrous targets are free from electrical runout concerns.

TRANSFORMATIVE SENSORS

Transformative sensors make use of the principles of transformer action, that is magnetic flux created by one coil links with the other coil to induce voltages. There are many different types, such as linear variable transformers, rotary variable differential transformers, and flux-gate magnetometers.

Linear Variable-Differential Transformer

The linear variable-differential transformer (LVDT) is a passive inductive transducer that has found many applications. It consists of a single primary winding positioned between two identical secondary windings wound on a tubular ferro-

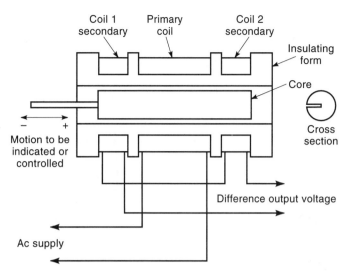

Figure 10. A linear variable-differential transformer is a passive inductive transducer consisting of a single primary winding positioned between two identical secondary windings wound on a tubular ferromagnetic former. As the core inside the former moves, the magnetic paths between primary and secondaries alter, thus giving secondary outputs proportional to the movement. The two secondaries are made as similar as possible by having equal sizes, shapes, and number of turns.

magnetic former, as shown in Fig. 10. The primary winding is energized by a high-frequency 50 Hz to 20 kHz ac voltage. The two secondaries are made identical by having an equal number of turns. They are connected in series opposition so that the induced output voltages oppose each other.

In many applications, the outputs are connected in opposing form, as shown in Fig. 11(a). The output voltages of individual secondaries v_1 and v_2 at null position are illustrated in Fig. 11(b). However, in opposing connection, any displacement in the core position x from the null point causes amplitude of the voltage output v_o and the phase difference α to change. The output waveform v_o in relation to core position is shown in Fig. 11(c). When the core is positioned in the middle, there is an equal coupling between primary and secondaries, thus giving a null point or reference point of the sensor. As long as the core remains near the center of the coil arrangement, output is very linear. The linear ranges of commercial differential transformers are clearly specified, and the devices are seldom used outside this linear range.

The ferromagnetic core or plunger moves freely inside the former; thus altering the mutual inductance between the primary and secondaries. With the core in the center, or at the reference position, the induced emfs in the secondaries are equal, and because they oppose each other, the output voltage is zero. When the core moves, say to the left, from the center, more magnetic flux links with the left-hand coil than with the right-hand coil. The voltage induced in the left-hand coil is therefore larger than the induced emf on the right-hand coil. The magnitude of the output voltage is then larger than at the null position and is equal to the difference between the two secondary voltages. The net output voltage is in phase with the voltage of the left-hand coil. The output of the device is then an indication of displacement of the core. Similarly, movement in the opposite direction to the right from the cen-

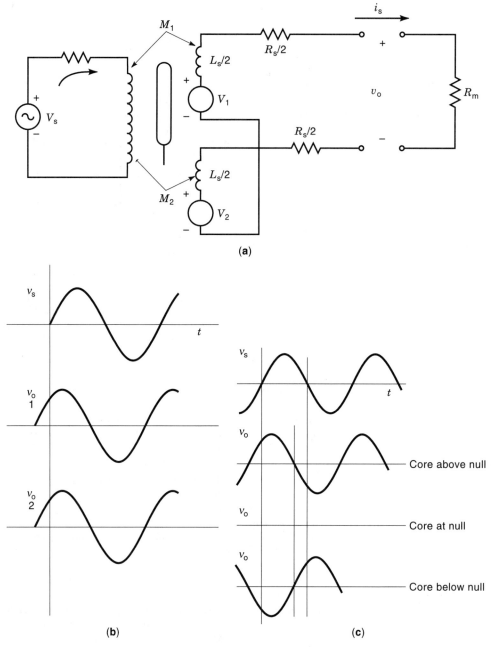

(a)

(b)

(c)

Figure 11. The voltages induced in the secondaries of a linear variable-differential transformer (a) may be processed in a number of ways. The output voltages of individual secondaries v_1 and v_2 at null position are illustrated in (b). In this case, the voltages of individual coils are equal and in phase with each other. Sometimes the outputs are connected opposing each other, and the output waveform v_o becomes a function of core position x and phase angle α as in (c). Note the phase shift of 180° as the core position changes above and below the null position.

ter reverses this effect, and the output voltage is now in phase with the emf of the right-hand coil.

For mathematical analysis of the operation of LVDTs Fig. 11(a) may be used. The voltages induced in the secondary coils are dependent on the mutual inductance between the primary and individual secondary coils. Assuming that there is no cross coupling between the secondaries, the induced voltages may be written as

$$v_1 = M_1 s i_p \quad \text{and} \quad v_2 = M_2 s i_p \qquad (15)$$

where M_1 and M_2 are the mutual inductances between primary and secondary coils for a fixed core position, s is the Laplace operator, and i_p is the primary current.

In the case of opposing connection, no load output voltage v_o without any secondary current may be written as

$$v_o = v_1 - v_2 = (M_1 - M_2) s i_p \qquad (16)$$

writing

$$v_s = i_p(R + s L_p) \qquad (17)$$

Substituting i_p in Eq. (16) gives the transfer function of the transducer as

$$\frac{v_o}{v_s} = \frac{(M_1 - M_2)s}{R + sL_p} \tag{18}$$

However, If there is a current resulting from output signal processing, then describing equations may be modified as

$$v_o = R_m i_s \tag{19}$$

where $i_s = (M_1 - M_2)si_p/(R_s + R_m + sL_s)$ and

$$v_s = i_p(R + sL_p) - (M_1 - M_2)si_s \tag{20}$$

Eliminating i_p and i_s from Eqs. (19) and (20) results in a transfer function

$$\frac{v_o}{v_s} = \frac{R_m(M_1 - M_2)s}{[(M_1 - M_2)^2 + L_sL_p]s^2 + [L_p(R + R_m) + RL_s]s + (R_s + R_m) + R} \tag{21}$$

This is a second-order system, which indicates that with the effect of the numerator the frequency of the system changes from $+90°$ at low frequencies to $-90°$ at high frequencies. In practical applications, the supply frequency is selected such that at null position of the core the phase angle of the system is $0°$.

The amplitudes of the output voltages of secondary coils are dependent on the position of the core. These outputs may directly be processed from each individual secondary coils for slow movements of the core, if the direction of the movement of the core does not bear any importance. However, for fast movements of the core, the signals may be converted to dc, and the direction of the movement from the null position may be detected. There are many options to do this; however, a *phase-sensitive demodulator* and filter are commonly used as shown in Fig. 12(a). A typical output of the phase-sensitive demodulator is illustrated in Fig. 12(b), for core positions as in Fig. 12(c), in relation to output voltage v_o, displacement x, and phase angle α.

The phase-sensitive demodulators are extensively used in differential-type inductive sensors. They basically convert the ac outputs to dc values and also indicate the direction of movement of the core from the null position. A typical phase-sensitive demodulation circuit may be constructed, based on diodes shown in Fig. 13(a). This arrangement is useful for very slow displacements, usually less than 1 or 2 Hz. In Fig. 13(a), bridge 1 acts as a rectification circuit for secondary 1, and bridge 2 acts as a rectifier for secondary 2. The net output voltage is the difference between the outputs of two bridges as in Fig. 13(b). The position of the core can be worked out from the amplitude of the dc output and the direction of the movement of the core can be determined from the polarity of the dc voltage. For rapid movements of the core, the output of the diode bridges need to be filtered, and this passes only the frequencies of the movement of the core and filters all the other frequencies produced by the modulation process. For this purpose, a suitably designed simple *RC* filter may be sufficient.

In the marketplace, there are phase-sensitive demodulator chips available, such as AD598 offered by Analog Devices,

Inc. These chips are highly versatile and flexible to suit particular application requirements. They offer many advantages over conventional phase-sensitive demodulation devices; for example, frequency of excitation may be adjusted to any value between 20 Hz and 20 kHz by connecting an external capacitor between two pins. The amplitude of the excitation voltage can be set up to 24 V. The internal filters may be set to required values by external capacitors. Connections to analog-to-digital converters are made easy by converting the bipolar output to a unipolar scale.

The frequency response of LVDTs is primarily limited by the inertia characteristics of the device. In general, the frequency of the applied voltage should be ten times the desired frequency response. Commercial LVDTs are available in a broad range of sizes, and they are widely used for displacement measurements in a variety of applications. The displacement sensors are available to cover ranges from ±0.25 mm to ±7.5 cm. They are sensitive enough to be used to respond to displacements well below 0.0005 mm. They can have operational temperature range from $-265°$ to $600°C$. They are also available in radiation-resistant designs for operation in nuclear reactors. For a typical sensor of range ±25 mm, the recommended supply voltage is 4 V to 6 V, with a nominal frequency of 5 kHz and a maximum nonlinearity of 1% full scale. Several commercial models, which can produce a voltage output of 300 mV for 1 mm displacement of the core, are available.

One important advantage of the LVDTs is that there is no physical contact between the core and the coil form; hence there is no friction or wear. Nevertheless, there are radial and longitudinal magnetic forces on the core at all times. These magnetic forces may be regarded as magnetic springs that try to displace the core from its null position. This may be a critical factor in some applications.

One problem with LVDTs is that it may not be easy to make the two halves of the secondary identical; their inductance, resistance, and capacitance may be different, causing a large unwanted quadrature output in the balance position. Precision coil-winding equipment may be required to reduce this problem to an acceptable value.

Another problem is associated with null position adjustments. The harmonics in the supply voltage and stray capacitances result in small null voltages. The null voltage may be reduced by proper grounding, which reduces the capacitive effects and the center-tapped voltage source arrangements. In center-tapped supplies, a potentiometer may be used to obtain a minimum null reading.

The LVDTs find a variety of applications, which include jet engines controls that are in close proximity to exhaust gases, and controls that measure roll positions in the thickness of materials in hot-slab steel mills. After some mechanical conversions, LVDTs may also make force and pressure measurements.

Rotary Variable-Differential Transformer

A variation from the linear variable-differential transformer is the rotary core differential transformer, as shown in Fig. 14. Here the primary winding is wound on the center leg of an E core; the secondary windings are wound on the outer legs of the E core. The armature is rotated by an externally applied force about a pivot point above the center leg of the

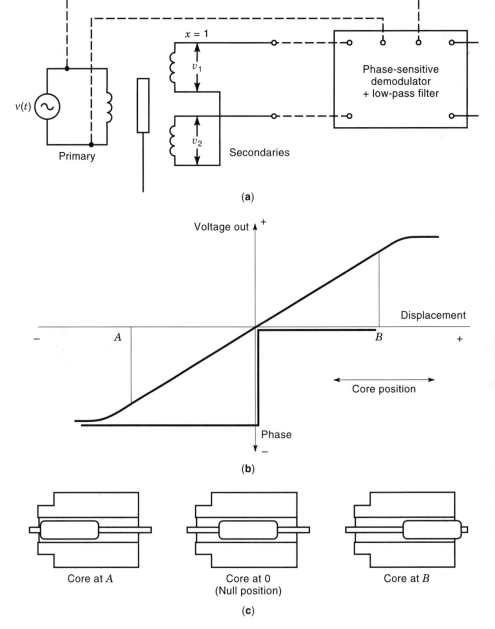

Figure 12. Phase-sensitive demodulator and filter (a) are commonly used to obtain displacement-proportional signals from LVDTs and other differential-type inductive sensors. They convert the ac outputs from the sensors into dc values and also indicate the direction of movement of the core from the null position. A typical output of the phase-sensitive demodulator is shown in (b). The relationship between output voltage v_o and phase angle α is also shown against core position x as sketched in (c).

core. When the armature is displaced from its reference or balance position, the reluctance of the magnetic circuit through one secondary coil is decreased; simultaneously the reluctance through the other coil is increased. The induced emfs in the secondary windings, which are equal in the reference position of the armature, are now different in magnitude and phase as a result of the applied displacement. The induced emfs in the secondary coils are made to oppose each other, and the transformer operates in the same manner as LVDTs. The rotating variable transformers may be sensitive to vibrations. If a dc output is required, a demodulator network may be used, as in the case of LVDTs.

In most rotary linear-variable differential transformers, the rotor mass is very small, usually less than 5 g. The nonlinearity in the output ranges between ±1% and ±3%, depending on the angle of rotation. The motion in the radial direction produces a small output signal that can affect the overall sensitivity. But this transverse sensitivity is usually kept less than 1% of the longitudinal sensitivity.

MOS MAGNETIC FIELD SENSORS

The technology of integrated magnetic field sensors, also called semiconductor magnetic microsensors, is well developed. This technology uses either high-permeability (e.g., ferromagnetic) or low-permeability (e.g., paramagnetic) materials. The integrated circuit magnetic techniques support many sensors, such as magnetometers, optoelectronics, Hall-effect sensors, magnetic semiconductor sensors, and superconductive sensors.

At the present, silicon offers an advantage of inexpensive batch fabrication for magnetic sensors. Most of the integrated circuit magnetic sensors are manufactured by following the

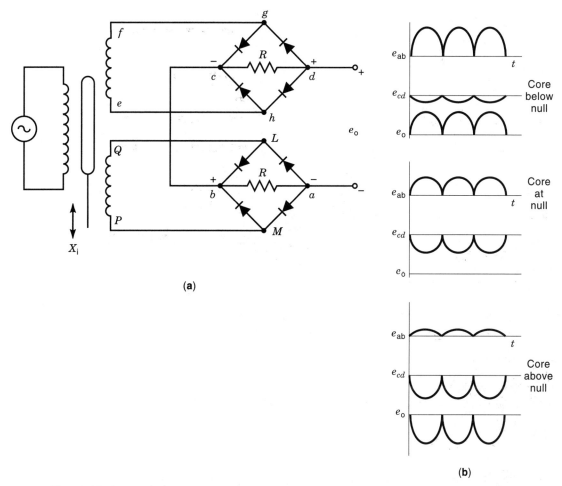

(a)

(b)

Figure 13. A typical phase-sensitive demodulation circuit based on diode bridges as in (a). The bridge 1 acts as a rectification circuit for secondary 1, and bridge 2 acts as a rectifier for secondary 2 where the net output voltage is the difference between the two bridges as in (b). The position of the core can be worked out from the amplitude of the dc output, and the direction of the movement of the core can be determined from the polarity of the voltage. For rapid movements of the core, the output of the diode bridges need to be filtered. For filters, a suitably designed simple RC filter may be sufficient.

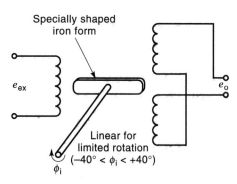

Rotational differential transformer

Figure 14. A rotary core differential transformer has an armature rotated by an externally applied force about a pivot point above the center leg of the core. When the armature is displaced from its reference or balance position, the reluctance of the magnetic circuit through one secondary coil is decreased; simultaneously the reluctance through the other coil is increased. The induced emfs in the secondary windings are different in magnitude and phase as a result of the applied displacement.

design rules of standard chip manufacturing. For example, MOS and complementary metal oxide semiconductor (CMOS) technologies are used to manufacture highly sensitive Hall-effect sensors, magnetotransistors, and other semiconductor sensors. Some of the magnetic semiconductor sensors are discussed in detail in the following sections.

MAGNETOMETERS

Magnetometers are devices that are produced to sense external magnetic fields mainly based on Faraday's Law of Induction. There are many different types of magnetometers including search coil magnetometers, SQUIDs, flux-gates, nuclear, and optical magnetometers. The two most commonly used types are the flux-gate and the search coil magnetometers.

Flux-Gate Magnetometers

Flux-gate magnetometers are made from two coils wound on a ferromagnetic material, as illustrated in Fig. 15. One of the coils is excited with a sinusoidal current, which drives the

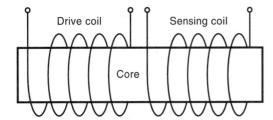

Figure 15. A flux-gate magnetometer consists of two coils wound on a ferromagnetic core. The driving coil is excited with a sinusoidal current, which drives the core into a saturation state. When saturated, the reluctance of the core to external magnetic field increases, thus repelling the external flux and hence reducing the effect of the external field on the second coil. The harmonics of the induced voltage in the sensing coil is an indication of the magnitude of the external magnetic field.

core into a saturation state. At this stage, the reluctance of the core to external magnetic field increases, thus repelling the external flux and so reducing the effect of the external field on the second coil. As the core becomes unsaturated, the effect of the external field increases. The increase and reduction of the effect of the external field on the second coil is sensed as the harmonics of the induced voltage. These harmonics are then directly related to the strength and variations in the external field. The sensitivity of these devices is dependent on the magnetic properties and the shape of the saturation curve of the core material. The measurement range can vary from 100 pT to 10 mT operating in the 0 kHz to 10 kHz frequency range.

Search Coil Magnetometers

Search coil magnetometers operate on the principle of Faraday's Law of Induction. A typical search coil magnetometer is shown in Fig. 16. The flux through the coil changes if the magnetic field varies in time or the coil is moved through the field. The sensitivity depends on the properties of the core material, dimensions of the coil, number of turns, and rate of change of flux through the coil.

Search coil magnetometers are manufactured from 4 cm in dimensions to 100 cm. They can sense weak fields as low as 100 pT within the frequency range of 1 Hz to 1 MHz. The upper limit is dependent on the relative magnitudes of resistance and inductance of the coil. During the signal processing, they can be used as part of a bridge or resonant circuits.

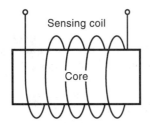

Figure 16. A search coil magnetometer uses Faraday's Law of Induction. The changing magnetic flux induces an emf in the coil. The sensitivity of the magnetometer depends on the properties of the core material, dimensions of the coil, number of turns, and rate of change of flux through the coil. They can sense weak fields as low as 100 pT within the frequency range of 1 Hz to 1 MHz.

SQUID Magnetometers

Superconducting quantum interference device sensors are used in many diverse applications from medicine, geophysics, and nuclear magnetic resonance to nondestructive testing of solid materials. The SQUID offers high sensitivity for the detection of weak magnetic fields and field gradients. They are made from conventional superconductors such as niobium operating at liquid helium temperatures. They are manufactured by using integrated circuit technology, as illustrated in Fig. 17.

The principle of operation is the Josephson effect. If a magnetic flux links to a ring-shaped superconducting material, a current is induced in the ring. This current flows forever due to the lack of any resistance. The intensity of current is proportional to the intensity of the field. The current in the ring is measured by using the Josephson effect wherein a weak link in the superconducting ring causes the superconducting current to oscillate as a function of magnetic field intensity. These oscillations can be sensed by many techniques such as coupling to a radio-frequency circuit and other resonance techniques.

The SQUIDs are extremely sensitive devices, with sensitivities ranging from 10 fT to 10 nT. The SQUID sensors can be arranged to measure magnetic fields in three dimensions in x, y, and z directions of Cartesian coordinates, as in the case of high-sensitivity gradiometers. One disadvantage of SQUIDs is that they need supercooling at very low temperatures. Much research is concentrated on materials exhibiting superconductivity properties at high temperatures. SQUIDs are used in gradiometers, voltmeters and amplifiers, displacement sensors, geophysics, gravity wave detection, and nondestructive testing, to name a few.

MAGNETORESISTIVE SENSORS

In magnetoresistive sensors, the magnetic field causes a change in resistance of some materials such as permalloys. In these materials, current passing through the material magnetizes the material in a particular magnetic orientation. An

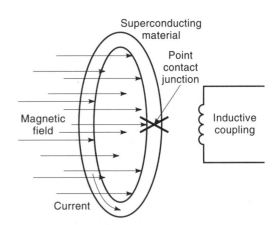

Figure 17. A SQUID consists of a superconducting ring. When subjected to an external field, the current is induced in the ring, which flows forever. The current in the ring is measured by using the Josephson effect and by creating a weak link in the ring. This weak link makes the superconducting current oscillate as a function of the external magnetic field intensity.

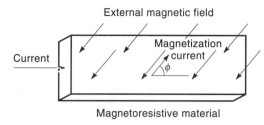

Figure 18. A magnetoresistive sensor's resistance changes in response to an external magnetic field. Magnetic domain orientation of the material is a function of external field and current flowing through it. The resistance is highest when the magnetization is parallel to the current and lowest when it is perpendicular to the current. These sensors are manufactured as thin films, and they have good linearity.

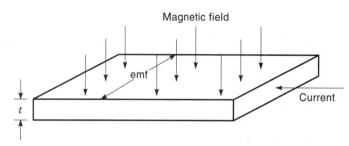

Figure 20. A Hall-effect sensor makes use of Lorentz force. The response of electrons to Lorentz force creates the Hall voltage, which is perpendicular to both the external magnetic field and the direction of current flow. If the current is dc, the voltage has the same frequency as magnetic flux.

external field perpendicular to the current, as illustrated in Fig. 18, causes the magnetic orientation to be disturbed. The resistance is highest when the magnetization is parallel to the current and lowest when it is perpendicular to the current. Hence, depending on the intensity of the external magnetic field, the resistance of the permalloy changes in proportion.

Magnetoresistive sensors are manufactured as thin films and usually integrated to be part of an appropriate bridge circuit. They have good linearity and low temperature coefficients. These devices have a sensitivity ranging from 1 μT to 50 mT. By improved electronics with suitable feedback circuits, the sensitivity can be as low as 100 pT. They can operate from dc to several gigahertz.

Magnetotransistor and Magnetodiode Sensors

Magnetotransistor and magnetodiode sensors are integrated silicon devices. They contain n-doped and p-doped regions forming pn, npn, or pnp junctions. In the case of magnetotransistors, there are two collectors, as shown in Fig. 19. Depending on the direction, an external magnetic field deflects electron flow between emitter and collector in favor of one of the collectors. The two collector voltages are sensed and related to the applied magnetic field. These devices are more sensitive than Hall-effect sensors.

In the case of magnetodiodes, p and n regions are separated by an area of undoped silicon containing the sensor. An external magnetic field perpendicular to the flow of charges deflects the holes and electrons in the opposite directions, re-

sulting effectively in a change in the resistance of the undoped silicon layer.

HALL-EFFECT SENSORS

In Hall-effect sensors, the voltage difference across a thin conductor carrying current depends on the intensity of the magnetic field applied perpendicular to the direction of current flow, as shown in Fig. 20. An electron moving through a magnetic field experiences Lorentz force perpendicular to the direction of motion and to the direction of the field. The response of electrons to Lorentz force creates a voltage known as the Hall voltage. If a current I flows through the sensor, the Hall voltage can mathematically be found by

$$V = R_\text{H} IB/t \tag{22}$$

where R_H is the Hall coefficient (cubic meters per degree Celsius), B is the flux density (tesla), and t is the thickness of the sensor (meters).

Therefore, for a specified current and temperature, the voltage is proportional to B. If the current is dc, the voltage has the same frequency as magnetic flux.

Hall-effect sensors can be made by using metals or silicon, but they are generally made from semiconductors with high electron mobility such as indium antimonide. They are usually manufactured in the form of probes with a sensitivity down to 100 μT. Silicon Hall-effect sensors can measure constant or varying magnetic flux having an operational frequency from dc to 1 MHz, within the range of 1 mT to 100 mT. They have good temperature characteristics from 200°C to near absolute zero.

MAGNETO-OPTICAL SENSORS

In recent years, highly sensitive magneto-optical sensors have been developed. These sensors are based on fiber-optics, polarization of light, Moire effect, and Zeeman effect, among others. This type of sensors leads to highly sensitive devices and is used in applications requiring high resolution such as human brain function mapping and magnetic anomaly detection. Here, because of the availability of space, only the polarization effect will be discussed briefly. Interested readers can find further information in the Refs. 1–6.

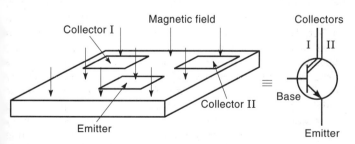

Figure 19. A magnetotransistor is an integrated silicon device that contains npn or pnp junctions. The electron flow between emitter and collector is influenced by the external magnetic field in favor of one of the collectors. The two collector voltages are sensed and related to the applied magnetic field.

In the polarization effect, a plane of polarized light in a strong magnetic field rotates its plane of vibration. The per-unit angular rotation is related to the per-unit magnetic field in a given length of material by the Verdet constant. For example, terbium gallium demonstrates a Verdet constant of 50 min/μT$\cdot$cm, whereas bismuth-substituted iron garnet can have up to 0.04 min/T$\cdot$cm. This polarization effect was first noticed by Faraday in 1845; hence, it is generally known as the Faraday effect. Recently, this principle was applied to semiconductors and crystals, which have different physical properties such as interband effects, intraband free career effects, and absorption of magnetism by impurities.

MAGNETIC THIN FILMS

Magnetic thin films are an important part of superconducting instrumentation, sensors and electronics in which active devices are made from deposited films. The thin films are usually made from amorphous alloys, amorphous gallium, and the like. As an example of this use of thin-film technology, a thin-film Josephson junction is given in Fig. 21. The deposition of thin films can be done by thermal evaporation, electroplating, sputter deposition, or chemical methods. The choice of technology depends on the characteristics of the sensors. For example, thin-film superconductors require low-temperature operations, whereas common semiconductors operate at room temperature.

The magnetic thin films find extensive applications in memory devices where high density and good sensitivities are required. In such applications, the magnetic properties of the coating are determined by the magnetic properties of the particles that can be controlled before coating. The choice of available materials for this purpose is extremely large. Thin-film technology is also developed in magneto-optics applications where erasable optical media for high-density magnetic storage is possible. The miniature magnetoresistive sensors for magnetic recording and pick-up heads, the Hall-effect sensors, and other magnetic semiconductors make use of thin-film technology extensively.

AMORPHOUS MAGNETIC MATERIALS

The amorphous magnetic materials can be classified as amorphous alloys (Fe, Co, Ni), amorphous rare earths, and amorphous superconductors. Amorphous alloys have good soft magnetic materials and are extensively used in magnetic heads. They have high-saturation magnetization, high permeability, and flat high-frequency dependence. Amorphous alloys are produced in the form of 20 μm to 50 μm thick ribbons using rapid solidification methods. Nevertheless, thermal stability of these alloys is the major drawback preventing the wider applications of these materials. Many of the amorphous metallic films are based on rare earth 3D transition metal alloys, such as Gd_xCo_x. They are used in bubble domain devices.

On the other hand, the amorphous superconductors are a class of superconducting materials such as bismuth and gallium. They are manufactured in the form of powders, ribbons, or flakes by using evaporation methods, ion mixing or ion implantation, or liquid quenching. The vapor deposition technique is used in semiconductor-type sensors. These materials are used in high field magnets, memory devices, and other computer applications.

SHIELDING AND SENSITIVITY TO ELECTROMAGNETIC INTERFERENCE

Magnetic fields are produced by currents in wires and more strongly by the coils. The fields produced by coils are important as a result of magnetic coupling, particularly when there are two or more coils in the circuit. The magnetic coupling between coils may be controlled by large spacing between coils, the orientation of coils, the shape of the coils, and shielding.

Inductive sensors come in different shapes and sizes. Even though some sensors have closed cores such as toroidal shapes, others have open cores and air gaps between cores and coils. Closed cores may have practically zero external fields, except small leakage fluxes. Even if the sensors do not have closed cores, most variable-inductor sensors have a rather limited external field, as a result of two neighboring sets of coils connected in opposite directions, thus minimizing the external fields. Because the inductive sensors are made from closed conductors, a current will flow, if the conductor moves in a magnetic field. Alternatively, a magnetic change produces current in stationary closed conductor. Unless adequate measures are taken, there may be external magnetic fields linking (interference) with the sensor coils, thus producing currents and unwanted responses.

Because of inherent operations, inductive sensors are designed to have a high sensitivity to magnetic flux changes. External electromagnetic interference and external fields can affect the performance of the sensors seriously. It is known that moderate magnetic fields are found near power transformers, electrical motors, and power lines. These small fields produce current in the inductive sensors elements. One way of eliminating external effects is accomplished by magnetic shielding of the sensors and by grounding appropriately. In magnetic shielding, one or more shells of high-permeability magnetic materials surround the part to be shielded. Multiple shells may be used to obtain a very complete shielding. Ends of each individual shell are separated by insulation so that the shell does not act as a single shorted turn, thus accommodating high current flows. Similarly, in the case of multiple shielding, shells are isolated from each other by proper insulation.

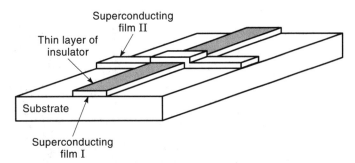

Figure 21. The magnetic thin films are made from deposited thin films from amorphous alloys such as gallium. This figure illustrates thin-film technology to form a Josephson. The deposition of thin films can be done by thermal evaporation, electroplating, sputter deposition, or chemical methods.

Alternating magnetic fields are also screened by interposing highly conductive metal sheets such as copper or aluminium on the path of the magnetic flux. The eddy currents induced in the shield give a counter mmf that tends to cancel the interfering magnetic field. This type of shielding is particularly effective at high frequencies. Nevertheless, appropriate grounding must be observed.

In many inductive sensors, stray capacitances may be a problem, especially at null position of the moving core. If the capacitive effect is greater than a certain value, say 1% of the full-scale output, this effect may be reduced by the use of a center-tapped supply and appropriate grounding.

BIBLIOGRAPHY

J. P. Bentley, *Principles of Measurement Systems,* 2nd ed., Burnt Mills, UK: Longman Scientific and Technical, 1988.

E. O. Doebelin, *Measurement Systems: Application and Design,* 4th ed., New York: McGraw-Hill, 1990.

J. P. Holman, *Experimental Methods for Engineers,* 5th ed., New York: McGraw-Hill, 1989.

J. E. Lenz, A review of magnetic sensors, *Proc. IEE,* **78** (6): 973–989, 1990.

W. Gopel, J. Hesse, and J. N. Zemel, *Sensors—A Comprehensive Survey,* Weinheim, Germany: WCH, 1989.

J. Evetts, *Concise Encyclopedia of Magnetic and Superconducting Materials,* New York: Pergamon, 1992.

HALIT EREN
Curtin University of Technology

MAGNETIC SHIELDING

Shielding is the use of specific materials in the form of enclosures or barriers to reduce field levels in some region of space. In traditional usage, *magnetic shielding* refers specifically to shields made of *magnetic* materials like iron and nickel. However, this article is more general because it covers not just traditional magnetic shielding but also shielding of alternating magnetic fields with conducting materials, such as copper and aluminum. In typical applications, shielding eliminates magnetic field interference with electron microscopes, computer displays (CRTs), sensitive electronics, or other devices affected by magnetic fields.

Although shielding of electric fields is relatively effective with any conducting material, shielding of magnetic fields is more difficult, especially at extremely low frequencies (ELF). The ELF range is defined as 3 Hz to 3 kHz (1). The selection of proper shield materials, shield geometry, and shield dimensions are all important factors in achieving a specified level of magnetic field reduction. Placing a shield around a magnetic field source, as shown in Fig. 1(a), reduces the field magnitude outside the shield, and placing a shield around sensitive equipment, as shown in Fig. 1(b), reduces the field magnitude inside the shield. These two options are often called shielding the source, or shielding the subject, respectively.

Both examples in Fig. 1 illustrate closed shield geometry. In many applications, it is impractical or impossible (due to physical constraints) to use an enclosure, and open shield geometries, also called partial shields, are required. Figure 2

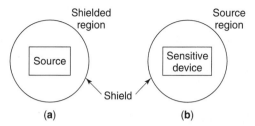

Figure 1. (a) Shielding the source. Placing a shield around a field source reduces the fields everywhere outside the shield. (b) Shielding the subject. A shield placed around a sensitive device reduces the fields from external sources.

shows two basic partial shield geometries, a flat plate shield (a), and a channel shield (b). For these configurations, the region where shielding occurs may be limited because the shield does not fully enclose the source or the subject, resulting in *edge effects*. A discussion of the geometrical aspects of shielding is contained in (2).

ELF SHIELDING VERSUS HIGH-FREQUENCY SHIELDING

Electric and magnetic fields radiate away from a source at the speed of light c. In the time it takes a source alternating with frequency f to complete one full cycle, these fields have traveled a distance λ, known as the electromagnetic wavelength:

$$\lambda = c/f \tag{1}$$

At distances from a field source on the order of one wavelength and larger, the dominant parts of the electric and magnetic fields are coupled as a propagating electromagnetic wave. If a shield is placed in this region, shielding involves the interaction of electromagnetic waves with the shield materials. Any mathematical description must be based on the full set of Maxwell's equations which involves calculating both electric and magnetic fields. Shielding of electromagnetic waves is often described in terms of reflection, absorption, and transmission (3). Because wavelength decreases with increasing frequency, shielding at radio frequencies in the FM band (88 MHz to 108 MHz) and higher typically involves the interaction of electromagnetic waves with shield materials.

At distances much less than one wavelength, the nonradiating portion of the fields is much larger than the radiating portion. In this region, called the reactive near-field region, the coupling between electric and magnetic fields can be ignored, and the fields may be calculated independently. This is called a quasistatic description. At 3 kHz (the upper end of the ELF band), a wavelength in air is 100 km. Thus, for ELF field sources, one is in the reactive near-field region in all practical cases, and a full electromagnetic solution is not re-

Figure 2. Examples of open shield geometries: (a) flat plate shield, (b) inverted channel shield.

quired. Instead, one need focus only on interaction of the magnetic field or the electric field with the shield material, depending on which field is being shielded. In some cases, shielding of the electric field with metallic enclosures is required. This article deals specifically with the shielding of dc and ELF magnetic fields.

MAGNETIC FIELDS

Moving electric charges, typically currents in electrical conductors, produce magnetic fields. Magnetic fields are defined by the Lorentz equation as the force acting on a test charge q, moving with velocity $\mathbf{v}$ at a point in space:

$$\mathbf{F} = q(\mathbf{v} \times \mu\mathbf{H}) \qquad (2)$$

in which $\mathbf{H}$ is the magnetic field strength with units of amperes per meter and μ is the permeability of the medium. By definition of the vector cross-product, the force on a moving charge is at right angles to both the velocity vector and the magnetic field vector. Lorentz forces produce torque in generators and motors and focus electron beams in imaging devices.

Unwanted, or stray magnetic fields deflect electron beams in the same imaging devices, often causing interference problems. Sources that use, distribute, or produce alternating currents, like the 60 Hz currents in a power system, produce magnetic fields that are time-varying at the same frequency.

Magnetic fields are vector fields with magnitude and direction that vary with position relative to their sources. This spatial variation or field *structure* depends on the distribution of sources. Equal and opposite currents produce a field structure that can be visualized by plotting lines of magnetic flux, as shown in Fig. 3. The spacing between flux lines, or line density, indicates relative field magnitudes, and the tangent to any flux line represents field direction. Another way to visualize field structure is through a vector plot, shown in Fig. 4. Lengths of the arrows represent relative field magnitudes, and the arrows indicate field direction.

Shield performance, or field reduction, is measured by comparing field magnitudes before shielding with the field magnitudes after shielding. In general, field reduction varies with

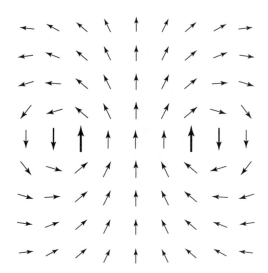

Figure 4. A vector plot graphically illustrates both field strength and direction as a function of position.

position relative to the source and shield. The shielding factor s is defined as the ratio of the shielded field magnitude $\mathbf{B}$ to the field magnitude $\mathbf{B_0}$ without the shield present at a point in space:

$$s = |\mathbf{B}|/|\mathbf{B_0}| \qquad (3)$$

The shielding factor represents the fraction of the original field magnitude that remains after the shield is in place. A shielding factor of zero represents perfect shielding. A shielding factor of one represents no shielding, and shielding factors greater than one occur at locations where the field is increased by the shield. It is incorrect to define the shielding factor as the ratio of the fields on opposite sides of a shield. Shielding factor is often called shielding effectiveness, expressed in units of decibels (dB):

$$\text{s.e. (dB)} = -20\log_{10}|\mathbf{B}|/|\mathbf{B_0}| \qquad (4)$$

Shielding effectiveness is sometimes alternatively defined as the inverse of the shielding factor, the ratio of unshielded to shielded fields at a point, but it is really a matter of preference. For example, a shielding effectiveness of two defined in this manner represents a twofold reduction, that is, the field is halved by the shield and the shielding factor is 0.5. When fields are time-varying, shielding is typically defined as the ratio of rms magnitudes.

SHIELDING MECHANISMS

Although shielding implies a *blocking* action, dc and ELF magnetic field shielding is more aptly described as altering or restructuring magnetic fields by the use of shielding materials. To illustrate this concept, Fig. 5(a) shows a flux plot of a uniform, horizontal, magnetic field altered (b) by the introduction of a ferromagnetic material.

There are two basic mechanisms by which shield materials alter the spatial distribution of magnetic fields, thus providing shielding. They are the flux-shunting mechanism and the induced-current mechanism (5).

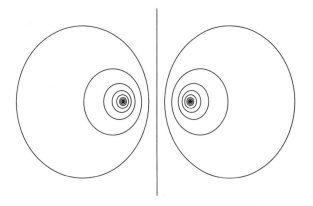

Figure 3. The lines of magnetic flux illustrate the field structure associated with one or more sources. The density of flux lines indicates the relative field strength and the tangent to any line indicates the field direction at that point.

Flux Shunting

An externally applied magnetic field induces magnetization in ferromagnetic materials. (All materials have magnetic properties, but in most materials these properties are insignificant. Only ferromagnetic materials have properties that provide shielding of magnetic fields.) Magnetization is the result of electrons acting as magnetic sources at the atomic level. In most matter, these sources cancel one another, but electrons in atoms with unfilled inner shells make a net contribution, giving the atoms a magnetic moment (6). These atoms spontaneously align into groups called domains. Without an external field, domains are randomly oriented and cancel each other. When an external field is applied, the Lorentz forces align some of the domains in the same direction, and together, the domains act as a macroscopic magnetic field source. A familiar magnetic field source is a bar magnet, which exhibits permanent magnetization even without an applied field. Unlike a permanent magnet, most of the magnetization in ferromagnetic shielding materials goes away when the external field is removed.

Basic ferromagnetic elements are iron, nickel, and cobalt, and the most typical ferromagnetic shielding materials are either iron-based or nickel-based alloys (metals). Less common as shielding materials are ferrites such as iron oxide.

Induced magnetization in ferromagnetic materials acts as a secondary magnetic field source, producing fields that add vectorially to the existing fields and change the spatial distribution of magnetic fields in some region of space. The term *flux-shunting* comes from the fact that a ferromagnetic shield alters the path of flux lines so that they appear to be *shunted* through the shield and away from the shielded region, as

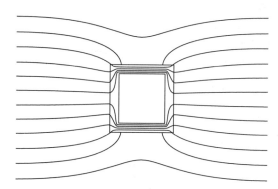

Figure 6. Example of the flux-shunting mechanism. The region inside a ferromagnetic duct is shielded from an external, horizontal magnetic field.

shown by the example in Fig. 6. Flux-shunting shielding is often described in terms of magnetic circuits as providing a low-reluctance path for magnetic flux.

Permeability μ is a measure of the induced magnetization in a material. Thus, permeability is the key property for flux-shunting shielding. The constitutive law

$$\mathbf{B} = \mu\mathbf{H} \tag{5}$$

relates magnetic flux density $\mathbf{B}$ to the magnetic field strength $\mathbf{H}$. More typically used, relative permeability is the ratio of permeability in any medium to the permeability of free space, $\mu_r = \mu/\mu_0$. Nonferrous materials have a relative permeability of one, and ferromagnetic materials have relative permeabilities much greater than one, ranging from hundreds to hundreds of thousands. In these materials, permeability is not constant but varies with the applied field $\mathbf{H}$.

The nonlinear properties of a ferromagnetic material can be seen by plotting flux density $\mathbf{B}$, as the applied field $\mathbf{H}$ is cycled. Figure 7 shows a generic $\mathbf{B}$–$\mathbf{H}$ plot that illustrates hysteresis. When the applied field is decreased from a maximum, the flux density does not return along the same curve, and plotting one full cycle forms a hysteretic loop. A whole family of hysteretic loops exists for any ferromagnetic material as the amplitude of field strength $\mathbf{H}$ is varied. The area of a hysteretic loop represents the energy required to rotate magnetic domains through one cycle. Known as hysteretic losses, this energy is dissipated as heat in the shield material.

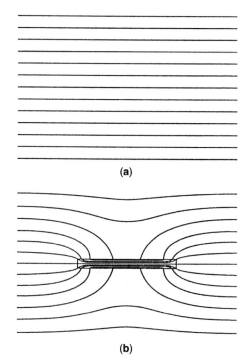

(a)

(b)

Figure 5. (a) Horizontal uniform field (b) altered by introduction of a ferromagnetic material; illustrates the concept that shielding is the result of induced sources in the shield material.

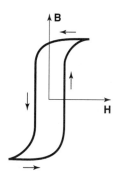

Figure 7. Typical $\mathbf{B}$–$\mathbf{H}$ curves showing how nonlinear properties of ferromagnetic materials result in a hysteretic loop as the applied field $\mathbf{H}$ is cycled.

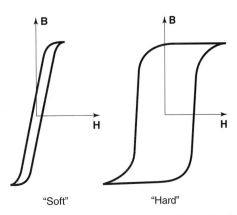

Figure 8. Examples of hysteretic loops for *soft* and *hard* ferromagnetic materials.

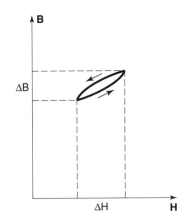

Figure 10. Hysteretic loop formed by a small ac field in the presence of a larger dc field.

For effective flux-shunting shielding, the flux density in a magnetic material should follow the applied field closely. However, it is obvious from the hysteretic loop of Fig. 7 that **B** does not track **H**. **B** lags **H**, as seen by the fact that there is a residual flux density (nonzero **B**) when **H** has returned to zero and that **B** does not return to zero until **H** increases in the opposite direction. Thus, *soft* ferromagnetic materials with narrow hysteretic loops are best for shielding, in contrast to *hard* ferromagnetic materials with wide hysteretic loops, typically used as permanent magnets and in applications such as data storage. Hysteretic curves illustrating "soft" and "hard" ferromagnetic materials are shown in Fig. 8.

At very low field levels relative permeability starts at some initial value (initial permeability) increases to a maximum as the applied field is increased, and then decreases, approaching a relative permeability of one as the material saturates, as shown in Fig. 9. Saturation occurs because there is a limit to the magnetization that can be induced in any magnetic material. In Fig. 7, the decreasing slope at the top and bottom of the curves occurs as the limit of total magnetization is reached. When a material saturates, it cannot provide additional shielding.

For shielding alternating magnetic fields via flux-shunting, the key property is ac permeability, $\Delta B/\Delta H$ through one cycle. Although Fig. 7 shows a hysteresis curve that swings from near saturation to near saturation in both directions, a hys-

teresis curve caused by a very small alternating field in the presence of a larger dc field might look like Fig. 10. In this case, the ac permeability is less than the dc permeability, **B/H**. In addition, the dc field creates a constant magnetization that affects the time-varying magnetization. Figure 11 shows how ac permeability for a small alternating field is reduced with increasing dc field. This plot, called a *butterfly* curve, is generated by measuring the ac permeability at different levels of dc field. The dc field is increased from zero to a maximum, reversed to the same maximum in the opposite direction, and then reduced to zero, and the ac permeability is measured at different points to generate the *butterfly* curve. The extent to which the ac permeability is affected depends on the properties of each ferromagnetic material. In general, the better ferromagnetic materials are more sensitive. This type of curve is relevant for shielding small ac fields in the presence of a larger dc field.

To gain an understanding of how flux-shunting varies with shield parameters, one can look at the analytical expression for the shielding provided by a ferromagnetic spherical shell with radius a, shield thickness Δ (that is much smaller than the radius), and relative permeability μ_r:

$$s = \frac{3a}{2\mu_r \Delta} \qquad (6)$$

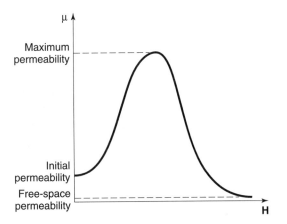

Figure 9. Permeability as a function of applied field strength.

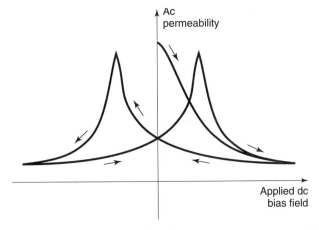

Figure 11. *Butterfly* curve illustrates how the ac permeability changes as a much larger dc field is applied and removed.

Table 1. Properties of Typical Shielding Materials

Name	Material Type	Max. Relative Permeability	Saturation Flux Density, T	Conductivity, S/m	Density, kg/m³
Cold-rolled steel	Basic steel	2,000	2.10	1.0×10^7	7880
Silicon iron	Electrical steel	7,000	1.97	1.7×10^6	7650
45 Permalloy	45% nickel alloy	50,000	1.60	2.2×10^6	8170
Mumetal	78% nickel alloy	100,000	0.65	1.6×10^6	8580
Copper	High conductivity	1	NA	5.8×10^7	8960
Aluminum	High conductivity	1	NA	3.7×10^7	2699

From Hoburt (8).

Equation (6) shows that shielding improves (shielding factor decreases) with increasing relative permeability and increasing shield thickness. It also shows that shielding gets worse with increasing shield radius. From the perspective of magnetic circuits, shielding improves as the reluctance of the flux path through the shield is lowered. Increasing permeability and thickness reduce the reluctance, improving shielding. Increasing shield radius increases reluctance by increasing the path length of the magnetic circuit, making shielding worse. In short, the flux-shunting mechanism works best in small, closed-geometry shields.

Flux-shunting shielding has been studied for a long time. A journal article (7) dating back to 1899 describes an effect whereby increased shielding is obtained using nested shells of ferromagnetic material with nonmagnetic materials or air gaps between the ferromagnetic shells. In other words, by changing the shield from a single thick layer to thinner double or triple layers, one can in some cases enhance the shielding effectiveness although using the same amount or even less ferromagnetic material. This effect occurs mainly with configurations where the total shield thickness is within an order of magnitude of the shield radius.

In some cases, a double layer shield is used to avoid saturation of the layer closest to the field source where the fields are strongest. For example, a steel material might be used as the first shield layer, whereas a high-performance nickel alloy is used as the second layer. The steel lowers the field enough that the nickel-alloy layer is not saturated. Saturation flux densities of typical shield materials are listed in Table 1.

Induced-Current Mechanism

Time-varying magnetic flux passing through a shield material induces an electric field in the material according to Faraday's law:

$$\nabla \times \mathbf{E} = -\frac{\partial \mathbf{B}}{\partial t} \tag{7}$$

In electrically conducting materials, the induced electric field results in circulating currents, or eddy currents, in the shield according to the constitutive relationship:

$$\mathbf{J} = \sigma \mathbf{E} \tag{8}$$

where $\mathbf{J}$ is the current density, σ is the material conductivity, and $\mathbf{E}$ is the electric field induced according to Eq. (7). The fields from these induced currents oppose the impinging fields, providing field reductions. Figure 12 shows a flat plate in a uniform field. Induced-current shielding appears to exclude flux lines from the shield, providing field reductions adjacent to the shield on both sides. Because the induced currents are proportional to the time rate-of-change of the magnetic fields, induced-current shielding improves with increasing frequency. Thus, at higher frequencies, magnetic fields are more easily shielded via the induced-current mechanism. In the limit of infinite conductivity or infinite frequency, flux lines do not penetrate the shield as shown in Fig. 13.

In a conducting shield, the magnetic field and induced-current magnitudes decrease exponentially in the direction of the shield's thickness with a decay length called the skin depth δ:

$$\delta = \sqrt{\frac{1}{\pi f \sigma \mu}} \tag{9}$$

which involves not only frequency f and conductivity σ but also permeability μ because it affects the flux density which induces the circulating currents. Because of exponential de-

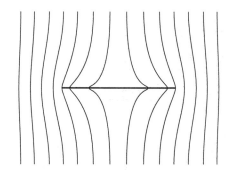

Figure 12. Conducting plate in an alternating vertical field tends to exclude flux from passing through the plate, thus providing shielding.

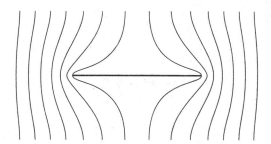

Figure 13. In the limit of zero resistivity or infinite frequency, a conducting shield totally excludes flux lines.

cay, shield enclosures with thickness on the order of a skin depth or thicker provide good shielding. For shield thicknesses much less than a skin depth, the induced current densities are constant across a shield thickness. However, significant shielding can still be obtained from thin conducting shields in some situations where the shield is sized properly. In these cases the shielding is a result of induced currents flowing over large loops.

The shielding factor equation for a nonferrous, conducting, spherical shield with radius a, thickness Δ, and conductivity σ provides insight into how these parameters affect the induced-current mechanism:

$$s = \frac{1}{\sqrt{1 + \left(\frac{2\pi f \mu_0 \sigma a \Delta}{3}\right)^2}} \qquad (10)$$

Because all parameters are in the denominator of Eq. (10), induced-current shielding improves (shielding factor decreases) with increasing frequency f, increasing shield thickness, and increasing shield radius. The effect of shield radius is opposite to that for flux-shunting shielding, and although flux-shunting shields static fields, the induced-current mechanism does not. In general, induced-current shielding is more effective for larger source-shield configurations whereas flux-shunting is more effective for smaller shield configurations.

Combined Shielding Mechanisms

Until now, the shielding mechanisms have been discussed separately. Equations (6) and (10) are the shielding factor equations for flux-shunting and induced-current shielding alone. In many shields, both mechanisms are involved. For example, most ferromagnetic materials, being metals, also have significant conductivity in addition to high permeability. Or a shield might be constructed using two materials, one layer of a high permeability material and one layer of a high conductivity material. In these cases both shielding mechanisms contribute to the shielding to an extent that depends on material properties, frequency of the fields, and details of the shield configuration.

To illustrate the combined effect of both shielding mechanisms, Fig. 14 shows the shielding factor, calculated by a method described in Ref. 8, as a function of shield radius for a spherical steel shield in a 60 Hz uniform field. For these calculations steel is assigned a conductivity of 6.76×10^6 S/m, a relative permeability of 180, and the shield thickness of 1 mm is held constant as the shield radius is varied. Flux-shunting dominates at the smaller radii, induced-current shielding dominates at the larger radii, and there is a worst case radius of about 0.4 m where a transition occurs between the dominant shielding mechanisms.

The combined effect of both flux-shunting and induced-current shielding can be exploited with multilayer shields made from alternating ferromagnetic and high-conductivity materials. Also using the method described in Ref. 8, one can explore this type of shield construction. Alternating thin layers of high permeability and high conductivity perform like a single-layer shield made with a material with enhanced properties.

SHIELDING MATERIALS

Basic magnetic field shielding materials can be grouped in two main categories: ferromagnetic materials and high conductivity materials. For dc magnetic fields, ferromagnetic materials are the only option. They provide shielding through the flux-shunting mechanism. For ac magnetic fields, both ferromagnetic and high conductivity materials may be useful as shielding materials, and both shielding mechanisms operate to an extent determined by the material properties, operating frequency, and shield configuration.

The practical high conductivity materials are those commonly used as electrical conductors, aluminum and copper. Copper is almost twice as conductive as aluminum, but aluminum is about 3.3 times lighter than copper and generally costs less than copper on a per pound basis. For shielding that depends on the induced-current mechanism, conductivity across a shield is paramount and copper has the advantage that it is easily soldered whereas aluminum is not—it should be welded. Mechanical fasteners can be used for connecting aluminum or copper sheets, but the longevity of these connections is questionable due to corrosion and oxidation.

Although there appears to be a large variety of ferromagnetic shielding materials, most fit into one of five basic types:

- Basic iron or steel—typically produced as coils and sheet for structural uses
- Electrical steels—engineered for good magnetic properties and low losses when used as cores for transformers, motors, etc.
- 40 to 50% nickel alloys—moderately expensive materials with very good magnetic properties
- 70 to 80% nickel alloys—highest cost materials with the best magnetic properties, often referred to generically as mumetal, although this was originally a trade name.
- Amorphous metals—noncrystalline metallic sheet formed by an ultrarapid quenching process that solidifies the molten metal; the noncrystalline form provides enhanced ferromagnetic properties.

Different manufacturers produce slightly different compositions of these basic materials, and they have different procedures for heat treating, but the percentages of the main elements, iron or nickel, are similar. There are only a few large producers of nickel-alloy materials. Shielding manufacturers

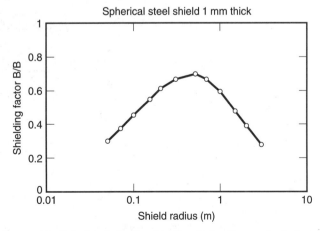

Figure 14. Calculated 60 Hz shielding factor for a spherical steel shell in a uniform magnetic field as a function of shield radius. The shield thickness of one millimeter is held constant.

typically purchase materials from a large producer, heat the materials in a hydrogen atmosphere (hydrogen annealing) to improve the ferromagnetic properties, and then utilize the metal to fabricate a shield enclosure or shield panels. Smaller shields are often annealed after fabrication because the fabrication process may degrade the magnetic properties.

Important properties for ferromagnetic shield materials are the initial permeability, the maximum permeability, and the magnetic field strength (or flux density) at which the material saturates and further shielding cannot be obtained. Because the ferromagnetic properties are nonlinear, the operating permeability depends on the magnitude of the magnetic field being shielded. In general, increasing magnetic properties go hand in hand with increasing cost, lower saturation levels, and lower conductivity. Table 1 shows nominal values of maximum permeability, saturation flux density, conductivity, and density for basic shielding materials including copper and aluminum (9). Note that the initial permeabilities of ferromagnetic materials are often one and two orders of magnitude smaller than the maximum permeabilities (see Table 1).

SHIELDING CALCULATIONS

Because there are an infinite variety of shield-source configurations and a wide variety of shield materials for building effective magnetic field shields, shielding calculations are a key part of practical shield design. Elaborate experiments need not be made to characterize the performance of each unique shield design. Extensive experiments are not only impractical but unnecessary. However, closed-form analytical expressions exist only for a limited set of ideal shield geometries, such as cylindrical shells, spherical shells, and infinite flat sheets. Even for these ideal shield geometries, the expressions can be quite complicated, especially solutions for shields with more than one material layer. For general shielding calculations, one must either select a simple approximation to obtain an order of magnitude shielding estimate or utilize more complex numerical methods to solve the shielding problem.

In high frequency shielding, calculations for plane waves propagating through infinite sheets are used to arrive at shielding estimates. Because the resulting equations are analogous to transmission line equations, this method is often called the transmission line approach (10). As described previously, this approach is not relevant to ELF shielding except for a limited set of conditions. Reference 8 describes a technique similar to the transmission line approach, but specifically tailored to ELF magnetic field shielding calculations for ideal shield geometries with multiple layers having different material properties. This method is well suited for calculations involving nested cylindrical or spherical shields or shields constructed from alternating layers of conducting and ferromagnetic materials.

Another technique found in literature is the circuit approach (11). In this method typically used to calculate ELF induced-current shielding, the shield enclosure is viewed as a shorted turn that can be characterized by an inductance and resistance. This method suffers from the assumption that significant details of field structure for the shielding problem are known a priori to properly set the circuit parameters. This severely limits application of the method.

General modeling of ELF magnetic field shielding amounts to calculating magnetic fields in the presence of conducting and ferromagnetic materials. The computation must account for induced currents and magnetization throughout the shield material. This involves solutions to the quasistatic form of Maxwell's equations for magnetic fields over a continuum that represents the problem region. In differential form the basic equations to be solved are the following:

$$\nabla \times \mathbf{H} = \mathbf{J} \tag{11}$$

$$\nabla \cdot \mathbf{B} = 0 \tag{12}$$

$$\nabla \times \mathbf{E} = -\frac{\partial \mathbf{B}}{\partial t} \tag{13}$$

along with the constitutive relationships for permeability, Eq. (5), and conductivity, Eq. (8), which describe the macroscopic properties of shield materials. This quasistatic description, which ignores the displacement current term $\partial \mathbf{D}/\partial t$, normally on the right-hand side of Eq. (11), is valid as long as an electromagnetic wavelength is much larger than the largest dimension of the shield. General solutions to these equations are often called *eddy current* or *magnetic diffusion* solutions. At zero frequency or zero conductivity in the shield, there are no induced currents. Only permeability restructures the magnetic field. This simplification is called the magnetostatic case, and solutions must satisfy only Eq. (11) and Eq. (12), along with the constitutive relationship that defines permeability, Eq. (5).

In finding exact solutions to the governing magnetic field equations previously described, one approach is to define a vector potential $\mathbf{A}$ that satisfies Eq. (12):

$$\nabla \times \mathbf{A} = \mathbf{B} \tag{14}$$

Substituting Eq. (14) in Eq. (13),

$$\mathbf{E} = -\frac{\partial \mathbf{A}}{\partial t} \tag{15}$$

Combining Eqs. (8), (11), (14), (15), and using a vector identity gives the following:

$$\nabla^2 \mathbf{A} - \mu\sigma\frac{\partial \mathbf{A}}{\partial t} = -\mu\mathbf{J_s} \tag{16}$$

in which $\mathbf{J_s}$ is the known distribution of source currents producing magnetic fields that require shielding.

When the source currents are sinusoidal, $\mathbf{A}$ and $\mathbf{J_s}$ can be represented as phasors, and the time derivative in Eq. (16) is replaced by $j\omega$:

$$\nabla^2 \mathbf{A} - j\omega\mu\sigma\mathbf{A} = -\mu\mathbf{J_s} \tag{17}$$

When the shield material has zero conductivity or the magnetic fields are constant (zero frequency), Eq. (17) becomes

$$\nabla^2 \mathbf{A} = -\mu\mathbf{J_s} \tag{18}$$

Equation (17) can be used for the general case where a shield provides field reduction through both flux-shunting and induced-current mechanisms. Equation (18) is only for flux-shunting. The shielding factor for a specific source-shield configuration is determined by first solving for the magnetic vector potential **A** without the shield in the problem and then solving for **A** with the shield. Using Eq. (14), one calculates the flux densities from both vector potential solutions. Ratios of the field magnitudes as in Eq. (2) define the field reduction provided by the shield as a function of position.

NUMERICAL SOLUTIONS FOR SHIELDING

Except for the ideal shield geometries mentioned previously, solving the governing equations requires numerical methods. Two common numerical techniques are the finite-element method and the boundary-integral method (12,13,14).

In the finite-element method, the problem region is subdivided into elements—typically triangles for two-dimensional problems and tetrahedra for three-dimensional problems—that form a *mesh*. The continuous variation of vector potential **A** over each element is approximated by a specified basis function. Then the unknowns become the coefficients of the basis function for each element. Variational concepts are used to obtain an approximate solution to the governing partial differential equation, for example, Eq. (17), across all elements. The net result is a system of algebraic equations that must be solved for the unknowns. Finite-element software is commercially available, and features that provide automatic meshing, graphical preprocessing, and visualization of results make it an accessible and useful general shield calculating tool for some shield problems, especially problems that can be modeled in two-dimensions or problems with symmetry about an axis. Figures 3, 5, 6, 12, and 13 were produced with finite-element software.

However, there are weaknesses to the finite-element method. Shield geometries typically involve very thin sheets of materials with much larger length and width dimensions. This, along with the need to accurately model significant changes in field magnitudes across the shield thickness, requires large numbers of elements in the shield region. Shielding problems are also characterized by large regions of air and complicated systems of conductors that are the field sources for the problem. In terms of energy density, the fields in the shielded region are negligible compared with fields near the sources, so one cannot rely on energy as the criterion for determining when an *adequate* solution has been obtained. Finally, solving the partial differential equations means that the problem region must be bounded and a boundary condition must be specified at the edges. The problem region must be made large enough that the boundary conditions do not affect the solution in the region where shielding is being calculated. This results in more unknowns and a larger problem to solve.

Instead of differential equations, it is also possible to use the integral form of the quasistatic equations. For determining magnetic fields in air due to some distribution of currents, one can derive an integral equation, often called the Biot–Savart law, that gives the magnetic field contribution at a point in space due to a differential *piece* of current density:

$$\mathbf{H} = \frac{1}{4\pi} \int_{V'} \frac{\mathbf{J}(\mathbf{r}') \times (\mathbf{r} - \mathbf{r}')}{|\mathbf{r} - \mathbf{r}'|^3} \, dv' \tag{19}$$

in which $\mathbf{J}(\mathbf{r}')$ is the current density in the problem as a function of position defined by the vector $\mathbf{r}'$ (from the origin to the integration point) and $\mathbf{r}$ defines the point where the magnetic field is being evaluated (vector from origin to the field evaluation point). Integrating over all of the currents in the problem gives the total field at one point in space. This equation is not valid when shield materials, that is, conducting and ferromagnetic materials, are introduced into the problem region. The boundary integral method overcomes this difficulty by replacing the effect of magnetization or induced-currents within the materials with equivalent sources at the surface of the materials where discontinuities in material properties occur. In contrast to the finite-element method, only the surfaces are divided into elements. Basis functions are used to approximate a continuous distribution of equivalent sources over these surfaces, and a system of equations is developed in which the unknowns are the coefficients for the basis functions. After solving for the unknown sources on the shield surface, one can then calculate the new magnetic field at any point by combining the contributions of all sources—the original field sources and the induced sources in the shield—to obtain the *shielded* magnetic field distribution.

The key advantages of the boundary-integral method are that only the surfaces of the shield need to be subdivided into elements and that the method is ideal for open boundary problems with a large air region. The method is also ideally suited for complex systems of currents. Thus, the boundary-element method is better suited for three-dimensional problems than the finite-element method. The main weakness of the boundary-integral method is that it results in a full system of equations that is more difficult to solve than the sparse system produced by the finite-element method. An integral method based on surface elements, developed expressly for solving three-dimensional quasistatic shielding problems, is described in (15).

The underlying theoretical basis for shield calculations is as old as electricity itself and goes back to Faraday and Maxwell. Although materials science is a rapidly changing area with developments in composite materials and materials processing, the basic materials for shielding of dc and ELF magnetic fields have, for the most part, remained unchanged. For basic shield configurations, calculations are straightforward. However, actual application of shielding requires practical expertise in addition to theoretical knowledge. For example, construction methods used to fabricate a shield from multiple sheets must ensure that conductivity and permeability are maintained across the entire shield surface, especially in critical directions. Edge effects and holes in shields for conduits, doors, windows, etc. degrade shield performance and must be accounted for early in the design process. With proper shield calculating tools and proper construction practices, shields can be designed that attenuate magnetic fields by factors ranging from 10 to 1000 (shielding factors ranging from 0.100 to 0.001), thus eliminating problems with stray or unwanted magnetic fields.

BIBLIOGRAPHY

1. *IEEE Standard Dictionary of Electrical and Electronics Terms* ANSI Std 100-1997, 6th ed., New York: IEEE, 1997.

2. L. Hasselgren and J. Luomi, Geometrical aspects of magnetic shielding at extremely low frequencies, *IEEE Trans. Electromagn. Compat.*, **37**: 409–420, 1995.

3. R. B. Schulz, V. C. Plantz, and D. R. Brush, Shielding theory and practice, *IEEE Trans. Electromagn. Compat.,* **30**: 187–201, 1988.

4. J. F. Hoburg, Principles of quasistatic magnetic shielding with cylindrical and spherical shields, *IEEE Trans. Electromagn. Compat.,* **37**: 547–579, 1995.

5. T. Rikitake, *Magnetic and Electromagnetic Shielding,* Boston: D. Reidel, 1987.

6. R. M. Bozorth, *Ferromagnetism.* Piscataway, NJ: IEEE Press, 1993 Reprint.

7. A. P. Wills, On the magnetic shielding effect of trilamellar spherical and cylindrical shells, *Phys. Rev.,* **IX** (4): 193–243, 1899.

8. J. F. Hoburt, A computational methodology and results for quasistatic multilayered magnetic shielding, *IEEE Trans. Electromagn. Compat.,* **38**: 92–103, 1996.

9. R. C. Weast, ed., *Handbook of Chemistry and Physics,* 56th ed., Boca Raton, FL: CRC Press, 1975–1976.

10. S. A. Schelkunoff, *Electromagnetic Waves,* New York: Van Nostrand, 1943.

11. D. A. Miller and J. E. Bridges, Review of circuit approach to calculate shielding effectiveness, *IEEE Trans. Electromagn. Compat.,* **EMC-10**: 52–62, 1968.

12. P. P. Silvester and R. L. Ferrari, *Finite Elements for Electrical Engineers,* Cambridge, UK: Cambridge University Press, 1983.

13. S. R. Hoole, *Computer-Aided Analysis and Design of Electromagnetic Devices,* New York: Elsevier Science, 1989.

14. R. F. Harrington, *Field Computation by Moment Methods,* New York: Macmillan, 1968.

15. K. C. Lim et al., Integral law descriptions of quasistatic magnetic field shielding by thin conducting plate, *IEEE Trans. Power Deliv.* **12**: 1642–1650, 1997.

DAVID W. FUGATE
Electric Research and Management, Inc.

FRANK S. YOUNG
Electric Power Research Institute

MAGNETIC SOURCE IMAGING

More than 200 years ago, it was discovered that biological processes are accompanied by electrical currents. Since then, measurements of bioelectric signals have become widespread procedures of great importance in both biophysical research and medical applications in clinical use. These studies include, for example, measurements of electric potential differences arising from human heart [the electrocardiogram (ECG)], brain [the electroencephalogram (EEG)], and other organs.

The same bioelectric activity that generates electrical potentials also generates weak magnetic fields. Because these biomagnetic fields measured outside the body are extremely low in magnitude ($\sim$10 fT to 100 pT), it was not until 1963 that the first successful detection of the magnetic field arising from human heart was performed (1). This was the beginning of magnetocardiography (MCG). Magnetoencephalography (MEG) was introduced in 1968 when magnetic signals due to the spontaneous α-rhythm in the brain were detected (2). However, it was only after the development of ultrasensitive superconducting quantum interference device (SQUID) detectors in the beginning of the the 1970s (3) that easier detection of biomagnetic signals became possible. In addition to the

magnetocardiogram and the magnetoencephalogram, various biomagnetic fields arising from the body have been studied since then.

Biomagnetic measurements offer information that is very difficult to obtain with other imaging methods (4–8). MEG and MCG are generated by the electric currents in neurons or myocardial cells, and therefore the measurements provide direct real-time functional information about the brain or the heart, respectively. The time scale of the detectable signals ranges from fractions of a millisecond to several seconds or even longer periods. The biomagnetic measurements are totally noninvasive and the body is not exposed to radiation or high magnetic fields. Mapping of biomagnetic signals at several locations simultaneously is easy and fast to perform with multichannel systems.

The metabolic processes associated with the neural or myocardial activity can be studied with positron emission tomography (PET), but the imaging times are several minutes, and the spatial resolution is about 5 mm. Better spatial resolution is obtained from functional magnetic resonance imaging.

Estimation of bioelectric current sources in the body from biomagnetic measurements is often called magnetic source imaging (MSI). To relate the functional information provided by MSI to the underlying individual anatomy, other imaging methods are employed, such as magnetic resonance imaging (MRI), computer tomography (CT), and X ray. In this article, we focus on MEG and MCG, followed by a brief discussion of other fields of biomagnetism. Furthermore, instead of a comprehensive review of MEG and MCG applications we provide a few illustrative examples of recent MSI studies.

MEG AND MCG STUDIES

During recent years, MEG and MCG have attained increasing interest. The ability of these methods to locate current sources combined with precise timing of events is valuable both in basic research and in clinical studies.

One common type of an MEG experiment is to record the magnetic field associated with a sensory stimulus or a movement. Since these fields are usually masked by the ongoing background activity, signal averaging is routinely employed to reveal the interesting signal component. Recordings of neuromagnetic fields have provided a wealth of new information about the organization of primary cortical areas (9).

Sensor arrays covering the whole head have made studies of complicated phenomena involving simultaneous or sequential processing in multiple cortical regions feasible. Because MEG is a unique tool to study information processing in healthy humans, several language-related studies have recently been conducted (10–12).

It is also possible to record the ongoing rhythmic spontaneous brain activity in real time and follow its changes under different conditions (13,14). In addition to the well-known 10 Hz α-rhythm originating in the vision-related cortical areas, similar spontaneous signals occur, for example, in the somatosensory system. MEG measurements have provided new information about both the generation sites of these rhythmic activities and their functional significance (15).

Both evoked responses and spontaneous activity recordings can be utilized in clinical studies (16). For example, the locations of the somatosensory and motor cortices deduced from evoked MEG signals can be superimposed on three-dimensional surface reconstructions of the brain, computed from MRI data. The resulting individual functional map can be a valuable aid in planning neurosurgical operations. Encouraging results have also been obtained in locating epileptic foci in candidates for epileptic surgery.

High-resolution MCG recordings have been applied both in basic cardiac research and in clinical studies. In the first MCG studies in the 1970s and 1980s, only single-channel devices were available, which limited the use of MCG to subjects and patients with normal sinus rhythm. Introduction of multichannel recording systems in the 1990s made the technique more suitable for routine clinical studies and for analysis of beat-to-beat variations. Currently, MCG is being used at some hospitals to test and further develop its clinical use.

Multichannel MCG studies are particularly promising in two clinically important problems: (1) in locating noninvasively abnormal cardiac activity critical for the arousal of life-threatening arrhythmias and (2) in evaluating the risk of such arrhythmias in different cardiac pathologies, especially after myocardial infarction. Successful MCG results have been reported, for example, in locating abnormal ventricular preexcitation sites associated with the Wolff–Parkinson–White syndrome, the origin of ventricular extrasystolic beats, and the origin of focal atrial tachycardias (17–21).

Furthermore, MCG localization accuracy has been tested with artificial sources, such as a pacing catheter in the heart (20,22). The localization accuracy reported so far, ranging from about 5 mm to 25 mm, is sufficient to provide valuable information for preablative evaluation of the patients. In addition to localization studies, MCG has been applied to retrospective identification of patients prone to malignant arrhythmias with about 90% sensitivity and specificity (23).

INSTRUMENTATION

Detection of Neuromagnetic Fields

The detector that offers the best sensitivity for the measurement of these tiny fields is the SQUID (24,25), which is a superconducting ring, interrupted by one or two Josephson junctions (26). These weak links limit the flow of the supercurrent, which is characterized by the maximum critical current I_c that can be sustained without loss of superconductivity. Direct-current (dc) SQUIDs, with two junctions, are preferred because the noise level is lower in them than in radio-frequency (RF) SQUIDs (27–29).

The magnetic signals from the body are extremely weak compared with ambient magnetic field variations (5) Thus, rejection of outside disturbances is of utmost importance. Significant magnetic noise is caused, for example, by fluctuations in the earth's geomagnetic field, by moving vehicles and elevators, and by the omnipresent powerline fields.

For rejection of external disturbances, biomagnetic measurements are usually performed in a magnetically shielded room. To make such an enclosure, four different methods exist: Ferromagnetic shielding, eddy-current shielding, active compensation, and the recently introduced high-T_c superconducting shielding. Many experimental rooms have been built for biomagnetic measurements utilizing combinations of these techniques (34–36). Commercially available rooms utilized in biomagnetic measurements usually employ two layers of aluminum and ferromagnetic shielding, possibly combined with active compensation. The inside floor area is usually 3 m by 4 m, and the height around 2.5 m.

In addition, the sensitivity of the SQUID measuring system to external magnetic noise can be greatly reduced by the proper design of the flux transformer, a device normally used for bringing the magnetic signal to the SQUID. For example, an axial first-order gradiometer consists of a pickup (lower) coil and a compensation coil with identical effective area and connected in series but wound in opposition [see Fig. 1(a)]. This system of coils is insensitive to a spatially uniform background field, but it responds to inhomogeneities. Therefore, a source near the lower coil, which will cause a much greater field in the pickup loop than in the more remote compensation coil, will thus produce a net output.

Most biomagnetic measurements have been performed with axial gradiometers. However, the off-diagonal planar configuration of Fig. 1(b) has some advantages over axial coils: The double-D construction (30) is compact in size, and it can be fabricated easily with thin-film techniques. The locating accuracies of planar and axial gradiometer arrays are essentially the same for superficial sources (31–33). The spatial sensitivity pattern, lead field, of off-diagonal gradiometers is narrower and shallower than that of axial gradiometers. These sensors thus collect their signals from a more restricted area near the sources of interest, and there is less overlap between lead fields of adjacent sensors in a multichannel array.

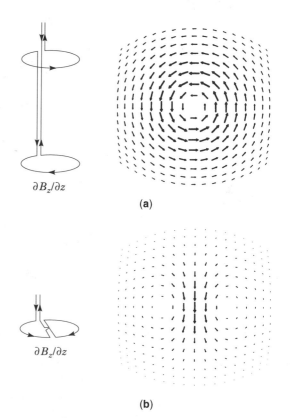

Figure 1. Left: Coil configurations for (a) an axial and (b) a planar gradiometer. Right: The corresponding sensitivity patterns, lead fields (see section entitled "Distributed Source Models."). The plots show the lead fields on a spherical surface. The gradiometer coil is located above the center of each pattern. The direction and size of the arrows indicate the magnitude and direction of the lead field at the center of the arrow.

Nevertheless, distant sources can often be detected more easily with axial gradiometer or magnetometer sensors. Therefore, many experimental and commercial systems include these coil configurations, possibly in combination with planar gradiometers.

Multichannel Magnetometers

The first biomagnetic measurements were performed with single-channel instruments. However, reliable localization of current sources requires mapping in several locations, and this is time-consuming with only one channel. Besides, unique spatial features present in, for example, brain rhythms cannot be studied. Fortunately, during the past 15 years, multichannel SQUID systems for biomagnetic measurements have been developed to provide reliable commercial products. A detailed account of this development can be found in Ref. 5.

A state-of-the-art multichannel MEG system comprises more than 100 channels in a helmet-shaped array to record the magnetic field distribution across the brain simultaneously. The latest MCG systems contain 60 to 80 detectors in a flat or slightly curved array to cover an area about 30 cm in diameter over the subject's chest or back. The dewar containing the sensors is attached to a gantry, which allows easy positioning of the dewar above the subject's head or chest. The position of the dewar with respect to the subject's head or torso is typically determined by measuring the magnetic field arising from an ac current fed into small marker coils attached to the skin (37,38) and by calculating their locations with respect to the sensor array. The locations of the marker coils with respect to an anatomical frame of reference are determined before the biomagnetic measurement by a three-dimensional digitizer.

As an example of an MEG installation we describe the Neuromag-122 system (Neuromag Ltd., Helsinki, Finland) (33). This device employs planar first-order two-gradiometer units to measure the two off-diagonal derivatives, $\partial B_z/\partial x$ and $\partial B_z/\partial y$, of B_z, the field component normal to the dewar bottom at 61 locations. The thin-film pickup coils are deposited on 28×28 mm^2 silicon chips; they are connected to 122 dc SQUIDs attached to the coil chip. The separation between two double-sensor units is about 43 mm. The system is depicted in Fig. 2.

The MCG system from the same company shown in Fig. 2 comprises 67 channels arranged on a slightly curved surface with diameter about 30 cm. The magnetic-field component (B_z) perpendicular to the sensor array surface is sensed by seven large-coil axial gradiometers and 30 two-channel planar gradiometer units identical to those described in the previous paragraph.

GENERATION OF BIOELECTROMAGNETIC FIELDS

Cellular Sources

To interpret the measured signals, one has to understand how electric and magnetic fields are generated by biological tissue. In this article, we consider biomagnetic signals generated by the electric currents in excitable tissue. These magnetic fields are linked to bioelectric potentials, and it is useful to consider both the magnetic field and the electric potential together.

Living cells sustain a potential difference between intra- and extracellular media. In a static situation, most cells are, as seen externally, electrically and magnetically silent. Excitable cells can produce electric surface potentials and external magnetic fields, which can be detected from outside.

The Quasi-Static Approximation

The total electric current density in the body, $\boldsymbol{J}$, is time-dependent, and the electric field ($\boldsymbol{E}$) and the magnetic field ($\boldsymbol{B}$) produced by $\boldsymbol{J}$ can be found from Maxwell's equations. However, the variations in time are relatively slow (below 1 kHz) (8,39), which allows treatment of the sources and the fields in a quasi-static approximation. This means that inductive, capacitive, and displacement effects can be neglected. In the quasi-static approximation Maxwell's equations thus read:

$$\nabla \cdot \boldsymbol{E} = \rho/\epsilon_0 \tag{1}$$

$$\nabla \times \boldsymbol{E} = 0 \tag{2}$$

$$\nabla \cdot \boldsymbol{B} = 0 \tag{3}$$

$$\nabla \times \boldsymbol{B} = \mu_0 \boldsymbol{J} \tag{4}$$

where μ_0 and ϵ_0 are the magnetic permeability and electric permittivity of the vacuum, respectively.

Primary Current

It is useful to divide the total current density in the body, $\boldsymbol{J}(\boldsymbol{r})$, into two components. The passive volume or return current is proportional to the conductivity $\sigma(\boldsymbol{r})$ and the electric field $\boldsymbol{E}$:

$$\boldsymbol{J}^{\mathrm{v}}(\boldsymbol{r}) = \sigma(\boldsymbol{r})\boldsymbol{E}(\boldsymbol{r}) \tag{5}$$

$\boldsymbol{J}^{\mathrm{v}}$ is the result of the macroscopic electric field on charge carriers in the conducting medium. Everything else is the primary current $\boldsymbol{J}^{\mathrm{p}}$:

$$\boldsymbol{J}(\boldsymbol{r}) = \boldsymbol{J}^{\mathrm{p}}(\boldsymbol{r}) + \sigma(\boldsymbol{r})\boldsymbol{E}(\boldsymbol{r}) \tag{6}$$

This definition would be meaningless without reference to the length scale. Here $\sigma(\boldsymbol{r})$ is the macroscopic conductivity; cellular-level details are left without explicit attention. The division in Eq. (6) is illustrative in that neural or cardiac activity gives rise to primary current mainly inside or in the vicinity of a cell, whereas the volume current flows passively everywhere in the medium.

It should be emphasized that $\boldsymbol{J}^{\mathrm{p}}$ is to be considered the driving "battery" in the macroscopic conductor; although the conversion of chemical gradients to current is due to diffusion, the primary current is largely determined by the cellular-level details of conductivity. In particular, the membranes, being good electrical insulators, guide the flow of both intracellular and extracellular currents.

If the events are considered on a cellular level, it is customary to speak about the impressed rather than primary current (39).

Neurons

Signals propagate in the brain along nerve fibers called axons as a series of action potentials. During an action potential, the primary current can be approximated by a pair of current

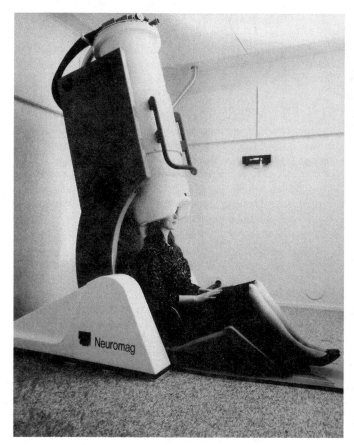
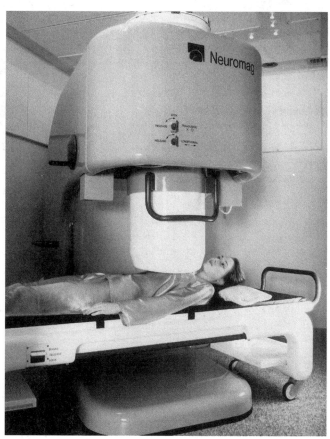

Figure 2. Left: The Neuromag-122™ MEG system. Right: The 67-channel MCG system. (Photographs courtesy of Neuromag, Ltd.)

dipoles corresponding to a local depolarization of the cell membrane, followed by repolarization. This source moves along the axon as the activation propagates. Although the model is a simplified one, the experimental magnetic findings are in reasonable agreement with this concept (40,41).

The axons connect to other neurons through synapses. In a synapse, transmittter molecules are released to the synaptic cleft and attach to the receptors on the postsynaptic cells. As a result, the ionic permeabilities of the postsynaptic membrane are modified and a postsynaptic potential is generated. The postsynaptic current can be adequately described by a single-current dipole.

The dipolar field produced by the postsynaptic current flow falls off with distance more slowly than the quadrupolar field associated with the action potentials. Furthermore, the postsynaptic currents last tens of milliseconds, whereas the duration of a typical action potential is only about 1 ms. On this basis, it is believed that that the electromagnetic signals observed outside and on the surface of the head are largely due to the synaptic current flow.

The two principal groups of neurons on the surface layer of the brain, the cortex, are the pyramidal and the stellate cells. The former are relatively large; their apical dendrites from above reach out parallel to each other, so that they tend to be perpendicular to the cortical surface. Since neurons guide the current flow, the resultant direction of the electrical current flowing in the dendrites is also perpendicular to the cortical sheet of gray matter.

Myocardium

In heart tissue, there are three main types of cells: pacemaker cells in the sinus and atrioventricular nodes, cells specialized for rapid conduction of the excitation along the bundle of His and Purkinje network, and, finally, muscle cells that perform mechanical work. Cardiac muscle consists of a large number of individual cells, each about 15 μm in diameter and 100 μm long. The intracellular spaces of adjacent muscle cells are interconnected, which makes the three-dimensional structure very complex.

An action potential in myocardial cells lasts 300 ms to 400 ms, which is over 100 times longer than a typical neural action potential. Provided that we observe a single myocardial cell at some distance from the membrane, the depolarization and repolarization can be modeled, respectively, by an equivalent depolarization and repolarization dipole.

Ventricular depolarization or repolarization propagates as about 1 mm thick wavefronts in the heart. A commonly used model to describe such propagating fronts is a uniform double layer (39). It consists of dipoles with equivalent strengths (assuming a constant dipole density), oriented perpendicular to the wavefront. The model is more suitable than a single current dipole in characterizing an excitation taking place simultaneously in a spatially large region, but it cannot account for possible holes in the wavefront (e.g., necrotic tissue). In addition, the classical concept of a uniform double layer is not valid if the anisotropic nature of myocardial tissue is to be included.

Calculation of the Bioelectromagnetic Fields

In the quasi-static approximation, the electric potential ϕ obeys Poisson's equation:

$$\nabla \cdot (\sigma \nabla \phi) = \nabla \cdot \boldsymbol{J}^{\mathrm{p}} \tag{7}$$

while the magnetic field due to the total current density, $\boldsymbol{J}$, is obtained from the Ampère–Laplace law:

$$\boldsymbol{B}(\boldsymbol{r}) = \frac{\mu_o}{4\pi} \int_V \frac{\boldsymbol{J}(\boldsymbol{r}') \times \boldsymbol{R}}{R^3} \, dv' \tag{8}$$

where the integration is performed over a volume V containing all active sources, $\boldsymbol{r}' \in V$, $\boldsymbol{R} = \boldsymbol{r} - \boldsymbol{r}'$.

It can be shown that the volume currents in an infinite homogeneous volume conductor give no contribution to the electric potential or the magnetic field, which are solely due to the primary currents, $\boldsymbol{J}^{\mathrm{p}}$ (42).

Next, we assume that the body consists of homogeneous subvolumes v'_k, $k = 1, 2, \ldots, M$, bounded by the surfaces S_k. The electrical conductivity within v'_k is constant, σ_k. Usually, the body is surrounded by air, and thus the conductivity outside the body surface is zero. In this case, the surface potential, ϕ_{S}, can be obtained from an integral equation (43)

$$(\sigma''_l + \sigma'_l)\phi_{\mathrm{S}}(\boldsymbol{r}) = 2\sigma_n \phi_\infty(\boldsymbol{r}) + \frac{1}{2\pi} \sum_{k=1}^M (\sigma''_k - \sigma'_k) \int_{S_k} \phi_{\mathrm{S}} \, d\boldsymbol{S}_k \cdot \frac{\boldsymbol{R}}{R^3} \tag{9}$$

where σ_n is the conductivity at the source location, σ'_k is the conductivity inside and σ''_k is the conductivity outside the surface S_k, and $d\boldsymbol{S}_k$ is the surface element vector perpendicular to the boundary. The term ϕ_∞ denotes the electric potential in an infinite homogeneous medium (in the absence of the boundaries S_k), and the surface integral accounts for the contribution of the counductivity change on the boundary S_k.

The external magnetic field is then evaluated by substituting the total current density, $\boldsymbol{J}$, into Eq. (8). It can be shown (44) that the result can be transformed to the form

$$\boldsymbol{B}(\boldsymbol{r}) = \boldsymbol{B}_\infty(\boldsymbol{r}) + \frac{\mu_o}{4\pi} \sum_{k=1}^M (\sigma''_k - \sigma'_k) \int_{S_k} \phi_{\mathrm{S}} \, d\boldsymbol{S}_k \times \frac{\boldsymbol{R}}{R^3} \tag{10}$$

where the term $\boldsymbol{B}_\infty$ is the magnetic field in the absence of the boundaries, S_k. Again, the surface integral accounts for the contribution of the counductivity change on the boundary S_k.

Analytic Solutions. Analytic solutions of Eqs. (9) and (10) exist only in a few simple symmetric geometries. If we approximate the head or the torso by a layered spherically symmetric conductor, it is possible to derive a simple analytic expression for the magnetic field of a current dipole (41):

$$\boldsymbol{B}(\boldsymbol{r}) = \frac{\mu_0}{4\pi} \frac{F \boldsymbol{Q} \times \boldsymbol{r}_Q - (\boldsymbol{Q} \times \boldsymbol{r}_Q \cdot \boldsymbol{r})\nabla F(\boldsymbol{r}, \boldsymbol{r}_Q)}{F(\boldsymbol{r}, \boldsymbol{r}_Q)^2} \tag{11}$$

where $\boldsymbol{r}_Q$ is the location of the current dipole, $\boldsymbol{Q}$ is the dipole moment vector, $F(\boldsymbol{r}, \boldsymbol{r}_Q) = a(ra + r^2 - \boldsymbol{r}_Q \cdot \boldsymbol{r})$, and $\nabla F(\boldsymbol{r}, \boldsymbol{r}_Q) = (r^{-1}a^2 + a^{-1}\boldsymbol{a} \cdot \boldsymbol{r} + 2a + 2r)\boldsymbol{r} - (a + 2r + a^{-1}\boldsymbol{a} \cdot \boldsymbol{r})\boldsymbol{r}_Q$, with $\boldsymbol{a} = (\boldsymbol{r} - \boldsymbol{r}_Q)$, $a = |\boldsymbol{a}|$, and $r = |\boldsymbol{r}|$.

An important feature of the sphere model is that the result is independent of the conductivities and thicknesses of the layers; it is sufficient to know the center of symmetry. The calculation of the electric potential is more complicated: The results can be expressed only as a series expansion of Legendre polynomials, and full conductivity data are required (45). Furthermore, radial currents do not produce any magnetic field outside a spherically symmetric conductor. Thus MEG is, to a great extent, selectively sensitive to tangential sources, and EEG data are required to recover all components of the current distribution.

The obvious advantage of a simple forward model is that a fast analytical solution is available. It has also been shown (46) that a sphere model fitted to the local curvature of the skull's inner surface (4) provides accurate enough estimates for many practical purposes. However, when the source areas are located deep within the brain or in the frontal lobes, it is necessary to use more accurate approaches.

In the first MCG localization studies the body was approximated as a homogeneous semi-infinite space, which can be regarded as a generalization of a spherical model with the radius extended to infinity (47,48).

However, later computer studies have shown that the semi-infinite approximation is oversimplified, and a more accurate description of the thorax shape is needed in the inverse studies (21,49). A slightly more accurate description of the thorax geometry can be obtained by using cylindrical or spheroidal models. However, the analytical expressions for arbitrary dipolar sources become substantially more complex than in the spherical case (50), and only a few studies to apply spheroids have been reported.

Numerical Approaches. When a realistic geometry of the head or the thorax is taken into account, numerical techniques are needed to solve the Maxwell equations. When applying the boundary-element method (BEM), electric potential and magnetic field are calculated from the (quasi-static) integral equations [Eqs. (9) and (10)], which can be discretized to linear matrix equations (46,49,51).

In most BEM applications to the bioelectromagnetic forward problem, the surfaces are tessellated with triangular elements, assuming either constant or linear variation for the electric potential on each triangle. However, the accuracy of the magnetic-field computation may suffer if a dipole source is located near a triangulated surface. The accuracy can be improved, for example, by applying Galerkin residual weighting instead of the standard collocation method and by approximating the surfaces with curved elements instead of plane triangles (52).

Realistically shaped geometries of each subject are usually extracted from MRI data. The regions of interest (e.g., the heart, the lungs, and the thorax; or the brain, the skull, and the scalp) need to be segmented from the data first (see section entitled MRI). The volumes or the surfaces are then discretized for numerical calculations. The segmentation and tessellation problems are still tedious and nontrivial (53).

The relatively low conductivity of the skull greatly facilitates the modeling of MEG data. In fact, a highly accurate model for MEG is obtained by considering only one homogeneous compartment bounded by the skull's inner surface (46). With suitable image processing techniques it is possible to isolate this surface from high-contrast MRI data with little or no user intervention.

The boundary-element model is more complex for EEG, because three compartments need to be considered: the scalp, the skull, and the brain. While the surface of the head can be easily extracted from the MRI data, it is difficult to construct a reliable algorithm to automatically isolate the scalp-skull boundary. In addition, special techniques are required to circumvent the numerical problems introduced by the high conductivity contrast due to the low-conductivity skull.

It is also possible to employ the finite-element method (FEM) or the finite-difference method (FDM) in the solution of the forward problem. The solution is then based on the discretization of Eq. (7). In this case, any three-dimensional conductivity distribution and even anisotropic conductivity can be incorporated (54). However, the solution is more time-consuming than with the BEM, and therefore the FEM or FDM has not been used in routine source modeling algorithms which require repeated calculation of the magnetic field from different source distributions.

SOURCE MODELING

The Inverse Problem

The goal of the bioelectric (EEG, ECG) and biomagnetic (MEG, MCG) inverse problems is to estimate the primary source current density underlying the electromagnetic signals measured outside or on the surface of the body. Unfortunately, the primary current distribution cannot be recovered uniquely, even if the electric potential and the magnetic field were known precisely everywhere at the surface and outside the body (55). However, it is often possible to use additional anatomical and physiological information to constrain the problem and facilitate the solution. One can also replace the actual current sources by equivalent generators that are characterized by a few parameters. The values of the parameters can then be uniquely determined from the measured data by a least-squares fit. The solution of the forward problem is a prerequisite for dealing with the inverse problem requiring repeated solution of the forward problem.

The Current Dipole Model. The simplest physiologically sound model for the neural or myocardial current distribution comprises one or several point sources, current dipoles. In the simplest case the field distribution, measured at one time instant, is modeled by that produced by one current dipole. The best-fitting *equivalent current dipole* (ECD) can be found by using standard least-squares optimization methods such as the Levenberg–Marquardt algorithm (56).

In the time-varying dipole model, introduced by Scherg and von Cramon (57,58), an epoch of data is modeled with a set of dipoles whose orientations and locations are fixed but whose amplitudes vary with time. Each dipole corresponds to a small patch of cerebral cortex or other structures activated simultaneously or in a sequence. The precise details of the current distribution within each patch cannot be revealed by the measurements, which are performed at a distance in excess of 3 cm from the sources.

As a result of the modeling, one obtains the locations of the sources and the orientation of the dipole component tangential to the inner surface of the overlying skull. In addition,

traces of the evolution of the source strengths are obtained. Again, the optimal source parameters are found by matching the measured data collected over a period of time with those predicted by the model using the least-squares criterion.

From a mathematical point of view, finding the best-fitting parameters for the time-varying multidipole model is a challenging task. Because the measured fields depend nonlinearly on the dipole position, the standard least-squares minimization routines may not yield the globally optimal estimates. Therefore, global optimization algorithms (59) and special fitting strategies (60), taking into account the physiological characteristics of particular experiments, have been suggested. For each candidate set of dipole positions and orientations it is, however, straightforward to calculate the optimal source amplitude waveforms using linear least-squares optimization methods (61).

In cardiac studies, an ECD is applicable for approximating the location and strength of the net primary current density confined in a small volume of tissue. Myocardial depolarization initiated at a single site spreads at a velocity of about 0.4 mm/ms to 0.8 mm/ms, and the ECD can be thought to be moving along the "center of mass" of the excitation. In practice, localization based on a single ECD is meaningful only during the first 10 ms to 20 ms of excitation.

Because both nonlinear fits for spatial coordinates and linear fits for dipole moment parameters need to be searched at every time instant, the use of even two ECDs becomes very complicated in cardiac studies. Alternatively, cardiac excitation can be modeled with a set of spatially fixed stationary or rotating dipoles, but attempts to define the time courses of the dipole magnitudes usually result in physiologically unacceptable results.

The Current Multipole Expansion. It is often convenient to present the electric potential and the magnetic field as multipole expansions. In the current multipole expansion the field due to the primary current, B_∞ in Eq. (10), is expressed as a Taylor series (62). Thus, more complex source current configurations can be described as higher-order multipole moments, such as quadrupole moments.

Different source models can be built by combining dipole and quadrupole moments. A current dipole is actually the lowest-order term in a general current multipole expansion (62); higher-order terms, such as quadrupoles and octupoles, can be used to account for more complex primary current configurations (48).

Distributed Source Models. Another approach often taken in source modeling is to relax the assumptions on the sources and use various estimation techniques to yield a distributed image of the sources. These methods include, for example, the minimum-norm estimates (63), magnetic-field tomography (MFT) (64), and low-resolution electromagnetic tomography (LORETA) (65).

The source images can provide reasonable estimates of complex source configurations without having to resort to complicated multidipole fitting strategies. However, one must keep in mind that even if the actual source is pointlike, its image is typically blurred, extending a few centimeters in each linear dimension. Therefore, the size of the "blobs" in the source images does not directly relate to the actual dimensions of the source but rather reflects an intrinsic limitation of the imaging method.

The basic concept relevant to all distributed source estimation methods is the lead field. The signal b_k detected by the kth sensor in the sensor array is a linear functional of the primary current distribution J^p and can be expressed as

$$b_k = \int_G L_k(r) \cdot J^p(r) \, dv \qquad (12)$$

where the integration extends over the source region G, which can be a curve, a surface, or a volume. The functions L_k are often called lead fields, which can be readily obtained by solving the forward problem for dipole sources.

The minimum-norm estimate (63) is the current distribution that has the smallest norm and is compatible with the measured data. Here, the norm is defined by

$$\|J\|^2 = \int_G J(r)^2 \, dv \qquad (13)$$

The minimum-norm estimate J^* can be expressed as a weighted sum of the lead fields, $J^* = \sum_{k=1}^{N} w_k L_k$. The weighting coefficients are found by fitting the data, computed from the minimum-norm estimate with those actually measured. Since the lead fields in a large array are almost linearly dependent, regularization techniques are needed to produce stable estimates.

Another type of a distributed source model was developed for reconstructing the sequence of ventricular depolarization by van Oosterom et al. (66). Their model is based on a uniform double layer of constant strength. Lead fields for MCG and ECG sensors are evaluated at each node on the endo- and epicardial surfaces of the heart. These transfer functions are weighted by the Heaviside time step function to define the onset of excitation at each surface node. Physiological constraints and regularization are then applied to limit the number of solutions.

Regularization and Constraints

The bioelectromagnetic forward problem can be written as $b = Lx + e$, where vector x represents the unknown (linear and nonlinear) source parameters, vector b consists of the measured (MEG/EEG or MCG/ECG) signals, vector e contains the contribution of measurement noise, and matrix L is effectively the transfer (lead field) function between the sources and the measurement sensors. Even small contributions of the noise e make the solution for x very ill-posed. Therefore, regularization techniques are needed to stabilize the solution (67).

In bioelectromagnetic studies dealing with source distributions, the most frequently applied techniques include the truncated-eigenvalue singular value decomposition (63) and the L-curve method (67). Another new approach is based on Wiener filtering and orthogonalized lead fields (68). In addition, spatial weighting can be applied to improve the solutions (21,69). Futher improvements are achieved by applying more than one constraint at the same time (70).

One can also make explicitly the additional assumption that the activated areas have a small spatial extent. For example, the MFT algorithm obtains the solution as a result of an iteration in which the probability weighting is based on the previous current estimate (64). According to the authors, this procedure produces more focal images than the traditional minimum-norm solutions. Another possibility is to use a MUSIC-type probability weighting (61) combined with cortical constraints to focus the image (71).

An approach that incorporates the desire to procure focal source images is to use the L^1 norm, that is, the sum of the absolute values of the current over the source space, as the criterion to select the best current distribution among those compatible with the measurement (72–74). In contrast to the traditional L^2-norm cost function [see Eq. (13)], the L^1-norm criterion yields estimates focused to a few small areas within the source space.

The most powerful way to constrain the bioelectromagnetic inverse problem is to apply anatomical and functional *a priori* information. For example, accurate reconstruction of the cortex surface or myocardial tissue from MRI data limits the spatial extent and orientation of the sources (75). Solutions can also be made more robust by requiring temporal smoothness. Invasively recorded signals such as intraoperative potential recordings from the cortex or from the heart can also be very valuable in developing proper physiological and temporal constraints for distributed sources.

The Relation between Bioelectric and Biomagnetic Signals

Both bioelectric and biomagnetic fields are generated by the same activity. As a consequence, there must be a correlation both in the temporal waveforms and in the spatial maps of the measured signals. Therefore, it is evident that bioelectric and biomagnetic measurements reveal partly redundant information. However, neither one can be used to uniquely reproduce the other; there are current configurations that produce either electric or magnetic field, but not both. A practical example of a magnetically silent source is a radial dipole in a spherically symmetric conductor. On the other hand, solenoidal currents do not produce any electric potential. This may become important, for example, in cardiac exercise studies (76). Therefore, a combination of magnetic and electric recordings seems appealing to obtain more complete information about the current distributions. Still, few attempts to combine electric and magnetic data have been reported (21,58,77,78).

In the previous considerations it was assumed that the volume conductor is homogeneous or piecewise homogeneous. However, many biological tissues are organized directionally, and the electrical conductivity depends on the direction of the fibers. For example, the conductivity in myocardial fibers is about three times higher in the main fiber direction than across the fibers. Colli-Franzone et al. (79) showed that a classical uniform dipole layer, as representing the myocardial wavefront, should be revised to take into account the anisotropic nature of the tissue. They were able to explain experimentally measured potential distributions with an oblique dipole layer, where the dipoles may also have tangential components in addition to the normal component.

Wikswo (80) studied isolated animal preparations and employed microSQUIDs and microelectrodes to measure magnetic and electric fields during and after applying a current stimulus. According to their results, the magnetic field is more sensitive to the underlying anisotropy than the electric potential. With such combined electric and magnetic recordings, it is at least in principle possible to determine the intra- and extracellular conductivity values.

The anisotropic properties of the heart are especially evident near the ventricular apex, where the spiral arrangements of the myocardial fibers can be observed on the epicardial surface. It has been argued that this kind of vortex geometry leads to electrically silent components in magnetic field. However, van Oosterom et al. (66) arrived at the conclusion that the anisotropy does not play a significant role in the ECG or MCG during the normal ventricular depolarization. On the other hand, the findings of Brockmeier et al. (76) in pharmacological MCG stress testing indicate that the anisotropy may cause larger repolarization changes in multichannel MCG signals than in the simultaneously recorded ECG maps.

The tissue is directionally oriented also in the brain. For example, the conductivity of the white matter in the direction of the fibers may be 10 times higher than the conductivity across the fibers. In the cerebral cortex, the corresponding factor is about two. In general, the anisotropy influences the body surface potentials and magnetic fields. However, in the sphere model a difference between the radial and the two tangential conductivities does not affect the magnetic field, while the influence on the electric potential is still substantial.

INTEGRATION WITH OTHER IMAGING MODALITIES

Fast high-field MRI devices provide precise anatomical data. Besides reconstruction of accurately shaped volume-conductor models, anatomic MRI data on the heart and the brain are necessary to combine the inverse solutions with the anatomy in a clinically useful presentation. Examples of source displays of both MRI slices and three-dimensional surface reconstructions are shown in the section entitled "Applications."

Segmentation of the structures of interest from image data is presently the most time-consuming part in constructing individualized boundary element models. In the medical imaging field, accurate extraction of anatomic structures from image data sequences is still an open problem. In practice, manual extraction of the objects of interest—for example, from MRI slices—is often considered the most reliable technique.

Recently, automated region-based and boundary-based segmentation and triangulation methods have been developed, for example, for extracting the lungs, heart, and thorax, or the brain and skull. In region-based methods, some features based on the intensity of the images are used to merge voxels. The boundary-based methods, in turn, rely on an intensity gradient detection. Both methods have limitations, but the utilization of prior geometrical knowledge, such as triangulated surfaces generated from data on other subjects, provides useful additional information. For example, a deformable pyramid model can then provide automatic segmentation and triangulation of the anatomic objects (53,81).

MRI is still fairly expensive, especially for large patient populations. Thus, methods are being developed to use other imaging methods for reconstructing individualized triangulated surfaces. In cardiac studies, two orthogonal thorax X-ray projections, or ultrasound images of the heart combined with three-dimensional (3D) digitization of the thorax surface, can be utilized to acquire patient-specific geometry models.

In principle, CT images could be used instead of MRI to construct a boundary-element model of the head. The skull is particularly easy to isolate from these data. However, the classification of soft tissues is often easier from MR images and the radiation load imposed by a CT scan is generally considered too high for healthy subjects.

To present the MEG/MCG inverse solutions accurately on the individual anatomy, special care needs to be taken with regard to combining the different coordinate frames. Prior to biomagnetic recordings, one has to fix some marker points, for example, with a 3-D digitization system. During MRI or X-ray imaging, specific markers clearly visible and identifiable, such as vitamin pills or tubes filled with MgCl solution, are attached on the reference points. Three or more markers are usually required to achieve sufficient accuracy in the data fusion.

Functional MRI perfusion studies of ischemic or infarcted heart are particularly valuable in developing physiological constraints and in validating the MCG/ECG localization results of ischemia or arrhythmogenic tissue. In brain studies, new possibilities are opened by combining the millimeter-level spatial resolution of functional MRI and the millisecond-scale temporal resolution of MEG and EEG. Weighting of minimum-norm solutions by functional MRI voxel information has been applied, for example, in visual stimulation studies (82).

However, it must be taken into account that fMRI and biomagnetic measurements are not always detecting common activity. Very clear changes of electric and magnetic signals can be easily missed by fMRI if they occur rarely or are very transient thus producing relatively small average changes in the metabolic level. Furethermore, all experimental setups cannot be easily used in both biomagnetic and fMRI studies. It may thus be necessary to often compare the final results of the analysis of each modality rather than aiming at a combination during the source reconstruction.

It is also possible to utilize positron emission tomography (PET) studies in combination with the electromagnetic methods. However, PET imposes a radiation load on the subject, and therefore the possibilities to perform multiple studies on a given subject are limited. Furthermore, PET is available only in a few centers, whereas MRI systems capable of functional imaging are generally available in modern hospitals.

APPLICATIONS

Brain Studies

Auditory Evoked Fields. Fig. 3 shows the results of a typical auditory evoked-response study performed with a whole-head MEG instrument (83). The responses were elicited by 50 ms tones delivered every 4 s to the subject's right ear. The data were averaged over about 100 repetitions with the stimulus onset as a trigger.

The signals were modeled with two current dipoles in a spherically symmetric conductor. The optimal locations, orientations, and time courses of the dipoles were determined with a least-squares search. Fig. 3 shows the averaged data, the distribution of the magnetic field component normal to the measurement surface at the peak signal value, the time courses of the source amplitudes, and the locations of the sources superimposed on a 3-D surface rendering computed

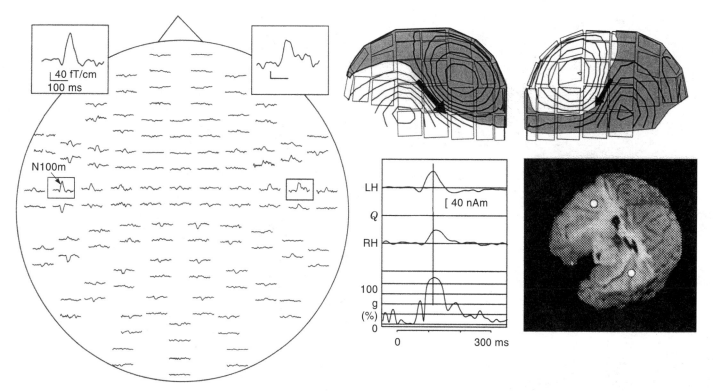

Figure 3. Left: Auditory evoked magnetic fields recorded with a 122-channel magnetometer (33) to 50 ms, 1 kHz tones presented to the subject's right ear once every 4 s. The head is viewed from above, and the helmet surface has been projected onto a plane; the nose points up. Right, above: The pattern of the field component normal to the helmet surface, B_z, shows the peak of the response. White indicates magnetic flux into and gray out of the head. The locations of the sensor units are indicated with squares. The positions and orientations of the two current dipoles modeling the data are projected to the helmet surface. Middle: Time dependence of the dipole strengths, indicating the time behavior of the active area in the left (LH) and right (RH) hemispheres. Q denotes the dipole moment; goodness-of-fit (g) indicates how well the model agrees with the measurement. Right, below: The locations of the dipoles projected on an MRI surface rendering, viewed from above. To show the supratemporal surface, frontal lobes have been removed from the images. (Modified from Ref. 83.)

from the subject's MRI data. The locations of the sources agree nicely with the known site of the auditory cortex on the supratemporal plane. Furthermore, the time courses of the source amplitudes show that the source in the left hemisphere, opposite to the stimulus, is stronger and peaks about 20 ms earlier than the source on the right.

Characterization of Cortical Rhythms and their Reactivity. Since the advent of EEG, the rhythmic oscillations of various cortical areas have been described and also utilized in clinical diagnosis, but their functional significance has remained unclear. With the whole-scalp neuromagnetometers, studies of cortical rhythms have become feasible. Because these rhythms do not repeat themselves, it is mandatory to record them simultaneously over the whole scalp.

The neuromagnetic brain rhythms in healthy adults have been recently characterized in (84) and their reactivity has been quantified during different situations. An efficient way to reveal task-related changes in the level of different frequency components is to filter the signal to the frequency passbands of interest, rectify it, and finally average the rectified signal with respect to the event of interest, like the onset of a voluntary movement (85).

Such an analysis has unraveled new features—for example, of the well-known mu rhythm, which is seen in the EEG records over the somatomotor cortices of an immobile subject. The comb shape of the mu rhythm already indicates the coexistence of two or three frequency components, strongest around 10 Hz and 20 Hz. The sources of the magnetic mu rhythm components cluster over the hand somatomotor cortex, with slightly more anterior dominance for the 20 Hz than for the 10 Hz cluster (see Fig. 4) (85). This difference suggests that the 20 Hz rhythm receives a major contribution from the precentral motor cortex, whereas the 10 Hz component seems mainly postcentral (somatosensory) in origin.

Further support for the functional segregation of these rhythms comes from their different reactivity to movements (15). The level of the 10 Hz rhythm starts to dampen 2 s before a voluntary movement and then returns back within 1 s after the movement. Suppression of the 20 Hz rhythm starts later and is relatively smaller, and the "rebound" after the movement is earlier and stronger than in the 10 Hz band.

Locating Epileptic Foci. Many patients with drug-resistant epilepsy suffer from seizures triggered by a small defective brain area. In preoperative evaluation of these patients, it is

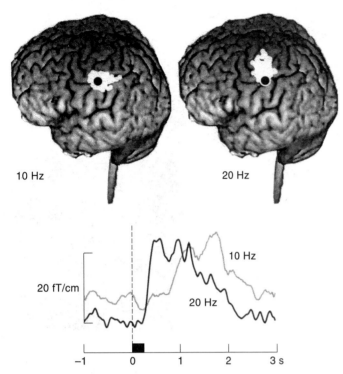

10 Hz   20 Hz

20 fT/cm

10 Hz

20 Hz

−1   0   1   2   3 s

Figure 4. Reactivity of spontaneous activity over the somatomotor hand region in association with voluntary right index finger movements. The time dependencies of the 10 Hz and 20 Hz activities are indicated by the two traces showing the temporal spectral evolution (84) of the signal recorded over the left somatomotor hand area. The locations of sources corresponding to 10 Hz and 20 Hz activity are indicated on the 3 D surface rendition of the subject's magnetic resonance images. The site of the source for electrical stimulation of the median nerve at the wrist is shown by the black dot.

important to know whether their epileptic discharges are focal and how many brain areas are involved, what is the relative timing between the foci, and how close they are to functionally irretrievable locations such as the motor and speech areas. MEG recordings have been able to answer some of these questions (86,87). The patients cannot be studied with MEG during major seizures, owing to movement artifacts, but in many cases the foci can be identified from interictal discharges occurring during the periods between the seizures.

As an example of a recording during an actual seizure Fig. 5 depicts MES signals from a patient who suffered from convulsions in the left side of his face (88). He was able to trigger the seizure by touching the left-side lower gum with his tongue.

The recordings show clear epileptic spikes which appear only in the right hemisphere at first, but later start to emerge in the corresponding areas of the left hemisphere as well. After the 14 s seizure, the epileptic discharges ended abruptly. Figure 5 also depicts locations of the spike ECDs, superimposed on the patient's MRI surface rendering. The sources are clustered along the anterior side of the central sulcus, extending 1 cm to 3 cm lateral to the SI hand area, as determined by somatosensory median-nerve evoked responses. The sources of the epileptic spikes thus agree with the face representation area in the precentral primary motor cortex and are in accord with the patient's clinical symptoms. Spikes generated by the focus in the left hemisphere lagged

behind the right-sided spikes by about 20 ms and probably reflected transfer of the discharges through corpus callosum from the primary to the secondary focus. Identification of secondary epileptogenesis is important for presurgical evaluation of patients because the secondary foci may with time become independent, and removal of the primary focus would then no longer be efficient in preventing the seizures.

Cardiac Studies

Ventricular Preexcitation. Ventricular preexcitation associated with the Wolff–Parkinson–White (WPW) syndrome is caused by an accessory pathway between the atria and the ventricles, which may lead to supraventricular tachycardias and life-threatening arrhythmias refractory to drug therapy. Intervention therapy, such as catheter ablation, is then needed, but a necessary condition for successful elimination of the premature conduction is the reliable localization of the accessory pathway.

Catheter ablation techniques have significantly decreased the need for cardiac surgery, but simultaneously increased the need for accurate noninvasive localization techniques. Noninvasively obtained prior knowledge of the site of the accessory pathway can improve the result and shorten the time needed in invasive catheter mapping, and thus diminish patient discomfort and surgical risk. In addition, shortening the time needed in invasive catheterization also reduces radiation exposure due to fluoroscopy monitoring of catheter positions.

Several MCG studies have been reported on localizing the ventricular preexcitation site in patients with the WPW syndrome (17,20,89,90). The reported accuracy of MCG localiza-

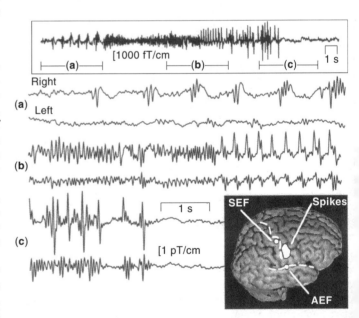

Figure 5. Epileptic discharges after voluntary triggering (88). The trace on the top illustrates MEG activity from the right hemisphere during the whole 14 s seizure. In the middle and lower parts of the figure, selected periods (a, b, c) are expanded and signals generated by the corresponding area in the left hemisphere are shown for comparison. Lower right corner: Locations of ECDs for ictal and interictal spikes (white cluster), and for auditory (AEF) and somatosensory (SEF) evoked fields, superimposed on the patient's MRI surface rendering. The course of the Rolandic and Sylvian fissures are indicated by the white dashed lines. (Adapted from Ref. 88.)

tions ranges from 5 mm to 25 mm, which is sufficient to be useful in preablative or presurgical consideration of the patients.

Ventricular Tachycardia. Generally, malignant ventricular tachycardia (VT) is much more difficult to locate for ablation treatment than the ventricular preexcitation. It is estimated that the lesion produced by the application of RF current is about 6 mm in diameter and about 3 mm in depth. Currently, clinical practice for precise localization introduces several catheters through arteries and veins into the ventricles for invasive recordings of cardiac activation sequences. This procedure can be very time-consuming, and noninvasively obtained information could shorten the procedure from several hours even to less than 1 hour.

VT patients include postmyocardial infarction patients, patients with different cardiomyopathies, and patients with monomorphic VT. MCG studies reported so far have attempted to locate the origin of ventricular extrasystoles or arrhythmias that have occurred spontaneously during the MCG recording (17,19,20). The results have been compared to the results of successful catheter ablations, presented over X-ray and magnetic resonance images when available. In such comparisons the average MCG locations were found to be within 2 cm from the invasively determined sites.

Examples of such studies are displayed in Fig. 6. These results were obtained with an individualized boundary-element torso model; an example is displayed in Fig. 7(a).

Cardiac Evoked Fields. Artificial dipole sources inserted in the heart with catheters (e.g., during routine electrophysiological studies) have been tested to verify the MCG localization accuracy (17,20,22). For example, Fenici et al. (22) studied five patients in whom a nonmagnetic pacing catheter (16) was used to stimulate the heart during MCG recordings. In general, the MCG localization results at the peak of 2 ms catheter stimuli and at the onset of paced myocardial depolarization were within 5 mm from each other. Because these are two physically different sources, the study provided further support for the good localization accuracy of the MCG method. In another recent study (91), simultaneous MCG and ECG mapping recordings were performed during pacing in 10 patients. The localizations were compared to catheter positions documented on fluoroscopic X-ray images. MCG results were, on the average, within 5 mm from the documented catheter position, while the ECG showed somewhat worse accuracy.

Reconstruction of Distributed Sources. For comparison of MCG and ECG mapping results, simultaneously recorded MCG and ECG data were applied in reconstructing ventricular depolarization isochrones on the endo- and epicardial surfaces of the heart of a healthy normal subject (92). The results showed almost identical isochrones from both magnetic and electric data. An example of the MCG isochrone reconstructions is shown in Fig. 6(d).

The minimum-norm estimates (MNE) have been applied in estimating the primary current distributions underlying measured MCG signals. An intrinsic problem associated with MNE is that it has a poor depth resolution of the sources without proper regularization and physiological constraints.

Various regularization and depth weighting methods have shown promising results (21,69).

Recently, depth-weighted MNE reconstruction has been applied in MCG data recorded in patients with chronical myocardial ischemia (18,93). Clinical validation for the results was provided by SPECT imaging. In general, the site of smallest current density, i.e., the missing depolarization component, was in good agreement with the SPECT result.

OTHER APPLICATIONS

Until now, magnetic source imaging studies have been focused in the brain or the heart. However, other applications are being developed as well.

Studies of compound action fields (CAF) from peripheral nerves require a very high sensitivity, because the signal amplitudes are below 10 fT (94–97). In addition, signal averaging of hundreds of stimulated sequences may be required to find the CAF waveforms. Analysis of the waveforms demonstrates the quadrupolar nature of neural activity, provided that the observations are performed at a distance from the depolarized segment (96). Multipole analysis has been applied to model the depolarization process; dipole terms reveal the location and intensity of the source, while octupolar terms are related to its longitudinal extension along the nerve fibers. In addition to studying propagation of the nerve impulses, multichannel measurements can reveal abnormalities such as proximal conduction blocks in the spinal nervous system (97).

High sensitivity and specific signal processing are also needed to detect low-frequency (0.05–0.15 Hz) magnetic signals from gastrointestinal system (98,99). Distinguishing between gastric and small bowel signals may provide a new tool to study abnormalities in the gastrointestinal system (98). Multichannel recordings allow feasible and continuous monitoring of magnetically marked capsules within the gastrointestinal tract with a temporal resolution on the order of milliseconds and a spatial resolution within a range of millimeters (99).

DISCUSSION

Modeling

Despite the inevitable ambiguities in source analysis, very useful information has been obtained by using relatively simple models. For example, the localization of functional landmarks in the brain using the current dipole model and a spherically symmetric conductor in the forward calculations has already been developed to the extent of being a reliable clinical tool (16). The results of these studies have also been often verified in direct intraoperative recordings.

The focal source analysis methods are sometimes criticized for being too extreme simplifications of the actual current distributions, which renders them rather useless in the study of complicated functions performed by the human brain. This intuitively appealing opinion is not well backed up by experimental data. Rather, recent fMRI data may be taken to indicate that the significant changes in metabolic activity associated even with complicated cognitive tasks might well be relatively focal.

Reconstruction methods to deal with source distributions are under development, but there are still difficulties in interpretation of the results obtained from measured data. Implementation of available physiological information and constraints is probably needed to obain a reasonable correlation with actual physiological events in the source regions. If the assumptions of the source model are not compatible with the characteristics of the actual electrophysiological sources, misleading estimates may ensue. As discussed in the section entitled "Distributed Source Models," the distributed source model may produce a distributed estimate even for a focal source. Only very recently have there been attempts to reliably estimate the actual extent of the current source using Bayesian parameter estimation (100) in conjunction with reasonable physiological and anatomical constraints.

Invasively recorded cardiac signals, such as potentials measured during electrophysiological studies on epi- and endocardial surfaces, provide the golden standard for validation of the MCG/ECG inverse solutions. Even though patient populations studied by MCG before or during invasive catheterization are still relatively small, the localization studies of various cardiac arrhythmias have shown encouraging results. Multichannel systems and accurate combination of the results with cardiac anatomy have improved the accuracy to the order of 5 mm to 10 mm, which is sufficient to aid in planning the curative therapy of arrhythmia patients. Further valida-

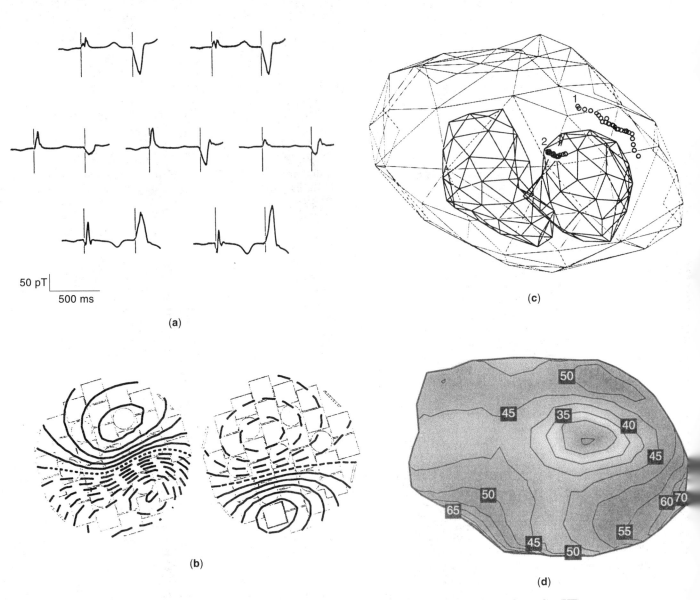

Figure 6. (a) MCG curves recorded from a patient suffering from ventricular tachycardia (VT). The seven axial gradiometers show (1) a normal sinus-rhythm beat and (2) an arrhythmogenic ventricular extrasystole (VES). (b) Isocontours of the magnetic-field component perpendicular to the sensor array (see Fig. 2). The field values were interpolated from the measured data with the minimum-norm estimation (68). Solid and dashed lines here indicate, respectively, magnetic flux toward or out of the chest. The step between adjacent contours is 1 pT. (c) MCG localization results obtained with a single moving ECD. (d) Ventricular activation sequence reconstructed from the VES by the method reported in Ref. 92.

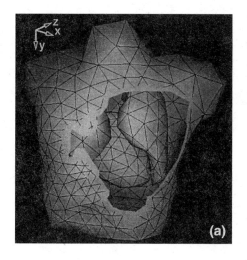
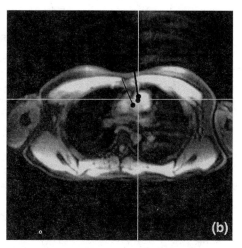

Figure 7. (a) An example of a boundary-element torso model constructed from MRI data. The surfaces of the body, the lungs, and the heart are tessellated into triangulated networks. The total number of triangles here is about 1500. (b) An example of MCG localization of tachycardias. The patient was suffering from continuous atrial tachycardia with the heart rate of over 140 beats per minute. ECD localization was performed from 67-channel MCG data at the onset of the P-wave, and the ECD locations were superimposed on the MRI data. Catheter ablation performed later at the location pinpointed by the MCG result terminated all arrhythmias.

tion for the MCG localization accuracy has been obtained by locating artificial dipole sources, such as pacing catheters inserted into the heart during electrophysiological studies.

Future Trends

The arrhythmogenic substrate is not manifested in all normal sinus rhythm recordings, and interventions may be needed during MCG to stimulate controlled arrhythmias to locate them. Thus, MCG should be available in a catheterization laboratory, but the demand of magnetical shielding and liquid helium is, in practice, limiting the use of MCG mapping in guiding invasive arrhythmia localization. For this purpose, compact-size higher-order magnetometer arrays operated without external shielding would be required.

MCG mapping under exercise is a promising tool for noninvasive characterization and localization of myocardial ischemia (76). Improved source modeling and localization methods are under test, especially in patients with coronary artery disease (93).

Despite over 20 years of MCG and MEG research, common standards of measurement techniques, data processing, and presentation are still lacking. Suggestions for such standards are emerging, but it is clear that there is and will be large differences between sensors and their arrangement in multichannel magnetometers. Fortunately, tools such as MNE (63,68) have been developed to interpolate signal morphologies and isocontour maps that are directly comparable to studies performed in other centers.

The field of magnetic source imaging may expand during the next few years with the implementation of low-noise high-T_c SQUID arrays that can be operated at the temperature of liquid nitrogen (101–103). The higher noise level of the high-T_c SQUIDs is, particularly in MEG studies, partly compensated by the smaller distance between the sensors and the body. At present, however, the low-T_c SQUIDs are easier to produce and thus cheaper than the high-T_c ones.

The future of MEG, with commercial whole-head instruments now available, looks promising. The capability to monitor activity of several cortical regions simultaneously in real time provides a unique window to study the neural basis of human cognitive functions. Important information can be obtained both from evoked responses and from spontaneous ongoing activity.

Effective signal processing and source modeling software is going to be increasingly important to extract all available functional data from the electromagnetic signals emerging from the brain and the heart. The widely discussed issue of whether the electric or magnetic technique is superior to the other is not of primary importance. Rather, one should apply a suitable alliance of different types of imaging methods, taking into account the characteristics and aims of the experiment being performed to yield optimal information about the functions of the biological system.

BIBLIOGRAPHY

1. G. Baule and R. McFee, Detection of the magnetic field of the heart, *Am. Heart J.,* **66**: 95–96, 1963.

2. D. Cohen, Magnetoencephalograpy: Evidence of magnetic fields produced by alpha-rhythm currents, *Science,* **161**: 784–786, 1968.

3. D. Cohen, E. A. Edelsack, and J. E. Zimmerman, Magnetocardiograms taken inside a shielded room with a superconducting point-contact magnetometer, *Appl. Phys. Lett.,* **16**: 278–280, 1970.

4. R. Hari and R. Ilmoniemi, Cerebral magnetic fields, *CRC Crit. Rev. Biomed. Eng.,* **14**: 93–126, 1986.

5. M. Hämäläinen et al., Magnetoencephalography—theory, instrumentation, and applications to noninvasive studies of the working human brain, *Rev. Mod. Phys.,* **65**: 413–497, 1993.

6. J. Nenonen, Solving the inverse problem in magnetocardiography, *IEEE Eng. Med. Biol.,* **13**: 487–496, 1994.

7. R. Näätänen, R. Ilmoniemi, and K. Alho, Magnetoencephalography in studies of human cognitive brain function, *Trends Neurosci.,* **17**: 389–395, 1994.

8. W. Andrae and H. Nowak (eds.), *Magnetism in Medicine,* Berlin: Wiley-VCH, 1998.

9. R. Hari, Magnetoencephalography as a Tool of Clinical Neurophysiology, in E. Niedermeyer and F. Lopes da Silva (eds.), *Electroencephalography: Basic Principles, Clinical Applications, and Related Fields,* 3rd ed., Baltimore, MD: Williams and Wilkins, 1993, pp. 1035–1061.

10. R. Salmelin et al., Dynamics of brain activation during picture naming, *Nature,* **368**: 463–465, 1994.

11. R. Salmelin et al., Impaired visual word processing in dyslexia revealed with magnetoencephalography, *Ann. Neurol.,* **40**: 157–162, 1996.

12. R. Näätänen et al., Language-specific phoneme representations revealed by electric and magnetic brain responses, *Nature,* **385**: 432–434, 1997.

13. R. Hari and R. Salmelin, Human cortical oscillations: A neuromagnetic view through the skull, *Trends Neurosci.,* **20**: 44–49, 1997.

14. C. Tesche et al., Characterizing the local oscillatory content of spontaneous cortical activity during mental imagery, *Cogn. Brain Res.,* **2**: 243–249, 1995.

15. R. Salmelin et al., Functional segregation of movement-related rhythmic activity in the human brain, *NeuroImage,* **2**: 237–243, 1995.

16. J. D. Lewine and W. W. Orrison, Magnetoencephalography, in W. G. Bradley and G. M. Bydder (eds.), *Advanced MR Imaging Techniques,* London: Martin Dunitz, 1997, pp. 333–354.

17. R. Fenici and G. Melillo, Magnetocardiography: Ventricular arrhythmias, *Eur. Heart J.,* **14** (Suppl. E): 53–60, 1993.

18. M. Mäkijärvi et al., New trends in clinical magnetocardiography, in C. Aine et al. (eds.), *Advances in Biomagnetism Research: Biomag96,* New York: Springer-Verlag, 1998 (in press).

19. M. Oeff and M. Burghoff, Magnetocardiographic localization of the origin of ventricular ectopic beats, *PACE,* **17**: 517–522, 1994.

20. W. Moshage et al., Evaluation of the non-invasive localization of cardiac arrhythmias by multichannel magnetocardiography (mcg), *Int. J. Cardiac Imaging,* **12**: 47–59, 1996.

21. G. Stroink, R. Lamothe, and M. Gardner, Magnetocardiographic and Electrocardiographic Mapping Studies, in H. Weinstock (ed.), *SQUID Sensors: Fundamentals, Fabrication and Applications,* Amsterdam: Kluwer Academic Publishers, 1996, NATO ASI Ser., pp. 413–444.

22. R. Fenici et al., Non-fluoroscopic localization of an amagnetic stimulation catheter by multichannel magnetocardiography, *PACE,* 1998, in press.

23. J. Montonen, Magnetocardiography in identification of patients prone to malignant arrhythmias, in C. Baumgartner et al. (eds.), *Biomagnetism: Fundamental Research and Clinical Applications,* New York: Springer-Verlag, 1995, pp. 606–611.

24. O. V. Lounasmaa, *Experimental Principles and Methods Below 1K,* London: Academic Press, 1974.

25. T. Ryhänen et al., SQUID magnetometers for low-frequency applications, *J. Low Temp. Phys.,* **76**: 287–386, 1989.

26. B. D. Josephson, Possible new effects in superconductive tunnelling, *Phys. Lett.,* **1**: 251–253, 1962.

27. J. Clarke, A superconducting galvanometer employing Josephson tunnelling, *Philos. Mag.,* **13**: 115, 1966.

28. J. Clarke, W. M. Goubau, and M. B. Ketchen, Tunnel junction dc SQUID fabrication, operation, and performance, *J. Low Temp. Phys.,* **25**: 99–144, 1976.

29. C. D. Tesche et al., Practical dc SQUIDs with extremely low 1/f noise, *IEEE Trans. Magn.,* **MAG-21**: 1032–1035, 1985.

30. D. Cohen, Magnetic measurement and display of current generators in the brain. Part I: The 2-D detector, *Dig. 12th Int. Conf. Med. Biol. Eng.,* Jerusalem, p. 15, 1979 (Petah Tikva, Israel: Beilinson Medical Center).

31. S. N. Erné and G. L. Romani, Performances of Higher Order Planar Gradiometers for Biomagnetic Source Localization, in H. D. Hahlbohm and H. Lubbig (eds.), *SQUID'85 Superconducting Quantum Interference Devices and their Applications,* Berlin: de Gruyter, 1985, pp. 951–961.

32. P. Carelli and R. Leoni, Localization of biological sources with arrays of superconducting gradiometers, *J. Appl. Phys.,* **59**: 645–650, 1986.

33. J. E. T. Knuutila et al., A 122-channel whole-cortex SQUID system for measuring the brain's magnetic fields, *IEEE Trans. Magn.,* **29**: 3315–3320, 1993.

34. D. Cohen, Low-field room built at high-field magnet lab, *Phys. Today,* **23**: 56–57, 1970.

35. S. N. Erné et al., The berlin magnetically shielded room (BMSR): Section B—performances, in S. N. Erné, H.-D. Hahlbohm, and H. Lubbig (eds.), *Biomagnetism,* Berlin: de Gruyter, 1981, pp. 79–87.

36. V. O. Kelhä et al., Design, construction, and performance of a large-volume magnetic shield, *IEEE Trans. Magn.,* **MAG-18**: 260–270, 1982.

37. J. Knuutila et al., Design considerations for multichannel SQUID magnetometers, in H. D. Hahlbohm and H. Lubbig (eds.), *SQUID'85: Superconducting Quantum Interference Devices and their Applications,* Berlin: de Gruyter, 1985, pp. 939–944.

38. S. N. Erné et al., The positioning problem in biomagnetic measurements: A solution for arrays of superconducting sensors, *IEEE Trans. Magn.,* **MAG-23**: 1319–1322, 1987.

39. R. Plonsey, *Bioelectric Phenomena,* New York: McGraw-Hill, 1969.

40. J. P. Wikswo, Jr., J. P. Barach, and J. A. Freeman, Magnetic field of a nerve impulse: First measurements, *Science,* **208**: 53–55, 1980.

41. J. P. Wikswo, Jr.,, Cellular magnetic fields: Fundamental and applied measurements on nerve axons, peripheral nerve bundles, and skeletal muscle, *J. Clin. Neurophysiol.,* **8**: 170–188, 1991.

42. J. Sarvas, Basic mathematical and electromagnetic concepts of the biomagnetic inverse problem, *Phys. Med. Biol.,* **32**: 11–22, 1987.

43. A. C. L. Barnard, I. M. Duck, and M. S. Lynn, The application of electromagnetic theory to electrocardiology. I. Derivation of the integral equations, *Biophys. J.,* **7**: 443–462, 1967.

44. D. B. Geselowitz, On the magnetic field generated outside an inhomogeneous volume conductor by internal current sources, *IEEE Trans. Magn.,* **MAG-6**: 346–347, 1970.

45. Z. Zhang, A fast method to compute surface potentials generated by dipoles within multilayer anisotropic spheres, *Phys. Med. Biol.,* **40**: 335–349, 1995.

46. M. S. Hämäläinen and J. Sarvas, Realistic conductivity geometry model of the human head for interpretation of neuromagnetic data, *IEEE Trans. Biomed. Eng.,* **36**: 165–171, 1989.

47. S. N. Erné et al., Modelling of the His–Purkinje Heart Conduction System, in H. Weinberg, G. Stroink, and T. Katila (eds.), *Biomagnetism: Applications & Theory,* New York: Pergamon, 1985, pp. 126–131.

48. J. Neonen et al., Magnetocardiographic functional localization using current multipole models, *IEEE Trans. Biomed. Eng.,* **38**: 648–657, 1991.

49. J. Nenonen et al., Magnetocardiographic functional localization using a current dipole in a realistic torso, *IEEE Trans. Biomed. Eng.,* **38**: 658–664, 1991.

50. B. N. Cuffin and D. Cohen, Magnetic fields of a dipole in speical volume conductor shapes, *IEEE Trans. Biomed. Eng.,* **BME-24**: 372–381, 1977.

51. B. M. Horacek, Digital model for studies in magnetocardiography, *IEEE Trans. Magn.,* **MAG-9**: 440–444, 1973.

52. C. Brebbia, J. Telles, and L. Wrobel, *Boundary Element Techniques—Theory and Applications in Engineering,* Berlin: Springer-Verlag, 1984.

53. J. Lötjönen et al., A triangulation method of an arbitrary point set selected from medical volume data, *IEEE Trans. Magn.*, **34**: 2228–2233, 1998.

54. H. Buchner et al., Inverse localization of electric dipole current sources in finite element models of the human head, *Electroencephalogr. Clin. Neurophysiol.*, **102**: 267–278, 1997.

55. H. Helmholtz, Ueber einige Gesetze der Vertheilung elektrischer Ströme in körperlichen Leitern, mit Anwendung auf die thierisch-elektrischen Versuche, *Ann. Phys. Chem.*, **89**: 211–233, 353–377, 1853.

56. D. W. Marquardt, An algorithm for least-squares estimation of nonlinear parameter, *J. Soc. Ind. Appl. Math.*, **11**: 431–441, 1963.

57. M. Scherg and D. von Cramon, Two bilateral sources of the late aep as identified by a spatiotemporal dipole model, *Electroencephalogr. Clin. Neurophysiol.*, **62**: 232–244, 1985.

58. M. Scherg, R. Hari, and M. Hämäläinen, Frequency-Specific Sources of the Auditory N19–P30–P50 Response Detected by a Multiple Source Analysis of Evoked Magnetic Fields and Potentials, in S. J. Williamson et al. (eds.), *Advances in Biomagnetism*, New York: Plenum, 1989, pp. 97–100.

59. K. Uutela, M. Hämäläinen, and R. Salmelin, Global optimization in the localization of neuromagnetic sources, *IEEE Trans. Biomed. Eng.*, **45**: 716–723, 1998.

60. P. Berg and M. Scherg, Sequential Brain Source Imaging: Evaluation of Localization Accuracy, in C. Ogura, Y. Coga, and M. Shimokochi (eds.), *Recent Advances in Event-Related Brain Potential Research*, Amsterdam: Elsevier, 1996.

61. J. C. Mosher, P. S. Lewis, and R. Leahy, Multiple dipole modeling and localization from spatiotemporal MEG data, *IEEE Trans. Biomed. Eng.*, **39**: 541–557, 1992.

62. T. E. Katila, On the current multipole presentation of the primary current distributions, *Nuovo Cimento*, **2D**: 660–664, 1983.

63. M. S. Hämäläinen and R. J. Ilmoniemi, Interpreting magnetic fields of the brain: Minimum-norm estimates, *Med. Biol. Eng. Comput.*, **32**: 35–42, 1994.

64. A. A. Ioannides, J. P. R. Bolton, and C. J. S. Clarke, Continuous probabilistic solutions to the biomagnetic inverse problem, *Inverse Problems*, **6**: 523–542, 1990.

65. R. D. Pascual-Marqui, C. M. Michel, and D. Lehmann, Low resolution electromagnetic tomography: A new method for localizing electrical activity in the brain, *Int. J. Physchopsysiol.*, **18**: 49–65, 1994.

66. A. van Oosterom et al., The magnetocardiogram as derived from electrocardiographic data, *Circ. Res.*, **67**: 1503–1509, 1990.

67. P. Hansen, Numerical tools for analysis and solution of fredholm integral equations of the first kind, *Inverse Problems*, **8**: 849–872, 1992.

68. J. Numminen et al., Transformation of multichannel magnetocardiographic signals to standard grid form, *IEEE Trans. Biomed. Eng.*, **42**: 72–77, 1995.

69. K. Pesola et al., Comparison of regularization methods when applied to epicardial minimum norm estimates, *Biomed. Tech.*, **42** (Suppl. 1): 273–276, 1997.

70. R. MacLeod and D. Brooks, Recent progress in inverse problems in electrocardiography, *IEEE Eng. Med. Biol.*, **17**: 73–83, 1998.

71. A. M. Dale and M. I. Sereno, Improved localization of cortical activity by combining EEG and MEG with MRI cortical surface reconstruction: A linear approach, *J. Cog. Neurosci.*, **5**: 162–176, 1993.

72. K. Matsuura and Y. Okabe, Selective minimum-norm solution of the biomagnetic inverse problem, *IEEE Trans. Biomed. Eng.*, **42**: 608–615, 1995.

73. K. Matsuura and Y. Okabe, A robust reconstruction of sparse biomagnetic sources, *IEEE Trans. Biomed. Eng.*, **44**: 720–726, 1997.

74. K. H. Uutela, M. S. Hämäläinen, and E. Somersalo, Spatial and temporal visualization of magnetoencephalographic data using minimum-current estimates, *NeuroImage*, **5**: S434, 1997.

75. M. Fuchs et al., Possibilities of functional brain imaging using a combination of MEG and MRT, in C. Pantev (ed.), *Oscillatory Event-Related Brain Dynamics*, New York: Plenum Press, 1994, pp. 435–457.

76. K. Brockmeier et al., Magnetocardiography and 32-lead potential mapping: The repolarization in normal subjects during pharmacologically induced stress, *J. Cardiovasc. Electrophysiol.*, **8**: 615–626, 1997.

77. D. Cohen and B. N. Cuffin, A method for combining MEG and EEG to determine the sources, *Phys. Med. Biol.*, **32**: 85–89, 1987.

78. M. Fuchs et al., Improving source reconstructions by combining bioelectric and biomagnetic data, *Electroenceph. Clin. Neurophysiol.*, **107**: 93–111, 1998.

79. P. Colli-Franzone et al., Potential fields generated by oblique dipole layers modeling excitation wavefronts in the anisotropic myocardium. Comparison with potential fields elicited by paced dog hearts in a volume conductor, *Circ. Res.*, **51**: 330–346, 1982.

80. J. P. Wikswo, Jr., Tissue Anisotropy, the Cardiac Bidomain, and the Virtual Cathode Effect, in D. Zipes and J. Jalife (eds.), *Cardiac Electrophysiology: From Cell to Bedside*, 2nd ed., Orlando, FL: 1994, Saunders, pp. 348–361.

81. P. Reissman and I. Magnin, Modeling 3d deformable object with the active pyramid, *Int. J. Pattern Recognition & Artif. Intell.*, **11**: 1129–1139, 1997.

82. G. Simpson et al., Spatiotemporal Mapping of Brain Activity Underlying Visual Attention Through Integrated MEG, EEG, FMRI and MRI, in C. Aine et al. (eds.), *Advances in Biomagnetism Research: Biomag96*, Springer-Verlag, New York, 1998.

83. J. P. Mäkelä, Functional differences between auditory cortices of the two hemispheres revealed by whole-head neuromagnetic recordings, *Hum. Brain Mapp.*, **1**: 48–56, 1993.

84. R. Salmelin and R. Hari, Characterization of spontaneous MEG rhythms in healthy adults, *Electroencephalogr. Clin. Neurophysiol.*, **91**: 237–248, 1994.

85. R. Salmelin and R. Hari, Spatiotemporal characteristics of sensorimotor neuromagnetic rhythms related to thumb movement, *Neuroscience*, **60**: 537–550, 1994.

86. W. W. Sutherling and D. S. Barth, Neocortical propagation in temporal lobe spike foci on mangetoencephalography and electroencephalography, *Ann. Neurol.*, **25**: 373–381, 1989.

87. R. Paetau et al., Magnetoencephalographic localization of epileptic cortex—impact on surgical treatment, *Ann. Neurol.*, **32**: 106–109, 1992.

88. N. Forss et al., Trigeminally triggered epileptic hemifacial convulsions, *NeuroReport*, **6**: 918–920, 1995.

89. J. Nenonen et al., Noninvasive magnetocardiographic localization of ventricular preexcitation in wolff-parkinson-white syndrome using a realistic torso model, *Eur. Heart J.*, **14**: 168–174, 1993.

90. M. Mäkijärvi et al., Magnetocardiography: Supraventricular arrhythmias and preexcitation syndromes, *Eur. Heart J.*, **14** (Suppl. E): 46–52, 1993.

91. R. Fenici et al., Clinical validation of three-dimensional cardiac magnetic source imaging accuracy with simultaneous magnetocardiographic mapping, monophasic action potential recordings, and amagnetic cardiac pacing, *11th Int. Conf. Biomagnetism, Biomag98*, Abstracts, Sendai, 1998, p. 119.

92. T. Oostendorp, J. Nenonen, and G. Huiskamp, Comparison of inverse solutions obtained from ECG and MCG maps, *Proc. 18th Annu. Int. Conf. IEEE Eng. Med. Biol. Soc.,* pp. CD-rom, 1996.

93. U. Leder et al., Non-invasive biomagnetic imaging in coronary artery disease based on individual current density maps of the heart, *Int. J. Cardiol.,* **64**: 83–92, 1998.

94. L. Trahms et al., Biomagnetic functional localization of a peripheral nerve in man, *Biophys. J.* **55**: 1145–1153, 1989.

95. R. Hari et al., Multichannel Detection of Magnetic Compound action Fields of Median and Ulnar Nerves, *Electroenceph. Clin. Neurophysiol.,* **72**: 277–280, 1989.

96. I. Hashimoto et al., Visualization of a moving quadrupole with magnetic measurements of peripheral nerve action fields, *Electroenceph. Clin. Neurophysiol.,* **93**: 459–467, 1994.

97. B.-M. Mackert et al., Mapping of tibial nerve evoked magnetic fields over the lower spine, *Electroenceph. Clin. Neurophysiol.,* **104**: 322–327, 1997.

98. W. O. Richards et al., Non-invasive magnetometer measurements of human gastric and small bowel electrical activity, in C. Baumgartner et al. (eds.), *Biomagnetism: Fundamental Research and Clinical Applications,* New York: Springer Verlag, 1995, pp. 743–747.

99. W. Weitschies et al., High-resolution monitoring of the gastrointestinal transit of a magnetically marked capsule, *J. Pharm. Sci.* **86**: 1218–1222, 1997.

100. D. M. Schmidt, J. S. George, and C. C. Wood, Bayesian inference applied to the electromagnetic inverse problem, Tech. Report, Los Alamos National Laboratory, Los Alamos, LA-UR-97-4813.

101. D. Drung et al., Integrated $YBa_2Cu_3O_{7-x}$ magnetometer for biomagnetic measurements, *Appl. Phys. Lett.,* **68**: 1421–1423, 1996.

102. M. Burghoff et al., Diagnostic application of high-temperature SQUIDs, *J. Clin. Eng.,* **21**: 62–66, 1996.

103. J. M. ter Brake et al., A seven-channel high-T_c SQUID-based heart scanner, *Meas. Sci. Techn.,* **8**: 927–931, 1997.

Reading List

H. Weinberg, G. Stroink and T. Katila, Biomagnetism, in J. G. Webster (ed.), *Encyclopedia of Medical Devices and Instrumentation, Vol. 1,* New York: Wiley, 1988, pp. 303–322.

S. J. Williamson et al., *Advances in Biomagnetism,* New York: Plenum, 1989.

M. Hoke et al., *Biomagnetism: Clinical Aspects,* Amsterdam: Elsevier, 1992.

C. Baumgartner et al., *Biomagnetism: Fundamental Research and Clinical Applications,* New York: Springer-Verlag, 1995.

C. Aine et al., *Advances in Biomagnetism Research: Biomag96,* New York: Springer-Verlag, 1998.

M. S. HÄMÄLÄINEN
J. T. NENONEN
Helsinki University of Technology

MAGNETIC STORAGE MEDIA

Magnetic recording of data relies on the controlled creation and reliable detection of regions of differing magnetization in a magnetic medium. Today's magnetic recording media are the result of impressively successful development since magnetic recording media were invented and first demonstrated 70 years ago. The development has been particularly fast in the last 15 to 20 years, during which many technical and commercial advances have been made. Media storage capacity has been dramatically raised from 5 to 10 Mbyte (megabytes) per disk platter in the early 1980s to 2 to 4 Gbyte (gigabytes) per disk platter in early 1998. (A decimal base is not used for numbers describing data stored. Arising from the computing usage of binary numbers, 1 Mbyte is the nearest power of 2 equivalent of 1 million, in other words, 1,048,576 bytes ≡ 2^{20} bytes; and 1 Gbyte is the nearest binary equivalent of 10^9, that is, 2^{30}, bytes, in other words, 1,073,741,824 bytes.) In the same period, the price of a disk platter has also fallen by more than a factor of 100. Despite these successes, progress in raising the areal density (number of data bits per unit area) of recording media has not stopped or slowed down, and the total data storage capacity of disk drives is still increasing rapidly. It is predicted that an areal density of 20 Gbit/in.^{2} will be demonstrated, and that media above 10 Gbit/in.^{2} should be available commercially by the turn of the century. (With respect to disk diameter and area, the units used in the United States are still the inch and square inch, respectively.)

The purpose of this article is to give an overview of magnetic recording media, especially of the thin-film media developed recently on rigid disks, since this is the dominant, continuing area of magnetic recording media research and commercial progress. We begin with a short look at the history of recording media, followed by detailed discussion of current media and reference to the future of these media.

Historical Perspective

The earliest magnetic recording medium was a wire made of stainless steel, containing nickel and chromium. The wire was annealed so that single-domain particles of the ferrite phase precipitated in an austenitic, nonmagnetic matrix (1). The resulting coercivity of this wire recording medium, that is, the applied magnetic field required to reduce the magnetization to zero, was only a few hundred oersteds. This type of medium served to establish the concept of magnetic recording. However, the medium was never commercialized for two practical reasons: the recording head could not reliably read all the previously recorded information, since the wires were easily twisted, and the read–write process was interrupted whenever the thin wires broke and needed to be "repaired" by knotting the broken ends together. To improve upon this type of magnetic recording medium and to avoid its mechanical problems, a spliceable tape coated with synthetic particles, including γ-Fe_2O_3, was developed in the early 1940s (2). The particles were aciculate (needle-shaped) and were held with their long axes parallel to a polymer tape backing. The particles were believed to be single domain and to undergo magnetization reversal by coherent rotation. Extensive research and development work since that time has led to the successful commercial applications of a wide variety of magnetic tape media.

The particulate-coating approach was extended to rigid disk media in the 1950s with IBM's development of the first rigid magnetic recording medium—a 24 in. disk coated with γ-Fe_2O_3. As the technology evolved, in the late 1970s the sputtering form of vacuum deposition became well established as the method for applying the magnetic layer in the magnetic recording industry.

Sputtering is far superior to the earlier coating methods for making films because it can create a much higher packing density of the magnetic constituents, and this is a prerequisite for higher recording densities. The magnetic materials used in the vacuum deposition were mostly CoCr alloys (3,4), which easily provided media coercivities of about 300 Oe to 500 Oe—a large improvement from the 200 Oe to 300 Oe coercivities of γ-Fe_2O_3–coated particulate media. By the early 1980s, ternary CoCrX alloys (X being a metallic element), such as CoCrNi and CrCrTa, had evolved with substantially improved recording media characteristics (5). With the addition of Ni or Ta, the anisotropy of the alloys was greatly increased, resulting in higher coercivities and better recording performance. The ternary alloys CoCrTa have been so successful so that they have been widely used in the past 10 years, until they reached a coercivity limit at about 2500 Oe to 2700 Oe. Exceeding that limit has involved a large effort throughout academia and industry. By the early 1990s, Pt was found to be an excellent candidate for increasing the magnetocrystalline anisotropy by forming CoCrPt- and CoCrPtTa-based alloys (6,7,24) and so enhancing magnetic film coercivities. Since the crystalline anisotropy is a key intrinsic parameter for increasing the film coercivity, quaternary alloys have therefore emerged as the means to further development of high-recording-density media by breaking the ternary coercivity limit. Coercivities up to 4000 Oe have reliably been obtained for the CoCrTaPt alloy system (8). At the same time, other high-anisotropy alloys, such as CoSm- and FeSm-based alloys, have also been studied but their recording characteristics, both from academia and from industry, have been exceeded by the quaternary alloy CoCrPtTa. Development of underlayer and seed-layer technologies and optimization of the manufacturing process has also assisted in obtaining dramatic improvements in the performance of media, including the media signal-to-noise ratio (SNR), as we discuss later.

Fundamentals of Magnetic Recording

Magnetic recording is based on storing information in a magnetic medium by controlled writing (creation) of regions with differing magnetization. Requirements for high-density longitudinal recording media include desirable bulk magnetic properties, appropriate magnetic domain structure, and stability of the magnetizations and the read–write process. In this section we briefly review the writing and reading processes and analyze their performance, which is correlated with bulk magnetic properties and the spacing between the head and the medium. The section concludes with a short discussion of thermal stability, which is also correlated with the micromagnetic properties. The magnetic domain and medium microstructure will be discussed in the later sections.

The writing process is illustrated in Fig. 1(a). The coil current creates a writing field in the medium that establishes the local magnetization. Reversal of the current creates a magnetization transition. The longitudinal writing field H_x generated by the head coil current in the Karlqvist model (9) is

$$H_x(x, y) = \frac{H_g}{\pi} \left(\arctan\frac{x+g/2}{y} - \arctan\frac{x-g/2}{y} \right) \quad (1)$$

where g is the gap width of head, H_g is the magnetic field inside the head gap, and x and y are the coordinates at which H_x is to be calculated, as shown in Fig. 1(a). The peak value

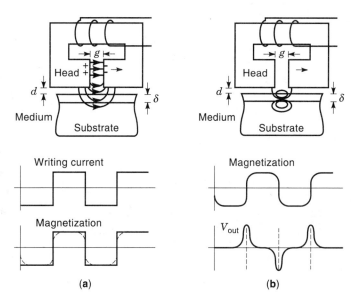

Figure 1. (a) Illustration of magnetic writing process. (b) Illustration of magnetic reading process.

of H_x is at $x = 0$ and can be expressed as

$$H_{x(\text{peak})} = H_x(0, y) = 2\frac{H_g}{\pi} \left(\arctan\frac{g/2}{y} \right) \quad (2)$$

A sufficient writing field must be applied to write magnetic transitions or to achieve a good overwrite of previously written data. It is commonly required that the writing field at the back of the medium is about 2.5 times the medium coercivity H_c. Taking $x = 0$, $y = d + \delta$ (where d is the spacing between head and medium and δ is the medium layer thickness) and $H_x(0, d + \delta) = 2.5H_c$, one can calculate the required H_g for a given H_c as

$$H_g = \frac{2.5\pi H_c}{2 \arctan\left(\dfrac{g}{2(d + \delta)} \right)} \quad (3)$$

Assuming the following reasonable parameter values for calculating a magnitude for H_g, $H_c = 2000$ Oe, $d = 40$ nm, $\delta = 25$ nm, and $g = 260$ nm, and substituting these values into Eq. (3), one finds $H_g = 5900$ Oe. The parameters that are projected for a 10 Gbit/in.^{2} medium are (10) to be $H_c = 3000$ Oe, $d = 20$ nm, $\delta = 10$ nm, and $g = 200$ nm, from which one may find $H_g = 9200$ Oe. Hence, permalloy, which is a common head material, cannot be used to write such a high coercivity medium, and new head materials with higher saturation flux densities are required.

The magnetization transition obtained by writing data can be described fairly well with a simple arctangent function as shown in Fig. 2,

$$M_x = \frac{2}{\pi}M_r \left(\arctan\frac{x}{a} \right) \quad (4)$$

where M_r is the remanent magnetization of the medium and a is known as the transition parameter. It is notable that only one parameter a is needed to specify the transition characteristics: a small value of a corresponds to a rapid transition,

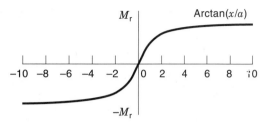

Figure 2. Arctangent form of magnetic transition.

and this favors high-density recording, as will be discussed in more detail in connection with the reading process.

The reading process is illustrated in Fig. 1(b). When a written magnetic transition in the medium passes under the head gap, the stray magnetic field from the medium will cause a change in magnetization in the head core and so will result in an induced output signal from the head coil. This signal is correlated with the head and medium parameters and the spacing between them. For an arctan magnetization transition, this output voltage can be expressed as follows (9,11,12):

$$V(x', d) = -2\mu_0 \left(\frac{\eta n v w}{\pi g} \right) (M_r \delta)$$
$$\times \left(\arctan \frac{x' + g/2}{a + d} - \arctan \frac{x' - g/2}{a + d} \right) \quad (5)$$

where η, n, v, w, and x' are the head efficiency, the number of coil turns in a head, the velocity of the medium relative to the head, and the recording track width in the medium, respectively. We note that Eq. (5) has the same form as Eq. (1) and the output voltage $V(x', d)$ is proportional to $M_r \delta$. Taking $x' = 0$ in Eq. (5), the peak voltage V_{p-p} is

$$V_{p-p} = 8\mu_0 \left(\frac{\eta n v w}{\pi g} \right) (M_r \delta) \left(\arctan \frac{g/2}{a + d} \right) \quad (6)$$

The width of the output pulse, measured at one-half of the maximum amplitude level, is called PW_{50} and is given by

$$PW_{50} = [g^2 + 4(a + d)(a + d + \delta)]^{1/2} \quad (7)$$

where a is the transition parameter introduced in Eq. (4). The derivation of an expression for a has to take into account the details of the writing process. We shall not go into this process but simply give the result (89):

$$a = 2\sqrt{\frac{M_r \delta (d + \delta/2)}{H_c}} \quad (8)$$

The desirable bulk magnetic properties of media for high-density recording can be understood by considering the pulse width of an output signal. A narrow pulse, that is, a small PW_{50}, will allow an increase in the linear recording density. Equation (7) suggests that small g, d, δ, and a favor a decrease in the value of PW_{50} and therefore an increase in the linear density. However, the values of g, d, and δ are limited by other considerations: the gap width g has to be large enough to produce a sufficient field [see Eqs. (1) and (2)]; the head–medium spacing d has to be large enough to prevent mechanical wear of both the head and medium, and so has a minimum value that is limited by the surface roughness of

the medium; and the medium thickness δ has to be great enough to provide sufficient signal [see Eqs. (5) and (6)]. To reduce δ and retain sufficient output signal, a large M_r of the medium is preferred. Therefore, an important approach to reducing PW_{50} is to shorten the transition parameter a, guided by Eq. (8), which indicates that this optimally requires decreasing $M_r \delta$ and increasing H_c.

Thermal stability is another important issue in high density recording as the crystalline grain size is reduced to attain increased areal recording density. The grain size should be as small as practical to satisfy the needs for a narrow transition, a high signal-to-noise ratio, and a smooth film surface. However, the grain size must be large enough to provide adequate thermal stability to guard against time-dependent magnetic effects (109,110). Using the thermal activation model, Sharrock pointed out that the measured coercivity H_c is time-dependent and can be expressed as

$$H_c(t) = H_0 \left\{ 1 - \left[\frac{k_B T}{K_u V} \ln(At) \right]^n \right\} \quad (9)$$

where k_B, T, K_u, and V are the Boltzmann constant, absolute temperature, magnetic anisotropy, and grain volume, respectively, A is a time-independent constant, and n varies from 1/2 to 2/3 depending on the orientation distribution of grain moments with respect to the field. Therefore, the coercivity H_c that is relevant to long-term storage (large t) can be significantly less than that which is relevant to high frequency writing (small t). To satisfy the thermal stability requirement, the following condition should be obeyed:

$$\frac{K_u V}{k_B T} \geq 60 \text{ to } 80 \quad (10)$$

This formula indicates that a medium with high anisotropy K_u and large grain volume V will have enhanced thermal stability. However, a large grain volume V will degrade the signal-to-noise ratio and transition width, both of which are undesirable.

Further analysis indicates that if the condition

$$\frac{K_u V}{k_B T} \leq 25 \quad (11)$$

holds, the medium will lose coercivity and become superparamagnetic because the grains are too small to retain the orientation of their magnetic moments, owing to thermal agitation. This is the so-called superparamagnetic limit of recording media.

In summary, high-density recording requires a high medium coercivity H_c and a small $M_r \delta$. Since $M_r \delta$ must be large enough to give sufficient output signal, the most appropriate way to reduce the transition a is by increasing coercivity H_c. However, one should notice the correlation between H_c and H_g in Eq. (3) and should choose a high value of H_c that is within the saturation flux density limitation of the head material. The grain size has to be large enough to provide thermal stability, but as small as practical to increase the signal-to-noise ratio and reduce the transition width. It should also be pointed out that high anisotropy K_u favors the enhancement of both H_c and thermal stability.

Units Used for Magnetic Recording Media

The units used in describing magnetic recording media have not been unique, although most in the field have adopted mks units. Researchers and disk makers in the United States in this field still use cgs units, partly because the electromagnetism mathematics description and derivations were so much simplified by using cgs units and partly because of a reluctance to change. Most disk makers in the United States now use cgs units, but a few use mks units or a combination of both units. Table 1 contains those parameters that are often used for magnetic recording media and their unit conversions.

Types of Magnetic Recording Media

Magnetic recording media technology has rapidly evolved in the following categories: (1) particulate media, (2) rigid continuous thin media, and (3) patterned media.

Particulate media comprise magnetic particle dispersions coated onto tapes or flexible disks to form a polymer–particle composite structure akin to paint. The common early magnetic materials were iron oxide and cobalt-surface-treated iron oxide. More recently aciculate iron particles have been adopted. Particulate media are widely used in audiotapes and videotapes, in removable floppy disks, and in plastic credit, bank, or other cards.

Rigid continuous media are made of magnetic films deposited onto a rigid substrate. There are two types of film media: longitudinal and perpendicular, in which the magnetic moment axis is parallel or perpendicular to the film surface, respectively. In principle, perpendicular media can permit higher recording areal density than that of longitudinal media because of the demagnetization character resulting from recording data bits with opposite polarities in perpendicular media. However a number of factors, including head writing concerns, have limited the attainments of perpendicular recording. Perpendicular recording is still in the testing stage and today's typical rigid continuous thin film media rely on longitudinal films.

Therefore, in this article, we limit the discussion to rigid longitudinal thin films. Most commonly, the rigid substrates are made of aluminum–magnesium alloys, glass, or glass ceramics. The magnetic films have been deposited on to the substrates originally by plating, later by physical deposition of vapor from a thermal evaporation source, and currently are deposited by vacuum sputtering. Generally, the magnetic layers are Co alloys, most commonly on a CoCr-alloy base matrix.

Patterned media are considered to be the media for the future, although they are still in their infancy. Much effort is being devoted to patterned media, which have been demonstrated to have a much larger area density capacity than continuous media (13). Processes for making patterned media include photolithography etching and optical-interference-controlled etching (14). Many other techniques are in the development stage (15–18).

MAGNETIC RECORDING MEDIA ON RIGID DISKS

The development of magnetic rigid recording disks has progressed considerably, both in the fabrication methods and in the areal density. IBM built the first rigid disk in 1957. It was 24 in. in diameter, coated with Fe_2O_3, and held 5 Mbyte of data with an areal density of 10 Mbit/in.^{2} and a coercivity about a few hundred oersteds. Three years later, 14 in. disks were made using the same process but the disk was improved to hold more data, with the areal density increased to 100 Mbit/in.^{2} and the coercivity doubled over that of its predecessor.

In the early 1970s, thermal evaporation and electroplating technologies became available for magnetic recording media and showed substantial advantages over the particulate coating technique. New media could then be manufactured with higher media areal density. Meanwhile, the disk size was reduced further to 8 in. In the early 1980s, sputtering technology was adopted by virtue of its superiority to all the earlier media fabrication methods. Since the resultant magnetic thin films were much more homogeneous and had much better film integrity, sputtered media could attain much higher areal recording densities. Therefore, from the mid-1980s to the mid-1990s, rigid media have advanced rapidly from media areal densities of hundreds of Mbit/in.^{2} to the current production level of several Gbit/in.^{2} At the same time, the media size has shrunk further, through 6, 5.25, 3.5, 3.0, and 2.5 in., all the way to 1.8 in. in diameter for mobile computer applications.

Magnetic materials for the rigid media of the last 10 years have mostly been Co–Cr–X–Y–Z alloys, where X, Y, and Z refer to different transition metal elements. At present, 3.5 in. and 2.5 in. disks are the dominant, popular sizes and commonly employ CoCrPtTaX media alloys. The cost of the media has dropped dramatically to about \$0.03 (US) per Mbyte from about \$5.00 per MB a decade ago. Progress in this technology continues and is expected to result in 10 Gbit/in.^{2} disks being commercially available at the turn of the century.

Studies of Magnetic Media on Rigid Disks

Magnetic recording media form one of the most successful research areas of transfer of research results into commercial application. Extensive studies of rigid media started in the early 1980s during the transition between electroplating and vacuum-sputtering technologies. Co, CoCr, CoCrNi alloys were the early materials studied in great detail (19–22). Fisher, Allan, and Pressesky proposed the CoCrTa magnetic alloy system (23), which lasted 10 years in commercial production until improvements in the performance of CoCrTa media became limited by coercivity limits of this alloy system. Lal and Eltoukhy (24) and other researchers (25) found that

Table 1. Parameters and Units Used in Magnetic Recording

Quantity	cgs	mks	Notes
Permeability of free space μ	1	$4\pi \times 10^{-7}$	
Intrinsic coercivity H_c	Oe	Oe	
Remanent coercivity H_{cr}	Oe	Oe	$H_{cr} \geq H_c$
Moment per volume M	emu/cm^3	A/m	
Remanent magnetization M_r	emu/cm^3	A/m	
Product of remanent moment and film thickness $M_r\delta$	emu/cm^2	A/m^2	
Product of remanent magnetic field flux and film thickness $B_r\delta$	G/cm^2		
Anisotropy constant K_u	ergs/cm^3	J/m^2	
Anisotropy field H_k	G	T	1 T = 10^4 G

the addition of Pt to CoCr or CoCrTa systems could dramatically increase film coercivity. With the quaternary alloy system, coercivity ranges as high as 4000 Oe have been achieved at higher Pt concentration (8).

Magnetic recording media can be classified into three categories according to the principles and schemes employed in the recording process. The first category comprises the longitudinal recording media, on which large numbers of studies have been reported around the world. Longitudinal recording is one of the most successful examples of academic research and commercial applications stimulating each other and of science leading directly to products. The second category is of the perpendicular recording media. Studies of these media have shown the high potential of perpendicular recording for ultrahigh area recording density. For practical reasons, this type of recording scheme is still not in use and there is no commercial product available. However, a large number of studies have been done, mostly in Japan, where the work originated and where there is still the most research and continuing academic discussion. The last category is of the longitudinal type of media that have an added "keeper layer." Keeper layer media have also attracted considerable attention, which has resulted in quite a number of studies. The rationale is that the resulting magnetic flux closure can increase the fundamental read-back signal. However, such media are less suited to high-areal-density recording because of the increased effective spacing between the head and the magnetic layers and higher media noise generated by the keeper layer. Such media are still in the experimental stage and are not used commercially.

Techniques of Fabricating Recording Media

Magnetic thin films can be fabricated by various techniques, but three major ones have been used commercially by the thin-film media industry.

Electroplating played an important early role and was the first technique to deposit thin films of magnetic material, such as Co or CoNi, onto disk substrates from solutions containing salts of Co and Ni and hypophosphite salts, along with buffers to maintain the solution pH. This technique is no longer considered suitable, however, for high-density media for reasons of film density, nonuniformity, and defects.

The second technique was thermal evaporation, which was first used in flexible media (29) and later used in rigid-disk media manufacturing. In this process, materials such as Co, FeCo, or FeCoCr alloys are inductively melted and evaporated on to a substrate that is constantly moving to ensure film uniformity. The evaporated material is deposited onto substrates at an angle away from normal incidence. This was reported to yield advantages such as higher in-plane anisotropy, and therefore higher coercivity and better hysteresis loop squareness, all of which are needed in longitudinal recording (30). This oblique deposition scheme was also helpful in that it allowed two sides of rigid disks to be deposited simultaneously. Figure 3 illustrates the oblique evaporation process of making a rigid media disk. The evaporation methods have been surpassed by sputtering.

Sputtering, the third technique, is now the most important and yields highly compact, dense film that is well suited to high-areal-density recording. Importantly, sputtered metal alloy deposits can closely copy the chemical composition of the

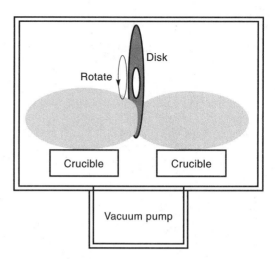

Figure 3. Illustration of the oblique evaporation process of making a rigid-disk media.

sputtering target material and the film magnetic properties can largely be controlled by alloy stoichiometry and by process conditions. Of the range of available sputtering techniques—dc and RF forms of diode sputtering and dc, ac, and RF forms of magnetron sputtering—dc magnetron sputtering has proven to be the best manufacturing process.

Given its importance in attaining the highest-density recording media, the sputtering process is described in detail in the following sections.

SPUTTERED THIN-FILM MEDIA

Sputtering Processes. Sputtering is an ion-bombardment process in which atoms are sputtered (removed) from a target (source of material) and deposited onto a substrate.

In a vacuum chamber at a low pressure of inert gas, a voltage difference is applied between a cathode and an anode to create a plasma, containing electrons and ions—Ar^+ in the case of argon, the most commonly used inert gas. The ions are accelerated towards the target (cathode) where they eject atoms of the target material, some of which, depending on the geometry of the system, can travel to be deposited on the substrate. This is a description of diode sputtering.

In magnetron sputtering, the plasma is additionally concentrated above the target by an enclosing magnetic field flux located at the sputtering sources. The magnetic field is imposed in such a way that the electrons are trapped in a region near the target surface, causing more intense ionization there. In standard sources, most of the sputtering occurs in an annular region around the center, as is shown in Fig. 4. The fierce ion bombardment does cause localized heating of the target, and this heating can cause some target materials to shatter if very large thermal gradients are allowed to develop and the target material cannot dissipate the heat fast enough. Sputtering targets and cathode systems are therefore cooled by flowing water to assist in dissipating the heat.

Electrical power can be supplied as RF, lower-frequency ac, or dc power, but dc power provides the highest sputtering rate, is more stable, and is best suited to manufacturing.

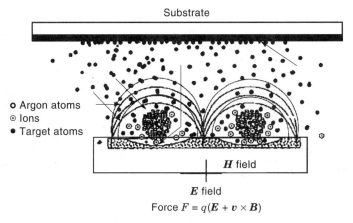

o Argon atoms
⊚ Ions
• Target atoms

H field

E field

Force $F = q(E + v \times B)$

Figure 4. Schematic diagram of sputtering from a circular target.

Place of Sputtering in the Disk-Manufacturing Process. Figure 5(a) shows a flow chart of the process for making disks and Fig. 5(b) shows a configuration of a pass-by-type sputtering machine for manufacturing rigid-disk media. Blank substrates are polished to a smoothness that will permit the head to fly at the required height without any impact with the disk during disk rotation. Disks are then washed and sputter-coated in a system of the type shown schematically in Fig. 6. The sputtering process itself is discussed in more detail in the following sections. After sputtering, disks are given post-sputtering treatments, including tape burnishing of the disk surfaces and dip lubrication. The disks are then subjected to head glide (flying) and certification tests.

Types of Sputtering. The main sputtering categories are diode sputtering, dc (constant-field) magnetron sputtering, rf (radio frequency) magnetron sputtering, and ac (alternating-field) magnetron sputtering.

Diode-sputtering systems are undemanding in design and construction and were the first to be used because the cathode can be quite simple and operation only requires application of an electric field between the cathode and anode. During diode sputtering, a plasma of positive inert-gas ions and electrons is generated, but the plasma is not closely confined. Many electrons therefore can bombard the substrate, which increases the substrate temperature in an uncontrolled manner and so affects film growth. It was found that the sputtering efficiency for this technique was low and that the film mechanical properties could not be kept within a narrow range. The diode technique was therefore quickly abandoned and was replaced for magnetic media by magnetron sputtering.

In magnetron sputtering, the cathode target design accommodates a magnet behind the target whose external field encloses the surface of the sputtering target. Under the influence of both the applied electric and magnetic fields, plasma electrons are strongly confined and so both the electron heating of the substrate and the consequent interference in film growth are minimal. The confinement of the electrons through the action of the Lorentz force is illustrated in Fig. 4. The nature of the applied electric field generates three important variants of magnetron sputtering.

Most popular in the media industry is dc magnetron sputtering, in which a constant voltage is applied between cathode and anode. This form of magnetron sputtering is easy and efficient, but demands that every part of the target surface must have good electrical conductivity to avoid both microarcing and the associated ejection of particles that can otherwise form detrimental defects in the disk. This "target spit" problem is commonly observed in dc sputtering of carbon targets but can easily be overcome by using RF magnetron sputtering. By preventing the accumulation of charge on the target, a radio-frequency oscillating electric field eliminates micro-arcs and so minimizes the introduction of defects in the disk surface. However, RF magnetron sputtering is not as straightforward as dc sputtering in its operation, because it is less stable and because it demands careful matching of the power source to the dynamic load that the sputtering system

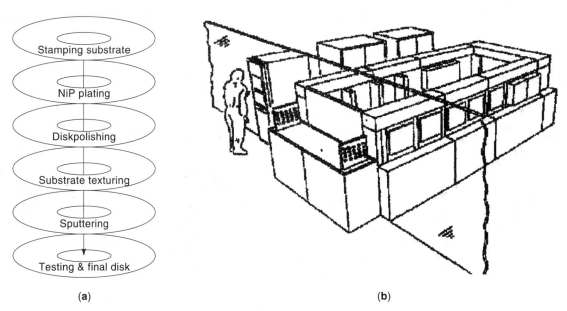

(a)

(b)

Figure 5. (a) The disk-manufacturing sequence. (b) A configuration of a pass-by type of sputter system (ULVAC sputtering system). [Fig. 5(b) courtesy of ULVAC.]

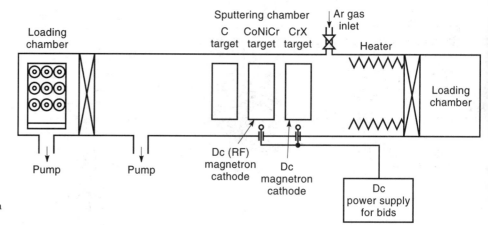

Figure 6. Schematic arrangement of a magnetron sputtering system.

presents. Few disk makers have therefore used this technique. To minimize instability problems and to permit use of less conductive targets without micro-arcing, the lower-frequency (<1 kHz) ac magnetron sputtering can be used with improved results. However, most disk makers currently use dc magnetron sputtering because it is simple and reliable.

Effects of Sputtering Process Control on Media Properties. The sputtering process has a great impact on thin-film mechanical properties, magnetic properties, and recording performance. For magnetic recording media, the sputtering process is as important as correct selection of the magnetic alloy composition and of the underlayer materials. Many parameters and factors affect the sputtering process. For example, changes in the sputtering power, the sputtering working pressure, the substrate temperature, and the substrate bias voltage all act to change the mobility of deposited adatoms.

Although the sputtering process is complex, it is the energy of adatoms that finally controls the film growth and film properties. It has been demonstrated that altering the mobility of the adatoms causes large changes in the mechanical and magnetic properties and in the morphology of the magnetic film. Figure 7 is the classic diagram that illustrates how the sputtering process affects film growth, which therefore alters the film properties (31,32). Films sputtered onto a substrate at high temperature, high substrate bias voltage, low argon pressure, and high sputtering power tend to be fully dense and to have few grain boundaries. The grain size of the films sputter deposited under these conditions therefore tends to be relatively large and the films have a well-defined grain morphology. On the other hand, films sputter deposited under conditions of lower substrate temperature, no bias, higher working pressure, or lower sputtering power tend to be porous, with a smaller grain size and wavy film surface morphology. Gao, Malmhall, and Chen studied and reported on the correlation between sputtering process conditions with film mechanical properties (32,91).

To make media that have good magnetic properties and good recording performance, the preferred sputtering parameters are high substrate temperature T_{sub} (for NiP/A1 substrate media, this means above $T_{sub} = 250°C$), about a 250 V substrate bias, low pressure, and relatively high sputtering power. Of course, making a medium with specified magnetic properties requires a combination of these sputter-deposition parameters. How to select a suitable combination depends on

the actual sputtering system, alloy selection, and media configuration that is going to be used. There is no absolute recipe for making a universally good magnetic recording medium; in industrial production, certain process parameters cannot be adjusted for optimal media performance because of constraints on production throughput and practicality. For example, low-mobility sputtering can usually result in a medium with a better signal-to-noise ratio (SNR), but the process has not been adopted commercially because of its limited throughput.

Sputtering System Configurations. An additional aspect of sputtering systems that is important to making good recording media is the configuration of the sputtering systems as *static* or *pass-by*. In pass-by systems, the substrates are conveyed past the sputtering targets; in static systems the substrates are stationary in front of target during film deposition. The static types of systems are usually smaller and have targets that can readily be exchanged. In these systems, it is

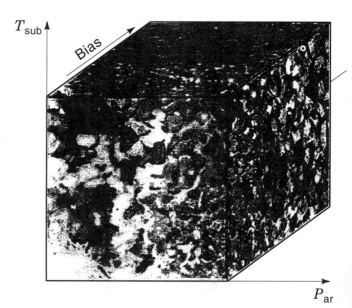

Figure 7. Schematic illustration of the effects of bias voltage, sputter chamber pressure, and substrate temperature on thin film morphology and grain structure.

also easier to develop and to control the sputtering process. Pass-by sputtering systems usually have the advantage in mass production if the sputtering yields can be properly maintained.

Disks sputtered in static sputter systems usually have good magnetic property uniformity around the circumferential direction but less so in the radial direction. Radial variations can, however, be minimized by optimizing both the magnetic flux distribution in front of the target surface and also the combination of sputtering power and duration. On the other hand, pass-by sputter systems usually yield disks with better magnetic property profiles in the radial direction. Studies have shown that pass-by sputtering systems tend to provide disks with poorer modulation in the circumferential direction (33,34) but that this problem can be minimized by depositing an appropriate seed layer underneath the underlayer and the magnetic layer (35–37).

Pass-by sputtering systems usually tend to have larger chambers than those of static sputtering systems; sufficiently good vacuum is therefore harder to attain and maintain. Studies have shown that vacuum integrity is also important to making high-density recording media disks of high quality and that improved vacuum correlates strongly with higher coercivity in the disk (38,39). Therefore it is common that magnetic alloys with higher anisotropy have to be used in pass-by systems to achieve the same media coercivity as is obtained in static sputtering systems with alloys of lower anisotropy. Constantly improving and optimizing the vacuum performance of sputtering system is an important and essential task for system makers.

Thin-Film Media Magnetics

Magnetic Properties of Thin Films. Magnetic properties that characterize media are most importantly the intrinsic anisotropy, the coercivity H_c, the saturation magnetization, the squareness S, and the coercivity squareness S^* [or the switching field distribution (SFD)]. Since the film coercivity encountered by a recording head is actually the remanent coercivity H_{rc}, this is the parameter more frequently used by disk makers. The relation between the intrinsic and remanent coercivity can be seen in Fig. 8.

The product $M_r\delta$ of remanent magnetization M_r and medium physical thickness δ is often used because of its direct relationship with the recording readback amplitude. As the media density is increased, the magnetic transition length parameter a must become shorter. According to the Williams–Comstock model [Eq. (8)], to reduce a and so shorten the tran-

sition, either H_c should be increased or $M_r\delta$ should be reduced, or both should occur.

Coercivity in thin-film media is predominantly controlled by the magnetocrystalline anisotropy of the target materials, by processing that affects the film microstructure and the magnetic interaction between grains, and by the extrinsic aspects of the microstructure of the magnetic media that influence the difficulty of domain-wall motion and of nucleating reverse domains. Both intrinsic anisotropy and media processing must be optimized to create media with high loop squareness and high coercivity.

Anisotropy and Its Origin. A magnetic recording medium exhibits a high coercivity and a high hysteresis loop squareness because of the anisotropy and the interactions among the grains in the film. The total, or effective, anisotropy of a medium is the sum, principally, of three parts: crystalline anisotropy, shape anisotropy, and elastic anisotropy, the latter being due to the film stress created during the sputtering process. This total anisotropy is therefore

$$K_{\text{total}} = K_u + K_{\text{shape}} + K_{\text{elastic}} \qquad (12)$$

Here, K_u is the magnetic anisotropy constant, which is an intrinsic parameter and is determined for each crystalline phase by its composition and structure. K_{shape} and K_{elastic} are extrinsic parameters and so their values are process dependent. K_{shape} includes the effect of the geometry of the thin film and the shape of the grains and depends on the sputtering process and on the surface morphology of the substrate. K_{elastic} includes strain and stress effects but contributes significantly if the medium possesses a nonzero magnetostriction (magnetostriction constant $\lambda \neq 0$). In the CoCrTa alloy system, $K_{\text{elastic}} = \frac{3}{2}\lambda\sigma$, where σ is the mechanical film stress, and can contribute up to one-quarter of the total medium coercivity (38).

There is still ambiguity about the role of the mechanical surface texture of the disk substrate in improving H_c and loop squareness. Results are scattered, but there are four main hypotheses: (1) the mechanical substrate texture induces grain shape anisotropy (39,40); (2) the mechanical texture induces crystalline easy-axis alignment along the texture line (41); (3) the substrate texture line induces anisotropic stress between the radial and circumferential directions (38,42), and (4) texture line geometry effects induce anisotropy (43).

This last hypothesis assumes that crystal c axes of all grains in the magnetic layer are epitaxially well aligned on the underlayer—parallel to the (002) plane of the underlayer in the case of a body-centered-cubic underlayer structure. In this case, therefore, the grain c axes lying along the texture groove lines do not tilt out of the film plane, whereas c axes perpendicular to the texture groove lines may be tilted out of the film plane.

Statistically, a predominant effect of c-axis alignment along texture grooves is observed. This last hypothesis is convincing and is confirmed by experiments with and without substrate heating (39). On the other hand, K_{elastic} may not depend only on one mechanism; two or three mechanisms may actually combine to change the total anisotropy K_{total}.

Magnetostatic and Exchange Interactions and the Loop Squareness. The origins of hysteresis squareness behavior

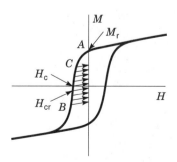

Figure 8. Hysteresis loop of a thin-film disk.

were explored theoretically and experimentally by Hughes, Bertram, and Chen. Hughes (44) considered magnetostatic energy and anisotropic energy and Zhu and Bertram (45) principally explored exchange energy aspects. The total energy density E_t describing the magnetization therefore is

$$E_t = E_a + E_m + E_e \qquad (13)$$

where E_a, E_m, and E_e are the anisotropic, magnetostatic, and exchange energy densities, respectively. The hysteresis squareness depends on their magnitudes and on their relative contributions to E_t.

For a medium that has high magnetostatic grain interactions among grains, the hysteresis loop is square but has lower H_c. When the exchange interaction is taken into account, the hysteresis loop of the medium obtained by computer modeling also showed higher loop squareness and smaller coercivity. Chen and Yamashita obtained experimental confirmation of this result. Figure 9 shows the media morphology and the hysteresis loops of three film media whose grains have different magnetostatic and exchange interactions (46).

Magnetization Reversal. Magnetic reversal in thin-film media can be classed into three different models: In the first, the grains are completely separate and reverse their magnetization independently; in the second, the grains are strongly exchange coupled and are grouped closely together as grain clusters, which reverse their magnetization cluster by cluster;

and in the third, the grains are not strongly exchange coupled but are coupled magnetostatically, over a longer range, in magnetic domains in which magnetic reversal occurs by movement of domain walls.

Understanding the mechanism of magnetization reversal is of great importance in making high-performance media, because the mechanism of the magnetization reversal directly relates to the media noise and the media signal-to-noise ratio. For example, the first reversal mode results in very low media noise but the second mode usually leads to high media noise. Media noise is a key limiting factor that we discuss in the later sections.

Magnetic Alloy Selection. Magnetic alloy selection predominantly depends on the bulk magnetic properties, specifically the film coercivity H_c required to satisfy hard drive specifications based on recording models such as that of Williams and Comstock. The next most important is the media signal-to-noise ratio performance attainable with alloys of suitable H_c. This selection criterion has become significantly more important as the media areal density has increased.

Early magnetic inductive media required only low coercivity (less than 2000 Oe) to deliver a high signal, and so the bulk magnetic properties were of greatest concern. Most of the alloys used were then CoCrTa, CoNiCr, CoCrX (where X represents a different element) or something close to these ternary systems. As the required coercivity is increased, however, alloys with higher magnetocrystalline anisotropy must be considered. For this purpose, Pt-containing alloys have

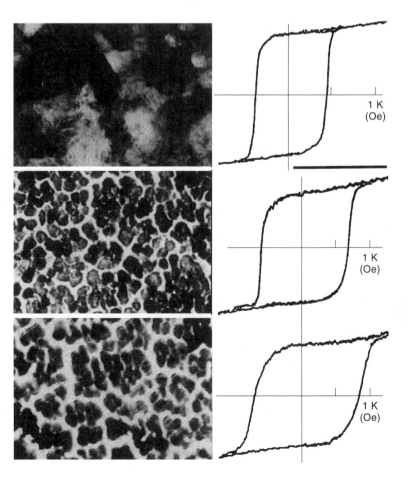

Figure 9. Morphology and hysteresis loops of films with different intergranular interactions.

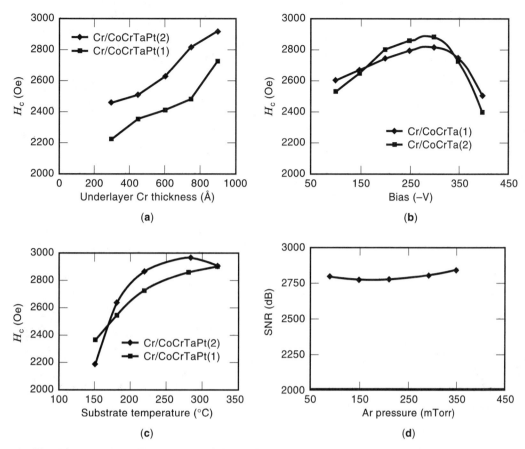

Figure 10. Process effects on H_c and the media signal-to-noise ratio for CoCrPtTa and CoCrTa media on Cr underlayers: (a) underlayer thickness, (b) bias effect, (c) substrate temperature, and (d) chamber pressure. Curves 1 and 2 refer to differing media compositions within the labeled alloy system.

been the most popular choice, although other alloy systems have been demonstrated to give higher film coercivity (47). Currently popular alloys with high crystalline anisotropy are CoCrPtTa, CoCrPt, CoCrPtB, or CoCrPtTaX where X can be Ni, B, Si, Zr, or Hf. Many studies have shown advantages in recording performance resulting from the addition of a fifth element X to CoCrPtTa. The performance enhancement includes improvements in media signal-to-noise ratio, and/or overwrite and PW$_{50}$ improvements (48,49). However, the physics behind these gains is still under active discussion.

Correlation of Magnetic Properties with Sputtering Process and Process Control. The magnetic properties of the media films closely depend on the film-making process. When the media configuration is defined—that is, when the underlayer, seed layer (which is sometimes used) and magnetic alloys are defined—variations in the process of fabricating the media still exhibit pronounced effects on the magnetic properties of the film medium. The coercivity depends strongly on the substrate temperature, the underlayer thickness, and the substrate bias, and sometimes also on the sputtering working pressure. However, sputtering under higher power density minimizes the pressure effect on film coercivity. Figure 10 shows the effects of these process parameters on the film coercivity of a CoCrPtTa medium.

Process control is an important aspect in fabricating media with magnetic properties lying within a narrow range about the designed values. In manufacturing, for a specified magnetic property specification, the coercivity H_c can often be controlled by the substrate temperature and the underlayer thickness, which in turn are set effectively by the heater power and the underlayer sputtering powers, respectively. The product of remanent magnetization M_r and magnetic layer physical thickness δ is controlled by the layer thickness and the squareness of the hysteresis loop. Depositing films under conditions of higher adatom mobility tends to form magnetic films with the c axes oriented in the film plane, which increases loop squareness and, therefore, increases the M_r of the medium. These deposition conditions also help establish a desirably small switching-field distribution (SFD), or equivalently large S^* (approximately $S^* = 1 -$ SFD), the coercivity squareness, which is a measure of the narrowness of the range of applied magnetic fields required to flip the magnetic moments in the film. Figure 11 shows the effect of process parameters on S^* of a CoCrPtTa alloy system.

Thin-Film Media Microstructures

Cobalt-based alloys with hexagonal close packed (hcp) structure have been developed for use as the current longitudinal recording media. However, to confer on them the required magnetic properties in the read–write recording process, an underlayer with a specific microstructure and crystallo-

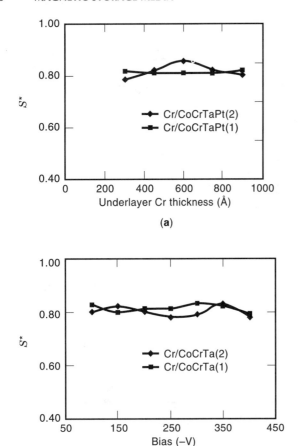

(a)

(b)

Figure 11. Process effects on coercivity squareness S^*. (a) Effect of the Cr underlayer thickness on CoCrTaPt media. (b) Effect of bias on the CoCrTa media on Cr underlayers. Curves 1 and 2 in each case refer to differing compositions within the same alloy systems.

graphic orientation is needed. Three decades ago, Lazzari, Melnick, and Randet (50) found that a Cr underlayer increased the in-plane coercivity of Co-based alloys.

It has been accepted for the last decade that the mechanism behind this coercivity increase is lattice-match-induced epitaxial growth of the magnetic film in a suitable crystallographic orientation. In a polycrystalline film with the easy magnetic axis of each grain oriented randomly, the magnetic anisotropy averages to zero and the film is isotropic. The only significant coercivity-enhancing mechanism then available is the shape anisotropy from the film itself and so it is difficult to obtain high coercivity and good hysteresis squareness. However, if an underlayer such as Cr, or a Cr alloy with B2 crystal structure in an appropriate orientation is deposited prior to the cobalt-based alloy, the crystalline orientation of Co- or Co-based alloys can preferentially be altered to enhance both the film coercivity and the hysteresis squareness. It is generally accepted that this underlayer effect is essential to ensuring alignment of the c axis of an hcp cobalt alloy structure in the film plane and so is critical in longitudinal recording.

Magnetic Layer Microstructure. To make a medium with square-hysteresis characteristic, which is highly desirable for better longitudinal recording performance, an appropriate

magnetic microstructure is essential. The squareness of the hysteresis loop originates both from the intergranular interactions and from the distribution of easy c axes. Calculation shows that a three-dimensional random distribution of the easy axes of a magnetic constituent, without counting the interactions among the grains, has a remanent M_r that is only half of the saturation magnetization M_s and that a two-dimensional random easy-axis distribution has M_r equal to $0.62M_s$ (51). Highly compact, sputtered grain structures with little grain separation usually result in strong magnetostatic interactions and so assist in achieving higher loop squareness. To obtain c-axis orientation in the film plane, the (11.0) or (10.0) planes of an hcp structure have to be grown parallel to the film plane. This can be done through control of the sputtering process but can most easily be achieved through epitaxial grain growth on top of an underlayer with the appropriate atomic structure, as discussed in the next section.

Crystalline Orientations of the Magnetic Layer and Underlayers. The most commonly used underlayer materials in current magnetic recording media are Cr, Cr-based alloys, or alloys with B2 crystal structure and similar atomic lattice constants (52). This structure is depicted in Fig. 12, which also lists atomic lattice constants. As we discussed in the preceding section, a specific orientation is needed for the desirable form of Co-based alloy film growth. Many studies have shown that having the (002) lattice orientation of the suitable B2 underlayer structure parallel to the film growth surface will induce epitaxial growth of an hcp Co alloy magnetic layer because of the close lattice match between the (11.0) plane of the Co alloys and the (002) plane of the CrX (X = V, Mo, . . .) underlayer (50). This (11.0) orientation corresponds to c-axis alignment, or texture, in the film plane. Recent work has also shown that c-axis alignment or in-plane texture, in the magnetic layer could be promoted by epitaxial film growth on Cr (112) and (110) orientations of the surface planes of the underlayer (53). Figure 13 shows the atomic structure and orientation relationships.

Film Morphology, Grain Size, and Grain-Size Distribution. For good magnetic recording media, besides growing films with

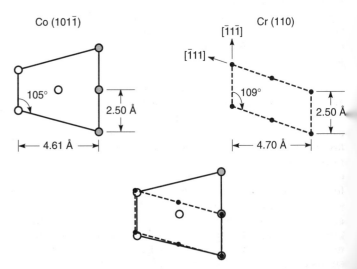

Figure 12. The crystal structure relationship of Cr and Co for Co(10.1) crystal planes parallel to Cr(110) planes.

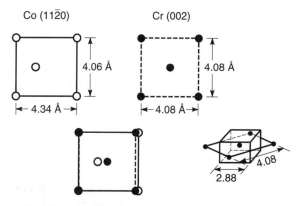

Co (11$\bar{2}$0)   Cr (002)

4.06 Å

4.34 Å

4.08 Å

4.08 Å

4.08

2.88

Figure 13. The crystal structures and relationship between Cr and Co, relevant to a Co(110)‖Cr(002) texture [i.e., with the Co(110) crystal planes parallel to the Cr(002) planes].

the correct atomic structure and crystalline orientation, it is also necessary to have uniform film morphology, small grain size, and narrow grain-size distribution. Studies have shown that, as the recording media layer is made thinner, the film growth technique becomes increasingly critical (54–57). Achieving two-dimensional film growth is preferable to three-dimensional growth—a surface and interfacial energy issue. A suitable substrate sublayer—for example, a NiP plated layer—can help establish a two-dimensional growth mode that improves the mechanical integrity of the subsequent layers. When the surface energy of a depositing layer has smaller surface energy than the existing surface material, the depositing adatoms tend to enlarge the existing surface layers by diffusing until they can be accommodated on the lower side of steps at the boundaries of the layers. This two-dimensional mode of growth therefore tends to form a continuous film by extension of the existing topmost layer in the plane of the surface. On the other hand, when the material of the depositing layer has a higher surface energy than the existing surface material, the depositing layer tends to grow in three dimensions by the formation of islands. For high-recording-density media, the layer-by-layer, two-dimensional film growth is particularly desirable since the magnetic film thickness must be very small and the mechanical integrity of the film is otherwise hard to maintain. Figure 14 illustrates the two types of film growth, which result in quite different recording performance (54).

The magnetic layer grain size is another key parameter that affects the film magnetic properties and recording performance in several ways. Figure 15 shows the dependence of H_c on grain size—behavior that is observed by many researchers (58–61). The coercivity initially increases with increasing grain size and then drops for grain sizes larger than 40 nm. The recording performance is also closely related to the grain size and to the grain size distribution. The smaller grain sizes and narrower grain-size distributions result in lower media noise and higher media signal-to-noise ratio, since the media noise originates in zigzag domain-edge irregularities, whose dimensions are correlated with the grain sizes. Therefore, controlling the grain size and minimizing the width of the grain-size distribution are important goals for media makers. Approaches to these goals include (1) selection of materials with different surface energies, (2) selection of the underlayer or seed layer, and (3) optimization of the film

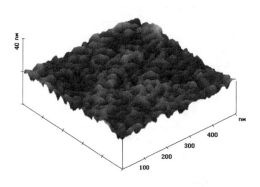

(a)

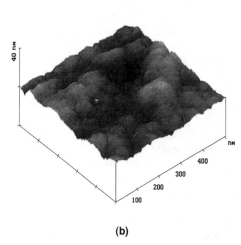

(b)

Figure 14. Two types of epitaxial growth of a Co-alloy magnetic layer on a Cr underlayer.

growth process. Grain-size control is closely related to the film morphology that we discussed earlier. When the grains are smaller, the film will of course present a smoother, more uniform morphology that can largely be retained throughout subsequent processing.

Magnetic Switching Volume. Studies have suggested that the magnetic film grain size and switching volume have im-

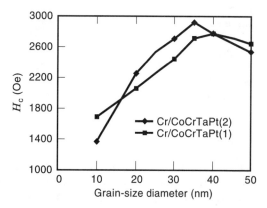

Figure 15. Grain-size effect on coercivity H_c.

portant effects on both the media performance and the thermal stability in high-density recording media (62–64). For low noise, as discussed previously, a small grain size is preferred. However, for sufficient thermal stability of the written information, the quantity $K_u V/(k_B T)$—the ratio of the energy associated with the magnetic anisotropy of the magnetic switching volume to the unit of thermal energy—must be sufficiently large (65). (In this ratio, K_u, k_B, T, and V are the magnetic anisotropy, Boltzmann's constant, the temperature, and the magnetic switching volume, respectively.) Thus, one would expect that there might be an optimal magnetic switching volume and so grain size, although these may not be identical quantities.

For a better understanding of the magnetization reversal process and the correlation between the reversal process and media noise, it is instructive to investigate the switching (activation) volumes and the physical grain sizes. It is known that these do not match for most materials, but the relationship is not fully understood at present. It is expected, however, that the medium noise should be related to the volume of the grains and/or the switching volume, since both quantities are relevant to the possible spatial resolution for recordings on these films.

Researchers have demonstrated that magnetization reversals do not imply switching volumes that are always identical with the physical grain size (66). In the simplest case, switching by homogeneous rotation of the magnetization, one would expect the physical grain and the switching volumes to coincide. The observed mismatch between physical grain and switching volumes has been known (67) and attributed to inhomogeneous magnetization reversals. Typically, for rather large particles, as occur in magnetic tape media, switching volumes are reported to be smaller than the particle volumes. Physically, this may be interpreted by assuming that the thermal energy $k_B T$ can cause a change in the magnetization in only a certain fraction of the particle volume (i.e., the switching volume) in order to reverse the magnetization in that region of the particle. An example for such an inhomogeneous thermally assisted magnetization-reversal process was theoretically analyzed in Ref. 68. Experimental measurements imply that the magnetization-reversal process in "large" single-domain grains is inhomogeneous—most likely initiated at defect locations. On the other hand, when grains are too small to support a domain structure, they will switch entirely. Thus films with larger single-domain grains are expected to be noisy.

An additional factor that must be considered is magnetic interaction. Since magnetic interactions between grains must have an influence on the magnetization-reversal process, by affecting the external field at each nearby location they will affect the switching volume as well. Experimentally, it is found, however, that decreasing the grain size does not change the switching volume significantly. On the other hand, grain size controls the nature of the interactions in the magnetic media. One may speculate that making the grain and switching volume the same will result in a good design of recording medium.

Process Control of Thin-Film Microstructure. It is clear that selections of a suitable magnetic alloy and a suitable combination of underlayer and magnetic layer are essential. However, the film-deposition process is equally important for making good magnetic recording media. As discussed in the sections entitled "Thin-Film Media Magnetics" and "Thin-Film Media Microstructures," the film microstructure is strongly dependent on having the sputtering process yield a magnetic layer with its c axes in-plane and it is clear that this requires preferably an underlayer of B2 structure with [200] texture. In practice, this [200] orientation requires sputtering conditions that provide high adatom mobility. These conditions involve (1) high sputtering power, (2) high substrate temperature, (3) high substrate bias, and (4) low sputtering pressure. The [200] crystalline orientation has been grown on glass substrates by varying the substrate temperature (69). Figure 16 illustrates how [200] texturing can be achieved by varying the underlayer thickness and substrate temperature. The process conditions are critical for the underlayer but less so for the subsequent magnetic layer. It is observed that, even although higher-energy adatoms are required for formation of [11.0] or [10.0] orientation films of hcp Co-based alloys, once the [200] structure of underlayer is laid down, the epitaxial influence on subsequent growth is stronger than the influence of the process conditions.

Correlation of Thin-Film Microstructure with the Film Fabrication Process. The correlations between the microstructures, the morphology of deposited media films, and the film-growth-process parameters are demonstrated in Fig. 17 for CoCrPtTa media. High substrate temperature, high substrate bias, high power-density sputtering, and low working-gas pressure sputtering result in a film with well-defined grain morphology and with a strong tendency for the c axes to be aligned in the film plane. As an example, Fig. 17 shows the transmission electron microscope (TEM) images of films for a range of sputtering pressures and Table 2 gives the corresponding atomic structure change and the ratio of the c axis in the plane of the film to that out of the plane of the film. Since this morphology, the lattice structure, and the orientation variations are all results of sputtered adatom mobility, other sputtering process parameters that similarly affect the adatom mobility can be expected to change the film morphology and lattice structure in a similar way.

Thin-Film Media Structures and Configurations

Although thin-film media structures have evolved into several different configurations, illustrated in Fig. 18, since they were first used in the early 1980s, the basic structure remains the same. Basically, an underlayer, typically of Cr or Cr alloy, is deposited, followed by the magnetic layer(s), and finally a protective overcoat layer, typically a form of carbon. The whole sputtering deposition process is carried out within one vacuum chamber.

Conventional Magnetic Media. Conventional media structures always have an underlayer, a magnetic layer, and an overcoat. Commonly, the underlayers are Cr, CrX alloys with X representing elements that primarily include V, Ti, Mo, Hf, Zr, and W in concentrations between 0 at.% and 30 at.%. The rationale for the X-element addition is first to alter the Cr lattice spacing while retaining the bcc crystal structure and second to inhibit development of larger underlain grains. The magnetic layer is now usually a quaternary CoCrXY alloy in which X and Y are Ta, Ni, Pt, Nb, and other elements. Ter-

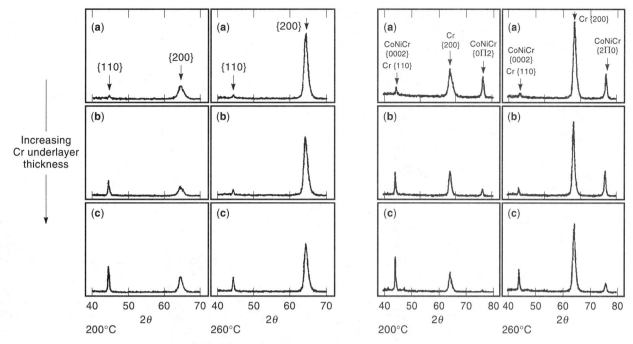

Figure 16. Influence of changes in Cr underlayer thickness and substrate temperature on film-growth texture observable through x-ray diffraction patterns, whose peaks reveal the major crystal planes parallel with the disk surface.

nary alloys without Pt were used in the early days for thin-film media with relatively low H_c ($H_c < 2.5$ kOe) but have been replaced by quaternary Pt-containing alloys for media requiring higher H_c. The addition of Pt linearly increases the coercivity at low concentrations and has resulted in coercivities as high as 4 kOe in thin film media. This dependence of coercivity on Pt concentration is quantified in Fig. 19. Carbon films have been used successfully as protective layers since the start of thin film media manufacturing. However, it has later been demonstrated that both hydrogenated and nitrogenated carbon films are even better than pure carbon films (70,71). Both CN_x and CH_y are currently in use by disk makers.

Seed-Layered or Double-Underlayered Media. It has been found that a double underlayer could result in better magnetic and recording performance for some alloys and certain underlayer materials (72–75). The first underlayer serves two purposes: It acts as a seed layer by initiating growth of smaller grains and by establishing a suitable epitaxial relationship in the deposition of the second underlayer. The second underlayer therefore grows both with the correct crystalline orientation and with a smaller grain size. This double underlayer approach has frequently been used for media, particularly in the pass-by type of sputtering system. The two underlayer materials need not be the same, but may be different, depending on the specific process, magnetic property requirement, and substrates used in the media design (57,76,77).

Flash Magnetic Layer and Double-Magnetic-Layer Media. Engineering of the working magnetic layer structures can also modify and improve the recording performance. The film coercivity can be increased substantially and the medium

signal-to-noise ratio can be enhanced significantly by several process modifications: (1) using a flash magnetic layer before depositing the working magnetic layer (78,79), (2) depositing the same magnetic materials under successively different sputtering condition (49), or (3) using two different magnetic alloys to build one working magnetic layer (80).

Sandwich-Layered Magnetic Media. A nonmagnetic spacer layer sandwiched between two magnetic layers can reduce the effects of intergranular exchange coupling and of demagnetizing fields. Once the sandwich-layered construction was demonstrated to confer lower media noise and higher media SNR (81–84), sandwich-configuration media were produced at the time when the media still had relatively high $M_r\delta$. These layer-spacing effects are illustrated in Fig. 20.

However, producing media with dramatically increased areal density necessitated lower $M_r\delta$ and higher H_c and so required a reduced magnetic film thickness. This is incompatible with inserting a nonmagnetic spacer between two magnetic layers, since the film coercivity then suffers a severe drop and the distance between the outermost surfaces of the magnetic layers is increased. The coercivity decrease is grain-size related and, in fact, at low enough $M_r\delta$ one could reach the grain sizes at which the grains are no longer ferromagnetic—the superparamagnetic size limit. Therefore, sandwich media are only useful in applications for higher-$M_r\delta$ media, unless a new alloy with yet higher crystalline anisotropy is found.

Keeper-Layered Magnetic Media. Keeper-layered media were another early invention in the field of inductive media. The concept was to enclose a larger magnetic transition flux and so to boost the signal that had not been used in any commercial product. Since high-saturation magnetization materi-

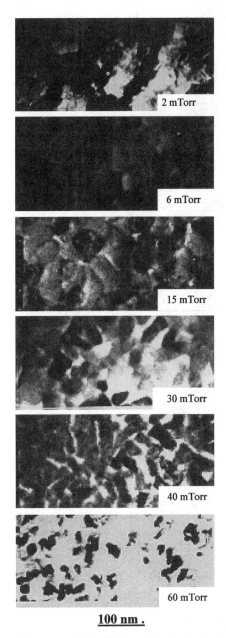

Figure 17. TEM micrographs showing the correlation between the film microstructures and sputter gas pressure for a thin film system.

als must be used for the keeper layer, the results are higher demagnetizing fields and greater exchange coupling. Consequently, for keeper-layer media to be useful, the gain in signal amplitude must be balanced against the loss in media signal-to-noise ratio, as Yen, Richter, and Coughlin have

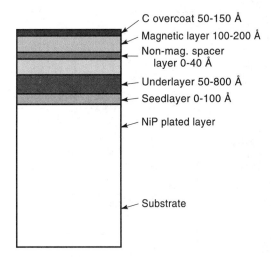

Figure 18. The configuration of a thin-film medium.

demonstrated (88). Keeper-layer media therefore remain in the research laboratories but not in commercial production.

Recording Performance

Magnetic Transition Wavelength. During a magnetic head writing process, magnetic transitions are generated in the magnetic medium. The minimum repeat distance between the magnetic transitions is called the magnetic transition wavelength—the shorter the wavelength, the higher the number of magnetic transitions per unit distance. This transition length is illustrated in Fig. 21. Suppose the medium is dc erased (that is, erased with a constant applied field) and the magnetization is in the positive direction, with a value of M_r in the hysteresis loop. As the field of the head reverses and increases in magnitude, the magnetization reduces, following the hysteresis loop curve. At $H = -H_c$, the minor hysteresis loop illustrates that there will still be a small remanent moment left. So, to take the remanence to zero the reversal field must be increased in magnitude to slightly more than H_c—to a field that is commonly called the remanent coercivity H_{cr}.

Assume that a medium is being written by a head that moves in the x direction—the direction of the field gradient in the gap in the recording head. Between media locations corresponding to points A and B in the hysteresis loop, there is a length a over which the transition occurs, and this can be estimated by

$$\frac{dM}{dx} = \frac{dM}{dH}\frac{dH}{dx} \qquad (14)$$

If we can assume the slope of the hysteresis loop and the head field gradient are constant, we can integrate this to obtain

Table 2. Easy-Axis Orientation of Magnetic Film versus Sputtering Pressure

Pressure (mTorr of Ar)	2	6	10	15	20	30	40	60
Angles (deg)[a]	90	85	70	77	60	20	17	9
Interpretation[b]	In	Out	Out	Out	Out	Out	Out	⊥

[a] Easy axis angles are measured with respect to the film plane normal. Completely in plane corresponds to 90° and fully perpendicular to the plane is 0°.
[b] The interpretation of the angular orientation of the easy axes are in-plane (In), out-of-plane (Out), and perpendicular-to-plane (⊥).

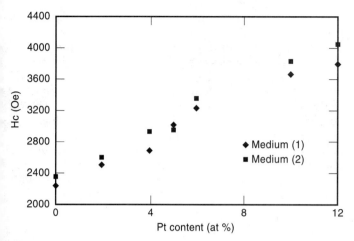

Figure 19. Pt concentration effect on coercivity H_c of CoCrTaPt media. Data sets 1 and 2 refer to differing underlayers.

$$a = M_r \bigg/ \frac{dM}{dH}\frac{dH}{dx} \qquad (15)$$

which gives the importance of the squareness (through the slope dM/dx) of the hysteresis loop of the medium. The magnetic transition is most popularly assumed to have the shape of an arctangent function (89,90), which leads for film media to

$$a_{min} = 2M_r\delta/H_c \qquad (16)$$

where δ is the magnetic medium thickness. Consequently, for the shorter magnetic transition lengths required for high lin-

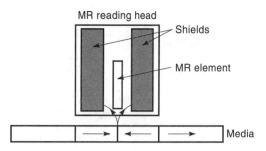

Figure 21. Transition length of a thin-film medium.

ear recording density, thin media with low magnetization and high coercivity are preferred.

Readback Amplitude, Fundamental Signal of Thin-Film Media. The presence of the magnetization distribution in the recording medium generates a magnetic field flux that extends over the surface of the medium. Based on Faraday's law, when a coil cuts through this flux, or, as in the case of an inductive readback head, a coil flies above a moving magnetic medium, a current will be generated in the coil. This current can be used as a signal and is proportional to the fluctuation of the magnetic flux changes over the surface of the media. This is essentially the readback signal. When this phenomenon was first studied, it was assumed that the magnetization had a longitudinal sinusoidal form. With the additional assumptions that the reproducing head consisted of a semi-infinite block of high permeability material with a flat face spaced a distance d above the medium and that the magnetic flux distribution along vertical direction obeyed Poisson's equation, the flux per unit width Φ of the reproducing head was calculated to be

$$\Phi = \Phi_x = \int_{d+\delta/2}^{\infty} B_x dy$$
$$= -[2\mu/(\mu+1)]2\pi\delta M \sin kx[(1-e^{-kd})/k\delta]e^{-kd} \qquad (17)$$

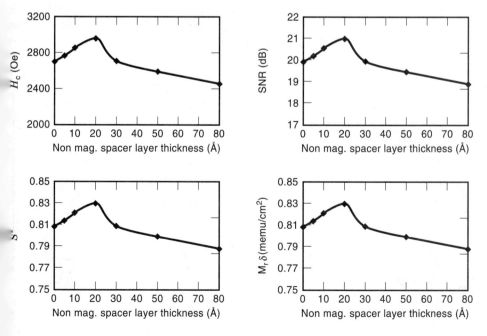

Figure 20. Effect of spacer layer thickness on H_c, SNR, S^*, and $M_r\delta$ in a double magnetic layer medium CoCrTa/Cr/CoCrPtTa system.

Using $x = vt$, where v is the linear velocity of the medium with respect to the head, t is time, and w is the width of the head, the time derivative of the flux Φ is

$$d\Phi/dt \propto 4\pi w v M[1 - \exp(-k\delta)]\exp(-kd)\cos \omega t \quad (18)$$

Now if the head has a pickup coil with N turns and the efficiency of the head is q, the voltage generated by the head should be

$$V(t) = N q \frac{d\Phi}{dt} \quad (19)$$

Since the voltage is proportional to the flux time derivative, which is proportional to $\exp(-kd)$, the conclusions are (1) that there is a spacing loss that is dependent on d as

$$20 \log_{10}[10 \exp(-kd)] = -54.6(d/\lambda) \, \text{dB}$$

and (2) that there is a dependence of the readback signal on the medium thickness through the term

$$(1 - e^{-kd}) = k\delta = 2\pi d/\lambda = \delta w/v$$

Figure 22 shows Wallace's data of spacing loss and readback signal dependence on frequency. He experimentally confirmed the behavior just described by inserting spacers between a ring head and a rotating magnetic disk during readback.

This discussion is given here for a medium read by an inductive head (inductive media) but for a medium read by a magnetoresistive head (MR media), the starting point will be different. For MR media the readback signal is not proportional to the rate of the magnetic flux change but rather is proportional to the total intensity of the flux change. Thus any effort towards increasing the rotational speed of drive, the drive revolutions per minute, in the case of drives with magnetoresistive heads, is only for the purpose of increasing the data rate and not for improving the readback signal amplitude. The reader is referred to the excellent, detailed discussion appropriate to the MR media, written by Bertram (89).

To characterize the readback signal in media studies, average amplitudes are usually measured with high- and low-frequency tracks—that is, tracks, respectively, with high and low linear densities of magnetization transitions. The output from a sequence of transitions is a superposition of their individual amplitudes. In any sequence, two sequential transitions always have opposite signs. At a very high (spatial) frequency, the pulses are close together and tend to cancel each other, whereas at low frequency the pulses are completely separate and there is no interference between them. The readback process therefore exhibits nonlinearity at high frequency. Only the low-frequency amplitude is correlated unambiguously with the media $M_r\delta$ values. The high-frequency amplitude reveals information that includes the bulk magnetic property $M_r\delta$ and the sharpness of pulses, pulse width PW_{50}, which will be discussed in the next section.

Pulse Width PW_{50}. The recording signal pulse width is very much of interest in studies of magnetic recording media. The measure that is often used is PW_{50}—the pulse width at 50% of pulse amplitude—which describes the suitability of a medium for achieving high linear recording density. The smaller the PW_{50}, the greater the number of magnetic transitions that can occur both in the space domain and in the time domain. Therefore, the narrower the PW_{50}, the better the readback resolution will be.

Williams and Comstock (112), as well as Middleton, studied the pulse width and derived the relation between PW_{50} and the head gap width g, the head flying height d, the medium thickness δ, and the transition length a that is given in Eq. (7). Different studies have shown how PW_{50} varies as a function of head gap and flying height; their results are contained in Fig. 23. The variation in PW_{50} from the recording media point of view is given in Fig. 24, which shows the effect of media thickness δ and flying height for a fixed head gap and transition length.

Overwrite. Overwrite is an important measure of the effectiveness of writing one frequency signal over a previously written pattern that is most probably of a different frequency. The practical way of assessing the overwrite characteristics is first to write a pattern at frequency f_1 in the medium and measure the readback signal $V_0(f)$, then to overwrite another pattern of frequency f_2 (e.g., $f_2 = 2f_1$ or $4f_1$) and measure the residual signal $V_{\text{Res}}(f_1)$ at frequency f_1 as well, and finally to

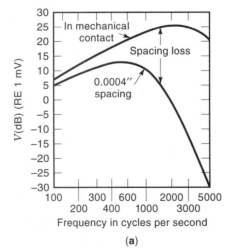

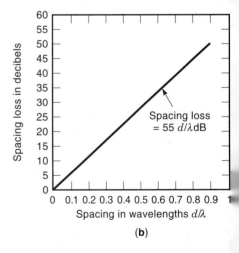

Figure 22. (a) Frequency effect on the read-back signal V (dB). (b) Dependence of the spacing loss (dB) on the ratio of the spacing d to recorded period of magnetization λ.

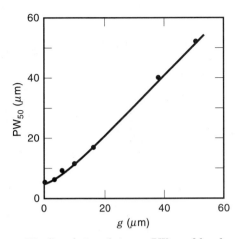

Figure 23. Correlations between PW_{50} and head gap g.

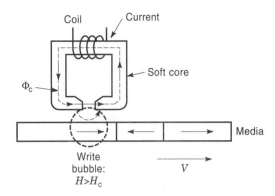

Figure 25. Writing a transition in a moving medium.

compute the overwrite OW as

$$OW = 20 \log \frac{V_0(f_1)}{V_{Res}(f_1)}$$

For high-performance thin films, this parameter should be at least over 30 dB.

In magnetic recording, the overwrite characteristics depend on both media and head properties. For example, increasing the writing field or decreasing the flying height of the head will improve overwrite performance. Here we discuss only the overwrite performance correlated to the media properties.

Overwrite is largely controlled by coercivity and $M_r\delta$. Overwrite can indicate whether the coercivity of a medium is excessively high. When the coercivity of a medium is high, the second set of magnetic transitions may not completely replace the old signals, thus reducing the overwrite. In high-density media, overwrite must be balanced against the need to have a short magnetic transition length and a small nonlinear transition shift (NLTS), which we discuss in the next section. This arises because a high-film coercivity is required for a short magnetic transition length and small NLTS, but lower coercivity is required for a good enough overwrite.

Nonlinear Transition Shift. At high recording density, magnetic transitions are nonlinearly distorted because the ratio

of PW_{50} to the transition separation is increased. When two transitions are written too closely together, the demagnetizing field from the previous transition affects the writing of the next transition. This results in a shift of the location of the transition and is called the nonlinear transition shift. This shift depends on the transitions that are already written. From the magnetic flux view point, the origin of NLTS can be derived as follows.

Since NLTS is a result of demagnetizing field H_{demag}, the medium sees the field from the head as $H_{eff} = H_{demag} + H_{head}$, in which H_{head} must exceed H_c in order to write a new transition.

Therefore, the transition is written not at a location x_0 but instead at the shifted location $x_0 + \Delta$, at which

$$H_{eff} = H_{demag} + H_{head}(x_0 + \Delta) > H_c \qquad (20)$$

Since the shift Δ is small, a Taylor-series expansion around the location x_0 enables the following expression to be derived (80):

$$\Delta X = \frac{4M_r\delta(d + \delta/2)^3}{\pi Q H_c B^3} \bigg| \qquad (21)$$

where Q is a head-dependent parameter and B is the intertransition spacing. Therefore, the NLTS is proportional to both $M_r\delta$ and d, is inversely proportional to medium coercivity, and competes with overwrite. A balance must therefore be found between NLTS and overwrite when developing a high-performance medium.

Media Noise and Media Signal-to-Noise Ratio. There has been concern that media noise will finally limit the attainable increases in areal recording density. Media noise is caused by the randomness and uncertainty of magnetic transitions and results from the characteristics of the magnetic constituents in the media. Any spatial variation and magnetic fluctuation will produce noise. As the packing fraction of grains in thin film media approaches unity and the related intergranular exchange coupling increases, the media noise increases at the higher recording densities. Figure 25 shows a medium with transitions produced by a writing head. As we discussed in the section entitled "Magnetization Reversal," when all the grains reverse their magnetization independently under the influence of the externally applied field, there will be no uncertainty in the transition location other than a geometric

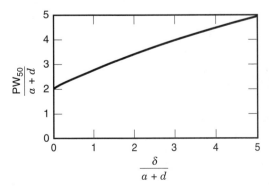

Figure 24. Correlation between PW_{50}, transition length, medium thickness, and head flying height.

grain-size effect and there will be less magnetization fluctuation. Therefore, the magnetization transition will be sharp, leading to low media noise and high media signal-to-noise ratio. On the other hand, if the magnetization reversal occurs in grains clustered and coupled tightly together, the magnetization transition will not be sharp, because its location conforms to the boundaries of the clusters of grains. This magnetic structure generates a so-called irregular or zigzag transition wall and results in media noise. The origin of the clustered grain reversal behavior lies mostly in exchange coupling but also in magnetostatic interaction.

As progressively higher areal densities are sought, media noise becomes more and more a problem. To make a medium with lower noise, different techniques and approaches have been reported. (1) *Physical grain separation:* Magnetic grains can be physically separated during the sputtering process by creating microvoids in the grain boundaries. This mechanism of physical isolation, which cuts the exchange coupling, is an effective way of reducing media noise (91). However, this technique has not really been applied in production, because, as discussed earlier, to make a porous medium, lower sputtering adatom mobility must be used and this slows down production throughput. (2) *Chemical segregation.* Studies have shown that the local chemical composition in the film could be varied during sputter fabrication. Surrounding or interlacing the magnetic grains with a nonmagnetic phase could also be effective in cutting down the exchange interactions and so lowering media noise (93). Techniques based on this idea have been widely used in the industry with quite remarkable results. However, new studies have shown that, at least for some media that exhibit low noise, neither physical separation nor chemical segregation could be detected. The only noticeable differences were the lower saturation magnetization and the extremely uniform and homogeneous film grain structure (94). This is, of course, of great current interest to disk makers.

Write Jitter. Peak jitter measurements are a valuable complement to the popular technique of integrating the frequency spectrum of media noise power. By using the standard peak jitter measurement algorithm described in Hewlett-Packard Application Note 358-3, the total peak jitter can be separated into write and read components. Under typical recording conditions, write jitter is known to be the dominant component of media noise. (Media noise also includes three variances associated with the readback waveform: peak jitter, PW_{50} fluctuations, and amplitude fluctuations.) By definition, write jitter is the standard deviation of the distribution of peak positions relative to their expected positions and is conveniently expressed in nanometers. If jitter measurements are performed at low recording density, in the absence of transition interference and NLTS, the peak write jitter is also the transition jitter. Transition jitter directly relates to the microstructure of the magnetic film.

Off-Track Capability. Unlike the other parameters we have discussed that characterize media performance, off-track capability (OTC), in units of length, refers to the performance change found when the readback head is positioned off the track center in a radial direction of the disk (94–96).

When a head is positioned over a specified track written on the magnetic disk, a certain positioning error is always

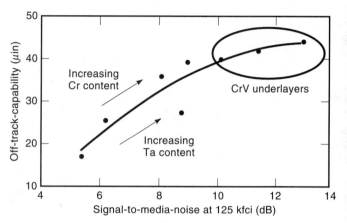

Figure 26. Off-track capability as a function of the media signal-to-noise ratio obtained by media composition and underlayer variation.

present because the tract is written in one location and is later read with the head slightly shifted off the track center. As the track density is increased (for high areal density), neighboring tracks start to interfere with each other by generating a signal in the read head, particularly when it is off the center of the track.

OTC characterizes how capable a medium is of allowing more tracks to be written in the fixed available radial dimension of a disk surface. OTC is mostly dependent media noise—the higher the OTC value, the greater the number of tracks that can be used per inch (TPI). The signal-to-noise ratio (SNR) and OTC give measures of a medium's recording density in two dimensions, the SNR revealing the range of linear recording densities along a track and OTC revealing the upper limits on the track number density. Figure 26 shows the correlations of OTC and media SNR. OTC always improves as SNR improves. Figure 27 shows the OTC curves of a group of samples that have different media SNRs (97). Interested readers are referred to the analysis by Taratorin (94).

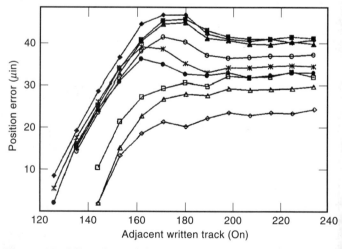

Figure 27. Off-track capability curves for a group of media of differing SNR as a consequence of differing underlayers and media compositions.

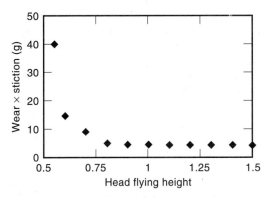

Figure 28. Head media products [(wear) × (stiction)] as a function of head flying height (in multiples of 1 μin., i.e., 25.4 nm).

Thin-Film Media Mechanics

Disk Tribology and the Head–Media Interface. Magnetic recording is realized by the relative motion between a recording head and a recording medium. To obtain a high signal at a high linear density with a narrow track pitch, according to the model of Williams and Comstock [Eq. (7)], the space between the head and the medium should be as small as possible. Therefore, the media surface needs to be made as flat and smooth as possible. However, smooth surfaces tend to lead to increased adhesion (stiction), friction, and thermal instability at the head–disk interface. A very small separation between the head and the media can indeed lead to greater friction and increased wear (98). Therefore, interface and media tribology is an important aspect of magnetic recording technology, recognized as critical in the magnetic recording media industry. Resolution of friction and wear issues involves appropriate selection of the substrate type and surfaces, the overcoat film, the lubricants, and the environment of the head–media interface and thus involves controlling the dynamics of the head and medium. The ultimate goal is to achieve contact recording with minimal friction and wear and with minimal adhesion between head and disk. Figure 28 shows the relationship between the product of wear and stiction as a function of recording head flying height. This behavior occurs because at a higher head-media flying height both wear and stiction can be very low, but as the head–media spacing is reduced both stiction and wear become large. The product of the two tends to infinity as the flying height tends towards zero, namely for contact recording.

Substrates. Conventional thin-film substrates are made from AlMg alloys containing about 5% Mg, since these have particularly high mechanical hardness. The substrates are stamp cut into circular plates and then electroplated with NiP film. This film further increases the surface hardness and also provides a base that can be highly polished. Aluminum substrates are widely used because they are good, easily made, and low in cost, but concerns have been raised about whether they can be made flat enough to permit ultralow head-flying heights and whether the hardness is high enough to withstand accidental head impacts without damage. For these reasons, glass, glass ceramic, and aluminum–boron-carbide have been considered as alternative substrate materials. Glass and glass ceramic substrates have now also been used in commercial products. However, advances in the technology

of the conventional substrates have repeatedly extended the possibilities for AlMg. The best AlMg substrates have been made with flatness better than 3 nm and with local substrate surface roughnesses of only about 0.2 nm. Therefore, conventional AlMg substrates are still preferred in the industry.

Overcoat and Lubricant. There are two aspects to disk tribology in addition to substrate texturing: the overcoat and the lubrication technologies. Disks require an overcoat after magnetic layer deposition for two important reasons—to protect the mechanical integrity of the magnetic layer and also to prevent or at least minimize corrosion of the magnetic layer. Carbon, or carbon modified by being hydrogenated or nitrogenated, has been widely and effectively used as an overcoat in the industry.

To allow the head to fly freely over the disk surface and to minimize wear resulting from contact between head and media, a lubricant must be applied over the protective carbon layer. The lubricant is usually applied by dipping the disk into a solution containing the lubricant or its precursor. The thickness of lubricant on the disk surfaces is controlled by the rate at which the disks are pulled out of the solution. A good bond between the lubricant and overcoat is needed to prevent the lubricant from being squeezed out from any contact between head and media and to prevent the lubricant flying off the disk surface during drive operation. Sometimes, double lubrication or baking the disk in an oven between two lubrication steps can improve the bonding.

Studies have also shown that the molecular weight of the lubricant has a strong effect on the tribological performance (99). The class of lubricants that has been used extensively in disk industry are the perfluoropolyethers, including Fomblin ZDOL or Fomblin AM2001 lubricants. Perfluoropolyethers are long-chain fluorocarbon compounds, a desirable combination of properties that include good thermal and chemical stability, low surface tension, and low vapor pressure.

In addition to dip lubrication, vapor lubrication has been investigated by disk makers for media intended for ultralow head flying height. If successful, vapor lubrication could result in still lower cost and better performance of magnetic recording disk drives.

Substrate Texturing. Applying a layer of lubricant does reduce wear, but the meniscus force associated with the lubricant can cause serious stiction problems, sometimes resulting in the recording head sticking to the surface of the media and causing drive failure. To solve this problem, technology has evolved for substrate "texturing"—controlled introduction of slight variations from surface flatness. (Note that, confusingly, this texturing is not the same as the crystallographic idea of texture referred to in discussing underlayer and magnetic layer microstructure, in which similar crystallographic planes and directions in a polycrystalline material are more or less closely aligned. However, texturing of a substrate surface can indeed influence the crystallographic alignment, and therefore the crystallographic texture, of subsequently deposited layers.) Several different techniques are used for texturing disk substrates: (1) mechanical texturing, (2) laser bump texturing, (3) sputter texturing, and (4) intrinsic texturing.

(1) In mechanical texturing, a tape coated with hard particles such as diamond or Al carbide is used to abrade

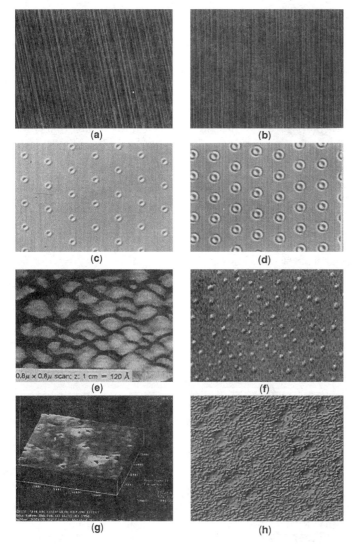

Figure 29. Images of different types of disk substrate texture; details given in the text.

the substrate surface mechanically in the circumferential direction. This technique has been very effective and widely used in the recording media industry for over 10 years. The circumferential texture not only minimizes head–media interface stiction problems, but also improves the recording performances because the circumferential texture induces anisotropic film magnetic properties (88). Figures 29(a) and 29(b) show two optical images of the circumferentially textured NiP/Al-Mg substrates. Figure 29(a) is of a pure circumferential texture and Fig. 29(b) is an optimized circumferential texture with cross-hatch markings that demonstrated to have greater tribological advantages.

(2) Laser bump texturing was first demonstrated by Ranjan et al. (100) and is now widely used in the magnetic recording media industry. A relatively high-energy laser beam is used to recrystallize the polished surface regions of the NiP layer and so to create bumps. Figures 29(c) and 29(d) contain images of two commonly used laser-created bump patterns. The bumps can be quite precisely controlled in their size, heights, and

number density. The accuracy of creating bumps by this technique has allowed zone texturing of substrates—namely only laying bumps in the head landing or take-off zone to serve the tribological purposes and so leaving the data zone without texture. This is demonstrably a good strategy and is now widely used in the disk-making industry.

(3) Sputter texturing was probably one of the best low-cost methods. In this method, before the underlayer and magnetic layer are deposited, a sputter texturing layer is deposited using a film-growth technique that provides a rough film surface morphology. The subsequent underlayer and magnetic layer(s) then replicate the surface morphology of the texture layer and thus serve the required mechanical purpose.

Two types of materials have been used to create sputter textures, one by Mirzamaani and co-workers using lower melting point materials and the other by Gao and co-workers (101–103).

The first method relied on the concept of surface energy mismatch between the sputtered metal layer and the substrate. The relative surface energies play a central role in determining the film growth mode at thermodynamic equilibrium. If the deposited material has a larger surface energy than the substrate, it will tend to form three-dimensional structures, imparting a greater surface roughness that serves as substrate texturing. This is related to the wetting phenomenon that takes place between two materials that have different surface energies.

The second method relied on the concept of film stress and stress release. The substrate was held at high temperature during film deposition and then cooled at a controlled rate after the deposition, thus creating and then releasing the stress. The result of the stress release is to form hillocks or bumps. These bumps could be used as a means of substrate texturing technique. Figures 29(e) and 29(f) illustrate these two types of sputter-created bumps. The bump size, height, and the density of the bumps can be controlled to a certain degree. This type of texturing is simple and of low cost, but is not adaptable to zone texturing. However, a head flying height of 15 nm has been demonstrated without significantly increased stiction.

Sputter texturing has mostly been used with alternative substrates, although the technique has been demonstrated on several different types of substrates. Sputter textured glass substrates have been used in commercial disks.

(4) Intrinsic substrate texturing applies to alternative substrates, such as glass ceramic. The substrate blanks are made with a mixture of amorphous and crystalline phases that provides hard crystallites embedded in the amorphous matrix. If the amorphous phase is more easily removed, polishing leaves the harder crystallites protruding as surface bumps. On the other hand, if the crystallites can be pulled from the amorphous matrix, polishing may leave holes. Both types of surface serve the necessary mechanical purpose and have been used to provide texturing in the

industry. Figures 29(g) and 29(h) illustrate types of intrinsic substrate texture.

THE FUTURE OF MAGNETIC RECORDING MEDIA

The future for magnetic recording media is strong—it is estimated to be possible to develop media with a further order of magnitude increase to 40 or even 100 Gbit/in.^{2} in the areal data recording density, based on development of currently known technologies and ideas. As yet no other data recording technology can compete in the combination of cost, capacity, and speed. However, like anything else, magnetic recording media must face limitations, both fundamental and technological.

Magnetic Recording Media Limitations

It is worth discussing the approaches available for increasing the areal density and trying to understand its limits. The key limiting properties are: (1) media magnetic properties, particularly H_c and $M_r\delta$, (2) media noise, and (3) thermal instability.

In the sections entitled "Fundamentals of Magnetic Recording" and "Recording Performance," we have pointed out that PW_{50} must be small for high-density recording media and that this requires the media to have certain macromagnetic properties—high H_c and small $M_r\delta$. The coercivity of current CoCrTaPt alloys is about 2500 Oe to 3500 Oe and the new generation of media should have an H_c of 3500 Oe to 5000 Oe. For still higher H_c, the new media will require both a high intrinsic anisotropy and a microstructure optimized as discussed previously. As $M_r\delta$ is decreased, the read-back signal decreases and increased head sensitivity becomes important. Therefore, the conventional read head—an inductive or magnetoresistive head—has to be replaced by a more sensitive head, based on giant magnetoresistance (GMR) or colossal magnetoresistance (CMR).

When the recording areal density is increased significantly, the transition length is reduced and zigzag magnetic transitions become dominant because of the demagnetizing field. This will generate substantial media noise and lower the media signal-to-noise ratio (94). Although the use of signal processing techniques can significantly increase the sensitivity and the signal from media, media noise eventually will be an unavoidable limitation for magnetic recording media.

The smaller magnetic grains required for higher areal densities are more susceptible to thermally activated effects that alter the recorded magnetizations. Thermal activation not only causes the so-called superparamagnetic limit (104) but also leads to the transition decay of coercivity (105). The ratio K_uV/k_BT of the magnetic anisotropy energy to the thermal energy must exceed 60 to satisfy the requirement of thermal stability. [K_u, V, k_B, and T are the (intrinsic) magnetic anisotropy, grain volume, Boltzmann constant, and temperature, respectively.] This means that materials with the highest anisotropy are required and that control of the grain-size distribution is critical.

The areal density of longitudinal magnetic recording media has increased at a compound rate of about 60% per year in the last decade. Media are commercially available today with an areal density of 3 Gbit/in.^{2} to 4 Gbit/in.^{2}; 10 Gbit/in.^{2} media are predicted to be available within a few years. Roughly speaking, from 1980 to 1990 more attention was paid to rais-ing the areal density by improving the macromagnetic properties H_c and $M_r\delta$ and since 1990 more attention has been devoted to reducing media noise by improving micromagnetic characteristics such as decoupling of the grains. Now, grain size and distribution control are becoming important aspects of media design. The areal density of recording media must have an upper limit, but Johnson (111) predicts that 100 Gbit/in.^{2} may be possible in future.

The Outlook for Magnetic Recording Media

The current development of magnetic recording media is now highly advanced, but still offers considerable scope for increased data storage per unit area, available at high data transfer rate. The closest competition is likely to come from near-field optical recording, which was recently announced (106,107). Near-field optical recording has been successfully demonstrated and is claimed to offer a capacity that is an order of magnitude higher than that of magnetic recording media. The near-field data rate is potentially high, since it uses a great deal of hard-drive technology. However, even before the demonstration of the first near-field magneto-optical recording drive, several drive makers have already demonstrated 10 Gbit/in.^{2} in magnetic recording media. It is not clear when near-field magneto-optical technology will be ready to compete with magnetic recording technology in three key areas: (1) capacity, (2) cost, and (3) data transfer rate.

For future ultrahigh recording density media, novel magnetic alloy materials probably need to be developed. Film coercivity needs to be increased even while $M_r\delta$ is reduced. A higher media SNR and lower bit error rate is also required. The magnetic media will need to have an improved microstructure, with even smaller grain size and a much narrower grain-size distribution (108). A thinner overcoat or zone carbon overcoat will need to be developed to enable the head flying height to be reduced. Less or no substrate texturing will need to be used and dynamic head load–unload technology will need to be adopted.

Although the superparamagnetic size barrier will eventually limit magnetic recording densities, 100 Gbit/in.^{2} recording has been predicted (111) and, within the magnetic recording community, there is consensus that 40 Gbit/in.^{2} at least is attainable (99). Already, by the turn of the century, a magnetic recording medium drive with an areal density of 10 Gbit/in.^{2} is expected to be commercially available.

BIBLIOGRAPHY

1. P. P. Zapponi, U.S. Patent No. 2,619,454, 1952.

2. R. L. Comstock and E. B. Moore, Ferrite film recording surfaces for disk recording, *IBM J. Res. Develop.,* **18**: 556, 1974.

3. W. T. Maloney, RF-sputtered chromium-cobalt films for high-density longitudinal magnetic recording, *IEEE Trans. Magn.,* **MAG-15**: 1135, 1979.

4. T. Yamada et al., CoNiCr/Cr sputtered thin film disks, *IEEE Trans. Magn.,* **21**: 1429, 1985.

5. M. Ishikawa et al., Film structure and magnetic properties of CoNiCr/Cr sputtered thin film, *IEEE Trans. Magn.,* **22**: 573, 1986.

6. B. Lal, H. Tsai, and A. Eltoukhy, A new series of quaternary co-alloys for rigid-disk applications, *IEEE Trans. Magn.,* **27**: 4739, 1991.

7. Y. Cheng et al., Evaluation of the CoCrTaPt alloy for longitudinal magnetic recording, *J. Appl. Phys.,* **75** (10): 6138, 1994.

8. C. Gao, Seagate Recording Media, 1996, unpublished result.

9. Karlqvist, Calculation of the magnetic field in the ferromagnetic layer of a magnetic drum, *Trans. R. Inst. Technol., Stockholm,* **86**: 3, 1954.

10. M. H. Kryder, J. A. Bain, and M. Xiao, Advanced materials and devices for high density magnetic recording and readback heads, *Proc. 3rd Int. Symp. Phys. Magn. Mater. (ISPMM '95),* Seoul, Korea, 1995, p. 457.

11. K. G. Ashar, *Magnetic Disk Drive Technology,* Piscataway, NJ: IEEE Press, 1997.

12. Y. C. Feng, Ph.D. dissertation, Carnegie Mellon Univ., Pittsburgh, PA, 1995.

13. E. S. Murdock, Roadmap for 10 Gbit/in² media: Challenges, *IEEE Trans. Magn.,* **28**: 3078, 1992.

14. S. Y. Zhou et al., Single-domain magnetic pillar array of 35 Nm diameter and 65 Gbits/in² density for ultrahigh density quantum magnetic storage, *J. Appl. Phys.,* **76**: 6673, 1994.

15. M. Farboud et al., Fabrication of large area nano-structured magnets by interferometric lithography, *7th Jt. (Magnetism and Magnetic Materials (MMM) Intermagn. Conf.,* San Francisco, 1998, Paper 04.

16. C. A. Grimes, Li Chun, and M. F. Shaw, The magnetic properties of permalloy thin films patterned into rectangular picture frame arrays, *7th Jt. MMM Intermagn. Conf.,* San Francisco, 1998, Paper 06.

17. S. Y. Chou, P. R. Krauss, and L. Kong, Nanolithographically defined magnetic structure and quantum magnetic disk, *J. Appl. Phys.,* **79**: 6101, 1996.

18. L. Kong, L. Zhuang, and S. Y. Chou, Writing and reading 7.5 Gbits/in² longitudinal quantized magnetic disk using magnetic force microscope tips, *IEEE Trans. Magn.,* **33**: 3019, 1997.

19. G. Bate et al., *Proc. Int. Conf. Magn.,* Nottingham, England, 1964, p. 816.

20. Y. Suganuma et al., Production process and high density recording characteristics of plated disks, *IEEE Trans. Magn.,* **MAG-18**: 1215, 1982.

21. K. Yoshida, T. Yamashita, and M. Saito, Magnetic properties and media noise of Co-Ni-P plated disks, *J. Appl. Phys.,* **64**: 270, 1988.

22. M. Ishikawa et al., Effects of thin Cr film thickness on CoNiCr/Cr sputtered hard disk, *IEEE Trans. Magn.,* **26**: 1602, 1990.

23. R. D. Fisher, J. C. Allan, and J. L. Pressesky, Magnetic properties and longitudinal recording performance of corrosion-resistant alloy films, *IEEE Trans. Magn.,* **22**: 352, 1986.

24. B. B. Lal and A. Eltoukhy, High-coercivity thin-film recording medium and method, U.S. Patent No. 5,004,652, 1991.

25. M. F. Doerner et al., Composition effects in high density CoPtCr media, *IEEE Trans. Magn.,* **29** (6, pt. 2): 3667–3669, 1990.

26. S. Iwasaki, Y. Nakamura, and K. Ouchi, Perpendicular magnetic recording with a composite anisotropy film, *IEEE Trans. Magn.,* **MAG-15**: 1456, 1979.

27. J. H. Judy, Thin film recording media, *MRS Bull.,* **15** (3): 63, March, 1990.

28. K. Ouchi and S. Iwasaki, Recent subjects and progress in research on Co-Cr perpendicular magnetic recording media, *IEEE Trans. Magn.,* **23**: 2443, 1987.

29. S. L. Zeder et al., Magnetic and structural properties of CoNi thin films prepared by oblique incidence deposition, *J. Appl. Phys.,* **61**: 3804, 1987.

30. J. S. Gau and W. E. Yetter, Structure and properties of oblique-deposited magnetic thin films, *J. Appl. Phys.,* **61**: 3807, 1987.

31. J. A. Thornton, The macrostructure of sputter-deposited coatings, *J. Vac. Sci. Technol. A,* **4**: 3059, 1986.

32. C. Gao, R. Malmhall, and G. Chen, Quantitative characterization of sputter process controlled nano-porous film state, *IEEE Trans. Magn.,* **33**: 3013, 1997.

33. M. R. Kim and S. Guruswamy, Microstructural origin of in-plane magnetic anisotropy in magnetron in-line sputtered CoPtCr thin film disks, *J. Appl. Phys.,* **74**: 4643, 1993.

34. H.-C. Tsai, M. S. Miller, and A. Eltoukhy, The effects of Ni₃P-sublayer on the properties of CoNiCr/Cr media using different substrate, *IEEE Trans. Magn.,* **28**: 3093, 1992.

35. E. Teng et al., Dispersed-oxide seedlayer induced isotropic media, *IEEE Trans. Magn.,* **33**: 2947, 1997.

36. J. L. Vossen and W. Kern (eds.), *Thin Film Processes II,* San Diego: Academic Press, 1991.

37. M. Takahashi, A. Kikuchi, and S. Kawakita, The ultra clean sputtering process and high density magnetic recording media, *IEEE Trans. Magn.,* **33**: 2938, 1997.

38. T. Chen et al., Isotropic vs. oriented thin film media: Choice for future high density magnetic recording, *Intermag'96,* Seattle, WA, 1996, Invited Talk.

39. R. Nishikawa et al., Texture-induced magnetic anisotropy of Co-Pt films, *IEEE Trans. Magn.,* **25**: 3890, 1989.

40. T. Coughlin et al., Effect of Cr underlayer thickness and texture on magnetic characteristics of CoCrTa media, *J. Appl. Phys.,* **67**: 4689, 1990.

41. A. Kawamoto and F. Hikami, Magnetic anisotropy of sputtered media induced by textured substrate, *J. Appl. Phys.,* **69**: 5151, 1991.

42. Y. Hsu, J. M. Sivertsen, and J. H. Judy, Texture formation and magnetic properties of RF sputtered CoCrTa/Cr longitudinal thin films, *IEEE Trans. Magn.,* **26**: 1599, 1990.

43. T. Lin, R. Alani, and D. N. Lambeth, Effects of underlayer and substrate texture on magnetic properties and microstructure of a recording medium, *J. Magn. Magn. Mater.,* **78**: 213, 1989.

44. G. F. Hughes, Magnetization reversal in cobalt-phosphorous films, *J. Appl. Phys.,* **54**: 5306, 1983.

45. J.-G. Zhu and H. N. Bertram, Micromagnetic studies of thin metallic films, *J. Appl. Phys.,* **63**: 3284, 1988.

46. T. Chen and T. Yamashita, Physical origins of limits in the performance of thin-film longitudinal recording media, *IEEE Trans. Magn.,* **24** (6): 2700, 1988.

47. T. Y. Shiroishi et al., Magnetic properties and read/write characteristics of CoCrPtTa/CrTa/ (Cr-Ti, Cr), *J. Appl. Phys.,* **73**: 5569, 1993.

48. L. W. Song et al., Magnetic properties and recording performance of multilayer films of CoCrTa, CoCrPtTa, and CoCrPtTa with CocrPtb, *IEEE Trans. Magn.,* **30**: 4011, 1994.

49. D. J. Rogers, Y. Maeda, and K. Takei, The dependence of compositional separation on film thickness for Co-Cr and Co-Cr-Ta magnetic recording media, *IEEE Trans. Magn.,* **30**: 3972, 1994.

50. J. P. Lazzari, I. Melnick, and D. Randet, *IEEE Trans. Magn.,* **MAG-3**: 205, 1967.

51. B. D. Cullity, *Introduction to Magnetic Materials,* Reading, MA: Addison-Wesley, 1972.

52. D. E. Laughlin and B. Y. Wong, The crystallography and texture of co-based thin film deposited on Cr underlayers, *IEEE Trans. Magn.,* **27**: 4713, 1991.

53. D. E. Laughlin et al., The control of microstructural features of thin films for magnetic recording, *Scr. Metall. Mater.,* **33**: 1525, 1995.

54. E. G. Bauer and B. W. Dodson, Fundamental issues in heteroepitaxy, *J. Mater. Res.,* **5**: 852, 1990.

55. C. Gao et al., Sub-micron scale morphology effects on media noise of CoCrTa thin films on smooth substrates, *J. Appl. Phys.,* **81**: 3958, 1997.

56. T. Yeh, J. M. Sivertsen, and J. H. Judy, Effect of very thin Cr underlayer on the micromagnetic structure of RF sputtered CoCrTa thin films, *IEEE Trans. Magn.*, **26**: 1590, 1990.

57. Q. Chen, J. Zhang, and G. Chen, Thin film media with and without biscrystal cluster structure on glass ceramic substrates, *IEEE Trans. Magn.*, **32**: 3599, 1996.

58. H. Sato et al., Effects of grain size and intergranular coupling on recording characteristics in CoCrTa media, *IEEE Trans. Magn.*, **32**: 3596, 1996.

59. H. Mitsuya et al., Effect of grain size on intergranular magnetostatic coupling in thin film media, *Magnetism and Magnetic Materials (MMM'96)*, Atlanta, GA, 1996, QQ-08.

60. H. S. Chang et al., Effects of Cr/Al underlayer on magnetic properties and crystallography in CoCrPtTa/Cr/Al thin films, *IEEE Trans. Magn.*, **32**: 3617, 1996.

61. Y. Nakatani and N. Hayashi, Effect of grain-size dispersion on read/write properties in thin film recording media affected by thermal fluctuation, *7th Jt. Magnetism and Magnetic Materials (MMM) Intermagn. Conf.*, San Francisco, 1998.

62. M. P. Sharrock, Time-dependent magnetic phenomena and particle-size effects in recording media, *IEEE Trans. Magn.*, **26**: 193, 1990.

63. P. I. Mayo et al., *J. Appl. Phys.*, **69**: 4733, 1991.

64. S. S. Malhatra et al., Magnetization reversal behavior in cobalt rare-earth thin films, *IEEE Trans. Magn.*, **81**: 1997.

65. P.-L. Lu and S. H. Charap, Thermal instability at 10 Gbit/in^2 magnetic recording, *IEEE Trans. Magn.*, **30**: 4230, 1994.

66. C. Gao et al., Correlation of switching volume with magnetic properties, microstructure, and media noise in CoCr (Pt)Ta thin films, *J. Appl. Phys.*, **81**: 3928–3930, 1997.

67. R. W. Chantrell, Magnetic viscosity of recording media, *J. Magn. Magn. Mater.*, **95**: 365, 1991.

68. H. B. Braun and H. N. Bertram, Nonuniform switching of single domain particles at finite temperatures, *J. Appl. Phys.*, **75**: 4609, 1994.

69. S. Duan et al., The dependence of the microstructure and magnetic properties of CoNiCr/Cr thin films on the substrate temperature, *IEEE Trans. Magn.*, **26**: 1587, 1990.

70. B. Marchon, N. Heiman, and M. R. Khan, Improvement of hydrogenated carbon overcoat performance, *IEEE Trans. Magn.*, **26**: 168, 1990.

71. B. B. Lal et al., Asymmetric DC-magnetron sputter carbon-nitrogen thin-film overcoat for rigid-disk applications, *IEEE Trans. Magn.*, **32**: 3774, 1996.

72. J. K. Howard, The effect of Cr and W nucleation layers on the magnetic properties of CoPt Films, *J. Appl. Phys.*, **63**: 3263, 1988.

73. L. L. Lee, D. E. Laughlin, and D. N. Lambeth, NiAl underlayer for CoCrTa magnetic thin films, *IEEE Trans. Magn.*, **30**: 3951, 1994.

74. Y. C. Feng, D. E. Laughlin, and D. N. Lambeth, Magnetic properties and crystallography of double-layer CoCrTa films with various interlayers, *IEEE Trans. Magn.*, **30**: 3960, 1994.

75. Y. Okuimura et al., Effects of CrNi Pr-layer on magnetic properties of CoCrTa longitudinal media, *IEEE Trans. Magn.*, **33**: 2974, 1997.

76. H. Mahvant, Oxidation of seed-layer for improved metal film performance, *IEEE Trans. Magn.*, **29**: 3691, 1993.

77. W. Xiong and H.-L. Hoo, Cr-Ta$_2$O$_5$ seedlayer for recording media on alternative substrates, *7th Jt. Magnetism and Magnetic Materials (MMM) Intermagn. Conf.*, San Francisco, CA, 1998.

78. L. Fang and D. N. Lambeth, CoCrPt media on hcp intermediate layers, *CMU Data Storage Syst. Cent. Rep.*, 1994.

79. N. Inaba et al., Magnetic and microstructural properties of CoCrPt/CoCrPtSi dual-layered magnetic recording media, *J. Appl. Phys.*, **75**: 6162, 1994.

80. B. Zhang et al., CoCrTa/CoCrPtTa double-layer films for magnetic recording, *IEEE Trans. Magn.*, **32**: 3590, 1996.

81. E. Teng and A. Eltoukhy, Flash chromium interlayer for high performance disks with superior noise and coercivity squareness, *IEEE Trans. Magn.*, **29**: 3679, 1993.

82. S. E. Lambert, J. K. Howard, and I. L. Sanders, Reduction of media noise in thin film metal media by lamination, *IEEE Trans. Magn.*, **26**: 2706, 1990.

83. E. S. Murdock, B. R. Natarajan, and R. G. Walmsley, Noise properties of multilayered co-alloy magnetic recording media, *IEEE Trans. Magn.*, **26**, 2700, 1990.

84. Y. C. Feng, D. E. Laughlin, and D. N. Lambeth, Magnetic properties and crystallography of double-layer CoCrTa thin films with various interlayers, *IEEE Trans. Magn.*, **30**: 3960, 1994.

85. B. Gooch et al., A high resolution flying magnetic disk recording system with zero reproduce spacing loss, *IEEE Trans. Magn.*, **27**: 4549, 1991.

86. T. M. Coughlin, B. R. Gooch, and B. Lairson, Keepered recording media, *Data Storage Mag.*, October 1996.

87. G. Mian and P.-K. Wang, Effects of a highly permeable keeper layer on transition shifts for thin film media, *IEEE Trans. Magn.*, **30**: 3984, 1994.

88. E. Yen, H. J. Richter, and T. Coughlin, Evaluation of keeper layer media, *Intermag'98*, San Francisco, CA, 1998.

89. N. Bertram, *Introduction to Magnetic Recording*, Cambridge, England: Cambridge Univ. Press, 1996.

90. R. M. White, *Introduction to Magnetic Recording*, New York: IEEE Press, 1985.

91. T. Yogi et al., Role of atomic mobility in the transition noise of longitudinal media, *IEEE Trans. Magn.*, **26**: 1578, 1990.

92. J. E. Wittig et al., Cr segregation in CoCrTa/Cr thin films for longitudinal recording media, *7th Jt. Magnetism and Magnetic Materials (MMM) Intermagn. Conf.*, San Francisco, CA, 1998, Paper CC-03.

93. C. Gao et al., Correlation of bulk saturation magnetization with magnetic properties and media signal-to-noise ratio of CoCrPtTa films, *J. MMM*, Sept. 1998 (in press).

94. A. Taratorin, *Characterization of Magnetic Recording Systems*, San Jose, CA: Guzik Technical Enterprises, 1996.

95. C. R. Paik et al., Magnetic properties and noise characteristics of high coercivity CoCrPtB/C media, *IEEE Trans. Magn.*, **28**: 3084, 1992.

96. P. I. Bonyhard and J. K. Lee, Magnetoresistive read magnetic recording head off-track performance assessment, *IEEE Trans. Magn.*, **26**: 2448, 1990.

97. C. Gao et al., Composition and underlayer effects on media noise and off-track capability in CoCrPtTa films, *IEEE Trans. Magn.*, **32**: 3569, 1996.

98. B. Bushan, *Tribology and Mechanics of Magnetic Storage Devices*, 2nd ed., New York: Springer-Verlag, 1996.

99. M. F. Doerner and R. L. White, Materials issues in magnetic-disk performance, *MRS Bullet.*, **21** (9): 28–34, 1996.

100. R. Ranjan et al., Laser texturing for low-flying-height media, *J. Appl. Phys.*, **69**: 5745, 1991.

101. M. Mizamaani, C. V. Jahnes, and M. A. Russak, Thin film disks with transient metal underlayers, *IEEE Trans. Magn.*, **28**: 3090, 1992.

102. C. Gao and D. Massey, *Thermally cycled metal layer deposition for substrate texturing*, U.S. Patent pending ª9450174, 1995.

103. C. Gao et al., Hillocks formation, a new approach for sputter texturing disk substrates, *IEEE Trans. Magn.*, **25**: 3965, 1996.

104. M. Kryder, Ultrahigh-density recording technologies, *MRS Bull.*, p. 17, September, 1996.

105. K. Johnson, Thin-film recording media: Challenges for physics and magnetism in the 1990s, *J. Appl. Phys.*, **69**: 4932, 1991.

106. G. Knight, Terastor Inc., Near-field recording, *Diskon'97, Tech. Symp.* 1997.

107. Near-field recording, technology update, *Data Storage Magazine*, 1997, p. 8.

108. E. G. Bauer, Fundamental issues in heteroepitaxy, *J. Mater. Res.*, **5**: 852, 1990.

109. M. P. Sharrock, Time dependence of switching fields in magnetic recording media, *J. Appl. Phys.* **76**: 6413, 1994.

110. R. W. Chantrell et al., Models of slow relaxation in particulate and thin film materials, *J. Appl. Phys.* **76**: 6407, 1994.

111. K. E. Johnson, Will media magnetics pace areal density?, in *Proc. DISCON USA '97, Int. Tech. Conf.*, Sept. 23–25, 1997, San Jose, CA.

112. M. L. Williams and R. L. Comstock, An analytical model of the write process in digital recording, *AIP Conf.*, Vol. 5, 1971, p. 738.

CHUAN GAO
Seagate Magnetic Recording

BRIAN W. ROBERTSON
University of Nebraska

Z. S. SHAN
HMT Technology

MAGNETIC STRUCTURE

The identification and characterization of the magnetic structure of a material depends on the viewpoint from which the characterization is required. At a strictly atomistic level, the relevant magnetic structure may be the relative orientation of the magnetic moment of one atom with respect to other atomic moments (these other moments may be made of the same or different elements), or it may be the relative orientation of the atomic magnetic moment with respect to the crystalline axes of the material. At the nanoscopic or microscopic level the magnetic structure may refer to the relative ordering of magnetic domains, magnetic particulates, or small magnetic structures which have formed within the material. For an artificially structured multilayer film, it may constitute the orientation of the magnetic directions of each film in the layered system, or it may refer to the correlation of magnetic domains between layers. In each case, from each of these distinct viewpoints, a knowledge of the magnetic structure of the film, and its evolution with external influences (like an applied magnetic field) would be very beneficial in understanding the overall behavior of the system.

SOURCE OF MAGNETISM

In an introduction to magnetic structure it is always useful to briefly outline the underlying source of the atomic magnetic moment. There are two contributions to the total magnetic moment of an atom, the spin-moment, m_S, and the orbital-moment, m_L. An intuitive view of the source of the atomic magnetic moment can be obtained by analogy with the magnetic properties of a loop of wire carrying a current, as de-

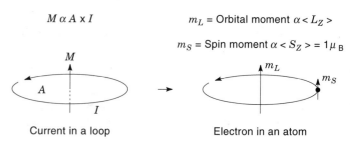

Figure 1. Source of the atomic magnetic moment. At left is the dipole moment generated by a current in a loop of wire. At right is the moment in an atom generated by the electron orbital motion and spin.

scribed in Fig. 1. As in an electromagnet, the current carried in a loop of wire creates a magnetic field which, at distances from the loop much larger than the loop diameter, can be adequately modeled by a magnetic dipole positioned at the center of the loop pointing upward (z-direction) according to the right-hand rule. In this mathematical simplification the moment generated by the current flowing in the loop is proportional to the product of the current in the loop, I, and the area enclosed by the loop, A. These two parameters characterize the strength of the magnetic dipole, and therefore the magnetization felt at a distance from the loop.

A simplistic view of an electron in an atom portrays the electron (charge) moving in an orbit around the nucleus. (A rigorous treatment can be found in Ref. 1.) This moving charge can be viewed as generating a magnetic field similar to the moving charge in a loop of wire, but in this case the area of the loop must be treated in quantum mechanical terms. The resulting magnetic dipole caused by the orbital motion of the electron, termed the orbital-moment, m_L, is proportional to the electron charge multiplied by the expectation value of the angular momentum of the electron (L) projected along the z-axis (z-component of L, labeled L_Z). But, unlike the current in a loop, the electron in an atom has a second contribution to the total magnetic moment. In addition to the intuitive orbital-moment, the individual electrons possess an intrinsic spin-moment which is proportional to the expectation value of the quantum mechanical spin of the electron (S), projected along the z-axis (S_Z component) which can only have a value of $+\frac{1}{2}$ or $-\frac{1}{2}$. The moment of a single electron spin is equal to 1 bohr magneton (1 μ_B). For a multielectron atom the total magnetic moment of the atom is the sum of the various orbital and spin-moment contributions of each electron of the atom.

The atomic moment forms the basic building block of magnetic structure in condensed materials. Interactions between the atomic moments, constrained by the imposed crystallographic architecture, form macroscopic spin structures. This article will not discuss the atomistic magnetic structures generated from these interactions between the individual magnetic moments. [The magnetic ordering can be incompletely classified in the standard way: diamagnetism, paramagnetism, antiferromagnetism, ferromagnetism, metamagnetism, superparamagnetism, ferrimagnetism, and parasitic ferromagnetism; (see Refs. 2 and 3)]. Instead, because of the tremendous advancements in thin film magnetism for application as magnetic devices, this report will focus on characterizing the nanoscale and microscale magnetic structures, especially those generated in thin film and multilayer magnetic structures.

DETERMINING MAGNETIC STRUCTURE

The complex and intricate interplay among material properties, thin film morphology, lithographic processing, and micromagnetic dynamics of the new classes of magnetic materials and devices lead us to anticipate a variety of barriers to their successful implementation as magnetic memories and magnetic based sensors (4,5). Since these devices are multilayered, multielement, and heteromagnetic (more than one magnetically active element) and because the vast majority of magnetic techniques cannot separate the magnetic signatures of the various materials, it is clear that an element-specific, layer-sensitive magnetic order probe to determine the magnetic structure is needed.

Most standard magnetic characterization tools are not sensitive to the atomic type but instead give the total magnetic moment of the sample within the probed volume. The few exceptions are usually only sensitive to one atomic species and are not broadly useful. A new direction in magnetic characterization that does have broad-based element selectivity exploits recent advancements in the use of synchrotron radiation facilities (6). These large experimental facilities serve as sources for the generation of intense X rays and are a powerful and proven tool for basic and applied studies in biology, chemistry, material science, medicine, and physics and their related subfields (6).

Magnetic Circular Dichroism

Whenever a charged particle undergoes an acceleration, it emits electromagnetic radiation. If high-energy electrons traveling at relativistic speeds (near the speed of light) are constrained to move in a circular orbit by the strong magnetic fields of a vacuum storage ring, the resulting emitted radiation, termed synchrotron radiation, is extremely intense and varies over a very broad range of wavelengths from infrared through visible and into soft and hard X rays. It turns out that this intense, highly collimated emitted radiation is also strongly polarized. Due to the dipole character of the synchrotron radiation, X rays emitted in the plane of the synchrotron (the plane-of-orbit of the circulating electrons) are linear polarized (6). Collecting X rays emitted out of the plane of the synchrotron (either above or below the plane-of-orbit) yields X rays with a high degree of left circular or right circular polarization (they are actually elliptical polarized X rays). Although many uses have been made of the linear polarized photons, it is the interaction of these left and right circular polarized X rays (also referred to as positive or negative helicity photons) with the spin-polarized electrons in magnetic materials that gives this technique element selectivity and magnetic sensitivity.

Magnetic circular dichroism (MCD) is a term used to describe a technique whose underlying mechanism is simply the difference between the absorption of left and right circular polarized photons by a magnetic material (7–13). MCD is complementary to the many, more familiar, spin-resolved electron techniques, but instead of resolving or selecting the electron spin, it uses the photon selection-rules associated with the absorption of circular polarized X rays to probe the wave-function character (spin + symmetry) of the unfilled electron states of the material. The MCD measurements are much more sensitive than electron spin-detection or spin-selection techniques [the figure-of-merit for MCD is 2 orders of magnitude higher than spin detection; see (8,14)] and can be acquired in the presence of magnetic fields. Also MCD measurements are element and even site specific with high sensitivity, enabling radically new experiments for unique insights into magnetic structures and interactions.

MCD is a technique that has element selectivity and magnetic sensitivity. The element specificity of MCD is generated from the same mechanism as that of X ray absorption spectroscopy. The electronic structure of a magnetic material (particularly the itinerant magnet materials of the transition metals) can be described in a picture where the available electron states are separated by their spin direction (up or down). In Fig. 2 is shown the number of available states for the electrons of a magnetic material as a function of the energy of the state, but separated by the electron spin. In this spin-resolved density-of-states picture, the right side represents the majority-spin (up-spin) states, and the left side the minority-spin (down-spin) states where the dotted line represents the fermi energy, E_f, separating the occupied and unoccupied states. (There are more up-spin electrons in the material, hence the term majority-spin electrons.) Shown at much deeper binding energies are the electrons of the spin-orbit split $2p_{3/2}$ core level (associated with the L_3 absorption edge) and the $2p_{1/2}$ core level (associated with the L_2 absorption edge).

For an incident photon, as the energy of the photon is increased, it reaches a value that is sufficient to promote a bound electron from the $2p_{3/2}$ core level up to the unfilled states above the fermi level, resulting in an increase in the photoabsorption by the sample. Since the binding energy of the $2p_{3/2}$ electron depends on the number of protons in the nucleus (and to a much lesser extent the chemical environment of the atom), the energy needed to promote the $2p_{3/2}$ electron, or the binding energy of the electron, will be different for different elements. Using photon energies near a particular absorption edge of an element will only probe the elec-

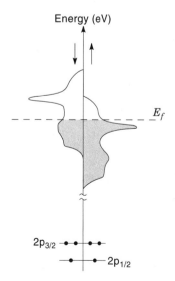

Figure 2. Electronic states available in a magnetic material separated by the electron spin direction. The fermi level, E_f, separates the occupied and unoccupied states.

tronic structure of that element, making this technique chemical specific.

The magnetic selectivity rises from the fact that for a circular polarized photon, the probability that the photon is absorbed by a magnetic material depends on the relative orientation of the spin of the electron and the spin of the photon. [The photon spin is defined by whether the photon rotation, or spin, is to the right (up spin) or the spin is to the left (down spin).] There is a large difference in the absorption of a magnetic sample depending on whether these two spin directions are aligned or anti-aligned. This is clearly demonstrated in Fig. 3 where is shown the X-ray absorption spectra of Fe for the two photon polarizations (helicities), corresponding to when the photon and electron spin directions are aligned and antialigned. The MCD (not shown) is just the difference of these two absorption spectra. Note that in Fig. 3, the L_3 and L_2 absorption edges have opposite MCD, characteristic of the transition metals.

EXAMPLES

To illustrate the utility of MCD in characterizing magnetic structures, three examples are described below to illustrate increasingly sophisticated characterization of the magnetic structure. Each of these measurements were conducted at the Naval Research Laboratory/National Synchrotron Light Source (NRL/NSLS) Magnetic Circular Dichroism Facility located at the NSLS Beamline U4B (15).

Elemental Selectivity: Cr/Fe(001)

The most straightforward application of MCD is to determine in which elements of a material are ferromagnetic and in which direction their moments point. This is extremely useful in examining alloy systems, overlayer systems, and buried layers. Therefore, in this first example, the presence of an MCD signal in the L-edge spectra of an element is used as a clear indication that the element is ferromagnetic and that it posses a net magnetic moment. What is more, the sign of the

MCD is indicative of the moment direction, allowing for the determination of the relative moment orientations in samples with more than one magnetic element. Used first in the hard X-ray (14) and later in the soft X-ray regions (9), this straightforward and simple variant of MCD is now used at NSLS and by a variety of other groups to unambiguously identify the presence of ferromagnetism or induced moments in an element-specific and site-specific manner. MCD in absorption has been used to identify induced moment and relative moment orientations for transition metals and rare earths deposited on ferromagnetic substrates where the magnetic signature of the substrate would completely mask the very small response of the overlayer (10).

To exploit the unique capabilities of MCD, namely its element specificity, reported here are MCD measurements for the determination of the magnetic structure of one monolayer (ML) of Cr deposited on single crystal body-centered cubic (bcc) Fe(001) (10). This system is of interest because it is the preliminary stage for the development of the Fe/Cr/Fe heterostructure which is representative of a class of layered systems that display aligned or antialigned coupling of the two ferromagnetic films through the interlying spacer layer (16–19), where the sign of the coupling is dependent on the thickness of the spacer layer (20).

To generate and maintain a clean single monolayer Cr overlayer, the sample is created in an ultra-high vacuum growth chamber with a base pressure of 1×10^{-10} torr. First, the Fe is deposited from a vacuum furnace on a GaAs(001) substrate which has been heat-cleaned in vacuum at 580°C to generate an oxide-free GaAs(001) surface. The Fe is deposited with the substrate held at 175°C and at a vacuum of 2×10^{-10} torr until a 150 Å Fe body-centered cubic (bcc) film is developed (10). This film is single crystal and displays the magnetic properties associated with the bcc Fe magnetic structure (21). A subsequent growth of Cr is carried out with the sample held at room temperature to eliminate interdiffusion and alloying between the Cr and the Fe.

The top panel of Fig. 4 shows the two room temperature X-ray absorption spectra (XAS) and the difference spectra (MCD) of the L_3 and L_2 absorption edges of the as-deposited first Fe film as a function of the soft X-ray energy. The solid line is taken with the spin direction of the photons parallel to that of the majority electrons of the Fe film, while the dashed line is taken with antiparallel spin directions. The dots are the difference between the parallel and antiparallel spectra which is the magnetic circular dichroism (MCD). After the 150 Å thick Fe film is measured, we deposit a fractional (0.25 ML) Cr overlayer. Unlike the layer-by-layer growth of Cr on unstressed Fe whiskers at elevated temperatures (19,22), we have found that the room temperature deposition of Cr on thin Fe films deposited on heat-cleaned GaAs(001) substrates form three-dimensional islands. The differences may be due to the rough nature of the surface of the Fe/GaAs(001) (23). Because the Cr grows in 3-D islands on the Fe/GaAs(001), submonolayer coverage down to 0.25 ML were recorded to maximize the amount of Cr in the first layer position and minimize the presence of Cr in second layer sites.

The bottom panel of Fig. 4 show XAS and MCD spectra of the lowest coverage, 0.25 ML Cr film. Because the 2p electron binding energies differ for different materials, these absorption L edges occur at different photon energies (see Fig. 4), yielding the elemental selectivity. At 0.25 ML coverage nearly

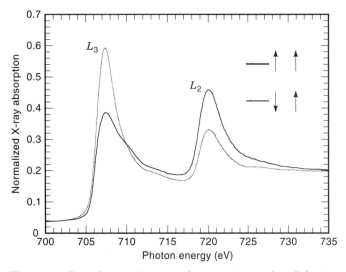

Figure 3. Typical transition metal magnetic circular dichroism (MCD) spectra. Solid line for aligned spins, dotted line for antialigned spins. Note that the L_3 and L_2 edges have opposite MCD behavior.

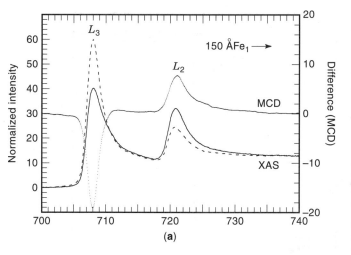

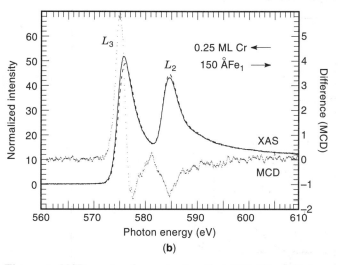

Figure 4. MCD spectra demonstrating that 0.25 ML of Cr has a magnetic moment aligned opposite to Fe moment. Top panel is absorption (XAS) and difference (MCD) scans for Fe; bottom panel is XAS and MCD for Cr with solid line for aligned spins, dashed for antialigned, and dots difference spectra.

all the deposited Cr should occupy the first layer with very little in second layer sites. Since MCD is element specific, it is immediately evident from the reversal of the Cr MCD intensity at both the L_3 and L_2 edges in comparison to the Fe MCD spectra, that submonolayer coverage of Cr are ferromagnetic and antialigned with the first Fe layer (as depicted in the inset of Fig. 4). If the Cr were not magnetic, no MCD signal would be detected. This alignment is in agreement with theoretical calculations for single monolayer Cr/Fe(001) structures which also predicted a ferromagnetic Cr layer adjoining to and antialigned with the Fe films (24–26). In addition to the reversal of the MCD signal, the Cr data show a strong differentiationlike lineshape at the L_3 edge which is due to a peak energy shift between the two XAS spectra, resembling the result of a recent atomic multiplet and crystal field calculation (27).

This example demonstrates that by simply inspecting the separate MCD signals of the overlayer and the magnetic substrate material, the relative orientation of the two moments created from the strong coupling of the moments at the inter-

face can be determined. This effect has been used in a wide variety of overlayer systems to determine the relative magnetic structure of induced moments of materials in close proximity to a ferromagnetic substrate. Similarly this inspection method has been used in multielement alloys to determine the magnetic role that each element plays (28) and to identify absorbate Curie temperatures (29).

Element-Specific Magnetometry

In the previous discussion of MCD, it was pointed out that it is the relative orientations of the spin of the electron and the spin of the photon that mediate the absorption strength of the material. By changing the photon spin direction, we can generate the MCD spectra and determine the magnetic state of the element. But, instead of reversing the photon spin direction, it is equivalent to reverse the electron spin direction (through the reversal of the magnetization of the sample by applying a magnetic field). In fact simultaneously reversing both the photon and electron spin direction MUST result in the same absorption and MCD spectra, since it is only the relative orientation of the two that is important. The benefit to reversing the electron spin directions is that we can now measure the MCD intensity, which is proportional to the net magnetization of that element within the material, *as a function of magnetic field*. These spectra now reflect the variation of the elemental moment as a function of applied magnetic field, and they serve as an element-specific magnetometer.

If the MCD is measured in transmission or by fluorescence detection by reversing the applied magnetic field direction while keeping the polarization of the incident photon beam fixed, this method permits for the measurement of element-specific magnetic information in an applied field intensity. In particular, by monitoring the peak height of a given elemental absorption edge (typically the L_3 white line) as a function of the applied magnetic field, element-specific magnetic hysteresis curves can be obtained (30).

The utility of element-specific magnetometry was demonstrated at NSLS where the element-specific hysteresis curves were measured and compared to the total hysteresis curve for a multilayer sample consisting of a Co (51 Å)/Cu (30 Å)/Fe (102 Å) trilayer structure deposited on a glass substrate and capped with an additional Cu film (40 Å) to prevent oxidation (see Fig. 5). Multilayer films of this type have important applications in magnetic read-heads and as elements in ultradense, nonvolatile memory. The top panel of Fig. 5 shows the measured MCD intensity for the Fe L_3 and Co L_3 absorption edge as a function of the applied magnetic field. These are the element-specific magnetic hysteresis loops. The bottom panel shows the total moment hysteresis curve as measured by vibrating sample magnetometry (shown as a solid line labeled VSM). Superimposed on the VSM hysteresis curve of the bottom panel is the scaled sum of the two individual Fe and Co element-specific magnetic hysteresis curves (shown as a dashed line labeled Fe + Co). The agreement is nearly perfect. From these scaling values, the known Co and Fe film thicknesses, and the total magnetic moment value as measured by VSM, the average elemental atomic magnetic moment of the Fe and the Co can be extracted. Our best fit determines the average moments to be 2.1 μ_B for the Fe and 1.2 μ_B for the Co.

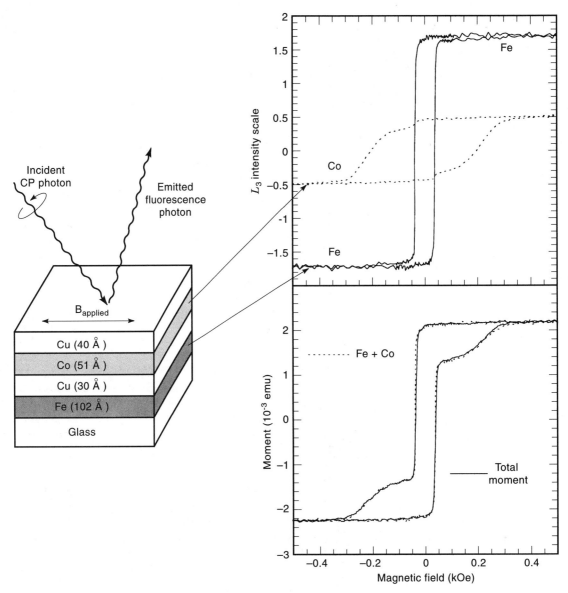

Figure 5. Element-specific hysteresis curve for Fe/Cu/Co multilayer. Left panel shows multilayer configuration. Top right panel shows L_3 intensity as a function of applied magnetic field for Fe and Co, these are the element-specific magnetic hysteresis curves. Bottom right panel shows total moment magnetometry curve (solid line) and sum of scaled Fe and Co curves (dotted line). The agreement is exact.

Figure 5 describes the dissection of a magnetometry curve into its chemically distinct parts for magnetic moment determination. Although this allows for the determination of individual moments or for the extraction of coefficients of magnetoresistance (31), this process was undertaken for only one component of the magnetic moment vector. By recording all three components of the magnetic moment vector sequentially, a coherent picture of the magnetic moment behavior as a function of applied field for each element is obtainable. This powerful technique of element-specific vector magnetometry (ESVM) is accomplished through the use of an in situ electromagnet which permits either independent or concerted rotation of the sample and the magnet with respect to the photon beam direction (32).

A powerful application of ESVM is for the characterization of buried magnetic layers. In a recent measurement of a Co/

Mn/Co trilayer system (33), the rotation and reversal of the Mn and Co magnetic moment vectors were determined as a function of applied field. In this system the two ferromagnetic films are coupled at an angle through the interlying Mn film (34), this coupling angle is dependent on the thickness, and therefore the magnetic structure of the Mn film. From the element-specific vector magnetometry curves of Co and Mn, the magnetic structure of the Mn was determined to be an antiferromagnetic helix, resembling the recent theoretical predictions for frustrated moment systems (35). This is the only method that could identify the applied field dependent variation of the extremely small Mn moment. All other total moment techniques would only measure the comparatively very large signature of the enclosing ferromagnetic films where the minuscule signal associated with the Mn would be lost.

Identifying Layer Switching

So far we have been investigating only the differential absorption of circular polarized photons by a magnetic heterostructure. Complimentary to this technique is the study of the differential reflectivity of circular polarized photons from a magnetic multilayer. Variants of XRMS have been used to identify layer thicknesses with high accuracy, characterize layer switching (36), and determine interfacial parameters, including separately measuring the chemical and magnetic roughness of the interface (37).

As a technologically relevant example of the importance of magnetic structure characterization at these larger scales, Fig. 6 shows a cartoon of a lithographically patterned trilayer magnetic device which can be used as an element in the next generation of magnetic memory. The structure consists of a conducting substrate onto which has been deposited two ferromagnetic films separated by a nonmagnetic spacer layer and capped with a metallization layer. The multilayer film can be structured to generate a small, all metal memory element, where the logical state depends on the relative orientation of the two magnetic moment directions. The magnetic element is interrogated by a 4-probe method and the resistance of the structure indicates the relative moment orientations (high resistance corresponds to antialigned moments, and low resistance corresponds to aligned moments) (17) which can be used as the memory configurations of a 2-level system. Although the illustration only depicts two magnetic layers, real magnetic devices typically are multiple repeats of this structure where the various layer orientations now represent the many memory configurations in this n-level system.

For a trilayer or multilayer structure of this type, it is clear that another area where determining magnetic structure is critical is in the determination of the order of magnetic layer switching as a function of applied filed. One technologically promising trilayer architecture, categorized as a spin valve (38), is composed of a hard magnetic layer (e.g., Co) and soft magnetic layer (e.g., NiFe alloy) separated by a nonmagnetic spacer (e.g., Cu). Since the soft magnetic layer has a lower coercive field, it should switch first leading to an antialigned state. Unfortunately, for device applications, effects such as interlayer and intralayer dipolar coupling, film micromorphology, and lithographic patterning are known to alter layer coercivities and may change the order of layer switching. Standard magnetometry techniques, which measure only the total magnetic response, can only infer the order of the

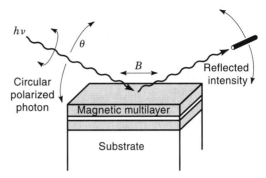

Figure 7. X-ray resonant magnetic scattering experimental geometry.

layer switching indirectly. To predict device performance and understand the fundamental dynamics of the system, it is necessary to directly determine the order in which the magnetic layers switch.

Such information is available using the technique of X-ray resonant magnetic scattering (XRMS) (39–43). XRMS is a soft X-ray reflectivity technique that involves a combination of X-ray resonant scattering and magnetic circular dichroism. In X-ray resonant scattering (XRS) (44,45), when the energy of an incident photon is equal to the binding energy of a core electron (i.e., resonantly tuned to an absorption edge), the reflectivity, or scattering factor, is greatly enhanced with respect to the nonresonant value. Using circular polarized X rays effectively combines XRS with MCD to yield a scattering technique with not only enhanced chemical selectivity but also with magnetic sensitivity and the ability to determine the magnetic depth profile of a layered system.

These reflectivity measurements were also conducted at the NRL/NSLS MCD facility located at beamline U4B of the NSLS (15). The experimental chamber consists of a vacuum compatible θ–2θ spectrometer (46) modified to incorporate a Si photo-diode detector to measure the reflected soft X-ray intensity, which is normalized to the incident photon flux as depicted in Fig. 7. Simultaneously changing the incident and detector directions allows for the acquisition of the specularly reflected intensity as a function of angle, photon energy, and applied magnetic field.

The magnetic trilayer to be studied is a NiFe(50 Å)/Cu(30 Å)/Co(20 Å) trilayer which was prepared on a Si substrate at room temperature by RF sputtering. To protect the film from oxidation, a thin Cu cap layer was deposited before the sample was removed. A standard magnetometry measurement of the NiFe/Cu/Co spin valve structure using vibrating sample magnetometry (VSM) is shown in the left panel of Fig. 8. The step in the hysteresis and the intermediate plateau regions are a clear indication of layer switching, giving rise to an antialigned state and a reduction in the total magnetization of the structure. From this measurement alone the order of the layer switching can only be indirectly inferred from knowledge of the behavior of thin films of the individual, unpatterned constituents. To directly measure the order of layer switching by XRMS, a comparison is made between the magnetic contributions to the reflectivity at the resonant energies for Co and Fe when the trilayer is in the aligned and antialigned states. The VSM data are used to determine the ap-

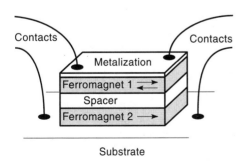

Figure 6. Prototypical magnetic-based memory element. Conductivity of all-metal device varies depending on relative orientation of two moment directions of magnetic films. Four contacts are for four-point probe interrogation.

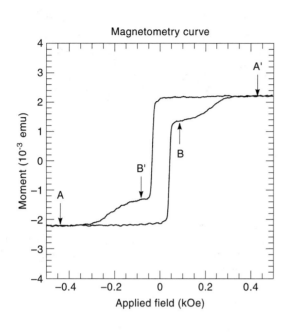

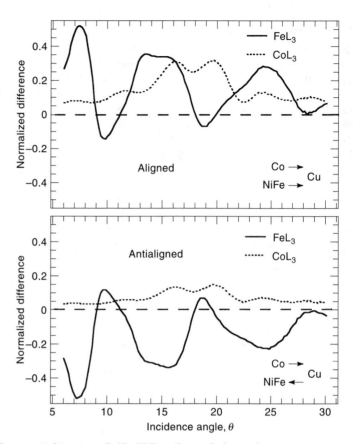

Figure 8. Determination of the order of layer switching in a Co/Cu/NiFe trilayer. Left panel shows magnetometry curve for trilayer. Top right panel shows the difference in reflectivity for Co and Fe taken at points A and A' of the magnetometry curve. Bottom right panel shows the difference in reflectivity for Co and Fe taken at points B and B' of the magnetometry curve and after one layer has flipped.

plied field values at which the aligned and antialigned states are realized.

For extraction of the magnetic information, it is necessary to measure the reflectivity for both left and right circular polarized light. Since there are difficulties associated with switching the helicity, it is equivalent to instead make measurements with the sample in two exactly opposite magnetic configurations while keeping the helicity fixed (9,11). For the aligned state case this is accomplished by measuring the reflectivity after applying a field large enough to saturate the sample in either magnetic field direction (marked A and A' in Fig. 8). Reflectivity measurements of the antialigned state involves a two-step process, where it is first necessary to magnetically saturate the sample at a high field (A in Fig. 8) and then reduce and reverse the field to move along the hysteresis loop until the intermediate plateau is reached (B in Fig. 8). The measurement of the mirrored antialigned magnetic configuration is achieved by repeating the sweep in the opposite field direction (following the A' to B' path).

Results of the two aligned state reflectivity scans at both the Fe L_3 edge (707.5 eV) and Co L_3 edge (778.0 eV) are presented in the top right panel of Fig. 8. This panel shows the normalized difference of the measured specular reflectivity as a function of the grazing angle, θ. The magnetic information is contained in this asymmetry ratio, R, which is the normal-

ized difference of the absorption spectra and is defined as

$$R = \frac{(I^+ - I^-)}{(I^+ + I^-)}$$

where I^+ and I^- denote the reflectivity with the magnetic structure either parallel or antiparallel to the photon helicity, respectively. The inset is a diagram of the resulting magnetization directions of the two films. The observed modulations in the difference spectra are not only due to interference effects from reflections at the various interfaces within the layered sample, they are also sensitive to the magnetic configuration of the multilayer. The lower right panel of Fig. 8 shows the normalized difference spectra for the antialigned case (B and B').

It is clearly evident that the Fe asymmetry in the antialigned case has reversed sign with respect to the aligned case, indicating that indeed the NiFe layer is the first layer to switch, as indicated by the inset. It is interesting to note, however, that even though the Co asymmetry is the same sign, it is reduced in magnitude at the lower applied field when compared to the high field value. This may indicate that the Co film is not completely antialigned to the NiFe layer but has broken into magnetic domains, in agreement with the VSM data. The lack of being in a completely antialigned state is possibly due to dipolar magnetic coupling and/or film micromorphology.

This example has demonstrated that the order of layer switching can be determined in a multilayer system when the layers have different constituent elements. The same type of characterization can be accomplished if the ferromagnetic layers within the multilayer have the same constituent materials (47), but in this case the depth selectivity of this technique is exploited. By varying the incidence direction, the majority of the scattering occurs for the first magnetic layer at grazing incidence angles but quickly includes deeper lying layers as the angle is increased.

SUMMARY

Characterization of the magnetic structure of a system has become a critical and necessary aspect of understanding magnetic behavior of multilayer systems. As magnetic systems become increasingly complex, utilizing multiple layers with more than one magnetically active element in each, magnetic structure characterization becomes increasingly important. The utility of element-specific probes, like magnetic circular dichroism, are now clear and well demonstrated.

BIBLIOGRAPHY

1. J. B. Goodenough, *Magnetism and the Chemical Bond*, New York: Wiley, 1963.
2. R. M. Bozorth, *Ferromagnetism*, Piscataway, NJ: IEEE Press, 1978.
3. S. Chikizuma and S. H. Charap, *Physics of Magnetism*, Malabar, FL: Krieger, 1964.
4. G. A. Prinz, *Science*, **250**: 1092, 1990.
5. L. M. Falicov et al., *J. Mater. Res.*, **5**: 1299–1340, 1990.
6. H. Winick, *Synchrotron Radiation Sources: A Primer*, Singapore: World Scientific, 1994.
7. C. T. Chen, *Nucl. Instrum. Methods Phys. Res. A*, **256**: 595–604, 1987.
8. C. T. Chen and F. Sette, *Rev. Sci. Instrum.*, **60**: 1616–1621, 1989.
9. C. T. Chen et al., *Phys. Rev. B*, **42**: 7262–7265, 1990.
10. Y. U. Idzerda et al., *Phys. Rev. B*, **48**: 4144–4147, 1993.
11. Y. U. Idzerda et al., *Nucl. Instr. Methods Phys. Res. A*, **347**: 134, 1994.
12. J. G. Tobin, G. D. Waddill, and D. P. Pappas, *Phys. Rev. Lett.*, **68**: 3642–3645, 1992.
13. Y. Wu et al., *Phys. Rev. Lett.*, **69**: 2307–2310, 1992.
14. C. T. Chen, *Rev. Sci. Instrum.*, **63**: 1229–1233, 1992.
15. S. Hulbert et al., *Nucl. Instrum. Methods Phys. Res. A*, **291**: 343, 1990.
16. P. Grunberg et al., *Phys. Rev. Lett.*, **57**: 2442, 1986.
17. M. N. Baibich et al., *Phys. Rev. Lett.*, **61**: 2472, 1988.
18. C. Carbone and S. F. Alvarado, *Phys. Rev. B, Condens. Matter*, **36**: 2433, 1987.
19. J. Unguris, R. J. Celotta, and D. T. Pierce, *Phys. Rev. Lett.*, **67**: 140, 1991.
20. S. S. P. Parkin, N. More, and K. P. Roche, *Phys. Rev. Lett.*, **64**: 2304, 1990.
21. G. A. Prinz, *Ultramicroscopy*, **47**: 346–354, 1992.
22. J. Unguris, R. J. Celotta, and D. T. Pierce, *Phys. Rev. Lett.*, **69**: 1125, 1992.
23. R. A. Dragoset et al., *Mater. Res. Soc. Symp. Proc.*, **151**: 193, 1989.
24. R. H. Victora and L. M. Falicov, *Phys. Rev. B, Condens. Matter*, **31**: 7335, 1985.
25. P. M. Levy et al., *J. Appl. Phys.*, **67**: 5914, 1990.
26. D. Stoeffler, K. Ounadjela, and F. Gautier, *J. Magn. Magn. Mater.*, **93**: 386, 1991.
27. G. van der Laan and B. T. Thole, *Phys. Rev. B, Condens. Matter*, **43**: 13401, 1991.
28. K. M. Kemner et al., *J. Appl. Phys.*, **81**: 1002, 1997.
29. L. H. Tjeng et al., *J. Magn. Magn. Mater.*, **109**: 288–292, 1992.
30. C. T. Chen et al., *Phys. Rev. B, Condens. Matter*, **48**: 642, 1993.
31. Y. U. Idzerda et al., *Appl. Phys. Lett.*, **64**: 3503, 1994.
32. V. Chakarian et al., *Appl. Phys. Lett.*, **66**: 3368, 1995.
33. V. Chakarian et al., *Phys. Rev. B, Condens. Matter*, **53**: 11313, 1996.
34. M. E. Filipkowski et al., *Phys. Rev. Lett.*, **75**: 1847–1850, 1995.
35. J. C. Slonczewski, *J. Magn. Magn. Mater.*, **150** (1): 13–14, 1995.
36. J. W. Freeland et al., *Appl. Phys. Lett.*, **71**: 276, 1997.
37. J. W. Freeland et al., *J. Appl. Phys.*, **83**: 6290–6292, 1998.
38. B. Dieny et al., *Phys. Rev. B, Condens. Matter*, **43**: 1297, 1991.
39. K. Namikawa et al., *J. Phys. Soc. Jpn.*, **54**: 4099, 1985.
40. C. Kao et al., *Phys. Rev. Lett.*, **65**: 373, 1990.
41. C.-C. Kao et al., *Phys. Rev. B, Condens. Matter*, **50**: 9599, 1994.
42. J. M. Tonnerre et al., *Phys. Rev. Lett.*, **75**: 740, 1995.
43. V. Chakarian et al., *Applications of Synchrotron Radiation in Industrial, Chemical, and Materials Science*, New York: Plenum Press, 1996, p. 187.
44. D. Gibbs et al., *Phys. Rev. Lett.*, **61**: 1241, 1988.
45. E. D. Isaacs et al., *Phys. Rev. Lett.*, **62**: 1671, 1989.
46. E. D. Johnson, C.-C. Kao, and J. B. Hastings, *Rev. Sci. Instrum.*, **63**: 1443, 1992.
47. J. W. Freeland et al., *J. Vac. Sci. Technol. A, Vac. Surf. Films*, **16**: 1355–1358, 1998.

Yves U. Idzerda
Naval Research Laboratory

MAGNETIC SWITCHING

Many applications of ferromagnetic materials are based on the fact that the magnetization can move from one stable state (direction) to another stable state. Ferromagnetic materials, like ferrimagnetic materials, possess spontaneous magnetic polarization in the absence of an applied field. The manner in which the magnetization changes, together with the time necessary for completing the magnetization state change, is of primary significance. A directional change in magnetization state can be thermally activated or result from applying an external magnetic field. The process of changing the magnetization direction is called magnetization switching or reversal. The magnetic switching process is extremely complex, because it is inherently nonlinear and depends nontrivially on material properties, magnetic structure, magnetic history, specimen geometry, amplitude and direction of the applied switching magnetic field, damping mechanisms, and switching-field time dependence. This article addresses the physical phenomenology involved in magnetic switching and is organized as follows:

- Magnetic moments (phenomenological and atomic models, interactions with applied static and frequency-dependent magnetic fields, magnetization equations of motion, resonance, magnetic susceptibility, dipole interaction energy, orbital and spin moments)
- Magnetic domains (general structure; magnetocrystalline anisotropy, exchange, magnetostatic, and magnetostriction energy; Bloch walls and Néel walls)
- Magnetic switching analysis (quasi-static and dynamic, single and multiple domains, coherent and incoherent rotation, damping mechanisms, hysteresis, giant magnetoresistance, spin valves)

MAGNETIC MOMENTS

Magnetic moments can be understood by phenomenological or atomic models. An elementary model is the magnetic dipole in which magnetic moments are described in terms of *equivalent* opposite magnetic charges separated by a distance d whose product yields the magnetic dipole moment $\boldsymbol{m}_d$. This model is analogous to that of an electric dipole moment of a pair of charges, $-q$ and $+q$, separated by a distance d. An equivalent phenomenological model is given in terms of a conventional current I flowing in a closed loop of area A [Fig. 1(a)], which has a magnetic dipole moment amplitude $m_d = IA$. The interaction of a magnetic dipole with an externally applied field $\boldsymbol{H}_0$ is given by the torque imposed on the dipole in the presence of the applied field:

$$\begin{aligned} \boldsymbol{T}_d &= \mu_0 \boldsymbol{m}_d \times \boldsymbol{H}_0 \\ &= -\mu_0 m_d H_0 \sin\theta \end{aligned} \qquad (1)$$

where μ_0 is the magnetic permeability of vacuum ($4\pi \times 10^{-7}$ H/m), θ is the angle between the dipole axis and the externally applied field, and the sign is chosen so that T_d is a restoring torque [Fig. 1(b)]. The angular gradient of the interaction (potential) energy between the magnetic dipole and the external field also represents the torque exerted on the dipole. The interaction energy E is given by

$$E = -\mu_0 m_d H_0 \cos\theta \qquad (2)$$

and

$$T_d = -\frac{\partial E}{\partial \theta} \qquad (3)$$

In an applied field, the dipole has minimum energy ($-\mu_0 m_d H_0$) at $\theta = 0$ and increases as the dipole moment is turned away from $\boldsymbol{H}_0$. At $\theta = \pi$, the dipole is in unstable equilibrium. The fundamental equation of motion of a magnetic dipole in an external field is derived from the torque equation because the torque is the gradient with respect to time of the angular momentum $\boldsymbol{L}_d$ of the dipole. The magnetic dipole moment $\boldsymbol{m}_d$ may be written as

$$\boldsymbol{m}_d = \gamma' \boldsymbol{L}_d \qquad (4)$$

where $\gamma' = m_d/L_d$. For a single electron moving in a circle, γ' is given by $e/(2m_e)$, where e and m_e are the charge and mass of the electron. The torque may be written

$$\begin{aligned} \boldsymbol{T}_d &= \frac{d\boldsymbol{L}_d}{dt} \\ &= \mu_0 \boldsymbol{m}_d \times \boldsymbol{H}_0 \end{aligned} \qquad (5)$$

or, equivalently, the equation of motion of the magnetic dipole in an applied field $\boldsymbol{H}_0$ is

$$\begin{aligned} \frac{d\boldsymbol{m}_d}{dt} &= \mu_0 \gamma' \boldsymbol{m}_d \times \boldsymbol{H}_0 \\ &= \gamma_c \boldsymbol{m}_d \times \boldsymbol{H}_0 \end{aligned} \qquad (6)$$

where $\gamma_c = \mu_0 \gamma'$ is the gyromagnetic ratio of a circulating charge (neglecting spin). Because of spin and angular momentum in a real magnetic ion, γ_{ion} can differ significantly from $\mu_0 e/(2m_e)$. In fact, a magnetic material may contain several magnetic ions so that its effective gyromagnetic ratio γ_{eff} differs from those of the individual ionic species present.

It is useful to review the effects of a static field and a frequency-dependent field on the equation of motion for a simple magnetic dipole. Time-dependent equations of magnetization motion, which will be addressed later in this chapter, can be derived from inverse Fourier transforms of frequency-domain expressions. In addition, specific details of individual magnetic fields comprising the net *effective* field acting on the time-dependent magnetization vector in a ferromagnetic medium may be better understood.

If the total external applied magnetic field is written as the sum of a static field $\boldsymbol{H}_0$ and a frequency-dependent field $\boldsymbol{h}e^{j\omega t}$, where $\omega = 2\pi f$ is the angular frequency,

$$\boldsymbol{H}(t) = \boldsymbol{H}_0 + \boldsymbol{h}e^{j\omega t} \qquad (7)$$

the time-harmonic case may be examined. Equation (7) may be substituted in Eq. (6) to give the *magnetic susceptibility* tensor relating the applied static external magnetic field, the components of magnetization, and the frequency-dependent field transverse to $\boldsymbol{H}_0$. Their dependence on the frequency ω,

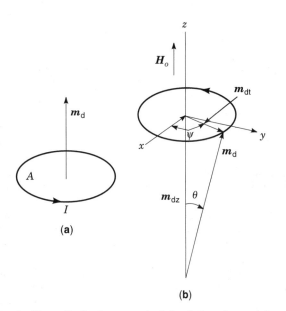

Figure 1. Magnetic dipole moment of circulating charge (a) and interaction of dipole with external magnetic field (b).

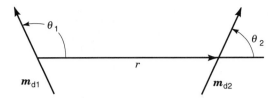

Figure 2. Interaction of two magnetic dipoles in the same plane oriented at arbitrary angles to a position vector separating them.

and the *Larmor* precessional frequency, $\omega_0 = d\psi/dt = |\gamma|H_0$, can also be derived. The phenomenological model is illustrated in Fig. 1(b). Allow the direction of $\boldsymbol{H}_0$ to be along the z-axis, so that

$$\boldsymbol{H} = H_0\hat{a}_z + (h_x\hat{a}_x + h_y\hat{a}_y + h_z\hat{a}_z)e^{j\omega t} \tag{8}$$

If $|\boldsymbol{h}| \ll |\boldsymbol{H}_0|$, then $|\boldsymbol{m}_{dt}| \ll |\boldsymbol{m}_{dz}|$ and

$$\boldsymbol{H} = \begin{bmatrix} h_x \\ h_y \\ H_0 \end{bmatrix}, \quad \boldsymbol{m}_d = \begin{bmatrix} m_{dx} \\ m_{dy} \\ m_{dz} \end{bmatrix} \tag{9}$$

where $\boldsymbol{m}_d = \boldsymbol{m}_{dt} + m_{dz}\hat{a}_z$, $\boldsymbol{m}_{dt} = m_{dx}\hat{a}_x + m_{dy}\hat{a}_y$ and $m_{dz} \simeq m_d$. The equation of motion of the magnetic dipole of moment m_d yields

$$\begin{bmatrix} m_{dx} \\ m_{dy} \\ m_{dz} \end{bmatrix} = \begin{bmatrix} \gamma^2 m_d H_0/(\gamma^2 H_0^2 - \omega^2) & -j\omega\gamma m_d/(\gamma^2 H_0^2 - \omega^2) & 0 \\ j\omega\gamma m_d/(\gamma^2 H_0^2 - \omega^2) & \gamma^2 m_d H_0/(\gamma^2 H_0^2 - \omega^2) & 0 \\ 0 & 0 & 0 \end{bmatrix} \begin{bmatrix} h_x \\ h_y \\ H_0 \end{bmatrix} \tag{10}$$

This may be simply written

$$\boldsymbol{m}_d = [\chi] \cdot \boldsymbol{H} \tag{11}$$

where $[\chi]$ is the classical Polder magnetic susceptibility tensor without damping. The magnetic permeability is $[\mu] = [1] + [\chi]$ where $[1]$ is the unit tensor. Equation (10) gives a resonant condition when $\omega = |\gamma| H_0$, and shows that the components of the magnetic dipole moment depend on the orientation of the field $\boldsymbol{H}$. Equation (10) represents a *steady-state* solution for the magnetic dipole in an applied field $\boldsymbol{H}$. Other frequency-domain expressions can be derived that correspond to various transient excitations.

Dipole–Dipole Interactions

The interaction of dipole moments influences the equations of motion of magnetization. One simple case is the general configuration of two dipoles that have magnetic moments $\boldsymbol{m}_{d1}$ and $\boldsymbol{m}_{d2}$ illustrated in Fig. 2. The dipole–dipole interaction energy is given by

$$E = \frac{\mu_0\boldsymbol{m}_{d1} \cdot \boldsymbol{m}_{d2}}{r^3} - \frac{3\mu_0(\boldsymbol{m}_{d1} \cdot \boldsymbol{r})(\boldsymbol{m}_{d2} \cdot \boldsymbol{r})}{r^5} \tag{12}$$

where $\boldsymbol{r}$ is the position vector from dipole 1 to dipole 2. In a ferromagnetic material, $\boldsymbol{m}_{d1} = \boldsymbol{m}_{d2} = \boldsymbol{m}_d$ and are parallel, so

the interaction energy becomes

$$E = \frac{\mu_0 m_d^2}{r^3}(1 - 3\cos^2\theta) \tag{13}$$

If the dipole moments are parallel to the position vector [Fig. 3(a)] and are rotated through an angle θ from the z-axis [Fig. 3(b)], there is a restoring torque in addition to the torque from any external field. The restoring torque is represented by

$$\begin{aligned} T &= -\frac{\partial E}{\partial \theta} \\ &= -2\mu_0 m_d\left(\frac{3m_d}{r^3}\cos\theta\right)\sin\theta \\ &= -\mu_0 M_d H_{Dz}\sin\theta \end{aligned} \tag{14}$$

where $M_d = 2m_d$ and H_{Dz} is an apparent field in the z-direction, that is,

$$H_{Dz} = \frac{3M_{dz}}{2r^3} = DM_{dz} \tag{15}$$

where $M_{dz} = M_d \cos\theta$.

Then the magnetization equation of motion of two aligned magnetic dipoles in an external field $\boldsymbol{H}_0$ becomes

$$\frac{d\boldsymbol{M}_d}{dt} = \gamma\boldsymbol{M}_d \times (\boldsymbol{H}_0 + \boldsymbol{H}_D) \tag{16}$$

If $\boldsymbol{H}_0$ is directed along the z-axis, the time derivatives of the transverse components of the total magnetization vector become

$$\begin{aligned} \frac{dM_{dx}}{dt} &= \gamma[M_{dy}H_0 - M_{dz}(-DM_{dy})] \\ \frac{dM_{dy}}{dt} &= \gamma[M_{dz}(-DM_{dx}) - M_{dx}H_0] \end{aligned} \tag{17}$$

Similar equations can be written when magnetic dipoles are parallel but normal to the position vector connecting them. In this case, for an applied field $\boldsymbol{H}_0$ in the y-direction

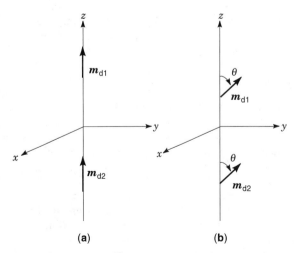

(a)   (b)

Figure 3. Aligned magnetic dipoles parallel to a position vector separating them (a) and rotational angle θ causing restoring torque (b).

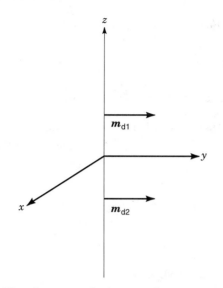

Figure 4. Aligned magnetic dipoles normal to a position vector separating them.

(see Fig. 4), the equations of motion of the resultant magnetization are

$$\frac{dM_{dx}}{dt} = \gamma[-M_{dz}(H_0 - DM_{dy})]$$

$$\frac{dM_{dz}}{dt} = \gamma[M_{dx}(H_0 - DM_{dy}) - M_{dy}(-DM_{dx})]$$

(18)

Both Eqs. (17) and (18) have the general form

$$\frac{d\boldsymbol{M}_d}{dt} = \gamma \boldsymbol{M}_d \times \left(\boldsymbol{H}_0 - \begin{bmatrix} D & 0 & 0 \\ 0 & D & 0 \\ 0 & 0 & 0 \end{bmatrix} \boldsymbol{M}_d \right)$$

$$= \gamma \boldsymbol{M}_d \times (\boldsymbol{H}_0 - [D]\boldsymbol{M}_d)$$

(19)

where $\boldsymbol{H}(t) = \boldsymbol{H}_0 - [D]\boldsymbol{M}_d$ is an internal *effective* field and $[D]$ is the two-dipole analog of the *demagnetization* tensor. Similar magnetization equations of motion for many-dipole arrays may be derived to gain insight into magnetic dipole alignment in an external magnetic field. However, dipole forces in solids are too weak to account for the large interaction energies that maintain alignment between a distribution of magnetic moments. These forces result from electron exchange, are quantum mechanical in origin, and must be understood in terms of the electron orbital and spin states of a magnetic ion.

Microscopic moments originate in very definite magnetic ions, elements of the periodic table. These ions are from one of two transition series that have an incomplete inner shell. The first transition series is that where the $3d$ electron shell is incomplete and consists of the elements ranging from calcium (atomic number 20) to zinc (atomic number 30). The second is the rare-earth series with an incomplete $4f$ shell. This contains elements 57 (lanthanum) through 71 (lutecium). The magnetic moments and angular momenta of all of the electrons occupying either the $3d$ or $4f$ shells combine to give a resultant magnetic moment of the ion.

The overall orbital angular momentum of a magnetic ion is specified by the quantum number L. In the presence of a

magnetic field, a free magnetic ion with orbital quantum number L may only have specific discrete values of angular momentum which are integral multiples of $\hbar = h/(2\pi)$, where h is Planck's constant (6.625×10^{-34} J·s). So the associated orbital magnetic moment is given by

$$m_{dL} = \gamma L \hbar$$
$$= \mu_0 e \hbar L/(2m_e)$$

(20)

Electron spin also gives rise to a magnetic moment. However, in this case, the magnetomechanical ratio $\gamma_{spin} = 2\gamma$, and the magnetic moment due to spin is given by

$$m_{dS} = 2\gamma S \hbar$$

(21)

where S is the spin quantum number. Neighboring ions affect the net orbital angular momentum. For elements of the first transition series, orbital angular momentum and associated magnetic moment are almost completely destroyed by electrostatic interactions with the lattice environment. The orbital contributions for the rare-earth series are also disturbed by neighboring ions but to a lesser extent. Therefore, for the first transition series of elements, the overall magnetic moment is close to that given by Eq. (21), so that the general form of the gyromagnetic ratio is

$$\gamma = \frac{g\mu_0 e}{(2m_e)},$$

(22)

where g is the Lande spectroscopic splitting factor ($g = 2.0023$ for a free electron). The value of $|\gamma|$ is 2.2×10^5 rad·m/(A·s).

MAGNETIC DOMAINS

Magnetic reversal, permeability, and hysteresis all depend on the structure of the magnetic domain and the variability of this structure under the effect of an applied external field. There are two general approaches for magnetization analysis. One approach, first introduced by Weiss (1) in 1907, is based on experimental evidence of the spatial distribution of magnetization in magnetic domains. Another approach, proposed by Brown (2), is based on the assumption that the magnetic polarization is a function of coordinates of the point at which it is being evaluated. In the Weiss model, each magnetic domain has a spontaneous, generally uniform, magnetic polarization density I_s oriented differently from those of its neighbors (see Fig. 5). The different *magnetic domains* or *Weiss domains* are separated by transition zones called *Bloch walls*. The atomic magnetic moments within each domain are aligned parallel to each other and across the Bloch wall the magnetic moments rotate to that in a neighboring domain. This rotation is illustrated for a 180° Bloch wall in Fig. 6. Thus the parallel orientation of magnetic moments is a *local* phenomenon. Several experimental techniques make it possible to show the distribution of magnetic polarization in a specimen.

One of the first experiments for observing magnetic domain walls in a bulk ferromagnetic specimen was performed by Barkhausen in 1919. In Barkhausen's experiment, a permanent magnet was rotated slowly with an approximate period of one second about a solenoidal winding holding a ferromagnetic rod specimen. The winding around the specimen detected variations in magnetic flux that were transmitted

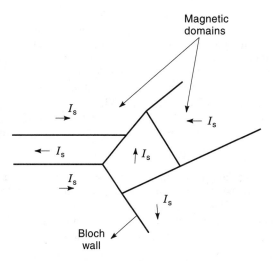

Figure 5. The magnetic domain structure of a specimen showing magnetic polarization I_s within each domain.

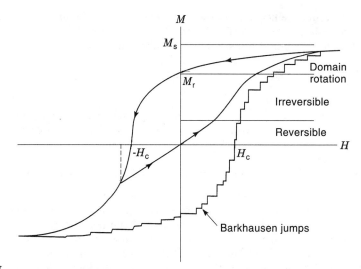

Figure 7. Magnetization loop characteristics showing Barkhausen's jumps from domain wall movements. M is magnetization as a function of applied magnetic field H. M_s, M_r are saturation and remanent magnetizations. H_c is intrinsic coercive field.

after amplification to a loudspeaker. Each time the polarization was reversed, audible sounds reflecting small magnetic grain deformation could be heard. The change in polarization of the ferromagnetic rod occurs in small successive steps, now known to correspond to the sudden change of *pinning points* on the Bloch walls (see Fig. 7). Pinning points, or points that fix or retain Bloch wall structure, are principally caused by various kinds of specimen inhomogeneity rather than specimen size or shape. The types of inhomogeneity affecting domain structure are voids, inclusions, fluctuations in alloy composition, local internal stresses, and the local directional order of crystal boundaries. Figure 7 illustrates the typical appearance of a *magnetic hysteretic* loop, which results from irreversible changes in magnetization that cause dissipation in the form of heat in the specimen. The *coercivity* or coercive field H_c is that magnetic field that must be applied in the opposite direction to restore zero magnetization. The area enclosed by the hysteretic loop is proportional to the energy dissipated into heat for each cycle around the loop. The coercive field is the ultimate determining factor for magnetic switch-

ing energy, which is approximately equal to the product of $2H_c$ and the remanent magnetization M_r. The coercive field also limits the minimum drive field required for domain switching. Another important parameter of the hysteretic loop for switching performance is the *squareness ratio* defined by M_r/M_s. Better switching performance is obtained when the squareness ratio has a value close to one. Irreversible changes in magnetization are addressed in the Preisach model for magnetic switching analysis of multiple-domain particles.

Another technique for observing magnetic domains directly is the Bitter method. The Bitter method consists of spreading a colloidal solution of magnetic particles on the surface of a specimen that has been electrolytically polished. Upon evaporation of the solvent, magnetic flux leakage occurs where the Bloch walls intersect the surface of the specimen. The leakage produces an accumulation of magnetic particles that can be observed with a microscope and which define the Bloch walls. Other methods that permit observing Bloch walls in movement are based on the rotation of the polarization of linearly polarized light during transmission (the *Faraday effect*) or during reflection (the *Kerr effect*). High-quality optical systems must be used to observe the small angles of polarization rotation in the last two methods.

Recently, magnetic force microscopy (MFM) has been used to image magnetic domain wall structures. The divergence of the magnetization in the walls is reflected by fringing magnetic fields outside the specimen. The magnetostatic interaction between the magnetic specimen and a magnetic sensor is measured. The sensor is a sharp tip coated with a magnetic thin film at the end of a flexible cantilever. When the tip is positioned a sufficient distance (typically 50 to 150 nm) away from the specimen surface, the cantilever responds to magnetic force gradients acting on the tip. MFM data permit high resolution, high sensitivity imaging of magnetic domains and other micromagnetic structures.

Magnetic domains can have highly variable shapes. A simple model for the modulus of the spontaneous polarization density vector I_s is that it has the same value in all domains

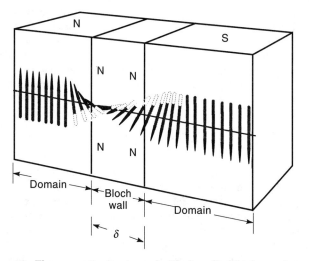

Figure 6. The magnetic structure of a Bloch wall of thickness δ separating $180°$ domains (6).

of a homogeneous material, provided that the temperature distribution is uniform. The magnetic moment $\boldsymbol{M}$ of a specimen of volume V must lie between the limits

$$0 \leq \boldsymbol{M} = \int_V \boldsymbol{I}_s dv \leq \boldsymbol{I}_s V \tag{23}$$

Hence the magnetic moment can vary from 0 to $\boldsymbol{I}_s V$ with an applied field H_0, which modifies the number, shape, and distribution of magnetic domains in the specimen.

General Structure of Magnetic Domains: Energy Considerations

Magnetic domain structures in either bulk or thin-film materials are expressed as statistical averages in terms of their size, shape, and internal orientation of the polarization vector $\boldsymbol{I}_s$. Only those structures are permissible in which a *local*, not *absolute* minimum in internal magnetic energy occurs. The internal energy E_i associated with magnetic domain structure is the sum of the magnetic anisotropy energy E_a, the exchange energy E_{ex}, the magnetostatic energy E_{ms}, and the magnetostriction energy E_{mt}.

Magnetocrystalline Anisotropy Energy. The interaction of atomic magnetic moments that leads to their parallel alignment is measured in terms of energy. In a crystal, a part of this energy, called the magnetocrystalline anisotropy energy, is a function of the angles that the magnetic moments have with respect to the crystal's axial directions. In a ferromagnetic crystal there are *easy* and *hard* directions of magnetization. In other words, the energy required to magnetize the crystal depends on the direction of the applied magnetic field relative to the crystalline axes. For single-crystal materials, such as iron, nickel, and their alloys, the anisotropy has cubic symmetry, provided they are prepared in the absence of a magnetic field. For crystals prepared or deposited in the presence of a magnetic field, a uniaxial symmetry is developed. A polycrystalline specimen of iron, nickel, or their alloys appears isotropic because of the random orientation of the constituent cubic crystallites.

The difference between the energies required to magnetize a crystal in the hard and easy directions is called the *anisotropy energy*. The magnetic anisotropy energy $E_a(\alpha_1, \alpha_2, \alpha_3)$, where $\alpha_1, \alpha_2, \alpha_3$ are the directional cosines that the magnetic polarization vector makes with the crystal axes, varies in different materials (4,5). If only magnetic anisotropy energy is taken into account, spontaneous magnetic polarization in a crystal specimen generally aligns itself parallel to principal crystallographic planes that can be described by their *Miller indices*. In these directions E_a has local minima in which there are *easy magnetization directions*. In general, the anisotropy energy of cubic metals, such as iron or nickel, may be expressed in terms of the anisotropy energy E that has the form (3)

$$E_a = E_0 + E_1 + E_2 \tag{24}$$

where E_0 is purely isotropic and $E_1 = K_1(\alpha_1^2\alpha_2^2 + \alpha_2^2\alpha_3^2 + \alpha_3^2\alpha_1^2)$ and $E_2 = K_2\alpha_1^2\alpha_2^2\alpha_3^2$ are anisotropic. The constants K_1 and K_2 are the first- and second-order anisotropy constants that arise from interaction between the orbital angular momentum and the crystalline lattice and are usually determined by fitting experimental data. In the case of a uniaxial material, the an-

isotropy energy E_{ua} is given, to the first order, by $E_{ua} = K_u(1 - \alpha_z^2) = K_u \sin^2 \theta$, where K_u is the first-order uniaxial anisotropy constant. E_{ua} is independent of the azimuthal angle.

Generally, coercive fields increase monotonically with $|K_1|$ because of higher domain-wall energies and stronger pinning of domain walls at voids, inclusions, and fluctuations in ferromagnetic material composition. Where frequent and rapid switching of the magnetic state is required, materials are chosen that have small first-order anisotropy constants and low domain-wall energies.

Exchange Energy. Exchange energy takes into account the interaction between the atomic magnetic moments. This interaction can be understood only in terms of quantum-mechanical exchange integrals, which have no classical analogs. The exchange energy of an assembly of spins S is given by

$$E_{ex} = -2 \sum_{i,j} J_{ij} \boldsymbol{S}_i \cdot \boldsymbol{S}_j \tag{25}$$

where the subscripts i,j distinguish two atoms with different spin quantum numbers S_i and S_j which are given in terms of $\hbar = h/(2\pi)$, J_{ij} is the exchange integral that has dimensions of energy, and the sum represents the total exchange energy of all atomic spins in the specimen. The spin quantum numbers S_i and S_j are always multiples of 1/2 and are related to the atomic magnetic moments by $m = 2Sm_B$, where m_B is the Bohr magneton, that is, the magnetic moment of electron spin, $9.273 \cdot 10^{-24}$ A $\cdot$ m^2. The ground state for a ferromagnetic assembly has all spins parallel, so that the exchange integral J_{ij} is positive. If the spins vary from point to point (as in a ferrimagnetic specimen), then J_{ij} is negative, and the exchange energy is increased above its maximum negative value. In general, both the magnetocrystalline anisotropy and exchange energies increase inside a domain wall. The sum of the anisotropy and exchange energies usually characterizes the energy associated with the domain wall itself and can be used to approximate the thickness of the domain wall. However, the domain-wall energy limits the area of the wall because the magnetostatic energy due to demagnetization decreases in direct proportion to the number of domains formed. The actual number of domains formed in a specimen stabilizes when the reduction of magnetostatic energy is compensated by the increase in domain-wall energy due to anisotropy and exchange. Again, the *total* energy must be a local minimum.

The domain-wall energy per unit area U_{wall} for a planar domain wall in a bulk specimen that has average anisotropy energy per unit volume $\overline{U}_a$, wall thickness δ, and n lattice planes in δ may be written approximately as

$$U_{wall} = \overline{U}_a\delta + \frac{JS^2\pi^2}{a\delta}, \tag{26}$$

where a is the atomic lattice spacing. This relationship is valid only when rotation is uniform within the wall (so the anisotropy energy is independent of wall thickness), when the exchange energy of two parallel magnetic moments in Eq. (25) is set equal to 0, and when the spin quantum numbers S_i and S_j are the same. The cosine of the angle θ between the spin quantum numbers S_i and S_j is also expanded in a power series in θ. Terms higher than the second term are neglected.

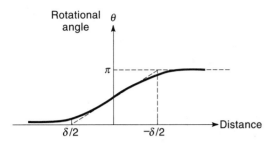

Figure 8. Nonuniform variation of rotational angle of magnetic moments across a Bloch domain wall as a function of distance from the center of the wall.

For a minimum in domain-wall energy, $\partial U_{\text{wall}}/\partial \delta = 0$, which yields the approximate relationship for evaluating domain wall thickness:

$$\delta = \left(\frac{JS^2 \pi^2}{a\overline{U}_a} \right)^{1/2} \tag{27}$$

For a body-centered cubic iron crystal, Eq. (27) yields a domain-wall thickness that is approximately 0.04 μm or 146 base vectors of the lattice. In fact, the rotation of magnetic moments across a Bloch wall is not uniform but has the form given in Fig. 8.

The minimum energy state of a bulk ferromagnetic specimen is generally one of multi-domain formation. For very thin films, whose easy axis is along the film normal, domain structure is approximately parallel to the surface of the film in the main body of the film, and closure domains form at the edges of the film (3). As film thickness increases, however, planar multidomains and domains whose planes tilt away from the film surface form. The direction of magnetization in the domain wall of a thin film (whose thickness is much less than the domain-wall thickness) also changes from that of a Bloch wall, normal to the plane of the film, to that of a Néel wall (6), parallel to the surface of the film (see Fig. 9). This is energetically favorable because the surface energy associated with demagnetization in such a film is minimized. This is particularly true for thin films of ferromagnetic materials that have negligible crystalline anisotropy, such as Permalloy. In general, the thickness of Bloch domain walls increases, whereas the thickness of Néel walls decreases with increasing film

thickness. The domain-wall energy of Bloch walls, on the other hand, decreases with increasing film thickness and that of Néel walls increases.

Magnetostatic Energy. The magnetostatic energy of a specimen results from the effects of demagnetization and is highly shape-dependent. Equivalent magnetic poles occur at the intersection of a domain wall with the surface of the material and give rise to stray fields external to the specimen. These create an opposing general *demagnetizing field* $\boldsymbol{H}_D$ with orientation opposite to the polarization vector internal to the specimen. The energy associated with the demagnetizing field may be written

$$E_{\text{ms}} = -\frac{1}{2} \int_V \boldsymbol{M} \cdot \boldsymbol{H}_D \, dv \tag{28}$$

Minimization of magnetostatic energy in bulk materials increases the number of domains until the decrease of magnetostatic energy associated with domain growth is compensated for by the increased energy of the Bloch walls. This gives a net minimum in energy. For magnetocrystalline anisotropy that has cubic symmetry, there are special domain configurations in which no external fields are created because adjacent domains may produce equal and opposite equivalent magnetic mass densities. In a thin film the configuration of the magnetization vector in the domain wall changes from that of a Bloch wall to a Néel wall or a *cross-tie* wall (3) to minimize surface regions that would have high demagnetization energies.

Magnetostriction Energy

Magnetostriction energy is the elastic energy associated with the dimensional change in a specimen resulting from magnetic polarization between neighboring domains. Magnetostriction occurs because of anisotropy. Magnetostrictive expansion or contraction of a crystalline lattice during the magnetization process is the reaction of the lattice to rotation of the magnetization vector from easy to hard directions. A material has positive magnetostriction if a specimen expands in the direction of the magnetization and negative magnetostriction if it contracts in the direction of magnetization. Hence, this energy varies according to the structure of the domains. Magnetostriction deforms domain configuration so as to minimize magnetostrictive energy. Magnetostriction energy is defined as the work necessary to reassemble domains to their initial states.

MAGNETIC SWITCHING ANALYSIS

Magnetic switching depends strongly on the dynamics, or the time dependence, of the process. The timescales required for magnetic switching cover a wide range. Magnetization reversal processes in some applications are on the order of seconds, while others are on the order of 10 ns or less. Switching speeds are limited by both eddy currents and by the intrinsic or relaxation damping mechanisms. Because of the wide timescales, the process of magnetic reversal can be studied either by seeking information about the final equilibrium position of the magnetization vector at a given point in the material without determining the details of the magnetization

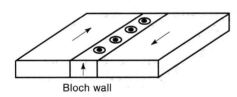

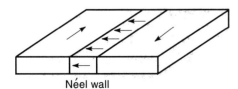

Figure 9. A Bloch wall and a Néel wall in a magnetic thin film (6).

process before equilibrium is attained or by following the position of the magnetization vector as a function of time during the magnetization reversal process. The approach followed depends on the application, but the second approach is more complex. The analysis for both approaches is outlined and discussed.

Thin metallic films represent the principal media for present-day high density magnetic recording. They are prominent examples of dynamic, fast magnetic switching applications. Hence we refer here mostly to thin films. The films of Permalloy ($Ni_{81}Fe_{19}$) are widely used in manufacturing more than one billion recording heads annually.

In the process of magnetic recording, information must first be stored on special magnetic media. Next, this information must be retrieved by reading the information stored on the media. Both the writing and reading are strongly tied to the magnetic switching process and to the relevant properties of the medium. During writing, the magnetization of the medium has to be locally switched from the well-defined orientation corresponding to the negative remanent state to the well-defined positive remanent state. These two magnetization states may be represented by "0" or "1" to store information on the binary system. The dimensions of the switched area are usually on the order of micrometers that have submicrometer separation between magnetization states. For example, the transition spacing between oppositely magnetized regions on present-day hard disk drives is about 0.2 μm, and track periods are on the order of 3 μm to 4 μm. Future applications will have even smaller transition spacings. To decrease noise, the switched area should have sharp boundaries and be stable with time, while allowing complete erasure. All of these parameters define the properties of the recording medium and the writing head. In reference to the switching process, they determine the amplitude of the bias field necessary to reverse the position of the magnetization, or the so-called *switching field,* and the *switching time.* The switching time is the time needed to reverse the magnetization position that uniquely defines the desired state.

During reading, the reading head moves above the recorded signal, and the response to the presence of the 0 or 1 state is monitored. Formerly, recording heads were inductive, and the response of the head corresponded to the voltage induced in a coil coupled to the magnetic structure. More recently, these heads have been replaced by *magnetoresistive* heads, whose response is proportional to the change in electrical resistivity of the head in a magnetic field. This change in resistivity occurs as the head moves over the recorded information. Other reasons for using magnetoresistive heads are not discussed here. What is important is the fact that the resistivity of the head depends on the strength and orientation of the magnetic field produced by the orientation of the magnetization on the recording medium. The first resistive reading heads were based on the effect of anisotropic magnetoresistance (AMR). The resistivity depends on the relative orientation of the current I through the film with respect to the direction of the magnetization M in the film. For Permalloy films deposited on silicon wafers, the resistivity at room temperature with M parallel to I is about 2.5% larger than when M is perpendicular to I.

Giant Magnetoresistance

The possibility for significantly enhancing the sensitivity and dynamic range of recording magnetoresistive heads resulted from the discovery of the giant magnetoresistance (GMR) in coupled iron-chromium layered structures (7,8). GMR is characteristic of electric current behavior in materials consisting of alternating ferromagnetic and nonmagnetic layers of ferromagnetic and nonmagnetic metals deposited on an insulating substrate. The resistance, measured with current flowing parallel to the layers, is greater when the magnetic moments in the alternating ferromagnetic layers are oppositely aligned

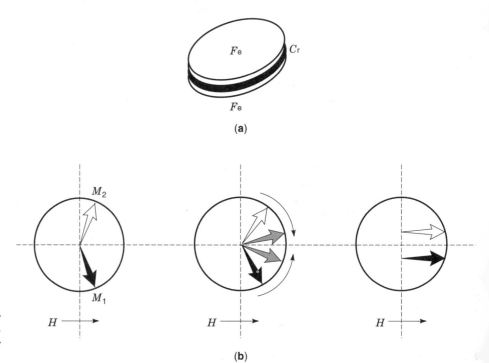

Figure 10. Giant magnetoresistive Fe/Cr layered structures (a) and possible equilibrium positions of magnetic moments for several applied fields $\boldsymbol{H}$ (b).

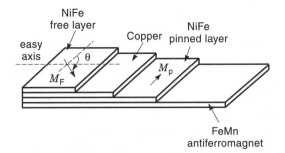

Figure 11. Structure of spin valve showing two uncoupled ferromagnetic films separated by a nonmagnetic metal layer. The bottom film is pinned.

and smallest when they are all parallel. The measured relative change in resistivity can be several hundred percent at low temperatures. The structure is shown for two single-domain layers in Fig. 10. Figure 10 also shows possible equilibrium positions of the layers' magnetic moments for several applied fields. In contrast with AMR, GMR depends only on the relative orientation θ of the magnetic moments of the layers. GMR results from spin-dependent scattering of the conductive electrons at the interfaces and may be expressed as

$$R = R_0 - \frac{\Delta R}{2}(\cos\theta + 1) \qquad (29)$$

where R_0 is the resistivity in zero field ($H = 0$), and ΔR is the maximum difference in resistivity corresponding to the parallel and antiparallel orientation of moments. Even larger effects are observed in a perpendicular geometry where the current flows perpendicular to the layers. GMR has also been observed in heterogeneous alloys and granular multilayers.

Spin Valve. The magnetoresistance of GMR structures increases with the number of alternating layers. This, however, complicates mass production and could decrease manufacturing yield. One question is whether it is possible to find a combination that allows GMR behavior and to make a more simple head structure. One of the solutions that fulfills these criteria is the *spin valve* (9). The schematic structure of the spin valve is shown in Fig. 11. A spin valve consists of just two uncoupled ferromagnetic films (Permalloy or cobalt) separated by a nonmagnetic metal layer (copper, silver, gold) spacer. The magnetization of one layer is pinned through exchange coupling with an underlying antiferromagnet (iron manganese) in a specific direction. Like GMR structures, the resistance of the spin valve is proportional to the angle between the magnetic moments in the magnetic layers. Because the ferromagnetic layers are only weakly coupled, the magnetization of the free layer is easy to rotate relative to the pinned layer, making the spin valve sensitive to low fields.

The previous examples demonstrate that a clear understanding of the magnetization reversal process is requisite to optimize heads for recording applications. In the following sections we outline the essential features of the magnetic switching process. The analysis is performed in two steps. First, only the quasi-static magnetization reversal is considered. Second, the more complex problems of dynamic switching are discussed.

Quasi-static Magnetization Reversal

Single-Domain Particle Behavior. Many models have been employed to explain hysteresis properties of single-domain particles (10). These models may be classified into those in which the particle reverses its magnetization in the single-domain state (coherent rotation) and those in which the magnetization does not remain uniformly parallel during the switch (incoherent rotation). The single-domain particle remains physically fixed during these rotations. Only the magnetization changes direction. The classical model of Stoner–Wohlfarth (11) involves only coherent rotation. Although this model does not completely apply to even single-domain particles, it provides an important understanding of magnetic switching and of more complicated models of the magnetization reversal process. It also permits systematic introduction of the necessary terms describing magnetic switching.

In static equilibrium, the torque on the magnetization exerted by the total magnetic field must vanish at every point in the medium, that is,

$$\boldsymbol{M} \times \boldsymbol{H}_{\text{eff}} = 0 \qquad (30)$$

where the effective magnetic field consists of the external applied magnetic field $\boldsymbol{H}_0$, a magnetic field due to the magnetization of the medium, the magnetostatic or demagnetizing field $\boldsymbol{H}_{\text{D}}$, the magnetic anisotropy field $\boldsymbol{H}_{\text{a}}$, and the magnetic exchange field $\boldsymbol{H}_{\text{ex}}$.

Assume a single-domain particle in the form of a thin film as portrayed in Fig. 12, where the saturation magnetization of the particle is M_{s} and where $\boldsymbol{M}$ is tilted from the z-axis by an angle θ_0 with an applied bias field $\boldsymbol{H}_0$ in the y-direction. If the particle has a preferred direction of magnetization along the z-axis, the minimum energy state is obtained in this *easy* direction. In such a uniaxial material, any deviation of $\boldsymbol{M}$ away from the easy axis by an angle θ leads to an increase of the energy. When $\boldsymbol{M}$ is inclined by an angle θ to the easy axis, the anisotropy field can be simply interpreted as the magnetic field H_{a} appearing perpendicular to the easy axis that is opposed to the component of $\boldsymbol{M}$ in this direction. This may be expressed mathematically by

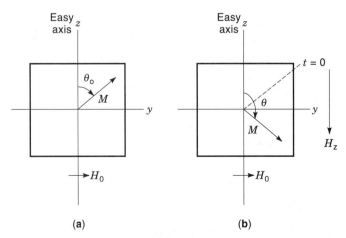

Figure 12. A single-domain particle in the form of a thin film. $\boldsymbol{M}$ is magnetization and $\boldsymbol{H}_0$ is the applied field.

$$H_a = -\frac{1}{M_s} \nabla_\alpha E_a |_{\alpha_{eq}} \qquad (31)$$

where M_s is the saturation magnetization of the material and the gradient of the anisotropy energy E_a is evaluated at the equilibrium position. The z-component of the anisotropy field in a uniaxial material may be written directly from Eq. (31) as

$$H_{az} = \frac{2K_u}{M_s} \cos\theta = H_a \cos\theta \qquad (32)$$

When a switching field H_z is applied (Fig. 12), the static equilibrium condition expressed by Eq. (30) gives the following relationship between θ and the magnetic field H_z:

$$H_z = -H_a(\cos\theta - \sin\theta_0 \cot\theta) \qquad (33)$$

A plot of M_z/M_s versus H_z/H_a in Fig. 13 shows the quasi-static magnetization characteristic of a single-domain particle when pure rotation of the magnetization takes place. Even for this simple case of magnetic reversal by uniform rotation, magnetic hysteresis is evident. In other words, decreasing or increasing the applied field does not necessarily lead to the same magnetic state. The field H_0 in Fig. 13 was chosen arbitrarily so that $\theta_0 = 30°$. Magnetic switching of the particle must occur when $\theta \geq \theta_c$, at which point $dH_z/d\theta = 0$. From Eq. (33)

$$\sin^3\theta_c = \sin\theta_0 \qquad (34)$$

and

$$\cos^3\theta_c = \frac{H_{zc}}{H_a} \qquad (35)$$

A plot of H_{zc} versus H_0 is usually known as the *switching asteroid* that has a minimum critical switching field for $\theta_0 = 45°$. The previous results are also obtained from energy considerations. The free energy density of the single-domain particle in Fig. 12(b) is given by

$$E_0 = K_u \sin^2\theta_0 + M_s H_0 \sin\theta_0 \qquad (36)$$

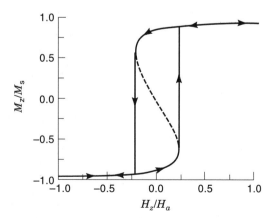

Figure 13. Hysteresis plot of M_z/M_s versus H_z/H_a showing the quasi-static magnetization characteristic of a single-domain particle in uniform rotation.

for the initial state and

$$E = K_u \sin^2\theta + M_s H_0 \sin\theta - M_s H_z \cos\theta \qquad (37)$$

for the final state. The equilibrium condition is obtained at a (local) minimum of the free energy with respect to the angles θ and θ_0, that is, $\partial E/\partial\theta = 0$ and $\partial E/\partial\theta_0 = 0$. The latter equilibrium conditions yield Eq. (33).

In general, for single-domain particles, the lowest energy eigenmodes of magnetic reversal are sought. For ellipsoids of revolution, these are either the *fanning* modes of coherent rotation or the *curling* and *buckling* modes of incoherent rotation (10). The fanning modes are named from the model of individual domains of equal volume spheres forming chains, where the magnetization vector in each sphere rotates in a direction alternate from its neighbor. Recent measurements of the angular dependence of the switching field in single-domain ferromagnetic particles (12), however, show that switching generally differs from the uniform rotational model. Both measurement data and analysis indicate that switching is represented by curling modes for $\theta \sim 0°$. For $\theta > 30°$, switching agrees closely with the coherent rotational model previously described. Incoherent reversal is usually thought to occur as a result of inhomogeneities in the particles, such as grain boundaries, voids, dislocations, cracks, and mechanical strains.

Multiple-Domain Particle Behavior. In multiple-domain particles the reversal process occurs through domain-wall motion that consists of both rotation in unison and "incoherent" rotation. The incoherent rotation depends on the magnitudes of both the switching and biasing fields. The quasi-static magnetization characteristics of reversal in this case are not unlike the hysteresis shown for the single-domain particle in Fig. 13. Ferromagnetic hysteresis has its origin, as illustrated by the previous simple analysis, in the multiplicity of metastable states in the system's free energy. This very general statement is difficult to quantify for predictive use. Just as difficult is the task of following the behavior of each domain and domain wall during the reversal process at each spatial position of the specimen—even for quasi-static treatment. Hence a more viable approach pursued by many researchers is based on phenomenological models in which all essential hysteresis characteristics are obtained as a direct consequence of a few, simple, initial assumptions. These assumptions are usually not concerned with the details of the magnetic state at a given point in the specimen nor are they concerned with the details of the domain walls and structure. Instead, they rely on a set of experimental data, collectively called an *identification process,* to determine a few constants relating a specific model to the specimen that permit predicting magnetic material behavior as a function of the applied magnetic field in a given direction.

Among the most widely known models are the Preisach (13) and the Williams–Comstock (14). The Preisach model was later extended by DelaTorre (15,16), Mayergoyz (17), and Jiles and Atherton (18). In the Preisach model, hysteresis is described by a collection of a large number of elementary reversible λ_α and irreversible $\gamma_{\alpha\beta}$ bistable units (see Fig. 14), each characterized by the values α and β, $\alpha > \beta$, at which the external field H_0 makes the unit switch up or down. The bistable unit $\gamma_{\alpha\beta}$ can be considered a quasi-domain that switches

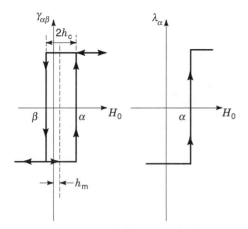

Figure 14. Preisach model of hysteresis described by a collection of elementary reversible λ_α and irreversible $\gamma_{\alpha\beta}$ bistable units.

with increasing applied field from the state $-M_s$ to the state $+M_s$ at the field α. With a decreasing applied field, the bistable unit reverses from the $+M_s$ to the $-M_s$ state at the field value β. For such a quasi-domain a local coercivity (switching field) $h_c = (\alpha - \beta)/2$ and an effective field from the neighboring quasi-domains $h_m = (\alpha + \beta)/2$ correspond. For $\beta < H_0 < \alpha$, $\gamma_{\alpha\beta}$ depends on the previous state of the specimen. Then a system of multidomain particles is described by the switching distributions $p(\alpha, \beta)$ and $\nu(\alpha)$, and the magnetization is expressed as

$$M(t) = \int_\alpha \int_\beta p[\alpha, \beta, H_0(t)]\gamma_{\alpha\beta}[H_0(t)]\,d\alpha\,d\beta$$
$$+ \int_{-\infty}^\infty v(\alpha)\lambda_\alpha[H_0(t)]\,d\alpha \tag{38}$$

Although Eq. (38) contains time-dependent variables, nonetheless the model is static and does not take into account the speed at which magnetization changes. The time-dependent magnetization corresponds to a step-by-step field change, but the output is still a sequence of equilibrium states that depend only on the value of the field H_0. Mayergoyz (17) proved that a system with hysteresis is equivalent to the Preisach model if it obeys the "wiping out or return-point memory" and "congruency or geometrical equality" properties. The first condition means that each local maximum H_{0m} field value wipes out the influence of the previous local maxima. The congruency property that defines the geometric equivalence of all minor loops delimited by the same peak field values has not been verified for real systems. Experimentally observed minor hysteresis loops become smaller when remanence is approached. However, this defect in the second condition is eliminated if a feedback correction to H_0 is added which produces an internal effective field H_i:

$$H_i = H_0 + CM \tag{39}$$

As the constant C becomes larger, the hysteresis loop becomes more rectangular or the squareness ratio defined by M_r/M_s in Fig. 7 becomes closer to one. The squareness ratio depends on the ratio of the average magnetostriction constant to the first-order anisotropy constant. A smaller ratio of the magnetostriction constant to the first-order anisotropy constant yields more hysteresis-loop squareness. The completely reversible magnetization process is given by the second term in Eq. (38).

Dynamic Magnetization Reversal

The quasi-static models describing magnetic reversal do not give the system's time response straightforwardly, especially for fast switching processes. A different approach is necessary to calculate the fast magnetization switching response at an arbitrary time and at each point of the recording medium. All forces acting on the magnetization at any point in space on the recording medium and at any time must be used to characterize the complete magnetization reversal. In general, it is assumed that the direction cosines and the magnitude of the magnetization vector vary continuously with position. Depending on the actual problem, changes may vary on a scale that is smaller or comparable to the domain size. This type of treatment forms the foundation of the micromagnetic modeling approach advocated by Brown (2).

Here a procedure based on the Landau–Lifshitz equation is followed (19). Landau and Lifshitz first introduced the phenomenological representation of the interactive forces with an effective internal field $\boldsymbol{H}_{\text{eff}}$. The action of this field is equivalent to the applied field but also includes other interactions in the ferromagnetic material. Generally, the magnetization becomes parallel to the effective field. Therefore Landau and Lifshitz next proposed an ad hoc damping term into the magnetization equation of motion which has the direction of $\boldsymbol{H}_{\text{eff}} - \boldsymbol{M}$ and an amplitude that approaches 0 when $\boldsymbol{H}_{\text{eff}}$ and $\boldsymbol{M}$ become parallel. The Landau–Lifshitz equation, which conserves total magnetization, can be written as

$$\frac{d\boldsymbol{M}}{dt} = -|\gamma|(\boldsymbol{M} \times \boldsymbol{H}_{\text{eff}}) + \boldsymbol{R} \tag{40}$$

where the specific forms of the damping term $\boldsymbol{R}$ most used are

$$\boldsymbol{R} = |\gamma|\lambda\left[\boldsymbol{H}_{\text{eff}} - (\boldsymbol{H}_{\text{eff}} \cdot \boldsymbol{M})\frac{\boldsymbol{M}}{M_s^2}\right] \tag{41}$$

$$\boldsymbol{R} = -\frac{\alpha|\gamma|}{M_s}(\boldsymbol{M} \times \boldsymbol{M} \times \boldsymbol{H}_{\text{eff}}) \tag{42}$$

and

$$\boldsymbol{R} = \frac{\alpha}{M_s}\left(\boldsymbol{M} \times \frac{d\boldsymbol{M}}{dt}\right) \tag{43}$$

Equation (43) is known as the Gilbert damping form. λ is the damping parameter that has the dimension of magnetization. The dimensionless Gilbert damping parameter $\alpha = \lambda/M_s$.

Now the effective field $\boldsymbol{H}_{\text{eff}}$, which represents all interactions, is introduced. If the free energy of the system is evaluated by the variational principle, the amplitude and direction of the internal field can be determined with good approximation. However, the resulting form is too general, and it is more practical to express the effective field as the power series:

$$\boldsymbol{H}_{\text{eff}}(\boldsymbol{M}) = -\frac{\partial E}{\partial \boldsymbol{M}} + \sum_i^3 \frac{\partial}{\partial x_i}\left[\frac{\partial E}{\partial(\partial \boldsymbol{M}/\partial x_i)}\right] \tag{44}$$
$$= \boldsymbol{H}_0(t) + \boldsymbol{H}_a(\boldsymbol{M}) + \boldsymbol{H}_D(\boldsymbol{M}) + \boldsymbol{H}_{\text{ex}}(\boldsymbol{M})$$

The $\boldsymbol{H}_0$ and $\boldsymbol{H}_a$ terms of Eq. (44) are the time-dependent applied field and the anisotropy field. The term $\boldsymbol{H}_D(\boldsymbol{M})$ represents the demagnetizing field of the specimen [see also Eqs. (15, 16)], which is solved by Maxwell's equations for the particular specimen shape of interest, that is,

$$\nabla \cdot [\boldsymbol{H}_D(\boldsymbol{r},\,t) + \boldsymbol{M}(\boldsymbol{r},\,t)] = 0$$
$$\nabla \times \boldsymbol{H}_D(\boldsymbol{r},\,t) = 0 \qquad (45)$$

For the quasi-static case, this set of equations leads to the Poisson equation, and, for a specimen that has the shape of an ellipsoid of revolution, the solution for $\boldsymbol{H}_D$ can be expressed in terms of the demagnetizing tensor $[N]$. If the specimen is magnetized along one of its principal axes, the tensor is diagonal with the individual components N_x, N_y, and N_z satisfying the condition $\sum_i^3 N_i = 1$. The ensuing solution for $\boldsymbol{H}_D$ is given by

$$\boldsymbol{H}_D(\boldsymbol{r}) = \frac{1}{4\pi} \nabla \left(\int_V \frac{\nabla \cdot \boldsymbol{M}}{r}\, dV + \int_S \frac{\boldsymbol{M} \cdot \boldsymbol{n}}{r}\, dS \right) \qquad (46)$$

The first integral is taken over the entire volume of the magnetized specimen, the second integral over the enclosing surface of the magnetized specimen, and the unit normal vector $\boldsymbol{n}$ to the surface is directed inward.

The last term in Eq. (44) is the exchange field which quantifies the quantum mechanical exchange interactions between neighboring spins:

$$\boldsymbol{H}_{ex}(\boldsymbol{M}) = \frac{G}{M_s} \nabla^2 \boldsymbol{M}(\boldsymbol{r},\,t) \qquad (47)$$

where $G = 4JS^2/(aM_s) = 2A/M_s$ is the exchange parameter, J is the exchange energy integral, S is the spin quantum number, a the lattice constant of the magnetic unit cell, and $A = 2JS^2/a$ ranges between 10^{-11} and 10^{-12} J/m.

One example that gives insight into the dynamic switching process is the uniform rotation of magnetization in a thin film. The procedure is outlined by Soohoo (20). For coherent rotation in a single-domain particle, magnetization does not depend on position in the specimen, so that the analysis is considerably simplified. However, the magnetization is a function of both azimuthal and polar angles (see Fig. 15). The effective field in Eq. (44) may be expressed in terms of the free energy and its derivatives. If the Gilbert modification of the Landau–Lifshitz equation is expressed in spherical coordinates,

$$M_s \frac{d\theta}{dt} = \frac{-|\gamma|}{\sin\theta} \frac{\partial E}{\partial\phi} - \alpha M_s \sin\theta \frac{d\phi}{dt}$$

and

$$M_s \sin\theta \frac{d\phi}{dt} = |\gamma| \frac{\partial E}{\partial\theta} + \alpha M_s \frac{d\theta}{dt} \qquad (48)$$

Now the dynamic behavior of the magnetization is obtained by numerically solving the two coupled differential Eqs. (48) as a function of θ and ϕ. Standard numerical techniques, such as the Euler and Runge–Kutta-type methods or predictor-corrector methods (21) may be employed. For sufficiently small switching speeds, $\partial\phi/dt \sim 0$ and the switching time $(\Delta t)_s$ can be written from the second of Eqs. (48) as

$$(\Delta t)_s = -\frac{\alpha M_s}{|\gamma|} \int_{\theta_i}^{\theta_f} \frac{d\theta}{\partial E/\partial\theta} \qquad (49)$$

where θ_i and θ_f are the initial and final orientations of the magnetization.

Two significant differences are apparent when comparing quasi-static reversal analysis to dynamic analysis. The first difference is that the switching time depends on the damping constant of the system and on the time-dependent torque acting on the magnetization during its entire reversal. The second difference is that even in the simplest case of single-domain uniform rotation, the motion of magnetization is more complicated and *is not confined to the plane of the film.* Because the torque $\boldsymbol{M} \times \boldsymbol{H}_{eff}$ for uniform rotation of a single-domain is directed out of the plane of the film, in the process of rotation the magnetization $\boldsymbol{M}$ tips slightly out of the plane of the film. This gives rise to an x-component ΔM_x and an accompanying demagnetizing field $H_{Dx} \simeq N_x\Delta M_x$. The demagnetizing field, in turn, exerts a torque on the magnetization in the plane of the film so as to rotate the magnetization toward the negative z-axis. The previous analysis is simplified and does not predict switching times observed experimentally. However, it gives proper insight into the general switching process. The discrepancy between calculated switching times from Eq. (49) and experimentally observed times is attributed to an incorrect assumption of a position-independent magnetization. In fact, thin films have initial magnetization ripples, and the magnetostatic field originating from $\nabla \cdot \boldsymbol{M}$ of the ripple distribution plays a significant part in the reversal process (22–24).

In this article we investigated the simple case of the dynamic behavior of a single-domain particle in a thin ferromagnetic film. Recent technological applications involve magnetic systems of much greater complexity. These systems possess complicated geometries and have nonuniform demagnetizing fields within multidomain particles, which leads to highly nonuniform distributions of the magnetization within the particles. The Landau–Lifshitz equation expressing damping is used almost exclusively to calculate the dynamics of polycrystalline, thin-film recording media and thin metallic films. It is also used to calculate the dynamic micromagnetics of submicrometer ferromagnetic particles.

The analysis of more complex systems can be the same as that outlined. In principle, the Landau–Lifshitz equation can

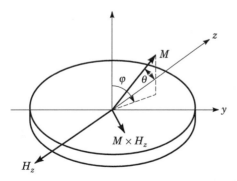

Figure 15. Rotation of magnetization as a function of azimuthal and polar angles in a thin film, where z represents the easy axis and the vertical line is x.

be solved for a continuous magnetic medium in three-dimensional regions of arbitrary geometry and arbitrary properties for nanosecond time-dependent processes. All interactions, based on calculating the effective field from free energy system considerations, are included in the micromagnetic approach. The magnetic specimen can be discretized and the magnetization as well as the acting effective fields can be taken as position-dependent (spatially inhomogeneous). Then appropriate boundary conditions may be applied. Computational times for the analysis depend on the number of cells used in the model. For some cases, computational time is reduced by assuming that the magnetization in each cell is uniform and that the switching process in the cell is described by pure rotation. Numerical simulations of the dynamic process reveal that the magnetization switching process is divided into three stages that are not significantly different from the micromagnetic calculations outlined previously or from experimental data: (1) precursor nucleation process (critical field or coercive field for magnetization reversal); (2) magnetization curling (incoherent rotation); and (3) actual switching and damping.

During the first stage, the motion of magnetization is almost stationary, and precession occurs along the equilibrium direction. If the switching field is less than critical, the magnetization settles in the equilibrium position without switching. In the curling stage, the magnetization along the applied magnetic field direction increases exponentially. For an ellipsoid of revolution, the curling starts from the center of the ellipsoid and extends from this center with time. Parts of the specimen switch first. This results in the appearance of domain walls and subsequent events that depend on both the damping and switching field, such as the generation of vortices (25–28). Additional complexity can be introduced in the model by taking into account possible random orientation of the crystalline anisotropic axes and interactions between neighboring cells. The same analytic procedures apply.

BIBLIOGRAPHY

1. P. Weiss, L'hypothese du champ moleculaire et la propriete ferromagnetique, *J. Phys. Theor. Appl.* 4E, **series 6**: 661, 1907.

2. W. F. Brown, Jr., *Micromagnetics,* New York: Interscience Publishers, 1963.

3. R. F. Soohoo, *Magnetic Thin Films,* New York: Harper and Row, 1965.

4. N. S. Akulov, Magnetostriction in iron crystals, *Z. Physik,* **52**: 389–405, 1928.

5. N. S. Akulov, *Ferromagnetism,* Moscow and Leningrad: ONTI, 1939.

6. C. Kittel, *Introduction to Solid State Physics,* New York: Wiley, 1986.

7. M. N. Baibich et al., Giant magnetoresistance of (001)Fe/(001)Cr magnetic superlattices, *Phys. Rev. Lett.,* **61**: 2472–2475, 1988.

8. P. Grünberg et al., Layered magnetic structures: Evidence for antiferromagnetic coupling of Fe layers across Cr interlayers, *Phys. Rev. Lett.,* **57**: 2442–2445, 1986.

9. B. Dieny et al., Magnetotransport properties of magnetically soft spin-valve structures, *J. Appl. Phys.,* **69**: 4774–4779, 1991.

10. J. C. Mallinson, *The Foundations of Magnetic Recording,* New York: Academic Press, 1993.

11. E. C. Stoner and E. P. Wohlfarth, A mechanism of magnetic hysteresis in heterogeneous alloys, *Phil. Trans. R. Soc. London A,* **240**: 599–642, 1948.

12. M. Lederman, S. Schultz, and M. Ozaki, Measurement of the dynamics of the magnetization reversal in individual single-domain ferromagnetic particles, *Phys. Rev. Lett.,* **73**: 1986–1989, 1994.

13. F. Preisach, Über magnetische Nachwirkung, *Z. Physik,* **94**: 277–302, 1935.

14. M. L. Williams and R. L. Comstock, An analytical model of the write process in digital magnetic recording, *17th Annu. AIP Conf. Proc.,* Chicago, IL, 1971, pp. 738–742.

15. E. Della Torre, Effect of interaction on the magnetization of single domain particles, *IEEE Trans. Audio Electroacoust.,* **14**: 86–93, 1966.

16. F. Vajda, E. Della Torre, and M. Pardavi-Horvath, Analysis of reversible magnetization-dependent Preisach models for recording media, *J. Magn. Magn. Materials,* **115**: 187–189, 1992.

17. I. D. Mayergoyz, Mathematical models of hysteresis, *Phys. Rev. Lett.,* **56**: 1518–1521, 1986.

18. D. C. Jiles and D. L. Atherton, Theory of ferromagnetic hysteresis, *J. Magn. Magn. Matter,* **61**: 48–60, 1986.

19. L. Landau and E. Lifhshitz, On the theory of the dispersion of magnetic permeability in ferromagnetic bodies, *Phys. Z. Sowjetunion,* **8** (2): 153–169, 1935.

20. R. F. Soohoo, *Theory and Application of Ferrites,* Englewood Cliffs, NJ: Prentice-Hall, 1960.

21. G. A. Korn and T. M. Korn, *Mathematical Handbook for Scientists and Engineers,* New York: McGraw-Hill, 1968.

22. M. H. Kryder and F. B. Humphrey, Dynamic Kerr observations of high-speed flux reversal and relaxation processes in permalloy thin films, *J. Appl. Phys.,* **40**: 2469–2474, 1969.

23. M. H. Kryder and F. B. Humphrey, Mechanisms of reversal with bias fields, deduced from dynamic magnetization configuration photographs of thin films, *J. Appl. Phys.,* **41**: 1130–1138, 1970.

24. H. Hoffmann and M. H. Kryder, Blocking and locking during rotational magnetization reversal in ferromagnetic thin films, *IEEE Trans. Magn.,* **9**: 554–558, 1973.

25. B. Yang and D. R. Fredkin, Dynamical magnetics of a ferromagnetic particle: Numerical studies, *J. Appl. Phys.,* **79**: 5755–5757, 1996.

26. Q. Peng and H. N. Bertram, Micromagnetic studies of switching speed in longitudinal and perpendicular polycrystalline thin film recording media, *J. Appl. Phys.,* **81**: 4384–4386, 1997.

27. J.-G. Zhu, Y. Zheng, and X. Lin, Micromagnetics of small size patterned exchange biased Permalloy film elements, *J. Appl. Phys.,* **81**: 4336–4341, 1997.

28. Y. Zheng and J.-G. Zhu, Switching field variation in patterned submicron magnetic film elements, *J. Appl. Phys.,* **81**: 5471–5473, 1997.

RICHARD G. GEYER
National Institute of Standards and Technology

PAVEL KABOS
Colorado State University

MAGNETIC TAPE EQUIPMENT

Since the Allies' discovery in 1944 of the "magnetophones" created in Germany during World War II (1), the technology of magnetic recording has evolved to the point that in the developed world there are few people without access to a mag-

Table 1. Some Magnetic Tape Equipment Formats and Applications

Name	Application	Recording Method	Tape Width	Digital Analog
VHS	Consumer video	H	0.5 in.	A
Beta	Consumer video	H	0.5 in.	A
Digital beta	ProVideo	H	0.5 in.	D
R DAT	ProAudio	H	4 mm	D
8 mm	Consumer and ProVideo	H	8 mm	A
8 mm	Data	H	8 mm	D
D2	ProVideo and data	H	19 mm	D
D3	ProVideo and data	H	0.5 in.	D
DLT	Data	L	0.5 in.	D
3480/90	Data	L	0.5 in.	D
QIC	Data	L	0.25 in.	D
QIC wide	Data	L	0.315 in.	D
4 mm	Data and audio	H	3.8 mm	D
Type C	ProVideo	H	1 in.	A
DCRSi	Data	T	1 in.	D
DST	Data	H	19 mm	D
DCT	ProVideo	H	19 mm	D
BetaSP	ProVideo	H	0.5 in.	D
Quad	ProVideo	T	2 in.	A
U Matic	ProVideo	H	0.75 in.	A
DDS	Data	H	4 mm	D
AIT	Data	H	8 mm	D
LTO/U	Data	L	0.5 in.	D
LTO/A	Data	L	8 mm	D
Travan	Data	L	0.315 in.	D
3590	Data	L	0.5 in.	D

L, Linear; H, helical; T, transverse; D, digital; A, analog.

netic tape recording/reproducing system. The prevalence of the consumer video cassette recorder (VCR) is the most obvious example, but the audiocassette recorder, whether stand-alone or in a vehicle, is equally ubiquitous. In addition, the magnetic tape drive is also the basis of the consumer video camera, the "Camcorder," that is now found in many homes. There are hosts of competing professional video, audio, and data recording systems used as video and data source devices, music recorders, and content editing systems. In business and industry, the use of magnetic tape drives as computer data repositories is also very common. These include the local backup device to the home computer, the server/network backup system, the instrumentation data recorder, and the data archive or warehouse prevalent in many corporate environments.

Table 1 lists the range of the magnetic recording tape equipment industry with a rough classification as to application and utility.

This article will first explain some of the basic principles of magnetic recording on tape, followed by a description of the three fundamental methodologies of helical recording, linear recording, and the older but still used transverse recording.

The formats and tape handling techniques that underlie these three methods will be illustrated by descriptions of various, but not by any means all, of the devices in current use.

BASIC ELEMENTS OF MAGNETIC TAPE DRIVES

Device Fundamentals

Figure 1 shows, at the most conceptual level, the basic elements of a magnetic tape drive. The components that are present in almost all modern drives include the following:

1. *A tape medium*. This varies in width generally from 4 mm to 2 in. and is used in a wide variety of thicknesses. Tape technology is addressed elsewhere in this encyclopedia; extensive technical reviews are available (2).

2. *Magnetic recording and reproducing heads*. The density of the recorded information is largely set by the transfer function of the record/read stylus as well as the

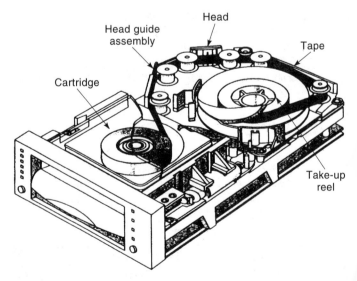

Figure 1. Basic elements of a magnetic tape drive. The example shown is a linear tape drive (DLT) where the drive has the take-up spool and the cartridge has the suppy spool.

properties of the medium. The critical parameters include the frequency response function, the physical wear properties, the record gap dimensions, and the resulting signal-to-noise ratio (SNR) achieved by the head–tape interface. Details of recording heads are given elsewhere in this encyclopedia and elsewhere (3).

3. *Ancillary magnetic heads.* These provide additional functions depending on the application, timing control, servo control marks, bulk or local erase functions and so on.

4. *Methods to hold the tape and heads in intimate contact while moving relative to each other.* Contoured tape head assemblies and/or heads incorporated into rotating drums are used. The air foil attributes of the interface is critical (4) as are the resulting wear properties on the drum and tape medium (5). The head drum assemblies can rotate at rates above 5000 rpm.

5. *A control method for the tape tension, as the drive is used in read/write and often fast seek modes.* Tension arms and or pinch rollers or capstans have been used to control tape movement as well as through the use of push–pull motor control.

6. *Drive motors to move the tape over the heads or, on occasion, lift the tape from the heads.* Speeds vary from a few inches per second to several hundred inches per second. Also used to control tape tension.

7. *Guides to control the tape position relative to the heads and spools.* Fixed guides, roller guides, and air-flow lubricated guides are used.

8. *Tape spools;* sometimes provided in the cartridge and sometimes split between cartridge and drive. Critical issues involve the ability of the tape-guiding and tension control system to ensure that the tape spools up uniformly and with good parallelism with the spool flanges so that tape damage is avoided.

9. *An automated loading and ejection system that accepts the cartridge and positions it appropriately over the drive spindles.* Associated with this is a mechanism that pulls the tape out and locates it over the drum or threads it ready for alignment.

10. *Sensing systems*—for example, to ensure that the cartridge is inserted correctly, to lock out unallowed write functions, to sense potential tape damage events, to detect potential harmful environments, and to feed back to servo systems as below.

11. *Servo systems*—for example, to drive the magnetic heads to follow the data tracks, to control tape tension, to control tape speed, to control the various timing elements in a drive, and so on.

12. *A preamplifier circuit to improve the signal picked up by the read head.* Critical is the frequency response and gain of the preamp together with its noise characteristics.

13. *A detector to detect the amplified signal.* Traditional peak detect systems have been used; but more recently, methods that permit more sophisticated analysis of the preshaped read pulse are in use. These are, in particular, maximum-likelihood type detectors as described below (6).

14. *A coding method, together with encoding and decoding circuitry.* This translates the detected signal into data bits that are conveniently and efficiently processed by the data channel. Code selection is very important and is discussed below (7).

15. *A read/write channel.* Circuitry that takes the processed signal and applies it to the write head, and/or accepts the read signal from the head in read mode.

16. *A data channel.* This is the circuitry and software that does the coding, processes the coded signal to provide error detection and correction functions, and provides data in a form that can be understood and interpreted by the interface and controller.

17. *Cache memory or buffer.* Embedded memory that can be used for internal drive functions, generally for smoothing data flow, and so on.

18. *Controller.* Usually a microprocessor that provides the logic controls for the above electronic functions. Associated with this is the firmware, or otherwise nonaccessible code that provides the housekeeping functions intrinsic to the drive.

19. *Interface.* The standardized connection to the external world that allows drive integration through universally understood links. Both the read and write data enter and exit through the interface, as do standard control functions.

Several of these basic elements are similar to those used in other data storage or data processing systems. The data channel and error correction systems have a strong resemblance to those used in telecommunication applications for example. Other elements are discussed in other parts of this text. In the following sections the descriptions of existing tape drive technology will illustrate some of these aspects. Other topics are treated separately. General references for tape recorder technology are given in Refs. 8 through 10.

Applications

The applications that tape drives have been put to can also be used as a method to classify them. The following is not intended to be fully comprehensive but does illustrate the major classes.

Analog Audio Recording. This technology uses classical alternating-current (ac) bias recording methods (11), but the equipment can vary from the standard "audio cassette deck" with a two-track format and a tape speed of 48 mm/s, to a 32-track 2-in.-wide tape format used by professional audio engineers. In this recording method the tape is moved over the three ring heads—erase, read, and record. The read signal is amplified, integrated, and equalized. The equalization corrects for the falloff in response at the higher frequencies. An ac bias of about 100 kHz is used to linearize the otherwise highly nonlinear recording process.

Digital Audio Recording. Rotating head audio recording systems that are based on conventional digital-bit-based recording have emerged (R-DAT) that complement the CD (nontape) music distribution technology. Very frequently the master recordings are done on 2 in. ac-bias recorders and then converted to digital data sometimes using multiple channels to achieve the necessary data rate of about 1 Mbit/s. The R-DAT format uses azimuth recording and a track following servo. Digital PCM encoding methods are also in use as well as FM pulse code modulation. Stationary head digital recording (DASH) formats are also used with lower tape speeds but more limited signal-to-noise ratios. A half inch tape drive typically has 24 tracks, uses ferrite heads, and operates at 30 in./s (12).

FM Video Recording. The basic FM recording principle can be simply illustrated by Fig. 2, where the video information is recorded by frequency modulation (8).

The large bandwidth requirement of video information placed significant burdens on the development of video tape recorders that were only solved with the advent of the transverse scanning four-headed "Quadruplex" recorder developed by Ampex in 1956 (13). This was later replaced by the angled scan (helical) system that used a rotating drum with the long tracks written across 1 in. tape by the use of the acute record angle. In both of these cases the drum speed allows very high effective tape-head speeds, 1600 in./s for example, while the actual tape speed is a stable, controllable, 5 in./s. This, among other factors, opens up the available bandwidth, although other issues such as time base stability are still significant. The SNR of analog video is proportional to the relative head to tape speed and to the square root of the track width. The bandwidth of the record process is between 1 MHz and 15 MHz, with the white carrier level at 10 MHz and the black level at 8 MHz. Cameras for video recording are based on very similar principles and in consumer products generally use the same formats as the playback devices. (VHS, 8 mm, Compact VHS, etc). In the professional camera arena, Super VHS is used and the Beta format is very successful (Beta SP). All these cameras are based on the use of helical scan technology, which is detailed below.

Digital Video Recording (14). The essence of digital video recording is in the coding method. Common is pulse code modulation where the differences in the zero crossing points in the signal shown in Fig. 2 is sampled, quantized into discrete bits made up of integer intervals, and then coded into a binary format. In order to achieve the total information content in a typical TV picture, a bit rate approaching 90 Mbit/s is

needed; when ECC bits and so on, are added, the bit rate is closer to 100 Mbit/s. Very-high-density recording is needed for this process. This, while readily achieved in the professional arena (the D2 Digital Video recorder for example), remains a challenge for consumer applications. The recording bandwidth for an 8-bit/sample quantizing system, and a sampling frequency of three times the color subcarrier, is 0.1 MHz to 86 MHz. The incompatibility between composite-signal-based NTSC, PAL, and SECAM broadcast standards is carried forward into the digital video arena, but the development of a component-based coding scheme now allows for compatibility. In this method the luminance signal and the two color difference signals are separately coded. With 8-bit quantization a total data rate of 216 Mbit/s is needed, and this is now available. Digital cameras have also been developed for the professional market with the Beta format predominating.

Digital Data Recording. Digital data are recorded on magnetic media using coded binary formats and mapped onto the medium in a "non-return-to-zero" (NRZ) form where the maximum and minimum number of zeros allowed between ones is constrained. These run-length-limited (RLL) codes are very similar to the code methods used by communication systems and magnetic and optical discs. These codes allow for effective high-density recording with simple error correction methods applicable. Digital data magnetic storage is characterized by the need for very low error rates after correction, 10^{-17} for example, and the hosts of competing technologies all have this in common. Originally based on an open-reel nine-track form, this application is now largely based on cartridge and cassette tape systems with linear recording, helical scan recording, and even transverse recording devices used. This is discussed below in more detail.

Instrumentation Data Recording. Originally developed from the technology of the audio recorder, this method of capturing long sequences of generally unstructured data, from telemetry or instrumentation data streams, now often uses pulse code modulation (PCM) methods. However, a base band linear method with ac bias is also in use utilizing a constant noise power ratio (NPR) method. Tape speeds can be very high, and very high densities can be achieved (3000 flux reversals/mm or fr/mm, for example). These recorders also incorporate error correction and detection methods.

FUNDAMENTALS OF MAGNETIC RECORDING ON TAPE

Recording Physics

Figure 3 shows a ring-type magnetic recording head and medium (10). The inductive ring head consists of a core of soft magnetic material (low coercive field, small magnetic remanence) with a coil of wire wrapped around it. For an inductive write and read head the coil of wire conducts a current around the soft magnetic core during write to generate a magnetomotive force in the head core. This magnetomotive force generates a magnetic field in the core which circulates through the core until it reaches a nonmagnetic gap in the core adjacent to the hard magnetic (high coercive field, large magnetic remanence) recording medium. The magnetic field fringes out across the gap with some of the field directed toward the hard magnetic recording medium layer.

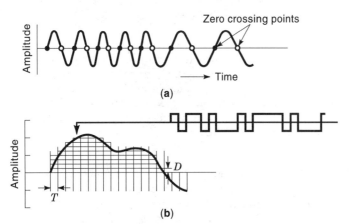

Figure 2. (a) The frequency modulation (FM) recording method and (b) the associated pulse code modulation (PCM) technique.

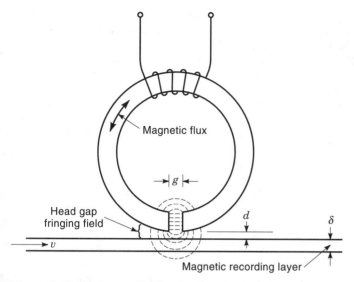

Figure 3. Magnetic recording head and media configuration. g is the gap size, d is the head-to-medium spacing, δ is the recording layer thickness, and v is the surface velocity.

When the field from the head penetrates the hard magnetic medium surface, it moves the magnetic particles in the medium through their magnetic hysteresis loop. The final field which the medium particles experience from the head field as the head moves across the medium determines the direction of remanent magnetization in the medium. As the medium (e.g., a magnetic tape) moves under the gap of the head, the direction of current flow through the wire coil wrapped around the magnetic head core changes direction as directed by the changing voltages from the preamplifier electronics driving the head coil. This changing current direction leaves the magnetic recording medium with a known remanent magnetization pattern as shown in Fig. 4. In this way, information is written into the magnetic recording medium which can be accessed during the readback process (10).

As the current in the coil around the write head core changes, the direction of the magnetization changes from one direction to the other. The recording mode we have been describing is called longitudinal recording in which the medium is magnetized in the plane of the medium along the direction of medium motion.

The width of this transition region between the two remanent magnetization region in the magnetic recording region is called the transition width (usually described by a transition width parameter, a). This transition width is determined primarily during the write process and only changes a few percent due to demagnetization reduction effects during the playback process. The primary determinants of the transition width are the magnetic properties of the magnetic recording media (in particular the magnetic coercivity and the slope or magnetic squareness of the magnetic hysteresis curve near the coercive field value) and the sharpness of the field from the magnetic recording head in the medium. The sharpness of the recording head field is a function of the geometry of the head (especially the head gap length, g) and the distance between the recording head and the magnetic recording medium.

Once the information is written in the magnetic recording media, it can be accessed using magnetic playback processes. Various sorts of magnetic playback transducers can be used in a magnetic recording system from an inductive sensor to a magnetoresistive (MR) or giant magnetoresistive sensor to a Hall or similar semiconductor magnetic playback sensor. These playback transducers experience loss mechanisms which can lower the signal level recovered from the medium and hence increase the susceptibility to errors in signal recovery due to recording system noise. Figure 4 shows readback signals which would be recovered from the series of written magnetic transitions (10).

An inductive playback head may use the same head core as used for the write process or may use a separate magnetic core with gap lengths for the write and read head optimized for each process. Although this increases the cost and complexity of the head, it can allow higher recording densities and a more robust signal recovery system. Such a multigap record/playback system is common in tape recording systems.

The heads used in tape recording may have several adjacent magnetic cores, each with its own magnetic write and read transducers. Each of these adjacent cores is referred to as a head track, and the head is called a multitrack head. Such a multitrack head allows several adjacent tracks of information to be written on and read off the recording medium in parallel, thus increasing the speed with which the recorded information is accessed.

If we assume in playback a sinusoidal recording pattern on the magnetic recording medium described by

$$M_x(x) = M_0 \sin(kx)$$

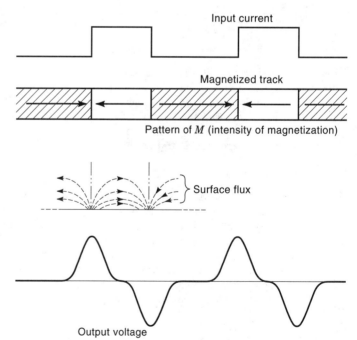

Figure 4. Pattern of magnetization in the medium recorded by the changing direction of input current through the record head. On the readback the surface flux from the medium is picked up by the read head, which generates voltage pulses for amplification.

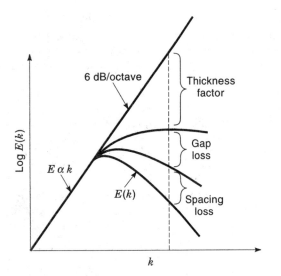

Figure 5. Significance of various reproduce loss terms as a function of wavenumber, k.

and if the playback transducer is an inductive head whose head field (which is the write head field and the read head sensitivity due to reciprocity theory), we get an inductive playback voltage generated by the flux from the magnetic transition picked up by the head core and passing through the wire coil wrapped around the head core. If we assume that the head fields are described by the Karlqvist field equations (15), this output voltage is described by

$$E(x) = \frac{\mu_0 vwM_0 H_g gk\delta}{NI} \exp(-kd)$$
$$\times \left(\frac{1 - \exp(-k\delta)}{k\delta} \right) \left[\frac{\sin\left(\frac{kg}{2}\right)}{\frac{kg}{2}} \right] \cos(kx)$$

This equation shows several terms that lower the signal output levels. Figure 5 shows that the most significant of these signal amplitude loss terms is the spacing loss (which is an exponential loss), followed in order of effect on the playback by the medium thickness loss (which may be significant for particulate coated tape media) and the gap loss (10).

Recording System Design

The basic magnetic recording system can be represented by the block diagram shown in Fig. 6.

The writing process is driven by a current driver in the preamplifier circuit. The input signal to the preamp is often preconditioned to account for known nonlinearities in the magnetic recording process. An example of this preconditioning is precompensation in which the time interval between adjacent transitions is modified to correct for expected

shifts in the transition location due to fields generated by adjacent transitions. The write process including input signal conditioning is shown in Fig. 7(a) (16).

In the readback process the voltage output from the head is fed into a amplifier which magnifies the signal and system noise up to that point in the circuit (including head, media, and circuit noise). The amplified signal can then be filtered to remove any systematic interference and equalized in a manner to make the signal easier to detect in the channel circuitry. Figure 7(b) shows the typical readback process prior to the channel detection.

Information is recorded onto the tape according to special codes which match the recorded transition characteristics to the data channel used for reading back the recorded information. Table 2 shows some of the modulation codes used in magnetic recording (7).

Encoded information is read out from the head passing over the tape. The signal enters the recording channel where it is detected and corrected (using error correction techniques) if needed. One of the most popular modern types of recording channels is the partial response class 4 (PR4) channel and its derivatives EPR4 and E²PR4. This channel requires shaping the pulses coming out of the playback head so they match a perfectly shaped pulse and can detect recorded information at very high densities if there are no nonlinear interference phenomena such as nonlinear superposition of pulses or nonlinear transition shift. This shaping technique is called equalization. Figure 8 shows a PR4 channel with a maximum like-

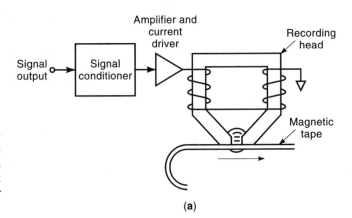

(a)

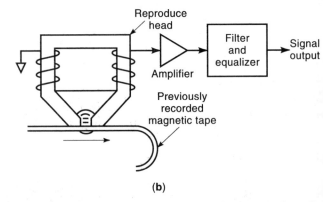

(b)

Figure 7. (a) Tape recording write process including signal conditioning and the current driver. (b) Tape recording read processes including amplifier and signal conditioning prior to the playback channel.

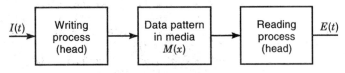

Figure 6. Block diagram of a basic magnetic recording system.

Table 2. Modulation Codes Used in Magnetic Recording

Code	RLL d,k	Rate	Range of Pulse Widths in Waveform[a]	PW Ratio (max/min)	Density Ratio	Detection Window	Maximum dc Charge	Usage
NRZI	0,∞	1		∞	1	1	∞	800 bpi tape
NRZI-S	0,8	8/9		9	0.89	0.89	∞	13xx disk
PE	0,1	1/2		2	0.5	0.5	T	24xx tape 23xx disk
MFM (Miller)	1,3	1/2		2	1	0.5	∞	33xx disk
GCR	0,2	4/5		3	0.8	0.8	∞	3420 tape
8/9	0,3	8/9		4	0.89	0.89	∞	3180 tape
ZM	1,3	1/2		2	1	0.5	$3T/2$	3850 MSS
Miller	1,5	1/2		3	1	0.5	$3T/2$	
3PM	2,11	1/2		4	1.5	0.5	∞	ISS-8470
2,7	2,7	1/2		2.67	1.5	0.5	∞	3370-80 disk
1,7	1,7	2/3		4	1.33	0.67	∞	

[a] The left-hand vertical bar denotes the pulse width of the clock.

lihood detector used to reduce the error in the read-back signals (17).

Figure 9 shows the effect of a PR4 equalizer on the frequency response of the channel (8). The equalizer boosts the high-frequency response so as to turn the Lorenzian-shaped isolated pulses from the recorded media into the more complex pulse shapes needed for sampling and decoding by the PRML channel.

Table 3 lists the user information density per recorded transition on the tape for two of the more popular codes used for PRML channels, (17) compared to the older peak detect channels (17).

SPECIFIC EQUIPMENT IMPLEMENTATIONS OF MAGNETIC TAPE RECORDING

The three basic mechanisms for magnetic recording will now be described in some detail; these are transverse, linear, and helical scan recordings.

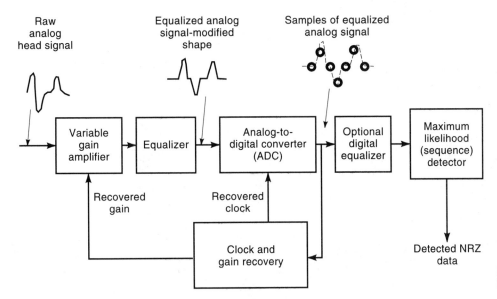

Figure 8. PR4 channel with maximum likelihood detector for error correction.

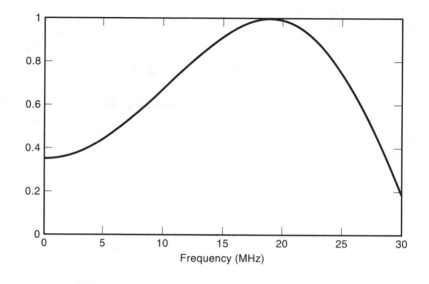

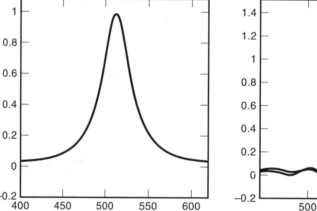
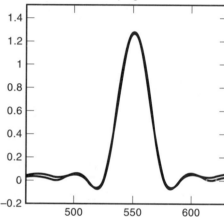

Figure 9. The equalizer circuit changes the frequency characteristics to shape isolated pulses into the complex shape required for accurate detection by the detector.

Transverse Recording

The origin of today's video recording industry lies in the 1956 introduction of the Ampex rotating head device. This technology allowed the use of a head to tape speed of 38 m/s (1500 in./s) while keeping the tape speed down to 38 cm/s and allowed a record time of 1 h. The recording was done by four heads with rotating drum writing transversely on the tape as shown in Fig. 10.

The limitations of this technology include the fact that each transverse track can only record a fraction of the one field, and complex switching is needed to assemble a complete field. Banding is produced by any amplitude or phase differences between the picture segments. Since it is necessary to use a 2 in. tape width, tape cost is very high. These limitations led to the helical scan technology detailed below. The standardized head and track format is shown in Fig. 11 (18).

This transverse method of recording has survived in a 1-in. form factor in the extremely high-performance digital data recorder called the DCRSi (Ampex Corporation). This data recorder used mostly for surveillance and related image capture

Table 3. User Densities for Various PRML Codes

Method	User Density (1, 7) Rate 2/3 Code	User Density (0, 4/4) Rate 8/9 Code
Peak detection (70% resolution)	1.33	0.88
Peak detection (80% resolution)	1	0.66
PR4	1.1	1.46
EPR4	1.33	1.77
E^2PR4	1.54	2.05

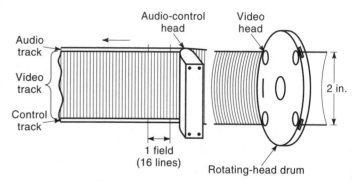

Figure 10. The transverse recording method. A rapidly rotating drum scans the tape surface vertically while the tape moves longitudinally at a much slower speed. The heads are embedded in the drum.

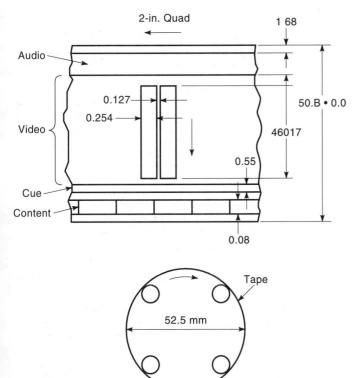

Figure 11. The 2 in. quad standard format. The universally accepted definitions of recorded data on tape for this format. The standard allows for interchange between differing versions of equipment that meet the standard.

applications has perhaps the highest data rate of any equivalent tape recorder (240 Mb/s). Of interest here is also the fact that the detector used maximum likelihood methods (Viterbi technology) (19) well before any other recording system.

Linear Recording

As shown in Fig. 12, the principle is the recording of a linear track of data produced when a tape is moved linearly over a stationary head. The recording can be simple parallel or serpentine; the latter allows bidirectional performance.

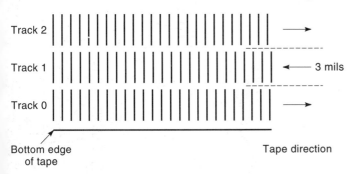

Figure 12. Linear tape recording formats. The information is written in a parallel data stream longitudinally along the entire tape length. It can also be interleaved in an alternating bidirectional form called serpentine.

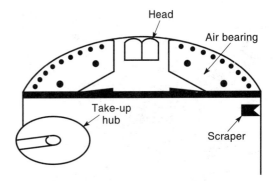

Figure 13. Typical tape head and handling configuration for the linear IBM 3480/3490 equipment family.

Nine-track open-reel methods originated this technique but have been supplanted by cartridge formats such as the 0.5 in. IBM 3480/3490 devices illustrated in Fig. 13.

Typically the tapes are chrome oxide, and there can be 36 tracks with tape speeds up to 80 in./s and data rates of 3.0 MB/s. More recently the 3590 technology has been introduced that uses metal particle tape and multitrack magnetoresistive heads. Native (uncompressed) sustained data rates of 6.6 MB/s and 144 tracks are achieved. The increased data rate is accomplished through the use of a track following servo which takes 8 steps over the 144 tracks (128 data tracks and 16 servo tracks). The head itself has 16 data tracks and 2 servo gaps.

An alternate solution is based on the DLT format; the cartridges have a supply reel but no take-up reel, which is permanently mounted in the drive. Figure 1 is an illustrative outline of a basic DLT drive. Up to 128 tracks are written in multiple linear passes down the entire length (1800 ft) of the tape. The inability to provide track to track cross-talk rejection, as will be discussed below, meant that a guard band had to be provided, as shown, this limits capacity. Hence symmetric phase recording (Fig. 14) was developed that allows 208 tracks on the 0.5 in. tape. A switchable head that pivots in the drive achieves the azimuth-type recording. The ability to electronically reject data derived from the opposing azimuth angle allows close packing on the tracks and eliminates the need for the guard band.

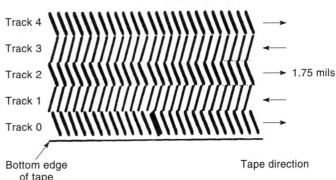

Figure 14. Symmetric phase recording—azimuth-syle linear recording. The opposing angle of the adjacent information simplifies the rejection of cross-talk between tracks.

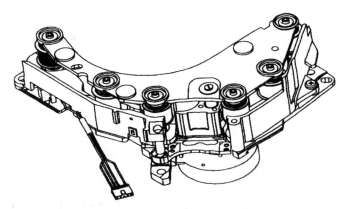

Figure 15. Linear (DLT) head guide and tape handling assembly. Note the six guiding rollers that control tape position.

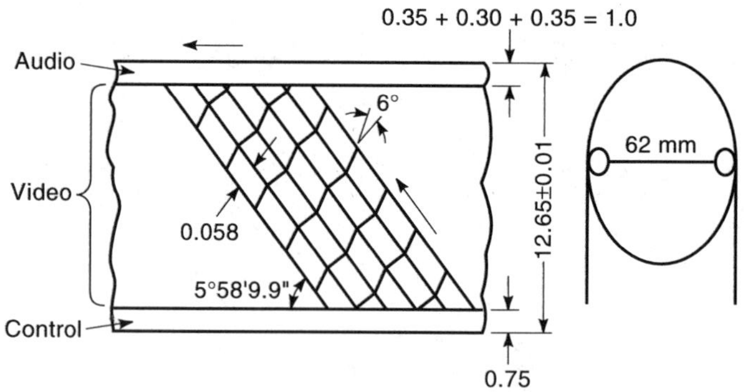

Figure 17. NTSC standard head and track format for helical scan as used in VHS recording for consumer applications.

The head stack switches to vertical when reading older formats ensuring backward compatibility. The four read/write channels are set up with a parallel channel architecture. This allows the drive to perform a read after write command and automatically re-record data on a parallel channel if it finds an error.

The heads in this drive format have been typically metal in gap (MIG) heads as discussed elsewhere. More recent versions utilize magnetoresistive (MR) heads. Additionally, to wipe accumulated debris off the surface of the tape, raised edges are fabricated on the edge of the head stack to act as scourers and to remove loose material. The head has to be contoured to the tape, this is to achieve intimate contact and is often achieved through a wearing process with special abrasive tapes.

Tape handling in these linear formats is critical. Since the take-up reel is internal and fixed, it is possible to use very low tape tension and achieve low head and tape wear. The head guide assembly is shown in Fig. 15. The drive head is the only contact point for the magnetic tape surface; and the six guides, or rollers, only touch the back surface. The drive motor is tachometer-controlled with one motor pulling the tape, while the other applies an opposing force to maintain optimal tension. To minimize stops and starts these drives, and others, use a form of memory cache buffering. Adaptive methods allow the system to adjust block sizes to match the host data rate. Data are compacted as they enter the cache buffer at a rate that matches the rate at which it is written to the tape. This helps keep the drive streaming as much as possible and reduces delays due to repositioning.

A further linear format of interest is the quarter-inch cartridge (QIC) form factor. In this device the cartridge itself performs several of the functions that otherwise are performed by the drive—for example, tape guidance and tension control. Bit densities of 106 kbpi and flux densities of 80 kfrpi are achieved. The coding is 1,7 RLL, and the track pitch up to 65 μm. In more recent versions, metal particle (MP) tapes are used with coercivities of 1659 Oe.

Helical Scan Recording

To achieve the recording of one video field continuously on tape, it is clearly desirable to lengthen the track: This is difficult in transverse formats; but if the tape can be recorded diagonally, it can be done. To do this the tape is physically wrapped around a very rapidly rotating drum in a helical-shaped tape path (20). This is shown in Fig. 16.

When this is combined with the concept of azimuth recording (21), this has proved the most effective way of achieving high-density recording of both digital and analog information.

Figure 17 shows the NTSC standard for head and track formats for VHS tape recorders, the most common consumer

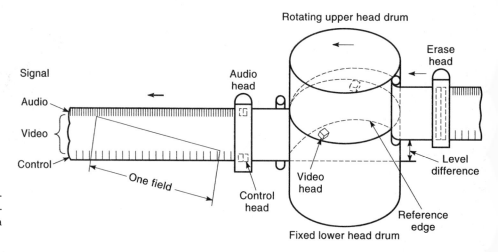

Figure 16. Basics of helical scan recording. The tape is wrapped around a rotating drum so that angled recoding can be achieved with consequent long tracks.

video tape recorder. Other helical scan standards include: type C—1 in.; U matic—0.75 in.; 8 mm, video and DDS; 4 mm; 19 mm (D2, DST, DCT), beta—0.5 in. and many other lesser used formats.

A critical design element in helical scan design is the ensuring of alignment of the horizontal sync pulse. This ensures machine-to-machine compatibility and stability.

To ensure this, it is necessary to link the tape speed V_m, the drum diameter d, and the helix angle θ_0.

The tape speed is then given by

$$V_m = \frac{\pi d \alpha F}{2(n_h \pm \alpha) \cos \theta_0}$$

where F is the field frequency, n_h is the number of horizontal sync pulses on one video track (262.5—NTSC), and α is the number of horizontal sync pulses between the adjacent track edges.

The actual recorded track angle is then given as the resultant of both head and tape movements:

$$\sin \theta = \frac{\sin \theta_0}{[1 \pm 2(2V_m/\pi Fd) \cos \theta_0 + (2V_m/\pi Fd)^2]^{1/2}}$$

The azimuth angle ϕ must be minimized in helical scan recording or the effective head gap length increases (by $1/\cos \phi$). Today the VHS standard azimuth angle is 6°. The ratio of the cross-talk, C, and the signal, S, can be calculated for azimuth recording from the following equation for azimuthal loss.

$$L_{az} = 20 \log_{10} \left(\frac{\pi w/\lambda \tan(2\phi)}{\sin[(\pi w/\lambda) \tan(2\phi)]} \right)$$

where λ is the recorded wavelength, w is the video track width, and 2ϕ is the angular difference in the head gaps.

If the video track and the play back head are misaligned by Δw, then the C/S ratio is given by

$$C/S = 20 \log_{10} \left(\frac{\sin[(2\pi \Delta w/\lambda) \tan(2\phi)] \Delta w}{(2\pi \Delta w/\lambda) \tan(2\phi)(w - \Delta w)} \right) \quad \text{dB}$$

A critical factor in helical scan recording is the maintaining of the proper spacing between the heads, drum, and tape. The heads are furthermore protruding from the plane of the drum as it rotates. While is necessary to maintain good head to tape contact, the drum to tape contact can be wearing to both components. Furthermore, variation in output signal can occur as the spacing varies. This problem has been studied using Reynolds equations and finite element analysis (5). When the tape is running over the drum, the displacement is determined by the shape of the head, the relative velocities, and the tape tension.

Numerous tape path configurations are used in helical scan. As an illustration Fig. 18 shows the tape path over the scanner and head in an 8 mm format. Note the use of a servo head in this format.

The use of complex servo systems has been an enabling technology for magnetic tape recording. Numerous schemes exist including pilot tones where the use of frequency components read by the head are used to control its position. Also used are embedded servos where there is information literally

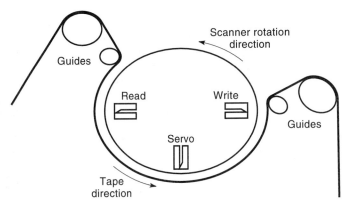

Figure 18. The 8 mm helical scan tape path. This well-established consumer and professional format uses the helical heads and a servo system to control head position.

embedded among with all the other data; these are used in helical audio formats. In the 8 mm technology where all the information on the tape is high frequency, the servoing is often accomplished by a piezoelectric bimorph that is driven by an error signal. Similar technology is used in 19 mm helical scan where the automatic scanning and tracking (AST) system uses heads driven by voice coil actuators (22).

BIBLIOGRAPHY

1. J. T. Mullin, Creating the craft of tape recording, *High Fidelity Mag.,* **April**: 62, 1976.

2. D. Jiles, *Magnetism and Magnetic Materials,* London: Chapman & Hall, 1991.

3. F. Jorgensen, *Magnetic Recording,* Blue Ridge Summit, PA: Tab Books, 1980.

4. K. Sookyung, M. Oakkey, and K. Haesung, Flying characteristics of tape above rotating drum with multiple heads, *IEEE Trans. Consum. Electron.,* **38** (3): 671, 1992.

5. C. Lacey and F. E. Talke, Measurement and simulation of the contact at the head/tape interface, *Trans ASME,* **114**: 646, 1992.

6. J. Hong, R. Wood, and D. Chan, An experimental PRML channel for magnetic recording, *IEEE Trans. Magn.,* **27**: 4532, 1991.

7. A. Patel, Signal and Error-Control Coding, in C. D. Mee and E. Daniel (eds.), *Magnetic Recording Handbook,* Part 2, New York: McGraw-Hill, 1989, p. 1115.

8. C. D. Mee and E. Daniel (eds.), *Magnetic Recording Handbook,* New York: McGraw-Hill, 1989.

9. J. C. Mallinson, *The Foundations of Magnetic Recording,* New York: Academic Press, 1993.

10. A. Hoagland and J. E. Monson, *Digital Magnetic Recording,* New York: Wiley, 1991.

11. K. B. Benson, *Audio Engineering Handbook,* New York: McGraw-Hill, 1988.

12. H. Nakajima and K. Okada, A rotary head digital audio tape recorder, *IEEE Trans. Consum. Electron.,* **CE-29**: 430, 1983.

13. C. P. Ginsberg, A new magnetic video recording system, *J. SMPTE,* **65**: 302, 1956.

14. SMPTE/EBU Digital Video Tape Recorder Specifications, 1984.

15. O. Karlqvist, Calculation of the magnetic field in the ferromagnetic layer of a magnetic drum, *Trans. R. Inst. Technol.,* Stockholm, Vol. 1, No. 86, 1954.

16. EMI, *Modern Instrumentation Tape Recording and Engineering Handbook,* 1978.

17. A. Taratorin, *Characterization of Magnetic Recording Systems,* San Jose: Guzik, 1996.

18. H. Sugaya and K. Yokoyama, Video Recording, in C. D. Mee and E. Daniel (eds.), *Magnetic Recording Handbook,* Part 2, New York: McGraw-Hill, 1989, p. 884.

19. R. Wood and D. Petersen, Viterbi detection of class IV partial response on a magnetic recording channel, *IEEE Trans. Commun.,* **34**: 1986, 454.

20. E. Schuller, German Patent No. 927,999, 1954.

21. S. Okamura, *Magnetic recording processing apparatus,* Japanese Patent No. 49-44535, 1964.

22. R. Revizza and J. R. Wheeler, *Automatic scan tracking using a magnetic head supported by a piezoelectric bender element,* U.S. Patent No. 4,151,570, 1979.

THOMAS M. COUGHLIN
SyQuest Inc.

DAVID H. DAVIES
Optitek Inc.

MAGNETIC TAPE RECORDING

HISTORY

In 1888, Oberlin Smith published an article, "Some possible forms of a phonograph." Oberlin Smith had already carried out the work in 1878, inspired by the telephone invented by Bell in 1877. In his paper, Smith described the basic principle of magnetic recording as it is used today. To record, he suggested leading small steel particles on a cotton thread through a coil. The coil carries a current proportional to a sound signal. For playback, the cotton thread is led through the coil again and an apparatus similar to Bell's telephone can reproduce the recorded sound. Oberlin Smith did not receive much recognition, though, because he had no functional device. A Danish engineer, Valdemar Poulsen, constructed the first recorder, the *Telegraphon,* in 1898. At the Paris World Exhibition in 1900, Valdemar Poulsen impressed the community with a recording of sound. Poulsen used a thin steel wire as recording medium.

Since there were no suitable electronics to amplify the signal, the Telegraphon did not gain importance for sound recording in the following years. After the invention of the vacuum tube, devices similar to the Telegraphon indeed recorded sound on steel tapes as Poulsen had suggested earlier. The steel tapes were heavy, clumsy to handle, and expensive. Fritz Pfleumer—who wanted to extend the use of his new cigarette paper manufacturing process—glued small iron particles onto his paper and created the first magnetic tape. (Fritz Pfleumer was actually not the first to have this idea: the American Joseph O'Neill had already applied for a patent in 1926, one year before Fritz Pfleumer.) The paper, however, tore readily, and Pfleumer decided to get help from the industry to commercialize his idea. An alliance between Allgemeine Elektrizitätswerke Gesellschaft (AEG) and Badische Anilin und Soda Fabrik (BASF) formed, in which AEG developed the recording device, and BASF developed the tape. In 1935, they presented the Magnetophon, which made use of a "ring head" that Eduard Schueller had invented during the development. BASF had replaced the paper substrate by a plastic base film. These two components improved the mechanics of the re-

corder significantly. Finally, the independent discovery of ac bias in Germany, Japan, and the United States in the late 1930s led to a large improvement of the sound quality of the recordings. The magnetic tape also improved. The carbonyl iron particles used for the Magnetophon tape were replaced by magnetite (Fe_3O_4) and later by γ-Fe_2O_3. Today needle-shaped γ-Fe_2O_3. particles are still used, but improved materials are needed for high-density recording. Apart from audio recording, magnetic tape has also found wide application in analog video recording, as well as digital data recording.

MANUFACTURING AND STRUCTURE OF TAPES AND FLOPPY DISKS

Particulate Tape: Structure and Manufacturing Process

Figure 1 shows a schematic cross-section for a *particulate tape.* A particulate tape consists of a base film, a back coating, and the magnetic layer. The base film is typically made of polyethyleneterephtalate (PET), with a thickness down to 7 μm. If a thinner base film is required, materials with a higher Young's modulus, such as polyethylene-2,6-naphtalate (PEN), have to be used. High-density recording is only possible if the magnetic surface is very smooth, which, in turn, presupposes a smooth base film. A tape wound with a lubricating film of air drawn in between the smooth front and back surfaces is impossible to handle, so the back coating is designed to provide a controlled surface relief. It should have an undulating rather than jagged structure so as not to damage the recording surface with which it comes into contact. In addition, the back coating is filled with conducting carbon black powder to counter the build-up of electrostatic charges. The relatively rough base film used for audio cassettes does not require a back coating. In addition to a magnetic powder, the typical recording layer contains carbon black, abrasive particles of Al_2O_3, dispersants, and lubricants, all held in a plastic binder consisting of a thermoplast and a polyurethane, together with an isocyanate cross-linking agent. The magnetic powder accounts for 20% to 50% in volume of the recording layer. The reader can obtain more information from Refs. 1–5.

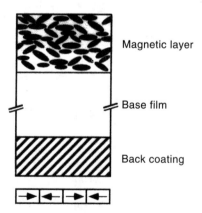

Figure 1. Cross-sectional view of the structure of a particulate tape. The magnetic layer typically contains needle-shaped particles together with a binder system. Particulate media are longitudinally magnetized.

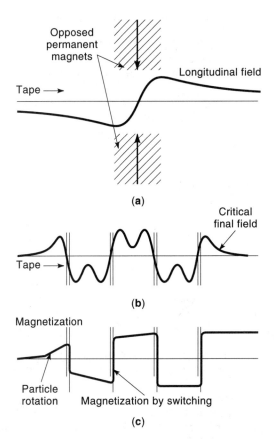

Figure 2. (a) A single orienting magnet consists of a pair of opposing permanent magnets, which are vertically oriented. An orientation with a multistage magnet is best if the field polarity changes sufficiently rapid for the magnetic moments to switch before the particles have time to rotate, (b) and (c).

The first stage in preparing the coating is to disperse the particles and so forth in a solution of the binder polymers. Vertical or horizontal bead mills, employed singly or in cascade, are standard dispersing equipment. Very fine particles, such as metal powder, can be better dispersed if a high-shear premix or kneading stage is added to the process. Just before coating, the cross-linking agent is added to the ink. Gravure, extrusion, or knife coating heads are used to put the ink onto the base film at web speeds between 100 m/min and 1000 m/min. Shortly after leaving the coating head, the still wet coating is magnetically oriented along the tape direction. Floppy disks are an exception: because they must be isotropic, it may be necessary to disorient them to remove any unwanted alignment due to shear in the coating head.

The most straightforward orienting magnet consists of two opposed permanent magnets [Fig. 2(a)], arranged symmetrically about the tape web, so that the wet coating sees only a longitudinal field. For better orientation, a sequence of such magnets may be installed. Figure 2(b) shows the resulting spatially alternating field. Orientation occurs if the polarity changes are sufficiently rapid for the magnetic moments of the particles to be switched around before the particles themselves have time to rotate, [see Fig. 2(c)]. This condition can be easily met for standard iron oxide and chromium dioxide tapes, but not for high-coercivity and high-moment metal particles (MP), which can be better oriented in the more unidirectional fields of large solenoids.

After the alignment, the coating has to dry until the chemical reactions in the binder system are (mostly) complete. The next step is to "calander" the tape. A calander presses the tape between polished rollers at a high temperature. This treatment results in a compressed magnetic film and, most important, in a very smooth surface finish. Finally, the tape is slit into the desired width and wound onto reels into cassettes. In the case of floppy disks, the "cookies" are punched out after calandering and mounted into their plastic housing.

Recording Particles

Only five magnetic materials are in use for particulate media:

1. *Maghemite* (γ-Fe$_2$O$_3$) is the light-brown material found on open-reel computer tapes, studio audiotape, and IEC I (International Electrotechnical Commission) audiocassettes. It is the lowest coercivity (24 kA/m to 30 kA/m) and oldest of the magnetic tape particles, suitable only for low-density storage. No new systems are designed to use γ-Fe$_2$O$_3$.

2. *Chromium dioxide* (CrO$_2$) was the first higher coercivity powder to improve on γ-Fe$_2$O$_3$ and established the "Chrome position" for audio cassettes. Most CrO$_2$ is now used in VHS (video home system) videotape, with smaller amounts going into IEC II audio and computer cartridge tape. Coercivities lie in between 40 kA/m and 60 kA/m, whereby CrO$_2$ tapes may have lower coercivities than equivalent Co modified γ-Fe$_2$O$_3$ tapes (see also the section titled Thermally Activated Magnetization Reversal Processes).

3. *Cobalt modified iron oxides* use the anisotropy of the cobalt ion to achieve coercivities of 30 kA/m to 80 kA/m. They are used in large quantities for consumer videotape, IEC II audio cassettes, QIC data cartridges (quarter inch cartridges), and floppy disks.

4. Doped *barium ferrite* (BaFe) has sometimes been hailed as the future high-density recording medium but, despite much development of experimental products, has so far appeared only in very small amounts in floppy disks.

5. *Metal particles* (MP) found their first significant application in high-output IEC IV audiotapes, but the main use is now in broadcast videotape as well as 8 mm consumer video, DAT (digital audio tape), data tape, and floppy disks. Since their introduction for video and DAT, the improvement in MP has been so spectacular that MP now dominates in new digital tape applications. The coercivity of video and digital MP ranges between 110 kA/m to 200 kA/m.

Table 1 summarizes magnetic properties of the powders used in tape and some other materials of interest.

Ferric Oxide (γ-Fe$_2$O$_3$). The essential problem in preparing ferric oxide, γ-Fe$_2$O$_3$, as well as Co-modified γ-Fe$_2$O$_3$ and metal particles, is to make uniaxially anisotropic particles from materials with cubic anisotropies and crystal habits. [As will be explained in the section "Magnetization Reversal in Fine Particles," elongated particles show the desired magnetic properties for tape.] Direct precipitation gives isometric particles, so a roundabout route, as illustrated in Fig. 3, via trigonal

Table 1. Material Properties for Some Magnetic Materials and Powders Used in Tape Manufacturing

Material	Magnetic Materials			Powders Used in Tape			
	σ_s (Am^2/kg)	M_s (kA/m)	ρ (g/cm^3)	σ_s (Am^2/kg)	H_c (kA/m)	Length (nm)	Diameter (nm)
γ-Fe$_2$O$_3$	76	350	4.6	73–75	24–30	250–500	30–50
Fe$_3$O$_4$	92	480	5.2				
CrO$_2$	106	490	4.6	75–85	35–60	250–320	30–45
CoFe$_2$O$_4$	80	425	5.3				
Co-doped γ-Fe$_2$O$_3$, (CoFe)				70–78	30–80	180–400	25–45
BaFe$_{12}$O$_{19}$[a]	72	380	5.3				
Doped BaFe$_{12}$O$_{19}$, (BaFe)[a]				55–65	50–120	10–40	40–100
Fe[b]	218	1710	7.8	120–150	90–200	60–250	20–40

[a] Platelet shaped.

[b] Tape particles here are oxide shell and may contain Co.

α-Fe$_2$O$_3$, must be followed to produce the desired acicular form. Three methods are in use to synthesize the α-Fe$_2$O$_3$, the most common being the dehydration of needles of FeOOH. Goethite, α-FeOOH, is formed by precipitation from a solution of FeSO$_4$ with NaOH. The second method is via lepidocrocite, γ-FeOOH precipitated from a solution of FeCl$_2$. In both cases, the next stage is the same, the dehydration to α-Fe$_2$O$_3$ at temperatures up to 800°C. To maintain the needle shape of the particles during dehydration, the FeOOH must be coated with some anti-sintering agent such as a phosphate. The third method is the direct hydrothermal synthesis of α-Fe$_2$O$_3$ from a suspension of Fe(OH)$_3$ using crystal modifiers to control the particle morphology. By avoiding the large density change on dehydration, the direct process introduces fewer of the defects and grain boundaries normally found in γ-Fe$_2$O$_3$ particles. Inhomogeneities give rise to internal magnetic poles which degrade the particle properties, and their absence has lent the direct-process particles the name *nonpolar* (NP). In hydrogen, or using a combination of hydrogen and organic compounds that can further hinder sintering, the α-Fe$_2$O$_3$ is reduced to

magnetite, Fe$_3$O$_4$, at temperatures below 500°C. Magnetite has a higher magnetization than γ-Fe$_2$O$_3$ and is, at first sight, an attractive recording material. It has, however, proved unsatisfactory, because in a finely divided form, it oxidizes naturally to γ-Fe$_2$O$_3$. It is also more susceptible to print-through (see the section titled Thermally Activated Magnetization Reversal Processes). γ-Fe$_2$O$_3$ is a metastable form, which reverts to hematite (α-Fe$_2$O$_3$) on heating to 400°C, so the final oxidation of Fe$_3$O$_4$ to γ-Fe$_2$O$_3$ must not exceed approximately 350°C. A densification process to improve the handling properties is commonly applied to the finished particles.

Typical particle lengths are 0.3 μm for the goethite and NP, and 0.4 μm for the lepidocrocite processes. The lepidocrocite particles tend to form bundles, which can be more easily oriented and packed, while the NP particles are very uniform and suitable for high-quality audio applications. About three-quarters of γ-Fe$_2$O$_3$ is prepared by the goethite process, which is also the basis for the manufacture of cobalt-modified oxides and metal particles.

Chromium Dioxide (CrO$_2$). In contrast to the iron oxides, CrO$_2$ is crystallized in a single-stage hydrothermal process. It has the rutile structure and forms smooth-faced acicular single-crystal particles. They tend to occur in parallel bundles and can be very well oriented. In addition to its shape anisotropy, CrO$_2$ has magnetocrystalline anisotropy. Chromium dioxide is an unusual material, being a ferromagnetic oxide and a good electrical conductor.

Synthesis of CrO$_2$ involves a reaction of an aqueous paste of CrO$_3$ and Cr$_2$O$_3$ under hydrothermal conditions. First, a thick mash is prepared and then heated in an autoclave to 300°C at a pressure of 350 MPa. The reacted product is a solid black mass, which must be drilled out of the reactor cans. It is then dispersed and treated with Na$_2$SO$_3$ or NaOH solution to topotactically convert the outer layer of the particles to β-CrOOH. This treatment is necessary to improve the stability of the powder in the presence of water but, as in the case of MP, it reduces the magnetization. The shape, size, and coercivity of the particles can be controlled by additives. Antimony and tellurium are used to vary the particle size, and iron to control the coercivity, although the iron doping also has an effect on the particle geometry. Up to about 3% of Fe^{3+} can be incorporated into the CrO$_2$ structure, increasing the magnetocrystalline anisotropy and the coercivity to over 80 kA/m. Between the Fe and Cr the exchange coupling is

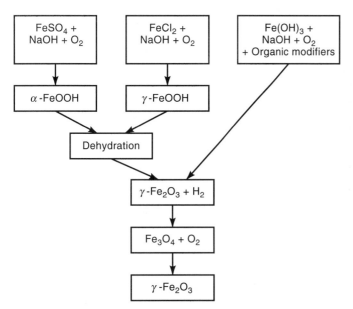

Figure 3. Three synthesis routes to produce elongated γ-Fe$_2$O$_3$ particles. The roundabout routes are necessary to grow elongated particles from materials with cubic crystal habits.

antiferromagnetic and stronger than the Cr–Cr exchange. Consequently, Fe doping decreases the saturation magnetization of CrO_2, while increasing the Curie point from 118°C for undoped material to about 170°C. Iridium is the most effective dopant for CrO_2, producing a spectacular rise of coercivity up to 220 kA/m. Although this material remains an expensive laboratory curiosity, mixed doping with very low levels of Ir can be used commercially for high-coercivity powders.

Cobalt-Modified Iron Oxides (CoFe). Cobalt-modified iron oxides (CoFe) utilize the high anisotropy of the Co^{2+} ion to increase the coercivity of the iron oxides described above. There are two classes of CoFe powders.

1. *Bulk Doped.* The most straightforward way to add cobalt is to deposit Co hydroxide onto either the γ-FeOOH or α-Fe_2O_3, and then to proceed with the normal heat treatments. The Co diffuses into the body of the particle. For lower quality applications of powders up to about 50 kA/m, this procedure is adequate, but it has certain weaknesses. The anisotropy and coercivity are strongly dependent on temperature, are subject to magnetostrictive losses, and demonstrate strong magnetic annealing effects. The latter is due to the well-known tendency of Co^{2+} ions to form pairs or groups ordered along the direction of magnetization. Therefore, the coercivity drifts with time, and because the particles adjust to local magnetic fields of the recorded signal, erasure is poor.

2. *Surface Modified.* Rather than treating the precursors, a 1 nm to 2 nm coating of $CoFe_2O_4$ is formed on the γ-Fe_2O_3 particle. No high temperatures are encountered and the Co remains at the surface. There are two methods of preparation. In the adsorption technique, cobalt hydroxide is precipitated onto γ-Fe_2O_3 and a portion of the cobalt is incorporated into the surface layer of the oxide. In the epitaxial method, cobalt ferrite from a mixture of Fe and Co solutions is precipitated directly onto γ-Fe_2O_3. Despite the anisotropy of $CoFe_2O_4$ being cubic, the dominant effect of the coating is to increase the uniaxial anisotropy of the particle. This is not properly understood, but could, for example, be due to stress or ordering effects at the boundary. The instabilities of bulk-doped material are roughly halved in importance by the surface modification, and such powders coat the majority of IEC II audio tapes, VHS, and S-VHS videotapes as well as floppy disks and QIC data cartridges.

Barium Ferrite (BaFe). Barium ferrite ($BaFe_{12}O_{19}$) has the M-type hexaferrite structure and a large uniaxial magnetocrystalline anisotropy ($H_A = 1350$ kA/m) directed along the hexagonal axis. Barium ferrite forms flat plates perpendicular to the easy axis in which the shape anisotropy, in contrast to other recording materials, works to reduce the coercivity. The pure form, with coercivities in the range of 300 kA/m, is used in credit card stripes and the like, but for tape applications, only doped material with lower coercivity has been used. Substituting elements such as Co^{2+} and Ti^{4+} for some of the Fe^{3+} adjusts the coercivity to the range of 50 kA/m to 200 kA/m and reduces the otherwise problematic temperature variation of the coercivity.

The ceramic method of firing and milling used to make barium permanent magnets is unsuitable for fine recording particles for which two methods are in use. In the *hydrothermal process,* the metal hydroxides are precipitated from salt solutions with an excess NaOH, and the resulting suspensions are heated in an autoclave to 200°C to 300°C. The washed-and-dried product is then annealed at 700°C to 800°C, to increase the magnetization. In the *glass crystallization process,* the components for the desired barium ferrite are dissolved in a borate glass melt. After rapid quenching, the amorphous glass flakes are annealed at temperatures up to 800°C. Last, the glass matrix is dissolved to separate the barium ferrite particles. Both methods can produce platelets of 50 nm and smaller in diameter. It can be observed that the magnetization decreases for very thin platelets, which is attributed to a 'dead layer' at the surface. One way to increase the magnetization is to deposit magnetite on the surface.

From its shape, barium ferrite appears to be the ideal powder to make vertically oriented media, especially floppy disks. However, the difficulty in producing a vertically oriented coating with a smooth surface, the tendency for the flat particles to stack together, and problems with the mechanical integrity of the coatings, have hindered its widespread introduction. The new, much smaller MP particles have also eroded barium ferrite's one-time advantage in theoretical signal-to-noise ratio. The main advantage remaining to barium ferrite is its stability against corrosion.

Metal Particles (MP). Metal particles (MP) are prepared from γ-Fe_2O_3 precipitated from $FeSO_4$. Three generic processes may be defined, although the actual manufacturing process may involve some combination of these: *basic process,* a solution of $FeSO_4$ is added to an excess of NaOH in solution, pH > 7; *acid process,* the $FeSO_4$ and NaOH are mixed in stoichiometric proportions, pH < 13 (in the range $7 <$ pH < 13 cubic magnetite is precipitated instead of γ-FeOOH); *carbonate process:* $FeSO_4$ is added to an excess of Na_2CO_3. The basic and acid processes both produce cylinder-shaped γ-FeOOH, while the carbonate process leads to so-called "spindle-shaped" particles. The spindle-shaped particles are actually fine fibers of FeOOH forming bundles with tapering ends and can be much smaller than the cylinders. Following precipitation and surface treatment, the next stages are similar to those for γ-Fe_2O_3, except that the reduction of the α-Fe_2O_3 in H_2 is carried through to the metal. The pure metal powder is pyrophoric and cannot be used, so a controlled oxidation of the surface is carried out at 80°C to 100°C to generate a passivation layer, approximately 4 nm thick. Fe_3O_4 or γ-Fe_2O_3 can be identified by X-ray and Mössbauer analysis of the passivation layer. The crystallites are superparamagnetic at room temperature and do not contribute to the magnetic properties of the particles. It appears that a polycrystalline/amorphous layer can better accommodate the lattice mismatch between metal and oxide and forms a less permeable protective layer. Doping with nickel up to 3% and cobalt up to 30% increase the magnetization of iron and/or facilitate the reduction process. A Co content of 30% is standard in advanced metal particles.

MP is increasingly used in high-density recording systems, in which a very smooth tape surface is essential. It is crucial, therefore, to apply surface treatments at different stages of manufacture to prevent sintering, not only to preserve the particle shape and coercivity, but also to improve the dispersing properties. To this end, a combination of SiO_2, $AlOOH$, or rare-earth oxides may be deposited on the γ-$FeOOH$ or α-Fe_2O_3. For further reading, see Ref. 5.

Particulate Tape: Double-Layer Coating

Although there have been early attempts, this technology did not receive much interest until very recently. Recent double-layer media have a very thin magnetic layer (MP), and a nonmagnetic underlayer that contains very small TiO_2 or α-Fe_2O_3 particles (6). The two layers are coated simultaneously, whereby their rheological properties need to be adjusted properly. The first commercial product was a Hi 8 videotape with considerably improved recording performance. The thinness of the magnetic layer itself—which is about 400 nm for this tape—is not responsible for the increased output. The manufacturing process requires a certain minimum coating thickness, regardless of whether it is magnetic or not, to achieve smooth surfaces. Depending on the recipes used, either double or (thick) single layer coatings can be made smoother. Magnetically, thin magnetic layers have advantages in overwrite behavior

Meanwhile it has been demonstrated that magnetic layers as thin as 120 nm can be achieved (7,8). The most advanced MP double-layer tapes can compete with Metal Evaporated (ME) tapes. The application for these tapes is the digital video cassette (DVC), which is a tape system intended for digital video recording.

Thin-Film Tape: Structure and Manufacturing Process

Metal Evaporated (ME) Tape. Figure 4 shows a schematic cross-section of ME tape. ME tape consists of a base film, a back coating, the magnetic layer, a carbon protection layer, and a lubrication layer. While the back coating is virtually identical to those used for particulate media, the base film shows a distinct difference. The surface of the base film, onto which the magnetic layer is deposited, carries a thin coating

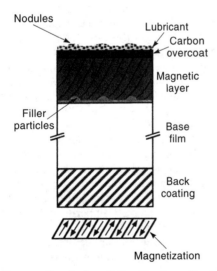

Figure 4. Cross-sectional view of the structure of a thin film tape (metal evaporated tape). A carbon overcoat and a lubrication layer protect the magnetic layer. The base film contains filler particles, which lead to nodules sticking out of the tape. The favored magnetization direction is oblique.

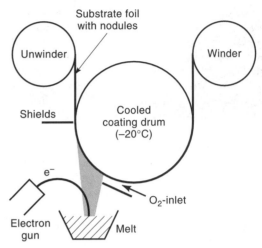

Figure 5. Sketch of a roll-coater used to manufacture ME tape. The magnetic alloy is heated with an electron gun and evaporated onto a base film that passes by. Oxygen is present during the evaporation process, which helps to isolate the magnetic grains.

with very small particles. After deposition of the magnetic film, the surface of the magnetic layer then shows 'nodules'. These nodules are about 10 nm to 20 nm high, and have a density of 10 to 50 per μm^2. The invention of the nodules has been the technological breakthrough for ME tape. The nodules improve tribology at the expense of an additional head to tape spacing loss, which sacrifices a bit of the recording performance.

Hi 8 ME tape is manufactured by oblique evaporation of a $Co_{0.8}Ni_{0.2}$ alloy in an oxygen atmosphere (9) (see Fig. 5). The typical composition of ME tape is about $(Co_{0.8}Ni_{0.2})_{0.8}O_{0.2}$ for the Hi 8 ME tape. Bulk $Co_{0.8}Ni_{0.2}$ has a 19% lower saturation magnetization than Co, and a comparable anisotropy field. Advanced ME tapes for DVC application contain no Ni (10). Due to its sufficiently high production speed (up to about 100 m/min) evaporation is—in contrast to sputtering—a suitable technology to produce videotapes. In the very beginning of the evaporation of the layer, the vapor arrives at the substrate at grazing incidence. A film grown at grazing incidence shows uniaxial anisotropy in an oblique direction (11,12). Roll-coaters operate in a wide range of evaporation angles [continuously varying incidence, (CVI) (9)]. The film deposition has to start at grazing incidence to preserve oblique anisotropy. A CVI process leads to a curved columnar microstructure, as sketched in Fig. 6. The magnetically easy axis is tilted out of the film plane. The tilt angle roughly coincides with the angle at which the columns start to grow on the base film. The columns themselves are not the relevant magnetic subunits for magnetization reversal and form a secondary structure (13). Individual crystals of ME tape are very small (about 5 nm in size). The magnetocrystalline anisotropy of the Co-alloy is the major source of anisotropy in ME tape.

Self-shadowing effects and low surface mobility (the substrate temperature is typically between $-20°C$ and $-30°C$), lead to a formation of a very porous layer, especially at grazing incidence (14). The columns do not grow in the direction of the incoming beam. Due to the effect of shadowing, the growth direction is closer to the film normal. This can be ap-

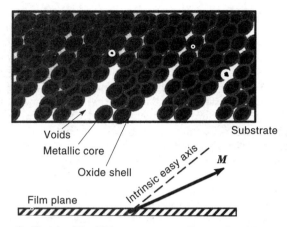

Figure 6. Sketch of the ME tape structure. Due to the oblique evaporation, the tape has a columnar structure. The magnetically easy axis is tilted out of the film plane. In zero field, the demagnetization energy pulls the magnetization closer to the film plane.

proximated by the "tangent rule":

$$\tan\alpha_c = 0.5\tan\alpha_B \qquad (1)$$

Here α_c and α_B are the angles that the column and the beam make with the film normal, respectively. At higher substrate temperatures, more continuous layers emerge, which are unsuitable for high-density recording. In case of ME tape, almost half of the volume of the magnetic layer consists of voids due to shadowing effects (13).

For improvement of particle separation and material yield, the evaporation is performed in the presence of oxygen. The oxidation of the magnetic material largely removes exchange coupling in ME tape, but there remains a magnetic correlation along the columns, which is also discussed in section titled Transition Models. Auger depth profiling shows the formation of oxide-rich surface and bottom layers. The upper oxide layer improves the mechanical performance of the tape, but lowers the output level due to the increased magnetic spacing between head and the active part of the medium. Another benefit of the upper oxide layer is corrosion protection of the tape. As indicated in Fig. 6, the "ME tape particles" are believed to have an oxide shell of CoO (and NiO) around a metallic core. Exchange anisotropy has been reported in ME tapes (15), which is consistent with this assumption.

The carbon protection layer is absent in some of the Hi 8 ME tapes. Advanced ME tapes utilize a Diamond Like Carbon (DLC) layer, to mechanically protect the tape. A layer thickness of 10 nm or even less is sufficient to improve wear resistance and provide additional corrosion protection for the metallic film (16). Since the protection layer is very thin, it can be sputtered in a separate station in the roll-coater. For improved runnability, the tape needs a lubricant layer. For particulate tape, it is believed that the lubricant forms a monomolecular layer on top of the coating. If worn off, the reservoirs inside the magnetic layer continuously replenish the lubricant. Since there are no pores on the surface of thin film media that can retain the lubricant, the lubricant must anchor itself to the thin-film surface.

Other Thin-Film Tapes. There have been many attempts to prepare thin-film media other than ME tape on flexible sub-

strates. Co–Cr thin films having an easy axis of magnetization perpendicular to the film plane have been investigated intensively. From a production point of view, ME tape is better suited for videotape, because a vertical Co–Cr medium requires an additional magnetically soft underlayer. (For a discussion of vertical recording, see the section titled 'Magnetization and Demagnetizing Fields'.) Oblique evaporation of single-layered Co-Cr media for use together with a ring head has been suggested (17).

The vapor pressures of Co and Cr—unlike Co and Ni—are considerably different, which makes control of an evaporation process more difficult than in the case of ME tape. The basic challenge in preparation of Co–Cr media is to break up the exchange coupling between the individual magnetic subunits as far as possible. Perpendicular Co–Cr layers have a columnar structure with grain sizes of about one-tenth of the layer thickness, which typically ranges from 0.1 μm to 0.3 μm. The underlying mechanisms that cause the formation of more or less magnetically independent subunits are still under discussion. Up to now, good recording results were only reported for films prepared at high substrate temperature (200°C to 250°C). For flexible substrates (videotape), these high temperatures require the use of the expensive polyimide (PI) film as a substrate, rather than the cheaper PET film, which can be used for ME tape. Changes in composition inside the grain of Co–Cr films show characteristic patterns that have been named 'chrysanthemum-like' (18,19). Thin-film media on flexible substrates other than ME tape gained virtually no practical importance. Apart from the aspects discussed above, the poor tribological properties, especially for Co–Cr-based media, prevented any practical implementations. The reader can find more information on Co–Cr and vertical recording in Refs. 20–22.

MAGNETIC PROPERTIES OF TAPES

Magnetic Parameters

The various types of tape differ in magnetic properties. Figure 7 illustrates the most important magnetic parameters. The magnetic properties of tapes are typically measured with vi-

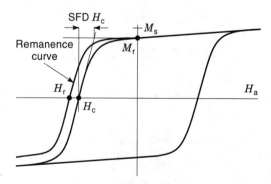

Figure 7. Hysteresis loop, remanence curve, and characteristic magnetic parameters. The saturation magnetization, M_s, is the maximum obtainable magnetization, the remanent magnetization, M_r, is the magnetization at zero field after saturation. The coercivity, H_c, is the field required to make the magnetization zero after saturation. The switching field distribution, SFD, defines how uniformly the medium switches. The remanent coercivity, H_r, gives the field, which makes the remanent magnetization zero after saturation.

brating sample magnetometers (VSM). These instruments measure the magnetization as a function of the applied field. The following properties characterize hysteresis loops:

- *Saturation Magnetization, M_s.* The saturation magnetization is the maximum attainable magnetization at very large applied field. The saturation magnetization for tape has to be distinguished from the saturation magnetization of the particles themselves. The tape magnetization is lower because the binder system (or voids) dilutes the magnetic material.

- *Remanent Magnetization, M_r.* The magnetization that remains after removal of a field. To avoid confusion, the remanent magnetization obtained after saturation with a large field is often referred to as *saturation remanence.*

- *Squareness.*

$$S_q = \frac{M_r}{M_s} \qquad (2)$$

In zero field, the magnetization of each particle will be on its easy axis. Since the particles in a tape are never aligned perfectly, and the measured magnetization is the magnetization component along on the field axis, it is smaller than the saturation magnetization, that is, $S_q \leq 1$. Therefore, the squareness reflects the degree of particle alignment. Magnetic interactions also tend to increase the squareness.

- *Orientation Ratio.* This is the ratio of the two saturation remanences in longitudinal (x) and transverse (y) direction. Since inhomogeneous magnetization within the particles at zero field can reduce the remanent magnetization, the orientation ratio captures the degree of particle alignment better than the squareness.

- *Coercivity.* The coercivity, H_c, gives the field at which the magnetization component along the applied field becomes zero.

- *Switching Field Distribution, SFD.* In magnetic tapes, there is a distribution of the switching fields of the particles. Köster has shown that the normalized slope of the hysteresis loop—which contains reversible and irreversible magnetization changes—is a convenient measure for the 'real' switching field distribution (23):

$$\text{SFD} = \frac{M_r}{H_c} \left(\frac{dM}{dH}\right)^{-1} \text{ at } H = H_c \qquad (3)$$

The SFD is related to the normalized slope, S^*, by SFD = $1 - S^*$.

The hysteresis loop contains reversible and irreversible magnetization changes. For information storage, only irreversible changes are of importance. *Remanence curves* can separate irreversible from irreversible magnetization changes. A point on the remanence curve is measured as follows: (1) magnetically saturate the sample, (2) apply a field in the opposite direction, (3) measure the magnetization with the field removed. The remanent magnetization plotted against the previously applied field value is the remanence curve (see Fig. 7). The field at which the remanent magnetization becomes zero is the *remanent coercivity, H_r.* In accordance

with the theory, the remanent coercivity is somewhat larger than the coercivity. There are different definitions for the SFD, but it is common practice to use Eq. (3) for convenience.

Table 2 summarizes typical tape parameters for some applications. The well-established oxide media (γ-Fe_2O_3, Co-doped γ-Fe_2O_3, and CrO_2) show the smallest tape magnetization and coercivities. Tape data for the same pigment type can deviate considerably depending on the application. Although videotapes made of CrO_2 and Co-doped γ-Fe_2O_3 are compatible with one another, their coercivities do not agree. CrO_2 shows a stronger time dependence of the coercivity than Co-doped γ-Fe_2O_3, which leads to the same coercivity at recording (see also the section titled "Thermally Activated Magnetization Reversal Processes"). Table 2 illustrates that the more advanced tapes—such as MP and ME tape—have considerably more particles per unit volume. For further reading, see Refs. 1 and 3.

Magnetic Parameters of ME Tape

ME tape is prepared by oblique evaporation. Therefore, the magnetically easy axis of ME tape is tilted out of the film plane. It was realized early that the tilt of the easy axis makes the recording performance in one direction different from that of the other (see also the section titled "Transition Models"). Little attention was paid to the demagnetization effects present in simple measurements of the magnetic properties. Magnetic properties of regular particulate tapes are measured in longitudinal direction, which coincides with the alignment direction (i.e., the easy axis of the tape). This standard procedure yields low values for the squareness, S_q, and large values for the SFD for ME tape. Magnetically, ME tapes did not seem to be very attractive.

If the tape magnetization has a perpendicular component, there is a perpendicular demagnetization field that has to be added vectorially to the applied field. The magnetic 'particles' in the tape, therefore, do not 'see' the external field alone, but rather the vectorial sum of the external field, H_a, and the demagnetizing field, H_d:

$$H_i = H_a + H_d \qquad (4)$$

If the external field is applied at an angle ϑ_E, with ϑ_E being the angle between film plane and applied field, the internal field is:

$$H_{i\parallel} = H_i \cos \vartheta_i = H_a \cos \vartheta_E \qquad (5)$$

$$H_{i\perp} = -H_i \sin \vartheta_i = -H_a \sin \vartheta_E - M_\perp \qquad (6)$$

Eqs. (5) and (6) assume that the demagnetization factor perpendicular to the film plane is equal to one and the other demagnetization factors are zero. Therefore, the demagnetizing field is always perpendicular to the film plane and equal in magnitude to the perpendicular magnetization component, $M_\perp$.

Consider the case of a very large external field applied in film plane, which is subsequently removed. With no external field present, the sample is lying in its own demagnetizing field. During the removal of the external field, the internal field has not only changed its magnitude, but also its *direction*, namely, from an in-plane orientation to a perpendicular orientation. Evidently, sweeping the external field—as it is

Table 2. Magnetic Parameters of Various Tapes and Floppy Disks

Application	Type	M_r (kA/m)	H_c (kA/m)	Coating (μm)	n^b (1000/μm^3)
Reel-to-reel tape	γ-Fe$_2$O$_3$	100–120	23–28	10	0.3
Audiotape IEC I	γ-Fe$_2$O$_3$	120–140	27–32	5	0.6
Audiotape IEC II	CrO$_2$	120–140	38–42	5	1.4
	Co-γ-Fe$_2$O$_3$	120–140	45–52	5	0.6–1.4
VHS tape	CrO$_2$	110	44–58	3–4	2
	Co-γ-Fe$_2$O$_3$	110	52–74	3–4	2
Hi 8 tape	MP	200	120–135	2–3	8
	ME^a	350	90	0.2	~125
DVC tape	MP	>300	180	0.15	50
	ME^a	450	135	0.15	~125
1.4 MB floppy	Co-γ-Fe$_2$O$_3$	50	50	1	1.4
100–120 MB floppy	MP	160	125	0.3	8
Data cartridge	MP	200	160	0.2–1.0	20

^{a} Denotes intrinsic properties.
^{b} Particle density.

done in magnetometers—leads to a rotation of the internal field whose magnitude depends on the tape magnetization itself. Figure 8 illustrates the complex process. The sketches in the upper row indicate some points on the magnetization curve. The middle row roughly indicates the magnitudes and the orientations of the external field, the demagnetizing field, and the internal field for the three points. In order to make a fair comparison between particulate media and ME tape, the direction of the internal field needs to be held fixed. The bottom row in Fig. 8 indicates that the sample has to rotate *during* the measurement of the hysteresis loop. In Fig. 8, the direction of the internal field is held fixed along the longitudinal direction; the little flags indicate the orientation of the film for the three points. Bernards et al. (24) and Richter (13)

have measured compensated hysteresis loops of ME tape. These measurements show that the 'intrinsic' hysteresis loop of ME tape is almost perfectly square, with a squareness larger than 0.9 and an SFD smaller than 0.1.

The compensated measurement shows symmetry around the easy axis as one expects from a magnetically uniaxial material. The 'intrinsic' easy axis forms an angle of 35° to 40° with the film plane. The angle dependence of the switching field, that is, remanent coercivity, is consistent with an incoherent magnetization process. The switching field is lowest along the easy axis and highest perpendicular to the easy axis with value close to the anisotropy field. Uncompensated measurements are not symmetrical, because the demagnetizing field distorts the magnitude and the direction of the internal field as outlined above. Further data show that ME tape is extremely well oriented, with an orientation ratio of about 10.

MAGNETIZATION REVERSAL OF FINE PARTICLES

A magnetic recording medium must consist of a magnetic material with a high enough coercivity to have sufficient safety margin against unwanted erasure. External as well as internal fields (demagnetization) can cause unwanted erasure. The information storage should use as little space as possible. Single-domain particles, that is, magnetic particles that are so small that they cannot break up into a multidomain structure, are consequently best suited for magnetic recording applications. All particles used in recording have uniaxial magnetic anisotropy. Magnetic particles with multiaxial anisotropy were also under discussion (1,25), but did not gain practical importance. To better understand the magnetic recording process, the fundamental switching behavior of single-domain particles is thus of primary interest.

Stoner–Wohlfarth Model (Coherent Rotation)

Stoner and Wohlfarth introduced a simple model for magnetization reversal in single-domain particles in 1948 (26). They assumed that the magnetization in these particles is always homogenous (model of coherent rotation). The model thus applies to elliptical particles only. Initially, it was argued that the strong (but short-ranged) exchange forces are strong

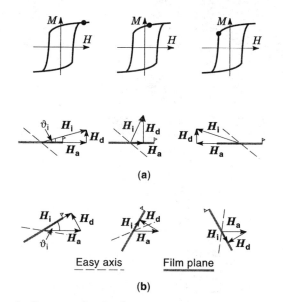

Figure 8. Compensation for demagnetization for metal evaporated tape. *Top row:* points on the hysteresis curve; *middle row:* external field, H_a, demagnetizing field, H_d, and internal field, H_i, for a standard hysteresis measurement; *bottom row:* external field, demagnetizing field, and internal field for a compensated hysteresis measurement. A proper compensation for demagnetization forces the internal field to stay on the same axis during hysteresis measurement.

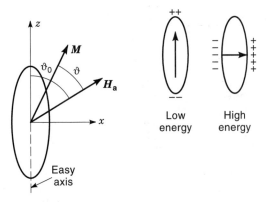

Figure 9. System of coordinates. In the state of lowest magnetostatic energy the magnetic poles are separated as much as possible.

enough to always ensure a homogeneous magnetization. In order to show a hysteresis, the magnetic material must have a magnetic anisotropy. In case of *shape anisotropy,* the magnetization of a single-domain particle seeks to orient itself such that it minimizes magnetostatic energy. As Fig. 9 illustrates, this occurs when the magnetic poles are separated as much as possible. In case of *magnetocrystalline anisotropy,* the crystal itself energetically favors certain magnetization orientations. As an example, consider a material with a hexagonal elementary cell such as cobalt. The c-axis of the elementary cell, which is the direction perpendicular to the hexagonal base plane, distinguishes itself from the directions in the base plane. For the particular case of cobalt, the c-axis is 'magnetically easy' and the magnetization likes to point along the easy axis.

For a Stoner–Wohlfarth particle with uniaxial anisotropy, the magnetic energy is:

$$E(\vartheta) = -\mu_0 M_s H_a V \cos\vartheta + \frac{1}{2}\mu_0 M_s H_A' V \sin^2(\vartheta - \vartheta_0) \quad (7)$$

Figure 9 illustrates the angle definitions. ϑ is the angle between the magnetization and the applied field, H_a, and ϑ_0 is the angle between the easy axis and the applied field. In Eq. (7), μ_0 is the permeability of free space, $4\pi\ 10^{-7}$ V·s/A·m, M_s is the saturation magnetization, H_A' is the effective anisotropy field, and V is the volume of the particle. The effective anisotropy field H_A' takes both the shape and the magnetocrystalline anisotropy into account:

$$H_A' = H_A + (N_\perp - N_\parallel)M_s \quad (8)$$

Here, H_A is the magnetocrystalline anisotropy field, and $N_\perp$ and $N_\parallel$ are the demagnetization factors perpendicular and parallel to the easy axis, respectively. Equation (8) assumes that the two easy axes coincide. In addition, Eq. (8) assumes that the shape of the specimen is an ellipsoid of revolution. In this case, the relation $N_\parallel + 2N_\perp = 1$ holds. Often the anisotropy constant, K_1, is used to describe the magnetocrystalline anisotropy energy. The anisotropy field, H_A, relates to K_1, as follows:

$$H_A = \frac{2K_1}{\mu_0 M_s} \quad (9)$$

Conceptually, the anisotropy field can be understood as a fictitious field that pulls the magnetization towards the easy axis.

The evaluation of Eq. (7) predicts magnetic hysteresis. The free parameter is the angle of the magnetization, ϑ. Figure 10 shows the energy according to Eq. (7) as function of ϑ. It is convenient to normalize Eq. (7) to $\mu_0 M_s H_A' V$ and to write:

$$h = \frac{H_a}{H_A'} \quad (10)$$

Depending on the field h, there can be either one or two energy minima (see Fig. 10). If there are two energy minima, two values for the magnetization can be assigned to one field value, that is, there is hysteresis. For higher field magnitudes, Fig. 10 illustrates that one of the minima becomes shallower until it disappears completely. At this point, the magnetization will switch irreversibly to the other energy minimum. Finding the energy minima of Eq. (7) thus determines the hysteresis loop. A necessary condition for an energy minimum is $dE/d\vartheta = 0$, which reads in normalized form:

$$2h\sin\vartheta + \sin[2(\vartheta - \vartheta_0)] = 0 \quad (11)$$

The stability of the magnetic state requires that $d^2E/d\vartheta^2 > 0$, which reads:

$$2h\cos\vartheta + 2\cos[2(\vartheta - \vartheta_0)] > 0 \quad (12)$$

The solution of Eq. (11) is not analytical, with the exceptions of the special cases $\vartheta_0 = 0°$ (easy axis parallel to the field), $\vartheta_0 = 90°$ (easy axis perpendicular to the field), and $\vartheta_0 = 45°$. Figure 11 shows the result for $\vartheta_0 = 0°$ and $\vartheta_0 = 90°$. For $\vartheta_0 = 0°$, starting from positive saturation, the magnetization remains on the easy axis until the applied field reaches the critical value. Then the magnetization reverses irreversibly to the opposite direction. The Stoner–Wohlfarth model predicts that the coercivity is equal to the effective anisotropy field. For the case $\vartheta_0 = 90°$ there are only reversible, that is, rotational, processes. For the intermediate cases $0° < \vartheta_0 < 90°$, the magnetization reversal process consists of both reversible and irreversible processes. There is an important difference between the switching field, h_s, and the coercivity, h_c. The coercivity is defined to be the magnetic field at which the projection of the magnetization on the field axis is zero. The

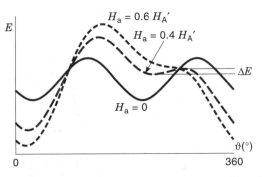

Figure 10. Magnetic energy for a single-domain particle as function of magnetization angle with the field, ϑ. Depending on the field H_a, there can be one or two energy minima. The energy barrier ΔE is required to switch the magnetization.

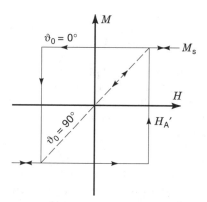

Figure 11. Magnetization reversal by coherent rotation in a single-domain particle: Hysteresis loops for the two cases easy axis aligned with the field ($\vartheta_0 = 0°$) and easy axis perpendicular to the field ($\vartheta_0 = 90°$).

switching field, or remanent coercivity, is the field at which the magnetization switches irreversibly. The coercivity can be equal or less than the switching field.

The magnetic recording process is vectorial in nature. Therefore, the angle dependence of the switching field is of importance. For the case of the Stoner–Wohlfarth model, the angle dependence of the switching field is:

$$h_s(\vartheta_0) = -\frac{1}{(\cos^{2/3}\vartheta_0 + \sin^{2/3}\vartheta_0)^{3/2}} \quad (13)$$

The negative prefix indicates that the switching occurs at negative field after the sample has been saturated in positive direction. The angle of the magnetization ϑ_c, with respect to the field axis just before switching, is:

$$\vartheta_c(\vartheta_0) = \vartheta_0 + \arctan\sqrt[3]{\tan\vartheta_0} \quad (14)$$

Real recording media consist of ensembles of single-domain particles. Therefore, the hysteresis loop of an ensemble is an average of the individual loops. For the case of a random distribution of the easy axes, the remanent magnetization is 0.5 M_s and the coercivity is $h_c = H_a/H_A' = 0.48$. This calculation assumes that there exists no magnetic interaction between the particles.

Chain of Spheres (Fanning)

While the Stoner–Wohlfarth model provides a good understanding of basic hysteresis phenomena, the predicted values for the switching fields are too high. The magnetization can also switch inhomogeneously. Rather than modeling elongated recording particles as prolate spheroids, Jacobs and Bean (27) suggested describing the shape anisotropy by a chain of spheres. γ-Fe$_2$O$_3$ particles, in particular, have shapes like peanuts. A chain of spheres has a lower shape anisotropy when compared with a prolate spheroid. The shape anisotropy field of a chain of $n \geq 2$ spheres is:

$$H_A^{shape} = \frac{M_s}{4} K_n \quad (15)$$

where

$$K_n = \sum_{j=1}^{n} \frac{n-j}{nj^3}$$

Jacobs and Bean discovered that the magnetization reversal is considerably facilitated if the magnetization vectors of adjacent spheres fan out rather than remaining parallel. Figure 12 illustrates that the additional magnetostatic energy partially cancels at magnetization reversal. This means that there is less resistance for the magnetization to overcome at reversal, making the magnitude of the switching field smaller.

As a simplification, the magnetization vectors are often assumed to fan out symmetrically. The switching field of *symmetric fanning* in a chain of n spheres ($n \geq 2$), with additional magnetocrystalline anisotropy along the chain axis, can be calculated analytically:

$$h_s(\vartheta_0) = \frac{1 - f_n}{f_n\sqrt{1 - f_n(2 - f_n)\sin^2\vartheta_0}} \quad (16)$$

$$\text{for} \quad \vartheta_0 \leq \arctan\left(\frac{1}{(f_n - 1)^{3/2}}\right)$$

where

$$f_n = \frac{3}{2}\frac{K_n + 4/\omega}{L_n}$$

$$L_n = \sum_{j=1}^{1/2(n-1) < j < 1/2(n+1)} \frac{n - (2j-1)}{n(2j-1)^3}$$

The parameter ω is proportional to the ratio between magnetocrystalline and shape anisotropy energy:

$$\omega = \frac{\frac{1}{2}\mu_0 M_s^2}{\frac{1}{2}\mu_0 M_s H_A} = \frac{M_s}{H_A} \quad (17)$$

For an infinitely long chain of spheres, Fig. 12 gives the result for the switching field as function of field angle. The switching field is normalized to its effective value, that is, the sum of the magnetocrystalline and shape anisotropy field. For dominating shape anisotropy, $\omega \to \infty$, the switching field is considerably reduced, especially for small angles ϑ_0 between the

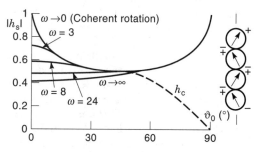

Figure 12. Reduced switching field ($h_s = H_s/H_A'$) as function of angle between applied field and easy axis, ϑ_0, for the fanning mechanism. In case of coherent rotation, the magnetization remains parallel, while they fan out otherwise. Increasing magnetocrystalline anisotropy ($\omega = M_s/H_A$) drives the reversal mechanism back to coherent rotation.

field and the easy axis. At larger angles ϑ_0, the fanning procedure does no longer efficiently lower the energy barrier for the magnetization reversal. Then the Stoner–Wohlfarth process (parallel magnetization vectors of the spheres) takes over. For strong magnetocrystalline anisotropy, $\omega \to 0$, the contribution of the shape anisotropy field to the total anisotropy field is small and the switching fields approach those of coherent rotation.

The chain-of-spheres model and variations thereof have been discussed in various papers, for example (28). It was noted that the magnetization inside these spheres does not remain homogenous as assumed before (29).

Nucleation Theory

While the fanning model successfully describes an inhomogeneous magnetization reversal process, it still fails to predict the size dependence of the switching field of fine particles. Using a micromagnetic approach, Brown (30) and Frei et al. (31) obtained more realistic switching fields than those predicted by Stoner and Wohlfarth. Micromagnetic theory works on a scale that is small enough to describe magnetization distributions in ferromagnetic bodies with sufficient accuracy, but large enough to replace the individual spins by a continuous magnetization (32–34). The total magnetic energy E of the particle is composed of exchange energy, magnetocrystalline energy, field energy, and magnetostatic energy:

$$E = \iiint \left\{ A[(\nabla m_x)^2 + (\nabla m_y)^2 + (\nabla m_z)^2] \right.$$
$$\left. + e_c - \mu_0 \boldsymbol{M} \cdot \boldsymbol{H_a} - \frac{1}{2} \mu_0 \boldsymbol{M} \cdot \boldsymbol{H_d} \right\} dV \quad (18)$$

where

A = exchange constant
m_i = direction cosines of the magnetization
e_c = magnetocrystalline anisotropy energy density
$\boldsymbol{H_d}$ = demagnetizing field
ΔV = volume element

Starting again with an ellipsoidal particle magnetized homogeneously by application of a very large positive field, the field is lowered slowly (in order to avoid dynamic effects) and, if required, reversed until the magnetization switches irreversibly. As long as the magnetization has not switched irreversibly, the equilibrium angle ϑ of the magnetization is still given by solving Eq. (11). The next step is to allow a small deviation from that equilibrium state (since any magnetization reversal must begin with a small change) and determine the total energy change associated with that deviation. It is important to allow the magnetization to leave its equilibrium state in an arbitrary manner, in order to find the *mode* that facilitates magnetization reversal the most. Mathematically, this corresponds to a linearization of the particle's magnetic energy around the current equilibrium state. The magnetization reversal mode is determined by minimization of the *second*-order energy change (the first-order energy change is zero). For the case discussed here, this energy change is

$$\Delta E^{(2)} = \iiint \{[(\nabla \xi)^2 + \lambda_\xi \xi]\epsilon^2 + [(\nabla \eta)^2 + \lambda_\eta \eta]\epsilon^2\} dV \quad (19)$$

Here, $\xi = \xi(x, y, z)$ and $\eta = \eta(x, y, z)$ are test functions, ϵ is a small quantity and the factors λ_ξ and λ_η have to be determined from the total energy. Applying a standard variational procedure to Eq. (19) leads to a set of differential equations known as *Brown's equations*. Any nontrivial solution of these equations indicates that either the state of homogeneous magnetization can be left or that coherent rotation occurs (Stoner–Wohlfarth switching). The largest external field strength at which this can happen is the *nucleation field*.

The most important nonuniform reversal process is the *curling* mode. When the magnetization leaves the uniform state, curling creates no additional poles in the plane perpendicular to the magnetization (see Fig. 13). This happens at the expense of exchange energy. Since the exchange energy is very strong and has a very short range, while the magnetostatic energy is weaker, but has a long range, the particle size has a strong influence on the nucleation field, H_n:

$$H_n = -H_A - \frac{k_c M_s}{2S^2} + N_\parallel M_s \quad (20)$$

where

$$S = \frac{R}{R_0}$$

$$R_0 = \frac{1}{M_s} \sqrt{\frac{4\pi A}{\mu_0}}$$

Here S is a reduced radius and k_c depends on the aspect ratio of the ellipsoid of revolution. The factor k_c varies between 1.08 (infinite cylinder) and 1.42 (very thin plate) (35,36). If the magnitude of the second term exceeds $N_\parallel M_s$, coherent rotation takes over. Equation (20) holds for alignment of both anisotropy axes with the external field.

The Euler–Lagrange equations deduced from Eq. (19) cannot be solved analytically for $\vartheta_0 \neq 0$ and $e_c \neq 0$. There is no curling in this case, because the magnetocrystalline anisotropy breaks the symmetry for $\vartheta_0 \neq 0$. Omitting the energy associated with the creation of additional poles at the beginning of magnetization reversal (which is strictly true only for curling), the following equation gives an upper bound for H_n (37):

$$H_n \cos \vartheta + H_A \cos 2(\vartheta - \vartheta_0) - M_s[N_\parallel \cos^2(\vartheta - \vartheta_0)$$
$$+ N_\perp \sin^2(\vartheta - \vartheta_0)] + \frac{k_c M_s}{2S^2} = 0 \quad (21)$$

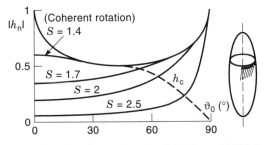

Figure 13. Reduced curling nucleation field ($h_n = H_n/H_A'$) for a prolate spheroid with aspect ratio 4 as function of angle between applied field and easy axis, ϑ_0. For small particle radius, S, the nucleation fields increase until they approach those for coherent rotation. No additional magnetic poles are created at curling.

Figure 13 shows the normalized nucleation fields as a function of field angle for elongated ellipsoidal particles for $e_c = 0$ (aspect ratio: 4). Similar as for fanning, increasing magnetocrystalline anisotropy drives the nucleation fields back to the Stoner-Wohlfarth solution. Figure 13 also shows that the switching fields of curling approach those of coherent rotation for small radius S. The general shapes of the angle dependence of the switching field of curling and fanning are similar (see Figs. 12 and 13). There exists always a reduced radius S which makes the curling solution agree with that of fanning. Therefore, a measurement of the angle dependence of the switching or the coercive field cannot identify the magnetization reversal mechanism.

In addition to rotation in unison and curling, *magnetization buckling* is another solution of Brown's equations for an infinite cylinder. Magnetization buckling is similar to the fanning mechanism. This mode introduces only very small changes in nucleation field, compared with the other two modes (31). It is of no practical interest, since it cannot occur in prolate spheroids of reasonable aspect ratio (38). In recent years, magnetization reversal has extensively been studied numerically; Schabes gives a review (39). Generally, the numerical calculations indicate that magnetization reversal starts at one end of the particle. The reader can obtain information on classical nucleation theory in (32–34,36,40,41).

Thermally Activated Magnetization Reversal Processes

For high-density recording, the particles in a recording medium should be as small as possible. As will be discussed in the section "Recording Physics," small particles lower the medium noise and potentially allow for smoother tape surfaces. On the other hand, extremely small particles—although magnetically ordered—lose their hysteresis. As shown in Fig. 10, stable magnetization states have local energy minima that are separated by energy barriers. If thermal energy can overcome these energy barriers, the critical fields discussed above will no longer be valid. In a noninteracting particle assembly with identical energy barriers, $\Delta E(h)$, for each particle, virtually any theory leads to (42):

$$\nu_{1,2} = f_0 \exp\left(-\frac{\Delta E(h)}{k_B T}\right) \qquad (22)$$

where

$k_B = 1.38 \times 10^{-23}$ J/K (Boltzmann's constant)
T = temperature in K
f_0 = "attempt" frequency

In Eq. (22) $\nu_{1,2}$ is the probability for one particle to switch from the magnetization state 1 to state 2 in the time interval dt. Within the model of coherent rotation, the energy barrier is identical to the particle volume. For very small particle volumes, $\nu_{1,2}$ will be so large that a particle assembly cannot remain magnetized after removal of a field (superparamagnetism). However, a stable remanence is not the criterion for a lower limit of particle size for recording media. The thermal energy seeks to completely randomize a magnetic system. The longer the time interval in which the thermal forces can operate, the more attempts they can make to successfully demagnetize the system. Similarly, if a magnetic field is applied, the

thermal forces will assist to switch the magnetization. For an assembly of Stoner–Wohlfarth particles with their easy axes aligned with the field, the energy barrier depends on the normalized field $h = H_a/H_A'$ as follows:

$$\Delta E(h) = \frac{1}{2}\mu_0 M_s H_A' V(1+h)^2 = \Delta E_0(1+h)^2 \qquad (23)$$

The time-dependent switching field can be calculated (43,44):

$$|H_s(t)| = H_A'\left\{1 - \sqrt{\frac{k_B T}{\Delta E_0}\ln\left(\frac{f_0 t}{\ln 2}\right)}\right\} \qquad (24)$$

Equation (24) imposes a practical limit on the minimum particle size useful for magnetic recording. Figure 14 shows some curves for the normalized switching field as a function of time. The height of the energy barrier ΔE_0 is varied and given in multiples of $k_B T$. For small values ΔE_0, the difference between the long-term coercivity, or "storage coercivity," and the short-term coercivity, or "writing coercivity" becomes large. A difference of about factor of two between the two coercivities may be tolerable.

Experimental data on the time dependence of the switching field follow the theoretical curves predicted by Eq. (24) well over many orders of magnitude (45). For very short times, the switching field increases sharply (46). This may indicate that Eq. (24) is no longer valid because the pulse length approaches $1/f_0$. Unfortunately, little is known about the escape frequency f_0. Brown gave an estimate in 1963 (47):

$$f_0 = \frac{\mu_0|\gamma|H_A'\alpha}{1+\alpha^2}\sqrt{\frac{\mu_0 M_s H_A' V}{2\pi k T}}(1-h^2)(1+h) \qquad (25)$$

In Eq. (25), $\gamma = -1.761\ 10^{11}$ 1/Ts is the gyromagnetic ratio, and α is the damping constant. Typical data for recording media yield 10^9 Hz for the order of magnitude for f_0. Using Mößbauer measurements a value of 10^{12} Hz has been reported (48), but the switching data fit better if the value of f_0 is 10^9 Hz (46). The reasonable agreement between theory and experiment is presumably fortuitous. It is well established that

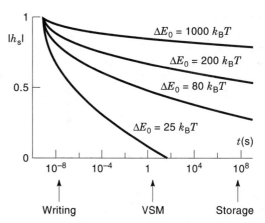

Figure 14. Reduced switching field ($h_s = H_s/H_A'$) as function of time scale for coherent rotation. The magnetic energy barrier at zero field, ΔE_0, is given in multiples of $k_B T$. For small energy barriers, the thermal energy reduces stability, which introduces a time-dependent coercivity.

magnetic recording particles do *not* reverse their magnetization coherently and the assumptions leading to Eq. (24) are therefore not valid. Equation (24) is not even valid for the Stoner–Wohlfarth model, since the field dependence of the energy barrier, Eq. (23) does not hold for arbitrary ϑ_0. A theoretical argument by Victora suggests that the exponent of 2 in Eq. (23) should be replaced by 1.5 (49).

An alternative approach to analyze thermally activated magnetization processes is the concept of a fluctuation field (50). The fluctuation field is best defined by equating two energies, namely, the thermal energy, k_BT, and the Zeeman energy of the magnetic moment M_sV_A in a fictitious fluctuation field, H_F:

$$H_F = \frac{k_BT}{\mu_0 M_s V_A} \quad (26)$$

Here V_A is the *activation volume*. Street and Woolley introduced time-dependent magnetization measurements (*magnetic viscosity*), which serve to determine fluctuation fields and activation volumes (51). For a Stoner–Wohlfarth particle, one expects the activation volume to be proportional to the particle volume. In most cases, fine magnetic particles have activation volumes smaller than the particle volume. The interpretation is that only a small portion of the particle is 'thermally activated' in the first instance. Once the small volume reverses magnetization, the rest of the particle follows. Using magnetic viscosity measurements, the fluctuation field of CrO_2 was found to be larger than that of Co doped γ-Fe_2O_3 (52). This is presumably due to the lower Curie temperature of CrO_2. Because of this large magnetic viscosity, CrO_2 tapes with lower nominal coercivity turn out to be equivalent in recording to their counterparts based on Co doped γ-Fe_2O_3. Magnetic viscosity measurements were also applied to standard and advanced MP and showed that, contrary to the expectation, the bigger particles were magnetically less stable than the smaller ones (53). The structure of the particles explains this result: the larger particles contain more (and smaller) crystal grains, which probably are the relevant units for magnetization reversal. For further study, the reader is referred to Refs. 42, 47, 50.

Print-through. A practical consequence of thermal activation is *print-through* (PT), which occurs in analog audio systems (1,54). PT is the unwanted copying of the recorded signal onto neighboring layers in the tape reel. Because of spacing and thickness loss factors, the wavelength with the maximum printing field is $2\pi t$, where t is the total tape thickness. For audiocassettes, this corresponds to a frequency of approximately 650 Hz. PT leads to audible echoes or even more disturbing foretastes of coming load passages. Since the human ear is sensitive to PT levels lower than −50 dB, only a small fraction of the particles need to have a small enough activation volume to be susceptible to the printing field.

Using larger particles reduces PT, but leads to an increased particle noise. Therefore, tightening the particle size distribution is a measure to improve PT. Another approach has been double-layer coating: the top layer uses finer particles for good noise performance, while the bottom layer uses larger particles to reduce PT. Digital systems, which are the only new systems being introduced, use only short wavelengths (less than a few micrometers). Then the spacing loss attenuates the stray fields and PT is not a problem.

RECORDING PHYSICS

When information is recorded magnetically on a tape, the recording head imprints magnetic patterns onto the medium. The magnetic layer of the tape serves to retain these patterns and to provide some means to retrieve the information. The stray field outside the tape reflects the magnetization of the tape and can thus be used for information retrieval. Since the field lines of the magnetic flux density must be closed, there is a magnetic field inside the magnetic material as well. This field—the demagnetizing field—opposes the magnetization. Since the demagnetizing field seeks to destabilize its own source, there has always been concern that excessive demagnetization may eventually destroy the recorded magnetization pattern. For tape recording, it is current understanding that, in particular, geometrical effects strongly influence the written magnetization patterns, which makes the question of demagnetization far less critical than initially thought.

Magnetic Recording Principle

Figure 15 sketches the basic recording principle. For writing information, a current is fed into the *recording head,* which moves relatively to the medium. Depending on the current direction, the stray field of the head switches the magnetization to the left or to the right. In case of digital recording, one always wishes to magnetize the medium up to saturation. Between the regions of saturated magnetization, there are *magnetization transitions.* For achieving a high recording density, the width of these transitions has to be kept as small as possible. The number of transitions (or flux changes) per length is the *linear density.*

Information is stored in digital magnetic recording by changing the distance between transitions. Since there is a clock in every digital system, the transitions are always separated by multiples of

$$B = vT_0 \quad (27)$$

where T_0 is the clock period and v is the head to medium velocity. Although not strictly correct, it is customary to call B the *bitlength.* Squarewave recording is used to study the fundamental recording behavior of heads and media. In this

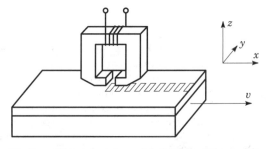

Figure 15. Recording principle and definition of the coordinate system. When the energized recording head is moved relatively to the medium, a magnetization pattern can be written onto the medium. Where the recorded medium is passed over the readback head, a readback voltage is induced.

case, the spacing between transitions is constant and one defines a wavelength λ as

$$\lambda = 2B \tag{28}$$

Typically, the width of the track is much wider than the transition spacing and the medium thickness. Therefore, track width dependencies are often neglected and the magnetic recording process becomes two-dimensional.

The magnetic flux, ϕ, emanating from the medium, is used to retrieve the written information. In case of inductive reading, any change of the stray flux that is sensed by the readback head, ψ, leads to an induced voltage:

$$\mathrm{emf} = -w \frac{d\psi}{dt} \tag{29}$$

Here w is the number of turns. Many recording systems use the same head for reading and writing. Magnetoresistive readback provides higher output voltages than inductive reading. While the bulk of the tape systems still use inductive readout, most of the newly introduced systems use a magnetoresistive transducer.

The recording principle for *perpendicular* recording is the same as that for longitudinal recording. The recording head shown in Fig. 15 is also termed *ring head*. Although a ring head can record on perpendicular media, it is more effective to use its magnetic counterpart, a *single-pole head*. The true magnetic counterpart to the ring head would be a single-pole head without a flux return path (55), but the efficiency of this structure is virtually zero. A magnetically soft layer underneath the recording layer provides a part of the required flux return path. This underlayer belongs magnetically to the head and physically to the medium. Such an underlayer has to be relatively thick [several times the thickness of the recording layer (22)] to prevent saturation at longer wavelengths. Unfortunately, the additional underlayer creates noise (56). ME tape has an easy axis tilted out of the film plane and is thus in between longitudinal and perpendicular media, but is identified as a longitudinal medium.

The recording process is most conveniently studied on a macroscopic scale, because a detailed discussion of the relevant magnetization processes on a micromagnetic level is too complicated. Even on a macroscopic scale, theoretical recording simulations require extensive computational efforts. For many purposes, simplified models are more appropriate than the more complex numerical models.

Magnetization and Demagnetizing Fields

In digital magnetic recording, the magnetization of a medium consists of magnetically saturated regions separated by magnetization transitions. The direction of the magnetization in neighboring "bits" points in opposite directions. Continuous magnetization distributions—which is a reasonable assumption if the particle size is much smaller than the bit pattern—create demagnetizing fields that can be calculated from:

$$\begin{aligned} \boldsymbol{H}_{\mathrm{d}} = &-\frac{1}{4\pi} \underset{\text{volume}}{\iiint} \frac{\nabla' \cdot \boldsymbol{M}'(\boldsymbol{r}')(\boldsymbol{r}-\boldsymbol{r}')}{|\boldsymbol{r}-\boldsymbol{r}'|^3} \, dV' \\ &+ \frac{1}{4\pi} \underset{\text{surface}}{\iint} \frac{\boldsymbol{n}' \cdot \boldsymbol{M}(\boldsymbol{r}')(\boldsymbol{r}-\boldsymbol{r}')}{|\boldsymbol{r}-\boldsymbol{r}'|^3} \, dA' \end{aligned} \tag{30}$$

Here the first expression gives the field due to volume charges $\rho_m = -\nabla \cdot \boldsymbol{M}$ and the second term that due to surface charges $\sigma_m = \boldsymbol{n} \cdot \boldsymbol{M}$, where $\boldsymbol{n}$ is a normal vector. For some special magnetization patterns, Eq. (30) can be evaluated analytically. For a tape of thickness δ and infinite width in the direction across the track, the volume charges correspond to a longitudinal magnetization M^{L} and the surface charges to a vertical magnetization M^{V}. Therefore, it is favorable to split Eq. (30) into longitudinal and vertical magnetization components. A sinusoidally magnetized tape with a constant magnetization through its depth (z-direction) creates sinusoidal demagnetizing fields in x- as well as in z-direction. Outside the tape, the stray flux (and the stray field) decays exponentially with distance (57,58):

$$\phi = \phi_0 \exp(-kz) \tag{31}$$

Here k is the wavenumber, $k = 2\pi/\lambda$. Eq. (31) is a direct consequence of the Laplace equation in two dimensions. Equation (31) can be converted to yield -54.6 dB z/λ. This *spacing loss* is the most important single factor for designing a recording system.

The decay of the stray flux in Eq. (31) does not depend on the orientation of the magnetization. Thus, there is no difference in stray field magnitude for longitudinal and perpendicular recording. If the tape magnetization rotates by an angle θ_0, the stray fields rotate by $-\theta_0$, (59). The fields *inside* the medium depend on the orientation of the magnetization and are evaluated here for two planes, $z' = 0$ (center plane of the medium) and $z' = \delta/2$ (surface plane of the medium) (55,60):

$$H_{\mathrm{d},x} = -M^{\mathrm{L}} \begin{cases} 1 - \exp\left(-\dfrac{k\delta}{2}\right) & \text{for } z' = 0 \\ \dfrac{1}{2}\left[1 - \exp(-k\delta)\right] & \text{for } z' = \dfrac{\delta}{2} \end{cases} \tag{32}$$

$$H_{\mathrm{d},z} = -M^{\mathrm{V}} \begin{cases} \exp\left(-\dfrac{k\delta}{2}\right) & \text{for } z' = 0 \\ \dfrac{1}{2}\left[1 + \exp(-k\delta)\right] & \text{for } z' = \dfrac{\delta}{2} \end{cases} \tag{33}$$

The primes indicate that the coordinate system is attached to the medium. Unprimed quantities will be used for the head system. If only the center plane of the medium, $z' = 0$, is considered, perpendicular recording seems to be much more favorable. When the recording density is increased, demagnetization increases for longitudinal recording, while it decreases for perpendicular recording. On the other hand, the demagnetizing field at high density always approaches 50% of the magnetization *at the medium surface*. This means that the external fields—which account for the reading signal—are not necessarily expected to be larger for perpendicular recording. A potential demagnetization advantage of perpendicular recording requires that the product $k\delta$ is large enough. In practice, perpendicular media must remain writable, which limits δ and, therefore, the product $k\delta$ (56). The presence of a soft magnetic underlayer does not change these conclusions (61).

Another useful field configuration that can be calculated analytically is an arctan-like magnetization transition. There is no physical justification for assuming that the transition shape is exactly like an arctan, but this approach has been widely used because of mathematical convenience:

$$M(x') = -\frac{2M}{\pi} \arctan \frac{x'}{a} \qquad (34)$$

Here, a is the *transition parameter,* which describes the transition sharpness. Equation (34) assumes that the magnetization transition does not change as a function of depth, that is, it is valid for thin media.

Several analyses have shown that the arctan transition has too long tails. The hyperbolic tangent or the error function more realistically describes the shape of a magnetization transition. Evaluation of Eq. (30) with Eq. (34) yields for the longitudinal demagnetizing field due to the longitudinal magnetization:

$$H_{d,x}^{L}(x',z') = \frac{M^{L}}{\pi} \left\{ \arctan\left(\frac{x'\left(\frac{\delta}{2}+z'\right)}{x'^2 + a^2 + \left|\frac{\delta}{2}+z'\right|a} \right) \right.$$
$$\left. + \arctan\left(\frac{x'\left(\frac{\delta}{2}-z'\right)}{x'^2 + a^2 + \left|\frac{\delta}{2}-z'\right|a} \right) \right\} \qquad (35)$$

The perpendicular demagnetizing field of the magnetization transition $M_x(x')$ is:

$$H_{d,z}^{L}(x',z') = \frac{M^{L}}{2\pi} \ln \left[\frac{x'^2 + \left(a + \left|\frac{\delta}{2}+z'\right|\right)^2}{x'^2 + \left(a + \left|\frac{\delta}{2}-z'\right|\right)^2} \right] \qquad (36)$$

For a vertical magnetization transition:

$$M_z(x') = -\frac{2M^{V}}{\pi} \arctan \frac{x'}{a} \qquad (37)$$

One obtains for the vertical demagnetizing field:

$$H_{d}^{V}(x',z') = \frac{M^{V}}{\pi} \left\{ \text{sgn}\left(\frac{\delta}{2}+z'\right) \arctan\left[\frac{x'}{a + \left|\frac{\delta}{2}+z'\right|} \right] \right.$$
$$\left. + \text{sgn}\left(\frac{\delta}{2}-z'\right) \arctan\left[\frac{x'}{a + \left|\frac{\delta}{2}-z'\right|} \right] \right\} \qquad (38)$$

The functional dependence of the longitudinal demagnetization field due to the vertical magnetization component is $H_{d,x}^{V}/M^{V} = -H_{d,z}^{L}/M^{L}$.

For a longitudinal magnetization transition, the demagnetization fields are strongest at some distance to the middle of the transition. For $M^{L} > H_c$, a perfectly sharp transition cannot occur because the demagnetizing fields will exceed coercivity and the transition must broaden. This imposes a limit on the transition parameter a (*demagnetization limit*). The demagnetization limit is of little practical importance. As outlined in the next section, typically, the transition length determined by the writing process itself is already larger than that imposed by the demagnetization limit.

For pure vertical magnetization, the strongest demagnetization occurs 'in the bit' which limits M^{V} to the coercivity. Since demagnetization does not hinder the vertical magnetization change at the transition, the magnetization can, in principle, increase up to its saturation value when the transition center is approached. From the considerations on sinusoidal magnetization, it follows, however, that the surface demagnetization fields limit the magnetization near the surface to $M_{max} = 2H_c$ at close transition spacing.

Readback

The reciprocity principle allows calculating the reading signal from a tape with a known magnetization pattern. The flux linkage ψ between the tape and a head having a field $\boldsymbol{h}_H$ when fed by a unit current is:

$$\psi = \mu_0 \iiint_{\text{medium}} \boldsymbol{M} \cdot \boldsymbol{h}_H \, dV \qquad (39)$$

Good approximations of the ring head fields are the formulas of Karlqvist (62) for distances larger than about 0.2 g from the head gap edges:

$$H_{H,x} = \frac{H_g}{\pi} \left[\arctan\left(\frac{g/2+x}{z}\right) + \arctan\left(\frac{g/2-x}{z}\right) \right] \qquad (40)$$

$$H_{H,z} = \frac{H_g}{2\pi} \ln \left[\frac{(g/2-x)^2 + z^2}{(g/2+x)^2 + z^2} \right] \qquad (41)$$

Here H_g is the deep gap field and g is the gap length. For distances closer to the gap edges, Szczech et al. give analytical formulas (63). Assuming a track width W being large compared with the other dimensions, a sinusoidal magnetization with peak value M_r homogeneous throughout the depth of the medium yields an induced voltage:

$$V_{0p} = \mu_0 M_r v \eta w W (1 - e^{-k\delta}) e^{-kd} \frac{\sin 1.13 \, g \cdot k/2}{1.13 \, g \cdot k/2} \qquad (42)$$

Here η gives the efficiency of the head, w the number of turns, d the head to medium separation, and v the relative velocity between head and medium. Equation (42) contains the spacing loss (e^{-kd}) discussed previously. Equation (42) also takes into account that the parts of the medium closer to the head suffer less from spacing loss than those, which are further away. If no losses occurred, one would expect that the output voltage is proportional to the medium thickness δ. The integration through the depth of the medium yields that the equivalent thickness δ_{eff} fully contributing to the output is:

$$\delta_{\text{eff}} = \frac{(1 - e^{-k\delta})}{k\delta} \delta \qquad (43)$$

The ratio $\delta_{\text{eff}}/\delta$ is called *thickness loss*. The name is somewhat misleading, since there is no loss due to the thickness rather than less gain than expected.

The *gap loss* [last term of Eq. (42)] takes into account that the output decreases if the wavelength λ approaches the (read) gap length g. For the Karlqvist approximation, the factor 1.13 in Eq. (42) is missing. This factor or similar ones appear for more accurate modeling of the head field and can be used for $\lambda < g$. The field components $H_{H,x}$ and $H_{H,z}$ form a

Hilbert pair, which means that, after a Fourier transformation, they have identical amplitude spectra, but their phase spectra are shifted by 90°. Note that Eq. (42), as it stands, holds for inductive heads. For magnetoresistive readout, the output signal no longer depends on the linear velocity.

For digital recording, the output of an isolated transition is of interest. Combining Eq. (29) and Eq. (39) for thin media, the induced voltage can be written:

$$V(x) = -\mu_0 v \eta w W \delta \int_{-\infty}^{+\infty} h_H(x'-x) \frac{\partial M(x')}{\partial x'} dx' \quad (44)$$

For a step-like magnetization transition, $a \to 0$ (i.e., dM/dx' is a delta function), Eq. (44) demonstrates that the voltage $V(x)$ simply follows the head field. The magnetization orientation determines whether the shape of the replay pulse samples the longitudinal or the vertical component of the head field, or a mixture thereof. If the transition sharpness is finite, $(a > 0)$, the readback pulse is approximately equal to that which would occur if a perfectly sharp transition would be written at spacing $d + a$. The equivalence of head to medium spacing and transition parameter is also seen from the Fourier transform of Eq. (34):

$$\mathcal{M}(k) = \frac{2jM_r}{k} e^{-|k|a} \quad (45)$$

Here $\mathcal{M}$ indicates the Fourier transform and $j = \sqrt{-1}$. Therefore, the effect of transition sharpness and head to medium spacing cannot be distinguished experimentally. The quantity $d + a$ is also called *generalized spacing*. For further study, the following literature is recommended: (60,61,64–67).

Recording Geometry

The *recording geometry* strongly controls the shape and width of the written transitions. For contact recording, omitting demagnetization is a reasonable assumption (68). Bertram et al. point out that the angle dependencies of the particle switching control the shape of the written transitions (69). It is convenient to plot the head fields in a form $|H_H|$ versus θ rather than using the conventional splitting into longitudinal and vertical field components ($|H_H|$ = magnitude of the head field and θ = angle of the head field with the film plane).

Figure 16 shows a parametric plot of the magnitude of the head field from a ring head. The field is plotted as function of

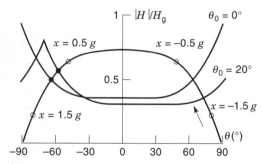

Figure 16. Polar plots: field magnitudes as function of angle. When the medium passes by the head, the field seen by the medium continuously changes magnitude and direction. The open circles indicate the magnitude and direction of the head field for some positions x in units of gap length g. The other two curves labeled $\theta_0 = 0°$ and $\theta_0 = 20°$ represent switching field curves for two different particle orientations. The filled circles indicate where the head field is larger than the medium switching field for the last time (writing location).

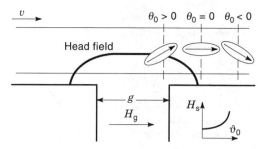

Figure 17. Recording geometry: the interplay between head field and angle dependence of the switching field creates semicircular magnetization patterns at writing.

angle that the tape sees when the head passes by. The hidden parameter is the location x as indicated on the curve. Here the angle $\theta = 0$ corresponds to a field pointing into the longitudinal direction. At large negative values x/g, the head field points into the positive z direction. With increasing x/g, the magnitude of the head field changes (and peaks at the gap edges for small z/g), while the field angle decreases continuously.

For incoherent magnetization processes, the switching field typically increases monotonically with the angle between easy axis and field (see the section titled "Nucleation Theory"). If such a particle is oriented longitudinally, the switching field curve has its lowest point at $\theta = 0$ when plotted into Fig. 16. If the particle is tilted in the xz plane ($\theta_0 > 0$), the switching curve shifts as illustrated in Fig. 16. The cusp for the switching field curve shown in Fig. 16 corresponds to the hard axis location. The state of magnetization "freezes" when the head field becomes smaller than the switching field, which is indicated by the filled circles in the figure. Figure 16 also illustrates the effect of different switching fields. For a larger switching field—which corresponds to a larger value for $h_r = H_r/H_g$—the freezing point will be closer to the gap. Therefore, the SFD has a direct effect on the transition width, even when there is *no demagnetization*. Since nonuniform particle orientations also result in a spread of writing locations, the effect of particle alignment is equivalent to an additional SFD (70).

These geometrical effects have a direct consequence for tape recording. Figure 17 sketches the writing locations for three identical particles with different orientation. The magnetization of each particle is assumed to lie on its easy axis. Evidently, the resulting magnetization forms a semicircular pattern. Tjaden and Leyten observed similar patterns as early as 1964 in a scaled up model system (71). Mallinson has shown theoretically, that a rotating magnetization can create a *one-sided flux* (72), which has double the field intensity on one side and none on the other. Due to the different magnetization directions and writing locations, the external stray fields of one-sided fluxes can either increase or decrease. For small wavelengths, this mechanism predicts nulls in the output curves as function of write current.

Transition Models

A very successful model for longitudinal recording on thin-film media is the *slope model* of Williams and Comstock (67). The main idea is that the head field gradient counterbalances the slope of the demagnetizing field at the center of a transition being written. The original slope model takes only longitudinal components into account, which implies that the medium switching field as function of field angle, H_s, is proportional to $1/\cos\theta$. The slope model discussed in the following is also valid for different angle dependencies of the switching field and easy axis orientations other than longitudinal (73).

The head fields and the mean demagnetizing fields are evaluated at a distance $z = d + \delta/2$ from the head surface (center plane of the medium). The first step is to determine the writing location x_w, which is the center of the magnetization transition being written. The slope of the total magnetization $d|\boldsymbol{M}|/dx$ is:

$$\frac{d|\boldsymbol{M}|}{dx} = \frac{d|\boldsymbol{M}|}{d|\boldsymbol{H}_{\text{tot}}|}\frac{d|\boldsymbol{H}_{\text{tot}}|}{dx} \qquad (46)$$

Assuming a transition shape according to an arctan function one obtains from Eq. (46):

$$a = -\frac{\delta}{4} - \frac{M_r^V \sin\theta_w}{\pi Q} \mp \frac{H_s}{\pi Q}\text{SFD}$$
$$\pm \sqrt{\left(\frac{\delta}{4} + \frac{M_r^V \sin\theta_w}{\pi Q}\right)^2 - \frac{M_r^L \delta \cos\theta_w}{\pi Q} \mp \frac{\delta H_s}{\pi Q}\text{SFD}} \qquad (47)$$

where $Q = \cos\theta_w\, dH_{\text{H}x}/dx + \sin\theta_w\, dH_{\text{H}z}/dx$. Here $dH_{\text{H}x}/dx$ and $dH_{\text{H}z}/dx$ give the gradients of the two head field components and θ_w is the field angle at the writing point x_w. The model can be extended to cover the shunting effect of the head by adding the magnetic images, but the solution is no longer analytical. Apart from the effects of magnetization, coercivity, and SFD on the a-parameter in longitudinal recording, the model explains essential features of recording on media with an inclined easy axis (ME tape). At positive θ_0 ($M_r^V > 0$)—easy axis orientation as in the left particle shown in Fig. 17—the model predicts a smaller a-parameter, that is, the transitions are sharper and the output is larger. In this case, a thicker layer can be written with a high field gradient as opposed to the other easy axis inclination. Inspection of Eq. (47) also shows that the perpendicular demagnetizing field narrows the transitions for $\theta_0 > 0$ while it broadens them for $\theta_0 < 0$. For ME tape, the optimum recording occurs when the writing takes place near the magnetically hard axis, that is, at writing the head field is *not* aligned with the easy axis. Typically in the range of an easy axis inclination of $\theta_0 = 20°$ to $50°$, the writing field direction crosses the hard axis, which means that the recorded magnetization changes direction. In such a case, the switching field curve shown in Fig. 16 would shift so far to the right, that the head field would intersect it on the left-hand side from the cusp. Since real ME tape is not perfectly oriented, the recording geometry can lead to double transitions, which means that there are two freezing points rather than one. One can also say that the trailing edge of the head can no longer overwrite entirely what the leading edge has written. These kinds of double transitions also occur in perpendicular media when written with a ring head (73).

Another type of writing interference occurs in isotropic media at low writing currents (25). In this case, the leading edge and the trailing edge of the head both record on the medium. The field at the trailing edge does not fully overwrite the recording of the leading edge any more (geometrical reasons

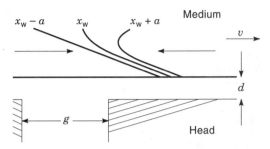

Figure 18. Sketch of transition center and width as function of depth for recording on acicular particulate media. The transition shape is not vertical due to geometrical effects.

prevent the head field from catching all of the switched particles again).

The slope model described so far holds for the thin-film approximation. Since tape, including ME tape, is a 'thick' recording medium, no one-dimensional model can describe tape recording well. Middleton et al. have suggested to decompose a thick medium into laminae, which themselves can be treated using the thin-film approximation (75). Other workers utilized the idea and several models have been developed, which yield results remarkably close the numerical models (70,76,77). Each sublayer is treated conventionally, but has a different magnetization, transition center and transition parameter. The magnetic interaction between the sublayers has to be calculated using Eqs. (35), (36), and (38).

Figure 18 illustrates a typical transition shape as it occurs for longitudinal MP tape. The three lines indicate how the transition shape varies as function of depth. The middle line gives the change of the transition center with depth, $x_w(z')$, and the other two lines show $x_w(z') \pm a(z')$ indicating the transition width. The demagnetization can broaden the transition by either increasing the transition tilt or increasing $a(z')$. Details depend on the angle dependence of the switching field, that is, the recording geometry. Figure 18 also helps to understand the recording behavior of thin-layer MP tapes. In a thin-layer MP tape, the parts of the medium far away from the head do not exist. The existence or nonexistence of the deeper layers does not have a first-order effect on what happens in the upper part of the medium, since geometric effects rather than magnetic effects dominate the recording. Theory and experiment indeed did not show significant changes in recording output for short wavelength for thick- and thin-layer MP tape (7,70). As expected, the thin MP tape shows higher overwrite and lower output at long wavelength. As a consequence of the reduced layer thickness, the saturation curves for thin-layer MP tape show a broader maximum, which is a second-order effect.

Modeling of ME tape is more difficult than modeling of longitudinal tape. Cramer has developed a numerical model for ME tape that is very successful (78). Stupp et al. report a simple multilayer slope model for ME tape, which agrees well with Cramer's numerical model (76). Although similar and successful for longitudinal tape, Richter's model failed for ME tape (70). The two slope models solve the nonlinear equations differently. This seemingly trivial detail directed the solution such that the transition shape follows the column inclination for the successful model. The unsuccessful model ended up with a different transition shape similar to that reported in

Ref. 69. In hindsight, the correct explanation is that there is a magnetic correlation in ME tape that forces the transition to follow the columns. The iteration used in the successful model implicitly assumes this correlation. The existence of such a correlation is consistent with noise investigations (79).

Noise

At playback, the reading head senses a limited volume of the tape. The average magnetization of all magnetic particles in that volume determines the signal. The deviation from the mean magnetization is noise. The more particles are contained in the sensing volume, the smaller the deviations of the mean, and the larger the signal-to-noise ratio, SNR, will be. Simple statistical arguments indicate that the (power) SNR is proportional to the number of particles per unit volume. Since the particles are much smaller than the recording dimensions, *particulate noise,* as described above, is approximately additive.

Particulate noise had been regarded as independent from the magnetization state for a long time. Today it is known that the particulate noise contribution should be largest when the medium is not magnetized, that is, the noise is maximum at the magnetization transitions. Therefore, the noise is not *stationary.* Tape media—including ME tape—have a relatively small volumetric packing fraction, typically smaller than 50%. According to Mallinson, the wide-band signal-to-noise ratio is (80):

$$\text{SNR} \approx \frac{\overline{m}^2 n W \lambda^2}{2\pi(1 - p\overline{m}^2)} \qquad \overline{m} = \overline{M}/M_r \qquad (48)$$

where λ is the minimum wavelength occurring in the recording system, $\overline{M}$ is the mean magnetization sensed by the head, p is the volumetric packing fraction of the tape, and n is the number of particles per unit volume. The tape volume sensed by the head is proportional to $W\lambda^2$, which directly appears in Eq. (48). Due to the relatively small packing fraction, the nonstationarity of the noise is removed, to some extent.

While Eq. (48) predicts the noise to be largest in an ac state, experiment shows that most tapes are noisiest when dc erased. Other noise sources cover the (uncorrelated) particle noise (64). Clusters of particles that have not been separated during the dispersing process and surface roughness effects are reasons for this additional noise.

Especially for particulate tape, *modulation noise* is of great interest. Modulation noise is multiplicative in nature: the noise becomes larger with larger signal. It occurs when the size of the noise sources is no longer small compared with the trackwidth. In practice, it is most difficult to achieve a very smooth surface that successfully eliminates modulation noise. Long wavelength modulation noise is easily recognized in a noise spectrum by the skirts around the signal, or *carrier* (64). In digital recording, modulation noise causes the error rate performance to deteriorate (81). In analog recording, the modulation noise interferes with the modulated signals to be recorded.

Analog Audio Recording (Ac Bias)

Analog audio recording aims directly to reproduce the sound waves of music or speech as a magnetization pattern on a tape. The human ear is sensitive to frequencies between 20 Hz and 20 kHz, and the faithful reproduction of music requires a bandwidth of 50 Hz to 15 kHz. This very large frequency ratio means that the recording conditions are very different from those for digital systems.

In the ubiquitous compact cassette (CC) format, the tape speed is 4.75 cm/s, and the wavelengths corresponding to 50 Hz and 15 kHz are 950 μm and 3.2 μm, respectively. Considering the 'thickness loss' term, Eq. (43), the appropriate depths of recording for these two signals would be ~300 μm and ~1 μm. A second distinguishing feature of the analog tape system is that the system needs to have a linear amplitude response. This requirement conflicts with the essentially nonlinear mechanism of magnetic remanence, so a linearizing process is necessary. In early tape recorders, a dc bias was added to the signal. In the magnetization versus field diagram, this serves to move the recording point to the nearly straight part of the isothermal remanence curve close to the coercivity (Fig. 19). The bottom of Fig. 19 illustrates that the zero signal level moves to a finite depth of recording. The great disadvantage of this procedure is that the upper layers of the tape are magnetically saturated, but have no signal contribution and high noise. The addition of a large high-frequency dither signal is therefore the method now universally used. Although the dither signal is symmetrical, it is, perhaps unfortunately, termed *ac bias,* in analogy to the dc bias described above. It will be seen that the application of the dither signal can be understood as an anhysteretic magnetization process. Figure 19 shows that the anhysteretic remanence curve (labeled M_{ar}) is more sensitive to signal than the dc biased IRM curve. Most importantly, the anhysteretic remanence is symmetrical about the origin, which assigns zero signal to a demagnetized tape, that is, zero signal is associated with low noise.

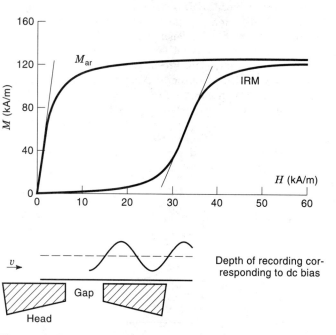

Figure 19. In contrast to the initial remanent magnetization (IRM) curve, the more linear anhysteretic remanence curve assigns zero magnetization to zero signal (or field). The bottom figure shows that dc bias, which would be required to linearize the IRM curve, results in a writing depth modulation.

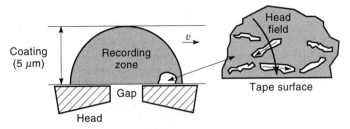

Figure 20. Recording zone for ac-bias recording. Due to the overbiasing of the short wavelengths, the head field relevant for the recording is almost vertical, which tends to demagnetize the tape.

Figure 20 shows the essential recording geometry of a standard cassette recorder. The recording gap is typically 2 μm and the coating thickness 5 μm. In order to use the full tape thickness for maximum long wavelength response, the bias is set to write through the entire magnetic coating. This means that the head current and depth of recording are set far too high for optimum recording of the highest frequencies. The enlargement of Fig. 20 illustrates one negative consequence of this overbiasing: near the surface, the head field relevant for the switching is almost vertical. Thus, there is a tendency to demagnetization, because differently inclined particles are magnetized in different directions. This mechanism is identical to that leading to the double transitions in ME tape, as previously described. Improved orientation of the particles diminishes this effect.

Two important performance parameters are the maximum output level (MOL) and the saturation output level (SOL). In the CC system, MOL is the 333 Hz output at which the third harmonic distortion reaches 3%, and the SOL is the maximum output attainable at 10 kHz. MOL and SOL are shown as function of the ac bias current in Fig. 21, confirming that at the specified bias, MOL is not far from its maximum and that the 10 kHz signal is grossly overbiased. Variations in tape coercivity mirror the dependence on bias current, so, because the bias current is set for the recorder, the coercivity must be tightly controlled. Presumably, this high sensitivity to coercivity has led to a general overestimation of the importance of coercivity in magnetic recording.

Applying the ac bias signal of 80 kHz to 100 kHz corresponds to an attempt to write a wavelength of about 0.5 μm onto the tape. An audio recorder is not designed to record such a short wavelength and the bias signal is heavily

damped. Instead of modeling a long sequence of blurred bias frequency transitions, it is usual to subsume the ac signal into the magnetic properties of the tape and treat the recording process as one of anhysteretic magnetization at the signal wavelengths. Anhysteretic magnetization is the process by which to a small offset field is added a large ac. The ac (bias) field is slowly reduced to zero from a large value sufficient to switch all the particles. The sample traverses many hysteresis cycles, destroying the memory of any previous magnetization and converging on a single-valued function of the offset field. [Anhysteretic magnetization does not quite correspond to ac bias recording: while the offset field remains constant for anhysteretic magnetization, it decays together with the ac field in a recording. This *modified anhysteretic process* is similar to the anhysteretic process (82).] Anhysteretic magnetization, M_a, with the offset field applied is always higher than anhysteretic remanence, M_{ar}, with the offset field removed. The difference between the two is small along the recording direction of well-oriented tapes, and M_{ar} will stand for both in the following discussion. The initial slope of the curve is called the *anhysteretic remanent susceptibility, χ_{ar},* and the main task is to explain its finite value. If the ac field is reduced sufficiently slowly, that is, the decrement ΔH_{ac} between subsequent half cycles is smaller than the magnitude of the signal field, all particles must be magnetized in the direction of the signal field. Therefore, one expects χ_{ar} to be infinite, which is in contradiction with experiment (see Fig. 19). One reason for χ_{ar} to be finite is that thermal activation introduces statistical uncertainty in particle switching, but the effect is rather small. For magnetic tapes, which are rather densely packed, the effects of interparticle magnetic interactions dominate. A graphical way to explain the effect of these interactions is the Preisach diagram (83), a version of which is shown in Fig. 22. The Preisach diagram is a particle density plot. In the representation used here, the strength of the Preisach function is indicated by the contours shown in

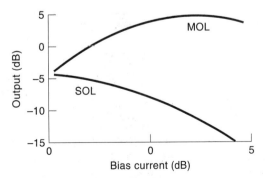

Figure 21. The ac bias current cannot be chosen to simultaneously optimize both the Maximum Output Level (MOL) and the Saturation Output Level (SOL).

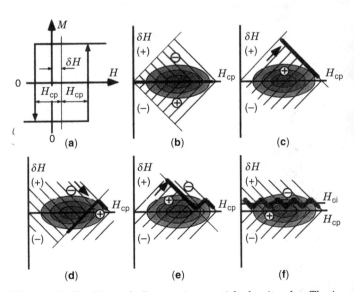

Figure 22. The Preisach diagram is a particle density plot. The intrinsic coercivity, H_{cp}, is on the abscissa and the interaction field, δH, is on the ordinate. The interaction field δH shifts the hysteresis loop (a). Depending on whether the applied field is increasing or decreasing, the boundary between the particles, which can and cannot be switched, is oriented at 45° or 135°.

Figs. 22(b)–22(f). The abscissa of the Preisach diagram gives the intrinsic particle coercivity, H_{cp}, without any interaction, while the ordinate represents the (magnetostatic) interaction field, δH, which the particle experiences. δH is created by all the other particles. As Fig. 22(a) illustrates, a positive interaction field δH shifts the hysteresis loop to the right. A decaying ac bias field will thus always negatively magnetize particles with positive interaction field, and vice versa. Figure 22(b) shows a representation of an ideally erased state ($M = 0$), with equal populations magnetized in either direction.

Consider now an increasing field applied to the (demagnetized) particle assembly. The negatively magnetized particles can switch to positive magnetization, if the applied field, H_a, is bigger than the coercivity augmented by the interaction field: $H_a > H_{cp} + \delta H$. Since $\delta H > 0$ for the negatively magnetized particles, the applied field can remagnetize all particles below $H_a = H_{cp} + \delta H$, which is the heavy line shown in Fig. 22(c). If the applied field is decreased again, those particles can switch back to negative magnetization, for which holds: $H_a > H_{cp} - \delta H$. In this case the terminating line rotates by 90°, as indicated in Fig. 22(d). Continuing the field sequence leaves a net magnetization without saturating the sample as in the interaction free discussion above [(Figs. 22(e) and (f)]. In small offset fields, H_{Si}, the magnetization is proportional to the density of particles near the H_{cp} axis. Thus, the Preisach representation qualitatively accounts for the finite χ_{ar}. The reader can find more detail in Refs. 58, 65, 84, 85.

While the Preisach model qualitatively describes magnetic interaction effects, it remains unsatisfactory, because the nature of the interaction is an assumption rather than a natural consequence of the model. In contrast to the Preisach model that merely assumes "positive" or "negative" interactions, Kneller (86) calculated the mean interaction field including structural information. He treated magnetic interaction as a *dynamic* interaction field, which is controlled by the magnetization. Using this concept, Kneller et al. calculated the anhysteretic susceptibility for an ensemble of infinite cylindrical particles (87):

$$\chi_{ar} = \frac{0.5p(1-p)M_s^2}{H_r^2(1-S_q^2)} \qquad (49)$$

After empirically replacing the factor 0.5 with 0.27 to account for SFD > 0, Eq. (49) is in good agreement with experiment. Köster (88) also found that to a good approximation, the M_{ar} curves of all oriented tape samples followed the universal relationship

$$\frac{M_{ar}}{M_r} = 0.569\frac{H}{H_{ar}} - 0.0756\left(\frac{H}{H_{ar}}\right)^3 + 0.0065\left(\frac{H}{H_{ar}}\right)^5 \qquad (50)$$

where H_{ar} is the offset field for which M_{ar} becomes 50% of the saturation remanence M_r. Using this expression, 3% MOL corresponds to a magnetization $M_r/2$; linearity has been bought at the expense of using only half of the potential magnetization and dynamic. The orientation and SFD have a subordinated but still important beneficial influence on the linearity and, of course, better orientation increases M_r.

Bertram (89) using the anhysteretic model calculated the magnetization profile in the tape coating and found, because at the tape surface the head field is almost perpendicular, the magnetization increased with depth into the tape. Details

depend on the angular switching behavior of the particles and the degree of orientation. Generally, higher orientation and M_r, together with narrower SFD, improve the output, while the correct balance between MOL and SOL is set by the coercivity.

Further Aspects of Recording

- *Erasure and Overwrite.* The hysteresis energy dissipated in recording a magnetic transition is sufficient to cause local heating of 0.1 K. Consequently, there is no physical or chemical damage to the tape and, from the magnetic standpoint, the tape may be rerecorded infinitely often. In order to reuse the tape, however, the old information must be erased. When the recording head itself is used for this purpose, several difficulties arise, owing to the small spatial extent of the head field. Clearly, if the head is incorrectly positioned or lifted from the tape surface by a speck of dirt, then the field will not reach all parts of the written track. A more subtle effect arises because the head field zone is so narrow that it is bridged by the field emanating from the unerased portion of the tape. In ac-bias recording, it has been shown (90) that even the tape noise can in this way be rerecorded and amplified leaving the so-called bias noise 3 dB above the true statistical particle noise. Analog recorders therefore always carry additional erase heads with gaps 5 to 10 times that of the recording head. Bulk erasure gives good results and is one of the last steps in tape manufacturing, where it is necessary to remove the magnetization induced by the orienting magnet as well as any control signal used by the processing equipment. The magnetic annealing effect in some cobalt-modified iron oxide tapes can hinder erasure [a 20 dB increase in residual signal was reported in (91)]. In digital recording, the old information is often overwritten directly. Overwriting is a very complex process: apart from incomplete erasure, the old information modulates the new information being written and complicated interference phenomena can occur. For improving overwrite behavior, a narrow SFD is beneficial. The different materials differ in overwrite behavior. In case of particulate media, Co-doped γ-Fe_2O_3 and BaFe are most difficult to overwrite, γ-Fe_2O_3 and MP are better to overwrite, and CrO_2 is best. ME tape outperforms all particulate media, which is due to the narrow SFD and the perpendicular magnetization component.

- *Writing Spacing Loss.* This adds to the reading spacing loss. The total spacing loss for acicular tape media is about 100 dB d/λ which agrees well with theory (69,70). The writing spacing loss is smaller in thin magnetic layers.

- *Length Loss.* If the length of the particles approaches the wavelength of interest, the output decreases:

$$LL = \frac{\sin(\pi L/\lambda)}{\pi L/\lambda} \qquad (51)$$

This analysis assumes the length loss to be independent of the recording process. Obviously, the transition parameter cannot become smaller than L/π, which has to be considered at the writing process. For investigations on Length Loss, see Refs. 92–94.

• At writing, transitions in longitudinal recording are shifted away from the gap due to the effect of the demagnetizing field of the previous transition. Similarly, transitions in vertical recording are drawn closer to the gap. Therefore Mallinson suggested that media with an oblique easy axis—ME tape—could eliminate the *nonlinear transition shift,* that is, the shift due to demagnetization (95). Later it was proven experimentally, that there is indeed an inclination angle of the easy axis that eliminates nonlinear transition shift (96).

RECORDING SYSTEMS

Linear Recorders

The number and variety of recording systems is huge. One can classify the recording systems into analog and digital or into linear and helical-scan systems. An example for an analog *linear recorder* is a tape deck. Figure 23 shows the basic mechanism of a linear tape recorder. After unwinding, the tape passes by an erase head, a recording head, and a replay head. Alternatively, one head can operate as both the recording and the replay head. All heads are stationary. The erase head has a large gap length that ensures full erasure of the tape. The tracks in such a recorder are parallel to the tape running direction, which gives the linear recorder its name. In audio recorders, the tracks 1 and 3 are used to run in one direction, and tracks 2 and 4 are used in the other. This minimizes crosstalk. Open-reel tape systems for audio recording operate in a linear mode. In 1963, Philips invented the compact cassette as an alternative for the bulky and difficult-to-handle open-reel tape recorders. The compact cassette systems have smaller dimensions and a lower head-to-medium speed, which sacrifices performance. Later, better tapes, heads, and electronics made up for the performance loss. Table 2 lists some magnetic properties of IEC I (standard compact cassette tape), and IEC II (improved compact cassette tape). The magnetic material in IEC II is CrO_2 or Co-doped γ-Fe_2O_3, while IEC I tape utilizes γ-Fe_2O_3 particles.

The tape speed and the minimum achievable wavelength determine the data rate or the maximum frequency of a linear recorder. Since high tape speed is undesirable because of high tape consumption, the frequency response of a linear recorder suffers. In order to increase the data rate, several heads can be operated in parallel. The operation of parallel channels requires a good head-to-tape contact over a large region, which is difficult to achieve. An example for a system that operates successfully nine parallel tracks is the digital compact cassette (DCC) system.

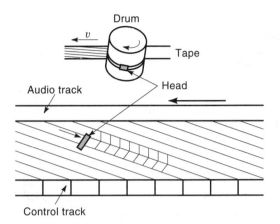

Figure 24. Helical scan principle: The tilted head drum records slanted tracks on the medium. The tape movement and the drum rotation are two independent, but simultaneous motions. The magnetization directions of adjacent tracks show an additional tilt (azimuth recording) to prevent side reading.

Helical-Scan Recorders

High-frequency applications such as video recorders use the helical-scan principle. In contrast to the linear recorders, the heads in helical-scan recorders are not stationary but mounted on a drum. There are two simultaneous motions in a helical-scan recorder: (1) the tape moves with a constant, relatively slow speed (few centimeters per second), and (2) the head drum rotates with a high speed (several meters per second). The axis of the head drum is not oriented parallel to the transverse direction, but it is tilted by a small angle. This results in slanted written tracks. Figure 24 shows a sketch of the recording configuration as well as a typical track format on the tape for a VHS recorder.

Analog video is always recorded using frequency modulation. Frequency modulation is chosen because it is insensitive to the amplitude changes caused by small surface variations at such short wavelengths. The modulated signal is amplitude limited and looks just like a digital signal. Demodulating the signal involves detecting the zero-crossings of the output signal, so again there is great similarity to digital techniques. Helical-scan became universal after several failed attempts to build mechanically sound high-velocity linear drives.

As Fig. 24 indicates, the head is wider than the spacing between the tracks (*track pitch*). Consequently, the head will always read a portion of the adjacent track(s). To prevent crosstalk, the gap orientations of the recordings of adjacent tracks are tilted against one another. This is called *azimuth recording.* Tilting the writing direction against the reading direction by an angle β reduces the output:

$$L = \frac{\sin\left(\dfrac{\pi W}{\lambda}\tan\beta\right)}{\dfrac{\pi W}{\lambda}\tan\beta} \tag{52}$$

Azimuth recording thus requires at least two heads on the head drum. The two heads with different azimuth angle record and read alternately. In analog video systems, the length of the track corresponds to one field or half picture. Therefore, the recorder can produce a picture (*still-frame*) without any tape motion. Achieving a reliable still-frame operation is one

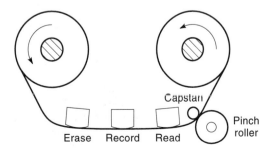

Figure 23. Sketch of a linear recorder with three heads.

of the most challenging tasks for a tape manufacturer. For tracking, there is often (a) control track(s), as indicated in Fig. 24.

The audio track shown in Fig. 24 is recorded using standard ac bias with a stationary head (see the section titled Analog Audio Recording). Due to the very low tape speed, the quality of the audio recording is poor. Recording the frequency modulated (FM) audio information with the rotating head improves the audio quality significantly. The audio information is recorded in a separate frequency band. In case of VHS, the audio information is first recorded using a large gap head and subsequently overwritten with the video information. The video information is recorded with a head with small gap length and occupies only a very shallow layer. This suffices for the short wavelength region relevant for the video signal. On the other hand, the audio signal remains essentially untouched.

The most important analog consumer video systems are VHS, S-VHS (Super-Video Home System), 8 mm and Hi 8. Other systems (e.g., Video 2000, Betamax) did not penetrate the market. The next consumer video system is DVC (digital video cassette), which is the counterpart to DVD (digital video disk). Professional recorders for broadcasting (such as D1 to D5, DVC-Pro, Betacam SP and digital Betacam) are similar in principle, but considerably more complex than the consumer devices with up to 32 different heads on one single drum. Most of these systems have NTSC (National Television System Committee) and PAL (phase alternating line) versions. The PAL and NTSC versions differ not only in the technology directly related to the broadcasting schemes, but also slightly in recording density and sometimes even in tape grade. Table 3 summarizes some data for the most important recording systems. Although several tape materials can be used for one system, switching between different types may not be advisable. The video heads stick out of the drum somewhat. Therefore, the head is pressed into the tape, which then forms a "tent." Mechanically, each head contours itself with time such that the head-to-medium contact is optimal. Changing the tape material may have the effect that the contouring process starts again from the beginning.

The new digital video systems show a new trend in storage philosophy. Not too long ago, it seemed mandatory that digital tape systems should store the full information. Digital Betacam and the DVC system utilize data-compression schemes. In contrast to audio or even data information, video information is highly redundant. Today much effort is directed toward identifying redundant information and to process the relevant data as economically as possible. Data compressions can reduce the amount of data by a factor of up to 50, without noticeable quality loss. This does not eliminate the need for higher recording density, since the playing times are still limited, even with data compression. It should be mentioned that helical scan recorders, derived from digital audio and video systems, are also used in digital data storage (DDS) systems.

Videotape Duplication. Most of the videotape produced enters the market as prerecorded tape. It is not attractive to record these tapes at the normal speed of a video recorder. There are two principles that allow to record videotape at much higher speed. In case of *thermal duplication,* the video information is recorded onto a CrO_2 tape. CrO_2 has a very low Curie temperature (120°C). At duplication, the CrO_2 tape is heated up to the Curie temperature and subsequently cooled down with a master tape in direct contact to it. At the cooling process, the magnetic pattern of the master—MP or Co-doped γ-Fe_2O_3 tape—is imprinted onto the slave. The duplication process can be very fast, if the tape is heated with a laser. CrO_2 is the only material suitable for thermal duplication. Another method for duplication is *anhysteretic duplication.* In this case, the master and the slave tape are subjected to an alternating transfer field while they are in contact. The master tape has about three times the coercivity as the slave and is recorded in a mirror image. The recording process is very similar to ac bias recording. The decaying alternating field takes the role of the ac bias and the stray field—which is added, as in ac bias recording—contains the information to be recorded. For further reading, see Refs. 85, 97–99.

MAGNETIC TAPE RECORDING: OUTLOOK

Magnetic recording and, in particular, magnetic tape recording, is a very old technology. The principles of magnetic

Table 3. Characteristic Data of Some Recording Systems

System		TW (mm)	TP (μm)	BL (μm)	AD (bit/μm^2)	Thickness (μm)
VHS	Analog video (PAL)	12.7	49	0.51	0.04	20
	Analog video (NTSC)	12.7	58	0.66	0.03	20
S-VHS	Analog video (PAL)	12.7	49	0.35	0.06	20
	Analog video (NTSC)	12.7	58	0.42	0.04	20
8 mm	Analog video (PAL)	8	34	0.29	0.10	10
	Analog video (NTSC)	8	20	0.35	0.14	10
Hi 8	Analog video (PAL)	8	34	0.21	0.14	10
	Analog video (NTSC)	8	20	0.25	0.20	10
R-DAT	Digital Audio	3.81	13.5	0.41	0.18	10
DVC	Digital video	6.35	10	0.24	0.42	7
HD, 1.4 MB	Floppy disk	—	188	1.46	0.0036	67
100–120 MB	Floppy disk	—	10	0.75	0.13	67
Data tape	Cartridge, 1998	8	69	0.32	0.046	~7
		6.35	34	0.5	0.06	~7

TW: tape width, TP: track pitch, BL: transition spacing, AD: areal density, and thickness of the tape/disk. Analog systems are converted for comparison, track pitch is given for short-play mode.

recording on particulate recording tape are well established, since the middle of 1940s. Since then, the technology has followed an evolutionary rather than a revolutionary path. Decreasing the particle size, narrowing the particle size distribution, and improving the manufacturing technology has resulted in considerably improved tape performance. The main path for future *particulate tape* is MP. To date, *thin-film* tape (ME tape) has shown the best recording performance. ME tape has been commercially available since the beginning of the 1990s. Present estimates indicate that both types of media, particulate and thin film, can support significantly higher areal storage density than those already demonstrated (7).

While the recording performance data indicate that the difference between thin-film media and particulate media is moderate, thin-film media may take advantage of their physical 'thinness'. The coating thickness of particulate tapes cannot be made smaller than 1 μm or 2 μm. This also holds for the double-layer tapes, which have a thin magnetic layer, but require a fairly thick (1 μm to 2 μm) underlayer. Base films can be as thin is 5 μm or even less. The thickness of the two layers adds much more to the total thickness of a tape than the thin metal film does. In the end, more tape fits into a cassette, which favors thin-film tape.

Magnetic tape recording offers very large storage capacity per volume at low cost. While the data rates are high, the access times are longer when compared with recording technologies that use disk-shaped media. These features make tape ideally suited for applications in which access time can be traded off against storage capacity.

BIBLIOGRAPHY

1. E. Köster, Particulate Media, in C. D. Mee and E. D. Daniel (eds.), *Magnetic Recording Handbook,* 2nd ed., New York: McGraw-Hill, 1996.

2. H. Jakusch and R. J. Veitch, Paticles for magnetic recording, *J. Inf. Rec. Mats,* **20**: 325–344, 1993.

3. M. P. Sharrock, Particulate magnetic recording media: A review, *IEEE Trans. Magn.,* **25**: 4374–4389, 1989.

4. G. Bate, Magnetic recording materials since 1975, *J. Magn. Magn. Mat.,* **100**: 413–424, 1991.

5. E. Schwab and H. Hibst, Magnetic recording materials, in R. Cahn, P. Haasen and E. Kramer (eds.), *Materials Science and Technology,* Vol. 3B, Weinheim: VCH, Verlagsgesellschaft, 1993.

6. H. Inaba et al., The advantages of the thin magnetic layer on a metal particulate tape, *IEEE Trans. Magn.,* **29**: 3607–3612, 1993.

7. H. J. Richter and R. J. Veitch, Advances in magnetic tapes for high density information storage, *IEEE Trans. Magn.,* **31**: 2883–2888, 1995.

8. S. Saitoh, H. Inaba, and A. Kashiwagi, Developments and advances in thin layer particulate recording media, *IEEE Trans. Magn.,* **31**: 2859–2864, 1995.

9. K. Shinohara et al., Columnar structure and some properties of metal evaporated tape, *IEEE Trans. Magn.,* **20**: 824–826, 1984.

10. T. Kawana, S. Onodera, and T. Samoto, Advanced metal evaporated tape, *IEEE Trans. Magn.,* **31**: 2865–2870, 1995.

11. M. S. Cohen, Anisotroy in Permalloy films evaporated at grazing incidence, *J. Appl. Phys.* Suppl., **32**: 87S–88S, 1961.

12. J. M. Alameda, M. Torres, and F. López, On the physical origin of in-plane anisotropy axis switch in oblique-deposited thin films, *J. Magn. Magn. Mat.,* **62**: 209–214, 1986.

13. H. J. Richter, An analysis of magnetization processes in metal evaporated tape, *IEEE Trans. Magn.,* **29**: 21–33, 1993.

14. A. G. Dirks and H. J. Leamy, Columnar microstructure in vapor-deposited thin films, *Thin Solid Films,* **47**: 219–233, 1977.

15. G. Bottoni, D. Gandolfo, and A. Ceccetti, Exchange anisotropy in metal evaporated tape, *IEEE Trans. Magn.,* **30**: 3945–3947, 1994.

16. Y. Kaneda, Tribology of metal evaporated tape for high density magnetic recording, *IEEE Trans. Magn.,* **33**: 1058–1068, 1997.

17. R. Sugita et al., Incident angle dependence of recording characteristics of vacuum deposited Co-Cr films, *IEEE Trans. Magn.,* **26**: 2286–2288, 1990.

18. Y. Maeda, S. Hirono, and M. Asahi, TEM observation of microstructure in sputtered Co-Cr film, *Jpn. J. Appl. Phys.,* **24**: L951–L953, 1985.

19. Y. Maeda and M. Asahi, Segregation in sputtered Co-Cr films, *IEEE Trans. Magn.,* **23**: 2061–2063, 1987.

20. K. Ishida and T. Nishizawa, The Co-Cr (cobalt-chromium) system, *Bull. Alloy Phase Diag.,* **11**: 357–369, 1990.

21. J. E. Snyder and M. H. Kryder, Quantitative thermomagnetic analysis of CoCr films and experimental determination of the CoCr phase diagram, *J. Appl. Phys.,* **73**: 5551–5553, 1993.

22. Y. Nakamura and S. Iwasaki, *Perpendicular magnetic recording method and materials,* Magnetic Materials in Japan: Research, Applications and Potential, Japan Technical Information Service, Elsevier, pp. 4–106, 1991.

23. E. Köster, Recommendation of a simple and universally applicable method for measuring the switching field distribution of magnetic recording media, *IEEE Trans. Magn.,* **20**: 81–83, 1984.

24. J. P. C. Bernards and H. A. J. Cramer, Vector magnetization of recording media: A new method to compensate for demagnetizing fields, *IEEE Trans. Magn.,* **27**: 4873–4875, 1991.

25. J. U. Lemke, An isotropic particulate medium with additive Hilbert and Fourier field components, *J. Appl. Phys.,* **53**: 2561–2566, 1982.

26. E. C. Stoner and E. P. Wohlfarth, A mechanism of magnetic hysteresis in heterogeneous alloys, *Trans. R. Soc.,* **A240**: 599–642, 1948.

27. I. S. Jacobs and C. P. Bean, An approach to elongated fine-particle magnets, *Phys. Rev.,* **100**: 1060–1067, 1955.

28. Y. Ishii and M. Sato, Magnetic behaviors of elongated single-domain particles by chain-of-spheres model, *J. Appl. Phys.,* **59**: 880–887, 1986.

29. A. Aharoni, Nucleation of magnetization reversal in ESD magnets, *IEEE Trans. Magn.,* **5**: 207–210, 1969.

30. W. F. Brown, Jr., Criterion for uniform micromagnetization, *Phys. Rev.,* **105**: 1479–1482, 1957.

31. E. H. Frei, S. Shtrikman, and D. Treves, Critical size and nucleation field of ideal ferromagnetic particles, *Phys. Rev.,* **106**: 446–455, 1957.

32. W. F. Brown, Jr.,, *Magnetostatic Principles in Ferromagnetism,* Amsterdam: North-Holland, 1962.

33. W. F. Brown, Jr., Micromagnetism, Recent Advances in Engineering Sciences, 5, of *Proc. 6th Tech. Meeting Soc. Eng. Sci.,* London: Gordon and Breach, 1970, pp. 217–228.

34. W. F. Brown, Jr., *Micromagnetics,* Huntington, NY: Krieger, 1978.

35. A. Aharoni, Some recent developments in micromagnetics at the Weizmann Institute of Science, *J. Appl. Phys. Suppl.,* **30**: 70S–78S, 1959.

36. A. Aharoni, Perfect and imperfect particles, *IEEE Trans. Magn.,* **22**: 478–483, 1986.

37. H. J. Richter, Media requirements and recording physics for high density magnetic recording, *IEEE Trans. Magn.,* **29**: 2185–2201, 1993.

38. A. Aharoni, Nucleation modes in ferromagnetic prolate spheroids, *J. Physics: Condensed Matter,* **9**: 10009–10021, 1997.

39. M. E. Schabes, Micromagnetic theory of non-uniform magnetization processes in magnetic recording particles, *J. Magn. Magn. Mat.,* **95**: 249–288, 1991.

40. S. Shtrikman and D. Treves, The coercive force and rotational hysteresis of elongated ferromagnetic particles, *J. Phys. Rad.,* **20**: 286–289, 1959.

41. A. Aharoni, Magnetization curling, *Phys. Stat. Sol.,* **16**: 3–42, 1966.

42. W. F. Brown, Jr., The Theory of Thermal and Imperfection Fluctuations in Ferromagnetic Solids, in R. E. Burgess (ed.), *Fluctuation Phenomena in Solids,* New York: Academic Press, 1965.

43. E. F. Kneller and F. E. Luborsky, Particle size dependence of coercivity and remanence of single-domain particles, *J. Appl. Phys.,* **34**: 656–658, 1963.

44. M. P. Sharrock and J T. McKinney, Kinetic effects in coercivity measurements, *IEEE Trans. Magn.,* **17**: 3020–3022, 1981.

45. L. He, D. Wang, and W. D. Doyle, Remanence coercivity of recording media in the high speed regime, *IEEE Trans. Magn.,* **31**: 2892–2894, 1995.

46. L. He et al., High-speed switching in magnetic recording media, *J. Magn. Magn. Mat.,* **155**: 6–12, 1996.

47. W. F. Brown, Thermal fluctuations of a single-domain particle, *Phys. Rev.,* **130**: 1677–1686, 1963.

48. D. P. E. Dickson et al., Determination of f_0 for fine magnetic particles, *J. Magn. Magn. Mat.,* **125**: 345–350, 1993.

49. R. H. Victorla, Predicted time dependence of the switching field for magnetic materials, *Phys. Rev. Lett.,* **63**: 457–460, 1989.

50. E. P. Wohlfarth, The coefficient of magnetic viscosity, *J. Phys. F,* **14**: L155–L159, 1984.

51. R. Street and J. C. Woolley, A study of magnetic viscosity, *Proc. Phys. Soc.,* **A62**: 562–572, 1949.

52. H. Jachow, E. Schwab, and R. J. Veitch, Viscous magnetization of chromium-dioxide particles: Effect of increased magnetocrystalline anisotrophy, *IEEE Trans. Magn.,* **27**: 4672–4674, 1991.

53. R. J. Veitch et al., Thermal effects in small metallic particles, *IEEE Trans. Magn.,* **30**: 4074–4076, 1994.

54. E. D. Daniel and P. E. Axon, Accidental printing in magnetic recording, *BBC Q.,* **5**: 214–256, 1951.

55. J. C. Mallinson and H. N. Bertram, A theoretical and experimental comparison of the longitudinal and vertical modes of magnetic recording, *IEEE Trans. Magn.,* **20**: 461–467, 1984.

56. J. R. Desserre, Crucial points in perpendicular recording, *IEEE Trans. Magn.,* **20**: 663–668, 1984.

57. R. L. Wallace, Jr., The reproduction of magnetically recorded signals, *Bell Syst. Tech. J.,* **30**: 1145–1173, 1951.

58. W. K. Westmijze, Studies on magnetic recording, parts 1–5, *Philips Res. Rep.,* **8**: 148–157; 161–183; 245–255; 255–269; 343–354; 1953.

59. J. C. Mallinson, On the properties of two-dimensional dipoles and magnetized bodies, *IEEE Trans. Magn.,* **17**: 2453–2460, 1981.

60. H. N. Bertram, Fundamentals of the magnetic recording process, *Proc. IEEE,* **74**: 1494–1512, 1986.

61. J. J. M. Ruigrok, *Short-wavelength magnetic recording,* Eindhoven: Elsevier, 1990.

62. O. Karlqvist, Calculations of the magnetic field in the ferromagnetic layer of a magnetic drum, *Trans. Roy. Soc. Techn. (Stockholm),* **86**: 1–27, 1954.

63. T. J. Szczech, D. M. Perry, and K. E. Palmquist, Improved field equations for ring heads, *IEEE Trans. Magn.,* **19**: 1740–1744, 1983.

64. H. N. Bertram, *Theory of Magnetic Recording,* Cambridge: Cambridge Univ. Press, 1994.

65. B. K. Middleton, Recording and reproducing processes, in C. D. Mee and E. D. Daniel, (eds.), *Magnetic Recording Handbook,* 2nd ed., New York: McGraw-Hill, 1996, Chap. II.

66. J. C. Mallinson, *The foundations of magnetic recording,* 2nd ed., San Diego: Academic Press, 1993.

67. M. L. Williams and R. L. Comstock, An analytic model of the write process in digital magnetic recording, *AIP Conf. Proc.,* **5**: 738–742, 1971.

68. H. N. Bertram and R. Niedermeyer, The effect of spacing on demagnetization in magnetic recording, *IEEE Trans. Magn.,* **18**: 1206–1208, 1982.

69. H. N. Bertram and I. A. Beardsley, The recording process in longitudinal particulate media, *IEEE Trans. Magn.,* **24**: 3234–3248, 1988.

70. H. J. Richter, A generalized slope model for magnetization transitions, *IEEE Trans. Magn.,* **33**: 1073–1084, 1997.

71. D. L. A. Tjaden and J. Leyten, A 5000:1 scale model of the magnetic recoding process, *Philips Tech. Rev.,* **25**: 319–329, 1964.

72. J. C. Mallinson, One-sided fluxes—a magnetic curiosity?, *IEEE Trans. Magn.,* **9**: 678–682, 1973.

73. H. J. Richter, An approach to recording on tilted media, *IEEE Trans. Magn.,* **29**: 2258–2265, 1993.

74. H. A. J. Cramer et al., Write-field interference in the recording process on Co–Cr media: An experimental study, *IEEE Trans. Magn.,* **26**: 100–102, 1990.

75. B. K. Middleton, A. K. Dinnis, and J. J. Miles, Digital recording theory for thick media, *IEEE Trans. Magn.,* **29**: 2286–2288, 1993.

76. S. E. Stupp, S. R. Cumpson, and B. K. Middleton, A quantitative multi-layer recording model for media with arbitrary easy axis orientation, *J. Magn. Soc. Jpn.,* **18** (Suppl. S1): 145—148, 1994.

77. D. Wei, H. N. Bertram, and F. Jeffers, A simplified model for high density tape recording, *IEEE Trans. Magn.,* **30**: 2739–2749, 1994.

78. H. A. J. Cramer, *On the Hysteresis and the Recording Properties in Magnetic Media,* Ph.D. Thesis, Den Haag: Twente University, 1993.

79. S. E. Stupp and A. B. Schrader, Ac-erased and modulation noise in metal evaporated tape, *IEEE Trans. Magn.,* **31**: 2848–2850, 1995.

80. J. C. Mallinson, A new theory of recording media noise, *IEEE Trans. Magn.,* **27**: 3519–3531, 1991.

81. E. Y. Wu, J. V. Peske, and D. C. Palmer, Texture-induced modulation noise and its impact on magnetic recording performance, *IEEE Trans. Magn.,* **30**: 3996–3998, 1994.

82. E. D. Daniel and I. Levine, Experimental and theoretical investigation of the magnetic properties of iron oxide recording tape, *J. Acoust. Soc. Amer.,* **32**: 1–15, 1960.

83. F. Preisach, Über die magnetische Nachwirkung, *Zeitschrift für Physik,* **94**: 277–302, 1935.

84. C. D. Mee, The physics of magnetic recording, in E. P. Wohlfarth (ed.), *Selected Topics in Solid State Physics,* vol. II, Amsterdam: North Holland, 1964.

85. F. Jørgensen, *The Complete Handbook of Magnetic Recording,* 4th ed., New York: McGraw-Hill, 1996.

86. E. F. Kneller, Magnetic-interaction effects in fine particle assemblies and thin films, *J. Appl. Phys.,* **39**: 945–955, 1968.

87. E. F. Kneller and E. Köster, Relation between anhysteretic and static magnetic tape parameters, *IEEE Trans. Magn.,* **13**: 1388–1390, 1977.

88. E. Köster, A contribution to anhysteretic remanence and AC bias recording, *IEEE Trans. Magn.,* **11**: 1185–1187, 1975.

89. H. N. Bertram, Long wavelength ac bias recording theory, *IEEE Trans. Magn.,* **10**: 1039–1048, 1974.

90. E. D. Daniel, Tape noise in audio recording, *J. Audio Eng. Soc.,* **20**: 92–99, 1972.

91. L. Lekawat, G. W. D. Spratt, and M. H. Kryder, Annealing study of the erasability of high energy tapes, *IEEE Trans. Magn.,* **29**: 3628–3630, 1993.

92. J. C. Mallinson, Maximum signal-to-noise ratio of a tape recorder, *IEEE Trans. Magn.,* **5**: 182–186, 1969.

93. T. Fujiwara, Magnetic properties and recording characteristics of barium ferrite media, *IEEE Trans. Magn.,* **23**: 3125–3130, 1987.

94. H. Auweter, et al., Experimental study of the influence of particle size and switching field distribution on video tape output, *IEEE Trans. Magn.,* **27**: 4669–4671, 1991.

95. J. C. Mallinson, Proposal concerning high-density digital recording, *IEEE Trans. Magn.,* **25**: 3168–3169, 1989.

96. S. E. Stupp and C. R. Cumpson, Experimental and theoretical studies of nonlinear transition shift in metal evaporated tape, *IEEE Trans. Magn.,* **33**: 2968–2970, 1997.

97. C. D. Mee and E. D. Daniel, *Magnetic Storage Handbook,* 2nd ed., New York: McGraw-Hill, Chapters 5–8, 1996.

98. J. R. Watkinson, *The Art of Digital Audio,* Boston/Oxford: Focal Press, 1989.

99. S. B. Luitjens, Magnetic recording trends: Media developments and future (video) recording systems, *IEEE Trans. Magn.,* **26**: 6–11, 1990.

HANS JÜRGEN RICHTER
Seagate Technology

RONALD J. VEITCH
BASF Aktiengesellschaft

MAGNETIC TAPES

Magnetic tapes can be described as multilayer composite materials consisting of a magnetic layer deposited on to a substrate. There are two basic types of magnetic tapes: (1) particulate tapes where magnetic particles are dispersed in a polymeric matrix and coated onto a polymeric substrate, and (2) thin-film tapes where continuous films of magnetic materials are deposited onto the substrate using vacuum techniques. Cross-sectional views of a particulate and a thin-film tape are shown in Fig. 1. The thin-film tape is commonly referred to as a metal-evaporated (ME) tape since it consists of a coating which is applied using evaporation techniques under vacuum. Currently, particulate tapes are more prevalent than ME tapes. However, as discussed by Bhushan (1–3) requirements for higher recording densities with low error rates have resulted in increased use of smoother, ultrathin ME tapes for digital recording.

DESCRIPTION OF TAPE MATERIALS

Standard and Advanced Substrates

The substrate (or base film) for magnetic tapes is typically a polyester material. Polyethylene terephthalate (PET) film is the most commonly used material, however a new polyester material called polyethylene naphthalate (PEN) is beginning to be used for advanced high-density tapes, such as the DC2120XL tape made by 3M/Imation. Ultrathin aromatic polyamide (ARAMID) substrates have been used for Sony

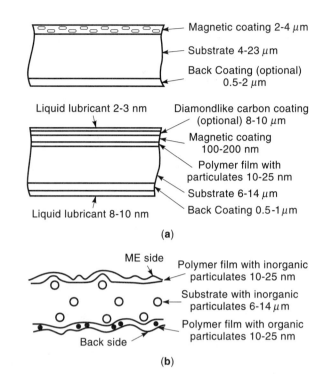

Figure 1. Sectional view of (a) A particulate and a metal-evaporated magnetic tape. (b) A coated PET substrate for ME tapes.

NTC-90 microcassettes, and advanced polymers such as polyimides (PI) and polybenzoxazole (PBO) have been studied for their potential as substrates. Table 1 provides a list of substrates studied by Weick and Bhushan (4–6), and a list of magnetic tapes that use ultrathin substrates and/or advanced magnetic coatings can be found in Table 2.

Magnetic Coatings

Particulate. The substrate is coated on one side of a tape with a magnetic coating that is typically 1 to 4 μm thick. This coating contains 70 to 80 wt.% (or 43–50 vol.%) submicron and acicular magnetic particles such as γ-Fe_2O_3, Co-modified γ-Fe_2O_3, CrO_2, or metal particles (MP) for longitudinal recording. Hexagonal platelets of barium ferrite ($BaO \cdot 6Fe_2O_3$) have been used for longitudinal recording, and they can be used for perpendicular recording. These magnetic particles are held in polymeric binders such as polyester-polyurethane, polyether-polyurethane, nitrocellulose, poly(vinyl chloride), poly(vinyl alcohol-vinyl acetate), poly(vinylidene chloride), VAGH, phenoxy, and epoxy. To reduce friction, the coating consists of 1 to 7 wt.% of lubricants (mostly fatty acid esters, e.g., tridecyl stearate, butyl stearate, butyl palmitate, butyl myristate, stearic acid, myrstic acid). Finally, the coating contains a cross linker or curing agent (such as functional isocyanates); a dispersant or wetting agent (such as lecithin); and solvents (such as tetrahydrofuran and methyl isobutyl-ketone). In some media, carbon black is added for antistatic protection if the magnetic particles are highly insulating, and abrasive particles (such as Al_2O_3 and Cr_2O_3) are added as a head cleaning agent and to improve wear resistance. The coating is calendered to a root-mean-square (rms) surface roughness of 8 to 15 nm.

For antistatic protection and for improved tracking, most magnetic tapes have a 1 to 3 μm thick backcoating of polyes-

Table 1. List of Substrate Materials for Magnetic Tapes

Material	Chemical Name[a]	Tradename	Manufacturing Method	Supplier	Thickness (μm)	Currently Used?
PET	Polyethylene terephthalate	Mylar A (57DB)	Drawing	Dupont	14.4	Yes
PEN	Polyethylene naphthalate	Teonex	Drawing	Teijin	4.5	Yes
ARAMID	Aromatic polyamide	Mictron TX-1	Casting	Toray	4.4	Yes
PI	Polyimide	Upilex	Casting	Ube	7.6	No
PBO	Polybenzoxazole		Casting	Dow	5.0	No

[a]Chemical structures of the substrate materials can be found in Ref. 6.

ter–polyurethane binder containing a conductive carbon black and TiO$_2$, typically 10% and 50% by weight, respectively. More information on particulates used for magnetic tapes can be found in Refs. 1–3.

Thin-Film (Metal-Evaporated). Thin-film (also called metal-evaporated or ME) flexible media consist of a polymer substrate (PET or ARAMID) with an evaporated film of Co–Ni (with about 18% Ni) and experimental evaporated/sputtered Co–Cr (with about 17% Cr) (for perpendicular recording, which is typically 100 to 200 nm thick). Electroplated Co and electroless-plated Co–P, Co–Ni–P, and Co–Ni–Re–P have also been explored but are not commercially used. Since the magnetic layer is very thin, the surface of the thin-film media is greatly influenced by the surface of the substrate film. Therefore, an ultrasmooth PET substrate film (rms roughness ~1.5 to 2 nm) is used to obtain a smooth tape surface. A 10 to 25 nm thick precoat composed of a polymer film with additives is generally applied to both sides of the PET substrate to provide controlled topography, Fig. 1(b). The film on the ME treated side generally contains inorganic particulates (typically SiO$_2$ with a particle size of 100 to 200 nm diameter and areal density of typically 10,000/mm^2). The film on the back side generally contains organic (typically cross-linked polystyrene) particles. The rms and peak-to-valley (P–V) distances of the ME treated side and the backside typically are 1.5 to 2 nm and 15 to 20 nm, and 3 to 5 and 50 to 75 nm, respectively. The polymer precoat is applied to reduce the roughness (mostly P–V distance) in a controlled manner from that of the PET surface, and to provide good adhesion with the ME films. Particles are added to the precoat to control the real area of contact and consequently the friction. A continuous magnetic coating is deposited on the polymer film. The polymer film is wrapped on a chill roll during deposition, which keeps the film at a temperature of 0 to −20°C. Co$_{80}$Ni$_{20}$ material is deposited on the film by a reactive evaporation process in the presence of oxygen; oxygen increases the hardness and corrosion resistance of the ME film. The deposited film, with a mean composition of (Co$_{80}$Ni$_{20}$)$_{80}$O$_{20}$ consists of very small Co and Co–Ni crystallites which are primarily intermixed with oxides of Co and Ni (Feurstein and Mayr (7); Harth et al. (8)). Various inorganic overcoats such as diamondlike carbon (DLC) in about 10 nm thicknesses are usually used to protect against corrosion and wear. A topical liquid lubricant (typically perfluoropolyether with reactive polar ends) is then applied to the magnetic coating and the backside by rolling. The topical lubricant enhances the durability of the magnetic coating, and if no DLC overcoat is present the lubricant also inhibits the highly reactive metal coating from reacting with ambient air and water vapor. A backcoating is also applied to balance stresses in the tape, and for antistatic protection. More information on thin films used for magnetic tapes can be found in Refs. 1–3.

DESIGN CONSIDERATIONS FOR DEVELOPMENT OF ADVANCED ULTRATHIN MAGNETIC TAPES

The thinner substrates shown in Table 1 allow for a higher volumetric density when a magnetic tape is wound onto a reel. To achieve this higher volumetric density, the product of the linear and track densities must also be high. This product is commonly referred to as the areal density, and as described by Wallace (9) this density is directly related to the signal amplitude reproduced after the tape is used for recording purposes. Therefore, to make an advanced storage device with a volumetric density of one terabyte per cubic inch the following characteristics are required: a substrate which is approximately 4 μm thick, a magnetic medium with a track density of about 9000 tracks per inch with a 64 head array and 8 head positions, and a linear density of about 160 kbits/inch. To make a magnetic tape with such high areal densities the substrate must be mechanically and environmentally stable with a high surface smoothness. For high track densities, lateral contraction of the substrates from thermal, hygroscopic, viscoelastic, and/or shrinkage effects must be minimal. To

Table 2. List of Magnetic Tapes That Use Ultra-Thin Substrates and/or Advanced Magnetic Coatings

Magnetic Coating	Substrate	Tradename	Supplier	Thickness (μm)
ME[a]	PET	Hi8 ME-180	Sony	7.5
BaFe[b]	PET	Hi8 BaFe-120	Toshiba	11.0
ME[a]	ARAMID	NTC-90	Sony	4.8
BaFe[b]	PEN	DC2120XL	3M/Imation	7.5

[a]Metal-Evaporated Tape
[b]Barium Ferrite (BaO · 6FE$_2$O$_3$) Particulate Coating

minimize stretching and damage during manufacturing and use of thin magnetic tapes, the substrate should be a high modulus, high strength material with low viscoelastic and shrinkage characteristics. As discussed by Bhushan (1), if the storage device is a linear tape drive, any linear deformations can be accounted for by a change in clocking speed. However, if the storage device is a rotary tape drive, anisotropic deformations of the substrate would be undesirable. To minimize stretching during use of thinner substrates, the modulus of elasticity, yield strength, and tensile strength should be high along the machine direction. Also, since high coercivity magnetic films on metal-evaporated tapes are deposited and heat

treated at elevated temperatures, a substrate with stable mechanical properties up to a temperature of 100 to 150°C or even higher is desirable. In addition to mechanical and environmental stability, the cost of the material is also a major factor in the selection of a suitable substrate. Lastly, Bhushan (1) has discussed the fact that various long-term reliability problems including uneven tape-stack profiles (or hardbands), mechanical print-through, instantaneous speed variations, and tape stagger problems can all be related to the substrate's viscoelastic characteristics.

TENSILE PROPERTIES

Mechanical properties of typical magnetic tapes and substrates are presented in Fig. 2. These properties were measured by Weick and Bhushan (4) using a Monsanto Tensometer T20 tensile test machine in accordance with ASTM Spec. 1708. Properties presented in Fig. 2 include modulus of elasticity, strain-at-yield/failure, breaking strength, and strain-at-break. The strain-at-break measurements correspond with the strain at which the substrates break, and the strain-at-yield/failure measurements correspond with the strain at which the substrates start to deform irreversibly. Figure 2(a) shows modulus of elasticity measurements for PET, two PET tapes, and the alternative substrates. PEN, ARAMID, PI, and PBO all offer improvements in elasticity when compared to 3 to 4 GPa for PET. Elastic moduli range from 4 to 5.4 GPa for PEN, 10 GPa for ARAMID, 4 to 5 GPa for PI, and 9 to 17 GPa for PBO. The alternative substrates also offer improvements in breaking strength when compared to PET and PET tapes. Therefore, based on a typical tape tension of 7.0 MPa (1000 psi), the alternative materials would be stressed to only a fraction of their breaking strength (typically 1/10 to 1/30). Strain-at-break measurements indicate that PET tends to be more ductile than the alternative substrates. However, for PET irreversible strain occurs at its yield point. This is indicated in Fig. 2(b), and strains of only 0.02 to 0.03 were measured when PET begins to yield. The PET tapes yield at slightly higher strains of 0.03 to 0.04. PI also fails at its yield point, but PEN, ARAMID, and PBO do not have yield points. Instead, they fail irreversibly only at their breaking points. The failure strains for the alternative substrates are typically higher than those measured for PET ranging from 0.03 to 0.08 for PEN, 0.035 to 0.05 for ARAMID, 0.04 to 0.045 for PI, and 0.03 to 0.04 for PBO.

Anisotropic characteristics of the tape substrate materials are shown in Fig. 3. Modulus of elasticity measurements along different material orientations are shown in this polar plot. Since ARAMID has a circular curve, it is a relatively isotropic material with an elastic modulus of approximately 10 GPa regardless of material orientation. PI is also relatively isotropic, but its modulus is significantly less than ARAMID's. PET and PEN not only have low moduli, but they are anisotropic materials as indicated by their elliptical curves. Figure 3 also shows that even though PBO has a high modulus, it is also anisotropic.

VISCOELASTIC CHARACTERISTICS (NONPERMANENT DEFORMATION)

Time-Dependent Creep Behavior

Viscoelasticity refers to the combined elastic and viscous deformation of a polymeric material when external forces are

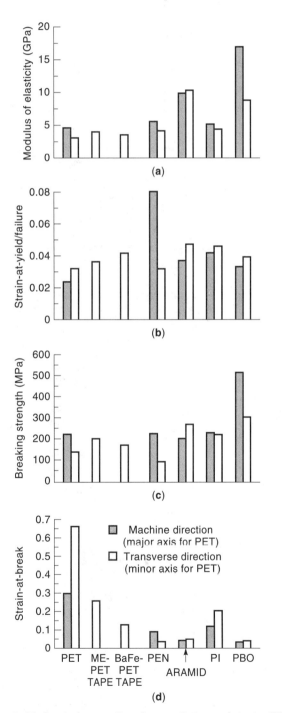

Figure 2. Mechanical properties of magnetic tape substrates [Weick and Bhushan (4)].

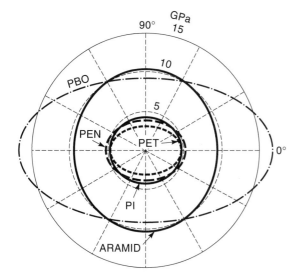

Figure 3. Anisotropy in modulus of elasticity for magnetic tape substrates. 0°—machine direction (major axis for PET); 90°—transverse direction (minor axis for PET) [Weick and Bhushan (4)].

Typical creep-compliance measurements for magnetic tapes and substrates are shown in Figs. 4(a) and (b) for the 50 °C temperature level, and the first derivative of the creep-compliance data for the substrates is shown in Fig. 5. This first derivative represents the creep velocity of the substrates, and is a more direct depiction of the rate at which the deformation occurs. More information about how the creep velocity curves were obtained can be found in Ref. 6.

Creep-compliance data presented in Fig. 4(a) show that there is an initial creep response which occurs in the first minute of each experiment due to immediate elastic and short term viscoelastic behavior of the materials. Throughout the rest of the experiments the materials creep (or stretch) due to the viscoelastic behavior of the particular polymer being evaluated. PET shows the largest amount of creep along its

applied. It is a function of time, temperature, and rate of deformation, and viscoelastic deformation of a magnetic tape can lead to the loss of information stored on the tape. For instance, various long-term reliability problems including uneven tape-stack profiles (or hardbands), mechanical print-through, instantaneous speed variations, and tape stagger problems can all be related to the substrate's viscoelastic characteristics. To minimize these reliability problems, it is not only important to minimize creep strain, but the rate of increase of total strain needs to be kept to a minimum to prevent stress relaxation in a wound reel.

A common method to measure time-dependent viscoelastic behavior at elevated temperatures is to perform creep experiments. Weick and Bhushan (5,6) have performed such experiments for magnetic tapes and substrates. During a creep experiment, a constant stress is applied to a strip of material (i.e., tape or substrate), and the change in length of this test sample is measured as a function of time at an elevated temperature. The amount of strain the material is subjected to can be calculated by normalizing the change in length of the specimen with respect to the original length. Creep compliance can then be calculated by dividing the time-dependent strain by the constant applied stress:

$$\epsilon(t) = \frac{\Delta l(t)}{l_0} \tag{1}$$

$$D(t) = \frac{\epsilon(t)}{\sigma_0} = \frac{\Delta l(t)}{\sigma_0 l_0} \tag{2}$$

where

$\Delta l(t) \equiv$ change in length of the test sample as a function of time

$l_0 \equiv$ original length of the test sample

$\epsilon(t) \equiv$ the amount of strain the test sample is subjected to

$\sigma_0 \equiv$ constant applied stress

$D(t) \equiv$ tensile creep-compliance of the test sample as a function of time

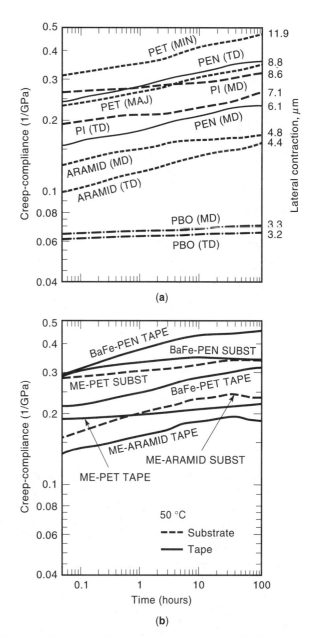

Figure 4. Creep-compliance measurements for (a) Magnetic tape substrates. The right axis shows lateral contraction calculations assuming a Poisson's ratio of 0.3 [Weick and Bhushan (5,6)]. (b) Actual magnetic tapes and their respective substrates.

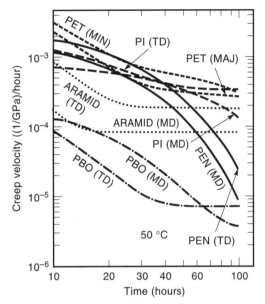

Figure 5. Creep velocity measurements for magnetic tape substrates [Weick and Bhushan (5,6)].

minor axis. PEN and PI show somewhat less creep; whereas ARAMID and PBO have total creep-compliances which are significantly lower. Creep-compliance measurements at ambient temperature show similar trends and are discussed by Weick and Bhushan (6) along with relationships between creep behavior and the molecular structure of each polymeric substrate.

From the total creep-compliance measurements at 50°C the amount of lateral contraction can be calculated as shown on the right-hand axis of Fig. 4(a). For a 12.7 mm wide tape, PET shows the largest amount of lateral contraction (approximately 12 μm along the minor axis), and PBO shows the least amount of contraction (a little more than 3 μm). PEN shows 6 to almost 10 μm of lateral contraction depending on the material orientation, whereas ARAMID shows only 4 to 5 μm.

Creep-compliance results for actual magnetic tapes are shown in Fig. 4(b). With the exception of BaFe-PET, measurements are presented for both the magnetic tapes and substrates. The substrates for the ME tapes were obtained by dipping the tapes in a 10% (vol.) HCl solution to dissolve the magnetic coating. The BaFe-PEN tape was scrubbed with methyl ethyl ketone to remove the magnetic layer and obtain the substrate. This method could not be used successfully for the BaFe-PET tape. From Fig. 4(b), the two particulate tapes (BaFe-PET and BaFe-PEN) have rates of creep which are relatively equivalent with BaFe-PEN showing a higher initial compliance. The ME-PET and ME-ARAMID tapes have lower total compliances than the particulate tapes, and the ME-ARAMID tape tends to have a higher creep velocity and lower initial compliance. When the creep behavior for each tape is compared to their respective substrate there is also an apparent difference between metal-evaporated and particulate tapes. Both ME-PET and ME-ARAMID have substrates which show a higher total creep compliance than the tape, and the creep velocity of these substrates is similar to that measured for the tapes. BaFe-PEN, on the other hand, has a substrate which tends to have a lower total creep compliance and creep velocity. This is possibly due to the presence of a

lower modulus (higher compliance) elastomeric binder coating on the BaFe-PEN tape which contributes to the overall creep behavior of the composite material. Note that the thickness of the magnetic layer/coating is 3 μm and the thickness of the PEN substrate is 4.5 μm.

Creep velocity data sets are shown in Fig. 5 on a log–log scale. PBO appears to creep at the lowest rate throughout the 100 hour experiments. When compared to PET, PI offers only a slight improvement in creep velocity; whereas the creep velocity for ARAMID is always lower than the velocity for PET. During the first part of the experiment PEN creeps at a rate which is nearly equal to the creep velocity for PET. However, at the end of the experiment the creep velocity for PEN is an order of magnitude lower than the velocity for PET. PEN also creeps at a lower rate than ARAMID after 100 h.

Figure 5 not only shows relative creep rates for the materials, additional information can be extracted from the slopes of the creep velocity curves. These slopes indicate acceleration (or deceleration) during the creep process. Typically, the materials show a decreasing creep velocity and a negative slope which indicates deceleration during the creep process. For ARAMID, the slope of the creep velocity curve remains constant after 100 h. This means that ARAMID continues to creep at the same rate without a change in velocity. In comparison, PEN not only creeps at a lower rate than ARAMID after 100 h, but the changing slope of the curves for PEN indicates that the creep velocity for PEN is decreasing. Recall from Figs. 4(a) and (b) that the total creep for ARAMID is actually less than the total creep for PEN. However, from this discussion PEN actually creeps at a lower rate than ARAMID, and this rate shows a decreasing trend after 100 h.

An analytical technique known as time-temperature superposition (TTS) has been used by Weick and Bhushan (5, 6) to predict long-term creep behavior of magnetic tape substrates at ambient temperature. Using this technique creep measurements at elevated temperatures are assembled to predict behavior at longer time periods. Results are presented in Fig. 6 for the machine direction (major optical axis for PET). The trend lines (or master curves) are assembled to predict the creep-compliance at 25°C over a 10^6 h time period. Shift factors tabulated above the figure show how much each curve was shifted (in hours) to enable a smooth fit at the 25°C reference temperature. Therefore, an indication of viscoelastic behavior at very short (<0.1 h) and very long time periods (>10^6 h) can be obtained. PET and PI show similar amounts of creep until the final decades when the amount of creep for PET exceeds that for PI. PEN shows somewhat less creep at all time periods, and ARAMID has creep characteristics which are always slightly lower than PEN's. The total creep for PBO is significantly lower than that measured for the other materials.

Frequency-Dependent Dynamic Mechanical Behavior

Weick and Bhushan (10,11) have shown that the dynamic mechanical response of a magnetic tape as it is unwound from a reel and travels over a head also depends on the elastic and viscoelastic characteristics of the magnetic tape. This elastic/viscoelastic recovery and subsequent conformity of the tape with the head occurs in just a few milliseconds, and requires optimization of the dynamic properties of the materials that comprise a magnetic tape. Note that a lack of tape-to-head conformity can lead to an increase in wear of the head as demonstrated by Hahn (12), and tape stiffness has been shown by Bhushan and Lowry (13) to be related to edge wear of a head.

Dynamic viscoelastic properties of magnetic tapes and substrates can be measured using dynamic mechanical analysis (DMA), and the information acquired from this analysis includes E' which is the storage or elastic modulus, and the lost tangent, $\tan(\delta)$, which is a measure of the amount of viscous or nonrecoverable deformation with respect to the elastic deformation. Both E' and $\tan(\delta)$ are measured as a function of temperature and deformation frequency, and results can be used to predict the dynamic response of tapes over several orders of magnitude. Equations used to calculate the storage modulus, E', and loss tangent, $\tan(\delta)$, are as follows:

$$E' = \cos(\delta)\left[\frac{\sigma}{\epsilon}\right] \tag{3a}$$

$$E'' = \sin(\delta)\left[\frac{\sigma}{\epsilon}\right] \tag{3b}$$

$$|E^*| = \sqrt{(E')^2 + (E'')^2} \tag{3c}$$

$$\tan(\delta) = \frac{E''}{E'} \tag{3d}$$

where

E' = storage (or elastic) modulus
E'' = viscous (or loss) modulus
$|E^*|$ = magnitude of the complex modulus
ϵ = applied strain
σ = measured stress
δ = phase angle shift between stress and strain

A Rheometrics RSA-II dynamic mechanical analyzer was used to measure the dynamic mechanical properties of magnetic tapes and substrates. At temperature levels ranging from -50 to $50°C$, the analyzer applies a sinusoidal strain on the sample at frequencies ranging from 0.016 to 16 Hz. The strain is measured by a displacement transducer, and the corresponding sinusoidal load on the sample is measured by a load cell. Since the polymeric tapes are viscoelastic there will be a phase lag between the applied strain and the measured load (or stress) on the specimen. The storage (or elastic) modulus, E', is therefore a measure of the component of the complex modulus which is in-phase with the applied strain, and the loss (or viscous) modulus, E'', is a measure of the component which is out-of-phase with the applied strain. The in-phase stress and strain results in elastically stored energy which is completely recoverable, whereas out-of-phase stress and strain results in the dissipation of energy which is nonrecoverable and is lost to the system. Therefore, the loss tangent, $\tan(\delta)$, is simply the ratio of the loss (or viscous) modulus to the storage (or elastic) modulus. Refer to the texts by Ferry (14), Tschoegl (15), and Aklonis and MacKnight (16) for more information about polymer viscoelasticity.

Representative E' and $\tan(\delta)$ data for five substrate materials are shown in Figs. 7(a) and (b), and E' and $\tan(\delta)$ data for representative magnetic tapes are shown in Figs. 8(a) and (b). Note that these are master curves which were generated by Weick and Bhushan (10,11) using the raw E' and $\tan(\delta)$ data that are functions of both frequency and temperature. This technique is known as frequency-temperature superposition, and is analogous to the time–temperature superposition technique described for time-dependent creep behavior. A 20°C reference temperature was used for the frequency-temperature superposition, and the shift factors can be found in Refs. 10 and 11.

At the 20°C reference temperature used to construct Figs. 7(a) and (b), PET and PI show the lowest storage moduli. Note that these data sets are for the machine direction (MD), and similar results have been found by Weick and Bhushan (10) for the transverse direction (TD). However, PET clearly has the lowest E' when measured in the transverse direction. PEN, ARAMID, and PBO have significantly higher storage

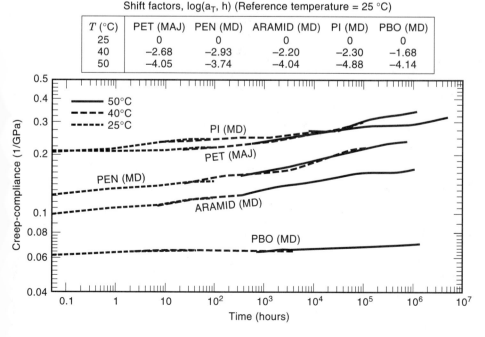

T (°C)	PET (MAJ)	PEN (MD)	ARAMID (MD)	PI (MD)	PBO (MD)
25	0	0	0	0	0
40	−2.68	−2.93	−2.20	−2.30	−1.68
50	−4.05	−3.74	−4.04	−4.88	−4.14

Figure 6. Creep-compliance master curves and shift factors for magnetic tape substrates. Machine direction data (major axis for PET) [Weick and Bhushan (5,6)].

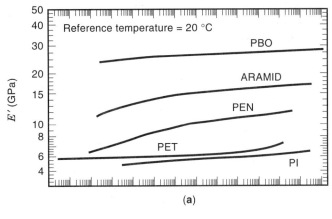

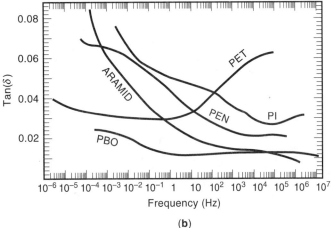

Figure 7. Typical dynamic mechanical analysis results for five magnetic tape substrates. Data sets are for PEN, ARAMID, PI, and PBO samples cut in the machine direction, and for PET samples cut along their major optical axis. [Weick and Bhushan (10)].

moduli than the other materials regardless of material orientation. Note that creep-compliance is inversely proportional to the storage modulus, and the general trends shown in Fig. 7(a) are in agreement with what is shown in Fig. 4(a). The storage modulus and tan(δ) data for PET are also in agreement with what was reported previously by Bhushan (1) for 23.4 μm thick PET substrates. The same trends have been observed but the storage moduli for the 14.4 μm thick PET are 2 to 2.5 GPa higher than what was reported by Bhushan (1), and the tan(δ) measurements are approximately 0.02 lower. These differences are not unexpected due to probable improvements in manufacturing the newer 14.4 μm thick PET material.

In general, the storage moduli for tapes made with PET substrates are lower than those measured for the tapes made with PEN and ARAMID substrates. This is shown in Fig. 8(a), and is primarily due to the fact that the more advanced PEN and ARAMID substrates have higher storage moduli than the PET substrates [Weick and Bhushan (10)]. The only exception to this is the MP-PET tape which has a somewhat higher modulus than the BaFe-PEN tape. Weick and Bhushan (11) have suggested that this could be due to the relatively thick MP film on the PET substrate which when applied to the substrate causes a more substantial increase in the modulus of the tape as a whole.

Loss tangent master curves in Figs. 7(b) and 8(b) show the relative amount of nonrecoverable, viscous deformation experienced by each substrate or tape. Recall that the loss tangent, tan(δ), is the ratio of the loss modulus, E'', to the storage modulus, E'. The storage modulus, E', is the elastic component of the modulus which responds in-phase with the applied strain, and E'' is the viscous component which lags the applied strain. Therefore, tan(δ) is the phase lag between the two components and is a relative measurement of nonrecoverable deformation. From Fig. 7(b) it can be seen that the loss tangent for all the substrates but PET shows a decreasing trend with increasing frequency. Therefore, at higher frequencies PI, PEN, ARAMID, and PBO do not dissipate as much nonrecoverable energy as PET, and PET is more likely to be deformed and stretched when it experiences high frequency transient strains in a tape drive. Similar results are shown in Fig. 8(b) for the tapes. The PET tapes show an increasing trend with increasing frequency, and the ME-ARAMID and BaFe-PEN tapes show a decreasing trend.

A complete discussion of the differences between DMA data for the tapes when compared to the substrates has been presented by Weick and Bhushan (11). To do this comparison accurately, DMA data sets were obtained for tapes and their actual constitutive substrates. To get these constitutive substrates, magnetic films and backcoatings were removed from magnetic tapes using suitable solvents [Weick and Bhushan

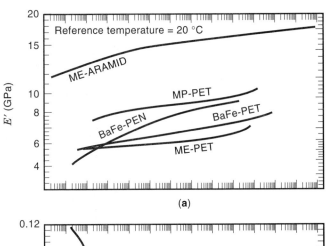

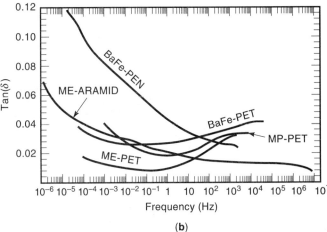

Figure 8. Dynamic mechanical analysis results for five magnetic tapes. (a) Storage modulus (E') master curves. (b) Loss tangent (Tan δ) master curves [Weick and Bhushan (11)].

(11)]. (Note that the substrates used to obtain the measurements shown in Fig. 7 were obtained directly from the polymer film manufacture, and had never been used in an actual tape.) In general, the storage modulus for a magnetic tape is typically higher than the storage modulus for its respective substrate. This is not surprising since magnetic coatings are comprised of either rigid ceramic or metal particles in an elastomeric film, or a continuous metal film. Since these magnetic films are likely to have a higher modulus than the polymeric substrates, when they are applied to the substrate the overall modulus of the composite magnetic tape is higher than the substrate alone. However, orientation, shape, and stiffness of the particles used in the magnetic coating could lead to exceptions to this general trend [Weick and Bhushan (11)].

SHRINKAGE, THERMAL, AND HYGROSCOPIC EXPANSION (PERMANENT DEFORMATION)

At elevated temperatures and humidities, polymeric materials (and therefore magnetic tapes) are susceptible to permanent deformation. One of these deformation mechanisms is known as shrinkage. Weick and Bhushan (5,6) have shown that when certain magnetic tapes (and substrates) are subjected to relatively small tensile stresses (<0.5 MPa), they tend to shrink or contract rather than creep. Shrinkage results for magnetic tape substrates are shown in Fig. 9(a). Only ARAMID (MD) and PEN (MD & TD) shrink at this temperature level. PEN shrinks as much as 0.035% after 100 h, and ARAMID (MD) shrinks 0.01% after 100 h. The relatively large amount of shrinkage for PEN could be reduced by stress stabilizing the material at 65°C. Creep appears to be a more dominant factor for PET since its change in length is positive rather than negative. The effect of polymeric structure and processing conditions on substrate shrinkage behavior can be found in Ref. 6.

Shrinkage measurements have also been obtained for the magnetic tapes. These results are presented in Fig. 9(b). BaFe-PET initially creeps and then shrinks in a manner which is similar to that observed for a 14.4 μm thick PET substrate. The ME-PET tape shrinks considerably more (0.03% after 100 h), and the ME-PET substrate shrinks 0.05%. This could mean that there are residual stresses present in the ME-PET substrate which are somewhat attenuated when the metal coating is applied. The ME-ARAMID tape and substrate shrink in a manner which was already observed for the ARAMID (MD) substrate evaluated separately (see Fig. 9(a)). Since the substrate shrinks less than the tape, residual stresses may have actually been added when the metal coating was applied.

Certain magnetic tape materials also undergo free expansion when subjected to elevated temperatures and humidities. Coefficient of thermal expansion (CTE) and coefficient of hygroscopic expansion (CHE) measurements have been reported by Weick and Bhushan (4), and are shown in Figs. 10(a) and (b). CTE measurements were made in accordance with ASTM D696-79 using a Zygo laser dimension sensor. Specimens cut into 13 × 1/2 inch strips were used for CTE, and the temperature was varied from 24 to 46°C. CHE measurements were made using a Neenah paper expansimeter. The humidity was varied from 27–95% using a salt solution, and 5 × 1/2 inch specimens were used. ARAMID and PEN have lower thermal

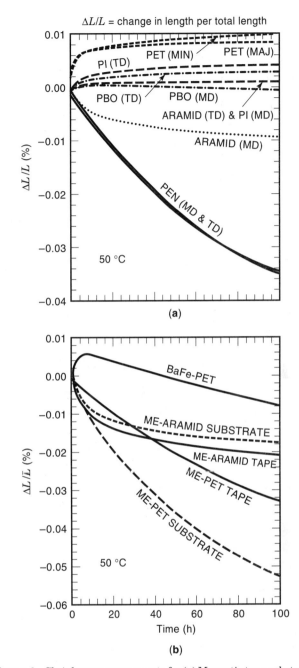

Figure 9. Shrinkage measurements for (a) Magnetic tape substrates. (b) Actual magnetic tapes and their respective substrates [Weick and Bhushan (5,6)].

expansion characteristics than PET. However, only ARAMID and PBO show a significant decrease in hygroscopic expansion. See Perettie et al. (17), Perettie and Pierini (18), and Perettie and Speliotis (19) for additional information.

A summary of the deformation characteristics measured for magnetic tape substrate materials by Bhushan (1), Weick and Bhushan (4), and Perettie and Pierini (18) is presented in Table 3, and CTE, CHE, creep, lateral contraction, and shrinkage measurements are presented on a percentage basis in Fig. 11. ARAMID and PEN both offer improved expansion and contraction characteristics when compared to PET. Although these materials shrink slightly more than PET at 50°C, their expansion and contraction characteristics are lower.

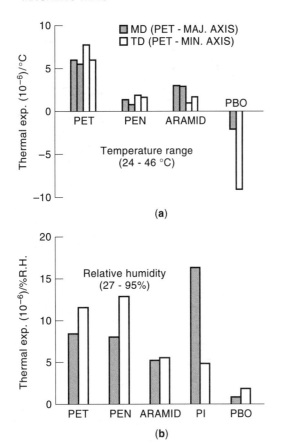

Figure 10. Thermal and hygroscopic expansion measurements for magnetic tape substrates. MD is the machine direction (major axis for PET), and TD is the transverse direction (minor axis for PET) [Weick and Bhushan (4)].

Weick and Bhushan (1,6) have reported that the shrinkage for PEN and polyesters in general could be reduced by stress-stabilizing the material at 65°C. ARAMID and PEN are also closer to the contraction criteria for an advanced magnetic tape with 256 tracks per inch to be read at a time. Based on this criteria lateral contraction should be less than 0.08% if a 1/10 track mismatch is tolerable and the head can be recentered. It is not entirely clear whether ARAMID or PEN should be used for the next generation of magnetic tapes. However, ARAMID has a

higher T_g than the polyester films, which makes it more suitable for metal evaporated tapes. Also, the elastic modulus is a factor of two higher for ARAMID when compared to PEN, and ARAMID is more isotropic which renders it more compatible with rotary tape drives. PEN on the other hand has a creep velocity shown in Fig. 5 which is lower than that for ARAMID, and the creep velocity for PEN continues to decrease after 100 h at 50°C. In comparison, ARAMID continues to deform viscoelastically at the same rate. See Refs. 5 and 6 for more information about creep velocity.

Although other substrates such as PBO and ARAMID clearly offer advantages over PET and PEN, it should be noted that PET is still the standard substrate used for magnetic tapes. PBO and PI have not been used due to their lack of availability and high cost. ARAMID is also a high cost material since it is manufactured using a casting technique rather than the drawing technique used for the polyester films (PET and PEN). PEN is now being used in place of PET for certain long-play video tapes as well as higher density data storage tapes such as the 3M DC2120XL, and ARAMID is used as the substrate for the Sony NTC-90 microcassette. A final ranking of substrates can be made based on the summaries presented in Table 3 and Fig. 11: 1st choice—PBO (unavailable), 2nd choice—PEN or ARAMID, 3rd choice—PI (availability in question), 4th choice—PET.

TRIBOLOGICAL CHARACTERISTICS

Surface Roughness

Roughness of the substrates is a concern when higher areal densities are required. The substrate must have a high surface smoothness in addition to being mechanically and environmentally stable. Furthermore, for the advanced magnetic tapes such as metal-evaporated (ME) tape, the surface of the substrate must be tailored to allow for the deposition and adhesion of thin metal coatings. Surface topography is also an important parameter which influences handling of the substrate. Throughout manufacture, conversion, and use Bhushan (1) has reported that the film must be able to be wound at high speeds without stagger (lateral slip) in the transverse direction. During winding it is necessary for each layer to tighten and cinch upon itself. To insure stability in the layer-

Table 3. Summary of Mechanical, Hygroscopic, Thermal, Viscoelastic, and Shrinkage Characteristics of Magnetic Tape Substrates

Material		Mod. of Elasticity (GPa)	Strain at Yield/ Failure	Breaking Strength (MPa)	Strain at Break	Density g/cm³	Moist. Absorb.[a] (%)	CHE (27–95%) (10⁻⁶)/%RH	CTE (24–46°C) (10⁻⁶)/°C	T_g (°C)	Melting Point (°C)	Creep-Compl. 100 hrs @ 50°C (GPa⁻¹)	Lat. Contract.[b] 100 hrs @ 50°C (μm)	Shrinkage 100 hrs @ 50°C (%)
PET	MAJ	4.3	0.02	221	0.29	1.395	0.4	8.5	6.0	116	263	0.35	9.3	neg.
	MIN	2.9	0.03	141	0.66			11.7	7.9			0.47	12.5	
PEN	MD	5.4	0.08	222	0.08	1.355	0.4	8.1	1.5	156	272	0.23	6.1	0.034
	TD	4.1	0.03	298	0.03			12.9	1.9			0.36	9.6	0.034
ARAMID	MD	9.8	0.04	200	0.04	1.500	1.5	5.3	3.1	277	None	0.18	4.8	0.011
	TD	10.2	0.05	271	0.05			5.5	1.0			0.16	4.3	neg.
PI	MD	4.8	0.04	227	0.12	1.420	2.9	13.0	—	360–410	None	0.33	8.6	neg.
	TD	4.2	0.05	223	0.21			—	—			0.27	7.1	
PBO	MD	16.8	0.03	511	0.03	1.540	0.8	1.0	-2.0	—	None	0.069	1.84	neg.
	TD	8.8	0.04	305	0.04			2.0	-9.0			0.067	1.79	

[a]24 hrs at 22 °C. [b]Calculated from creep-compliance data for a 12.7 mm wide substrate using a Poisson's ratio of 0.3 and applied stress of 7.0 MPa.

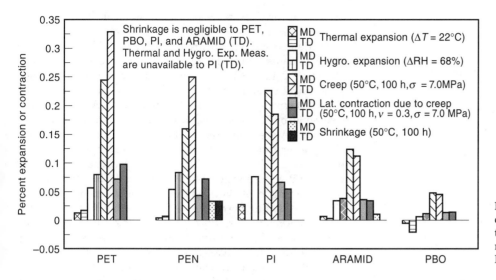

Figure 11. Summary of expansion and contraction characteristics for magnetic tape substrates where the data has been reduced to a percentage scale [Weick and Bhushan (4)].

to-layer contact, Bhushan (1), Bhushan and Koinkar (20), and Oden et al. (21) have reported that particles known as "anti-slip" agents are commonly dispersed in PET before extrusion. The presence of these particles also affects the abrasion of the PET film (unbackcoated tape surface) when it is in sliding contact with tape drive components or during winding or spooling.

Surface roughness profiles are shown in Fig. 12 for the magnetic tape substrates. These profiles were taken using a commercially available atomic force microscope (AFM), and the scan size was set to 100 × 100 μm. Root-mean-square (rms) roughness values are tabulated in this figure along with peak-to-valley (P–V) and peak-to-mean (P–M) distance measurements for each substrate. The roughness profile for PET indicates that anti-slip particles are present in the polymer. These particles ensure that the substrate can be wound tightly on to itself, and prevent any layer-to-layer slippage between wraps. Typically, two size distributions of particles are used: submicron particulates with a height of approx. 0.5 μm to reduce interlayer friction, and larger particles with a height of approximately 2 to 3 μm to control the air film (Bhushan (1) and Bhushan and Koinkar (20)). The larger particles are typically composed of ceramics such as silica, titania, bentonite, calcium carbonate, or clays of different kinds. Bhushan (1) has discussed the fact that these particles are known to affect the abrasion resistance of the PET film (unbackcoated tape surface) when used in sliding contact with components in a tape drive or during winding on the spool.

The surface of the alternative substrate PEN is substantially smoother than the surface of PET. Furthermore, the P–V and P–M distance values for PEN are less than half that for PET indicating that anti-slip agents are not present in PEN. ARAMID also has a substantially lower rms than PET and PEN. The rms for ARAMID is 10.6 nm compared to 37.2 nm for PET and 15.8 nm for PEN. P–V and P–M distance values for ARAMID are less than 1/3 that measured for PET and half that measured for PEN. PI offers the smoothest topography with an rms of 3.20 nm; however the P–V and P–M values for PI are higher than that for ARAMID possibly due to imperfections in the surface. PBO has an rms of 16.5 nm which is similar to that measured for PEN, but the P–V and P–M values for PBO are much lower indicating that PBO has fewer surface imperfections.

In addition to the 100 × 100 μm scans, 10 × 10 μm AFM scans were also performed. Results are summarized for both scan sizes in Table 4. With the exception of ARAMID and PI, rms roughness decreases with scan size as reported by Bhushan and Ruan (22). This is also true for P–V and P–M with the exception of ARAMID. Therefore, in general, smaller regions of the substrates appear to be smoother, and the probability of hitting a significantly high peak decreases with scan size. ARAMID is the exception, rms doubles, and P–V and P–M increase slightly. This increase in rms is most likely indicative of a surface modification performed on ARAMID to accommodate metal-evaporated coatings.

Friction

Since magnetic recording devices require low motor torque and high magnetic reliability, the finished magnetic medium must also exhibit low friction/stiction and high durability. Interlayer friction between a substrate and itself is important since the substrate must be wound onto itself during manufacturing. Similarly, interlayer friction between an unbackcoated tape (the substrate) and a coated magnetic tape is important during actual tape drive operation since an unbackcoated side of a tape will be in contact with a coated tape surface during reel winding. In addition, for unbackcoated tapes the substrate can contact tape path components directly. Therefore, friction between the substrate and these components is important, and ferrite is typically chosen as the countersurface for tribological studies [Bhushan and Koinkar (20)].

Friction measurements have been made by Weick and Bhushan (10) using a reciprocating friction tester (REFT). For these experiments a 12.7 mm wide by 270 mm long substrate strip was drawn back and forth over a countersurface at a constant velocity of 25 mm/s for each pass. The applied stress was 7.0 MPa and the duration of the experiments was 15 min. Three different countersurfaces were utilized for the friction experiments: (1) a Ni–Zn ferrite tape head (IBM 3480/3490-type, rms = 2.2 nm, P–V distance = 21 nm), (2) a sample of the substrate itself, and (3) a metal particle tape (MP-tape) with an rms roughness of 5.7 nm and peak-to-valley distance of 56 nm. Three to four repeats were performed for each sub-

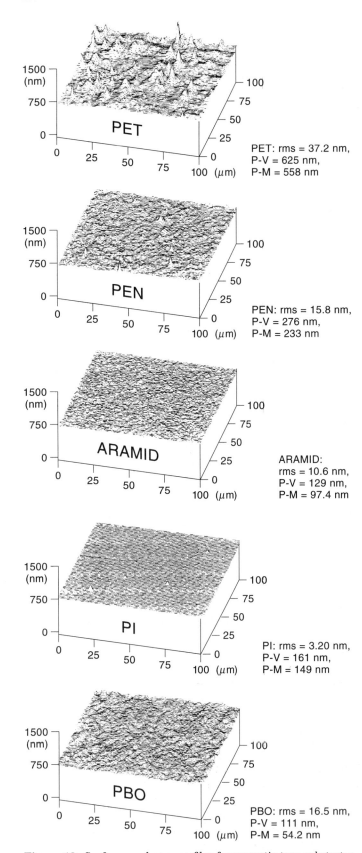

PET: rms = 37.2 nm,
P–V = 625 nm,
P–M = 558 nm

PEN: rms = 15.8 nm,
P–V = 276 nm,
P–M = 233 nm

ARAMID:
rms = 10.6 nm,
P–V = 129 nm,
P–M = 97.4 nm

PI: rms = 3.20 nm,
P–V = 161 nm,
P–M = 149 nm

PBO: rms = 16.5 nm,
P–V = 111 nm,
P–M = 54.2 nm

Figure 12. Surface roughness profiles for magnetic tape substrates (100 × 100 μm² AFM scans). Root mean square (rms) roughness values are shown along with peak-to-valley (P–V) and peak-to-mean (P–M) roughness values [Weick and Bhushan (10)].

strate, and the average peak friction was calculated for each pass along with the 95% confidence intervals.

Friction measurements are tabulated in Table 4 along with 95% confidence intervals. Friction traces are shown in Fig. 13. For PET, higher friction coefficients were measured when the substrate was rubbed against itself or an MP tape than when it was rubbed against a ferrite surface. A similar trend was found by Bhushan and Koinkar (20), but lower friction coefficients were measured in the present study possibly due to the newer 14.4 μm PET (Mylar A DB grade) substrate used in this research. The higher friction coefficient measured when PET was rubbed against itself could be attributed to higher plowing contributions from the interaction of the antislip particles. Furthermore, the MP tape used in this study has an rms roughness of 5.7 nm versus 2.2 nm for the ferrite surface. Therefore, plowing interactions between antislip agents in the PET substrate and asperities on the relatively rough MP tape surface could lead to higher friction. When PET is rubbed against the ferrite surface there is less plowing, and the harder, higher modulus ceramic antislip particles could lead to a lower real area of contact, lower frictional forces, and a lower adhesive friction component.

For PEN, higher friction coefficients were measured when it was rubbed against a ferrite surface than when it was rubbed against itself or an MP tape countersurface. This is the opposite trend to what was observed for PET. Note that PEN has a lower rms roughness and P–V distance than PET due to a lack of hard, ceramic, antislip particles. Therefore, due to the lack of these particles, the adhesive friction component will be lower and the friction coefficient for PEN rubbing against ferrite will be higher. The PEN substrate could also form larger real areas of contact with the head leading to higher adhesive friction forces. Furthermore, in keeping with concepts presented by Fowkes (23,24), *acidic* groups in the polyester backbone of PEN can interact and adhere more readily with the *basic* ferrite surface which is comprised of various oxides (11% NiO, 22% ZnO, 67% Fe_2O_3). Friction coefficients measured against itself and MP tape are also higher for PEN versus PET although the confidence intervals are rather high. This could be attributed to more asperities interlocking with the countersurface leading to an increase in the plowing component. Although the particles on a PET surface are large and will indeed interact, the total summation of the areas which are interlocking could be smaller for PET versus PEN.

The friction trend for ARAMID is similar to that observed for PEN: friction against ferrite is higher than against itself or MP tape. Since the structure for ARAMID contains the relatively acidic amide linkages, acid-base interactions with the basic ferrite surface is again more likely. Although this is not indicated in Table 4 by the lower friction coefficient of 0.49 ± 0.06 for ARAMID versus 0.59 ± 0.06 for PEN, the trend line in Fig. 13 shows that the confidence interval for ARAMID is large at the beginning and the end of the experiment. Electrostatic attraction between ARAMID and the head was also more pronounced than for the other materials, and could play a role along with acid-base interactions. Frictional effects for ARAMID against itself and the MP tape are also lower when compared to the same measurements for PET. Since the rms, P–V, and P–M are significantly lower for ARAMID than for

Table 4. Coefficient of Friction and Surface Roughness Measurements for Magnetic Tape Substrates

| | Surface Roughness—Atomic Force Microscope Measurements | | | | | | Coefficient of Friction for Substrate | | |
| | Scan Size: $100 \times 100 \ \mu m$ | | | Scan Size: $10 \times 10 \ \mu m$ | | | | | |
Substrate	Rms (nm)	Pk-Valley (nm)	Pk-Mean (nm)	Rms (nm)	Pk-Valley (nm)	Pk-Mean (nm)	Against Ni–Zn Ferrite	Against Itself	Against MP Tape
PET	37.2	625.	558.	16.1	147.	116.	0.29 ± 0.02	0.39 ± 0.02	0.41 ± 0.10
PEN	15.8	276.	233.	10.8	126.	105.	0.59 ± 0.06	0.47 ± 0.02	0.47 ± 0.09
ARAMID	10.6	129.	97.4	20.2	142.	101.	0.49 ± 0.06	0.21 ± 0.05	0.27 ± 0.10
PI	3.20	161.	149.	3.76	35.8	24.8	—	—	—
PBO	16.5	111.	54.2	9.99	66.1	32.2	0.43 ± 0.01	0.43 ± 0.07	0.45 ± 0.18

PET (based on $100 \times 100 \ \mu m$ AFM scans), frictional effects due to plowing could be lower for ARAMID. Although the rms roughness for ARAMID increases to 20.2 nm (versus 16.1 nm for PET) when a $10 \times 10 \ \mu m$ scan size is used, a reduction in plowing is still felt to be a feasible explanation for the lower ARAMID - ARAMID and ARAMID - MP tape friction. This is due to the fact the friction measurements are macroscopic in nature and should be compared to large scale roughness measurements such as the $100 \times 100 \ \mu m$ AFM scans.

There is no significant difference in friction measurements for PBO when it is rubbed against ferrite, itself, or an MP tape. Although PBO has an rms which is approximately equal to PEN's, its P–V and P–M values are significantly lower. Therefore, plowing contributions from interlocking asperities are not likely to be present when PBO is rubbed against itself or an MP tape. Furthermore, since PBO does not contain the acidic groups present in ARAMID and PEN, there is no measurable increase in friction when it is rubbed against a basic ferrite countersurface. However, friction is slightly higher for PBO when compared to PET. No concrete explanation for this can be made, but PBO does have a significantly higher tensile strength than PET, and it is likely to have a higher shear strength, which would lead to an increase in the friction based on a simple adhesion model.

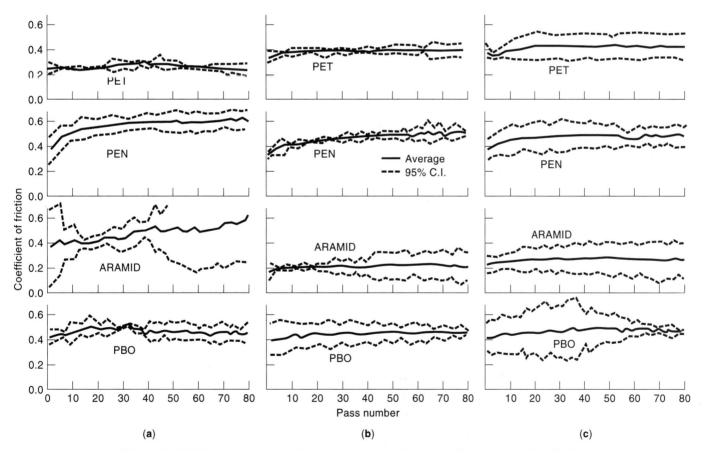

Figure 13. Friction measurements for magnetic tape substrates rubbing against (a) A Ni–Zn ferrite head, (b) Themselves, and (c) An MP tape. The velocity = 25.4 mm s^{-1}, sample length = 270 mm, applied stress = 7.0 MPa, and the duration of the experiments = 15 min [Weick and Bhushan (10)].

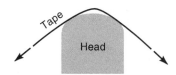

Figure 14. Schematic drawing of a tape traveling over a head and forming two distinct contact regions due to lack of tape-to-head conformity [Weick and Bhushan (10)].

MECHANICAL DESIGN CONSIDERATIONS

Tape-to-Head Conformity and Dynamic Tape–Head Interactions

Typically, a high modulus substrate is desirable for a magnetic tape during fabrication and use of the tape. Bhushan (1) and Weick and Bhushan (4,5) have shown that high modulus, high strength films such as PEN, ARAMID, or PBO are more desirable to minimize stretching and damage of the tape and subsequent loss of information stored on the tape. However, Hahn (12) has found that there is a direct relationship between head wear and bending stiffness of a magnetic tape. Since bending stiffness is proportional to the product of modulus and moment of inertia, it is clear that higher modulus substrates can lead to increased head wear. This increasing relationship is attributed to a lack of tape-to-head conformity. Stiffer, higher modulus tapes will form contact zones as shown schematically in Fig. 14.

From the creep and DMA studies performed by Weick and Bhushan (4–6,10,11) using magnetic tapes and substrates, it is clear that a viscoelastic tape substrate material has a temperature and frequency-dependent modulus. Therefore, under certain circumstances in a tape drive the modulus of a tape substrate will increase and affect the tape-to-head conformity. Two of these circumstances are discussed by Weick and Bhushan (10) using a PEN substrate as an example. The first circumstance is referred to as Case 1, and it considers the deformation frequency for a perturbation in the tape as it comes off a roll (or bearing) near the head. When this occurs the segment can experience a perturbation which propagates at a velocity ΔV. This propagation velocity is related to the tape tension and mass density of the tape [Stahl et al. (25)].

$$\Delta V = \sqrt{\frac{T}{m}} \tag{4}$$

where

T = the applied tape tension in the drive
m = the linear mass density of the tape in kg/m

From Fig. 15 the frequency of this velocity disturbance, f_{d1}, is inversely related to the time it takes for a tape segment to travel from the roll at $x = 0$ to the head at $x = L$. Therefore, f_{d1} is equal to ΔV divided by the distance L:

$$\text{Case 1:} \qquad f_{d1} = \frac{\Delta V}{L} = \frac{1}{L}\sqrt{\frac{T}{m}} \tag{5}$$

For a 3480/3490-type head, the distance L is approximately equal to 20 mm. The typical axial stress applied to a tape is 7.0 MPa. Using a 4.5 μm thick PEN substrate as an example, the tension is 0.40 N for a 12.7 mm wide tape substrate. Since the linear mass density for PEN is 7.3(10^{-5}) kg/m, f_{d1} is 3.71 kHz from Eq. (5).

The other circumstance, Case 2, considers the deformation frequency when a tape substrate encounters and travels over the head. This second deformation frequency is directly related to the fact that the tape does not immediately form a stable hydrodynamic air film. The sudden change in curvature associated with the tape bending over the head together with such factors as the air film viscosity, velocity of the tape, and tension per unit width all contribute to the deformation of the tape over a finite arc distance along the head. The coordinate for this distance is s, and a tape segment encountering the head will undergo a change in height, Δh_0, which is a maximum at the air film entrance region, and will become zero when a stable air film of thickness h_0 is formed. This change in height is due to the increase in pressure from ambient to the pressure associated with the hydrodynamic air film. Gross (26) has presented the following equations to calculate the ratio $\Delta h(s)/h_0$:

$$\frac{\Delta h(s)}{h_0} = Ae^{-\xi} \tag{6a}$$

$$\xi = \frac{s\epsilon^{1/3}}{h_0} \tag{6b}$$

$$\epsilon = \frac{6\mu V}{T/w} \tag{6c}$$

where

$\Delta h(s)$ = change in the thickness of the hydrodynamic air film as a function of s
h_0 = thickness of the stable air film
s = longitudinal coordinate along the head
A = a constant which is typically exceedingly small (23)
μ = the dynamic viscosity in N $\cdot$ s/m^2
V = the average tape velocity in m/s
T/w = the tape tension per unit width of tape in N/m

Equations (6a–c) are only true at the entrance region to the head and can be used to predict the distance s_1 through which the tape is subjected to the transient deformation. Since the change in thickness of the air film is a function of h_0 as well as the other variables already listed, curves can be drawn for various h_0 values which show $\Delta h/h_0$ as a function of s. These curves are shown in Fig. 16, and $\Delta h/h_0$ has been divided by the unknown constant A. The initial height of the deformation is governed by the magnitude of this constant. The distance s_1 can be found from the point at which each of the curves reaches a $\Delta h/Ah_0$ of zero. However, since the functions shown in Eq. 6(a–c) are exponential functions that decrease asymptotically and therefore never truly reach zero, s_1 will be se-

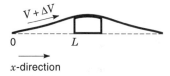

Figure 15. Diagram of a tape traveling from a roll or bearing at $x = 0$ to the head $x = L$ [Weick and Bhushan (10)].

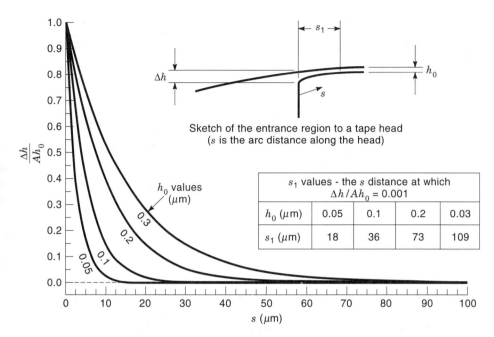

Figure 16. Variation in air-bearing thickness at the entrance region (Δh) as a function of stable film thickness values h_0. Curves are for a PEN substrate, similar curves can be developed for other substrates and tapes [Weick and Bhushan (10)].

lected at the point where $\Delta h/Ah_0$ is 0.1% of its maximum value of 1. Once s_1 is found Eq. (7) can be used to calculate a second deformation frequency using the average tape velocity, V:

$$\text{Case 2}: \quad f_{d2} = \frac{V}{s_1} \tag{7}$$

From Fig. 16, s_1 is equal to 36 μm for an h_0 of 0.1 μm which is a typical value for the thickness of the stable air film between a tape and an IBM 3480/3490-type head (Bhushan (1)). For an average velocity, V, of 2 m/s the frequency, f_{d2}, is equal to 56 kHz.

To determine the deformation strain, the frequency-dependent storage modulus from DMA master curves such as the those shown in Figs. 7(a) or 8(a) can be used together with the maximum stress in the tape, $\sigma_{\max}$. The following equation gives an estimate of the strain as a function of deformation frequency where f_d is equal to either f_{d1} or f_{d2}:

$$\epsilon(f_d) = \frac{\sigma_{\max}}{E'(f_d)} \tag{8}$$

Although Eq. (8) shows that ϵ and E' are functions of frequency, it should be understood that these parameters are also functions of temperature.

In Eq. (8) $\sigma_{\max}$ is equal to either $\sigma_{\max1}$ or $\sigma_{\max2}$ depending on whether Case 1 or 2 is being used to determine how the tape is deformed. $\sigma_{\max1}$ is the axial stress on the tape as it travels from the roll to the edge of the head, and $\sigma_{\max2}$ is the stress on the tape as it travels over the head. Therefore, the equation for $\sigma_{\max2}$ considers the fact that the tape is not only subjected to an axial tensile stress, it is also subjected to a bending stress from the head. Although the tape is flexible and can be slowly wrapped around objects such as the head without exceeding its stress limit, Bhushan (1) and Hahn (12) have reported that when the tape is used in a high-speed tape drive it will act like a stiff, rigid member at high deformation rates.

Equations and derivations for $\sigma_{\max1}$ and $\sigma_{\max2}$ can be found in Ref. 10.

For both Cases 1 and 2 the storage modulus will increase as a function of frequency as shown in Figs. 7(a) and 8(a). These frequency-dependent modulus values are shown in Table 5 for magnetic tape substrates as calculated using the techniques discussed by Weick and Bhushan (10). For both cases, PET and PI clearly have the lower storage moduli followed by PEN, ARAMID, and PBO. Based on the direct relationship between head wear and bending stiffness measured by Hahn (12), tapes manufactured with the lower modulus substrates such as PET or PI will potentially conform to the head more readily than PEN, ARAMID, or PBO, and wear of the head will be less for PET or PI. However, this assumes that the substrates are all of equal thickness. Since bending stiffness is directly proportional to the product of elastic modulus and the moment of inertia, thinner tapes will have a substantially lower bending stiffness even though their modulus is higher. This is because the moment of inertia for a rectangular cross-section is proportional to the cube of the thickness.

Transverse Curvature Due to Anisotropy

Recent studies by Bhushan and Lowry (13) have shown that under certain conditions there can be a higher amount of wear at the edges of a tape head when compared to the center. More specifically, the amount of edge wear relative to the wear at the center of the head is higher. This higher relative edge wear has also been shown to be related to tape stiffness, and could correspond with the edges of the tape contacting the head. Furthermore, due to the multilayer composite structure of the tape, it is likely that the tape will show transverse curvature when an axial load is applied. This transverse curvature results in a lack of transverse conformity as shown in Fig. 17. The amount of this curvature depends on the relative thickness of the layers as well as the material properties of each layer. To evaluate the extent of this transverse curva-

Table 5. Calculated Stresses and Strains Imposed on a Tape Substrate When a "Loose" Tape Segment Comes Off a Roll (Case 1), and Stresses and Strains on the Substrate Due to Deformations from Encountering the Tape Head (Case 2).[a]

Substrate	Thickness. (μm)	Mass Density (10^{-4} kg/m)	Deformation Frequencies and Stress–Strain Calculations for Case 1.				Deformation Frequencies and Stress–Strain Calculations for Case 2[b]					Tensile Test Measurements	
			f_{d1} (kHz)	E' @ f_{d1} (GPa)	σ_{max1} (MPa)	ϵ @ f_{d1}	s_1 (μm)	f_{d2} (kHz)	E' @ f_{d2} (GPa)	σ_{max2} (MPa)	ϵ @ f_{d2}	Strength (MPa)	Strain at Yield/failure
PET	23.4	4.21	3.51	6.75	7.0	0.001	63	32	7.1	22.2	0.003	220	0.05
PET	14.4	2.56	3.54	6.75	7.0	0.001	53	38	7.2	27.9	0.004	221	0.02
PEN	4.5	0.73	3.71	11.0	7.0	0.0006	36	56	11.3	52.9	0.005	222	0.08
ARAMID	4.4	0.83	3.43	16.2	7.0	0.0004	36	56	16.9	53.8	0.003	200	0.04
PI	7.6	1.40	3.48	6.25	7.0	0.001	43	47	6.4	39.5	0.006	227	0.04
PBO	5.0	1.01	3.29	27.5	7.0	0.0003	37	54	28.0	49.0	0.002	511	0.03

[a]Applied stress = 7.0 Mpa; width of substrates = 12.7 mm. Tensile test measurements are included for comparison. Calculations shown below are for a PEN substrate and an IBM 3480/3490-type head.
[b]h_o = 0.1 μm.

ture, classical lamination theory (CLT) has been used by Weick and Bhushan (11) to determine stress-strain and stiffness relationships for each layer and the composite magnetic tape as a whole. See Jones (27) for more information about CLT and mechanics of composite materials.

A material with such properties as PET is usually referred to as an orthotropic material, and the orthotropy ratio, α, is defined as the modulus along the major (or stiff) axis with respect to the modulus along the minor (or compliant) axis. PEN also has a tendency to have orthotropic characteristics; whereas a substrate such as ARAMID tends to have isotropic characteristics for which the modulus is the same regardless of material orientation. As already demonstrated in this paper, the magnetic layer will have different modulus characteristics depending on if it is a metal-evaporated (ME) film or a particulate coating (BaFe or MP). The ME film can be considered as a continuous film with isotropic characteristics. Although the particulate MP or BaFe coatings are themselves composites comprised of an elastomeric binder with hard metal or ceramic particles, Weick and Bhushan (11) modeled them as being macroscopically isotropic.

Using CLT, it can be shown that when a load per unit width, N_x, is applied to the tape, the tape will stretch, contract, and curve in the transverse direction. The analytical expression that describes this phenomenon is defined as follows:

$$N_x = A_{11}\epsilon_x + A_{12}\epsilon_y + B_{11}\kappa_x + B_{12}\kappa_y \qquad (9)$$

where

$N_x \equiv$ axial force per unit width
A_{11} and $A_{12} \equiv$ stiffness terms for the axial and transverse strains, respectively

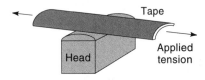

Figure 17. Lack of tape-to-head transverse conformity [Weick and Bhushan (11)].

B_{11} and $B_{12} \equiv$ stiffness terms for the axial and transverse curvatures, respectively
ϵ_x and $\epsilon_y \equiv$ strains in the x and y directions
κ_x and $\kappa_y \equiv$ curvatures in the x and y directions

Note that B_{12} is the transverse curvature stiffness, which is a measure of the tapes resistance to curvature when a load per unit width, N_x, is applied to the tape. Although it would be desirable to calculate κ_y and determine explicit numbers for the amount of transverse curvature a tape is subjected to under an axial load, this is not plausible since it would require the determination of all the stiffnesses, and a knowledge of the other strains and curvatures. Therefore, the simple approach is to calculate B_{12} from equations defined by the classical lamination theory for a two- layer composite (i.e., magnetic tape). From Eq. (9) it should be clear that higher B_{12} values correspond with lower curvatures, κ_y. Therefore, B_{12} as calculated using the equation below can be thought of as a measure of the tapes resistance to transverse curvature. Or, stated another way, tapes with higher B_{12} values will transversely conform to the head, and the edge wear phenomenon observed by Bhushan and Lowry (13) will be minimal. Since it is desirable to develop a nondimensionalized measurement of a tapes resistance to transverse curvature, B_{12} will be divided by B_0, which is defined as the B_{12} value for the tape when the magnetic film thickness is zero:

$$\frac{B_{12}}{B_0} = 1 - \left(\frac{v_a(\alpha - v_{b12}^2)}{v_{b12}(1 - v_a^2)}\right)\left(\frac{E_a}{E_{b1}}\right)\left(\frac{a}{b}\right)^2 \qquad (10)$$

where

v_a and $v_{b12} \equiv$ Poisson's Ratios for the coating and substrate, respectively
E_{b1} and $E_{b2} \equiv$ elastic moduli for the major and minor axes of the substrate, respectively
$E_a \equiv$ elastic modulus for the magnetic coating
a and $b \equiv$ thicknesses of the magnetic coatings and substrates, respectively
$\alpha \equiv$ orthotropy ratio for the substrate, (E_{b1}/E_{b2})

From Eq. (10), as the thickness ratio a/b increases, the value of B_{12}/B_0 decreases. Similarly, B_{12}/B_0 will also decrease with

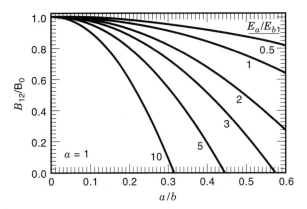

Figure 18. Transverse stiffness of magnetic tapes as a function of the thickness ratio, a/b (Poisson's ratios are assumed to be equal to 0.3). [Weick and Bhushan (11)].

an increasing modulus ratio E_a/E_{b1}, or an increasing orthotropy ratio α. Note that even if $\alpha = 1$, there will still be some curvature which will increase as a/b increases. Also, B_0 itself is a function of the orthotropy ratio. Therefore, a substrate with a high α will have a low B_0 value, and will experience some transverse curvature when an axial tension per unit width is applied.

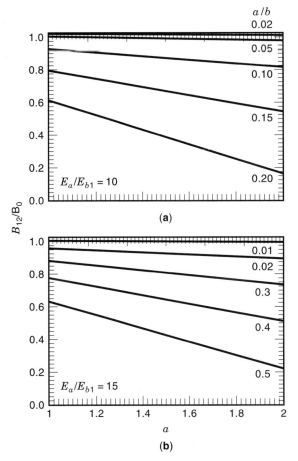

Figure 19. Transverse stiffness of magnetic tapes as a function of substrate orthotropy, α (Poisson's ratios are assumed to be equal to 0.3). [Weick and Bhushan (11)].

Figures 18 and 19 show the transverse stiffness trends for magnetic tapes in a nondimensional, graphical form. In Fig. 18, B_{12}/B_0 is shown as a function of the thickness ratio for various values of the modulus ratio. A higher thickness or modulus ratio will cause the transverse stiffness to decrease. In other words, as the thickness and/or modulus of the magnetic coating increases relative to the thickness or modulus of the substrate, the transverse curvature of the tape will increase. Note that the effect of orthotropy is not considered in Fig. 18 since the orthotropy ratio for the substrate, α, is assumed to be 1.

Figure 19 shows the effect of orthotropy on transverse stiffness for two modulus ratios. The E_a/E_{b1} ratio of 10 was used to generate the top graph in Fig. 19 and is a typical value for ME tapes; whereas the modulus ratio of 1.5 used for the bottom graph is indicative of particulate tapes such as the MP and BaFe tapes. As expected, Fig. 19 shows that as the orthotropy of the substrate increases, the transverse curvature of the tape will increase. The extent of this curvature will be greater for particulate tapes than ME tapes since the thickness ratios are typically larger for particulate tapes. Thickness ratios for the particulate tapes used in this study typically range from 0.4 to 0.6, and a/b ratios for the ME tapes are substantially smaller ranging from 0.02 to 0.04. Therefore, from Fig. 19 ME tapes with modulus ratios on the order of 10 will have B_{12}/B_0 values approaching 1. In comparison, particulate tapes will have lower transverse stiffness values ranging from only 0.2 to 0.6. As a result, tapes like ME-ARA-MID will have minimal transverse curvature, and will transversely conform to the head. This will lead to minimal edge grooving of the head such as that observed by Bhushan and Lowry (13).

Critical Tension for Tape Flyability

Due to the projected use of advanced substrates for magnetic tapes, Weick and Bhushan (10) have used an experimental

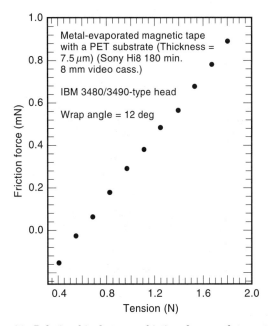

Figure 20. Relationship between friction force and tape tension. Measurements were made using a modified Honeywell Ninety Six vacuum controlled tape drive. The tape speed was 1.5 m s^{-1}, and friction forces are averages for 600 m of tape [Weick and Bhushan (10)].

technique to measure the critical tension required to maintain tape flyability and prevent overstressing of the tape. In this work, a 7.5 μm thick, 8 mm wide, metal-evaporated magnetic tape (Sony Hi-8 ME-180) was used in a Honeywell "Ninety-Six" vacuum controlled tape drive which was modified by Bhushan and Lowry (13) to accept 3480/3490-type heads. Friction forces could be measured using a split I-beam, and tension could be controlled by adjusting the vacuum pressure. Experiments were performed at different tensions, and friction forces were measured as shown in Fig. 20. Although it was expected that the friction would reach a low threshold value indicative of a complete loss of the tape-to-head air bearing, this was not observed. Instead, as tension was reduced, frictional forces between the tape and head were also reduced in a linear fashion. The limit of 0.42 N (1.5 oz) corresponded to a loss of vacuum pressure, and no visible or measurable translation of the tape, or catastrophic loss of the air bearing was observed. Furthermore, it is interesting to note that the 0.42 N limit is equivalent to a 7.0 MPa applied stress on the 7.5 μm thick magnetic tape. Although the results shown in Fig. 20 do not provide a direct solution to the question of a critical tension for the alternative substrates, they do provide an indication that frictional forces do tend to decrease with tension, and lower tensions are potentially feasible in advanced tape drives.

BIBLIOGRAPHY

1. B. Bhushan, *Mechanics and Reliability of Flexible Magnetic Media.* New York: Springer-Verlag, 1992.

2. B. Bhushan, *Tribology and Mechanics of Magnetic Storage Devices.* 2nd ed., New York: Springer-Verlag, 1996.

3. B. Bhushan, Tribology of the head-medium interface, in C. Denis Mee and Eric D. Daniel (eds.), *Magnetic Recording Technology,* 2nd ed., New York: McGraw-Hill, 1996.

4. B. L. Weick and B. Bhushan, Characterization of magnetic tapes and substrates, *IEEE Trans. Magn., 32*: 3319–3323, 1996.

5. B. L. Weick and B. Bhushan, Shrinkage and viscoelastic behavior of alternative substrates for magnetic tapes, *IEEE Trans. Magn., 31*: 2937–2939, 1995.

6. B. L. Weick and B. Bhushan, Viscoelastic behavior and shrinkage of ultra-thin polymeric films, *J. Appl. Polymer Sci., 58*: 2381–2398, 1995.

7. A. Feurstein and M. Mayr, High vacuum evaporation of ferromagnetic materials—a new production technology for magnetic tapes. *IEEE Trans. Mag.,* **MAG-20**: 51–56, 1984.

8. K. H. Harth et al., *J. Mag. Soc. Jpn.,* 13(S-1): 69–72, 1989.

9. R. L. Wallace, The reproduction of magnetically recorded signal, *Bell Syst. Tech. J., 30*: 1145–1173, 1951.

10. B. L. Weick and B. Bhushan, The tribological and dynamic behavior of alternative magnetic tape substrates, *Wear,* **190**: 28–43, 1995.

11. B. L. Weick and B. Bhushan, The relationship between mechanical behavior, transverse curvature, and wear of magnetic tapes, *Wear, 202*: 17–29, 1996.

12. F. W. Hahn, Wear of recording heads by magnetic tape, in B. Bhushan, et al. (eds.), *Tribology and Mechanics of Magnetic Storage Systems,* Park Ridge, IL: ASLE, **1** (SP-16): 41–48, 1984.

13. B. Bhushan and J. A. Lowry, Friction and wear studies of various head materials and magnetic tapes in a linear mode accelerated test using a new nano-scratch wear measurements technique, *Wear, 190*: 1–15, 1995.

14. J. D. Ferry, *Viscoelastic Properties of Polymers,* 3rd. ed., New York: Wiley, 1980.

15. N. W. Tschoegl, *The Phenomenological Theory of Linear Viscoelastic Behavior: an Introduction,* New York: Springer-Verlag, 1989.

16. J. J. Aklonis and W. J. MacKnight, *Introduction to Polymer Viscoelasticity,* New York: Wiley, 1983.

17. D. Perettie et al., *J. Mag. Magn. Mat., 120*: 334–337, 1993.

18. D. Perettie and P. Pierini, Advanced substrates for high performance flexible media, in D. Speliotis (ed.), *Advances in Applied Magnetism, Vol I: Barium Ferrite and Advanced Magnetic Recordings.* Amsterdam: Elsevier, October 1997.

19. D. Perettie and D. Speliotis, *J. Mag. Soc. Jpn., 18* (S1): 279–282, 1994.

20. B. Bhushan and V. N. Koinkar, Microtribology of PET films, *Trib. Trans., 38*: 119–127, 1995.

21. P. I. Oden et al., AFM imaging, roughness analysis and contact mechanics of magnetic tape and head surfaces, *ASME J. Tribol., 114*: 666–674, 1992.

22. B. Bhushan and J. Ruan, Atomic-scale friction measurements using friction force microscopy: Part II—application to magnetic media, *ASME J. Tribol., 116*: 389–396, 1994.

23. F. M. Fowkes, Role of acid-base interfacial bonding in adhesion, *J. Adhes. Sci. Tech.,* **1**: 7–27, 1987.

24. F. M. Fowkes, Acid-Base interactions in polymer adhesion, in J. M. Georges (ed.), *Microscopic Aspects of Adhesion and Lubrication, Tribology Series 7.* Amsterdam: Elsevier, 119–137, 1982.

25. K. J. Stahl, J. W. White, and K. L. Deckert, Dynamic response of self-acting foil bearings, *IBM J. Res. Develop., 18*: 513–520, 1974.

26. W. A. Gross, *Fluid Film Lubrication.* New York: Wiley, 1980.

27. R. M. Jones, *Mechanics of Composite Materials.* New York: Hemisphere, 1975.

BRIAN L. WEICK
University of the Pacific

BHARAT BHUSHAN
Ohio State University

MAGNETIC THIN-FILM DEVICES.   See MAGNETIC MICROWAVE DEVICES.

MAGNETIC VARIABLES CONTROL

Applications of variable-speed electric drive systems are constantly diversifying and growing in numbers. Such a trend is primarily caused by ever-increasing levels of factory automation during the past few decades. It is difficult today to find a single manufacturing process that does not involve at least one variable-speed electric drive. Electric drives power elevators, overhead cranes, machine tools, robots, conveyer belts, steel production lines, paper mills, pumps, compressors, and so on. The other important and expanding area of application is transportation systems. Electric machines drive trains, electric vehicles, and electric forklifts and are indispensable in various motion control systems of airplanes, ships, satellites, and space craft. Last but not least, the application area correlated with domestic appliances has experienced considerable development during the last two decades as well. Electric machines power washing machines and tumble dryers, hand tools, lawn movers, and hair dryers, to name a few. What all these very different applications have in common is that variable-speed operation is required.

Different applications of variable-speed electric drives impose differing requirements on accuracy of the speed control. In a number of cases the rotating speed of a drive has to be close to but not necessarily equal to the reference speed. Typical examples are fans, pumps, and compressors, for which it is frequently sufficient to have only an approximate control of speed as even a small change in speed causes considerable variation in the output power. Electric drives that are utilized in these applications today are usually called general-purpose ac drives. Their characteristics are lack of closed-loop speed control, low cost, low maintenance requirements, and high reliability. The second type of variable-speed electric drive is the one that is used when rough speed control does not suffice. Closed-loop speed control is then necessary so that the shaft of the machine is equipped with a speed-measuring device. Speed control is accurate in the steady state only and no attempt is made to control the transition from one reference speed to the other. The transient response of such a drive is therefore poor, but the speed holding is good in the steady state. In contrast to this, the third type of a variable-speed drive, usually called high-performance drive, is capable of providing both accurate steady-state speed control and a controlled transition from one reference speed to the other. Applications that necessitate use of a high-performance drive include robotics, machine tools, elevators, rolling mills, paper mills, spindles, mine winders, electric traction, and electric vehicles. As the transition from one speed to another has to be controllable, it is necessary to control not only speed but transient torque as well. (If a drive is used for positioning, rotor position control is also required.) Such high-performance applications typically require steady-state speed control accuracy better than 0.5%, a wide range of speed control (typically at least $20:1$), and very fast and accurate transient response [1].

A separately excited dc motor was until recently the only available electric machine that could be used in a high-performance drive. A dc motor is, by virtue of its construction, ideally suited to meeting control specifications for high performance. However, dc motor construction is at the same time the reason why dc drives are today replaced with ac drives wherever possible. A separately excited dc motor has two windings: One is on the stator and the other one is on the rotor. Stator winding is supplied with dc current and provides excitation flux. The rotor (armature) winding is supplied with dc current as well, through the commutator assembly that encompasses stationary brushes that move along the commutator surface as the rotor rotates. The commutator with brushes is the major weakness of a dc motor. It requires maintenance, and the brushes have to be replaced on a regular basis. Because the brushes slip along the commutator during rotation, the maximum current that can be supplied during transient operation is limited, as is the allowable rate of change of current. The commutator limits the maximum operating speed as well. The cost, size, and weight of a dc machine are all higher than for an ac machine of the induction motor type with an equivalent power rating. The inertia of a dc machine is higher than the inertia of an ac machine. This means that under equal developed torque conditions, a dc motor will take longer to reach the desired speed than an ac motor. The efficiency and overall power factor of an ac drive are, in general, better than for the equivalent dc drive [1]. These are the reasons behind the general trend of replacing dc motors with ac motors. Nevertheless, the idea behind the concept of a high-performance drive is most easily explained by taking a separately excited dc motor as an example.

High performance requires a controlled transition from one steady-state operating speed to another. Because the motion of the rotor takes place due to the developed electromagnetic torque of the machine, the motor torque has to be controlled during the transient. The electromagnetic torque of a separately excited dc motor is determined with the product of the armature current and the excitation flux. The existence of the commutator with brushes, which are positioned in an axis perpendicular to the axis along which excitation flux acts, makes it possible to control excitation flux and developed torque independently by means of two dc currents. Excitation flux is determined solely by the value of the excitation current and does not change if armature current varies. Suppose that the machine is excited with constant excitation current, so that excitation flux is constant as well. Developed electromagnetic torque is then controllable solely by armature current. The desired torque value in both steady-state and transient operation is achieved by supplying the armature winding with an appropriate current value. Hence the existence of the commutator assembly inherently enables independent flux and torque control. It is usually said that flux and torque control are decoupled as flux is controlled by one (excitation winding) current while torque is, in constant flux operation, controlled by another (armature winding) current.

The important conclusion that results from the discussion of torque production in a dc machine is that decoupled flux and torque control requires control of armature current. This means that a current-controlled dc source must feed the machine's armature winding. In other words, the output voltage of the dc source is controlled in a closed-loop manner in such a way that the required armature current is produced. Current-controlled sources that are used to supply a dc machine are power electronic converters of an ac/dc type that come in various configurations.

The preceding discussion can be summarized in five statements: (1) High-performance operation requires that electromagnetic torque of a motor is controllable in real time; (2) instantaneous torque of a separately excited dc motor is directly controllable by armature current as flux and torque control are inherently decoupled; (3) independent flux and torque control are possible in a dc machine due to its specific construction, which involves a commutator with brushes whose position is fixed in space; (4) instantaneous flux and torque control require that the machine windings are fed from current-controlled dc sources; and (5) current and speed sensing is necessary in order to obtain feedback signals for real-time control.

Substitution of dc drives with ac drives in high-performance applications has become possible only recently. From the control point of view, it is necessary to convert an ac machine into its equivalent dc counterpart so that independent control of two currents yields decoupled flux and torque control. What the commutator with brushes does in a dc machine physically (statements 2 and 3) has to be done in an ac machine mathematically. This is unfortunately far from being a trivial task. The most frequently utilized types of ac machines are of brushless construction, so the physical structure of the machine is not any help in establishing means for independent flux and torque control. Fundamental principles that enable mathematical conversion of an ac three-phase machine into an equivalent dc machine were established in the early 1970s for both induction and synchronous machines (2) and are known as vector control or field-oriented control (1,3–5). What remains common for both dc and ac high-performance drives is that the supply sources are current controlled (statement 4), current feedback and speed feedback are required (statement 5), and torque is controlled in real time (statement 1). Stator winding of three-phase ac machines is supplied with ac currents, which are characterized by amplitude, frequency, and phase rather than just by amplitude, as in dc case. Thus an ac machine has to be fed from a source of variable output voltage, variable-output frequency type. Power electronic converters of dc/ac type (inverters) are the most frequent source of power in high-performance ac drives. Application of vector-controlled ac machines in high-performance drives became a reality in the early 1980s and has been enabled by developments in the areas of power electronics and microprocessors. Control systems that enable realization of decoupled flux and torque control in ac motor drives are complex and involve a coordinate transformation that has to be executed in real time. Application of microprocessors or digital signal processors is therefore mandatory.

The frequency of the stator winding supply uniquely determines the speed of rotation of a synchronous machine. Permanent magnets or dc excitation current in the rotor winding provide excitation flux. The rotor carries with it the excitation flux as it rotates, and the instantaneous spatial position of the rotor flux is always fixed to the rotor. Hence, if rotor position is measured, the position of the excitation flux is known. Such a situation leads to relatively simple vector control algorithms for both permanent magnet and wound rotor synchronous motors (1,3,4). The situation is more involved in synchronous reluctance machines. The rotor is of salient pole structure but without either magnets or excitation winding, so that excitation flux stems from the ac supply of the three-phase stator winding. By far the most complex situation results in induction machines where not only the excitation flux stems from the stator winding supply but the rotor rotates asynchronously with the rotating field. This means that even if the rotor position is measured, the position of the rotating field in the machine remains unknown. Vector control of induction machines is thus the most complicated case (1,3–5).

Vector control of ac machines has reached a mature stage and is today widely applied when high performance is required. A squirrel-cage induction machine is the most frequently used type of electric machines and is found in applications that cover the entire power range. The main advantages of an induction motor over the other types of ac motors are low cost and very rugged construction that requires virtually no maintenance. Application of synchronous motors takes place either in relatively low-power regions (permanent magnet machines) or in very high-power regions (wound rotor synchronous machines). Although permanent magnet machines are brushless as well, their cost is at present considerably higher than the cost of an induction motor of the same power rating. Wound rotor synchronous machines are of the brushed type and are applied as motors in very high-power regions only, where their higher cost is offset by some advantages over the induction motors. It follows from this consideration that induction machines are used whenever possible, and it is for this reason that only vector control of induction machines is dealt with in this article.

Vector control of ac machines enables decoupled torque and flux control in much the same way as it is achieved in dc machines. To explain vector control principles, it is necessary to perform at first transformation of the model of an ac machine from the original phase domain into a fictitious, so-called arbitrary reference frame domain.

MATHEMATICAL MODELING OF AN INDUCTION MACHINE

Vector control provides instantaneous control of the machine's flux and torque, which is to be realized via instantaneous current control. Thus the time domain mathematical model, in terms of original phase variables, has to be utilized as a starting point. Unfortunately, principles of vector control cannot be explained and understood from this time domain model. Instead, this model has to be mathematically transformed into a new model, which describes a fictitious induction machine equivalent to the original one, by a suitably chosen mathematical transformation.

The procedure of mathematical modeling of an induction machine is subject to a number of assumptions (4,6,7). Those that are relevant for subsequent considerations are that winding resistances and leakage inductances are constant parameters, iron losses and higher spatial harmonics of magnetomotive force (m.m.f.) are neglected, and it is assumed that the magnetizing curve of the machine is linear (flux saturation is neglected). By performing mathematical transformation of the model given in terms of existing phase quantities, the physical induction machine is substituted with an equivalent induction machine that does not exist in reality. Both stator and rotor three-phase windings are transformed, using different mathematical transformations for stator and rotor. The resulting mathematical model may be given in terms of either real (6) or complex (7) variables. The main feature of the transformed model is that it describes a fictitious machine whose equivalent stator and rotor windings all rotate at the same, in general arbitrary, angular speed. Relative motion between stator and rotor phase windings is substituted with fictitious electro-motive forces. The transformation is therefore referred to as transformation to the common arbitrary reference frame. The following two subsections review the real, so-called d-q-axis model and the complex, so-called space vector model of a three-phase induction machine.

Real d-q-Axis Model

Let the phases of the original three-phase windings on stator and rotor be denoted with indices a, b, c and A, B, C, respectively. Symbols v, i, and ψ stand for instantaneous voltage, current, and flux linkage of any of the windings. Stator and rotor three-phase windings are to be transformed into a common reference frame, which rotates at an arbitrary angular speed ω_a, so that resulting new stator and rotor windings (ds, qs, and dr, qr) will all rotate with the same arbitrary angular speed. Hence the transformation enables substitution of a six-winding induction machine with an equivalent four-winding machine.

The magnetic axis of the stator phase winding a is taken as a stationary axis with respect to which all the angular positions are measured. As the rotor rotates at an electrical angular speed ω, instantaneous position of the rotor winding A magnetic axis with respect to the stator phase a magnetic axis is determined with $\theta = \int \omega \, dt$. All the windings are transformed to the arbitrary reference frame rotating at speed ω_a, so that instantaneous position of d-axis with respect to the stationary phase a axis is determined with $\theta_s = \int \omega_a dt$. The angle between d-axis and the rotor A axis is hence $\theta_r = \theta_s - \theta$. The axes d and q are mutually perpendicular. Let f denote either voltage, current, or flux linkage of any of the windings of stator and rotor. The transformation of original to equivalent d-q-axis windings is governed with transformation angles θ_s and θ_r for stator and rotor quantities, respectively. In particular,

$$
\begin{aligned}
f_{ds} &= \frac{2}{3}\left[f_a \cos\theta_s + f_b \cos(\theta_s - 2\pi/3) + f_c \cos(\theta_s - 4\pi/3) \right] \\
f_{qs} &= -\frac{2}{3}\left[f_a \sin\theta_s + f_b \sin(\theta_s - 2\pi/3) + f_c \sin(\theta_s - 4\pi/3) \right]
\end{aligned}
\tag{1}
$$

$$
\begin{aligned}
f_{dr} &= \frac{2}{3}\left[f_A \cos\theta_r + f_B \cos(\theta_r - 2\pi/3) + f_C \cos(\theta_r - 4\pi/3) \right] \\
f_{qr} &= -\frac{2}{3}\left[f_A \sin\theta_r + f_B \sin(\theta_r - 2\pi/3) + f_C \sin(\theta_r - 4\pi/3) \right]
\end{aligned}
\tag{2}
$$

where indices s and r stand for stator and rotor, respectively. Equations (1) and (2) define transformation from the original phase domain into the common d-q-axis reference frame as a single-step transformation. The transformation may be looked at as being composed of two transformations. The first one replaces original three-phase windings with equivalent two-phase windings that still rotate with the speed of rotation of the original windings, while the second transformation replaces two-phase machine with d-q-axis windings that all rotate at the same speed.

The factor $\frac{2}{3}$ in Eqs. (1) and (2) is correlated with powers in the original and the equivalent induction machine. While the actual machine is three phase, the equivalent one is two-phase. Factor $\frac{2}{3}$ preserves the equality of power per phase in the original and the equivalent machine. As the rotor winding is short-circuited (either squirrel-cage machine or slip-ring machine with short-circuited rotor winding is assumed), instantaneous power in terms of transformed quantities equals

$$
P_e = (3/2)(v_{ds} i_{ds} + v_{qs} i_{qs})
\tag{3}
$$

The mathematical model of a three-phase induction machine is, upon completion of the transformation, obtained in the following form:

$$
v_{ds} = R_s i_{ds} + \frac{d\psi_{ds}}{dt} - \omega_a \psi_{qs}
$$
$$
v_{qs} = R_s i_{qs} + \frac{d\psi_{qs}}{dt} + \omega_a \psi_{ds}
\tag{4}
$$

$$
v_{dr} = 0 = R_r i_{dr} + \frac{d\psi_{dr}}{dt} - (\omega_a - \omega)\psi_{qr}
$$
$$
v_{qr} = 0 = R_r i_{qr} + \frac{d\psi_{qr}}{dt} + (\omega_a - \omega)\psi_{dr}
\tag{5}
$$

where d-q-axis flux linkages are given by

$$
\psi_{ds} = L_s i_{ds} + L_m i_{dr} = L_\sigma s i_{ds} + L_m(i_{ds} + i_{dr})
$$
$$
\psi_{qs} = L_s i_{qs} + L_m i_{qr} = L_\sigma s i_{qs} + L_m(i_{qs} + i_{qr})
\tag{6}
$$

$$
\psi_{dr} = L_r i_{dr} + L_m i_{ds} = L_\sigma r i_{dr} + L_m(i_{ds} + i_{dr})
$$
$$
\psi_{qr} = L_r i_{qr} + L_m i_{qs} = L_\sigma r i_{qr} + L_m(i_{qs} + i_{qr})
\tag{7}
$$

and magnetizing flux, magnetizing current, and magnetizing inductance (all denoted with an index m) are defined as

$$
\begin{aligned}
i_{dm} &= i_{ds} + i_{dr} & i_{qm} &= i_{qs} + i_{qr} \\
\psi_{dm} &= L_m i_{dm} & \psi_{qm} &= L_m i_{qm} \\
i_m &= \sqrt{i_{dm}^2 + i_{qm}^2} & \psi_m &= \sqrt{\psi_{dm}^2 + \psi_{qm}^2} \\
L_m &= \psi_m / i_m
\end{aligned}
\tag{8}
$$

The magnetizing inductance in Eq. (8) is constant, due to assumed linearity of the magnetic circuit. Index σ in Eqs. (6) and (7) denotes leakage inductances of stator and rotor.

Terms of the form $\omega\psi$ in Eqs. (4) and (5) are already mentioned fictitious electromotive forces that represent relative motion between stator and rotor phase windings, as well as the relative motion between all the phase windings and the common d-q reference frame.

Taking mechanical power as positive in motoring, the equation of mechanical motion is

$$
T_e - T_L = \frac{J}{P}\frac{d\omega}{dt}
\tag{9}
$$

where J, P, and T_L stand for inertia, number of pole pairs, and load torque, respectively; mechanical angular speed is ω/P; friction is neglected; and electromagnetic torque is given by

$$
\begin{aligned}
T_e &= \frac{3}{2}P(\psi_{ds} i_{qs} - \psi_{qs} i_{ds}) = \frac{3}{2}P\frac{L_m}{L_r}(\psi_{dr} i_{qs} - \psi_{qr} i_{ds}) \\
&= \frac{3}{2}P(\psi_{dm} i_{qs} - \psi_{qm} i_{ds})
\end{aligned}
\tag{10}
$$

Coefficient $\frac{3}{2}$ in the torque expression is a consequence of the power correlation between the original and the equivalent induction machine. Inverse correlation between d-q-axis variables and original phase domain variables is established in the same way for voltages, currents, and flux linkages. For example, for stator quantities

$$
\begin{aligned}
f_a &= f_{ds} \cos\theta_s - f_{qs} \sin\theta_s \\
f_b &= f_{ds} \cos(\theta_s - 2\pi/3) - f_{qs} \sin(\theta_s - 2\pi/3) \\
f_c &= f_{ds} \cos(\theta_s - 4\pi/3) - f_{qs} \sin(\theta_s - 4\pi/3)
\end{aligned}
\tag{11}
$$

Transformation expressions of the same form, with appropriate change of indices, apply to rotor variables.

Equations (1), (2), and (4) through (11), together with definitions of various spatial angles, completely describe an induction machine in an arbitrary common reference frame. All the variables and parameters of the rotor winding are referred to the stator by means of turns ratio. The model in an arbitrary reference frame yields an appropriate model for any specified value of the common reference frame angular speed. In the special case when $\omega_a = 0$, equations in a stationary α,

β reference frame result:

$$v_{\alpha s} = R_s i_{\alpha s} + \frac{d\psi_{\alpha s}}{dt}$$
$$v_{\beta s} = R_s i_{\beta s} + \frac{d\psi_{\beta s}}{dt} \tag{12}$$

$$v_{\alpha r} = 0 = R_r i_{\alpha r} + \frac{d\psi_{\alpha r}}{dt} + \omega\psi_{\beta r}$$
$$v_{\beta r} = 0 = R_r i_{\beta r} + \frac{d\psi_{\beta r}}{dt} - \omega\psi_{\alpha r} \tag{13}$$

Equations (1), (2), and (11) remain valid, with change of indices, $d \to \alpha$ and $q \to \beta$, and provided that it is recognized that $\omega_a = 0$ means $\theta_s = 0$ and $\theta_r = -\theta$. Equations (3), (6)–(8), and (10) remain unaltered, with change of indices $d \to \alpha$ and $q \to \beta$. Equation (9) is unchanged.

Complex Space Vector Model

A mathematical model of an induction machine in terms of real d-q-axis variables, expressed in an arbitrary reference frame, contains for each variable (voltage, current, flux linkage of either stator or rotor) two components that are mutually perpendicular. This is a consequence of the 90° displacement between the d-axis and the q-axis. It is possible to regard the d-q reference frame as an orthogonal rotating system of axes in which a variable along one axis represents a real part of a complex variable, while the same variable along the other axis represents the imaginary part of the complex variable. So defined complex numbers are called space vectors (4,7).

It is convenient to define space vectors initially in a stationary reference frame. Components along the α, β axes of both stator voltage and current are correlated with phase quantities by Eq. (1), with angle θ_s set to zero. The stator voltage space vector and stator current space vector are defined in the stationary reference frame as follows (symbols for space vectors are in boldface italic, while the superscript s denotes the stationary reference frame):

$$\boldsymbol{v}_s^s = v_{\alpha s} + j v_{\beta s} = v_s e^{j\beta_s}$$
$$\boldsymbol{i}_s^s = i_{\alpha s} + j i_{\beta s} = i_s e^{j\epsilon_s} \tag{14}$$

where both, being complex numbers, are expressed in polar form as well. As α, β components of both voltage and current are time-varying quantities, the phase of both complex numbers is a time-varying quantity. If transformation expressions for α, β components, Eq. (1) with θ_s set to zero, are substituted in Eq. (14), correlation between space vectors in stationary reference frame and actual phase variables is obtained

$$\boldsymbol{v}_s^s = \frac{2}{3}(v_a + \boldsymbol{a}v_b + \boldsymbol{a}^2 v_c)$$
$$\boldsymbol{i}_s^s = \frac{2}{3}(i_a + \boldsymbol{a}i_b + \boldsymbol{a}^2 i_c) \tag{15}$$

where $\boldsymbol{a} = e^{j2\pi/3}$ is a spatial operator. To express a space vector in an arbitrary reference frame, it is necessary to rotate the space vectors of Eq. (15) for an angle that defines an instantaneous position of the d-axis of the common reference frame with respect to the stationary phase a-axis. Stator voltage and current space vectors are given in an arbitrary reference

frame with

$$\boldsymbol{v}_s = \boldsymbol{v}_s^s e^{-j\theta_s}$$
$$\boldsymbol{i}_s = \boldsymbol{i}_s^s e^{-j\theta_s} \tag{16}$$

Substitution of Eq. (14) in Eq. (16) yields

$$\boldsymbol{v}_s = v_{ds} + j v_{qs} = v_s e^{j(\beta_s - \theta_s)}$$
$$\boldsymbol{i}_s = i_{ds} + j i_{qs} = i_s e^{j(\epsilon_s - \theta_s)} \tag{17}$$

Manipulation of Eqs. (4)–(8) enables creation of the complex, space vector model of an induction machine (7):

$$\boldsymbol{v}_s = R_s \boldsymbol{i}_s + \frac{d\boldsymbol{\psi}_s}{dt} + j\omega_a \boldsymbol{\psi}_s$$
$$0 = R_r \boldsymbol{i}_r + \frac{d\boldsymbol{\psi}_r}{dt} + j(\omega_a - \omega)\boldsymbol{\psi}_r \tag{18}$$

$$\boldsymbol{\psi}_s = L_s \boldsymbol{i}_s + L_m \boldsymbol{i}_r = L_\sigma \boldsymbol{i}_s + \boldsymbol{\psi}_m$$
$$\boldsymbol{\psi}_r = L_r \boldsymbol{i}_r + L_m \boldsymbol{i}_s = L_\sigma \boldsymbol{i}_r + \boldsymbol{\psi}_m$$
$$\boldsymbol{\psi}_m = L_m \boldsymbol{i}_m \tag{19}$$
$$\boldsymbol{i}_m = \boldsymbol{i}_s + \boldsymbol{i}_r$$

where the remaining space vectors are defined as

$$\boldsymbol{\psi}_s = \psi_s e^{j(\phi_s - \theta_s)}; \quad \boldsymbol{\psi}_r = \psi_r e^{j(\phi_r - \theta_s)}; \quad \boldsymbol{i}_r = i_r e^{j(\epsilon_r - \theta_s)}$$
$$\boldsymbol{\psi}_s^s = \psi_s e^{j\phi_s}; \quad \boldsymbol{\psi}_r^s = \psi_r e^{j\phi_r}; \quad \boldsymbol{i}_r^s = i_r e^{j\epsilon_r}$$
$$\boldsymbol{\psi}_m = \psi_m e^{j(\phi_m - \theta_s)}; \quad \boldsymbol{i}_m = i_m e^{j(\phi_m - \theta_s)}$$
$$\boldsymbol{\psi}_m^s = \psi_m e^{j\phi_m}; \quad \boldsymbol{i}_m^s = i_m e^{j\phi_m} \tag{20}$$

Equation (9) remains unchanged. Electromagnetic torque and input power, Eqs. (10) and (3), can be given as

$$T_e = \frac{3}{2}P\,\mathrm{Im}\{\boldsymbol{i}_s\boldsymbol{\psi}_s^*\} = \frac{3}{2}P\frac{L_m}{L_r}\,\mathrm{Im}\{\boldsymbol{i}_s\boldsymbol{\psi}_r^*\} = \frac{3}{2}P\,\mathrm{Im}\{\boldsymbol{i}_s\boldsymbol{\psi}_m^*\}$$
$$P_e = \frac{3}{2}\mathrm{Re}(\boldsymbol{v}_s\boldsymbol{i}_s^*) \tag{21}$$

where the symbol $*$ denotes complex conjugation. Equations (9), (18), (19) and (21) constitute, together with the appropriate transformation expressions, the complete complex space vector model of an induction machine.

PRINCIPLES OF VECTOR CONTROL OF INDUCTION MACHINES

The clue for decoupled flux and torque control in an induction machine lies in the choice of the common reference frame in which the machine is represented. The reference frame can be selected as firmly fixed to any of the three flux space vectors (stator, air-gap, and rotor flux) in the machine. The idea of field-oriented control requires that instantaneous values of the magnitude and spatial position of the stator current space vector with respect to the selected flux space vector in the machine can be controlled [i.e., that the stator current space vector is oriented with respect to the chosen flux space vector (4)]. Because in an induction machine there exist three flux space vectors [Eq. (20)], it is possible to realize stator flux oriented, rotor flux oriented and air-gap flux oriented control

(1,4,5). The requirement for instantaneous control of the magnitude and spatial position of the stator current space vector translates itself into the requirement that amplitude, frequency, and phase of stator phase currents have to be instantaneously controllable (8).

The induction machine is fed from a power electronic converter with closed-loop current control in vector-controlled drives. The converter is usually a voltage source inverter operated in the pulse-width modulated (PWM) mode. Stator voltages are obtained on the basis of the closed-loop current control, so that the machine is fed from a current regulated PWM (CRPWM) voltage source inverter. It is possible to view an induction machine in two different ways, depending on the selected current control method. If closed-loop current control is performed using actual phase currents (current control in a stationary reference frame), it is possible to regard an induction machine as current fed. Reference stator phase currents are equal to actual phase currents under ideal conditions, and it may be assumed that the currents rather than the voltages are impressed into the machine's stator winding. Stator currents are thus known and stator voltage equations may be omitted from consideration. In contrast to this, if closed-loop current control is performed using transformed d-q-axis stator current components (current control in rotational reference frame), the machine cannot be regarded as current fed and stator voltage equations have to be considered. In this case a voltage-fed machine results (3,5,9). (It should, however, be noted that the voltage source is still current-controlled.)

Rotor flux oriented control yields the simplest configuration of the control system and is therefore the most frequently applied method. Although rotor flux oriented control is utilized in conjunction with both current-fed and voltage-fed machines, the analysis is here restricted to the simpler of the two (i.e., to rotor flux oriented control of a current-fed induction machine).

Rotor Flux Oriented Control of a Current-Fed Induction Machine

Consider the space vector model of an induction machine, [Eqs. (18)–(21)]. The machine is assumed to be current fed and stator voltage equation is therefore omitted. Let the common reference frame be fixed to the rotor flux space vector and, moreover, let the d-axis (real axis) of this common reference frame coincide with the rotor flux space vector. Then

$$\begin{aligned}
\theta_s &= \phi_r \\
\theta_r &= \phi_r - \theta \\
\omega_a &= \omega_r \\
\omega_r &= d\phi_r/dt
\end{aligned} \tag{22}$$

The rotor flux space vector of Eq. (20) is a real variable in this reference frame:

$$\begin{aligned}
\boldsymbol{\psi}_r &= \psi_{dr} + j\psi_{qr} = \psi_r \\
\psi_{qr} &= 0 \\
d\psi_{qr}/dt &= 0 \\
\psi_{dr} &= \psi_r
\end{aligned} \tag{23}$$

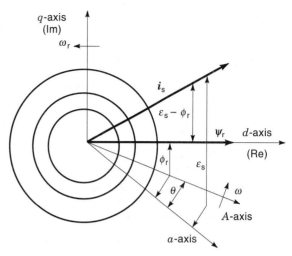

Figure 1. Rotor flux and stator current space vectors in the common d-q reference frame fixed to the rotor flux space vector. Rotor flux space vector is aligned with d-axis of the reference frame at all times, and speed of rotation of the two is the same.

Figure 1 illustrates rotor flux and stator current space vectors in this special common reference frame.

Consider now electromagnetic torque expressed in terms of the rotor flux space vector, Eq. (21). Taking into account Eq. (23), Eq. (21) yields

$$T_e = \frac{3}{2}P\frac{L_m}{L_r}\psi_r\,\mathrm{Im}\{i_s\} = \frac{3}{2}P\frac{L_m}{L_r}\psi_r i_{qs} \tag{24}$$

The torque equation is of the same form as in a separately excited dc machine and, if magnitude of the rotor flux is kept constant, torque can be controlled solely by stator q-axis current.

To accommodate the rotor voltage equation of Eq. (18) to the chosen reference frame, rotor current space vector has to be expressed from Eq. (19) using Eq. (23):

$$i_r = (\psi_r - L_m i_s)/L_r \tag{25}$$

Substitution of Eq. (25) into Eq. (18), with Eq. (22) accounted for, results in the following complex rotor voltage equation (rotor time constant is introduced as $T_r = L_r/R_r$):

$$0 = \frac{1}{T_r}\psi_r + \frac{d\psi_r}{dt} + j(\omega_r - \omega)\psi_r - \frac{1}{T_r}L_m i_s \tag{26}$$

Separation of Eq. (26) into real and imaginary parts yields

$$\psi_r + T_r\frac{d\psi_r}{dt} = L_m i_{ds} \tag{27}$$

$$(\omega_r - \omega)\psi_r T_r = L_m i_{qs} \tag{28}$$

Equation (27) reveals that the magnitude of the rotor flux can be controlled by the stator d-axis current and that it is constant if the stator d-axis current is constant. According to Eq. (28), the angular slip frequency $\omega_{sl} = \omega_r - \omega$ linearly depends on the stator q-axis current when the magnitude of rotor flux is constant. Developed torque is then proportional to slip frequency. If the stator d-axis current is constant, rotor flux is constant and torque can be instantaneously altered if it is possible to change the stator q-axis current instantaneously. If the machine is current fed, Eqs. (24), (27), and (28) constitute the complete model. Thus decoupled torque and flux control can be obtained with a current-fed machine, provided that the control system operates in the rotor flux oriented reference frame. The machine is, however, supplied with three-phase ac currents. It is therefore necessary to include a coordinate transformation between the controller and the power supply.

Coordinate transformation requires information on the instantaneous position of the rotor flux space vector. The schemes of rotor flux oriented control may be subdivided into two groups, depending on how this information is obtained. In indirect schemes the position of the rotor flux space vector is calculated without the use of measured electromagnetic variables. In direct control schemes some measured electromagnetic variables are used for rotor flux position calculation.

Direct Rotor Flux Oriented Control of an Induction Machine

Figure 2 illustrates control system of a current-fed direct rotor flux oriented induction machine, which comprises a rotor flux (PI) controller, speed (PI) controller, and torque (PI) controller. The information regarding rotor flux position and amplitude (and torque) is obtained from measured signals, as discussed shortly. The two control paths that generate stator d-q-axis current references operate in parallel, independently one of the other. An asterisk denotes reference (commanded) quantities, while a superscript e stands for estimated variables. Feedback signals include rotor speed (which is mea-

sured or calculated from measured rotor position) and measured stator currents. The CRPWM inverter is assumed to be ideal, so that reference and actual stator phase currents are equal. The two transformation blocks between d-q-axis currents and phase current references describe the coordinate transformation of Eq. (11), performed in two steps as discussed in conjunction with Eq. (1). The outputs of the speed and rotor flux controllers are limited and the provision for field weakening is included. The field-weakening block keeps rotor flux at a constant rated value in the base speed region (up to the rated speed) and reduces rotor flux reference inversely proportionally to the speed above base speed. Such a change of rotor flux reference is frequently applied although it is simplified with respect to the optimal rotor flux reference change in the field weakening region (1,10).

Rotor flux estimation in direct schemes of rotor flux oriented control can be performed in various ways. Air-gap (main, magnetizing) flux, stator currents, stator voltages, and rotor speed (position) are measurable quantities, and different combinations of these signals can be used (11). The method based on stator current and air-gap flux measurement performs calculations using Eqs. (6) and (7) in a stationary reference frame. It requires installation of flux sensors or tapping of stator windings (11) and is rarely applied today. The second method asks for measurement of stator voltages and stator currents (12), and the magnitude and position of the rotor flux space vector are calculated from Eqs. (12), (6), and (7) in the stationary reference frame. The method involves integration, and estimation becomes inaccurate at low speeds.

A frequently utilized method of rotor flux space vector estimation, shown in the upper part of Fig. 2, uses measured stator currents and rotor speed (9) (it will be denoted as $i_s - \omega$ estimator). The main advantages of this scheme are that there is no need for special construction or modification of the machine, integration of voltages is avoided, and estimation is operational at zero speed. It is used in the vector control system of Fig. 2 and in rotor flux oriented voltage-fed induction machines with current control in the rotational reference frame (9). The estimator performs calculations on the basis of the model of an induction machine in a rotor flux oriented reference frame [Eqs. (24), (27), and (28)]. Rotor flux position is calculated by integrating the sum of the measured rotor speed and estimated angular slip frequency, as shown in Fig. 2, where symbol s denotes the Laplace operator. Measured stator currents have to be transformed into a rotor flux oriented reference frame. The major shortcoming of this method is the strong dependence of estimation accuracy on parameter variation effects. The same remark applies to indirect rotor flux oriented control, which is discussed next.

Indirect Rotor Flux Oriented Control of an Induction Machine

If the rotor flux space vector position is calculated in a feedforward manner, using measured speed (position) and references rather than measured electromagnetic variables, indirect rotor flux oriented control results (current measurement remains necessary in order to establish closed-loop current control). Equations (24), (27), and (28) enable calculation of the reference stator d-q-axis current components, reference angular slip speed, and desired rotor flux spatial position as

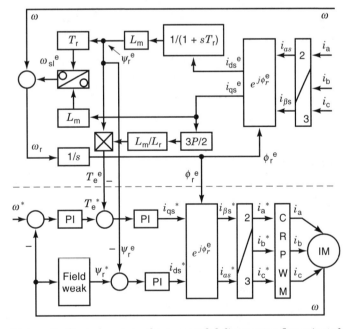

Figure 2. Control system of a current-fed direct rotor flux oriented induction machine. Stator d-q-axis current references are created by two independent control loops operating in parallel and are converted into phase current references by means of a coordinate transformation. Estimation of the rotor flux space vector is performed using measured stator currents and rotor speed. The estimator operates in the rotor flux oriented reference frame, and measured stator currents have to be transformed using inverse coordinate transformation.

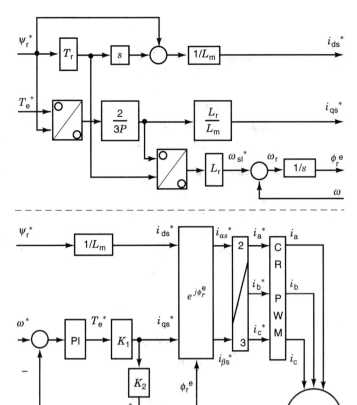

Figure 3. Outlay of an indirect rotor flux oriented controller and its implementation in conjunction with a current-fed induction machine for operation in the base speed (constant flux) region [$K_1 = (2/3P)(L_r/L_m^2)/i_{ds}^*$, $K_2 = 1/(T_r i_{ds}^*)$].

follows (1):

$$i_{qs}^* = \frac{2}{3P} \frac{T_e^*}{\psi_r^*} \frac{L_r}{L_m}$$

$$i_{ds}^* = \frac{1}{L_m}\left(\psi_r^* + T_r \frac{d\psi_r^*}{dt}\right) \quad (29)$$

$$\omega_{sl}^* = \frac{L_m}{T_r} \frac{i_{qs}^*}{\psi_r^*}$$

$$\phi_r^e = \int (\omega_{sl}^* + \omega)\, dt \quad (30)$$

Figure 3 shows an indirect rotor flux controller, based on Eqs. (29) and (30) (an asterisk again denotes reference quantities). Prevailing applications are for drives that require operation in the base speed region only (where rotor flux reference is constant and rated) and it is then possible to simplify the control scheme. Such a drive is shown in the lower part of Fig. 3, where due to $\psi_r^* = $ const., $i_{ds}^* = \psi_r^*/L_m$ is constant as well. Torque command is obtained as output from the speed PI controller. Indirect vector control is a frequent choice in practical realizations as the control system is significantly simpler, compared to direct orientation schemes.

Performance of a Rotor Flux Oriented Induction Machine

Dynamic performance of a rotor flux oriented machine is most easily examined using simulations. Simulation programs must include representation of the control system and an appropriate model of the induction machine. A power electronic converter can be omitted from the simulation if performance is analyzed under ideal supply conditions. The most appropriate model of the induction machine is then the one given by Eqs. (4)–(10), formed in the reference frame fixed to the rotor flux space vector angular speed determined by the control system. Such an approach enables omission of the coordinate transformation blocks, as outputs of the control system become directly inputs into the machine model. Stator voltage equations are not required when a current-fed machine is analyzed. Note that the motor model must include an equation for the time derivative of the rotor flux q-axis component. If rotor flux oriented control is achieved, simulation will give rotor flux along the q-axis as equal to zero.

If a current-controlled PWM voltage source inverter is included in the simulation model, it is most convenient to represent the induction machine in the stationary reference frame. Stator voltage equations now have to be included in the model.

Figure 4 presents a simulation illustration of dynamics of a rotor flux oriented induction machine. The results apply to the direct rotor flux oriented induction machine of Fig. 2. All the parameters of the estimator are taken as equal to those in the machine model and the operation is simulated assuming ideal current feeding (i.e., the model does not include inverter representation, the machine is represented with d-q-axis model in the reference frame fixed to the estimated rotor flux position, and stator current commands are inputs into the machine model so that stator voltage equations are omitted). All the three controllers of Fig. 2 are of PI type, so that in any steady-state operation reference and actual speed are equal, as are the estimated and reference torque, and rotor flux. The machine initially operates in steady state with rated rotor flux command, zero load torque, and speed equal to 40% of the rated in negative direction of rotation. Speed reversal is then initiated with a ramplike speed reference change, from −40% to +40% of the rated speed. Rotor flux reference remains rated. Change in speed command leads to fast buildup of the stator q-axis current, leading to a corresponding buildup of torque of the same profile. Actual and estimated torque values coincide. Maximum torque is limited to seven times rated. No change in rotor flux takes place, stator d-axis current remains the same as before the transient, and decoupled control of flux and torque is achieved. The final steady state is identical to the original one, except that the machine rotates in the positive direction.

Another important feature of the drive is its response to a sudden application and removal of the load torque. Figure 5 shows transients that now take place. It is an experimental recording of the operation of an indirect rotor flux oriented induction machine of the structure illustrated in Fig. 3 (index n stands for rated values). The machine initially operates with 70% of the rated rotor flux (70% of the rated stator d-axis current) at a speed of 600 rpm under no-load conditions. Step load torque is at first applied and then removed. Stator q-axis current command and the measured speed are shown. Speed initially drops following the application of the load

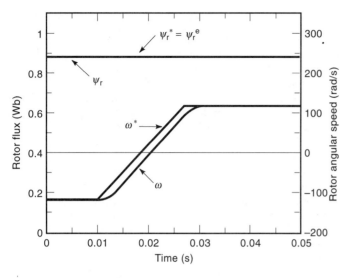

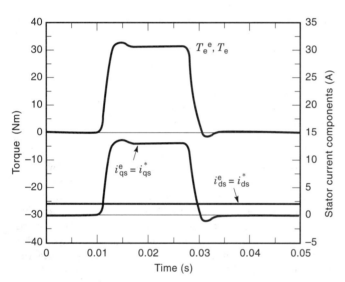

Figure 4. Speed reversal of a current-fed direct rotor flux oriented induction machine in the base speed region (simulation, 0.75 kW machine). Torque builds up rapidly, and interaction between d- and q-axes does not take place, as witnessed by the unchanged value of the rotor flux in the machine. Decoupled rotor flux and torque control is thus achieved.

torque. Torque quickly builds up and returns speed to the reference value. Removal of the load torque has the opposite effect. Speed initially exceeds reference, causing rapid reduction in the torque, which leads to return of the speed to the reference value.

Simulation and experimental results in Figs. 4 and 5 show that rotor flux oriented control is indeed characterized by very quick torque response. Torque response is smooth, without any unwanted oscillations, so that speed change is uniform and as rapid as possible.

Other Orientation Possibilities

As already noted, rotor flux oriented control can be realized with closed-loop current control of stator d-q-axis current components, resulting in a voltage-fed machine. The model derived for a rotor flux oriented current-fed induction ma-

chine remains valid. However, stator voltage equations now have to be considered and outputs of the control system are now references for stator voltage d-q-axis components rather than references for stator current components. Correlation between stator d-q-axis voltages and stator d-q-axis currents is not decoupled, and it is necessary to include a decoupling circuit in the control system, which decouples stator voltages and currents along d-q-axes (9). The resultant control structure is more complex than the one of a current-fed machine.

It is possible to realize vector control with stator current orientation along stator flux and along air-gap flux space vectors (4). If the derivation procedure for a current-fed machine is done for stator flux and air-gap flux oriented control, the main reason for predominant use of rotor flux oriented control becomes obvious. The models of a current-fed machine in these two reference frames do not possess the main feature of rotor flux oriented control—namely, that flux and torque can be independently controlled by two stator current components in a decoupled manner. The equations are coupled and it is therefore necessary to introduce decoupling circuits into the control system even when the machine is current fed (13). Decoupling circuits significantly increase the complexity of the drive.

As a result of existence of three types of vector control (orientation along rotor, air-gap and stator flux), which each can be realized using either indirect or direct vector control, so-called "universal field oriented controller" has been introduced (14). This controller can operate in any of the three reference frames, with either indirect or direct type of control, and various operating modes can be used in different speed regions, with smooth changeover during operation of the machine (15).

PARAMETER VARIATION EFFECTS IN ROTOR FLUX ORIENTED INDUCTION MACHINES

The constant parameter model of an induction machine is used for calculation of rotor flux space vector position in both direct and indirect vector control schemes. If the value of any of the parameters in the control part differs from the corre-

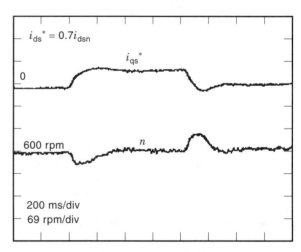

Figure 5. Step loading and unloading of an indirect rotor flux oriented induction machine (experiment, 0.75 kW machine). Application and removal of the load torque initiate rapid torque response, which returns the speed to the reference value.

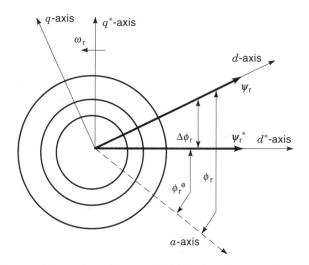

Figure 6. Illustration of commanded (d^*-q^*) and actual (d-q) rotor flux oriented reference frames in detuned operation. Because the commanded reference frame does not coincide with the actual one, decoupled rotor flux and torque control does not take place.

sponding actual value in the machine, so-called detuning occurs. This means that the estimated rotor flux position, calculated by the controller and used for coordinate transformation, does not correspond to the actual rotor flux position in the machine, so that field orientation is not achieved. Figure 6 illustrates detuned operation; the d-q-axis reference frame determined by the controller is denoted with an asterisk. The actual rotor flux space vector is displaced from the d-axis of this reference frame, so that the actual rotor flux oriented d-q reference frame does not coincide with the frame assumed by the controller. The consequence of detuning is that decoupled rotor flux and torque control does not take place, and this leads to unwanted transients in torque response and to steady-state errors in both rotor flux and torque.

Induction machine parameters are subject to variation due to various reasons. Stator and rotor resistance change with operating temperature. Stator and rotor leakage inductance vary due to different levels of saturation of the leakage flux paths. Magnetizing inductance varies with changes in the saturation level of the main flux path. Rotor resistance and rotor leakage inductance may change with rotor frequency due to skin effect. Iron losses are often neglected in the model of the induction machine from which vector control principles are derived.

A variety of schemes of rotor flux oriented control make a unified analysis of performance deterioration due to parameter variation effects impossible. Different schemes rely on the use of different parameters in a different way. The analysis of parameter variation effects is here restricted to a current-fed machine with control schemes of Figs. 2 and 3. The reason for this specific selection is twofold. First, these two types of rotor flux orientation are most frequently utilized. Second, as both of these schemes rely on the same equations for achieving field orientation [Eqs. (29) and (30) for the indirect scheme and Eqs. (24), (27), and (28) for the direct scheme], their steady-state behavior under detuned conditions is identical. Thus the following steady-state analysis of detuned operation applies to the both schemes. Stator resistance and stator leakage inductance are not involved in the rotor flux

position calculation. Hence their variations have no impact on operation of the drive and can be excluded from further analysis.

Rotor resistance and rotor leakage inductance vary with rotor frequency only in deep-bar and double-cage induction machines (16). Variation of rotor leakage inductance due to saturation of the rotor leakage flux path is a secondary order effect (17), as rotor leakage inductance enters all the controller equations summed with the magnetizing inductance, which is 10 to 100 times greater. Omission of iron loss representation in the vector controller leads to detuning that is relatively small (18,19). These considerations leave two sources of detuning as most relevant: main flux saturation and temperature-related variation in rotor resistance.

Main Flux Saturation

Induction machines are designed to operate around the knee of the nonlinear magnetizing curve. Magnetizing flux and magnetizing inductance have rated values in the rated operating point on this curve. If the magnetizing flux is changed from rated value in either direction, the value of the magnetizing inductance will change. Nonlinearity of the magnetizing curve has numerous consequences on the operation of a vector-controlled induction machine. It affects available torque per ampere and needs to be accounted for if this ratio is to be maximized in steady-state and/or transient operation (1,20). Developed torque and its linearity are affected as well (1,21). Next, there are numerous situations when rotor flux reference, and hence the main flux saturation level as well, are altered dynamically during operation of the drive. When speed exceeds rated value, the machine operates in the field-weakening region and rotor flux reference is reduced below rated value. Since a reduction of flux reference leads to an increase in the magnetizing inductance, nonlinearity of the magnetizing curve has to be taken into account (22). Rotor flux oriented control is well suited to operation of an induction machine with optimal efficiency. In this case rotor flux reference is varied until input power consumption reaches minimum for a given load. When the load is light, optimal efficiency is achieved with reduced flux (23). Thus, depending on the operating cycle of the machine, continuous variation of the reference rotor flux takes place although the machine may operate in the base speed region only. Variation of rotor flux reference changes magnetizing inductance, so that accuracy of field orientation is affected.

Steady-state analysis of detuning due to main flux saturation requires incorporation of the magnetizing curve into the steady-state model (1,21). The analysis is done in the d-q reference frame determined by the controller (d^*-q^* reference frame of Fig. 6). Steady-state operation of the indirect vector controller (Fig. 3) is described by

$$i_{ds}^* = \psi_r^*/L_m^* \qquad i_{qs}^* = T_e^*/\psi_r^*(L_r^*/L_m^*)(1/K)$$
$$\omega_{sl}^* = (L_m^* i_{qs}^*)/(T_r^* \psi_r^*) \quad \phi_r^e = \int(\omega_{sl}^* + \omega)\,dt = \int \omega_r\,dt \quad (31)$$
$$\omega^* = \omega \qquad T_e = T_L$$

Equality of reference and actual speed is a consequence of the action of the PI speed controller, which forces steady-state speed error to zero. Actual torque developed by the machine equals load torque. In Eq. (31) and in what follows, apart from reference values, an asterisk denotes parameter values used in the controller. Constant K stands for $(\frac{3}{2})P$.

Consider next Eqs. (5) and (10). Elimination of rotor current components by means of Eq. (7), formulation of the equations in the reference frame dictated by the controller, and application of the steady-state constraint $d/dt = 0$ lead to the following equations:

$$(1/T_r)\psi_{dr} = (L_m/T_r)i_{ds}^* + \omega_{sl}^*\psi_{qr}$$
$$(1/T_r)\psi_{qr} = (L_m/T_r)i_{qs}^* - \omega_{sl}^*\psi_{dr} \qquad (32)$$
$$T_e = K(L_m/L_r)(\psi_{dr}i_{qs}^* - \psi_{qr}i_{ds}^*)$$

Stator current d-q-axis components and slip frequency in Eq. (32) equal reference values due to the choice of the reference frame and idealized treatment of the inverter. Magnetizing inductance in Eq. (32) in general differs from the one in Eq. (31). Hence rotor time constants differ as well. Let the ratio of magnetizing inductances be $\beta = L_m/L_m^*$. Then $\alpha = T_r/T_r^* = L_r/L_r^* = (\beta + \sigma_r)/(1 + \sigma_r)$, where $\sigma_r = L_{\sigma r}^*/L_m^* = \mathrm{constant}$. Steady-state detuning is characterized by three quantities: ratio of actual to commanded rotor flux, ratio of actual to commanded torque, and error in orientation angle. Magnitude of the actual rotor flux and error in the orientation angle are defined as

$$\psi_r = \sqrt{\psi_{dr}^2 + \psi_{qr}^2}$$
$$\Delta\phi_r = \tan^{-1}(\psi_{qr}/\psi_{dr}) \qquad (33)$$

where actual rotor flux d-q-axis components are projections of the actual rotor flux space vector onto the d^*-q^* system of axes in Fig. 6. If Eq. (32) is solved for rotor flux d-q-axis components, these are further inserted into the torque equation of Eq. (32) and the controller equations [Eq. (31)] are then accounted for, detuning characteristics are obtained in the form

$$\frac{\psi_r}{\psi_r^*} = \beta\sqrt{\frac{1 + (T_r^*\omega_{sl}^*)^2}{1 + \alpha^2(T_r^*\omega_{sl}^*)^2}}$$
$$\frac{T_e}{T_e^*} = \beta^2\frac{1 + (T_r^*\omega_{sl}^*)^2}{1 + \alpha^2(T_r^*\omega_{sl}^*)^2} = \left(\frac{\psi_r}{\psi_r^*}\right)^2 \qquad (34)$$
$$\Delta\phi_r = \tan^{-1}\left(\frac{(1-\alpha)T_r^*\omega_{sl}^*}{1 + \alpha(T_r^*\omega_{sl}^*)^2}\right)$$

Equation (34) can be used to assess trends in detuning due to saturation, by taking the ratio of magnetizing inductances β as an independent variable. However, to predict behavior of a given motor quantitatively, it is necessary to account for the actual magnetizing curve of the machine. Indeed, for the given rotor flux reference, torque reference, and magnetizing inductance in the controller, coefficient β is a dependent rather than independent variable, whose value is determined with the magnetizing curve:

$$\psi_m = f(i_m); \quad L_m = \psi_m/i_m = f_1(i_m) = f_2(\psi_m) \qquad (35)$$

The procedure is iterative and encompasses Eqs. (7), (8), (34), and (35). The model derived so far is sufficient for characterization of the torque mode of operation (i.e., when speed control loop is open) as torque reference is taken as an inde-

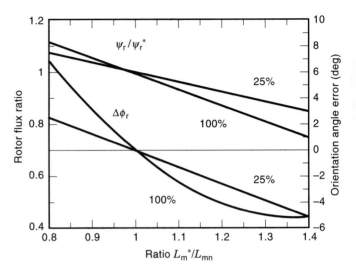

Figure 7. Orientation angle error and rotor flux ratio for incorrect setting of the magnetizing inductance in the controller (operation in the base speed region with rated rotor flux command, load torque as parameter, 4 kW machine). Error in the magnetizing inductance setting can lead to severe detuning.

pendent input. In operation with closed-loop speed control, the equality of the load torque and machine's torque, given in Eq. (31), has to be satisfied. Due to detuning, reference torque does not equal actual torque and is, in general, unknown. If the product of the controller rotor time constant and reference slip frequency is expressed from Eq. (31) as $\omega_{sl}^*T_r^* = (1/K)(L_r^*/\psi_r^{*2})T_e^* = hT_e^*$ and is then substituted into the torque ratio equation of Eq. (34), a third-order equation is obtained:

$$T_e^{*3} + T_e^{*2}\left(-\frac{\alpha^2}{\beta^2}T_L\right) + T_e^*\frac{1}{h^2} + \left(-T_L\frac{1}{\beta^2h^2}\right) = 0 \qquad (36)$$

whose solution determines reference torque value. [Condition $T_e = T_L$ is accounted for in Eq. (36).] Coefficient h is a constant for given rotor flux reference and given value of the magnetizing inductance in the controller. Load torque is the independent variable in Eq. (36).

Figure 7 shows the steady-state detuning characteristics for the speed mode of operation (i.e., closed-loop speed control), for operation in the base speed region with constant rated rotor flux command. Load torque is the parameter, and characteristics are plotted against the ratio of magnetizing inductance in the controller to the rated magnetizing inductance value. Torque ratio, being equal to the rotor flux ratio squared, Eq. (34), is omitted. Detuning is independent of the speed, and characteristics show that incorrect setting of the magnetizing inductance in the controller can lead to severe orientation angle errors.

To investigate the dynamics of a saturated rotor flux oriented induction machine, a convenient dynamic saturated induction machine model is required. It is therefore necessary to modify the d-q-axis model given by Eqs. (4)–(7) by accounting for nonlinear correlation between magnetizing flux and magnetizing current in Eq. (8). State-space models of a saturated induction machine can be formed in various ways, depending on which variables are selected as state-space variables (24). Any of the dynamic saturated machine models may be used for simulation purposes. Furthermore, these models can be used to assess qualitatively the impact of main flux saturation on rotor flux oriented control. The net effect of main flux saturation is loss of decoupled rotor flux and torque control during transients and in steady states other than rated, even when the machine operates with rated constant rotor flux command and the value of magnetizing inductance in the controller equals rated (25). This is due to the fact that, according to Eq. (8), variation of stator q-axis current leads to variation of q-axis magnetizing current, so that the q-axis component of the magnetizing flux changes and causes alteration in the total magnetizing flux. Hence the magnetizing inductance varies as well. This effect is usually termed cross saturation and is insignificant in practice when torque is limited to at most twice the rated value (25). When the current-fed machine is analyzed, stator voltage equations can be omitted and stator d-q-axis current components and their derivatives then act as inputs to the model of the machine, which is formed in the reference frame fixed to the rotor flux position calculated by the controller.

Transient behavior is investigated for the direct rotor flux oriented current-fed induction machine of Fig. 2. The value of the magnetizing inductance in the estimator is set to rated and operation is simulated for the same reversing transient already depicted in Fig. 4. Figure 4 applies to an idealized situation when saturation is neglected, while the results, shown in Fig. 8, are obtained with saturation accounted for in the machine's model. As acceleration torque attains seven times the rated torque value, the q-axis component of the magnetizing current becomes significant and drives the machine into deep saturation. The flux estimator does not recognize this change in magnetic conditions, so that the rotor flux is erroneously estimated as remaining constant and equal to the reference value during the transient and the stator d-axis current command is kept unchanged. Actual torque is smaller than the estimated value. Speed response differs insignificantly from the one in Fig. 4 and is therefore not shown.

Much the same behavior is observed if a change in saturation level takes place due to a change in rotor flux reference, which causes a change in stator d-axis current command (26). Thus it follows that even when the magnetizing inductance in the estimator (or indirect controller) is set to the rated value, there will be detuning if either significant cross saturation occurs or if the machine is operated with variable rotor flux command. If the magnetizing inductance value in the controller does not correspond to the rated (this situation is illustrated for steady states in Fig. 7), transient response deteriorates further.

Rotor Resistance Variation

Rotor resistance enters equations of both the indirect vector controller of Fig. 3 and the rotor flux estimator of Fig. 2 through the rotor time constant. The rotor time constant determines the accuracy of slip frequency calculation [Eqs. (28) and (30)]. As slip frequency is summed with rotor speed in order to calculate the position of the rotor flux space vector, any error in the value of the rotor resistance directly leads to detuned operation. The effects of variation in rotor resistance are analyzed in considerable depth in Refs. 27–29. An investigation of detuning due to rotor resistance variation has to be done in such a way that main flux saturation is included in the model of the machine (27–29). Thus the complete modeling procedure described in the previous subsection fully applies to the analysis of detuning due to rotor resistance variation. It is only necessary to set L_m^* to L_{mn}, define the ratio between the actual rotor resistance and the rotor resistance used in the controller as $r = R_r/R_r^*$, and take this coefficient as an independent variable. Introduction of this coefficient modifies detuning equations, so that detuning in the steady state with closed-loop speed control is described by

$$\frac{\psi_r}{\psi_r^*} = \beta \sqrt{\frac{1 + \left(\omega_{sl}^* T_r^*\right)^2}{1 + \left(\dfrac{\alpha}{r}\right)^2 \left(\omega_{sl}^* T_r^*\right)^2}}$$

$$\frac{T_e}{T_e^*} = \frac{\beta^2}{r} \frac{1 + \left(\omega_{sl}^* T_r^*\right)^2}{1 + \left(\dfrac{\alpha}{r}\right)^2 \left(\omega_{sl}^* T_r^*\right)^2} = \frac{1}{r}\left(\frac{\psi_r}{\psi_r^*}\right)^2$$

$$\Delta\phi_r = \tan^{-1}\left(\frac{\omega_{sl}^* T_r^*\left(1 - \dfrac{\alpha}{r}\right)}{1 + \omega_{sl}^{*2} T_r^{*2}\dfrac{\alpha}{r}}\right) \tag{37}$$

$$T_e^{*3} + T_e^{*2}\left[-\left(\frac{\alpha}{\beta}\right)^2 \frac{1}{r} T_L\right] + T_e^* \frac{1}{h^2} + \left(-\frac{T_L}{h^2}\frac{r}{\beta^2}\right) = 0$$

Figure 9 displays steady-state detuning characteristics, which are again speed independent. The orientation angle error becomes significant for large discrepancies between actual and reference rotor resistance values.

Figure 10 illustrates transient operation with detuned rotor resistance. The scheme of Fig. 2 is simulated. The initial and final steady state correspond to operation with rated rotor flux, 80% of the rated speed, and zero load torque. The simulated transient is a step application and removal of the rated load torque. Two cases are shown: rotor resistance in the machine equal to 150% and 66%, respectively, of the value

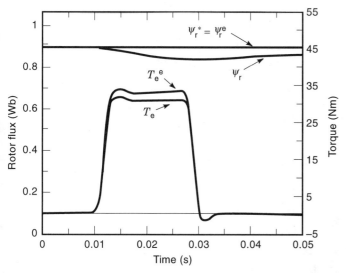

Figure 8. Influence of cross saturation on reversing transient (simulation, 0.75 kW machine). Actual rotor flux decreases, while rotor flux estimator erroneously judges flux as remaining constant. Developed transient torque is therefore smaller than the estimated torque. Speed response insignificantly differs from the one of Fig. 4, due to filtering action of the drive's inertia, and is therefore not shown.

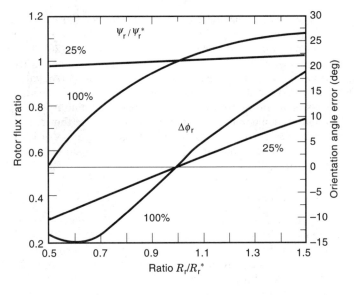

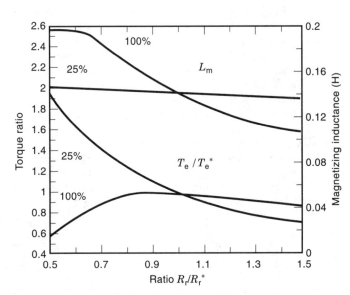

Figure 9. Steady-state detuning due to rotor resistance variation in the base speed region (rated rotor flux command, load torque as parameter, 4 kW machine). Detuning is speed independent but load dependent and is severe for large discrepancies between actual and reference rotor resistance values. Magnetizing inductance in the controller equals rated (0.141 H). Rotor resistance variation can cause significant change in the saturation level in the machine, as confirmed by magnetizing inductance variation in the machine for rated load torque operation.

used in the controller. As the rotor flux estimator is unaware of the change in rotor resistance, estimated rotor flux in both cases equals commanded rotor flux. Actual torque developed by the machine equals applied load torque in steady state. However, estimated torque (which almost equals commanded torque in both transient and steady-state operation as the torque controller is very fast) significantly differs from the actual one. The same applies to the actual rotor flux. The actual rotor flux and estimated torque responses are oscillatory when rotor resistance in the machine is smaller than the one used in the controller. It should be noted that effects of rotor resistance detuning on transient performance are usually

simulated in the torque mode of operation (i.e., with open speed control loop, Refs. 1, 27, 28), rather than with closed-loop speed control, as it is done here. A step torque command is then applied as the input, and the behavior of the actual torque is observed. If such an approach is utilized, then actual torque in the machine essentially has the response obtained here for the estimated torque.

Results of presented detuning studies indicate the importance of initial correct setting of all the parameters used in the controller. Methods of experimental parameter identification are numerous (11) and are beyond the scope of interest here. Only one simple experimental method (30) of rotor time constant tuning in indirect vector control scheme of Fig. 3 is

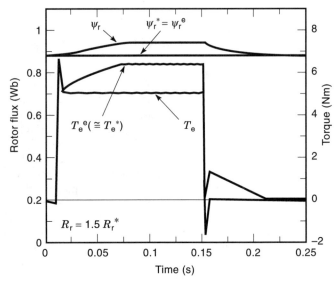

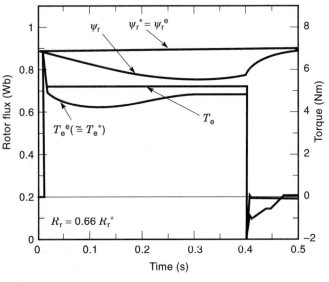

Figure 10. Illustration of detuning effects in transient operation caused by rotor resistance variation: responses to step loading and unloading with rated load torque for $R_r = 1.5R_r^*$ and $R_r = 0.66R_r^*$ (0.75 kW machine). Actual and estimated rotor flux and torque coincide when rotor resistance in the controller equals the one in the machine. However, with detuned rotor resistance only estimated rotor flux equals reference, while actual rotor flux deviates, causing a corresponding discrepancy between estimated and actual torques.

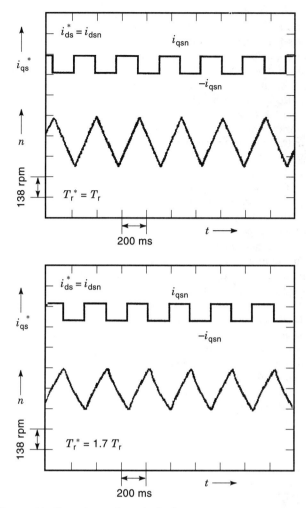

Figure 11. Experimental method of rotor time constant tuning in indirect vector controller: speed response to alternating square-wave torque command with correct rotor time constant and with 1.7 times correct rotor time constant (0.75 kW machine). Speed response is a triangular function of time when rotor time constant is correctly set in the controller. Speed response deviates from triangular when the rotor time constant value is incorrect.

described. Rotor flux command is set to rated value, so that the stator d-axis current command is rated. The machine is accelerated to a certain speed under no-load conditions and the speed loop is then opened. The machine further operates in torque mode and an alternating square-wave torque command, leading to the alternating square-wave q-axis stator current command, is applied. If the rotor time constant value in the controller is correct, the actual torque response is square wave. Square-wave torque, according to Eq. (9), causes a triangular variation of speed. If the rotor time constant value is incorrect, the actual torque is not a square wave and the speed response deviates from triangular. Figure 11 depicts experimentally recorded speed response obtained with correct and incorrect rotor time constant setting. The machine operates with rated rotor flux command and with a square-wave rated torque (rated stator q-axis current) command. The deviation of speed response from triangular is evident when the value of rotor time constant is incorrect.

COMPENSATION OF PARAMETER VARIATION EFFECTS

Sources of parameter changes in an induction machine differ in nature. Stator and rotor resistance variation with temperature is thermal and is inherently slow, as the thermal time constant of the machine is much bigger than the electromagnetic time constants. Parameter variations caused by main flux saturation, leakage flux saturation, and skin effect, as well as the iron loss, are of electromagnetic nature. Change of parameters due to electromagnetic phenomena is much quicker than thermally caused variations, as it is governed by electromagnetic time constants. This principal difference in both the nature of parameter variations and in the rate at which the variations take place has led to the development of two different approaches to compensation of parameter variation effects. Parameter variations due to electromagnetic phenomena are most appropriately compensated if the standard, constant parameter controller is substituted with a modified one that takes into account given parameter variation. This method of compensation is of the open-loop type and it provides compensation in both transient and steady-state operation. Compensation of temperature-dependent variation of resistances is most adequately provided by on-line identification of the resistance, which is usually operational in the steady state only.

Compensation of the skin effect related parameter variations in a vector-controlled drive can be accomplished by substituting the rotor circuit with two equivalent rotor circuits (16,31). As there are now two rotor flux space vectors, it is impossible to define a unique rotor flux space vector and the orientation of the stator current space vector is performed with respect to the air-gap flux space vector (16,31). Rotor leakage flux saturation can be included in the model of the machine by making rotor leakage inductance a variable parameter, dependent on the rotor current (31). Similarly, iron loss representation can be included in the induction machine model, and it is possible to design both an indirect rotor flux oriented controller and the $i_s - \omega$ estimator that fully compensate for the iron loss if such a modified model is used for development of vector control schemes (18,19). As already noted, all these parameter variation effects are of minor importance when compared to main flux saturation and rotor resistance variation. The following discussion therefore concentrates on compensation of the two major sources of detuning.

Compensation of Main Flux Saturation

The approach to main flux saturation compensation, which is to be discussed, is based on modification of the standard d-q-axis model, given by Eqs. (18), (19), and (21) (equivalent T-circuit approach). It is possible to deal with compensation of main flux saturation using other approaches, such as the equivalent π circuit approach (32) or equivalent inverse Γ circuit (equivalent rotor magnetizing current) approach (1,9), and to develop again appropriate modified vector control schemes.

Equations (18) and (19) are the starting point in derivation of a modified $i_s - \omega$ estimator that fully accounts for main flux saturation. If rotor current space vector is eliminated from Eq. (18), one obtains

$$T_{\sigma r}d\boldsymbol{\psi}_r/dt + j(\omega_a - \omega)\boldsymbol{\psi}_r T_{\sigma r} + \boldsymbol{\psi}_r = \boldsymbol{\psi}_m \qquad (38)$$

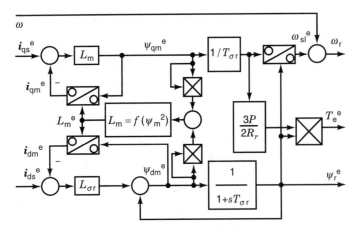

Figure 12. A modified rotor flux estimator with compensation of main flux saturation. The estimator recognizes changes in saturation level due to both variation of the rotor flux reference and cross saturation and thus provides instantaneous adaptation to the actual saturation level in the machine.

where rotor leakage time constant is defined as $T_{\sigma r} = L_{\sigma r}/R_r$. Application of the rotor flux orientation constraints, Eqs. (22) and (23), followed by resolution into d-q-axis components, yields the first two equations:

$$T_{\sigma r} d\psi_r/dt + \psi_r = \psi_{dm}$$
$$\omega_r - \omega = \omega_{sl} = \psi_{qm}/(T_{\sigma r}\psi_r) \tag{39}$$

As the dependence of the magnetizing flux on magnetizing current is a known nonlinear function, then from Eq. (8) it follows that the magnetizing current components are

$$i_{dm} = \psi_{dm}/L_m = \frac{\psi_{dm}}{\psi_m} i_m(\psi_m) = i_{dm}(\psi_m)$$
$$i_{qm} = \psi_{qm}/L_m = \frac{\psi_{qm}}{\psi_m} i_m(\psi_m) = i_{qm}(\psi_m) \tag{40}$$

where $i_m = i_m(\psi_m)$ represents inverse magnetizing curve. Application of Eq. (23) in conjunction with Eq. (7) and subsequent substitution of Eq. (40) yields two additional equations of the estimator:

$$\psi_{dm} = \psi_r + L_{\sigma r}[i_{ds} - i_{dm}(\psi_m)]$$
$$\psi_{qm} = L_{\sigma r}[i_{qs} - i_{qm}(\psi_m)] \tag{41}$$

Electromagnetic torque, Eq. (10), can be expressed in rotor flux oriented reference frame as

$$T_e = K\psi_r\psi_{qm}/L_{\sigma r} \tag{42}$$

Equations (39)–(42) describe a saturated current-fed rotor flux oriented induction machine and enable design of a modified rotor flux estimator (Fig. 12) that includes nonlinear function $L_m = f(\psi_m^2)$, obtainable from the no-load magnetizing curve. The modified estimator compensates for a change of saturation level due to both change in reference flux setting and due to the cross-saturation effect. Operation of the scheme of Fig. 2, when the estimator of Fig. 12 is used, is examined in Ref. 26 for the same reversing transient already discussed in conjunction with Fig. 8. The estimator of Fig. 12

correctly detects a tendency of the rotor flux to decrease during the reversal. As a consequence, the PI flux controller increases the reference stator d-axis current, and the rotor flux in the machine remains at almost constant value. Estimated torque and actual torque are in good agreement as well. Similarly, if a change in the saturation level in the machine takes place due to variation of the stator current d-axis command, the estimator correctly detects this change and again provides full compensation (26).

Modified indirect vector controller can be designed by utilizing Eqs. (39)–(42) as the starting point. However, if one recalls that the main reason for widespread application of the indirect rotor flux oriented control is its simplicity, then it is desirable to minimize the increase in complexity due to addition of main flux saturation compensation. A possibility is to provide only partial compensation, by ignoring two phenomena (33). First, the impact of cross saturation is pronounced only if the machine operates with very high transient torques. If this is not the case, the q-axis component of the magnetizing flux is small and its contribution to the total magnetizing flux can be neglected. Second, calculation of reference stator q-axis current reference and calculation of the reference angular frequency in Eqs. (29) and (30) involve the ratio of magnetizing to rotor inductance. Change in this ratio is always small and may be neglected. These two approximations enable development of the controller equations separately for the d-axis and q-axis. From Eqs. (39) and (41), taking the first approximation into account, one gets

$$\psi_m \approx \psi_{dm} = T_{\sigma r} d\psi_r^*/dt + \psi_r^*$$
$$i_{ds}^* = i_{dm}(\psi_m) + (1/L_{\sigma r})T_{\sigma r} d\psi_r^*/dt \tag{43}$$

while equations for stator q-axis current command and for the angular slip frequency command of Eqs. (29) and (30) remain unchanged:

$$\omega_{sl}^* = K_1 i_{qs}^*/\psi_r^* \quad K_1 = R_r L_{mn}/(L_{\sigma r} + L_{mn})$$
$$i_{qs}^* = K_2 T_e^*/\psi_r^* \quad K_2 = (1/K)(L_{\sigma r} + L_{mn})/L_{mn} \tag{44}$$

Figure 13 shows indirect rotor flux oriented controller with partial compensation of main flux saturation (33). It provides compensation when the machine operates with variable d-axis current command.

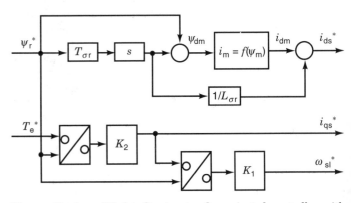

Figure 13. A modified indirect rotor flux oriented controller with partial compensation of main flux saturation. The controller compensates for change in saturation level due to change in rotor flux reference and ignores the cross-saturation effect.

Compensation of Rotor Resistance Variation

Rotor resistance plays a crucial role in establishing the accurate rotor flux oriented control in both schemes considered here, and most of the research in the area of compensation of parameter variation effects has been devoted to rotor resistance tuning. A number of various methods are available (34). Due to the thermal nature of rotor resistance variation, compensation consists of on-line rotor resistance identification. The prevailing method is based on principles of model reference adaptive control and is dominant due to its relatively simple implementation. The idea is that one quantity can be calculated in two different ways. The first value is calculated from references inside the control system. The second value is calculated from measured signals. The difference between the two is an error signal, whose existence is assigned entirely to the error in rotor resistance. The error signal drives an adaptive mechanism (PI or I controller) that provides correction of the rotor resistance. Many methods belong to this group (35–39) and they primarily differ with respect to which quantity is selected for adaptation purposes. The reactive power method does not involve stator resistance and is therefore frequently applied (35,36). Other possibilities include methods based on the special criterion function (37), air-gap power (38), torque (36,39), rotor flux magnitude (36), stator voltage d- or q-axis components (36), etc. There are a couple of common features that all these methods share. First, rotor resistance adaptation is usually operational in steady states only and is therefore based on the steady-state model of the machine. Second, stator voltages are usually required for calculation of the adaptive quantity and they have to be either measured or reconstructed from the inverter firing signals and measured dc link voltage (11). Third, identification usually does not work at zero speed and at zero load torque. Finally, identification relies on the model of the machine, in which, most frequently, all the other parameters are treated as constants. This is at the same time the major shortcoming of this group of methods. An analysis of the parameter variation influence on accuracy of rotor resistance adaptation (40) shows that, if other parameters vary, performance of the drive with resistance identifier can be worse than performance without the identifier. The other drawback, impossibility of adaptation at zero speed and zero load torque, is sometimes successfully eliminated (37).

The principle of model reference adaptive control approach is elaborated next, using reactive power as the reference and adaptive quantity. Rotor resistance adaptation utilizes a steady-state model of the machine. The input reactive power is given in terms of d-q-axis quantities by

$$Q_e = (3/2)(v_{qs}i_{ds} - v_{ds}i_{qs}) \tag{45}$$

Stator voltage equations in rotor flux oriented reference frame are in the steady state (9):

$$
\begin{aligned}
v_{ds} &= R_s i_{ds} - \omega_r \sigma L_s i_{qs} \\
v_{qs} &= R_s i_{qs} + \omega_r \sigma L_s i_{ds} + \omega_r (L_m/L_r)\psi_r
\end{aligned} \tag{46}
$$

where $\sigma = 1 - L_m^2/L_s L_r$. In steady-state operation with correct rotor flux orientation, $\psi_r = L_m i_{ds}$. If this condition is taken into account and Eq. (46) is substituted into Eq. (45), the following

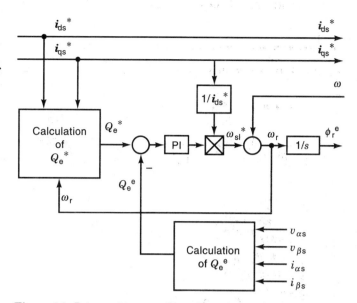

Figure 14. Rotor resistance on-line adaptation mechanism using approach based on model reference adaptive control. The difference between the two reactive powers is an error signal used to drive a PI controller, whose output is shown here as inverse of the rotor time constant.

correlation is obtained:

$$v_{qs}i_{ds} - v_{ds}i_{qs} = \omega_r L_s i_{ds}^2 + \omega_r \sigma L_s i_{qs}^2 \tag{47}$$

The rotor resistance adaptation mechanism is constructed using Eq. (47). The reference quantity is defined in terms of reference stator d-q-axis currents using the right-hand side of Eq. (47), while the estimated value of the same quantity is calculated on the basis of the measured stator voltages and currents, using the left-hand side of Eq. (47). As power is invariant with respect to the reference frame, the left-hand side is calculated using voltages and currents in the stationary reference frame. Thus finally

$$
\begin{aligned}
Q_e^* &= (3/2)(\omega_r L_s i_{ds}^{*2} + \omega_r \sigma L_s i_{qs}^{*2}) \\
Q_e^e &= (3/2)(v_{\beta s}i_{\alpha s} - v_{\alpha s}i_{\beta s})
\end{aligned} \tag{48}
$$

The adaptation mechanism is independent of the stator resistance, which is a good feature of this method. All the inductances are assumed to be constant. The difference between the two reactive powers of Eq. (48) is assigned to discrepancies between rotor resistance value used in the controller and the actual one. This error signal is processed through a PI controller, whose output is an updated rotor resistance value, which is subsequently used in calculation of the reference slip command. Figure 14 illustrates the adaptation mechanism, where operation in the constant rotor flux region is assumed and the output of the corrective PI controller is shown as the inverse of the rotor time constant.

The reactive power method is sensitive to inductance variation. A convenient way of eliminating this drawback is to combine vector control schemes that include compensation of main flux saturation (Figs. 12 or 13) with the rotor resistance identifier of Fig. 14. Information on magnetizing inductance is then passed from the modified indirect vector controller (or

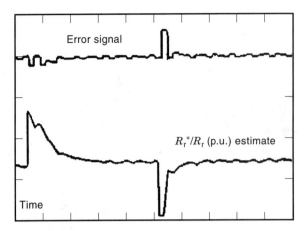

Figure 15. Experimentally recorded operation of the rotor resistance adaptation in an indirect rotor flux oriented induction machine. Rotor resistance in the controller is deliberately detuned by ±50% in a stepwise manner. The rotor resistance identifier always returns the value to the correct one. (Scales: time—10 s/div, error function—0.5 p.u./div, rotor resistance estimate—0.4 p.u./div; 0.75 kW machine.) Figure provided courtesy of Dr. S. N. Vukosavic, Department of Electrical Engineering, University of Belgrade, Belgrade, Yugoslavia.

rotor flux estimator) to the resistance identifier, while the resistance identifier supplies the indirect controller (or flux estimator) with updated values of the rotor resistance. It is thus possible to compensate for both effects by combining the two methods of parameter variation compensation (37). Figure 15 illustrates operation of such a rotor resistance adaptation scheme by means of experimentally recorded traces. The error function, which serves as the input to the PI controller, is shown together with the rotor resistance estimate in per unit (i.e., ratio of rotor resistance in the controller to the actual one in the machine). The drive operates at zero speed with 0.2 per unit load torque. The adaptation mechanism operation is illustrated for step variation of rotor resistance used in the controller of ±50%. The adaptation mechanism always returns rotor resistance in the controller to the previous value (i.e., to $R_r^*/R_r = 1$).

TRENDS IN VECTOR CONTROL OF INDUCTION MACHINES

The problem of parameter variations is well understood, and many solutions are available. However, none of them are capable of completely solving the problems, and viable alternatives are continuously being sought. One possibility of improving the accuracy of rotor flux position estimation consists of the use of observers (5,41) or extended Kalman filters (42). Rotor flux estimation can in both cases be combined with on-line parameter identification (11,43,44). These approaches have already been thoroughly investigated and proved to yield improved performance. However, their complexity is likely to preclude implementations in commercially available drives.

Two emerging trends that seem to be capable of offering improved performance while still retaining relatively simple realization requirements appear to be fuzzy logic (FL) control and artificial neural networks (ANNs). Fuzzy logic controllers can be used in various ways in vector-controlled drives. A speed PI controller can be substituted with an FL controller

or an FL controller can be used to tune continuously the parameters of the speed PI controller (45). A fuzzy logic controller can be used instead of a PI controller for correction of detuning due to parameter variation effects in schemes similar to Fig. 14 (46). An optimal efficiency controller that changes flux-producing current command can be based on an FL controller (46). Similarly, an ANN can be used as a rotor flux estimator (47), as a speed controller (48), or as part of the rotor resistance adaptation mechanism (48). The system of Ref. 48 uses two ANNs, one as a speed controller and the other as a rotor resistance identifier.

It has been assumed throughout this article that a vector-controlled induction machine is equipped with a speed or position sensor. This is the Achilles tendon of the drive as it considerably increases cost, requires space for mounting, and has to be electrically connected to the controller. This leads to reduction in reliability of the drive with an increase in cost; thus it is highly desirable to eliminate the speed (position) sensor. Such a situation has initiated development of numerous so-called sensorless vector-controlled induction motor drives, which rely on speed estimation rather than on speed (position) measurement. Approaches to speed estimation vary to a great extent but are almost exclusively based on measurement of either stator currents only, or stator currents and stator voltages (5,9,11,49). Although considerable developments in this area have already taken place during the last decade, exciting new achievements may be expected.

BIBLIOGRAPHY

1. D. W. Novotny and T. A. Lipo, *Vector Control and Dynamics of AC Drives,* Oxford: Clarendon Press, 1996.

2. F. Blaschke, Das Prinzip der Feldorientierung, die Grundlage für die TRANSVECTOR—Regelung von Drehfeldmaschinen, *Siemens—Zeitschrift,* **45**: 757–760, 1971.

3. I. Boldea and S. A. Nasar, *Vector Control of AC Drives,* Boca Raton, FL: CRC Press, 1992.

4. P. Vas, *Vector Control of AC Machines,* Oxford: Clarendon Press, 1990.

5. A. M. Trzynadlowski, *The Field Orientation Principle in Control of Induction Motors,* Norwell, MA: Kluwer Academic Publishers, 1994.

6. P. C. Krause, O. Wasynczuk, and S. D. Sudhoff, *Analysis of Electric Machinery,* Piscataway, NJ: IEEE Press, 1995.

7. P. Vas, *Electric Machines and Drives—A Space Vector Theory Approach,* Oxford: Clarendon Press, 1992.

8. A. Hughes, J. Corda, and D. A. Andrade, Vector control of cage induction motors: a physical insight, *IEE Proc. Electr. Power Appl.,* **143**: 59–68, 1996.

9. W. Leonhard, *Control of Electrical Drives,* 2nd ed., Berlin: Springer-Verlag, 1996.

10. X. Xu and D. W. Novotny, Selection of the flux reference for induction machine drives in the field weakening region, *IEEE Trans. Ind. Appl.,* **28**: 1353–1358, 1992.

11. P. Vas, *Parameter Estimation, Condition Monitoring, and Diagnosis of Electrical Machines,* Oxford: Clarendon Press, 1993.

12. R. Joetten and G. Maeder, Control methods for good dynamic performance induction motor drives based on current and voltage as measured quantities, *IEEE Trans. Ind. Appl.,* **IA-19**: 356–363, 1983.

13. E. Y. Ho and P. C. Sen, Decoupling control of induction motor drives, *IEEE Trans. Ind. Electron.,* **35**: 253–262, 1988.

14. R. W. De Doncker and D. W. Novotny, The universal field oriented controller, *IEEE Trans. Ind. Appl.*, **30**: 92–100, 1994.

15. R. W. De Doncker et al., Comparison of universal field oriented (UFO) controllers in different reference frames, *IEEE Trans. Power Electron.*, **10**: 205–213, 1995.

16. R. W. De Doncker, Field-oriented controllers with rotor deep bar compensation circuits, *IEEE Trans. Ind. Appl.*, **28**: 1062–1071, 1992.

17. R. W. De Doncker, Parameter sensitivity of indirect universal field-oriented controllers, *IEEE Trans. Power Electron.*, **9**: 367–376, 1994.

18. E. Levi, Impact of iron loss on behaviour of vector controlled induction machines, *IEEE Trans. Ind. Appl.*, **31**: 1287–1296, 1995.

19. E. Levi et al., Iron loss in rotor flux oriented induction machines: identification, assessment of detuning and compensation, *IEEE Trans. Power Electron.*, **11**: 698–709, 1996.

20. I. T. Wallace et al., Verification of enhanced dynamic torque per ampere capability in saturated induction machines, *IEEE Trans. Ind. Appl.*, **30**: 1193–1201, 1994.

21. R. D. Lorenz and D. W. Novotny, Saturation effects in field-oriented induction machines, *IEEE Trans. Ind. Appl.*, **26**: 283–289, 1990.

22. H. Grotstollen and J. Wiesing, Torque capability and control of a saturated induction motor over a wide range of flux weakening, *IEEE Trans. Ind. Electron.*, **42**: 374–381, 1995.

23. D. S. Kirschen, D. W. Novotny, and T. A. Lipo, On-line efficiency optimization of a variable frequency induction motor drive, *IEEE Trans. Ind. Appl.*, **IA-21**: 610–616, 1985.

24. E. Levi, A unified approach to main flux saturation modelling in d-q axis models of induction machines, *IEEE Trans. Energy Convers.*, **10**: 455–461, 1995.

25. E. Levi and V. Vuckovic, Field-oriented control of induction machines in the presence of magnetic saturation, *Electric Machines and Power Systems*, **16**: 133–147, 1989.

26. E. Levi, Magnetic saturation in rotor flux oriented induction motor drives: operating regimes, consequences and open-loop compensation, *European Transactions on Electrical Power Engineering*, **4**: 277–286, 1994.

27. R. Krishnan and F. C. Doran, Study of parameter sensitivity in high-performance inverter-fed induction motor drive systems, *IEEE Trans. Ind. Appl.*, **IA-23**: 623–635, 1987.

28. K. B. Nordin, D. W. Novotny, and D. S. Zinger, The influence of motor parameter deviations in feedforward field orientation drive systems, *IEEE Trans. Ind. Appl.*, **IA-21**: 1009–1015, 1985.

29. O. Ojo, M. Vipin, and I. Bhat, Steady-state performance evaluation of saturated field-oriented induction machines, *IEEE Trans. Ind. Appl.*, **30**: 1638–1647, 1994.

30. R. D. Lorenz, Tuning of field oriented induction motor controllers for high performance applications, *Proc. Annu. Meet. IEEE Ind. Appl. Soc.*, **20**: 607–612, 1985.

31. S. Williamson and R. C. Healey, Space vector representation of advanced motor models for vector controlled induction motors, *IEE Proc. Electr. Power Appl.*, **143**: 69–77, 1996.

32. C. R. Sullivan et al., Control systems for induction machines with magnetic saturation, *IEEE Trans. Ind. Electron.*, **43**: 142–152, 1996.

33. E. Levi, S. Vukosavic, and V. Vuckovic, Saturation compensation schemes for vector controlled induction motor drives, *Proc. Annu. IEEE Power Electron. Spec. Conf.*, **21**: 591–598, 1990.

34. R. Krishnan and A. S. Bharadwaj, A review of parameter sensitivity and adaptation in indirect vector controlled induction motor drive systems, *IEEE Trans. Power Electron.*, **6**: 695–703, 1991.

35. L. J. Garces, Parameter adaption for the speed-controlled static ac drive with a squirrel-cage induction motor, *IEEE Trans. Ind. Appl.*, **IA-16**: 173–178, 1980.

36. T. M. Rowan, R. J. Kerkman, and D. Leggate, A simple on-line adaption for indirect field orientation of an induction machine, *IEEE Trans. Ind. Appl.*, **27**: 720–727, 1991.

37. S. N. Vukosavic and M. R. Stojic, On-line tuning of the rotor time constant for vector-controlled induction motor in position control applications, *IEEE Trans. Ind. Electron.*, **40**: 130–138, 1993.

38. D. Dalal and R. Krishnan, Parameter compensation of indirect vector controlled induction motor drive using estimated airgap power, *Proc. Annu. Meet. IEEE Ind. Appl. Soc.*, **22**: 170–176, 1987.

39. R. D. Lorenz and D. B. Lawson, A simplified approach to continuous, on-line tuning of field-oriented induction machine drives, *IEEE Trans. Ind. Appl.*, **26**: 420–424, 1990.

40. A. Dittrich, Parameter sensitivity of procedures for on-line adaptation of the rotor time constant of induction machines with field oriented control, *IEE Proc. Electr. Power Appl.*, **141**: 353–359, 1994.

41. G. C. Verghese and S. R. Sanders, Observers for flux estimation in induction machines, *IEEE Trans. Ind. Electron.*, **35**: 85–94, 1988.

42. T. Du, P. Vas, and F. Stronach, Design and application of extended observers for joint state and parameter estimation in high-performance ac drives, *IEE Proc. Electr. Power Appl.*, **142**: 71–78, 1995.

43. D. J. Atkinson, P. P. Acarnley, and J. W. Finch, Observers for induction motor state and parameter estimation, *IEEE Trans. Ind. Appl.*, **27**: 1119–1127, 1991.

44. L. Salvatore, S. Stasi, and L. Tarchioni, A new EKF-based algorithm for flux estimation in induction machines, *IEEE Trans. Ind. Electron.*, **40**: 496–504, 1993.

45. P. Vas, F. Stronach, and M. Neuroth, Design and DSP implementation of fuzzy controllers for servo drives, *Electrical Eng.*, **79**: 265–276, 1996.

46. J. B. Wang and C. M. Liaw, Indirect field-oriented induction motor drive with fuzzy detuning correction and efficiency optimisation controls, *IEE Proc. Electr. Power Appl.*, **144**: pp. 37–45, 1997.

47. M. G. Simoes and B. K. Bose, Neural network based estimation of feedback signals for a vector controlled induction motor drive, *IEEE Trans. Ind. Appl.*, **31**: 620–629, 1995.

48. H. T. Yang, K. Y. Huang, and C. L. Huang, An artificial neural network based identification and control approach for the field-oriented induction motor, *Electric Power Systems Res.*, **30**: 35–45, 1995.

49. K. Rajashekara, A. Kawamura, and K. Matsuse (eds.), *Sensorless Control of ac Motor Drives*, Piscataway, NJ: IEEE Press, 1996.

Reading List

M. P. Kazmierkowski and H. Tunia, *Automatic Control of Converter-fed Drives*, Amsterdam: Elsevier, 1994. (Good textbook with wide coverage of electric drive control; contains excellent reference section with literature published until 1990.)

S. A. Nasar and I. Boldea, *Electric Machines: Dynamics and Control*, Boca Raton, FL: CRC Press, 1993. (Somewhat similar in approach and content to Ref. 1, but less detailed).

B. K. Bose (ed.), *Power Electronics and Variable Frequency Drives*, Piscataway, NJ: IEEE Press, 1997. (A collection of exceptionally fine and in-depth articles, some of them at an advanced level.)

D. W. Novotny and R. D. Lorenz (eds.), *Introduction to Field Orientation and High Performance ac Drives*, IEEE Ind. Appl. Soc. Tutorial Course, Presented at IEEE Ind. Appl. Soc. Annu. Meet., Toronto, ON, 1985. (Good starting point for a beginner.)

V. R. Stefanovic and R. M. Nelms (eds.), *Microprocessor Control of Motor Drives and Power Converters,* IEEE Ind. Appl. Soc. Tutorial Course, Presented at IEEE Ind. Appl. Soc. Annu. Meet., Dearborn, MI, 1991. (Covers many realisation related issues in great depth.)

P. Pillay (ed.), *Performance and Design of Permanent Magnet ac Motor Drives,* IEEE Ind. Appl. Soc. Tutorial Course, Presented at IEEE Ind. Appl. Soc. Annu. Meet., San Diego, CA, 1989. (Detailed treatment of various issues relevant to vector control of permanent magnet synchronous machines.)

References 5, 7, 9, and 11 contain exhaustive bibliographies. Reference 49 contains an exceptional collection of most important papers in the area of sensorless vector control.

EMIL LEVI
Liverpool John Moores University

MAGNETIZATION REVERSAL.   See MAGNETIC MEDIA,
MAGNETIZATION REVERSAL.
MAGNETOCARDIOGRAPHY.   See BIOMAGNETISM;
MAGNETIC SOURCE IMAGING.
MAGNETOENCEPHALOGRAPHY.   See BIOMAGNETISM;
MAGNETIC SOURCE IMAGING.

MAGNETOHYDRODYNAMIC POWER PLANTS

Magnetohydrodynamic (MHD) power generation refers to a technique in which a conducting gas is moved through a magnetic field to produce an electric field, which is utilized to produce power. A conventional generator is illustrated in Fig. 1, in which an electrical conductor is moved through a magnetic field and an electric field is generated in the wire. In the MHD

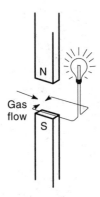

Figure 2. Illustration of MHD generator concept in showing electrical conducting gas moving through magnetic flux lines to generate an electric field. Note that electrodes must be placed in contact with the gas to permit the current to flow through an external circuit.

case, as illustrated in Fig. 2, the electrical field is generated in the conducting gas which is passing through a magnetic field. In order to utilize this power, electrodes must be in physical contact with the gas to permit current flow. The cathode emits electrons and the anode collects them, the same convention as in an electron tube.

The conducting gas used in power generation is a partially ionized gas. In the regime of central power generation, it is described as slightly ionized, having free electron densities on the order of 10^{16} electrons/m^3. The method of producing the conducting gas, or plasma, as it is more correctly called, leads to a classification of type of MHD generators: equilibrium and nonequilibrium. In the equilibrium case, the gas is heated to a temperature such that a small fraction (around 1 in 10,000 of the atoms in the gas) are thermally ionized. In nonequilibrium ionization, energy is selectively added to the electrons, resulting in a fraction of ionization that exceeds that due to thermal excitation alone. In the latter case, the physics of the gas (especially collision cross section of electrons with heavier particles and the mean time between collisions) must be such that the ionization rate is greater than the recombination rate. The conditions that meet this requirement are very low pressures for combustion gases or the use of noble gases that have low collision cross sections, preferably also at reasonably low pressures. The low-pressure condition severely limits application of combustion gases for nonequilibrium applications. The use of a noble gas, such as argon, forces the constraint that the gas be reused and, hence, is called closed-cycle MHD. By contrast, the equilibrium ionization technique has been applied primarily to combustion products which are utilized once and then exhausted to the atmosphere. Thus, it is classified as open-cycle MHD.

To produce the plasma for open-cycle MHD power generation, a fossil fuel such as coal, char, oil, or gas is burned at a high enough temperature to produce the required ionization. This temperature required for ionization is too high for a practical power system if just the normal combustion products are present. The lowest possible temperature is about 4500 K, the temperature at which nitrogen oxide (NO) begins to ionize. However, an element that has a lower ionization potential can be added to reduce this temperature to more manageable levels. Potassium, with an ionization potential of 4.34 eV, is the most practical choice [cesium (with an ioniza-

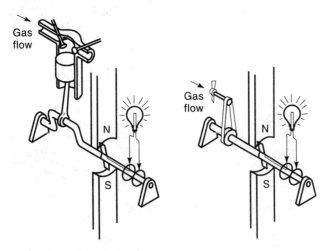

Piston engine generator   Turbo-generator

Figure 1. Illustration of conventional generator concepts showing a copper conductor moving through magnetic flux lines to generate an electric field.

tion potential of 3.89 eV) and rubidium (with an ionization potential of 4.16 eV) are the only elements that are lower, but both are very expensive]. With 1% potassium in the flow, efficient MHD power production requires a flame temperature of about 2800 K. To achieve this high a flame temperature with fossil fuel requires significant preheating of the combustion air or oxygen enrichment of the air or some combination of the two techniques.

Fuels that have been considered for MHD power plants include coal, natural gas, oil, bitumen, char, and refinery coke. Fuels that have a high carbon-to-hydrogen ratio produce better electrical conductivity at the same conditions. This is because the water that is formed when hydrogen burns disassociates into monatomic hydrogen and an OH radical. The latter has an affinity for electrons. When it joins with an electron to become a negative ion, its mass is so much larger than an electron that it effectively removes that electron from the conductivity process. However, fuels that are high in hydrogen, such as natural gas, can be used but require a few tens of degrees Kelvin higher temperature to achieve the same electrical conductivity. Calculated electrical conductivity and flame temperature for coal with 1% potassium added as potassium carbonate is shown in Fig. 3(a) and (b), respectively, for a range of air preheat temperatures and pressures. These calculations are all for a fuel-rich condition, 85% of theoretical oxygen. The reasons for choosing this fuel-rich condition are twofold: It optimizes the electrical conductivity, and it is essential for the control of nitrogen oxide emissions to a very low level.

The actual configuration of the MHD generator, in its simplest form, is shown in Fig. 4(a). This arrangement is called a continuous electrode Faraday generator because the electrodes are continuous along the generator length. This configuration is used in some short-duty time applications, such as the Russian PAMIR, which generates a pulse of several megawatts for several seconds. However, it is inefficient because there is a completed path for circulating axial currents to flow through the plasma and back along the length of the electrode, causing electrical losses. This problem is circumvented in the segmented Faraday generator as shown in Fig. 4(b). In this design, the electrodes are narrow strips with insulators between the electrodes, preventing a return path for circulating currents through the electrodes in the axial direction. The segmented Faraday generator is the most efficient configuration, but it suffers from one disadvantage for application to commercial power generation: It requires a separate loading circuit for each pair of electrodes, as shown in Fig. 4(b).

The diagonally conducting generator, as illustrated in Fig. 4(c), avoids this problem and its efficiency approaches that of the segmented Faraday generator when properly designed and loaded. In an MHD generator, when current flows in the direction mutually perpendicular to the velocity and the magnetic field, it is called the Faraday current and is in the y-direction on the coordinate system shown in Fig. 4. When current flows in a magnetic field, there is a force on moving charged particles that tends to separate the electrons from the positive ions. This effect is called the Hall effect. The result is an electric field along the length of the generator, called the Hall field. The diagonal generator loads the vector sum of the Faraday field and the Hall field, and the wall angle is determined as $\tan^{-1} \boldsymbol{E}_\mathrm{F}/\boldsymbol{E}_\mathrm{H}$, where $\boldsymbol{E}_\mathrm{F}$ is the Faraday electric

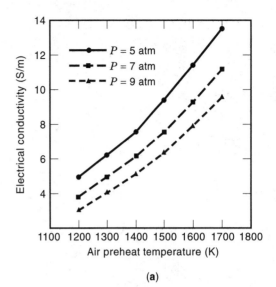

(a)

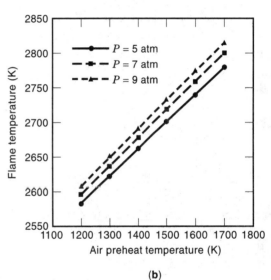

(b)

Figure 3. (a) Electrical conductivity versus air preheat temperature for Illinois #6 coal dried to 2% moisture with 1% potassium added as potassium carbonate. (b) Flame temperature versus air preheat temperature for Illinois #6 coal dried to 2% potassium. Note the preheat temperature required to achieve 2800 K flame temperature.

field and $\boldsymbol{E}_\mathrm{H}$ is the Hall electric field. Electrodes at this wall angle lie along equipotential lines at design conditions because the electric field vector is perpendicular to the electrode. This generator can be loaded from one end to the other, or it can be loaded with more than one load connection along the length of the generator. For a generator designed for central power application, where high efficiency and high reliability are desired, it would normally be loaded with multiple loads to approximate the optimum current. This would result in something like four to six separate load circuits versus several hundred for the segmented Faraday configuration. Insofar as known to the author, the diagonal connection is the only configuration considered for commercial power generation in a linear open-cycle system.

A principal attraction of the MHD generator for central power generation is its high conversion efficiency, as mea-

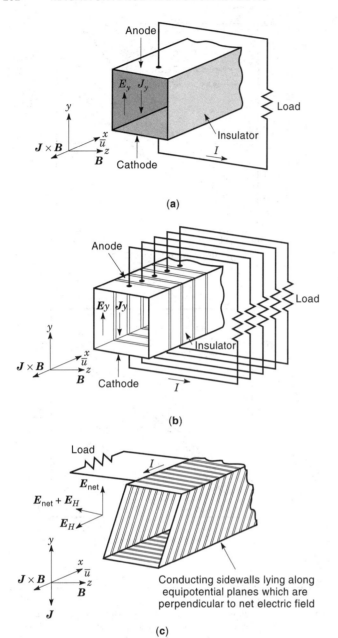

Figure 4. (a) Continuous electrode Faraday MHD generator. (This is the simplest MHD generator.) (b) Segmented electrode Faraday MHD generator. The electrodes are segmented in the X-direction to avoid a return path for circulating currents. (c) Diagonal conducting wall MHD generator. The sidewalls are perpendicular to the net electrical field and lie along equipotential lines.

sured on the basis of the first law of thermodynamics (energy conservation). That efficiency is defined as the electrical power output divided by the energy lost by the gas stream while passing through the generator. Alternatively, it may be calculated as the electrical power generated divided by the sum of the electrical power generated plus the heat loss to the walls. Efficiency as so defined can typically be around 90%, depending on the size and pressure ratio (thus length) of the generator. The MHD generator is only useful in the temperature range in which there is sufficient electrical conductivity. At about 2200 K and at exhaust pressures typically slightly

less than 100 kPa, the conductivity has decreased to around 1 S/m and MHD generation is no longer attractive. However, the working gas still contains considerable sensible energy.

This leads to the concept of a combined cycle MHD steam power plant in which a MHD topping cycle is used on a steam Rankine bottoming cycle. This concept is shown schematically in Fig. 5. Note that there are numerous opportunities to integrate the topping and bottoming cycles for improved efficiency and resource allocation. The requirement for heating the combustion air can be incorporated into the bottoming cycle furnace and result in a large efficiency improvement because the energy is recycled from the less efficient bottoming cycle to the more efficient topping cycle. Also, the mechanical power to run the compressors for the topping cycle combustion can be met with a steam turbine economically, especially in large plants. The cooling water from the topping cycle can be incorporated into the boiler feedwater heating design to utilize that energy. The potassium used as seed in the topping cycle needs to be recovered in the bottoming cycle and processed for reuse.

The low levels of environmental intrusion from the MHD steam combined cycle power plant are one of its attractive features. Department of Energy studies (1) have shown that such plants have the lowest levels of any of the possible coal-fired combined cycles. Sulfur dioxide emissions are completely eliminated by combination of the sulfur in the fuel with potassium seed, which is collected as solid potassium sulfate in the particulate control system. Nitrogen oxides are formed in large amounts (10 kppm to 12 kppm) in the high-temperature combustors of MHD plants, even at fuel-rich stoichiometries. However, by cooling the gases in a radiant furnace before completing the combustion, the nitrogen oxides (actually, almost completely nitric oxide, NO, at fuel-rich conditions) relax toward equilibrium at lower temperatures and at a secondary combustion temperature of about 1000 K (1350°F), very little NO remains (2). The secondary combustion must be made in a way that avoids much temperature rise to avoid the formation of additional nitrogen oxides.

The MHD plant has a higher particle loading in the exhaust because of the potassium seed added. Furthermore, the size distribution tends to be smaller due to the higher temperatures having vaporized some of the solids. However, the resistivity of the ash is lower due to the high potassium content, and electrostatic precipitators (ESPs) are very efficient in removing the solids, although more capability is required than would be required for a normal coal plant. In addition to needing to remove the particulate to meet Clean Air Act standards, it needs to be removed in the MHD case in order to recycle the potassium seed. MHD pilot plants have operated with a completely invisible stack plume. Volatile organic compound (VOC) emissions, which are essential components for formation of fine particulate in the atmosphere, are almost completely eliminated in the MHD plant because combustion efficient enough to produce electrically conducting plasma has been shown to contain no unburned elemental or organic compound carbon (3). This is an important advantage because power plants seek to control particulates in the atmosphere having mean diameters of 2.5 μm and less because the troublesome part of these particulates are formed in the atmosphere from unburned VOCs, sulfur oxides, and nitrogen oxides.

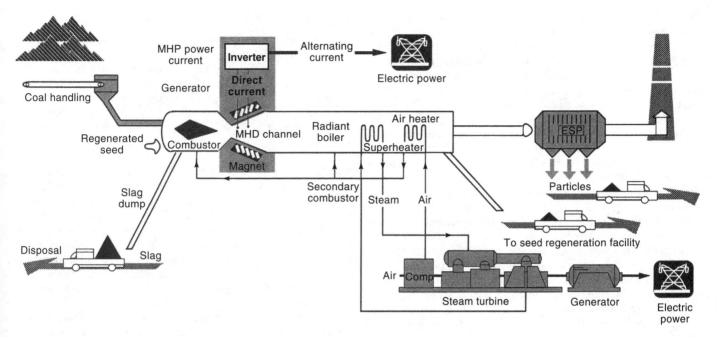

Figure 5. Schematic of MHD/steam combined cycle power plant.

THE MHD TOPPING CYCLE

The MHD topping cycle consists of a means of generating the plasma (generally a combustor), a nozzle to accelerate the plasma, an MHD generator with associated control, power take off system, invertors to convert the generated direct current power to alternating current, a magnet to provide a magnetic field in the generator volume, and a diffuser to reduce the plasma velocity and recover static pressure before it enters the radiant furnace. As previously noted, there are several streams that are integrated with the bottoming cycle including air preheat and/or oxygen production, recovery and reprocessing of the potassium seed, and use of boiler feedwater for cooling of all topping cycle components.

Combustor

The combustor used in a MHD power plant faces some extreme conditions compared with the burners in other fossil plants. It must generate a plasma at about 2800 K and deliver it at pressures of perhaps 0.2 MPa to 1.0 MPa. In order to reach this temperature, it is necessary to have a small volume, with good mixing of fuel and oxidant and extremely rapid heat release. When burning coal, the volume must be a compromise (optimization) whereby time is allowed for carbon burnout while keeping the combustor small enough to limit the heat losses to the walls. There have been two successful coal combustors demonstrated that differ substantially in design.

One of these is a single-stage, full-slag-carryover combustor modeled somewhat after rocket engine technology (4). A sketch of this combustor is shown in Fig. 6. In this design, the pulverized coal is injected in a conical pattern from the middle of the injector plate. Air is fed through a number of small holes in the injector plate where a pressure drop is taken to accelerate the air. These jets of air mix very rapidly with the coal and produce high heat release concentrations,

typically over 120 MW/m³. The walls of the combustor rapidly become covered with coal slag, solidified next to the wall and molten on the flame side. In spite of insulation effects of this slag layer, heat fluxes to the walls average about 100 W/cm². The potassium seed is pulverized with the coal and fed together in this configuration. The solids, consisting of coal ash and seed, are molten at the exit of the combustor and are blown over with the gas for separation later in the steam bottoming plant. Scale-up of this combustor concept is feasible by using clusters of these burners, firing into a common chamber.

The advantages of this type of combustor are simplicity, low cost, and higher, more uniform electrical conductivity because of the time available for the seed to reach chemical and thermal equilibrium with the gas.

The other type of combustor that has been developed is the two-stage, slag-rejecting combustor (5). The concept for this

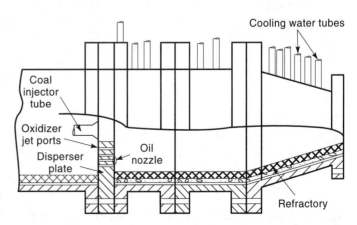

Figure 6. Single stage MHD coal combustor concept. This design is optimized for rapid turbulent mixing and for minimum heat loss to the walls.

design is to produce mixing in the first stage by injecting the air tangentially to produce swirl for mixing. The air preheat temperature and stoichiometry are chosen so that the coal ash is liquid in this stage and it is tapped out at the bottom of this first stage. The gases are taken off tangentially so as to get a deswirl effect going into the second stage. The seed and additional air are added in the second stage. The objective of this design is to remove as much coal slag as possible before adding the potassium seed to minimize the amount of potassium seed that chemically combines with the slag and may be more difficult to recover. Slag removal efficiencies of up to 50% were demonstrated experimentally with coal ground to 70% through 200 mesh (6). Additional slag removal can be achieved with larger particles of coal, but the carbon burnout suffers as the coal particle size increases. Also, this combustor has problems in scale-up because as the size of the combustor increases, slag removal decreases unless the air injection velocity is increased accordingly. As noted, the advantage of this design is that it removes some of the coal ash before the potassium seed is injected. The disadvantages are larger size, more heat loss, and possibly lower electrical conductivity due to insufficient residence time for the potassium seed to reach thermal and chemical equilibrium.

Regardless of which type of combustor is used, the primary stoichiometry should be chosen to meet two objectives: to maximize the electrical conductivity and minimize the nitrogen oxides emitted by the plant. A plot of flame temperature and electrical conductivity versus primary stoichiometry is shown in Fig. 7. It shows that the primary combustor should be operated under fuel-rich conditions at about 85% to 90% of theoretical oxygen to maximize temperature and electrical conductivity. This happily turns out to be compatible with very low nitrogen oxide emissions as well.

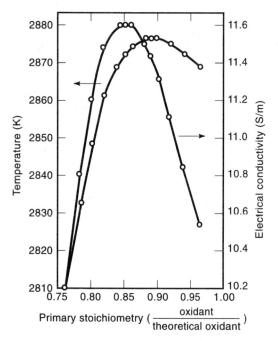

Figure 7. Flame temperature and electrical conductivity versus primary stoichiometry; coal with oxygen-enriched air, 1% potassium, pressure = 7 atm. Note that the electrical conductivity peaks at a stoichiometry of ~0.85.

The Nozzle

A gas dynamic nozzle is needed between the primary combustor and the MHD generator to accelerate the flow to the desired entrance condition. The voltage generated in the MHD generator is the product of gas velocity, magnetic field flux density, and the distance over which it acts. From this viewpoint alone, the velocity should be high. Other considerations such as convective heat transfer coefficient, size of generator, and total amount of power that can be generated from a unit of fuel, however, place upper limits on the desired entrance velocity. In practice, a high subsonic flow (Mach number 0.8 to 0.9) or low supersonic flow (Mach number 1.1 to 1.3) is normally chosen for central power generation. It is desirable to choose the velocity and area lofting to avoid the formation of a shock in the generator. Thus, a subsonic generator should be lofted to remain subsonic throughout its length, and a supersonic generator should be designed to remain supersonic throughout its length and under all operating conditions. This places some difficult restrictions on turn-down of the generator to operate at part load conditions.

The isentropic one-dimensional gas dynamic relations can be used to design the nozzle. The supersonic nozzle cross-sectional area converges to a throat, and then it diverges to the area required for the desired entrance Mach number. The subsonic nozzle must be lofted to provide a throat (a short supersonic region) and then it converges to produce the desired subsonic entrance Mach number. Heat fluxes to the walls are normally higher in the nozzle throat than in any other part of the plant. Design of the cooling for the nozzle, especially at the throat, is a challenge. The best success has been demonstrated with channeled cooling water passages with high-velocity water to ensure good convective heat transfer between the inside of the cooling water passage and the water.

The MHD Generator

Analysis and characterization of the MHD generator involves the consideration of both Maxwell's equations for electromagnetic fields and the conservation of mass, energy, and momentum equations for fluid flow (7). In addition, of course, an equation of state is required for the fluid. In MHD calculations, these sets of equations are coupled. The electrical effect, $\boldsymbol{J} \times \boldsymbol{B}$, is a body force which must be considered in the fluid flow. Similarly, the electric field produced by the motion of the electrically conducting gas in the magnetic field must be considered in the electromagnetic solution. This is illustrated in Fig. 8. There are no cases of practical interest for which the equations can be solved in closed form. Thus, the equations must be solved numerically for analysis or design of the MHD generator. In cases where the interaction is low, it may be a satisfactory approximation to neglect the effect of the body force, $\boldsymbol{J} \times \boldsymbol{B}$, in the flow calculations. In those cases where the body force is neglected in the calculation of flow quantities, the flow is said to be uncoupled. We consider here only the regime of MHD generation for central power plants. In this regime, the $\boldsymbol{J} \times \boldsymbol{B}$ body force is the major sink for momentum and cannot be neglected in the design calculations for the MHD generator. Full solutions to this set of equations are available. See, for example, Ref. 8, which is a public domain code published by Argonne National Laboratory.

The problem can be simplified for preliminary design calculations by considering the problem as one-dimensional (9).

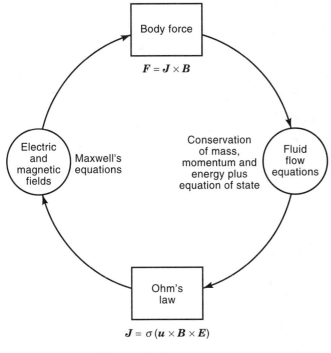

$$J \text{ is current density}$$
$$B \text{ is magnetic field flux density}$$
$$F \text{ is force}$$
$$u \text{ is velocity}$$
$$E \text{ is electrical field}$$
$$\sigma \text{ is electrical conductivity}$$

Figure 8. Illustration of coupling of the gas-dynamic and electromagnetic relations.

This simplification involves the assumption that the magnetic field is aligned with one axis—for example, the z-axis as shown in Fig. 4. This implies that induced magnetic fields due to currents in the generator are negligible. This is certainly the case in the central power generation regime. We also need to assume that the current density and flow quantities are uniform in a plane perpendicular to the direction of flow, the yz-plane, and are represented by the average value in the plane. With such assumptions, the simplified equations become:

Conservation of Momentum Equation:

$$\rho u \frac{du}{dx} + \frac{dP}{dx} = -J_y B = 4 \frac{P_w}{D} \tag{1}$$

where

τ_W = shear stress
D = hydraulic diameter
ρ = density of the plasma

Conservation of Energy Equation:

$$\rho u \frac{dh}{dx} + \rho u^2 \frac{du}{dx} = J \cdot E + 4 \frac{q_w}{D} \tag{2}$$

where q_w = heat flux to walls.

Conservation of Mass or Continuity Equation:

$$\rho u A = \dot{m} \tag{3}$$

where A = cross-sectional area of duct and

$$\dot{m} = \text{mass flow rate through duct}$$

Ohm's Law (including only significant terms for this case):

$$J = \sigma(E + u \times B + E_d) - \frac{\Omega}{B}(J \times B) \tag{4}$$

where E_d is an electrical field loss due to plasma to electrode voltage drop. The electrical field corresponding to a potential drop, Δ, defined as

$$\Delta = \frac{E_d}{uB} = \frac{V_d}{uBd}$$

Ω = Hall parameter = electron cyclotron frequency times mean time between collision of electron with heavy particles

From Ohm's law, an equation for $J \cdot E$ can be derived for use in the energy equation (10)

$$J \cdot E = \frac{J^2}{\sigma} - J_y B(1 - \Delta) \tag{5}$$

An equation of state is required. That is, given two thermodynamic variables such as T and P, a relationship is required such that the other thermodynamic variables can be calculated. This is typically a table of values calculated from a chemical equilibrium code such as the NASA SP-273 Code or one of its variants. An alternative that is now feasible due to the greatly expanded capability of modern computers is to use a chemical equilibrium code to calculate the other variables at each point.

The electrical conductivity, σ, must also be available from the table or calculation. The method used by the author was originally described by Frost (11).

Equations (1), (2), (3), and (5) can be simplified to three independent differential equations in four unknowns, P, T, U, and A. If one of these variables is fixed or known, a solution for the other variables can be computed along the length of the generator. Once the flow variables are calculated, the electrical variables can be computed via the relations tabulated in Table 1 of Ref. 10.

For a practical design calculation, constant velocity is the case of most interest because this is very near to the maximum power generation point for a given gas. The equations can be modified to perform a constant Mach number case which is useful in designing for the absence of shocks in the generator. Either case then gives all the electrical and thermodynamic variables along the generator and the cross-sectional area at each point along the length. A loading factor, which relates the impedance of the load to that of the generator, is also required. This is defined as

$$K = \frac{R_L}{R_L + R_g} \quad \text{or} \quad \frac{V_L}{V_G}$$

where

R_L is load resistance
R_g is internal generator resistance
V_L is load voltage
V_G is generated voltage

There is an elementary theorem in electrical engineering that says the maximum power transfer from a circuit to a load occurs when the load resistance equals the generator internal resistance. Thus, one is tempted to set $K = 0.5$. This is not a good choice, however, because it means that half the generated power is dissipated by the current flowing through the gas. If a higher K is chosen, the generator becomes longer and more efficient. The correct choice, in general, will depend on the overall plant economics in terms of levelized cost of electricity being minimized. The MHD generator tends to be relatively cheap, but making the magnet longer adds considerable expense. A loading factor, K, of 0.88 to 0.90 is typical.

For a given load factor and gas conditions, the amount of power that can be generated depends on the pressure ratio across the generator. The exit pressure must be determined by the conditions needed at the diffuser entrance in order to recover to the pressure needed at the furnace inlet (it may be slightly less than atmospheric for induced draft designs).

Some common efficiency definitions that apply to the MHD generator are given below

$$\text{Enthalpy extraction} = \frac{\text{Electrical power output}}{\text{Sensible enthalpy of gas}}$$
$$= \frac{P_{\text{out}}}{(h_{0_{\text{in}}} - h_{\text{ref}})\dot{m}}$$

where

$h_{0_{\text{in}}}$ is total enthalpy at generator entrance
h_{ref} is static enthalpy at reference condition of 1 atm, 298 K (77°F)

Note that since the only energy loss from the gas is electrical power output and heat transfer to the walls, we obtain

$$\text{Generator efficiency} = \frac{\text{Electrical power output}}{\text{Decrease in enthalpy across generator}}$$
$$= \frac{P_0}{(h_{0_{\text{in}}} - h_{0_{\text{out}}})\dot{m}}$$

where $h_{0_{\text{out}}}$ = total enthalpy at generator exit. Also, since the only energy loss from the gas is electrical power output and heat transfer to the walls, we obtain

$$\text{Generator efficiency} = \frac{P_{\text{out}}}{P_{\text{out}} + Q_w}$$

when Q_w = total heat transferred from gas to walls.

The Diffuser

The function of the diffuser is to slow the gas down and recover (convert) as much of the dynamic pressure as possible to static pressure. The diffuser for a subsonic generator is just a diverging duct, with (a) the wall angles chosen to be small enough to avoid flow separation at the walls and (b) the length sufficient to recover as much static pressure as possible but no longer so as to minimize the heat loss to the walls. For a supersonic generator, a constant area section is placed at the generator outlet to cause the flow to shock to subsonic, then the same type diffuser is used as for the subsonic diffuser. A common measure of performance of the diffuser is called the pressure recovery coefficient, C_p, which is the ratio of the recovered static pressure to the dynamic pressure available. It may be expressed as

$$C_p = \frac{P_2 - P_1}{\frac{1}{2}\rho u^2}$$

where

P_2 is static pressure at the outlet
P_1 is static presure at the inlet
ρ is fluid density at the inlet
u is fluid velocity at the inlet

The coefficient of performance for practical diffusers in MHD topping cycles are expected to be 0.5 to 0.6. Heat loss in the diffuser walls is incorporated into the boiler feedwater heating chain and, in some aggressive designs, may be used for boiling surface.

THE STEAM BOTTOMING CYCLE

The steam bottoming cycle takes the hot gases from the MHD diffuser and uses the sensible heat in them to generate electrical power by a Rankine cycle in a manner that is similar to a conventional coal-fired steam power plant. There are some differences in the MHD bottoming cycle application, and these will be emphasized in this treatment.

Boiler Design

The first difference from a conventional steam plant is that the hot gases enter from a single duct rather than distributed burners. The gas is at a higher temperature (2000 K to 2200 K, 3140°F to 3500°F) and is reducing. The gas is reducing because the gases exiting the MHD cycle must be cooled before completing the combustion. At this point they contain carbon monoxide, hydrogen, and hydrogen sulfide, which are known to be corrosive to power plant materials. The basic structure can be a boiler fabricated of membrane boiler tube, but the interior metal surface must be protected from corrosion.

A standard technique is to weld studs to the interior surface of the boiler tube and refractory coat all metal surfaces that will come into contact with the combustion gases (12). This technique is well established commercially because it is used in cyclone-fired boilers, in some gasifier applications, and in the black liquor boilers of paper plants. The refractory selected must be rated to withstand the expected gas temperatures. It can be anticipated that the refractory will burn back to an equilibrium thickness, thinner in the lower, hotter regions and thicker in the upper, lower-temperature regions. The effect of the total flow entering the boiler at a single point primarily affects the desirability of distributing the flow of hot combustion gases uniformly over the interior surface of the boiler so as to utilize all the boiling surface efficiently.

Air Heating

As noted previously, preheating combustion air is one of the techniques used to attain a flame temperature sufficient to get adequate equilibrium ionization of the potassium seed. This temperature needs to be around 2800 K (4580°F). With standard atmospheric air, a preheat temperature of 1900 K (2960°F) to 2000 K (3140°F) is needed to achieve this flame temperature. A method of heating air to this temperature economically is one of the biggest remaining development problems for the MHD steam combined cycle power plant. (It should also be noted that there are a myriad of other power plant applications of such an air heater if it were developed.) It has generally been presumed that a metallic tube air heater is not applicable to heating air to these temperatures, although it may be the most economical solution for heating the air to an intermediate temperature. The original MHD plant concept envisioned the use of recuperative heat exchangers such as those used to preheat the air for blast furnaces. Clearly these are technically feasible, but the huge investments and operational problems inherent in these type heaters for a power plant make them unlikely final choices. More recently, ceramic materials for heat transfer tubes have been developed. Tests of individual tubes have shown satisfactory results for this application, but the problems of manifolding, treatment of thermal stress, and so on, remain to be solved. In the absence of an air heater that will preheat air to these temperatures, the design solution for the plant is to heat the air to as high as feasible with metallic air heaters and use some oxygen enrichment to achieve the required flame temperatures (13). A typical solution to this problem is to add oxygen so that the fraction by weight is about 36% and preheat the resulting mixture to 922 K (1200°F) in a metallic air heater.

Secondary Combustion

As previously noted, the primary combustor must operate fuel rich for optimum conductivity and low NO_x emissions. After the combustion products are cooled to the point that nitrogen oxide has decomposed to the desired level, the combustion must be completed to recover the remaining fuel energy and prevent emission of unburned and corrosive compounds, especially CO, H_2, and H_2S. The design goals are to achieve as complete a combustion as possible and to avoid a temperature increase that will increase the formation of nitrogen oxides. These goals are contradictory, of course, and a compromise must be chosen that is acceptable in both respects. The completeness of combustion is largely a function of the efficiency of mixing the exhaust gases with the air. This can be modeled numerically or by a laboratory simulation. In order to avoid a significant temperature increase with the accompanying increase in the formation of NO_x, the air can be introduced over a time period during which heat is removed from the gas. An alternative technique is to use exhaust gas recirculation, which involves taking exhaust gases from a lower-temperature region in the plant and, by use of a blower or fan, reinjecting them into the secondary combustor region so as to keep the mixture temperature from rising. From a combustion kinetics viewpoint and an overall steam plant efficiency

viewpoint, it is desirable to preheat the secondary combustion air to some modest temperature such as 589 K (600°F). This can be done as an alternative to economizer surface in the plant. Another design issue is the overall plant stoichiometry, or the amount of excess air added. The impact of increasing excess air is increased stack losses. Experimental work with a pilot plant showed that 5% to 10% excess air is a reasonable design point with control by measurement of CO emissions in the stack. Conventional coal plants typically operate at much higher excess air levels. Staged combustion and stack gas monitoring permit operating at this low level of excess air to gain the benefit of increased efficiency.

Particulate Loading

The particulate loading in the steam plant portion of the MHD steam combined cycle power plant is inherently higher than in a conventional coal-fired power plant. This is because of the potassium seed added to permit equilibrium ionization at reasonable temperatures. Potassium is needed to the extent of something like 1% of the total flow. If the potassium is added as a compound such as potassium carbonate or potassium formate, the total weight percentage of compound added is 1.77% and 2.15%, respectively, of the total flow. The carbonate or formate disassociates in the combustor of course, but it re-forms potassium sulfate or potassium carbonate in the lower-temperature regions of the steam plant. This turns out to result in typically a doubling of solids content in the plant components compared to operation with just the coal ash particulate. In addition, these potassium compound particulates tend to be much smaller in size than the coal ash particles. It should be noted that potassium sulfate forms preferentially as long as any sulfur is available. After all the sulfur is used, the remaining potassium forms carbonate. The significance of this observation here is that the carbonate has a lower melting point (1170 K, 1646°F) than the sulfate (1342 K, 1955°F). Any heat transfer surface having a gas side temperature higher than this will encounter liquid deposits that are harder to remove by soot-blowing. Since the larger particles are more likely to be removed in the furnace or elsewhere in the plant, the fraction of the total particulate that is composed of small potassium compounds in the flue gas toward the lower temperature part of the plant often approaches 80% to 90%. The design implications of this are threefold: (i) The tendency to collect on convective heat transfer surfaces and block the gas passages is greatly increased, (ii) heat transfer efficiency of convective sections is reduced (higher fouling factors), and (iii) the removal of sufficient particles to meet stack gas emission requirements may be more difficult. Early experiments with potassium seed resulted in convective sections that were completely plugged to gas flow with deposits. The boiler must be designed with wide tube spacing, at least as wide as those in used for the worst fouling coals and lignites. If the plant will ever burn low sulfur coal, the spacings should be further increased to provide a margin of safety from the expected potassium carbonate deposits. In any case, adequate soot-blowing capability must be provided. The amount of soot-blowing provided must be balanced against the design fouling factor used in calculating the amount of heat transfer surface required, but it should be noted that fouling factors tend to be higher in the MHD case due to the potassium compounds present.

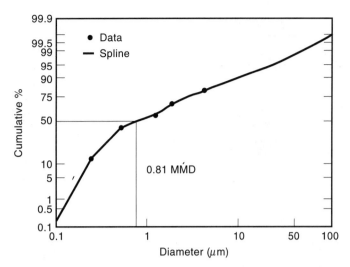

Figure 9. Typical particle size distribution in particulate from MHD pilot plant, upstream of the ESP/baghouse. Note the bimodal distribution with the smaller particles being potassium compounds and the larger ones coal ash.

Particulate Collection

As previously noted, the principal differences between the particulate in the MHD steam plant and that in a conventional coal-fired steam plant are the higher mass loading, smaller particulate, and different chemical composition in that it is high in potassium compounds.

In pilot plant tests designed to simulate a commercial plant (14), the particle loading was measured to be 14 g/m³ to 16 g/m³ (6 grains/ft³ to 7 grains/ft³). A typical particle size distribution is shown in Fig. 9. The mean mass diameter was about 0.6 μm. The size distribution is bimodal, with the larger particles tending to be coal ash and the smaller ones potassium sulfate. The resistivity of this mixture for two similar cases is shown in Fig. 10. The resistivity for temperatures of interest are in the range of 10^9 Ω·cm to 10^{10} Ω·cm. This is a desirable range for good ESP operation and tends to partially compensate for the very heavy mass loading and small size. A four-field dry ESP with a design specific collection area (SCA) of 475 ft²/1000 acfm performed in the pilot plant with

99.5% removal efficiency. With such a fine particulate, a major problem with greatly improved efficiency is reintrainment of small particles when the plates are rapped to remove the collected particles. An attractive alternative to the dry ESP seems to be the wet ESP, in which the plates are washed continuously with water. A baghouse is also a feasible choice. The bag fabric must be chosen to avoid blinding. A Gore-Tex fabric, manufactured by W. L. Gore & Co., Knoxville, TN, was used successfully in the pilot plant.

If a dry ESP is used, it has been suggested that if should be designed to operate at a high enough gas temperature so that it can be placed before the economizer/low temperature air heater surface. Since the temperature difference between flue gas and feedwater or air is low in this region, a large amount of convective surface is required and substantial savings result from having a cleaner gas and lower fouling factor.

Seed Recovery and Regeneration

The economics of the MHD steam power plant depend on recovery and reuse of the potassium seed. If the potassium is recovered as a sulfur or chlorine compound, chemical processing to convert it to a sulfur and chlorine free compound is required. Potassium is largely recovered from the exhaust by the ESP or baghouse, but significant quantities may also be recovered from hoppers under heat transfer surfaces.

If the plant has a slag rejecting combustor or wet bottom furnace, it can be anticipated that some potassium will be present with the coal slag from these effluent streams. The chemical form of potassium in these streams may be as compounds with the coal slag components, especially silicates. A method has been devised to economically recover this potassium by leaching with calcium hydroxide to recover potassium hydroxide, which integrates well with the formate seed regeneration process (15,16). A number of possible processes were studied during the US Department of Energy MHD development program with the conclusion that the formate process, which was used in Germany during World War II, is the best choice for converting the potassium sulfate recovered to potassium formate for reuse. The recovered potassium carbonate can be separated from the sulfate and chloride by differences in their solubility in water. Separation of sulfate from carbonate was performed well in the pilot plant program by a

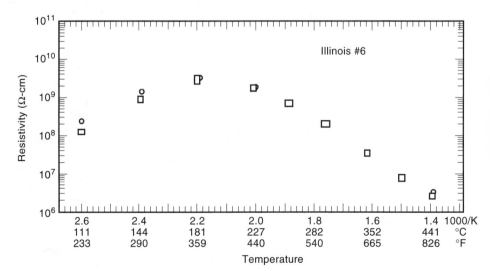

Figure 10. Ash resistivity from MHD pilot plant as a function of temperatures. Resistivities from 10^9 to 10^{10} Ω·cm are desirable from ESP operation.

rotary vacuum filter working on the water from the wet ESP. The system was saturated with carbonate, and the sulfate was removed by the filter.

Economic Plant Design

In order to adequately optimize the performance of the MHD steam combined cycle power plant, a model of the cost and performance is needed. The Electric Power Research Institute (EPRI) has developed a methodology for calculating the levelized cost of electricity over the planned lifetime of the plant (17). The complexities and trade-offs in the MHD plant require a model that calculates performance and cost for use in the EPRI methodology. Potential designers should develop such a model or use one of those listed in the references (9,18–20).

BIBLIOGRAPHY

1. G. Seikel and L. P. Harris, A summary of the ECAS MHD power plant results, *NASA Technical Memorandum X-73491*, 1976.

2. L. W. Crawford and R. A. Attig, Impact of process variables on NO_x emissions for Illinois #6 and rosebud coal fuels in MHD power generation, *Proc. 30th Symp. Eng. Aspects MHD*, Baltimore, MD, 1992, pp. VIII, 1.1–1.7.

3. K. D. Parks, Priority pollutant analysis of MHD-derived combustion products, *Proc. 28th Symp. Eng. Aspects MHD*, Chicago, IL, 1980.

4. R. C. Attig et al., Design report for the combustor for the low mass flow coal fired flow facility, Report No. DR-2.1.2-79-02, Tullahoma, TN: The University of Tennessee Space Institute, 1979.

5. TRW Applied Technology Division, *10th Q. Progress Report*, Report No. MHD-ITC-90-311, 1990.

6. A. Grove et al., 50-Mwt combustor performance summary, *Proc. 28th Symp. Eng. Aspects MHD*, Chicago, IL, 1990.

7. R. J. Rosa, *Magnetohydrodynamic Energy Conversion*, New York: McGraw-Hill, 1968.

8. J. X. Bouillard et al., *User's manual for MG MHD: A multigrid three-dimensional computer code for the analysis of MHD generators and diffusers, ANL/MHD-89/1*, Argonne, IL: Argonne National Laboratory, 1988.

9. J. N. Chapman, On the economic optimization of the magnetohydrodynamic-steam power plant, PhD dissertation, Eng. Sci. Mechan. Dept. University Tennessee, Knoxville, TN, 1977.

10. Y. C. L. Wu, Performance theory of diagonal conducting wall MHD generators, *AIAA J.*, **14** (10): 1362–1368, 1976.

11. L. S. Frost, Conductivity of seeded atmospheric pressure plasmas, *J. Appl. Phys.*, **32** (10): 1961.

12. Anonymous, *Steam: Its Generation and Use*, 39th ed., New York: Babcock & Wilcox, 1978.

13. J. N. Chapman and W. H. Boss, An economic analysis of the optimum stoichiometry for an early commercial MHD-steam plant, *Proc. 27th Symp. Eng. Aspects MHD*, Reno, NV, 1989.

14. R. C. Attig, *Illinois No. 6 coal 2000 Hour Proof-Of-Concept Report*, University Tennessee Space Ins., Tullahoma, TN, 1994.

15. J. K. Holt et al., Recovery of insoluble potassium by base extraction from MHD spent seed and slag, *10th Int. Conf. MHD Electr. Power Generation*, Tiruchirapalli, India, 1989.

16. A. C. Sheth et al., (UTSI) and R. L. Solomon et al. (RCC), Technical and economical consideration of the formate process for MHD seed recovery and regeneration, *10th Int. Conf. MHD Electr. Power Generation*, Tiruchirapalli, India, 1989.

17. *Technical assessment guide (TAG), EPRI Report No. P-2410-SR*, 1981 ed., Palo Alto, CA: The Electric Power Research Institute, 1982.

18. J. N. Chapman and W. H. Boss, *User's Manual For The UTSI MHD/Steam Power Plant Systems Code, DOE/ET/10815-98*, University of Tennessee Space Institute, 1985.

19. J. N. Chapman and J. H. Hollis, The UTSI performance–cost model of combined cycle MHD steam power plants, *Proc. Int. Liaison Group Symp. Modelling MHD Power Stations*, Eindhoven University, The Netherlands, 1986.

20. L. E. Van Biber and J. F. Pierce, A conceptual design of the Scholz MHD retrofit plant, *Proc. 26th Symp. Eng. Aspects MHD*, Nashville, TN, 1988.

JAMES N. CHAPMAN
Tennessee University Space
Institute

MAGNETOOPTIC EFFECT. See FARADAY EFFECT.

MAGNETORESISTANCE

Perovskite materials continue to be focal points for materials research and development efforts around the world. Perovskites such as barium titanate were extensively studied and used in the 1940s as active elements in communications systems. More recently, the confluence of novel thin film deposition techniques and the observation of superconductivity at temperatures above that of liquid nitrogen have triggered a flurry of activity in a broad class of perovskite and related materials. Another parallel area of considerable technological and scientific interest is ferroelectric and dielectric perovskites. Thin films of these materials, integrated on Si/GaAs, are strong candidates for solid-state memories, sensors, actuators, monolithic microwave elements, infrared sensors, and so on. Focusing specifically on the magnetic perovskites, considerable work has already been carried out on magnetic perovskites based on Fe, Co, Ni, and Mn. A very detailed review of the structural chemistry, magnetism, and transport in these systems is available in a review by Goodenough (1). We begin this review by summarizing some of the initial results that have led to the "rebirth" of this field. Because manganese-based rare earth perovskites with the typical chemical formulation of $La_{1-x}M_xMnO_{3\pm y}$ (where M is a divalent Group II cation such as Ca, Sr, Ba) exhibit the most interesting magnetotransport effects, we focus specifically on this family of compounds. Figure 1 schematically illustrates the crystal structure of this perovskite phase. What is noteworthy is that very rarely do these compounds adhere to the basic cubic perovskite structure; a variety of structural distortions are possible and have been observed as a function of composition and processing. Almost simultaneously, Chahara et al. (2) and Von Helmholt et al. (3) presented results of large magnetoresistance in epitaxial thin films of two of the prototypical compounds within this family. Almost immediately afterward, two different groups, one led by Jin et al. (4) and the other from this laboratory (5), published data showing magnetoresistance values much higher than that published by the two original papers. Figure 2 shows one typical plot of the resistance as a function of magnetic field; the colossal magnetore-

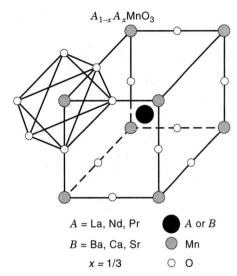

$A_{1-x}A_xMnO_3$

A = La, Nd, Pr	● A or B
B = Ba, Ca, Sr	◒ Mn
x = 1/3	○ O

Figure 1. A schematic of the perovskite crystal structure of substituted rare earth manganites.

sistance (CMR) value in this case was much larger than $10^5\%$ at a temperature of 77 K and a magnetic field of 5 T. Note that this value of MR is larger than 100% because it is defined as $(R_0 - R_H/R_H$, where R_0 is the resistance at zero field and R_H is the value in an applied field of H. When one puts this into a technological perspective, the field values needed to get useful MR are much too high. In many research programs, this was the trigger point that led to detailed studies of various factors that have an impact on the temperature and field dependence of magnetoresistance. Among them, the effect of lattice strain, oxygen stoichiometry, and the effects of layering and heterostructuring are being studied. The thin films and heterostructures are typically deposited by pulsed laser deposition. Refer to one of the many articles in current literature for a detailed description of the thin film processing procedures as well as the details of the various experimental

techniques that are being employed to characterize the materials.

Perovskite materials, even in the bulk, are very susceptible to lattice distortions, which may occur either spontaneously (e.g., ferroelectrics) or as a consequence of process-induced variations of, for example, the oxygen stoichiometry. When grown in thin film form, they are even more susceptible to lattice strains, especially when constrained by heteroepitaxy to the substrate lattice. For example, the bulk lattice parameter of $La_{0.7}Sr_{0.3}MnO_3$ (LBMO) is 3.88 Å whereas that of one common single crystal oxide substrate $LaAlO_3$ (LAO) is 3.79 Å, leading to a 2.4% compressive strain in the plane of the film (6). On the other hand, if the same material is grown on $SrTiO_3$ (STO), the lattice mismatch is smaller (1.0%) and is tensile. This leads to out-of-plane strains in the film, as illustrated in Fig. 3 where we see that 1000 Å thick films grown on these two types of surfaces show different out-of-plane lattice parameters. What is interesting to note is that, in both cases, the films do not seem to have relaxed into their bulk lattice dimensions, even after growth of 1000 Å. Because these materials do not have any other source of magnetic anisotropy, they strongly exhibit stress (or conversely strain) induced magnetic anisotropy. For example, in the absence of any magnetocrystalline anisotropy, one would expect the easy direction of magnetization to be controlled by the magnetostatic energy, thereby inducing an in-plane easy magnetization direction (or easy magnetization plane). This is indeed seen in magnetic hysteresis, measured in the plane of the film, of the films grown on STO, Fig. 4. However, the films grown on LAO are actually preferentially magnetized out of the plane (see Fig. 4), with the magnetization vector possibly lying at an angle to the substrate surface. Furthermore, this conclusion is authenticated by magnetic force microscopy (MFM) studies of the film surface, which are illustrated in Fig. 5. The films grown on STO show a "feathery" surface structure (reflecting the fact that there is no magnetization vector normal to the film surface), whereas the films on LAO show the characteristic maze domain structure typical of uniaxial materials. Is this a possible ingredient in obtaining large MR values? The magnetotransport measurements do not indicate dramatic differences in MR values for these two types of films. Figure 6 shows MR as a function of applied magnetic field for the films on STO and LAO, and although there is a small increase in MR for the films on STO, it is not dramatic.

Several approaches to reduce or eliminate this lattice strain are currently being studied. Among them, it has been found that changing the thickness of the film during deposition can progressively relax the mismatch strain, as illustrated through the X-ray diffraction spectra in Fig. 7. The progressive decrease in the c-axis lattice parameter can be interpreted as a consequence of a reduction in the compressive in-plane stresses that lead to an increase in the c-axis lattice parameter in the first place. It is worth noting that this does not lead to a completely relaxed film. Annealing a 1000 Å thick film in oxygen at 900°C also produces some partial relaxation, but the c-axis lattice parameter is still larger than the bulk value, as shown in Fig. 8. However, this does have a significant effect on the magnetic hysteresis loops. Figure 9 compares hysteresis loops for the same sample, before and after the oxygen annealing. The initial susceptibility has increased by almost an order of magnitude with an almost

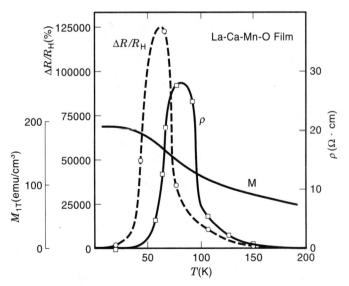

Figure 2. Temperature dependence of saturation magnetization, resistivity, and magnetoresistance for an epitaxial La–Ca–Mn–O film.

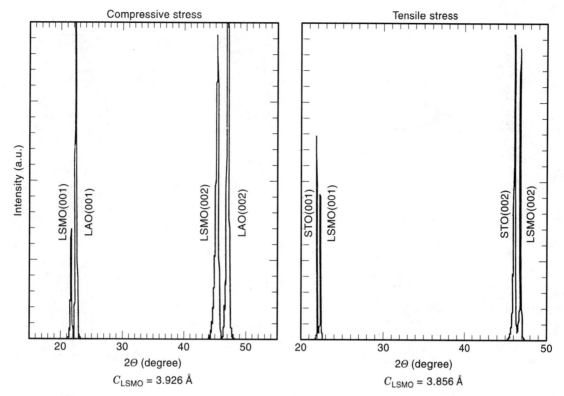

Figure 3. X-ray Bragg scans from epitaxial LSMO films grown on LAO and STO substrates illustrating the influence of substrate-film lattice mismatch on the film lattice parameters.

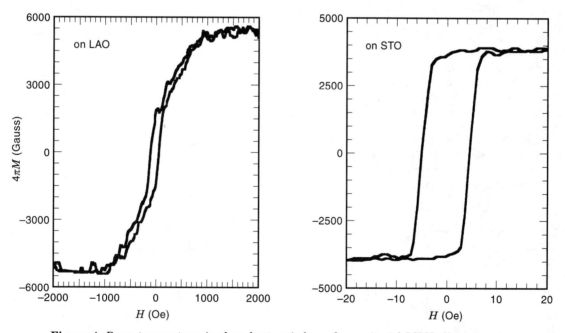

Figure 4. Room temperature, in-plane hysteresis loops from epitaxial LSMO films grown on LAO and STO substrates showing the influence of strain on the preferred magnetization direction.

AFM   MFM

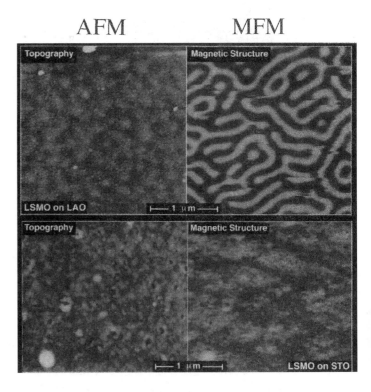

LaAlO$_3$

SrTiO$_3$

Figure 5. Magnetic force microscopy images of the film surface in the case of STO and LAO substrates illustrating the effect of mismatch strain on the preferred magnetization direction and consequently on the domain structure.

doubling of the saturation magnetization. Transmission electron microscopy of the as-grown samples indicate that the film has ordered domains with sizes in the range of 10 nm to 30 nm, as illustrated in the dark field micrograph in Fig. 10. This image was obtained using the superlattice reflection which is circled in the diffraction pattern shown in the inset. The regions in bright contrast correspond to one set of ordered domains, whereas the regions in dark contrast correspond to the other. We also observe that the domain size increases considerably with annealing in oxygen. The interfaces between these domains (i.e., the antiphase boundaries) can act as scattering sites during transport. The effect of these types of defects on the magnetotransport properties needs to be understood through further systematic experiments.

One of the most effective ways to alleviate the effects of lattice mismatch strain is to use buffer layers. A very suitable buffer layer is epitaxially grown SrTiO$_3$ on a [001] LaAlO$_3$ substrate. STO has a lattice parameter of 3.905 Å, almost

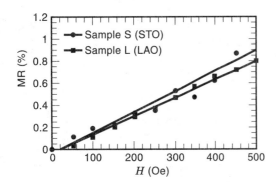

Figure 6. Magnetoresistance as a function of applied field for the epitaxial films grown on STO and LAO showing that the mismatch strain has only a minimal influence on the MR properties.

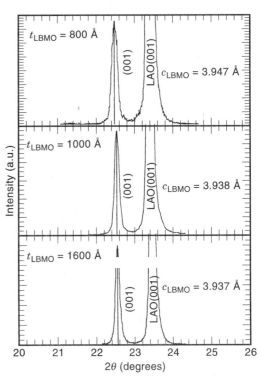

Figure 7. X-ray diffraction spectra as a function of thickness of the LSMO layer, showing the partial relaxation of lattice mismatch strain with increasing thickness.

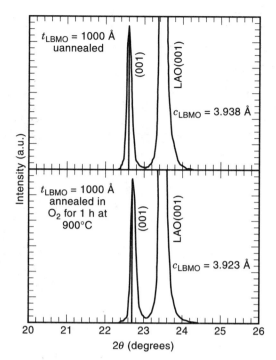

Figure 8. X-ray diffraction spectra of an epitaxial LSMO film before and after annealing at 900°C showing the reduction in the out-of-plane lattice parameter resulting from relaxation of the lattice mismatch strain.

identical to that of LBMO (3.9 Å). A thin layer (about 700 Å to 1000 Å) of STO is typically used as a buffer layer on which the LBMO layer is then grown. The X-ray diffraction patterns in Fig. 11 clearly show the effectiveness of the STO layer. The LBMO layer now has an out-of-plane lattice parameter of 3.907 Å, which is very close to the bulk value. As expected, this is also reflected in the magnetic properties, primarily the initial susceptibility and the saturation magnetization, which are shown in Fig. 12.

The obvious question now arises: is this sufficient to improve the MR properties of the CMR layer, especially at low fields? Direct current (dc) and microwave MR measurements

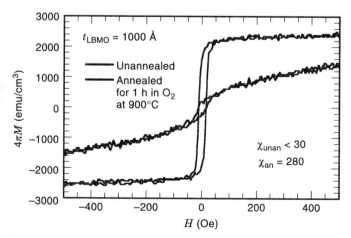

Figure 9. In-plane hysteresis loops of the LSMO film before and after annealing at 900°C showing the significant improvement in the squareness as well as the low-field susceptibility.

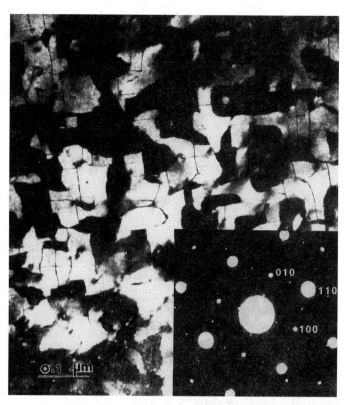

Figure 10. A dark-field transmission electron micrograph of the as-grown LCMO film showing the existence of ordered antiphase domain structure. The inset shows the electron diffraction pattern and the superlattice reflection that was used to form this dark-field image.

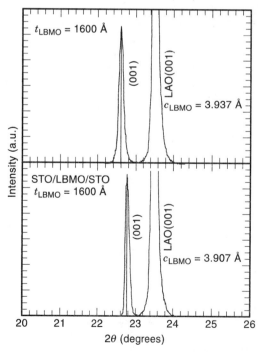

Figure 11. X-ray diffraction patterns from an epitaxial LBMO thin film grown on a [001] LAO substrate with and without a STO buffer layer, showing the beneficial effect of the STO layer in relaxing the mismatch strain.

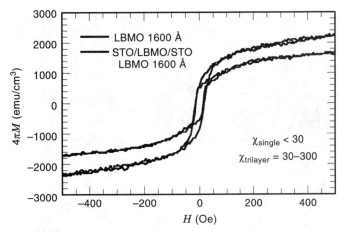

Figure 12. In-plane hysteresis loops of the LBMO film grown with and without the STO buffer layer, showing the strong enhancement of low-field susceptibility.

indicate that there is some improvement in the low-field MR values, but it is not significant enough to have an impact on an actual implementation in low-field sensing applications. Figure 13 compares the microwave and dc MR values, measured at 1800 Oe, and clearly demonstrates that there is very little difference between these two types of measurements. An interesting, but perhaps obvious, point that arises from this result is that the magnetotransport and the consequent MR effect are directly associated with the electronic properties of the manganite. Further supporting evidence for this is also available from optical measurements (7). It is also of interest to note that in these epitaxial films, the highest MR is obtained very close to the metal-insulator transition, although there is still considerable debate over the actual location of the peak in the MR value vis-à-vis the ferromagnetic Curie temperature (8). Contrast this with data that will be presented later on nonepitaxial films that show a very different MR behavior. Finally, note that at this field level, the MR values are still only of the order of a few percent, which is still not sufficient for low-field magnetic sensing applications

such as in read heads. Obtaining large MR values at low fields clearly requires further ingenuity and engineering of the thin films and their microstructures.

Two different directions are currently being explored by many of the groups that are intensively involved in this field. Hwang et al. (9), Ju et al. (10), and Gupta et al. (11) made a very interesting observation that in polycrystalline CMR materials, the MR values actually increase as the temperature is progressively decreased below the peak resistance temperature. This is in direct contrast to the epitaxial films, which show the largest MR values very close to the peak resistance temperature and drop to very small values well below the ferromagnetic transition. This was the first evidence for magnetoresistance behavior that is distinctly different from the CMR effect and that is not closely associated with the metal-insulator transition. This effect is associated with the granularity of the system (either bulk or thin film) and has been described as a spin-polarized tunneling phenomenon. Interestingly enough, a similar effect was also observed in epitaxial LBMO/STO/LBMO heterostructures by Lu et al. (12) and Sun et al. (13), in which the STO tunnel barrier layer thickness was on the order of 50 Å. The authors have suggested that the STO layer behaves like a spin-polarized tunnel barrier. Although the detailed physics behind this is still being resolved, what has emerged is that there is another source of MR, associated with granularity or alternatively structural interfaces. Two important types of structural interfaces can be visualized, namely, natural interfaces such as grain boundaries and antiphase boundaries and those created artificially by heterostructuring. We discuss the structural aspects of these two types next and then conclude with some results and comments on the magnetotransport behavior across such interfaces.

Four distinct classes of films can be generated, based on the type of substrate structure, surface chemistry, and deposition conditions. For example, by systematically changing the nature of the substrate surface, we can obtain these four distinct microstructures. For example, a fully epitaxial film of the CMR material (e.g., LSMO) on an almost lattice matched substrate such as LAO or STO can be obtained, in which case there is full in-plane orientational locking (i.e., $[100]_{LSMO}\|[100]_{LAO}$). A highly c-axis oriented film is obtained with in-plane crystallographic grain boundaries using a substrate with a large lattice mismatch (e.g., MgO) and a suitable template layer [e.g., Y–Ba–Cu–O (YBCO)]. On a substrate such as SiO_2/Si with a thin layer of bismuth titanate as a structural template, highly c-axis out-of-plane oriented films with random in-plane grain boundaries are obtained. Finally, when there is no structural or chemical similarity between the film and the substrate, typically polycrystalline films result. Figure 14 shows representative examples of in-plane phi-scans of the first three types and illustrates the difference in in-plane structural coherence between them. On an LAO substrate [Fig. 14(a)], the observation of intensity maxima spaced 90° apart reflects the fourfold symmetry of the film in response to the same for the substrate and clearly reflects orientational locking. On the [001] MgO substrate with the YBCO layer, the YBCO layer grows highly c-axis oriented with in-plane grain boundaries which are dictated by specific geometrical relationships, also known as the coincidence site lattice (CSL) model (or in the case of YBCO on MgO, the near model (14). Figure 14(b) shows the phi-scan for this case and

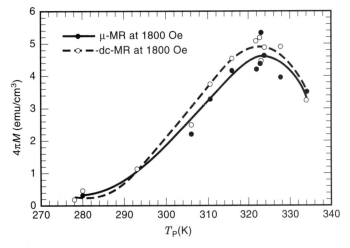

Figure 13. Comparison of the dc and microwave MR of an epitaxial LBMO film at an applied field of 1800 Oe, clearly demonstrating that the two types of measurements yield the same MR value.

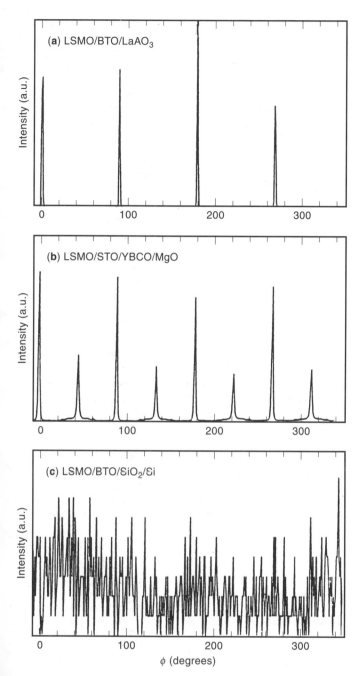

Figure 14. X-ray phi-scans from three types of films: *Top:* epitaxial film on LAO substrate showing the full in-plane orientation locking; *center:* highly [001] oriented film on MgO with a predominance of 45° in-plane grain boundaries; *bottom:* highly [001] oriented film on SiO₂/Si with no in-plane orientation locking.

clearly shows the presence of intensity maxima spaced 45° apart, indicating the existence of 45° rotation grain boundaries in the plane of the film. Indeed, these types of grain boundaries are observed in planar section TEM images, shown for example in Fig. 15. Figure 15(a) shows the actual image of this grain boundary, whereas Fig. 15(b) shows the same image after being digitally filtered. Figures 15(c) and (d) show light optical diffraction patterns obtained from the two grains showing that they are of a single orientation, whereas the composite optical diffractogram in Fig. 15(e) shows clearly

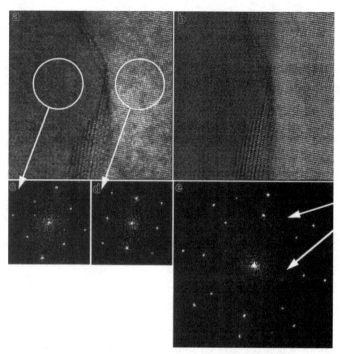

Figure 15. Planar section transmission electron micrographs of the film grown on MgO showing the characteristic 45° grain boundaries, identified from the optical diffraction patterns.

the 45° rotation between them. Returning to Fig. 14(c), this X-ray phi-scan shows the case of a LSMO film grown on BTO/SiO₂/Si where the film is c-axis oriented but there is no in-plane orientational locking.

Figure 16 shows the effect of these grain boundaries on the low-field magnetotransport properties, at a temperature (77 K) well below the Curie temperature or the maximum resistance temperature. The epitaxial film shows almost no MR; however, the films with grain boundaries show a much larger, negative magnetoresistance compared to the epitaxial films. This is one of the interesting aspects of transport in these

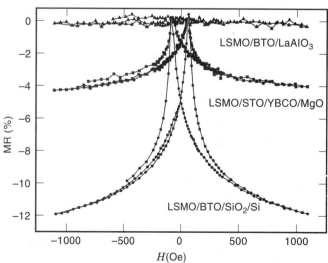

Figure 16. Low-field MR for the three types of films described in Fig. 14, illustrating the beneficial effects of the in-plane grain boundaries.

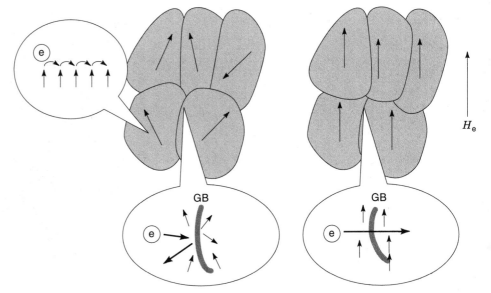

Figure 17. A schematic illustration of the possible mechanism by which the large MR effects are observed in the granular films. At zero field, the magnetization vectors in each grain is randomly oriented in addition to the intrinsic spin disorder at the grain boundaries; when a field is applied the spins within the grains reorient and thus reduce the scattering within the grain, leading to a lower resistance.

materials that still needs detailed elucidation. Figure 17 shows schematically a possible model to explain the large field-dependent MR in the films with grain boundaries. We suspect that the intrinsic structural disorder at the grain boundaries (the degree of which depends on the nature of the grain boundary, namely whether it is a CSL-type semicoherent grain boundary or a fully random grain boundary) leads to spin disorder very close to the grain boundaries. Upon application of a small magnetic field, the ferromagnetic material inside the grain becomes fully magnetized while the grain boundary regions are still not fully aligned in the applied field direction (15). The effective field at the grain boundaries is of the order of $H_e + 4\pi M$, and even though the external field H_e is only a few hundred Oersteds, the total field is of the order of a few thousand Oersteds. It is still unclear if this effective internal field is sufficient to cause the alignment of spins in the grain boundary regions, and further systematic studies are required.

A second approach to low-field MR in these oxides is through heterostructuring. Two types of heterostructure devices, namely spin valves (16–18) and spin tunnel devices, are being explored in many laboratories. A typical example of a spin valve is shown schematically in Fig. 18. Such a device typically consists of two ferromagnetic (FM) layers (significantly different coercive fields) separated by a metallic, nonferromagnetic layer. Case 1 in this figure shows the condition where the two FM layers are polarized in the same direction. Under these conditions, if a spin-polarized electron is launched from the bottom FM layer, then it can pass into the upper FM layer through the normal metal layer without any hinderance leading to a low resistance. However, if Case 2 is true, the spin-polarized electron launched from the bottom FM layer will get scattered at the upper FM/metal interface, and the resistance will be higher. In some sense, these two different resistance values correspond to two different logic states (as in a memory element). Control of the spin orientation is through the coercive fields of the two layers. Typically, one of the layers has a much higher coercive field or is pinned to one magnetization state by exchange biasing through a layer underneath. This is illustrated in Fig. 19 through sche-

matic hysteresis loops. The hysteresis loop in the bottom of this figure shows the characteristic split indicative of the two distinctly different coercive fields. The corresponding MR plot is shown in Fig. 20. Large values of spin-dependent MR values are obtained in the intermediate field regime where the spins in the two FM layers are antiparallel.

Using these concepts, we have been exploring approaches to obtain large MR values in heterostructures that use the CMR perovskites as one of the FM layers, as schematically illustrated in Fig. 21. The intrinsic similarity in crystal structure and chemistry between these materials (i.e., they are all based on oxygen octahedra) means that heteroepitaxy is feasible. Indeed, X-ray diffraction patterns from prototypical het-

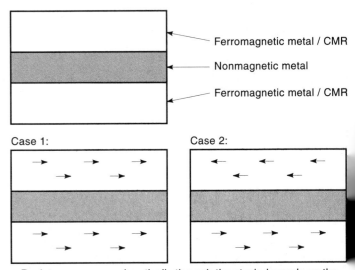

Resistance measured vertically through the stack depends on the relative spin orientation of the top and bottom ferromagnetic layers. If they are parallel (Case 1) there is no scattering and the resistance will be lower than when they are antiparallel (Case 2) and there is scattering.

Figure 18. A schematic illustration of the trilayer structure of a spin valve illustrating the origin of scattering that is dependent on the relative orientations of the spin polarizations in the two FM layers.

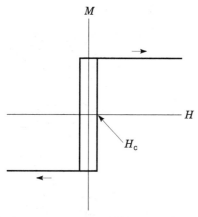

H_c is a parameter of the material that is controlled by the composition of the material and the preparation of the material.

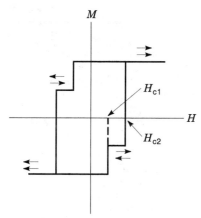

R_J is related to the relative spin orientation, and the relative spin orientation depends on the magnetic field, so $R_J = R_J(H)$.

Figure 19. A schematic diagram of the method by which the spin polarizations in the two layers are controlled through the coercivities of the two FM layers.

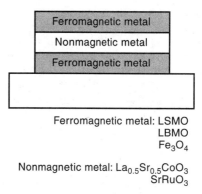

Ferromagnetic metal: LSMO
LBMO
Fe_3O_4

Nonmagnetic metal: $La_{0.5}Sr_{0.5}CoO_3$
$SrRuO_3$

Figure 21. A schematic diagram of the all-oxide spin valve that is being explored in our program.

erostructures grown using LSMO as the FM top and bottom layers and LSCO as the nonferromagnetic metallic layer show excellent in-plane as well as out-of-plane structural coherence. Ferromagnetic hysteresis loops obtained from these heterostructures show clear proof of the existence of two distinct coercive fields (Fig. 22), which is one prime requirement for a spin valve to operate. Transport measurements are currently under way, and the details of magnetotransport in these spin valves should become clearer over the next year.

CONCLUDING REMARKS

Colossal magnetoresistance in perovskite manganates is not a new phenomenon in the same way that the materials themselves are not new. What has happened over the past few years is that new materials processing approaches and the knowledge that has been obtained from the high-temperature superconductivity research arena have transferred over to magnetic oxides. For example, the ability to grow epitaxial perovskites has been and continues to be an important prerequisite to the explosion of research in the CMR area. Although it appears doubtful that large MR values will be obtained at low magnetic fields in single thin-film layers of the CMR perovskites, the recent results of magnetotransport measurements in polycrystalline films appear to be very

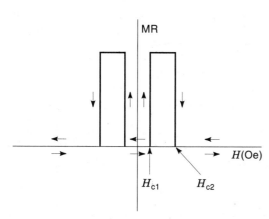

Figure 20. The MR corresponding to the hysteresis loop shown in Fig. 19 for the spin valve illustrated in Fig. 18.

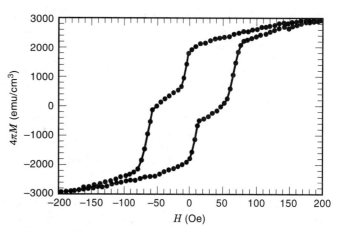

Figure 22. Magnetic hysteresis loop of a LSMO/LSCO/LSMO trilayer spin valve showing the two distinct coercive fields that are a basic requirement for a spin valve operation.

promising in this regard. More detailed studies are required to understand the origins of the behavior and to see if large MR values can be obtained at room temperature and low fields. What has become more exciting is the natural evolution to multilayered heterostructures, such as the spin valve or the spin tunnel junctions. The next few years should bring considerable excitement to this area of research.

BIBLIOGRAPHY

1. J. Goodenough, *Prog. Solid State Chem.,* **5**: 145, 1972.

2. K. Chahara et al., *Appl. Phys. Lett.,* **63**: 1990, 1993.

3. R. Von Helmholt et al., *Phys. Rev. Lett.,* **71**: 2331, 1993.

4. S. Jin et al., *Science,* **264**: 413, 1994.

5. G C. Xiaong et al., *Appl. Phys. Lett.,* **66**: 1427, 1995.

6. C. Kwon et al., *J. Magn. Mag. Matls.,* in press.

7. S. Kaplan et al., *Phys. Rev. Lett.,* **77**: 2081, 1996.

8. K. Ghosh et al., *Phys. Rev. B, Condens. Matter,* in press.

9. H. Y. Hwang et al., *Phys. Rev. Lett.,* **77**: 2041, 1996.

10. H. L. Ju et al., *Phys. Rev. B, Condens. Matter,* **51**: 6143, 1995.

11. A. Gupta et al., *Phys. Rev. B, Condens. Matter,* **54**: 15629, 1996.

12. Y. Lu et al., *Phys. Rev. B, Condens. Matter,* **54**: 8357, 1996.

13. J. Z. Sun et al., *Appl. Phys. Lett.,* **69**: 3266, 1996.

14. T. S. Ravi et al., *Phys. Rev. B, Condens. Matter,* **42**: 10141, 1990.

15. N. D. Mathur et al., *Nature,* in press.

16. M. Johnson and R. H. Silsbee, Interfacial charge-spin coupling: Injection and detection of spin magnetization in metals, *Phys. Rev. Lett.,* **55**: 1790, 1985.

17. M. Johnson, Bipolar spin switch, *Science,* **260**: 320, 1993.

18. M. Johnson, The all-metal spin transistor, *IEEE Spectrum,* **31**: 47, 1994.

R. RAMESH
T. VENKATESAN
University of Maryland

MAGNETORESISTANCE, THERMAL.   See THERMAL
MAGNETORESISTANCE.

MAGNETOSTRICTIVE DEVICES

Magnetostrictive materials transduce or convert magnetic energy to mechanical energy and vice versa. As a magnetostrictive material is magnetized, it strains; that is, it exhibits a change in length per unit length. Conversely, if an external force produces a strain in a magnetostrictive material, the material's magnetic state will change. This bidirectional coupling between the magnetic and mechanical states of a magnetostrictive material provides a transduction capability that is used for both actuation and sensing devices. Magnetostriction is an inherent material property that will not degrade with time.

The history of magnetostriction began in the early 1840s when James Prescott Joule (1818–1889) positively identified the change in length of an iron sample as its magnetization changed. This effect, known as the Joule effect, is the most common magnetostrictive mechanism employed in magneto-

strictive actuators. A transverse change in dimensions accompanies the length change produced by the Joule effect. The reciprocal effect, in which applying a stress to the material causes a change in its magnetization, is known as the Villari effect (also referred to as the magnetostrictive effect and magnetomechanical effect). The Villari effect is commonly used in magnetostrictive sensors.

An additional magnetostrictive effect used in devices is the Wiedemann effect, a twisting that results from a helical magnetic field, often generated by passing a current through the magnetostrictive sample. The inverse Wiedemann effect, also known as the Matteuci effect, is used for magnetoelastic torque sensors (1,2).

The existence of both direct and reciprocal Joule and Wiedemann effects leads to two modes of operation for magnetostrictive transducers: (1) transferring magnetic energy to mechanical energy and (2) transferring mechanical energy to magnetic energy. The first mode is used in the design of actuators for generating motion and/or force and in the design of sensors for detecting magnetic field states. The second mode is used in the design of sensors for detecting motion and/or force; in the design of passive damping devices, which dissipate mechanical energy as magnetically and/or electrically induced thermal losses; and in the design of devices for inducing change in a material's magnetic state.

In many devices, conversion between electrical and magnetic energies facilitates device use. This is most often accomplished by sending a current through a wire conductor to generate a magnetic field or measuring current induced by a magnetic field in a wire conductor to sense the magnetic field strength. Hence, most magnetostrictive devices are in fact electromagnetomechanical transducers.

Some of the earliest uses of magnetostrictive materials during the first half of this century include telephone receivers, hydrophones, magnetostrictive oscillators, torque-meters, and scanning sonar. These applications were developed with nickel and other magnetostrictive materials that exhibit bulk saturation strains of up to 100 μL/L (units of microlength per unit length). In fact, the first telephonic receiver, tested by Philipp Reis in 1861, was based on magnetostriction (3).

The discovery of "giant" magnetostrictive alloys (materials capable of over 1000 μL/L) in the 1970s renewed interest in magnetostrictive transducer technologies. Many uses for magnetostrictive actuators, sensors, and dampers have surfaced in the past two decades as more reliable and larger strain and force giant magnetostrictive materials (e.g., Terfenol-D manufactured by Etrema Products, Inc. and Metglas manufactured by Allied Corp.) have become commercially available (in the mid- to late 1980s). Current applications for magnetostrictive devices include ultrasonic cleaners, high-force linear motors, positioners for adaptive optics, active vibration or noise control systems, medical and industrial ultrasonics, pumps, and sonar. In addition, magnetostrictive linear motors, reaction mass actuators, and tuned vibration absorbers have been designed, whereas less obvious applications include high-cycle accelerated fatigue test stands, mine detection, hearing aids, razor blade sharpeners, and seismic sources. Ultrasonic magnetostrictive transducers have been developed for surgical tools, ultrasonic cleaners, and chemical and material processing.

The state of the art for commercially available magnetostrictive ultrasonic devices is a motor rated at 6 kW, with development of a prototype 25 kW ultrasonic motor in progress (4). These motors are rated at resonance (20 kHz). Flextensional underwater sonar magnetostrictive devices have been driven to a source level of 212 dB (Ref. 1 μPa at 1 m) (5). Also, a recently designed high force test rig for use by the US Air Force in fatigue testing of jet engine turbine blades produces broadband forces from 200 Hz to 2000 Hz with peak to peak forces of 14 kN at 2 kHz (6).

MAGNETOSTRICTIVE MATERIALS

All ferromagnetic materials exhibit magnetostriction; however, in many materials, its magnitude is too small to be of consequence. In the early 1970s, the search for a material that exhibited large magnetostriction at room temperature grew out of the discovery that the rare earth elements terbium and dysprosium exhibit magnetostriction (basal plane strains) on the order of 1% at cryogenic temperatures. From 1971 to 1972, Clark and Belson of the NSWC (Naval Surface Warfare Center) (7), and Koon, Schindler, and Carter of the NRL (Naval Research Laboratory) (8), independently and almost simultaneously discovered the extremely large room temperature magnetostriction of the rare earth–iron, RFe$_2$, compounds.

Partial substitution of other rare earths, such as dysprosium, for terbium in the Tb–Fe compound resulted in improvements in magnetic and mechanical properties. The stoichiometry of Tb–Dy–Fe that became known as the giant magnetostrictor Terfenol-D (terbium: Ter; iron: Fe; Naval Ordnance Laboratory: NOL; dysprosium: D) is given by Tb$_x$Dy$_{1-x}$Fe$_y$, where $x = 0.27$ to 0.3 and $y = 1.92$ to 2.0. Small changes in x and y can result in significant changes in the magnetic and magnetostrictive properties of the material. Decreasing y below 2.0 reduces the brittleness of the compound dramatically but reduces the strain capability. Increasing x above 0.27 does not significantly effect the strain capability, but it reduces the magnetocrystalline anisotropy, which effectively allows for increased magnetostriction at lower fields and more efficient energy transduction.

Nominal longitudinal strains for various materials are shown in Table 1. Note that nickel has a negative magnetostrictive constant, and its length shortens in the presence of a magnetic field, whereas other materials including Terfenol-D have a positive magnetostrictive constant. In the case of iron, the magnetostrictive constant changes from positive to negative, known as the Villari reversal, as the field is increased (9).

In 1978 Clark and co-workers (10) introduced a second new magnetostrictive material based on amorphous metal, produced by rapid cooling of magnetic alloys consisting of iron, nickel, and cobalt together with one or more of the following elements: silicon, boron, and phosphorus. These alloys, known commercially as Metglass (metallic glass), have been processed to achieve very high coupling coefficients, on the order of 0.95, and are commonly produced in thin-ribbon geometries. Metglass is typically used for sensing applications and for converting mechanical or acoustical energy into electrical energy. The ease with which they can be magnetized and demagnetized and low core losses of these alloys constitute a distinct set of magnetic properties that have triggered high interest in research on metallic glasses (11).

Giant magnetostrictive materials are currently available in a variety of forms, including thin films, powder composites, and monolithic solid samples. Reviews of giant thin film magnetostrictive material are available (2,12,13). Giant magnetostrictive particle composites (GMPCs) and commercially available (Midé Technology Corp.) magnetostrictive elastomers are powdered magnetostrictive material (often Terfenol-D) solidified in a composite or rubber matrix (14). This form has the advantages that the nonelectrically conducting matrix material can be used reducing eddy current losses associated with ac operation, varied geometries are readily achieved, and the elastic properties, in particular ductility and machinability, can be greatly improved. Furthermore, devices using a driver element made with GMPCs as opposed monolithic material may avoid the need for prestress mechanisms and the bulky permanent magnets needed for biased ac operation. The major disadvantage with GMPCs is a reduction in output strain and strain rates related to the matrix–Terfenol-D composition ratio (14–17).

Some of the latest research on new magnetostrictive materials includes demonstration of a sputtered, amorphous thin film version of Terfenol-D and the doping of TbFeDy alloys with additional rare earths including holmium to minimize magnetic hysteresis losses (18). One of the most promising new research directions is work on magnetomemory materials, materials that combine the magnetostrictive behavior of ferroelectric materials with the high-strain attributes found in shape memory alloys. Currently, the FePdCo magnetomemory system exhibits magnetostriction several times that of the best giant magnetostrictive materials and is the focus of theoretical work by a number of researchers. Devices based on magnetomemory may provide significant advances in magnetostrictive sensor capabilities (19,20).

For this survey of magnetostrictive devices, applications using nickel and the monolithic forms of Terfenol-D will be emphasized, as these are the most common commercially available materials used in magnetostrictive transducer applications and are well suited for discussing common transducer design and operation issues.

MAGNETOSTRICTIVE DEVICE TRANSDUCTION BASICS

Typical magnetostrictive transducers employ both electromagnetic energy conversion and magnetomechanical energy conversion. A wound wire solenoid is used for conversion between electric and magnetic energies, and a magnetostrictive material is used as the transducer driver to convert between magnetic and mechanical energies. For actuation, passing a current through the solenoid converts electrical energy into magnetic energy. Maxwell's equations can be used to show that the magnetic field seen by the magnetostrictive driver is

Table 1. Nominal Strain of Magnetostrictive Materials

Material	Magnetostriction (μL/L)
Iron	20
Nickel	−40
Alfenol 13	40
NiCo	186
Terfenol-D	2000

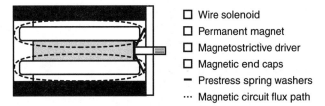

☐ Wire solenoid
☐ Permanent magnet
☐ Magnetostrictive driver
☐ Magnetic end caps
– Prestress spring washers
··· Magnetic circuit flux path

Figure 1. Components of basic magnetostrictive devices: magnetostrictive driver, magnetic end caps, permanent magnet, wound wire solenoid, magnetic circuit, and prestress mechanism.

proportional to the current in the solenoid. The magnetic energy generated by the solenoid simultaneously magnetizes and produces strain in the magnetostrictive core. This same device can also be used as a sensor. Application of a sufficient force to strain the transducer magnetostrictive driver will produce a change in the magnetization of the driver. In turn, this changing magnetic energy can be measured directly with a Hall probe or, as can be shown by using the Faraday–Lenz law, it can be converted to electrical energy in the form of an induced voltage in the solenoid and then measured. One of many common magnetostrictive device configurations is sketched in Fig. 1.

Figure 2(a) shows an output displacement versus applied magnetic field " butterfly" for a typical commercially available unbiased magnetostrictive actuator (Etrema Products Inc., Model AD-140j). For ac operation, either a dc current or a permanent magnet having a critical field strength of H_c is used to offset the initial actuator length, centering ac operation in the near-linear strain region. This is shown in Fig. 2(b) for a commercially available biased magnetstrictive actuator (Etrema Products, Inc., Model AA140j).

The efficiency with which a magnetostrictive material converts magnetic energy to mechanical and vice versa is often used as a key attribute to quantify the performance of a magnetostrictive device. It is noteworthy, however, that the design of a highly efficient magnetostrictive transducer also requires significant attention be paid to the efficiency of conversion between electrical energy and magnetic energy that actually goes into magnetization of the magnetostrictive driver. This requires careful design of the transducer magnetic circuit, including incorporation of a closed magnetic flux return path. This is a task that can be readily undertaken with modern finite element analysis tools. (See MAGNETIC CIRCUITS.)

Furthermore, proper matching of the system power supply to the input impedance of the transducer needs to be considered when optimizing overall system efficiency (just as a home stereo output impedance is matched to a standard 8 Ω loud speaker input impedance). The power supply rating in watts is equal to voltage times amperes and is given based on use with a specific load. Competing electrostrictive transducer technologies, based on piezoelectric and ferroelectric transduction, are capacitive devices that for generating moderate to high force output typically require power supplies capable of providing high voltages (on the order of 1000 V) at low currents (microamperes). Magnetostrictive transducers, on the other hand, are inductive devices that require power supplies capable of providing moderate voltages (less than 100 V to 200 V) at moderate currents (milliamperes up to several amperes) for similar force outputs. Depending on the power requirements and transducer design, the power supply for a magnetostrictive system could be similar to those used to drive common commercial electromagnetic devices, such as mechanical shakers and voice coil loudspeakers.

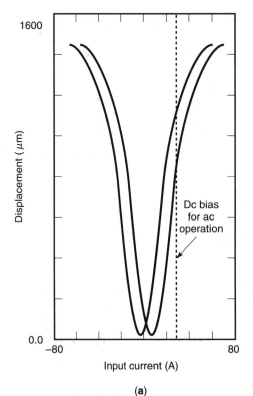

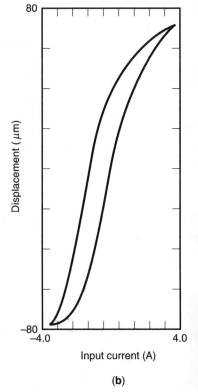

Figure 2. Commercially available magnetostrictive Terfenol-D actuator output: (a) unbiased Model AD-140j; (b) biased Model AA-140j. (Courtesy of ETREMA Products, Inc.)

(a)

(b)

The magnetostrictive effect itself is not frequency limited; however, in practice, ac losses (due to magnetic hysteresis and eddy currents) and device-specific mechanical and electrical resonances tend to restrict the output bandwidth capability to low ultrasonics (<100 kHz). Magnetostrictive devices do operate down to dc; a step input magnetic field will cause a step change in device length.

THE MAGNETOELASTIC EFFECT

The magnetostrictive process relating the magnetic and mechanical material states can be described with the two coupled linear equations given in Eq. (1). These equations neglect temperature effects and have been reduced from a three-dimensional vector form to reflect only axial behaviors. These magnetostrictive equations of state are expressed in terms of mechanical parameters (strain ϵ, stress σ, Young's modulus at constant applied magnetic field E_y^H), magnetic parameters (applied magnetic field H, magnetic induction B, permeability at constant stress μ^σ), and two magnetomechanical coefficients (the axial strain coefficient $d_{33} = d\epsilon/dH|_\sigma$ and $d_{33}{}^* = dB/d\sigma|_H$),

$$\epsilon = \sigma/E_y^H + d_{33}H \tag{1a}$$
$$B = d_{33}^*\sigma + \mu^\sigma H \tag{1b}$$

These equations will be discussed in detail in a following section but for now should be considered as a basis for capturing the coupled mechanical and magnetic nature of magnetostriction. Equation (1a) indicates that the strain of a magnetostrictive element changes with stress and applied magnetic field. First, looking at the applied stress σ, when a stress is applied to a magnetostrictive sample, it will strain. If the field H is held constant and the stress varied, the sample will get shorter if the stress is compressive and longer if the stress is tensile. The magnitude of the effect is scaled by the inverse of the material's elastic modulus E_y^H. Next, an applied magnetic field H can also change the sample's length. If the stress is held constant, the effect of H in the strain is proportional to the applied field scaled by the piezomagnetic coefficient d_{33}. The elastic modulus and the piezomagnetic coefficients vary from one magnetostrictive material to the next and often vary with operating conditions. They need to be experimentally determined and are usually provided by the material manufacturer. Similarly, the magnetic induction B will also vary with stress and applied field, only scaled by the coefficients of Eq. (1b) as indicated.

Figure 3 shows a cartoon to help depict the strain and magnetic induction observed in magnetostrictive materials as they are magnetized and as described by these equations. In Fig. 3, the essence of a magnetostrictive device is lumped into discrete mechanical and magnetic attributes that are coupled in their effect on the magnetostrictive core strain and magnetic induction (21). As shown, the external stress is set to a constant compressive value ($\sigma = \sigma_0$) provided by the mass resting on a stiff spring on top of the magnetostrictive core.

Looking first at the case of no applied field ($H = 0$), the sample has an initial length ($\epsilon = 0$), and no net axial magnetic induction ($B = 0$), as depicted in Fig. 3(c). As the magnitude of the applied field H increases to its saturation limits, $\pm H_s$, the elliptical magnets rotate to align with the applied

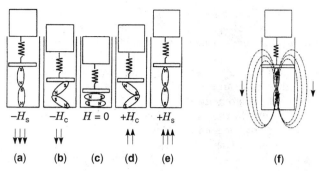

Figure 3. Cartoon of changing strain ϵ and magnetic induction B in a magnetostrictive element subjected to a constant compressive stress σ_0. The applied magnetic field H increases from (a) $-H_s$ through (c) $H = 0$ to (e) H_s. Figure (f) indicates the magnetic flux lines associated with magnetization of the driver shown in (e).

field; the axial strain increases to ϵ_s and magnetization of the element in the axial direction increases to $+B_s$ [(Fig. 3(e)] or decreases to $-B_s$ [(Fig. 3(a)]. At an applied field strength of H_s, the saturation magnitudes of strain and magnetic induction have been reached; that is, the strain and magnetization of the sample will not change with further increases in the applied field. Thus, both the magnetically induced strain and the magnetic induction magnitudes increase moving from the center figure outward as the magnitude of the applied field increases to its saturation values.

Alternatively, picture the applied field being set to a constant, like H_s, and then placing an increasing mass load or compressive force on the magnetostrictive element from the outermost figures to the center figure. Both the axial strain and axial magnetization magnitude in the element will decrease with increasingly negative (compressive) stress.

Figure 3(f) shows the flux lines associated with alignment of the magnetic domains in the magnetostrictive driver shown in Fig. 3(e). These lines are technically the superposition of the contributions caused by the magnetization of the magnetostrictive driver element and the applied field H_s. Flux lines for Fig. 3(a) would be similar in shape, but flux would be flowing in the opposite direction. Similarly, Figs. 3(b,d) would exhibit the same shapes in the appropriate direction but would be lower in magnitude. Figure 3(c) would not exhibit external flux lines, as the driver as shown is demagnetized. The flux field is used in sensors to measure the magnetization, strain, and/or force on the magnetostrictive driver by monitoring the flux either with a Hall effect probe or by detecting the voltage induced in a wire conductor perpendicular to the flux lines.

Even though this cartoon is not to be taken literally, it conveys the essence of the behaviors observed in magnetostrictive materials. For instance, the cartoon suggests that for ac applications, a dc magnetic bias can be applied to strain the material to half its saturation length and cycle between initial and saturated lengths. While this does convey the idea behind biasing, in practice the dc bias needed for ac operation, H_c, is based on operation centered in the steepest region of the curve shown in Fig. 2 where the strain–field slope is a maximum. This region, called the "burst" region, arises as a result of reorientation of magnetic moments (produced at the atomic level by electron spins) from an "easy" crystallographic axis perpendicular to the applied field to one more closely aligned with the applied field. Easy axes correspond to orientations for the magnetic moment vectors that satisfy local crystallographic energy minimization states in the magnetostrictive material as the applied stress and magnetic field vary.

Furthermore, an initial compressive stress is often used to increase the alignment of magnetic moments along easy axes perpendicular to the applied field. Although a given sample cannot get any longer than its saturation length [Figs. 3a,e], under a prestress the sample's zero field length can decrease [get shorter than shown in Fig. 3(c)], thereby maximizing the net achievable strain.

MAGNETOSTRICTIVE ACTUATORS

Numerous types of magnetostrictive transducers are distinguished by their configuration and use of the magnetostrictive driver. Common transducer classifications can be found in the literature (22,23). In most cases, the Joule effect in a magnetostrictive element is employed to produce a longitudinal vibration in rods or radial vibrations of tubes or rings. The most common design is the longitudinal vibrating Tonpilz in which a cylindrical or rectangular sample provides axial output (22). The tube-and-plate or tube-and-cone transducers use 1/4 wavelength magnetostrictive tubes to excite blocks tuned to transfer mechanical power to the surrounding medium efficiently. Other designs employ a number of magnetostrictive elements fitted end to end to form a ring or polygon (24). The simultaneous longitudinal excitation of the elements causes the ring's diameter to oscillate. The flextensional design (23) employs a longitudinal vibrating magnetostrictive element to excite an arched membrane, commonly an oval-shaped shell. Transducer designs have also been based on other magnetostrictive effects, such as the Wiedeman-effect twisting pendulum (2). Many variations in design on these transducers are available including some uses in noncontact activation, where, for instance in rotating environment, a nonrotating magnetic field source can be used to produce and/or sense shape changes in a rotating element.

Prior to the 1940s, the materials conventionally used for many transducer applications were the magnetostrictive elements nickel, cobalt, iron, and their alloys Permendur, Permalloy, and so on. In the 1940s and 1950s, the emergence of piezoelectric ceramics barium titanate and lead zirconate titnate (PZT) led to higher efficiency transducer designs. The twofold increase in coupling coefficient of PZT over that of nickel provided higher transducer acoustic efficiencies and larger output powers in a number of applications. With the development of giant magnetostrictive materials in the 1970s, even higher transducer efficiencies are now achievable. The superiority of Terfenol-D over piezoelectrics in high-power applications has been extensively demonstrated. Moffett, Powers, and Clark (25) present a theoretical comparison of the power handling capability of Terfenol-D and PZTs. A more recent paper by Moffett and Clay (26) address the superiority of Terfenol over PZT in a barrel-stave flextensional transducer in which the active motor was either Terfenol-D or PZT, with the rest of the device design otherwise unchanged. The Terfenol-D transducer had an FOM (figure of merit) of 60 W/kg · kHz · Q, about four times that of the PZT transducer, and an 8 dB higher source level throughout a wider bandwidth. Akuta (27) reports a comparison of displacement versus output force relationships between PZT and Terfenol-D.

This should not be taken to imply that nickel-based actuators and sensors are outdated. Nickel still costs considerably less than giant magnetostrictive materials and shares the magnetostrictive attribute that (unlike their piezoelectric and ferroelectric counterparts) the magnetostrictive shape change capability will not degrade with use. Companies such as Blue Wave Ultrasonics offer commercial nickel-based ultrasonic cleaners with "the only lifetime guarantee in the industry."

Performance Evaluation

Many techniques to characterize a transducer's performance are available. The technique selected usually depends on several factors, including the intended use of the transducer and allowable resources. Performance quantification in many cases boils down to analysis of the material properties under as-run conditions, that is, material properties of the active driver element based on measurements taken in the transducer under loads typical of a given application.

Magnetostrictive material properties are intrinsically related to the material magnetization, stress, and temperature distribution in the transducer. In fact, this relationship is strongly coupled so changes in operative parameters affect material properties and vice versa (28). Typical material properties considered when analyzing magnetostrictive transducers include Young's modulus, mechanical quality factor, magnetic permeability, saturation magnetization and magnetostriction, magnetomechanical or piezomagnetic coefficient, and magnetomechanical coupling factor. Measurements of a transducer performance provide magnetoelastic parameters that are characteristic of both the magnetostrictive driver and the transducer itself. For instance, the magnetomechanical coupling gauges the energy conversion between the elastic and magnetic sides of the device magnetostrictive material. However, in a transducer, the effectiveness of the energy conversion is highly dependent upon the magnetic circuit that routes the field applied by current through a solenoid into the driver. The magnetomechanical coupling then represents, when considered *in the transducer,* a property of both the driver and the specific transducer design. On the other hand, the saturation magnetization and magnetostriction are properties of the driver material itself, not the transducer. Further treatment on material property analysis can be found in Refs. 21,29–31.

To illustrate how these elastomagnetic properties relate to some important transducer performance indicators, we discuss six very distinct measures of transducer performance: (1) the strain-applied field (ϵ–H) characteristic curve, (2) transducer electrical impedance mobility loops, (3) transducer blocked force, (4) output energy and energy densities, (5) output power limitations, and (6) figures of merit.

ϵ–H **Characteristic Curve.** The idea of measuring quasi-static transducer displacement response for performance characterization has been explored extensively, mostly because of the simplicity of the procedure. The magnetic field intensity is, for a given input current, determined by knowledge of the solenoid characteristics, and the mechanical output is measured with displacement sensors such as the LVDT (linear variable differential transformer) or accelerometers. Typical strain measurements at different mechanical preloads are shown in Fig. 4. A strain-applied field characteristic curve aids in choosing the optimum polarization field for achieving quasi-linear performance and for avoiding frequency doubling. Additionally, it provides information about some of the longitudinal elastomagnetic coefficients.

Returning to the coupled linear equations given in Eq. (1), which are known as the piezomagnetic constitutive equations, we see that, for fixed stress, the strain and applied magnetic field are related by the piezomagnetic coefficient d_{33}, with $d_{33} = (\partial/\partial H)\epsilon|_\sigma$, the slope of the ϵ–H characteristic curve. This is a very common measure of performance as a first look at a transducer's output capability. A value for d_{33} can be obtained by estimating the slope of a line drawn between the minimum and maximum strains for the rated output based on transducer data such as that shown in Fig. 2. Note that this may not be the same as placing a line on the tip and tail of the hysteresis loop shown in Fig. 2(b).

By forgoing the last 10% of achievable strain at the saturation and low field ends of the curve, much higher efficiency performance can be achieved. Trade-offs is quantifying a d_{33} based on measured data arise, whereby if additional displacement is required (assuming space and cost constraints allow for it), one may opt for using a longer sample in the steep region of the strain-field characteristic curve rather than driving a shorter sample to saturation. This would provide two device designs using drivers of the same magnetostrictive material with potentially quite different d_{33} coefficients.

This method for characterization of d_{33}, however, should be interpreted for use as measured (i.e., a dc measure of the transducer output). The differential values of d_{00} vary not only with H and σ (see Figs. 2 and 4), but also with the magnetic bias, stress distribution in the material, and operating frequency (31,32). Hence the strain-field characteristic curve is of little use in assessing transducer dynamic operation, where the preceding factors play a significant role. In such cases, methods based on dynamic excitation of the transducer are preferred (31).

Transducer Electrical Impedance Mobility Loops. The total transducer electrical impedance, the electrical voltage measured per input current as a function of frequency, contains a plethora of information on transducer behavior. As discussed in detail by Hunt (3) and Rossi (33) among others, the electrical impedance is the sum of the blocked and the mobility impedances. The blocked impedance is the voltage per current associated with a transducer in which the output motion is restricted to zero, or "blocked." The mobility impedance is the difference in the total (measured) and blocked impedances and is a result of "mobility" or the force and velocity in the mechanical side of the vibrating transducer. (Although discussed here in the context of performance evaluation, understanding how to predict a transducer impedance function and having the ability to tailor its attributes for specific applications are also important with regard to power supply matching and optimization of total system performance.)

One use of impedance function information is to obtain fundamental resonant and antiresonant frequencies and the two associated elastic moduli for the magnetostrictive transducer driver. The resonant condition occurs when the magnitude of the mobility component of the measured impedance is its maximum and is represented by the diameter of the mobility impedance loop in the Nyquist representation (34). This resonant frequency f_r is associated with the open circuit (constant field) elastic modulus of the material E_Y^H so that $f_r \propto \sqrt{E_Y^H}$. Similarly, the maximum-diameter point of the electrical admittance loop ($Y_e = 1/Z_e$) represents the antiresonant frequency f_a associated with the short circuit (constant induction) elastic modulus E_Y^B. This relationship is such that $f_a \propto \sqrt{E_Y^B}$.

The short circuit modulus E_Y^B represents the stiffest material condition, which occurs when all available magnetic energy has been transduced into elastic potential energy. When energy is transferred from the elastic to the magnetic side, the effective modulus decreases to the value E_Y^H. The two elastic moduli are related by the magnetomechanical coupling factor k ($0 \le k \le 1$), as follows (35):

$$E_Y^H = (1 - k^2)E_Y^B \qquad (2)$$

In a similar fashion, the intrinsic or uncoupled magnetic permeability μ^σ of Eq. (1b) is reduced to a value corresponding to the constant strain permeability μ^ϵ because of the energy conversion from the magnetic to the elastic side. Clark (35) provides further detail and proves, by means of energy considerations, that both permeabilities are also related by k

$$\mu^\epsilon = (1 - k^2)\mu^\sigma \qquad (3)$$

By analogy to the cartoon in Fig. 3, holding the magnetic induction B constant while allowing σ and H to vary implies that the ellipses will not rotate from a given orientation. Hence the transducer will appear stiffer to an external force

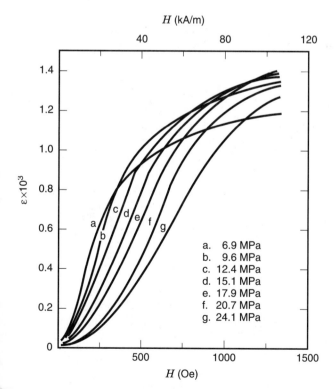

Figure 4. Strain-applied magnetic field relationship at 0.7 Hz, no magnetic bias, prestress: 6.9, 9.6, 12.4, 15.1, 17.9, 20.7, 24.1 MPa (32).

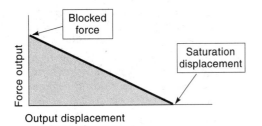

Figure 5. Cartoon of typical transducer blocked force–saturation displacement relationship.

than if H was held constant but B and σ were allowed to vary. Alternatively, holding strain ϵ constant while varying H limits the potential change in B; thus the material looks less permeable under operating conditions of constant strain than constant stress.

Ultimately, the electrical impedance provides information on how the energy is transferred between the magnetic and mechanical sides of the transducer. Knowledge of resonances f_r and f_a permits calculation of elastic moduli, magnetic permeabilities, and magnetomechanical coupling factor. Discussion on the latter will be continued in the "Figures of Merit" section.

Blocked Force. Figure 5 shows a sketch of a load line, a curve that represents the output force and displacement capabilities of a transducer. As the transducer tries to do work against an increasing load, its output displacement capability decreases. The system's blocked force F_B is the load against which the transducer can no longer do work. At this load, the transducer displacement is reduced to zero, and the transducer is virtually "clamped" or "blocked."

For a quasi-static loading case, a simplified transducer model can be developed using Newton's Second Law which equates the force output of the transducer to an elongating elastic spring element,

$$F_T = k_m u_H \qquad (4)$$

where the magnetostrictive driver material is assumed to have stiffness k_m and the transducer output displacement u_H is found as the magnetostriction [Eq. (1a)], times the driver length: $u_H = \epsilon \times L$. The effective stiffness of the driver, based on axial extension of a slender rod having a cross-sectional area A, length L, and short circuit elastic modulus E_Y^B, can be modeled as

$$k_m = \frac{E_Y^B A}{L} \qquad (5)$$

Thus, $F_T = E_Y^B A \epsilon$. The transducer will elongate to its saturation length with a maximum force potential $F_{Tmax} = E_Y^B A \epsilon_s$ under no external load. At the other extreme of the load line, an external load prevents strain of the magnetostrictive driver (hence the use of E_Y^B instead of E_Y^H). Equating maximum external and internal force capabilities for no-strain and no-load, respectively, the blocked force rating for a transducer can be found as

$$F_B = E_Y^B A \epsilon_s \qquad (6)$$

This force represents the maximum force that the material is capable of producing under quasistatic and lossless conditions, and it is a function of the rod geometry and the material properties E_Y^B and ϵ_s. Based on published values of the short circuit elastic modulus $E_Y^B = 50$ GPa and of the saturation magnetostriction $\epsilon_s = 1600$ μL/L, the blocked force rating for a device with a 24.5 mm diameter Terfenol-D driver is 38 kN.

Output Energy and Energy Densities. Let us assume a simple linear elastic load of stiffness k_L acting upon the Terfenol-D driver of stiffness k_m, as depicted in Fig. 6. Such a model represents, for instance, the actuator acting upon an acoustic load such as water and assumes no losses. The mechanical energy acting upon the load is

$$E_M = \frac{k_L u^2}{2} \qquad (7)$$

where u is the displacement at the rod end.

One can separate the total displacement of the magnetostrictive driver into an active component and a passive component. The active component is the displacement of the driver free end u_H under no external load [Eq. (1a)]. The elastic or passive component is the response u_{el}, which opposes u_H as a result of the transducer doing work against an external load (possibly including work against a prestress mechanism). The total driver free end displacement is $u = u_H - u_{el}$.

The free end of the driver is subjected to a total force F, where the passive displacement is $u_{el} = F/k_m$, and the total end displacement is

$$u = u_H - \frac{F}{k_m} \qquad (8)$$

For a linearly elastic external load, the force acting upon the driver end is $F = k_L u$. Substituting this expression for F into Eq. (8) yields

$$u = u_H - \frac{k_L u}{k_m} \qquad (9)$$

or equivalently after defining $t = k_L/k_m$

$$u = \frac{1}{1+t} u_H \qquad (10)$$

Finally, substituting Eq. (10) into Eq. (7) and letting $k_L = t k_m$ yields

$$E_M = \frac{t}{(1+t)^2} \left(\frac{k_m u_H^2}{2} \right) \qquad (11)$$

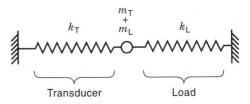

Figure 6. Schematic representation of an actuator of stiffness k_T loaded with an elastic load of spring constant k_L.

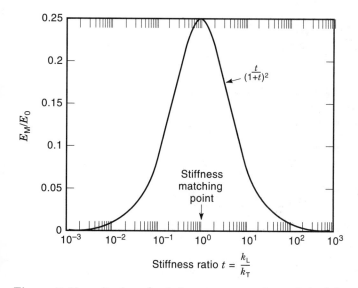

Figure 7. Normalized mechanical energy output is maximized for stiffness matching between the magnetostrictive driver and the combined transducer and external load stiffness characteristics.

This expression is plotted in Fig. 7 as a function of t, where it is readily apparent that the energy is maximum for $t = 1$. This situation indicates that stiffness matching is desired to achieve maximum output energy. It is also important to note that in some prestress mechanism designs, a spring or a set of washers is placed in series with the magnetostrictive driver. The spring stiffness has to be accounted for in calculations of stiffness matching. As noted previously, the driver stiffness is directly related to the elastic modulus E_Y, which can be measured in different ways (28,30).

The output energy of a transducer is often scaled by the transducer volume, mass, and price. These energy "densities" facilitate the fair comparison between transducers. However, they do not give information on efficiency of energy conversion in the transducer. This issue is discussed further later.

Projector Power Limitations. As discussed in Refs. 36 and 37, the power output near resonance may be stress- or field-limited, depending on whether the mechanical quality factor Q is greater or smaller than the optimum value. This optimum Q hovers around unity for Terfenol-D (38). Hence, a low Q raises the maximum power limit of the transducer near resonance. In addition to the ability to handle more power output, a low Q means a broader operational bandwidth. This situation may or may not be desirable because at low Q, the transducer is less able to filter out unwanted modes of vibration that might lead to harmonic distortion. This highlights some of the compromises present in designing the transducer's mechanical quality factor. Figure 8(a) depicts a typical radiated power limit-Q curve indicating typical performance tradeoffs at frequencies near transducer resonance for a Terfenol-D transducer, and Fig. 8(b) presented an equivalent circuit representation of the transducer based on linear behavior.

Figures of Merit. The single most important transducer FOM is the magnetomechanical coupling factor k. This is particularly true in the low-signal linear regime, where k^2 quan-

tifies the fraction of magnetic energy that can be converted to mechanical energy per cycle, and vice versa. Improvements in manufacturing techniques over the past decade have helped increase the magnetomechanical coupling factors reported in bulk samples of giant magnetostrictives like Terfenol-D. Today, values of $k \approx 0.7$ are commonly reported for Terfenol-D ($k \approx 0.3$ for nickel). Note that a transducer's coupling factor will be lower than that of its magnetostrictive core due to magnetic, mechanical, and thermal losses inherent to other aspects of the device design and its components.

As discussed in detail in Ref. 35, rearrangement of the piezomagnetic equations (1a,b) leads to expressions for the magnetomechanical coupling on the basis that the energy transfer between elastic and magnetic sides in the Terfenol produces a change in the effective material elastic modulus and magnetic permeability. In a lossless scenario, the resonance and antiresonance frequencies of the electrical impedance function in Fig. 9(a–c) represent f_r and f_a as defined by the mobility impedance. In Fig. 9b, the prestress is constant at 1.0 ksi (kilopound per square inch) and in 9c the magnetic field is constant at 75 Oe. In the presence of losses, f_r and f_a shift slightly from the resonance and antiresonance points. Based on electroacoustic principles (3), it can be demonstrated that k is directly related to the system resonant and antiresonant fre-

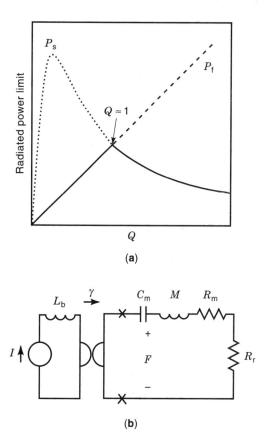

Figure 8. (a) Radiated power versus quality factor Q shown as the minimum of the stress limited output P_s and the field limited output P_f. (b) Van Dyke transducer equivalent circuit representation for the magnetostrictive device used to predict (a), where I = current, L_b = blocked inductance, γ = gyrator ratio, C_m = open circuit compliance, F = force, M = mass, R_m = mechanical resistance, and R_r = radiation resistance. Reprinted with permission from Ref. 36. Copyright © 1993 Acoustical Society of America.

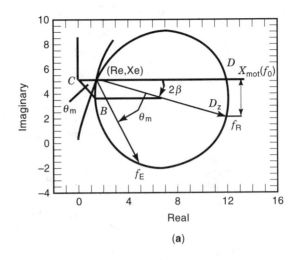

(a)

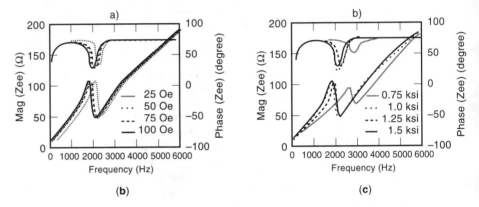

Figure 9. Measured Terfenol-D transducer electrical impedance functions: (a) Nyquist representation indicating diameter of mobility loop for use in determination of f_r; (b) impedance for 1.0 ksi prestress at varied ac drive levels; and (c) impedance functions for 25 Oe drive level at varied initial prestress settings (39).

quencies, f_r and f_a. Assuming no losses,

$$k^2 = 1 - \left(\frac{f_r}{f_a}\right)^2 \tag{12}$$

The k dependence on ac field intensity and magnetic bias condition is clearly demonstrated in Refs. 28 and 39. In a lossy magnetic circuit, however, this relationship is somewhat different, although the essential aspects are preserved (28). The effect of losses can be minimized by carefully designing the magnetic path, and by using laminated material for the driver and all components of the magnetic circuit flux path by using either radial or longitudinal slots as appropriate. The penalty is typically increased complexity in manufacture and modeling.

An alternative formulation for k is derived from energy formulations, known as the three parameter method (35),

$$k^2 = \frac{d^2}{s^H \mu^\sigma} \tag{13}$$

Transducers intended for underwater use are often rated according to the desired characteristics of these devices: high power P, low resonant frequency f_0, low mass M (or volume V), and low quality factor Q. Hence, in the literature FOM = P/Qf_0V is commonly found. Claeyssen et al. (40) present a comparison of several Terfenol-D and PZT devices with FOM

ranging between 10 J/m³ and 655 J/m³, where a Terfenol-D flextensional has a remarkable 655 J/m³ figure of merit. A similar PZT device reached only 15 J/m³; the difference is caused exclusively by the much smaller strain output of the PZT (the acoustical power output is proportional to the strain squared).

Types of Devices

In order to illustrate the needs and issues of different transducer applications, we classify magnetostrictive transducers in three broad categories as follows:

1. High-power, low-frequency applications. These applications are typically associated with underwater acoustic generation and communications. The designs discussed here are the flextensional transducer, the piston-type transducer, and the ring-type transducer.

2. Motion generation. In this category, we include transducers designed to do work against external structural loads attached to them. The motion may be linear, such as in the inchworm or Kiesewetter motors, or rotational.

3. Ultrasonic applications. This category involves a fairly broad range of actuators, whose final use may actually vary from surgical applications to cleaning devices used in high-speed machining.

Despite the differences among the three categories, there is intrinsic commonality to all magnetostrictive devices based upon the magnetomechanical nature of the device. First of all, it is often desirable to achieve linearity in the performance. Because magnetostrictive materials respond nonlinearly to applied magnetic fields (hysteresis, quadratic response at low field, saturation at high fields), it is common practice to subject the material to mechanical and magnetic bias (known as bias conditions). A magnetic bias is supplied with permanent magnets, located in series or parallel with the drive motor, and/or with dc. The mechanical bias is applied by structural compression of the driver with either bolts or via the transducer structure itself, such as in flextensional transducers.

Second, maximum effectiveness is obtained at mechanical resonance, where dynamic strains of peak-to-peak amplitudes higher than the static saturation magnetostrain can be achieved. Dynamic strains are of primary importance in low-frequency, high-power transducers, namely those for sonar and underwater communications.

Among the other technological issues that affect the design and operation of magnetostrictive transducers are thermal effects, transducer directionality (determined by ratio between size and acoustic wavelength), resonant operation, and power source.

Low-Frequency Sonar Applications

Research efforts searching for smaller sonar transducer systems capable of delivering increasingly higher acoustic power output have existed since the inception of sonar technology. Terfenol-D has proven attractive for underwater sound projection given its output strain, force, and impedance-matching characteristics. Not surprisingly, much of the research effort on magnetostrictive underwater devices has been performed by the US Navy, where Terfenol-D was originally developed.

The quest for more powerful sonar units forced researchers to increase either the size of the radiating surface or the vibration amplitude of the device. As a consequence, the former issue brought the re-emergence of a transducer principle first developed in the 1930s, the flextensional transducer. Magnetostrictive flextensional transducers offer high power at low frequencies. The power output of Terfenol-D flextensionals is about 25 times larger than that of PZT flextensionals because the dynamic strains are approximately five times larger and the power output is approximately proportional to the square of the strain. By comparison with the Tonpilz transducer, flextensionals present the advantages of providing a larger radiating surface per volume, using less active material, and requiring lower voltages. Technological challenges inherent to flextensionals are stress-induced fatigue, hydrostatic compensation, and prestressing for use in deep submersion. The Tonpilz-type transducer typically packs high power in a compact size. Moreover, because these transducers use an oversized piston for acoustic field generation, the diameter of the active element and magnetic circuit components can be increased without change in overall transducer diameter. A variation of the Tonpilz design is that with two end radiating surfaces, which have an impact on the radiated acoustic field's directivity pattern. Square-ring transducers are used in either omnidirectional or directional mode. They provide low-frequency acoustic emissions.

A discussion of actuator devices for sonar applications is presented next. The discussion addresses the three main types of devices: the flextensional, the ring-type, and the piston-type.

Flextensional Actuators. Flextensional transducers radiate acoustic energy by flexing a shell, usually oval-shaped, caused by the longitudinal extension and contraction of a cylindrical drive motor mounted in compression inside the shell. These transducers are capable of producing high power at fairly low frequencies. Their history dates back to 1929–1936, when the first flextensional device for use as a foghorn was first built and patented by Hayes at the NRL (41).

Several factors limited the effectiveness of the early flextensional devices. Hayes's original design lacked a preload mechanism, which limited the strain and force capability of the transducer and lead to fatigue-related failure. The parallel development in the late 1920s of the ring transducer, a relatively cheaper and more rugged device with similar electroacoustic efficiency, severely undermined the budget destined for research on flextensionals. The flextensional concept was a high-power, high-efficiency transducer being used in the wrong field. It was not until the 1950s that flextensionals started to be considered for underwater applications, leading to a 1966 patent by Toulis (42). This device was nearly identical to Hayes's, except for the explicit intention for underwater projection and the use of a piezoelectric stack for driving the device. Interestingly enough, the flextensional transducer is usually credited to Toulis (43).

Flextensional transducers are identified by classes. This is one aspect of flextensionals that often leads to confusion among transducer designers. The confusion arises as to which geometry defines which class. To make matters worse, there are currently two different classification schemes and at least two proposed variations of those. With the exception of the Class I and IV, which are common to both schemes, the use of classification numbering must be accompanied by declaration of the scheme employed.

The following is a brief review of both classification schemes, as illustrated in Figs. 10(a,b). The reader is pointed to the references for further details.

In the classification scheme of Brigham and Royster (43,44), the radiating surfaces of Class I flextensionals are formed by revolving an ellipse around its long axis. In other words, Class I flextensionals have a football-shaped shell with the magnetostrictive motor mounted in compression inside, along the major axis. A variant of this design known as the University of Miami flextensional has flat surfaces as described in Ref. 43. The Class II flextensional is essentially a modified Class I in which, in order to accommodate more active material, the major axis is extended in both directions. The extra amount of active material allows more power handling without changing the fundamental resonance of the oval radiator. One strong limitation of the flextensional design is the lack of true broadband capability. To overcome this limitation, a Class III design has been proposed. The shell of the Class III flextensional consists of two Class I shells of slightly different sizes grated together. The two closely spaced resonances help to broaden the acoustic bandwidth beyond that of the Class I design. Unquestionably, the most common shell geometry is that of the Class IV flextensional. Typically, the shell is convex in shape, although there are variants, such as

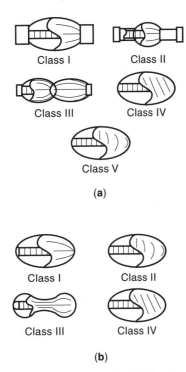

(a)

(b)

Figure 10. (a) Brigham and Royster's classification scheme for flextensional transducers. (b) Pagliarini and White's classification scheme for flextensional transducers.

the flat-oval shell or the concave shell of Merchant (classified type VII by Rynne) (45,46). Finally, the Class V consists of two shallow spherical caps attached to either side of a vibrating ring or disk (sometimes called clam shell). This design is credited to Abbott [however, this design had been described by Pallas in 1937 (23)], who in a 1959 patent (47), shows a ceramic-driven unit with either concave or convex spherical shells. Note that in the Brigham–Royster scheme, Classes I, IV, and V are basically distinguished by shape. On the other hand, Classes I, II, and III are differentiated by end use (i.e., Class II is a high-power version of Class I and Class III is a broadband version of Class I).

The Pagliarini–White scheme classifies flextensionals in four classes (I to IV), based exclusively on shell geometry (48–50). The Class I and Class II are similar to Brigham and Royster's Classes I and II. The Class III is known as the barrel-stave flextensional, and the Class IV is the typical oval-shaped shell. Jones (49) and Rynne (46) provide their own classification schemes by subtly modifying the previous schemes.

It is characteristic of flextensional transducers to have two types of radiating modes: a low-frequency flexing mode [or modes (51)] associated with the bending of the shell, and a higher-frequency breathing mode, in which the whole shell expands and contracts in unison. As the names implies, the desired mode of vibration is a flexural one. However, the breathing mode is acoustically more like a monopole; thus, it is usually of higher efficiency and improves the effective system magnetomechanical coupling factor. This higher mode has a strong effect on the parameters used to describe the flexural modes and, hence, it cannot be overlooked. Under hydrostatic pressure, the shell geometry determines whether the flexing mode occurs in phase or out of phase with respect

to the longitudinal vibration of the driver. The former case is characteristic of the Class IV flextensional, whereas the latter case is characteristic of Merchant's barrel-stave design.

In the Class IV design, the hydrostatic pressure acting on the shell tends to unload the magnetostrictive driver. This presents a technological challenge that defines the compromise between acoustic power and submersion depth. This problem has been addressed in different ways, including depth-compensation, mechanical filtering, and change of the shell geometry to concave instead of convex (Merchant).

In summary, flextensional devices designed for use with magnetostrictive drivers offer several advantages over other transducer materials such as PZTs, but successful demonstrations of their practicality for actuation are fairly recent. One such work reports a Terfenol-D-powered acoustic projector operated at a depth of 122 m (400 ft) driven to a source level of 212 dB (Ref. 1 μPa at 1 m) (5). Recent work (25,40) demonstrates that good magnetomechanical coupling and efficiency characteristics are possible in flextensional devices. Technological issues such as the magnetic circuit design, unloading of the magnetostrictive driver caused by submersion depths, the effects of cavitation at shallow depths, and stress-induced fatigue in the shell have been addressed effectively, even though trade-offs among these issues are inevitable (5).

Piston-Type Actuators. High dynamic strains are instrumental in achieving high acoustic power radiation. As with many other transducer designs, the availability of giant magnetostrictive materials such as Terfenol makes it possible to design more powerful underwater piston transducers with little or no bulk penalty. It has been suggested (40) that because of the need for much less active material, the cost of a Terfenol-driven piston transducer would be lower than, for instance, that of an equivalently rated piezoelectric flextensional transducer. The simpler design and the lack of bending parts likely to suffer fatigue-induced failure also indicate some of the advantages of piston-type designs over conventional flextensional transducers.

The Tonpilz transducer (Tonpilz is German for "sound mushroom") is quite simple in principle, as illustrated in Fig. 11. The magnetostrictive rod is surrounded by a drive solenoid, which provides the magnetic field excitation. The magnetostrictive element actuates upon an inertial mass ("tail"

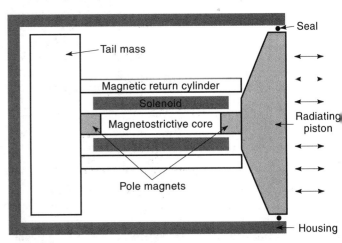

Figure 11. Typical Tonpilz magnetostrictive transducer.

mass) and has a front radiating surface which generates the desired sound waves. The magnetic path is completed by magnetic couplers and a magnetic return path cylinder. In fact, the magnetic circuit can have many possible configurations. For instance, some designs use cylindrical permanent magnets and ferromagnetic end pieces known as a "barrel magnet" configuration instead of the "stacked magnet" configuration shown in Fig. 11. Each configuration has a specific set of merits, which means that the optimum magnetic circuit configuration should be identified based upon the particular application requirements (52). As a rule of thumb, rods shorter than 20 cm and with diameters under 2.5 cm have increased magnetic coupling, in the order of 5%, in the barrel magnet configuration over what can be achieved using a stacked magnet configuration (reference: T. T. Hansen of Extrema Products, Inc., personal communications). However, when longer and thicker rods are used, significant phenomena such as saturation and end effects are likely to occur in the barrel magnet configuration. The stacked magnet configuration does have the potential for significantly increasing the systems' resonant frequency due to stiff pole magnets located in series with the magnetostrictive element. Typical permanent magnet materials currently in use are Alnico V, samarium-cobalt, and neodimium-iron-boron, while the magnetic return materials currently used are laminated steel and ferrites.

Recently, researchers have been able to develop innovative magnetostrictive Tonpilz radiators capable of excellent source level outputs and very high FOMs. Steel describes (53) a Tonpilz transducer in which six hexagonal cylinders of Terfenol-D are arrayed forming a tube inside which there is a prestress rod. The transducer itself is mounted on its central node in such a way that the magnetostrictive tube and the head mass form a so-called balanced piston design. The vibration of head and tail masses is almost symmetrical (with the same velocity but in opposite directions). The advantage of this design is that the Terfenol-D tube is isolated from hydrostatic pressure and freed of undesirable shifts in resonant frequency as a result of changing preloads. However, balanced piston operation occurs only near resonance, so the device is a narrow bandwidth transducer.

Conversely, a novel hybrid magnetostrictive/piezoelectric transducer designed at the NUWC (Naval Underwater Warfare Center) (54) is capable of broadband operation by virtue of its double resonance configuration. The transducer consists of a series arrangement of a tail mass, an active driver, a center mass, a second active driver, and a head mass. Because the velocities of the piezoelectric and magnetostrictive drivers are 90° out of phase with each other, self-tuning similar to that in the balanced piston design is obtained. In addition, the improved coupling coefficient of this device translates into better acoustic response at even lower frequencies than in conventional Tonpilz transducers of similar size and weight.

Meeks and Timme (55) present a Tonpilz vibrator consisting of three Terfenol rods spaced 120° apart, head and tail masses, and a center rod for mechanical and magnetic biasing. The transducer was not as powerful as more recent units, but it was useful for demonstration of the potential of Terfenol-D for this type of application.

Claeyssen and Boucher (56) developed a set of magnetostrictive transducers capable of very high acoustic power output. An initial design, called the Quadripode, uses four rods

to vibrate the radiating piston at an in-water resonant frequency of 1200 Hz. A follow-up design, named Quadripode II, has better effective coupling and broader operational bandwidth, thanks to modifications in the head mass and prestress rod. Both transducers feature a forced cooling device. A transducer based on three rods instead of four, called Tripode, was also developed. It better uses the giant magnetostriction effect to yield twice the power output of the four-rod transducers (3.8 kW versus 1.6 kW), packed in less than half the volume (22 dm³ versus 51 dm³). Hydrostatic compensation permits deep submersion up to 300 m. The source level of the Tripode at resonance (1.2 kHz) is 209 dB (Ref. 1 μPa at 1 m), and the FOM is 24 J/m³ (40). Another interesting design is the double-ended vibrator; a radiating piston replaces the tail mass so that the transducer is symmetric around its midpoint. This transducer competes favorably against the flextensional in the 300 Hz range because of its simpler, cheaper design, the lack of fatigue-related problems, and the lower voltages required for operation. More details on this device can be found in Refs. 40 and 56.

Ring Actuators. In a general sense, ring transducers employ the radial vibrations of tubes or plates. This concept was first developed during the late 1920s (23). The interest in ring transducers during those early days was based on the ruggedness and lower cost compared to other available transducer technologies. One example of those early magnetostrictive devices is the radially vibrating cylinder patented by Hayes in 1935 (57).

The ring transducer concept was extensively researched, not only with magnetostrictive materials as the active driver but also with piezoelectric ceramics (58,59). Today, the better coupling factor and lower elastic modulus of the rare earth iron alloys gives these materials an edge over conventional piezoelectrics.

For instance, the square ring transducer in Ref. 60 is a device capable of providing either omni- or directional sound propagation at low frequencies. Four Terfenol rods are arranged forming a square; the radiating surface consists of four curved pistons attached to each corner of the square, as shown in Fig. 12. A similar principle was utilized in the octagonal ring transducer discussed in Ref. 61.

One interesting feature of the square-ring transducer is its ability to generate either dipole or omnidirectional sound ra-

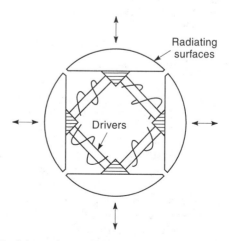

Figure 12. Schematic representation of a square-ring transducer.

diation. The dipole mode is accomplished by switching the magnetic bias on two of the rods and maintaining a constant direction on the ac magnetic field on all four rods. This feature translates into a transducer having two effective resonant frequencies and Q values, one for each mode of operation.

Motion/Force Generation Devices

There is growing interest in the use of magnetostrictive devices as a source for motion and/or structural forces. A variety of magnetostrictive motors have been designed, with the objective of providing accurate actuation at a given force rating over a range of operating frequencies. Conventional electromagnetic and hydraulic devices are commonly used in combination with gear boxes for motion conversion, and as a result are prone to mechanical play and bulkiness. Magnetostrictive motors produce outputs comparable to many conventional systems without the need for a gearbox interface, thereby avoiding many of the added system design issues (mass, volume, wear, play, backlash, etc.).

In this section we classify magnetostrictive motors as linear and rotational, depending on the type of motion provided by the device.

Linear Motors. The simplest type of linear motion device, a piston actuator, was presented in Fig. 1. Piston devices output mechanical strains and forces over bandwidths from dc to over 5 kHz. Piston actuators have been employed in a variety of applications, ranging from linear positioners and control of valves to achieving active vibration control in structures in single-ended configurations (62–65) and as active struts in double-ended configurations (66). The stroke of these devices is limited to the strain of the magnetostrictive driver element. The stroke can be increased by going to a longer device or using any of a variety of simple motion amplification devices, with the drawback of introducing play.

Magnetostrictive motors capable of displacement strokes greater than the core saturation strain include the elastic wave motor (EWM), or Kiesewetter motor (67). The EWM consists of a cylindrical Terfenol-D rod moving in inchwormlike fashion inside a stator tube. The magnetostrictive core is snugly placed inside the stator tube; the stator is in turn surrounded by several short coils located along its length. As the first coil is energized, the magnetostrictive element lengthens locally, while its diameter decreases by virtue of the volume invariance of the magnetostrictive effect. When the field is removed, the rod clamps itself again inside the tube. By energizing the coils sequentially along the length of the stator, it is possible to induce a displacement d_s per cycling of the coils as indicated in Fig. 13. The length of the step d_s can be adjusted by varying the magnetic field intensity generated by the coils. The speed of the motor is regulated by the speed at which the coils are energized and by the shielding effects of eddy currents. Normally the coils are excited with a single traveling pulse, but increased speed can be achieved by multiple-pulse excitation. Reversal of the direction of motion is achieved by reversing either the bias polarization or the direction of the traveling pulse. The device load handling rating is constrained by either the interference forces generated between the core and the stator tube or the blocked force capability associated with core elongation. A

prototype designed for use in the paper industry develops 1000 N of force, 200 mm of stroke, and 20 mm/s speeds (68). Other areas of application for this device are valve control and precision positioners. The EWM has also been utilized for active control of the profile of aircraft wings. The proof-of-concept device in Ref. 69 facilitates fuel savings in aircrafts of 3–6%, by changing the profile of the "smart trailing edge" at speeds of 0.4 mm/s and strokes of 25.4 mm.

A variant of the inchworm principle uses fixed-position and translating transducer elements to simultaneously clamp and unclamp adjacent sections of a load shaft. When the translating clamps are engaged, pusher transducers are actuated to move the translating clamps from their resting position. When the fixed position clamps are engaged, the pusher elements return the (now unclamped) translating elements to their resting position. The fixed position clamps maintain the load shaft position during the return portion of the translating clamps' motion cycle. By sequential clamping and unclamping, bidirectional linear motion of the load shaft is in-

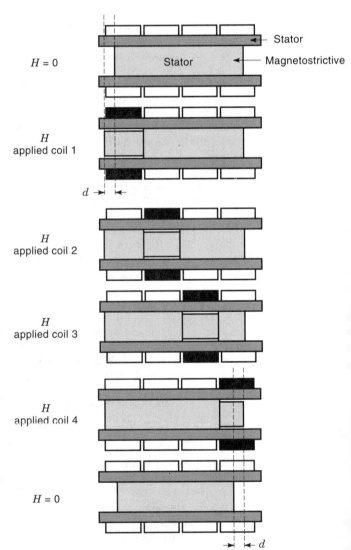

Figure 13. Principle of operation of the Kiesewetter inchworm motor. Energized solenoids are marked in black; passive solenoids are white.

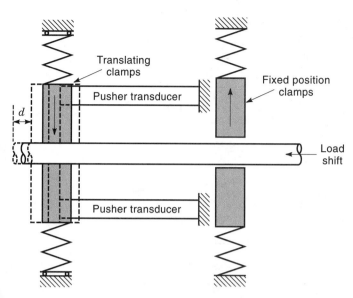

Figure 14. Principle of operation of the hybrid piezoelectric/magnetostrictive inchworm actuator.

duced as illustrated in Fig. 14. The device load handling rating is constrained by either the clamping forces generated between the load shaft and clamping transducer elements or by the blocked force capability associated with the pusher transducers.

Some researchers have demonstrated hybrid devices, in which both piezoelectric and magnetostrictive devices are employed. In Fig. 14 the clamping is done with piezoelectric stacks while translation is provided by Terfenol-D devices. A prototype motor of this kind is presented in Ref. 70. Designs employing an inchworm design using magnetostrictive clamps and piezoelectric pusher elements are also being studied (71). The attraction for design of hybrid devices arises when the motor is driven at its electrical resonance, characterized by the inductive and capacitive nature of the magnetostrictive and piezoelectric materials. The 180° phase lag between the piezoelectric stacks, and the 90° phase lag between the inductor current and the capacitor voltage, provide natural timing for the clamping, unclamping, and linear motion actions. In addition, the amplifier design is greatly simplified because of the significant reduction of the reactive power load (transducer impedance is almost purely resistive). The Fig. 14 prototype achieves a stall load of 115 N and no-load speeds of 25.4 mm/s. A similar idea (72) uses a magnetostrictive bar as a load shaft, allowing a very compact design. When used in conjunction with a dedicated switching power drive, the predicted speed of the motor is 7.8 mm/s, at a frequency of 650 Hz and force of 130 N. Other examples of linear magnetostrictive motors are the push-pull actuator by Kvarnsjo and Engdahl (73), the oscillating level actuator by Cedell et al. (15), and the broadband shaker by Hall (21).

Rotational Motors. The magnetostrictive principle has been lately employed in rotational motors as well. Several areas such as the aerospace and the automotive industries benefit from the higher controllability of the "smart material" motors compared to conventional hydraulic or electromagnetic devices. In particular, there is also a niche for high torque, low speed motors, in which the inchworm technique is particu-

larly suitable. A device having these characteristics, capable of a maximum torque of 3 N · m and speed of 3°/s, is presented by Akuta in Ref. 74. Another working prototype of the inchworm type is that presented by Vranish et al. (75), a quite sophisticated design capable of very high torque of 12 N · m at speeds of 0.5 rpm and precision microsteps of 800 microradians. Despite the great positioning accuracy and high holding torques, to date the inchworm-type devices sometimes lack in efficiency, which can represent a severe limitation in some applications. To overcome this, a resonant rotational motor based on a multimode stator has been proposed by Claeyssen et al. (76). Two linear Terfenol-D actuators are used to induce elliptic vibrations on a circular ring (stator), which in turn transmits rotational motion to two rotors pressed against the ring. The prototype did not achieve the expected deficiencies, but at 2 N · m of torque and speeds of 100°/s, the potential of this idea has been clearly demonstrated. The vibrations of the stator are illustrated in Fig. 15.

A proof-of-concept hybrid magnetostrictive/piezoelectric rotational motor is illustrated in Fig. 16, following the concept presented in Ref. 77. A piezoelectric stack clamps a piece of friction material onto the rotating disk, while two magnetostrictive rods move the clamp tangentially to the disk to achieve rotational motion. The direction of induced motion is bi-directional, determined by the direction of excitation of the oppositely connected driving solenoids. Once again, the inductive and capacitive nature of the transducer's electrical impedance is used to minimize the reactive power requirement on the power amplifiers, and the timing is naturally determined by the time lag between inductors and capacitors. Thus operation near electrical resonance is ultimately desired. The prototype shown in Ref. 77 achieves near 4 rpm at excitation frequencies between 650 Hz and 750 Hz and voltages of between 30 V and 40 V.

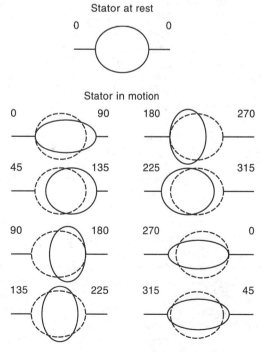

Figure 15. Modes of vibration of the rotational motor of Claeyssen et al. (76).

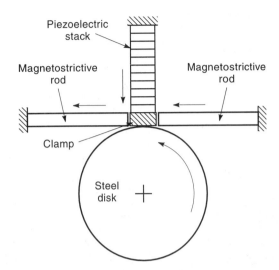

Figure 16. Hybrid piezoelectric/magnetostrictive rotational motor [after (77)].

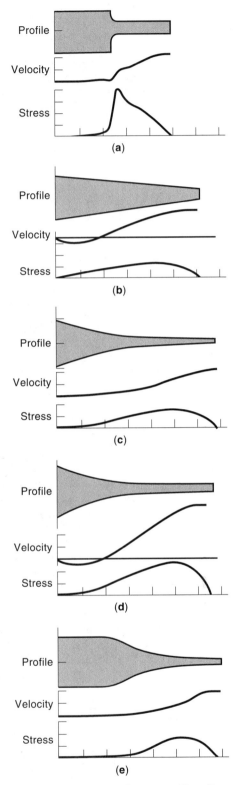

Figure 17. Geometry, velocity, and stress profiles of horns for amplifying the displacement output of ultrasonic transducers. (a) Stepped (b) conical, (c) exponential, (d) catenoidal, (e) Fourier. (78) Reproduced with permission, copyright © John Wiley & Sons.

Ultrasonic Devices

Some of the rotational motors presented in this section lend themselves very well to ultrasonic frequencies. These devices, along with other types of ultrasonic transducers, are discussed in the next section.

Ultrasonics is the science of acoustic waves with frequency above the human audible range, nominally above about 20 kHz. This definition is somewhat arbitrary, since most sound propagation phenomena behave the same at sonic and ultrasonic frequencies. However, the distinct features of ultrasonic devices merit a separate study from other types of transducers. For instance, one major problem in the generation of airborne ultrasound is the difficulty associated with coupling the energy generated by the electroacoustic device into the medium due to the large impedance ratio mismatches between the transducer and the medium (on the order of 10^{-5}). At very high frequencies in the megahertz range, the ultrasonic energy can be concentrated by shaping a thin film transducer into a bowl shape. The center of curvature represents the energy focus. An alternative way, usually employed at frequencies in the tens of kHz range, is by using a half-wavelength metal horn (the length of the horn is one-half the wavelength of elastic waves at the transducer's resonant frequency). The tapered shape of the horn produces an increase in the displacement, velocity, and pressure amplitudes, as illustrated in Fig. 17 (78). Note each profile has its own modal characteristics, thus the right shape must be carefully chosen for the application in hand.

Traditionally, the most common magnetostrictive materials used for ultrasonic applications were pure nickel and ironcobalt alloys, such as Permendur (49% iron, 49% cobalt, 2% vanadium). The magnetostrictive material is shaped so that good closure of the magnetic flux is achieved, and several laminas of the material are stacked together to minimize eddy current losses. The power losses due to eddy currents are proportional to the square of the operating frequency, so at ultrasonic regimes their significance is high. The material is as usual driven with solenoids. These transducers are normally operated at the half-wave fundamental frequency, which means there is a node in the center plane of the magnetostrictive element. This plane, where no motion is produced, constitutes a good mounting point for the transducer. Alternatively, if concentration of acoustic energy is sought through the use of a horn, the mounting point can be moved to the horn's node. This situation is illustrated in Fig. 18 (79).

Devices similar to that of Fig. 18, with or without the coupling horn, are being used in diverse applications such as cleaning of intricate or hard-to-reach parts, emulsification and homogenization of immiscible liquids, atomization of liquids or inks, particle agglomeration, degassing of liquids, catalysis of chemical reactions, machining, welding of plastics and metals, medical therapy and surgery, metal casting, dispersion of solids in liquids, and foam control (80–83). The underlying principle behind each of these processes can be related to cavitation effects, heat generation, mechanical effects, diffusion, stirring, and chemical effects. Of these effects, cavitation is perhaps the one that plays an integral role in most processes. As the sound wave propagates within the medium, individual particles are alternatively subjected to compression and refraction, although their relative positions do not change. For large sound wave amplitudes the magnitude of the negative pressure in the rarefraction areas, associated with the low-pressure half-cycle of vibration of the transducer, produces micron-size gaseous cavities to be formed. These cavitation bubbles grow until an unstable state is reached during the high-pressure half-cycle of operation, in which the bubbles collapse giving rise to huge pressure gradients as the liquid rushing from opposite sides of the bubble collides. The implosion event generates a microjet which impinges at high velocity on the surface to be cleaned. It has

been calculated that locally the pressure rises to about 10 ksi, with an accompanying temperature rise of about 10,000 K (80). The combination of high speeds, pressure, and temperature, together with conventional cleaning agents, frees contaminants from their bonds with the surface.

Magnetostrictive transducers utilizing nickel as the active element are currently used for industrial cleaning and degassing of liquids, at operating frequencies of 20 kHz to 50 kHz (Blue Wave Ultrasonics, PMR Systems, Pillar Power Sonics). For dental and jewelry applications the operating frequencies exceed 50 kHz. Although piezoelectric transducers are sometimes preferred for megahertz-range ultrasonic generation, the ruggedness and durability of magnetostrictive devices constitutes a very desirable characteristic. In addition, magnetostrictive materials do not need to be repolarized when accidentally heated beyond the Curie point, as is the case with piezoelectrics.

Details on ultrasonic half- and quarter-wave prototype transducers using a GMPC (giant magnetostrictive powder composite) material can be found in Ref. 15. The half-wavelength unit was tested for liquid atomization, whereas the quarter-wavelength unit was used for ultrasonic washing and, when coupled to a 10:1 horn, for welding of plastics and cutting of wood.

Both nickel and PZT materials are "high-Q" materials, i.e., their characteristic response exhibits a sharp peak at resonance, providing a very narrow range of frequencies over which the material is capable of transducing large amounts of energy. The output and efficiency decrease drastically at operating frequencies away from the fundamental resonant frequency. This effect can prove very limiting in applications where the system resonance is prone to varying, such as due to load changes. On the other hand, Terfenol-D is a low-Q ($Q \leq 20$) material and hence it overcomes much of the bandwidth limitations of nickel and PZTs. Hansen reports (81) current progress in Terfenol-driven ultrasonic transducers for surgical applications (348 μm displacement at 1 MHz), and for degassing of liquids for the bottling industry. Ultrasonic welding and engraving using Terfenol are also being studied. A 15 kW, 20 kHz ultrasonic projector is currently being developed in an Etrema-NIST research effort (4). The projected areas of applicability of the device are devulcanization of rubber for recycling, catalysis of chemical reactions, and ultrasonic treatment of seeds to improve yield and germination efficiency.

Another field of high research and commercial interest is that of ultrasonic motors. The use of ultrasonic motors is being extensively explored in robotics, aerospace, automotive, and consumer applications. One visible example at the consumer level is the ring-type ultrasonic motor used for autofocusing camera lenses. Many kinds of ultrasonic motors have been developed, but two main categories can be distinguished (84): the vibrating driver type, and the progressive wave type. The vibrating driver uses the elliptical motion of a PZT driver, whereas the progressive wave motor uses the frictional force between the PZT driver and translator. Either type presents attractive features such as fine controllability, good efficiency, and compact size. However, the driving force is restricted by the friction generated between the moving surfaces, and most importantly, by the shearing strength of the PZT. To overcome these limitations, Akuta (74) developed a rotational actuator using Terfernol-D to achieve large am-

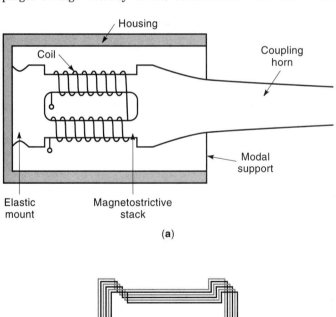

Figure 18. (a) Ultrasonic transducer with coupling exponential horn (79). (b) Detail of magnetostrictive stack. The stack can be attached (by silver brazing for instance) directly to the resonating diaphragm in the washing tank (79).

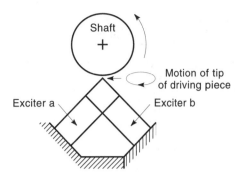

Figure 19. Terfenol-D ultrasonic rotational motor, after Akuta (74).

plitude elliptical vibrations, high torque, and fine angular controllability. The motor, illustrates in Fig. 19, has a speed of 13.1 rpm, and maximum torque of 0.29 N · m at a rotational speed of 1300 Hz.

Recent research (85) indicates it is feasible to produce remotely powered and controlled standing wave ultrasonic motors (SWUMs) using thin film rare earth–iron alloys. The motor is completely wireless, and the excitation coil can be placed away from the active rotor. This represents a unique feature, not possible to achieve with piezoelectric technology, which can be used for intrabody distribution of medication, micropositioning of optical components, and actuation of small systems such as valves, electrical switches, and relays. A linear motor with these characteristics producing speeds of 10 to 20 mm/s and a rotational unit producing 1.6 μN · m of torque at 30 rpm were built.

Other Actuator Applications

We include a list of miscellaneous applications which take advantage of the large strains and force characteristics of magnetostrictive materials: wire bonding clamp for the semiconductor packaging industry (86); dexterous force reflection device for telemanipulation applications (87); hearing aid device based on the generation of low amplitude vibrations to bone, teeth, and similar hard tissue using Terfenol-D (88); magnetostrictive borehole seismic source (89); system for high-frequency, high-cycle fatigue testing of advanced materials using a magnetostrictive actuator (6,90); magnetostrictive fuel injection valve (15). An interesting review of some additional actuators is given by Restorff in Ref. 91.

MAGNETOSTRICTIVE SENSOR APPLICATIONS

The focus of this section will be magnetostrictive sensors, devices that rely on a material's magnetoelastic properties to convert a physical dimension into an electrical signal that can be processed and transmitted. The sensor system includes the sensor components themselves, the target, and supporting electronics. Some sensor systems use the inherent magnetoelastic properties of the target to effect the detection of the property of interest. For example, magnetoelastically induced elastic waves can be used to monitor a ferromagnetic specimen, such as pipes or the reinforcement bars (rebar) used in concrete structures. In other sensor configurations, the magnetostrictive properties of part of the sensor allow the measurement of the property of interest. In particular, we focus

on sensor designs and configurations that have some experimental verification.

SENSING EFFECTS

Magnetostrictive sensors can be divided into three groups based on how the magnetomechanical properties of the system components are used to measure the parameters of interest: (1) passive sensors, (2) active sensors, and (3) combined sensors. Passive sensors rely on the material's ability to change as a result of environmental stimuli to make measurements of interest. Passive sensors use the magnetomechanical effect such as the Villari effect to measure external load, force, pressure, vibration, and flow rates. Active sensors use an internal excitation of the magnetostrictive core to facilitate some measurement of core attributes that change in response to the external property of interest. For example, the magnetic permeability of Terfenol-D is sensitive to temperature. With proper calibration, temperature can be determined by measuring changes in the strain produced through excitation of a Terfenol-D sample with a known applied magnetic field.

Designs that employ two coils, one to excite the magnetostrictive component and one for measurement, are known as transformer-type sensors. The most common active sensor design mentioned in literature is the noncontact torque sensor. This employs variations on a general theme of using a magnetostrictive wire, thin film, or ribbon wrapped around or near the specimen that is subject to a torque. The change in the magnetic induction can then be related to the torque on the specimen.

Finally, combined sensors that use Terfenol-D as an active element to excite or change another material have been developed. They will allow measurement of the property of interest. For example, a fiber optic magnetic field sensor uses the change in length of a magnetostrictive element in the presence of a magnetic field to change the optical path length of a fiber optic sensor. There are numerous examples of combined sensors, including those to measure current, shock (percussion) and stress, frost, proximity, and touch. Stress can be measured using photoelastic material, and highly accurate displacement measurements can be made with the help of a magnetostrictive guide. Fiber optics and diode lasers have been used with magnetostrictive elements to measure magnetic flux density (magnetometers) (92). NDE (nondestructive evaluation) applications have also been developed, such as a corrosion sensor for surveying insulated pipes. Applications are discussed later in terms of the measured property or quantity: torque, magnetic field, material characterization, motion, force, and miscellaneous characteristics.

Torque Sensors

Torque measurement has great benefit in a variety of applications and industries, including the automotive industry and high-speed machining. Magnetostrictive torque sensors are traditionally based on the inverse magnetostrictive effect (Villari effect), where a torque-induced change in stress in the target causes a change in the magnetization of a magnetostrictive element in the sensor-target system. This change in magnetization can be measured directly (passive) or as a change in permeability with active excitation (active). A thorough overview of torque sensor technology, focusing on magnetostrictive torque sensors, is given in papers by Fleming (93).

Noncontact Torque Sensors. One of the most common torque sensor configurations is the noncontact torque sensor. The change in the stress in a ferromagnetic shaft is measured by detecting the change in permeability of a ferromagnetic element flux linked to the shaft. Figure 20 shows the basic design components, which include a C-shaped ferromagnetic core with an excitation and detection coil. Two variables in the sensor system are the excitation coil current I and the air gap between the shaft target and the sensor core. Noncontact torque sensors are advantageous because implementation is simple and fast. The performance is highly stable, accurate, and sensitive. Fleming investigated several issues with these sensors, including nonlinearities caused by sensor element properties, the effects of magnetic saturation, and excitation frequency (94).

Thin-Film Torque Sensors. An alternative torque sensor configuration is to apply the magnetostrictive material directly to the target. This idea was developed by Yamasaki, Mohri, and their collaborators, who used a wire explosion spraying technique to adhere thin layers of Ni, Fe–Ni, and Fe–Co–Ni to shafts (95). In this method, the conductive wire is exploded at a high temperature into fine particles that adhere strongly to the shaft. The wire explosion technique results in a sufficiently strong adhesion that they report increased sensor reliability relative to competing technologies that rely on epoxies to fix magnetostrictive material to a surface. When the shaft is twisted, stress in the film causes a change in its magnetization-applied magnetic field hysteresis loop. When two such regions of magnetostrictive material are surrounded by coils connected in a multivibrator bridge circuit (96) a change in voltage caused by a change in the torque can be detected. A linear, nonhysteretic relationship between the bridge circuit output voltage and the torque was obtained, with Ni and $Fe_{42}–Ni_{58}$ providing the greatest sensitivity.

A second system using a 300 μm thick Ni layer applied by plasma jet spraying was investigated by Sasada et al. (97). The instantaneous torque on a rotating shaft was related to

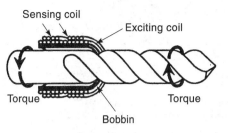

Figure 21. Noncontact measurement of torque on drill employing one excitation coil and two sensing coils one over the shank and one over the flutes (98) © IEEE 1994.

the magnetic circuit permeability measured using a pair of U-shaped magnetic heads positioned at ±45° from the shaft axis. The sensitivity of the measurements to the air gap between the shaft and the magnetic heads was examined, and a self-compensating method was presented. A relatively linear sensitivity of 500 V/N · m with little hysteresis was measured. In addition, the output response of the sensor was found to be nearly independent of the rotational frequency.

Sensor Shafts. These applications are examples of taking advantage of the magnetostrictive effects of the target material itself. In the first example, the in-process detection of the working torque on a drill bit is related to the drill permeability. An excitation coil surrounds part of the drill including the shank and the flutes, as in Fig. 21 (98). Two sensing coils, one positioned over the flutes and one over the shank, are connected in series opposition allowing the measurement of the permeability. The permeability of the shank is less sensitive to changes in torque than the flutes, and the difference in the output voltages of the two coils changes in proportion to the applied torque. In the second example, a sensor shaft was made of Cr–Mo steel suitable for automobile transmission applications. Two grooved sections are surrounded by coils, which are configured in an ac bridge circuit. When torque is applied to the shaft, the Villari effect results in a change in impedance measured by the bridge. According to Shimada et al., this sensor design is robust with respect to temperature (99).

Motion Sensors

Bidirectional Magnetostrictive Transducers. The existence of the Villari effects makes it possible for a magnetostrictive transducer, such as those described in the sonar and motion/force device sections to have two modes of operation, transferring magnetic energy to mechanical energy (actuation) and transferring mechanical energy to magnetic energy (sensing). As with many other transducer technologies such as electromagnetic (moving coils) and piezoelectricity, magnetostrictive transduction is reciprocal, and a transducer has the ability to both actuate and sense simultaneously. Applications such as the telephone and scanning sonar use this dual mode. For example, Terfenol-D sonar transducers can be used as either a transmitter or receiver or both at the same time. Another potential use of dual-mode operation is in active vibration and noise control. One transducer can be used to sense deleterious structural vibrations and provide the actuation force to suppress them. This approach to active control provides what is called colocated sensing and actuation, which aids in ensuring

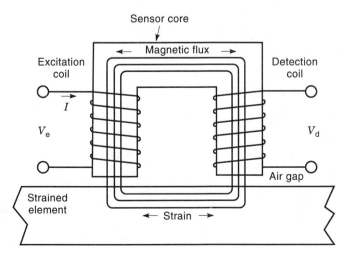

Figure 20. Noncontact torque sensor with excitation and detection coils around legs of C-shaped ferromagnetic core (32).

control system stability. Self-sensing control uses the sensed signal in a feedback loop to drive the transducer. Numerous papers have described systems based on this effect and shown its effectiveness (66,100,101).

Fenn and Gerver have developed, tested, and modeled a velocity sensor based on a permanent magnet biased Terfenol-D actuator (102). The output from the device was a voltage induced in a coil surrounding the monolithic Terfenol-D core, which was proportional to the time rate of change of the strain in the Terfenol-D core and ultimately the velocity of the connected target. Peak sensitivities of 183 V/m/s were seen when the coil was left open. In addition, the coil could be shunted to provide passive damping capability, and the voltage proportional to the target velocity could be monitored across the shunt resistor.

Noncontact Magnetostrictive Strain Sensors. Noncontact magnetostrictive strain sensors (NMASSs) use the magnetic field to couple the straining target to the sensing element. A noncontact system has several advantages (103). First, it is a noninvasive technique; that is, it does not require mechanical bonding to the target. This is a significant advantage for measuring strain of rotating targets. Second, the sensor can be moved to measure strain at different points easily and quickly, perhaps providing a three-dimensional strain-mapping capability. Finally, this sensor is rugged, with good sensitivity and overload capacity. Figure 20 shows the general configuration of the sensor (103) with the C-shaped ferromagnetic sensor core wound with an excitation and sensing coil. The flux path crosses the air gap to the target. Because of the magnetoelastic effect, strain in the target causes a change in the magnetic circuit permeability, which will be seen as a change in the sensed voltage. In general, the sensor will detect changes in strain of ferromagnetic materials; however, strains in nonferromagnetic materials can detected by applying a ferromagnetic layer on the surface. This sensor configuration has also been used to measure torque in shafts because the principal strains are 45° from the shaft axis.

Magnetostrictive Waveguide Position Sensors. The magnetostrictive waveguide position sensors detect the position of a permanent magnet connected to the target, which is free to move along the length of a magnetostrictive waveguide. An emitter is used to send a current pulse continuously through the waveguide, which produces a circumferential magnetic field. This combined with the longitudinal magnetic field produced by the permanent magnet results in a helical magnetic field.

The Wiedemann effect described earlier then results in a torsional strain pulse; the triggered torsional acoustic wave travels at the speed of sound in both direction away from the permanent magnet along the waveguide. One end of the waveguide is fitted with a receiver; at the other end a damper attenuates the acoustic wave so that it will not reflect back corrupting the signal at the receiver and avoiding the development of standing waves. The receiver measures the time lapse between the current pulse and the acoustic wave, which is related to the distance between the receiver and the permanent magnetic/target. The acoustic wave can be measured by the change in permeability resulting from the strain pulse in

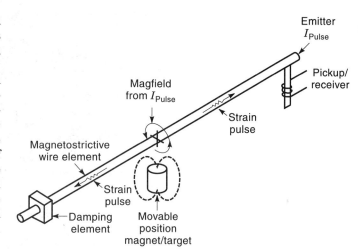

Figure 22. Magnetostrictive waveguide position sensor (93).

the waveguide. In Fig. 22, the receiver, or pick-up element, is shown as a magnetostrictive ribbon welded to the waveguide that converts the torsional pulse to a longitudinal elastic pulse. The permeability of the ribbon, which changes as a result of the elastic pulse, is monitored via Faraday's Law with a coil wrapped around the ribbon. A piezoelectric element can also be used as the receiver to measure the acoustic wave.

Several versions of this sensor are available commercially and discussed in the literature. Nyce describes the operation of MTS model LP in detail (104). Current pulses with frequencies between 10 Hz and 10 kHz provide the excitation. The sensor performance is considered as a function of the magnetostrictive waveguide material (high magnetostriction, low attenuation, and temperature stability are desired), and geometry. Lucas Control System Products (United Kingdom and United States) has developed the MagneRule Plus, a compact position sensor for measurements up to 120 in. (305 cm) with high linearity and repeatability. In addition, it can measure fluid levels by connecting the permanent magnet to a float, which is then placed in the fluid to be monitored. Finally, Equipel (France) manufactures and sells Captosonic position sensors capable of measuring distances up to 50 m with ±1 mm accuracy (105).

Magnetoelastic Strain Gages. Using the fact that the permeability of many magnetostrictive materials is stress sensitive, a strain gage made from strips of Metglass 2605SC has been developed by Wun-Fogle and associates (106). The permeability of the ribbons decreased in tension and increased in compression. The ribbons were prepared by annealing in a transverse 2.6 kOe magnetic field at 390°C for 10 min and then rapidly cooled in a saturation magnetic field. To maintain their high sensitivity, the ribbons must be strongly bonded to the target in an initially stress-free condition. A highly viscous liquid bond was found to bond the ribbons to the target adequately, although it did result in a loss of dc response. For experimental verification, two ribbons were placed on the top and bottom of a beam and surrounded by two coils connected in opposition. The ribbons were excited by an external magnetic field of 1 kHz. The net voltage in the coil system was related to changes in permeability and, hence, changes in strain of the ribbons. The strain measurement system resulted in a figure of merit, given by $F_\mu = (\partial\mu/\partial\epsilon)/\mu$, of 4×10^5, which compares favorably with conventional strain gages. It was found that strains on the order of 10^{-9} could be measured at 0.05 Hz.

Magnetostrictive Delay Lines (MDLs). An acoustic pulse is propagated through the magnetostrictive delay line and detected by a receiving unit. A current pulse through a conductor (PCC) orthogonal to the magnetostrictive delay line (MDL) generates a pulsed magnetic field in the MDL, which generates an elastic wave, see Fig. 23. An active core (AC) of soft magnetic material, placed near the PCC, is connected to the target and is free to move relative to the MDL. The magnetic pulse and hence elastic wave generated in the MDL is sensitive to the magnetic coupling between the AC and MDL. As the target and AC move away from the MDL, the magnetic coupling between the PCC and MDL increases so that the magnetic pulse and elastic wave in the MDL increases in strength. The output to the sensor is the pulse generated in the receiving unit coil (RC) as described by Faraday's Law. The output voltage induced in the RC is sensitive to the gap distance between the MDL and AC; for MDL-AC, displacement is less than 2 mm. Most importantly, the output is fairly linear and anhysteretic with respect to the MDL-AC displacement. Sensitivities of 10 μV/μm have been reported using 24-μm thick Metglas 2605SC amorphous ribbon as the MDL (79). Permanent magnets were also used to maximize the generation of the acoustic pulse and the measured voltage in the RC. Multiple AC-PCC elements can be used with one MDL to form an integrated array. This novel aspect of the sensor system and several AC-PCC-MDL configurations are discussed by Hristoforou and Reilly (107). Applications for this sensor include tactile arrays, digitizers, and structural deformations.

Force Sensors

Magnetostrictive Delay Lines. The magnetostrictive delay line displacement sensor configuration has been modified to produce a force distribution sensor (108,109). A force applied directly to the MDL will distort the acoustic signal by the emitter as described previously. The change in the acoustic wave measured by a receiver coil is related to the force applied to the delay line. An experimental device tested by Hristoforou and Reilly (108) used a Metglas 2605SC FeSiBC amorphous ribbon as the delay line embedded in a fiberglass channel. The channel bends under an applied force, stressing

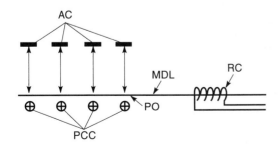

Figure 23. Magnetostrictive delay line with active core connected to target, pulsed current conductor, receiving coil, and point of origin (PO) for acoustic stress (79) © IEEE 1994.

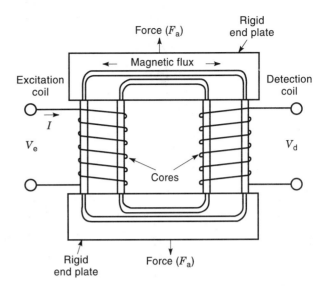

Figure 24. Magnetostrictive sensor for measuring force, composed of two ferromagnetic cores, one with an excitation coil and the other with a sensing coil, between two rigid end plates (85).

the MDL. In addition, the channel ensures that the PCC oriented perpendicular to the MDL ribbon on the bottom of the channel does not move relative to the MDL. For a given current, the voltage detected by the receiver coil caused by a force F is proportional to the e^{-cF}, with calibration constant c. Integrated arrays for measuring force can be constructed with multiple MDL (each with a receiver) and PCC oriented perpendicularly. The values of multiple forces on the two-dimensional array can be backed out from the voltages measured by the receivers.

Magnetostrictive Force Sensor. Kleinke and Uras describe a force sensor that employs the change in electrical impedance of a coil that is flux linked to a magnetic circuit to measure the stress or force acting on a magnetostrictive component in the magnetic circuit (110). It is similar in construction to the noncontact magnetostrictive strain sensor discussed earlier. However, rather than having a C shape, two magnetostrictive cores are held in place by rigid end pieces, see Fig. 24. A coil surrounds each core, one of which is used for excitation and the other for sensing. A constant amplitude ac is impressed in the excitation coil generating an oscillating magnetic field. This results in a voltage in the detection coil with a magnitude proportional to the time rate of change of flux linking the detection coil. An applied force on the sensor will cause a strain change in the magnetostrictive cores resulting in a change in the core magnetization. In this mode, where the magnetomotive force is kept constant, a change in the output voltage from the detection coil is linearly related to the change in force. In a constant flux operation mode, the excitation current is allowed to vary in order to maintain a constant detection coil output voltage. In this case the change in excitation current is related to the change in force. Compared with conventional force transducers such as those that employ strain gages, this force sensor is simpler, more rugged, relatively inexpensive, and employs simpler electronics.

Amorphous Ribbon Sensors. A tensile force sensor based on the strong Villari effect of amorphous ribbons, such as Ni-,

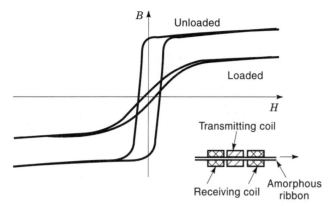

Figure 25. Amorphous ribbon force sensor with one excitation coil and two receiving coils so that changes in permeability caused by a force are detected (129).

Fe-, and Co-based alloys, has been described by Seekircher and Hoffman (111). A transmitting coil excites the ribbon while a pair of detection coils measure the maximum induction, which is dependent on the stress as shown in Fig. 25. Loads below 4 N were measured with Co alloy ribbon, 25 μm thick by 3 mm wide with negligible hysteresis. The high Young's modulus of the ribbons results in low-displacement sensors, which can be load-bearing elements. In some cases, temperature compensation is required.

Numerous torque sensor designs, patents, and literature are also available. Highly sensitive shock-stress sensors employing iron-rich amorphous ribbons are described in detail by Mohri and Takeuchi (112). A combined torque-force sensor has also been developed (113). The change in permeability (or magnetic flux) in a magnetic circuit resulting from strain in an element of the circuit can be used to measure both torque as described previously and force.

Material Characterization Magnetostrictive Sensor

Figure 26 shows a noncontact sensor that uses the magnetostrictive properties of the target material to excite elastic waves that can be measured and monitored for use in characterizing the target material properties (114–116). The system can be used only with ferromagnetic material that has a magnetoelastic response. The sensor consists of a transmitting coil (pulse generator, power amplifier, and bias magnet) surrounding the object, which generates the mechanical wave via

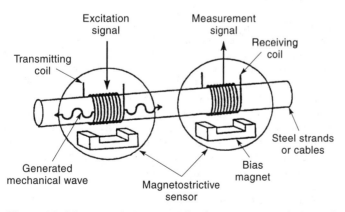

Figure 26. Magnetostrictive sensor for characterizing corrosion and monitoring ferromagnetic materials such as pipes and strands (130).

magnetostrictive excitation. A receiving coil (signal preamplifier, data acquisition hardware, and permanent magnet) located at a distance from the excitation coil measures the signal produced by the waves. These signals can be used to characterize the material for corrosion, measure stress in strands, and so on. Signals are generated by changes in the material geometry. Experimentation has shown that the wave attenuation increases with the degree of corrosion. This method has been used successfully to identify corrosion in strands, reinforced bars (including those embedded in cement), water pipes, and other systems where noninvasive monitoring techniques are preferred.

Magnetometer

There are numerous designs for magnetic field sensors, including many which rely on magnetostrictive properties of component materials. These sensors vary considerably, in part because they are designed to detect magnetic fields of different strengths and frequencies (117).

The first magnetometer design, developed by Chung et al., employs a Terfenol-D sample to convert a magnetic field into a measurable quantity (118–120). A Terfenol-D rod strains in the presence of an ac magnetic field. This displacement can be measured accurately with a laser interferometer calibrated to output a signal related to the magnetic field. A dc magnetic bias is used to optimize the sensitivity of the Terfenol-D strain with magnetic field, resulting in values of up to 10 μL/L/gauss. In addition, the sensitivity was found to be a function of the mechanical prestress.

Other magnetic field sensors are also available. In 1979 Yariv and Winsor proposed a now common configuration that uses a magnetostrictive film coating applied to an optic fiber (121). The magnetic field causes the magnetostrictive film to deform, straining the optic fiber. This causes changes in the optical path length and the phase of the laser light passing through the optic fiber, which can be detected by an interferometer. Mermelstein shows that the resolution limit at dc to low frequency (less than 1 Hz) of such a sensor is approximately 3×10^{-11} Oe (122). Magnetostrictive amorphous metals, often in the form of ribbons, are extremely sensitive to external magnetic fields and have been used as the active element of a magnetometer (96). Finally, a magnetic field sensor has been developed based on magnetostrictive delay line technology (123).

Miscellaneous

Magnetostrictive sensors have been used to measure or monitor a number of other properties and characteristics. Some examples found in the literature include hearing aids (88), magnetoelastic delay line digitizers (124), magnetoacoustic keyboards (125), thermometers (126), biomedical monitors for lung ventilation (127) and spine movement (128), bimorphs for two-dimensional scanning elements (129), and composite cure monitors. An interesting review of additional sensors is given by Restorff in Ref. 91.

MAGNETOSTRICTIVE DEVICE MODELING

The increased use of magnetostrictive materials has not been accompanied by the availability of complete and accurate transducer models. Complete models must incorporate the different operating regimes present in magnetostrictive transducers in a manner compatible with the requirements of specific applications. The regimes present in magnetostrictive transducers are electric (which involves routing of magnetic field through magnetic circuit for magnetization of the magnetostrictive core), mechanical (elastic state of the materials present in the transducer), and thermal (the temperature distribution in the transducer materials). Extensive experimental evidence (21,28,31) demonstrates that it is crucial to consider the interaction between these regimes; therefore, models for the various regimes need to be coupled to provide a complete description of a transducer system.

Most of the currently available models of magnetostrictive transducers fail to incorporate all regimes simultaneously. This is in part because transducer models are still predominantly based on the linear constitutive piezomagnetic equations [Eq. (1)] (14). These equations neglect the effects of coupling between stress, magnetization, and temperature in the material, which have a bearing on the mechanical performance as indicated in Ref. 35, among others.

Carman and Mitrovic (130) and Kannan and Dasgupta (131) extended the scope of the linear constitutive equations by including specific nonlinear effects. The nonlinear dynamics of magnetostrictive transducers has been also modeled following a phenomenological approach. One such approach is obtained through the use of generalized Preisach operators. Preisach models that specifically target giant magnetostrictive materials are discussed by Restorff et al. (132), Adly and Mayergoyz (133), and Smith (134). These models characterize different operating regimes but are cumbersome to implement in magnetostrictive transducers where performance changes significantly during operation.

Other modeling efforts have combined physical laws with phenomenological observation. Engdahl and Berqvist (135) present a model capable of dealing with nonlinearities and various losses, such as magnetomechanical hysteresis, eddy currents, ohmic heating, and mechanical losses. Physically based Preisach models combined the empirical robustness of nonphysical laws with some understanding of the physical processes that govern magnetic and magnetostrictive hysteresis. Basso and Bertotti (136), for instance, developed a Preisach-based hysteresis model for materials where the magnetization is dominated by domain wall motion.

Nevertheless, there does not exist a fundamental model applicable to the complete performance space of magnetostrictive transducers. The development of physics-based models that incorporate the different operating regimes and the issues inherent to them will certainly extend the utility of these transducers. In this section, some of the fundamental issues in magnetostrictive transducer modeling are identified, and some of the current state-of-the-art modeling techniques that handle these different issues are reviewed.

Linear Transduction

The linear transduction equations provide a "black box" model of the relationship between electrical and mechanical sides of the transducer through the transducer's total electrical impedance. These equations can be written

$$V = Z_e I + T_{em} v \tag{14a}$$

$$F = T_{me} I + z_m v \tag{14b}$$

where V and I are voltage and current across transducer leads, Z_e is the blocked electrical impedance, v is velocity, T_{em} and T_{me} are electrical due to mechanical and the mechanical due to electrical transduction coefficients, and z_m is mechanical impedance. Hunt, Rossi, and many others have authored authoritative texts on the subject of transduction covering the use of equivalent circuit models such as that presented in Fig. 8(b) and the transduction equations as applied to modeling of magnetostrictive devices.

Magnetization

Magnetization and the magnetoelastic interactions taking place inside the magnetostrictive core and other transducer components have been modeled in different manners. Transducer magnetic models can be formulated using standard techniques (see MAGNETIC CIRCUITS) (i.e., using equivalent circuit representations and/or finite element modeling approaches). The least well-defined component of such models is inevitably the magnetic state of the magnetostrictive core, which, referring back to Fig. 3, changes significantly with operating conditions.

One approach for modeling a magnetostrictive material is to consider the material at its micromagnetics level. Classical micromagnetics work by Brown (137) assumes that the material is simultaneously magnetizable and deformable with its energy state defined by elasticity, thermodynamics, and electromagnetic principles. Recent work in this field deals specifically with Terfenol-D (138). Although the modeling results are useful at the domain level, the extension of micromagnetic models to describe the performance of transducers is a major undertaking.

The Preisach model has been used extensively for characterization of ferromagnetic materials, and more recently for characterization of magnetostrictives as well. Much research effort has been devoted to accommodate physical aspects into the Preisach approach, which in its original form, cannot be traced back to fundamental principles. Reimers and Della Torre implement in Ref. 139 a fast inverse hysteresis model amenable to control applications and with applicability to magnetostrictive materials.

Following the classical anisotropy domain rotation model by Stoner and Wohlfarth (140), Lee and Bishop (141) extended their idea to a random assembly of domain particles having cubic anisotropy. Clark et al. (142) included compressive loading along the [111] easy axis direction in a two-dimensional scheme. Jiles and Thoelke (143) generalized this model to three dimensions, by considering the anisotropy, magnetoelastic, and field energies along all three directions. Although model results are in good agreement with measured data, the identification of the fractional occupancies, which define the participation of different easy axes in the total magnetization, is by no means trivial.

Another modeling approach that yields significant results is the ferromagnetic hysteresis model. Theory by Jiles and Atherton (144) predicts quasi-static magnetization-applied field (M–H) loops in ferromagnetic materials by considering the energy of domain walls as they bow and translate during magnetization. Later extensions include models of magnetostriction, eddy current losses, minor loops, and mechanical prestress (see Ref. 39). The appeal of the Jiles–Atherton model stems from the physical basis of the model parameters and the fact that only five parameters are needed for complete description of the magnetic state of the material. However, this model is purely magnetic in nature, which highlights the need for accurate and general magnetomechanical characterization laws to describe the transduction process in the transducer completely.

Stress

It is common practice to place giant magnetostrictive samples under a mechanical compressive prestress for operation. The ability of the material to survive high accelerations and shock conditions improves under compressive stress because, for example, Terfenol-D is much more brittle in tension (tensile strength $\approx$ 28 MPa) than in compression (compressive strength $\approx$ 700 MPa) (32). Under dynamic operating conditions in systems using the more brittle magnetostrictives such as monolithic Terfenol-D, the rated system output force is usually limited by the magnitude of the device prestress according to Newton's Second Law, where $F = ma = \sigma_o A$. Above the prestress, the output dynamic load will accelerate away from the magnetostrictive driver, only to return with potentially damaging results. In addition, adequate preloading is capable of improving the magnetic state in the material as a consequence of the coupling between the magnetic and mechanical states. Figure 9(c) shows transducer electrical impedance function sensitivity to changes in prestress.

Considering an energy balance in the material, the mechanical prestress is an additional source of anisotropy energy that competes against the magnetocrystalline anisotropy, strain, and applied field energies. The application of the compressive preload forces a larger population of magnetization vectors to align perpendicular to the direction of application of the preload, where a state of local minimum energy is reached. This translates into both a smaller demagnetized length and increased saturation magnetostriction. However, for compressive preloads larger than a certain value, the prestress energy overpowers the elastic energy produced by the material and the magnetostriction decreases.

The relationship $\epsilon-M$ in Terfenol-D has been reported to be nearly independent of prestress except for very low prestresses and high magnetizations. Clark et al. (145) report a range of about 15 MPa to 55 MPa for the $\epsilon-M$ stress independence in Bridgman-grown samples. In addition, work by Kvarnsjö (146) indicates that at high prestresses, the primary magnetization mechanism is domain magnetization rotation; as a consequence, the quadratic law for magnetostriction discussed earlier is appropriate. Clark et al. (145) found that the quadratic law is incorrect in the low prestress regime (below 30 MPa for statically loaded, single-crystal material) where it was found that strain depends directly on stress and that the magnetostriction cannot be accounted for by a single-valued magnetization law (i.e., hysteresis is observed).

Thermal Effects

Temperature effects can be incorporated in the linear piezomagnetic model given in Eq. (1). Furthermore, by considering higher-order interactions between magnetization, stress, and temperature, Carman and Mitovic (130) developed a model capable of producing results in good agreement with experimental data at high preloads. However, the model is not capable of predicting saturation effects. Following the lead of Hom and Shankar (147), Duenas, Hsu, and Carman (14) developed

an analogous set of consititutive equations for magnetostriction. These equations are given by

$$\epsilon_{ij} = s_{ijkl}^{M,T}\sigma_{kl} + Q_{ijkl}^{T}M_kM_l + \alpha_{ij}^{M}\Delta T \tag{15a}$$

$$H_k = -2Q_{ijkl}^{T}M_l\sigma_{ij} + \frac{M_k}{k|M|}\text{arctanh}\left(\frac{|M|}{M_s}\right) + P_k^{\sigma}\Delta T \tag{15b}$$

where ϵ is strain, s is compliance, Q is the magnetostrictive parameter, M and M_s are magnetization and saturation magnetization, a is coefficient of thermal expansion, T is temperature, H is magnetic field, P is the pyromagnetic coefficient, and k is a constant. Superscripts indicate the constant physical condition under which the parameter is measured, and subscripts indicate tensor order following conventional notation. This model successfully captures the quadratic nature of magnetostriction for low to medium field levels, but it does not completely describe the saturation characteristics. The model allows the incorporation of magnetostrictive hysteresis via complex model parameters.

Heat not only affects the magnetization processes in the magnetostrictive core but also has bearing on the overall transducer design. Together with magnetomechanical hysteresis and eddy current losses, ohmic heating in the excitation coil is perhaps one of the most significant sources of losses. Other thermal effects to consider in transducer design are thermal expansion of the magnetostrictive driver itself [for Terfenol-D, $\alpha \approx 12 \times 10^{-6}/^{\circ}\text{C}$ (32)], dependence of material properties and performance of Terfenol-D on temperature, thermal expansion of the coil (coils used in prototype transducers showed expansion above 1.5% when heated from room temperature to temperatures of around 100°C), thermal expansion of other transducer components, and the thermal range of the coil insulation. Maintaining controlled temperatures during operation is critical for certain transducer applications. One clear example of this is precision machining. Transducers for this type of application need to use efficient thermal sinks, either active (cooling fluids), passive (thermally conducting materials, superconducting solenoids), or a combination of the two.

Ac or Eddy Current Losses

Dynamic operation leads to additional complications in the performance of magnetostrictive transducers. One important loss factor to consider is the eddy (or Foucault) currents. As modeled by the Faraday–Lenz Law, eddy currents are set up in the transducer's conducting materials to resist magnetic flux changes. These currents produce a magnetic flux that resists the externally applied magnetic field and simultaneously cause a nonuniform distribution of current density often known as skin effect.

Classical eddy current power loss formulations assume complete magnetic flux penetration and homogeneous permeability throughout the material. This assumption is valid only for small material thickness (e.g., laminas); hence, it is invalid for solid, thick cylindrical transducer cores. The characteristic frequency above which the homogeneity of penetration of the magnetic flux is compromised, for cylindrical samples, is given by

$$f_C = \frac{2\rho}{\pi D^2 \mu^{\epsilon}} \tag{16}$$

where ρ is the resistivity of the material [$\rho \approx 0.6 \times 10^{-6}\ \Omega \cdot m$ for Terfenol-D (120)], D is the rod diameter, and μ^ϵ is the clamped permeability. For a 6.35 mm (0.25 in.) diameter rod, the characteristic frequency is about 5 kHz. Laminations in the magnetostrictive core, low operating frequencies, silicon steel end caps, and slit permanent magnets help to mitigate the effects of eddy currents.

A modeling approach based on energy considerations is shown by Jiles (148). This approach considers eddy currents as a perturbation to the quasi-static hysteresis. The simplicity of this model is, however, offset by the limitations imposed by its assumptions. Because uniform flux penetration is assumed, its applicability is limited to thinly laminated material or low operating frequencies.

Another classical approach to the eddy currents issue is the one presented by Bozorth (9) among others. In this model, a so-called eddy current factor X is used to account for the reduced inductance caused by the oppositely induced magnetic field. The complex quantity $X = X_r + jX_i$ can be written for cylindrical current-carrying conductors in terms of Kelvin ber and bei functions and is dependent upon frequency of operation f and the characteristic frequency f_c as follows:

$$X_r = \frac{2}{\sqrt{p}}\left(\frac{\mathrm{ber}\ \sqrt{p}\ \mathrm{bei}'\ \sqrt{p} - \mathrm{bei}\ \sqrt{p}\ \mathrm{ber}'\ \sqrt{p}}{\mathrm{ber}^2\ \sqrt{p} + \mathrm{bei}^2\ \sqrt{p}}\right) \tag{17a}$$

$$X_i = \frac{2}{\sqrt{p}}\left(\frac{\mathrm{ber}\ \sqrt{p}\ \mathrm{ber}'\ \sqrt{p} - \mathrm{bei}\ \sqrt{p}\ \mathrm{bei}'\ \sqrt{p}}{\mathrm{ber}^2\ \sqrt{p} + \mathrm{bei}^2\ \sqrt{p}}\right) \tag{17b}$$

where $p = f/f_c$ is a dimensionless frequency parameter. An alternative, simpler presentation of the same formulation is shown by Butler and Lizza in Ref. 149.

Dynamic Effects

As can be seen in Figs. 9(b,c), the resonant frequency of a Terfenol-D transducer varies significantly with operating conditions. The transducer mechanical resonance varies as a result of a number of factors. Some of these factors are the magnetostrictive core geometry, bias condition, delta E effect, ac magnetic field, external load, springs stiffness, and operating temperature. Other factors intrinsic to the specific design such as damping of internal components can also be significant.

The first mode axial resonance f_0 of a Terfenol-D transducer such as that sketched in Fig. 1 is given by

$$f_0 = \frac{1}{2\pi}\sqrt{\frac{k_m + k_{ps}}{m_{eff}}} \tag{18}$$

where k_m is the magnetostrictive core stiffness given by Eq. (5), k_{ps} is the prestress mechanism stiffness, and m_{eff} is the system effective dynamic mass. The effective mass in the equation is formed by one-third of the mass of the rod plus the external load plus components of the prestress mechanism. Recognizing that the magnetostrictive material's compliance is strongly dependent upon operating conditions, the resonance frequencies of the device will be strongly dependent on operating conditions. Increasing ac magnetic field intensity increases the system compliance; hence, it decreases transducer resonant frequencies [Fig. 9(b)], while increasing prestress increases the system stiffness and hence resonant fre-

quencies [Fig. 9(c)]. Control of the transducer resonant frequency over a range of 1200 Hz to 1800 Hz using the delta E effect has been used in design of magnetostrictive moving-resonance or tunable vibration absorbers (150).

The dynamics of Terfenol-D transducers coupled to external loads are studied in Ref. 151. An in-plane force-balancing model is used to consider both passive forces (those associated with the transducer and external load structural dynamics) and active forces (derived from the magnetostriction effect) to yield a PDE (partial differential equation) transducer model. The active component of the force is characterized by the strains predicted by the Jiles–Atherton ferromagnetic hysteresis model, combined with a magnetostriction quadratic law. The passive component is characterized by a PDE model of the transducer and external load structural mechanics. The model is capable of predicting the quasi-static magnetization state of the transducer core, and of accurately predicting strains and displacements output by the transducer. The parameter requirements of this model are minimal [experimental identification of six physically based parameters from quasi-static hysteresis loop measurements (e.g., saturation strain and magnetization)] and the model is computationally nondemanding given the relative simplicity of the underlying PDE system. Work is in progress to incorporate the effects of ac losses.

CONCLUDING REMARKS

The high strains and forces achievable with magnetostrictive transducers, high-coupling coefficients, and high-energy density of magnetostrictive materials have justified their use in an ever-increasing number of sensor and actuator applications, ranging from sonar projectors to vibration control, ultrasonics, and sensors designed to find corrosion in pipes. Magnetostrictive device performance capabilities are sometimes overlooked in the interest of avoiding nonlinear and hysteretic input–output transducer behaviors. The unique electromagnetomechanical coupling that takes place in these devices poses rigorous engineering challenges to those who would undertake magnetostrictive transducer design. As evidenced by the resurgence in patented devices based on magnetostriction, designers continue to overcome these challenges and make advances in modeling and controlling magnetostrictive device performance attributes. Furthermore, shrewd transducer designers are recognizing that many of these unique nonlinear behaviors afford the engineer who understands the material attributes a uniquely rich performance space. As material advances continue, it is expected that magnetostrictive device design engineers will find new magnetostrictive solutions to an ever-growing variety of transducer applications.

BIBLIOGRAPHY

1. E. Lee, Magnetostriction and magnetomechanical effects, *Rep. Prog. Phys.*, **18**: 184–229, 1955.

2. E. du Tremolet de Lacheisserie, *Magnetostriction Theory and Applications of Magnetoelasticity*, Boca Raton, FL: CRC Press, 1993.

3. F. V. Hunt, *Electroacoustics: The Analysis of Transduction, and its Historical Background*, Amer. Inst. Phys. Acoust. Soc. Amer., 1982.

4. Nist Advanced Technol. Program, *Terfenol-D high power ultrasonic transducer,* Nist Grant No. 97-01-0023.

5. M. Moffett, R. Porzio, and G. Bernier, High-power Terfenol-D flextensional transducer, *NUWC-NPT Tech. Doc.,* **10**: 883-A, 1995.

6. G. Hartman and J. Sebastian, Magnetostrictive system for high-frequency high-cycle fatigue testing, Univ. Dayton Res. Inst., UDR-TR-97-116, 1997.

7. A. E. Clark and H. S. Belson, Giant room-temperature magnetostrictions in $TbFe_2$ and $DyFe_2$, *Phys. Rev. B,* **5**: 3642–3644, 1972.

8. N. C. Koon, A. I. Schindler, and F. L. Carter, Giant magnetostriction in cubic rare earth-iron compounds of the type RFe_2, *Phys. Lett.,* **37A**: 413–414, 1971.

9. R. M. Bozorth, *Ferromagnetism,* Princeton, NJ: Van Nostrand, 1968.

10. J. L. Butler, *Magnetostrictive transducer design and analysis,* Image Acoustics, Inc., short course notes, Aug. 1997.

11. Y. Lu and A. Nathan, Metglass thin film with as-deposited domain alignment for smart sensor and actuator applications, *Appl. Phys. Lett.,* **70** (4): 526–528, 1997.

12. C. Body et al., Finite element modeling of a magnetostrictive micromembrane, *Proc. Actuator '96, 5th Int. Conf. New Actuators,* VTI-VDE, Bremen, Germany, 1996, pp. 308–311.

13. H. Uchida et al., Effects of the preparation method and condition on the magnetic and giant magnetostrictive properties of the (Tb,Dy)Fe$_2$ thin films, *Proc. Actuator 96, Intern. Conf. New Actuators,* VDI-VDE, Bremen, Germany 1996, pp. 275–278.

14. T. A. Duenas, L. Hsu, and G. P. Carman, Magnetostrictive composite material systems analytical/experimental, *Symp. Advances Smart Materials-Fundamentals Applications,* Boston, MA, 1996.

15. T. Cedell, Magnetostrictive materials and selected applications. Magnetoelastically induced vibrations in manufacturing processes, Dissertation, Lund Univ., Sweden, 1995.

16. Y. Wu and M Appa, Modeling of embeded magnetostrictive particulate actuators, *Proc. Symp. Smart Structures Materials,* Vol. 2717, 1996, pp. 517–529.

17. P. P. Pulvirenti et al., Enhancement of the piezomagnetic response of highly magnetostrictive rare earth-iron alloys at kHz frequencies, *J. Appl. Phys.,* **79**: 6219–6221, 1996.

18. M. Wun-Fugle et al., Hysteresis reduction and magnetostriction of dendritic [112] Tb$_{-x}$Dy$_{-y}$Ho$_{-z}$Fe$_{-1.95}$ $(x+y+z=1)$ rods under compressive stress, *1998 ONR Transducer Materials Transducers Workshop Proc.,* May 1998.

19. R. James and M. Wuttig, Ferromagnetic shape memory alloys, *Proc. SPIE's 5th Annu. Int. Symp. Smart Structures Materials,* Vol. 3324, San Diego, CA, 1998.

20. R. James and M. Wuttig, Magnetostriction of martensite, submitted to *Phil. Mag. A.,* 1997.

21. D. Hall, *Dynamics and vibrations of magnetostrictive transducers,* Ph.D. dissertation, Iowa State Univ., Ames, 1994.

22. F. A. Fischer, *Fundamentals of Electroacoustics,* New York: Interscience, 1955.

23. K. Rolt, History of the flextensional electroacoustic transducer, *J. Acoust. Soc. Amer.,* **87** (3): 1340–1349, 1990.

24. Nat. Defense Res. Committee, *The design and construction of magnetostriction transducers,* Office scientific res. develop., Summary Tech. Rep. Division 6, NDRC, Vol. 13, Washington, DC, 1946.

25. M. Moffett, J. Powers, and A. Clark, Comparison of Terfenol-D and PZT-4 power limitations, *J. Acoust. Soc. Amer.,* **90** (2): 1184–1185, 1991.

26. M. Moffett and W. Clay, Demonstration of the power-handling capability of Terfenol-D, *J. Acoust. Soc. Amer.,* **93** (3): 1653–1654, 1993.

27. T. Akuta, An application of giant magnetostrictive material to high power actuators, in *Workshop on Rare-Earth Magnets and their Applications,* Tokyo: Soc. Non-Traditional Technol., 1989, p. 359.

28. M. Dapino, A. Flatau, and F. Calkins, Statistical analysis of Terfenol-D material properties, *Proc. SPIE Smart Struct. Mater.,* **3041**: 256–267, 1997.

29. M. Dapino et al., Measured Terfenol-D material properties under varied applied magnetic field levels, *Proc. SPIE Smart Struct. Mater.,* **2717**: 697–708, 1996.

30. M. B. Moffett et al., Characterization of Terfenol-D for magnetostrictive transducers, *J. Acoust. Soc. Amer.,* **89**: 1448–1455, 1991.

31. F. Calkins, M. Dapino, and A. Flatau, Effect of prestress on the dynamic performance of a Terfenol-D transducer, *Proc. SPIE Smart Struct. Mater.,* **3041**: 293–304, 1997.

32. J. L. Butler, *Application Manual for the Design of ETREMA Terfenol-D Magnetostrictive Transducers,* Ames, IA: Edge Technol., 1998.

33. M. Rossi, *Acoustics and Electroacoustics,* Norwood, MA: Artech House, 1988.

34. F. T. Calkins and A. B. Flatau, Transducer based measurements of Terfenol-D material properties, *Proc. SPIE Smart Struct. Mater.,* **2717**: 709–719, 1996.

35. A. E. Clark, Magnetostrictive rare earth-Fe$_2$ compounds, in E. P. Wohlfarth (ed.), *Ferromagnetic Materials,* Amsterdam: North-Holland, 1980, Vol. 1, Chap. 7.

36. M. B. Moffett, On the power limitations of sonic transducers, *J. Acoust. Soc. Amer.,* **94** (6): 3503–3505, 1993.

37. R. S. Woolleett, Power limitations of sonic transducers, *IEEE Trans. Sonics Ultrasonics,* **SU-15**: 218–229, 1968.

38. M. B. Moffett and J. M. Powers, Comparison of terfenol-D and PZT-4 power limitations, *J. Acoust. Soc. Amer.,* **90** (2): 1184–85, 1991.

39. F. T. Calkins, *Design, analysis and modeling of giant magnetostrictive transducers,* Ph.D. dissertation, Iowa State Univ., Ames, 1997.

40. F. Claeyssen et al., Progress in magnetostrictive sonar transducers, *Conf. Proc. UDT,* Cannes, France, 1993, pp. 246–250.

41. H. C. Hayes, *Sound generating and directing apparatus,* U.S. Patent No. 2,064,911, 1936.

42. W. Toulis, *Flexural-extensional electromechanical transducer,* U.S. Patents No. 3,274,537 and 3,277,433, 1966.

43. L. H. Royster, The flextensional concept: A new approach to the design of underwater acoustic transducers, *Applied Acoustics,* London: Elsevier, 1970, pp. 117–125.

44. G. A. Brigham and L. H. Royster, Present status in the design of flextensional underwater acoustic transducers, *J. Acoust. Soc. Am.,* **46**: 92, 1969.

45. H. C. Merchant, *Underwater transducer apparatus,* U.S. Patent No. 3,258,738, 1966.

46. E. F. Rynne, Innovative approaches for generating high power, low frequency sound, in *Transducers for Sonics and Ultrasonics,* Lancaster, PA: Technomic, 1993, pp. 38–49.

47. F. A. Abbott, *Broadband electroacoustic transducer,* U.S. Patent No. 2,895,062, 1959; reviewed in *J. Acoust. Soc. Amer.* **32**: 310, 1960.

48. C. J. Purcell, Terfenol driver for the barrel-stave projectors, in *Transducers for Sonics and Ultrasonics,* Lancaster, PA: Technomic, 1993, pp. 160–169.

49. D. F. Jones, Flextensional barrel-stave projectors, in *Transducers for Sonics and Ultrasonics,* Lancaster, PA: Technomic, 1993, pp. 150–159.

50. J. A. Pagliarini and R. P. White, A small, wide-band, low frequency, high-power sound source utilizing the flextensional transducer concept, *Oceans 78. The Ocean Challenge,* New York: IEEE, 1978, pp. 333–338.

51. J. C. Debus, J. N. Decarpigny, and B. Hamonic, Analysis of a Class IV flextensional transducer using piece-part equivalent circuit models, in *Transducers for Sonics and Ultrasonics,* Lancaster, PA: Technomic, 1993, pp. 181–1197.

52. F. Claeyssen, Giant magnetostrictive alloy actuators, *J. Appl. Electromagn. Mater.,* **5**: 67–73, 1994.

53. G. A. Steel, A 2-kHz magnetostrictive transducer, in *Transducers for Sonics and Ultrasonics,* Lancaster, PA: Technomic, 1993, pp. 250–258.

54. S. C. Butler and J. F. Lindberg, A broadband hybrid magnetostrictive tonpilz transducer, *Proc. UDT,* Sydney, Australia, 1998.

55. S. W. Meeks and R. W. Timme, Rare earth iron magnetostrictive underwater sound transducer, *J. Acoust. Soc. Amer.,* **62**: 1158–1164, 1977.

56. F. Claeyssen and D. Boucher, Design of lanthanide magnetostrictive sonar projectors, *Proc. UDT,* Paris, 1991, pp. 1059–1065.

57. H. C. Hayes, U.S. Patent No. 2,005,741, 1935.

58. S. L. Ehrlich and P. D. Frelich, *Directional sonar system,* U.S. Patent No. 3,176,262, 1962.

59. S. L. Ehrlich and P. D. Frelich, *Sonar transducer,* U.S. Patent No. 3,290,646, 1966.

60. S. M. Cohick and J. L. Butler, Rare-earth iron 'square-ring' dipole transducer, *J. Acoust. Soc. Amer.,* **72** (2): 313–315, 1982.

61. J. L. Butler and S. J. Ciosek, Rare earth iron octagonal transducer, *J. Acoust. Soc. Amer.,* **67** (5): 1809–1811, 1990.

62. R. L. Zrostlick, D. H. Hall, and A. B. Flatau, On analog and digital feedback for magnetostrictive transducer linearization, *Proc. SPIE's Smart Structures and Materials,* Vol. 2715, San Diego, CA, 1996, pp. 578–599.

63. M. D. Bryant et al., Active vibration control in structures using magnetostrictive Terfenol-D with feedback and/or neural network controllers, *Proc. Conf. Recent Advances Adaptive Sensory Materials Applications,* Technomic, 1992.

64. J. R. Pratt and A. H. Nayfeh, Smart structures and chatter control, *Proc. 5th Annu. Intern. Symp. Smart Structures Materials,* Vol. 3329, San Diego, 1998, pp. 161–172.

65. H. Nonaka et al., Active vibration control of frame structures with smart structure using magnetostrictive actuators, *Proc. 5th Annu. Int. Symp. Smart Structures Materials,* Vol. 3329, San Diego, 1998, pp. 584–595.

66. L. Jones and E. Garcia, Application of the self-sensing principle to a magnetostrictive structural element for vibration suppression, *ASME Proc. 1994 Int. Mech. Eng. Congr. Expos.,* Chicago, 1994, Vol. 45, pp. 155–165.

67. L. Kiesewetter, Terfenol in linear motors, *Proc. 2nd Inter. Conf. Giant Magnetostrictive Alloys,* 1988.

68. R. C. Roth, The elastic wave motor-a versatile Terfenol driven, linear actuator with high force and great precision, *Proc. 3rd Int. Conf. New Actuators,* AXON Tech., Bremen, German, 1992, pp. 138–141.

69. F. Austin, Smart Terfenol-D powered trailing-edge experiment, *Proc. 5th Annu. Int. Symp. Smart Structures Materials,* Vol. 3326, San Diego, CA, 1998.

70. J. E. Miesner and J. P. Teter, Piezoelectric/magnetostrictive resonant inchworm motor, *Proc. SPIE's Smart Structures Materials,* Vol. 2190, Orlando, FL, 1994, pp. 520–527.

71. B. J. Lund, L. E. Faidley, and A. B. Flatau, Hybrid linear motor design issue analysis, submitted to *SPIE Smart Structures and Integrated Systems Conf.,* Newport Beach, 1999.

72. B. Clephas and H. Janocha, New linear motor with hybrid actuator, *Proc. 4th Annu. Int. Symp. Smart Structures Materials,* Vol. 3041, San Diego, CA, 1997, pp. 316–327.

73. L. Kvarnsjo and G. Engdahl, A new general purpose actuator based on Terfenol-D, *Proc. 3rd Int. Conf. New Actuators,* AXON Tech., Bremen, Germany, 1992, pp. 142–146.

74. T. Akuta, Rotational type actuators with Terfenol-D rods, *Proc. 3rd Int. Conf. New Actuators,* VDI-VDE, Bremen, Germany, 1992, pp. 244–248.

75. J. M. Vranish et al., Magnetostrictive direct drive rotary motor development, *IEEE Trans. Magn.,* **27**: 5355–5357, 1991.

76. F. Claeyssen et al., A new resonant magnetostrictive rotating motor, *Proc. 5th Intern. Conf. New Actuators,* AXON Tech., Bremen, Germany, 1996, pp. 272–274.

77. R. Venkataraman, A hybrid actuator, M.S. Thesis, Univ. of Maryland, 1995.

78. J. R. Frederick, *Ultrasonic Engineering,* New York: Wiley, 1965.

79. G. L. Gooberman, *Ultrasonics Theory and Applications,* London: English Univ. Press, 1968.

80. J. Szilard, Ultrasonics, *Encyclopedia of Physical Science and Technology,* Vol. 17, New York: Academic Press, 1992, pp. 153–173.

81. T. T. Hansen, Magnetostrictive materials and ultrasonics, *Chemtech,* 56–69, 1996.

82. A. I. Markov, *Ultrasonic Machining of Intractable Materials,* London: Iliffe Books, 1966.

83. H. E. Bass et al., Ultrasonics, *Encyclopedia of Science and Technology,* 8th ed., New York: McGraw-Hill, 1977.

84. S. Ueha et al., *Ultrasonic Motors Theory and Applications,* Oxford, UK: Clarendon Press, 1993.

85. F. Claeyssen et al., Micromotors using magnetostrictive thin films, *Proc. 5th Annu. Int. Symp. Smart Structures Materials,* Vol. 3329, San Diego, CA, 1998.

86. D. M. Dozor, Magnetostrictive wire bonding clamp for semiconductor packaging; initial prototype design, modeling, and experiments, *Proc. 5th Annu. Int. Symp. Smart Structures Materials,* Vol. 3326, San Diego, CA, 1998.

87. O. D. Brimhall and C. J. Hasser, Magnetostrictive linear devices for force reflection in dexterous telemanipulation, *Proc. SPIE's Smart Structures Materials,* Vol. 2190, Orlando, FL, 1994, pp. 508–519.

88. Barry Mersky and Van P. Thompson, *Method and apparatus for imparting low amplitude vibrations to bone and similar hard tissue,* U.S. Patent 5,460,593, 1995.

89. R. P. Cutler, G. E. Sleefe, and R. G. Keefe, Development of a magnetostrictive borehole seismic source, Sandia Report SAN97-0944, Sandia Laboratories, 1977.

90. G. Hartman and J. Sebastian (UDT 1997), *Magnetostrictive system for high-frequency high-cycle fatigue testing,* U.S. Patent No. 5,719,339, 1998.

91. J. B. Restorff, Magnetostrictive materials and devices, *Encyclopedia of Applied Physics,* Vol. 9, VCH, 1994, pp. 229–244.

92. R. P. Culter, G. E. Sleefe, and R. G. Keefe, Development of a magnetorestrictive borehole seismic source, Sandia National Labs, Rep. No. SAND97-0944, 1997.

93. W. Fleming, Magnetostrictive torque sensors—Comparison of branch, cross, and solenoidal designs, *Int. Congress and Exposition,* Detroit, 1990, SAE Tech. paper series, 900264, pp. 51–78.

94. W. Fleming, Magnetostrictive torque sensor performance—Nonlinear analysis, *IEEE Trans. Veh. Technol.,* **38**: 159–167, 1989.

95. J. Yamasaki et al., Torque sensors using wire explosion magnetostrictive alloy layers, *IEEE Trans. Magn.*, **MAG-22**: 403–405, 1986.

96. K. Mohri, Review on recent advances in the field of amorphousmetal sensors and transducers, *IEEE Trans. Magn.*, **MAG-20**: 942, 1984.

97. I. Sasada, S. Uramoto, and K. Harada, Noncontact torque sensors using magnetic heads and a magnetostrictive layer on the shaft surface—Application of plasma jet spraying process, *IEEE Trans. Magn.*, **MAG-22**: 406–408, 1986.

98. I. Sasada et al., In-process detection of torque on a drill using magnetostrictive effect, *IEEE Trans. Magn.*, **30**: 4632–4634, 1994.

99. M. Shimada, Magnetostrictive torque sensor and its output characteristics, *J. Appl. Phys.*, **73** (10): 6872–6874, 1993.

100. J. Pratt and A. B. Flatau, Development and analysis of a self-sensing magnetostrictive actuator design, *Proc. SPIE*, **1917**: 952, 1993.

101. J. Pratt, *Design and analysis of a self-sensing Terfenol-D magnetostrictive actuator,* MS. thesis, Iowa State Univ., Ames, 1993.

102. R. C. Fenn and M. J. Gerver, Passive damping and velocity sensing using magnetostrictive transduction, *Proc. SPIE*, **2190**: 216–227, 1994.

103. D. K. Kleinke and H. M. Uras, A noncontact magnetostrictive strain sensor, *Rev. Sci. Instrum.*, **64** (8): 2361–2367, 1993.

104. D. Nyce, Magnetostriction-based linear position sensors, *Sensors,* April, pp. 22–26, 1994.

105. J. Peyrucat, Des distances de 50 metres connués 1 mm pres: C'est l'effect Wiedemann, *Mesures*, **43**: 16, juin 1986.

106. M. Wun-Fogle, H. T. Savage, and M. L. Spano, Enhancement of magnetostrictive effects for sensor applications, *J. Mater. Eng.*, **11** (1): 103–107, 1989.

107. E. Hristoforou and R. E. Reilly, Displacement sensors using soft magnetostrictive alloys, *IEEE Trans. Magn.*, **30**: 2728–2733, 1994.

108. E. Hristoforou and R. E. Reilly, Force sensors based on distortion in delay lines, *IEEE Trans. Magn.*, **28**: 1974–1977, 1992.

109. R. Reilly, Force transducers for use in arrays, U.S. Patent No. 4,924,711, 1990.

110. D. K. Kleinke and H. M. Uras, A magnetostrictive force sensor, *Rev. Sci. Instrum.*, **65** (5): 1699–1710, 1994.

111. J. Seekircher and B. Hoffmann, New magnetoelastic force sensor using amorphous alloys, *Sensors Actuators*, **A21–A23**: 401–405, 1990.

112. K. Mohri and S. Takeuchi, Stress-magnetic effects in iron-rich amorphous alloys and shock-stress sensors with no power, *IEEE Trans. Magn.*, **MAG-17**: 3379–3381, 1981.

113. J. Zakrzewski, Combined magnetoelastic Transducer for Torque and Force measurement, *IEEE Trans. Instrum. Meas.*, **46**: 807–810, 1997.

114. K. A. Bartels, H. Kwun, and J. J. Hanley, Magnetostrictive sensors for the characterization of corrosion in rebars and prestressing strands, *SPIE Proc.*, Nondestructive Evaluation of Bridges and Highways, Vol. 2946, Scottsdale, AZ, December 1996.

115. H. Kwun and C. M. Teller, II, *Non destructive evaluation of pipes and tubes using magnetostrictive sensors*, U.S. Patent No. 5,581,037, 1995.

116. J. W. Brophy and C. R. Brett, Guided UT wave inspection of insulated feedwater piping using magnetostrictive sensors, *SPIE Proc.*, **2947**: 205–209, 1996.

117. S. Foner, Review of magnetometry, *IEEE Trans. Magn.*, **MAG-17**: 3358–3363, 1981.

118. R. J. Weber and D. C. Jiles, A Terfenol-D based magnetostrictive diode laser magnetometer, U.S. Dept. Commerce, grant ITA 87-02, 1992.

119. R. Chung, R. Weber, and D. Jiles, A Terfenol-D based magnetostrictive diode laser magnetometer, *IEEE Trans. Magn.*, **27**: 5358–5243, 1991.

120. J. Doherty, S. Arigapudi, and R. Weber, Spectral estimation for a magnetostrictive magnetic field sensor, *IEEE Trans. Magn.*, **30**: 1274–1290, 1994.

121. A. Yariv and H. Windsor, Proposal for detection of magnetic field through magnetostrictive perturbation of optical fibers, *Opt. Lett.*, **5**: 87, 1980.

122. M. Mermelstein, Fundamental limit to the performance of fibreoptic metallic glass DC magnetometers, *Electron. Lett.*, **21**: 1178, 1985.

123. E. Hristoforou, H. Chiriac, and M. Neagu, A new magnetic field sensor based on magnetostrictive delay lines, *IEEE Trans. Instrum. Meas.*, **46**: 632–635, 1997.

124. T. Meydan and M. Elshebani, A magnetostrictive delay line digitizer by using amorphous ribbon materials, *IEEE Trans. Magn.*, **27**: 5250–5252, 1991.

125. T. Worthington et al., A magnetoacoustic keyboard, *IEEE Trans. Magn.*, **MAG-15**: 1797, 1979.

126. K. Shirae and A. Honda, *IEEE Trans. Magn.*, **MAG-17**: 3151, 1981.

127. T. Klinger et al., Magnetostrictive amorphous sensor for biomedical monitoring, *IEEE Trans. Magn.*, **28**: 2400–2402, 1992.

128. T. Klinger et al., 3D-CAD of an amorphous magnetostrictive sensor for monitoring the movements of the human spine, *IEEE Trans. Magn.*, **28**: 2397–2399, 1992.

129. E. Orsier et al., Contactless actuation of giant magnetostriction thin film alloy bimorphs for two-dimensional scanning application, *SPIE Proc.*, **3224**: 98–108, 1997.

130. G. P. Carman and M. Mitrovic, Nonlinear constitutive relations for magnetostrictive materials with applications to 1-D problems, *J. Intell. Materials Syst. Structures*, **6**: 673–684, 1995.

131. S. Kannan and A. Dasgupta, Continuum magnetoelastic properties of Terfenol-D; what is available and what is needed, *Adaptive Materials Symposium*, Summer Meeting of ASME-AMD-MD, Univ. of California, Los Angeles, 1995.

132. B. Restorff et al., Preisach modeling of hysteresis in Terfenol-D, *J. Appl. Phys.*, **67**: 5016–5018, 1990.

133. A. A. Adly and I. D. Mayergoyz, Magnetostriction stimulation using anisotropic vector Preisach-type models, *IEEE Trans. Magn.*, **32**: 4773–4775, 1996.

134. R. C. Smith, Modeling techniques for magnetostrictive actuators, *Proc. SPIE Symp. Smart Structures Materials*, Vol. 3041, San Diego, CA, 1997, pp. 243–253.

135. G. Engdahl and A. Berqvist, Loss simulations in magnetostrictive actuators, *J. Appl. Phys.*, **79**: 4689–4691, 1997.

136. V. Basso and G. Bertotti, Hysteresis models for the description of domain wall motion, *IEEE Trans. Magn.*, **32**: 4210–4212, 1996.

137. W. F. Brown, *Magnetoelastic interactions,* Berlin: Springer-Verlag, 1966.

138. R. D. James and D. Kinderlehrer, Theory of magnetostriction with applications to $TB_xDy_{1-x}Fe_2$, *Philos. Mag. B,* **68**: 237–274, 1993.

139. E. Della Toree, Fast Preisach based magnetization model and fast inverse hysteresis model, *IEEE Trans. Magn.* submitted.

140. E. C. Stoner and E. P. Wohlfarth, A mechanism of magnetic hysteresis in heterogeneous alloys, *Phil. Trans. Roy. Soc.*, **A240**: 599–642, 1948.

141. E. W. Lee and J. E. Bishop, Magnetic behavior of single-domain particles, *Proc. Phys. Soc.,* **89**: 661, 1966.

142. A. E. Clark, H. T. Savage, and M. L. Spano, Effect of stress on the magnetostriction and magnetization of single crystal $Tb_{0.27}Dy_{0.73}Fe_2$, *IEEE Trans. Magn.,* **MAG-20**: 1443–1445, 1984.

143. D. C. Jiles and J. B. Thoelke, Theoretical modeling of the effects of anisotropy and stress on the magnetization and magnetostriction of $Tb_{0.3}Dy_{0.7}Fe_2$, *J. Magn. Magn. Mater.,* **134**: 143–160, 1994.

144. D. C. Jiles and D. L. Atherton, Theory of ferromagnetic hysteresis, *J. Magn. Magn. Mater.,* **61**: 48, 1986.

145. A. E. Clark et al., Magnetomechanical coupling in Bridgman-grown $Tb_{0.3}Dy_{0.7}Fe_{1.9}$ at high drive levels, *J. Appl. Phys.* **67**: 5007–5009, 1990.

146. K. Kvarnsjö, On characterization, modeling and application of highly magnetostrictive materials, PhD dissertation, Royal Institute of Technology, TRITA-EEA-9301, Stockholm, Sweden, 1993.

147. C. L. Hom and N. Shankar, A fully coupled constitutive model for electrostrictive ceramic materials, *2nd Int. Conf. Intell. Materials,* ICIM, 1994, pp. 623–624.

148. D. C. Jiles, Frequency dependence of hysteresis curves in conducting magnetic materials, *J. Appl. Phys.,* **76**: 5849–5855, 1994.

149. J. L. Butler and N. L. Lizza, Eddy current loss factor series for magnetostrictive rods, *J. Acoust. Soc. Am.,* **82**: 1997.

150. A. B. Flatau, M. J. Dapino, and F. Calkins, High-bandwidth tunability in a smart passive vibration absorber, *Proc. SPIE Smart Structures Materials,* #3329-19, San Diego, CA, 1998.

151. M. Dapino, R. Smith, and A. Flatau, An active and structural strain model for magnetostrictive transducers, *Proc. SPIE Smart Structures Materials,* #3329-24, San Diego, CA, 1998.

Marcelo J. Dapino
Iowa State University

Frederick T. Calkins
The Boeing Company

Alison B. Flatau
Iowa State University

MAGNETS FOR MAGNETIC RESONANCE ANALYSIS AND IMAGING

Nuclear magnetic resonance (NMR) was discovered in 1946 by Purcell (1) and Bloch (2). Classically, it is the precession of the spins of nuclei with magnetic moment, subjected to a transverse radio frequency (RF) field in the presence of a longitudinal magnetic field. Nuclear species of biological interest having nonzero magnetic moment are listed in Table 1 together with their Larmor precession frequency-field dependence.

Table 1. The Larmor Precession Constant for Various Nuclides

Nuclide	Atomic Number	NMR Frequency (MHz/Tesla)
Hydrogen	1	42.5759
Deuterium	2	6.5357
Carbon	13	10.705
Oxygen	17	5.772
Sodium	23	11.262
Phosphorus	31	17.236

Experimentally, NMR is performed as follows (3,4): nuclei are immersed in a static field B_0, which results in the development of a net polarization along the field direction, occurring with a time constant τ_1. Transitions among their spin states are excited by a high-frequency field B_1, oriented perpendicular to the static field. In a rotating frame of reference (rotating at the Larmor precession frequency), the magnetization vector of the spin is tilted from a longitudinal direction (along B_0) toward the transverse plane. The angle of precession depends on the strength and duration of the applied RF field and is given by

$$\psi = \gamma B_1 \tau_p \qquad (1)$$

where τ_p is the pulse width and γ is the gyromagnetic ratio.

Following the pulse, the magnetization decays transversely with a time constant τ_2, the spin–spin relaxation time. The polarization develops (or decays) along the field with a time constant τ_1, the spin–lattice relaxation time.

NMR is the preeminent method for the identification of chemical species in weak solution. It also has useful applications in solid materials. The most exacting specifications for an NMR magnet are imposed by high-resolution NMR. The resonant frequency of a nucleus depends not only on B_0 but also, to a small extent, on the shielding provided by the electronic structure of the chemical compound. This effect is the chemical shift and is distinctive for each chemical species. Thus the resonant frequency of the ^1H nucleus in water is different from that in benzene (C_6H_6) or in the methyl or methylene group in alcohol (CH_3CH_2OH). These small differences in frequency are typically a few parts per million and provide a means to identify the components of a complex molecule.

Early NMR spectrometers used continuous wave (CW) methods in which the frequency of the B_1 field would be changed slowly and the absorption of a tank circuit enclosing the sample would be recorded as a spectrum of power absorption versus frequency. At the resonant frequency a sharp increase in absorption would be observed. The width of the peak depended, among other things, on the magnification Q of the tank circuit.

In modern NMR, a pulse of RF of sufficient strength and duration is applied to the sample so that all the spins are excited. The pulse is then switched off, and the signals emitted at various frequencies by the sample during relaxation of the spins are monitored. A Fourier analysis of the signal then transforms the time-dependent spectrum into a frequency-dependent signal, thus revealing the resonance peaks associated with the chemical shifts (3,4).

The uniformity of the static field B_0 is the key to high-resolution NMR and to sharp images in magnetic resonance imaging (MRI). A uniform field allows large numbers of nuclei to precess at exactly the same frequency, thus generating a strong signal of narrow bandwidth. The underlying theory and practice of high-homogeneity superconducting magnets is described in this section. MRI magnets differ from those used for NMR analysis in that spatial distribution of either signal strength or τ_1 or τ_2 relaxation times are measured over a volume far greater than that of a sample for chemical analysis. In MRI, the predominant nuclear species examined is hydrogen in water. Density or τ_1 or τ_2 is measured on planes throughout a body and reconstructed as two-dimensional maps.

The field strengths of NMR magnets are higher than those of MRI magnets. From the discovery of NMR, in 1946, to 1967, magnetic fields were limited to 2 T that could be generated by electromagnets. A 5 T superconducting magnet was introduced in 1967, and slow improvements led to the 20 T magnetic field available in NMR magnets today. The driving forces for the increase in field strength are the chemical shift (separation of the nuclear species which is linearly proportional to field strength), signal strength (which is proportional to the square of the field strength), and signal-to-noise ratio (which is proportional to the 3/2 power of the field strength). Although various experimental techniques have been applied to improve signal-to-noise ratio, including tailored pulse sequences, signal averaging, cooled conventional receiver coils and superconducting receiver coils, increased field strength is still desirable for increased chemical shift. Magnets for high-resolution NMR are now almost exclusively superconducting, and it is only that type that is described here.

The optimal field strength for MRI is determined by a number of factors, including reduction in imaging time, reduction in chemical shift artifacts, and reduction in cost, and by limits to the exposure of a patient to electromagnetic radiation, as set by regulations. Even though most MRI magnets have field strengths up to 1.5 T, a few experimental magnets have been built or designed for functional MRI studies with field strengths up to 8 T. In order to achieve the desired combination of field strength, working volume, and stability, superconducting magnets represent the principal type employed. However, both water-cooled resistive magnets, iron-cored electromagnets, and rare earth permanent magnet MRI systems have been used or are in use for special applications. The superconducting magnet is here considered exclusively.

DESIGN PRINCIPLES

NMR Magnets

The analysis of weak solutions imposes several requirements on the magnet. In order to obtain usable signal-to-noise ratios, a large volume of sample must be used. This immediately demands good field uniformity so that variations in background field strength do not give rise to different frequencies, which would, of course, mask the small chemical shifts being sought. High field strength is desired as detailed above. Even if these requirements are met, the dilution of the sample may be such that repeated pulses are required. The final signal-to-noise that can be obtained from a number of pulses is proportional to $\sqrt{N}$. A run may take many hours or even days to accomplish. During that time not only must the spatial homogeneity of the background field be excellent, but the magnitude of the field also must be constant, or at least must change only very slightly. (The reason that any change is permitted is that a frequency lock can be used to adjust the frequency of the RF to match a slow and slight change in the background field.)

To summarize, the NMR magnet should have high field strength and great uniformity, and the field must be stable.

MRI Magnets

The essential principles for MRI magnets are identical to those for NMR magnets, but the volumes of homogeneity are much greater, whereas the homogeneity is somewhat lower. Despite field strengths lower than those of NMR magnets, the stored magnetic energies of MRI magnets are greater by reason of the large bore, which must be sufficient to house correction coils, pulsed gradient coils, and a patient. The MRI system differs significantly from that of the NMR system by including means to superimpose linear field gradients on the background homogeneous field. These pulsed gradients define thin planes in which the field is known but different from that elsewhere. Thus the frequency of nuclear magnetic resonance is spatially encoded so that the signals generated by the relaxing nuclei have frequencies which define their position. As in NMR for chemical analysis, the MRI signals may interrogate either the density of nuclei or the τ_1 or τ_2 relaxation times.

THEORETICAL DESIGN

Almost all superconducting NMR and MRI magnets are solenoids. The reason for that is the relative simplicity and ease of manufacture and design of solenoids, compared with, for instance, extended dipoles. Although the generation of the RF field could be simpler with a transverse background field, the difficulty of manufacture of a high-background field magnet would far outweigh any advantage in the RF coil. The construction of a high-homogeneity solenoid proceeds in three parts: a winding array is designed, based solely on the analysis of the axial variation of the field of a solenoid; the magnet is wound and the spatial variation of its actual field is measured; and the unwanted errors in the field arising from manufacturing imperfections are removed by shimming.

The center field of a solenoid is given by

$$B_0 = \mu_0 J a_0 \ln\{[\alpha + (\alpha^2 + \beta^2)^{1/2}]/[1 + (1 + \beta^2)^{1/2}]\} \quad (2)$$

where J is the overall winding current density, a_0 is the inner radius, α is the ratio of outer to inner radii, and β is the ratio of length to inner diameter (5). Because SI units are used throughout, μ_0 is the permeability of free space, $4\pi \times 10^{-7}$ H/m, a_0 is in meters, and B_0 is in Tesla.

The field strength decreases at points on the axis away from the center of the solenoid. The axial variation of field strength on the z axis is expressible as a Taylor's series

$$\begin{aligned} B(z) = B_0 &+ (d^2B/dz^2)z^2/2 + (d^4B/dz^4)z^4/4! \\ &+ (d^6B/dz^6)z^6/6! + \cdots \end{aligned} \quad (3)$$

Only even terms appear because the center of the solenoid coincides with the origin.

Figure 1 illustrates the geometry of a thin solenoid, of radius a_0 and extending a length z_0 to the right of the origin. For such a thin solenoid, the derivatives of the field at the origin are as follows:

$$\begin{aligned} B_0 &= \tfrac{1}{2}\mu_0 i z_0 (a_0^2 + z_0^2)^{-1/2} \\ dB/dz &= -\tfrac{1}{2}\mu_0 i a_0^2 (a_0^2 + z_0^2)^{-3/2} \\ d^2B/dz^2 &= -\tfrac{1}{2}\mu_0 i 3 z_0 a_0^2 (a_0^2 + z_0^2)^{-5/2} \\ d^3B/dz^3 &= -\tfrac{1}{2}\mu_0 i a_0^2 (3a_0^2 - 12z_0^2)(a_0^2 + z_0^2)^{-7/2} \end{aligned} \quad (4)$$

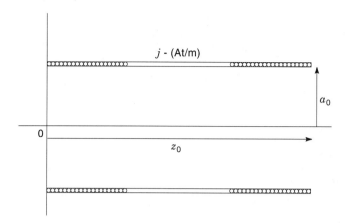

Figure 1. Geometry of a thin solenoid showing the coordinate system used to define current geometries.

where i is the sheet current density in amp-turns per meter and a_0 and z_0 are as illustrated in Fig. 1.

The field of a solenoid symmetric about the center plane has even symmetry and no odd derivatives. So, by evaluating the even derivatives of the field at the center, the axial variation of field generated by a solenoid can be calculated to an accuracy determined by the number of derivatives used and by the distance from the center. The derivatives can be treated as coefficients of a Cartesian harmonic series so that

$$B_z = B_0 + b_2 z^2 + b_4 z^4 + b_6 z^6 + \cdots$$

in which

$$
\begin{aligned}
b_2 &= \mu_0 i [3 z_0 a_0^2 (a_0^2 + z_0^2)^{-5/2}]/4 \\
b_4 &= \mu_0 i [(45 a_0^2 z - 60 z^3)(a_0^2 + z_0^2)^{-9/2}]/48 \\
b_6 &= -\mu_0 i [(5 a_0^4 z - 20 a_0^2 z^3 + 8 z^5)(a_0^2 + z_0^2)^{-13/2}]/1440
\end{aligned}
\tag{5}
$$

For coils of odd symmetry, such as shim coils described later, the corresponding harmonics are

$$
\begin{aligned}
b_1 &= \mu_0 i [(a_0^2 + z_0^2)^{-3/2}]/2 \\
b_3 &= \mu_0 i [(3 a_0^2 - 12 z^2)(a_0^2 + z_0^2)^{-7/2}]/12
\end{aligned}
\tag{6}
$$

Notice that the magnitude of any harmonic coefficient is mediated by the denominator of the expressions that each include the term $(a_0^2 + z_0^2)^{(n+1/2)}$, where n is the order of the harmonic. Thus, the generation of high-order harmonics requires coils with large values of current (ampere-turns) or small radius. This is significant in the construction of shim coils, as is noted later.

Associated with an axial variation of field is a radial variation, arising from radial terms in the solution of the Laplace scalar potential equation. For instance, even-order axial variations are accompanied by axisymmetric radial variations (6) of the form

$$
\begin{aligned}
B_2(z, x, y) &= b_2 [z^2 - \tfrac{1}{2}(x^2 + y^2)] \\
B_4(z, x, y) &= b_4 [z^4 - 3(x^2 + y^2) + \tfrac{3}{8}(x^2 + y^2)^2]
\end{aligned}
\tag{7}
$$

These equations show that if b_2 or b_4 are zero there will be no axisymmetric radial variation of field.

Figure 2 illustrates a set of nested solenoids. Solenoid 1 gives rise to nonzero values of the harmonic coefficients b_2, b_4, b_6, etc. If dimensioned correctly, solenoid 2 by contrast can produce equal values for some or all these coefficients but with opposite polarity. Then at least b_2 and b_4 will have net zero values, and the first uncompensated harmonic to appear in the expression for axial field variation will be the sixth order. A minimum, but not necessarily sufficient, condition is that as many degrees of freedom are needed in the parameters of the coils as there are coefficients to be zeroed.

This method can be extended to as many orders as desired. In most high-resolution NMR magnets, the required uniformity of the field at the center is achieved by nulling all orders up to and including the sixth. That is, the solenoid is of eighth order. In the design of the solenoids, no odd order appears, of course. The first residual harmonic will have a very small value close to the center, although at greater axial distances, the field will begin to vary rapidly. Thus, the design of a high-homogeneity solenoid requires only the calculation of the field or the field harmonics on axis, and those harmonics may be easily calculated using only Cartesian coordinates.

MANUFACTURING ERRORS

The theoretical design of a high-homogeneity magnet can be simple because only axial terms in the z field need to be considered. However, the manufacturing process introduces errors in conductor placement which generate both even- and odd-order axial and, most significantly, radial field gradients. Further, the materials of the coil forms, the nonisotropic contraction of the forms and windings during cool-down to helium temperature and the effects of the large forces between the windings may also introduce inhomogeneity. Typically, the homogeneity of an as-wound set of NMR solenoids is not better than 10^{-5} over a 5 mm diameter spherical volume (dsv) at the center. For high-resolution NMR, an effective homogeneity of 10^{-9} over at least 5 mm dsv is required. The improvement of the raw homogeneity to this level is achieved by three steps, superconducting shim coils, room temperature shim coils and, in NMR magnets only, sample spinning. (Additionally, in cases of poor raw homogeneity, ferromagnetic shims may be used occasionally in NMR magnets and routinely in MRI magnets to compensate for large errors or significant high-order harmonics.) The presence of radial field gradients necessitates a more comprehensive field analysis than is convenient with Cartesian coordinates.

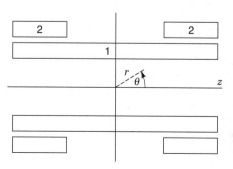

Figure 2. Principle of harmonic compensation. Coaxial solenoids generating field harmonics of opposite sign.

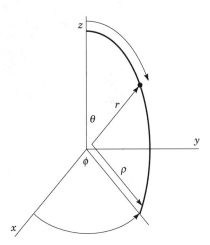

Figure 3. The system of spherical coordinates specifying field points and current sources.

LEGENDRE FUNCTIONS

The expression of the harmonics of the field in terms of Cartesian coordinates provides a simple insight into the source of the harmonics. However, as the order of the harmonic increases, the complexity of the Cartesian expressions renders manipulation very cumbersome, and an alternative method is needed. The Laplace equation for the magnetic field in free space is conveniently solved in spherical coordinates. These solutions are spherical harmonics, and they are valid only in the spherical region around the center of the solenoid, extending as far as, but not including, the nearest current element. Figure 3 illustrates the coordinate system for spherical harmonics. The convention followed here is that dimensions and angles without subscripts refer to a field point, and with subscripts they refer to a current source.

The axisymmetric z field generated by a coaxial circular current loop can be expressed in the form of a Legendre polynomial, thus,

$$B_z = \sum_{n=0}^{\infty} g_n r^n P_n (\cos \theta) \tag{8}$$

where r and θ define the azimuth of the field point in spherical coordinates, and u is $\cos(\theta)$. $P_n(u)$ is the zonal Legendre polynomial of order n and g_n is a generation function given by

$$g_n = \mu_0 i P_{n+1} \cos(\theta_0) \sin(\theta_0)/(2\rho^{n+1}) \tag{9}$$

where θ_0 and ρ_0 define the position of the current loop in spherical coordinates. In this text, it is the convention that $n = 0$ represents a uniform field. The field strength given by Eqs. (8) and (9) is constant with azimuth at constant radius r.

Equations (8) and (9) are equivalent in spherical coordinates to those of Eqs. (4), (5), and (7) on the z axis but additionally predict the z field off axis. In the design of the main coils Eqs. (8) and (9) offer no more information than Eq. (5). However, in the calculation of the off-axis z fields, they provide important additional information that can be used in the optimization of coil design when fringing fields must be considered.

The harmonic components of the z field can also be expressed in the form of associated Legendre functions of order n, m (7). Those functions define the variation of the local z field strength at points around the center of the magnet and include variation of the field with azimuth φ. Thus,

$$\begin{aligned} B_z(n, m) = r^n (n + m + 1) P_{n,m}(u) \\ \times [C_{n,m} \cos(m\varphi) + S_{n,m} \sin(m\varphi)] \end{aligned} \tag{10}$$

where $C_{n,m}$ and $S_{n,m}$ are the harmonic field constants in tesla per meter^n, $P_{n,m}(u)$ is the associated Legendre function of order n and degree m, and u is $\cos(\theta)$. The order n is zonal, describing the axial variation of z field. The degree m is tesseral, describing the variation of the z field in what would be the x–y plane in Cartesian coordinates. φ is the azimuth to the point at radius r from an x–z plane. θ is the elevation of the point from the z axis. Tables of the values of the Legendre polynomials can be found in standard texts on mathematical functions (8).

In Eq. (10), m can never be greater than n. For example, if $n = m = 0$, $B_z(0,0)$ is a uniform field independent of position. If $n = 2$ and $m = 0$, $B_z(2,0)$ is a field whose strength varies as the square of the axial distance [i.e., B_2 of Eq. (7)]. If $n = 2$ and $m = 2$, $B_z(2,2)$ is a field that is constant in the axial direction but increases linearly in two of the orthogonal radial directions and decreases linearly in the other two. Figure 4 shows a map of the contours of constant field strength of a $B_z(2,2)$ field harmonic for which $S_{2,2} = 0$. The $B_z(2,2)$ field has zero magnitude at the origin and along the x and y coordinate axes. Of course, the direction of the zero values of the $B_z(2,2)$ harmonic will not generally lie in the Cartesian x and y planes. Depending on the relative values of $C_{n,m}$ and $S_{n,m}$ in Eq. (10), the zero harmonic planes will lie at an angle other than $\phi = 0$ or $m\pi/2$. The constant field contours of $B_z(2,2)$ extend to infinity along the z axis and represent, arbitrarily in this figure, values for $B_z(2,2)$ of 10^{-4}, 10^{-6}, and 10^{-8}, for example. Within the indicated cylinder centered on the z axis, the value of the harmonic is everywhere less than 10^{-6}. For

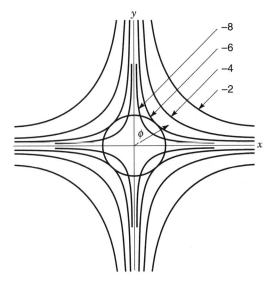

Figure 4. Surfaces of constant magnitude of a $B(2,2)$ harmonic field, showing that the tesseral harmonic is zero when the azimuth ϕ is a multiple of $\pi/2$.

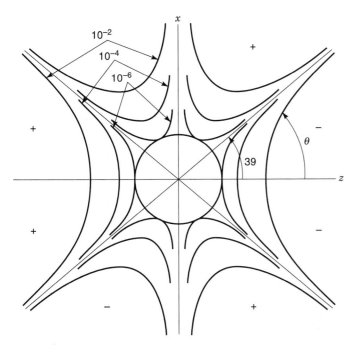

Figure 5. Surfaces of constant magnitude of a $B(3,0)$ harmonic field, showing that the zonal harmonic is zero when the elevation μ is $39°$ or $90°$.

higher values of m, there are more planes of zero value. Thus, $B_z(4,4)$ has eight planes of zero value, $B_z(8,8)$ has 16, and so forth. A harmonic $B_z(4,2)$ defines a field in which the second-degree azimuthal variation itself varies in second order with axial distance.

The zonal harmonics $B_z(2,0)$, $B_z(3,0)$, $B_z(4,0)$ have conical surfaces on which the value of the field is zero. Thus, for instance, $B_z(3,0)$ has contours of zero value such as are shown in Fig. 5 to lie at $\theta = 39°$ and $90°$.

The four hyperbolas are actually surfaces of rotation about the z axis, and each represents a constant value for $B_z(3,0)$ of, say, 10^{-6} [the uniform field, $B_z(0,0)$ at the origin having unity value]. Within the indicated ellipsoid, centered on the origin, the value of the $B_z(3,0)$ harmonic is therefore everywhere less than 10^{-6}.

The contours of the zero values of the spherical harmonics are analogous to combinations of Figs. 4 and Fig. 5. The zero values now lie on straight lines radiating from the origin. The surfaces of constant value look like the spines of sea urchins. As for the zonal harmonics, ellipsoidal surfaces roughly describe boundaries within which the magnitudes of the spherical harmonics do not exceed a given value. These error surface diagrams are often used in the design of an MRI magnet to identify the maximum calculated field error within a central volume caused by the highest uncompensated harmonic.

Thus, in general, the deviations from the ideal uniform solenoidal field can be expressed as the sum of a large number of harmonics each described by the associated Legendre function of order n and degree m. Although the Cartesian expressions of Eq. (4) can be used for the design of a coil system to generate a uniform field, the associated Legendre functions must generally be used for the analysis of the measured field and the design of shim coils or of ferromagnetic shims to compensate for harmonics with nonzero values of m (9).

Optimization Methods

With the recent rapid increase in the speed and size of computers, an alternative technique for the design of uniform field magnets has been developed. Not only is a uniform field of specific magnitude required but that should be combined with other criteria. For instance it could be accompanied by the smallest magnet, that is, the minimum of conductor, or by a specified small fringing field. To achieve these ideal solutions, an optimization technique is now generally used. The field strength of a set of coils is computed at points along the axis, and, if fringing field is a consideration, at points outside the immediate vicinity of the system. The starting point may be a coil set determined by a harmonic analysis as described earlier. Now however, mathematical programming methods are employed to minimize the volume of the windings satisfying the requirement that the field should not vary by more than the target homogeneity for each of the chosen points. Again, for purposes of homogeneity, only field on axis is considered because the radial variation of axisymmetric components of field is zero if the axial component is zero. The field strengths at points outside the magnet will be minimized by inclusion of a set of coils of much larger diameter than the main coils but carrying current of reverse polarity.

All design techniques, but particularly that of optimization, are complicated by the highly nonlinear relationship between the harmonic components generated by a coil and the characteristics of the coil. Thus the reversal in sign of the harmonic components occurs rapidly as the dimensions or position of a coil are changed. In the example of an NMR magnet shown in Fig. 6, the value of the second harmonic changes by 4 ppm for an increase in the diameter of the wire in the small coil "l" of only 0.1 mm. The optimization of the ampere turns, shape and position of a coil thus affects the various harmonics in highly nonlinear and often conflicting ways.

Design optimization involves the computation of an objective function which contains all the elements that have to be minimized, subject to a set of constraints (10). For example, it may be required to minimize some combination of winding volume or magnet length subject to constraints on the field error at a number of points within the bore and on the fringing field at some point outside the magnet. The objective function would then be of the form

$$\sum_{i=1}^{N} pV_i + L \qquad (11)$$

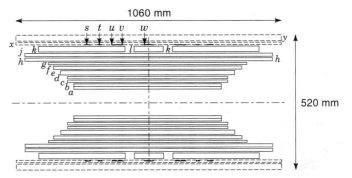

Figure 6. Coil profiles of an actual 8th order compensated NMR solenoid. The graded sections a through j produce axial harmonics of which orders 2, 4, and 6 are compensated by sections k and l. Layers x and y are shim coils.

where V_i is the volume of a coil, N is the number of coils, and L is the length of the magnet. The factor p weights the relative importance of the volume and of the length. This objective function is then minimized subject to the following constraints:

$$\left[\sum_{i=1}^{N} B_{i,p} - B_0\right]^2 < \Delta B^2 \qquad (12)$$

$$\sum_{i=1}^{N} B_{i,f} < B_f \qquad (13)$$

In Eq. (12) $B_{i,p}$ is the field at point p due to coil i. The equation represents the constraint on uniformity of field. It could also be expressed in terms of harmonic terms; for example, each even term up to $P_{10,0}$ being less than $10^{-6} B_0$, the center field. [The inclusion of the squared terms in Eq. (12) allows for either positive or negative error field components.] Equation (13) expresses the condition that the fringing field should be less than, say, 1 mT (10 gauss) at a point, outside the magnet system. The 10 gauss criterion frequently represents the maximum field to which the public may be exposed in accessible areas around an MRI system.

The minimization of the objective function is performed by a mathematical programming algorithm, whereas the solution of the constraining Eqs. (12) and (13) will require a nonlinear technique (such as Newton–Raphson), in order to deal with the extremely nonlinear variation of the harmonics as they change with coil geometry (11).

SHIELDING

The minimization of the external fringing field is becoming increasingly important for the siting of MRI systems, so the active shielding of MRI magnets with center fields up to 2 T is now almost universal. (Active shielding of MRI magnets with center fields above 2 T is uneconomical and is not generally attempted.) Active shielding is generally achieved by the inclusion in the coil array of two reverse polarity coils at diameters typically twice that of the main coils. Because of the large dipole moment of an MRI magnet, the unshielded fringing field will extend several meters from the boundary of the cryostat. Consequently, active shielding is applied to many MRI magnets with central fields of over 0.5 T (12). The effect of the shielding on the harmonics of the center field must, of course, be included in the design of the compensation coils.

SHIMMING

The harmonic errors in the field of an as-built magnet divide into purely axial variations (axisymmetric zonal harmonics, which are accompanied by radial variations dependent on the elevation θ from the z axis, but independent of ϕ) and radial variations (tesseral harmonics, which depend on ϕ, where ϕ is the angle of azimuth in the x–y plane).

In order to compensate for the presence of various unwanted harmonic errors in the center field of the as-built coils, additional coils capable of generating the opposite harmonics are applied to the magnet. For each set of n and m in the associated Legendre functions, a current array can be

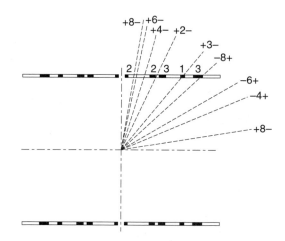

Figure 7. A set of axial shim coils for harmonic correction up to $B(3,0)$. These coils generate small harmonics of 4th order and higher.

designed in the form of a set of arcs of varying azimuthal extent and symmetry and with various positions and extents along the z axis. The magnitude of the harmonic field that an array generates can be controlled by the current. This is the principle of variable harmonic correction for both MRI and NMR magnets. (Correction by means of ferromagnetic shims is not variable.)

The shimming of the unwanted harmonics is a process in two independent parts. First, there is the design of as many sets of coils as are needed to generate the compensating harmonics. Second is the measurement of the actual field errors to determine the magnitudes of the various harmonic components and the application of currents to the previously designed coils to provide the compensation. In fact, because superconducting shims must be built into the magnet prior to installation in the cryostat and cooldown, the range of harmonic errors in the field of the as-built magnet must be largely anticipated. Typically it might be assumed that the level of harmonic error decreases by a factor of three for each unit increase in n or m. Therefore, as a rough guide it has been found that compensation of up to $B(3,0)$ for the zonal harmonics and up to $B(2,2)$ for the tesseral harmonics is satisfactory in most cases for the superconducting shims of small bore NMR magnets. There will also be a set of room temperature shims in a high resolution NMR system. Those will compensate for errors typically up to $B(6,0)$ and $B(3,3)$ in many cases. Typically there may be up to 28, but exceptionally up to 45 independent shims in all. They will be constructed according to a different principle from the superconducting shims. The shimming of MRI magnets is accomplished by current shims, typically up to $n = 3$ and $m = 2$, and by ferromagnetic shims.

Superconducting Axial Shims

These will be simple circular coils combined in groups so as to generate a single harmonic only (13). Thus, a coil to generate $B(3,0)$ must generate no $B(1,0)$ nor $B(5,0)$. Because the superconducting shim coils need to generate only a small fraction of the field due to the main coil, they generally need only comprise one to three layers of conductor. For that reason the harmonic sensitivities can be calculated directly from Eqs. (4) and (5). A set of axial shims providing correction of $B(n,0)$ harmonics for $n = 1$ through 3 are shown in Fig. 7. Note that, for a fixed linear current density, only the angles defining the start and end of each coil are needed, together, of course, with the current polarities, either side of the center plane of the magnet, odd for $n = 1, 3, 5, . . .$ and even for $2, 4, 6, . . .$.

The set of coils illustrated in Fig. 7 generate negligible harmonics above the third order, $B(3,0)$. The individual coils of each harmonic group are connected in series in sets, there being in each set enough coils to generate the required axial harmonic but excluding, as far as is practical, those harmonics that are unwanted. Thus, in the figure, coils labeled 2 generate second-order $B(2,0)$ but no fourth order. However, they do generate higher orders. The first unwanted order is $B(6,0)$ but that is small enough that it may be neglected. So also with all higher orders because the denominator in the expressions of Eqs. (4) and (5) strongly controls the magnitude of the harmonic. Also illustrated in the figure is the effect on harmonic generation of the angular position of a circular current loop. Each of the dashed lines lies at the zero position of an axial harmonic. Thus, at an angle of $70.1°$ from the z axis, the $B(4,0)$ harmonic of a single loop is zero. Two loops carrying currents of the same polarity and suitable magnitude may be located on either side of the $70.1°$ line to generate no fourth-order harmonic yet generate a significant second order harmonic. Similarly, a coil for the generation of only a first order axial harmonic is located on the line for zero third order. The zero first-order harmonic line is at $90°$, the plane of symmetry. In order therefore to generate a third order with no first, two coils must be used, with opposing polarities. The coils are all mirrored about the plane of symmetry, but the current symmetries are odd for the odd harmonics and even for the even harmonics. The loops may be extended axially as multiturn coils while retaining the property of generating no axial harmonic of a chosen order, if the start and end angles subtended by the coils at the origin are suitably chosen.

The principles described earlier can be applied both in the design of shim coils and in the selection of main coil sets. A further observation from the zero harmonic lines of Fig. 7 is that the higher harmonics reverse sign at angles close to the plane of symmetry of the system. This implies that, to produce single, high-order harmonics, coil positions close to the plane of symmetry must be chosen because the other coil locations where the sign of the harmonic reverses are too far from the plane of symmetry to be usable; the coils lying a long way from the plane of symmetry generate weak high-order harmonics.

Superconducting Radial Shims

The radial shims are more complex than those for purely axial harmonics because the finite value of m requires a $2m$-fold symmetry in the azimuthal distribution of current arcs, the polarity of current always reversing between juxtaposed arcs in one z plane (6,9). For instance, $m = 2$ requires four arcs, as shown in Fig. 8. However, as for $m = 0$, the set of current arcs shown in Fig. 8 will generate $B(n,m)$, where n is $2, 4, 6$, etc., or $1, 3, 5$, etc., depending on even or odd current symmetry about the $z = 0$ plane. So, the positioning of the arcs along the z axis is again crucial to the elimination of at least one unwanted order, n. Fortunately, the azimuthal symmetry generates unique values of the fundamental radial harmonic m. (Eight equal arcs cannot generate an $m = 2$ harmonic.) However, depending on the length of the arc, higher

radial harmonics may be generated. For the shim coil configuration of Fig. 8, the first unwanted radial harmonic is $m = 6$. The higher tesseral harmonics are much smaller than the fundamental because of the presence in the expression for the field of a term $(r/r_0)^n$. Generally, the arc length is chosen to eliminate the first higher-degree radial harmonic. As an example, if the arc length of each shim coil shown in Fig. 8 is $90°$ the $B(6,6)$ harmonic disappears. The $B(10,10)$ harmonic is negligible.

The superconducting shims are almost invariably placed around the outside of the main windings. Although the large radius reduces the effective strength of the harmonics they generate, the shim windings cannot usually be placed nearer to the center of the coil because of the value of winding space near the inner parts of the coil and because of the low critical current density of wires in that region due to the high field. A comprehensive treatment of shim coil design may be found in Refs. 6 and 9. Those references also include details of superconducting coil construction. It should be noted, however, that some expressions in Ref. 6 contain errors.

Ferromagnetic Shims

Ferromagnetic shimming is occasionally used in high field, small bore NMR magnets, but its principal use is in MRI magnets. It is in that application that it will be described. The principle invoked in this kind of shimming is different from that of shim coils. The shims now take the form of discrete pieces of ferromagnetic material placed in the bore of the magnet. Each piece of steel is subjected to an axial magnetizing field at its position sufficient to saturate it. It then generates a field at a point in space that is a function of the mass of the shim and its saturation magnetization B_s with little dependence on its shape. For ease of example, a solid cylinder of steel will be assumed. The axis of the cylinder is in line with the field, as shown in Fig. 9. (In Fig. 9 the axis labeled z is that of the shim, not that of the MRI magnet itself. In fact, the shim will usually be placed at the inside surface of the bore of the MRI magnet.)

The field B, caused by the ferromagnetic shim, contains both axial and radial components. The axial component B_z is the correcting field required, and it adds arithmetically to the field of the magnet. The radial component adds vectorially to the field and produces negligible change in the magnitude of

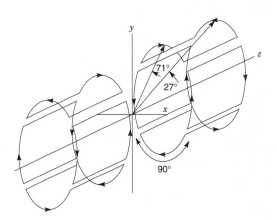

Figure 8. Schematic of a set of radial shim coils for correction of a $B(2,2)$ harmonic showing the positioning necessary to eliminate $B(4,2)$ and $B(4,4)$.

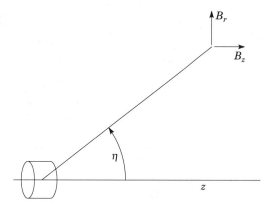

Figure 9. Field vectors generated by a ferromagnetic shim in the bore of an MRI magnet. B_z adds arithmetically to the main field; B_r adds vectorially and so has negligible influence on the field.

the axial field. Therefore, only the axial component of the shim field must be calculated. If the saturation flux density of the shim is B_s, the axial shim field is given by

$$B_z = B_s V[(2 - \tan^2 \eta)/(\tan^2 \eta + 1)^{5/2}]/(4\pi z^3) \qquad (14)$$

where V is the volume of the shim and z and η are as shown in Fig. 9.

The practical application of ferromagnetic shims involves the measurement of the error fields at a number of points, and the computation of an influence matrix of the shim fields at the same points. The required volumes (or masses) of the shims are then determined by the inversion of a U, W matrix, where U is the number of field points and W is the number of shims. In an MRI magnet, the shims are steel washers (or equivalent) bolted to rails on the inside of the room temperature bore of the cryostat. In the occasional ferromagnetic shimming of an NMR magnet, the shims are coupons of a magnetic foil pasted over the surface of a nonmagnetic tube inserted into the room temperature bore or, if the cryogenic arrangements allow, onto the thermal shield or helium bore tube. As in the design of the magnet, linear programming can be used to optimize the mass and positions of the ferromagnetic shims (e.g., to minimize the mass of material).

Resistive Electrical Shims

The field of an NMR magnet for high resolution spectroscopy must be shimmed to at least 10^{-9} over volumes as large as a 10 mm diameter cylinder of 20 mm length. If, as is usually the case, substantial inhomogeneity arises from high-order harmonics (n and m greater than 3), superconducting shims are of barely sufficient strength. This arises because of the large radius at which they are located, at least in NMR magnets (e.g., in the regions x and y of Fig. 6). In general, the magnitude of a harmonic component of field generated by a current element is given by

$$B_n \propto r^{n+1}/r_0^{n+1} \qquad (15)$$

where r is the radius vector of the field point, and ρ_0 is the radius vector of the source. Thus, the effectiveness of a remote source is small for large n.

In order to generate useful harmonic corrections in NMR magnets for large n and m, electrical shims are located in the warm bore of the cryostat. Although in older systems those electrical shims took the form of coils tailored to specific harmonics, modern systems use matrix shims. Essentially, the matrix shim set consists of a large number of small saddle coils mounted on the surface of a cylinder. The fields generated by unit current in each of these coils form an influence matrix, similar to that of a set of steel shims. The influence matrix may be either the fields produced at a set of points within the magnet bore, or it may be the set of spherical harmonics produced by appropriate sets of the coils.

FIELD MEASUREMENT

The accurate measurement of the spatial distribution of field in the as-wound magnet is essential to shimming to high homogeneity. Sometimes, measurement of the field is possible at very low field strengths with tiny currents flowing in the windings at room temperature. That may allow mechanical adjustment of the positions of the main compensations coils ($k - k$ in Fig. 6) to reduce the $B(1,0)$, $B(2,0)$ and $B(1,1)$ harmonics. Major field measurement is made with the magnet at design field strength and in persistent mode. The methods of measurement in NMR and MRI magnets are generally different.

In NMR magnets, because of the small bore, the field is measured by a small NMR probe on the surface of a cylindrical region about 8 mm diameter and over a length of up to 10 mm. The measurements are made at typically 20° azimuthal intervals. From these field measurements, the predominant harmonics can be deduced, using a least-squares fit, and shimmed by means of the superconducting shim coils, both axial and radial. With subsequent measurements, as the harmonic content becomes smaller, the higher harmonics become evident and in turn can be shimmed. The field measurements are usually reduced to harmonic values because the shim sets are designed to generate specific harmonics. The correcting current required in any particular shim set is then immediately determined. Measurement and shimming is always an iterative process, generally requiring several iterations to achieve homogeneities of better than 10^{-9} over 5 mm dsv.

Field measurement in an MRI magnet is usually performed differently because much more space is available and because knowledge of the magnitudes of the harmonics in associated Legendre polynomial form is an advantage in the shimming process. In this case, the measuring points will lie on the surface of a sphere. Typically, the diameter of this sphere may be 500 mm. The field is measured at intervals of ϕ, often 30°, around each of the circles of intersection with this spherical surface of several $z = $ const planes, called Gauss planes. From these measurements and by the property of orthogonality of the associated Legendre functions, the values of the constants $C_{n,m}$ and $S_{n,m}$ in Eq. (10) can be deduced by the following methodology.

The double integral

$$\int_{-1}^{+1} \int_0^{2\pi} P(n,m)(u)[\cos(m\phi)]P_{i,j}(u)[\cos(j\phi)] \, du.d\phi$$

is nonzero only if $i = n$ and $j = m$. Then, for $m > 0$, its value is

$$2\pi (n + m)!/(2n + 1)(n - m)!$$

So, if both sides of Eq. (10) are multiplied by $P_{i,j}(u)[\cos(j\phi)]$ or $P_{i,j}(u)[\sin(j\phi)]$ and double integrated, the right-hand side will be nonzero only if $i = n$ and $j = m$. Then

$$\int_{-1}^{+1} \int_{0}^{2\pi} B_z(n, m) P_{i,j}(u)[\cos(j\phi)] \, du.d\phi$$
$$= C_{n,m} r^n (n + m + 1) 2\pi (n + m)!/(2n + 1)(n - m)!$$

and

$$\int_{-1}^{+1} \int_{0}^{2\pi} B_z(n, m) P_{i,j}(u)[\sin(j\phi)] \, du.d\phi$$
$$= S_{n,m} r^n (n + m + 1) 2\pi (n + m)!/(2n + 1)(n - m)! \quad (16)$$

Equation (16) are realized in practice by the measurement of the field at each of 60 points (for example) and the multiplication of each value by the spherical harmonics, $P_{i,j}(u) \cos(j\phi)$ and $P_{i,j}(u) \sin(j\phi)$ at that point. The integration is numerical. The method usually employed is Gaussian quadrature, similar in principle to Simpson's rule for numerical integration in Cartesian coordinates, but in which the $z = $ const planes are the roots of the Legendre polynomial and the weights assigned to the values measured on each of these planes are derived from the Lagrangian. Tables of the roots and weights are found in standard texts on numerical analysis (14). For the purposes of example, assume the number of planes $p = 5$, and the number of azimuthal points per plane $q = 12$, for a total of 60 points on the surface. The planes are at $z/r = 0$, $z/r = \pm0.5385$, and $z/r = \pm0.9062$; r is the radius of the spherical surface. The corresponding weights are 0.5689, 0.4786, and 0.2369. The measurements are made on the circles of intersection and two numerical integrations of Eq. (16) performed, one for the $\cos(m\phi_\theta)$ terms and the other for the $\sin(m\phi_\theta)$ terms. Then the values of $C_{n,m}$ and $S_{n,m}$ are obtained from

$$C_{n,m} = \left[\sum_{p}^{p} \sum_{q}^{q} w_q B(u_p, \phi_q) P_{n,m}(u_p) \cos(m\phi_\theta) \right] \quad (17)$$
$$\times [(2n + 1)(n - m)!]/[2\pi r^n (n + m + 1)!]$$

$$S_{n,m} = \left[\sum \sum w_q B(u_p, \phi_q) P_{n,m}(u_p) \sin(m\phi_\theta) \right] \quad (18)$$
$$\times [(2n + 1)(n - m)!]/[2\pi r^n (n + m + 1)!]$$

where $B(u_p, \phi_q)$ is the field at the point p, q, the subscripts p and q denote each of the 60 points, and w_q is the Gaussian weighting for the plane q.

NMR MAGNET DESIGN AND CONSTRUCTION

Practical issues peculiar to the design and construction of NMR magnets include the following:

> The wire diameter must be such that layers of windings near the inner radius of the solenoids do not generate discrete field fluctuation of a size comparable to the desired homogeneity. For example, if the wire diameter is very large, say, greater than 3 mm, the field will develop a fine structure away from the z axis. If the winding lay of a large diameter wire is helical in each layer, a helical structure may arise in the amplitude of the field with consequent problems in the correction of the resulting high harmonics.

> Nonmagnetic coil forms must be used because the presence of discrete regions of ferro- or strong paramagnetism will generate large harmonics of high order (large n and possibly also m), which would be very difficult to shim.

The index of the wire must be high. All high-resolution NMR requires high field stability, with a decay not exceeding about 10^{-8} per hour. To achieve that, the magnets operate in persistent mode. A superconducting switch is closed across the winding after energization so that the current flows without loss in a resistanceless circuit. The superconducting switch consists of a small coil (usually noninductive) of a superconducting wire equipped with a resistance heater. When the heater is energized, the temperature of the coil is raised above the critical value, and the coil becomes resisitive. The charging voltage applied to the magnet then causes only a small current to flow in the switch. When the magnet has been charged, the switch heater is turned off, the coil cools, and the magnet current can then flow through the switch without loss. If a magnet does not need frequent resetting, its rate of field decay must be small. The joints between wire lengths and the switch and the magnet must be superconducting, and the wire must be without resistance. Although the joints can indeed be made so that their critical currents exceed the operating current, the effective resistance of the wire, owing to its index, may be high enough that decay in persistent mode exceeds acceptable levels for NMR. The resistance of the wire, manifest as a low value of the index, arises from variation in the critical current along the length of the wire. If a short region exists where the superconducting filaments are thin or have low pinning strength, a fraction of the current transfers between superconducting filaments through the copper (or bronze) matrix, giving rise to the resistance. The voltage associated with this resistance appears in critical current measurements on samples of the wire.

Figure 10 shows the typical trace of voltage gradient along a superconducting wire in a fixed field as a function of cur-

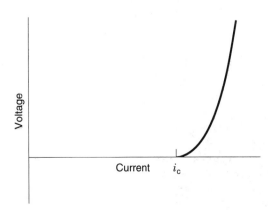

Figure 10. Voltage gradient along a composite superconductor as a function of steady current, showing the effect of index.

rent. The defined critical current is that at which a voltage gradient of, typically, 0.1 μV/cm is measured. As the current is increased beyond the critical value, that voltage gradient increases. An approximation to the gradient near to the critical current is

$$v \sim (i/i_c)^N \qquad (19)$$

where i/i_c is the ratio of actual current to critical current and N is the index of the wire. The higher the value of N, the sharper is the superconducting to normal transition. Clearly, for values of i below the critical value, the voltage gradient will be small; the larger the index, the smaller the gradient. So, for NMR magnets, an appropriate combination of index and the ratio of i/i_c must be chosen. The index of most niobium–titanium (NbTi) wires suitable for MRI and NMR is typically 50. However, for niobium–tin (Nb_3Sn) wires, the index is lower, typically around 30, and the matrix is the more resistive bronze. So, for high field NMR magnets using Nb_3Sn inner sections, lower ratios of i/i_c are necessary. The concept of the index is only an approximation to the behavior of voltage as a function of current. In fact, theory and measurement indicate that the effective index increases as i/i_c decreases below 1 (15,16). Field decay arising from the index is constant and is distinguished from that caused by flux creep. The latter is a transient effect. It dies away with a logarithmic time dependence after a magnet has been set in persistent mode.

Protection of the magnet from the consequences of quenching must be compatible with the electrical and thermal isolation of the magnet from room temperature systems. Quenching is an spreading irreversible transition from the superconducting to the normal resistive state in the winding. The energy released during quenching in an NMR or MRI magnet must be dissipated as heat in the winding. In order to limit the energy and hence heat dissipated in any part of the winding, the magnet must be electrically divided into sections, each of which is provided with a shunt, often in the form of diodes. This subdivision limits the energy that can be transferred between sections and thereby minimizes the temperature rise and voltage generated within a section during quenching (17).

NMR Magnet Design

Figure 6 illustrates the winding array of a typical 750 MHz NMR magnet (18). Table 2 specifies the dimensions and winding specifications of the sections.

At a current of 307.86 A, these windings generate 17.616 T at the center; that corresponds to 750 MHz proton resonance frequency. The total inductance is 109.2 H, and the stored energy is 5.17 MJ. The first nonzero harmonic of the design is the 12th. The coils s, t, u, v, and w and their mirror images are the axial shim coils located in the annular space x. The radial shim coils are located in the space y.

The winding of the Nb_3Sn sections of high field NMR magnets presents particular problems. The wire is wound in the unreacted state after which it must be heated at about 700°C for up to 200 h to transform the separate niobium and tin components into the superconducting compound. The wire is insulated with S-glass braid, with a softening temperature of about 1000°C. (An alternative is E-glass. Although the E-glass may start to soften during the heat treatment, it is stronger in the prefired state than S-glass and therefore better survives the exigencies of winding.) After the heat treatment the winding is consolidated by impregnation with epoxy resin.

The forms on which the Nb_3Sn wire is wound must also endure the heat treatment without distortion. Stainless steel is the universal choice for the coil forms although titanium alloys have been used. The alloy 316 L is generally preferred because of its very small magnetic susceptibility. If the form is assembled with welds, those must be made with nonmagnetic filler, if used. The inner bore of the form must be quite thick if distortion is not to occur. The reason for that lies in the expansion coefficients of the wire and of stainless steel. The unreacted Nb_3Sn wire consists of bronze, niobium, tin, tantalum, and copper. During reaction, the copper and bronze have negligible strength, and the mechanical properties of the niobium and tantalum dominate. Their coefficients of thermal expansion are smaller than that of stainless steel with the consequence that, as the temperature rises during the heat treatment, the bore of the form will expand faster than the inner diameter of the winding. If the bore tube is thin, it can buckle against the constraint of the winding.

The need for thick bore tubes leads to windings of several wire diameters on one form. The thick bore tube occupies space that could otherwise be used by field-generating winding. In order to minimize the diluting effect of the bore tube, large winding builds are used. However, in the high field regions of Nb_3Sn windings, the wire diameter must be graded to optimize the cross section of Nb_3Sn corresponding to the local field. An alternative to the thick stainless steel bore tube is the transfer of the reacted winding to an aluminum form before impregnation with epoxy resin. This has been used occasionally, as in the example of Fig. 6 (18).

MRI Magnet Design

The design construction of MRI magnets follows the principles involved in the construction of NMR magnets (19). The forces and energies are generally greater. For example, the force tending to center each large end coil of the MRI magnet illustrated in Fig. 11 is 1,339,000 N (150 tons). NbTi conductor is used exclusively in MRI magnets because, to date, center fields of no more than 5 T are used. The NbTi filaments in the copper matrix of composite NbTi wires are twisted to approximate transposition. Because of the low fields in MRI magnets, wires with few NbTi filaments can be used. Those filaments can then be arrayed as a single circular layer within a copper matrix. The filaments are then fully transposed and are magnetothermally very stable. Mechanical perturbation is nevertheless a problem, and attention has to be paid to the interface between a winding and the coil form against which it presses. Because of the high stored energies, large copper cross sections are needed in the conductor to avoid over heating during quenching. Currents are typically up to 500 A. A common form of conductor is a composite wire embedded in a copper carrier. The latter frequently has a grooved rectangular cross section into which the composite wire is pressed or soldered. Insulation may be cotton or kapton wrap instead of enamel, and wax may be used as an impregnant as an alternative to epoxy.

Table 2. Dimensions of the Windings of a 700 MHz NMR Magnet

Section Number	Peak Field (T)	Wire Type	Wire Diameter (mm)	Inner Diameter (mm)	Outer Diameter (mm)	Winding Length (mm)	Number of Turns
a	17.62	Nb$_3$Sn	2.4	43	52.2	600	1000
b	16.97	Nb$_3$Sn	2.22	52.2	65.3	600	1620
c	15.93	Nb$_3$Sn	1.84	68.8	79.9	650	1770
d	14.8	Nb$_3$Sn	1.83	83.4	92.5	650	1780
e	13.83	Nb$_3$Sn	1.63	99.0	115.2	700	3870
f	11.80	Nb$_3$Sn	1.61	121.7	134.8	750	4194
g	9.83	NbTi	1.41	141.3	149.5	800	3396
h	8.31	NbTi	1.30	158.0	171.0	1000	6948
j	5.70	NbTi	1.14	176.0	186.2	1000	6992
k	3.13	NbTi	1.30	194.2	212.5	88.7	1020
l / l	3.14	NbTi	1.30	194.2	212.5	377.1	4110 (each)

At a current of 307.86 A these windings generate 17.616 T at the center; that corresponds to 750 MHz proton resonance frequency. The total inductance is 109.2 H and the stored energy is 5.17 MJ. The first nonzero harmonic of the design is the twelfth.

Most whole-body MRI magnets used in clinical applications have room temperature bores of between 1 and 1.3 m, with fields up to 2 T. An example of the profile of the windings of a whole-body MRI magnet is illustrated by the simple five-coil system of Fig. 11.

The center field is 1.5 T and the dimensions of the windings are listed in Table 3. The compensation is to tenth order (10 ppm over a 500 mm sphere). The current is 394 A for a 1.5 T center field. The inductance is 78 H and the stored energy 6 MJ.

The fringing field of this magnet extends a long way from the cryostat in which the coils are housed. The 1 mT (10 gauss) line is at an axial distance of 11.3 m and at a radial distance of 8.8 m from the center. Access to the space within these limits must be restricted because of the dangers to the wearers of pacemakers, the attraction of ferromagnetic objects, and the distortion of video monitors.

This may be an expensive restriction in a crowded hospital. Therefore, methods of shielding the space from the fringing fields are frequently used. Three methods are generally available: close iron, remote iron, and active shielding. The use of iron close to the coils has been used in a few instances. However, the iron must be at room temperature, to avoid otherwise severe cryogenic penalties. That leads to difficulties in balancing the forces between the coils and the iron in order to minimize the loads that the cryogenic supports must resist.

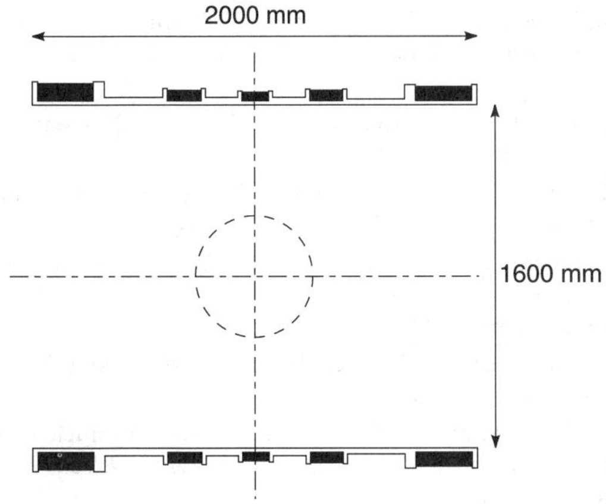

Remote iron takes the form of sheet, typically several millimeters thick, placed against the walls of the MRI room. This involves rather awkward architectural problems but is used frequently where the restricted space can still extend several meters from the cryostat.

The third form of shielding is by superconducting coils, built around the main coils, operating in series with the main coils as part of the persistent circuit, and in the same cryogenic environment. Those shield coils generate a reversed field to cancel, or reduce, the external fringing field. Typical of the resulting magnet is the eight-coil configuration shown schematically in Fig. 12.

Particular aspects of the illustration follow. The outwardly directed body forces in unshielded MRI windings are supported in tension in the conductor. However, in the shielded version, those forces are too large to be supported by the conductor alone, and a shell is applied to the outside of the winding against which the accumulated body forces act. Thus, the body forces on coils 3 and 4 are supported on their outer surfaces by a structural cylinder. Coil 4 provides the compensation of the dipole moment of the three inner windings so that the fringing fields of the magnet are much reduced from those of the unshielded magnet. The reduction in the volume of the restricted space is about 93%. The magnet is much heavier (and more expensive) than the simple unshielded type and the structural design of the cryostat and the suspension system is accordingly stronger. The highest fields generated at the center of shielded whole-body MRI magnets is 2 T. See also Ref. 12.

Figure 11. Coil profile of a 1.5 T unshielded MRI magnet illustrating typical coil placement.

Table 3. Example of a 1.5 T Whole-Body MRI Magnet

Coil Number	Inner Radius (mm)	Outer Radius (mm)	Left End (mm)	Right End (mm)	Number of Turns
1	741.9	807.1	−977.0	−724.9	3030
2	742.2	785.7	−391.9	−244.6	1180
3	742.1	777.0	−74.9	74.9	960
4	742.2	785.7	244.6	391.9	1180
5	741.9	807.1	724.9	977.0	3030

CRYOGENICS

As for NMR magnet systems, the economic operation of super-conducting MRI magnets demands cryogenic systems with low heat in-leak. The evolution of MRI cryostats has been significant over the past 15 years. They have changed from simple liquid helium, liquid nitrogen shielded reservoirs with relatively high cryogen evaporation rates to single or multi-cryocooled cryostats. In one embodiment, no refrigerant is used in some types of cryocooled MRI magnets; in other examples, a combination of cryocoolers and refrigerants provide a zero evaporation rate. Demountable current leads are an essential feature of any magnet system with low refrigerant evaporation rate, and have been a standard feature of MRI magnet systems since 1974. An implication of demountable current leads is the need for the MRI magnet to be self-protecting during quenching, just as an NMR magnet must be.

PULSED GRADIENT COILS

In addition to the uniform background field, which it is the function of the MRI magnet to generate, pulsed gradient fields must be superimposed on that field in order to create the spatial encoding of the resonant frequencies of the protons (or other species) within the body. Those pulsed gradient fields are generated by three sets of room temperature coils, each set being driven by a powerful ramped current source. The pulsed field gradients are linear (dB/dz, dB/dx, dB/dy), as far as it is possible to design pure first-order gradient coils. Two problems arise in the overall design as a consequence of these pulsed gradient fields. The first is the effect on the superconductor of the periodic incident fields. Although the thermal shields and coil forms lie between the gradient coils and the superconductor, the incident pulsed field would still be significant there. Those small fields would cause a loss within the conductor through the mechanisms of hysteresis and coupling.

The second is the distortion of the gradients arising from currents induced in adjacent structures, such as the thermal

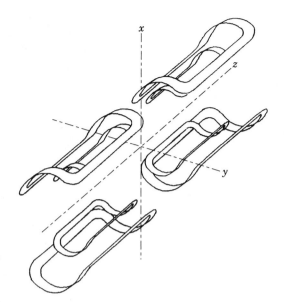

Figure 13. Schematic of actively shielded pulsed gradient coils for the dB_z/dx gradient showing the spacing between the main and shield coils.

shields and helium vessel, also sometimes the room temperature bore tube, if that is metallic. This distortion is minimized by locating sets of shield coils near the room temperature bore tube. These active shield coils are energized in opposition to the main pulsed gradient coils. They serve to confine the return flux of the gradient coils to flow in the space between the main gradient coils and the shield coils. The eddy currents induced in the surrounding structures are thereby minimized. The shield coils reduce the efficiency of the pulsed gradient system, that effect becoming more pronounced as the diameter of the main coils becomes a large fraction of that of the shield coils. At a diameter ratio greater than about 0.85, the efficiency is so reduced that the driving power required for useful gradient fields becomes prohibitively large. Figure 13 illustrates the form of the shielded dB/dx or dB/dy pulse coils. The dB/dZ coils are simple solenoids surrounded by shielding solenoids.

BIBLIOGRAPHY

1. E. M. Purcell, H. C. Torrey, and R. V. Pound, Resonance absorption by nuclear magnetic moments in a solid, *Phys. Rev.*, **69**: 37, 1946.

2. F. Bloch, W. W. Hansen, and M. Packard, Nuclear induction, *Phys. Rev.*, **69**: 127, 1946.

3. E. Becker, *High resolution NMR, Theory and Applications*, New York: Academic Press, 1980.

4. R. R. Ernst, G. Bodenhausen, and A. Wokaun, *Principles of Nuclear Magnetic Resonance in One and Two Dimensions*, Oxford: Clarendon Press, 1987.

5. D. B. Montgomery, *Solenoid Magnet Design*, New York: Wiley-Interscience, 1969, p. 4.

6. M. D. Sauzade and S. K. Kan, High resolution nuclear magnetic spectroscopy in high magnetic fields, *Adv. Electron. Electron Phys.*, **34**: 1–93, 1973.

7. W. R. Smythe, *Static and Dynamic Electricity*, New York: McGraw-Hill, 1950, pp. 147–148.

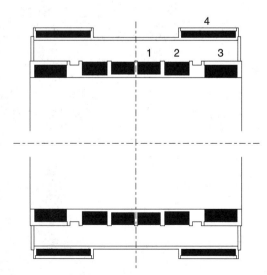

Figure 12. Schematic of an actively shielded MRI magnet showing the large coils needed to generate a main field while the shield coils generate an opposing field.

8. M. Abramowitz and I. A. Stegun (eds.), *Handbook of Mathematical Functions,* Washington, DC: US Dept. of Commerce, Natl. Bur. of Standards, 1964.

9. F. Romeo and D. I. Hoult, Magnetic field profiling: Analysis and correcting coil design, *Magn. Resonance Med.,* **1**: 44–65, 1984.

10. M. R. Thompson, R. W. Brown, and V. C. Srivastava, An inverse approach to the design of MRI main magnetics, *IEEE Trans. Magn.,* **MAG-30**: 108–112, 1994.

11. W. H. Press et al., *Numerical Recipes: The Art of Scientific Computing,* Cambridge, UK: Cambridge Univ. Press, 1987.

12. F. J. Davies, R. T. Elliott, and D. G. Hawksworth, A 2 Tesla active shield magnet for whole body imaging and spectroscopy, *IEEE Trans. Magn.,* **MAG-27**: 1677–1680, 1991.

13. E. S. Bobrov and W. F. B. Punchard, A general method of design of axial and radial shim coils for NMR and MRI magnets, *IEEE Trans. Magn.,* **MAG-24**: 533–536, 1988.

14. P. Davis and P. Rabinowitz, Abscissas and weights for Gaussian quadrature of high order, *J. Res. NBS,* (RP2645), AMS (55): 35–37, 1956.

15. Y. Iwasa, *Case studies in Superconducting Magnets,* New York: Plenum, 1994, pp. 306–307.

16. J. E. C. Williams et al., NMR magnet technology at MIT, *IEEE Trans. Magn.,* **MAG-28**: 627–630, 1992.

17. B. J. Maddock and G. B. James, Protection and stabilisation of large superconducting coils, *Proc. Inst. Electr. Eng.,* **115**: 543–546, 1968.

18. A. Zhukovsky et al., 750 MHz NMR magnet development, *IEEE Trans. Magn.,* **MAG-28**: 644–647, 1992.

19. D. G. Hawksworth, Superconducting magnets systems for MRI, *Int. Symp. New Develop. in Appl. Superconductivity,* Singapore: World Scientific, 1989, pp. 731–744.

JOHN E. C. WILLIAMS
Massachusetts Institute of
Technology

MAGNETS FOR NMR.   See MAGNETS FOR MAGNETIC RESONANCE ANALYSIS AND IMAGING.

MAGNETS, PERMANENT.   See PERMANENT MAGNETS.

MAGNETS, SUPERCONDUCTING.   See SUPERCONDUCTING CRITICAL CURRENT; SUPERCONDUCTING MAGNETS FOR PARTICLE ACCELERATORS AND STORAGE RINGS; SUPERCONDUCTING MAGNETS, QUENCH PROTECTION.

MAGNETS, SUPERCONDUCTING FOR NUCLEAR FUSION.   See SUPERCONDUCTING MAGNETS FOR FUSION REACTORS.

MAINTENANCE, AIRCRAFT.   See AIRCRAFT MAINTENANCE.

MAINTENANCE, JET TRANSPORT AIRCRAFT.   See JET TRANSPORT MAINTENANCE.

MAINTENANCE, SOFTWARE.   See SOFTWARE MAINTENANCE.

MAJORITY LOGIC

Majority logic is used to implement a majority decision mechanism. In particular, a majority logic block (MLB) can be regarded as a black box that receives different data values at its inputs and gives, as its output, the data value present on the majority of its inputs. For instance, if an MLB receives the three input data "1," "1," and "0," it gives the output data "1."

MLBs are generally used as component blocks of digital electronic systems devoted to critical operations, like control systems of nuclear plants, flight control systems of aircraft, speed control systems of trains, or onboard control systems of satellites, etc. The correct uninterrupted operation of these systems is mandatory because their malfunction could cause catastrophic consequences, such as loss of human life or huge economic loss. To avoid such losses, the system must be designed to guarantee its correct behavior (that is, the behavior expected in the fault-free case) despite the occurrence of internal faults.

In fact, electronic systems are prone to faults. Faults occur during the system's manufacturing process and during its operation. For instance, in the presence of a high electrical field, high current density within a circuit metal line might cause the line to break. This might make the faulty circuit provide an output datum different from the correct one, for example, a "0" rather than a "1." If the faulty signal, for instance, is a signal that activates the alarm device of a train's speed control system when equal to "1," it is obvious that a faulty "0" may cause the whole system to become ineffective with possible catastrophic consequences.

Unfortunately, faults of this kind (and faults of a different kind) might occur within a digital electronic system. The likelihood of their occurrence is reduced by using proper electronic components, but ensuring that they never occur is impractical.

Hence, if the correct operation of a system is crucially important, some precautions must be taken to avoid the catastrophic consequences that faults might produce. The techniques used to reach this goal are generally called fault-tolerant techniques.

In particular, to guarantee that a system tolerates its possible internal faults (that is, to ensure that no system malfunction occurs because of such faults), redundancy is used. As an example, to guarantee that a train's speed control system tolerates faults in the circuit activating the alarm device, redundant copies of such a circuit are used. If these redundant copies are properly isolated from one another, a single fault during the system operation affects only one copy, hence only this copy provides incorrect (or faulty) output data, whereas the other copy (or copies) gives correct output data.

However, it should be obvious that redundancy alone is not sufficient to ensure that a system tolerates its possible internal faults. To reach this goal, we need some decision criterion that allows us to determine which data value, among those at the output of the redundant copies, can be regarded as correct and which one as incorrect. This decision criterion normally is the majority criterion, implemented by an MLB.

Of course, three is the minimum number of copies of the same circuit (or, more generally, module) that we must have to allow the MLB's selection of the majority data value.

The use of MLBs and redundant copies of the same module characterize the fault-tolerant *N*-modular redundancy (NMR) systems where, as previously introduced, *N* must be ≥3. The idea to use *N*-modular redundancy and majority logic blocks (also called majority "voting blocks," or "voters") to

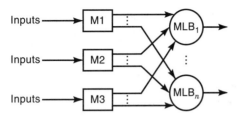

Figure 1. Representation of a triple modular redundancy (TMR) system.

achieve fault-tolerance was first introduced in Ref. 2 and has been adopted for several critical applications, such as space exploration missions and nuclear reactor protection (3,4).

In the particular case where $N = 3$, these systems are called triple modular redundancy (TMR) systems. Hence a TMR system consists of (1) three copies of the original, non-fault-tolerant module, that concurrently process the same information (where a module is simply a circuit, a gate, or even a microprocessor, depending on system's related choices), (2) n MLBs (where n is the number of outputs of the replicated module), each comparing bit-by-bit the corresponding outputs of the three replicated modules, and producing at its output the majority data value among those at its inputs (Fig. 1). Moreover, suitable techniques are generally adopted at the system level to avoid exhausting all of the system's redundancy as modules become successively faulty.

Obviously, MLBs play a critical role in the fault-tolerant systems described. In fact, the correct operation of the whole system is compromised if an MLB becomes faulty. For instance, it is obvious that, if the output of an MLB is affected by a stuck-at "1" fault (for instance, because of a short between an output line and the circuit power supply), the faulty MLB always gives an output "1," regardless of the data values at its inputs (that is, also when the data value on the majority of its inputs is a "0").

In the remainder of this article we consider the problem of designing MLBs for TMR and NMR systems and the case of faulty MLBs.

MAJORITY LOGIC FOR TRIPLE MODULAR REDUNDANCY SYSTEMS

As previously introduced, an MLB can be regarded as a black box that must produce at its output the datum value present on the majority of its inputs. To distinguish such a majority datum value, an MLB must have at least three inputs. When this is the case, the MLB must satisfy the truth table shown in Table 1, where Z denotes the output of the MLB, and $X = (X_1, X_2, X_3)$ is the MLB input vector.

Table 1. Truth Table of a Three-Input Majority Logic Block

$X_1X_2X_3$	Z
000	0
001	0
010	0
011	1
100	0
101	1
110	1
111	1

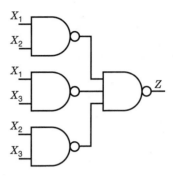

Figure 2. NAND-NAND logical implementation of a majority logic block for a TMR system.

Hence, the MLB can be implemented (at the logic level) by a two-level NAND-NAND (Fig. 2) or NOR-NOR (Fig. 3) circuit. Equivalently, a two level AND-OR or OR-AND implementation can be considered (5).

Of course, the electrical implementations of these MLBs depend on the technology adopted. As a significant example, if the complementary MOS (CMOS) technology is used, the NAND-NAND majority logic block can be implemented by the circuit shown in Fig. 4. Alternatively, the circuit shown in Fig. 5, for instance, can be used (6). Another possible electrical implementation of MLBs for TMR systems (Fig. 6) was proposed in (6). Different from the conventional MLB realizations considered up to now, this MLB produces an output error message in case of internal faults that make it give an incorrect majority output datum value. The behavior of this circuit is described later while dealing with the problems due to faults affecting MLBs.

As regards the number of errors in X tolerated by a general TMR system, it is evident that, if only one of the MLB input data is incorrect (that is either X_1, X_2, or X_3), the MLB produces the correct output datum value. Instead, if two input data are incorrect, the MLB produces the incorrect output datum value. Hence, a TMR system tolerates only a single error on the corresponding outputs of the replicated modules. If a higher number of errors must be tolerated (because of system level requirements), an NMR system with n MLBs (where n is the number of outputs of the replicated module), each with N inputs ($N > 3$), must be used.

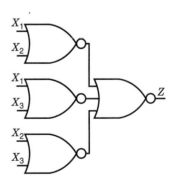

Figure 3. NOR-NOR logical implementation of a majority logic block for a TMR system.

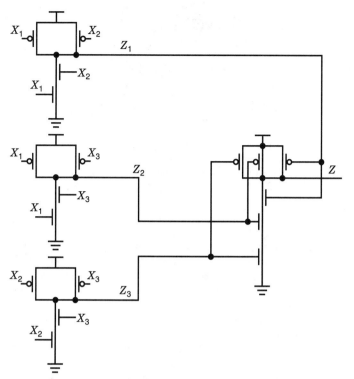

Figure 4. Example 1 of a possible CMOS implementation of the NAND-NAND majority logic block in Fig. 2.

MAJORITY LOGIC FOR *N*-MODULAR REDUNDANCY SYSTEMS

To tolerate E errors in the vector $\boldsymbol{X} = (X_1, X_2, \ldots, X_N)$ given to the input of an MLB of an NMR system, N and E must satisfy the following equation (7):

$$N \geq 2E + 1 \qquad (1)$$

In fact, when this is the case, the number of erroneous bits E is smaller than the number of correct bits $(N - E)$.

Barbour and Wojcik (5) added an upper bound to the value of N of an NMR system tolerating E errors. In particular, they

transformed Eq. (1) into

$$2E + 1 \leq N \leq (E + 1)^2 \qquad (2)$$

Moreover, they proposed possible two-level implementations of MLBs for NMR systems, whose values of N and E satisfy Eq. (2). These are the implementations most widely used for MLBs of general NMR systems.

In particular, the first level of the proposed MLBs consists of

$$\binom{N}{K}$$

AND (or OR, or NAND, or NOR) gates, where K is a parameter [called a "grouping parameter" (5)] that must satisfy the following condition:

$$E + 1 \leq K \leq N - E \qquad (3)$$

The inputs to these first level gates are the

$$\binom{N}{K}$$

combinations of K elements out of the N elements of the MLB input vector $\boldsymbol{X}$.

The second level of the proposed MLBs consists of a simple OR (or AND, or NAND, or NOR) gate, depending on whether AND (or OR, or NAND, or NOR, respectively) gates are used at the first level.

As an example, the general NAND-NAND implementation of such an MLB, with generic "grouping parameter" K, is

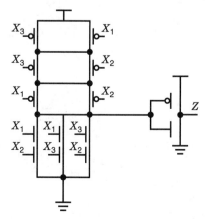

Figure 5. Example 2 of a possible CMOS implementation of the NAND-NAND majority logic block in Fig. 2.

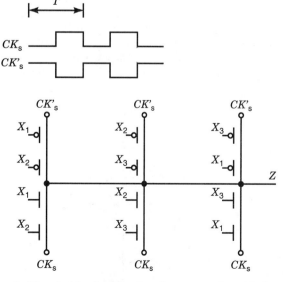

Figure 6. Electrical implementation of a majority logic block giving an output error message in case of internal faults affecting its correct operation.

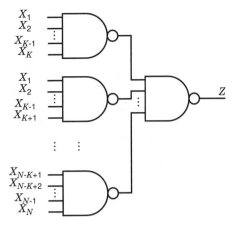

Figure 7. NAND-NAND logical implementation of a majority logic block for an NMR system with "grouping parameter" = K.

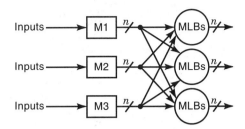

Figure 9. Triplicated MLB scheme proposed in Ref. 9. In principle, triplication of the MLB allows faults that could possibly affect one of the replicated MLBs to be tolerated. In practice, problems arise in correctly discriminating the value given by the majority of the replicated MLBs.

shown in Fig. 7. We can easily verify that, if $N = 3$ and $K = 2$, this implementation equals that reported in Fig. 2.

The derived electrical implementations of these MLBs are straightforward.

An alternative implementation of MLBs for NMR systems was proposed in Ref. 8. These MLBs (Fig. 8) are suitable for NMR dynamic systems, that is, in systems where, differing from the conventional cases considered until now, the number of processed replicated modules (hence the number of inputs of the MLBs) can be dynamically changed during the system's operation.

PROBLEMS IN CASE OF FAULTY MAJORITY LOGIC

As previously mentioned, if the MLB of a TMR (NMR) system becomes faulty, it might produce at its output an incorrect majority datum value, hence making the adoption of the TMR (NMR) fault-tolerant technique useless.

To deal with this problem, MLBs can themselves be replicated (9), as schematically shown in Fig. 9 for a TMR system. Note that in this and in the following figures, we use a common symbol to represent n MLBs and the n outputs of the replicated modules, respectively. It is obvious that, if this scheme is adopted, faults affecting one of the replicated MLBs that make it produce incorrect output data can be tolerated. However, similar to the case of replicated modules only, in

this case the problem of distinguishing the datum value provided by the majority of the replicated MLBs also arises. Of course, if other MLBs (receiving the outputs of the replicated MLBs) are used to fulfill this purpose, the problem discussed here is simply translated to these final MLBs.

An alternative strategy for dealing with this problem is to renounce the fault-tolerance requirement of the MLBs and to require only that the MLBs give an output error message in case of internal faults that make them produce incorrect output data. Such an error message can be exploited at the system level to avoid the dangerous consequences possibly resulting from incorrect MLB operation. For instance, if an error message is received by one of the MLBs of a train's speed control system, a system level recovery procedure can be automatically started, eventually making the train stop.

MLBs of this kind can be found in Refs. 1, 6, and 10.

In particular, in Ref. 1, the MLBs are duplicated (Fig. 10), and their outputs are given to the inputs of a comparator that verifies whether or not such outputs are equal to one another. In case of a disagreement, the comparator gives an output error message. It is obvious that this strategy guarantees that, in case of faults affecting an MLB that make it produce erroneous majority output data, the comparator gives an output error message.

However, compared with the case where single MLBs are used, this solution implies an inevitable increase of area overhead and power consumption costs of the fault-tolerant system.

To reduce these costs in TMR systems, the MLB introduced in Ref. 6 can be adopted. The electrical structure of this MLB is shown in Fig. 6, where CK_S and CK_S' denote a periodic signal whose period is T and an opposite waveform, respectively.

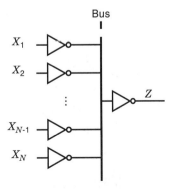

Figure 8. Majority logic block suitable for NMR dynamic systems introduced in Ref. 8.

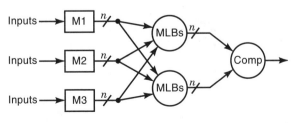

Figure 10. Duplicated MLB scheme presented in Ref. 1. Faults affecting one of the duplicated MLBs cannot be tolerated but can be detected by the output comparator.

This MLB provides an output error message when internal faults occur that make it give an output incorrect majority datum value. This behavior is guaranteed for all MLB node stuck-at, transistor stuck-on, transistor stuck-open, and resistive bridging faults (whose values of the connecting resistance are in the range 0 Ω to 6 kΩ).

In the fault-free case, this MLB gives as its output (1) a signal equal to CK_S' if the majority of its input signals is equal to "0;" (2) a signal equal to CK_S if the majority of its input signals is equal to "1" (Table 2). Hence, in the fault-free case, this MLB gives at its output a signal assuming both high and low logic values within the same period T. In particular, the logic value produced when $CK_S = 1$ is equal to the majority datum value (MV in Table 2) among those given to its input during such a period T.

Instead, in case of an internal fault of the kind previously reported, either the MLB is not affected by the fault (that is, it continues providing the correct majority output data), or it produces an output error message (in particular a signal that does not change logic value within T). If a fault does not affect the MLB and following internal faults occur, either the MLB is not affected by these internal faults or it provides an output error message. This condition holds for all possible sequences of internal faults under the general assumptions that internal faults occur one at a time during the operation of an integrated circuit and that the time interval elapsing between two succeeding faults is long enough to allow the fault-free modules to produce both high and low output data values.

The presence of an error message at the output of this MLB can be simply revealed. For instance a flip-flop sampling the MLB output on both the CK_S rising and falling edges (11) can be used.

This MLB implementation can also be extended to NMR systems (10). In this case, compared with the electrical structure shown in Fig. 6, (1) the number of parallel pull-up/pull-down branches changes from

$$\binom{3}{2}$$

to

$$\binom{N}{(N+1)/2}$$

(2) each branch consists of $(N + 1)/2$ series transistors, rather than of 2. Obviously these conditions may limit the maximal

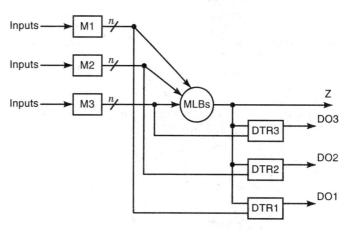

Figure 11. Detecting scheme possibly adopted to avoid the exhaustion of a TMR system redundancy. The detectors (DTRi, $i = 1, 2, 3$) reveal the disagreement between the outputs of the replicated modules and the majority output data given by the MLBs.

value of N for which this implementation can be conveniently used.

As previously introduced, suitable techniques must be also adopted to avoid exhausting the TMR (NMR) system's redundancy because modules become successively faulty. To fulfill this purpose, detectors that reveal the disagreement between the outputs of the MLBs and of the replicated modules can be used (12), as shown schematically in Fig. 11. A possible electrical implementation of detectors of this kind, suitable for use together with the MLBs described, can be found in Ref. 10. Finally, several critical applications require that the MLBs of the used TMR (NMR) systems provide fail-safe outputs; that is, signals whose value is either correct or safe (where, as an example, the red color is the "safe" value for traffic control lights). Possible implementations of this kind of MLBs can be found in Ref. 13.

BIBLIOGRAPHY

1. K. G. Shin and H. Kim, A time redundancy approach to TMR failures using fault-state likelihoods, *IEEE Trans. Comput.*, **43**: 1151–1162, 1994.

2. J. V. Neumann, Probabilistic logics and the synthesis of reliable organisms from unreliable components, *Automata Studies, Ann. Math. Studies,* **34**: 43–98, 1956.

3. C. E. Stroud and A. E. Barbour, Testability and test generation for majority voting fault-tolerant circuits, *J. Electron. Test.: Theory Appl.*, **4**: 201–214, 1993.

4. D. Audet, N. Gagnon, and Y. Savaria, Quantitative comparisons of TMR implementations in a multiprocessor system, *Proc. 2nd IEEE Int. On-Line Testing Work,* Biarritz, France, 1996, pp. 196–199.

5. A. E. Barbour and A. S. Wojcik, A general, constructive approach to fault-tolerant design using redundancy, *IEEE Trans. Comput.*, **38**: 15–29, 1989.

6. C. Metra, M. Favalli, and B. Riccò, Compact and low power self-checking voting scheme, *Proc. IEEE Int. Symp. Defect Fault Tolerance VLSI Syst.,* Paris, France, 1997, pp. 137–145.

7. W. H. Pierce, *Failure-Tolerant Computer Design,* New York: Academic, 1965.

8. N.-E. Belabbes, A. J. Guterman, and Y. Savaria, Ratioed voter circuit for testing and fault-tolerance in VLSI processing arrays,

Table 2. Truth Table of the Majority Logic Block in Ref. 6 and Corresponding Majority Datum Value

$X_1X_2X_3$	Z	MV
000	CK_S'	0
001	CK_S'	0
010	CK_S'	0
011	CK_S	1
100	CK_S'	0
101	CK_S	1
110	CK_S	1
111	CK_S	1

IEEE Trans. Circuits Syst. I, Fundam. Theory Appl., **43**: 143–152, 1996.

9. P. K. Lala, *Fault Tolerant and Fault Testable Hardware Design,* Englewood Cliffs, NJ: Prentice-Hall, 1985.

10. C. Metra, M. Favalli, and B. Riccò, On-line self-testing voting and detecting schemes for TMR systems, *J. Microelectron. Syst. Integration,* **5** (4): 261–273, 1997.

11. M. Afghahi and J. Yuan, Double edge triggered D flip-flop for high speed CMOS circuits, *IEEE J. Solid State Circuit,* **SC-26**: 1168–1170, 1991.

12. N. Gaitanis, The design of totally self-checking TMR fault-tolerant systems, *IEEE Trans. Comput.,* **37**: 1450–1454, 1988.

13. M. Nicolaidis, Fail-safe interfaces for VLSI: Theoretical foundations and implementation, *IEEE Trans. Comput.,* **47** (1): 62–77, 1998.

CECILIA METRA
University of Bologna

MANAGEMENT EDUCATION

Engineers have been candidates for managerial positions in industrial and government organizations for most of the twentieth century. During the latter half of the century, the percentage of engineers moving into managerial positions has increased dramatically. About two-thirds of all individuals with engineering degrees now pursue a management career track, as opposed to design, research, or other career tracks, and about 85% of all engineers will, at some point in their careers, have managerial responsibility. With the current emphasis on integrating sophisticated information, communication, and automation technologies into engineered systems as well as into the processes for designing, manufacturing, and marketing those systems, adequate technological knowledge to understand the intricacies of these systems and processes is recognized as an important attribute of the managers whose task is to coordinate the processes. Hence, the percentage of individuals being promoted into managerial positions who have some form of technical background (engineers and applied scientists being prominent among that group) has been increasing over the past two decades. With the trend toward wider distribution of management functions, and in particular the emergence of the project form of organization in the high-tech industries, these managerial positions tend to have titles like project manager (or leader), program manager, product manager, systems manager, operations (or production) manager, and field office manager or assistant manager. Both the redistribution of management functions and their redefinition within a framework of integration has led to special needs for the management education of engineers.

A BRIEF HISTORY

During the first half of the twentieth century, most organizations relied heavily on on-the-job training and internal management training programs to prepare engineers and other employees for positions of management. While large organizations continue to staff training departments and organizations of all types contract for internal management training services, there emerged after World War II an awareness of the value of more formal management education. For those with engineering degrees, a master's degree in management, most notably the master of business administration (MBA), became the program of choice. Many employers selected certain individuals to enroll in these programs full time and set up reimbursement schemes to encourage others to enroll part time. By the late 1950s and early 1960s, the approaches and techniques being developed to manage new and complex military and space technologies, in particular, suggested a form of management education different from what was being offered through MBA programs. The new form needed to be designed specifically for those who were to become program and systems managers. For engineers and applied scientists, some early responses to this need included master's degree programs in engineering administration at George Washington University and in engineering management at the University of Missouri–Rolla, each of which evolved into its own department. Other programs began sprouting within existing departments (e.g., a master's degree in engineering management within the University of Pittsburgh's department of industrial engineering) and, some departments changed their name when an engineering management program was added (e.g., the department of industrial engineering and engineering management at Stanford University and the department of operations research and engineering management at Southern Methodist University).

There are now, in the United States, more than one hundred master's degree programs in management that cater to engineers and applied scientists, as well as many technology and systems management programs for both engineers and nonengineers. Each of these programs has its own flavor, representing a variety of curricula, forms of staffing, and delivery modes. The importance of employing engineers with a managerial perspective has also led to the development of undergraduate programs, both majors and minors, and undergraduate management courses (required of engineers at some institutions) to provide that perspective. Nondegree certificate programs have recently become popular alternatives to master's degrees as a way to deliver management education in a shorter timeframe and in nontraditional formats.

The history that has created the current variety of management education programs presents, for the engineer, the problem of discerning the differences between the options. For the providers of management education, the absence of standardized curricula presents a problem of identity, which is important in developing employer credibility. The flavor of each type of program is influenced by the clientele to be served within the geographical region of the institution, and by the nature of that institution and its other academic programs. Some management education programs have a manufacturing orientation, others a project management orientation; some cater to the private sector, others to the public sector; some reflect a traditional business administration curriculum, others an industrial engineering curriculum, and still others a mix of the two. The American Assembly of Collegiate Schools of Business (AACSB) has, over the years, provided curricular guidance for management programs offered through schools of business. The Accreditation Board for Engineering and Technology (ABET) focuses its attention on undergraduate programs and has not served to provide curricular guidance for graduate management programs in schools of engineering. The rapidly changing needs of management education in a high-tech world may make this a desirable state of affairs; that is, perhaps an identity for these new forms of management education that focuses on the value of a diversity of perspectives (and programs) and on continuous change (in curricula) will emerge.

There is general agreement that management education has been and needs to continue to be multidisciplinary and interdisciplinary. This, by itself, presents special problems in staffing and delivering nontraditional management programs within the traditional university. There is also agreement regarding the high priority that should be placed in management education on continuing to develop communication skills, thinking and problem-solving skills, and interpersonal and team-building skills. This presents special problems when using distance learning technology, a common and increasingly used mode for delivering management education. There is disagreement on the extent to which the philosophy of management appropriate for the type of managerial position into which engineers are likely to move differs from the philosophy of management predominant in traditional education programs. Or, to put the same issue into the form of a question, Does the emergence of the project form of organization imply a need for a form of management education quite different from the traditional forms in order to reflect the new philosophy? And, if so, how can it be programmed, staffed, and delivered so that it creates an identity readily recognized and valued by students, teachers, educational administrators, and employers?

MANAGEMENT PHILOSOPHY

The term *management philosophy* is used here to refer to a system of motivating concepts, principles, and values, and hence a viewpoint on or way of thinking about the practice of management. Over time, discernible shifts in philosophies of management tend to be aligned with shifts in socioeconomic factors, organizational structures and processes, and/or technological breakthroughs. The currently emerging, although not yet well defined, philosophy of management can be linked to the role of information and communication technologies in facilitating new, streamlined organizational structures, and of automation and simulation technologies in coordinating and customizing the processes of design, production, and delivery of an organization's goods and services, which themselves are increasingly incorporating those technologies. It is in this context that engineers are attractive prospects for managerial positions due to their familiarity and comfort level with the technology. The diversity of management education programs for engineers is a consequence of differences in perception of or approach to the emerging philosophy of management.

Organizational Trends

Although ad hoc project teams (work groups, task forces, program offices, etc.) have been a part of organizations for a long time, the matrix form of organization was the first formaliza-

tion of a project-oriented structure. Companies in the military, aerospace, electronics, chemical, and other high-tech industries needed a way to free project managers from the usual chain of command, which involved reporting along functional lines of authority (e.g., engineering, manufacturing, marketing, and finance). The complexity of the technologies being employed required that all functions be considered simultaneously and without bias. By creating a line of reporting separate from these functional lines, this integration could occur. While functional management serves to maintain stability in organizations, project management serves as a change agent.

The matrix organization is not without its problems. Members of project teams typically have homes in functional departments, creating opportunities for conflict between functional and project managers. Also, projects are temporary, so when a project is completed there may or may not be a new project to which the project manager can move. Many organizations have lost some of their top management talent as a result of an uneven flow of projects. The multiple chains of command in the matrix organization also present a special problem of coordination. Computerized information systems, even in the early years, provided a means to facilitate the coordination of projects. With further advances in information and communication technology, a project form of organization that is significantly different from the matrix organization is becoming possible. The trend involves a redefinition and redistribution of traditional functions of management—both in the sense of functions like production, marketing, and finance and in the sense of functions like planning, organizing, and controlling. For example, management control is being redefined in terms like *dynamic coordination of teams* rather than in terms like *chain of command* and *span of control*. The redistribution of functions suggests a greater need for individuals with some management education, rather than less, as every member of a project team must develop an appreciation for all the functions and take responsibility for the success of the team. This is demanding a shift in thinking about management.

Management Paradigms

The current popularity of the term *paradigm* can be traced to Thomas Kuhn and his book *The Structure of Scientific Revolutions* (1). As the use of the term spread, it was picked up by writers and thinkers in nonscientific disciplines, including management consultants and educators, who began using it to address the dramatic changes occurring in the marketplace and in society and the need for managers to embrace new ways of thinking in order for their organizations to survive. The word has now taken on the status of a buzzword, devaluing some of the original impact intended by Kuhn. The word *paradigm* can denote a model, a pattern, or an example. Kuhn preferred to use the word *exemplar* for example and saved *paradigm* for talking about a predominant pattern of thought or worldview within a profession, discipline, or sector of society. He specified two conditions for qualifying a change in thinking as a paradigm shift: First, the change has to be linked to an accomplishment of sufficient magnitude to attract a group of adherents away from competing patterns, and second, the new pattern must possess an open-endedness sufficient to give these adherents something to do.

Paradigm shifts are not, and cannot, be planned in the usual sense. They occur when they are needed, and they become needed when desirable concepts, tools, systems, and processes cannot be implemented without them. Implementation of modern information, communication, and automation technologies, competition from international and multinational corporations (particularly in Pacific Rim countries), and socioeconomic changes in North America are cited as the factors that provide the impetus for recent shifts in management thinking and practice. Whether these changes qualify as a transformation of management paradigm remains for history to record.

The number of new management concepts, tools, and systems cited by educators, writers, and consultants is so great that it would not be possible or fruitful to list them here, and more are being created daily. The management section of one of the large chain bookstores typically contains 40 to 50 titles on management tools and philosophy, leadership style, and organizational change, with each claiming new ways of thinking. There are, however, a few writers whose ideas and language have become so widely known and used in both the rhetoric of everyday management life and actual management practice, that they deserve mention. The first three items in the following list offer a societal context in which new thinking is being stimulated; the remainder focus specifically on management thinking.

1. *The Third Wave.* With the publication of his books *Future Shock* (2) and *The Third Wave* (3), Alvin Toffler has acquired the status of social visionary. Focusing on the impact of information, communication, and automation technologies, Toffler chronicles structural shifts in economic, political, and social systems worldwide that are having consequences on and creating new possibilities for national priorities, organizational transformation, individual work, and everyday life. The first wave was agricultural, the second industrial, and the third informational.

2. *The Knowledge Society.* As Toffler is widely regarded as America's social visionary, Peter Drucker is widely regarded as its management visionary. Through a long history of books, Drucker has popularized certain ideas on management. In *Post-Capitalist Society* (4), he introduces the idea of the knowledge society and prophesies that information and knowledge will soon replace labor, land, and capital as the most important (even the only important) resource (and product) of economic organizations.

3. *The Age of Unreason/Paradox.* The British author Charles Handy has gone a step further in his books *The Age of Unreason* (5) and *The Age of Paradox* (6), examining the contradictions created by sudden and dramatic change and the implications of that on organizations of all types. As does Toffler and Drucker, Handy addresses educational institutions as well as economic and political ones, and the changes needed to manage these for the social good. The reader is led to conclude a need for new logic, a logic of change.

4. *Systems Thinking.* The early history of systems thinking in management begins at the Tavistock Institute in London in the 1960s. It was here that experiments with a variety of corporations were conducted, employing ideas in democratic management and participative decision making. Emery and Trist (7) are credited with developing the sociotechnical systems approach to organizational change, stimulating their colleagues and sponsors to follow with books like *Towards a New Philosophy of Management* (8) and *Alternatives to Hierarchies* (9). In the United States, Russell Ackoff became a dominant advocate of systems thinking in management, building on his work with Emery, *On Purposeful Systems* (10), and popularizing it in books like *Redesigning the Future* (11), *Creating the Corporate Future* (12), and *The Democratic Corporation* (13). He developed idealized design as a tool for participative and consensus decision making and applied it to many types of organization. The concepts of the circular organization, the internal market economy, and the multidimensional approach to organizational design represent ways to implement democratic management. He offers a challenge to total quality management (see list item 6).

5. *Change Management.* The first management book to reach the status of a national bestseller in the modern era was Peters and Waterman's *In Search of Excellence* (14). Based on case studies of successful and unsuccessful firms, the authors identify eight attributes of successful organizations that can serve as guidelines for change. Peters followed this with some sequels, including *A Passion for Excellence* (15) and *Thriving on Chaos* (16). Bradford and Cohen in *Managing for Excellence* (17) offered a practical approach to thinking about these changes, including the transformation of the role of the manager from that of technician and conductor to that of developer. They suggest that the role of developer requires a team approach to work and its organization. The idea of self-directed work teams has received substantial attention in recent years. (See Ref. 18.) Rosabeth Moss Kantor, also relying on a set of case studies, declared in *The Change Masters* (19) the importance of flat, team-based, entrepreneurial-style organizations for success with innovation and change.

6. *Total Quality Management.* The apparent success of Japanese companies in the 1970s and 1980s has been attributed in part to the implementation of ideas developed by W. Edwards Deming of the United States. Deming was hired by the Japanese in the 1950s to apply statistical concepts of quality to design, production, and other processes in Japanese firms. The design quality and reliability of many of the products of those firms made them competitive in the global marketplace. As a result of his experiences in Japan, Deming extended his ideas to general management thinking. His 14 points of management presented in his *Out of the Crisis* (20) and in Scherkenbach's *The Deming Route to Quality and Productivity* (21) form the core concepts of total quality management (TQM). TQM has received attention in many corporate and government organizations in the United States and Europe. Implementation of TQM involves both the introduction of new tools (e.g., statistical process control, robust de-

sign, quality function deployment) and the development of a new culture oriented toward the customer and continuous improvement of products and processes. U.S. companies have attempted to use Deming's 14 points of management as a way to accomplish the latter, but few have been successful in implementing all of them. The primary obstacle appears to be the degree to which the reliance on numbers-based management systems inhibits a customer orientation and continuous improvement. These systems—namely, systems for work standards, training, purchasing, performance appraisal, production control, and financial management—are deeply embedded in the thinking about accepted ways to do business. Irrespective of the difficulties, Deming and TQM have influenced management thinking by raising awareness of the importance of satisfying customers, encouraging employee creativity, maintaining the flexibility to change quickly, building quality into the processes of design and production (rather than inspecting for quality in the end products), and focusing attention on variation in those processes.

7. *Process Reengineering.* With Hammer and Champy's *Reengineering the Corporation* (22) and Champy's *Reengineering Management* (23), the idea of continuous improvement has met its greatest challenge. While the focus is still on building quality into the fundamental processes of a business and its organization, the thesis is that creating quality in a rapidly changing and turbulent world requires fundamental transformation of these processes and discontinuous thinking. These radical and dramatic transformations rely on breakthroughs, not on ideas for incremental improvements (or Kaizen). Considered one of the major process innovations in recent years is that of concurrent or simultaneous engineering—an approach to product development that emphasizes integration of the design, production, marketing, distribution, and other aspects of a product by considering them concurrently rather than sequentially. The idea is not only that interrelationships among these aspects are better addressed, but also that the time to market is less and the responsiveness to the customer greater.

8. *The Learning Organization.* Argyris and Schon introduced the concept of organizational learning in their 1978 book by that title (24). It has recently been picked up by a number of authors, most notably Peter Senge in his book *The Fifth Discipline* (25). The learning organization is one in which learning holds a position of the highest priority in the strategic mission, as well as the daily operations, of the organization. The five disciplines of effective learning organizations are shared vision, mental models, team learning, personal mastery, and systems thinking. A related concept is that of the virtual organization, one that transcends the boundaries of any single organization by taking advantage of the collaborative possibilities offered by multiple organizations. The resulting organizational arrangement is not a physically defined entity, but a virtual one with the ability to respond more quickly, more frequently, and more innovatively than any of its component organizations could individually. This attribute has been given the name *agility.* (See Ref. 26.)

9. *Nonlinear Dynamics and Chaos.* The theoretical foundations for the ideas just presented have not been well articulated. Systems theory is a recurring theme, but only in its system dynamics version does it reflect a rigor deserving of scientific status. The science of nonlinear dynamics and chaos has provided an additional foundation for management theory. Notable books that have explicated this theory include Wheatley's *Leadership and the New Science* (27), Priesmeyer's *Organizations and Chaos* (28), and Goldstein's *The Unshackled Organization* (29). Common themes include advocacy of self-organization (as opposed to planned change), flexibility as a criterion for decision making (as opposed to optimality), and variety generation (as opposed to efficiency and predictability). The shift in thinking from productivity to agility is consistent with a shift from a goal orientation to a process orientation, the latter requiring special attention to the dynamics of the organization and its operations. Strategic planning becomes an ongoing process of reevaluating and resetting goals, rather than an occasional exercise in evaluating strategies for achieving fixed goals. Paradoxes no longer need to be treated as aberrations; they become sources of creativity and transformation.

10. *The Knowledge-Creating Company.* In their book by this title, Nonaka and Takeuchi (30) build on the ideas in nonlinear dynamics and make a case for an alternative view of the paradigm shift taking place. Rather than treating middle management as superfluous and advocating significant reductions in middle manager ranks, they regard middle management as acquiring a central role in the management and operations of the firm. They contend that the primary product of modern organizations is knowledge, and to think otherwise is to put the firm at a competitive disadvantage. Middle managers are the knowledge engineers of the organization. What is particularly noteworthy is that these authors are both professors in a Japanese university who have worked in industry and did their graduate education in the United States. Their contention that the success of Japanese companies in the 1970s and 1980s was based on skills and expertise in organizational knowledge creation is in contrast to the commonly held belief that success was attributable to the implementation of quality control, quality circles (bottom-up decision making), and lean management. While not in contradiction with these principles, knowledge creation does represent a perspective that perhaps has more potential as a way of thinking about change in Western organizations than do ideas in books like Ouchi's *Theory Z* (31) and Pascale and Athos's *The Art of Japanese Management* (32).

While many of the ideas discussed in the preceding list are contradictory, there are also some common themes. These include a trend toward semiautonomous (self-directed) project teams or work groups; a focus on flexibility/agility in planning, design, production, and marketing; treatment of the entire life cycle of a product/service concurrently throughout its development; and recognition of the importance of information sharing, knowledge building, and learning in everyday activities. Organizations that embrace these notions tend to exhibit distributed, dynamic, networked, and multidimensional structures, and parallel, integrated, nonlinear, and circular processes. Where these organizations exist, they rely heavily on information, communication, and automation technologies to help maintain coordination. Intelligent systems are often used and will be increasingly used in the future.

Implications for Management Education

A case has been made by many (the aforementioned authors among them) that if shifts in management paradigms are occurring in industrial and government organizations, then shifts in the philosophy and delivery of management education should follow. These shifts, it is argued, should not be limited to changes in what is taught, but should include changes in how it is taught as well. What can be said at this time is that there are multiple strands of change occurring simultaneously, some of which could be viewed as contradictory. For example, one shift is being driven by information and communication technologies (particularly television and Internet-based technologies) and is directed at the ability to deliver courses to a greater number of geographically distributed students. While this mode of delivery has been heavily content oriented, the pressures from industry are for a more process-oriented form of education, directed at developing the new thinking, interpersonal, and communication skills suggested by the paradigm shift(s) discussed previously. Master's degrees and certificates in engineering management and the management of technology have been prime candidates for experiments with both content-oriented, televised delivery and process-oriented, weekend/evening programs that make extensive use of case studies and group projects. The question that remains is whether these divergent philosophies of management education will continue to specify two quite different categories of program, or whether they will merge into an integrated concept and approach.

EDUCATIONAL MODELS

There are at least three ways to distinguish different approaches to management education for engineers. There are different program types corresponding to educational level; different curricula depending on the student and employer market targeted; and different pedagogical approaches reflecting teaching styles and modes of delivery supported by different institutional types.

Program Types

Management education programs designed specifically for engineers and applied scientists began with master's degrees, including MBAs with concentrations in areas like industrial management and technology strategy, master's degrees in engineering management or engineering administration, management concentrations within industrial engineering master's programs, and master's degrees in the management of technology. MBAs are offered by business schools; master's degrees in engineering management/administration tend to be offered by engineering schools either as stand-alone programs or as programs within an engineering department, and may or may not involve participation of a business school; industrial engineering concentrations in management tend to favor a more quantitative approach than do the other programs; and management of technology programs tend to be offered as collaborative efforts between business and engineering schools. In all of these categories, there are some programs that are designed for full-time students and others designed for part-time, working professionals. While some of these programs do not require work experience for admission, others require two years or more of full-time work experience in an engineering environment since receiving the bachelor's degree, and virtually all regard such work experience as highly desirable. MBAs emphasize functional areas of business—particularly production, marketing, human resources, and finance—and integrate these through a policy perspective. Engineering management/administration programs emphasize the management of technical projects, programs, and operations, integrating functional areas of business throughout the curriculum. Management of technology programs tend to emphasize the strategic management of technology and technical resources.

There are three ABET accredited undergraduate degrees in engineering management in the United States: University of Missouri–Rolla, Stevens Institute of Technology, and the U.S. Military Academy. These programs emphasize a curriculum in engineering fundamentals, a concentration in an engineering discipline, and a broad set of courses in management topics. A recent trend in undergraduate programs is toward a minor in engineering management. The undergraduate minor at Old Dominion University is a four-course sequence emphasizing decision making, quality control, economic analysis, project management, and team building. In contrast, the undergraduate minor in technology and management at the University of Illinois at Urbana–Champaign is offered jointly by the colleges of commerce/business and engineering and is open to students from both colleges. The curriculum consists of six courses, three of which combine the students of both colleges, and an industrially sponsored capstone project. The use of industrially sponsored projects in undergraduate engineering curricula is becoming quite popular; such projects offer opportunities to expose students to the management and business aspects of the engineering profession.

Certificate programs as alternatives to degree programs are being introduced at both the postbachelor's and postmaster's levels. These programs allow greater flexibility in curriculum and format than do many degree programs, and employers are becoming increasingly supportive of programs with educational content that meets their needs, even if such programs cannot be accomplished within an official degree-granting structure. Case Western Reserve University's certificate in the management of technology has received strong support from industry. It can be anticipated that as the demand for continuing education in management beyond the master's degree continues to grow, professional doctoral programs (as opposed to research-oriented doctoral programs) that can be delivered in part-time and nontraditional formats will begin to emerge. Such programs will not require a research dissertation.

Curricular Content

The focus of the curricula of management programs varies substantially from one institution to another. There is general agreement that an introduction to the functional areas of finance, marketing, and production is important; however, in some programs this is accomplished in separate courses, while in others it is accomplished by integrating these subjects across the curriculum. Programs at institutions that reside in regions with a high concentration of federal government employers and government contractors tend to focus on program and contract management. Programs at institutions that reside in regions with high concentrations of industrial employers tend to focus more on manufacturing management. Programs associated with industrial engineering departments tend to favor either a manufacturing or an operations research focus. Programs associated with business schools tend to favor a strategic management focus. Programs within schools of engineering tend to get pressured to be more technical and quantitative, although some programs have maintained a strong behavioral component in their curricula. There is an ongoing debate about the degree to which management curricula for engineers should be standardized, and that debate is likely to continue.

Pedagogical Approaches

Just as there is a range of curricular content in management programs that target engineers and applied scientists, there is also a range of pedagogical approaches to the teaching and delivery of that curriculum. While there has been a move toward more project-based education in undergraduate engineering programs, the demand for master's programs in management that are accessible to the part-time student has prevented a similar trend from occurring at the graduate level. Many master's programs do require a capstone project or thesis; and some on-campus programs, including those with weekend or other nontraditional formats, may permit a greater concentration of project work throughout the curriculum. Capstone projects conducted as individual study have a history in some programs of taking multiple semesters to complete.

At many institutions, master's degrees in engineering management have been targeted for delivery via distance learning technology, particularly video. These programs have favored lecture and question/answer modes of discussion in their approach to teaching. Leaders in this form of delivery, including the University of Missouri–Rolla, Old Dominion University, National Technological University, and the University of Maryland, have pioneered the types of management education that work well through the video medium; they also recognize that it is a different form of education from that of the traditional classroom.

The latest trend is toward Internet-based delivery. This approach is primarily asynchronous, requiring a high level of motivation on the part of the student for successful completion to be realized. Asynchronous learning has been around for many years in the form of correspondence courses. Curricula developed for the World Wide Web offer many advantages over the traditional correspondence course. It remains to be seen if this form of delivery is able to capture the population of students needed to make it cost efficient, or if it can be integrated with other modes of delivery to circumvent its drawbacks.

ISSUES IN MANAGEMENT EDUCATION

The changes that are occurring in the structure and management of economic organizations, and that will certainly continue to occur, are forcing institutions of higher education to reconsider the way they deliver educational products in general. The demand for management education, in particular, is such that a variety of innovative programs have been forthcoming in recent years. The growing political pressure to hold public institutions accountable, through demonstrations of the value of their services to the public, and the resultant tightening of budgets have raised questions about the future roles of the university in our society. The rise of the "virtual university" certainly suggests some profound changes. The predominant opinion at this time is that there is much of management education that can be delivered through a virtual university, much of it better (particularly when the visual advantages of the various media are utilized) than what could be delivered through the traditional university; but there are other needs of management education (like face-to-face group exercises and team-based projects) that cannot be met in this way.

The following trends in the content of management education for engineers and applied scientists are regarded as significant:

1. The team-based project form of organization
2. The management of parallel and distributed (as opposed to sequential) processes
3. Global and multicultural perspectives on business
4. Entrepreneurship/intrapreneurship skills

The following trends in the process of management education for engineers and applied scientists are regarded as significant:

1. Industrial involvement in courses/projects
2. Interdisciplinary approaches to teaching
3. The use of educational technology
4. A mix of scheduled and asynchronous learning environments

The extent to which management education continues to embrace these trends will depend in great part on how institutions of higher education respond to their changing roles.

BIBLIOGRAPHY

1. T. S. Kuhn, *The Structure of Scientific Revolutions,* 2nd ed., Chicago: The University of Chicago Press, 1970.
2. A. Toffler, *Future Shock,* New York: Bantam Books, 1980.
3. A. Toffler, *The Third Wave,* New York: Bantam Books, 1980.
4. P. F. Drucker, *Post-Capitalist Society,* New York: HarperBusiness, 1993.
5. C. Handy, *The Age of Unreason,* Boston: Harvard Business School Press, 1989.
6. C. Handy, *The Age of Paradox,* Boston: Harvard Business School Press, 1994.
7. F. E. Emery and E. L. Trist, Socio-technical systems. In C. W. Churchman and M. Verhulst (eds.), *Management Sciences: Models and Techniques,* Oxford: Pergamon, 1960.

8. P. Hill, *Towards a New Philosophy of Management,* New York: Barnes & Noble, 1971.

9. Ph. G. Herbst, *Alternatives to Hierarchies,* Leiden, The Netherlands: Martinus Nijoff Social Sciences Division, 1976.

10. R. L. Ackoff and F. E. Emery, *On Purposeful Systems,* Chicago: Aldine-Atherton, 1972.

11. R. L. Ackoff, *Redesigning the Future,* New York: Wiley, 1974.

12. R. L. Ackoff, *Creating the Corporate Future,* New York: Wiley, 1981.

13. R. L. Ackoff, *The Democratic Corporation,* New York: Oxford University Press, 1993.

14. T. J. Peters and R. H. Waterman, Jr., *In Search of Excellence,* New York: Harper & Row, 1982.

15. T. Peters and N. Austin, *A Passion for Excellence,* New York: Random House, 1985.

16. T. Peters, *Thriving on Chaos,* New York: Alfred A. Knopf, 1987.

17. D. L. Bradford and A. R. Cohen, *Managing for Excellence,* New York: Wiley, 1984.

18. J. D. Orsburn, L. Moran, E. Musselwhite, and J. H. Zenger, *Self-Directed Work Teams,* Burr Ridge, IL: Irwin, 1990.

19. R. M. Kanter, *The Change Masters,* New York: Simon & Schuster, 1983.

20. W. E. Deming, *Out of the Crisis,* Cambridge, MA: Massachusetts Institute of Technology, Center for Advanced Engineering Study, 1982.

21. W. W. Scherkenbach, *The Deming Route to Quality and Productivity,* Washington, DC: CEEPress Books, 1992.

22. M. Hammer and J. Champy, *Reengineering the Corporation,* New York: HarperBusiness, 1993.

23. J. Champy, *Reengineering Management,* New York: HarperBusiness, 1995.

24. C. Argyris and D. A. Schon, *Organizational Learning: A Theory of Action Perspective,* Reading, MA: Addison-Wesley, 1978.

25. P. M. Senge, *The Fifth Discipline: The Art and Practice of the Learning Organization,* New York: Doubleday Currency, 1990.

26. S. L. Goldman, R. N. Nagel, and K. Preiss, *Agile Competitors and Virtual Organizations,* New York: Van Nostrand Reinhold, 1995.

27. M. J. Wheatley, *Leadership and the New Science,* San Francisco: Berrett-Koehler, 1991.

28. H. R. Priesmeyer, *Organizations and Chaos: Defining the Methods of Nonlinear Management,* Westport, CT: Quorum Books, 1992.

29. J. Goldstein, *The Unshackled Organization,* Portland, OR: Productivity Press, 1994.

30. I. Nonaka and H. Takeuchi, *The Knowledge-Creating Company,* New York: Oxford University Press, 1995.

31. W. G. Ouchi, *Theory Z,* New York: Avon Books, 1981.

32. R. T. Pascale and A. G. Athos, *The Art of Japanese Management,* New York: Warner Books, 1981.

LAURENCE D. RICHARDS
Bridgewater State College

MANAGEMENT INFORMATION SYSTEMS

Advances in technology have been occurring at an increasing rate, and the rate of technological advances is expected to continue to increase for the foreseeable future. These advances have enhanced the flow of information around the globe in a time frame that has accelerated decision processes and time frames. Now managers can influence ongoing business processes in any part of the world with competitive speed. The potential scope of business operations, markets, and competition is truly global in nature. The term "world class" has real meaning because of the application of technology. This is evident by the fact that 25% of the US economy consists of imports and exports. The global potential of business operations places a premium on the effective and efficient use of technology because the use of management information systems is nearly always required for companies to compete successfully within the global framework. Technology has permanently altered the nature of business operations and competition. Corporations not only can operate globally but can collaborate with other corporations to gain economies of scale and to capitalize on new expertise by outsourcing parts of their operations to other companies. As an example, data entry can be outsourced to locations where labor costs are lower, and systems development can be outsourced to countries like India and Ireland where there are highly skilled labor forces. The end result of all of this is that the corporate infrastructure (which includes skilled MIS people), information technology, and management information systems play a major role in the corporate world. The speed with which corporations can incorporate new technologies and adapt to the changing technological environment is a major corporate concern.

This article discusses the competitive uses of technology in business environments and how changes in technology effect a paradigm shift in management information systems; namely that companies increase their operations in the global environment as communication technology advances, such as the World Wide Web. Next, the information architectures that support the ability of companies to do business in the global environment and support commerce and decision making are discussed: complex systems such as centralized systems, LAN-based systems, client–server systems, and distributed cooperative systems. The next section outlines decision making, decision support systems architecture, group decision support systems, and advances in decision support systems. The article concludes with a discussion of the qualifications required of the MIS personnel who develop and maintain the information systems.

COMPETITIVE USE OF INFORMATION TECHNOLOGY

Corporations are very sensitive to the bottom line and the important role technology plays in achieving lower costs, higher product quality, and improved customer service. One of the models used to evaluate an organization's competitive standing is the Porter Competitive Force model (1,2), depicted in Fig. 1. The model provides a systematic way for managers to evaluate their corporation with respect to their competitors and other competitive forces. Following is a systematic discussion of the competitive forces addressed by the model.

The first force and logical starting point for assessing a corporation's use of technology is for the corporation to understand its purpose for existence and the role it can realistically play within its environment. The corporation must realistically determine its overall goal and market strategy in the context of its competition before any analysis. The corporation must also be prepared to sustain its efforts to capitalize on technology to gain competitive advantage because competitive advantage would otherwise be short term. It takes sustained improvement and innovation to sustain competitive advan-

Figure 1. An analytical framework by which a company understands its relationship to its environment.

tage. Then the corporation must understand and analyze its competitors and the industry. The appropriate strategy can then be determined. Alternative strategies usually involve either achieving the status of low-cost producer or trying to differentiate its products or services from other competitors. The second consideration concerns defining its market scope which can be local, regional, national, or global. Corporations need to reassess their strategy and market scope periodically because of the dynamic changes in technology and the market they are in or are targeting for the future. Our interest concerns the role that information technology (IT) can play in corporate strategy and analysis. Some IT services that may support the initial analysis include news clipping services, government databases, commercial databases, electronic data interchange (EDI) data, and data warehousing coupled with data mining.

The second of the five major forces that must be addressed is the threat of new entrants. A new entrant into the market would certainly add new capacity and competition to the market. A new entrant is also likely to increase price pressure and may also put upward pressure on the factors of production and promotion. The question with respect to this force is whether a corporation can cover the capital cost and compete on a cost-effective basis to thwart the entry of potential competitors. A second consideration would be to address the potential of using IT to differentiate its product or to provide customer service that would be difficult to replicate or expensive to develop. The final consideration for our purposes is whether the corporate infrastructure, information architecture, and applications can be used to provide more effective market access than its competitors. The objective is to build barriers so that a competitor finds it unprofitable to enter the market. One method of achieving this is accomplished by gaining economies of scale internally or through an outsourcing arrangement. A second method is to increase the switching costs of customers to make it more difficult or expensive to switch products or services. Other methods include blocking access to distribution channels, using financial and market strength to intimidate, and eliminating competition through an outright purchase.

The third major force is the bargaining power of buyers. The power of a buyer increases with the percentage of sales.

A powerful buyer has the potential to force prices down, demand higher quality, demand more services, and force competition. One obvious method is to reduce the percentage of sales to any one buyer by increasing the number of buyers. A second method is to increase the loyalty of the buyer by establishing a close relationship between the two companies. A closer relationship can be facilitated through electronic mail, point-to-point communications, and a promotional presence and linkage on the World Wide Web (WWW). This technique basically increases the switching costs. The linking of respective order, distribution, and inventory systems is widely used to achieve this objective. This approach will no doubt be less effective as more standards are established. Other approaches include blocking out the competition and differentiating products and services.

The fourth major force is supplier bargaining power. There is a relationship between the power of suppliers and the upward pressure on the cost of materials, parts, and services. Supplier power can also result in downward pressure on quality and service. One approach to reducing supplier bargaining power is in some respects similar to reducing buyer power. There is greater safety from higher prices, lower quality, and poor customer service as supplier competition is increased. In essence, there is safety in numbers. The power of buyers can be reduced by selecting and developing close relationships with multiple suppliers. The communications outages in the Chicago and New York areas caused many firms to rethink their sole dependence on a single data and voice communication system supplier. A second method of reducing supplier power is the threat of backward integration. A corporation can also combat high supplier power through backward integration. The power of the supplier is certainly diminished if the supplier knows that the corporation can and potentially would produce its products or service. These countermeasures encourage fair prices, higher quality, and better service with reasonable profits.

The fifth force is the impact of substitute products and services on the competitive stature of the corporation. The impact is similar to new entrants into the market. The availability of substitute products puts downward pressure on prices and profit margins and upward pressure on quality and service. IT can be used to assist managers in improving their price/performance, and the potential exists to allow them to differentiate their product by enhancing product features through the use of microprocessors or enhancing their customer services through communication systems.

CORPORATE MANAGEMENT INFORMATION SYSTEMS

If anything is true in the world of corporate management information systems, it is that the rate of change has increased over time and that corporations are critically dependent on computer-based management information systems. Advances have tremendously increased a corporation's capability to receive, process, store, and communicate information. Data communications has been a major facilitator for expanding corporate management information systems beyond the traditional boundaries of the firm. The focus of the application of technology and management information systems has undergone a transition from an internal corporate focus to a collaborative focus that uses interorganizational systems. Interorganizational systems enable the coordination of the corporation's daily operations with those of its supporting suppliers, distributors, and customers. The corporate data communications architecture has been enlarged to include one-to-one and one-to-many links to enable the integration of the supporting operations of suppliers and distributors. Interorganizational systems have greatly increased the speed of information flow which in turn increases the capability of managers to speed up decisions. The end result is faster, more accurate decisions, the minimization of physical document flows, lower inventories, less storage space, and more responsive and accurate delivery of products and services. The day of the self-sufficient corporation is a relic of the past. Computer processors provide a good illustration of a category of products produced through the cooperative efforts of multiple international suppliers. The processor chips are manufactured in the United States. Computer primary memory is manufactured in Korea, Japan, and the United States. The hard drive's secondary storage devices are manufactured for US companies in southeast Asia.

EDI systems have expanded the knowledge domain of managers through the acquisition and use of high volumes of point-of-sale data for their own and competitor products. The amount, accuracy, and timeliness of this EDI data provides a wealth of product performance knowledge with respect to geography, customer profiles, promotions, and sales patterns. This has been enabled by dramatic increases in storage, multidimensional databases, data warehouses, neural net technologies, and analytical software advances. Technology, such as Lotus Notes, has increasingly been employed to bring greater expertise, greater collaboration, and broader participation to bear on problem solving, planning, and other creative activities. This type of collaboration within the company and among interorganizational systems mentioned earlier is prevalent today, and some theorists suggest that it will lead to virtual corporations tomorrow in which a corporation may consist only of the guiding intelligence that contracts out all of the nonexecutive operations.

The New Management Information Systems Paradigm

Corporate management information systems are constrained by the human ability to understand, apply, and adapt new developments in technology. Corporate information technology is typically available for several years before corporations can put it into general use. The delay is also caused by the higher expense associated with new technology, early defects, and the higher risks associated with using a new technology before standards are established. The differing characteristics of information at the operational, middle management, and executive levels are fairly well known. Information at the operational level is detailed, and the scope of the data is usually a subfunction. As we move to the middle management level, the information represents the whole functional area, and at the executive level it encompasses the corporation in an integrated, summarized manner. MIS professionals can design very fine operational systems that have user-friendly graphical user interfaces. Systems design is much easier when the functionality is well defined and decisions are repetitive in nature. Systems design is much more difficult at the execu-

tive level because the decisions are frequently unstructured and nonrepetitive. The decision time frame is very short at the operational level and quite long at the executive level. The complexity of the decision is much simpler at the operational level and usually very complex at the executive level. We know that as we move from the operational to the executive level, the information mix changes from predominantly internal to predominantly external.

Transaction Processing Systems (TPS). As depicted in Fig. 2, operational applications are often called transaction processing systems. Transaction processing systems support the normal business activities of an organization by supporting the operational level activities of an organization. They provide fast and efficient processing of large volumes of data that are input or output by the organization. For example, flight reservations for an airline is an example of a transaction processing system. When data are input and output, the system verifies that the data is accurate, free from errors, and ensures that the data are kept up to date. Transaction processing systems form the basis for most of the organizational data.

Applications, such as accounts receivable, accounts payable, payroll, and inventory, were typically the first business applications. TPS applications shared common characteristics. The applications were event-driven, human-intensive, high-volume, highly structured, dealt with internal data, and were used at the operational level of the corporation. They were also designed for one subfunction within a functional area and were used as stand-alone systems. The result was a one-to-one relationship between the application program and the data file. A reduction in labor was frequently used as justification for implementation, and these systems did in fact eliminate significant numbers of employees. Transaction processing systems also greatly improved the accuracy of processing and greatly facilitated the generation of periodic reports. Early computers were often housed in the accounting area and usually relied on batch processing. Transaction data were stored primarily in tape files because disk capacity was severely limited and expensive. The data were merged and summarized for daily, weekly, monthly, quarterly, and annual reports. The reports were most suitable for operational and second level managers to use for control purposes internally within the firm. As technology decreased in cost and

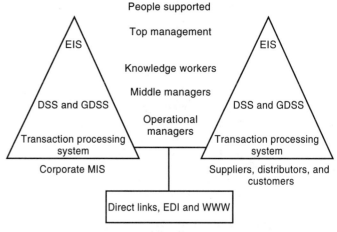

Figure 2. Advances in IT such as EDI and the WWW have dramatically changed the IS organizational boundaries and architecture.

increased in capability, the benefit of additional applications was greater than their marginal cost, and they were implemented in functional areas across corporations at the operational level. TPS generate much of the internal data stored in the corporate database.

Management information systems use the data from transaction processing systems and provide reports to business managers to enable them to gain insight into the company's operations. This helps managers plan, organize, and control their business activities more effectively and efficiently. The reports may be generated periodically or on the demand of managers. These reports may also provide for exception-based reporting (when a situation is unusual) and allow for drill-down when the reports are on-line. MIS reports can support managerial decision making and support managers at more levels of the organization than those supported by transaction processing systems. The input data for an MIS system are primarily internal and are generated by the various transaction processing systems within an organization. The traditional management information system in the past had an inward focus. Textbooks often represented MIS by a single pyramid in which a varying taxonomy of information systems was used to characterize the various levels and types of information systems. The management information of past years usually respected the boundaries of the corporation. Today, the management information system no longer has neatly defined boundaries. State of the art database, client server, and communications technologies have dramatically changed the focus of MIS to be more process oriented, collaborative, and interorganizational. The high-capacity, direct communication links, switched networks, EDI, and the WWW have facilitated the development of a more complex management information system that links transaction level application systems between different corporate management information systems. The effective operation of these interorganizational systems requires interorganizational associations to coordinate, standardize, and control these interfaces. Now corporations have integrated parts of the management information systems, allowing routine daily access so as to become more competitive within their markets.

Decision Support Systems (DSSs). Decision support systems are used primarily by managers or professionals [or groups of managers (GDSS)] at middle management levels to make operational and planning decisions. The decisions are semi-structured, the problems are less repetitive, and the user–interface is graphical and intuitive. Decision support systems are used to resolve operational problems, interorganizational system problems, and problems assigned by higher levels of management. The problems may address decisions that are solely internal or relate to external relationships. The internal and external mix of data varies with the nature of the decision at hand. A GDSS/DSS is typically capable of accessing both internal and external databases. The DSS typically can easily enter data for analytical purposes, create graphics, and generate custom reports. The tools may include spreadsheets, mathematical models, expert systems, neural net systems, financial models, high-level programming languages, and a wide variety of other tools. DSSs are discussed in further detail in a later section.

Executive Information Systems (EISs). EISs are used by high-level executives for planning and control. The executive makes decisions that have a longer term impact than middle or operational managers. These executives make the strategic directional decisions that guide the corporation in terms of people, products, and markets. The decisions are unstructured and based on substantial amounts of external data. Executives must have access to data covering longer historical and projected future time periods. Top executives are interested in issues critical to the success of the corporation. The concept of management by exception is important to the effective use of their time. They often need to access information by product line, vendor, geographic location, competitor, and other similar categories. The ad hoc ability to look at overall corporate performance indicators and drill down into more detail is an important capability within the EIS. EISs are very difficult to design and relatively expensive to operate because of their dynamic nature.

EISs reduce information overload on the executive and include user-friendly versions of the DSS used by middle level managers. In general, the interface is graphical and easy to use. Executives do not enter queries, but rather use preselected models with options that enable them to customize the information needed to respond to their request. The information is provided in a summary format so that the executive can get a broad picture. Then the executive can choose to drill down for the details or do a pivot to view the data from a different perspective so as to identify problems and seek solutions. The internal databases that an EIS uses often are in the form of data warehouses. EISs also link to external databases and must provide access to both qualitative and quantitative information.

THE GLOBAL MARKET

The international nature of the corporation and its management information systems are currently going through a paradigm shift with the advent of the WWW which is making a dramatic impact on businesses worldwide. In 1996 sales reached $2.6 billion and were estimated to be more than $220 billion in 2002. Although the Web is still essentially unregulated, has security deficiencies, and lacks standards, the information available is doubling on a yearly basis. The WWW has dramatically increased the access of even small businesses to local, regional, national, and international markets. The use of the WWW is widely accepted for providing information about companies, products, services, and employment opportunities. The Web's rapid acceptance is driven by the potential for increasing profit through electronic commerce and also by the danger that a given firm will become less competitive because of its failure to stay current with the electronic marketplace. The WWW increases competition by equally facilitating the access of other foreign and domestic corporations into US markets. The price of technological obsolescence can be high and might be life-threatening. The increased competition puts pressure on corporations to improve product quality, improve customer service, and decrease price, while simultaneously improving efficiency. To succeed in this increasingly competitive environment, corporations must effectively use information technology to create, promote, distribute, and support products and services which are truly "world class." Companies operating in this environment must marshal the resources necessary to create and maintain the infrastructure and corporate IT architecture that enables these operations.

INFORMATION ARCHITECTURE

The decreasing cost and increasing power of IT have continually reduced the marginal cost of employing IT in an increasing range and intensity of business applications. Over the years the use of the technology has steadily increased and the computer-based applications have permeated competitively critical operations. In most cases, corporations could not function for any significant period of time without their data and application portfolio. Corporations have had to invest consistently in MIS to remain competitive in products, services, and distribution. Information architecture has evolved with advances in technology.

Information architecture is the form that information technology takes in an organization so as to achieve organizational goals. The organization's architecture consists of the way the components, hardware, software, networks, and databases are organized. The manner in which these components are organized for many organizations has changed from pure centralized systems, to client-server systems, and to distributed cooperative processing systems. Much of this changeover has taken place because of the emergence of powerful personal computers and the growth of sophisticated knowledge users who demand easy-to-navigate graphical interfaces and use powerful software to accomplish their work. The cost of these systems has also meant that organizations have pushed more of their processing to personal computers and networks of personal computers, as opposed to handling the processing centrally. In addition, when connected cooperatively, new organizational structures, such as teams and work groups, can be supported effectively, and they provide organizational flexibility.

Centralized Systems. Traditionally, prior to the mid-1980s, almost all processing was done centrally in a mainframe, and users were provided with terminals and sharing access to the mainframe. This architecture provided control and security to organizational data. Centralized systems started with stand-alone systems and later evolved to integrated functional systems.

Stand-Alone Systems. These systems were developed for well-defined applications, for example, accounts payable, accounts receivable, payroll, and inventory processing. Being transaction-based systems, they usually processed large volumes of data. The outputs of these systems were periodic, printed reports responding to predefined queries usually written in a low level language. These systems were under the control of centralized management, predominantly batch processed, and with relatively little end user access. Hence, they were not very useful for executive decision making. These centralized systems gave way to integrated systems.

Integrated functional systems are what we might today call "suites," for example, an accounting, marketing, or production suite. These functional suites were predominantly transaction-based and stored data in early database management formats that supported more summarized reports and presented somewhat integrated data reports to middle and upper management. These suites were more customizable and allowed for ad hoc queries and on demand reports. However, these systems were all centralized, and users were clam-

oring for additional functionality in applications that was often not provided by the information systems staff.

LAN-Based Systems. Because of the time lag for applications to be developed on mainframe systems, many departments purchased personal computers and connected them via local area networks (LANs) to meet their needs. These systems were characterized by network servers, to which users connected by using LANs. The network file server stored common office productivity applications, such as spreadsheets and word processing, and provided other services, such as print and email services. Users connected to the server and could use these applications and communicate with other members of the network via email.

Client-Server Computers. A logical extension of the LAN occurred when users were connected to the network server and also to the centralized mainframe where corporate data existed. Users could use their computers as dumb terminals to the mainframe and also use their personal computers to do additional processing, such as formatting the presentation of data and error checking. This led to client-server processing. In client-server applications, the processing is done on more than one computer. Typically the applications are broken down into presentation, business logic, and data. The data layer is typically a database management system stored on the main-frame that allows the user to query and perform functions on the centralized database. The business logic layer checks and makes sure that integrity considerations for the application are not violated. For example, salary raises in a particular year may not be greater than 5% is an integrity rule that may be stored separately in the business layer. The presentation layer formats the data for the user. When all these three layers are in separate computers, as in Fig. 3, then we have a three-tier, client-server system. More commonly, two of these, such as the business layer and presentation layer, may be in the client system leading to the two-tier, client-server system. Three-tier, client-server systems are better because they provide more flexibility and scalability (i.e., we can add additional clients without lowering performance), and changes made in one layer do not affect other layers.

Distributed Cooperative Systems. When client-server systems first came into use, they were generally constrained by proprietary interfaces. These have slowly given way to open or industry standard interfaces allowing organizations to mix and match computers and software more easily. The emer-

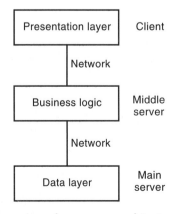

Figure 3. An overview of a common architecture used to design client–server applications.

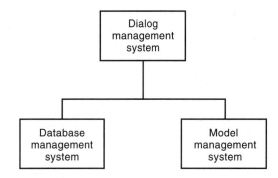

Figure 4. An illustration of the main components in a decision support system.

gence of the Web browser as a universal client and the internet as the underlying networking mechanism has helped to promote open systems. Cooperative systems take client-server processing to its logical end, wherein different (*n*-tier as opposed to two- or three-tier) computers work together to solve a common problem. These computers can be connected via a LAN or a wide area network. For example, in a cooperative processing system, the order processing system may log into a supplier's information system to check the status of an order. Standards are developing wherein users would not have to know where on the network of computers a service is being performed, and all cooperative processing takes place transparently to the user.

DECISION SUPPORT SYSTEMS

As discussed previously, a DSS supports the decision making activities of managers. DSSs help managers identify and study alternative solutions to problems and choose among them.

Decision Making

Managers make decisions on a variety of problems. Simon stated that managers make decisions that are either programmed or nonprogrammed (3). Programmed decisions are fairly structured and can be made via rules or standard procedures. For example, the decision to reorder products in an inventory can probably be automated based on the product concerned. Nonprogrammed decisions are more difficult to quantify and inherently lend less structure to the problem. For example, the decision to open a new plant is an unstructured problem.

Simon also described the following three phases that managers go through when making a decision or solving a problem:

Intelligence. In this phase the manager scans the environment for problems calling for a solution. This is akin to the intelligence activities undertaken by the military. The problem is identified, constraints specified, and the problem formulated.

Design. The manager designs, develops, and analyzes alternative courses of action. Here the manager identifies and evaluates alternatives for feasibility.

Choice. In the choice phase the manager decides on one of the alternative courses of action and then implements the decision. Then a continuous review of choice activities takes place.

Decision makers do not necessarily always optimize and find the best solution. They may find a solution that is good but not necessarily optimal. This is reasonable in situations where finding the optimal solution proves too costly or is time consuming. Information systems, such as a decision support system, can help the manager make programmed or nonprogrammed decisions, provide support in all three phases of decision making, and help in identifying either optimal or satisficing solutions.

DSS Architecture

A DSS architecture, shown in Fig. 4, consists of a database management system, model management system, and a dia-

log management system. A DSS uses a database system that provides access to data. These allow the manager to query data. These data may reside in a traditional database system, client-server environment, or in a data warehouse. In addition to traditional data, DSS typically contain links to external databases, such as those available from different sources on the Web.

Models are built using model base systems. These help the manager analyze the data. A model base includes (1) formulas and equations, (2) simulation models, (3) linear integral programming models, (4) statistical models, and (5) financial formulas, such as cash flow. Typically these models may be created using spreadsheets or other special purpose software. For example, Taco Bell combines three models—a forecasting model, a simulation model, and an integer model—to schedule its employees. The *forecasting model* is used to predict customer arrival so that managers can predict the sales in fifteen-minute intervals. A discrete event *simulation model* then helps the manager develop labor or staffing tables. These are then fed to an *integer-programming model* that allows the manager to decide what the exact employee schedule should look like. The output specifies how many people are needed, their positions throughout the day, and what their shifts would look like including breaks. These are provided in both graphical and tabular formats to the manager. For engineering purposes, a model might contain a simulation of an integrated chip, before it is physically produced. The simulation may provide valuable information on layout, thermal considerations, performance, and other factors. Then these simulations enable the design engineers to arrive at a suitable design before spending large amounts of money to produce a chip.

The dialog management system provides the user interface to the DSS. Using the dialog management system the user interacts with both the database and the model base. Interactions include performing queries and entering parameters into the system. The dialog management system also returns the results in a format selected by the user, such as tabular, graphical, or animation. The dialog management system allows the user to switch views easily between a tabular or graphical display.

Group Decision Support Systems

Businesses are now moving toward an environment where teams and groups work together to complete projects. Group decision support systems (GDSSs) support the decision making activities of groups of managers or teams. GDSS enhance decisions by removing group process impediments, such as groupthink or pressures to conform in making a decision. Typically, a GDSS is installed in a conference room and contains hardware and software that facilitate meetings.

Some types of software that are installed include brainstorming and idea generation software, wherein members of the group enter ideas anonymously. Others can comment on these ideas and participants can rank, consolidate, and vote on these ideas. Alternatives can be evaluated based on these ideas, and they can be ranked using different weighting factors. In addition, software for stakeholder analysis, problem formulation, and contingency planning aids are included. Collaborative tools, such as a private scratchpad and a group scratchpad, are typically included.

GDSSs are typically run by using a meeting facilitator and giving anonymity to all participant actions. This provides the advantage that some users do not dominate meetings and improves the decision making capacity of the group as a whole. In addition, GDSSs provide group memory facilities by keeping records of the meeting.

Groupware. In contrast to GDSSs that provide support for decision making, groupware, such as Lotus Notes, supports more collaborative work. Groupware provides bulletin boards, e-mail, calendar facilities, etc. to support teams and groups. It facilitates the movement of messages or documents to enhance communication among team members. Groupware also provides access to shared databases, work-flow management, and conferencing. Using these, team members across the globe can keep in touch with each other and keep every other member of the group posted on their activities and the project's progress. The Internet and its discussion groups, e-mail, and browsing software provide some of the same capabilities using open standards and nonproprietary software. The discussion groups on the Internet can be limited to users within a corporation by limiting access to specified individuals.

Advances in Decision Support Systems

Organizations have started to realize a new purpose for gathering information, that is, as a measurement on which to base future action rather than as a postmortem and a record of what has already occurred. On-line analytical processing (OLAP) tries to support this new desire of organizations. OLAP refers to the dynamic enterprise analysis required to create, manipulate, and synthesize information from enterprise data models. OLAP includes the ability to discern new or unanticipated relationships among variables and the ability to

- Identify the parameters necessary to handle the large volume of data
- Create an unlimited number of dimensions (consolidation paths)
- Specify cross-dimensional conditions and expressions
- Analyze data according to these multiple dimensions

Multidimensionality is the key requirement for OLAP, which handles large volumes of data, typically resident in a data warehouse. An OLAP server typically sits between the data warehouse and the client software that the decision

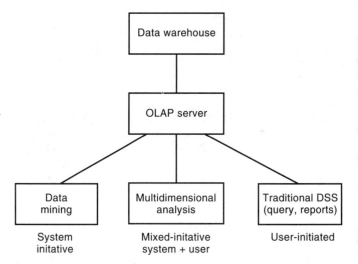

Figure 5. An illustration of the role that data warehouses, OLAP, data mining, and multidimensional analysis plays in augmenting the traditional DSS in supporting decision making.

maker uses. The client software used for analysis can be user-driven (e.g., traditional queries and reports), system-driven as in data mining, or driven jointly by both system and user, as in multidimensional analysis. This new DSS architecture is shown in Fig. 5.

Data Warehouse. Regular databases do not meet the needs of a DSS because they do not include historical data, in general lack some data integrity (not free from errors), and also are organized for ease of application rather than for decision making. A data warehouse is a database created specifically to support decision making. Hence, the data in a data warehouse are highly integrated (from many different databases in the organization), scrubbed (free of errors), contain time-variant data (historical data), and are organized by subject rather than by application. In general, data in a data warehouse are not deleted.

Multidimensional Database. Multidimensional databases optimize storage and manipulate output to help users investigate patterns of data in the data warehouse. The dimensions represent the user's perception of the data. For example, the set of all products or sales regions is a dimension, the quarterly year may be a dimension, etc. The objective of multidimensional analysis is to help decision makers perceive the meaning contained in the data. The user can visualize the interrelationships in the data more easily than otherwise. Multidimensional databases allow users to look at all possible combinations of entities and their interrelationships by providing the following capabilities:

Pivoting. Users can pivot or rotate across tabs by moving dimensions displayed in columns and rows. This changes the orientation of the object that is displayed allowing the user to investigate the different relationships.

Sorting and Collapsing. Users can sort by any dimension or collapse two dimensions into one.

Aggregation and Drill-Down. Users can see the data at the desired level of detail. They can receive a summary report or can choose to drill down and investigate further.

Time-Dimension. Support for time dimension is built into multidimensional analyses, as most data warehouses have historical data stored in them. Users can query for averages for last year, sales by month, etc.

Data Mining. In multidimensional analyses, users search for the patterns in the data. However, in data mining the computer system searches for patterns of information in data. Data mining is the computer-assisted process of searching through and analyzing enormous sets of data and extracting the meaning of the data. Basically two approaches are used: (1) predictive modeling and (2) automatic discovery. If the search is based on a predetermined idea of the patterns or some hypothesis about what the patterns might be, it is called predictive modeling. In the absence of any predetermined hypothesis, the search is called automatic discovery. Ideally, a combination of the two methods is used in conjunction with the user to discover the patterns. The user's role is to guide the search process, using a visualization system.

The process is usually iterative, so that the user can review the output to help refine the search process or to form a narrower or more elaborate search. Data mining software should be transparent (the system can explain why it performed certain operations), so that it can help the user understand and guide the system. Once the search process is complete, it is still the responsibility of the user to interpret the results so as to ensure that useful knowledge is derived from the data.

Generally data mining tools solve problems that

- Partition data into two sets, based on the presence or absence of data. This is useful in direct mail campaigns, fraud detection, and bankruptcy prediction.
- Partition data into multiple predetermined sets of classes. This is useful in medical diagnosis, and in establishing a credit rating and bond rating.
- Perform function reconstruction, such as time-series forecasting. This is useful in areas, such as forecasting financial data and sales data at the level of the individual.

QUALIFIED IS PERSONNEL

The information system is organized to optimize the information system services so as to meet the organizational goals and culture. In general, the structure of the IS organization reflects its informational architecture. However, it can be either centralized or decentralized. A chief information officer (CIO) typically heads the information systems organization. John Whitmarsh, editor of *CIO* magazine, defines the CIO as one who designs a "technology blueprint" and develops technology strategies that advance corporate goals.

A CIO's role is to

- Guide and unify the entire IT resources
- Coordinate all resources
- Be business oriented, not technology oriented
- See the advantage of technology and where to apply it broadly in the business
- Engineer technology organizations and infrastructure

A growing role for the CIO is to add value to their customers and also make sure that their customers recognize the value added. In general, the CIO has to ensure that the right things are done and is responsible for management of information technology and technical issues in information systems. In a centralized organization, the CIO supervises all IS professionals. In a decentralized setup, the CIO supervises the corporate unit. Each division has its own divisional information leader. The CIO relies on database administrators (DBAs), a telecommunications manager, programmers, and systems analysts to help in carrying out the various information system tasks. Figure 6 shows a typical organization chart for the information systems group.

Programmers and Systems Analysts. Most IS professionals start their careers either as programmers or systems analysts. They analyze the needs of the users and design and write codes to develop custom solutions. They also help maintain and update existing systems. To succeed, these professionals must have a technical background and also must also understand the business functions and possess excellent communication skills. They advance to become project leaders and are in charge of other analysts, allocate resources, such as hardware and software used in the development process, and use project management skills to ensure that the project is developed and delivered to the customer on time and within budget.

Database Administrator. The information stored in the corporate database is one of the most valuable resources within the organization. A separate database administrator (DBA) is responsible for managing all aspects of the database. Some responsibilities include planning, design, maintenance, and organization of the database, determining access to the database, user support and training with regard to database, maintaining a secure database environment, and providing for recovery from failures. To perform these functions, the DBA must be able to work with programmers and systems analysts and also with end users who have varying levels of computing skills. The person must be skilled in design and administration of databases while simultaneously having the

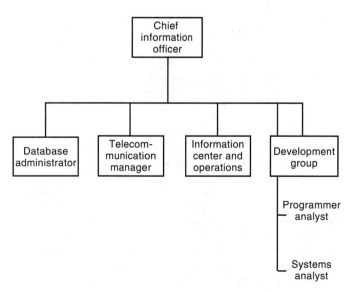

Figure 6. A typical organizational structure for an MIS department.

ability to work with end users and management. In many organizations, the data administration functions may be spread over a group of people rather than an individual.

Telecommunications Manager. The telecommunications manager is responsible for designing, implementing, and maintaining the corporate networks (both local and wide area) and ensuring that users can get to the computing resources that they need. The manager must be able to assess current and future telecommunication needs, the direction of communication technology, plan the network topologies, recommend hardware and software, design the system and implement it. In addition, the manager must implement security measures for authentication, authorization, and prevent against any unauthorized break-in and use of the corporate network. With the emergence of the Internet as a feasible commercial medium, these functions have taken on additional importance within the IS organization. Communication audits must be undertaken to check that the networks are being operated and used as intended.

Information Center. The information center is responsible for providing help to end users and assisting users with any problems that arise as they use the system. The center provides functions, such as training, documentation, equipment selection and setup, standards, and trouble shooting.

Operations. This group focuses on the efficiency of computer operations in the organization and consists of systems operators. Systems operators are responsible for ensuring that computers, networks, printers, etc., work efficiently. They generally hold certification from the industry vendors and work directly with hardware and software.

FUTURE DEVELOPMENTS

Information systems provide value to their business units by (1) allowing the organization to respond rapidly to changing market conditions or customer requests, (2) improving quality, and fostering innovation, and (3) competing and serving customers on a global basis. They also create value indirectly when they improve user interfaces and make it easier for users to respond to customer queries or when information systems are developed using different techniques to allow them to respond rapidly to competitive pressures. Advances are taking place in new user interfaces. For example, speech recognition is starting to become available for issuing commands to the computer. Systems development is changing to using components off the shelf, as opposed to developing them in-house. In addition, developments in electronic commerce are making interorganizational systems commonplace and provide organizations with another mechanism to reach their customers and end users. Videoconferencing technology and collaborative technologies are also starting to emerge. These allow organizations to support group meetings and project teams globally. They provide organizations with flexibility and the ability to compete worldwide. Technological advances make possible business reengineering using information technology. All of these advances imply that organizations, customers, and end users can expect to perform or have their work performed quicker with better quality and at low cost.

BIBLIOGRAPHY

1. M. Porter, *Competitive Strategy: Techniques for Analyzing Industries and Competitors,* New York: Free Press, 1980.
2. M. Porter, *Competitive Advantage: Creating and Sustaining Superior Performance,* New York: Free Press, 1985.
3. H. A. Simon, *The New Science of Management Decisions,* rev. ed., Englewood Cliffs, NJ: Prentice-Hall, 1977.

Additional Reading

Books

L. M. Applegate, F. Warren McFarlan, and James L. McKenney, *Corporate Information Systems Management: Text and Cases,* Fourth Edition, Chicago: Richard D. Irwin, 1996.

James Champy, *Reengineering Management: The Mandate for New Leadership,* New York: Harper Business, 1995.

V. Dhar and R. Stein, *Intelligent Decision Support Methods,* Upper Saddle River, NJ: Prentice-Hall, 1997.

L. M. Jessup and J. S. Valacich, *Group Support Systems: New Perspectives,* New York: Macmillian, 1993.

Kenneth C. Laudon and Jane P. Laudon, *Management Information Systems: Organizations and Technology,* Fifth Edition, Saddle River, NJ: Prentice-Hall, 1998.

T. M. Rajkumar and J. Domet, Databases for decision support, in B. Thuraisingham (ed.), *Handbook of Data Management 1996–1997 Yearbook,* Boston: RIA Group, 1996.

V. Sauter, *Decision Support Systems,* New York: Wiley, 1997.

Articles

Nancy Bistritz, Taco Bell finds recipe for success, *ORMS Today,* 20–21, October 1997.

E. K. Clemons, Evaluation of strategic investments in information technology, *Commun. ACM,* 23–36, January 1991.

Kevin P. Coyne and Renee Dye, The competitive dynamics of network-based businesses, *Harvard Bus. Rev.,* 99–109, January–February 1998.

G. DeSanctis and R. B. Gallupe, A foundation for the study of group decision support systems, *Manage. Sci.,* 589–609, May 1987.

Shihar Ghosh, Making sense of the internet, *Harvard Bus. Rev.,* 126–135, March–April 1998.

R. Grohowski et al., Implementing electronic meeting systems at IBM: Lessons learned and success factors, *Manage. Inf. Syst. Q.,* 369–383, December 1990.

J. R. Nunamaker, Jr. et al., Electronic meeting systems to support group work, *Commun. ACM,* 40–61, July 1991.

Stephen Pass, Digging for value in a mountain of data, *ORMS Today,* 24–27, October 1997.

Michael Porter, How information can help you compete, *Harvard Bus. Rev.,* 149–160, July–August 1985.

Ziff Davis, E-commerce and the internet grow together [online], Nov. 10, 1997. Available http://www.cyberatlas.com/segments/retail/market_forecast.html

Jeanette Borzo, E-commerce to total $333 billion by 2002 [online], May 11, 1998. Available http://www.infoworld.com/cgi-bin/displayStory.pl?980511.eiecomm.htm

DONALD L. DAWLEY
T. M. RAJKUMAR
Miami University

MANAGEMENT OF CHANGE

In response to changes in industrial conditions and other competitive challenges, firms commonly adopt new practices, policies, and technologies to sustain their competitiveness. In many cases, the implementation of these changes within the workplace fails to deliver the expected benefits of improved work quality and increased productivity. Instead, the results are often a disrupted work flow, unanticipated downtime, worker dissatisfaction, and loss of productivity. Some authors argue that most firms incur a substantial "adjustment cost" in implementing new technology that prevents them from initially realizing any net benefits from their investment (1). This period of adjustment arises because of a mismatch between the new technology, existing processes, and the organization itself. It can be a long and costly process for any firm. Yet the phrase "no pain, no gain" may aptly characterize this adjustment period, since ultimately organizational change may be essential for the long-term survival of the firm.

The adoption of a new technology highlights some of the issues associated with managing change because it requires the institution of new policies and procedures, the restructuring of technical work, and the acquisition of new skills. Firms considering a major technological change need to recognize that implementation is itself a learning process. Clearly, there is a need to understand and account for learning curve effects in the use of the new technology. More importantly, organizations need to learn how to integrate human resources into implementation strategies in order to minimize the *resistance to change* exhibited by workers and managers. Resistance to change can manifest itself in a wide range of behavioral problems from absenteeism to all-out sabotage of the technology. In general, worker resistance is caused by the fear of being replaced or the uncertainty about developing required new skills. Managerial resistance often stems from perceived changes in status or power base (2). A plan for managing technological change helps alleviate resistance (3). Computer-aided design (CAD) is an example of a technology adoption prevalent in the electronic/telecommunications industries that requires a plan for managing technological change.

Introducing a new technology requires the development of a change management strategy addressing two stages: preadoption and implementation. The change strategy, as a component of the technological adoption plan, should be an interactive and dynamic process of integration between the technological resources and the organizational environment. Communication and training are key elements in managing the technological change. This article proposes a number of methods for managing technological change. It examines the development of a plan for the management of technological change that aligns the organizational structure with the selection of training methods. It also discusses the effects of technology on worker resistance and how a change management plan can overcome this resistance. The introduction of a new CAD system is used as an example to illustrate the idea of technological change and demonstrate the methods of change management.

COMPUTER-AIDED DESIGN

CAD is widely used in many industries today, including the automotive/transportation industry, electronics/telecommuni-

cations, and architecture and engineering services. As new applications of CAD are developed and add-on software becomes popular, the use of CAD will become even more widespread. Many manufacturers use CAD in product development. Given the complex designs used in many products today, it would be difficult to produce and revise the required designs without the use of a CAD system. Most firms that use CAD have reported substantial time savings in their design work and improved quality (4). However, CAD is more than a drafting tool; it can serve as a managerial tool as well. The use of CAD can help integrate a product development team by allowing the design process to proceed in parallel. Different engineers can retrieve the design data from the CAD database and work on complementary features of the product in synchronized fashion. The result can be a dramatic reduction in the engineering time needed to complete a customer's order or to develop a new product.

Although firms may adopt CAD for many different reasons, a firm using CAD can no longer assume that it holds an advantage over its competitors. With so many firms using CAD, the technology has been reduced to a *qualifying requirement* for sustained competitiveness in design work. While failure to adopt CAD could result in a substantial loss of competitiveness, the use of CAD only implies that the firm is at parity with its competitors. However, CAD use does not guarantee success. Studies suggest that a majority of U.S. firms experience some type of failure when CAD is implemented (5). Yet firms have achieved a variety of strategic benefits, including cost savings, improved design flexibility, and better coordination with partners, suppliers, and customers, through the development of a technology strategy that addresses both human resources and organizational issues. Human resources includes the development of all CAD users—technical support staff, engineers, and managers—in order to smooth the implementation process, improve the skills and abilities of all users, and decrease worker resistance.

BACKGROUND

Resistance to Change

The adoption of a new technology often requires firms to overcome the obstacle of employee resistance in order to realize the full benefits of that technology. Technological changes create uncertainty for employees by disrupting organizational factors such as job task, role, internal relationships, and decision-making processes. Changes in these factors can produce either positive or negative psychological effects, which, in turn, can lead to positive or negative outcomes. Figure 1 illustrates the many factors creating resistance and the possible psychological effects and associated outcomes (6). The primary motive for resistance is change in the job task. Technology often reshapes the nature of the job task and requires new knowledge, skills, and abilities to perform these new tasks. Individual employees may interact with CAD in vastly different ways and thus require different levels of understanding of the operation of the new system. Since implementation and training may be staggered throughout the organization, the employee who does not yet understand the system may feel disadvantaged with respect to the ability to perform. Fear of a negative performance review can manifest itself in a negative view of the new technology and feelings of stress,

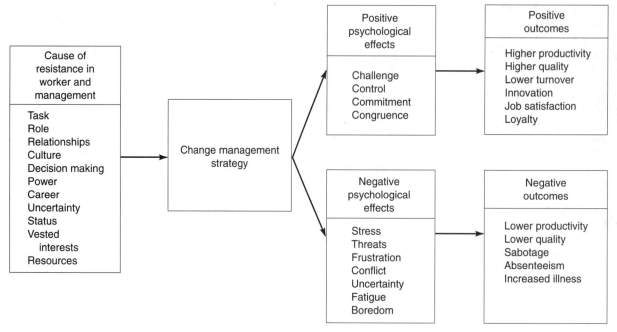

Figure 1. Causes of resistance to change.

frustration, and anger. Employees become less motivated and express these feelings through lower productivity and poor quality. Organizations may also experience an increase in absenteeism as well as extreme sabotage. For example, draftspeople have been known to pour coffee into their CAD station or stick paper into the computer's disk drive.

Unfortunately, management can also resist change. Managers may believe that the new technology will negatively impact their decision-making power and control over employees. In the case of a CAD adoption, managers lose some control to computer system administrators, who often dictate how designs and programs are stored and accessed. Managers may suffer during the "period of adjustment." The inevitable productivity downturn during implementation may be blamed on the manager, who is often evaluated on the basis of departmental performance (1). Management resistance to change creates a work atmosphere in which uncertainty is higher. The result can be a decrease in office morale, and a higher turnover can occur (3).

Another problem that contributes to resistance during the new technology implementation stage is the misalignment between the technology and the user environment. The misalignment usually occurs when there is conflict among the technical specifications of the work, the computer delivery system, and the performance criteria by which the user is evaluated (1). For example, CAD may be adopted on the basis of automating routine design tasks, ease of designing in three dimensions, allowing the electronic transfer of designs to manufacturing, or reducing paper in the design function. CAD's potential to be a tool of integration and communication between different work groups is often overlooked, and in many cases CAD is used only as an electronic drawing board. The lack of change within the user environment, in either the work structure and process, stymies the expected benefits of CAD. As Forslin and Thulestedt (Ref. 4, p. 201) state, "In hierarchic organizations, competition between functions and de-

fending of territories are inherent. This creates barriers for the necessary cooperation in technical changes." Thus, until the user environment becomes more cooperative, the integrative capabilities of CAD can rarely be realized.

Successful management of technological change attempts to reduce organizational barriers and produces a number of positive outcomes by boosting employee morale. Employees become challenged and committed to the adoption of new technology, which leads to higher job satisfaction. This, in turn, limits the duration of the period of adjustment, reduces the productivity downturn, and results in higher quality. The successful management of technological change also creates a congruence between management and employee expectations, which reduces employee fear and uncertainties.

Change Strategies

There are a variety of methods for dealing with resistance to change. The key elements in any plan for technological change management are knowledge and communication (7). To develop a change strategy effectively, management must first understand the impact of the organizational structure on workers and work flow.

Organizational Structure. Organizational structure can effect the adoption of a technology while at the same time the technology can influence and change the organizational structure. Understanding the organizational structure's impact on the adoption of technology can help the manager choose an appropriate change management strategy. Burns and Stalker (8) have examined the importance of organizational structure to successful business ventures. Their studies show that organic firms experience more success in innovative environments than mechanistic organizations.

Organic firms exhibit characteristics of flexibility and pushing decision making down the chain of command. They are environments in which the individual worker is given a great deal of independence, and thus relationships between manager and worker tend to be less formal. Organic environments are more team oriented, more professional, and less formal than mechanistic structures (8,9). Davis and Wilkof (10) define the organic system as a professional organization that draws its cohesiveness through formal and informal norms derived from a communality of interest. It is this common interest that keeps the organization together. The rules tend to be less rigid, and the manager's spans of control are smaller. In organic environments, task accomplishment and innovation are moved from management to the most knowledgeable parties.

In contrast, mechanistic environments are more formal and structured than organic systems. They have more authoritative and hierarchical relationships between managers and worker. Managers are the decision makers and resolve work-related problems. Workers have little control of their own environment and the way they do their work (8).

Research indicates that organic and mechanistic environments are linked to different organizational activities and competitive conditions. For example, Link and Zmud (11) discovered that organic structures are the preferred environments in research and development (R&D) activities. Their study indicates that organic structures encourage greater R&D efficiency. Covin and Slevin's (12) study of small firms found that organic structures are more successful in hostile, competitive environments, while mechanistic structures are more successful in situations in which there is no hostility and a less competitive environment. Davis and Wilkof (8), while researching the transfer of scientific and technical information, observed the relationship between organizational structure and efficiency of information transfer. They concluded that one of the best ways to improve communication and information transfer is to alter the organizational structure to a more open, organic system. Clearly, research has shown that the work environment is distinctly different for organic and mechanistic structures.

Organizational Structure and CAD. In practice, design departments exhibit different characteristics in the use of CAD.

Organic firms allow employees more autonomy. This feature of an organic environment is visible when workers are allowed to have influence over the decision process concerning work issues. Employees in organic environments exhibit more independence from their managers (8). The level of autonomy can be measured by the amount of contact the worker has with his or her manager. Workers in independent, organic structures have less contact with their managers and receive less direction on work methods. In organic structures, the workers are assigned work and allowed the freedom and responsibility to complete it in a way they deem appropriate (8,9). Another indicator of a firm's organic nature is the workers' involvement in the decision process of purchasing a CAD system. In organic structures, consideration is given to the employees' opinions about which system to use. Although the level of organic characteristics may seem difficult to measure, answers to simple questions such as "Where is the manager's office in relation to the employees?" and "How frequently do employees interact or speak with their manager or coworkers?" tend to reveal the work environment and structure quite quickly.

CAD firms with mechanistic structures have a different work approach. They are organized in a top-down fashion. Unlike organic structures, workers in mechanistic structures have little say in the day-to-day decisions that affect their occupations. Instead, management decides what CAD system to purchase and how it is used. In general, management in mechanistic firms makes itself more visible to the worker than in organic structures. Management is more likely to dictate rules and policy then to ask for the opinion and input of the workers. For example, when implementing a CAD system, management will choose the menus, naming conventions and drawing management procedures without consulting the workers.

Change Methods

There are a number of approaches for dealing with resistance to change. Table 1 lists six approaches and describes the advantages and disadvantages for each method. Kotter and Schlesinger (3) in their research found that the most fre-

Table 1. Method for Dealing with Resistance to Change

Method	Tactic Used in the Case	Organizational Structure	Advantages	Disadvantages
Training and communication	Where there is a low level of information	Mechanistic and organic	People will not hinder the change in the short term	Can be time consuming
Participation	When groups have the power to resist	Organic	People can become very innovative, gives higher quality	An unsuitable program may be designed
Facilitation and support	When work restructuring causes adjustment problems	Organic	Works best with adjustment problems	Very expensive and can still fail
Negotiation and agreement	Where there are status issues and a power struggle—also across department changes	Mechanistic	An easy way to smooth resistance between interdepartmental groups	Is expensive and cause inefficiencies
Coercion and cooptation	Where speed is essential	Mechanistic	Quick and easy solution to change in the short term	Leads to future problems
Explicit and tacit force	Where the initiator possesses great power	Mechanistic	Quick and overcomes any kind of resistance	Lower productivity and less integration

Adapted from J. P. Kotter and L. A. Schlesinger, Choosing strategies for change, *Harvard Business Rev.* March/April, 1979.

quently used methods are (1) explicit and tacit force and (2) training and communication. Explicit and tacit force is most frequently used in situations in which speed is a necessity. This method is more common in mechanistic organizations because managers are more authoritarian and employees tend to have less control over their work environment. Although this method has the advantage of being inexpensive, employees can become angry over being forced to perform tasks. In contrast, training and communication persuade employees to participate in the implementation. Because many people are involved, this approach can be very time consuming. However, it promotes innovative behavior and increases long-term productivity. While training and communication are used in both mechanistic and organic organizational structures most methods are more commonly used in either one or the other.

When implementing new technologies, organic organizations tend to solicit the involvement and opinions of the employees. These firms tend to use strategies in which communication is an important component (6). Beatty and Gordon (13), in an in-depth study of CAD implementation of ten companies, found that teams are among the best means of integrating technology into an organization. They also found that successful implementation has a technology "champion," a worker who has a great deal of political sensitivity and good communication skills that can be used to persuade and motivate employees. A technology champion and teams are often present in the participation and facilitation procedures. Organic firms often have communication channels and internal mechanisms that promote easy development of these strategies. Both the participation and facilitation methods for managing change bring full employee commitment and control adjustment problems to work restructuring. However, these methods are liable to take more time and sometimes fail (3).

Unfortunately, mechanistic structures are unable to implement participation and facilitation methods. Their internal structure tends to inhibit group planning. As can be seen in Table 1, mechanistic organizations tend to choose strategies that use power or coercion to gain employee acceptance. Such strategies are commonly used in situations where the job tasks are more structured and the employee has little voice in the implementation process. Firms using these methods observe a quicker implementation time and a lower cost. However, these methods run the risk of causing worker dissatisfaction. As long as time is not an issue, the best option to smooth implementation and maximize technology benefits is training and communication (14).

Training Issues. The implementation of a new technology often alters the occupational structure and changes the required skills. Training, when used properly, can do much to enhance the change process and the adoption of new technology. When implementing new technologies, companies must make decisions regarding user training so that the new skills required by the technology may be obtained by the organization in an effective manner. This makes training a perfect tool for implementing a change strategy. It becomes a vehicle that allows managers the opportunity to communicate and involve the employees (7,15).

Training issues can be broken into two stages: preadoption and implementation. Preadoption training issues are those in which training is considered prior to a firm's final decision

regarding purchase of the new technology. These issues are critical because unless a firm is cognizant of the important issues (such as the extent to which the entire work force *needs* to be trained, or whether the present work force is willing and/or able to *be* trained for the new technology) prior to the purchase of the technology, there exists a potential danger that the adoption will not be successful. Implementation issues are those in which the firm has *already committed* to the adoption of the system and is designing the implementation process in order to maximize the usefulness of the system while keeping costs in check. Examples of implementation issues include such decisions as choice of the method of training and the use of lower skilled workers to use the new technology.

Preadoption Issues. One question that needs to be considered in adoption decisions is whether or not it is possible to retrain the present work force to perform the tasks that will be required under the new technology. This question encompasses two possibilities. It is possible that the potential users might not possess the necessary skills or education to be able to learn the new system within a reasonable time period. In other situations, the existing workers decide that they are not interested in learning and using the new technology; for example, such nonadoptive behavior was found in certain instances of CAD implementation (16). Without worker support, firms have found it extremely difficult to implement the new technology successfully. It is critical, therefore, that the firm has the ability to assess the workers' ability and desire to use the new system accurately.

A firm considering the adoption of new technology also needs to estimate accurately the total expense of educating the work force. For example, the expense of purchasing a CAD system is much more than just the price of the computer hardware and software. A more complete estimate of expenses will include both training program expenses and a "cushion" for loss of productivity while the trainee masters the new equipment. A price estimate that does not include these costs can cause financial problems for the firm. Underestimation of training and education expenses is a major reason for failure in the implementation of CAD systems (17). The underestimation of the costs of CAD has caused firms to overextend themselves to the point where they could no longer remain competitive, resulting, in some instances, in the ultimate demise of the business (15).

Another closely related issue that needs to be considered initially is the extent to which training needs to be integrated throughout the firm. It is often critical to train employees besides those who come in direct contact with the technology in order that the equipment can be used at a level closer to its full potential. Examples of such employees might be CAD supervisors, secretarial staff, and manufacturing employees. Some new technologies are so radically different from existing systems that anyone who will have the slightest contact with the new system should receive training. For example, a recent survey of firms that have CAD systems showed that almost 50% of all companies with CAD/CAM training provided training to *all* occupational groups related to design, manufacturing, and materials management within the whole company (5). CAD systems were found to be so different from traditional design methods that workers who did not receive any training in CAD were at a tremendous disadvantage when they were needed to use the system in any way. Thus, with the implementation of CAD technology, many firms appear to have found a need to educate anyone who might come into contact with the system. It is therefore critical that management is aware of these needs (and costs) before choosing to adopt.

Implementation Issues. Implementation issues are those confronting the firm when it is already committed to the new technology. The main concern of these issues is how to develop a work force most efficiently and effectively that is sufficiently skilled at operating the new technology so that the system can achieve its potential. Some implementation issues are similar to preadoption issues, while many are completely different.

When a new technology is brought into a firm, management needs to decide the extent to which they are going to train the work force. Specifically, they must decide which workers *within* specific functional groups are to receive training. Although much is being done in universities to provide future workers with specialized technological skills (18), almost all firms find the need to retrain a group of their existing employees to operate and work with the new technology. From a financial standpoint, it might sound appealing for management to set up a specialized group within the firm to be trained while the rest of the employees are left alone. By doing this, the firm would only be spending training money on a percentage of the workers within the functional group. This might be possible, for example, if only *part* of the work in design utilizes a new CAD system, while the remaining work is done using traditional methods. But this does not work when the output designs must be CAD drawings.

The literature on electronic data interchange (EDI), another emerging technology, provides insight into another training implementation issue. The firm bringing in a new technology must decide how the workers will be trained. The firm must decide whether the training will be done in-house by their own personnel or if it will be done externally by either a vendor or an outside consultant. The firm also needs to decide if the training program it offers will be formal (classroom type) or a more tutor-oriented system. Internal training programs can often be more tailor-made for the specific system and can allow for a more informal and open exchange of ideas than their external counterparts. External education, on the other hand, is frequently more formal and has the advantage of being less expensive and is often the only option for smaller firms that may not have the money, facilities, or personnel to have internal training (19,20). Externally provided training programs have the disadvantage of being more generic and therefore frequently less useful.

Although researchers appear to agree upon the list of advantages and disadvantages, there is no consensus about which type of training is better. Engleke (21), for example, recommends that the training be done on site, without vendors, while Hubbard (22) feels that the best training sessions "are highly organized, relatively formal, classes taught by professional instructors at off-site locations." One study examined this issue with respect to the firm's managerial style and firm environment. It found that the mechanistic firms chose more formal methods of training while organic firms took a hands-on approach. Size of the firm may also influence the method of training (23). Majchrzak (Ref. 5, p. 200) found that 45% surveyed had in-house company-sponsored training programs. She also found that bigger firms generally chose to use the in-house, more focused training, while the smaller firms did not.

Another tactic used for keeping training costs down limits the training to those who are new hires. These policies may sound appealing, but in practice they have often led to disastrous results (10,24,25). These studies on deliberate separation of a portion of workers from a training program show that these policies often lead to feelings of exclusion. The workers feel that there is segregation into an "elite" group of workers and a group of workers that are being "put out to pasture." These feelings frequently lead to bad morale and poor worker/manager relationships. Some firms have also tried to hire employees from other companies that already have been trained in the new technology (21). In larger firms with substantial, established design teams, this has led to similar bad feelings and yielded similar results to situations in which workers were excluded from training. It is interesting, but not surprising, to note that when companies choose to offer training to some employees and not others, there are distinct patterns concerning which workers receive the offers to be trained. Liker and Fleisher (25) found that, although managers would not *say* that age entered into their decision process, the probability of being chosen to be a CAD user drops by 2% for each year of your age. For example, a difference in age of six years would correspond to a 12% difference in the probability of being selected. In their study, the average age of users was 39, while the average age of nonusers was 48. This apparent bias may not be unfounded—a study that analyzed the ability of a worker to learn new computer software found that younger workers (under 45 years of age) did significantly better than their older counterparts on comprehensive exams given after the training session (26).

An important issue in implementing a training program is determining when to train workers, before or during technology implementation. Managers should be particularly concerned with the transfer of training, which is the application of material learned in the classroom to the workplace. The length of time that passes between the time of training and the time of actual hands-on usage of the new technology greatly affects the transfer of training. Beatty (16) tells of a company that trained its employees six months in advance of the receipt of a new CAD system. The results were disastrous. Most of the information taught in the training sessions was forgotten in the period between training and actual usage. Ideally, little time should elapse between training and routine usage. Engelke (21) has found that the half-life of advanced package training is about two weeks if not applied immediately.

Many firms face a dilemma in training. If training is given before installation of the new system, the worker is allowed the benefit of learning the system prior to installation and will help to reduce productivity losses that might occur if the worker were trying to learn the new system after installation. The problem with prior training is that the purchased equipment often arrives later than scheduled or does not run properly immediately after installation, and the worker quickly forgets the training that was given. To avoid this problem, some firms opt to wait to train the worker until after the system has arrived and is functional, even though there will be productivity losses from downtime while learning (6).

Another critical concern for managers is the development of tacit skills. Tacit skills, which come from experience, tend to diminish in relative importance when compared to implicit skills, which are generally acquired through some form of a training program (27). It is not necessarily true that the tacit skills are actually reduced with the introduction of a new technology (as the previous examples of failed attempts to use computer operators clearly illustrate), but the number of new machines or technology skills that must be learned reduce tacit skills to a partial role in the education process. A fully trained worker must possess both tacit and implicit skills for the system to work at its fullest potential (6).

Implementation issues involving training for the multiple layers of the firm must also be considered. One might approach this subject by deciding to train only those workers who work directly with the new technology. In CAD, this would mean that training would only be given to the design group and not to any of the peripheral workers who might come in contact with but would not use the system routinely. The literature, however, suggests that training multiple layers of the firm is beneficial, if not necessary, in many applications (16,18). Brooks and Wells (28) found that managers who are not familiar with the new CAD system frequently experience difficulties. One common problem is the loss of status for a supervisor, especially if a skill differential develops between worker and supervisor. The literature in EDI agrees with the previously noted findings of CAD experiences. Carter et al. (Ref. 29, p. 14) notes that management training in EDI is "a key to increasing the likelihood that managers assigned the task of implementing EDI will succeed." Another problem that an untrained supervisor faces is the difficulty of effectively planning and controlling the work flow if he or she has no basis from which to estimate drawing and alteration times. The supervisor may also encounter great difficulty in evaluating a worker's progress and assessing performance of an individual. The combination of these problems often leads to the untrained supervisor "losing track" of the workers, thus causing a strained relationship and loss of productivity (15).

One mechanism to decrease worker resistance during the training period is to choose a pilot project that will help develop tacit skills and implement the technology. For example, in CAD implementation the pilot project should include the following characteristics: (1) a great deal of drafting and design work, (2) extensive design revision, (3) designs that are used by other functions and work groups, and (4) the need for extensive visual demonstrations and presentations of the design to customers and other projects. Workers involved in the pilot project would be among the first to receive CAD training. Such a pilot project would showcase CAD's capabilities in improving design/drawing productivity, integrating diverse users of design work, and communicating the output of design work. The successful use of CAD can be documented by an evaluation of the time savings and the quality improvements that were achieved. Often a single demonstration of success is enough to mitigate possible worker resistance. An illustration of this implementation strategy, often called the "quick slice" approach, occurred in a firm that implemented EDI in only one distribution center. Soon the workers in the distribution center were telling others in the organization of the benefits that were being achieved by using the technology. In no time at all, workers in other areas decided that they wanted the technology as well and were clamoring for EDI training (30). Thus, minimizing worker resistance to a technology can be achieved through the prudent choice of a pilot project, the development of a training program suited to the project needs and the organizational structure, and the demonstration of a successful use of the technology.

RESTRUCTURING DESIGN WORK THROUGH CAD

Traditionally, design work has encompassed both high-value-added activities (e.g., creative thought, problem solving, and design innovation) and low-value-added activities [e.g., producing a hard copy of an existing design, making minor design changes, and executing engineering change notices (ECNs)] that have *all* been performed by engineers and other technical professionals. In the *work restructuring* process, the low-value-added activities associated with design work are made routine and automated via CAD, and thus less-skilled workers who may not have design experience are capable of performing these tasks, which were previously performed only by engineers and draftspeople (31). Engineers and other technical professionals then have more time to devote to the higher-value design activities. The results are more opportunities for intellectual enrichment of their work, increased job effectiveness, and more time to integrate design activities with other areas of the organization (4). The work restructuring process in CAD reallocates tasks that were once considered *design work* across a broader spectrum of personnel. The work process is changed to incorporate more collaboration and coordination between engineers and other employees. Since work restructuring promotes both a team approach to design and better time utilization the result is designs that are completed sooner at a lower cost.

The use of technology as a facilitator in restructuring work has created debate with respect to the long-term implications for the work force. Two different effects have been hypothesized: deskilling and intellectual specialization. *Deskilling* refers to the devaluation of workers' intellectual skills when technology assumes the tasks previously performed by those workers, thus rendering their skills unnecessary. The deskilling effects of technology on workers has been debated with some arguing that technology-induced deskilling will lead to both a fragmentation of work content and an erosion of required work skills and ultimately will create a large class of unskilled workers (32). However, the deskilling proposition has not been empirically supported in studies of the effects of numerical control machining in the metal working industry and office automation on Canadian clerical workers (33,23). Although some skills were rendered unnecessary in these cases, workers were required to develop new skills to use the technology effectively, and a wide variety of new jobs was spawned by the technology.

A different impact of the new technology may be intellectual specialization, in which the knowledge domain of the design engineer or technical professional is profoundly changed from that of a generalist to a specialist. The shortening of product life cycles and the rapid pace of technological obsolescence so prevalent in the electronics and telecommunications industries imply that the design engineer must simultaneously increase his or her overall level of technical knowledge as well as his or her familiarity with increasingly sophisticated technology to remain at a state-of-the-art level. With more and more routine and low-value-added design activities performed through CAD, the design engineer may "intellectually specialize" by devoting the majority of his or her time to those activities that require the highest level of cognitive skills with which the engineer is qualified to perform (31). Thus, the use of CAD may promote increased specialization in design work—not because of work simplification or a degradation in the skill requirements—but because the task breadth and the knowledge intensity of the remaining design work create such a level of technical complexity that only a staff of narrowly focused or "specialized" professionals can perform it (20). Carried to an extreme, intellectual specialization is sometimes hypothesized as creating a division in the work force with an increasingly larger staff of professional workers performing highly specialized yet cognitively challenging work, while the other workers are relegated to performing relatively lower-level tasks. In these cases, intellectual specialization does not necessarily imply cost savings, since the organization must increase the size of its professional staff, many members of which represent some of the most highly paid individuals in the organization.

However, the success of work restructuring and the balance between deskilling and intellectual specialization are highly dependent on the effectiveness of the training program. In a study of the German mechanical engineering CAD industry (34), emphasis on *computer literacy* at the expense of actual design techniques during the training process limited the scope of tasks that could be relegated to less-skilled employees. Two other failures were documented when workers trained only as computer operators lacked sufficient design background to perform design work successfully (31). Many of these failures have been attributed to a conventional mindset in CAD training that tends to equate humans with machinery. The Swedish project UTOPIA represented a major departure from this mindset by designing training programs that focused on the development of both design and negotiation skills on the part of the nontechnical worker and on advocating the restructuring of design work so as to use CAD in order to enhance, not replace, the skills of the design worker.

DISCUSSION

Research provides general support for the idea that there are relationships among change management strategies, organizational structure, the formality of the training program, and the restructuring of design work. Managing the technological change process depicted in Fig. 2 suggests a very proactive approach in deciding what kinds of programs to establish when implementing a new technology. Figure 2 implies that for an effective technological change strategy, selection of an approach should be based on the organizational structure.

In the case of implementing a new CAD system the type of training program utilized by a firm is closely related to the organizational structure of the firm. For many companies, the choice of training program is more a matter of finding the best "fit" to the specific organizational structure of the particular firm than just a decision based on firm size alone. However, it appears that if firms are particularly concerned with benefits associated with the deskilling process, then it might be in their best interest to use more informal methods to train their workers.

In a recent study on CAD adoption (23), a mechanistic firm was observed using *informal* training methods specifically to

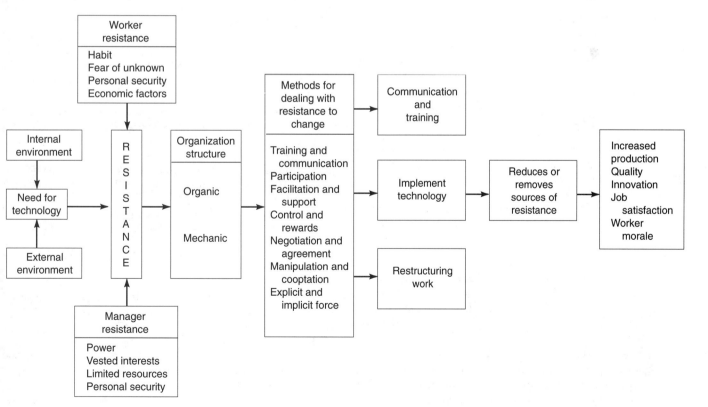

Figure 2. Managing the technological change process.

take advantage of the deskilling process. It might be suggested that other mechanistic firms alter their training format to do the same. By allowing for more informal, loose flows of information, these mechanistic firms might be able to achieve the same quality of CAD education. Organic firms concerned with management training, on the other hand, might be advised to pursue more formal methods of educating CAD managers. As noted previously, the literature strongly suggests that this is highly beneficial for making the management of the systems run smoothly.

In the firms involved in the study noted previously, only those using informal, tutorlike training programs were found to have lower-skilled workers contributing through the CAD system. Those firms relying exclusively on formal methods had no such work restructuring. This difference could be significant, for work restructuring holds the potential of providing firms with financial savings in the use of CAD by allowing for the use of lower-paid workers in routine design activities and the use of more highly paid professional workers for the most intellectually challenging design tasks.

In the telecommunications and electronics industries, the penalty for not adopting or successfully implementing CAD is prohibitive—total noncompetitiveness to the point at which the firm must essentially abandon its markets and leave the industry. Adoption of other technologies such as third party design software, groupware, email, project management software, and processing technologies all result in changing the work environment. To successfully adopt these technologies, managers must match the change in the management program with the goals (work restructuring) and organizational structure.

Suggested Areas of Future Research

Although the observations of this article are strongly related to CAD adoption, it is important to recognize that other technological adoptions may have a wider impact on an organization's coordination ability and work processes. For example, incorporating new processing technologies can often have wide ranging consequence which effect the management and work structure of multiple departments (1). However this article does give a general overview on the major considerations involved when managing a technological change. In addition, there are many different factors besides organizational structure and training that can influence a CAD implementation. The development of a contingency model is clearly beyond the scope of this article. Future studies may explore these contingencies and look at other technological adoptions. This observation raises a number of interesting research questions.

Questions

1. Does the implementation of technology or change within organic structures take longer than those in mechanistic structures? Is the long-term outcome of technology adoption better with one structure than the other?

2. What are the dynamics when adopting a technology across a number of departments? What are the coordinating mechanisms that help to implement change within multiple departments? Are these mechanisms the same for the two organizational structures?

BIBLIOGRAPHY

1. W. B. Chew, D. Leonard-Barton, and R. E. Bohn, Beating Murphy's law. *Sloan Manage. Rev.,* **32** (3): 5–16, 1991.

2. D. Twig, C. Voss, and G. Winch, Implementing integrating technologies: Developing managerial integration for CAD/CAM. *Int. J. Prod. Manage.,* **12** (7): 76–91, 1992.

3. J. P. Kotter and L. A. Schlesinger, Choosing strategies for change. *Harvard Business Rev.,* March/April, 1979.

4. J. Forslin and B. M. Thulestedt, Computer aided design: A case strategy in implementing a new technology. *IEEE Trans. Eng. Manage.,* **36**: 191–201, 1989.

5. A. Majchrzak, A national probability survey on education and training in CAD/CAM. *IEEE Trans. Eng. Manage.,* **33**: 197–206, 1986.

6. H. Noori, *Managing the Dynamics of New Technology: Issues in Manufacturing Management.* Englewood Cliffs, NJ: Prentice Hall, 1990.

7. L. A. Berger, *The Change Management Handbook: A Road Map to Corporate Transformation.* Burr Ridge, IL: Irwin Professional Publishers, 1994.

8. T. Burns and G. M. Stalker, *The Management of Innovation.* London: Tavistock Press, 1961.

9. J. A. Courtright, G. T. Fairhurst, and L. E. Rogers, Interaction patterns in organic and mechanistic systems. *Acad. Manage. J.,* **32**: 773–802, 1989.

10. P. Davis and M. Wilkof, Scientific and technical information transfer for high technology: keeping the figure on its ground. *R&D Manage.,* **18**: 45–58, 1988.

11. A. N. Link and R. W. Zmud, Organizational structure and R&D efficiency. *R&D Manage.,* **16**: 317–323, 1986.

12. G. C. Covin and D. P. Slevin, Strategic management of small firms in hostile and benign environments. *Strategic Manage. J.,* **10**: 75–87, 1989.

13. C. A. Beatty and J. R. M. Gordon, Barriers to the implementation of CAD/CAM systems. *Sloan Manage. Rev.,* **29** (4): 25–33, 1988.

14. A. Berger, Towards a framework for aligning implementation change strategies to a situation-specific context. *Int. J. Operations & Production,* **12**: 32–45, 1992.

15. B. M. Bouldin, *Agents Of Change: Managing the Introduction of Automated Tools.* Englewood Cliffs, NJ: Yourdon Press, 1989.

16. C. A. Beatty, Tall tales and real results: implementing a new technology for productivity. *Business Quart.,* **51** (3): 70–74, 1986.

17. P. S. Adler, New technologies, new skills. *California Manage. Rev.,* **29** (1): 9–28, 1986.

18. S. A. Abbas and A. Coultas, Skills and knowledge requirements for CAD/CAM. In P. Arthur (ed.), *CAD/CAM Edu. Training: Proc. CAD ED 83 Conf.* Garden City, NY: Anchor Press, 1984.

19. P. S. Goodman and S. M. Miller, Designing effective training through the technological life cycle. *Natl. Productivity Rev.,* **9** (2): 169–177, 1990.

20. H. Rolfe, In the name of progress? Skill and attitudes towards technological change. *New Technol., Work and Employment,* **5** (2): 107–121, 1990.

21. W. D. Engelke, *How to Integrate CAD/CAM Systems: Management and Technology.* New York: Marcel Dekker, 1987.

22. S. W. Hubbard, *Applications for Business.* Phoenix, AZ: Oryx Press, 1985.

23. C. McDermott and A. Marucheck, Training in CAD: An exploratory study of methods and benefits. *IEEE Trans. Eng. Manage.,* **42**: 410–418, 1995.

24. I. L. Goldstein, Training in work organizations. *Annu. Rev. Psychol.,* **31**: 229–272, 1980.

25. J. K. Liker and M. Fleisher, Implementing computer aided design: the transition of nonusers. *IEEE Trans. Eng. Manage.* **36**: 180–190, 1989.

26. M. Gist, B. Rosen, and C. Schwoerer, The influence of training method and trainee age on the acquisition of computer skills. *Personnel Psychol.*, **41**: 255–265, 1988.

27. P. S. Adler, CAD/CAM: Managerial challenges and research issues. *IEEE Trans. Eng. Manage.*, **36**: 202–215, 1989.

28. L. S. Brooks and C. S. Wells, Role conflict in design supervision. *IEEE Trans. Eng. Manage.*, **36**: 271–282, 1989.

29. J. R. Carter et al., Education and training for successful EDI implementation. *J. Purchasing Mater. Manage.*, **23**: 13–19, 1987.

30. R. B. Handfield et al., International purchasing through electronic data interchange systems. In P. C. Deans and K. R. Karwan (eds.), *Global Information Systems and Technology.* Harrisburg, PA: Idea Group Publishing, 1994.

31. H. Salzman, Computer-aided design: Limitations in automating design and drafting. *IEEE Trans. Eng. Manage.*, **36**: 252–261, 1989.

32. H. Braverman, *Labor and Monopoly Capital: The Degradation of Work in the 20th Century.* New York: Monthly Review Press, 1974.

33. G. Zicklin, Numerical control machining and the issue of deskilling. *Work and Occupations*, **14**: 452–466, 1987.

34. F. Manske and H. Wolfe, Design work in change: Social conditions and results of CAD use in mechanical engineering. *IEEE Trans. Eng. Manage.*, **36**: 282–297, 1989.

Recommended Readings

G. George, Employee technophobia: understanding, managing and rewarding change. *J. of Compensation & Benefits*, **11**: 37–41, 1996.

S. L. Herndon, Theory and practice: Implications for the implementation of communication technology in organizations. *J. Business Communication*, **34**: 121–129, 1997.

A. Judson, *Changing Behavior in Organizations: Minimizing Resistance to Change.* Cambridge, MA: B. Blackwell, 1991.

CHRISTOPHER MCDERMOTT
Rensselaer Polytechnic Institute

ANN MARUCHECK
University of North Carolina-Chapel Hill

THERESA COATES
Rensselaer Polytechnic Institute

MANAGEMENT OF DOCUMENTATION PROJECTS

Three key practices mark the successful management of information projects—planning, estimating, and tracking. The purpose of these practices is to ensure that the final information products meet the needs of the intended users, and that they are accurate, complete, accessible, and usable. All management activities should thus be directed to achieving these goals. If information is not accurate, users will make mistakes in performing tasks and interpreting results. If information is not complete from the users' perspective, they will not have the information they need to perform tasks or to solve problems. If information is not accessible, users will not be able to find the information they need quickly and easily. And, if information is not usable, that is, clear and understandable, users will not be able to achieve their goals.

Unfortunately, many people who are responsible for preparing technical information believe that their task is finished once they have written down what they know. They focus on recording their knowledge of a subject rather than considering the needs of those who will read and use the information. We have all had the experience of trying to use a technical manual to assemble and use a product we have purchased, only to discover that the information provided is not what we need to know. We have read the results of someone's research, only to discover that the methods are unclear and cannot be duplicated or that the conclusions do not appear to follow from the results. Information we need for our own work is simply not available.

Professionally prepared technical information requires that we take into account from the very first who will be using the information and what they will want to do with it. This article presents a five-phase process that will assist in conducting and managing an information-development project to ensure the delivery of user-focused information products at the end.

Information planning is the first of the five phases of the information-development life cycle. The five phases to be discussed in detail below are

- Phase 1: Information planning
- Phase 2: Content specification and prototyping
- Phase 3. Implementation
- Phase 4: Testing and production
- Phase 5: Evaluation

This life cycle is discussed in detail in Ref. 1. The discussion below focuses on explaining the key elements of the information development life cycle as it relates to the activities of the entire information-development team, in addition to the project manager.

PHASE 1. INFORMATION PLANNING—IDENTIFYING THE USERS' INFORMATION NEEDS

The Information Planning phase marks the beginning of the information-development life cycle. The planning phase allows the information designers to gather data that assist them in answering four key questions:

- Who are the potential users of the product or process and its documentation, including users of the interface, the online help, and others, such as those training and supporting the users?
- What is their range of knowledge, skills, and experience in the subject matter covered by the documentation? What are their range of skills and experience in the tools used in the product?
- What goals do they want to achieve using the product and the documentation? How do they achieve those same goals today?
- What information styles will be most effective in helping the users to perform and learn?

To answer these questions about the users, information developers consider it critical to interact directly with the information users. This interaction includes observing them in the environment in which they will use the information. This direct observation of users is especially critical when the information is intended to support users in performing specific tasks, for example, operating a machine, using software to operate equipment or perform tasks, maintaining equipment, using information for design and development, and so on.

As the goal of information planning is to develop information products that will meet user needs and enable them to achieve their knowledge and performance goals, the greater the understanding of the users and their requirements, the better the information plan is likely to be.

For example, the information developers may discover that the needs of the assembly line personnel are best served by the development of job aids displayed in the workplace rather than lengthy and complex manuals. The information plan would establish the case for producing job aids rather than manuals. The information developers might discover that their design engineering audience is best served by conceptual information about the relationship between the design tools and their design goals. Similarly, they will make their case for this information design in the Information Plan.

The Information Plan serves as a proposal for the information strategy devised for the project. The strategy ought to be established using first-hand knowledge of users and their needs, not on unexamined assumptions about the users. Several years ago, a team at one of the major semiconductor manufacturers investigated the information needs of the design engineers who purchased microchips from the company. The designers were most likely to be electrical engineers, who often had advanced degrees in the field. The internal design engineers had always assumed that the information needs of their professional customers were identical to their own information needs. However, after the team surveyed the customers and conducted site visits, they learned that the customers wanted information that explained how the chip might be effectively used in their designs rather than information about how the chip had been designed in the first place. The investigation demonstrated that the existing information design for the manuals was not meeting the customers' needs. To quote a customer interviewed for the study, "We need to know how to use the chip, not how to design it."

Creating the Information Plan

At the end of the investigation, after the users' information needs are well understood, the team of information developers prepares an Information Plan. The Information Plan summarizes the results of the investigation and presents the strategy for meeting user needs. This strategy outlines the role of all the layers of information delivery available. From the information designed into the interface, through context-sensitive help and more detailed conceptual, procedural, and instructional text and graphics, the information designers should explain how each piece of information will meet the needs of a particular segment of the user population. Included in the strategy should also be discussions of training requirements and ongoing support of users with information provided by field engineers, telephone support, or through continually updated information accessed electronically through Web pages, bulletin boards, fax lines, and more.

The Information Plan presents the architecture of the information solution for the users. In the next phase, Content Specification, the broad brush of the Information Plan strategy is translated into the details of the many types of media that will be designed.

Figure 1 illustrates part of a typical information plan, including a summary of the proposed strategy and predicted costs.

How Long Should Information Planning Take?

In the standard five-phase information development life cycle, it is estimated that Phase 1 should take between 10 and 20% of the total information project time. For example, if you have estimated that the information development project should take 6 person-months or approximately 800 development hours, then the information planning phase should take about 80 h to complete. The calendar duration of the 80 h may be more than 2 weeks because of the logistics in scheduling site visits and other information-gathering activities. For projects that have many unknowns in terms of users and their goals and activities, a higher percentage of project time may need to be allocated to information planning.

Estimating the Cost of the Information Project. The initial estimate of the information project is best made following the development of the Information Plan. If an estimate must be made before any information planning has occurred, it is best made in reference to previous similar projects. For example, if the last project took 750 h over 6 weeks to complete and was similar in scope to the current project, and if the evaluation of the information delivered showed that it met user needs, then the next project is likely to take a similar amount of time. It would be wise, however, to conduct a preliminary analysis of project dependencies to ensure that they also are the same from the previous project to the next project. Conducting a dependencies analysis is explained below.

To estimate the cost of a documentation project requires that

1. The scope of the proposed project has been determined
2. The level of quality required to meet user needs has been determined
3. A history of previous similar projects has been compiled
4. The dependencies of the proposed project have been evaluated
5. The resources available to complete the project have been examined
6. The deadline and milestone schedule requirements of the project have been determined.

Each of these estimating requirements is described in the following sections.

Determining the Scope of the Information-Development Project

The scope of an information-development project reflects the amount of information to be developed. The scope should be derived from the Information Plan and from previous experience providing information for similar users. Note that estimating the information scope is often difficult for those inexperienced in information design and development. It may be

Information plan template (first page only)

Purpose of the Project

In this section, explain the purpose of the technical project. Provide some background explaining the company's motivation for producing the product and its information and training. For example, is it because they have introduced a new product that requires a new approach? In this section it is appropriate to include information about the company's marketing strategies for the product. What is this product's niche? What is special about it that the information should highlight from a sales angle?

Goals for the Information

In this section, describe the goals of learning and use that the proposed information addresses. For example, will you have a mix of information and training, or will information stand alone as a user's sole information resource? How will the deliverables you are proposing help to achieve these goals?

Quality/Usability Goals for the Information

In this section describe the quality/usability goals you have established for the information. How will you ensure that users will be successful in learning and using the product or system? For example, upon completing training and using manuals or online help can you ensure that users are able to perform their tasks within a specified time with as few errors or calls for help as possible?

Product Description

In this section, include a brief description of the product and its basic functions. This information should be kept brief because more detailed information about product functionality is included in the content specifications for the individual deliverables.

User Profile

Describe the background and experience of the user(s) for the information. What are their expectations about the product? In what kind of environment do they work? Are there any special circumstances surrounding their use of the product?

Include a brief description of the individual user segments. In this description, include only information that will have an impact on the final design and organization of the information for these users. For example, if one user segment contains people with very similar educational backgrounds, this fact may be less important to the information project than the fact that some have no typing experience but will be expected to use a keyboard.

Figure 1. A representative example of an Information Plan.

useful to compare the project being estimated with previous projects, competitors' projects, or similar projects with which you may be familiar from other industries. For example, you may discover that a project similar in scope developed the previous year resulted in a 150-page manual, a 50-page set of error messages and solutions, and 250 online help topics. Then you might initially assume that the new project will have a similar scope.

For a completely new project, you may need to make a rough estimate of tasks to be supported in the documentation and to multiply by a standard such as 5 or 10 pages for each task. It may be necessary to estimate the number of help topics to be included with a software package by calculating the number of help topics produced for a similar project and correlating these with the number of screen displays, dialog boxes, menu items, check boxes, choice buttons, and so on. The number of choices the user has may provide a rough correlation with the user tasks to be documented, as well as the number of context-sensitive help topics.

Depending on the nature of the information media chosen, different units are employed as indicators of the scope of work. Table 1 lists units that are traditionally related to different types of information media.

Experts in each media type typically use units of this sort as scope definitions for estimating purposes. Added to the scope definition is the level of quality and complexity to be achieved. For example, a simple graphic image may take only

Table 1. Traditional Information Units for Purposes of Scope Estimates

Information Media	Unit
Printed or electronic text	Pages (approx. 400–500 words)
Context-sensitive online help	Topics (approx. 250 words)
Graphics	Image type
Video	Minutes of video
Classroom instruction	Hours of instruction
Interactive multimedia	Events (page turn, popup, animation, voiceover unit, and so on)
Sound	Seconds
Quick reference	Impressions (print images)

Table 2. Typical Scope Definition from an Information Plan

Information Unit	Scope Definition
Users' guide	60 pages
Online help	200 help topics
Conceptual guide	45 pages
Training class	6 h of instruction
Quick reference card	2 impressions

an hour or so to create, while an exploded view of a complex mechanism may require a 100 hours or more.

A typical estimate of scope for an information-development project might look like Table 2.

Determining the Quality Level Needed to Meet User Needs

In most hardware and software development projects, the quality level is determined by the effort made to reduce the number of defects in the final product delivered to the customer. In a software project, for example, the product specification may require that no Level 1 defects remain in the product before it is shipped, defining a Level 1 defect as one that causes loss of data for the user. Following typical testing algorithms, sufficient testing is conducted to estimate with reasonable certainty that no Level 1 defects exist.

The quality level of information products can also be set by the level of effort taken to ensure that no defects exist and that the information meets the usability requirements set by the developers and the users. In some organizations, for example, information-development standards require that all task-oriented documentation be tested with the product to ensure that no errors exist in the instructions. Other organizations require that documents pass the criteria set for usability testing of the documentation with the customers. If the test subjects can perform the selected tasks within a specified amount of time and at an acceptable level of performance (no unrecoverable errors, a stated level of user satisfaction, and so on), then an acceptable level of quality has been achieved.

The quality level set for an information-development project will determine how much effort to devote to ensuring that the resulting products are defect-free and usable. It will take more time to complete a project that requires a high level of quality with regard to user requirements and usability than to complete a project in which defects in the documentation go unchecked and uncorrected. You must determine if lower levels of quality are acceptable to your organization, team, and customers you choose to reduce documentation effort below a standard of quality. Remember that lower levels of information quality may result in greater customer dissatisfaction, higher training costs, higher support, and lower product sales. Organizations that tolerate lower quality standards for documentation often also tolerate lower quality standards for their products in general. One general manager of a development organization reported that his company's quality standard was to "ship it because the customer will find the defects before we do."

Other organizations require high levels of quality in documentation going to customers or employees because they recognize the impact of poor quality on their organizations' reputation. They also may be interested in reducing training costs or reducing the cost of customer support by providing infor-

mation that permits customers to learn and act independently.

Compiling a History of Similar Projects

The histories compiled of similar projects completed in your organization will provide a base line of data for evaluating the cost of a new project. For example, you may have determined that previous projects completed in the organization resembled those in Table 3.

Hours to complete includes all development time (including planning), information gathering, writing, editing, capturing screen images, creating simple graphics, and managing the project. The hours do not typically include time for reviews by product developers and others not part of the information-development team.

After histories of similar projects have been collected, you must determine if these projects represented the quality standard sought for the current project. The following questions must be considered. Have you surveyed user satisfaction with the information products? Are you aware of the number of support calls attributable to incorrect or unusable documentation? Are you aware of the time required by your customers to train end users of your product? Do you know if your customers are rewriting your documentation because it is inadequate to meet their needs? Are you providing more information than your customers find necessary? Review the section entitled "Phase 5: Evaluation" for ideas about studying the customers and evaluating documentation quality.

Once you have obtained a project history and decided that the project achieved an appropriate standard of quality, the data collected to help in estimating the next project. If you find that the quality level is inadequate, you may want to increase the metric to allow more time for quality assurance activities and quality improvement, including customer observations, usability testing, and functional testing.

In Table 3, note that the smallest users' guide also required the most time to create in terms of the unit metric (5.7 h per page). Frequently, efforts made to minimize information and make it more accessible and usable increase the unit metric at least temporarily until the new standard is well understood by the development team. Also note that the average of the four projects is 4.17 h per page. In estimating the cost of a new project, you may want to begin your calculation with 4.17 h per page.

If project histories are not available, you may choose to use industry averages as a starting point or a benchmark with other information-development organizations in the industry. Be certain, however, that you understand exactly what the managers are counting toward total development time and what level of quality they are working toward. One organization I studied produced high-quality minimalist manuals for a hardware product and averaged 9.75 h per page in doing so.

Table 3. History of Typical Information-Development Projects

Project Type	Hours to Complete*
Users' guide of 150 pages	562 h (3.75 h/page)
User's guide of 234 pages	983 h (4.2 h/page)
Users' guide of 567 pages	2336 h (4.12 h/page)
Users' guide of 56 pages	319 h (5.7 h/page)

Multimedia project

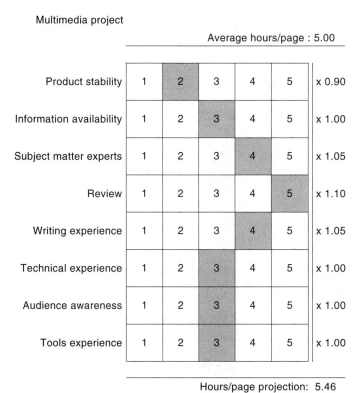

Average hours/page : 5.00

	1	2	3	4	5	
Product stability	1	**2**	3	4	5	x 0.90
Information availability	1	2	**3**	4	5	x 1.00
Subject matter experts	1	2	3	**4**	5	x 1.05
Review	1	2	3	4	**5**	x 1.10
Writing experience	1	2	3	**4**	5	x 1.05
Technical experience	1	2	**3**	4	5	x 1.00
Audience awareness	1	2	**3**	4	5	x 1.00
Tools experience	1	2	**3**	4	5	x 1.00

Hours/page projection: 5.46

Figure 2. Dependencies calculator.

Another organization averages about 4 h per page for task-oriented software documentation. I have found that the development of usable context-sensitive online help typically requires about 2.5 h to 3.5 h per help topic, depending on the experience of the developers and other project dependencies. Classroom-based instructional design and course development is often quoted at between 20 h and 40 h per deliverable hours of instruction. Refer to others in the same industry who are known to track their projects conscientiously to help provide a starting point for project estimates.

Evaluating the Dependencies of the Proposed Project

Even if you have developed an adequate history of previous projects in your own organization, using the metrics of an individual project or even the metrics of an average project may not produce a good estimate of the cost of a new project. Not all projects are equal nor are all projects average. I recommend using the Dependencies Calculator illustrated in Fig. 2 to calculate the effects of several potentially significant factors on a particular project. The Dependencies Calculator weighs the effects of ten typical project dependencies, using the midpoint (a factor of 3) as 1.00 (the neutral point) and assigning each rating above 3 a factor 5% over the previous rating and each rating below 3 a factor 5% below the previous rating.

The ten dependencies on the sample calculator in Fig. 2 represent the typical dependencies we have seen in documentation projects. Dependencies that affect your projects can be added or dependencies that do not apply may be omitted. Nine of the dependencies have increments of 5% above and below the center except for the first dependency, which has a 10% increment above and below the center. This dependency

for project stability applies to the stability of the larger development project of which information development is a part. For example, a project would be considered highly unstable if you anticipated many functional changes throughout the development life cycle. Such product changes will affect the number of information changes that will have to be made during the information development life cycle. A stable product with few changes will take fewer hours and cost less to document.

Consider the dependencies calculation for a specific project. Based on an analysis of previous projects with the same team and new information about the current situation, the project manager creates the following set of dependencies. On each five-point scale, 3 represents the average case in the organization, 4 and 5 represent worse than average scenarios, and 1 and 2 represent better than average scenarios, as illustrated in Fig. 3.

On the basis of the calculation and with the starting point of 4.17 hour/page, the project manager calculates the hours per page for this project to be 4.55.

Once the average hours per unit have been determined and the dependencies for a specific project have been calculated, multiply the hours per unit by the number of units to calculate the hours required to complete the project of the scope and quality specified. For example, a project manager knows that previous online help projects have averaged 3.1 h per help topic. The dependencies calculation for the new project results in 3.5 h per help topic. The manager estimates they will need to write 250 help topics based on a comparison with previous projects of similar scope. Therefore, the total hours required to complete the projects are 250 times 3.5 or 875 hours. At an average charge per fully burdened hour of

Online help project

Average hours/page : 4.17

	1	2	3	4	5	
Product stability	1	**2**	3	4	5	x 0.90
Information availability	1	2	**3**	4	5	x 1.00
Subject matter experts	1	2	3	**4**	5	x 1.05
Review	1	2	3	4	**5**	x 1.10
Writing experience	1	2	3	**4**	5	x 1.05
Technical experience	1	2	**3**	4	5	x 1.00
Audience awareness	1	2	**3**	4	5	x 1.00
Tools experience	1	2	**3**	4	5	x 1.00

Hours/page projection: 4.55

Figure 3. Sample dependencies calculation for a particular project.

Table 4. A Typical Information-Development Team

Team Members	Percentage of Project Time
Project manager	15%
Writer	50%
Editor	15%
Graphic artist or illustrator	15%
Production specialist	5%

employee time of $65/h, the manager then calculates that developing the online help will cost nearly $57,000 of development time.

Examining the Resources Available to Conduct the Project

The primary resources required for an information-development project are people—writers, editors, graphic artists, production specialists, project managers, and other specialists depending on the nature of the media selected. The information project manager must evaluate the team resources needed for the project as well as the resources available. In the dependencies calculation, the experience of the team members has already been accounted for in part. If team members change, the dependencies for the project may also change.

Table 4 illustrates a typical team and the percentages of time required per team member for an information-development project. If you decide to assign the project to a single individual, consider carefully if that decision will be cost effective. Can the individual do his or her own quality assurance? We often find that writers are not the best editors of their own work. Can the individual do the graphics needed for the project? The project may require photographs of equipment. Does the information developer have adequate skills to take the photographs, taking into account lighting, correct exposures, editing, and digitizing? Is final production better done by a production specialist or by the individual? I believe that using a highly skilled communicator to do production tasks is not cost effective. As information projects become increasingly complex, with many media to choose from, assuming that a single individual can do everything effectively is a mistake.

In addition to evaluating the skills needed for the project, consider the availability of people for the project. Are all the experienced staff members already dedicated to several other projects? What hours are allocated to those projects? Are they available to work full-time on the new project, or only part time?

If the cost of the rest of the projects has not been estimated, it is likely that team members have been overassigned to too many projects. Overassignment usually means that quality is compromised to meet deadlines. Underassignment of time does not guarantee quality either. People tend to fill the available time doing something but not always work that advances the quality of the product. In fact, underassignment may result in lower quality if too much unnecessary information is provided. Unnecessary information clutters the document, making it more difficult for users to find the information they actually need. Manuals that are too long are often not used because they are intimidating to certain users. Online information that is voluminous often means online searches that result in hundreds of topics selected, making the critical information difficult to separate from the "nice to know" or the completely irrelevant.

The key to maintaining quality at the specified level is to staff projects correctly, based on the estimates you have made. You need to ensure that sufficient staff be assigned from the beginning to take into account information changes that are likely to occur, but you must guard against using more staff than is necessary.

Determining the Project Milestones and Deadlines. Project milestones and deadlines for information-development projects are usually established in response to external factors such as product launch dates and customer requirements. However, to some extent, milestones should be viewed as dependent on your ability to conduct the project phases in a manner that produces a high-quality result. Based on experience with hundreds of projects, I suggest using the following, illustrated in Table 5, to schedule internal milestones.

To estimate Phase 4 requires taking into account the production requirements of the project. For example, a report that is simply reproduced on a copier machine will have an almost negligible Phase 4. But a manual of several hundred pages that will be offset printed and bound may take two or three weeks to complete. In evaluating the Phase 4 milestone, discuss the production techniques with people expert in the media that will be used. Remember that even information delivered electronically takes time to prepare and debug.

Developing a Project Spreadsheet

One of the best and simplest ways of representing your initial view of an information-development project is a spreadsheet. The spreadsheet allows you to staff the project appropriately to meet the interim milestones and the deadline while maintaining the quality level demanded by the users. Figure 4 illustrates a typical project spreadsheet for a project that includes a user manual and an online help system.

Each column in the spreadsheet represents a month or a week of the project. Each row represents the hours allocated to each person working on the project. The total hours per person adds up to the total hours required to complete the project as defined in the Information Plan.

The spreadsheet can also be used to define the due dates of project milestones; for example, to calculate when Phase 2 is expected to be completed. If Phase 2 represents approximately 30% of the total project time, then the spreadsheet can be used to calculate when 30% of the total hours has been expended.

In addition to using the project spreadsheet to estimate and schedule the project, you can use it to track project prog-

Table 5. Milestone Definitions

Phase	Name	Percentage of Total Project Time
Phase 1	Information planning	10
Phase 2	Content specification	20–25
Phase 3	Implementation	approximately 50, depending on Phase 4 requirements
Phase 4	Testing and production	18
Phase 5	Evaluation	2

Test project
Projected hours worksheet

Book Name Factor	Skill Level	Projected Sept-97	Projected Oct-97	Projected Nov-97	Projected Dec-97	Projected Total Hours	Projected Hours/Page
User's Guide		60		Page Count:			
0.12	Project manager	8	8	10	3	29	0.48
	Writer	60	60	40		160	2.67
0.15	Editor	9	9	11	3	32	0.53
	Graphic artist			30		30	0.50
	Production coordinator				20	20	0.33
	Subtotal	**77**	**77**	**91**	**26**	**271**	**4.52**
Online Help		150		Page Count:			
0.12	Project manager	9	20	18	4	51	0.34
	Writer	60	76	60		196	1.31
0.15	Editor	9	11	9	0	29	0.19
	Online specialist	10	80	80	30	200	1.33
	Subtotal	**88**	**187**	**167**	**34**	**476**	**3.17**
	Total Hours	**165**	**264**	**258**	**60**	**747**	**3.56**
	Hours/Month	120	136	120	136		
	Full-Time Equivalent/Month	1.38	1.94	2.15	0.44		

Figure 4. Sample project spreadsheet.

ress, measuring hours expended against the percent complete of the project.

PHASE 2. CONTENT SPECIFICATION—PRESENTING THE DETAILS OF THE DESIGN

In Phase 2 of the information-development process, information developers move from the general strategy for meeting user needs sketched in the Information Plan to detailed design plans for the media they intend to produce. The Content Specifications should demonstrate how the results of the user and task analyses will be played out in the design of technical information of all types, including context-sensitive help systems, a series of paper and electronic manuals, online or self-paced tutorials, computer-based training, and even classroom training and other online mechanisms for ongoing user support.

Some information developers may argue that detailed specifications for the design of information deliverables are premature, especially when aspects of the product functionality, the user interface, or the information content may not have yet been determined. This view reflects a mistaken notion of the concept of user- and task-oriented design. In a task-oriented information design, the users' need for information and the tasks the users need to perform should be the focus of the design, not the information content or the tasks performed by the product. Too often, information developers create superficially user- or task-oriented instruction that simply states information out of context or describes the product tasks, rather than providing information that reflects the users' goals and objectives. If the users' goals and objectives have been analyzed in Phase 1, then the results of the analysis should inform the detailed design process of Phase 2. An information design that is based upon user tasks is unlikely

to change in a substantial way during the course of the project, while an information design based upon subject matter, system tasks, or software design is likely to change whenever the subjects change or the underlying functionality is rethought.

A detailed content specification of the elements of the information design has several significant advantages over a vague plan in the head of the individual writer:

- Information developers are likely to consider the design implications of their user and task analyses more thoroughly if they are required to write a detailed specification rather than a simple heading-level outline.

- Reviewers are more likely to understand the intent of the design and its relationship to user needs from a detailed specification than they are from a vague, high-level outline.

- Implementation of the design should not begin until the overall approach to the information is thought through. Remember that many information developers are likely to "just start writing without a well organized plan in place."

- In case of personnel changes, a detailed plan enables a new information developer more easily and quickly to pick up where the former developer left off.

- Information managers can estimate resources required and plan a detailed schedule of milestone deliverables more effectively around a detailed plan than a vague outline.

A detailed Content Specification should include the following:

- A description of the purpose of each information media that will be delivered
- Measurable usability objectives for each deliverable
- Brief summaries of user characteristics and tasks
- A discussion of the design rationale for each deliverable
- A detailed annotated outline of each module to be designed, including behavioral objectives and a breakdown of information into two or three heading levels.

The Content Specifications for documents, manuals, and some online help designs include a table of contents for the information with annotations, as shown in the following example.

Example of an Annotated Outline. Section A. Setting up the oscilloscope

In this section, engineers and scientists will learn the setup steps that they must perform so that the instrument can be used. The setup steps will include options that the user can exercise and indicate how those options will affect the types of measurements that can be taken.

A table of contents-like outline may be the best way to display the organization of information in a book or a traditional help application. However, other organizational structures may be more effective for planning the details of Web sites, help systems, tutorials, computer-assisted training, and others that have a hypertext rather than a linear structure. These structures are more effectively specified through hierarchical or Web-like models (hierarchy charts, Web maps) with which the designer can show the relationships among the modules more easily than in a linear outline.

A Web map can be used to show the relationship between a help topic and the interface objects, facilitating the development of context-sensitive links. Each interface object in the Web becomes a starting point for access into the help system. Each help topic is shown as an object that can be linked to and from, accessed through a browse sequence or through a table of contents, keyword search, or full-text search. A Web map makes all the relationships clear, although it can become quite complex if there are several hundred or thousands of help topics or if the links are random rather than systematic. Systematic links help the users to predict the kind of information they will receive when they pursue a particular hyperlinked path. Random links often confuse the users, leading them to abandon references to the help system.

Prototyping—A Phase 2 Opportunity for Rapid Information Design

During Content Specification, the information developers should begin producing prototypes of their design ideas. Design prototypes serve the same function in information design as they do in product design. They permit more effective feedback by other members of the design team and by potential users. Prototypes of manuals, help systems, computer-based training, and others can be reviewed using cognitive walkthrough or heuristic evaluation techniques, or they can be subjected to usability assessments. For example, if an information developer is contemplating a new structure for a Web-based information library, it is possible to test the structure simply by prototyping the sequence of hypertext links. Users might be asked to find information and their responses

recorded to help decide if the design is usable before the information is finished.

Early prototypes of online help or user manuals might also be created to accompany early product prototypes to provide supporting information to assist learners who bring fewer resources to the performance and learning problem. In one early test of interface and documentation, we not only learned that the interface created an flawed conceptual model of the product functions in the users' mind but that the prototype documentation did not help the users correct their misconceptions. Both product and documentation needed to be redesigned.

Prototyping also gives the information developers an opportunity to show concretely how the information will look to the users. The prototype designs should be fully formatted according to the requirements of the Information Plan and the Content Specifications so that the prototypes approximate the final look and feel of the design. In this way, users and team members can judge if the information designs are both usable and attractive. For most people unskilled in Web or book design or the layout of help screens, an abstract description of the intended design fails to communicate effectively the full intent of the design. Only with a concrete representation is the user able to construct a comprehensive mental model of the design and provide useful feedback as to its effectiveness.

Combining Phase 1 and Phase 2 for Revision Projects

For most information-design projects, it is important to keep Phases 1 and 2 of the information development life cycle separate. Phase 1, Information Planning, encourages designers to look at the broad issues involved with satisfying user needs. In the Information Planning phase, media are selected, strategies developed, and usability goals established in keeping with the information learned from the users and other subject-matter experts. In Phase 2, Content Specification, the broad outline of goals and objectives is translated into specific information deliverables, each of which is itself carefully specified. Without both phases in place, information developers are more likely to recreate the status quo, creating the same dull, unusable information year after year without regard to performance issues or changes in the make-up of the user community.

On some projects, however, the information planning may have already been done for a previous version of the product. As a consequence, only Phase 2 specifications may need to be written to add new functionality to an existing structure or to make minor organizational adjustments based on feedback from users, trainers, and customer support. However, care should be taken to avoid endlessly maintaining an existing structure after it has lost its effectiveness. Too often, both product and information developers continue to make changes to an existing structure without ever examining its effectiveness or are afraid to make drastic changes to a structure that they have learned is ineffective. For that reason, it is useful to reconsider Phase 1, Information Planning, on a regular and frequent schedule throughout the life of a product.

PHASE 3: IMPLEMENTATION—TURNING DESIGN INTO AN INFORMATION PRODUCT

Some aspects of implementation of the information design begin during the prototyping activities of Phase 2. However, primary implementation work in Phase 3 should not begin in earnest until Phase 2 detailed planning, prototyping, and early usability studies are complete. We recommend that at least 30% of total project hours be devoted to Phases 1 and 2. That is, if a project is projected to take 1200 h, or two people working full-time for six months, then Phases 1 and 2 should consume at least 400 of the 1200 h. In that time, an information-design strategy will be completed and detailed.

During Phase 2, information developers are likely to begin assembling some of the technical content needed for the information deliverables. However, most of that content development should take place in Phase 3, Implementation. Phase 3 will take from 50% to nearly 60% of the total project time, depending upon the amount of time needed for testing and production during Phase 4. The percentage of time required for Phase 4, Production, will increase depending on such factors as printing, translation, packaging, and distribution.

During the Implementation Phase, information developers begin to produce the information types outlined during planning. Typically, information deliverables go through at least two formal review cycles, following first draft (alpha) and second draft (beta) development. Often, information deliverables go through at least one more informal review cycle early in Phase 3. In this review, small sections of a document or help system are circulated among informed people for feedback on content, style, layout, and so on.

During Phase 3, the information developers are learning more about how to present the content and how to relate the content to the users' goals. Writers, instructional designers, illustrators, graphic designers, layout specialists, online specialists, video producers, animators, and other individuals representing the wide variety of media we can include will be involved in contributing their expertise to the emerging information.

As the information is created, it should all be reviewed before every scheduled phase by a developmental editor, an individual skilled in heuristic evaluation techniques and alert for potential problems in organization, level of detail, completeness, tone, format, and more. The developmental editor should be a senior member of the information-development team with considerable experience working with information developers to assist them in ensuring that the project goals are being met. The developmental editor often assumes a teaching role, especially when some members of the team are inexperienced in the information design techniques used by the organization. The editor fulfills a significant quality assurance role by ensuring that goals and standards are met and that best practices are consistently followed. For detailed information on developmental editing, see Ref. 2.

In addition to developmental editing, the information-development team may include individuals expert in copyediting. The copyeditor ensures that the text, graphics, and layout conform to company and industry standards. Copywriters generally check documents for spelling, grammar, consistency, adherence to regulations, and more. Early copyediting ensures that information does not contain errors that distract reviewers from their primary task of ensuring the accuracy of the information.

Technical reviews are a standard part of the Implementation Phase, but are frequently unsuccessful in helping to ensure the quality of the information delivered. In fact, many information developers consider the current technical-review process to be broken. The primary reason for review problems is a lack of commitment to the review process by the reviewers themselves. A thorough technical review of technical information takes time, on the average 5 to 10 min a page. It also takes careful attention to detail to ensure that the conceptual, procedural, and instructional text includes the correct information. It takes even more careful attention to ensure that no information that might help the users learn and perform has been omitted. Too often, those with review responsibilities do not allow sufficient time in their schedules for thorough reviews. As a consequence, incorrect information often finds its way into final information products.

Usability Testing of Documentation and Help

As soon as documentation and help are prototyped, usability testing can begin with actual users. Although the most complete assessment of the usability of the documentation will take place once draft software and the draft information are complete, early tests can take place with early prototypes of interface and information. We might want to learn, for example, if a minimalist approach we have selected for procedural information is sufficient to assist novice users in performing new tasks. With basic procedures in place and an early view of the interface, including paper prototypes, we can ask potential users to perform tasks following the instructions.

More extensive usability testing is performed at the first, or alpha, draft phase. Generally, we divided Phase 3, Implementation, into three subphases: first draft; second draft; and production draft. We define each draft by the percentage complete of the information. For example, we might expect the first draft of the information to be 90% complete, with draft graphics in place, a complete table of contents, and a rudimentary index. The second draft may be 99% complete in terms of text and graphics with nothing left to add except the final corrections, formatting details, and complete index. At the second draft, sections of the documents are often released for translation. The final production draft is ready for shipment to printing or implementation on CD-ROM or as part of a Web site.

Usability assessment becomes a most serious activity at the first draft subphase. At this point, you can ask potential users to perform complete tasks with a product and provide the online help and paper documentation to answer their questions. During usability assessments at this point, however, a significant decision must be made. You can ask users to perform tasks with the product and simply make the technical information available without mention, you can remind users that help and paper documentation is available for their use, or you can constrain the task and ask users explicitly to use the help and the paper documentation to complete the assigned tasks. No technique is any more valid than any other and only the last will provide an explicit test of the documentation. It is entirely possible that the first two choices will give little or no information about the effectiveness of the documentation.

It is a good idea to pursue all three techniques. The first technique, not referring to the documentation explicitly, might be used before any help or paper documentation exists. The second technique will suggest at what point the users turn to the documentation for assistance. The third technique might indicate if the technical information adequately supports learning, especially for novice users.

Tracking the Project

Throughout the course of the project, but especially during Phase 3, you will need to track progress. Tracking progress includes knowing how much time has been allocated to each task, how much time has been used to date, how much time is remaining, and how much work is left to complete the task. To gather the information, you must know if the project is on track. Each member of the information-development team should be asked to track their own progress. They need to know their allocated hours for each task and report how many hours they have expended and how many are remaining. It is best to ask everyone to report their hours weekly, before they lose track.

The most difficult aspect of project tracking is, however, not the hours allocated but the progress of each task toward completion. People are often quite optimistic in reporting how much they have done and how much they have left to do; if they are performing a task they have never done before, they will have little sense of what remains to be done for completion. The project manager will need to assist them in evaluating their progress. The more you know about the nature of the tasks, the more assistance you can give in assessing progress.

You must also be alert for changes to the original scope defined for the project in the Information Plan and the Content Specification. If a help system with 150 topics has been specified, remaining on track will be difficult if team members add topics. They need to be aware of the original estimates and how they were made so that they track carefully any changes to project scope. No one should feel free to increase the project scope without approval of the project manager; the project manager must be prepared to estimate the affect of changes on the team's ability to complete the project on time and maintain the level of quality required by the user.

Reporting Project Progress

As the project is tracked, be prepared to report progress to team members and management. Progress reporting should include both oral reports and periodic written reports. In the written reports, include a summary of the progress and plans for the next period. Review the hours used and the hours remaining and estimate the overall progress toward completion. Finally, discuss any problems that have occurred or are anticipated. By dealing with problems immediately, you are more likely to solve them while they are still small. By anticipating problems, you will be able to deal with them effectively and quickly. Always keep the goals of the project in mind—to meet the users' information needs.

PHASE 4. PRODUCTION—ENSURING ACCURATE, COMPLETE INFORMATION FOR THE USER

For print documentation, including materials for use in classroom training, the details of production are best left to the specialists. It should be noted, however, that print production, even when advanced electronic production techniques are used, takes time that cannot be truncated. Large print runs of multicolor documents, including the binding and prepara-

tion for distribution, will take between three and four weeks to complete. During this production period, no further changes can be made to the information without incurring enormous expense. As a result, it is very important that no product or content changes occur. If they do, it is likely that users will be disappointed to learn that something they believe they should be able to do is not possible, or something is possible for which they have no information.

Even if information is distributed electronically, the final electronic files must be prepared. The preparation includes adding hypertext links, creating indexes to facilitate keyword searches, and testing functionality. Context-sensitive help systems, while requiring the same testing before distribution to users, also must have the links between software and help tested to ensure that the correct information appears.

Functional testing of electronic information, while it can begin in Phase 3, should become part of the functional testing of the product to ensure accuracy and completeness. Web-based information also requires functional testing to ensure that internal and external links operate as intended.

In addition to functional testing, Phase 4 often includes the most intense period of activity for translation and localization. A lack of discipline in the processes for developing product and information will have a considerable adverse affect on the cost and effectiveness of translation and localization efforts. To facilitate translation and localization, all product and information developers should work with standardized vocabularies, consistent structures for information, and a minimalist approach that reduces the verbiage to what the user needs to know. If the developmental and copyediting functions described earlier have been applied to all the information deliverables, including the product interface, translation and localization will proceed more smoothly and result in more accurate information delivery to a range of user communities worldwide. Once again, the need for a careful and consistent approach to all the information that touches the user will reap benefits for both users and developers.

PHASE 5. EVALUATION—REVIEWING THE PROJECT SUCCESSES AND FAILURES

Since no one likes to review a string of failures, let us hope that following some of the steps outlined in this article will result in a successful project, one in which all information delivered to users is carefully planned and well integrated. Even if the project has been successful, there are always many opportunities to improve. Few projects are completed without communication challenges, especially when a diverse team of information specialists finds itself working together for the first time. Whenever a project team consists of people with different experience, training, perspectives, and personalities, there will always be opportunities for improvement.

In Phase 5, two sorts of evaluation are recommended—written and oral. Each member of the team may want to write a project wrap-up for the part of the project under his or her responsibility. The wrap-up reports might then be combined by the project manager into a single report. The wrap-up report should include quantitative information about project activities, including hours used to complete each phase in comparison to the original hours predicted to be used in Phase 1. If the total hours were different than originally predicted, the project manager should account in the narrative for the differences, especially if the project has taken considerably longer than first estimated. Such an analysis will help the project team to better estimate its time on future projects.

In addition to the wrap-up report, the information-development team, including the product developers, should meet to review the report and discuss ways in which they might be able to work together more effectively in the future. It is better to discuss problems soon after they have occurred rather than allow bad feelings and resentment to linger into the next projects.

Information design and development promises to become an integral part of product development. Even so, the processes need to be improved so that no member of the team is made to feel like a second-class participant. To develop information products that meet the complex needs of a wide variety of users throughout the life of a product, you must work together effectively. As the responsibility of our organizations for meeting user needs increases, so must our teamwork and our ability to listen and to respect the perspectives of development professionals from diverse disciplines. The development professionals must also learn that teams mean collaboration not competition. Through a sound information-development process such collaboration will be enhanced.

BIBLIOGRAPHY

1. J. T. Hackos, *Managing Your Documentation Projects,* New York: Wiley, 1994.
2. J. Tarutz, *Technical Editing: The Practical Guide for Editors and Writers,* Reading, MA: Addison-Wesley, 1992.

JOANN T. HACKOS
Comtech Services, Inc.

MANAGEMENT OF SOFTWARE COMPLEXITY. See SOFTWARE MANAGEMENT VIA LAW GOVERNED REGULARITIES.

MANAGEMENT RISK. See RISK MANAGEMENT.

MANIPULATORS

FUNCTIONS AND MECHANISMS OF MANIPULATORS

Manipulators and Tasks

A manipulator is defined as a mechanical system that executes meaningful tasks by manual control, automatic control, or a combination of both. From that viewpoint, most conventional robots fall in the category of manipulators.

The tasks which the manipulator is commanded to perform can be diverse, e.g., performing hazardous tasks that humans cannot do, simply imitating human behavior, and performing tasks that humans can perform, but much more efficiently and with much improved reliability. We exclude manipulators, such as those that merely imitate human motions, those that function in special environments, e.g., underwater, space, vacuum, and so on, and those used for entertainment. Therefore we focus on manipulators that are actually used for practical task execution.

In a broad sense, manipulators that perform meaningful tasks in the domain of production activities, range widely in variety, as shown in Fig. 1. The main correspondences of those examples with tasks performed are as follows (Fig. 2):

1. Hydraulic excavator. A construction machine is not usually dealt with as a manipulator. A hydraulic excavator, however, is an exception. An excavator is one of the most representative construction machines which is mainly used to excavate earth [Fig. 2(a)]. It is also used to crush rocks, to grab timbers with a special attachment, and to transfer objects.

2. Machining center. The most representative machine tool used mainly for cutting metals [Fig. 2(b)]. When the attachment tool is replaced to cope with a required process, various operations can be carried out by replacing at least several tens of tools.

3. Multijoint articulated robot. Robots take part mostly in simple, repetitive motions with improved efficiency and reasonable accuracy. The main tasks involved in such motions are changing the position and the orientation of an object, that is, machine parts, materials, electronic components, and so on. These motions consist of a series of elementary motions, such as grasping [Fig. 2(c)], releasing, transferring, changing configuration, etc. The articulated robot is also used widely for painting [Fig. 2(e)] and welding [Fig. 2(f)] mostly in the automobile industry.

4. Industrial orthogonal robot. This type of robot is mainly used for assembly tasks in manufacturing whose typical motions include accurate positioning and insertion of mating parts [Fig. 2(d)].

5. Self-propelled mobile vehicle. This robot distributes and collects many parts and products to and from different work cells in a factory [Fig. 2(g)]. It performs elementary motions, such as loading, unloading, and transferring. There are cases where these loading and unloading motions are undertaken separately by a manipulator associated with each respective work cell.

Functions of a Manipulator and Their Constraints

Generally when designing a new product, we first determine the desired functions and then elaborate them in the order of functions, mechanisms, and structure (construction) while satisfying different constraints to which the product is subject (1). This stepwise elaboration process can be applied to the design of a specific mechanical device and also to the design of larger more abstract systems. Therefore, we describe those functions required of the manipulator from the viewpoint of theoretical design, those mechanisms used to realize respective elementary functions, and the design constraints associated with the functions.

As previously mentioned, various manipulators perform a variety of tasks, but common to all these tasks is that each manipulator generates relative motions between the work tool (end effector) and the subject item to be worked or between the item gripped and other subjects. In other words, this is the sole function of the manipulator. And, incidental thereto, the realization of motion, the conveyance of force, the transmission of information and transfer of materials, are required as functions, although they are lower functions. The

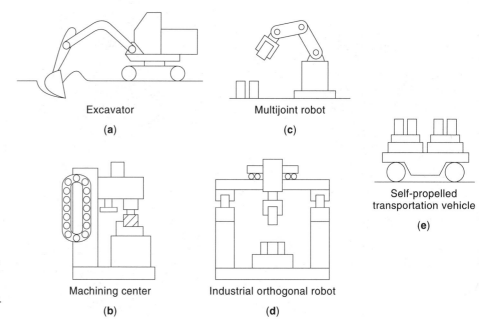

Figure 1. Manipulators working in different production fields.

mechanisms to realize these functions are explained in detail in the next section.

A very important function usually overlooked or underestimated when a manipulator is designed is the so-called "fail-safe function" that aims to avoid catastrophic destruction of the system in an unexpected situation. This function is better realized by using a structural device, not by relying on a software-based emergency handling procedure. At present, no manipulator sufficiently meets fail-safe need. However, it is strongly suggested that this function be given higher precedence when designing manipulators in the future.

To realize the desired functions, it is necessary to approach design from the viewpoints of both hardware (mechanism/ construction) and software (control), but as software is discussed in other sections of this encyclopedia, this section deals mainly with mechanisms.

When looking at the specific composition of a manipulator (Fig. 1), there are common items for many of the different manipulators regardless of which tasks they are intended to perform. As shown in Fig. 3, whatever the physical structure and configuration may be, attached tools or workpieces exert an action on another tool or workpiece, and the manipulator is the device that performs the action and establishes physical interactions between the two objects. As shown in Fig. 3(a), the manipulator has a base, a drive source, an actuator with a transmission mechanism, an arm, a sensor, and a gripping

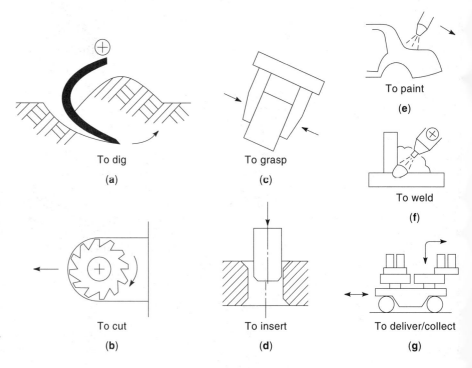

Figure 2. Various tasks which manipulators execute.

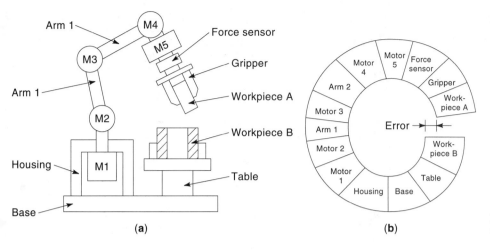

Figure 3. Basic structure of a manipulator (a) and C-circle of a manipulator (b).

tool connected nearly in series and forming one structural body. If the whole body is abstracted, the entire structure forms a C-shape beginning from the tools or the workpiece and reaching to the other end which is the object to be worked. This abstraction is called the C-circle of a manipulator (hereafter abbreviated as C-circle). The C-circle is the most important factor that determines a task's accuracy (in other words task tolerance) and required execution time, which are the fundamental performance criteria of a manipulator.

Constraints related to the design of a manipulator include operability, manipulability, maintainability, cost, etc., and the constraints on the manipulator itself are weight, operational accuracy, motion characteristics (rise time, settling time), tact time, etc. (Fig. 4). Of these, constraints on operability are limited by the mechanical characteristics of the hardware and the control characteristics of the software. These mechanical characteristics are coped with by considering rigidity, shape, resonance, thermal deformation, etc., of the C-circle as a whole. For example, although the harmonic drive reduction gear widely used in robots is light, small, and con-

venient, it essentially lessens the rigidity of the C-circle and therefore is recommended most where there is a strong demand for high accuracy and high-speed operation. Although there are several difficulties from the design viewpoint, there is no other approach but to simplify the C-circle using a direct-drive motor and improving rigidity.

Mechanism To Materialize Functions

An actual manipulator is an integration of mechanisms corresponding to a set of elementary functions designed to fulfill desired specifications under the aforesaid constraints. The mechanisms involved can be diverse, depending on their use. The next section in this article explains the mechanisms used for a manipulator in normal environments in terms of dimensions, and the section following explains the use of a manipulator in the microworld. Here we introduce some mechanisms that are considered essential to the future development of robots and manipulators although the world of robotics under present conditions does not pay much attention.

Fail-Safe Mechanism. In the process of operating a manipulator, if it collides with human operators or other machines, a part of the manipulator or the other machine may be damaged due to the excessive force of the collision, or it may lead to a fatal accident at worst. In the conventional design of a robot, a sensor-based control system is used to sense the failure and take appropriate action before a serious problem occurs. But the desired functions often cannot be fulfilled because of the time lag in the control system. Therefore, it is preferable that the fail-safe mechanism be realized by part of the structure, which should have the characteristics shown in Fig. 5(a). To produce this, for example, two pairs of springs and stoppers as shown in Fig. 5(b) are combined in tandem (2).

Variable Compliance Mechanism. The manipulator that fulfills desired functions by its motions necessarily forms a C circle. For this reason, it becomes difficult to fulfill two desired functions, a large work space and high positioning accuracy, at the same time. To solve this problem, a function is called for that changes the rigidity of individual parts and joints, which compose the C-circle. For this purpose, for example, increasing the apparent geometrical moment of inertia

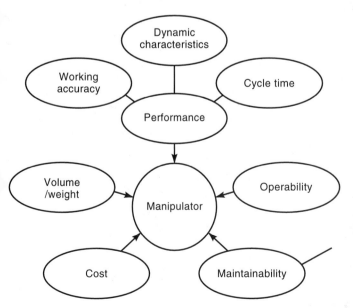

Figure 4. Constraints imposed on a manipulator.

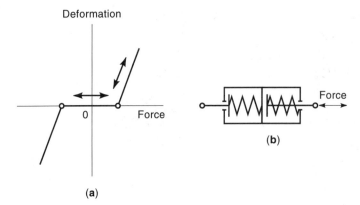

Figure 5. Fail-safe mechanism: (a) force-deformation characteristic of fail-safe mechanism; (b) concrete example of fail-safe mechanism realized by series connection of spring/stopper couples.

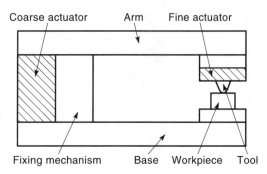

Figure 7. Realization of fine operation through piggyback structure.

(Fig. 6) by adsorption with an electromagnetic sheet and fixation of joints are effective.

Macro/Micro Structure. To substantiate the same desired functions mentioned previously, segmentation of role sharing of the moving parts (actuator, arm, etc.) is effective, that is, combining a coarse movement mechanism and a fine movement mechanism, as in the case of a direct-drive motor and a piezo actuator, to produce a large work space and high positioning accuracy. Even here, a combination of actuators alone is not sufficient, and at least a mechanism for fixing the interface between the two should be incorporated in combination to produce operations (Fig. 7).

Local C-Circle Structure. When fine motion tasks and small force operations are to be performed at the tip of the arm, it is difficult to maintain the relative position between tools and gripping parts and the object parts, no matter how accurately position control is performed at the base of the arm or how rigidly the arm and the actuator are raised. Particularly where reactive force is received by operations, deviations in position and orientation are caused by the reactive force no matter how small. Although a controller is often used to increase the apparent rigidity of the C-circle in solving the problem, its contribution is not so significant. Under such circum-

stances, incorporating small local C-circles in parallel with the large C-circle is extremely effective (Fig. 8).

Junction. When tasks performed by one manipulator are diverse, a variety of end effectors is necessary. In this case, more than a simple mechanical connection/disconnection is required at the junction between the manipulator and end effectors. What should be exchanged through the junction includes mechanical relative position, transmission of force and torque, transmission of information, and transfer of substances, such as cutting fluid, air, and gas. The interface carries out secure transmission and transfer through the connector ports. As of now, the most serious technical difficulty resides with the transmission of information. Sensory information about the interaction between the end effector (gripper or tool) and the object can be transmitted to the manipulator, but in many cases the signal level is very low. It is also common to have actuators on the tool side which provide necessary motion at the tool end. Therefore, sensor signals and also energy must be transferred from one end to the other (Fig. 9). An example of an application to manipulators is schematized in Fig. 10. In the future, technological efforts should be made for more secure and efficient transmission. For this purpose, magnetic and electromagnetic coupling are promis-

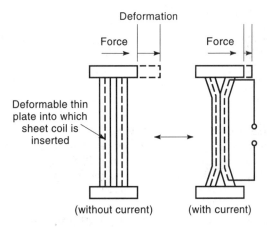

Figure 6. Variable compliance mechanism.

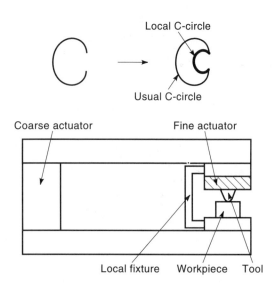

Figure 8. Improvement of working accuracy by local C-circle structure.

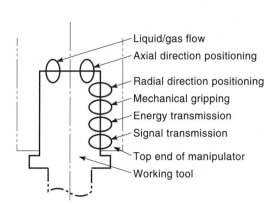

Liquid/gas flow
Axial direction positioning
Radial direction positioning
Mechanical gripping
Energy transmission
Signal transmission
Top end of manipulator
Working tool

Figure 9. Basic performance required for smart connectors of a manipulator.

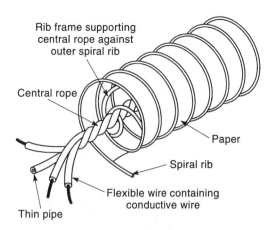

Rib frame supporting central rope against outer spiral rib
Central rope
Paper
Spiral rib
Flexible wire containing conductive wire
Thin pipe

Figure 11. Example of an arm structure which does not make humans afraid.

ing because they are not susceptible to contamination at the joint surface (3).

Soft Appearance. Although functions called for by the industrial manipulators are as described previously, manipulators in the future will, inevitably, have more occasion to interact with humans. Therefore a tender appearance which gives human users a sense of ease, but not a sense of fear or anxiety, will become vitally important. Humans will always remain the master over manipulators under any circumstances and they require that the manipulators not be in a position to make them afraid or to inflict injury. Manipulators that physically interact with humans should be weaker than humans. For this reason, manipulators must be soft and light enough to give humans a feeling of ease, but nonetheless rigid enough to execute given tasks. The functions considered contradictory in the engineering sense will be justified when higher precedence is given to safety issues. To fulfill the desired functions, adopting special structures and materials such as very light synthesized materials or paper materials, may be useful (Fig. 11).

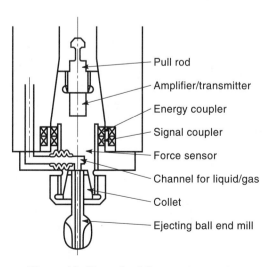

Pull rod
Amplifier/transmitter
Energy coupler
Signal coupler
Force sensor
Channel for liquid/gas
Collet
Ejecting ball end mill

Figure 10. Example of the smart connector.

COMPOSITION OF NORMAL SIZE MANIPULATORS

Hardware Design of Manipulators

Task Elements Called for by Manipulator. A programmable, automatically controlled, multijoint manipulator is called a robot in industry. This section introduces this kind of robotic manipulator, which we call a normal-size manipulator whereas a small-size robot, termed a micromanipulator, is treated later. There are many introductory books about robotics in general (4–9) that deal mainly with three major aspects of robotics, kinematics, dynamics, and control. This section views manipulators from a design perspective.

Industrial production processes that involve robots are divided mainly into three processes, such as assembling informational equipment, welding automobiles, and handling injection parts. In Japan particularly, these three large production processes account for approximately 70% of all processes in which robots are used. Analysis of specific tasks performed by the robots in these three processes tells us what kinds of task elements are required of manipulators. Two kinds of tasks appear commonly, picking and placing objects and tracing object surfaces.

Although robots are used in processes other than these three, a careful task analysis reveals that many of them consist of a series of the two previously mentioned task elements. There is only a minor difference in the overall task objectives, such as in heavy-load lifting or painting. These three processes are widely used because the tasks performed are suitable for a robot and because developing task-dependent robots dedicated to a specific task is too costly, resulting in increased employment of general-purpose robots.

Next, we shall examine the desired operations of robots in the future. It may safely be said that current production processes that can benefit from the investment in robots have already been developed almost to their fullest extent. Processes expected in the next generation have been spreading out of manufacturing industry, for example, operations in hazardous environments (e.g., nuclear reactor, minefield, active volcano, site of disaster), operations to care for the physically handicapped, operations where humans cannot see with the naked eye or cannot touch with fingers because the ob-

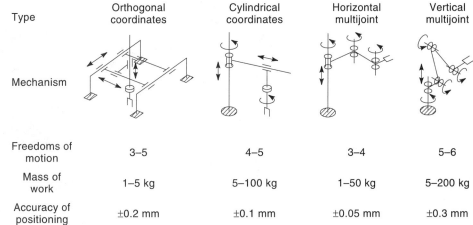

Type	Orthogonal coordinates	Cylindrical coordinates	Horizontal multijoint	Vertical multijoint
Mechanism				
Freedoms of motion	3–5	4–5	3–4	5–6
Mass of work	1–5 kg	5–100 kg	1–50 kg	5–200 kg
Accuracy of positioning	±0.2 mm	±0.1 mm	±0.05 mm	±0.3 mm

Figure 12. Specifications of different robotic structures.

jects are extremely small or are inside the human body (e.g., dissection of a single cell or keyhole surgery), and operations performed in an environment of clean air, vacuum, or a reducing atmosphere, where humans cannot share the environment.

Within their environment all of these robots more or less mimic operations performed by humans in the normal world. Similarly, when task elements involved in these operations are considered, in fact the two major task elements previously mentioned account for the greater part. For example, the task element specially called for by a nursing manipulator is to lift physically handicapped persons and move them, and the microsurgery manipulator is called on to move surgery tools as if tracing the affected part. Of course constraints associated with the task elements differ from those for industrial manipulator presently in use. Softness and tenderness, so that the handicapped person is not afraid of nursing manipulator, preciseness that allows highly accurate movement, and softness that leaves the affected part minimally invaded by the microsurgery manipulator are required.

Mechanism of Normal-Size Manipulators. To accomplish these two task elements, the manipulator has to undergo spatial movements within the given work space. As an example of its mechanism, let us introduce the mechanism of an industrial manipulator in Fig. 12.

Four mechanisms are shown in the figure, but choices of mechanisms are made by analyzing task elements. Three conditions are of primary concern: required degree of freedom of motion, weight of objects, and required accuracy of positioning. In the aforementioned three processes as examples, two require high positioning accuracy although the object is relatively light. An orthogonal coordinate type and horizontal multijoint type (SCARA) for assemblying informational equipment (e.g., circuit board) and an orthogonal coordinate type and cylinder coordinate type for handling of injection parts are used, respectively. On the other hand, when the task requires tracing curved surfaces, such as welding automobile frames, an articulated multijoint type is used. The degree of freedom of a respective mechanism is either rotational or prismatic. When rotation is used for a degree of freedom, the configuration of tools can be arbitrarily set. Although the rotational rate can be very fast, the rigidity of the actuator is rather low, hence positioning accuracy is worsened. On the

other hand, when translational motion is used, both the rigidity of the actuator and the accuracy of positioning improve whereas the tool orientation becomes fixed and the tool motion is rather slow.

Some constraints should be relaxed to overcome these shortcomings. First to be tackled is the rigidity of the driving system. In the case of a robot manipulator, generally, no substantial bending of arms, no distortion of joints nor deformation of structure occurs unless a flexible structure is intentionally used. What is crucial is the deformation and chattering of the drive mechanism caused by external forces and moments from actuators, couplings, gear trains, etc. Next, because the motion of tools is generally constrained by acceleration rather than velocity, the acceleration to which the drive system is subjected can be considered a constraint. To circumvent this constraint, there is no other way but to make the mass of moving parts including the object lighter to increase rigidity to suppress unwanted vibration, or to make the drive power of the actuator larger, although positioning accuracy is governed by how well thermal expansion and inertial force are compensated for by control. The more cumbersome problem would be the thermal expansion of the structure because the inertial force generally is repeatable and can be compensated for rather easily if the weight of the object and the dynamics of the drive system are estimated. However, the thermal expansion (determined by the average temperature of the whole and also by the internal temperature distribution) is hard to compensate for because the interior temperature of the structure generally cannot be measured. The final constraint is the limitation of tool configuration. In many cases this can be resolved quickly if design change on the side of the object is accommodated. The constraint in this case is the design of the object, which requires the flexibility and willingness of the designer to accept modification of objects.

The next section describes how these constraints are overcome.

Design Constraints on Manipulators. Four kinds of constraints for designing manipulators have been mentioned: rigidity of the drive system, acceleration of the drive system, thermal expansion, and design modification of the object. These are reviewed in depth in this section.

First we discuss the rigidity of the drive system. For the manipulator to trace a trajectory under the influence of external forces and inertial force, the more rigid the actuator becomes, the better the accuracy achieved. Ideally speaking, a robot with position feedback should retain infinite rigidity. In other words the same position should be maintained whether the robot is pushed or pulled from its extremities. At least the resonance frequency should be approximately on the order of several hundred Hz at the tip of tool. However, because the position sensor does not provide the precise position of the object, things do not work as expected in reality. For example, when the position is sensed by an encoder attached to the motor and drives the tools via reduction gears, like a harmonic drive, the resonance frequency at the end of the tools drastically decreases, for example, to 10 Hz.

However, as far as actual task executions are concerned, in many cases, especially in an emergency, the smaller the rigidity, the better the performance. Particularly, when an object placed at inaccurate position is to be handled, for example, inserting a peg in a hole or grasping a ball with the gripper, it can be resolved by highly precise servocontrol or by no position feedback at all, if artificial compliance is employed. When dealing with unexpected sudden moves or collisions with its environment, one method suggested is to incorporate a mechanical fuse in the manipulator which absorbs collision impact or unwanted excessive force to keep the damage minimal. The margin of such misoperation is immensely narrowed when humans are nursed or operated on by a manipulator. Under such circumstances it is desirable to insert a mechanical element with variable rigidity which automatically becomes softer in an emergency or makes the exterior skin of the manipulator softer like a car air bag when a collision occurs or an excess force is exerted.

Second we discuss the acceleration of the drive system. As long as the actuator size is comparable to a human, the positioning accuracy is reduced rather easily, that is, smaller than 0.1 mm readily using any actuator unless too sudden a start or a stop is involved. However, as the acceleration exceeds $2g$, vibration remains even after arrival at the target position, arms get twisted and the posture of tools changes, and the driving torque becomes insufficient, causing the motor to generate excessive heat. Recently, a direct-drive motors to increase the acceleration of rotation and linear motors to increase the acceleration of translation have frequently been employed. The latter, especially, makes it possible to attain an acceleration as high as $10g$, but heat generation also becomes significant requiring compensation for thermal deformation, the next issue to be discussed.

Third we discuss thermal expansion of the structure. If an arm one meter long is made of aluminum and its temperature is raised by one degree, then the thermal expansion at the tip is 0.2 mm. In reality, the temperature rise is easily several tens of degrees when the actuator is not appropriately cooled or when heat is generated by a welding torch or heat-emitting device. Then accuracy of positioning is even worsened, for example, by 0.1 mm. Thermal expansion is often forgotten by many robot designers. In manufacturing semiconductors and precision machining, however, compensation for thermal expansion is unavoidable. An easy solution to the problem is to keep the ambient temperature constant because it is quite difficult to achieve accurate compensation.

With the rapid advance of different kinds of sensors, including image sensors, force sensors, and noncontact position sensors, which use laser, eddy current, electrostatic capacity, and so on, the relative position between the end effectors (tools) and the object can be accurately measured. This allows accomplishing tasks even when the accurate, absolute position of the tip with respect to the distal base is not known. In the future, a combination of these sensors and fine motion mechanisms mounted on the tool, for example, piezoelectric or friction-drive, will be subject to frequent use to compensate for relative positional error at high speed.

The last issue is design modification of the object. To simplify the motion of a manipulator, it is often beneficial from the viewpoint of task execution to change the design of the object to a simpler one. For example, the shape of an object in which parts are inserted from more than one direction may be changed so that all parts are accessed from one direction, thus greatly simplifying the assembly task. Two parallel surfaces may be fabricated in the object with which handling of the object can be made secure and easier, or fabricating a large chamfer on the edge of the pin to be inserted produces self-aligning capability and makes an insertion task easier. In addition, although this may be a rather extreme example, assembly of two different parts which requires a robot may be replaced by a design change which integrates the two parts as one part. This kind of decision should be made carefully by considering several factors: cost for redesigning, necessary change in manufacturing process, matching with other parts, and so on.

Robot Hand

It was said a while ago that a robot is only as good as its hand or end effector. Although the situation may be slightly different after the emergence of a variety of locomotive mechanisms, such as legs, wheels, crawlers and so on, still many tasks largely depend on the end effector and any tool that actually performs the tasks. Nowadays the robot hand has been diversified in wide application of robotic systems from deep underwater to outer space, from construction machinery to microsurgery, from handling of toxic chemicals or explosives to tender care of elder people or handicapped. There is, however, no doubt that the robot hand plays a vital role in every one of the applications.

Generally speaking, the arm, wrist, and legs are versatile because they are used primarily for positioning the hand and any tool that it carries. The hand, on the other hand, is usually task-specific. The hand for one application is unlikely to be useful for another because most robot hands and end effectors are designed on an ad hoc basis to perform specific tasks with specific parts. Although the vast majority of hands are simple, grippers, pincers, tongs, or in some cases remote compliance devices, some applications may need more dexterity and versatility than these conventional hands can deliver. A multifingered hand offers some solutions to the problem. An excellent review of multifingered hands and their related issues can be seen in (10). One recent example of a multifingered hand, called the Barrett hand, is shown in Fig. 13.

This section gives a general description of robot hands from the practical point of view rather than introducing the state of the art of the field. In industrial applications, robot hands do not necessarily have fingers. As a matter of fact,

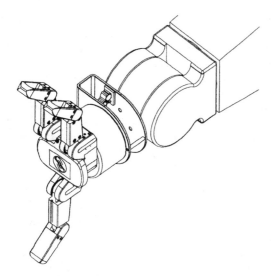

Figure 13. Three-fingered robot hand. Reprinted with permission of Barrett Technology Inc.

those which utilize magnetic force or vacuum suction force do not require fingers which may be added to prevent the object from falling off. Here we focus on grippers which have fingers (typically two or three).

When a robot hand is designed or selected, the following points need to be considered:

- Gripping force and workpiece weight
- Robot load capacity
- Power source
- Work space
- Environmental conditions

These five points are discussed in detail here.

1. *Relation of gripping force and workpiece weight:* It is suggested that the gripping force be about 5 to 20 times the workpiece weight. Its value may be affected by the shape of the finger attachment, the friction condition of the contact surface, the gripping method, and the transfer speed. The faster the transfer speed, the larger the gripping force required. Typical gripping methods are illustrated in Fig. 14. In the top figure, frictional force is induced by the rubber pad mounted on the attachment surface whose frictional coefficient is about 0.5 as opposed to 0.1 to 0.2 for steel material. The bottom figure has a hook for drop prevention. In case of low transfer speed without vibration or shock, for example, less than 0.1 m/s, the gripping force is chosen in the range of 5 to 10 times the workpiece weight. On the other hand, if the workpiece is subjected to high acting speed, vibration, or shock (high acceleration), for example, more than 1 m/s, or if the center of gravity of the workpiece is distant from the gripping point, a gripping force more than 20 times as large as the workpiece weight is required.

2. *Robot load capacity:* The total weight of hand, workpiece, mounting plate, and finger attachment must be smaller than the payload capacity of the manipulator. Attention should also be paid to the definition of the payload capacity, for example, worst case scenario or other specific configuration.

3. *Power source:* Most conventional robot hands use an electrical, pneumatic, or hydraulic power source. Each method has its pros and cons. Electric motors, such as dc motors or step motors, are commonly used to drive a robot hand and the joints of a manipulator because they are easy to use and are a good compromise of response, accuracy, and cost. Hydraulic motors are common in applications requiring large power, such as construction machinery, for lifting heavy objects or crushing rocks. One of the attractive features of the hydraulic drive is that it runs more smoothly at low speed than electric motors. However, a hydraulic drive may exhibit dynamic behavior that is quite oscillatory. The main drawback is possible leaks in the closed oil-flow system. Pneumatic drive systems are quite effective when a large force is not required and a simple open/close operation suffices. They also do not require return circuits as hydraulic drive systems do. The main limitations are low positioning accuracy, low rigidity, and exhaust noise.

Another concern closely related with choosing the power source is cables or hoses. Any cables or hoses

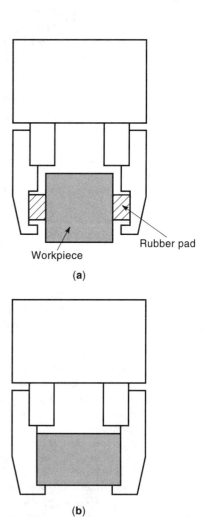

Figure 14. Examples of the finger attachment of a robot hand: (a) rubber pad mounted on the finger attachment; (b) hook on the finger tip for drop prevention.

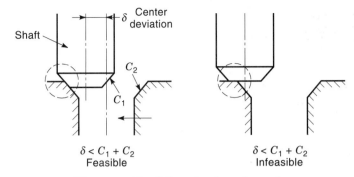

$$\delta < C_1 + C_2 \qquad\qquad \delta < C_1 + C_2$$
Feasible   Infeasible

Figure 15. Feasibility of an insertion task.

routed from the base to the end tip of a manipulator experience enormous flexure. An inappropriate choice may cause an early breakdown of the power system or severe hindrance to the smooth operation of the manipulator. Also, no excessive force should be exerted on connectors, cables, or hoses.

4. Work space: If picking and placing a workpiece is the task to be performed by the hand, an appropriate work space has to be provided for each action. The hand needs a certain space for approaching the work piece, and it also requires a space for picking and releasing it. The detection method for finger opening, closing, and catching the workpiece is also related to work space.

5. Environmental conditions: If a robot hand is used in a clean room, for eaxmple, handling semiconductor wafers or magnetic memory disks, special attention needs to be paid, especially to the power source, transmission mechanisms, and sealing mechanisms, so that no contamination is emitted from the hand. Other potential concerns include coolant, grinding dust, water, and ambient temperature.

Beside the points mentioned, there are peripheral technologies which are closely related to robot hands which are used widely in industrial applications.

- Automatic tool changer (or automatic hand changer): Unless the host robot is occupied with the same task, it is convenient to have a supply of different grippers and the ability to switch between them as tasks change. A robot arm exchanges grippers by using a turret or a tool changer. A turret is limited to switching between only two or three grippers whereas a tool changer handles a large number of grippers. In an automatic tool changer, each tool has a common mechanical interface through which the power source and the sensor signals are transmitted. Potential drawbacks of such tool changers are increased cycle time, reduced arm payload capacity, added expense, and reduced system reliability.

- Remote Center Compliance (11): For inserting a shaft into a hole, the relationship between the chamfers and the feasibility of the task is as illustrated in Fig. 15. The tolerable center deviation between the two parts is determined by the following equation (12).

$$\delta < C_1 + C_2 \qquad\qquad (1)$$

where δ is the center deviation and C_1 and C_2 are the chamfering dimensions of the shaft and the hole, respectively. If this equation is satisfied, then the shaft is passively guided into the hole by the chamfers. Otherwise, it is impossible to accomplish the insertion tasks unless the center deviation is actively compensated for until the equation is satisfied

Even when the equation is satisfied, some proximal part has to tolerate such an alignment motion of the parts. A Remote Center Compliance (RCC) device was specifically developed for this purpose (11). The device is equipped with two compliant elements. One allows lateral motions, and the other tolerates rotational motions about the tool tip. Its structure is schematized in Fig. 16(a). The device is often substituted by a simpler version of RCC which roughly realizes the lateral motions and the rotations [Fig. 16(b)].

This section concludes with pictures from more recent applications of parallel jaw grippers in agile robotic systems (Fig. 17).

Multiaxis Force Sensor

Object Manipulation and Force Sensing Information. In most cases, the robots that have become familiar for practical use up to the present are taught motions one by one, and the robots simply repeat them. In contrast, future robots will be self-supporting by taking in the information from outside, judging it, and acting. At that time, the intelligence that judges the information and the sensors that take in the information will play important roles.

Because robotic motions are diverse, we will consider the handling operations that become the center of motion, in other words, the handling of the object by the manipulator or the end effector attached thereto. What immediately comes to mind is the assembly robot in a factory production line, where object mechanical parts to be handled are hard and can be handled roughly. In the future, however, there will be soft or fragile, delicate objects or there will be the need to handle objects of unknown nature. In the cases of the nursing robot and the surgery robot, the objects are humans themselves.

What kind of processes are needed when an unknown object is to be handled? First of all, the position, size, and shape of the object must be estimated by visual inspection, followed by identification of the hardness or softness and contact condition with the object by touching it with fingers. Then by lifting it, the weight and the position of center of the gravity can be determined. In other words, humans grasp the characteristics of an object by using information derived from their sense of sight and by force sense (including tactile sensing) information from hands and arms. For fragile objects, force and tactile sensing become essential. Human hands are a highly complex and integrated system having many degrees of freedom and different sensing capabilities, which enables a very high level of dexterity and flexibility.

The functionality possessed by robotic hands still falls far short of that of human hands. However, those used in industrial applications are furnished with the minimal functions. Necessary information for handling the object can be obtained as visual information from a CCD camera and as force information from a force sensor attached to the manipulator or the

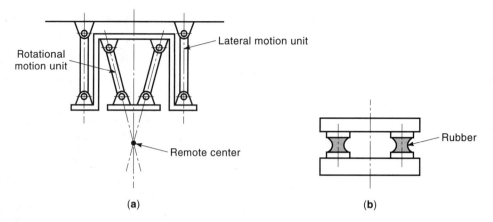

Figure 16. Remote center compliance (RCC): (a) remote center compliance (RCC); (b) quasi-RCC device.

Figure 17. Robot hands in agile robotic systems: (a) wheeled mobile robot picking up a tennis ball. Reprinted with the permission of Real World Interface Inc. (b) visually guided mobile robot approaching a coffee can. Reprinted with the permission of Real World Interface Inc.

end effector. The former is explained in detail in other sections. The rest of this section describes force-sensing issues.

Force Sensing and Force Sensors. Force sensory information means information about forces and moments observed at the interface between the end effector and the object when performing handling tasks. These forces and moments are fully described by six components: forces in three orthogonal axes and moments around the same axes.

Based on the force information, a lot of useful information is obtained, such as contact point, contact force, object configuration, object shape, and so on. Knowing this allows us to perform considerably complicated operations with force information alone. Currently, however, robots are used merely for deburring mechanical parts or fitting mating parts, but in the future, force control will be an indispensable technology when handling more delicate objects.

The techniques of sensing forces and moments are roughly divided into two methods, one that uses a force sensor and one that does not. In the latter case, the load of the object is calculated from the driving torque of the motor that drives the manipulator joints or end effector. Specifically, a dynamic model is described, and the load is calculated as a disturbance from the difference between the actual motion and the nominal motion derived from the mathematical model. To obtain force sense information with high accuracy, however, it is preferable to employ a force sensor.

Now, what kind of sensing mechanism is used in force sensors? In machine tools, piezoelectric elements are frequently used as pick-up devices. This utilizes the fact that because of the piezoelectric effect an electric charge is induced when a force is applied. The piezoelectric element has advantages that it has high rigidity and therefore it does not reduce the rigidity of the original machining system. There are also disadvantages, such as a large sensor, arbitrary setting of sensitivity is difficult, potential leakage of electric charge gives trouble in long-term stability, integration of multiaxis sensing is difficult, and so on.

The force sensors widely used for robotic applications use strain gauges as sensing elements. A strain gauge is placed at a position where a specific force or moment is effectively measured. The less cross talk, the better the sensor. Because the sensing resolution is inversely proportional to structural rigidity, increasing the sensitivity leads to a decrease in rigidity. Therefore, it is important to choose a force sensor to execute desired tasks which is sensitive and is also rigid. The major advantages of strain gauges are that they can be made

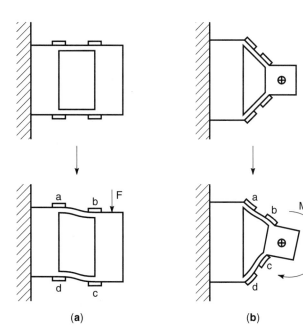

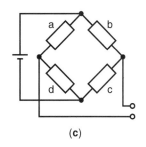

(c)

Figure 18. Sensing force and moment with parallel plates and radial plates: (a) sensing force with parallel-plate structure; (b) sensing moment with radial-plate structure; (c) bridge circuit with strain gauges.

smaller, sensitivity can be chosen at one's disposal, the operation is stable, integration of multiple axes is easy, and they are generally much cheaper than the piezoelectric type.

Multiaxis Force Sensor Using Parallel-Plate Structure. In this subsection, a multiaxis force sensor which utilizes a parallel-plate structure and strain gauge is introduced as a practical example. Figure 18 shows the schematics of a parallel-plate structure and a radial-plate structure that is a variation of the parallel-plate structure. The parallel-plate structure, Fig. 18(a), has its movable part connected to the fixed part with a pair of thin parallel plates. As a force is exerted on the movable part, it exhibits a translational displacement in parallel with the fixed part. The elastic strain on the plate surface is measured by the strain gauge. By forming a Wheatstone bridge, the force is detected as an electric signal. The characteristics of the parallel-plate structure are that its major displacement is permitted only in one direction and its high rigidity in the other directions yields an excellent sensing separativity. Also the sensitivity can be arbitrarily chosen according to the design of the parallel plate. The radial plate shown in Fig. 18(b) is employed to measure a moment applied to the movable part.

Figure 19 shows examples of variations and combinations of parallel-plate and radial-plate structures. Combination of these can provide many different types of force sensors. Fig-ure 20 shows a six-axis force sensor composed of three sets of the parallel plates and three sets of radial plates. In the construction the parallel plate is used for force measurement, and the radial plate is used for moment measurement. Six linearly independent force components have to be generated from the six sensing signals. Although the original signals contain some level of cross-talk signals, applying a decoupling matrix to the original signals suppresses such cross talk to as little as approximately 0.1%. Figure 20 shows a ring-shaped, six-axis force sensor composed of a parallel-plate structure alone (13). It has twelve detection parts from which the six components, three forces and three moments are computed. A force sensor is typically installed near an operational end, for example, the wrist of a robot, and, with this design, the hollow center is utilized for transmitting electrical power and sensor signals from the arm to the end effector or vice versa.

Figure 21 illustrates the primary and the secondary modes of deformation in parallel-plate structures. By using the sec-

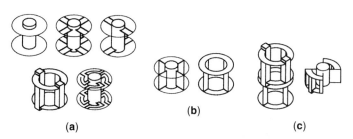

Figure 19. Examples of parallel-plate and radial-plate structures: (a) parallel-plate structure; (b) radial-plate structure; (c) combination.

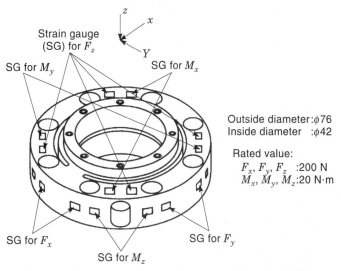

Outside diameter : ϕ76
Inside diameter : ϕ42

Rated value:
F_x, F_y, F_z :200 N
M_x, M_y, M_z:20 N·m

Figure 20. Structure of a ring-shaped, six-axis, force sensor.

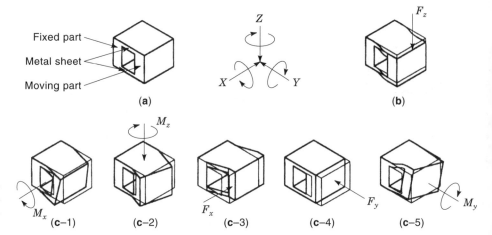

Figure 21. Sensing directions and corresponding deformation modes: (a) structure; (b) primary deformation; (c) secondary deformation.

ondary mode, a six-axis force sensor of a simpler construction can be composed (14). Figure 22 depicts a cylindrical six-axis force sensor using the secondary mode for sensing moments. This sensor can also be used as it is as a finger of a robotic hand.

Micromanipulation

Description of Operations. Micromanipulation (hereinafter to be called microoperation) includes assembling microparts and surgical operations conducted under an optical microscope or an electron microscope. Although the size of the manipulator is large, operations performed in handling the micro-objects, are called microoperations. The term microoperation is not affected by the size of the manipulator involved. It is rather defined by the size of the object to be handled.

Application Domains. In the microscale world where a wide range of microoperations are performed, the following are distinguishing characteristics:

1. Microbody effect: As an object becomes smaller, its weight becomes lighter, and its rigidity decreases accordingly. With respect to mechanical properties, as the mass becomes smaller the characteristic frequency increases (microstructural effect). It is also more likely that, with an excess force, the object is damaged or pushed out of sight. Effects due to crystallization of material or oxidized films cannot be disregarded, and the

assumption that the surface condition is uniform can no longer be taken for granted (microsurface effect). Also, a small object cannot be seen with the naked eye, and the behavior of such an object in many cases is far from our daily intuition (micro–macro gap).

2. Microfield effect: As the size of an object becomes extremely small, the scale effect prevails. In the microscale world, surface forces, such as electrostatic force, intermolecular forces, surface tension, etc., in proportion to the second power of the size become more dominant than the volume force, such as gravity, the force of inertia, etc., in proportion to third power of the size. Therefore, the handling method using gravity is no longer useful (microbody dynamic effect). In addition, the reaction of the object to the environment is sensitive. For example, many properties related to heat transfer are faster in inverse proportion to the third power of the dimensional ratio as the volume becomes small, and the thermal capacity similarly becomes smaller (micro phenomena effect).

Components To Implement Microscale Operations. To produce microscale operations, the following components must be realized:

1. Microscale operating system: To implement microscale operations that go beyond human hand operations, a special device which handles the microscale object is necessary. This system provides a tool for experiencing what is going on in the microscale world. For this purpose, it is suggested that the system retain a certain degree of universality, so that it can be utilized in different applications. Such universality is typically fulfilled by a manipulator which has several degrees of freedom, including rotational motion with sufficient range of motion. In other words, first, the desired functions for handling the microscale object should be made clear. Secondly, methodologies to substantiate individual functions should be considered, and thirdly, a total system design should be conceived of which satisfies the design constraints imposed by its environment.

2. Microbody dynamics: When handling a micro object, the adsorptive force imposed on the microbody raises problems. It is impossible to fabricate tools and handling

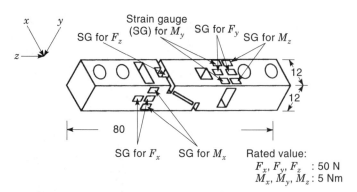

Figure 22. Structure of a bar-shaped, six-axis force sensor.

methods by trial and error which are suitable for a particular object in a particular environment. First of all, analyses of dynamic systems (microbody dynamics) involved in tools and handling techniques are needed, although there are many cases where different kinds of forces become dominant. In most cases, however, electrostatic force, intermolecular force, and surface tension are of primary concern. To know the magnitude of these adsorptive forces during an operation, it is important, to examine them theoretically and also to take actual measurements.

3. Micro-object handling technique: When handling a microobject in which adsorptive force is substantial, control of the force is necessary. One way to get around this problem is to devise a special effector for the manipulator that generates or eliminates the absorptive force at one's instance. However, this type of control can be used only in limited cases, and in many cases such control is difficult to implement. Hence handling the object becomes insecure. To make the microhandling task more reliable, it is necessary to handle the object even in the presence of these adsorptive forces. Handling difficulties similar to the current problem are observed in everyday life when handling sticky objects, and they may shed light on micro-object handling. An appropriate account of the magnitude of absorptive forces and knowledge of microparticle dynamics are mandatory for developing successful micro-operations.

Desired Functions of a Micromanipulator. A micromanipulator that implements operations in the world of such special characteristics, must have the following functions:

1. Monitoring function: Observation by microscopy is a powerful means of measuring the fine movement of an object. To obtain visual sensory information to the fullest extent, multidirectional observability and an ability to change the direction of view, if possible, are required.

2. Visible field operation: To obtain task information from the monitoring system at any time, it is essential that the task always be executed within the visible field of the microscope.

3. Position and orientation change of the object: From the standpoint of task execution, it is highly convenient if the position and posture can be changed independently.

4. Object gripping function: The gripping and releasing function is necessary, taking into account the dominant physical principles of the microobject.

5. Reactive force monitoring function: It is necessary that microforce sensing ability be installed to prevent the object or the probe from being deformed, damaged, or lost.

6. Inconsistency rectification function: Inconsistency caused by misjudgment due to a human operator's incorrect intuition, especially in terms of the physical size of the object, must be rectified.

Basic Components of a Micromanipulation System. A microscale manipulation system is composed of a monitoring system, a manipulator, and a workbench. This subsection describes how to build these essential components.

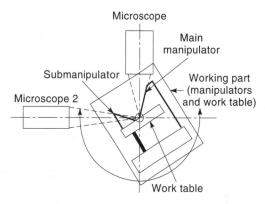

Figure 23. Concentrated visual field configuration.

1. Vision system: There is no doubt that visual information is vital in handling tasks. It is also important to achieve a system configuration so that the overall system is built around the field of view or a specific point within the view (Fig. 23). A multidirectional viewing system is composed of more than one microscope each of which monitors the same work space from different directions. These fields of view should contain the work space provided by the manipulation system. To achieve a variable monitoring direction without affecting task execution, a rotational degree of freedom should be given which rotates about the center of the field of view.

2. Manipulation system: To implement the task functions within the visual field and to attain a position and attitude control relative to an object, it is essential that all of the manipulator motions take place with respect to a specific point inside the visual field. This is typically a certain point on the object (Fig. 24). This means that all of the rotational axes intersect at the tip of the tool and the degrees of freedom are constructed sequentially in the order of the tool, the rotational degree of freedom, and the translational degree of rotation. The use of translational motions provided at the bottom of the ma-

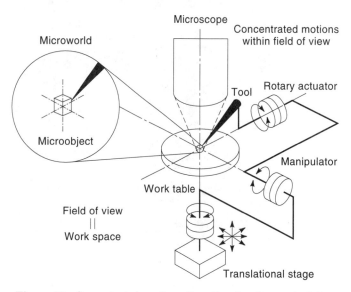

Figure 24. Concentrated configuration of motion for manipulator.

Figure 25. Photo of a microobject-handling system II.

nipulation system enables bringing the tip of the tool into the field of view of a microscope.

3. Components of the workbench: The workbench is responsible for the position and attitude control to compensate for discrepancies in position and orientation which supplement the motion of the manipulator. To achieve compensation, it is necessary to have a sufficient range of motion while assuring containment of the object within the field of view with position accuracy and without damaging the object or the tool.

Mechanism of a Micromanipulator. Following is an example of a microoperating system (Fig. 25) which consists mainly of two manipulators and two eyes all of which are coordinated to realize microoperations inside an electron microscope.

- Electron microscope: An electron microscope is employed as a main microscope, and an optical microscope is used as a supplement. These two microscopes are configured

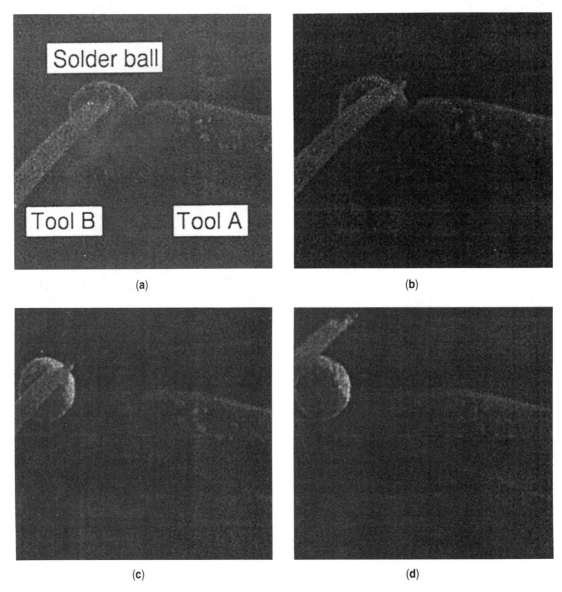

Figure 26. Placing a solder ball by holding the object still: (a) placing the object with Tool A; (b) Holding the object with Tool B; (c) Removing Tool A from the object; (d) Removing Tool B from the object.

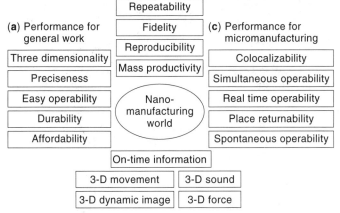

(b) Performance for general manufacturing

| Repeatability |
| Fidelity |
| Reproducibility |
| Mass productivity |

(a) Performance for general work

| Three dimensionality |
| Preciseness |
| Easy operability |
| Durability |
| Affordability |

Nano-manufacturing world

(c) Performance for micromanufacturing

| Colocalizability |
| Simultaneous operability |
| Real time operability |
| Place returnability |
| Spontaneous operability |

| On-time information |
| 3-D movement | 3-D sound |
| 3-D dynamic image | 3-D force |

(d) Performance for transfer of real feeling during manufacturing

Figure 27. Groupwise functions for the Nano-Manufacturing World.

so that they provide orthogonal views and also so that their focal points coincide with each other. By the choice of such a configuration, the task space is set around the focal point of the microscopes. This system has a rotational degree of freedom about the focal point which rotates both the manipulators and the workbench relative to the microscopes and which enables varying one's gaze angle during the operation.

- Manipulators: As described earlier there are two types of manipulator, the main operating arm and an assisting operating arm. The main operating arm has two rotational degrees of freedom and three translational degrees of freedom. The axes of the rotations intersect with each other at one point, and a translation mechanism aligns the tip of the tool with that point. An ultrasonic motor (operation range: 180°; resolution: 0.1°) is used for the rotational degree of freedom, and a piezoelectric element (operation range: 17 μm; resolution: 10 nm) is utilized for the translation. Because the second arm is required to move against the main operating arm together with the workbench while gripping the object on the workbench, it is mounted on the workbench so as to maintain the position relative to the workbench. The arm has one translational degree of freedom using a piezoelectric element with a displacement-magnifying mechanism.

- Workbench: The workbench has three translational degrees of freedom in orthogonal directions. The micromotion of the workbench is realized by a piezoelectric element (operation range: 17 μm; resolution: 10 nm), and the coarse motion is taken care of by an ultrasonic motor (operation range: 10 mm; resolution: 10 μm).

Force That Works on a Microobject. A variety of adsorptive forces are observed when working on a microobject. An electrostatic force, van der Waals force, surface tension, etc. These forces may cause many problems in manipulating microobjects. In a microoperation dealing with an object several tens of micrometers in size, the effect of surface tension is small because there is no liquid-bridge formation on the surface in the normal atmospheric environment. When the surface of the object is sufficiently clean and a certain surface roughness exists, the effect of the van der Waals force also

becomes small. As a consequence, electrostatic force is a dominant adsorptive force. Because an object that is not electrically grounded is always considered to be charged electrically through contact or friction, it also is necessary to take into account an electrostatic force due to charges.

Handling Techniques in Microoperation. When considering a pick-and-place operation performed by a manipulator on an object on a workbench, the adsorptive force F_t and F_w work between the effector and the object and the workbench and the object, respectively. When $F_t > F_w$, the object may be picked up by the tool, but the object picked up cannot be replaced on the workbench again. To replace the object, it is necessary to make F_t smaller than F_w. The manipulation of a microobject is produced by controlling one adsorptive force against another.

Example of Microoperation. Considering the special characteristics of surface force previously mentioned and taking advantage of the point that the adsorptive force is affected largely by contact area, a microoperating system under the electron microscope has succeeded in accomplishing a pick-and-place operation of a solder ball with two arms 20 μm in size by mechanically changing the contact area between the object and the effector (Fig. 26).

Nano-Manufacturing World and An Example of Its Operation. Fabricating a minute, three-dimensional structure involves many processes such as forming, transferring, assembling, joining, and inspection, which take place one after another. Because the object cannot be seen with the naked eye nor can be felt by hands, it is practically impossible to perform each individual operation with a separate system. To circumvent this problem, a system that performs operations from forming to assembling must be created.

The group of the functions, called for by a system that can fabricate three-dimensional microstructures, as shown in Fig. 27, submerges the operator in the world of microsubstances.

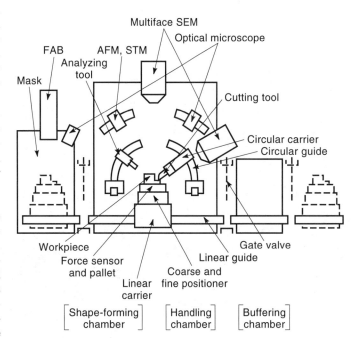

Figure 28. Construction of the Nano-Manufacturing World.

Figure 29. Front view of the manipulation system in the handling chamber.

The functions include forming using a removal process with no reactive force and assembling performed by two manipulators while observing from two orthogonal directions. A transferring mechanism also connects more than one operating site, a working space that is also used as a workbench is assured, and a parts shelf and a monitoring facility are provided to observe these situations from the external, accurate transmission of the operator's intention.

What has truly substantiated these functions, is the so-called Nano-Manufacturing World (NMW) shown in Fig. 28. Mounted on the vibration isolation table are a shape-forming chamber, handling chamber, and buffering chamber, and under the vibration isolation table are two vacuum pumps. In the forming chamber, a removal process with no reactive force takes place by a mask and fast atomic beam (FAB) forming micro parts. Then the parts formed are transferred to the handling chamber by a handling manipulator and pallet. A variety of operations are performed at the center of the operating chamber by two manipulators on the right and left. The progress of operations is monitored by a multiple-viewing scanning electron microscope (SEM) and is displayed on two

CRTs. Forces and sound generated during the operations are detected by a force sensor and a microphone and are transmitted to the operator after appropriate signal processing. Using a system like this, vivid sensations and feelings are delivered to the operator as if submerged in the micro world. Operations are carried out as if the operator were dealing with an object of normal size, not a tiny object which the operator can actually see or feel.

Figure 29 shows an example of the manipulation task in the handling chamber, and Fig. 30 is a microtorii (Japanese temple gate), an example of a micro three-dimensional structure fabricated by NMW. This microtorii is made of a crystal of GaAs and was formed by FAB. It is 80 μm high and 120 μm wide. The base is made of polyimide resin. The holes supporting the columns of the torii are 20 μm square and were fabricated using a YAG laser of the fourth harmonics. The two manipulators performed the transfer and assembly to complete the micro three-dimensional structure.

BIBLIOGRAPHY

1. Y. Hatamura, *Practice of Machine Design.* Oxford, UK: Oxford Univ. Press, 1998.

2. Y. Hatamura et al., Actual conceptual design process for an intelligent machining center, *Ann. CIRP,* **44**: 123, 1995.

3. T. Nagao and Y. Hatamura, Investigation into drilling laminated printed circuit board using a torque-thrust-temperature sensor, *Ann. CIRP,* **37**: 79, 1988.

4. R. P. Paul, *Robot Manipulators: Mathematics, Programming, and Control,* Cambridge, MA: MIT Press, 1981.

5. M. Spong and M. Vidyasagar, *Robot Dynamics and Control,* New York: Wiley, 1989.

6. H. Asada and J. J. E. Slotine, *Robot Analysis and Control,* New York: Wiley, 1986.

7. J. J. Craig, *Introduction to Robotics: Mechanics and Control,* 2nd ed., Reading, MA: Addison-Wesley, 1989.

8. A. J. Koivo, *Fundamentals for Control of Robotic Manipulators,* New York: Wiley, 1989.

9. T. Yoshikawa, *Foundations of Robotics: Analysis and Control,* Cambridge, MA: MIT Press, 1990.

10. R. M. Murray, Z. Li, and S. S. Sastry, *A Mathematical Introduction to Robotic Manipulation,* Boca Raton, FL: CRC Press, 1994.

11. D. E. Whitney, Quasi-static assembly of compliantly supported rigid parts, *Trans. ASME, J. Dyn. Syst. Meas. Control,* **104** (1): 65–77, 1982.

12. M. Fuchigami, *Robot-Based Production Systems,* Tokyo: Nikkan Kogyou Press, 1994.

13. Y. Hatamura, A ring-shaped 6-axis force sensor and its applications. *Proc. Int. Conf. Advanced Mechatronics,* Tokyo, Japan, 1989, p. 647.

14. Y. Hatamura, K. Matsumoto, and H. Morishita, A miniature 6-axis force sensor of multilayer parallel plate structure. *Proc. 11th Triennial World Cong. Int. Meas. Confed.,* Houston, TX, 1988, p. 621.

Y. Hatamura
T. Sato
M. Nakao
K. Matsumoto
University of Tokyo

Y. Yamamoto
Tokai University

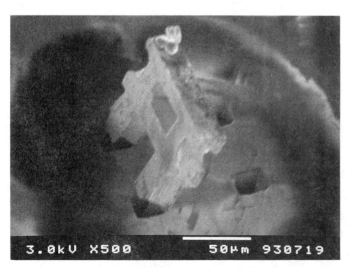

`3.0kV  X500          50µm  930719`

Figure 30. Photo of the microtorii fabricated by the NMW.

MANUFACTURABILITY USING CAD. See CAD FOR MANUFACTURABILITY.

MANUFACTURING, COMPUTER INTEGRATED. See COMPUTER INTEGRATED MANUFACTURING.

MANUFACTURING FLEXIBILITY, SEMICONDUCTOR. See FLEXIBLE SEMICONDUCTOR MANUFACTURING.

MANUFACTURING, LASERS. See LASER BEAM MACHINING.

MANUFACTURING OF SEMICONDUCTORS. See LITHOGRAPHY.

MANUFACTURING PROCESSES

Manufacturing processes can be broadly classified as follows:

1. *Casting, Foundry, or Molding Processes.* A permanent or a nonpermanent mold is prepared, and molten metal is poured into this prepared "cavity." The metal later solidifies and retains the shape of the mold cavity. In many cases the castings may have to be machined to conform to desired specifications and tolerances. Plastics and composites utilize molding processes, such as *injection molding.*

2. *Forming or Metalworking Processes.* This could be the next step after a casting or molding process is completed. The basic purpose is to modify the shape and size of the material. Examples are rolling, forging, extrusion, and bending.

3. *Machining or Material-Removal Processes.* In some cases, this could be the final process before commencing assembly operations. The objective is to remove certain designated, unwanted areas from a given part to yield the final desired "finished" shape. For example, a cutting tool may be used to cut a thread in a bolt. Traditionally, there are seven basic machining processes: shaping, drilling, turning, milling, sawing, broaching, and abrasive machining. Different machine tools have been developed to accomplish these tasks. For example, a lathe accomplishes the process of turning, and a drill press is used to drill holes. It is obviously convenient to perform several of these operations using a single workpiece setup. Equipment to accomplish such tasks are called *machining centers.*

4. *Nontraditional Machining Processes.* The traditional cutting tool in a lathe for example, removes a certain amount of material from the product and this results in a chip. There are several nontraditional chipless machining processes: chemical machining (as in photoengraving), chemical etching, physical vapor deposition (PVD), electropolishing or electroplating, thermochemical machining (TCM), electrochemical machining (ECM), electrodischarge machining (EDM), electron-beam machining (EBM), laser-beam machining (LBM), ultrasonic machining (USM), water-jet machining, plasma-arc welding (PAW), plasma-arc cutting (PAC), and plasma-jet machining.

5. *Assembly or Joining and Fastening Processes.* These include mechanical fastening (with bolts, nuts, rivets, screws, etc.), soldering and brazing, welding, adhesive bonding, and press, snap, and shrink fittings.

6. *Finishing or Surface Treatment Processes.* Some products may need this operation, whereas others totally skip this operation. Finishing ensures that all the burrs left by machining are removed so that the product is prepared appropriately and is ready for shipment or assembly. Sometimes a protective or decorative coating may be used in a finishing operation. Surface treatment may include chemical cleaning with solvents, painting, plating, buffing, and galvanizing.

7. *Heat Treatment Processes.* These processes are included only in certain cases, where the mechanical and metallurgical properties must be altered to meet specific needs. These processes subject metal parts to heating and cooling at predetermined rates.

8. *Miscellaneous Processes.* These include many processes that may not have been covered under the previous categories: inspection, testing, palletizing, packaging, material handling, storage, and shipping fall under this heading.

UNCONVENTIONAL OR NONTRADITIONAL MANUFACTURING PROCESSES

Conventional manufacturing processes always utilize a tool, which is harder than the workpiece, to remove unwanted material. A function box as shown in Fig. 1 can represent a manufacturing operation. The box has an input and an output. The "controlling factor" and the "manufacturing process" are supposed to represent how the operation is accomplished. Because harder tools are required to process tougher workpieces, a variety of new technologies have been developed. Some are discussed here.

Electrodischarge Machining (EDM)

The principle used is creating a high-frequency electric spark to erode the workpiece. A simple R–C circuit or a high-frequency impulse voltage generator provides high frequencies (250 kHz, for example) and high voltages to cause an electron avalanche in a dielectric medium. Figure 2 is a schematic representation of the EDM process. The idea is to create an ionized path and a rapid increase in the plasma temperature to something like 20,000°F. Transient heat transfer is subsequently accomplished to the tool and to the work piece. A thin surface layer of the metal is ionized. Surface finishes as thin as 2 μin. can be accomplished in this manner. EDM erodes both the tool and the workpiece. Therefore, it requires trial and error to optimize the tool–workpiece combination. Material removal rates of 2 cubic inches per hour have been achieved using a supply-input power of 10 kW. Electrodischarge grinding (EDG) is another modern manufacturing process that follows the same principle.

Electrochemical Machining

The electrochemical machining (ECM) process is based on Faraday's electrolytic law relating to anodic dissolution of metals. In this case the workpiece is connected to the positive of a dc source and made the anode. The tool is the cathode and is connected to the negative terminal. Electronic current

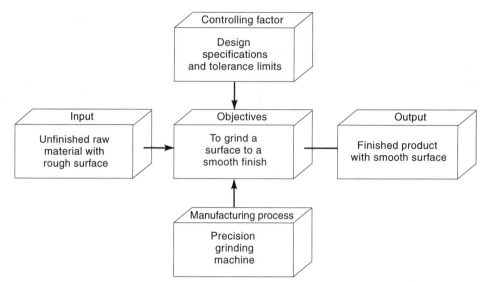

Figure 1. Representation of a manufacturing operation.

flows from cathode to anode and this effect is advantageously utilized to remove burrs or to drill a hole. It is estimated that 10,000 A of current remove steel at the rate of 1 in.³ per minute.

Ultrasonic Machining

In ultrasonic machining (USM) a transducer is used in conjunction with a tool that vibrates at low amplitudes but at high frequencies (25 kHz, for example). The principle is to remove material by microchipping and erosion. This is facilitated by an abrasive slurry contained between the workpiece and tool. The vibration of the tip of the tool results in imparting a very high velocity to the fine abrasive grains. Thus, the motion of the grinding grits is normal to the work surface,

as shown in Fig. 3. In a conventional operation, the motion of grinding grits is tangential to the work surface. This produces a cutting type of miniature chips. Sometimes this process is also called ultrasonic grinding (USG). Ultrasonic welding (USW) is another manufacturing process that utilizes the same principle for welding.

Powder Metallurgy

A crude form of powder metallurgy may have existed in Egypt as early as 3000 B.C. This process gained popularity during the late nineteenth century. Powder metallurgy is a manufacturing process wherein finely-powdered materials are blended and pressed into a desired shape by a process known as compacting. Then the compacted mass is heated at a controlled temperature to bond the contacting surfaces of the particles, a process called sintering. Thus the final product is manufactured in the shape required. In addition, it also possesses the desired properties and characteristics. Further, the product often needs no machining or finishing. There is almost no wastage of material. Porosity and permeability of the product are easily controlled. To be cost effective, powder metallurgy (abbreviated P/M) requires high production volume. Quality-grade metal powders, precision punches, compacting dies, and specialized sintering equipment are all very expensive, and

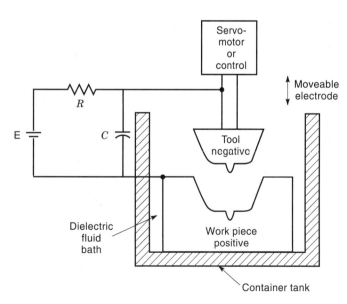

Figure 2. In the EDM process a powerful spark erodes the workpiece to a desired configuration.

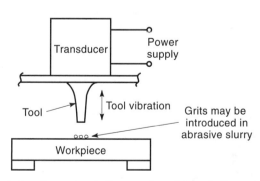

Figure 3. In ultrasonic grinding, motion of the grinding grits is perpendicular to the surface of the workpiece.

these contribute to high costs. Nevertheless, casting by traditional means may not be feasible for some high temperature alloys. Forging or hot extrusion, on the other hand, may result in poor tolerances and cause unnecessary die wear. Machining obviously generates waste material during processing. Powder metallurgy produces a wide range and variety of goods of diverse shape and size with good and acceptable tolerance limits. Modern manufacturing methods utilizing highly automated equipment produce several types of consumer items, toys, and automotive parts by the million. Because labor cost per part is low, powder metallurgy offers a viable alternative and is often preferred. Powder metallurgy is also preferred in cases where parts are produced in small quantities. Stringent specifications, tolerances, and desired metallurgical properties sometimes dictate the use of powder metallurgy. Small parts made from nickel-based super alloys, beryllium processing, certain types of self-lubricating bearings, and metallic filters are some examples where powder metallurgy is applied.

Laser Beam Machining (LBM), Welding, and Cutting

The principle used here is focusing the high-density energy of a laser (Light Amplification by Stimulated Emission of Radiation) beam to melt and evaporate portions of the workpiece in a controlled manner. A schematic representation is shown in Fig. 4. Extreme caution has to be exercised while using lasers because they can cause permanent retinal damage to the eyes. LBM is widely used in electronics and automotive parts manufacturing where precision drilling of holes (0.005 mm or 0.0002 in.) is required. Reflectivity and thermal conductivity of the workpiece surface play a major role in LBM. Excimer lasers are very popular for drilling holes and marking plastics and ceramics. Pulsed carbon dioxide lasers are commonly used for cutting ceramics and metals. Neodymium:Yttrium-aluminum-garnet (Nd:YAG) lasers and ruby lasers are used for welding metals.

Electron Beam Machining (EBM), Welding, and Cutting

Unlike laser beam machining, electron beam machining requires a vacuum. Dc voltages as high as 200 kV are used to accelerate electrons to speeds comparable to the speed of light. These high-speed electrons impinge on the surface of

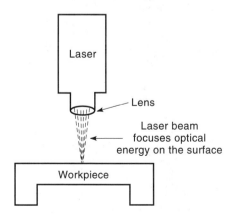

Figure 4. The principle behind the laser-beam-machining (LBM) process. The reflectivity and thermal conductivity of the workpiece influence LBM effectiveness.

the workpiece and generate heat which accomplishes the desired manufacturing operation, say, for example, drilling a hole or cutting a pattern in a precious metal. Caution must be exercised while using EBM because electrons, high voltages, vacuum, and metal surface all combine to generate hazardous X rays. Higher material removal rates (compared with EDM or LBM) are achieved by plasma-arc cutting (PAC). In this case ionized gases (plasma beams) are used which are particularly useful when cutting materials like stainless steel, and where very high temperatures (17,000°F) are needed.

Injection Molding

This type of fabrication is very popular for manufacturing complex-shaped plastic components. This method is similar to "die casting." In both, molten thermoplastic resin or some low melting point alloy is injected into a die. Then it is allowed to cool and harden. Modern day industry uses a wide variety of plastics and polymers. The idea implemented is to convert the plastic raw material directly into a finished product in a single operation. The process selected for fabricating these modern day plastics depends mainly on one criterion, whether the polymer is thermoplastic or thermosetting. Thermoplastic resins and polymers can be heated to a fluid state, so that they can be poured into a die or injected into a mold. In the case of thermosetting polymers, the polymerization process and the shape-forming process are achieved simultaneously because, once the polymerization has taken place, no further deformation is possible. Some of the methods available are casting, extrusion, thermoforming, etc. In addition, a variety of molding techniques are extensively used with plastics and polymers. Injection molding, reaction injection molding (RIM), rotational molding, foam molding, transfer molding, cold molding, compression molding, and hot-compression molding are some of the manufacturing processes most commonly used by the plastics industry.

Rapid Prototyping

Space age technologies and the computer revolution have required the manufacturing industry to produce prototypes of parts and components economically at a faster pace. This has resulted in the development of *rapid prototyping* techniques, also called *desktop manufacturing* or *free-form fabrication*. The idea is to manufacture an initial full-scale model of a product. The part is made directly from a three-dimensional CAD drawing. One of the methods is called *stereolithography*, a process based on curing and hardening a photocurable liquid polymer to the desired shape, using an ultraviolet laser source. Some of this equipment costs as much as half-a-million dollars. However, in many instances, this method is much cheaper than conventional prototyping, and the manufacturing industry has quickly recognized the importance and economic impact of these new technologies. Some of the other techniques used are selective laser sintering, three-dimensional printing, ballistic particle manufacturing, photochemical etching, and laminated object manufacturing. Almost all of these methods use CAD systems, and many cost in the region of hundreds of thousands of dollars. Some metals are used, but rapid prototyping with plastics and polymers, such as polystyrene, epoxy, polyester, PVC, and nylon, is more common.

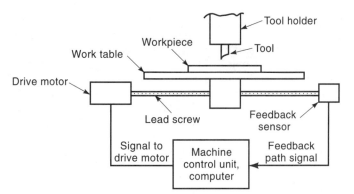

Figure 5. Improved quality and reduced manufacturing time are some of the advantages of using computers and closed-loop feedback systems.

Assembly or Joining and Fastening Processes

In most cases, this joining process is inevitable because the product cannot be manufactured in one single piece or one single operation. For example, a pressure cooker lid may be made from aluminum, but it has a plastic handle. In addition, the replaceable sealing ring is made from rubber. In other words, selected products may have to be replaced frequently, according to a routine maintenance schedule. In some cases it might be more economical to manufacture, transport, and assemble individual components at the customer's site. The functionality of different components may dictate that the desired properties be different. Besides traditional mechanical fastening, a variety of joining and fastening processes are available. If the material is "weldable," then the engineer has

a wide selection depending on the application and needs. Under the category of welding one can list shielded metal arc welding, submerged arc welding, gas metal arc welding, flux-cored arc welding, electrogas welding, electrosag welding, gas tungsten arc welding, plasma-arc welding, laser-beam welding, electron-beam welding, inertia friction welding, linear friction welding, resistance spot welding, resistance seam welding, resistance projection welding, flash butt welding, stud arc welding, percussion welding, explosion welding, and diffusion welding.

In many cases welding may not be the proper choice. For example, alloys containing zinc or copper are considered unweldable. Aluminum alloys are weldable only at a very high temperature. Brazing and soldering processes use much lower temperatures compared to welding. Further, soldering temperatures are lower than those used for brazing. Brazing is a joining operation wherein a filler material is placed between the surfaces to be joined and the temperature is raised to melt the filler material but not the workpieces. As such, a brazed joint possesses higher strength. It is believed that brazing dates as far back as 3000 B.C. Brazing methods are identified by the various heating methods employed. Torch brazing, furnace brazing, induction brazing, resistance brazing, dip brazing, infrared brazing, and diffusion brazing are noteworthy. Brazing is conducted at relatively high temperatures. For example, stainless steel and nickel-copper alloys need high brazing temperatures on the order of 1120°C. At the other extreme, titanium can be brazed at 730°C, using silver alloys.

Soldering is similar to brazing but requires lower temperatures. In this case, the filler material melts below 450°C and again, as in brazing, the base metal does not melt. A general purpose soldering alloy widely used in electronics assembly

Figure 6. Water-jet cutting machine. Courtesy of Stäubli Unimation, Duncan S.C.

operations melts at 188°C and is made of 60% tin and 40% lead. However, special soldering alloys are made of silver-lead, silver-tin, or silver-bismuth. Silver–lead and silver–cadmium soldering alloys are used when strength at higher temperatures is required. Gold, silver, and copper are easy to solder. Stainless steel and aluminum are difficult to solder and need special fluxes that modify the surfaces. Automated soldering of electronic components to printed circuit boards at high speeds is accomplished with wave soldering equipment. Other methods include torch soldering, furnace soldering, iron soldering with a soldering iron, induction soldering, resistance soldering, dip soldering, infrared soldering, and ultrasonic soldering. Soldering is commonly associated with electronics assembly operations, such as printed circuit assembly.

Adhesive bonding has been gaining increased acceptance by manufacturing engineers since World War II. Common examples are bookbinding, labeling, packaging, and plywood manufacturing. The three basic types of adhesives are natural adhesives (examples, starch, soya flour), inorganic adhesives (example, sodium silicate) and synthetic organic adhesives (examples, thermoplastics and thermosetting polymers).

AUTOMATED MANUFACTURING PROCESSES

Manufacturing operations have been carried out on traditional machines, such as lathes and drill presses, for a long time. This lacked flexibility and required skilled craftsmanship, trained mechanics, and was labor-intensive. Besides, "repeatability" of operations or production of exactly identical parts was extremely difficult because of human involvement.

Figure 7. The Merlin® Gantry Robot in an industrial setting Courtesy of American Robot Corporation, Oakdale, PA.

Automation is derived from the Greek word *automatos,* which means self-acting. Automation is broadly defined as the process of performing certain preselected sequences of operations with very little labor or human intervention. Numerically controlled machine tools were developed only recently in the 1950s. This breakthrough came almost two centuries after the industrial revolution! The postwar era gave automated manufacturing processes a gigantic boost that was long overdue. The latter part of the twentieth century saw some of the outstanding technological developments, such as integrated circuits, high-speed computers, programmable controllers, lasers, robots, vision systems, artificial intelligence, and expert systems. Figure 5 shows how sensors, feedback control systems, computers and computer numerically controlled (CNC) machine tools help in automating a manufacturing process.

Manufacturing processes are cost effective only when they are designed, planned, and executed efficiently. Process plan-

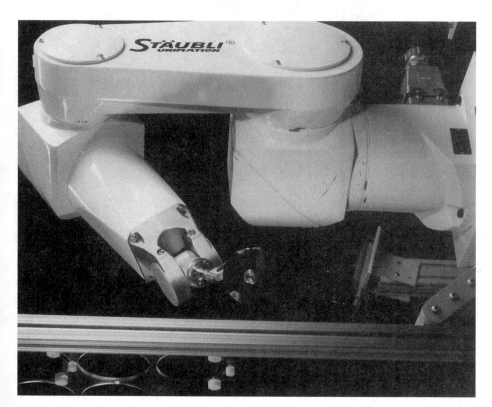

Figure 8. Clean room applications include wafer handling and sputtering. Courtesy of Stäubli Unimation, Duncan S.C.

ning is labor-intensive and time-consuming because the "process planner" has to selectively choose the methods and sequences required for the production and assembly operations. The planner also selects the necessary machine tools, fixtures, and dies. This tedious task is made simple by using computer-aided process planning (CAPP), a powerful tool that views the complete manufacturing operation as one integrated system. There are two types of CAPP systems, *the derivative system* (wherein the idea is to follow a standard process stored in the computer files) and *the generative system* (wherein the process is automatically generated based on some sort of "logic"). CAPP obviously requires expensive, sophisticated software that works appropriately with CAD/CAM systems. Some of the benefits include reduced planning costs, decreased "lead times," and improved product quality. Computers have helped in inventory management and other areas. Group technology (GT), cellular material-requirements planning (MRP), manufacturing resource planning (MRP-II) are some of the areas destined to gain wider acceptance and usage during the twenty-first century. Coordinate measuring machines (CMM), lasers, vision systems, ultrasonics, and other noncontact measurement techniques are helping to streamline inspection and quality control.

Programmable automation has several advantages. Some are listed here:

1. improves product quality
2. eliminates dangerous jobs and hazardous working conditions
3. increases the safety of operating personnel
4. eliminates human error by reducing human involvement.
5. minimizes cycle times
6. minimizes effort
7. enhances productivity
8. relieves skilled labor shortages
9. relieves boredom
10. stabilizes production
11. reduces labor costs
12. reduces waste of material
13. reduces manufacturing costs
14. maintains consistency of product uniformity
15. increases product diversification and product flexibility
16. designs more repeatable processes, just-in-time (J.I.T.)
17. increases punctuality and conformity to stipulated delivery dates
18. improves management of in stock material and improves inventory control
19. motivates work force whose capabilities are more challenged
20. improves compliance with OSHA regulations and reduces accidents

Environmentally safe manufacturing processes are obviously very desirable, and *Water-jet machining* (WJM) falls under this category. The force resulting from the momentum change of a stream of water can be advantageously utilized to create an efficient and clean cutting operation. Pressures ranging between 500 and 1200 MPa (1 Pascal = 1 Newton/meter^2 and 1 pound per square inch = 6891 Pa) are used to direct a jet of water to act like a saw. Water-jet machining, which is also called *hydrodynamic machinging,* can be very conveniently used to effectively cut plastics and composites. The food processing industry uses. WJM for slicing a variety of food products. Whether it is a strong and solid material like brick or wood, or a soft and flexible material such as vinyl or foam, hydrodynamic machining offers the engineer an advantageous choice for the selected manufacturing operation, because WJM eliminates the need for certain requirements, such as, for example, pre-drilled holes. A water-jet cutting machine can be seen in Fig. 6.

Robots have made a significant impact on the manufacturing shop floor, relieving humans from dull, dirty, and dangerous environments. They have been manufacturing high quality goods with minimal waste and at reduced costs. Robots are continuing to play a dominant role in streamlining several manufacturing processes. An example of a gantry robot installation is shown in Fig. 7.

Robots have helped the electronics manufacturing industry in a variety of ways. An example of a Robot being used in a semiconductor manufacturing processes is shown in Fig. 8.

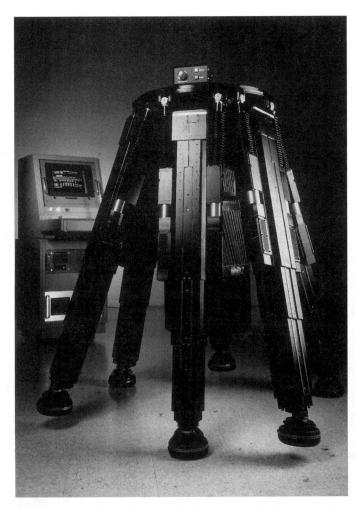

Figure 9. Odex-III with telescoping leg design. Courtesy of Odetics, Inc., Anaheim, CA.

Here, the robot helps in a *clean room* application handling a silicon wafer. Silicon processing for the electronics industry may include the following: epitaxial growth, cleaning, deposition, and lithography using masks, inspection, measurement, etching and doping.

Industry is currently using *manufacturing cells* that operate without direct human intervention. Remotely controlled robots have been designed to work in hazardous environments or locations that are inaccessible to conventional wheeled or tracked vehicles. "Walking" robots, such as the Odex-III with its telescoping leg design, can maneuver in confined spaces and "climb" steep stairs (see Fig. 9). The twenty-first century will see great progress in the area of manufacturing processes.

The factory of the future will be a fully automated facility wherein several processes such as material handling, machining, assembly, inspection, and packaging, will all be accomplished using sophisticated sensor/vision equipped robots and computer controlled machine tools. The focus is on re-directing an unskilled direct labor force toward more creative jobs, such as robot–computer programming or information processing. The highly competitive global marketplace demands the development and successful implementation of sophisticated manufacturing processes that can be termed world class.

BIBLIOGRAPHY

W. D. Compton (ed.), *Design and Analysis of Integrated Manufacturing Systems,* Washington, DC: Natl. Acad. Press, 1988.

N. H. Cook, *Manufacturing Analysis,* Reading, MA: Addison-Wesley, 1966.

E. P. DeGramo, J. T. Black, and R. A. Kohser, *Materials and Processes in Manufacturing,* Englewood Cliffs, NJ: Prentice-Hall, 1997.

M. P. Groover, *Fundamentals of Modern Manufacturing,* Englewood Cliffs, NJ: Prentice-Hall, 1996.

J. Harrington, Jr., *Understanding the Manufacturing Process,* New York: Dekker, 1984.

S. Kalpakjian, *Manufacturing Processes for Engineering Materials,* Menlo Park, CA: Addison-Wesley, 1997.

MYSORE NARAYANAN
Miami University

MANUFACTURING RESOURCE PLANNING

Manufacturing Resource Planning (MRP II) is essentially a business planning system. It is an integration of information systems across departments. In an enterprise implementing MRP II manufacturing, finance and engineering managers are linked to a company-wide information system. Thus, managers have access to information relating to their functional area of management as well as to information pertaining to other aspects of business. In reality, to reduce cost and provide good customer service, this integration is clearly mandatory. For example, the sales department has to have the production schedule to promise realistic delivery dates to customers, and the finance department needs the shipment schedule to project cash flow.

The concept of integrating information systems across departments is one that is basically common sense. Once integration across departments is achieved and its value experienced, it seems hard to believe that this has not always been the case. However, it is relatively recent. It is only since the 1970s that the integration of business functions and sharing of information across departments are being practiced. Several businesses still exist where business functions are working in isolation from each other, each focusing on their narrowly defined operational area with their own information system.

To fully comprehend and appreciate MRP II, one needs to understand the evolution of manufacturing planning. The questions of what, how much, and when to produce are the three basic questions in manufacturing planning. Over the years, different approaches to answer these questions have been proposed. The latest approach to answering these questions and, in fact, placing the answers in context within the whole business practice is MRP II. Although MRP II is largely borne out of the batch production and assembly environment, it is applicable in almost any facility.

BEFORE MANUFACTURING RESOURCE PLANNING

Until the 1970s the aforementioned three basic questions were typically answered by classic inventory control models. All these methods were based on the concept of stock replenishment where the depletion of each item in inventory is monitored and a replenishment order is released periodically, or when inventory reaches a predetermined level, or a hybrid of the two. Order quantities are determined by considering the tradeoff among related costs, based on the forecast demand and the level of fluctuations in demand. This approach fails to recognize the dependence between the components and end-items. Furthermore, it does not take into consideration the difference in demand characteristics between a manufacturing environment and a distribution environment. While demand in a distribution environment needs to be forecast for each item and does have fluctuations, in a manufacturing environment the demand needs to be forecast only for the end-product and not for the component items, in general. In addition, in a manufacturing environment the questions of what, when, and how much to order cannot be answered independent of production schedule. The production schedule states how much to produce of each product, and based on that the demand for each component item can be calculated since the usage of each item to build the end-product is exactly known.

The difference in the nature of demand in a manufacturing environment brought the development of Material Requirements Planning (MRP) systems, which translate the production schedule for the end-item referred to as the Master Production Schedule (MPS) into time-phased net requirements for each component item. This translation, however, involved large volume of transaction processing and thus warranted computing power. MRP systems found widespread acceptance once computers became available for commercial use starting in the late 1970s. It is not appropriate, however, to view MRP systems in isolation. As previously stated, material planning cannot be viewed in isolation of production and capacity planning. Each is a part of a broader system which is commonly referred to as Manufacturing Planning and Control (MPC) system. MPC includes sales and operations planning as well as detailed materials planning and ties up these plans with corresponding levels of capacity planning. A typical illustration of manufacturing resource planning is given in Fig. 1, where the hierarchy of MPC activities with corresponding levels of capacity planning are shown.

MANUFACTURING PLANNING AND CONTROL SYSTEMS

Sales and Operations Planning

Sales and Operations Planning is commonly known in the literature as Aggregate Production Planning and Resource Planning. Figure 1 shows the hierarchy of MPC activities. The highest level is basically concerned with matching capacity to estimated demand in the intermediate future, typically about 12 to 18 months, through the aggregate production plan. As the name implies, the aggregate production plan is usually prepared for product families or product lines as opposed to being prepared for individual end-products. This aggregated plan may be expressed in total labor hours, units, or dollars, or a combination of these. Likewise, time is also aggregated such that the plan is expressed on a monthly or quarterly basis. Typically, the time periods are monthly for the initial 3 to 6 months, and quarterly for periods thereafter. Because of the possible conflicts among the objectives of minimizing costs, keeping adequate inventory levels, and maintaining a stable rate of production, aggregate production planning is a complex task. Several costs such as those of inventory holding, hiring/firing, overtime/undertime, subcontracting, and backordering are considered in its preparation. During the preparation of the production plan, capacity is not considered a "given." This means that capacity may be increased or decreased based on the projected demand and the various costs. This could be through adding/deleting shifts, overtime/undertime, or expansion/closure of facilities. Capacity planning at this level is often referred to as *resource planning*.

The academic literature on aggregate production planning problem contains several different approaches to providing the solution. One approach is modeling it as a mathematical programming problem. Various mathematical programming models such as the transportation method of linear programming (1), linear programming (2,3), mixed-integer programming (4), and goal programming (5) have been applied to solve the problem. In a mathematical formulation, usually the objective is to minimize the total cost subject to demand and capacity constraints, by adjusting the above listed variables. A variation of this approach is the Linear Decision Rule (6) model where the assumption of linear costs (except for the labor cost) is relaxed. In addition to these optimizing methodologies, heuristic search procedures (7,8) and regression of past managerial decisions (9) have also been applied to the problem.

The aggregate production plan guides and constraints the scope of short-term decisions and needs to be disaggregated into detailed production schedules for individual end-products for short-term planning. In other words, the sum of individual end-product short-term production schedules must be consistent with the aggregate production plan. The disaggregation process provides the link between longer-term aggregate plans and shorter-term planning decisions. In the research literature the so-called "hierarchical production planning models" attempt to provide this link. These models utilize not just one but a series of mathematical models. Decisions made at one level constitute the constraints at lower levels where short-term decisions are made (10–17).

Master Planning and Material Requirements Planning

The next level of planning, shown as Master Production Scheduling in Fig. 1, is the result of disaggregating the aggre-

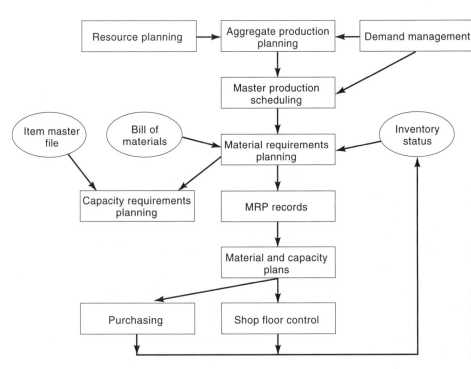

Figure 1. Manufacturing planning and control system schematic.

Table 1. Master Production Schedule for Item A

Weeks:	1	2	3	4	5	6
MPS:						100

gate production plan into production schedules for individual end-products. The Master Production Schedule (MPS) is usually expressed on a weekly basis and can be of varying lengths. The planning horizon for the MPS ranges from three months to one year. When several variations of the end-product are offered, the master production schedule is accompanied by a Final Assembly Schedule (FAS). Where FAS is maintained for a specific product configuration, MPS is maintained at the common subassembly level for options.

The master production schedule is a major input to the detailed planning of material requirements. The thrust of material planning is to determine component item requirements based on the master production schedule over the planning horizon. This obviously requires information about which components are needed, how many are needed, how they are assembled to build the end-product, and how much time is needed to obtain each component. This information is given in a product structure file referred to as the Bill of Materials (BOM). BOM is thus another major input for material planning in addition to the MPS. Note, however, that the determination of material requirements cannot be divorced from the information on how many units of each component item is already on hand and how many on order. This information is maintained in the inventory record for each component item. In addition, information on the routing and processing times for manufactured components is maintained in the so-called Item Master File (IMF). The Material Requirements Planning (MRP) system takes these three inputs—MPS, BOM, and inventory records—and calculates the exact time-phased net requirements for all component items. This, in turn, serves as the basis for authorizing the commencement of production for manufactured parts and release of purchase orders for purchased parts. The following simple example serves to illustrate the mechanics of how an MPR system processes the three inputs to obtain the time-phased net requirements for manufactured component items and purchased parts. Table 1 shows the MPS for end-product A. Figure 2 shows the BOM and the inventory record for end-product A, component item C, and purchased parts B and D. Table 2 shows the MRP records for all items.

The BOM shows that end-product A is made by assembling one unit of item B and two units of item C. Each unit of item C is made from two units of item D. Items B and D are pur-

Table 2. MRP Records

MRP Record for Item A

Periods	1	2	3	4	5	6
Gross Requirements						100
Scheduled Receipts						
On Hand	20	20	20	20	20	
Net Requirements						80
Planned Order Release					80	

MRP Record for Item B

Periods	1	2	3	4	5	6
Gross Requirements					80	
Scheduled Receipts						
On Hand	50	50	50	50	0	0
Net Requirements					30	
Planned Order Release			30			

MRP Record for Item C

Periods	1	2	3	4	5	6
Gross Requirements					160	
Scheduled Receipts						
On Hand	0	0	0	0	0	0
Net Requirements					160	
Planned Order Release			160			

MRP Record for Item D

Periods	1	2	3	4	5	6
Gross Requirements			320			
Scheduled Receipts			400			
On Hand	0	0	80	80	80	80
Net Requirements						
Planned Order Release						

chased parts. Lead times for each item is also presented (in parentheses) in the BOM. The inventory records show the on-hand and on-order quantities and their due dates. The MPS for item A indicates that 100 units of product A is planned to be completed in week 6. Note that the gross requirements for the end-item constitute the MPS.

Since 20 units of item A are on hand, and will remain on hand until week 6, 80 more units are needed in week 6. This information is shown in the MRP record for item A in the rows titled Gross Requirements (100 units in week 6), On Hand (20 units), and Net Requirements (80 units). A work order for 80 units of A, due at the beginning of week 6, needs to be released to the shop floor. Since the lead time is estimated as 1 week for item A, the order needs to be released in week 5. This is reflected in the row titled Planned Order Release. Since assembly of 80 units of A needs to start in week 5, and 1 unit of B and 2 units of C are required to make 1 unit of A, 80 units of B and 160 units of C are needed at the beginning of week 5 before the assembly of A can start. Thus, the planned order release for item A constitutes the gross requirements for its immediate components items B and C.

In the MRP records for items B and C, gross requirements are reflected in week 5 as 80 and 160, respectively. There is an on-hand quantity of 50 for B. Hence, the net requirement is only 30 units. Since the purchasing lead time for B is estimated to be 2 weeks, a purchase order for the remaining 30 units needs to be released 2 weeks ahead of the date of need, that is, in week 3, as shown in the MRP record. Similarly,

| Bill of material | Inventory records |
Item	On hand	On order
A	20	0
B	50	0
C	0	0
D	0	400	Due in wk. 3

Figure 2. (Left) The Bill of Materials showing all the components of end-item A, their relationships and usage quantities. The lead times for each component are given in parenthesis. (Right) Inventory on hand and on Order for all items in the Bill of Materials.

160 units of C are needed and there is no on-hand quantity. Therefore, a work order needs to be released to the shop to start making 160 units of C in week 3, since the lead time is 2 weeks and the need date is week 5.

The planned release date of the work order for making item C is week 3. Since 2 units of item D are used in each unit of item C, 320 units of item D needs to be withdrawn from stock in week 3 to start making the 160 units of item C. Therefore, again, the planned order release for item C determines the gross requirements of its immediate component D. As shown in the MRP record for item D, the gross requirement for item D is 320 units in week 3. Item D has a 3-week purchasing lead time. Thus, beginning of the current period is too late to release an order for item D. However, an order for item D has apparently been released in the amount of 400 units in the previous planning cycle. It is scheduled to be received in period 3. Since the order has been released in the past, it is an open order referred to as "Scheduled Receipt." This meets the requirement of 320 units in week 3. In addition, 80 units will remain on hand after period 3.

This example serves to demonstrate the two aspects of MRP: (1) netting of requirements for each item over on hand and on order quantities and (2) time phasing order releases by the estimated lead time for each item to meet the net requirements. It also demonstrates the coordination between order release date and order quantities of an item and the gross requirements of its immediate components. This process is referred to as the BOM explosion. Thus, as a result of the BOM explosion the MRP system produces (1) the planned order release schedule for manufactured and purchased items, (2) shop work orders, (3) purchase orders, and (4) reschedule notices, if necessary, for open orders.

The MRP records are processed (i.e., the BOM explosion is performed) on a periodic basis. The periodicity is influenced by the dynamism of the operating environment and by the computer processing power. With the ever-increasing processor speed it has become easier to update MRP records frequently. In the research literature the replanning of the MPS and the consequent BOM explosion (i.e., updating of the MRP records) is modeled via a rolling horizon procedure. Once the MRP records are processed for the planning horizon, it is assumed that the first period's decision is implemented and then the horizon is rolled to the beginning of the next period (or more than one period depending on the replanning frequency). The planning horizon length is fixed. Therefore, new periods' requirements are added at the end of the horizon, and the MRP records are updated based on the new information. Frequent replanning keeps MRP records updated. However, it is not necessarily desirable because it often results in changes in production schedules. Changes in demand and consequently the master production schedule, as well as the addition of the new periods' requirements at the end of the horizon, result in changes in (a) the due dates for open orders and (b) the quantity and timing of planned orders for the end item. Since end-item planned orders constitute the gross requirements for component items, the components' due dates for open orders and planned orders (timing or quantity) also change. This phenomenon is referred to as *system nervousness* and is identified as a major obstacle to the successful implementation of MRP systems (18–21). Several authors have investigated the impact of replanning frequency and the issue of system nervousness (22–26).

Lot-Sizing. An important issue in the BOM explosion is the order size determination. As item net requirements within the planning horizon are determined, order releases are planned to meet these requirements. In the example above, planned order quantities are equal to net requirements. However, ordering policy is not always "order as much as needed in each period." In fact, the order quantity may be quite different than the net requirements, such that a few periods' net requirements may be combined in one order. In that case, as the order is received, some of it goes to stock and is carried until it is consumed in the following periods whose net requirements are included in the order. How many periods' requirements should be combined in one order constitutes the issue of lot-sizing. The lot sizes are usually determined based on the tradeoff between the inventory carrying and ordering costs. Sometimes, the order quantity may be fixed, especially for purchased parts since the supplier may have control over the order quantity due to packaging and shipping requirements or because of quantity discounts, and so on. The lot-sizing procedure used has quite an impact on the system. As net requirements are consolidated into fewer orders, the pattern of gross requirements for components tend to be such that a period with a high requirement is followed by a number of periods with zero requirements. In other words, requirements tend to get more and more "lumped" for lower level items. This results in violent swings in capacity requirements from period to period and, in turn, causes implementation problems. In addition, lot size could amplify the impact of schedule changes and system nervousness.

The academic literature on lot-sizing is very rich. Several approaches have been proposed. One approach assumes that lot-sizing is performed for each item independent of other items and ignores the coordination between multiple levels of BOM. This is referred to as *single-level lot sizing*. Two methods have been proposed to obtain the optimum solution to the single-level lot-sizing problem assuming a finite horizon. One is a dynamic programming-based procedure (27) and the other is an efficient branch and bound procedure (28). Several heuristic procedures have also been developed to achieve the balance between the cost of carrying inventory versus the cost of ordering for the single-level lot-sizing problem (29–34). Some of the well-known heuristic procedures are:

Economic Order Quantity (EOQ). Order quantity is determined using the basic EOQ model with the average demand per period set-up cost and per period unit holding cost.

Periodic Order Quantity (POQ). A variant of EOQ. Order periodicity suggested by the EOQ model is used. Order quantity is equal to the number of periods' requirements within the periodicity.

Fixed Order Quantity (FOQ). Order quantity is a fixed quantity determined by an external constraint or preference.

Part Period Balancing (PPB). The order quantity is determined such that the cost of carrying inventory does not exceed the cost of placing a new order over the periods that the order covers.

Silver and Meal (SM). Cost is minimized over the number of periods that the order covers.

Several studies evaluating the performance of the heuristic procedures under a wide range of operating conditions have been reported in the literature (35–42). Some authors have evaluated the performance of these single-level heuristics, level by level, in a multilevel MRP system (43–49). Results from these studies show that under rolling horizons and demand uncertainty, conditions encountered in practice, none of the lot-sizing procedures provide the optimum solution and that the difference in the performance of lot-sizing rules tend to disappear (41,42).

Another approach to solving the lot-sizing problem is to take into consideration the dependency between the timing and quantity of the parent item order and component item requirements, as reflected in BOM. This is referred to as the *multilevel lot-sizing problem*. Several researchers have developed optimizing (50–56) as well as heuristic procedures (57–68) for the multilevel lot-sizing problem. Some authors also proposed capacitated lot-sizing procedures (68a,b). However, these procedures are not easily applicable to large size problems. The number of items and levels in BOM found in practice are often much too large for these methods to be useful. Furthermore, practical applications of such multilevel procedures have not been reported. The usual practice is to apply single-level heuristic lot-sizing procedures, on a level-by-level basis (69). Among such heuristics, only a few—LFL, EOQ, FOQ, and POQ—are reported as used by practitioners (70). Excellent reviews of lot-sizing research can be found in Refs. 71–73.

Safety Stock and Safety Lead Time. Safety stock is inventory that is kept in addition to the item requirements. Safety stocks exist in several different forms and may be needed for several different reasons. Extra inventory of the end-product may be kept as a protection against the uncertainty in demand—that is, forecast errors. At the component level, safety stock may be kept to protect against the uncertainties in the manufacturing process such as process yields. Safety stock of purchased items may be kept to protect against unreliable vendor deliveries. Ideally, there should not be any need for safety stock. However, since both demand and supply are uncertain in many manufacturing environments, safety stocks are commonly used in practice. They can be incorporated into the MRP system by adjusting the net requirements and, thus, the order quantities. Several research studies investigate the use of safety stocks in MRP systems (74–77). One of the reasons for safety stock is to reduce nervousness which results from the uncertainty in demand. However, this is a costly strategy and may not work as intended. Therefore, care should be taken in the determination of safety stock levels (76,77).

Safety lead time is a procedure where the shop or purchase orders are released and scheduled to arrive one or more periods earlier than the actual need date. It is used more against uncertainty in the timing rather than quantity. Both safety stock and safety lead time increase the amount of inventory in the system and inflate capacity requirements. Therefore, the decision to use either one has to be made with a proper understanding of their financial and physical implications on the system (78).

Capacity Planning

While translating the MPS into time-phased requirements for all the items in the BOM, the MRP system is capacity-insensitive. It implicitly assumes that sufficient capacity is available. This makes it necessary to determine the capacity requirements warranted by the MPS as well as by the detailed material plans sequentially, as shown in Fig. 1. First, a rough estimation of capacity requirements is made subsequent to the preparation of the MPS. This is used to ensure the validity of the MPS. Validation of the MPS is important since an unrealistic MPS may create problems in the execution of the production plan. Next, a more detailed determination of capacity requirements is made after the BOM explosion to produce work load profiles for all (or some critical) work centers which serves to confirm the feasibility of the material plan.

Rough Cut Capacity Planning. The viability of the master production schedule is checked by means of rough-cut capacity planning which may be as "rough" as using historical work center work loads or as detailed as using the routing and lead times for the individual products. Techniques available for rough-cut capacity planning include Capacity Planning Using Overall Factors (CPOF), Capacity Bills (CB) and Resource Profiles (RP).

Capacity Planning Using Overall Factors is the least detailed of the three methods. CPOF uses the MPS and historical work loads at work centers as inputs to obtain a rough estimate of capacity requirements at various work centers. Continuing from the above example, assume that one unit of item A requires 1.05 standard labor hours. Also, based on past data, assume that historical percentage of loads (labor hours) in Work Centers (WC) 1, 2, and 3 are 41%, 35%, and 24%, respectively. Based on the CPOF method the total capacity requirements would be (1.05*100) 105 total hours, distributed as (105*0.41) 43.05 hours for WC 1, (105*0.35), 36.75 hours for WC 2, and (105*.24) 25.20 hours for WC 3, all in period 6. The CPOF method is attractive because of its simplicity. However, it would be useful only to the extent the historical work center loads reflect the current requirements. Any change in the product mix or in the processing requirements due to product or process design change may easily outdate the historical figures and, thus, should be taken into consideration prior to the use of the CPOF method. Furthermore, this method shows the capacity requirements in the same MPS time periods where the end-product requirements are located—that is, this method does not time-phase capacity requirements by the estimated component lead times.

In addition to the MPS, CB requires BOM information, shop floor routings, and operation standard times for each item at each work center. From the BOM file, it retrieves the information concerning which components, and how many of each (usages), are needed to build the end-product. The component usages are multiplied by the MPS quantity to determine the total component requirements to build the MPS. Each component requirement is then multiplied by per unit operation standard times for each work center indicated on its shop floor routing. The capacity requirements are summarized by work center.

In CB, BOM information, routing, and operation standard times replace the historical work center load percentages used in CPOF. Therefore, any changes in the product mix, product, or process design (reflected in operation standard times and routings) will be incorporated in the determination of capacity requirements. This makes the CB method more attractive for those environments where such changes may occur frequently.

CB, like CPOF, shows work center capacity requirements (accumulated from all items in BOM) in the same MPS time period where the end-product requirements are located. This may not be an issue for those cases where the manufacturing lead time is short relative to the time bucket used in the MPS. However, when manufacturing lead time extends over multiple MPS periods, aggregating the capacity requirements into the same period may be far from reflecting the real capacity requirements. The Resource Profiles method uses the same information from the BOM, shop floor routings, and operation standard times as does the CB method. In addition, RP time-phases the capacity requirements by component lead times. The resulting output shows work center loads spread over the total manufacturing lead time, for each work center reflected in those time periods when the work is actually expected to be performed. Thus, RP is the most sophisticated of the three rough-cut capacity planning techniques described here.

Rough-cut capacity planning techniques are used to validate the MPS. If capacity requirements exceed available capacity, either the MPS or the capacity availability has to be altered. Thus, the preparation of MPS and its validation by checking capacity availability is an iterative process, where ultimately the correspondence between the MPS and capacity availability is to be achieved.

Capacity Requirements Planning. The next level of capacity planning is performed subsequent to the detailed planning of material requirements. MRP explosion provides the netting of gross requirements over on-hand and on-order quantities and reflects the actual lot-sizes for each component in the planned orders. Also, any additional requirements for components not included in the MPS (e.g., service parts) are also included in the calculations. The time-phased material plans produced by the MRP system are translated into detailed capacity requirements through Capacity Requirements Planning (CRP).

CRP uses the information on shop floor routings and operation time standards (setup and processing times) like sophisticated rough-cut procedures. However, instead of determining capacity requirements based on MPS quantities, CRP translates planned order quantities, reflecting actual lot sizes time-phased during the MRP process, into labor/machine hours. These hours are added to the labor/machine hours translated from open order quantities (work-in-process). This produces time-phased load profiles for work centers over the planning horizon. Calculating detailed capacity requirements enables the validation of material plans by checking for feasibility. Again, the correspondence between hours required and hours available needs to be achieved for successful execution of the plans at the shop floor.

Capacity insensitivity of the MRP approach has been an early source of criticism. An alternative to the infinite loading of CRP is finite loading. Finite loading also uses the planned orders as input. However, it also requires the orders to be prioritized—that is, placed in the sequence in which they will be processed (79). After prioritizing, it loads the orders to work centers until available capacity is reached. Because of its reflection of the relationship between capacity and scheduling, it is viewed more as a shop scheduling technique (80).

Shop Floor Control

The lowest level in the hierarchical MPC model presented in Fig. 1 is concerned with the execution of plans. Note that MRP output only specifies the release of orders for component item production. Each work order is comprised of several individual processing steps often performed at various work centers/machines. When the work order is released to the shop floor, the material needed to make the parts is withdrawn from the stock room and moved to the work center or machine where the first operation is to be performed. Typically, there may be a large number of work orders often competing for the same set of resources. Therefore, in each work center/machine there needs to be a mechanism to schedule the competing work orders. Operations scheduling is a major element of a manufacturing planning and control system due to its impact on customer service, in terms of on-time delivery performance.

Various scheduling rules exist for prioritizing jobs at each machine. These rules may be as simple as first come first served or based on some other more complicated criterion. Numerous rules have been developed and discussed in the literature (81–84). Some of the well-known rules are:

Shortest Operations First, also known as the Shortest Processing Time (SPT) rule. The jobs are prioritized in the ascending order of the processing times at the current work center.

Earliest Due Date (EDD) rule. Jobs are prioritized in the ascending order of their due dates.

Operations Due Date (OPNDD). Jobs are prioritized in the ascending order of their due dates for the current operation.

Critical Rule (CR). The ratio of time until due date to lead time remaining (in the current and subsequent operations until the completion of the job) is used to prioritize jobs. The job with the smallest CR is the most urgent and is thus given the highest priority.

Total Slack (TS). The difference between time until due date and lead time remaining is used to prioritize jobs. The job with the smallest slack is given the highest priority.

Slack per Remaining Operations (S/RO). The ratio of total slack to the number of operations is used to prioritize jobs. The job with the smallest S/RO is given the highest priority.

The effectiveness of scheduling rules differs based on the performance criterion used such as flow time (time from the arrival of the order until its completion), earliness, tardiness, inventory, number of tardy jobs, shop utilization, and so on. Research shows that SPT tends to minimize average flow time; due-date-based rules tend to perform well in terms of due date related criteria. The literature on scheduling is extensive. Several job shop (85) and flow shop (86) simulations compare the performance of dispatching rules under various operating conditions (87). In these simulations, usually scheduling of a machine is done using a specified dispatching rule, regardless of the scheduling in other work centers. Simultaneous scheduling of all machines is very difficult to model and is rarely done in job-shop settings. Recently, methods such as tabu search, genetic algorithms, and simulated annealing have also been applied to the scheduling problem (88).

Much of the existing literature focuses on problems in which all jobs are assumed available for processing at the beginning of the horizon. It also advocates schedules in which no machine is ever kept idle in the presence of waiting jobs. However, in a typical job shop (e.g., tool room, die shop, small component manufacturing shop) jobs arrive continuously. Thus, superior schedules may involve deliberately keeping a machine idle, in the presence of waiting jobs, in order to process an anticipated "hot" job that is yet to arrive. In addition to dynamic environments, deliberate idle times may also be necessary when both early and tardy completion of jobs is undesirable, or when there are multiple machines. For a review of the literature dealing with the issue of schedules with deliberate machine idle times, an interested reader may refer to Refs. 89–91.

Plossl and Wight (92) distinguish between loading and scheduling. Load is the amount of work waiting in the shop (or at the machine) to be performed and can be computed as the amount of total work. Load can be controlled by monitoring the work input into the shop. Shop loading is said to be balanced when the flow of work into the shop equals the output from the shop. By adjusting the input, one is able to control the amount of work backlog, machine utilization, and the shop throughput. Several authors have studied the issue of controlling the release of jobs to the shop by means of order review/release policies. In general, the results appear to be mixed in that it is not clear if and when such policies are effective in improving the overall system performance (93–99). It appears that controlling the release of orders to balance the load between the machines (100,101) may be a superior approach when compared to basing the order release and control decision on other objectives. See Ref. 102 for a framework for a manufacturing system where an order review/release policy is implemented.

Implementation of MRP Systems

Successful MRP system implementation requires more than just the information system. One of the major success factors is management commitment. First, a commitment needs to be made to provide accurate information that is input to the system. This requires cleaning and integrating the databases and their continuous maintenance as well as timely data entry. Companies successfully implementing MRP systems deal with accurate BOM, MPS, inventory, and lead time data in making inventory and scheduling decisions. Second, a commitment is needed to train the people who will use the system. These are prerequisites to successful MRP system implementation. Providing the prerequisites clearly have costs, and the extent of costs depend on the initial condition of the company. Therefore, a commitment of resources is also needed. Challenges in the implementation of MRP II systems include period-size resolution (short-term planning), data transaction intensity (and resulting accuracy challenges), iterative capacity planning versus finite), and non-intuitive knowledge requirements (extensive training), among the more general. A thorough discussion of these challenges can be found in (80).

Successful MRP system implementation brings several benefits. MRP systems bring a good match between demand and supply by making the need date for items coincide with their due date. Because of the closer match between demand and supply, finished goods inventories are reduced. Improving planning of priorities and scheduling reduces work-in-process, and improving timing for vendor deliveries reduces raw material inventories. Altogether, inventory turnover increases and obsolescence decreases (103,104).

In general, the extent of benefits derived from the system depends on how a company uses the MRP system. Users of MRP systems are classified into four classes: Class A to D. Those companies using it to its fullest capacity (with full support of the top management) for priority planning and capacity planning, with a realistic and stable MPS, are referred to as "Class A" users. At the other end of the spectrum there are Class D users for whom the MRP system exists only in data processing and does not reflect the physical realities of the organization. For a detailed discussion of the MRP users classification see Ref. 103.

Several empirical studies dealing with the practical issues surrounding efficient and effective implementation of MPC systems, in particular MRP-based systems, have appeared in the literature. See, for example, Refs. 105–107. Kochhar and McGarrie (107) report seven case studies and face-to-face meetings with senior managers and identify key characteristics for the selection and implementation of MPC systems. They conclude that (1) the operating environment significantly impacts the choice of the system and (2) the existing framework for an objective assessment of the need for individual control system functions is largely inadequate in serving the needs of managers. This result demonstrates the need for a modular design and a decentralized architecture for MPC systems, thus providing individual companies the maximum flexibility in tailoring the system to meet their needs within a common framework. Such an architecture and design, in our view, should automatically preserve the best features in all variants of the system and, thus, be able to guarantee efficiency and effectiveness (108).

Problems with MRP

There are a number of fundamental flaws in the MRP-based approach to production planning and control. Central weakness is MRP's modus operandi of sequential, independent processing of information. The approach attempts to "divide and conquer" by first planning material at one level and then utilization of manpower and machines at another level. The result is production plans which are often found to be infeasible at a point too late in the process to afford the system the opportunity to recover. Second, MRP-based systems do not provide a well-designed formal feedback procedure instead depend on ad hoc, off-line, and manual procedures. When a problem occurs on the shop floor, or raw material is delayed, there is no well-defined methodology for the system to recover. Thus, the firm depends on and actively promotes safety buffers, leading to increased chances for missing strategic marketing opportunities.

A third flaw concerns the use of planned lead times. Planned lead times are management parameters which are provided prior to the planning process and represent the amount of time budgeted for orders to flow through the factory. This can result in a tremendous amount of waste in terms of work-in-process inventory. For example, consider four single operation jobs A, B, C, and D with processing time requirements 5, 4, 7, 9 and all four due at time 25. Under an MRP system, the planned lead time is prespecified and fixed. Let the planned lead time for each of these jobs be 25. Thus, the material for all four would be made available at time 0 by the MRP system. Suppose, the jobs are processed in the order A-B-C-D. Since we know from Little's law that inventory is proportional to flow time, we shall focus on flow time as the performance measure. It is easy to verify that the average flow time for the given sequence is 25, assuming early delivery is not permitted, consistent with the just-in-time manufacturing philosophy. Suppose the material arrival dates for the four jobs are planned to coincide with their planned start dates. Then, the average flow time would be 17.5. This translates into a saving of 30% in inventory costs. Note that this is only possible if a complete schedule can be constructed and the information is used to plan material procurement and delivery. See Ref. 109 for a detailed report of how substantial reduction in inventory costs can be obtained by first constructing a complete schedule and then using the schedule information to plan material.

The fourth problem with MRP systems is that often schedules are extremely nervous, which, in turn, leads to increased costs, reduced productivity, low morale, and lower customer service leads (110). The following numerical example demonstrates the problem of nervousness in detail. Figure 3 shows the BOM. Table 3 shows the MRP records in subsequent planning cycles. The BOM includes two components below the end-item level: One unit of end-item A comprises one unit of component B and two units of component C.

The planning horizon is assumed to be 12 periods long, and the lead time is three periods for each item. End-item lot size is determined by using the periodic order quantity (POQ) method, with a periodicity of four periods. A lot-for-lot regime is employed for lot-sizing the component requirements. The MRP records for the first planning cycle, which covers periods 1 through 12, appear in Table 3a. The beginning inventory is 163 units for end-item A, 27 units for component B, and 54 units for component C. An order for 341 units of item A is scheduled to be received in period 2, and orders for 304 units of item B and 608 units of item C are scheduled to be received in period 3.

In period 1, the demand forecast for item A is 75 units, and the projected inventory balance (i.e., inventory on-hand) at the end of the period is 88 units (Table 3a). An order for 348 units of item A is planned for release in period 3 to cover the demand forecast in periods 6 through 9. Therefore, at the beginning of period 1, *expedited orders* are released for 17 units of item B and 34 units of item C, both of which are due at the beginning of period 3. The actual demand for item A in period 1 is 106 units, as opposed to the 75 units forecast. Consequently, the actual inventory balance at the end of period 1 is 57 units, instead of the anticipated 88 units. The

effect of this sudden spike in demand is evident in the MRP records presented in Table 3b.

At the beginning of period 2, demand forecast for period 13 becomes available and is added to the horizon. Since only 57 units of item A are on hand and 341 units are scheduled to be received in period 2, the total will not be sufficient to cover the anticipated demand during periods 2 through 5. Therefore, an *unplanned* order for 314 units of item A is released in period 2, and it is due in period 5. Consequently, the previously planned order for 348 units of A in period 3 is *canceled*. In turn, the due dates of open orders for items B and C are *expedited* from period 3 to period 2 (Table 3b). Furthermore, the expedited component orders released in period 1 are *rescheduled to later periods* to avoid inventory buildup.

Note that the cumulative lead time for end-item A is six periods. Between planning cycles 1 and 2, the following changes occurred in item A's schedule: (1) An unplanned order for 314 units is released in period 2, necessitating an emergency setup; (2) new planned orders are made for 335 units in period 6 and 110 units in period 10; and (3) the previous plans for producing 348 units in period 3 and 270 units in period 7 are canceled. Together, these changes cause a ripple effect, leading to a complete revision of the material plans for items B and C: Open orders for 304 units of items B and 608 units of item C are expedited from period 3 to period 2, and open orders for 17 units of item B and 34 units of item C are postponed from period 3 to period 6. Also, the new plan calls for order releases in periods 3 and 7 for both items whereas the previous plan did not. Likewise, planned orders in period 4 are canceled.

These types of changes to the material plan directly impact the capacity plan. In particular, changes within the cumulative lead time (periods 2 through 6) may not be feasible. The new and unplanned order for 314 units of product A and the expedited orders for the components (for 304 units of B and 608 units of C) may necessitate overtime and, thus, lead to an increase in cost. Such changes may also cause other jobs to become tardy. For a detailed discussion of the issue of nervousness see Ref. 111.

As mentioned earlier, uncertainties about supply and/or demand and dynamic lot-sizing combined with rolling planning are major causes of nervousness in schedules. Many strategies have been recommended to dampen nervousness, including freezing a portion of the master production schedule (112–120), time-fencing (80), using lot-sizing procedures selectively (121), forecasting beyond the planning horizon (122), incorporating the cost of changing the schedule into the lot-sizing process (123–125), using lot-for-lot ordering below level 0 (120), and using buffer stock at the end-item level (76,77,120,121). Freezing the MPS appears to be the most effective method for reducing nervousness (119). However, research is still ongoing to find ways to compensate for the likely reduction in service level when the MPS is frozen (126).

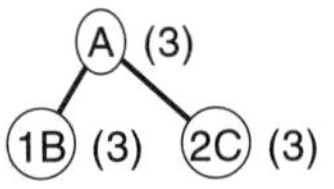

Figure 3. The Bill of Materials showing all the components of end-item A, their relationships and usage quantities. The lead times for each component are given in parenthesis.

MPC in Different Environments

The MPC system design, especially the activities in levels 2, 3, and 4, to a great extent depend on the nature of demand that a company is facing. Three principal environments where the approaches to MPC system design will differ are defined as Make-to-Stock, Make-to-Order, and Assemble-to-Order.

Table 3. MRP Records in Subsequent Planning Cycles

a. MRP Records in the Beginning of the First Planning Cycle

Item A

Periods	1	2	3	4	5	6	7	8	9	10	11	12
Gross Requirements	75	146	87	92	95	70	111	111	65	99	85	86
Scheduled Receipts		341										
On Hand	88	283	196	104	9							
Planned Order Release			348					270				

Item B

Periods	1	2	3	4	5	6	7	8	9	10	11	12
Gross Requirements			348				270					
Scheduled Receipts			304									
On Hand	27	27										
Planned Order Release	17			270								

Item C

Periods	1	2	3	4	5	6	7	8	9	10	11	12
Gross Requirements			696				540					
Scheduled Receipts			608									
On Hand	54	54										
Planned Order Release	34			540								

b. MRP Records in the Beginning of the Second Planning Cycle

Item A

Periods	2	3	4	5	6	7	8	9	10	11	12	13
Gross Requirements	146	87	92	95	70	111	111	65	99	85	86	110
Scheduled Receipts	341											
On Hand	252	165	73									
Planned Order Release	314				335				110			

Item B

Periods	2	3	4	5	6	7	8	9	10	11	12	13
Gross Requirements	314				335				110			
Scheduled Receipts	304				17							
On Hand	17	17	17	17								
Planned Order Release		301				110						

Item C

Periods	2	3	4	5	6	7	8	9	10	11	12	13
Gross Requirements	628				670				220			
Scheduled Receipts	608				34							
On Hand	34	34	34	34								
Planned Order Release		602				220						

Make-to-Order. When a company builds its products according to customer specifications, then MPS is expressed in terms of each customer order. Capacity requirements are based on the current backlog of customer orders. Bills of material are specific to each customer order; and since each order is unique, manufacturing lead time has a large degree of uncertainty.

Assemble-to-Order. When the products offered by the company have large variety, then it is not practical to stock each and every possible end-product. However, customers may expect delivery faster than the time it would take to manufacture the product after the order is received. Therefore, the MPS is maintained in terms of major subassemblies (options) level. When a customer order is received, the final assembly is made according to the desired end-item configuration. The specific cus-

tomer orders are maintained in the Final Assembly Schedules. In the assemble-to-order environment, Planning Bill of Material represent the major product options. Figure 4 shows the Planning Bill of material for a fictitious automobile.

Figure 4 shows that 40% of the cars made are Model A, 30% are Model B, 25% are Model C, and 5% are the Limited Model. Seventy-five percent of all cars have automatic transmission, and 25% have stick shift transmission. Engines can be V6 (75%) or V8 (25%). Also, cars can have two-wheel drive (60% of all cars) or four-wheel drive (40% of all cars). With these options there are $4 \times 2 \times 2 \times 2 = 32$ end-product configurations. Instead of building all possible configurations to stock, MPS is kept at the options level; that is, there are 13 MPS (1 for common items) and up to 32 FAS where only

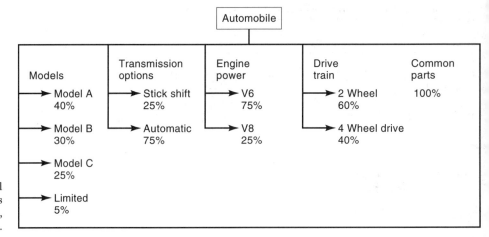

Figure 4. The Planning Bill of Material for the automobile showing the options available in building the end-item: Model, Transmission, Engine Power, Drive Train.

the record of actual customer orders are maintained. Keeping the MPS at options level reduces the delivery lead time and facilitates the forecasting of demand. The major uncertainty is in the product mix. The total of options can be more than 100% to buffer the uncertainty in the product mix.

Make-to-Stock. When the company is building standard products that the customers buy off-the-shelf, then the schedule is based on the forecast demand. Items are built to stock, and demand is satisfied instantaneously from stock. In this environment, MPS is stated in terms of end-products. Customer order promising is based on available-to-promise quantity. The available-to-promise values are calculated for the end-product for those periods where there is an order quantity (these order quantities constitute the MPS). For the first period, available-to-promise is the on-hand plus first-period order quantity (if any) minus the sum of all customer orders until the next period where there is an order quantity. For later periods, available-to-promise is the order quantity minus all customer orders in that and subsequent periods until the next period where there is an order. Since MPS is based on forecast information, customer orders consume the forecast. The forecast errors are monitored, and forecasts and the MPS are updated if needed. The available-to-promise logic facilitates the effective coordination between marketing/sales and production functions. The concept of available-to-promise is demonstrated in the example shown in Table 4.

Note that the MPS row shows production of 40 units of the end-product in weeks 1, 3, and 5. In period 1 the sum of the on-hand quantity and the MPS order quantity is 50. In periods 1 and 2 (until period 3 where there is the next MPS order quantity) the total of customer orders is 25. Therefore, up to

a total of 25 units are available-to-promise within periods 1 or 2. In periods 3 and 4, the sum of customer orders is 26. Thus, 16 units are still available-to-promise within periods 3 or 4. In periods 5 and 6, there are no actual customer orders. So the MPS quantity of 40 units in period 5 can be used to promise to customers in period 5 or 6.

In environments where the production process involves repetitive manufacturing and flow systems such as assembly lines, the production schedule is typically based on a rate of production and is stable over some period of time. Thus, material planning becomes much less sophisticated. Since item routing on the shop floor is determined by the flow of the line, and components need not wait or go in and out of stock between subsequent operations, tracking material on the shop floor is not needed. This reduces the number of levels in the BOM as well as the number of transactions on the shop floor. Lead times becomes shorter, and material flow on the shop floor can easily be controlled by kanbans. In this kind of environment, Just-in-Time manufacturing techniques can be applied to manage the shop floor operations. The design of the MPC system is thus determined by the market characteristics that the company is facing. See Ref. 80 for a detailed discussion of different MPC environments.

MANUFACTURING RESOURCE PLANNING (MRP II)

It is easy to see that manufacturing planning and control activities are closely related to the activities of other functional areas such as marketing and sales, product/process engineering and design, purchasing, and materials management. The quality of the major inputs to manufacturing planning—namely, the MPS, BOM, and inventory record information—is not determined solely by manufacturing. These inputs are prepared, shared, and updated by other functions within the organization as well. For example, consider the following.

While marketing creates the demand, manufacturing is responsible for producing the parts and products necessary to meet the demand. Therefore, any marketing activity that may influence future demand needs to be confirmed by manufacturing. Thus, as a statement of planned production, MPS provides the basis for making delivery promises via the ATP logic. It is valuable for coordinating the activities of sales and

Table 4. Order Promising for the End-Product

Periods	1	2	3	4	5	6
Forecast	20	20	20	20	20	20
Customer Orders	18	7	22	4		
On Hand	30	10	28	8	28	8
Available-to-Promise	25		14		40	
MPS Order Quantity	40		40		40	

production departments. Any change or update by sales needs to be approved by manufacturing and vice versa.

Changes in BOM impacts product routings and lead times which are used in material and capacity planning. Proper material and capacity planning, therefore, warrants close coordination between manufacturing and engineering so as to maintain valid bills of materials. Any changes in the BOM will have to be agreed upon by both engineering and manufacturing to assure (1) the feasibility of tolerances and (2) the impact of product revisions and new product introductions (where marketing also is involved) on the shop floor system.

Likewise, accounting/finance functions should also use the same data as manufacturing, for making revenue and cost projections. MPS converted to dollars depicts the revenue stream, purchase orders converted to dollars represent the cost of materials, and shop floor activities represented in work orders converted to dollars reflect the labor and overhead costs. In other words, production schedule converted to dollars reflects the cash flow schedule. Discrepancy in the information used by manufacturing and finance/accounting should not be acceptable.

Traditionally, however, each function within an organization had its own way of doing things, with unique databases. Furthermore, communication among the various functional areas has not always been perfect. However, such separation of the activities across functions is artificial. In any business, all activities are interrelated and constitute the whole rather than a collection of different functions. Therefore, the next logical step was to combine the manufacturing activities with those of finance, marketing, purchasing, and engineering through a common database. This recognition led to the evolution of MRP to what is called Manufacturing Resource Planning or MRP II (127).

It is easy to realize that since there is one physical system in operation in a company, there is no justification for having more than one information system representing different dimensions of this physical system. The information system should also be unique and reflect the actual physical system. Thus, MRP systems evolved into MRP II when a common database became available for use by all functions, and any change or update by one functional area would immediately become visible to the rest of the organization.

In addition to integrating the various functional areas within the business, MRP II systems also provide a "what if" capability. It can be used to simulate what would happen if various decisions were implemented, without changing the actual database. This makes it possible to see, for example, the impact on capacity and material requirements of changing the schedule and the impact on customer responsiveness of product design/engineering changes leading to BOM changes.

CONCLUSION

A vast majority of small and large manufacturing companies, around the world, have made significant and substantial investment in MRP II systems and, hence, continue to use MRP II-based systems for manufacturing planning and control (128). A recent survey of U.S. companies covering a wide spectrum of manufacturing industries (ranging from machine tools, automobile components, furniture, plastics, and medical equipment to computers and defense electronics) shows that MRP is the most widely used system (56% of the firms reported using an MRP system) for manufacturing planning and control (129). Furthermore, the American Production and Inventory Control Society (APICS) has listed "improved MRP" systems as one of the top 10 topics of concern to their 80,000 plus members in 1995. Just one MRP system software, MAPICS, has an installed base of an estimated 13000 sites worldwide (130). Recent evidence indicates that there are more than 100 MRP II software products available in the market (131). The dominance of MRP-II systems is further substantiated by a recently completed survey conducted by Advanced Manufacturing Research, Inc. The results suggest that the size of the market for MRP-based production planning and control software in 1993 alone has been over US $2 billion. Thus, it is clear that MRP systems not only continue to dominate the manufacturing planning and control (MPC) in practice but may continue to do so for several years to come (132,133).

BIBLIOGRAPHY

1. S. H. Bowman, Production scheduling by the transportation method of linear programming, *Oper. Res.,* **4**: 100–103, 1956.

2. F. Hansmann and S. W. Hess, A linear programming approach to production and employment scheduling, *Manage. Technol. Monogr.,* **1**: 46–51, 1960.

3. S. Eilon, Five approaches to aggregate production planning, *AIIE Trans.,* **7**: 118–131, 1973.

4. C. Chung and L. J. Krajewski, Planning horizons for master production scheduling, *J. Oper. Manage.,* 389–406, 1984.

5. U. A. Goodman, A goal programming approach to aggregate planning of production and work force, *Manage. Sci.,* **20**: 1569–1575, 1974.

6. C. C. Holt, F. Modigliani, and H. A. Simon, A linear decision rule for production and employment scheduling, *Manage. Sci.,* **2**: 1–30, 1953.

7. C. H. Jones, Parametric production planning, *Manage. Sci.,* **13**: 843–866, 1967.

8. W. H. Taubert, A search decision rule for the aggregate scheduling problem, *Manage. Sci.,* **14**: 1343–1359, 1968.

9. B. H. Bowman, Consistency and optimality in managerial decision making, *Manage. Sci.,* **9**: 310–321, 1963.

10. A. C. Hax and H. C. Meal, Hierarchical integration of production planning and scheduling, In M. A. Geisler (ed.), *Logistics,* New York: North-Holland/American Elsevier, 53–69, 1975.

11. G. R. Bitran, E. A. Haas, and A. C. Hax, Hierarchical production planning: A single stage system, *Oper. Res.* **29**: 717–743, 1981.

12. G. L. Bitran, E. A. Haas, and A. C. Hax, Hierarchical production planning: A two stage system, *Oper. Res.,* **30**: 232–251, 1982.

13. G. L. Bitran and A. C. Hax, On the design of hierarchical production planning systems, *Decis. Sci.,* **8**: 28–55, 1977.

14. G. R. Bitran and A. C. Hax, Disaggregation and resource allocation using convex knapsack problems with bounded variables, *Manage. Sci.,* **27**: 431–441, 1981.

15. M. D. Oliff and G. K. Leong, A discrete production switching rule for aggregate planning, *Decis. Sci.,* **18**: 382–397, 1987.

16. G. K. Leong, M. D. Oliff, and R. E. Markland, Improved hierarchical production planning, *J. Oper. Manage.,* **8**: 90–114, 1989.

17. M. R. Bowers and J. P. Jarvis, A hierarchical production planning and scheduling model, *Decis. Sci.,* **23** (1): 144–159, 1992.

18. K. L. Campbell, Scheduling is not the problem, *J. Prod. Inventory Manage.,* **12** (3): 53–77, 1971.

19. H. Mather, Reschedule the schedules you just scheduled—Way of life or MRP? *J. Prod. Inventory Manage.,* **18** (1): 60–79, 1977.

20. D. C. Steele, The nervous MRP system: How to do battle. *J. Prod. Inventory Manage.,* **16**: 83–89, 1971.

21. D. C. Whybark and J. G. Williams, Material requirements planning under uncertainty, *Decis. Sci.,* **7**: 595–606, 1976.

22. R. T. Barrett and R. L. LaForge, A study of replanning frequencies in a material requirements planning system, *Comput. Oper. Res.,* **18** (6): 569–578, 1991.

23. C. A. Yano and R. C. Carlson, An analysis of scheduling policies in multiechelon production systems, *IIE Trans.,* **17** (4): 370–377, 1985.

24. C. A. Yano and R. C. Carlson, Interaction between frequency of scheduling and the role of safety stock in material requirements planning systems, *Int. J. Prod. Res.,* **25** (2): 221–232, 1987.

25. R. Venkataraman, Frequency of replanning in a rolling horizon master production schedule for a process industry environment, *Proc. Annu. Decis. Sci. Conf.,* **3**: 1547–1549, 1994.

26. C. Chung and L. J. Krajewski, Replanning frequencies for master production schedules, *Decis. Sci.,* **17** (2): 263–273, 1986.

27. H. M. Wagner and T. Whitin, Dynamic version of the economic lot-size model, *Manage. Sci.,* **5**: 89–96, 1958.

28. F. R. Jacobs and B. M. Khumawala, A simplified procedure on optimal single-level lot-sizing, *J. Prod. Inventory Manage.,* **28** (3): 39–42, 1987.

29. J. J. DeMatteis, An economic lot-sizing technique I—The part period algorithm. *IBM Syst. J.,* **7** (1): 30–38, 1968.

30. N. Gaither, A near optimal lot-sizing model for material requirements planning systems, *J. Prod. Inventory Manage.,* (4): 75–89, 1981.

31. G. K. Groff, A lot-sizing rule for time-passed component demand, *J. Prod. Inventory Manage.,* (1): 47–53, 1979.

32. R. Karni, Maximum part-period gain (MPG)—A lot-sizing procedure for unconstrained and constrained requirements planning systems, *J. Prod. Inventory Manage.,* (2): 91–98, 1981.

33. E. A. Silver and H. C. Meal, A heuristic for selecting lot size quantities for the case of a deterministic, time-varying demand rate and discrete opportunities for replenishment, *J. Prod. Inventory Manage.,* (2): 64–74, 1973.

34. R. L. LaForge, A decision rule for creating planned orders in MRP, *J. Prod. Inventory Manage.,* **26** (4): 115–125, 1985.

35. W. L. Berry, Lot-sizing procedures for requirements planning systems: A framework for analysis, *J. Prod. Inventory Manage.,* **13** (2): 19–34, 1972.

36. J. D. Blackburn and R. A. Millen, Selecting a lot-sizing technique for a single-level assembly process, Part I: Analytical results, *J. Prod. Inventory Manage.,* (3): 42–47, 1979.

37. J. D. Blackburn and R. A. Millen, Selecting a lot-sizing technique for a single-level assembly process, Part II: Empirical results, *J. Prod. Inventory Manage.,* (4): 41–52, 1979.

38. J. D. Blackburn and R. A. Millen, Heuristic lot-sizing performance in a rolling schedule environment, *Decis. Sci.,* **11** (4): 691–701, 1980.

39. K. R. Baker, An experimental study of the effectiveness of rolling schedules in production planning, *Decis. Sci.,* **8** (1): 19–27, 1977.

40. M. A. DeBodt and L. N. Wassenhove, Cost increases due to demand uncertainty in MRP lot-sizing, *Decis. Sci.,* **14** (3): 345–361, 1983.

41. U. Wemmerlov and D. C. Whybark, Lot-sizing under uncertainty in a rolling schedule environment, *Int. J. Prod. Res.,* **22** (3): 467–484, 1984.

42. U. Wemmerlov, The behavior of lot-sizing procedures in the presence of forecast errors, *J. Oper. Manage.,* **8** (1): 37–47, 1989.

43. L. E. Yelle, Lot-sizing for the MRP multi-level problem, *Ind. Manage.,* **July–August**: 4–7, 1978.

44. L. E. Yelle, Materials requirements lot-sizing: A multi-level approach, *Int. J. Prod. Res.,* **17** (3): 223–232, 1979.

44a. E. A. Veral and R. L. LaForge, Incremental lot-sizing rule in a multi-level inventory environment, *Decis. Sciences,* **16** (1): 57–72, 1985.

44b. R. L. LaForge, J. W. Patterson, and E. A. Veral, A comparison of two part-period balancing rules, *J. of Purchasing and Materials Management,* **22** (2): 30–36, 1986.

45. F. R. Jacobs and B. M. Khumawala, Multi-level lot-sizing in material requirements planning: An empirical investigation, *Comput. Oper. Res.,* **9** (2): 139–144, 1982.

46. J. R. Biggs, Heuristic lot-sizing and sequencing rules in a multistage production inventory system, *Decis. Sci.,* **10**: 96–115, 1979.

47. J. R. Biggs, S. M. Goodman, and S. T. Hardy, Lot-sizing rules in hierarchical multistage inventory systems, *J. Prod. Inventory Manage.,* (1): 104–115, 1979.

48. H. Choi, E. M. Malstrom, and R. J. Classen, Computer simulation of lot-sizing algorithms in three stage multi-echelon inventory systems, *J. Oper. Manage.,* **4** (3): 259–277, 1984.

49. D. Collier, The interaction of single-stage lot-size models in a material requirements planning system, *J. Prod. Inventory Manage.,* (4): 11–17, 1980.

50. W. I. Zangwill, A backlogging model and a multi-echelon model of a dynamic economic lot-size production system—A network approach, *Manage. Sci.,* **15** (9): 506–527, 1969.

51. S. F. Love, A facilities-in-series inventory model with nested schedules, *Manage. Sci.,* **18** (5): 327–338, 1972.

52. W. B. Crowston, M. H. Wagner, and J. F. Williams, Economic lot-size determination in multi-stage assembly systems, *Manage. Sci.,* **19** (5): 517–527, 1973.

53. E. Steinberg and H. A. Napier, Optimal multi-level lot-sizing for requirements planning systems, *Manage. Sci.,* **26** (12): 1258–1271, 1980.

54. P. Afentakis, B. Gavish, and U. Karmakar, Computationally efficient optimal solutions to the lot-sizing problem in multi-level assembly systems, *Manage. Sci.,* **30** (2): 222–239, 1984.

55. E. L. Prentis and B. M. Khumawala, MRP lot-sizing with variable production/purchasing costs: Formulation and solution, *Int. J. Prod. Res.,* **27** (6): 965–984, 1989.

56. M. A. Mcknew, C. Saydam, and B. J. Coleman, An efficient zero–one formulation of the multilevel lot-sizing problem, *Decis. Sci.,* **22** (2), 280–295, 1991.

57. B. J. McLaren, A study of multiple-level lot-sizing procedures for materials requirements planning system, Ph.D. dissertation, Purdue University, 1977.

58. J. D. Blackburn and R. A. Millen, Improved heuristics for multi-echelon requirements planning systems, *Manage. Sci.,* **28** (1): 44–56, 1982.

59. J. D. Blackburn and R. A. Millen, The impact of a rolling schedule in a multi-level MRP system, *J. Oper. Manage.,* **2** (2): 125–135, 1982.

60. Q. Rehmani and E. Steinberg, Simple single pass multi-level lot-sizing heuristics, *Proc. Am. Inst. Decision Sci.,* **2**: 124–126, 1982.

61. S. C. Graves, Multi-stage lot-sizing: An iterative procedure, *TIMS Stud. Manage. Sci.,* **16**: 95–110, 1981.

62. P. Afentakis, A parallel heuristic algorithm for lot-sizing in multistage production systems, *IEEE Trans.,* (3): 34–41, 1987.

63. E. L. Prentis and B. M. Khumawala, Efficient heuristics for MRP lot-sizing with variable production/purchasing costs, *Decis. Sci.,* **20**: 439–450, 1989.

64. C. Canel, B. M. Khumawala, and J. S. Law, An efficient heuristic procedure for the single-item, discrete lot-sizing problem, *Int. J. Prod. Econ.,* **43** (3): 139–148, 1996.

65. O. Kirca and M. Kokten, A new heuristic approach for the multi-item dynamic lot-sizing problem, *Eur. J. Oper. Res.,* **75** (2): 332–341, 1994.

66. R. Sikora, A generic algorithm for integrating lot-sizing and sequencing in scheduling a capacitated flow line, *Comput. Ind. Eng.,* **30** (4): 969–981, 1996.

67. K. S. Hindi, Solving the CLSP by a tabu search heuristic, *J. Oper. Res. Soc.,* **47** (1): 151–161, 1996.

68. K. S. Hindi, Solving the single-item, capacitated dynamic lot-sizing problem with startup and reservation costs by tabu search, *Comput. Ind. Eng.,* **28** (40): 701–707, 1995.

68a. E. A. Veral, Integrating the Master Production Scheduling and Bills of Material Explosion Functions in Materials Requirements Planning Systems, *Unpublished Doctoral Dissertation,* Clemson University, 1986.

68b. E. A. Veral and R. L. LaForge, The integration of cost and capacity considerations in material requirements planning systems, *Decis. Sciences,* **21** (3): 507–520, 1990.

69. J. Haddock and D. E. Hubicki, Which lot-sizing techniques are used in material requirements planning? *J. Prod. Oper. Manage.,* **30** (3): 53–56, 1989.

70. U. Wemmerlov, Design factors in MRP systems: A limited survey, *J. Prod. Invent. Manage.,* **20** (4): 15–35, 1979.

71. H. C. Bahl, L. P. Ritzman, and J. N. D. Gupta, Determining lot size and resource requirements: A review, *Int. J. Prod. Res.,* **22** (5): 791–800, 1984.

72. E. Ritchie and A. K. Tsado, A review of lot-sizing techniques for deterministic time-varying demand, *J. Prod. Invent. Manage.,* **27** (3): 65–79, 1986.

73. R. Eftekharzadeh, A comprehensive review of production lot-sizing, *Int. J. Phys. Distribution Logist. Manage.,* **23** (1), 30–44, 1993.

74. C-H Chu and J. C. Hayya, Buffering Decisions Under MRP Environment: A Review. *Omega,* **16** (4): 325–331, 1988.

75. M. K. McClelland and H. M. Wagner, Location of inventories in an MRP environment, *Decis. Sci.,* **19**: 535–553.

76. V. Sridharan and R. L. LaForge, The impact of safety stock on schedule instability, cost and service, *J. Oper. Manage.,* **8** (4): 327–347, 1989.

77. V. Sridharan and R. L. LaForge, On Using Buffer Stock to Combat Schedule Instability. *Int. J. Oper. Prod. Manage.,* **20** (7): 37–46, 1990.

78. J. A. Buzacott and J. G. Shanthikumar, Safety stock versus safety time in MRP controlled production systems, *Manage. Sci.,* **40** (12): 1678–1689, 1994.

79. J. H. Blackstone, *Capacity Management,* Cincinnati OH: South-Western Publishing Co., 1989.

80. T. E. Vollmann, W. L. Berry, and D. C. Whybark, *Manufacturing Planning and Control System,* Dow Jones Irwin, 1997.

81. R. W. Conway, W. L. Maxwell, and L. W. Miller, *Theory of Scheduling,* Reading, MA: Addison-Wesley, 1967.

82. K. R. Baker, *Introduction to Sequencing and Scheduling,* New York: Wiley, 1974.

83. A. Kiran and M. Smith, Simulation Studies in Job Shop Scheduling—I, *Comput. Ind. Eng.,* 8 (2): 87–93, 1984.

84. A. Kiran and M. Smith, Simulation studies in job shop scheduling—II, *Comput. Ind. Eng.,* 8 (2): 95–105, 1984.

85. J. H. Blackstone, D. T. Phillips, and G. L. Hogg, A state-of-the-art survey of dispatching rules for manufacturing job shop operations, *Int. J. Prod. Res.,* **20** (1): 27–45, 1982.

86. J. R. King and A. S. Spachis, Heuristics for flow-shop scheduling, *Int. J. Prod. Res.,* **18** (3): 345–357, 1980.

87. S. C. Graves, A review of production scheduling, *Oper. Res.,* **29** (4): 646–675, 1981.

88. T. E. Morton and D. W. Pentico, *Heuristic Scheduling Systems,* New York: Wiley, 1993.

89. V. Sridharan and Z. Zhou, Dynamic non-preemptive single machine scheduling, *Comput. Oper. Res.,* **23** (12): 1183–1190, 1996.

90. V. Sridharan and Z. Zhou, A decision theory based scheduling procedure for single machine weighted earliness and tardiness problems, *Eur. J. Oper. Res.,* **94**: 292–301, 1996.

91. J. J. Kanet and V. Sridharan, Scheduling with inserted idle time: Problem taxonomy and literature review, Working Paper, Clemson University, 1997.

92. G. W. Plossl and O. W. Wight, Capacity planning and control, *Prod. Invent. Manage.,* **14** (3): 31–67, 1973.

93. W. Bechte, Controlling manufacturing lead time and work-in-process inventory by means of load-oriented order release, In *APICS Conference Proceedings,* 1982, pp. 67–72.

94. W. Bechte, Theory and practice of load oriented manufacturing control, *Int. J. Prod. Res.,* **26** (3): 375–395, 1988.

95. J. W. M. Bertrand, The use of workoad information to control job lateness in controlled and uncontrolled release production systems, *J. Oper. Manage.,* **3** (2): 79–92, 1983.

96. S. K. Shimoyashiro, K. Isoda, and H. Awane, Input scheduling and load balance control for a job shop, *Int. J. Prod. Res.,* **22** (4): 597–605, 1984.

97. J. J. Kanet, Load-limited order releases in job shop scheduling systems, *J. Oper. Manage.,* **7** (3). 44–58, 1988.

98. G. L. Ragatz and V. A. Mabert, An evaluation of order release mechanisms in a job shop environment, *Decis. Sci.,* **19** (1): 167–189, 1988.

99. P. M. Bobrowski and P. S. Park, Work release strategies in a dual resource constrained job shop, *Omega,* **17** (2): 177–188, 1989.

100. J. C. Irastorza and R. H. Deane, A loading and balancing methodology for job shop control, *IEEE Trans.,* **6** (4): 302–307, 1974.

101. J. C. Irastorza and R. H. Deane, Starve the shop—reduce work-in-process, *Prod. Invent. Manage.,* 105–123, 1989.

102. S. A. Melnyk and G. L. Ragatz, Order review/release research issues and perspectives, *Int. J. Prod. Res.,* **27** (7): 1081–1096, 1989.

103. J. C. Anderson et al., Material requirements planning: The state of the art, *Prod. Invent. Manage.,* **23** (4): 51–66, 1982.

104. J. Orlicky, *Material Requirements Planning,* New York: McGraw-Hill, 1975.

105. J. P. Monniot et al., A study of computer aided production management in UK batch manufacturing, *Int. J. Oper. Prod. Manage.,* **7** (2), 7–30, 1987.

106. P. Duchessi, C. M. Schaninger, and D. R. Hobbs, Implementing a manufacturing planning and control information system, *Calif. Manage. Rev.,* **31** (3), 1988.

107. A. Kochhar and B. McGarrie, Identification of the requirements of manufacturing control systems: A key characteristics approach, *Integrated Manuf. Syst.,* **3** (4), 1992.

108. V. Sridharan and J. J. Kanet, Planning and control system: State of the art and new directions, In Adelsberger, Marik, and Lazansky (eds.), *Information Management in Computer Integrated Manufacturing,* Berlin: Springer-Verlag, 1995.

109. J. J. Kanet and V. Sridharan, The value of using scheduling information for planning material requirements, Decis. Sci., (forthcoming, 1998).

110. R. H. Hayes and K. B. Clark, Explaining observed productivity differentials between plants: Implications for operations research, *Interfaces,* 15 (6): 3–14, 1985.

111. S. N. Kadipasaoglu and V. Sridharan, Measurement of instability in multi-level MRP systems, *Int. J. Prod. Res.,* 35 (3): 713–737, 1996.

112. V. Sridharan, W. L. Berry, and V. Udayabanu, Freezing the master production schedule under rolling planning horizons, *Manage. Sci.,* 33 (9): 1137–1149, 1987.

113. V. Sridharan, W. L. Berry, and V. Udayabanu, Measuring master production schedule stability under rolling planning horizons, *Decis. Sci.,* 19 (1): 147–166, 1988.

114. V. Sridharan and W. L. Berry, Freezing the master production schedule under demand uncertainty, *Decis. Sci.,* 21 (1): 97–121, 1990.

115. V. Sridharan and W. L. Berry, Master production scheduling, make-to-stock products: A framework for analysis, *Int. J. Prod. Res.,* 28 (3): 541–558, 1990.

116. V. Sridharan and R. L. LaForge, An analysis of alternative policies to achieve schedule stability, *J. Manuf. Oper. Manage.,* 3 (1): 53–73, 1990.

117. N. P. Lin and L. Krajewski, A model for master production scheduling in uncertain environments, *Decis. Sci.,* 23 (4): 839–861, 1992.

118. X. Zhao and T. S. Lee, Freezing the master production schedule for material requirements planning systems under demand uncertainty, *J. Oper. Manage.,* 11 (2): 185–205, 1993.

119. S. N. Kadipasaoglu and V. Sridharan, Alternative approaches for reducing schedule instability in multistage manufacturing under demand uncertainty, *J. Oper. Manage.,* 13: 193–211, 1995.

120. V. Sridharan and L. R. LaForge, Freezing the master production schedule: Implications for fill rate, *Decis. Sci.,* 25 (3): 461–469, 1994.

121. D. Blackburn, D. H. Kropp, and R. A. Millen, A comparison of strategies to dampen nervousness in MRP systems, *Manage. Sci.,* 33 (4): 413–429, 1986.

122. R. C. Carlson, S. L. Beckman, and D. H. Kropp, The effectiveness of extending the horizon in rolling production scheduling, *Decis. Sci.,* 13 (1): 129–146, 1982.

123. D. H. Kropp and R. C. Carlson, A lot-sizing algorithm for reducing nervousness in MRP systems, *Manage. Sci.,* 30 (2): 240–244, 1984.

124. D. H. Kropp, R. C. Carlson, and J. V. Jucker, Use of dynamic lot-sizing to avoid nervousness in material requirements planning systems, *J. Prod. Invent. Manage.,* 20 (3): 40–58, 1979.

125. D. H. Kropp, R. C. Carlson, and J. V. Jucker, Heuristic log-sizing approaches for dealing with MRP system nervousness, *Decis. Sci.,* 14 (2): 156–169, 1983.

126. S. N. Kadipasaoglu, V. Sridharan, and L. R. LaForge, An investigation and comparison of safety stock and freezing policies in master production scheduling, Working Paper, University of Houston, 1997.

127. O. W. Wight, *Manufacturing Resource Planning: MRP II Unlocking America's Productivity Potential,* Essex Junction, VT: Oliver Wight Limited Publications, 1984.

128. H. Kumar and R. Rachamadugu, Is MRP dead? *APICS Performance Advantage,* 5 (9), 24–27, 1995.

129. E. Newman and V. Sridharan, Manufacturing planning and control: Is there one definitive answer? *J. Prod. Invent. Manage.,* 33 (1), 1992.

130. D. Turbide, Alive and kicking. *APICS Performance Advantage,* 5 (6): 52–54, 1995.

131. MRP-II software buyer's guide, *IIE Solutions,* 36–41, July 1995.

132. W. Chamberlain and G. Thomas, The future of MRP-II: Headed for the scrap heap or rising from the ashes, *IIE Solutions,* 27: 32–35, July 1995.

133. D. Turbide, MRP-II still number one! *IIE Solutions,* 28–31, July 1995.

SUKRAN N. KADIPASAOGLU
University of Houston

V. SRIDHARAN
Clemson University

MANUFACTURING, SCHEDULING OF SEMICONDUCTOR. See SEMICONDUCTOR MANUFACTURING SCHEDULING.

MANUFACTURING SEMICONDUCTORS. See FUZZY LOGIC FOR SEMICONDUCTOR MANUFACTURING.

MANUFACTURING SYSTEMS, AUTOMATIC. See AUTOMATION.

MANUFACTURING TECHNOLOGY. See COMPUTER INTEGRATED MANUFACTURING.

MANUFACTURING WITH LASERS. See LASER DESKTOP MACHINING.

MARCONI. See ANTENNAS.

MARINE SYSTEMS. See UNDERWATER VEHICLES.

MARKETING AND SALES MANAGEMENT. See SALES AND MARKETING MANAGEMENT.

MARKETING HIGH TECHNOLOGY. See HIGH TECHNOLOGY MARKETING.

MARK SCANNING EQUIPMENT

As computer technology rapidly propelled us into the information age, it has become evident that the bottleneck of information exchange is the interface between the human operators and their computers. Barcodes and two-dimensional codes are technologies that help to ease this bottleneck.

Barcodes are the zebra-striped patterns that one sees on product packaging in retail environments. Far beyond retail points-of-sale, barcodes are also widely used in many industrial applications including manufacturing process control, inventory control, transportation, identification, and blood banks. Two-dimensional codes (2-D codes, sometimes referred to as 2-D barcodes) are extensions of barcodes, carrying more information in the same printed area. Although most people have seen barcodes, only a select few may know the intricacies of barcodes and barcode scanning.

There are many ways through which a computer can output information for human consumption: Display it on a screen, print it on paper, or even synthesize it into voice. All of these methods are simple, accurate, and relatively fast. To input data into a computer is a different matter. Often a keyboard is used, but it is slow and inaccurate. Optical character recognition (OCR) (see OPTICAL CHARACTER RECOGNITION—OCR) for print and handwriting is becoming more sophisticated, but is still not accurate enough for most business applications. It has been realized that in many situations the information to be input into the computer is printed, and the same printing process can produce information in two ways at the same time, one for machine reading and one for human reading. Barcodes and barcode scanners are simply the marks for machine reading and the readers that read these marks.

A barcode records a short string of text, and usually it is used as the index value that represents an item in a database. Figure 1 illustrates a system employing barcode scanners as input devices. Processing a barcoded item involves a barcode scanner decoding the barcode and transmitting the result to a terminal, which requests information with this index value from a main computer hosting the database, which in turn looks up the requested information and transmits the result back to the terminal. The whole process usually takes a small fraction of a second and appears to be instantaneous.

In the following sections, after a brief historical overview, we will discuss first barcodes, followed by barcode scanners, 2-D codes, and, finally, 2-D code capable scanners.

HISTORICAL OVERVIEW

The first US patent related to barcodes was issued in 1949 to J. Woodland and B. Silver (1). Interestingly, the patent did not cover a *bar* code, but rather a *ring* code [see Fig. 2(a)]. Reportedly, Woodland and Silver first thought of using bars of different thickness to record information as well, but then decided to make the code isotropic. By bending the bars into concentric circles, the ring code looks the same from all directions, and the scanner does not have to be lined up with the code to be scanned. While not of much commercial significance, the work of Woodland and Silver demonstrated that one of the most important design criteria for a machine-readable code is the ease with which the code can be scanned.

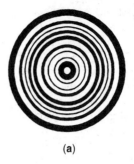

(a)

(b)

(c)

(d)

Figure 2. Examples of ring code and barcodes: a ring code illustration (no ring code standard exists today) (a), a UPC (b), a Code 39 (c) and a Code 128 (d) barcode. In (b), the UPC code is composed of two separately decodable parts, with each part being taller than wide. The start and stop patterns, as well as the center guard bars (the shared bars between the two parts), are usually extended as shown. All these measures help to ensure that a scan line misaligned at up to 45° can still cross all bars in each part completely.

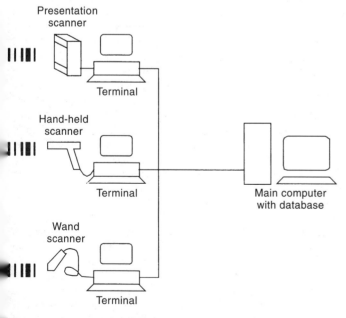

Figure 1. A system employing barcode scanners as one of its input means. Terminals, using index values encoded in barcodes, request information from the main computer hosting the database. Different types of barcode scanners can be mixed in the same application.

Indeed, the barcode scanning equipment which best demonstrates the possible productivity gain brought about by this technology could not have been invented in the time of the Woodland and Silver patent. As we shall discuss below, only two categories of scanners (i.e., laser scanners and imaging scanners) can generate many scan lines per second and thus demonstrate the highest possible throughput that barcode technology can offer. While lasers were invented in the early 1960s, the most popular electronic imager, the charge coupled device (CCD), was invented in the early 1970s. It is little wonder that barcode applications did not emerge until the late 1960s.

The earliest successful large-scale implementation of barcodes is probably the Universal Product Code [UPC, shown in Fig. 2(b)] (2), the type of barcode currently used in supermarkets in the United States. A superset of UPC is the European Article Numbering code (EAN), which, despite the origin of its name, is now a standard adopted worldwide. Outside of supermarkets, different types of barcodes are also used by transportation, warehousing, healthcare, and other industrial sectors. The most widely used types, other than UPC/EAN, are Code 39 and Code 128 [Fig. 2(c) and 2(d)]. Each type of barcode is formed under different rules, which define different *symbologies*. A symbology is standardized in at least two ways: a standard-setting organization accepts and maintains its specification, while an industry association coordinates which symbology to use and precisely how it should be used in the particular applications pertaining to that industry. A list of organizations that participate in barcode standardization is included in Appendix A of Ref. 1.

Today barcodes are used in more and more applications in industrial, governmental, and educational institutions. By using barcodes, repetitive labor is reduced, and so is error rate, while productivity is enhanced and more data become available for real-time tracking, analysis, and control. Barcode printing and reading represents a multibillion dollar industry. The use of 2-D codes opens up even more new opportunities to applications where the input of data into computers can be automated.

BARCODES

In this section, we will first explore how information is recorded in barcodes, and then we will cover the mathematical methods utilized in barcode design.

Barcode Fundamentals

Barcodes carry their information in the relative widths of their elements (bars and spaces). Most symbologies use both the bars and spaces between the bars to record information. A barcode scanner takes samples along a line (the *scan line*) crossing all elements, measuring their widths, and decodes from the measurement the recorded message. Since no information is carried in the exact value of the element widths, a barcode can be printed at any magnification (within the capability of the printer and the intended scanner) and read at any distance.

UPC/EAN barcodes record fixed-length numerical values only. Other types of barcodes may record both numbers and letters, or even the complete ASCII code, and may also record variable-length messages. For example, Code 39, one of the most widely used symbologies, can encode all digits, all uppercase English letters, and a few punctuation marks, and it can encode variable-length messages.

Compared to ring codes, barcodes are not isotropic, but they are shift-invariant, in the sense that two parallel scanlines crossing it completely are certain to get the same information. Furthermore, barcodes are angle-insensitive. A slanted scan through a barcode yields a longer sequence of data, and possibly an electrical signal that is less well-defined, but all transitions recorded in the barcode are reflected

in precise proportion. The maximum scanning angle is simply given by

$$|\theta| \leq \tan^{-1} H/W \tag{1}$$

where H and W are the width and height of the complete barcode, respectively. In many applications, this angular misalignment allowance is sufficient.

In some applications it is desirable that a barcode be read regardless of its orientation. Although a *barcode* is not *isotropic*, a *barcode scanning system* can be *omnidirectional*. To construct an omnidirectional barcode scanning system using line-scanning technology, a star pattern of N or more evenly distributed scan lines can be used, where

$$N = \left\lceil \frac{\pi}{2 \cdot \tan^{-1} H/W} \right\rceil \tag{2}$$

For example, an original omnidirectional design of UPC barcode system includes a UPC barcode as presently used, along with a laser-scanner with two scan lines perpendicular to each other. The UPC barcode can be decoded in parts (the most common UPC code has two parts that share a few endbars, as shown in Fig. 2(b); and for each part, $H > W$ (referred to as *over-square*). Thus for $N = 2$, it is guaranteed that one of the scan lines can pass through each part of the barcode completely, in a direction that crosses all the bars [Fig. 3(a)].

Barcode Design Considerations

The barcode being scanned is not ideal, nor is the scanning process. The main factors affecting the performance of a barcode scanner are *signal distortion* and *noise*, both of which we will elaborate on later. To facilitate accurate decoding, barcode symbology designs employ mathematical tools to make the symbology less sensitive to signal distortion and to noise.

Generally, a character (from the alphabet for the particular symbology) is recorded as a fixed number of bars and spaces, and its recorded version is called a *codeword*. Usually in a given barcode symbology a codeword also has a fixed total width. For example, in UPC/EAN every digit is represented by two bars and two spaces taking up the total width of seven times the width of the narrowest bar (the X *dimension*, or simply the X). When scanned, the barcode so designed shows a marked periodicity in the electronic signal: Every codeword is represented by one cycle, consisting the same number of peaks and valleys and taking up the same amount of time. This periodicity is designed for ease of decoding. Once a periodicity is detected in a scan line, the decoder can identify which symbology is used. Different codewords are also separated by the periodicity. Symbologies that exhibit this periodicity are said to be *self-clocking*.

Common barcode symbologies are categorized according to the number of element widths allowed. Some allow only two widths for the bars and spaces, thus they are commonly referred to as *binary* barcode symbologies (or binary codes). In these barcodes, a wide element is typically more than twice as wide as a narrow element, so they can remain distinct even with some distortion and noise. Other barcode symbologies encode the information with more allowable width values (usually more than 3) and are sometimes referred to as *delta* barcode symbologies (or delta codes); the origin of the word

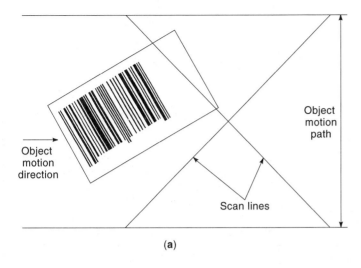

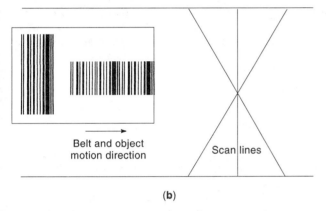

Figure 3. Two omnidirectional barcode scanning systems: (a) A UPC scanner with two crossed scan lines, each at 45° from the object motion direction and both extending across the complete motion path, and an over-square UPC barcode. (b) An overhead scanner with three scan lines (at 30° from each other) and a package bearing two identical and perpendicular barcodes.

delta is that the different widths are usually integer multiples of X. Table 1 lists the characteristics of several common barcodes, including binary codes and delta codes. The statistics shown in the table are calculated with the formula given below.

The maximum number of distinct codewords of a symbology and the recorded information density are discussed in Ref. 3. A binary code with e elements (bars and spaces) per

codeword, out of which w are wide, can be referred to as a w/e code. The maximum number of distinct codewords is given by

$$S_{\mathrm{B}}(e, w) = \binom{e}{w} \tag{3}$$

and the related information density is

$$H_{\mathrm{B}}(e, w) = \frac{1}{[e + (r - 1)w]} \log_2 S_{\mathrm{B}}(e, w) \quad \text{bits}/X \tag{4}$$

where r is the ratio of the wide elements' width to that of the narrow elements. In many cases a smaller number of codewords is used, and the achieved information density is less than calculated here.

A delta code is called an (n, k) code if each codeword is nX wide and contains k pairs of bars and spaces. For example, UPC/EAN is a $(7, 2)$ code. The maximum number of distinct codewords in an (n, k) code is given by

$$S_{\mathrm{D}}(n, k) = \binom{n - 1}{2k - 1} \tag{5}$$

and the related information density, is

$$H_{\mathrm{D}}(n, k) = \frac{1}{n} \log_2 S_{\mathrm{D}}(n, k) \quad \text{bits}/X \tag{6}$$

For any n, maximum H_{D} is achieved by *symmetric* codes, which obey

$$n = 4k - 1$$

and

$$H_{\mathrm{D}}\left(n, \frac{n + 1}{4}\right) \approx 1 - \frac{\log_2[2\pi(n - 1)]}{2n} \quad \text{bits}/X \tag{7}$$

Thus it can be seen that the larger the value n is, the larger the maximum H_{D} becomes. The trade-off is that the larger n value means a longer self-clocking period and therefore higher susceptibility to scan speed variations.

Sometimes the codewords with very wide elements are eliminated from an (n, k) code. An (n, k, m) code is the subset of an (n, k) code where no codeword has an element wider

Table 1. Characteristics of Several Symbologies Which Have Matching Ideal Models

Name	Code Type	S	H	Comment
Interleaved 2 of 5[a]	2/5	10	0.415	Codewords are interleaved: each codeword carrying information in bars is interleaved with another carrying information in spaces.
Code 39[a]	3/9	84	0.474	Intercodeword gap does not carry information.
UPC/EAN	(7, 2)	20	0.617	Only 10 distinct codewords are used in UPC-A, the most popular subtype of UPC.
Code 93	(9, 3)	56	0.645	Forty-six distinct codewords are used.
Code 128	(11, 3, 4)	216	0.705	One hundred two distinct codewords are used.
PDF417[b]	(17, 4, 6)	10480	0.786	Three clusters are used, each containing 929 codewords.

[a] For these binary codes, $r = 2.5$ is assumed.
[b] PDF417 is a two-dimensional code with a regular barcode codeword structure. We will cover two-dimensional codes later in this article.

than mX. The size of the alphabet of an (n, k, m) code is

$$S_D(n, k, m) = \binom{n-1}{2k-1} - 2k \sum_{u=m+1}^{n-2k+1} \binom{n-u-1}{2k-2} \quad (8)$$

And, as can be seen, when

$$m \approx n - 2k$$

the number of distinct codewords reduction is not significant.

The information density calculation used here does not take into consideration the non-information-carrying parts of a barcode, including *the start and stop patterns,* the special codewords or patterns used at the two ends of a barcode, and the *quite zones,* which are the required white space bordering the barcode. In addition, for ease of printing by certain specialized printing equipment, some symbologies do not use the width of the spaces (or bars) to record information. Some others leave a space between adjacent codewords, so this space can be printed with looser tolerance (these symbologies are called *discrete* symbologies). All these variations reduce the information density as calculated above.

Many delta codes can be decoded using the *t-sequence.* This is a feature designed to counter ink-spread, a phenomenon where all bar widths grow (or shrink) in such a way that each bar edge is shifted by the same distance, dx. Given a sequence of element widths (the *x-sequence*) from a scan line,

$$x_1, x_2, x_3, x_4, \ldots$$

the *t-sequence* is defined as

$$t_1 = x_1 + x_2, \, t_2 = x_2 + x_3, \, t_3 = x_3 + x_4, \ldots$$

When ink-spread is introduced, the *x-sequence* becomes

$$x_1' = x_1 + 2dx, \qquad x_2' = x_2 - 2dx, \qquad x_3' = x_3 + 2dx,$$
$$x_4' = x_4 - 2dx, \ldots$$

where dx is the amount of ink-spread per bar-space edge, but the *t-sequence* remains unchanged:

$$t_1' = x_1 + x_2 = t_1, \qquad t_2' = x_2 + x_3 = t_2, \qquad t_3' = x_3 + x_4 = t_3, \ldots$$

Thus *t-sequence* decoding is not affected by ink-spread.

Some symbologies are *self-checking.* This is the feature where if one width measurement or one edge location measurement is incorrect, the codeword under consideration becomes invalid. When that happens, a potential *misdecode* (i.e., decoding a barcode into incorrect message) becomes a nondecode, which is considered a more tolerable outcome. Self-checking is realized through selection of codewords such that no two codewords are too similar in their element composition. Such a selection process reduces the number of distinct codewords and the information density from those calculated by Eqs. (3) to (7).

Barcodes may also use checksums to avoid misdecode. Some symbologies dictate how the checksums are used, while others may allow user selection. A common single-character

checksum formula is

$$\text{checksum} = \left(\sum_i a_i C_i \right) \bmod S \quad (9)$$

where a_i are nonzero constants, C_i are values the symbology under consideration assigns to the codewords in the barcode, and S is the size of the alphabet of the symbology. The checksum is stored in the barcode, and the calculation is repeated in the scanner. If a codeword is misdecoded, no matter whether it is a user-data codeword or the checksum codeword, calculation with Eq. (9) will not agree with the decoded checksum value, invalidating the decode.

Some symbologies use multiple codebooks to enlarge the allowed alphabet. The idea is similar to using the Latin alphabet but with a method to specify whether the letters should convey words in English, French, or German. With three different interpretations, we effectively obtained an alphabet three times as large. As an example, the size of the Code 128 alphabet is only 102, excluding the start and stop patterns. But by using three codebooks, it has an effective alphabet of all ASCII characters (128 in total) plus all double digits (i.e., 00, 01 to 99). Specially designated codewords are used to switch between codebooks.

BARCODE SCANNERS

Requirements for barcode readers are stringent and diverse. The ideal scanner should be easy and natural to use, fast, accurate, inexpensive, rugged, and durable. In more technical terms, it should read all symbologies one possibly needs and decide automatically which symbology describes a barcode (referred to as *autodiscrimination*), cover a large area, scan barcodes in many angular configurations (rotation, pitch, and yaw) and at very different distances, and read barcodes of low quality (either low print quality or partially damaged). Usually all these are not achieved simultaneously, and the choice of scanners is based on the application's priorities and compromises.

There are at least two ways to categorize barcode scanners, namely, by their scanning mechanism and by the embodiment. By the scanning mechanism, scanners can be divided into laser scanners, where a moving laser spot does the scanning, and imaging scanners, where the scanning is done virtually in electronics. Wand scanners have no scanning mechanism, and the hand motion of a human operator does the scanning. By the embodiment, scanners can be separated into handheld, slot, presentation, wand, and overhead varieties.

In the remaining part of this section we will first discuss the basic steps of barcode signal processing. We will then investigate the signal degrading factors affecting the performance of scanners, which are distortions and noise. We will cover different barcode scanners first according to scanning technology—namely, laser scanning, electronic scanning (imaging scanners), and manual scanning (wands)—and then according to scanner embodiments.

Barcode Signal Processing

A general block diagram of a barcode scanner is shown in Fig. 4. We discuss the parts relating to signal processing in this

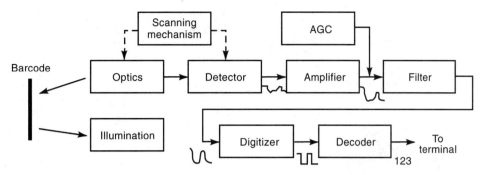

Figure 4. General block diagram of a barcode scanner. Light from the illumination source is scattered by the barcode, and part of the scattered light is focused by the optics onto the detector. Components in the optics may be scanned (in laser scanners), or virtual scan could be performed on the detector (in imaging scanners). Electronic signal conditioning and processing is performed to obtain the text information encoded in the barcode. Not all scanners contain all parts illustrated here, and some parts could be arranged differently.

subsection, while deferring the discussion on the remaining parts.

A barcode scanner scans the field in a scan line, in search of edges between areas of different reflectivity (i.e., bars and spaces). Part of the light scattered by the barcode is collected onto a detector, which converts it into an electronic signal. Specular reflection is avoided when possible to stabilize the signal level.

Signal amplification is often aided by an automatic gain control (AGC) circuit. Electronic filtering is first performed on the signal to block out high-frequency noise. A digitizer finds the bar-space edges represented in the signal. These edges are estimated with either (a) the locations where the signal crosses a particular signal (4), or (b) the zero-crossings of the second derivative of the waveform (5). The intervals between adjacent bar-space edges are measured, which results in the x-sequence.

A decoder translates the x-sequence (or t-sequence) into text message. Autodiscrimination is achieved by finding a symbology that best describes the barcode signal.

Distortion and Noise

The quality of a barcode scanner depends largely on its performance when the input data are not perfect. The imperfections a scanner must contend with include ink-spread, time-scale distortion, convolution distortion, and noise. Distortion and noise are the causes of misdecode. Most symbology-scanner systems can achieve misdecode rates lower than 10^{-6} in normal circumstances.

As mentioned, ink-spread widens all bars wide and narrow by the same amount. The amount of widening depends on the paper and ink used, and therefore it could be different from print to print. Ink-spread is also used in a more general sense to cover all causes of uniform bar width growth (or shrinkage) in the barcode label.

Time-scale distortion occurs when the signal, when mapped onto the scan line, is not sampled at uniform intervals. In wand scanners, this is due to the fact that the velocity of hand motion is not constant. In laser scanners, the reasons may be the nonlinear mapping between the angular scan speed and the linear scan speed, the variation in angular scan speed, the yaw of the barcode, and so on. In imaging scanners,

the contributing factors may include the yaw of the barcode and the distortion of the optical system.

Overall, barcode scanners exhibit a low-pass filtering behavior. A low-pass filter in the frequency domain is equivalent to a convolution in the spatial domain, hence the name convolution distortion. The effect is that the *depth of modulation* (DOM) decreases. DOM is defined as the ratio of signal level change caused by adjacent narrow elements versus that caused by adjacent wide elements. At small DOM, estimated edge locations tend to shift from their ideal locations in the x-sequence (6) (Fig. 5).

Both barcode printing and scanning processes introduce noise. Printing noise comes from the print head, the ribbon in a dot-matrix printer, and the paper (e.g., egg cartons). Scanning noise will be discussed later with scanners. Noise causes random error in edge-location measurements and, with a lower probability, introduces false elements.

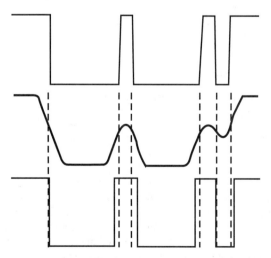

Figure 5. When an ideal barcode waveform (top) is convolved with the kernel of a low-pass filter, in the blurred waveform the depth of modulation decreases, especially for edges between narrow elements (middle). When the rectangular barcode waveform is recreated (bottom), in the form of x-sequence, the estimated edges are shifted from the original locations.

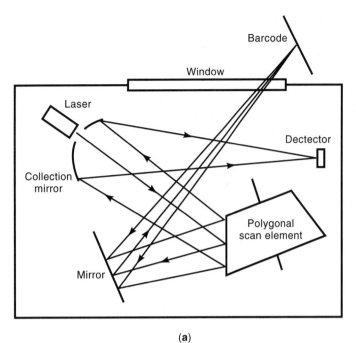

(a)

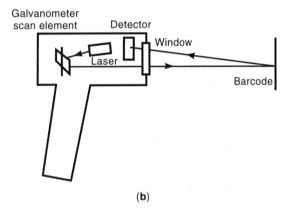

(b)

Figure 6. Two configurations of laser scanners: (a) A slot scanner with a polygonal scan element and a retroreflective collection optical system. (b) A hand-held scanner with a galvanometer-type scan element and a staring detector.

Scanning Technologies

In this subsection we will discuss scanners according to the means of scanning employed—that is, laser scanning (laser scanners), electronic scanning (CCD scanners), and hand scanning (wand scanners).

Laser Scanners. As illustrated in Fig. 6, a laser scanner consists of a laser source, a scan element, a window for light to exit and reenter the scanner, and one or more detectors. Some scanners may also have several mirrors in the optical path, as well as a collector element, which is either a lens or a concave mirror. We shall detail some of these parts below.

A laser beam illuminates a small spot that is scanned across the barcode. The two special qualities of a focused laser beam, namely high intensity and low divergence, are both used effectively. The high intensity of the laser illumination differentiates the laser spot from the surrounding area, which is illuminated by ambient light. The low divergence of the

laser beam provides laser scanners with large working range. First-generation laser scanners used He–Ne lasers, while most current laser scanners use red laser diodes emitting at 650 nm to 675 nm.

The working range achievable by a laser scanner depends primarily on signal strength and the DOM. Depending on the X dimension of the barcode, the working range achievable by a laser scanner can be tens of cm to multiple meters. Barcodes with large X dimensions may be scanned from a large distance, and the limiting factor is likely the signal strength (the maximum laser power emitted from scanners is regulated by government agencies). Barcodes with smaller X dimensions are more likely limited by DOM.

The DOM is determined partly by the laser beam size and the electronic filtering applied to the received signal. Often the laser beam from a scanner is Gaussian, or close to it. The transverse amplitude field distribution of a TEM_{00} Gaussian beam is given by (7)

$$E(x, y, z) = A(z) \exp\left[-\frac{x^2 + y^2}{\omega^2(z)} - j\Phi(x^2 + y^2, z)\right] \quad (10)$$

where z is along the beam propagation direction, A is a real-valued amplitude, and Φ is a real-valued phase factor. The beam size (here we refer to the beam radius, instead of diameter, to simplify the notation), $\omega(z)$, is given by

$$\omega(z) = \omega_0 \left[1 + \left(\frac{z - z_0}{z_R}\right)^2\right]^{1/2} \quad (11)$$

where ω_0 is the beam waist size (minimum value of ω), z_0 is the beam waist location, and z_R is the confocal parameter, related to ω_0 by

$$z_R = \frac{\pi \omega_0^2}{\lambda} \quad (12)$$

where λ is the laser wavelength. Usually the beam waist is optimized according to the X dimension of the intended barcode so that maximum working range can be achieved. For example, to get the maximum range in which

$$\omega(z) \leq aX$$

where a is a constant related to the design of the scanner, the beam waist size should be set to

$$\omega_0 = \frac{aX}{\sqrt{2}} \quad (13)$$

Different optimization merit functions may be used to include linear scan-speed and electronic filtering, as well as other factors, and the optimization solution is likely not analytical.

An effective beam size can be defined to include both the Gaussian beam size and the effect of low-pass electronic filtering (4). The system impulse response can be expressed as a convolution:

$$h_s(t) = h(t)^* s(t) \quad (14)$$

where h is the impulse response of the electronic filter, and s is the linear impulse response of the laser beam:

$$s(t) = \frac{\int E^2(vt, y, z)\, dy}{\iint E^2(x, y, z)\, dx\, dy} \qquad (15)$$

where v is the linear velocity of the laser spot, and x is the direction of the scan (parallel to v).

The transfer function of a typical electronic filter, which is the Fourier transform of h, may be expressed as

$$H(f) = \prod_k \frac{1}{1 + i2\pi\tau_k f} \qquad (16)$$

for the low-pass filter used in the electronics. The effective beam size can then be approximated as

$$\omega_s(z) = \sqrt{\omega^2(z) + 4\sum_k \tau_k^2} \qquad (17)$$

Laser scanners are frequently supersampled; that is, the spacing between samples is smaller than the laser spot size. Supersampling helps to achieve an actual resolution that is smaller than the laser beam size (the related topic of super-sampling imager pixel arrays is discussed in Ref. 8). Because of the availability of supersampled data, laser scanners often perform all signal processing in the analog domain before dig-itizing the data into x-sequences.

Often a laser scanner uses either a polygon or a galvanom-eter to deflect the laser beam and produce the scan line (scan-ners that use moving holograms to generate the scan line will be discussed with slot scanners). The scanning is usually not at constant speed, causing systematic time-scale distortion. When the scanning surface or the laser beam is not perpen-dicular to the rotational axis, the output beam often does not stay in a plane, producing a scan bow (9,10).

The scattered light from the barcode is collected by either one or more staring detectors [Fig. 6(b)] or, more likely, a ret-roreflective system (described below). An example is shown in Fig. 6(a). Staring collection systems are simple, but they re-ceive all ambient light in the field-of-view, which increases noise (see discussion below). A retroreflective collection sys-tem shares the scanning element between producing the out-put beam and collecting scattered light, and therefore it can use a small collection field-of-view that follows the scanning beam. The reduced field-of-view provides an increased signal-to-noise ratio.

Many retroreflective laser scanners allow the blurred im-age of the laser spot to overfill the detector when the object distance is a short, which is referred to as optical AGC (11). An optical AGC is sometimes preferred over an electronic AGC because of its instantaneous response. Following Fig. 7, we can see that

$$r = \left(1 - \frac{s}{f} + \frac{s}{L}\right)R \qquad (18)$$

where f is the focal length of the collection lens, and all other parameters are defined in Fig. 7. An ideal optical AGC is achieved when $s = f$, or the detector is at the back focal plane of the collection lens. If the detector is a circular one with radius r_0, then the received power (to the approximation that

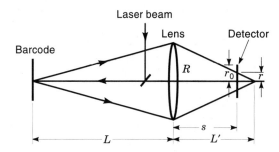

Figure 7. The equivalent retroreflective collection optical system of a laser scanner. Optical AGC is realized through the choice of detector location s and the detector size r_0.

the solid angle subtended by the collection lens is the same as the projected solid angle) is given by

$$P_o = \begin{cases} P_i \left(\dfrac{r_0}{s}\right)^2, & L < \dfrac{sR}{r_0} \\[2mm] P_i \left(\dfrac{R}{L}\right)^2, & L \geq \dfrac{sR}{r_0} \end{cases} \qquad (19)$$

where P_i is the laser beam's optical power. This calculation also assumes that the laser beam produces an ideal nondiver-gent spot throughout the working range. Other issues deviat-ing from the ideal optical AGC include (1) the collection lens not being focused at infinity and (2) shape mismatch between the detector and the aperture.

Ambient light adds noise to a laser scanner, because shot noise is proportional to the total light intensity at the detector (7, Chapters 10 and 11):

$$S_{\text{shot}}(f) = 2q\bar{I} \qquad (20)$$

where $\bar{I}$ is the direct-current (dc) component of the photocur-rent, and q is the charge of an electron. For a laser scanner to be able to operate under ambient light as bright as sunlight, a narrow-band optical filter matching the laser wavelength is usually used. Artificial light sources, even though not compa-rable in intensity to sunlight, also cause concern because they may be modulated at a frequency that interferes with the bar-code reading. Laser speckle noise is another unique noise for this type of scanner. The noise power spectral density is given by (12)

$$S_{\text{speckle}}(f) = \frac{(\lambda z\langle i\rangle)^2}{\sqrt{\pi}v\omega(z)A_d} \exp\left[-\left(\frac{\pi\omega(z)f}{v}\right)^2\right] \qquad (21)$$

where $\langle i \rangle$ is the ensemble average instantaneous photocur-rent, z is measured from the receiver (either a collection lens or a bare staring photodetector), v is the spot velocity, and A_d the size of the receiver. As shown in Eq. (21), the speckle noise is proportional to the photocurrent and hence the laser power (often referred to as a multiplicative noise), and the signal-to-noise ratio cannot be improved by increasing the la-ser power.

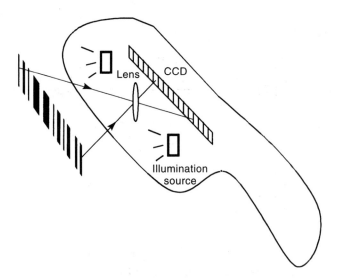

Figure 8. Illustration of the important components in an imaging scanner, including a linear CCD, a lens, and two illumination sources.

CCD Scanners. The major components of an imaging scanner are illustrated in Fig. 8 and are discussed in the following. Imaging scanners for one-dimensional (1-D) barcodes usually use linear CCD arrays (13) as the virtual scanning device. The CCD may have a few thousand pixels, and the barcode is imaged on the CCD by a single lens or by a lenslet array. For CCD scanners, the number of pixels in the imager array and the field-of-view determine the available resolution (the minimum resolvable feature size). A/D conversion is commonly performed on the waveform to preserve the resolution.

The working range of an imaging scanner is predicted by (14)

$$WR = \frac{2X\rho L}{R} \qquad (22)$$

where X and R are as defined before, L is the object distance, and ρ is a constant related to the required minimum DOM. A requirement of 20% DOM leads to $\rho \approx 0.83$. The working range of a CCD scanner, given by Eq. (22), is limited compared to that achievable by a laser scanner. Equation (22) is based purely on geometrical optics, but some CCD scanners use very small apertures which require diffraction-related analysis.

For CCD scanners, ambient light contributes to barcode illumination. With insufficient ambient light, the scanner may need to lengthen the exposure time or turn on its own illumination. Longer exposure time subjects the scanner to motion-induced blur, as usually either the scanner or the object bearing the barcode is in motion.

The transfer function and noise of CCD scanners are common to those of CCD imagers (13) and are not discussed here.

Wand Scanners. Wand scanners are the simplest and least expensive scanners, with the lowest scanning performance. They do not contain any scanning mechanism—the human operator does the scanning. A strict wand scanner usually transmits the scanned signal to a separate box for decoding. Newer types of hand-scanned scanners (e.g., credit-card-shaped) may incorporate the decoding electronics in the same hand-held physical unit.

Wand scanners work in contact or very close proximity with the barcode. They use an incoherent light source for illumination. The light collection optical train shares the same optical opening on the unit with the illumination optics. The self-clocking characteristic of symbologies is most important for wand scanners, as the hand-scanned scan line has a velocity that varies significantly over a barcode.

Without mechanical or electrical scanning, wand scanners do not get as much repeated data as laser or imaging scanners, and its decode speed (expressed in decodes per second) suffers as a consequence. The contact requirement may also damage the barcode, making this type of scanner less suited for environments where the barcode is reused multiple times.

Scanner Embodiments

In this subsection we discuss several common scanner embodiments. These include hand-held scanners, slot scanners, presentation scanners, and overhead scanners. We also present scan engines, which are miniature barcode scan modules that can be integrated into other devices. Wand scanners, also a unique scanner embodiment, have been discussed above.

Hand-Held Scanners. Nonwand hand-held scanners include mechanical or electronic scanning mechanisms. Most applications where the distance between the scanner and the barcode is highly variable use laser-based scanners. In applications where the scanner can be in contact with the barcode, CCD-based hand-held scanners can be used. A hand-held scanner can scan multiple times while the scanner's trigger is pulled, and this repetition helps to boost the scanner's decode speed.

Hand-held laser scanners use a mechanical scanning mechanism, which is usually a galvanometer driven by a miniature motor. The motor drives either a small mirror or the laser itself. The former is generally preferred because that the output beam is scanned at twice the angular velocity as the mirror. The laser scan line also serves as a visual feedback to the operator, indicating which barcode is scanned, how the scan line is aligned with the barcode, how long the scan line is, and so on. Sometimes an elliptical beam profile is used to average over possible noise in the barcode (Fig. 9).

Figure 9. An elliptical beam averages over noise in the barcode. Noise in barcodes is probably caused by the printer (print-head defect, ink shortage, worn-out ribbon, etc.) or the media (paper quality, watermark, lamination, shrink-wrapping, etc.).

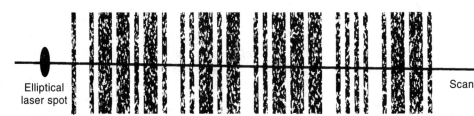

Elliptical laser spot   Scan

Some of hand-held imaging scanners have working ranges of tens of centimeters, while others only work when the barcode is nearly in contact with the scanner.

Slot Scanners. Slot scanners are most often used in supermarkets. The name is derived from the design of first-generation devices, which had a pair of crossed slots where laser beams come out to scan the merchandise from the bottom. Required to perform with high scan speed (see below), all slot scanners are laser scanners. Most slot scanners use an asymmetrical polygon, in which all surfaces have different slope angles in relation to the rotational axis, to move the laser beam. Some slot scanners use a holographic plate to move the laser beam, which we will discuss later.

The most important performance factor in a slot scanner is its "scan speed," a composite measure that includes human factors. The achievable scan speed relates to the size of the "scan zone" and the number of sides the scanner covers. To translate these terms into technical language, one can imagine a package being moved over the scanner (which can be modeled as linear motion at constant speed). The scan zone is the region where a barcode on the package is scanned. Ideally the scan zone should be wide enough to cover the entire belt, but should be more compact in the direction along the belt. To the scanner manufacturer, the problem is that barcodes facing different directions may be scanned at different locations along the belt. If these different locations are too far from each other along the belt direction, it may be difficult for the operator to distinguish whether the item scanned is the intended one or one still on the belt. The number of sides a scanner covers is another ergonomics-related issue. If a scanner can cover more sides of a package, the cashier needs to align the package less frequently. Most slot scanners can cover at least the leading and bottom sides, while some newer ones can cover the far and trailing sides, and even partially cover the near and top sides.

The fact that a slot scanner looks at an object from different sides means that it also has an extremely large overall field-of-view. All slot scanners, therefore, use the retroreflective collection system. Because of vignetting, the effective collection area, and hence the signal power, can change significantly along the scan line. This puts special requirements to the electronics, especially the AGC.

Laser spot ellipticity due to the angle between the laser beam and the barcode causes an additional problem for slot scanners. As shown in Fig. 10, the elliptical spot can partly act as an averaging filter, similar to that shown in Fig. 9, in a horizontal scan line. But in a vertical scan line, the elongated spot acts as an additional low-pass filter, which has the adverse effect of reducing the DOM.

Holographic scanners open new opportunities to optical design (15). This is because the beam-angle variations do not have to come at the cost of particular internal beam paths, and different scan lines can have different focusing powers. The former is helpful in producing more varied scan lines to cover different sides of the object, while the latter is useful in producing larger-than-usual scan zones. Furthermore, the collection lens can be built on the same rotating plate that generates the scan lines. This allows for more flexible optical AGC. For example, the value of R in Eq. (18) can now be tailored for scan lines with different L.

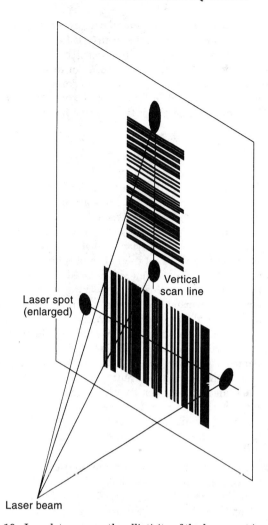

Figure 10. In a slot scanner, the ellipticity of the laser spot is caused by the angle between the laser beam and the barcode plane. This is not very critical in a horizontal scan line, but is a serious concern in a vertical scan line.

Presentation Scanners. These are mostly used in department stores, drug stores, libraries, and so on, in applications where it is preferable that the scanner not be maneuvered by the human operator and where the checkout space is limited. The barcode to be decoded is *presented* (hence the name) to the scanner and left in its field-of-view until decoded.

For presentation scanners, working range is less important. Laser scanners allow more samples on one scan line and thus can have a larger field-of-view than imagers at the same resolution, and therefore they are a common choice. Frequently an alignment requirement on the barcode is not desired, even though the barcode is not over-square. A star-shaped pattern of scan lines, discussed earlier in this article, is commonly used. Furthermore, to alleviate requirements on translational alignment, each scan line in the star pattern is duplicated into a group of parallel scan lines.

Overhead Scanners. High-speed overhead scanners are used to scan barcode-bearing packages traveling on conveyer belts. The width of the belt can be up to 1 m, while the speed of the belt can be up to 200 m/min. These requirements put high data rate demand on the scanner.

Laser-based overhead scanners usually employ multiple scan lines, and omnidirectional scanning is helped by printing the barcode in two orthogonal directions [see Fig. 3(b)]. With two identical barcodes, the number of scan lines of the scanner is effectively doubled. For example, if three scan lines are used, we can use $N = 6$ in Eq. (2), and the relationship between the width and height of the barcode becomes

$$\frac{W}{H} \leq \frac{1}{\tan(\pi/12)} \tag{23}$$

This allows the width of the barcode to be almost four times the height. In reality, more scan lines are needed to account for the objects' motion. Nevertheless, this method of achieving omnidirectionality is preferable because a scan line that is nearly parallel to the conveyer-belt motion direction cannot effectively cover the width of the belt and therefore has to be duplicated. Holographic scanners are used in some applications—for example, to increase the working range.

Imager-based overhead scanners use high-speed parallel processing, which allows omnidirectional decoding of barcodes and even 2-D codes. As we will discuss shortly, the high-speed reading of 2-D codes requires imager-based scanners.

Sometimes a package on the belt may not carry the barcode on the top, but on a different side. If other sides of the package are also to be scanned, the speed and angular requirement for the scanner becomes much more stringent. Usually several scanners are used to form such a high-throughput system.

Scan Engines. Scan engines are miniature barcode scan modules, which are mostly laser-based. Because of the ease of integration they provide, they are widely used in hand-held mobile computers and checkout terminals. Some hand-held scanners also employ scan engines inside.

With a volume of only several cubic centimeters, a scan engine adds the fast and accurate data input method of barcode scanning to a normal computer. Some scan engines contain integrated decoders as well, and the communication between the scan engines and the host computer is through a standard serial communication port. Other scan engines do not contain an integrated decoder; thus the system is simpler and more economical, and the host computer performs the decoding.

TWO-DIMENSIONAL CODES

Two-dimensional codes are generally used as portable data files, where the complete data file related to an item is recorded in the code. This contrasts with the short string of text recorded in a barcode which serves as an index value to a database. By using 2-D codes, the reliance on the network and database server is eliminated. To record more than tens of characters in a barcode is not practical, so 2-D codes are invented to record more data in less area, which facilitates printing and scanning.

In 2-D codes, data are recorded in both directions. Although the most direct way to use both directions to record data is to use square packing, other packing methods have also been used. Particularly, a class of 2-D codes called *stacked barcodes* are built with stacks of 1-D codewords, and

their modules are usually taller than wide. Calculating information density for 2-D codes is more involved than that for 1-D barcodes, because different 2-D codes have very different amount of overhead, which includes finder patterns, support structures, codeword overhead, and error-correction overhead (discussed later). To the first order of approximation the reading performance (such as working range) does not depend on the linear size, but depends instead on the area of the smallest module (16).

Dozens of 2-D symbologies have been invented (see Table 2 for a partial listing), but only a few of them are standardized and widely adopted. In this section we introduce three of these that have published standardized specifications and have been adopted by some industries as the standard symbology: PDF417, MaxiCode, and DataMatrix [Fig. 11(a–c)]. In addition, we also cover postal codes [Fig. 11(d) and 11(e)]. Because postal services around the world are all government-owned monopolies, postal codes are published and maintained by individual postal services, who also regulate their use.

PDF417

PDF417 is the most widely adopted 2-D code, and it can store over one kilobyte of user information in one barcode. True to its stacked barcode nature, PDF417 symbology exploits many ideas developed in 1-D barcodes, often to a greater extent. The PDF417 name refers to a portable data file with a (17, 4) 1-D codeword structure. These 1-D codewords are arranged in rows, each row having *row indicator* codewords at the left and right, next to the start and stop patterns. As can be seen from Fig. 11(a), the start and stop patterns of PDF417 are continuous throughout the height of the barcode.

PDF417 uses three codebooks (called *clusters* in PDF417), each containing an exclusive set of codewords to encode the same data (this contrasts with the practice of using the same codewords to encode different data in some 1-D barcode sym-

Table 2. A Partial Listing of 2-D Codes, Excluding Postal Codes

Name	Standardized	Public Domain
2-DI	No	No
Array Tag	No	No
Aztec Code	Yes	Yes
Codablock	Yes	Yes
Code 16K	Yes	Yes
Code One	Yes	Yes
CP Code	No	No
DataGlyph	No	No
Dot Code	No	No
HueCode	No	No
LEB Code	No	No
MaxiCode	Yes	Yes
MicroPDF417	Yes	Yes
MMC	No	No
PDF417	Yes	Yes
QR Code	Yes	Yes
SmartCode	No	No
Snowflake Code	No	No
SoftStrip	No	No
Supercode	No	Yes
Vericode	No	No

Code in public domain can be used without fee.

(a)

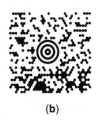

(b)

(c)

|..||..||...|.|..|.|.|..||.||.||..||......|||

(d)

|||.|.|.||.|||.||.||||.|||.||.|||.|.|.|.|.|.|.|.||.|.|.||.|.||.|||.|

(e)

Figure 11. Several two-dimensional codes, including (a) a PDF417 symbol, encoding 151 numeric and mixed-case alphabetical characters, (b) a MaxiCode symbol, encoding 93 numeric and uppercase alphabetical characters, (c) a DataMatrix symbol, encoding 10 mixed-case letters, (d) a PostNet code (US postal code), and (e) a Japanese postal bar code.

bologies). The cluster number of a particular codeword is defined as:

$$c = (x_1 - x_3 + x_5 - x_7) \bmod 9 \qquad (24)$$

where x_i is the width of ith element of the codeword. Only codewords in clusters 0, 3, and 6 are used in PDF417, making the codewords self-checking.

Codewords in each row of a PDF417 barcode have the same cluster number. Counting from the top of a PDF417 barcode, the cluster number of a codeword in the ith row is

$$c = (3i - 3) \bmod 9 \qquad (25)$$

This feature provides vertical tracking information for a tilted scan line. For example, if after a series of cluster 0 codewords a cluster 3 codeword is observed, the decoder may conclude that the scan line is tilting downward.

2-D codes contain more data than their 1-D counterpart, and therefore they require more robust data protection than simple checksums. The error-control method that PDF417 utilizes is the Reed–Solomon Code (hereafter RS Code, see CHANNEL CODING), which allows not only error detection, but also error correction of multiple codewords. PDF417 permits

user-selectable error-correction levels. Usually an error-correction capability of about 20% is used, but more (or less) can be selected if the application is more (or less) demanding.

The error-correction capability is manifested in the number of added error-control codewords added (readers familiar with or interested in error-control theory should note that the use of codeword here does not agree with its usage in error-control theory). These error-control codewords are generated with equations that parallel Eq. (9), but in a manner such that all the user-data related and error-control codewords are mutually interconnected. Error-correction codes such as RS Code can correct errors, where a codeword is misdecoded from the barcode, or erasures, where the codeword in the barcode is not decoded, or a combination of both, up to the maximum level of error-correction capability. As for the checksum calculation, it does not matter whether the error or erasure occurs at a data codeword or an error-control codeword. The maximum error correction capability of an RS Code with κ error-control codewords is

$$\epsilon + 2h \le \kappa \qquad (26)$$

where ϵ is the number of erasures, and h is the number of errors. A detailed description of RS Code can be found in ALGEBRAIC CODING THEORY.

The concept of error-detection budget is also introduced to PDF417 to further reduce its misdecode probability (17). This budget, b, is a number of error-control codewords reserved from being used for error correction, so that Eq. (26) is revised as

$$\epsilon + 2h \le \kappa - b \qquad (27)$$

This reduction reduces significantly the misdecode probability, at the cost of slightly reducing the error-correction capability (18).

High-level encode/decode is an additional layer of translation in PDF417. This concept is a direct extension to the codebook concept as practiced in some 1-D barcodes. With this concept, the PDF417 codewords, each recording a value between 0 and 928, does not record the user data directly. Instead, translation is done to facilitate data compression/compaction. Nowadays a common practice in 2-D codes, the data compression/compaction method used in these codes is different from those employed in general data-compression schemes. General data compression are most effective when the data model is adapted to the data. 2-D codes use prior knowledge of the likely data types [e.g., numeric, alphanumeric, electronic data interchange character sets (19), etc.] to encode with preselected schemes. The benefit is reduced overhead and more efficiency, especially suited for the typical amount of data encoded, which is much smaller than that treated by general data compression schemes.

MaxiCode

MaxiCode has been adopted as a standard 2-D code for high-speed sortation by several standard bodies. A MaxiCode symbol contains a hexagonal matrix of hexagonal elements surrounding the finder pattern [Fig. 11(b)], and it can encode up to 84 characters from a 6-bit alphabet.

Designed for high-speed over-the-belt scanning application, MaxiCode has an isotropic finder pattern and a fixed size. Commonly referred to as the "bull's eye," the finder pattern follows the ring code tradition. Both isotropic finder pattern and fixed size facilitate high-speed image processing by hardware.

As PDF417, MaxiCode also uses RS Code and high-level encoding/decoding. Two error-correction levels are available for user selection.

DataMatrix

DataMatrix [Fig. 11(c)] has been adopted as a standard 2-D code for small item marking. Applications include the marking of integrated circuit and parts. A typical symbol encodes less than 100 bytes of user data, although the specification allows much more.

A DataMatrix barcode consists of square modules arranged in a square grid pattern. The X dimension and number of modules are variable. The outer enclosure is the finder pattern, which contains two solid sides and two dotted sides. From an error-correction point of view, two major varieties of DataMatrix exist: One uses a convolutional error-correction code, whereas the other uses an RS Code.

Postal Codes

Postal codes [Fig. 11(d) and 11(e)] are a special kind of 2-D code because only the vertical dimension encodes data, while the horizontal direction is used only to produce a periodical signal which helps to maintain reading synchronization. The pioneer of postal codes, the United States Postal Service PostNet, uses two types of bar heights and is referred to as 2-state codes. Many other postal services in the world, such as those in Australia, Canada, Japan,and the United Kingdom, have designed 4-state codes where each bar can record up to 2 bits of information. The amount of data stored in postal codes is similar to or slightly more than that possible in 1-D barcodes.

TWO-DIMENSIONAL CODE SCANNERS

As mentioned earlier, PDF417 belongs to stacked barcodes, which can be scanned by a special class of scanners. On the other hand, reading most other 2-D barcodes requires 2-D imaging scanners.

PDF417/1-D Barcode Scanners

Stacked barcodes can be scanned by scanners that produce parallel scan lines. For an application requiring frequent PDF417 barcode scanning, or high scanning throughput, scanners which generate raster patterns can be employed. The scan pattern of these laser scanners mimics the raster pattern of television sets. These scanners also autodiscriminate between PDF417 and 1-D barcodes. Some of these can even scan postal codes. In addition, there are imager-based PDF417/1-D barcode scanners specifically designed to read barcode-bearing cards (such as driver's licenses).

Some PDF417/1-D scanners produce only 1-D scan lines. Either laser- or imager-based, these 1-D scanners are the most economical PDF417 scanners, and they can autodiscriminate among PDF417 and 1-D symbologies. The drawback is that the time needed to scan a PDF417 barcode is relatively long. The operator has to swipe the scan line up and down the barcode, while the scan line itself is swept from left to right (and back from right to left for laser scanners) automatically by the scanner.

General Two-Dimensional Code Scanners

Imager-based scanners are required for general scanning applications involving 2-D and 1-D barcodes. Hand-held 2-D barcode scanners usually employ 2-D CCDs.

If the 2-D/1-D barcodes are carried by sheet paper, then regular flatbed scanners can be used to image the paper, and the computer connected to the scanner can perform the decode. For high-speed over-the-belt applications, such as using MaxiCode for sortation, linear CCD-based scanners are used (as discussed earlier). DataMatrix codes are often so small that machine-vision equipment is required to scan them. But because of the special barcode features designed in these codes, especially the error-correction code employed, the read-rate and the accuracy achievable are still much higher than reading regular text with the same machine-vision equipment.

CONCLUDING REMARKS

We live in the age of information. Barcodes and 2-D codes, together with radio-frequency identification (RF-ID), magnetic-strip cards, smart cards, and contact memory devices, have become some of the preferred ways to input information quickly and accurately into computers and computer networks. With easy, quick, and accurate access to information, the computers and networks can then better help us to complete our work better and more efficiently. And we are still far from realizing the full potential that these technologies can bring.

BIBLIOGRAPHY

1. R. C. Palmer, *The Bar Code Book: Reading, Printing, Specification, and Application of Bar Code and Other Machine Readable Symbols,* 3rd ed., Peterborough, NH: Helmers, 1995.

2. D. Savir and G. Laurer, The characteristics and decodability of the Universal Product Code, *IBM Syst. J.,* **14**: 16–33, 1975.

3. T. Pavlidis, J. Swartz, and Y. P. Wang, Fundamentals of bar code information theory, *Computer,* **23** (4): 74–86, 1990.

4. E. Barkan and J. Swartz, System design considerations in barcode laser scanning, *Opt. Eng.,* **23** (4): 413–420, 1984.

5. N. Normand and C. Viard-Gaudin, A two-dimensional bar code reader, *Proc. 12th IAPR Int. Conf. Pattern Rec.,* **3**: 201–203, 1994.

6. S. J. Shellhammer, Distortion of Bar Code Signals in Laser Scanning, in G. A. Lampropoulos, J. Chrostowski, and R. M. Measures (eds.), *Applications of Photonic Technology,* New York: Plenum, 1995.

7. A. Yariv, *Optical Electronics,* 3rd ed., New York: Holt, Rinehart and Winston, 1985, Chap. 2.

8. K. M. Hock, Effect of oversampling in pixel arrays, *Opt. Eng.,* **34** (5): 1281–1288, 1995.

9. Y. Li and J. Katz, Laser beam scanning by rotary mirrors. I. Modeling mirror-scanning devices, *Appl. Opt.,* **34** (28): 6403–6416, 1995.

10. Y. Li, Laser beam scanning by rotary mirrors. II. Conic-section scan patterns, *Appl. Opt.,* **34** (28): 6417–6430, 1995.

11. J. Wang, Z. Chen, and Z. Lu, System analysis of bar code laser scanner, *Proc. SPIE,* **2899**: 32–40, 1996.

12. D. Yu, M. Stern, and J. Katz, Speckle noise in laser bar-code-scanner systems, *Appl. Opt.,* **35** (19): 3687–3694, 1996.

13. G. C. Holst, *CCD Arrays, Cameras, and Displays,* Bellingham, WA: SPIE Press, 1996.

14. D. Tsi et al., System analysis of CCD-based bar code readers, *Appl. Opt.,* **32** (19): 3504–3512, 1993.

15. L. D. Dickson and G. T. Sincerbox, Holographic Scanners for bar Code Readers, in G. F. Marshall (ed.), *Optical Scanning,* New York: Marcel Dekker, 1991.

16. A. Longacre, The resolvability of linear vs. polygonal barcode features, *Workshop Autom. Identification Advanced Technol.,* Stony Brook, NY, 1997, pp. 57–60.

17. Uniform Symbology Specification, PDF417, Pittsburgh, PA: AIM USA, 1994.

18. J. D. He et al., Performance analysis of 2D-barcode enhanced documents, *Proc. SPIE,* **2422**: 328–333, 1995.

19. Electronic Data Interchange X12 Draft Version 3 Release 4 Standards, Alexandria, VA: Data Interchange Standards Assoc., Inc., 1993.

Reading List

R. C. Palmer, *The Bar Code Book: Reading, Printing, Specification, and Application of Bar Code and Other Machine Readable Symbols,* 3rd ed., Peterborough, NH: Helmers, 1995. Covers all phases of barcode usage.

C. K. Harmon, *Lines of Communication: Bar Code and Data Collection Technology for the 90s,* Peterborough, NH: Helmers, 1994. Talks about barcodes, 2-dimensional codes and other automatic data-collection technologies.

D. J. Collins and N. N. Whipple, *Using Bar Code: Why it's Taking Over,* Duxbury, MA: Data Capture Inst., 1994. Covers barcode technology and applications. Easy to read.

The Source Book, Pittsburgh, PA: AIM USA, 1997. Introduces subfields of automatic identification and data collection technologies (AIDC) and lists participating companies.

Various AIM symbology standards.

ANSI standards on barcode printing and barcode usage.

Workshop on Automatic Identification Advanced Technologies, Stony Brook, NY, 1997. Contains recent developments of barcode related technologies.

JACKSON DUANFENG HE
Symbol Technologies, Inc.

MASSES. See MASS SPECTROMETERS.

MASS MEASUREMENT. See WEIGHING.

MASS SPECTROMETERS

APPLICATIONS OF MASS SPECTROMETRY

Mass spectrometers precisely determine the masses of ionized atoms or molecules with extreme sensitivity (1,2). Such investigations allow the following measurements.

The Identification of Atoms or Molecules

From a reasonably resolved mass spectrum, the original atoms or molecules (see Fig. 1) are identified by determining their atomic or molecular weight precisely. From such a mass spectrum the binding energies of different isotopes of one or of several elements are also determined (3,4) if the mass-resolving power $m/\Delta m$ of the mass analyzer is large enough (see also the section "The Magnetic Ion Trap Mass Spectrometer").

The Investigation of Macromolecules

The mass analysis of macromolecules is helpful for investigating very large biomolecules (see Fig. 2) and also atomic or molecular clusters. The goal of such investigations is to precisely determine the molecular weight.

The Determination of the Structure of Molecules

To understand the structure of complex organic molecules one can isolate ions of one specific mass in a first-stage mass analyzer, fragment this ion by gas collisions in some intermediate gas cell, and analyze the mass spectrum of the molecule fragments in a second-stage mass analyzer. This (MS/MS) technique (5) is used, for instance, in the amino acid sequencing of proteins (6).

The Measurement of Isotopic Distributions

When some isotopes of a certain element decay radioactively, the isotopic intensity distribution reveals the age of the sample. Some of the isotopes exist only in very small quantities while others are abundantly available. For this reason it is very important to use a mass analyzer that provides for a rather small mass cross-contamination from one mass to the next. The system must be designed properly for this purpose, and it must have a very good vacuum to reduce residual gas scattering of the most abundant ion species (see Fig. 3). Especially low mass cross-contamination is achieved by using high-energy ions accelerated in tandem accelerators (7). In

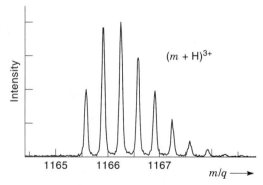

Figure 1. Mass spectrum of β-chain insulin of mass $m \approx 3494.9$ u, recorded as $(m + H)^{3+}$, that is, with one proton attached. This mass spectrum was recorded in a time-of-flight mass spectrometer (see Fig. 9) of mass resolving power $m/\Delta m = 17000$ (FWHM), that is, measured as the full line width at half maximum. Since carbohydrate molecules all contain about one ^{13}C-atom for every one hundred ^{12}C-atoms there is a mass multiplet of ions for every molecule. When the molecule is large, the probability is high that it contains one or several ^{13}C-atoms. Thus, the most abundant molecule is usually not the one that contains only ^{12}C-atoms.

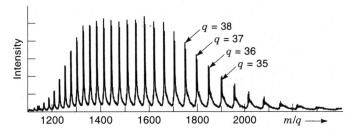

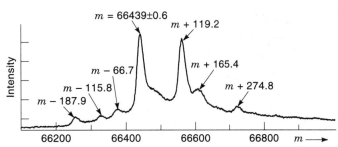

Figure 2. The mass spectrum of ≈ 30 to ≈ 55 times charged ions of albumin. This molecule has a molecular weight of about (66,439 $\pm$ 0.6) u. The time-of-flight mass spectrometer used for this investigation had only a mass resolving power $m/\Delta m \approx 12{,}000$ so that individual ion masses could not be separated from each other. Also shown is the mathematical combination of all of these mass distributions, from which one determines the molecular mass of albumin with certain adducts with a precision of only 10 ppm. The main adducts here had a mass of $m_1 = 119.2$ u.

such systems, for instance, the isotopic ratio $^{14}C/^{12}C$ can be investigated. This is important because ^{14}C, which has a half-life of 5730 years, is constantly produced in the earth's atmosphere by the sun, so that the $^{14}C/^{12}C$ ratio is a good measure for the age of old natural products, such as wood or bones.

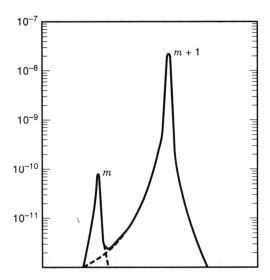

Figure 3. Mass cross-contamination caused by the tails of an intense neighboring mass line. The shown spectrum was recorded by a rather good sector field mass analyzer that had a (FWHM) mass-resolving power of $m/\Delta m \approx 5000$ and was operated at a pressure of only a few times 10^{-8} mbar.

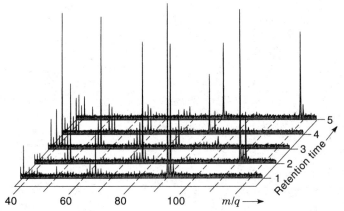

Figure 4. The record of a GC/MS investigation of traces of about 10^{-12} g of (1) toluene, (2) chlorobenzene, (3) nonane, (4) chloroheptane, and (5) t butylbenzene. For such a measurement several thousand or 10,000 mass spectra are recorded of the chromatograph effluent, and each mass spectrum is characteristic of the effluent at a specific time. For clarity, however, here only one mass spectrum is shown for each GC peak. A typical record, as shown here, requires several hundred seconds and in some cases up to a few thousand seconds because one must wait until the substance with the longest retention time leaves the gas chromatograph.

The Detection of Small Amounts of Specific Atoms of Molecules

To investigate the pollution, for instance, of water or air samples one can use the high sensitivity and specificity of mass spectrometers and combine them with the selectivity of gas chromatographs (GC/MS), liquid chromatographs (LC/MS), or of complex but powerful capillary zone electrophoresis (CZE) systems. In such chromatographs the molecules are separated according to their chemical adsorption properties (8), that is, simultaneously injected chemically different molecules leave the chromatograph successively. A specific substance is identified in such a system by the mass spectrum of the chromatograph's effluent recorded at a specific time (see Fig. 4).

FUNDAMENTALS OF MASS SPECTROMETRY

The very diverse applications described are all united by the use of the same instrumentation. Thus it is necessary to describe the mass spectrometric techniques in some detail.

An atomic or a molecular ion to be mass analyzed is characterized by

1. its mass m measured in *mass units* u, that is, the mass of one-twelfth of a ^{12}C-atom;
2. its charge q measured in *charge units,* that is, the negative value of the charge of one electron or $\approx 1.6022 \times 10^{-19}$ C;
3. its kinetic energy $K = qV$ measured in electron volts where V is the potential difference by which the ion has been accelerated; and
4. its velocity v measured in kilometers per second or in millimeters per microsecond (9) is given by

$$v \approx 9.82269 \sqrt{\frac{2K}{m}} \qquad (1)$$

The numerical multiplier is found from $c/\sqrt{m_u} \approx$ 299,792,458/$\sqrt{931,494,300}$ where c is the velocity of light in millimeters per microsecond and m_u is the energy equivalent of one mass unit in electron volts.

MASS ANALYZERS

To distinguish ions of different masses, the electromagnetic fields used can either provide a lateral dispersion, that is, a mass-dependent beam deflection or a longitudinal dispersion, that is, a mass-dependent flight-time difference. In the second case one must use a chopped and bunched ion beam, whereas in the first case bunched ion beams and dc beams can be used.

Laterally Dispersive Mass Analyzers

Laterally dispersive mass analyzers consist mainly of a magnetic deflecting field of flux density B measured in Tesla (T). In a magnetic flux density B, nonrelativistically fast ions ($v \ll c$) of mass m, energy K, and charge q move along radii ρ measured in millimeters according to the expression

$$B\rho \approx \frac{\sqrt{2Km}}{9.82269q} \qquad (2)$$

Usually the ions produced are all accelerated by the same potential difference V_0 so that their kinetic energies are all $K_0 = qV_0$. Thus their velocities $v_0 \approx 10\sqrt{2K_0/m}$ are also the same. However, because of the ionization process, the ions always have an energy spread $\pm\Delta K$ which is usually smaller than 1 eV.

When ions of different masses $m - m_0 \pm \Delta_m$ and of a range of energies $K = K_0 \pm \Delta K$ enter into a magnetic sector field, the ions are finally separated from the beam axis by $\pm [(x_B|m)(\Delta m/m_0) + (x_B|K)(\Delta K/K_0)]$ with $(x_B|m) = (x_B|K)$ for a given magnetic field. Here $\pm(x_B|m)(\Delta m/m_0)$ determines the desired mass separation between ions of masses $m_0 + \Delta m$ and $m_0 - \Delta m$, and $(x_B|K)(\Delta K/K_0)$ causes a detrimental beam widening because of the continuous distribution of ion energies from $K_0 - \Delta K$ to $K_0 + \Delta K$. This energy spread also causes the ions to diverge from the ion source at angles of divergence given by

$$\alpha_0 = \pm\frac{v_i}{v_0} = \pm\sqrt{\frac{\Delta K}{K_0}} \qquad (3)$$

because the velocity distribution of the unaccelerated ions is isotropic.

For a given magnetic field this widening of a mass line caused by the ions' energy distribution is unavoidable. However, for a combination of two or more sector fields, one can arrange the fields such that the energy dispersions compensate for each other, and the overall mass dispersion is $(\bar{x}|m)\Delta m \neq 0$. For a two-sector field system there are two possible arrangements (see Fig. 5):

1. The first solution uses an electrostatic sector field in addition to the magnetic sector which is dimensioned so that the forwardly calculated energy dispersion $(\bar{x}_E|K)(\Delta K/K_0)$ of the electrostatic sector field is equal to the backward calculated energy dispersion $(\bar{x}_B|K)(\Delta K/K_0)$ of the magnetic sector field (9,10), as indi-

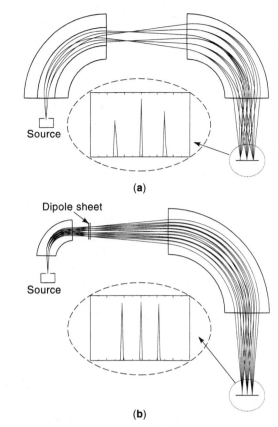

Figure 5(a) and (b). Ion trajectories are shown in two types of angle and energy focusing mass spectrometers which in the example shown both use the same geometry for the final magnetic sector-field mass analyzer. In Fig. 5(a) this magnetic sector field is preceded by (9,10) an electrostatic sector field and in Fig. 5(b) by a magnetic sector field placed at a different potential (10). It is assumed here that the ions are accelerated in a dipole sheet between the two stages. Note that in both systems 18 ion trajectories are shown characterized by two energies and three masses which leave the ion source at three different angles of inclination. For both systems there are also only three beams at the end (characterized by the three ion masses) independent of the angles at which the ions left the ion source and independent of the energy of these ions.

cated in Fig. 5(a). Because the electrostatic sector field has no mass dispersion, the mass dispersion of the magnetic sector field is also the mass dispersion of the full system.

2. The second solution uses a small magnetic sector field to compensate for the energy dispersion of the main magnetic sector (11), as indicated in Fig. 5(b). When the two sector magnets operate at different potentials and the energies K_1 and K_2 denote the ion energies in the two fields, one must postulate that the forwardly calculated energy dispersion $(\bar{x}_{B1}|K)(\Delta K/K_1)$ of the first magnetic sector field equals the backward calculated energy dispersion $(\bar{x}_{B2}|K)(\Delta K/K_2)$ of the second sector field (11). However, such a combination of two sector fields is useful only for $K_1 \neq K_2$ because only in this case the mass dispersions of the two sector magnets do not cancel each other for the case of a vanishing energy dispersion.

Comparing the two solutions, one sees that the overall mass dispersion is a little larger for the first case than for the second. The mass cross-contamination, however, is lower for the second solution (11) because the two momentum analyzers are used in series whereas only one momentum analyzer exists for the first solution (10).

Longitudinally Dispersive Mass Analyzers

Laterally dispersive mass analyzers have proven to be effective and powerful tools exhibiting the lowest mass cross-contamination of all known mass analyzers. However, longitudinally dispersive systems are becoming more and more popular because they are mechanically simpler. There are three types of such systems:

1. high-frequency mass analyzers (12,13,14) in which low mass ions pick up higher speeds than high mass ions within one frequency cycle. Thus, low mass ions swing in this field with larger amplitudes that can become larger than the electrode separation.

2. mass spectrometers in which ions circulate in a homogeneous magnetic field (15–17) or in race tracks (3,9).

3. time-of-flight mass analyzers (18–23) in which the mass-dependent flight times are observed directly for ions that have started simultaneously.

High-Frequency Mass Analyzers. High-frequency mass analyzers cause ions of energies of a few electron volts to swing in electric ac fields of some 100 V/mm and frequencies of $\nu \approx$ 1 MHz. Because low mass ions swing with larger amplitudes than high mass ions, ions of too low mass are intercepted by the electrodes and only high mass ions survive. Adding a dc potential to the electrodes (see Figs. 6 and 7) also eliminates the heaviest ions (12,13) so that only ions within a small range of masses move along stable trajectories. There are two basic configurations of such mass analyzers.

The Quadrupole Ion Trap. The *quadrupole ion trap* uses rotationally symmetrical, hyperbolically shaped electrodes (see

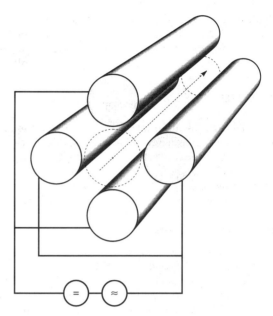

Figure 7. The quadrupole mass filter. Under the action of ac and dc fields, low energy ions swing laterally between the electrodes shown, and they move with constant velocity in the direction of the z-axis. Analogously to the ion motion in an ion trap, here also a pure ac field would allow only ions above a certain mass to move along stable orbits, whereas an added dc field would make the trajectories of the heavier ones unstable.

Fig. 6). Into such a *trap* ≈ 10 eV ions are introduced which then under the action of the high-frequency field mainly swing up and down between the two electrodes of the two-surface hyperboloids of Fig. 6. If the ac voltage is increased over time, the amplitudes of these swings also increase, and ions of higher and higher masses impinge on the electrodes or if appropriate holes are provided (14), leave the trap and can be recorded on some sensitive ion detector. By varying the voltages on these electrodes appropriately over time or during different time intervals, ions over a wide mass range or over a deliberately narrowed one can be made to move along stable orbits. One very useful MS/MS sequence requires that during the time interval Δt_1 the voltages are chosen so that stable ion motion is guaranteed only for ions of one particular molecule mass m_0, whereas during the time interval Δt_2 the voltages are chosen so that ions of a range of masses all perform stable motions. Though, at the beginning of this time interval Δt_2, only ions of mass m_0 are in the trap, these ions fragment because of collisions with residual gas atoms. With properly chosen ac and dc voltages, these fragments are all stored in the trap. Then during the time interval Δt_3, the voltages are scanned so that the mass range for the ions that perform stable motion is constantly decreased. Thus a mass spectrum is recorded of the fragment ions which become unstable successively according to their mass values.

The Quadrupole Mass Filter. A second well-established high-frequency mass analyzer consists of four rod-like electrodes (see Fig. 7) with ≈ 10 eV ions injected along the quadrupole axis. For appropriate ac and AD voltages of several 100 V, only ions within a very small mass range pass this *mass filter* (12,13), that is, move in the z-direction about 0.1 mm in each high-frequency period and at the same time swing transver-

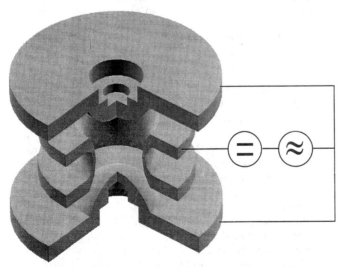

Figure 6. In the ion trap shown the ions can swing in the rotationally symmetrical quadrupolar ac field. If the dc power supply is left off, only ions above a certain mass move along stable orbits. The addition of a dc power supply also provides a limit for the high mass ions. In all cases the ion motion is mainly up and down in the figure shown here.

sally between the electrodes with amplitudes of about 5 mm. At the end, a mass spectrum is recorded in an ion detector if the ac and dc amplitudes are scanned appropriately over time.

The Magnetic Ion Trap Mass Spectrometer. A mass spectrometer of very high mass-resolving power $R = m/\Delta m$ is a system in which ions move in a magnetic flux density B along radii ρ according to Eq. (2) with a velocity v according to Eq. (1). Thus the flight time per turn in microseconds is given by

$$\bar{t} \approx 2\pi \frac{\rho}{v} \approx 0.0651212 \frac{m}{B} \qquad (4)$$

where B is in Tesla and m is in mass units. This flight time is independent of the ion energy K. There are several ways to determine this flight time per turn.

1. If the radius ρ is large enough, one can determine the time $\bar{t}$ directly for one turn by small pulsed beam deflectors (15). In principle, this method is also applicable for many turns (9).

2. One can amplify the potentials induced on electrodes close to the ion path (see Fig. 8). After a Fourier analysis these induced voltages reveal mass specific frequencies $\nu = 1/\bar{t}$ and thus the desired mass spectrum (3,16).

3. If the radius ρ is small enough, one can superimpose a high frequency electrostatic field to the magnetic one and register the finally left ions in the system by accelerating them axially out of the magnetic field. In this case the ions' azimuthal velocities are transformed into axial velocities that can be measured by a time-of-flight technique (17) which identifies the ions in resonance.

All these systems deliver rather high mass resolving powers of $m/\Delta m \geq 100,000$ or more which can be used for molecule mass analysis (16) or for the determination of nuclear mass defects (24) and other basic information like the CPT invariance (25). Note here that the mass resolving power of such systems is proportional to the number of rotations of the ion cloud and that the mass resolving power per turn can be rather small (16,17). Thus for a given experimental time it is advantageous to increase the magnetic flux density as much as possible. Note also that for $B = 7$ T, singly charged 100 eV ions of 100 u would move along circles of radius $\rho \approx 2.057$ mm with $\bar{t} \approx 0.930\ \mu s$.

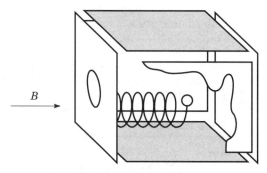

Figure 8. A Fourier transform mass analyzer that uses a high field magnetic solenoid (14). The mass-resolving power increases with an increased magnetic flux density B.

Instead of having ions move along circles in a homogeneous magnetic field, one can also arrange a number of magnetic and/or electrostatic sector fields into a *race track* with intermediate field-free regions (3,9,18) and then determine the flight time $\bar{t}$ per turn via Fourier transform techniques. In principle this flight time $\bar{t}$ depends on the ion mass and also on the ion energy. However, there are still two ways to achieve high mass-resolving power:

1. reduce the energy spread in the ion beam either by an electron cooler (3) for high energy ions or by gas collisions (26) for low energy ions.

2. introduce magnetic and/or electrostatic quadrupole lenses into the ring and excite them so that the overall system is energetically isochronous (9,18), that is, faster ions are sent around the ring along properly elongated trajectories.

Both race track systems have been used successfully for mass measurements of stable (14) and of short-lived nuclei (4).

Time-of-Flight Mass Analyzers

A different approach is a time-of-flight (TOF) mass spectrometer (MS) into which a bunch of ions of different masses is injected and the ion arrival is recorded at some downstream ion detector (see Fig. 9). Differing from all scanning mass analyzers, there is no limit in a TOF-MS on the mass values of the ions under investigation. Therefore TOF-MS systems are powerful tools for the investigation of large biomolecules or cluster ions.

The ion source for a TOF system requires that all ions of equal mass start from some point at the same time. This is guaranteed by

1. a pulsed ion acceleration in which a pulsed electric field acts on some cloud of ions. If all of these ions are stationary initially and then are accelerated in the z-direction, they all reach a properly placed ion detector simultaneously if their masses are the same. In other words, *they will be bunched* because the ions that start from $z_0 + \Delta z$ receive a little less energy than those that start from $z_0 - \Delta z$, but the first ones must also travel a little further (18,19). There are three ways to implement this method:

   a. introduce ions at a specific z_0, and wait until they move apart to different z-values because of their thermal energies (19). In this case the final ion positions z are correlated to their initial energies which improves the bunching properties considerably.

   b. store (20) the ions in the potential well caused by a beam of electrons of energy K_0. Depending on the electron beam current I_e, the potential in the middle of the beam is $V_e \approx 15200(I/\sqrt{U_0})$. For $U_0 = 70$ eV and $I = 0.0005$ A, one thus finds that $V_e \approx 0.75$ V. In this case the final bunch length is usually determined by the "turnaround time," that is the time in which an ion that moved initially in the $-z$-direction has reversed its velocity.

   c. introduce a low energy ion beam perpendicularly to the z-direction, that is, the direction of the pulsed ion acceleration (21). In this arrangement some "ion stor-

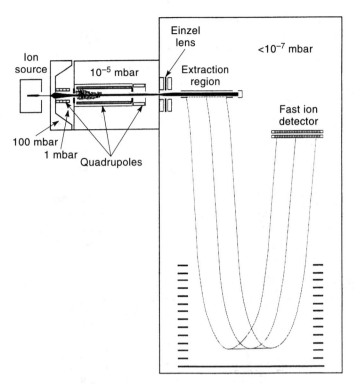

Figure 9. A time-of-flight mass analyzer for heavy molecules. Note the folded flight path in the z-direction and the large ion mirror used (23). Also note the orthogonal ion injection in the x-direction and the pulsed ion acceleration in the z-direction (21). The electrospray ion source (31) forms charged droplets and sends them into the vacuum system where in flight they evaporate all the solvent, so that highly charged ions remain. Then these ions are passed through three quadrupoles the last of which is powered in a dc mode and focuses the 10eV ions into the extraction region. The other two are two gas-filled ac only quadrupoles the first of which *heats* the ions to boil off any molecule adducts or to fragment the main molecule (6), and the second of which because of its different gas pressure *cools* the ions by gas collisions. This procedure works continuously so that finally the ions all have more or less the same velocity in the x-direction before being pulse accelerated orthogonally in the z-direction.

age" is also achieved because it takes a relatively long time for the ions to perpendicularly traverse the region from which they are then accelerated. Most importantly, however, there is almost no ion motion in the z-direction initially. Thus the "turnaround time" is reduced compared to 1(b).

For such a pulsed ion acceleration, the final energy distribution in the bunched ion beam is large, for instance, $\Delta K \approx 100$ eV.

2. a pulsed ion generation which is, for instance, the case for the laser ionization of a sample. A very good method is (20) to embed some organic material into a matrix, for instance one of glycerol, so that each molecule to be investigated is surrounded only by matrix molecules. Irradiating this sample by a short intense laser pulse releases such molecules as ions. This method is called *matrix-assisted laser desorption and ionization* (MALDI). The performance of this technique is improved by accelerating these ions in an electrostatic field that is switched on shortly after the laser pulse (19).

The time focus (19,21) achieved by pulsed ion acceleration is a good technique for achieving reasonable mass-resolving powers for a so-called *linear TOF-MS*. One can improve the simultaneous arrival of ions of equal mass by using this time focus as the source for a folded-flight-path TOF-MS (18,20). In such a system the ions are reflected by some *electrostatic ion mirror* in an energy isochronous manner, that is, so that the more energetic and thus faster ions reach the final ion detector only via a properly dimensioned detour (18,23) achieved by their deeper penetration into the reflector field. One can also say: *to first order the overall flight time becomes independent of the ions' energy spread* ΔK. Investigating q-times charged ions of energy $(K_0 \pm \Delta K)$ that enter a homogeneous repeller field E this condition of isochronicity postulates that

$$L_1 + L_2 = \frac{4qE}{K_0} \tag{5}$$

where L_1 and L_2 denote the lengths of the field-free regions between the ion source and the repeller field and between the repeller field and the final ion detector, respectively. In this case the overall flight time is $t + \Delta t$ with $\Delta t/t = (\Delta K/K_0)^2/8 + \ldots$. Using an ion reflector composed of two regions (23) of properly dimensioned field strengths or a grid-free ion reflector (18,20) that produces a properly dimensioned field region, the performance of such a TOF-MS can even be improved, so that finally only much smaller effects proportional to $(\Delta K/K_0)^3$ remain.

ION SOURCES

Many different ion sources have been developed for different mass spectrometers and different applications. Only in a few cases this ion formation can occur inside a mass analyzer, as for instance inside of an "ion trap." In most cases an external ion source is used that consists of two parts: the ionization device and the ion accelerating and beam-forming device.

Electron Impact Ion Sources

Energetic electrons ionize vaporized atoms or molecules by collisional impact. The electron energy here must be large enough that at least one electron in the shell of the atom or molecule in question is removed. Optimally, (2,27) beams of electrons of 70 eV or 80 eV are used with currents of usually a few 100 μA. In many cases these beams are held together by magnetic fields of perhaps 0.01 T produced easily by permanent magnets. This magnetic field usually confines the electron beam to diameters ≤ 2 mm and at the same time elongates the electron path thus increasing the overall ionization probability for a given electron current. Usually the ions are pulled from the ionization region through some narrow orifice by an accelerating electrode at a potential of some 100 V or even 1000 V. However, the ions often are also pushed toward this orifice by some relatively large pusher electrode at a potential of a few volts.

Plasma Ion Sources

In a plasma ions already exist and only need to be extracted. However, electrostatic fields cannot penetrate deeply into a conductive plasma and thus ions can be extracted only from the plasma surface which consequently is depleted of ions to some depth after a very short time. Thus a pulsed ion acceleration that extracts ions from a replenished *plasma surface* is especially effective (28). As one should expect, the extracting field strength shapes the plasma surface and thus greatly influences the ion optical properties of the extracted ion beam.

Thermal Ion Sources

If the ionization energy I_p for some atom under consideration is smaller than the work function W of a substrate, the atoms leave this substrate partially as ions. The ratio between ionized and neutral atoms evaporating from a filament heated to a temperature T is given by

$$\frac{N^+}{N^0} = \exp \frac{W - I_p}{kT} \tag{6}$$

where $k = 8.62 \times 10^{-5}$ is the Boltzmann constant measured in electron volts per K. Re and also W or Ta substrates have large work functions which cause them to efficiently remove one electron from evaporating alkali atoms, that is Li, Na, K, Rb, Cs, Fr and from some earth alkali atoms during evaporation. With Rb or Sr on 2500 K hot Re, and $I_p(\text{Rb}) = 4.18$ eV or $I_p(\text{Sr}) = 5.7$ eV one finds that $N^+/(N^0 + N^+)$ equals 100% or 85% if the work function of the Re surface is assumed to be 6.1 eV. However, other atoms, for instance, U are also ionized with some efficiency (2,27). Similarly, one can also attach one electron to atoms with large electron affinities (for instance, Br) if they are evaporated from a substrate that has a very low work function, for instance, LaB_6.

This simple ionization technique is improved considerably if the substrate forms the inner surface of a hot cavity (29). In this case most elements are ionized with efficiencies of about 1%. The reason for this enhancement is that the ions are extracted through a small hole in this cavity whereas the atoms stay in the cavity and thus get another chance to be ionized.

Laser Ion Sources

As one might expect, one can directly ionize specific isotopes by high-power lasers of some resonant frequencies. However, often one uses only lasers of broader frequency bands and thus ionizes all isotopes of a specific element simultaneously (30). In some cases one also uses the property of a finely focused laser beam to produce a small plasma at some surface as, for instance, for the previously mentioned MALDI sources (22).

Electrospray Ion Sources

A very interesting ion source for high mass biomolecules uses the electrospray technique (31). In one of these ion sources a very dilute solution of the molecules of interest is sprayed in fine droplets into some gas at about 1 bar. Then the gas and these droplets are sucked through a small orifice into a vacuum vessel where the droplets will evaporate most of the solvent and thus rapidly decrease in size. If there were only one molecule in the droplet there is only one molecule left at the end when all the solvent has evaporated. Because the droplets were multiply charged from the beginning, as is the case in the spray from any waterfall, the final molecule is multiply charged (see Figs. 1 and 4).

CONCLUSION

Mass spectrometry began as a special technique to determine the isotopic distribution of all elements (9), but it has grown into a very general, sensitive, and specific analytical technique for ionized atoms or molecules. For this reason mass spectrometers have been used in very diverse applications from nuclear physics to pharmacology. The fastest growing application certainly is the analysis of biomolecules and the trace detection of organic molecules.

BIBLIOGRAPHY

1. F. White and G. Wood, *Mass Spectrometry: Applications in Science and Engineering,* New York: Wiley, 1981.
2. M. Gross, *Mass Spectrometry in the Biological Sciences: A Tutorial,* Dordrecht: Kluwer, 1992.
3. B. Schlitt et al., *Hyperfine Interactions,* **99**: 117, 1996.
4. H. Wollnik et al., *Nucl. Phys.,* A616: 346, 1997.
5. K. L. Busch, G. L. Glish, and S. A. McLuckey, *Mass Spectrometry / Mass Spectrometry: Techniques and Applications of Tandem Mass Spectrometry,* New York: VCH, 1988.
6. A. Dodonov et al., *Rapid Commun. Mass Spectrom.,* **11**: 1649, 1997.
7. R. C. Finkel and M. Suter, "AMS in the Earth Sciences: Techniques and Applications," in *Adv. Anal. Geochem.* **1**: 211, 1990.
8. F. Brunner, *The Science of Chromatography,* Amsterdam: Elsevier, 1985.
9. H. Wollnik, *Optics of Charge Particles,* Orlando: Academic Press, 1987.
10. F. W. Aston, *Philos. Mag.,* **38**: 709, 1919.
11. H. Wollnik, *Nucl. Instr. Methods,* in press, 1998
12. W. Paul and H. Steinwedel, *Z. Naturf.,* **A8**: 448, 1953.
13. P. H. Dawson, *Quadrupole Mass Spectrometry and its Applications,* Amsterdam: Elsevier, 1976.
14. G. C. Stafford et al., *Int. J. Mass Spectrom. Ion Proc.,* **60**: 85, 1984.
15. A. G. Marshall and F. R. Verdun, *Fourier Transform in NMR, Optical and Mass Spectrometry,* New York: Elsevier, 1990.
16. G. Wendt, *Metrologia,* **22**: 174, 1986.
17. M. St. Simon et al., *Physica Script,* **T59**: 406, 1995.
18. H. Wollnik, in Mass spectrometry in biomolecular science, M. Caprioli, A. Malorni and G. Sindona, (eds.), Dordrecht: Kluwer, 1996, p. 111.
19. W. C. Wiley and I. H. McLaren, *Rev. Sci. Instr.,* **26**: 1150, 1955.
20. R. Grix et al., *Int. J. Mass Spectrom. Ion Proc.,* **93**: 323, 1989.
21. G. J. O'Halloran et al., Tech. Doc. ASD TDR Report 62-644, Bendix Co., 1964.
22. M. Karas, U. Bahr, and U. Giessmann, *Mass Spectrom. Rev.,* **10**: 335, 1991.
23. B. A. Mamyrin et al., *Sov. Phys. JETP,* **37**: 45, 1973.
24. D. Beck et al., *Nucl. Phys. A,* **626**: 343c, 1997.
25. G. Gabrielse et al., *Phys. Rev. Lett.* **74**: 3544, 1995.
26. G. Savard et al., *Phys. Lett. A,* **158**: 247, 1991.
27. B. Wolf, *Handbook of Ion Sources,* Boca Raton: CRC Press, 1995.
28. Y. Shirakabe et al., *Nucl. Instr and Methods,* **A 337**: 11, 1993.
29. A. Latushinsky and V. Raiko, *Nucl. Instr. Methods,* **125**: 61, 1975.
30. P. v. Duppen, *Nucl. Instr. Methods,* **B 126**: 66, 1997.
31. J. B. Fenn et al., *Science,* **246**: 64, 1989.

HERMANN WOLLNIK
Universität Giessen

MATCHED FILTERS

The history of the matched filter can be traced back to more than half a century ago. In 1940s, due to World War II, radar became a very important detecting device. To enhance the performance of radar, D. O. North proposed an optimum filter for picking up signal in the case of white-noise interference (1). A little bit later, this technique was called matched filter by Van Vleck and Middleton (2). Dwork (3) and George (4) also pursued similar work. The filter has a frequency response function given by the conjugate of the Fourier transform of a received pulse divided by the spectral density of noise. However, the Dwork–George filter is only optimum for the case of unlimited observation time. It is not optimum if observations are restricted to a finite time interval. In 1952, Zadeh and Ragazzini published the work"Optimum filters for the detection of signals in noise" (5), where they described a causal filter for maximizing the signal-to-noise ratio (SNR) with respect to noise with an arbitrary spectrum for the case of unlimited observation time, and second for the case of a finite observation interval. Since then, extensive works on matched filters were done in 1950s. A thorough tutorial review paper called "An introduction to matched filters" (6) was given by Turin.

In the 1960s, due to rapid developments of digital electronics and digital computers, the digital matched filter has appeared (7–9). Turin gave another very useful tutorial paper in 1976, entitled "An introduction to digital matched filters" (10), in which the class of noncoherent digital matched filters that were matched to AM signals was analyzed.

At this time, matched filters have become a standard technique for optimal detection of signals embedded in steady-state random Gaussian noise. The theory of matched filter can be found in many textbooks (11–13).

In this article, we will briefly discuss the theory and application of matched filters. We will start with a continuous input signal case. Then, we will look at the discrete input signal case. Finally, we will provide some major applications of matched filters.

THE MATCHED FILTER FOR CONTINUOUS-TIME INPUT SIGNALS

As mentioned previously, the matched filter is a linear filter that minimizes the effect of noise while maximizing the signal. Thus, a maximal SNR can be achieved in the output. A general block diagram of matched-filter system is described in Fig. 1. To obtain the matched filter, the following conditions and restrictions are required in the system:

1. The input signal consists of a known signal $s_i(t)$ and an additive random noise process $n_i(t)$ with continuous pa-

rameter t. The corresponding output signal and noise are $s_o(t)$ and $n_o(t)$, respectively.
2. The system is linear and time invariant.
3. The criterion of optimization is to maximize the output signal-to-noise power ratio. Since noise $n_o(t)$ is random, its mean squared value $E\{n_o^2(t)\}$ is used as the output noise power.

Mathematically, this criterion can be written as

$$\text{SNR}_o = \frac{s_o^2(t)}{E\{n_o^2(t)\}} = \text{maximum} \tag{1}$$

at a given time t. The form of the matched filter can be derived by finding the linear time-invariant system impulse response function $h(t)$ that achieves the maximization of Eq. (1). The mathematical derivation process can be described as follows.

Since the system is assumed to be linear and time invariant, the relationship between input signal $s_i(t)$ and output signal $s_o(t)$ could be written as

$$s_o(t) = \int_{-\infty}^{t} s_i(\tau)h(t-\tau)\,d\tau \tag{2}$$

Similarly, the relationship between input noise $n_i(t)$ and output noise $n_o(t)$ could also be expressed as

$$n_o(t) = \int_{-\infty}^{t} n_i(\tau)h(t-\tau)\,d\tau \tag{3}$$

Substituting Eqs. (2) and (3) into Eq. (1), the output power SNR can be shown to be

$$\text{SNR}_o = \frac{\left|\int_{-\infty}^{t} s_i(\tau)h(t-\tau)\,d\tau\right|^2}{\int_{-\infty}^{t}\int_{-\infty}^{t} R_n(\tau,\sigma)h(t-\tau)h(t-\sigma)\,d\tau\,d\sigma} \tag{4}$$

where $R_n(\tau, \sigma)$ is the autocorrelation function of the input noise $n_i(t)$ and is given by

$$R_n(\tau,\sigma) = E\{n_i(\tau)n_o(\sigma)\} \tag{5}$$

Now the unknown function $h(t)$ can be found by maximizing Eq. (4). To achieve this goal from Eqs. (4) and (5), one can see that the optimum $h(t)$ (i.e., the matched-filter case) will depend on the noise covariance $R_n(\tau, \sigma)$. Since $h(t)$ is required to be time invariance [i.e., $h(t - \tau)$ instead of $h(t, \tau)$], the noise at least has to be wide-sense stationary [i.e., $R_n(t, \tau) = R_n(t - \tau)$]. To obtain the optimum filter, based on the linear system theory (13), Eq. (4) can be rewritten as

$$\text{SNR}_o = \frac{\left|\int_{-\infty}^{\infty} H(f)S(f)e^{i\omega t_0}df\right|^2}{\int_{-\infty}^{\infty} |H(f)|^2 P_n(f)\,df} \tag{6}$$

where $H(f) = \mathscr{F}[h(t)]$ is the Fourier transform of the impulse response function $h(t)$ (i.e., the transfer function of the sys-

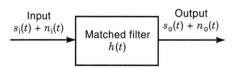

Figure 1. The block diagram of the matched filter in continuous time.

tem), $S(f) = \mathcal{F}[s(t)]$ is the Fourier transform of the known input signal $s(t)$, $\omega = 2\pi f$ is angular frequency, t_0 is the sampling time when SNR$_o$ is evaluated, and $P_n(f)$ is the noise power spectrum density function. To find the particular $H(f)$ that maximizes SNR$_o$, we can use the well-known *Schwarz inequality*, which is

$$\left| \int_{-\infty}^{\infty} A(f)B(f)\,df \right|^2 \leq \int_{-\infty}^{\infty} |A(f)|^2\,df \int_{-\infty}^{\infty} |B(f)|^2\,df \quad (7)$$

where $A(f)$ and $B(f)$ may be complex functions of the real variable f. Furthermore, equality is obtained only when

$$A(f) = kB^*(f) \quad (8)$$

where k is any arbitrary constant and $B^*(f)$ is the complex conjugate of $B(f)$. By using the Schwarz inequality to replace the numerator on the right-hand side of Eq. (6) and letting $A(f) = H(f)\sqrt{P_n(f)}$ and $B(f) = S(f)e^{i\omega t_0}/\sqrt{P_n(f)}$, Eq. (6) becomes

$$\text{SNR}_o \leq \frac{\int_{-\infty}^{\infty} |H(f)|^2 P_n(f)\,df \int_{-\infty}^{\infty} \frac{|S(f)|^2}{P_n(f)}\,df}{\int_{-\infty}^{\infty} |H(f)|^2 P_n(f)\,df} \quad (9)$$

In addition, because $P_n(f)$ is a non-negative real function, Eq. (9) can be further simplified into

$$\text{SNR}_o \leq \int_{-\infty}^{\infty} \frac{|S(f)|^2}{P_n(f)}\,df \quad (10)$$

The maximum SNR$_o$ is achieved when $H(f)$ is chosen such that equality is attained. This occurs when $A(f) = kB^*(f)$, that is,

$$H(f)\sqrt{P_n(f)} = \frac{kS^*(f)e^{-i\omega t_0}}{\sqrt{P_n(f)}} \quad (11)$$

Based on Eq. (11), the transfer function of the matched filter $H(f)$ can be derived as

$$H(f) = k\frac{S^*(f)}{P_n(f)}e^{-i\omega t_0} \quad (12)$$

The corresponding impulse response function $h(t)$ can be easily obtained by taking the inverse Fourier transform of Eq. (12), that is,

$$h(t) = \int_{-\infty}^{\infty} H(f)e^{i\omega t}\,df = \int_{-\infty}^{\infty} k\frac{S^*(f)}{P_n(f)}e^{iw(t-t_0)}\,df \quad (13)$$

In the matched-filter case, the output SNR$_o$ is simply expressed as

$$\max\{\text{SNR}_o\} = \int_{-\infty}^{\infty} \frac{|S(f)|^2}{P_n(f)}\,df \quad (14)$$

For the case of *white noise*, the $P_n(f) = N_0/2$ becomes a constant. Substituting this constant into Eq. (13), the impulse response of the matched filter has a very simple form

$$h(t) = Cs_i(t_0 - t), \quad (15)$$

where C is an arbitrary real positive constant, t_0 is the time of the peak signal output. This last result is one of the reasons why $h(t)$ is called a matched filter since *the impulse response is "matched" to the input signal in the white-noise case.*

Based on the preceding discussion, the matched filter theorem can be summarized as follows: The *matched filter* is the linear filter that maximizes the output signal-to-noise power ratio and has a transfer function given by Eq. (12).

In the previous discussion, the problem of the physical realizability is ignored. To make this issue easier, we will start with the white-noise case. In this case, the matched filter is physically realizable if its impulse response vanishes for negative time. In terms of Eq. (15), this condition becomes

$$h(t) = \begin{cases} 0, & t < 0 \\ s_i(t_0 - t), & t \geq 0 \end{cases} \quad (16)$$

where t_0 indicates the filter delay, or the time between when the filter begins receiving the input signal and when the maximum response occurs. Equation 16 also implies that $s(t) = 0$, $t > t_0$, i.e., the filter delay must be greater than the duration of the input signal. As an example, let us consider the following signal corrupted by additive white noise. The known input signal has the form

$$s_i(t) = \begin{cases} Be^{bt}, & t < 0, \ B, b > 0 \\ 0, & t \geq 0 \end{cases} \quad (17)$$

Substituting Eq. (17) into Eq. (16), the impulse response of matched filter $h(t)$ is

$$h(t) = \begin{cases} Be^{b(t_0-t)}, & t \geq t_0 \\ 0, & t < t_0 \end{cases} \quad (18)$$

The physical realizability requirement can be simply satisfied by letting $t_0 \geq 0$. The simplest choice is $t_0 = 0$ so that $h(t)$ has a very simple form

$$h(t) = \begin{cases} Be^{-bt}, & t \geq 0 \\ 0, & t < 0 \end{cases} \quad (19)$$

The output signal $s_o(t)$ of the system can be obtained by substituting Eqs. (17) and (19) into Eq. (2). The calculated result of $s_o(t)$ is

$$s_o(t) = \begin{cases} \dfrac{B^2}{2b}e^{bt}, & t < 0 \\ \dfrac{B^2}{2b}e^{-bt}, & t > 0 \end{cases} \quad (20)$$

To give an intuitive feeling about the results above, Figs. 2(a)–2(c) illustrate the input signal $s_i(t)$, matched filter $h(t)$, and output signal $s_o(t)$. From Fig. 2(c), indeed, one can get the maximum signal at time $t = t_0 = 0$. Note that, in Fig. 2, we have assumed the following parameters: $B = b = 1$. The physical implementation of this simple matched filter can be achieved by using a simple *RC* circuit as illustrated in Fig. 3, in which the time constant of the *RC* circuit is $RC = 1$.

In many real cases, the input noise may not be white noise and the designed matched filter may be physically unrealizable. Now, let us look at another example with color noise (11). We assume that the input signal $s_i(t)$ has a form of

$$s_i(t) = \begin{cases} e^{-t/2} - e^{-3t/2}, & t > 0 \\ 0, & t < 0 \end{cases} \tag{21}$$

and the input noise is wide-sense stationary with power spectral density

$$P_n(f) = \frac{4}{1 + 4(2\pi f)^2} \tag{22}$$

To obtain the matched filter, first, we take the Fourier transform of input signal $s_i(t)$. Based on Eq. (21), the spectrum of

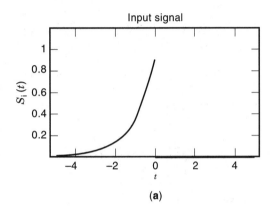

(a)

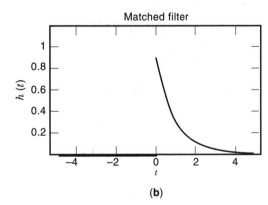

(b)

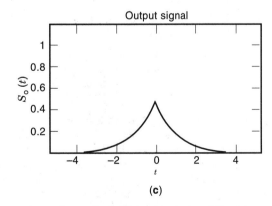

(c)

Figure 2. A simple matched-filter example for white noise and continuous time. (a) Input signal. (b) Matched filter. (c) Output signal. This figure provides an intuitive feeling about using matched filter for continuous time signal processing.

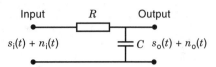

Figure 3. Implementation of the discussed example in text using a RC circuit. This figure shows that the continuous time matched filter can be physically implemented by using a simple RC circuit.

input signal can be shown to be

$$S_i(2\pi f) = \frac{4}{(1 + i4\pi f)(3 + i4\pi f)} \tag{23}$$

Substituting Eqs. (22) and (23) into Eq. (12), the transfer function of matched filter $H(f)$ can be derived as

$$\begin{aligned} H(f) &= k\frac{S^*(f)}{P_n(f)} e^{-i\omega t_0} \\ &= k\frac{1 + i4\pi f}{3 - i4\pi f} e^{-i\omega t_0} \end{aligned} \tag{24}$$

To simplify the expression, we let the arbitrary constant $k = 1$ for the later derivations. By taking the inverse Fourier transform of Eq. (24), the impulse response of the matched filter is

$$h(t) = -\delta(t - t_0) + 2e^{(t-t_0)3/2}u(t_0 - t) \tag{25}$$

where $u(t)$ is the unit step function. Note that this filter is not physically realizable because it has a nonzero value for $t < 0$. To solve this problem, one method is to take a realizable approximation by letting $h(t) = 0$ for $t < 0$. In this case, the approximated matched filter $h_a(t)$ can be expressed as

$$\begin{aligned} h_a(t) &= h(t)u(t) \\ &= -\delta(t - t_0) + 2e^{(t-t_0)3/2}u(t_0 - t)u(t) \end{aligned} \tag{26}$$

Then, the output spectrum $S_o(f)$ of the output signal $s_o(t)$ due to this approximated matched filter is

$$S_o(f) = S_i(f)H_a(f) \tag{27}$$

where $H_a(f)$ is the Fourier transform of $h_a(t)$. Again, by taking the inverse Fourier transform of Eq. (27), the output signal $s_o(t)$ can be derived as

$$\begin{aligned} s_o(t) = &-e^{-3t_0/2}e^{-t/2}u(t) + \tfrac{2}{3}e^{-3(t_0+t)/2}u(t) \\ &- \tfrac{1}{3}e^{-3(t_0-t)/2}u(-t) + \tfrac{1}{3}e^{-3(t_0-t)/2} \end{aligned} \tag{28}$$

Again, to have an intuitive feeling about this example, Fig. 4 illustrates the input signal $s_i(t)$, the ideal physically unrealizable matched filter $h(t)$, approximated realizable matched filter $h_a(t)$, and the output signal $s_o(t)$ obtained with the approximated filter.

For the purpose of convenience, we assume that $t_0 = 1$ for these plots. From Fig. 4(d), one can see that, indeed, the output signal has a maximum value at $t = t_0 = 1$. However, there is no guarantee that this approximated filter is the optimum filter. In fact, it is shown that, a better output SNR can be

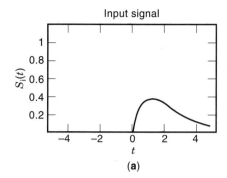

(a)

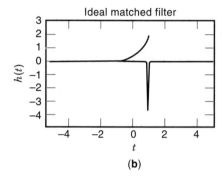

(b)

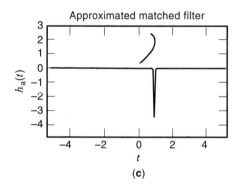

(c)

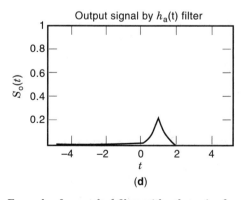

(d)

Figure 4. Example of a matched filter with color noise for a continuous-time signal. (a) Input signal. (b) Ideal matched filter. (c) Approximated matched filter. (d) Output signal with approximated matched filter. This figure illustrates how to deal with color noise with matched filter.

achieved for this problem if the prewhitening technique is employed for the signal detection (11).

Before the end of this section, we want to point out that, in practical terms, it is impossible to design an optimal matched filter for any signal which has an infinite time duration because it requires infinite delay time. However, the above examples are very fast exponential decaying signal, for which one can make the delay time long enough so that optimality can be approached to any desired degree. In other words, the practically "realizable" matched filter only exists for a time limited function. From this point of view, mathematically speaking, the above two examples both have infinite time duration. Thus, even for the second example, it becomes unrealizable. However, since there are extremely fast exponentially decaying signals, optimality can be achieved to any desired degree. In this sense, example 2 can be treated as "realizable" matched filter. Finally, since t_0 represents the delay time of the filter in the above examples, in practice, it must be selected longer than the time duration of the target signal. For the sole purpose of simplicity, in the above examples, the simple values (that are not strict in the mathematical sense) of t_0 are selected.

THE MATCHED FILTER FOR DISCRETE-TIME INPUT SIGNALS

In recent years, with rapid developments of the digital computers, digital signal processing becomes more and more powerful. Some major advantages of using digital signals as compared to their analog forms are the high accuracy, high flexibility, and high robustness. Right now, the matched filter can be easily implemented with the digital computer in real time. To implement the filter with digital computer, one has to deal with the discrete signal instead of continuous signal. In this case, for the same linear time invariant system as described in Fig. 1, the relationship between the output signal $s_o(t)$ and input signal $s_i(t)$ has changed from the continuous-time form Eq. (2) to the following discrete time form (11):

$$s_{oj} = \sum_{k=-\infty}^{j} h_{j-k} s_{ik} \qquad (29)$$

where s_{ik} represents the input signal at time k ($k = 0, \pm 1, \pm 2, \ldots$), h_k is the discrete impulse response function of the linear, time-invariant matched filter, and s_{oj} is the corresponding discrete output signal at time j. In other words, the integration in Eq. (2) has been replaced by the summation in Eq. (29). Similarly, in the discrete-time case, the Eq. (3) is rewritten as

$$n_{oj} = \sum_{k=-\infty}^{j} h_{j-k} n_{ik} \qquad (30)$$

Again, our objective is to find the optimum form of matched filter so that the output signal-to-noise power ratio will be maximum at some time q. Mathematically, it can be written as

$$\mathrm{SNR_o} = \frac{s_{oq}^2}{E\{n_{oq}^2\}} = \mathrm{maximum} \qquad (31)$$

To find h_k, we let maximim SNR, symbolized as SNR_{omax}, equal a constant $1/\alpha$. Since SNR_{omax} represents the maximum *power* ratio, it has to be larger than 0, i.e., $\alpha > 0$. Substituting this assumption into Eq. (31), one can obtain

$$\mathrm{SNR}_o = \frac{s_{oq}^2}{E\{n_{oq}^2\}} \le \mathrm{SNR}_{omax} = \frac{1}{\alpha} \tag{32}$$

Equation (32) can be rewritten as

$$E\{n_{oq}^2\} - \alpha s_{oq}^2 = C \ge 0 \tag{33}$$

where C is a positive real constant and the equality holds only for the optimum matched filter. To find this matched filter, one can substitute Eqs. (29) and (30) into Eq. (33). Then, one can get

$$\sum_{k=-\infty}^{q} \sum_{j=-\infty}^{q} R_n(k-j)h_{q-k}h_{q-j} - \alpha \left| \sum_{k=-\infty}^{q} s_{ik}h_{q-k} \right|^2 = C' \ge 0 \tag{34}$$

where $R_n(k-j)$ is the autocorrelation function of the input noise n_i and C' is another positive constant. Note that, in the process of deriving Eq. (34), we already assume that the input noise is at least wide-sense stationary. Under this assumption, the following condition holds:

$$R_n(k-j) = R_n(j-k) = E\{n_k n_j\} \tag{35}$$

Since the equality holds in Eq. (34) when h_k is an optimum matched filter regardless of the detail forms of input signal and noise, it can be shown that the following equation can be derived under this condition (11):

$$\sum_{j=-\infty}^{q} R_n(k-j)h_{q-j} = s_{ik} \tag{36}$$

To make our discussion easy to be understood, we start with the simple white-noise case. In this case, the autocorrelation function can be simply written as

$$R_n(k) = \begin{cases} N_0/2, & k=0 \\ 0, & k \ne 0 \end{cases} \tag{37}$$

Substituting Eq. (37) into Eq. (36), we obtain

$$\frac{N_0}{2}h_{q-k} = s_{ik} \tag{38}$$

To get a simpler expression of h_k, we let $l = q - k$. Then, Eq. (38) can be rewritten as

$$h_1 = \frac{2}{N_0}s_{i(q-1)} \tag{39}$$

Comparing Eq. (39) with Eq. (15), one can see that Eq. (39) is exactly the discrete form of Eq. (19).

As an example, let us consider a discrete input signal s_{ik} to be given by

$$s_{ik} = \begin{cases} e^k, & k \le 0 \\ 0, & k > 0 \end{cases} \tag{40}$$

The input noise is additive white noise with autocorrelation function

$$R_n(k) = \begin{cases} \dfrac{N_o}{2} = 1, & k=0 \\ 0, & k \ne 0 \end{cases} \tag{41}$$

Substituting Eqs. (40) and (41) into Eq. (39), we have

$$h_k = s_{i(q-k)} = \begin{cases} e^{q-k}, & k \ge q \\ 0, & k < q \end{cases} \tag{42}$$

Substituting Eqs. (40) and (42) into Eq. (29), the output signal s_{oj} can be derived as

$$s_{oj} = \frac{1}{1-e^2}e^{-|j|}, \quad j = 0, \pm 1, \pm 2, \ldots \tag{43}$$

Again, to have an intuitive feeling about this result, Figs. 5(a), 5(b), and 5(c) illustrate the discrete input signal s_{ik}, the discrete matched filter h_k, and the discrete output signal s_{oj}. To get the simplest form in Fig. 5, we assumed $q = 0$.

Equation (36) only deals with the physically realizable case. In general, Eq. (36) will be written as (11)

$$\sum_{j=-\infty}^{\infty} R_n(k-j)h_{q-j} = s_{ik} \tag{44}$$

In Eq. (44), we have replaced the limit q by ∞. To get the general form of a discrete matched filter for nonwhite noise, we take the z transform on both size of Eq. (44) and use the convolution theorem for z transforms. Then, Eq. (44) can be shown to be (14)

$$z^{-q}P_n(z)H(1/z) = S_i(z) \tag{45}$$

where

$$P_n(z) = \sum_{k=-\infty}^{\infty} R_n(k)z^{-k}$$

$$H(z) = \sum_{k=-\infty}^{\infty} h_k z^{-k} \tag{46}$$

$$S_i(z) = \sum_{k=-\infty}^{\infty} s_{ik}z^{-k}$$

represent the power density spectrum, z transform of a discrete matched filter, and z transform of discrete input signal. To obtain the z transform of a discrete matched filter $H(z)$, Eq. (45) is rewritten as

$$H(z) = \frac{S_i(1/z)}{P_n(z)}z^{-q} \tag{47}$$

In deriving Eq. (47), we have used a property of power density spectrum, that is, $P_n(z) = P_n(1/z)$. Theoretically speaking, the discrete matched filter in the time domain (that is, the impulse response function of discrete matched filter) can be obtained by taking the inverse z transform of Eq. (47) (14), that is

$$h_k = \frac{1}{2\pi i} \oint_\Gamma H(z) z^{k-1} \, dz = \frac{1}{2\pi i} \oint_\Gamma \frac{S_i(1/z)}{P_n(z)} z^{-q} z^{k-1} \, dz \quad (48)$$

where Γ represents a counterclockwise contour in the region of convergence of $H(z)$ enclosing the origin. Note that, similar to the continuous-time case, the discrete matched filter defined by Eqs. (47) and (48) may not be realizable for arbitrary input signal and noise because h_k will not vanish for negative values of the index k.

To implement the color noise effectively, the prewhitening technique is used (11). In this approach, the input power density spectrum $P_n(z)$ is written as the multiplication of two facts $P_n^+(z)$ and $P_n^-(z)$, that is,

$$P_n(z) = P_n^+(z) P_n^-(z) \quad (49)$$

where $P_n^+(z)$ has all of the poles and zeros of $P_n(z)$ that are inside the unit circle and $P_n^-(z)$ has all the poles and zeroes of $P_n(z)$ that are outside the unit circle. By this definition, it is easy to show that

$$P_n^+(z) = P_n^-(1/z) \quad (50)$$

Note that, in the time domain, $P_n^+(z)$ corresponds to a discrete-time input signal that vanishes for all times $t < 0$. Similarly, $P_n^-(z)$ corresponds to a discrete-time input signal that vanishes for all time $t > 0$. This property can be easily proven in the following way. Assume that n_k is a discrete-time function that vanishes on the negative half-line; that is,

$$n_k = 0, \quad k < 0 \quad (51)$$

If n_k is absolutely summable, that is, if

$$\sum_{k=-\infty}^{\infty} |f_k| = \sum_{k=0}^{\infty} |f_k| < \infty \quad (52)$$

then, the z transform of this discrete function n_k becomes

$$N(z) = \sum_{k=-\infty}^{\infty} n_k z^{-k} = \sum_{k=0}^{\infty} n_k z^{-k} \quad (53)$$

From Eq. (53), one can see that the function $N(z)$ exists everywhere when $|z| \geq 1$. Hence, the poles of $N(z)$ will all be inside the unit circle. Thus, $P_n^+(z)$ corresponds to a discrete-time input signal that vanishes for all time $t < 0$. Similarly, it can be shown that $P_n^-(z)$ corresponds to a discrete-time input signal that vanishes for all time $t > 0$. Assume $H_{pw}(z)$ is the prewhitening filter. Based on the definition of prewhitening filter, $H_{pw}(z)$ for the noise power spectrum $P_n(z)$ must satisfy (11)

$$[P_n^+(z) H_{pw}(z)][P_n^-(z) H_{pw}(1/z)] = 1 \quad (54)$$

From Eq. (54), one can conclude that the prewhitening filter is

$$H_{pw}(z) = \frac{1}{P_n^+(z)} \quad (55)$$

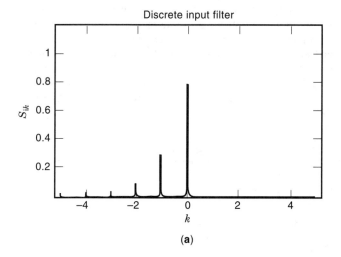

(a)

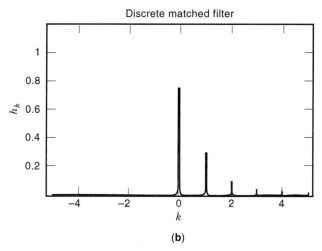

(b)

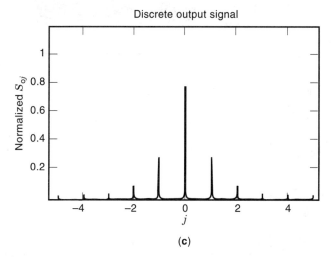

(c)

Figure 5. An example of matched filter for the white noise in discrete time. (a) Discrete input signal. (b) Discrete matched filter. (c) Discrete output signal. This figure gives an intuitive feeling about using matched filter for discrete time signal processing.

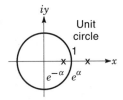

Figure 6. Pole locations of the power spectrum density.

Because $P_n^+(z)$ corresponds to a discrete-time input signal that vanishes for all time $t < 0$, the impulse response h_{pwk} of this prewhitening filter will vanish for $k < 0$. Hence, the prewhitening filter $H_{pw}(z)$ is physically realizable. For example, let us consider a color noise with power density spectrum

$$P_n(z) = \frac{N_0}{2} \frac{e^{2\alpha}}{(e^\alpha - z^{-1})(e^\alpha - z)}, \quad \alpha > 0 \quad (56)$$

Equation (56) shows that $P_n(z)$ contains poles both inside and outside the unit circle. As discussed in the early part of this section, this $P_n(z)$ can be written as the multiplication of $P_n^+(z)$ and $P_n^-(z)$. For the purpose of convenience and symmetry, we let

$$P_n^+(z) = \sqrt{\frac{N_o}{2}} \frac{e^\alpha}{e^\alpha - z^{-1}}$$
$$P_n^-(z) = \sqrt{\frac{N_o}{2}} \frac{e^\alpha}{e^\alpha - z} \quad (57)$$

Based on Eq. (57), it is easy to show that $P_n^+(z)$ has a pole at $z = e^{-\alpha}$ (11). Since $\alpha > 0$, $z = e^{-\alpha} < 1$. In other words, this pole is inside the unit circle. Similarly, $P_n^-(z)$ has a pole at $z = e^\alpha$ that is a real number greater than unity. Figure 6 illustrates these pole locations of above power spectral density $P_n(z)$ in the complex plane. In the figure, we assume that $z = x + iy$. For this particular example, the poles are on the real axis.

When applying this prewhitening technique to the discrete matched filter, Eq. (47) will be rewritten as

$$H(z) = \frac{1}{P_n^+(z)} \left(\frac{S_i(1/z)}{P_n^-(z)} z^{-q} \right) \quad (58)$$

Equation (58) is the multiplication of two terms. The first term is the prewhitening filter and the second term is the remainder of the unrealizable matched filter. Note that this multiplication is equivalent to put two linear systems in tandem. Similar to the continuous-time case, this remaining unrealizable filter can be made realizable by throwing away the part that does not vanish for negative time.

APPLICATIONS OF A MATCHED FILTER

As mentioned in the first part of this article, the major application of the matched filter is to pick up the signal in a noisy background. As long as the noise is additive, wide-sense stationary, and the system is linear and time invariant, the matched filter can provide a maximum output signal-to-noise power ratio. The signal can be a time signal (e.g., radar signal) or spatial signals (e.g., images). To have an intuitive feel-

ing about the time signal detection by a matched filter, let us consider the following simple example. For the purpose of convenience, the ideal input signal is assumed to be a normalized sinc function, that is, $\text{sinc}(t) = \sin(\pi t)/\pi t$, as shown in Fig. 7(a). This ideal signal is embedded into an additive broadband white noise. The corrupted signal is shown in Fig. 7(b). Figure 7(c) shows the system output when this corrupted

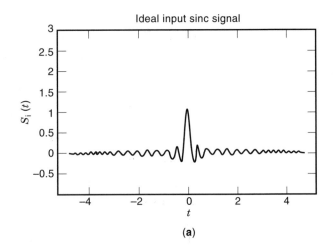

(a)

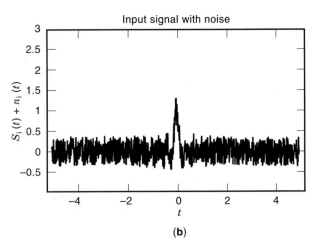

(b)

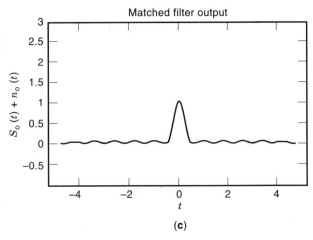

(c)

Figure 7. Results of the matched filter acting on an input signal with sinc function embedded in white noise. (a) Ideal input signal. (b) Signal with noise. (c) Matched-filter output.

signal passes through the matched filter. From Fig. 7(c), one can see that the much better signal-to-noise power ratio can be achieved by applying matched filter for the signal detection as long as the noise is additive at least wide-sense stationary noise.

Besides applying matched filters for the time-signal detection (such as the radar signal previously mentioned), they can also be used for spatial signal detection (15–17). In other words, we can use a matched filter to identify specific targets under the noisy background. Thousands of papers have been published in this field. To save space, here, we just want to provide some basic principles and simple examples of it. Since spatial targets, in general, are two-dimensional signals, the equations developed for the one-dimensional time signal needs to be extended into the two-dimensional spatial signal. Note that when matched filter is applied to the 2-D spatial (or image) identification, this filtering process can be described simply as a cross-correlation of a larger target image (including the noisy background) with a smaller filter kernel. To keep the consistency of the mathematical description, a similar derivation process (used for the 1-D time-signal case) is employed for the 2-D spatial signal. Assume that the target image is a two-dimensional function $s(x, y)$ and this target image is embedded into a noisy background with noise distribution $n(x, y)$. Thus, the total detected signal $f(x, y)$ is

$$f(x,y) = s(x,y) + n(x,y) \tag{59}$$

Similar to the one-dimensional time signal case, if $f(x, y)$ is a Fourier-transformable function of space coordinates (x, y) and $n(x, y)$ is an additive wide-sense stationary noise, the matched filter exists. It can be shown that $H(p, q)$ has a form of (15)

$$H(p,q) = k\frac{S^*(p,q)}{N(p,q)} \tag{60}$$

where $S^*(p, q)$ is the complex conjugate of the signal spectrum, $N(p, q)$ is the spectral density of the background noise, k is a complex constant, and (p, q) are corresponding spatial angular frequencies. Mathematically, $S(p, q)$ and $N(p, q)$ are expressed as

$$S(p,q) = \int_{-\infty}^{+\infty}\int_{-\infty}^{\infty} s(x,y)e^{-i(px+qy)}\,dx\,dy$$
$$N(p,q) = \int_{-\infty}^{+\infty}\int_{-\infty}^{\infty} n(x,y)e^{-i(px+qy)}\,dx\,dy$$
$$F(p,q) = \int_{-\infty}^{+\infty}\int_{-\infty}^{\infty} [s(x,y)+n(x,y)]e^{-i(px+qy)}\,dx\,dy$$
$$= S(p,q) + N(p,q) \tag{61}$$

For the purpose of simplicity, we assume that the input noise $n(x, y)$ is white noise. In this case, Eq. (60) is reduced to the simpler form

$$H(p,q) = k'S^*(p,q) \tag{62}$$

where k' is another constant. Now, assume that there is an input unknown target $t(x, y)$. Then, the corresponding spectrum is $T(p, q)$. When this input target passes through the matched filter $H(p, q)$ described by Eq. (62), the system output in the spectrum domain, that is, (p, q) domain, becomes

$$T(p,q)H(p,q) = K'T(p,q)S^*(p,q) \tag{63}$$

Assume that the final system output is $g(x', y')$, where (x', y') are the spatial coordinates in the output spatial domain. Based on the discussion in the section titled "The Matched Filter for Continuous-Time Input Signals," $g(x', y')$ can be obtained by taking the inverse Fourier transform of Eq. (63), that is,

$$g(x',y') = \int_{-\infty}^{+\infty}\int_{-\infty}^{\infty} T(p,q)S^*(p,q)e^{i(px'+qy')}\,dp\,dq \tag{64}$$

In Eq. (64), if the input unknown function $t(x, y)$ is the same as the prestored function $s(x, y)$, Eq. (64) becomes

$$g(x',y') = \int_{-\infty}^{+\infty}\int_{-\infty}^{\infty} S(p,q)S^*(p,q)e^{i(px'+qy')}\,dp\,dq$$
$$= \int_{-\infty}^{+\infty}\int_{-\infty}^{\infty} |S(p,q)|^2 e^{i(px'+qy')}\,dp\,dq \tag{65}$$

In this case, the system output $g(x', y')$ is the Fourier transform of the power spectrum $|S(p, q)|^2$, which is an entirely positive real number so that it will generate a big output at the original point $(0, 0)$. Notice that, in recent years, due to the rapid development of digital computers, most 2-D filtering can be carried out digitally at relatively fast speed. However, to have an intuitive feeling about 2-D filtering, an optical description about this filtering process that was widely used in the earlier stage of image identification (15) is provided. Optically speaking, the result described by Eq. (65) can be explained in the following way. When the input target $t(x, y)$ is same as the stored target $s(x, y)$, all the curvatures of the incident target wave are exactly canceled by the matched filter. Thus, the transmitted field, that is, $T(p, q)S^*(p, q)$, in the frequency domain, is a plane wave (generally of nonuniform intensity). In the final output spatial domain, this plane wave is brought to a bright focus spot $g(0, 0)$ by the inverse Fourier transform as described in Eq. (65). However, when the input signal $t(x, y)$ is not $s(x, y)$, the wavefront curvature will in general not be canceled by the matched filter $H(p, q)$ in the frequency domain. Thus, the transmitted light will not be brought to a bright focus spot in the final output spatial domain. *Thus, the presence of the signal $s(x, y)$ can conceivably be detected by measuring the intensity of the light at the focal point of the output plane.* If the input target $s(x, y)$ is not located at the center, the output bright spot simply shifts by a distance equal to the distance shifted by $s(x, y)$. Note that this is the shift-invariant property of the matched filter. The preceding description can also mathematically be shown by Schwarz's inequality. Based on the cross-correlation theorem of Fourier transform (17), Eq. (64) can also be written in the spatial domain as

$$g(x',y') = \int_{-\infty}^{+\infty}\int_{-\infty}^{\infty} t(x,y)s^*(x-x',y-y')\,dx\,dy \tag{66}$$

which is recognized to be the cross-correlation between the stored target $s(x, y)$ and the unknown input target $t(x, y)$. By

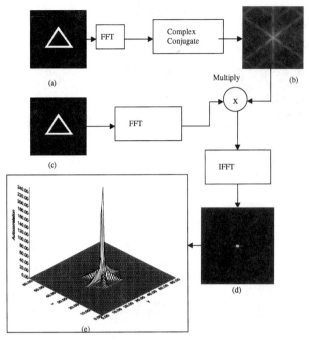

FFT: Fast Fourier Transform
IFFT: Inverse Fast Fourier Transform

Figure 8. Autocorrelation results of the matched filter application to pattern recognition. (a) Stored training image. (b) Absolute value of the matched filter. (c) Unknown input target. (d) Autocorrelation intensity distribution. (e) Three-dimensional surface profile of autocorrelation intensity distribution. This figure shows that there is a sharp correlation peak for autocorrelation.

applying Schwarz's inequality into Eq. (66), we have

$$\left| \int_{-\infty}^{+\infty} \int_{-\infty}^{\infty} t(x,y)s^*(x-x',y-y')\,dx\,dy \right|^2$$
$$\leq \int_{-\infty}^{+\infty} \int_{-\infty}^{\infty} |t(x,y)|^2\,dx\,dy \int_{-\infty}^{+\infty} \int_{-\infty}^{\infty} |s(x-x',y-y')|^2\,dx\,dy \quad (67)$$

with the equality if and only if $t(x,y) = s(x,y)$. Because the integral limit is $\pm\infty$ in Eq. (67), by letting $x = x - x'$, $y = y - y'$, we have

$$\int_{-\infty}^{+\infty} \int_{-\infty}^{\infty} |s(x-x',y-y')|^2\,dx\,dy = \int_{-\infty}^{+\infty} \int_{-\infty}^{\infty} |s(x,y)|^2\,dx\,dy \quad (68)$$

Substituting Eq. (68) into Eq. (67), we have

$$\left| \int_{-\infty}^{+\infty} \int_{-\infty}^{\infty} t(x,y)s^*(x-x',y-y')\,dx\,dy \right|^2$$
$$\leq \int_{-\infty}^{+\infty} \int_{-\infty}^{\infty} |t(x,y)|^2\,dx\,dy \int_{-\infty}^{+\infty} \int_{-\infty}^{\infty} |s(x,y)|^2\,dx\,dy \quad (69)$$

with the equality if and only if $t(x,y) = s(x,y)$. To recognize the input target, we can use the normalized correlation intensity function as the *similarity criterion* between the unknown input target and the stored target. The normalized correlation

intensity function is defined as

$$\frac{\left| \int_{-\infty}^{+\infty} \int_{-\infty}^{\infty} t(x,y)s^*(x-x',y-y')\,dx\,dy \right|^2}{\int_{-\infty}^{+\infty} \int_{-\infty}^{\infty} |t(x,y)|^2\,dx\,dy \int_{-\infty}^{+\infty} \int_{-\infty}^{\infty} |s(x,y)|^2\,dx\,dy} \quad (70)$$

Based on Eq. (69), we obtain

$$\frac{\left| \int_{-\infty}^{+\infty} \int_{-\infty}^{\infty} t(x,y)s^*(x-x',y-y')\,dx\,dy \right|^2}{\int_{-\infty}^{+\infty} \int_{-\infty}^{\infty} |t(x,y)|^2\,dx\,dy \int_{-\infty}^{+\infty} \int_{-\infty}^{\infty} |s(x,y)|^2\,dx\,dy} \leq 1 \quad (71)$$

with the equality if and only if $s(x,y) = t(x,y)$. Thus, one can conclude that the normalized correlation intensity function has a maximum value 1 when the unknown input target $t(x, y)$ is same as the stored target $s(x, y)$. In other words, if there is a 1 detected in the normalized correlation intensity function, we know that the unknown input target is just our stored target. Therefore, this unknown target is recognized.

Again, to have an intuitive feeling about the pattern recognition with matched filter, let us look at the following example. Figure 8(a) shows a triangle image that is used to construct the matched filter. Mathematically speaking, this image is $s(x, y)$. Then, the matched filter $S^*(p, q)$ is synthe-

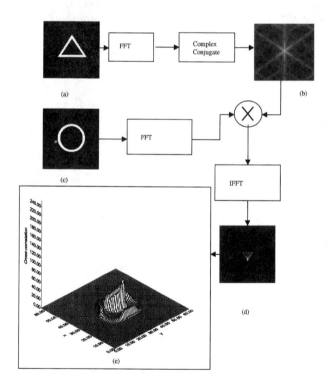

FFT: Fast Fourier Transform
IFFT: Inverse Fast Fourier Transform

Figure 9. Cross-correlation results of the matched filter applied to pattern recognition. (a) Stored training image. (b) Absolute value of the matched filter. (c) Unknown input target. (d) Cross-correlation intensity distribution. (e) Three-dimensional surface profile of cross-correlation intensity distribution. This figure shows that there is no sharp correlation peak for cross correlation.

sized based on this image. Figure 8(b) shows the absolute value of this matched filter. When the unknown input target $t(x, y)$ is the same triangle image as shown in Fig. 8(c), Fig. 8(d) shows the corresponding autocorrelation intensity distribution on the output plane. Figure 8(e) depicts the corresponding three-dimensional surface profile of the autocorrelation intensity distribution. From this figure, one can see that, indeed, there is a sharp correlation peak in the correlation plane. However, if the unknown input target $t(x, y)$ is not the same image used for the matched-filter construction, the correlation result is totally different. As an example, Figs. 9(a) and 9(b) show the same stored image and matched filter. Figure 9(c) shows a circular image used as the unknown input target. Figures 9(d) and 8(e) illustrate the cross-correlation intensity distribution and corresponding three-dimensional surface profile. In this case, there is no sharp correlation peak. Therefore, from the correlation peak intensity, one can recognize the input targets. In other words, one can tell whether the unknown input target is the stored image or not. Before the end of this section, we would like to point out that, besides the 2-D matched filter, in recent years, 3-D (spatial-spectral) matched filters were also developed. Due to space limitations, we can not provide a detail description about this work. Interested readers are directed to papers such as the one written by Yu et al. (19).

CONCLUSION

In this article, we have briefly introduced some basic concepts of the matched filter. We started our discussion with the continuous-time matched filter. Then we extended our discussion to the discrete-time input signals. After that, some major applications of the matched filters such as the signal detection and pattern recognition were addressed.

BIBLIOGRAPHY

1. D. O. North, Analysis of the factors which determines signal/noise discrimination in radar, *Technical Report No. PtR-6C,* Princeton, NJ: RCA Laboratories, 1943.

2. J. H. Van Vleck and D. Middleton, A theoretical comparison of visual, aural and meter reception of pulsed signals in the presence of noise. *J. Appl. Phys.,* **17**: 940–971, 1946.

3. B. M. Dwork, Detection of a pulse superimposed on fluctuation noise, *Proc. IRE,* **38**: 771–774, 1950.

4. T. S. George, Fluctuations of ground clutter return in air-borne radar equipment, *J. IEE,* **99**: 92–99, 1952.

5. L. Z. Zadeh and J. R. Ragazzini, Optimum filters for the detection of signals in noise, *Proc. IRE,* **40**: 1123–1131, 1952.

6. G. L. Turin, An introduction to matched filters, *IRE Trans. Inf. Theory,* **IT-6**: 311–329, 1960.

7. C. R. Cahn, Performance of digital matched filter correlation with unknown interference, *IEEE Trans. Commun. Technol.* **COM-19**: Part II, 1163–1172, 1971.

8. D. J. Gooding, *A digital matched filter for signals of large time-bandwidth product. Technical Report No. 16,* Waltham, MA: Sylvania Communications System Laboratory, February 1969.

9. K. Y. Chang and A. D. Moore, Modified digital correlator and its estimation errors, *IEEE Trans. Inf. Theory,* **IT-16**: 699–700, 1970.

10. G. L. Turin, An introduction to digital matched filters, *Proc. IEEE,* **64**: 1092–1112, 1976.

11. J. B. Thomas, *An Introduction To Communication Theory and System,* New York: Springer-Verlag, 1988, p. 202.

12. J. H. Karl, *An Introduction to Digital Signal Processing,* New York: Academic Press, 1989, p. 217.

13. L. W. Couch, *Digital and Analog Communication Systems,* New York: Macmillan, 1990, p. 497.

14. L. B. Jackson, *Digital Filters and Signal Processing,* Boston: Kluwer Academic Publishers, 1989, p. 34.

15. A. Vander Lugt, Signal detection by complex spatial filtering, *IEEE Trans. Inf. Theory,* **IT-10**, 139–145, 1964.

16. *SPIE Milestone Series on Coherent Optical Processing,* edited by F. T. S. Yu and S. Yin (eds.), Bellingham, WA: SPIE Optical Engineering Press, 1992.

17. S. Yin, et al., Design of a bipolar composite filter using simulated annealing algorithm, *Opt. Lett.* **20**: 1409–1411, 1995.

18. F. T. S. Yu, *Optical Information Processing,* New York: Wiley-Interscience, 1983, p. 10.

19. X. Yu, I. Reed, and A. Stocker, Comparative performance analysis of adaptive multispectral detectors, *IEEE Trans. Signal Process.,* **41**, 2639, 1993.

SHIZHUO YIN
FRANCIS T. S. YU
University Park, PA

MATCHING.   See BROADBAND NETWORKS.

MATERIALS.   See FUNCTIONAL AND SMART MATERIALS.

MATERIALS, CONDUCTIVE.   See CONDUCTING MATERIALS.

MATERIALS EVALUATION.   See EDDY CURRENT NONDESTRUCTIVE EVALUATION.

MATERIALS, FUNCTIONAL.   See FUNCTIONAL AND SMART MATERIALS.

MATHEMATICAL LINGUISTICS.   See COMPUTATIONAL LINGUISTICS.

MATHEMATICAL OPTIMIZATION.   See MATHEMATICAL PROGRAMMING.

MATHEMATICAL PROGRAMMING

Mathematical programming is an interdisciplinary branch of mathematical science, computational science, and operations research that seeks to answer the question, "What is best?," for problems in which the quality of any answer can be expressed as a computable value. In the context of *mathematical programming,* the term *programming* does not denote a particular type of computer programming, but is synonymous with the word *optimization* contemporarily or *optimal planning* originally. In the 1940s, the term programming was used to describe the planning or scheduling of activities within some large organization. Programmers found that they could represent the amount or level of each activity as a variable whose value was to be determined. They then could mathematically describe the restrictions inherent in the planning or scheduling problem as a set of equations or inequalities involving the variables. A solution to all of these constraints would be considered an acceptable or feasible plan or schedule.

It was soon found that modeling a complex operation simply by specifying constraints is hard. If there were too few constraints, many inferior solutions could satisfy them; if there were too many constraints, desirable solutions were ruled out, or in the worst case no solutions were possible. The success of programming ultimately depends on a key insight that provided a way around this difficulty. One could specify, in additional to the constraints, a measure of performance or objective that is a function of the variables or activities, such as cost or profit, that could be used to decide whether one solution was better than another. Then, a best or an optimal solution is the one that gives the best possible value, that is minimum or maximum value, of the objective function while satisfying all constraints. The term *mathematical programming,* which is interchangeable with *mathematical optimization,* came to be used to describe the minimization or maximization of an objective function of many variables, subject to constraints on the variables.

Not much was known about this field before 1940. For one thing, computers are necessary since applications usually require extensive numerical computation. However, there were some very early theoretical contributions; for instance, in the last century Cauchy described the method of steepest ascent (up a mountain) in connection with a system of equations (derivatives equated to zero). The field began to flourish in the 1940s and 1950s with the introduction and development of the very important branch of the subject known as *linear programming*—the case where all the costs, requirements, and other quantities of interest are terms strictly proportional to the levels of the activities or sums of such terms. In mathematical terminology, the objective function is a linear function, and the constraints are linear equations and inequalities. Such a problem is called a linear program. The term linear programming is referred to as the process of setting up a linear program and solving it.

Linear programming is without doubt the most natural mechanism for formulating a vast array of problems with modest effort and today a standard tool that has saved many thousands or millions of dollars for most companies or businesses of even moderate size in the various industrialized countries of the world. Its huge success is partially due to the facts that the mathematics involved is simple and easy to understand and that its first computational method, known as the simplex method, has been extremely successful in practice. But it seems that the popularity of linear programming lies primarily with the formulation phase of analysis rather than the solution phase. In fact, most existing optimization applications can be categorized more or less as some sort of optimal allocation of scarce resources in which a great number of constraints and objectives that arise in the real world are indisputably linear, especially in the area of managerial economics.

In spite of the broad applicability of linear programming, the linearity assumption is sometimes too unrealistic since many physical and economic phenomena in the world surrounding us are nonlinear. For many real-world problems in which functional relationships are nonlinear and that involve interactions between the problem variables, linear programming models are not sufficient to describe the relevant problem complexities. If instead some nonlinear functions of the variables are used in the objective or constraints, the problem is then called a nonlinear program. Most optimization problems encountered in engineering are of this nature. Solving such problems is harder, but in practice often achievable. Al-though the study of computational methods for solving nonlinear programs began in the 1960s, many effective algorithms that are able to solve problems with thousands of variables have been developed.

The solution of a linear or nonlinear program can be fractional. For some applications a fractional solution makes perfect sense; a financial investment decision expressed as a fraction of a large unit, say $1.3 million in portfolio selections is such an example. There are applications, however, in which fractional solutions do not make much sense. For example, the optimal solution of an airline scheduling model may be to fly 1.3 airplanes from city A to city B, which, while mathematically correct, is in reality utter nonsense. When we impose the extra restriction on a linear or nonlinear program that some or all variables must take on integral values, we obtain a mixed or pure integer linear or nonlinear program that in general is much harder to solve than its continuous counterpart. Nevertheless, a combination of faster computer technology and more sophisticated methods has made large integer programs increasingly tractable in recent years.

Mathematical programming has undergone rapid development in recent years and grown into a subject of many branches. This article aims at providing general background information in mathematical programming and presenting some basic notions and results in this subject. The emphases are linear, nonlinear, and integer programs. We shall begin in the first section with the so-called diet problem, a classical real-world optimization problem and discuss various ways to model this problem. The second section is concerned with the classification of optimization problems and their standard forms. The section entitled "Applications and Practicalities" gives a brief discussion on some aspects of solving optimization problems in practice and provides some information necessary to those who want to use software to solve optimization problems. A brief overview of important, well-known, and heavily used results in two typical classes of optimization problems, the smooth nonlinear programming problem and integer linear programming problem, is given in the section titled "Basic Theory." In the next section, key algorithms for solving these two classes of problems are reviewed. Finally, in the last section standard textbooks and references on further reading in this subject are provided, including several informative Internet resources related to this subject.

THE DIET PROBLEM

In this section we shall discuss the diet problem, a classical real-world problem that falls into the category of optimal allocation of scarce resources. In order to be solved by computerized optimization algorithms, a real-world problem must be stated in a very rigid algebraic form. We shall analyze the various characteristics of problem situations to formulate the diet problem in several different forms.

Problem Definition

Suppose that prepared foods of the kinds shown in Table 1 at the market are available at the prices per package indicated. These foods provide percentages, per package, of the minimum daily requirements for vitamins A, B, and C for an average individual, as displayed in Table 2. The diet problem is to find the most economical combination of packages that will

Table 1. Food Prices Per Package

Identifier	Food	Price
B	Beef	$3.49
F	Fish	$2.99
C	Cheese	$1.49
M	Meatloaf	$1.99
S	Spaghetti	$2.49

meet the basic minimum nutritional requirements for good health on a weekly basis—that is, at least 700% of the daily requirement for each nutrient. Such a problem might, for example, be faced by the dietician of a large army.

Linear Programming Model

The diet problem can be formulated as a linear program. Let us denote by x_B the number of packages of beef to be purchased, x_F the number of packages of fish, and so forth. Then the total cost of diet is

$$\text{Total cost} = 3.49x_B + 2.99x_F + 1.49x_C + 1.99x_M + 2.49x_S$$

The total percentage of requirement of vitamin A is given by a similar formula, except that x_B, x_F, and so forth are multiplied by the percentage instead of the cost per package:

Total percentage of vitamin A weekly requirement met
$$= 60x_B + 40x_F + 5x_C + 70x_M + 25x_S$$

This amount needs to be greater than or equal to 700%. Similar formulas are needed for other vitamins, and each of these is required to be greater than or equal to 700%. Putting these all together, we have the following linear program for the diet problem:

Minimize total cost

Subject to

Total percentage of vitamin A weekly requirement met
$\geq 700\%$

Total percentage of vitamin B weekly requirement met
$\geq 700\%$

Total percentage of vitamin C weekly requirement met
$\geq 700\%$

or in mathematical terms,

Table 2. Vitamin Requirements

	A	B	C
Beef	60%	25%	20%
Fish	40%	45%	40%
Cheese	5%	20%	20%
Meatloaf	70%	29%	30%
Spaghetti	25%	40%	49%

Minimize $3.49x_B + 2.99x_F + 1.49x_C + 1.99x_M + 2.49x_S$

Subject to

$$60x_B + 40x_F + 5x_C + 70x_M + 25x_S \geq 700$$
$$25x_B + 45x_F + 20x_C + 29x_M + 40x_S \geq 700$$
$$20x_B + 40x_F + 20x_C + 30x_M + 49x_S \geq 700$$
$$x_B \geq 0, x_F \geq 0, x_C \geq 0, x_M \geq 0, x_S \geq 0$$

(1)

Note that we have added lower bounds on the variables at the end in order to have an adequate description of the problem since it does not make sense to purchase fewer than zero packages of a food.

Solving a linear program with less than thousands of variables is considered rather trivial nowadays. The solution of the above linear program is $x_B = 0$, $x_F = 0$, $x_C = 0$, $x_M \approx 5.06$, $x_S \approx 13.83$ with the total cost $\approx \$44.51$, found in less than 0.01 s on a personal computer (PC) using a commercial linear programming software package. Thus the cost is minimized by a diet of about 5.06 packages of meatloaf and 13.83 packages of spaghetti. But this solution does not seem very much balanced. You can check that it neatly provides about 700% of the requirement for vitamins A and B, but about 830% for vitamin C, a bit more than necessary.

Alternative Linear Programming Models

One might guess that a solution for a more balanced diet would be generated by improving the model Eq. (1). There are at least two quick tricks. The first one is to limit the total percentage of requirement of all vitamins, say, in the range of 700% to 770%, which results in the following model with additional upper bounds on the constraints:

Minimize $3.49x_B + 2.99x_F + 1.49x_C + 1.99x_M + 2.49x_S$

Subject to

$$770 \geq 60x_B + 40x_F + 5x_C + 70x_M + 25x_S \geq 700$$
$$770 \geq 25x_B + 45x_F + 20x_C + 29x_M + 40x_S \geq 700$$
$$770 \geq 20x_B + 40x_F + 20x_C + 30x_M + 49x_S \geq 700$$
$$x_B \geq 0, x_F \geq 0, x_C \geq 0, x_M \geq 0, x_S \geq 0$$

(2)

The second is to require the amount of each vitamin to equal 700% exactly. The resultant model is simply to replace each $\geq$ sign with an $=$ sign in the constraints of Eq. (1), that is,

Minimize $3.49x_B + 2.99x_F + 1.49x_C + 1.99x_M + 2.49x_S$

Subject to

$$60x_B + 40x_F + 5x_C + 70x_M + 25x_S = 700$$
$$25x_B + 45x_F + 20x_C + 29x_M + 40x_S = 700$$
$$20x_B + 40x_F + 20x_C + 30x_M + 49x_S = 700$$
$$x_B \geq 0, x_F \geq 0, x_C > 0, x_M \geq 0, x_S \geq 0$$

(3)

Solving Eqs. (2) and (3), respectively, gives the following solutions. For Eq. (2), the total cost $\approx \$45.32$, and the solution $x = (x_B = 0, x_F \approx 4.30, x_C = 0, x_M \approx 4.08, x_S \approx 9.71)$; for Eq. (3), the total cost $\approx \$45.88$, and the solution $x = (x_B = 0, x_F \approx 9.35, x_C = 0, x_M \approx 2.92, x_S \approx 4.87)$. The optimal solutions for diet do become more balanced now as one can see the requirements for vitamins A, B, and C provided by the solution of Eq. (2) are at the level of 700%, 700%, and 770%, respectively. As for the solution of Eq. (3), it is in fact required by the constraints to be an exactly balanced diet. One can check that, however, since the constraints became more and more restrictive, the total cost went up from \$44.51 for model Eq. (1) to \$45.32 for model Eq. (2) and further to \$45.88 for model Eq. (3).

Integer Programming Model

To be really rigorous, one might insist that solutions of the diet problem must be integer-valued as foods are sold in the unit of one package. A straightforward way to obtain integral solutions is to round off the fractional variables of solutions obtained to their nearest integers. While this might be satisfactory in certain situations, a better alternative is to make the use of integer programming technique. To this end, the problem must be formulated as an integer program. For the diet program, this can be easily done by imposing one *additional* constraint that all variables must be integer-valued. The corresponding integer program for linear program Eq. (1) looks like

Minimize $3.49x_B + 2.99x_F + 1.49x_C + 1.99x_M + 2.49x_S$

Subject to

$$60x_B + 40x_F + 5x_C + 70x_M + 25x_S \geq 700$$
$$25x_B + 45x_F + 20x_C + 29x_M + 40x_S \geq 700 \qquad (4)$$
$$20x_B + 40x_F + 20x_C + 30x_M + 49x_S \geq 700$$
$$x_B \geq 0, x_F \geq 0, x_C \geq 0, x_M \geq 0, x_S \geq 0$$

x_B, x_F, x_C, x_M, x_S are integers

Solving this integer program using an integer programming software yields the total cost $\approx$ \$44.81, and the solution is $x = (x_B = 0, x_F = 0, x_C = 0, x_M = 5, x_S = 14)$. Note that this solution is exactly the result of rounding off the fractional variables (recall that $x_M \approx 5.06$, $x_S \approx 13.83$) of the solution of Eq. (1) to the nearest integers.

The solution for the integer program counterpart of linear program Eq. (2) is the total cost = \$45.32 and $x = (x_B = 0, x_F = 5, x_C = 0, x_M = 4, x_S = 9)$. Note, however, that this solution cannot be obtained by rounding off the fractional variables of solution of Eq. (2) to the nearest integers.

In continuing to solve the corresponding integer program for Eq. (3), we found that this time, however, our software reported an infeasibility of the model, meaning there are no integer-valued variables that would satisfy all the constraints.

Probabilistic Model

Modeling many real-world problems is complicated by the fact that the problem data cannot be known accurately for a variety of reasons. The simplest reason is due to measurement error. Other reasons could be that problem data are stochastic in nature as in many engineering systems and that some data represent information about the uncertain future as in many planning problems. Undoubtedly, the quality of solutions of a model depends not only on the accuracy of functional relationships involved but on the quality of the data in the model. As an example let us revisit the diet problem. One might agree that the amount of vitamin content in food is not a constant but some sort of random variable. The percentage of the minimum vitamin daily requirement per package of a food for an

average individual is therefore also a random variable. So is the total percentage of a vitamin weekly requirement met. Then what does constraint satisfaction mean in this instance? The constraints of linear programming formulation Eq. (1) in this context simply say that the means of the total percentages of vitamin requirements met must be greater than or equal to 700%, which, from a statistical point of view, is not adequate. In what follows we will use this example to illustrate briefly how a branch of mathematical programming, so-called *stochastic programming*, can be utilized to deal with the randomness involved in the problem data.

We shall deal with the vitamin A nutrient content first. Other nutrient contents can be treated in the same way. Let p_B, p_F, p_C, p_M, and p_S be random variables representing the percentages, per package, of the minimum weekly requirements for vitamin A of foods beef, fish, cheese, meatloaf, and spaghetti, respectively, for an average individual. Assume that they are normally distributed with means α_B, α_F, α_C, α_M, and α_S and variances σ_B^2, σ_F^2, σ_C^2, σ_M^2, and σ_S^2, respectively; further assume that they are independently distributed for the sake of simplicity of discussion. Now let us denote by x_B the number of packages of beef to be purchased, and so forth, as we did before. Then the total percentage of vitamin A weekly requirement met, denoted by $u_{A,x}$, is

$$u_{A,x}(p) = p_B x_B + p_F x_F + p_C x_C + p_M x_M + p_S x_S$$

which is also a normal random variable with mean

$$\alpha_{A,x} = \alpha_B x_B + \alpha_F x_F + \alpha_C x_C + \alpha_M x_M + \alpha_S x_S$$

and variance

$$\sigma_{A,x}^2 = \sigma_B^2 x_B^2 + \sigma_F^2 x_F^2 + \sigma_C^2 x_C^2 + \sigma_M^2 x_M^2 + \sigma_S^2 x_S^2$$

Obviously, it is too demanding to ask for a solution x to satisfy

$$u_{A,x} \geq 700 \text{ for all possible value of } u_{A,x}$$

In fact, this is impossible as $u_{A,x}$ is a normal distribution that ranges over all possible real values. However, one might soon realize that what is really required practically is that

$$\Pr(u_{A,x} \geq 700) \approx 1$$

where $\Pr(u_{A,x} \geq 700)$ denotes the probability of the event $(u_{A,x} \geq 700)$. This is the basic idea behind what is called the *chance-constrained programming*. In general, the chance-constrained approach is to require that

$$\Pr(u_{A,x} \geq 700) \geq l_A, \qquad (5)$$

where l_A is the desired probability that the nutrient constraint be satisfied and often called the *acceptance level*. The quantity l_A is a parameter of the model chosen by the modeler and reflects the modeler's attitude towards how often the nutrient constraint should be satisfied. It should always be, of course, less than 1. The value of 0.95 is often a reasonable pick. It is essential to note that, from a computational point of view, the uncertainty introduced by the randomness of problem data is removed in the above formulation Eq. (5) provided that the distribution function of problem data is known and computable.

Similarly, the chance constraints for vitamins B and C could be formulated as

$$\Pr(u_{B,x} \geq 700) \geq l_B,$$
$$\Pr(u_{C,x} \geq 700) \geq l_C,$$

respectively. Then based on the chance-constrained programming approach, a *certainty* or *deterministic* counterpart of the earlier diet linear model Eq. (1) while the problem data have uncertainty involved is

Minimize $3.49x_B + 2.99x_F + 1.49x_C + 1.99x_M + 2.49x_S$

Subject to

$$
\begin{aligned}
&\Pr(u_{A,x} \geq 700) \geq l_A \\
&\Pr(u_{B,x} \geq 700) \geq l_B \\
&\Pr(u_{C,x} \geq 700) \geq l_C \\
&x_B \geq 0, \ x_F \geq 0, \ x_C \geq 0, \ x_M \geq 0, \ x_S \geq 0
\end{aligned}
\tag{6}
$$

In general, an optimization model involving uncertainty can be converted to a deterministic nonlinear model, the explicit algebraic form of which can sometimes be obtained without much analytical effort. For instance, assume that the means and variances of the vitamin A contents of different foods are given in Table 3.

Then with some analytical manipulation, the chance-constrained constraint for vitamin A

$$\Pr(u_{A,x} \geq 700) \geq l_A$$

with acceptance level $l_A = 0.95$ is equivalent to the following nonlinear constraint:

$$60x_B + 40x_F + 5x_C + 70x_M + 25x_S - 1.645$$
$$(1.24x_B^2 + 0.19x_F^2 + 0.45x_C^2 + 10.2x_M^2 + 4.25x_S^2)^{1/2} \geq 700$$

Similar equivalent forms can be obtained for other vitamins. Thus Eq. (6) can in fact be converted to a deterministic nonlinear program.

Applicability

The diet problem is one of the first optimization problems studied back in the 1930s and 1940s and was first motivated by the Army's desire to meet the nutrient requirements while minimizing the cost. The original diet problem is essentially the same as the version given here and had 77 foods and 9 nutrients. It was first solved to optimality in the National Bureau of Standards in 1947 using then the newly created simplex method for the linear program. The solution process took nine clerks using hand-operated desk calculators 120 work days although it can be solved now in a few seconds on a PC.

Table 3. Vitamin A Contents

	Beef	Fish	Cheese	Meatloaf	Spaghetti
Mean	60%	40%	5%	70%	25%
Variance	1.24	0.19	0.45	10.2	4.25

The optimal solution turned out to consist of wheat flour, corn meal, evaporated milk, peanut butter, lard, beef liver, cabbage, potatoes, and spinach and does not seem to be tasty at all. This is not surprising as taste is in fact not a concern in the problem definition. It is not expected that people actually choose their foods by solving this model. However, similar models might be of practical use as a way for providing feed for animals. More sophisticated and practical versions of the diet problem taking into account color, taste, and variety as well as the frequency of food consumption have been proposed by dieticians and nutritionists since the original diet problem was published.

CLASSIFICATION, MATHEMATICAL FORMULATIONS, AND STRUCTURE OF OPTIMIZATION PROBLEMS

Although optimization problems arise in all areas of science and engineering, at root they have a remarkably similar form. In general, optimization problems are made up of three basic ingredients: an objective function that we want to minimize or maximize; a set of decision variables that affect the value of the objective function; and a set of constraints that allow the variables to take on certain values but exclude others. In mathematical terms, the most general form of optimization problem may be expressed as follows [*Mathematical program* (MP)]:

Minimize $f(x)$

Subject to

$$
\begin{aligned}
&h_i(x) = 0, \ i = 1, \ldots, p \\
&g_j(x) \leq 0, \ j = 1, \ldots, r \\
&x \in D
\end{aligned}
\tag{7}
$$

where the decision variable x is a n-component vector $x = (x_1, x_2, \ldots, x_n)$; the objective function f, equation constraint functions h_i and inequality constraint functions g_j are real functions; and D is the domain space where x can take values. A point $x \in D$ that satisfies all constraints is called a feasible point and the set consisting of all feasible points is called a feasible set or feasible region. In the rest of the article, when D is not specified it is always assumed that $D = R^n$, and in such case the last inclusion $x \in D$ is omitted.

The formulation [Eq. (7)] is a bit too general and further classification is possible based on problem characteristics and structure. The next section is an overview of classification of optimization problems and their standard forms. The subsequent section is concerned with some classes of optimization problems that have special mathematical structure.

Classification and Formulations

Optimization problems are classified into subclasses based on their intrinsic characteristics and the structure of problem functions. Each subclass has its own standard form and has been studied separately in order to develop the most effective algorithms for solving this subclass of problems. Although there is no unified taxonomy for optimization problems, the following considerations seem to lead to a reasonable classification scheme.

Constrainedness. Constrainedness means whether or not a MP has constraints present. Unconstrained programs are those without constraints and $D = R^n$. Constrained programs are those having at least one constraint.

Nonlinearity. This features the nonlinearity of problem functions. If any of the objective function f and constraints h_i and g_j of a MP is not linear, then it is said to be a nonlinear program. Otherwise, it is a linear program. A linear program has to be constrained or it is trivial since in this case, either it has no solution or it has the whole domain R^n as its solutions. An unconstrained nonlinear program is referred to as an unconstrained optimization program. Constrained nonlinear programs can be further classified according to their increasing nonlinearity as quadratic programs, bound-constrained programs, linearly constrained programs, and general nonlinear programs.

Dimensionality. Based on their dimensionality, optimization problems are classified into one-dimensional and multidimensional problems. Unconstrained optimization problems in which the decision variable x is a single-component vector are called one-dimensional optimization problems and form the simplest but nonetheless a very important subclass of optimization problems.

Integrality. Most discrete optimization problems impose integrality on decision variables. If the variables of a MP are required to take integer values, it is called an integer program. In such a case, we often write $D = Z^n$, where Z^n is the set of integral n-dimensional vectors. If some of the variables must be integers but the others can be real numbers, it is called a mixed-integer program. In many models, the integer variables are used to represent logical relationships and therefore are constrained to equal either 0 or 1. Then we obtain a restricted 0-1 or binary integer program. In a binary integer program, we write $D = B^n$, where B^n is the set of n-dimensional binary vectors. Although most integer programs are NP-complete [see Nemhauser and Wolsey (1)], many of them from the real world can be solved, at least close, to optimality by exploiting problem-specific structures.

Size and Sparsity. The size of a MP is measured in terms of the number of variables (components) of x and the number of constraints and is often, though not always, proportional to the difficulty of solving the problem. Traditionally, mathematical programs are grouped into small-scale, intermediate-scale, and large-scale problems. Today, with present computing power small-scale, intermediate-scale, and large-scale linear programs usually mean having from a few to a thousand variables and constraints, a thousand to a few hundred thousands variables and constraints, and more than a million variables and constraints, respectively. For the much harder nonlinear programs, small scale, intermediate scale, and large scale mean having from a few to a dozen variables and constraints, a few hundred to a thousand variables and constraints, and more than a thousand to tens of thousands of variables and constraints, respectively. Data sparsity is also one of the measures of the problem complexity. For most real-world optimization problems, the sparsity increases as the size gets large.

Data Uncertainty. Most practical optimization models include some level of uncertainty about the data or functional relationships involved. In many cases not much is lost by assuming that these "uncertain" quantities are deterministic either because the level of uncertainty is low or because these quantities play an insignificant role in the model. However, there are cases where these uncertain quantities play a substantial role in the analysis and cannot be ignored by the model builder. To deal with the uncertainty involved in optimization problems, stochastic programming has been developed. Stochastic programs are written in the forms of mathematical programs with the extension that the coefficients that are not known with certainty are given a probabilistic representation that could be a distribution function. To solve them with a computer, stochastic programs are in general converted to some certainty equivalents. Much of the study of stochastic programs lies in the phase of uncertainty modeling and how to convert them to deterministic equivalents.

The considerations previously noted can be used to classify optimization problems into subclasses and standard forms of these subclasses have been used to communicate the problem structure to general optimization software packages. We shall list some of the standard forms as follows.

A standard form of a *linear program* (LP) is

$$\text{Minimize } c_1 x_1 + c_2 x_2 + \cdots + c_n x_n$$

Subject to

$$a_{11} x_1 + a_{12} x_2 + \cdots + a_{1n} x_n = b_1$$
$$a_{21} x_1 + a_{22} x_2 + \cdots + a_{2n} x_n = b_2 \qquad (8)$$
$$\vdots$$
$$a_{m1} x_1 + a_{m2} x_2 + \cdots + a_{mn} x_n = b_m$$
$$x_1 \geq 0, \; x_2 \geq 0, \; \ldots, \; x_n \geq 0$$

where the objective and all constraints are linear. In more compact vector notation, this standard form becomes

$$\text{Minimize } c^T x$$

Subject to

$$Ax = b, \qquad (9)$$
$$x \geq 0$$

Note that an inequality constraint such as

$$a_{i1} x_1 + a_{i2} x_2 + \cdots + a_{in} x_n \leq b_i$$

can be converted into an equivalent equality constraint below by introducing a slack variable x_{n+1}

$$a_{i1} x_1 + a_{i2} x_2 + \cdots + a_{in} x_n + x_{n+1} = b_i, x_{n+1} \geq 0$$

Quadratic programs (QP) have linear constraints and quadratic objective functions:

$$\text{Minimize } c^T x + \frac{1}{2} x^T G x$$

Subject to

$$a_i^T x = b_i, \; i = 1, \ldots, p \qquad (10)$$
$$a_j^T x \leq b_j, \; j = 1, \ldots, r$$

where G is a symmetric matrix.

The *unconstrained optimization problem* (UOP) is a nonlinear program without constraints:

$$\text{Minimize } f(x) \tag{11}$$

When there are some simple bounds on the components of x, it is then called a *bound-constrained problem* (BCP)

$$\begin{aligned} &\text{Minimize } f(x) \\ &\text{Subject to} \\ &\quad l_i \leq x \leq u_i, \, i = 1, \ldots, n \end{aligned} \tag{12}$$

Problems with nonlinear objective and linear constraints are called *linearly constrained nonlinear programs* (LCNP)

$$\begin{aligned} &\text{Minimize } f(x) \\ &\text{Subject to} \\ &\quad a_i^T x = b_i, \, i = 1, \ldots, p \\ &\quad a_j^T x \leq b_j, \, j = 1, \ldots, r \end{aligned} \tag{13}$$

If at least one of the constraints of a MP is nonlinear and no specific structure can be detected, then it falls into the category of general *nonlinear programs* (NLPs), which has the following form:

$$\begin{aligned} &\text{Minimize } f(x) \\ &\text{Subject to} \\ &\quad h_i(x) = 0, \, i = 1, \ldots, p \\ &\quad g_j(x) \leq 0, \, j = 1, \, \ldots, r \end{aligned} \tag{14}$$

These standard forms for continuous optimization problems readily extend to their corresponding *integer programs* by adding one additional constraint $x \in Z^n$ for pure integer programs, or $x \in B^n$ for binary integer programs, etc.

The general formulation of a *stochastic program* (SP) is as follows:

$$\begin{aligned} &\text{Minimize } E\{f(x, \xi)\} \\ &\text{Subject to} \\ &\quad E\{h_i(x, \xi)\} = 0, \, i = 1, \ldots, p \\ &\quad E\{g_j(x, \xi)\} \leq 0, \, j = 1, \ldots, r \end{aligned} \tag{15}$$

where ξ is a random vector and E is the expectation functional. This model is rich enough to include a wide range of applications, and in fact, has been further classified.

The above taxonomy is not unique. For instance, a linear program is also a quadratic program, which is also a nonlinear program. In general, an optimization problem should be put in the most restricted class for which it qualities. This helps in accurately communicating the problem structure to software used for solving the problem.

Structures of Optimization Problems

The mathematical structure of an optimization problem has implications for the existence and behavior of solutions, the difficulty of solving the problem, and the speed of convergence of algorithms. The basic mathematical properties of optimization problems are continuity and smoothness. In this article we shall always assume that the problem functions of interest are continuous and smooth as most of them are in the real world. For problems involving nonsmooth functions, we simply comment that they do arise in practical situations and the study of them has in fact formed a branch of mathematical programming, called *nonsmooth optimization* [see Neittaanmaki (2)]. In this subsection we shall present two classes of optimization problems that have very desirable special structure, the convex program and least-square problem.

Convex Programs. Convexity is a very important structure for mathematical programs. A set C in R^n is *convex* if the line segment joining any two points in C is contained in C. A function f defined over C is said to be a *convex* function if the following inequality holds for any two points x_1 and x_2 in C:

$$f(\lambda x_1 + (1 - \lambda)x_2) \leq \lambda f(x_1) + (1 - \lambda)f(x_2) \quad \text{for } 0 \leq \lambda \leq 1$$

A function g defined over C is *concave* if and only if $-g$ is convex. A mathematical program is called a convex (concave) program if the feasible region is a convex set and the objective function is a convex (concave) function. A fundamental property of a convex program is that local solutions are also global solutions. Note that a linear program is a convex program by definition. Detailed information on convexity can be found in Rockafellar (3).

Least-Square Problems. Least-square problems arise from fitting mathematical models to data. Specifically, the assumption is made that the functional relationship between the variable x and function value y is

$$y = f(x, t)$$

where $t \in R^n$ is a vector of parameters that are to be determined, and the form of f is known. Assume that data

$$(x_i, y_i), \, i = 1, \ldots, m$$

have been collected, and we want to select the parameters t in the model $f(x, t)$ so that

$$f(x_i, t) \cong y_i, \, i = 1, \ldots, m$$

It makes sense to choose the "best" estimate of parameters t by solving

$$\text{Minimize}_t \sum_{i=1}^{m} [y_i - f(x_i, t)]^2$$

This unconstrained optimization problem is called a least-square problem. In some situations, it might be necessary for the parameters t of a least-square problem to be subject to certain constraints. Least-square problems can be solved by specifically designed algorithms that take advantage of the structure, namely, the objective is the sum of squares.

APPLICATIONS AND PRACTICALITIES

The practical applications of mathematical programming are incredibly vast, reaching into almost every activity in which numerical information is processed. To provide a comprehensive account of all the applications would therefore be impossible, but a selection of primary areas in engineering might include the following.

1. *Operations management.* Applications in this area are often related to allocation of scare resources to maximize some measure of profit, quality, efficiency, effectiveness, etc. Other types of applications are the analysis and tuning of existing operations and development of production plans for multiproduct processes. Representatives are airline crew scheduling problems [Hoffman and Padberg (4)] and inventory control problems [Hillier and Lieberman (5)]. Applications of this sort are often modeled by linear and integer programs. Planning problems concerning the future are usually handled by the stochastic programming technique [Murray (6)].

2. *Design of engineering systems.* Applications in engineering design range from the design of small systems such as trusses [McCormick (7)] and oxygen supply systems [Reklaitis, Ravindran, and Ragsdell (8)] to large systems such as aero-engine and bridges, from the design of individual structural members to the design of separate pieces of equipment to the preliminary design of entire production facilities. Nonlinear programming is often the choice of modeling device in engineering design problems.

3. *Regression and data fitting.* A common problem arising in engineering model development is the need to determine the parameters of some semitheoretical model given a set of experimental data. The regression and data fitting problems can be transformed to nonlinear optimization problems.

It should be noted that in considering the application of optimization methods in design and operations, the optimization step is but one step in the overall process of arriving at a good design or an efficient operation. As a powerful tool, optimization technique has to be well understood by the user in order to be employed effectively.

The process of implementing an optimization application generally consists of the following three major steps:

1. Problem definition and model development.
2. Use of software to solve the model.
3. Assessment of the result.

These steps might have to be repeated several times until a desirable result is obtained. In what follows, we shall comment on each of the three steps.

In step 1, several decisions have to be made, including defining the decision variables, creation of a single criterion (or objective) function, determination of the function forms and constraints representing the underlying cause and effect relationships among the variables, collection and quantification of the data involved, etc.

In step 2, the appropriate optimization software needs to be chosen, information about the problem must be communicated to the software through its user interface, and the solution found needs to be interpreted and understood in the context of the problem domain. Since most optimization packages are developed for solving a particular problem category and

there is no universal applicable optimization software existing as of today, selecting the appropriate software that is designed for solving the kind of problem in question is important in this step. The book *Optimization Software Guide* by Moré and Wright (9) published in 1993 provides information on about 75 optimization software packages available then covering many categories of optimization problems. The user interface of most optimization software expects the user to provide two kinds of information. The first kind is required and concerned with the problem description such as the algebraic forms and coefficients of objective and constraint functions and the type of the problem. The second kind is usually optional and related to certain algorithmic controlling parameters that are associated with the implementation of the algorithm. Since optimization algorithms involve many decisions that are inherently problem dependent, the controlling parameters allow the user to tune the algorithm in order to make it most effective for the problem being solved. To ease the use of software for the inexperienced or uninterested user, optimization software tends to provide default settings for these parameters.

Step 3 consists of the validation of solutions found and post-optimality analysis (often called sensitivity analysis [Fiacco (10)]). It is a fact of life that even a good optimization algorithm may claim that a solution is found when it is not. One intrinsic reason is that most optimization algorithms are designed to find points that satisfy necessary conditions that do not guarantee optimality. What is worse is that the problem model itself may be ill-posed. For ill-posed problem models, numerical errors such as round-off errors might drive the computed solution far away from the real one, and in such case model modification and data refinement are necessary.

There have been some serious efforts to define *standard input formats* (SIFs) for describing optimization problems of certain category, for example, the MPS format [see Nazareth (11)] that has become the de facto input format for linear and integer programs, the SMPS format proposed by Birge et al. (12) as a standard input format for multiperiod stochastic linear programs, the LANCELOT specification file developed by Conn, Gould, and Toint (13), and the MINOS specification file used by MINOS [Murtagh and Saunders (14)] for general nonlinear programs. These SIFs tend to be very specific and lengthy and are easy to be understood by computer programs but hard to generate and costly to maintain by humans. To get around this difficulty, algebraic modeling languages for mathematical programming, for example, the powerful AMPL developed by Fourer, Gay, and Kernighan (15), began to surge in the recent years. They provide computer-readable equivalents of notation such as $x_i + y_i$, $\sum_{j=1}^{n} a_{ij}x_j \leq b_i$, $i \in S$, etc., that are commonly seen in algebra and calculus and allow the optimization modeler to use traditional mathematical notation to describe objective and constraint functions. These algebraic forms of problem description will then be converted by computer to formats that are understood by optimization algorithms. The use of modeling language has made optimization model prototyping and development much easier and less error-prone than before.

BASIC THEORY

The theory of mathematical programming is incredibly rich. Specialized theories and algorithms have been developed for all problem categories. In the subject of linear programming alone, there have been thousands of research papers published and dozens of textbooks available. In this section we shall first give a brief overview of different types of optimal solutions and a fundamental existence result. Since it is impossible to expose even the very basic theoretical results for all problem categories here, we have decided to focus on two important and heavily used problem categories, namely, the smooth nonlinear programming problem and integer linear programming problem.

A general deterministic mathematical programming problem can be written in the following format [General mathematical program (GMP)]:

$$\text{Minimize } f(x)$$
$$\text{Subject to } x \in \Omega \tag{16}$$

where f is a real function and $\Omega \subset R^n$ is the feasible region. We shall briefly introduce in rigorous terms what we mean by an optimal solution of Eq. 16, and then present an existence result.

Type of Solutions

In general, there are several kinds of optimal solutions to Eq. 16. A point $x_0 \in \Omega$ is said to be a *strict local* (optimal) solution or a *weak local* (optimal) solution if there exists a neighborhood N of x_0 such that $f(x_0) < f(x)$ or $f(x_0) \leq f(x)$ for all $x_0 \neq x \in \Omega \cap N$. We say that x_0 is a *strict global* (optimal) solution if $f(x_0) < f(x)$ for all $x_0 \neq x \in \Omega$, and a *weak global* (optimal) solution if $f(x_0) \leq f(x)$ for all $x_0 \neq x \in \Omega$. A global solution is also a local solution by definition but not vice versa. For some mathematical programs with special structure, for example, convex programs, a local solution is also a global solution. For general nonlinear programs, however, a local solution might not be a global solution, and finding a local solution is usually much easier than a global one.

Existence of Solutions

A mathematical program may or may not have a global solution. When there does not exist a global solution, it could be due to the fact that the program is infeasible, that is, the feasible set is empty, or that the program is unbounded, that is, the feasible set is not empty but the objective function value is unbounded from below in the feasible set. A basic result of a mathematical program is the well-known theorem of Weierstrass, which states that if the objective function f is continuous and the feasible set Ω is nonempty and compact, then a global solution exists.

The rest of this section will be devoted to two typical classes of optimization problems, the smooth nonlinear programming problems in next subsection and integer linear programming problems in the subsequent subsection.

Smooth Nonlinear Programming Problems

The emphasis of this subsection is the optimality conditions of solutions of optimization problems. We shall confine ourselves mainly to the consideration of local solutions as they are simpler and more fundamental than global solutions. In fact, for nonlinear optimization problems local solutions are often, though not always, satisfactory enough in practical situa-

tions. For simplicity of presentation, we always assume in the article that the problem functions in question have the necessary smoothness required in the context.

Optimality Conditions. As we can see from the definition of optimal solutions, optimality of a local solution point is defined by its relationship with other feasible points in contrast to, say, seeking a point where $f(x) = 0$. The verification of optimality directly by the definition cannot be carried out by computers since it would be necessary to evaluate infinitely many neighboring feasible points of a proposed local solution. Fortunately, if the problem functions are smooth enough, it is possible to derive some practical optimality conditions that can characterize local solutions and involve analytical information only at a proposed solution point.

Optimality conditions have fundamental importance in optimization theory and algorithms since they are essential in understanding solution behavior conceptually and optimization algorithms are often motivated by attempts to find points satisfying them. Optimality conditions are of two types: necessary conditions, which must hold at a local solution, and sufficient conditions, conditions that, if satisfied at a feasible point, guarantee that point to be a local solution. To explain these seemingly abstract conditions, we shall begin with a simple case.

Optimality Conditions in the Unconstrained Case. We shall consider the unconstrained optimization problem:

$$\text{Minimize } f(x)$$

The key to the derivation of optimality conditions is to use models that are simple and easy to manipulate to approximate complicated ones. The mathematical ground of the approximation is the basic Taylor-series expansion of problem functions at a point of interest. When f is once differentiable at a point x_0, it can be expanded in its Taylor series about x_0 up to first-order, which gives

$$f(x) = f(x_0) + \nabla f(x_0)^{\mathrm{T}}(x - x_0) + o(\|x - x_0\|)$$

and when f is twice differentiable, the Taylor series up to second-order is

$$f(x) = f(x_0) + \nabla f(x_0)^{\mathrm{T}}(x - x_0) + \frac{1}{2}(x - x_0)^{\mathrm{T}}\nabla^2 f(x_0)(x - x_0)$$
$$+ o(\|x - x_0\|^2)$$

Another technique often used in the derivation of optimality conditions is to consider movement away from a proposed solution point in some given direction or curve that falls in the feasible region and to examine the behavior of problem functions along this direction or curve. Given a direction d, we say that it is a *descent* or *ascent* direction of f at x_0 if

$$\nabla f(x_0)^{\mathrm{T}}d < 0 \text{ or } \nabla f(x_0)^{\mathrm{T}}d > 0$$

And we say that f has negative or positive curvature in d at x_0 if

$$d^{\mathrm{T}}\nabla^2 f(x_0)d < 0 \text{ or } d^{\mathrm{T}}\nabla^2 f(x_0)d > 0$$

Consider the following Taylor series of f along d at x_0:

$$f(x_0 + td) = f(x_0) + t\nabla f(x_0)^T d + \frac{1}{2}t^2 d^T \nabla^2 f(x_0)d + o(t^2)$$

We can see that whether the sign of $\nabla f(x_0)^T d$ is positive or negative (or equivalently, whether d is a descent or ascent direction) determines whether the value of f increases or decreases initially when x moves away from x_0 along d. When $\nabla f(x_0)^T d = 0$, the curvature of f in d at x_0, that is, $d^T \nabla^2 f(x_0)d$, governs the initial behavior of f at x_0 along d. This observation leads to the following optimality conditions.

First-Order Necessary Conditions. If x_0 is a local solution, then $\nabla f(x_0)^T d \geq 0$ for all d, or equivalently, $\nabla f(x_0) = 0$.

Second-Order Necessary Conditions. If x_0 is a local solution, then

1. $\nabla f(x_0) = 0$
2. $d^T \nabla^2 f(x_0)d \geq 0$ for all d, that is, $\nabla^2 f(x_0)$ is positive semidefinite

Second-Order Sufficient Conditions. If a point x_0 satisfies $\nabla f(x_0) = 0$ and $d^T \nabla^2 f(x_0)d > 0$ for all $d \neq 0$, that is $\nabla^2 f(x_0)$ is positive definite, then x_0 is a local solution.

Approximations to the Feasible Region and Nonlinear Programs. Having derived the optimality conditions in the unconstrained case, we now turn our attention to the constrained nonlinear program Eq. (14), where the feasible region Ω is defined by

$$\Omega = \{x \in R^n: \quad h_i(x) = 0, \, i = 1, \ldots, p,$$
$$g_j(x) \leq 0, \, j = 1, \ldots, r\}$$

Note first that at a point x_0 of interest there are two types of inequality constrains: *active* constraints if $g_j(x_0) = 0$, *inactive* constraints otherwise. For a continuous inactive inequality constraint g_j, if $g_j(x_0) < 0$, then g_j stays that way at least locally, that is, $g_j(x) < 0$ for x near x_0 by virtue of continuity. Therefore, the inequality constraint $g_j(x) \leq 0$ is always satisfied for x near x_0 and so can be ignored locally. Thus it is of some importance to know which inequality constraint is active and which is not at a point of interest. Let $A(x)$ denote the index set of active constraints at x.

Given a feasible point x_0, it appears that linearizing the constraint functions by replacing them with their respective first-order approximations would give a good approximation to Ω around x_0:

$$F(x_0) := \{x \in R^n: \quad h_i(x_0) + \nabla h_i(x_0)^T(x - x_0) = 0, \, i = 1, \ldots, p$$
$$g_j(x_0) + \nabla g_j(x_0)^T(x - x_0) \leq 0, \, j \in A(x_0)\}$$
$$(17)$$

where the inactive constraints are ignored, and $h_i(x_0) = 0$ for $i = 1, \ldots, p$, and $g_j(x_0) = 0$ for $j \in A(x_0)$ by the feasibility of x_0. However, this is not always true on account of the fact that the boundary of the feasible region may be curved. To ensure the geometry of Ω around x_0 is adequately captured by $F(x_0)$, a *constraint qualification* is required at x_0. A standard constraint qualification requires that the set $\{\nabla h_i(x_0), i = 1, \ldots, p; \nabla g_j(x_0), j \in A(x_0)\}$ be linearly independent.

Better approximations of Ω around x_0 than $F(x_0)$ can be obtained by using higher-order approximations to the problem functions. For example, by replacing the constraint functions with their respective second-order approximations yields a better one:

$$F^2(x_0) := \{x \in R^n: \quad h_i(x_0) + \nabla h_i(x_0)^T(x - x_0)$$
$$+ \frac{1}{2}(x - x_0)^T \nabla^2 h_i(x_0)(x - x_0) = 0, \, i = 1, \ldots, p$$
$$g_j(x_0) + \nabla g_j(x_0)^T(x - x_0)$$
$$+ \frac{1}{2}(x - x_0)^T \nabla^2 g_j(x_0)(x - x_0) \leq 0, \, j \in A(x_0)\}$$
$$(18)$$

There are also ways to approximate the nonlinear program Eq. (14) at x_0. Two readily available approximations to Eq. (14) making the use of the approximations $F(x_0)$ and $F^2(x_0)$ are the following: *First-Order Approximation to NLP*

$$\text{Minimize } f^1(x) := f(x_0) + \nabla f(x_0)^T(x - x_0)$$
$$\text{Subject to } x \in F(x_0)$$
$$(19)$$

and *Second-Order Approximation to NLP*

$$\text{Minimize } f^2(x) := f(x_0) + \nabla f(x_0)^T(x - x_0)$$
$$+ \frac{1}{2}(x - x_0)^T \nabla^2 f(x_0)(x - x_0) \quad (20)$$
$$\text{Subject to } x \in F^2(x_0)$$

Necessary Optimality Conditions of Nonlinear Programs. Intuitively, if a point x_0 is a local location to Eq. (14), then along any feasible smooth curve $c(t): [0, 1] \to R^n$ emanating from x_0, that is, $c(0) = x_0$ and $c(t) \in \Omega$ for any $t \in [0, 1]$, the objective function f cannot decrease initially, which in mathematical terms means that

$$f'_+(c(0)) \geq 0 \quad (21)$$

since otherwise it would contradict the assumption that x_0 is a local solution. Without getting into detailed mathematics, we simply say that Eq. (21) leads to the following.

First-Order Necessary Conditions. If x_0 is a local solution to Eq. (14), and a constraint qualification holds at x_0, then we have

$$\nabla f(x_0)^T d \geq 0 \text{ for any } d = (x - x_0) \text{ such that } x \in F(x_0) \quad (22)$$

These necessary conditions are intuitive, but inconvenient to manipulate. Among the equivalents of Eq. (22), the following system is important:

$$\nabla L(x_0, u, v) = \nabla f(x_0) + \sum_{i=1}^{p} u_i \nabla h_i(x_0) + \sum_{j=1}^{r} v_j \nabla g_j(x_0) = 0 \quad (23)$$

$$h_i(x_0) = 0, \, i = 1, \ldots, p \quad (24)$$

$$g_j(x_0) \leq 0, \, j = 1, \ldots, r \quad (25)$$

$$v_j g_j(x_0) = 0, \, v_j \geq 0, \qquad j = 1, \ldots, r \quad (26)$$

where $L(x, u, v) = f(x) + \sum_{i=1}^{p} u_i h_i(x) + \sum_{j=1}^{r} v_j g_j(x)$ is the famous Lagrange function, u_i's and v_j's are Lagrange multipliers, and

Eqs. (23)–(26) are called the Karush-Kuhn-Tucker (KKT) conditions, which have a fundamental importance in optimization theory. Note that Eqs. (24) and inequalities (25) are actually the feasibility requirement, and Eq. (26) is the so-called complementarity condition. A triple (x, u, v) satisfying Eqs. (23)–(26) is sometimes referred to as a KKT point and x as a *stationary* point.

The complementarity condition Eq. (26) might look a bit strange at the first glimpse. It states that both v_j and $g_j(x_0)$ cannot be nonzero, or equivalently that inactive constraints have a zero multiplier. Note that when g_j is active, v_j could be either positive (in such a case g_j is said to be strongly active) or zero, the intermediate state between being strongly active and inactive. If there is no such j that $g_j(x_0) = v_j = 0$, then strict complementarity is said to hold, and in such a case, dropping all the inactive constraints and forcing all strongly active constraints to equation constraints will not change the behavior of the KKT system locally.

Second- or higher-order necessary conditions are also derivable by taking into account of second- or higher-order derivative information when available and will not be presented here as they are much less useful than the KKT conditions in practice.

Sufficient Optimality Conditions for General Nonlinear Programs. For convex programs, the first-order necessary conditions Eq. (22) are also sufficient for optimality, but for general nonconvex nonlinear programs, a gap exists between the sufficiency and necessity of Eq. (22). Note that, however, sufficient conditions can be obtained by strengthening Eq. (22) by replacing $\geq$ with $>$.

First-Order Sufficient Conditions. Assume x_0 is a feasible point for Eq. (14). If we have

$$\nabla f(x_0)^T d > 0 \text{ for any } d = (x - x_0) \text{ such that } x \in F(x_0) \quad (27)$$

then x_0 is a strict local solution.

Denote $F^-(x_0) = \{x \neq x_0 : \nabla f(x_0)^T(x - x_0) \leq 0\}$. Then, Eq. (27) can be formulated as $F(x_0) \cap F^-(x_0) = \emptyset$. Unfortunately, first-order sufficient conditions Eq. (27) are rather weak since in general $F(x_0) \cap F^-(x_0)$ is not empty and in such a case first-order derivative information is not sufficient to characterize the optimality. To complement this, second-order sufficient conditions have been developed that will be presented later.

Assume that x_0 is a stationary point and $x \in F^2(x_0)$. Multiplying the equations in Eq. (18) by u_i and the inequalities in Eq. (18) by v_j and adding them to the objective function of Eq. (20), we then obtain an interesting inequality

$$f^2(x) \geq L(x_0, u, v) + \nabla L(x_0, u, v)^T(x - x_0)$$
$$+ \frac{1}{2}(x - x_0)^T \nabla^2 L(x_0, u, v)(x - x_0)$$

Using the above inequality and the facts that $f(x) = f^2(x) + o(\|x - x_0\|^2)$ and that $L(x_0, u, v) = f(x_0)$ and $\nabla L(x_0, u, v) = 0$, we can conclude the following.

Second-Order Sufficient Conditions. Assume x_0 is a stationary point. If there exist Lagrange multipliers u and v such that for every $x \in F(x_0) \cap F^-(x_0)$ we have

$$(x - x_0)^T \nabla^2 L(x_0, u, v)(x - x_0) > 0 \quad (28)$$

then x_0 is a strict local solution. In such a case, we say that (x_0, u, v) satisfies the second-order sufficient conditions.

Integer Linear Programming Problems

In this subsection we shall consider the pure integer linear program

$$\text{Minimize } c^T x$$
$$\text{Subject to } x \in S = \{x \in Z^n : Ax \leq b, \, x \geq 0\} \quad (29)$$

Let P denote the polyhedron $\{x \in R^n : Ax \leq b, \, x \geq 0\}$ and z_{IP} the optimal value of Eq. (29). Then the feasible set can be rewritten as $S = P \cap Z^n$. For simplicity we shall always assume P is bounded and thus S consists of finitely many points. The focus of this subsection is the relationship between an integer program and its relaxations. The basic concepts and results covered here are those, such as valid inequalities and facets of polyhedron, that are concerned with using continuous objects to describe the discrete feasible set S and how to generate them.

We should stress that for integer programs only global solutions are of interest in general. The primary way of establishing the global optimality of a feasible solution x is to compare $c^T x$ with z_{IP} to check if $c^T x - z_{IP} = 0$ or more practically $c^T x - z_{IP} \leq \epsilon$ for some small $\epsilon > 0$. In the latter case, x is a near-optimal solution within the ϵ threshold. One might wonder how the previous verification of optimality can be carried out numerically as z_{IP} is usually unknown in the solution process. The trick is to establish a close enough lower bound w on z_{IP} since $c^T x - w \leq \epsilon$ would imply $c^T x - z_{IP} \leq c^T x - w \leq \epsilon$. A typical technique for finding a lower bound is to use relaxation. The idea is to replace Eq. (29) by an easier problem that can be solved and whose optimal value is then used as a lower bound. Frequently, it is necessary to refine these problems iteratively to obtain successively tighter bounds.

Relaxation. A relaxation of Eq. (29) is any optimization problem

$$\text{Minimize } z_{RP}(x)$$
$$\text{Subject to } x \in S_{RP}$$

where the subscript RP stands for *Relaxed Problem,* with the following two properties:

1. $S \subseteq S_{RP}$
2. $c^T x \geq z_{RP}(x)$ for $x \in S$

If the above relaxation has a solution x^* with optimal value z_{RP}, obviously we have $z_{IP} \geq z_{RP}$, that is, z_{RP} is a lower bound of z_{IP}. Furthermore, if x^* happens to be feasible for the original integer program, then it is also a solution of the original integer program.

An obvious way to obtain a relaxation is to satisfy property 1 by dropping one or more of the constraints that define S and to satisfy property 2 by setting $z_{RP}(x) = c^T x$. The *linear programming relaxation* of Eq. (29) is obtained by deleting the integrality constraints $x \in Z^n$ and thus is given by

$$z_{LP} = \text{Minimize } \{c^T x : x \in P\}$$

Solving this linear program results in a lower bound z_{LP} of z_{IP}. Unfortunately, this lower bound is usually not good enough for difficult integer programs and successive improvement is often needed.

Since the solutions of a linear program lie on its vertices, it is not hard to see that extending the feasible set S to its convex hull will result in a relaxation that is equivalent to Eq. (29),

$$\text{Minimize} \quad c^T x$$
$$\text{Subject to } x \in \text{conv}(S) \tag{30}$$

where $\text{conv}(S)$ is the convex hull of S, that is, the set of points that can be written as a convex combination of points in S, that is,

$$\text{conv}(S) = \left\{ x: x = \sum_{i=1} \lambda_i x^i, \ \sum_{i=1}^m \lambda_i = 1, \ \lambda_i \geq 0, \right.$$
$$\left. \text{where } x^1, \ldots, x^m \text{ is any set of points in } S \right\}$$

This observation, however, does not help us much since it is in general expensive to find a linear inequality description of $\text{conv}(S)$. The focus has largely been on the representation and construction of a weaker relaxation

$$\text{Minimize} \quad c^T x$$
$$\text{Subject to } x \in Q \tag{31}$$

where Q is a polyhedron satisfying $\text{conv}(S) \subseteq Q \subseteq P$ such that the linear program Eq. (31) gives an optimal or near-optimal solution to Eq. (29). To this end, the following concept is useful.

Valid Inequality. An inequality $\pi^T x \leq \pi_0$, where π is a vector, is valid for S, or equivalently $\text{conv}(S)$, if it is satisfied by all points in S. Given two valid inequalities $\pi^T x \leq \pi_0$ and $\gamma^T x \leq \gamma_0$ that are not scalar multiples of each other, we say that $\pi^T x \leq \pi_0$ is stronger than or dominates $\gamma^T x \leq \gamma_0$ if $\{x: \pi^T x \leq \pi_0, x \geq 0\} \subset \{x \ \gamma^T x \leq \gamma_0, x \geq 0\}$. A maximal valid inequality of S is the one that is not dominated by any other valid inequality of S.

Obviously the set of maximal valid inequalities for S describes $\text{conv}(S)$. Thus it would be of considerable interest to know how the valid inequalities, especially, maximal valid inequalities can be generated.

Generating Valid Inequalities. Note that $\text{conv}(S) \subseteq P$ since $S = P \cap Z^n \subseteq P$, and in general $\text{conv}(S) \neq P$. So there might exist valid inequalities for S that are not valid for P. Therefore, the valid inequalities for S cannot be derived only from information about P and have to be obtained using the additional integrality constraint $S \subseteq Z^n$. There are several general methods for generating valid inequalities and the one we shall present here is the so-called *Chvatal-Gomory* (GC) *rounding method*. This approach is based on the simple principle that if a is an integer and $a \leq b$, then $a \leq \lfloor b \rfloor$, where $\lfloor b \rfloor$ is the largest integer less than or equal to b. For $S = \{x: Ax \leq b, x \geq 0\} \cap Z^n$, where $A = (a_1, a_2, \ldots, a_n)$ and $N = (1, \ldots, n)$, the method is a three-step procedure:

1. Choose a nonnegative vector $u = (u_1, \ldots, u_n) \geq 0$, and take a linear combination of the constraints with weights u_i for all i and obtain the following valid inequality

$$\sum_{j \in N} (ua_j) x_j \leq ub$$

2. Since $x \geq 0$ implies $\sum_{j \in N} (ua_j - \lfloor ua_j \rfloor) x_j \geq 0$, subtracting it from the left-hand side of the preceding inequality yields the valid inequality

$$\sum_{j \in N} (\lfloor ua_j \rfloor) x_j \leq ub$$

3. Since $x \in Z^n$ implies $\sum_{j \in N} (\lfloor ua_j \rfloor) x_j$ is an integer, we invoke integrality to round down the right-hand side of the above inequality and obtain the valid inequality

$$\sum_{j \in N} (\lfloor ua_j \rfloor) x_j \leq \lfloor ub \rfloor \tag{32}$$

The valid inequality Eq. (32) can be added to $Ax \leq b$, and then the procedure can be repeated by combining generated inequalities and/or original ones. It can be proved that by repeating the CG procedure a finite number of times, all of the valid inequalities for S can be generated.

In fact, some of the maximal valid inequalities are necessary in the description of $\text{conv}(S)$ and others are not and thus can be dropped. To find out which are necessary and which are not, the following notion from the theory of polyhedra is useful.

Facets of Polyhedron. If $\pi^T x \leq \pi_0$ is a valid inequality for the polyhedron $\text{conv}(S)$ and $F = \{x \in \text{conv}(S): \pi^T x = \pi_0\}$, F is called a face of $\text{conv}(S)$, and we say that $\pi^T x \leq \pi_0$ represents F. If a face $F \neq \text{conv}(S)$, then $\dim(F)$, the dimension of F, must be less than $\dim(\text{conv}(S))$. A face F of $\text{conv}(S)$ is a facet of $\text{conv}(S)$ if $\dim(F) = \dim(\text{conv}(S)) - 1$.

It can be shown that for each facet F of $\text{conv}(S)$, one of the inequalities representing F is necessary in the description of $\text{conv}(S)$. For this reason techniques for finding facets are important in solving integer programs effectively. General methods for generating all valid inequalities such as the CG procedure can be quite inefficient in obtaining facets. The best-known technique for finding facet-defining inequalities of integer programs is to make the use of problem structure and is quite problem-specific. It is indeed more of an art than a formal methodology. Considerable efforts have been devoted to the determination of families of facet-defining inequalities or strong valid inequalities for specific problem classes, and there are many interesting problems for which facet-defining inequalities or strong valid inequalities have been obtained. Interested readers may consult Nemhauser and Wolsey (1) for more information.

ALGORITHMS

An algorithm is in our context a numerical procedure for starting with given initial conditions and calculating a sequence of steps or iterations until some stopping rule is satisfied. A variety of algorithms have been developed for each class of optimization problems. Similar to what we did in the section entitled "Basic Theory," to give the reader a sense of what optimization algorithms look like we shall mainly focus our attention on two typical classes of optimization problems discussed in the previous section, the smooth nonlinear program and integer linear program, and discuss algorithms for solving these two problem classes. The following subsection covers the Newton-type methods for solving smooth nonlinear programs, while in the subsection thereafter, two general methods for solving integer programs, the branch-and-bound method and cutting-plane method, will be presented.

Solving Smooth Nonlinear Programming Problems

Almost all algorithms for smooth optimization are iterative in the sense that they generate a series of points, each point being calculated on the basis of the points preceding it. An iterative algorithm is initiated by specifying a starting point. If an algorithm is guaranteed to generate a sequence of points converging to a local solution for starting points that are sufficiently close to the solution, then this algorithm is said to be *locally convergent*. If the generated sequence of points is guaranteed to converge to a local solution for *arbitrary* starting points, the algorithm is then said to be *globally convergent*. The focus of this subsection is Newton-type methods. We shall begin with the basic Newton method for solving unconstrained optimization problems, which is known to be only locally convergent, and then briefly review how we can globalize Newton's method so that it converges for any starting point. Finally, a generalization of the basic Newton method to constrained problems is presented. Interestingly enough, it has been noticed that almost all iterative algorithms for smooth nonlinear programming that perform exceptionally well in practice are some variants of Newton's method.

Before introducing Newton's method, we must stress that an algorithm being theoretically convergent does not mean it always converges to a solution in a practically allowed time period. The consensus in nonlinear optimization is that to be considered as practically convergent, an algorithm has to be at least superlinearly convergent, a notion related to the speed of convergence, which we shall briefly present next.

Speed of Convergence. Assume that the sequence $\{x_k\}$ generated by an algorithm converges to x^*. If we have

$$\lim_{k \to \infty} \frac{\|x_{k+1} - x^*\|}{\|x_k - x^*\|} = \beta < 1$$

the sequence is said to converge *linearly* to x^* and the rate of convergence is linear. The case for which $\beta = 0$ is referred to as *superlinear* convergence. If

$$\lim_{k \to \infty} \frac{\|x_{k+1} - x^*\|}{\|x_k - x^*\|^2} = \beta > 0$$

then the rate of convergence is quadratic. The algorithm is said to be linear, superlinear, or quadratic, according to the convergence rate of the sequence it generates. It is easy to see that quadratic convergence is faster than superlinear convergence, which is faster than linear convergence. A rich theory on speed of convergence, or convergence rates, for measuring the effectiveness of algorithms has been developed [see Ortega and Rheinboldt (16)].

We shall first consider the unconstrained optimization problem, which is central to the development of optimization algorithm. Constrained optimization algorithms are often extensions of unconstrained ones.

Newton's Method. The underlying principle in most iterative algorithms for smooth optimization is to build, at each iteration, a local model of the problem that is valid near the current solution estimate. The next, often improved, solution estimate is obtained at least in part from solving this local model problem. At the current iteration, the basic Newton method solves the local model that is obtained by replacing the original function with its quadratic approximation around the current iterate x_k

$$\underset{s}{\text{Minimize}}\, q_k(s) := f(x_k) + \nabla f(x_k)^\mathrm{T} s + \frac{1}{2} s^\mathrm{T} \nabla^2 f(x_k) s$$

where $s = x - x_k$. When the Hessian matrix $\nabla^2 f(x_k)$ is positive definite, q_k has a unique minimizer that can be obtained by solving the linear system $\nabla q_k(s) = 0$, that is,

$$\nabla f(x_k) + \nabla^2 f(x_k) s_k = 0 \text{ or } s_k = -\nabla^2 f(x_k)^{-1} \nabla f(x_k)$$

The next iterate is then $x_{k+1} = x_k + s_k$. Convergence is guaranteed if the starting point x_0 is sufficiently close to a local solution. The most notable feature of Newton's method is that the rate of convergence is quadratic.

Globalization of Newton's Method. When the starting point x_0 is far away from a local solution, the iterates generated by the basic Newton method may not even converge. A common approach is to use a line search to globalize the basic Newton method so that it converges from any starting point.

Given a descent search direction d_k, a line-search method generates the iterates by setting $x_{k+1} = x_k + \alpha_k d_k$, where α_k is chosen so that $f(x_{k+1}) < f(x_k)$. A practical criterion for a suitable α_k is to require α_k to satisfy the so-called sufficient decrease condition

$$f(x_k + \alpha_k d_k) \le f(x_k) + \mu \alpha_k \nabla f(x_k)^\mathrm{T} d_k$$

where μ is a constant with $0 < \mu < 1$.

Most line-search versions of the basic Newton method generate the search direction $d_k = s_k = -\nabla^2 f(x_k)^{-1} \nabla f(x_k)$ by occasionally replacing the Hessian matrix $\nabla^2 f(x_k)$ with $\nabla^2 f(x_k) + E_k$ such that the resultant matrix is sufficiently positive definite. This guarantees that the search direction s_k defined by Newton's method is a descent direction since $\nabla f(x_k)^\mathrm{T} s_k = -\nabla f(x_k)^\mathrm{T} [\nabla^2 f(x_k) + E_k]^{-1} \nabla f(x_k) < 0$.

Constrained Optimization. Many techniques have been proposed for solving the constrained nonlinear program Eq. (14). One of them is the *sequential quadratic programming* (SQP) method, which is a generalization of Newton's method for unconstrained optimization. At the current solution estimate, this method uses a linearly constrained quadratic local model to approximate the original problem. In its purest form, replacing the objective function with its quadratic approximation

$$q_k(s) := f(x_k) + \nabla f(x_k)^\mathrm{T} s + \frac{1}{2} s^\mathrm{T} \nabla^2 L(x_k, u_k, v_k) s$$

and the constraint functions with their respective linear approximations, SQP solves

$$\text{Minimize } q_k(s)$$

Subject to

$$h_i(x_k) + \nabla h_i(x_k)^{\text{T}} s = 0, \ i = 1, \ldots, p$$
$$g_j(x_k) + \nabla g_j(x_k)^{\text{T}} s \le 0, \ j = 1, \ldots, r$$

where $s = x - x_k$, and sets the new solution estimate $x_{k+1} = x_k + s_k$. As a variant of Newton's method, SQP inherits excellent local convergence property. Given a KKT point (x^*, u^*, v^*) satisfying the second-order sufficient conditions, when SQP starts at a point x_0 sufficiently close to x^* and all the Lagrange multiplier estimates (u_k, v_k) remain sufficiently close to (u^*, v^*), the sequence it generates converges to x^* at a quadratic rate. One complexity of SQP is that the Lagrange multiplier estimates are needed to set up the second-order term in q_k and so must be updated from iteration to iteration. A direct treatment is simply to use the optimal multipliers for the quadratic local problem at the previous iteration. The interested reader may consult Fletcher (17) for details.

Similar to the basic Newton method for unconstrained optimization, the SQP method in its pure form given earlier is not guaranteed to converge for a starting point that is far away from a local location. Again, a line search along the search direction s_k can be used to globalize SQP. Of course, we now want the next iterate not only to decrease the value of the objective function but also to come closer to the feasible region. But often these two aims conflict and so it is necessary to weight their relative importance and consider their joint effect in reaching optimality. One commonly used technique to achieve this is to use a *merit* or *penalty function* to measure the closeness of a point to optimality

$$m(x; c) := f(x) + \sum_{i=1}^{p} c_i |h_i(x)| + \sum_{j=1}^{r} c_{p+j} \max (g_j(x), 0)$$

where $c_k > 0$ are penalty parameters. Then a line search aiming at achieving sufficient decrease of the merit function can be used to choose an α_k for $x_{k+1} = x_k + \alpha_k d_k$, where $d_k = s_k$. The interested reader might consult Fletcher (17) for more information.

Solving Integer Linear Programming Problems

In the section titled "Integer Linear Programming Problems" we have addressed some basic properties of integer programs and discussed the relationship between an integer program and its relaxations and how to generate valid inequalities to improve the relaxations. In general, integer programs are much more complicated and expensive to solve than their continuous relaxations on account of the discrete nature of the variables. A simple-minded way to deal with an integer program is to form its corresponding continuous relaxation by dropping the integrality constraint, and then to solve the relaxation and round off the solution to its nearby integers in certain manner. In fact, this is how many integer programs are handled in practice. It is important to realize that there is no guarantee that a good solution can be obtained in this way, even by examining all integer points in some neighborhood of the solution of a relaxation. General techniques for

solving integer programs do exist, though they often need to be customized in order to be most effective. We shall present two general methodologies for solving integer programs, namely, the branch-and-bound method and cutting-plane method. For simplicity, we shall confine ourselves to the integer linear programming problem.

Branch-and-Bound Method. The branch-and-bound method solves an integer program by solving a series of related continuous programs in a systematic way. The basic idea behind it is the familiar *divide and conquer*. In other words, if it is too hard to optimize over the feasible set S, perhaps the problem can be solved by optimizing over smaller sets and then putting the results together. More precisely, we can partition the feasible set S into a set of subsets $\{S^i: i = 1, \ldots, k\}$ such that $\cup_{i=1}^{k} S^i = S$ and $S^i \cap S^j = \emptyset$ for $i, j = 1, \ldots, k, i \ne j$, and solve the problem over each of the subsets, i.e., solve (IP^i)

$$\text{Minimize } c^T x$$

$$\text{Subject to } x \in S^i$$

for $i = 1, \ldots, k$. Assume their respective solutions are x^i with optimal value z_{IP}^i for $i = 1, \ldots, k$. Then we can easily put the results together since it is obvious that the optimal value of the original problem $z_{\text{IP}} = \min_{i=1,\ldots,k} z_{\text{IP}}^i$. Let j be the one such that $z_{\text{IP}} = z_{\text{IP}}^j$. Then x^j is a solution of the original problem. Note that this scheme can be applied recursively, that is, if a particular subproblem IP^i cannot be easily solved, the divide-and-conquer process can be carried out for the subproblem IP^i, meaning the subset S^i can be further partitioned and the problem can be solved over the furthered partitioned subsets.

In general, partitioning is done by imposing additional bounds on certain components of x. For instance, the original problem can be partitioned into two subproblems by "branching" on some component, say x_1, yielding IP^1

$$\text{Minimize } c^T x$$

$$\text{Subject to } x \in S, \ x_1 \le 10$$

and IP^2

$$\text{Minimize } c^T x$$

$$\text{Subject to } x \in S, \ x_1 \ge 10 + 1$$

It is possible to repeat the branching process on IP^1 and IP^2, and again on the resulting problems. However, the total number of resultant subproblems increases *exponentially* with the number of levels of branching done, and it is unrealistic to solve all these subproblems when the total gets too high. The branch-and-bound method takes advantage of the fact that many of these subproblems can actually be "pruned" based on information about bounds on the optimal value. Specifically, since the subproblems are solved sequentially, at any stage we can keep track of the best feasible solution obtained so far and its objective function value, which we denote by $\bar{x}_{\text{IP}}$ and $\bar{z}_{\text{IP}}$, respectively. Assume IP^i is the current subproblem we are dealing with. We form a continuous relaxation RP^i of IP^i, solve it, and obtain its global solution x_{R}^i with optimal value z_{R}^i. Now if x_{R}^i is feasible for IP^i, it is then a solution of IP^i and so IP^i is already solved and can be pruned. If $z_{\text{R}}^i < \bar{z}_{\text{IP}}$, x_{R}^i is then a better feasible solution than $\bar{x}_{\text{IP}}$. Thus, $\bar{x}_{\text{IP}}$ and $\bar{z}_{\text{IP}}$

should be updated by setting $\bar{x}_{IP} = x_R^i$ and $\bar{z}_{IP} = z_R^i$. Otherwise, x_R^i is not better than $\bar{x}_{IP}$ and so can be ignored. If x_R^i is not feasible for IP^i, we also compare z_R^i with $\bar{z}_{IP}$. If $z_R^i \geq \bar{z}_{IP}$, we then conclude that there is no hope of finding a better solution than $\bar{x}_{IP}$ by solving subproblem IP^i. The reason is that z_R^i is a lower bound of the optimal value of IP^i due to the fact that RP^i is a relaxation of IP^i and $\bar{z}_{IP}$ is already as good as z_R^i. Thus, in such a case IP^i can be pruned. However, in the case for which $z_R^i < \bar{z}_{IP}$, we cannot rule out the possibility that IP^i could have a solution that is better than $\bar{z}_{IP}$, and so the branching process needs to be carried out further on IP^i.

Many strategies are known in the implementation of the branch-and-bound method with respect to how to branch a subproblem, how to pick the next subproblem to consider when the current subproblem is pruned, etc. For the interested reader, Nemhauser and Wolsey (1) is a good book for details. It is easy to see that the quality of produced bounds ($\bar{z}_{IP}$ and z_R^i) is crucial in pruning out subproblems effectively and in fact the primary factor in the efficiency of a branch-and-bound algorithm.

Cutting-Plane Algorithm. The cutting-plane algorithm works with a sequence of successively tighter continuous relaxations of the integer program Eq. (29) until, hopefully, an integer optimal solution is found. The basic idea is simple. Assume that at the current iteration a solution x^* to the current continuous relaxation is found. If x^* is an integer solution, then it is a solution to the integer program and the problem is solved. Otherwise, we try to find a valid inequality for S that is not satisfied by x^* by solving, often approximately, a so-called separation problem. Since this valid inequality cuts off x^* from S, or more appropriately from conv(S), we then add it to the current relaxation to form a tighter relaxation and proceed to the next iteration. In order to have a sufficiently tighter relaxation, it is desirable to generate a facet-defining inequality that cuts of x^* from conv(S). Generating good cuts is often problem specific and details can be found in Nemhauser and Wolsey (1).

Recently, the cutting-plane algorithm has been incorporated into the general branch-and-bound scheme for solving subproblems or at least improving the bounds. The combined method, called the branch-and-cut method, has proved to be quite effective in solving some hard integer programs [see Hoffman and Padberg (4)].

FURTHER READING

In the previous sections we have sketched some basic results in the subject of mathematical programming, which is now on its way to maturity. There exists a vast literature on this subject. In fact, all topics mentioned in the paper have been explored in great detail in the past several decades. In what follows, we shall suggest some general references based on our limited knowledge on this subject. A collection of articles on the historical accounts of many branches of mathematical programming is the interesting book edited by Lenstra, Rinnooy Kan, and Schrijver (18); excellent state-of-the-art expository articles on the most important topics in mathematical programming by leading experts in the field can be found in the handbook edited by Nemhauser, Rinnooy Kan and Todd (19). Standard textbooks or references in the subject are the

following: Dantzig (20) in linear programming; Bazaraa, Jarvis, and Sherali (21) in linear programs and network flows; Luenberger (22) in linear and nonlinear programming; McCormick (7) and Fletcher (17) in nonlinear programming; Fiacco (10) in sensitivity analysis; Gill, Murray, and Wright (23) in numerical methods and implementation; Rosen (24) in large-scale optimization; Fiacco and McCormick (25) and Megiddo (26) in interior point and related methods; Nemhauser and Wolsey (1) and Schrijver (27) in integer programming; Ahuja, Magnanti, and Orlin (28) in network flows; Hall and Wallace (29) in stochastic programming; Neittaanmaki (2) in nonsmooth optimization; Anandalingam (30) in multilevel programming; Sawaragi, Nakayama and Tanino (31) in multiobjective optimization; Moré and Wright (9) in evaluation and comparison of optimization software packages; Hocking (32) in optimal control. The introductory books in operations research by Hillier and Lieberman (5), Winston (33) and Winston and Albright (34) also cover many branches of mathematical programming. The *Mathematical Programming Society* has published several volumes of selective tutorial lectures given by leading experts covering many branches of mathematical programming at its triennial international symposiums, and the latest ones are the volume "Mathematical Programming: State of the Art 1994" (35) edited by Birge and Murty and the special issue "Lectures on Mathematical Programming, ISMP97" (36) edited by Liebling and Werra.

Many journals contain articles in mathematical programming. The ones devoted to this subject are *Mathematical Programming, Optimization, Journal of Optimization Theory and Applications, SIAM Journal on Optimization,* and *Journal of Global Optimization.* Some of the most relevant ones are *Mathematics of Operations Research, SIAM Journal on Control and Optimization, Operations Research, Management Science, The European Journal of Operational Research,* and *Operations Research Letters.*

There is also a tremendous amount of information relevant to the subject on the Internet. The Operations Research Page (http://mat.gsia.cmu.edu) of Professor Michael Trick at Carnegie Mellon University is a page for pointers to all aspects of Operations Research. The Optimization Technology Center founded jointly by Argonne National Laboratory and Northwestern University has a home page (http://www.mcs.anl.gov/home/otc) that has a lot information on optimization techniques and also implements the so-called network-enabled optimization system designed for solving optimization problems remotely over the Internet. The Mathematical Programming Glossary page (http://www-math.cudenver.edu/~hgreenbe/glossary/glossary.html) maintained by Professor Harvey Greenberg at University of Colorado at Denver contains many technical terms and links specific to mathematical programming.

BIBLIOGRAPHY

1. G. L. Nemhauser and L. A. Wolsey, *Integer and Combinatorial Optimization,* New York: Wiley, 1988.

2. M. Neittaanmaki, *Nonsmooth Optimization,* London: World Scientific Publishing, 1992.

3. R. T. Rockafellar, *Convex Analysis,* Princeton, NJ: Princeton University Press, 1970.

4. K. L. Hoffman and M. Padberg, Solving airline crew scheduling problems by branch-and-cut, *Management Sci.,* **39** (4): 657–682, 1993.

5. F. S. Hillier and G. J. Lieberman, *Introduction to Operations Research,* New York: McGraw-Hill Publishing, Inc., 1980.

6. W. Murray, Financial planning via multi-stage stochastic programs, In J. R. Berge and K. G. Murty (eds.), *Mathematical Programming: State of the Art 1994,* Ann Arbor, MI: 1994.

7. G. P. McCormick, *Nonlinear Programming: Theory, Algorithms and Applications,* New York: Wiley, 1983.

8. G. V. Reklaitis, A. Ravindran, and K. M. Ragsdell, *Engineering Optimization: Methods and Applications,* New York: Wiley, 1983.

9. J. J. Moré and S. J. Wright, *Optimization Software Guide,* Vol. 14 of Frontiers in Applied Mathematics, Philadelphia: SIAM, 1993.

10. A. V. Fiacco, *Introduction to Sensitivity and Stability Analysis in Nonlinear Programming,* New York: Academic Press, 1983.

11. J. L. Nazareth, *Computer Solution of Linear Programs,* New York: Oxford University Press, Inc., 1987.

12. J. B. Berge et al., A standard input format for multiperiod stochastic linear programs, *COAL Newsletter* **17**: 1–19, 1987.

13. A. R. Conn, N. I. M. Gould, and Ph. L. Toint, *LANCELOT,* Berlin: Springer-Verlag, 1992.

14. B. A. Murtagh and M. A. Saunders, MINOS 5.1 User's Guide, Technical Report No. SOL 83-20R, System Optimization Laboratory, Standard University, Standard, 1983.

15. R. Fourer, D. M. Gay, and B. W. Kernighan, *AMPL, A Modeling Language for Mathematical Programming,* San Francisco, CA: The Scientific Press, 1993.

16. J. M. Ortega and W. C. Rheinboldt, *Iterative Solution of Nonlinear Equations in Several Variables,* New York: Academic Press, Inc., 1970.

17. R. Fletcher, *Practical Methods of Optimization,* New York: Wiley, 1987.

18. J .K. Lenstra, A. H. G. Rinnooy Kan, and A. Schrijver (eds.), *History of Mathematical Programming,* Amsterdam: CWI, 1991.

19. G. L. Nemhauser, A. H. G. Rinnooy Kan, and M. J. Todd (eds.), *Optimization,* Amsterdam: North-Holland, 1989.

20. G. B. Dantzig, *Linear Programming and Extensions,* Princeton, NJ: Princeton University Press, 1963.

21. M. S. Bazaraa, J. J. Jarvis, and H. D. Sherali, *Linear Programming and Network Flows,* New York: Wiley, 1990.

22. D. G. Luenberger, *Linear and Nonlinear Programming,* Reading, MA: Addison-Wesley, 1984.

23. P. E. Gill, W. Murray, and M. H. Wright, *Practical Optimization,* London: Academic Press, 1981.

24. J. B. Rosen (ed.), *Supercomputers and Large-Scale Optimization: Algorithms, Software, Applications,* Annals of Operations Research, Vol. 22, Switzerland: J. C. Baltzer AG, Science Publishers, 1990.

25. A. V. Fiacco and G. P. McCormick, *Nonlinear Programming, Sequential Unconstrained Minimization Techniques,* New York: Wiley, 1968.

26. N. Megiddo (ed.), *Progress in Mathematical Programming— Interior Point and Related Methods,* New York: Springer, 1989.

27. A. Schrijver, *Theory of Linear and Integer Programming,* New York: Wiley, 1986.

28. R. K. Ahuja, T. L. Magnanti, and J. B. Orlin, *Network Flows,* Englewood Cliffs, NJ: Prentice Hall, 1993.

29. P. Hall and S. W. Wallace, *Stochastic Programming,* New York: Wiley, 1994.

30. G. Anandalingam (ed.), *Hierarchical Optimization,* Annals of Operations Research 34, Switzerland: J. C. Baltzer AG, Science Publishers, 1992.

31. Y. Sawaragi, H. Nakayama, and T. Tanino, *Theory of Multiobjective Optimization,* New York: Academic Press, 1985.

32. L. M. Hocking, *Optimal Control: An Introduction to the Theory with Applications,* New York: Oxford University Press, 1991.

33. W. L. Winston, *Operations Research: Applications and Algorithms,* Boston: PWI-Kent Publishing Co., 1991.

34. W. L. Winston and S. C. Albright, *Practical Management Science: Spreadsheet Modeling and Applications,* Belmont, CA: Duxbury Press, 1997.

35. J. B. Birge and K. G. Murty (eds.), *Mathematical Programming: State of the Art 1994,* Ann Arbor, MI: The University of Michigan, 1994.

36. T. M. Liebling and D. de Werra (eds.), *Lectures on Mathematical Programming, ISMP97, Mathematical Programming, Vol. 79,* New York: Elsevier Science, 1997.

JIMING LIU
Lucent Technologies

MATHEMATICAL PROGRAMMING. See GEOMETRIC PROGRAMMING.

MATHEMATICAL THEORY OF COMMUNICATIONS. See INFORMATION THEORY.

MATHEMATICS. See GEOMETRY.

MATLAB. See CIRCUIT STABILITY.

MATRIX PROPERTY. See EIGENVALUES AND EIGENFUNCTIONS.

MATRIX RICCATI EQUATIONS. See KALMAN FILTERS AND OBSERVERS.

MAXIMALLY FLAT GAIN FILTERS. See BUTTERWORTH FILTERS.

MAXIMUM ENTROPY. See MAXIMUM LIKELIHOOD DETECTION.

MAXIMUM LIKELIHOOD DETECTION

The task involved in pattern detection or Recognition is that of making a decision about the unknown, yet constant, nature of an observation. In this context, an observation could be a single scalar or a multidimensional vector, and the nature of such observations is related to their classification according to some criteria specific to the application. For instance, in a face detection scenario, the observations are images, and the overall goal of a system is to select those containing human faces.

The maximum likelihood principle states that in a given object classification scenario, one should pick the object class for which the observation in question is most likely to happen. For instance, if we knew that in some place most summer days are sunny and most winter days are cloudy and we are asked to guess the season based solely on the fact that one of its days is sunny, our best guess should be that it is summer.

For the purpose of object detection, we use as much of any available information as we can about the underlying pattern structure of the observations. In most cases, all available information comes in the form of examples whose classification is known beforehand. We refer to them as the training set. Although the basic idea of the maximum likelihood principle is simple, the estimation of the probability distributions of the observations from the training set could be rather complex. Therefore, two different approaches have been taken to deal with this, parametric versus nonparametric probability estimators.

In this article we deal mainly with object detection in the context of computer vision and image understanding. However, maximum likelihood detection and many other approaches in pattern recognition have much wider scope and applicability in a number of different scenarios. In the following sections we describe a visual object detection setup, the aforementioned approaches for maximum likelihood detection, and an automatic face detection system based on non-parametric probability models.

VISUAL OBJECT DETECTION

Most object detection techniques by themselves are not invariant to rotation, scale, illumination changes, object pose, and so on. To overcome this limitation, the training examples are normalized in illumination, scale, rotation, and position before they are used in the learning procedure. The result of this learning procedure is a pattern recognition module capable of detecting the objects in question within a limited range of variation in scale, rotation, and illumination.

Let us assume that a test image is given and that we are to detect objects on it. In the detection procedure, a collection of rescaled and rotated images is computed from the test image according to the desired range of detection capability. Then, each subwindow within these images is normalized for illumination and tested with the aforementioned pattern recognition module to decide whether the desired object is in this subwindow. As a result, a new collection of images is obtained. In these, the pixel value of each position is the result of the pattern recognition module for the corresponding subwindow position. For example, each pixel value could be proportional to the likelihood that this subwindow contains the object. Further analysis of these images is carried out to produce a robust list of candidates of the object being detected. Figure 1 illustrates the case in which faces of different sizes are detected using a pattern recognition module that computes the likelihood that there is a face in a subwindow of size 17×14 pixels.

The overall performance of the detection system depends on the choice of scale factors, rotation angles, illumination normalization algorithms, and the size of the detection subwindow. The narrower these ranges are set, the more consistent the patterns that are fed to the detection module in the learning procedure. However, a larger search space is also required in the detection procedure to cover a similar range of detection capability.

Different techniques can be used in the recognition module (1,2). The approaches of particular interest here are those based on the Bayes decision rule (3). These approaches take each subwindow and try to estimate the probability that it belongs to each of the object classes in question. Then, using the value of the probability as a confidence level, the class with highest probability is selected to describe the object in the subwindow. These approaches are known as probabilistic reasoning techniques and include both maximum likelihood and maximum a posteriori detection setups.

BAYES DECISION RULE

Assuming that the nature of the observations is well known and therefore that the conditional probability densities of the observations for each object class are given, then the Bayes Decision Rule yields the minimum error (4). This error, known as the Bayes error, is a measure of the class separability.

Let $\omega_1, \omega_2, \ldots, \omega_L$ be the object classes and O be the observation variable. Then, the a posteriori probability function of ω_i given O, is obtained using the Bayes Formula as

$$P(\omega_i \mid O) = \frac{p(O \mid \omega_i)P(\omega_i)}{p(O)}$$

where $p(O|\omega_i)$ is the conditional probability density of the observation and $p(\omega_i)$ is the a priori probability for the ith object class and $p(O)$ is the probability of the observation.

The Bayes decision rule states that we should pick the object class ω_i^* with maximum probability, given the observation. If we pick

$$\omega_i = \arg \max_{\omega_i} \{p(O \mid \omega_i)P(\omega_i)\}$$

for the observation in question, we obtain the maximum a posteriori (MAP) decision rule. However, in most cases, the a priori probabilities for the classes $p(\omega_i)$ are unknown, and therefore, for practical purposes, they are set equal. Then, the obtained rule

$$\omega_i^* = \arg \max_{\omega_i} \{p(O \mid \omega_i)\}$$

is known as the maximum likelihood (ML) decision rule.

PROBABILITY MODELS

In reality, the most serious limitation of the Bayes decision rule is the difficulty of estimating the probability distributions needed for its application. In most cases, the information or knowledge about the object classes is available in the form of examples, that is, a set of observations that have been classified beforehand, usually called the training set, are given, and the goal of the learning procedure is to find a discriminant function capable of classifying these observations and also others not available for training.

In this general approach, the training set is used to estimate the probability functions for each object class. The goal of the learning technique is determining the best set of parameters for the probability estimators. As is usual in data fitting problems, estimating the probability from a set of examples faces a number of issues, such as the completeness of the training set, the generalization properties of the models, the optimization criteria, etc.

Probability distributions are usually modeled with parametric functions, for instance, Gaussian mixture densities. Another approach, based on the assumption that the observations are of a discrete nature, is to model the probability functions using the statistical averages. We call the former parametric probability models and the latter nonparametric probability models. Once the probability functions are ob-

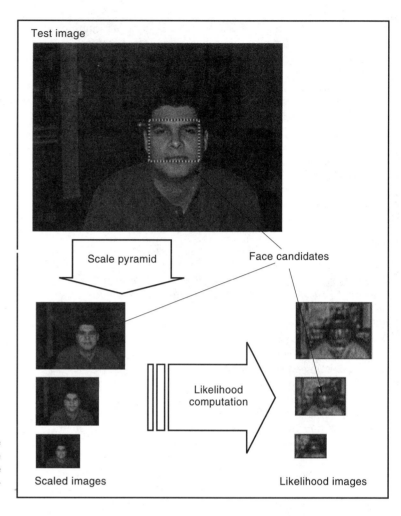

Figure 1. Scheme of a multiscale, maximum likelihood face detection setup. Each subwindow of the scaled version of the test image is tested, and the likelihood that it contains a face is displayed in the likelihood images from which the face candidates are obtained.

tained, they are used in a ML or MAP setup for object detection. In the following section we briefly describe an example of a parametric ML face detection system. However, in the rest of this article we concentrate our effort on a nonparametric ML face detection system.

Parametric Probability Models

In modeling probability functions or distributions of multidimensional random variables, one encounters a difficult issue. There is a compromise between the complexity of the model and the procedure used to fit the model to the given data. Extremely complex models would have to be used to consider all of the underlying dependency among all of the variables and to fit the model well to the training data.

The Karhunen–Loeve transform (5) is often used to reduce the dimensionality of the data and to overcome the limitation imposed by the dependency among the variables. Rather than estimating the probability densities on the original space, the observations are projected to the eigenspace in which the energy is packed to a subset of the components and the components are uncorrelated. Then, in this new space, the probability of the observation is often estimated using a Gaussian distribution or a mixture of Gaussian densities with diagonal covariance matrices (6).

Moghaddam and Pentland reported an example of a maximum likelihood detection system using this approach (7). In

their work, the parameter estimation is carried out using the expectation-maximization (EM) algorithm.

INFORMATION-BASED MAXIMUM DISCRIMINATION

The detection process described in this article is carried out as a classification using the Bayes decision rule. We mainly compute the likelihood ratio of an observation using the probability models obtained from the learning procedure and compare it to a fixed threshold to make the decision. We use statistical averages to construct nonparametric probability models, and the learning procedure is turned into an optimization whose goal is to find the best model for the given training data. From information theory we borrow the concept of Kullback relative information and use it as the optimization criteria that measure the class separability of two probability models.

Let the observed image subwindow be the vector $X \in I^N$, where I is a discrete set of pixel values. Let $P(X)$ be the probability of the observation X given that we know that it belongs to the class of objects we want to detect, and let $M(X)$ be the probability of the observation X, given that we know that it belongs to other classes.

We use the likelihood ratio $L(X) = P(X)/M(X)$ to decide whether the observation X belongs to the object class in question by comparing it to a threshold value. Setting this threshold to 1 leads to the Bayes decision rule. However, different values are used depending on the desired correct-answer-to-false-alarm ratio of the detection system.

Kullback Relative Information

Kullback relative information, also known as Kullback divergence or cross-entropy, measures the "distance" between two probability functions, and therefore it measures the discriminatory power of the likelihood ratio of these probability functions under the Bayes decision rule (8,9).

The divergence of the probability function P with respect to the probability function M is defined as

$$H_{P\|M} = \sum_{\boldsymbol{X}} P(\boldsymbol{X}) \ln \frac{P(\boldsymbol{X})}{M(\boldsymbol{X})}$$

Although it does not satisfy triangular inequality, this divergence is a nonnegative measure of the difference between the two probability functions that equals zero only when they are identical. In our context, we use the Kullback divergence as the optimization criteria in our learning procedure. Basically, we set up a family of probability models and find the model that maximizes the divergence for the given training data.

Modified Markov Model

Dealing with probability models that take full advantage of the dependency of all of the variables is limited by the dimensionality of the problem. On the other hand, assuming complete independence of the variables makes the model rather useless. In between these extremes, we use a modified Markov model. This family of models is well suited for modeling our random processes and also easy to handle mathematically.

We compute the probability of the modified kth order Markov model as

$$P(\boldsymbol{X}) = \prod_{i=1,\ldots,T} P(X_{S_i} \mid X_{S_{i-1}}, \ldots, X_{S_{i-k}})$$

where $\boldsymbol{S} = \{S_1, \ldots, S_T\}$ is a list of indices (e.g., each S_i denotes the pixel location), and the Kullback divergence between the probability functions $P(\boldsymbol{X})$ and $M(\boldsymbol{X})$ of such random processes as

$$H_{P\|M}(\boldsymbol{S}) = \sum_{i=1,\ldots,T} H_{P\|M}(X_{S_i} \mid X_{S_{i-1}}, \ldots, X_{S_i-k})$$

Information-Based Learning

The key idea behind this learning technique is to restate the learning problem as an optimization in which the goal is to find the list $\boldsymbol{S}^* = \{S_1^*, \ldots, S_T^*\}$ that maximizes the Kullback divergence $H_{P\|M}(\boldsymbol{S})$ for a given training set. It is clear that the computational requirements of such an optimization problem are prohibitive. However, we make some simplifications to find a practical solution to this problem.

First, we requantize the observation vector as part of the image preprocessing step so that each pixel has only a few possible values, for instance, four gray levels $X_i = \{0, 1, 2, 3\}$ for $i = 1, \ldots, N$. Then, using a first-order Markov model, the divergence of the two probability functions for a given list of indices $\boldsymbol{S} = \{S_1, \ldots, S_T\}$ is obtained from

$$H_{P\|M}(\boldsymbol{S}) = \sum_{i=1,\ldots,T} H_{P\|M}(X_{S_i} \mid X_{S_{i-1}})$$

where

$$H_{P\|M}(X_j \mid X_k) = \sum_{X_j, X_k} P(X_j, X_k) \ln \frac{P(X_j \mid X_k)}{M(X_j \mid X_k)}$$

is the divergence of each pair of pixels within the image subwindow, and is obtained from the training set using histogram counts and statistical averages.

Then, we treat our optimization as a minimum-weight spanning-tree problem in which the goal is to find the sequence of pairs of pixels that maximizes the sum of $H_{P\|M}(\boldsymbol{S})$. Finally, we use a modified version of Kruskal's algorithm to obtain suboptimal results (10).

Once a solution is obtained, it is used to precompute a three-dimensional lookup table with the log likelihood ratio for fast implementation of the detection test. Given an image subwindow $\boldsymbol{X} \in \boldsymbol{I}^N$, the computation of its log likelihood is carried out as $\log L(\boldsymbol{X}) = \sum_{i=1,\ldots,T} L'[i][X_{S_i}][X_{S_{i-1}}]$, where

$$L'[i][X_{S_i}][X_{S_{i-1}}] = \log \frac{P(X_{S_i} \mid X_{S_{i-1}})}{M(X_{S_i} \mid X_{S_{i-1}})}$$

It is worth noting that such an implementation results in very fast, highly parallelizable algorithms for visual pattern detection. This is particularly important when we consider that the likelihood ratio is computed for each of the image subwindows obtained from the tested image.

FACE AND FACIAL FEATURE DETECTION AND TRACKING

We tested the previously described learning technique in the context of face and facial feature detection. Examples of faces were obtained from a collection of "mug shots" from the FERET database (11) using the locations of the outer eye corners as a reference to normalize the face size and position within the image subwindows. As negative examples, we also used a collection of images of a wide variety of scenes with no frontal-view faces.

We used the likelihood model obtained with the training set in a ML detection setup to locate face candidates in the test images. Several scaled and rotated images are obtained from the input image and tested with this face detection module according to the desired range of detection capability. In addition to locating the face candidates, the system further tests those candidates with likelihood models for the right and left eyes so that the algorithm can accurately locate these facial features. A detailed description of this implementation, testing procedure, error criteria, performance description, etc. can be obtained from Refs. (12,13).

A real-time, automatic face and facial feature tracking system was implemented on an SGI-ONYX with 12 R10000 processors and a SIRIUS video acquisition board. Real-time video is grabbed from a camera to the computer memory for processing and sent back out to a monitor with additional labeling information, such as the position of the face and the facial

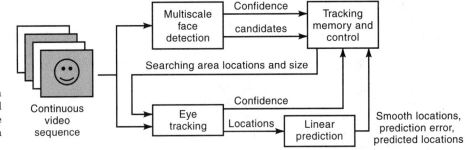

Figure 2. Block diagram of a face detection and eye tracking system. Both the initial face detection and the continuous eye tracking are implemented using maximum likelihood detection setups.

features. As illustrated in the block diagram in Fig. 2, the system handles continuous video sequences. Faces and their eyes are first detected in an upright frontal view using our ML setup. Then, the eyes are tracked accurately over the video sequence under face translation, rotation, and zooming by applying a similar ML detection setup. In the tracking setup, the predicted position of the eyes is used to limit the search for eye detection.

The system operates in two modes. In the detection mode, the system constantly carries out an exhaustive search to find faces and their outer eye corners. Set up to detect faces in a range of sizes (actually, the distance between the outer eye corners) between 100 and 400 pixels, this detection loop runs on 10 processors at about 3 frames per second. Once a face is successfully detected, that is, when the confidence level of the detection is above a fixed threshold, the system switches to the tracking mode.

In the tracking mode, the predicted positions of the outer eye corners are used to normalize the incoming video frames. A normalized image is obtained for each frame so that the tracked face lays in an upright position and with the appropriate size. Then, the locations of the eyes are continuously updated by applying the eye detector in these normalized images. The eye detection module is based on the likelihood models obtained with the aforementioned visual learning technique but at a much higher resolution than that used in face detection. As a result, there is no error accumulation or inaccuracy over long video sequences, and a wide range of rotation and zooming can be handled successfully. Whenever, the confidence level of the eye tracking falls below a predefined threshold, the system switches back to the detection mode, and this cycle starts over.

BIBLIOGRAPHY

1. R. Schalkoff, *Pattern Recognition: Statistical, Structural and Neural Approaches,* New York: Wiley, 1992.

2. K. Fukunaga, *Introduction to Statistical Pattern Recognition,* 2nd ed., New York: Academic Press, 1991.

3. H. Stark and J. Woods, *Probability, Random Processes, and Estimation Theory for Engineers,* Englewood Cliffs, NJ: Prentice-Hall, 1986.

4. L. Devroye, L. Gyërfi, and G. Lugosi, *A Probabilistic Theory of Pattern Recognition,* New York: Springer-Verlag, 1996.

5. I. T. Jolliffe, *Principal Component Analysis,* New York: Springer-Verlag, 1986.

6. R. A. Redner and H. F. Walker, Mixture densities, maximum likelihood and the EM algorithm, *SIAM Rev.,* **26** (2): 195–239, 1984.

7. B. Moghaddam and A. Pentland, Probabilistic visual learning for object representation, *IEEE Trans. Pattern Anal. Mach. Intell.,* **19**: 696–710, 1997.

8. R. M. Gray, *Entropy and InformationTheory,* New York: Springer-Verlag, 1990.

9. J. N. Kapur and H. K. Kesavan, *The Generalized Maximum Entropy Principle,* Waterloo, Canada: Sandford Educational Press, 1987.

10. T. H. Cormen, C. E. Leirserson, and R. L. Rivest, *Introduction to Algorithms,* New York: McGraw-Hill, 1990.

11. P. J. Phillips et al., The FERET September 1996 database and evaluation procedure, *Proc. 1st Int. Conf. Audio Video-Based Biometric Person Authentication,* Crans-Montana, Switzerland, March 12–14, 1997.

12. A. Colmenarez and T.S. Huang, Maximum Likelihood Face Detection, *Int. Conf. Automatic Face Gesture Recognition,* Vermont, USA, 1996.

13. A. Colmenarez and T. S. Huang, Face detection with information-based maximum discrimination, *CVPR,* San Jose, Puerto Rico, 1997.

Antonio J. Colmenarez
Thomas S. Huang
University of Illinois at Urbana-Champaign

MAXIMUM LIKELIHOOD IMAGING

Imaging science is a rich and vital branch of engineering in which electromagnetic or acoustic signals are measured, processed, analyzed, and interpreted in the form of multidimensional images. Because these images often contain information about the physical, biological, or operational properties of remote objects, scenes, or materials, imaging science is justly considered to be a fundamental component of that branch of engineering and science known as remote sensing. Many subjects benefit directly from advances in imaging science—these range from astronomy and the study of very large and distance objects to microscopy and the study of very small and nearby objects.

The photographic camera is probably the most widely known imaging system in use today. The familiar imagery recorded by this device usually encodes the spectral reflectance properties of an object or scene onto a two-dimensional plane. The familiarity of this form of imagery has led to a common definition of an image as "an optical reproduction of an object by a mirror or lens." There are, however, many other imaging systems in use and the object or scene properties encoded in their imagery can be very different from those recorded by a photographic camera. Temperature variations can, for instance, be "imaged" with infrared sensing, velocity variations can be "imaged" with radar, geological formations can be "imaged" with sonar, and the physiological function of the human brain can be "imaged" with positron emission tomography (PET).

A photographic camera forms images in a manner very similar to the human eye, and, because of this, photographic images are easily interpreted by humans. The imagery recorded by an infrared camera might contain many of the features common to visible imagery; however, the phenomena being sensed are different and some practice is required before most people can faithfully interpret raw infrared imagery. For both of these modalities, though, the sensor data is often displayed as an image without the need for significant signal processing. The data acquired by an X-ray tomograph or synthetic aperture radio telescope, however, are not easily interpreted, and substantial signal processing is required to form an "image." In these situations, the processing of raw sensor data to form imagery is often referred to as image reconstruction or image synthesis (1), and the importance of signal processing in these applications is great. To confirm this importance, the 1979 Nobel prize in physiology and medicine was awarded to Alan M. Cormack and Sir Godfrey N. Hounsfield for the development and application of the signal processing methods used for X-ray computed tomography, and the 1974 Nobel prize in physics was awarded to Sir Martin Ryle for the development of aperture synthesis techniques used to form imagery with radio telescope arrays. For both of these modalities the resulting images are usually very different from the visible images formed by photographic cameras, and significant training is required for their interpretation.

Imagery formed by photographic cameras, and similar instruments such as telescopes and microscopes, can also be difficult to interpret in their raw form. Focusing errors, for example, can make imagery appear blurred and distorted, as can significant flaws in the optical instrumentation. In these situations, a type of signal processing known as image restoration (2,3) can be used to remove the distortions and restore fidelity to the imagery. Processing such as this received national attention after the discovery of the Hubble Space Telescope aberrated primary mirror in 1990, and one of the most successful and widely used algorithms for restoring resolution to Hubble imagery was based on the maximum-likelihood estimation method (4). The motivation for and derivation of this image-restoration algorithm will be discussed in great detail later in this article.

When signal processing is required for the formation or improvement of imagery, the imaging problem can usually be posed as one of statistical inference. A large number of estimation-theoretic methods are available for solving statistical-inference problems (5–9), and the method to be used for a particular application depends largely on three factors: (1) the structure imposed on the processing; (2) the quantitative criteria used to define image quality; and (3) the physical and statistical information available about the data collection process.

Structure can be imposed on processing schemes for a variety of reasons, but the most common is the need for fast and inexpensive processing. The most common structure imposed for this reason is linear processing, whereby imagery is formed or improved through linear combinations of the measured data. In some situations structural restrictions such as these are acceptable, but in many others they are not and the advent of faster and more sophisticated computing resources has served to greatly lessen the need for and use of structural constraints in imaging problems.

Many criteria can be used to quantify image quality and induce optimal signal-processing algorithms. One might ask, for example, that the processed imagery produce the "correct" image on average. This leads to an unbiased estimator, but such an estimator may not exist, may not be unique, or may result in imagery whose quality is far from adequate. By requiring that the estimated image also have, in some sense, the smallest deviations from the correct image this criterion could be modified to induce the minimum variance, unbiased estimator (MVUE), whose imagery may have desirable qualities, but whose processing structure can be difficult or impossible to derive and implement. The maximum-likelihood method for estimation leads to an alternative criterion whereby an image is selected to optimize a mathematical cost function that is induced by the physical and statistical model for the acquired data. The relative simplicity of the maximum-likelihood estimation method, along with the fact that maximum-likelihood estimates are often asymptotically unbiased with minimum variance, makes this a popular and widely studied method for statistical inference. It is largely for this reason that the development and utilization of maximum-likelihood estimation methods for imaging are the focus of this article.

One of the most important steps in the utilization of the maximum-likelihood method for imaging is the development of a practical and faithful model that represents the relationship between the object or scene being sensed and the data recorded by the sensor. This modeling step usually requires a solid understanding of the physical and statistical characteristics of electromagnetic- or acoustic-wave propagation, along with an appreciation for the statistical characteristics of the data acquired by real-world sensors. For these reasons, a strong background in the fields of Fourier optics (10,11), statistical optics (12–14), basic probability and random-process theory (15,16), and estimation theory (5–9) is essential for one wishing to apply maximum-likelihood methods to the field of imaging science.

Statistical inference problems such as those encountered in imaging applications are frequently classified as ill-posed problems (17). An image-recovery or -restoration problem is ill posed if it is not well posed, and a problem is well posed in the classical sense of Hadamard if the problem has a unique solution and the solution varies continuously with the data. Abstract formulations of image recovery and restoration problems on infinite-dimensional measurement and parameter spaces are almost always ill posed, and their ill-posed nature is usually due to the discontinuity of the solution. Problems that are formulated on finite-dimensional spaces are frequently well-posed in the classical sense—they have a unique solution and the solution is continuous in the data. These problems, however, are often ill conditioned or badly behaved and are frequently classified as ill posed even though they are technically well posed.

For problems that are ill posed or practically ill posed, the original problem's solution is often replaced by the solution to a well-posed (or well-behaved) problem. This process is referred to as regularization and the basic idea is to change the problem in a manner such that the solution is still meaningful but no longer badly behaved (18). The consequence for imaging problems is that we do not seek to form a "perfect" image, but instead settle for a more stable—but inherently biased—image. Many methods are available for regularizing maximum-likelihood estimation problems, and these include: penalty methods, whereby the mathematical optimization problem is modified to include a term that penalizes unwanted behavior in the object parameters (19); sieve methods, whereby the allowable class of object parameters is reduced in some manner to exclude those with unwanted characteristics (20); and stopping methods, whereby the numerical algorithms used to solve a particular optimization problem are prematurely terminated before convergence and before the object estimate has obtained the unwanted features that are characteristic of the unconstrained solution obtained at convergence (21). Penalty methods can be mathematically, but not always philosophically, equivalent to the maximum a posteriori (MAP) method, whereby an a priori statistical model for the object is incorporated into the estimation procedure. The MAP method is appealing and sound provided that a physically justified model is available for the object parameters. Each of these regularization methods is effective at times, and the method used for a particular problem is often a matter of personal taste.

SCALAR FIELDS AND COHERENCE

Because most imaging problems involve the processing of electromagnetic or acoustic fields that have been measured after propagation from a remote object or scene, a good place to begin our technical discussion is with a review of scalar waves and the concept of coherence. The scalar-wave theory is widely used for two reasons: (1) acoustic wave propagation is well-modeled as a scalar phenomenon; and (2) although electromagnetic wave propagation is a vector phenomenon, the scalar theory is often appropriate, particularly when the dimensions of interest in a particular problem are large in comparison to the electromagnetic field wavelength.

A scalar field is in general described by a function in four dimensions $s(x, y, z; t)$, where x, y, and z are coordinates in three-dimensional space, and t is a coordinate in time. In many situations, the field fluctuations in time are concentrated about some center frequency f_0, so that the field can be conveniently expressed as

$$s(x, y, z; t) = a(x, y, z; t) \cos \left[2\pi f_0 t + \theta(x, y, z; t) \right] \quad (1)$$

or, in complex notation, as

$$s(x, y, z; t) = \mathrm{Re}\{u(x, y, z; t) e^{j2\pi f_0 t}\} \quad (2)$$

where

$$u(x, y, z; t) = a(x, y, z; t) e^{j\theta(x, y, z; t)} \quad (3)$$

is the complex envelope for the field. Properties of the field amplitude a, phase θ, or both are often linked to physical or operational characteristics of a remote object or scene, and the processing of remotely sensed data to determine these properties is the main goal in most imaging applications.

Coherence is an important concept in imaging that is used to describe properties of waveforms, sensors, and processing algorithms. Roughly speaking, coherence of a waveform refers to the degree to which a deterministic relationship exists between the complex envelope phase $\theta(x, y, z; t)$ at different time instances or spatial locations. Temporal coherence at time delay τ quantifies the relationship between $\theta(x, y, z; t)$ and $\theta(x, y, z; t + \tau)$, whereas the spatial coherence at spatial shift $(\Delta_x, \Delta_y, \Delta_z)$ quantifies the relationship between $\theta(x, y, z; t)$ and $\theta(x + \Delta_x, y + \Delta_y, z + \Delta_z; t)$. A coherent sensor is one that records information about the complex-envelope phase of a waveform, and a coherent signal-processing algorithm is one that processes this information. Waveforms that are coherent only over vanishingly small time delays are called temporally incoherent; waveforms that are coherent only over vanishingly small spatial shifts are called spatially incoherent. Sensors and algorithms that neither record nor process phase information are called incoherent.

Many phenomena in nature are difficult, if not impossible within our current understanding, to model in a deterministic manner, and the statistical properties of acoustic and electromagnetic fields play a fundamental role in modeling the outcome of most remote sensing and imaging experiments. For most applications an adequate description of the fields involved is captured through second-order averages known as coherence functions. The most general of these is the mutual coherence function, which is defined mathematically in terms of the complex envelope for a field as

$$\Gamma_{12}(\tau) = E[u(x_1, y_1, z_1, t + \tau)u^*(x_2, y_2, z_2, t)] \quad (4)$$

The proper interpretation for the expectation in this definition depends largely on the application, and much care must be taken in forming this interpretation. For some applications a definition involving time averages will be adequate, whereas other applications will call for a definition involving ensemble averages.

The mutual coherence function is often normalized to form the complex degree of coherence as

$$\gamma_{12}(\tau) = \frac{\Gamma_{12}(\tau)}{[\Gamma_{11}(0)\Gamma_{22}(0)]^{1/2}} \quad (5)$$

and it is tempting to define a coherent field as one for which $|\gamma_{12}(\tau)| = 1$ for all pairs of spatial locations, (x_1, y_1, z_1) and (x_2, y_2, z_2), and for all time delays, τ. Such a definition is overly restrictive and a less restrictive condition, as discussed by Mandel and Wolf (22), is that

$$\max_{\tau} |\gamma_{12}(\tau)| = 1 \quad (6)$$

for all pairs of spatial locations, (x_1, y_1, z_1) and (x_2, y_2, z_2). Although partial degrees of coherence are possible, fields that are not coherent are usually called incoherent. In some situations a field is referred to as being fully incoherent over a particular region and its mutual coherence function is modeled over this region as

$$\Gamma_{12}(\tau) \simeq \kappa I(x_1, y_1, z_1)\delta_3(x_1 - x_2, y_1 - y_2, z_1 - z_2)\delta_1(t - \tau) \quad (7)$$

where $I(\cdot)$ is the incoherent intensity for the field, $\delta_3(\cdot, \cdot, \cdot)$ is the three-dimensional Dirac impulse, $\delta_1(\cdot)$ is the one-dimensional Dirac impulse, and κ is a constant with appropriate units. Most visible light used by the human eye to form images is fully incoherent and fits this model. Goodman (13) and Mandel and Wolf (22) provide detailed discussions of the coherence properties of electromagnetic fields.

INCOHERENT IMAGING

Astronomical telescopes, computer assisted tomography (CAT) scanners, PET scanners, and many forms of light microscopes are all examples of incoherent imaging systems; the waveforms, sensors, and algorithms used in these situations are all incoherent. The desired image for these systems is typically related to the intensity distribution of a field that is transmitted through, reflected by, or emitted from an object or scene of interest. For many of these modalities it is common to acquire data over a variety of observing scenarios, and the mathematical model for the signal acquired by these systems is of the form

$$I_k(y) = \int h_k(y, x) I(x)\, dx, \quad k = 1, 2, \ldots, K \quad (8)$$

where $I(\cdot)$ is the object incoherent intensity function—usually related directly to the emissive, reflective, or transmissive properties of the object, $h_k(\cdot, \cdot)$ is the measurement kernel or system point-spread function for the kth observation, $I_k(\cdot)$ is the incoherent measurement signal for the kth observation, x is a spatial variable in two- or three-dimensions, and y is usually a spatial variable in one-, two-, or three-dimensions. The mathematical forms for the system point-spread functions $\{h_k(\cdot, \cdot)\}$ are induced by the physical properties of the measurement system, and much care should be taken in their determination. In telescope and microscope imaging, for example, the instrument point-spread functions model the effects of diffraction, optical aberrations, and inhomogeneities in the propagation medium; whereas for transmission or emission tomographs, geometrical optics approximations are often used and the point-spread functions model the system geometry and detector uncertainties.

For situations such as astronomical imaging with ground-based telescopes, each measurement is in the form of a two-dimensional image, whereas for tomographic systems each measurement may be in the form of a one-dimensional projection of a two-dimensional transmittance or emittance function. In either situation, the imaging task is to reconstruct the intensity function $I(\cdot)$ from noisy measurements of $I_k(\cdot)$, $k = 1, 2, \ldots, K$.

Quantum Noise in Incoherent Imagery

Light and other forms of electromagnetic radiation interact with matter in a fundamentally random manner, and, because of this, statistical models are often used to describe the detection of optical waves. Quantum electrodynamics (QED) is the most sophisticated theory available for describing this phenomenon; however, a semiclassical theory for the detection of electromagnetic radiation is often sufficient for the development of sound and practical models for imaging applications. When using the semiclassical theory, electromagnetic energy is transported according to the classical theory of wave propagation—it is only during the detection process that the field energy is quantized.

When optical fields interact with a photodetector, the absorption of a quantum of energy—a photon—results in the release of an excited electron. This interaction is referred to as a photoevent, and the number of photoevents occurring over a particular spatial region and time interval are referred to as photocounts. Most detectors of light record photocounts, and although the recorded data depend directly on the image intensity, the actual number of photocounts recorded is a fundamentally random quantity. The images shown in Fig. 1 help to illustrate this effect. Here, an image of Simeon Poisson (for whom the Poisson random variable is named) is shown as it might be acquired by a detector when 1 million, 10 million, and 100 million total photocounts are recorded.

Statistical Model

For many applications involving charge coupled devices (CCD) and other detectors of optical radiation, the semiclassical theory leads to models for which the photocounts recorded by each detector element are modeled as Poisson random variables whose means are determined by the measurement intensity $I_k(\cdot)$. That is, the expected number of photocounts acquired by the nth photodetector during the kth observation interval is

$$I_k[n] = \gamma \int_{\mathscr{Y}_n} I_k(y)\, dy \quad (9)$$

where n is a two-dimensional discrete index to the elements of the detector array, $\mathscr{Y}_n$ is the spatial region over which the nth detector element integrates the image intensity, and γ is a nonnegative scale factor that accounts for overall detector efficiency and integration time. Furthermore, the number of photocounts acquired by different detector elements are usually statistically independent, and the detector regions are often small in size relative to the fluctuations in the image intensity so that the integrating operation can be well-modeled by the sampling operation

$$I_k[n] \simeq \gamma |\mathscr{Y}_n| I_k(y_n) \quad (10)$$

where y_n is the location of the nth detector element and $|\mathscr{Y}_n|$ is its integration area.

Other Detector Effects

In addition to the quantum noise, imaging detectors introduce other nonideal effects into the imagery that they record. The efficiency with which detectors convert electromagnetic energy into photoevents can vary across elements within a detector array, and this nonuniform efficiency can be captured by attaching a gain function to the photocount mean

$$I_k[n] = a[n] \gamma |\mathscr{Y}_n| I_k(y_n) \quad (11)$$

Seriously flawed detector elements that fail to record data are also accommodated with this model by simply setting the gain to zero at the appropriate location. If different detectors are used for each observation the gain function may need to vary with each frame and, therefore, be indexed by k.

Because of internal shot noise, many detectors record photoevents even when the external light intensity is zero. The resulting photocounts are usually modeled as independent

Simeon Poisson

1 million total photons

10 million total photons

100 million total photons

Figure 1. Image of Simeon Poisson as it might be acquired by a detector when 1 million, 10 million, and 100 million total photocounts are recorded.

Poisson random variables, and this phenomenon is accommodated by inserting a background term into the imaging equation

$$I_k[n] \simeq a[n]\gamma|\mathscr{V}_n|I_k(y_n) + I_b[n] \tag{12}$$

As with the gain function, if different detectors are used for each observation this background term may need to vary with each frame and, therefore, be indexed by k. With the inclusion of these background counts, the number of photocounts acquired by detector element n is a Poisson random variable with mean $I_k[n]$ and is denoted by $N_k[n]$.

The data recorded by many detectors are also corrupted by another form of noise that is induced by the electronics used for the data acquisition. For CCD detectors, this is *read-out* noise and is often approximated as additive, zero-mean Gaussian random variables so that the recorded data are modeled as

$$d_k[n] = N_k[n] + g_k[n] \tag{13}$$

where $g_k[n]$ models the read-out noise at the nth detector for the kth observation. The variance of the read-out noise $\sigma^2[\cdot]$ may vary with each detector element, and the read-out noise for different detectors is usually modeled as statistically independent.

The appropriate values for the gain function $a[\cdot]$, background function $I_b[\cdot]$, and read noise variance $\sigma^2[\cdot]$ are usually selected through a controlled study of the data acquisition system. A detailed discussion of these and other camera effects for optical imaging is given in Ref. 23.

Maximum-Likelihood Image Restoration

Consistent with the noise models developed in the previous sections, the data recorded by each detector element in a photon-counting camera are a mixture of Poisson and Gaussian random variables. Accordingly, the probability of receiving N photocounts in the nth detector element is

$$\Pr\{N_k[n] = N; I\} = \exp(-I_k[n])(I_k[n])^N/N! \tag{14}$$

where

$$\begin{aligned} I_k[n] &= a[n]\gamma|\mathscr{V}_n|I_k(y_n) + I_b[n] \\ &= a[n]\gamma|\mathscr{V}_n| \int h_k(y_n, x)I(x)\,dx + I_b[n] \end{aligned} \tag{15}$$

contains the dependence on the unknown intensity function $I(\cdot)$. Furthermore, the probability density for the read-out noise is

$$p_{g_k[n]}(g) = (2\pi\sigma^2[n])^{-1/2}\exp[-g^2/(2\sigma[n])] \tag{16}$$

so that the density for the measured data is

$$\begin{aligned} p_{d_k[n]}(d; I) &= \sum_{N=0}^{\infty} p_{g_k[n]}(d-N)\Pr\{N_k[n] = N; I\} \\ &= \frac{(2\pi\sigma^2[n])^{-1/2}}{N} \sum_{N=0}^{\infty} \exp[-(d-N)^2/(2\sigma[n])] \\ &\quad \exp(-I_k[n])(I_k[n])^N \end{aligned} \tag{17}$$

For a given data set $\{d_k[\,\cdot\,]\}$, the maximum-likelihood estimate of $I(\,\cdot\,)$ is the intensity function that maximizes the likelihood

$$l(I) = \prod_{k=1}^{K} \prod_{n} p_{d_k[n]}(d_k[n]; I) \qquad (18)$$

or, as is commonly done, its logarithm (the log-likelihood)

$$\mathcal{L}(I) = \ln l(I)$$
$$= \sum_{k=1}^{K} \sum_{n} \ln p_{d_k[n]}(d_k[n]; I) \qquad (19)$$

The complicated form for the measurement density $p_{d_k[n]}(\,\cdot\,; I)$ makes this an overly complicated optimization. When the read-out noise variance is large (greater than 50 or so), however, $\sigma^2[n]$ can be added to the measured data to form the modified data

$$\tilde{d}_k[n] = d_k[n] + \sigma^2[n]$$
$$= N_k[n] + g_k[n] + \sigma^2[n] \qquad (20)$$
$$\simeq N_k[n] + M_k[n]$$

where $M_k[n]$ is a Poisson-distributed random variable whose mean value is $\sigma^2[n]$. The modified data at each detector element are then similar (in distribution) to the sum of two Poisson-distributed random variables $N_k[n]$ and $M_k[n]$ and, as such, are also Poisson-distributed with the mean value $I_k[n] + \sigma^2[n]$. This approximation is discussed by Snyder et al. in Refs. 23 and 24. The probability mass function for the modified data is then modeled as

$$\mathrm{Pr}[\tilde{d}_k[n] = D; I] = \exp\{-(I_k[n] + \sigma^2[n])\}(I_k[n] + \sigma^2[n])^D/D! \qquad (21)$$

so that the log-likelihood is

$$\mathcal{L}(I) = \sum_{k=1}^{K} \sum_{n} \{-(I_k[n] + \sigma^2[n]) \qquad (22)$$
$$+ \tilde{d}_k[n]\ln(I_k[n] + \sigma^2[n]) - \ln d_k[n]!\}$$

Two difficulties are encountered when attempting to find the intensity function $I(\,\cdot\,)$ that maximizes the log-likelihood $\mathcal{L}(I)$: (1) the recovery of an infinite-dimensional function $I(\,\cdot\,)$ from finite data is a terribly ill-conditioned problem; and (2) the functional form of the log-likelihood does not admit a closed form, analytic solution for the maximizer even after the dimension of the parameter function has been reduced.

To address the dimensionality problem, it is common to approximate the parameter function in terms of a finite-dimensional basis set

$$I(x) \simeq \sum_{m} I[m]\psi_m(x) \qquad (23)$$

where the basis functions $\{\psi_m(\,\cdot\,)\}$ are chosen in an appropriate manner. When expressing the object function with a predetermined grid of image pixels, for example, $\psi_m(\,\cdot\,)$ might be an indicator function that denotes the location of the mth pixel. For the same situation, the basis functions might alternatively be chosen as two-dimensional impulses co-located with

the center of each pixel. Many other basis sets are possible and a clever choice here can greatly affect estimator performance, but the grid of two-dimensional impulses is probably the most common. Using this basis, the data mean is expressed as

$$I_k[n] = a[n]\gamma|\mathscr{Y}_n|I_k(y_n) + I_b[n]$$
$$= a[n]\gamma|\mathscr{Y}_n| \int h_k(y_n, x) \sum_{m} I[m]\delta_2(x - x_m)\,dx + I_b[n] \qquad (24)$$
$$= a[n]\gamma|\mathscr{Y}_n| \sum_{m} h_k(y_n, x_m)I[m] + I_b[n]$$

where y_n denotes the location of the nth measurement, x_m denotes the location of the mth object pixel, and $\delta_2(\,\cdot\,)$ is the two-dimensional Dirac impulse. The estimation problem, then, is one of estimating the discrete samples $I[\,\cdot\,]$ of the intensity function from the noisy data $\{d_k[\,\cdot\,]\}$. Because $I[\,\cdot\,]$ represents samples of an intensity function, this function is physically constrained to be nonnegative.

Ignoring terms in the log-likelihood that do not depend upon the unknown object intensity, the optimization problem required to solve for the maximum-likelihood object estimate is

$$\hat{I}[n] = \arg\max_{I \geq 0} \left\{ -\sum_{k=1}^{K} \sum_{n} (I_k[n] + \sigma^2[n]) \right. \qquad (25)$$
$$\left. + \sum_{k=1}^{K} \sum_{n} \tilde{d}_k[n]\ln(I_k[n] + \sigma^2[n]) \right\}$$

where $\tilde{d}_k[n] = d_k[n] + \sigma^2[n]$ is the modified data and

$$I_k[n] = a[n]\gamma|\mathscr{Y}_n| \sum_{m} h_k(y_n, x_m)I[m] + I_b[n]$$

is the photocount mean. The solution to this problem generally requires the use of a numerical method, and a great number of techniques are available for this purpose. General-purpose techniques such as those described in popular texts on optimization theory (25,26) can be applied. In addition, specialized numerical methods devised specifically for the solution of maximum-likelihood and related problems can be applied (27,28)—a specific example is discussed in the following section.

The Expectation-Maximization Method. The expectation-maximization (EM) method is a numerical technique devised specifically for maximum-likelihood estimation problems. As described in Ref. 27, the classical formulation of the EM procedure requires one to augment the measured data—commonly referred to as the *incomplete data*—with a set of *complete data* which, if measured, would facilitate direct estimation of the unknown parameters. The application of this procedure then requires one to alternately apply an *E-step*, wherein the conditional expectation of the complete-data log-likelihood is determined, and an *M-step*, wherein all parameters are simultaneously updated by maximizing the expectation of the complete-data log-likelihood with respect to all of the unknown parameters. In general, the application of the EM procedure results in an iterative algorithm that produces a sequence of parameter estimates that monotonically increases the measured data likelihood.

The application of the EM procedure to the incoherent imaging problems has been proposed and described for numerous applications (29–32). The general application of this method is outlined as follows. First, recall that the measured (or incomplete) data $\tilde{d}_k[n]$ for each observation k and detector element n are independent Poisson variables with the expected value

$$E\{\tilde{d}_k[n]\} = a[n]\gamma\,|\mathscr{Y}_n|\sum_m h_k(y_n, x_m)I[m] + I_b[n] + \sigma^2[n] \quad (26)$$

Because the sum of Poisson random variables is still a Poisson random variable (and the expected value is the sum of the individual expected values), the incomplete data can be statistically modeled as

$$\tilde{d}_k[n] = \sum_m N_k^c[n, m] + M_k^c[n] \quad (27)$$

where for all frames k, detector locations n, and object pixels m, the data $N_k^c[n, m]$ are Poisson random variables, each with the expected value

$$E\{N_k^c[n, m]\} = a[n]\gamma\,|\mathscr{Y}_n|h_k(y_n, x_m)I[m] \quad (28)$$

and for all frames k and detector locations n, the data $M_k^c[n]$ are Poisson random variables, each with the expected value

$$E\{M_k^c[n]\} = I_b[n] + \sigma^2[n] \quad (29)$$

In the terminology of the EM method, these data $\{N_k^c[\,\cdot\,,\,\cdot\,], M_k^c[\,\cdot\,]\}$ are the complete data, and although they cannot be observed directly, their measurement, if possible, would greatly facilitate direct estimation of the underlying object intensity.

Because the complete data are independent, Poisson random variables, the complete-data log-likelihood is

$$\begin{aligned}
\mathscr{L}^c(I) = &-\sum_k \sum_n \sum_m a[n]\gamma\,|\mathscr{Y}_n|h_k(y_n, x_m)I[m] \\
&+ \sum_k \sum_n \sum_m N_k^c[n, m]\ln(a[n]\gamma\,|\mathscr{Y}_n|h_k(y_n, x_m)I[m])
\end{aligned}$$
$$(30)$$

where terms not dependent upon the unknown object intensity $I[\,\cdot\,]$ have been omitted. Given an estimate for the object intensity $I^{\text{old}}[\,\cdot\,]$, the EM procedure makes use of the complete data and their corresponding log-likelihood to update the object intensity estimate in such a way that $I^{\text{new}}[\,\cdot\,]$ increases the measured data log-likelihood. The E-step of the EM procedure requires the expectation of the complete-data log-likelihood, conditional on the measured (or incomplete) data and using the old object intensity estimate $I^{\text{old}}[\,\cdot\,]$

$$\begin{aligned}
Q(I; I^{\text{old}}) = &\,E[\mathscr{L}^c(I)|\{\tilde{d}_k[n]\}; I^{\text{old}}] \\
= &-\sum_k \sum_n \sum_m a[n]\gamma\,|\mathscr{Y}_n|h_k(y_n, x_m)I[m] \\
&+ \sum_k \sum_n \sum_m E[N_k^c[n, m]|\{\tilde{d}_k[n]\}; I^{\text{old}}] \\
&\quad \ln(a[n]\gamma\,|\mathscr{Y}_n|h_k(y_n, x_m)I[m])
\end{aligned}$$
$$(31)$$

The intensity estimate is then updated in the M-step by maximizing this conditional expectation over I

$$I^{\text{new}} = \arg\max_{I \geq 0} Q(I; I^{\text{old}}) \quad (32)$$

It is straightforward to show that the object estimate is then updated according to

$$I^{\text{new}}[m] = \frac{\sum_k \sum_n E[N_k^c[n, m]|\{\tilde{d}_k[n]\}; I^{\text{old}}]}{\sum_k \sum_n a[n]\gamma\,|\mathscr{Y}_n|h_k(y_n, x_m)} \quad (33)$$

As described in Ref. 29, the conditional expectation is evaluated as

$$\begin{aligned}
&E[N_k^c[n, m]|\{\tilde{d}_k[n]\}; I^{\text{old}}] \\
&= \frac{a[n]\gamma\,|\mathscr{Y}_n|h_k(y_n, x_m)I^{\text{old}}[m]}{\sum_{m'} a[n]\gamma\,|\mathscr{Y}_n|h_k(y_n, x_{m'})I^{\text{old}}[m'] + I_b[n] + \sigma^2[n]}\tilde{d}_k[n] \quad (34)
\end{aligned}$$

so that the iterative formula for updating the object estimate is

$$I^{\text{new}}[m] = I^{\text{old}}[m]$$
$$\frac{\sum_k \sum_n h_k(y_n, x_m)\left[\dfrac{a[n]\gamma\,|\mathscr{Y}_n|\tilde{d}_k[n]}{\sum_{m'} a[n]\gamma\,|\mathscr{Y}_n|h_k(y_n, x_{m'})I^{\text{old}}[m'] + I_b[n] + \sigma^2[n]}\right]}{\sum_k \sum_n a[n]\gamma\,|\mathscr{Y}_n|h_k(y_n, x_m)} \quad (35)$$

For the special case of uniform gain with no background or detector noise, the iterative algorithm proposed by Richardson (33) and Lucy (34) has the same form as these iterations. An excellent historical perspective of the application of the EM method to imaging problems is presented in Ref. 35, and detailed discussions of the convergence properties of this algorithm along with the pioneering derivations for applications in emission tomography can be found in Ref. 36.

Figures 2 and 3 illustrate the use of this technique on imagery acquired by the Hubble Space Telescope (HST). Shortly after the launch of the HST with its aberrated primary mirror in 1990, the imagery acquired by this satellite became a focus of national attention. Whereas microscopic flaws in the telescope's mirror resulted in the severely distorted imagery, image restoration methods were successful in restoring much of the lost resolution (4). Figure 2, for example, shows imagery of the star cluster R136 in a star formation called 30 Doradus as acquired by the telescope and as restored using the methods described in this article. Also shown in this figure are imagery acquired by the telescope after its aberrated mirror was corrected, along with a processed image showing the potential advantage of applying image restoration methods to imagery acquired after the correction. Figure 3 contains an image of Saturn along with restorations formed by simple inverse filtering, Wiener filtering, and by the maximum-likelihood method. According to scientific staff at the Space Telescope Science Institute, the maximum-likelihood restoration

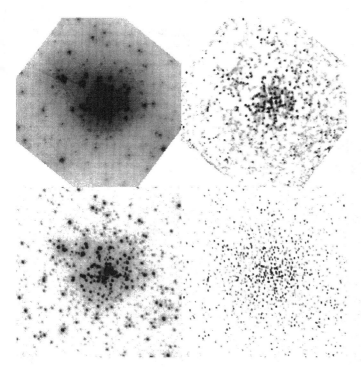

Figure 2. Imagery of the star cluster R136 in the star formation 30 Doradus as acquired by the Hubble Space Telescope both before and after its aberrated primary mirror was corrected. Upper left: raw data acquired with the aberrated primary mirror; upper right: restored image obtained from imagery acquired with the aberrated primary mirror; lower left: raw data acquired after correction; lower right: restored image obtained from imagery acquired after the correction. (Courtesy of R. J. Hanisch and R. L. White, Space Telescope Science Institute and NASA.)

provides the best trade-off between resolution and noise amplification.

Regularization. Under reasonably unrestrictive conditions, the EM method described in the previous section produces a sequences of images that converges to a maximum-likelihood solution (36). Imaging problems for which this method is applicable are often ill-conditioned or practically ill-posed, how-

ever, and because of this the maximum-likelihood image estimates frequently exhibit severe noise artifacts. Common methods for addressing this problem are discussed briefly in this section.

Stopping Rules. Probably the simplest method to implement for overcoming the noise artifacts seen in maximum-likelihood image estimates obtained by numerical procedures is to terminate the iterative process before convergence. Implementation of such a procedure is straightforward; however, the construction of optimal "stopping rules" can be challenging. Criteria for developing these rules for problems in coherent imaging are discussed in Refs. 21, 37, 38.

Sieve Methods. The basic idea behind the method of sieves is to constrain the set of allowable image estimates to be in a smooth subset called a sieve. The sieve is selected in a manner that depends upon the degree to which the problem is ill-conditioned and upon the noise level. Badly ill-conditioned problems and noisy data require a "small" sieve set containing only very smooth functions. Problems that are better conditioned with little noise can accommodate "large" sieve sets, and the sieve is ideally selected so that its "size" grows with decreasing noise levels in such a manner that the constrained image estimate converges to the true image as the noise level shrinks to zero. Establishing this consistency property for a sieve can, however, be a difficult task.

The general method of sieves as a statistical inference tool was introduced by Grenander (20). The application of this method to problems in incoherent imaging was proposed and investigated by Snyder et al. (39,40). The method is based on a kernel sieve defined according to

$$\mathscr{S} = \left\{ I : I[m] = \sum_p s[m, p]\alpha[p] \right\} \qquad (36)$$

where intensity functions within the sieve set $\mathscr{S}$ are determined by the nonnegative parameters $\{\alpha[p]\}$. The sieve-constrained optimization problem then becomes one of maximizing the likelihood subject to the additional constraint $I \in \mathscr{S}$. The smoothness properties of the sieve are induced by the sieve kernel $s[\cdot, \cdot]$. With a Gaussian kernel, for instance, the smoothness of the sieve set is determined by the variance parameter σ

$$s[m, p] = \frac{1}{\sqrt{2\pi\sigma^2}} \exp\left(-\frac{(m-p)^2}{2\sigma^2} \right) \qquad (37)$$

This Gaussian kernel was investigated in Refs. 39, 40, but kernels with other mathematical forms can be used. The EM method can, with straightforward modifications, be applied to problems in which kernel sieves are used for regularization.

Penalty and MAP Methods. Another method for regularizing maximum-likelihood estimation problems is to augment the likelihood with a penalty function

$$\mathscr{C}(I) = \mathscr{L}(I) - \gamma \Phi(I) \qquad (38)$$

where Φ is a function that penalizes undesirable qualities (or rewards desirable ones) of the image estimate, and γ is a nonnegative scale factor that determines the relative contribution of the penalty to the optimization problem. The penalized image estimate is then selected to maximize the cost function $\mathscr{C}$, which involves a trade between maximizing the likelihood $\mathscr{L}$

Figure 3. Raw imagery and restorations of Saturn as acquired by the Hubble Space Telescope. From left to right: telescope imagery; restoration produced by simple inverse filtering; restoration produced by Wiener filtering; restoration produced by the maximum-likelihood method. (Courtesy of R. J. Hanisch and R. L. White, Space Telescope Science Institute and NASA.

and minimizing the penalty Φ. The choice of the penalty can greatly influence the resulting image estimate, as can the selection of the scale factor γ. A commonly used penalty is the quadratic smoothness penalty

$$\Phi(I) = \sum_n \sum_{m \in \mathcal{N}_n} w_{nm}(I[n] - I[m])^2 \qquad (39)$$

where $\mathcal{N}_n$ denotes a neighborhood of pixel locations about the nth object pixel, and the coefficients w_{nm} control the link between pixel n and m. This penalty can also be induced by using a MAP formulation with Gaussian Markov random field (GMRF) prior model for the object. However, because the use of this penalty often results in excessive smoothing of the object edges, alternative penalties have been developed and investigated (41–43). A particularly interesting penalty is induced by using a MAP formulation with the generalized Gaussian Markov random field (GGMRF) model (43). The use of this prior results in a penalty function of the form

$$\Phi(I) = \gamma^q \sum_n \sum_{m \in \mathcal{N}_n} w_{nm}|I[n] - I[m]|^q \qquad (40)$$

where $q \in [1, 2]$ is a parameter that controls the smoothness of the reconstruction. For $q = 2$ this is the common quadratic smoothing penalty, whereas smaller values of q will, in general, allow for sharper edges in the object estimates.

Although the EM method is directly applicable to problems in which stopping rules or kernel sieves are used, the EM approach is less simple to use when penalty or MAP methods are employed. The major difficulty arises because the maximization step usually has no closed-form solution; however, approximations and modifications can be used (41,44) to address this problem.

Alternative Numerical Approaches

A major difficulty encountered when using the EM method for incoherent-imaging problems is its slow convergence (45). Many methods have been proposed to overcome this problem, and a few of these are summarized briefly here. Because of the similarities of the EM method to gradient ascent, line-search methods can be used to accelerate convergence (45), as can other gradient-based optimization methods (46,47). Substantial improvements in convergence can also be obtained by using a generalization of the EM method—the space-alternating generalized expectation-maximization (SAGE) method (28,48)—whereby convergence is accelerated through a novel choice for the complete data at each iteration. In addition, a coordinate descent (or ascent) optimization method has been shown to provide for greatly reduced computational time (49).

COHERENT IMAGING

For synthetic aperture radar (SAR), ultrasound, and other forms of coherent array imaging, an object or scene is illuminated by a highly coherent source (such as a radar transmitter, laser, or acoustic transducer), and heterodyne, homodyne, or holographic methods are used to record amplitude and phase information about the reflected field. The basic signal model for these problems is of the form:

$$u_p = \int h_p(x)u(x)\,dx + w_p, \quad p = 1, 2, \ldots, P \qquad (41)$$

where p is an index to sensor locations (either real or synthetic), u_p is the complex-amplitude measured by the pth sensor, $u(x)$ is the complex-amplitude of the field that is reflected from an object or scene of interest, $h_p(x)$ is a sensor response function for the pth sensor measurement, and w_p accounts for additive sensor noise. The response function accounts for both the sensor characteristics and for wave propagation from the object or scene to the sensor; in the Fraunhofer approximation for wave propagation, these functions take on the form of a Fourier-transform kernel (10).

When the object or scene gives rise to diffuse reflections, the Gaussian speckle model (50) is often used as a statistical model for the reflected field $u(\cdot)$. That is, $u(\cdot)$ is modeled as a complex Gaussian random process (13,51,52) with zero-mean and the covariance

$$E[u(x)u^*(x')] \simeq s(x)\delta_2(x - x') \qquad (42)$$

where $s(\cdot)$ is the object incoherent scattering function. The sensor noise is often modeled as zero-mean, independent complex Gaussian variables with variance σ^2 so that the recorded data are complex Gaussian random variables with zero-mean and the covariance

$$E[u_p u_{p'}^*] = \int h_p(x)h_{p'}^*(x)s(x)\,dx + \sigma^2\delta[p - p'] \qquad (43)$$

where $\delta[\cdot]$ is the Kronecker delta function. The maximum-likelihood estimation of the object scattering function $s(\cdot)$ then becomes a problem of covariance estimation subject to the linear structure constraint of Eq. (43).

Using vector-matrix notation the data covariance is, as a function of the unknown object scattering function

$$\begin{aligned} \boldsymbol{R}(s) &= E[\boldsymbol{u}\boldsymbol{u}^\dagger] \\ &= \int \boldsymbol{h}(x)\boldsymbol{h}^\dagger(x)s(x)\,dx + \sigma^2\boldsymbol{I} \end{aligned} \qquad (44)$$

where $\boldsymbol{u} = [u_1 u_2 \cdots u_P]^T$ is the data vector, $\boldsymbol{h}(x) = [h_1(x)h_2(x) \cdots h_P(x)]^T$ is the system response vector, $[\cdot]^T$ denotes matrix transposition, $[\cdot]^\dagger$ denotes Hermitian matrix transposition, and $\boldsymbol{I}$ is the $P \times P$ identity matrix. Accordingly, the data log-likelihood is

$$L(s) = -\ln\det[\boldsymbol{R}(s)] - \text{tr}[\boldsymbol{R}^{-1}(s)\boldsymbol{S}] \qquad (45)$$

where $\boldsymbol{S} = \boldsymbol{u}\boldsymbol{u}^\dagger$ is the data sample-covariance. Parameterization of the parameter function as in Eq. (23) is a natural step before attempting to solve this problem, but direct maximization of the likelihood is still a difficult problem. Because of this, the EM method has been proposed and discussed in Refs. 53–55 for addressing this problem, and the resulting algorithm has been shown to produce parameter estimates with lower bias and variance than alternative methods (56). A major problem with this method, though, is the high computational cost; however, the application of the SAGE method (28) to this problem has shown great promise for reducing the computational burden (57). The development and application of regularization methods for problems in coherent imaging is an area of active research.

SUMMARY

Imaging science is a rich and vital area of science and technology in which information-theoretic methods can be and have been applied with great benefit. Maximum-likelihood methods can be applied to a variety of problems in image restoration and synthesis, and their application to the restoration problem for incoherent imaging has been discussed in great detail in this article. To conclude, the future of this field is best summarized by the following quote from Bracewell (58):

> The study of imaging now embraces many major areas of modern technology, especially the several disciplines within electrical engineering, and will be both the stimulus for, and recipient of, new advances in information science, computer science, environmental science, device and materials science, and just plain high-speed computing. It can be confidently recommended as a fertile subject area for students entering upon a career in engineering.

BIBLIOGRAPHY

1. H. Stark (ed.), *Image Recovery: Theory and Application,* Orlando, FL: Academic Press, 1987.

2. A. K. Katsaggelos (ed.), *Digital Image Restoration,* Heidelberg: Springer-Verlag, 1991.

3. R. L. Lagendijk and J. Biemond, *Iterative Identification and Restoration of Images,* Boston: Kluwer, 1991.

4. R. J. Hanisch and R. L. White (eds.), *The Restoration of HST Images and Spectra—II,* Baltimore, MD: Space Telescope Science Institute, 1993.

5. H. L. Van Trees, *Detection, Estimation, and Modulation Theory, Part I,* New York: Wiley, 1968.

6. H. V. Poor, *An Introduction to Signal Detection and Estimation,* 2nd od., New York: Springer Verlag, 1994.

7. B. Porat, *Digital Processing of Random Signals: Theory and Methods,* Englewood Cliffs, NJ: Prentice-Hall, 1993.

8. L. L. Scharf, *Statistical Signal Processing: Detection, Estimation, and Time Series Analysis,* Reading, MA: Addison-Wesley, 1991.

9. S. M. Kay, *Modern Spectral Estimation: Theory and Applications,* Englewood Cliffs, NJ: Prentice-Hall, 1988.

10. J. W. Goodman, Introduction to Fourier Optics, 2nd edition, New York: McGraw-Hill, 1996.

11. J. D. Gaskill, *Linear Systems, Fourier Transforms, and Optics,* New York: Wiley, 1978.

12. M. Born and E. Wolf, *Principles of Optics,* 6th edition, Elmsford, NY: Pergamon, 1980.

13. J. W. Goodman, *Statistical Optics,* New York: Wiley, 1985.

14. B. E. A. Saleh and M. C. Teich, *Fundamentals of Photonics,* New York: Wiley, 1991.

15. A. Papoulis, *Probability, Random Variables, and Stochastic Processes,* 3rd edition, New York: McGraw-Hill, 1991.

16. R. M. Gray and L. D. Davisson, *Random Processes: An Introduction for Engineers,* Englewood Cliffs: Prentice-Hall, 1986.

17. A. Tikhonov and V. Arsenin, *Solutions of Ill-Posed Problems,* Washington, DC: Winston, 1977.

18. W. L. Root, Ill-posedness and precision in object-field reconstruction problems, *J. Opt. Soc. Am., A,* **4** (1): 171–179, 1987.

19. J. R. Thompson and R. A. Tapia, *Nonparametric Function Estimation, Modeling, and Simulation,* Philadelphia: SIAM, 1990.

20. U. Grenander, *Abstract Inference,* New York: Wiley, 1981.

21. E. Veklerov and J. Llacer, Stopping rule for the MLE algorithm based on statistical hypothesis testing, *IEEE Trans. Med. Imaging,* **6**: 313–319, 1987.

22. L. Mandel and E. Wolf, *Optical Coherence and Quantum Optics,* New York: Cambridge University Press, 1995.

23. D. L. Snyder, A. M. Hammoud, and R. L. White, Image recovery from data acquired with a charge-coupled-device camera, *J. Opt. Soc. Am., A,* **10** (5): 1014–1023, 1993.

24. D. L. Snyder et al., Compensation for readout noise in CCD images, *J. Opt. Soc. Am., A,* **12** (2): 272–283, 1995.

25. D. G. Luenberger, *Linear and Nonlinear Programming,* Reading, MA: Addison-Wesley, 1984.

26. R. Fletcher, *Practical Methods of Optimization,* New York: Wiley, 1987.

27. A. P. Dempster, N. M. Laird, and D. B. Rubin, Maximum likelihood from incomplete data via the EM algorithm, *J. R. Stat. Soc., B,* **39**: 1–37, 1977.

28. J. A. Fessler and A. O. Hero, Space-alternating generalized expectation-maximization algorithm, *IEEE Trans. Signal Process.* **42**: 2664–2677, 1994.

29. L. A. Shepp and Y. Vardi, Maximum-likelihood reconstruction for emission tomography, *IEEE Trans. Med. Imaging,* **MI-1**: 113–121, 1982.

30. D. L. Snyder and D. G. Politte, Image reconstruction from list-mode data in an emission tomography system having time-of-flight measurements, *IEEE Trans. Nucl. Sci.,* **NS-30**: 1843–1849, 1983.

31. K. Lange and R. Carson, EM reconstruction algorithms for emission and transmission tomography, *J. Comput. Assisted Tomography,* **8**: 306–316, 1984.

32. T. J. Holmes, Maximum-likelihood image restoration adapted for noncoherent optical imaging, *J. Opt. Soc. Am., A,* **6**: 666–673, 1989.

33. W. H. Richardson, Bayesian-based iterative method of image restoration, *J. Opt. Soc. Am.,* **62** (1): 55–59, 1972.

34. L. B. Lucy, An iterative technique for the rectification of observed distributions, *Astronom. J.,* **79** (6): 745–754, 1974.

35. Y. Vardi and D. Lee, From image deblurring to optimal investments: Maximum likelihood solutions for positive linear inverse problems, *J. R. Stat. Soc. B,* **55** (3): 569–612, 1993.

36. Y. Vardi, L. A. Shepp, and L. Kaufman, A statistical model for positron emission tomography, *J. Amer. Stat. Assoc.,* **80**: 8–37, 1985.

37. T. Hebert, R. Leahy, and M. Singh, Fast MLE for SPECT using an intermediate polar representation and a stopping criterion, *IEEE Trans. Nucl. Sci.,* **NS-34**: 615–619, 1988.

38. J. Llacer and E. Veklerov, Feasible images and practicle stopping rules for iterative algorithms in emission tomography, *IEEE Trans. Med. Imaging,* **MI-8**: 186–193, 1989.

39. D. L. Snyder and M. I. Miller, The use of sieves to stabilize images produced with the EM algorithm for emission tomography, *IEEE Trans. Nucl. Sci.,* **NS-32**: 3864–3871, 1985.

40. D. L. Snyder et al., Noise and edge artifacts in maximum-likelihood reconstructions for emission tomography, *IEEE Trans. Med. Imaging,* **MI-6**: 228–238, 1987.

41. P. J. Green, Bayesian reconstructions from emission tomography data using a modified EM algorithm, *IEEE Trans. Med. Imaging,* **9**: 84–93, 1990.

42. K. Lange, Convergence of EM image reconstruction algorithms with Gibbs priors, *IEEE Trans. Med. Imaging,* **MI-9**: 439–446, 1990.

43. C. A. Bouman and K. Sauer, A generalized Gaussian image model for edge-preserving MAP estimation, *IEEE Trans. Image Process.,* **2**: 296–310, 1993.

44. T. Hebert and R. Leahy, A generalized EM algorithm for 3-D Bayesian reconstruction from Poisson data using Gibbs priors, *IEEE Trans. Med. Imaging,* **MI-8**: 194–202, 1989.

45. L. Kaufman, Implementing and accelerating the EM algorithm for positron emission tomography, *IEEE Trans. Med. Imaging,* **MI-6**: 37–51, 1987.

46. E. U. Mumcuoglu et al., Fast gradient-based methods for Bayesian reconstruction of transmission and emission PET images, *IEEE Trans. Med. Imag.,* **MI-13**: 687–701, 1994.

47. K. Lange and J. A. Fessler, Globally convergent algorithm for maximum a posteriori transmission tomography, *IEEE Trans. Image Process.,* **4**: 1430–1438, 1995.

48. J. A. Fessler and A. O. Hero, Penalized maximum-likelihood image reconstruction using space-alternating generalized EM algorithms, *IEEE Trans. Image Process.,* **4**: 1417–1429, 1995.

49. C. A. Bouman and K. Sauer, A unified approach to statistical tomography using coordinate descent optimization, *IEEE Trans. Image Process.,* **5**: 480–492, 1996.

50. J. C. Dainty (ed.), Laser speckle and related phenomena. *Topics in Applied Physics,* vol. 9, 2nd ed., Berlin: Springer-Verlag, 1984.

51. F. D. Neeser and J. L. Massey, Proper complex random processes with applications to information theory, *IEEE Trans. Inf. Theory,* **39**: 1293–1302, 1993.

52. K. S. Miller, *Complex Stochastic Processes: An Introduction to Theory and Applications,* Reading, MA: Addison-Wesley, 1974.

53. D. B. Rubin and T. H. Szatrowski, Finding maximum likelihood estimates of patterned covariance matrices by the EM algorithm, *Biometrika,* **69** (3): 657–660, 1982.

54. M. I. Miller and D. L. Snyder, The role of likelihood and entropy in incomplete-data problems: Applications to estimating point-process intensities and Toeplitz constrained covariances, *Proc. IEEE,* **75**: 892–907, 1987.

55. D. L. Snyder, J. A. O'Sullivan, and M. I. Miller, The use of maximum-likelihood estimation for forming images of diffuse radar-targets from delay-doppler data, *IEEE Trans. Inf. Theory,* **35**: 536–548, 1989.

56. M. J. Turmon and M. I. Miller, Maximum-likelihood estimation of complex sinusoids and Toeplitz covariances, *IEEE Trans. Signal Process.,* **42**: 1074–1086, 1994.

57. T. J. Schulz, Penalized maximum-likelihood estimation of structured covariance matrices with linear structure, *IEEE Trans. Signal Process.,* **45**: 3027–3038, 1997.

58. R. N. Bracewell, *Two-Dimensional Imaging,* Upper Saddle River, NJ: Prentice-Hall, 1995.

TIMOTHY J. SCHULZ
Michigan Technological University

MEASUREMENT.
See ACCELERATION MEASUREMENT; DENSITY MEASUREMENT; DISPLACEMENT MEASUREMENT; MAGNETIC FIELD MEASUREMENT; MILLIMETER WAVE MEASUREMENT; Q-FACTOR MEASUREMENT.

MEASUREMENT, ATTENUATION.
See ATTENUATION MEASUREMENT.

MEASUREMENT, C-V.
See C-V PROFILES.

MEASUREMENT ERRORS

In applied science and engineering, it is agreed that all observations contain errors. This article discusses how these errors are described today. It also discusses how this description is used to compute the effect of the errors upon the measurement result, and how this effect is reduced or even minimized by rational use of the observations and by experimental design.

For the description of observations, use is made of a mathematical model. This is a mathematical expression intended to describe the observations fully. It will be supposed throughout this article that the mathematical model is parametric, and that the parameters are the quantities to be measured. For example, the parametric model may be a sinusoidal function of time with unknown amplitude, frequency, and phase. Then, these three quantities are the parameters of this model. Yet the model of the observations is incomplete without including errors. If there is reason to assume that the errors in the observations are nonsystematic and additive, they are taken into consideration by adding a term representing them to the expression for the sinusoidal function. Then, the resulting sum is the mathematical model of the observations. It is supposed to fully describe the observations. Nonsystematic errors may loosely be defined as errors that vary if the experiment is repeated under the same conditions and are equal to zero if averaged over many experiments. They are modeled as stochastic variables with an expectation equal to zero. The term expectation is the abstract mathematical term for the mean value. It will be used throughout to avoid confusion with averages, such as time averages, which are measurements. If, in the example, the errors are stochastic variables, so are the observations. Since each observation is equal to the sum of the sinusoidal function and the stochastic error, the expectation of an observation is the value of the function at the time instant concerned. Therefore, the expectation or, equivalently, the function value represents the hypothetical errorless observation.

Since the observations are stochastic variables, they are described by probability density functions. These define for discrete stochastic variables, such as counting results, the probability of occurrence of a particular discrete outcome. For continuous stochastic variables, the probability density function defines the probability of occurrence of an observation within a particular range of values. The probability density function also determines the expectation of the observations. Since this expectation is equal to the function value, the probability density function of the observations depends on the parameters of the function, and thus, the measurement problem has become a statistical parameter estimation problem. This observation has important consequences. It implies that for measurement, use can be made of the extensive theory and methods of statistics. It will be seen that this offers a number of exceptional advantages. A description of these advantages requires some familiarity with a number of notions from statistics. Therefore, these will first be introduced. References 1–4 are useful general texts on statistics.

In statistics, the function of the observations with which a parameter is estimated is called an estimator. Using the same observations, for a particular parameter, different estimators can be defined. Since the observations are stochastic variables, so is the estimator. Therefore, the estimator has a probability density function, an expectation, and a standard deviation. If the expectation of the estimator is equal to the hypothetical true value of the parameter to be estimated, the estimator is called unbiased. Otherwise, it is biased. The deviation of the true value from the expectation is called bias.

Bias is the systematic error. It is, therefore, equivalent to the concept accuracy. There are two essentially different sources of bias. In the first place, the expectation of the observations may be different from the function model assumed. In the above mentioned sinusoidal example, a trend may be present in addition to the sinusoid, while the model assumed and fitted to the observations consists of the sinusoidal function only. This will, of course, always result in a systematic deviation of the estimated parameters, even in the hypothetical complete absence of nonsystematic errors. The remedy is to include the trend in the model fitted. The inclusion of the trend has the effect that two additional parameters, the slope and the intercept of the trend, have to be estimated. It will be discussed later that there are various reasons to keep in measurement the number of parameters to be estimated as small as possible. Therefore, classical measurement measures to avoid errors such as trends, day-and-night cycles, and background are always preferable to including these contributions in the model. The second source of bias is of a completely different nature. It is produced by and is characteristic of the estimator, itself. It may, therefore, also occur if the assumed model of the observations and that of the expectations are the same. Two estimators of the same parameters from the same observations may have different bias. If the bias vanishes as the number of observations increases, the estimator is called asymptotically unbiased. An effective method to remove bias of this kind is described by Stuart and Ord (4).

The standard deviation of an estimator represents precision. It is the spread of the measurement result if the experiment is repeated under the same conditions. In the example of the estimation of the parameters of the sinusoid, the amplitude, frequency, and phase have to be estimated simultaneously. This estimator is, therefore, vector valued. A vector estimator has a covariance matrix associated with it. The diagonal elements of this matrix are the variances of the estimators of each of the elements of the vector of parameters. The off-diagonal elements represent the covariances of the estimators of different elements. Bias and standard deviation are statistical key properties of estimators for practical measurement purposes. They demonstrate the practical feasibility, clarity, and generality of the model based statistical parameter estimation approach to the treatment of errors in observations. In addition to these desirable properties, model based parameter estimation has a number of advantages which will now be discussed.

It has been mentioned earlier that for the measurement of the parameters of the same model from the same set of observations, use may be made of different estimators. These estimators will generally have different standard deviations, that is, have different precision. The question may, therefore, be asked which estimator is most precise? Or put somewhat differently, what precision is attainable if any estimator may be used? The answer to this question may be given using the concept of Fisher information. The Fisher information with respect to the model parameters is computed from the probability density function of the observations. If the model has one parameter, the quantity computed is called the Fisher information amount. If more than one parameter is measured, such as in the sinusoidal example, it assumes the form of a matrix, the Fisher information matrix. This is a symmetrical matrix of the same order as the number of parameters. The elements of the Fisher information matrix are dependent on the probability density of the observations and on the model of the expectations. They are independent of any method of estimation. With any set of observations, considered as stochastic variables, a Fisher information matrix with respect to the unknown parameters is associated. The inverse of the Fisher information matrix is called the Cramér Rao lower bound. It can be shown that the diagonal elements of the covariance matrix of any unbiased vector estimator, the variances of the elements, cannot be smaller than the corresponding diagonal elements of the Cramér Rao lower bound. Therefore, any unbiased estimator is at best as precise as a hypothetical estimator of the same parameters having the Cramér Rao lower bound as its covariance matrix. Thus, the Cramér Rao lower bound is a standard to which the precision of an unbiased estimator may be compared. This is the reason why the ratio of a diagonal element of the Cramér Rao lower bound to the variance of an estimator is called the efficiency of the estimator. Cramér Rao theory also extends to functions of the estimated parameters. For example, suppose that the height, width, and location of a Gaussian pulse or spectral peak are estimated, but that the quantities to be measured ultimately are the location and the area. Then, the Cramér Rao lower bound for the height, width, and location combined with the expression for the area in terms of the width and height may be used to compute the Cramér Rao lower bound for the location and area. The resulting expressions exactly describe the propagation of the Cramér Rao lower bound for the original parameters to that for location and area. Thus, they show exactly the sensitivity of the Cramér Rao lower bound for the area parameter to the various elements of the Cramér Rao lower bound for the original parameters. This means that an instrument has been found to compute the propagation of errors in the observations to errors in the parameters and, subsequently, to errors in functions of the parameters.

For the engineer and applied scientist, an important question is how to make the influence of errors in the observations upon the measurement result as small as possible. This is equivalent to the question of how to find the method that produces the most precise measurement result from the available observations. For calibration purposes, precision itself may be the ultimate purpose. In other applications, precision is often pursued not for its own sake but to make the conclusions drawn from the measurement result more reliable. The extent to which it is possible to find an efficient estimation method depends on the available a priori knowledge of the probability density function of the observations. As has been discussed earlier, this probability density function is parametric in the hypothetical exact values of the unknown parameters. This dependence of the probability density function of the observations on the parameters may be used to derive a so-called maximum likelihood estimator of the parameters. First, the numerical values of the available observations are substituted for the corresponding independent variables of the probability density function. Next, the true values of the parameters are considered to be variables. The function thus obtained is called the likelihood function of the parameters. Finally, the likelihood function is maximized with respect to the parameters. The values of the parameters at the maximum are the maximum likelihood estimates of the parameters. This procedure shows the first advantage of the maximum likelihood estimator: it is easily found. In addition, the maximum likelihood estimator has a number of favorable statistical properties. The most important of these is that under general conditions, it attains asymptotically the Cramér Rao lower bound. This means that for a large number of observations, it is most precise. The elements of the Cramér Rao lower bound depend on experimental variables. For example, in the estimation of the parameters of the sinusoid and those of the Gaussian pulse, these are the number of observations and their location. If experimental variables such as these may to a certain extent be freely chosen, this freedom may be used to minimize the Cramér Rao lower bound and, thus, the asymptotic variance. This manipulation of the covariance matrix using experimental variables is called experimental design. From a practical point of view, experimental design may be very attractive since it may lead to a more precise measurement result with the same effort or even less. For practical measurement, the invariance property of maximum likelihood estimators is also important: functions of maximum likelihood estimators are maximum likelihood estimators themselves. Generally, maximizing the likelihood function with respect to the parameters is a nonlinear optimization problem which can only be solved using an iterative numerical optimization method. For a long time, this has been a serious impediment to the application of maximum likelihood, but today, excellent optimization methods and software are available which make the method accessible to any user.

If the observations are normally distributed, the maximum likelihood estimator of the parameters can be shown to be the weighted least squares estimator with the inverse of the covariance matrix of the observations as weighting matrix. If the observations are linear in all parameters to be estimated, the least squares estimator is a relatively simple closed form expression linear in the observations. In addition, if the observations are not normally distributed, the weighted least squares method with the inverse covariance matrix as weighting matrix still has the smallest variance among all estimators that are both linear in the observations and unbiased. If the observations are nonlinear in one or more of the parameters to be estimated, the least squares estimator is, as a rule, no longer a closed form and has to be evaluated using an iterative numerical method. However, effective, specialized, and reliable numerical methods and software are available that make the use of nonlinear least squares straightforwardly. As a result, least squares has become a major tool in the handling of observations subject to error in general and not only of normally distributed observations.

EXPECTATIONS OF OBSERVATIONS

The reduction or minimization of the effect of errors in the observations upon the measurement result requires a mathematical model of the observations. In this article, additive nonsystematic errors in the observations will be modeled as stochastic variables with an expectation equal to zero. This implies that the observations are also stochastic variables, and that the expectations of the observations are the hypothetical exact or errorless observations. Thus, these expectations constitute the model underlying the observations. It will be assumed throughout that this model is a parametric function, and that parameters of this model are the quantities to be measured, or that the quantity to be measured can be computed from these parameters.

Example 1. The multiexponential model. Multiexponential observations are observations with expectations

$$y_n(\gamma) = \alpha_1 \exp(-\beta_1 x_n) + \ldots + \alpha_L \exp(-\beta_L x_n) \tag{1}$$

with $n = 1, \ldots, N$ where N is the number of observations, and the $2L \times 1$ vector γ is defined as $(\alpha_1 \ldots \alpha_L \beta_1 \ldots \beta_L)^T$, where the amplitudes α_ℓ and the decay constants β_ℓ are the parameters to be measured, and the superscript T denotes transposition. The measurement points $x_n, n = 1, \ldots, N$ are supposed known. If, different from Eq. (1), there is a linear trend in the observations, this deterministic contribution has to included in the model of the expectations of the observations

$$y_n(\eta) = \alpha_1 \exp(-\beta_1 x_n) + \ldots + \alpha_L \exp(-\beta_L x_n) + \lambda x_n + \mu$$

where $\eta = (\alpha_1 \ldots \alpha_L \beta_1 \ldots \beta_L \lambda \mu)^T$. In this expression, λ and μ are the slope and the intercept of the trend, respectively. These parameters have to be estimated along with the parameters α_ℓ and β_ℓ. This means that the number of parameters to be estimated has increased by two. It will be shown below that this is not only disadvantageous from a computational point of view, but it also unfavorably influences the precision with which the α_ℓ and β_ℓ can be measured. Therefore, it is worthwhile to keep the number of parameters as small as possible. As a consequence, changing the experimental conditions to remove the trend is always preferable to including it in the model of the expectations. On the other hand, if trends cannot be avoided, they have to be included since otherwise, the model of the expectations is wrong. Then, values for the amplitudes and decay constants are found systematically deviating from the α_ℓ and β_ℓ, even in the hypothetical case that nonsystematic errors in the observations are absent.

THE DISTRIBUTION OF THE OBSERVATIONS

The mathematical model of the observations is completed by a description of how the observations are distributed about their expectations. This is done in the form of the joint probability density function of the observations. If $w = (w_1 \ldots w_N)^T$ is the vector of the N available observations, their probability density function may be described as $p(w)$. Then, the expectation $E[w] = (E[w_1] \ldots E[w_N])^T$ is defined as

$$E[w] = \int \ldots \int w p(w) dw \tag{2}$$

where $dw = dw_1 dw_2 \ldots dw_N$, and the integrations are carried out over all possible values of w. Then,

$$E[w] = y(\theta) \tag{3}$$

where $y(\theta) = [y_1(\theta) \ldots y_N(\theta)]^T$, and $y_n(\theta)$ is the function parametric in the unknown parameters θ defining the errorless observations such as the exponential model described by Eq. (1). For engineering and applied science, two probability density functions are particularly important. These are the normal probability density function and the Poisson probability density function.

Example 2. The normal probability density function. The observations $w_1, \ldots, w_N$ are said to be normally distributed if their probability density function is described by

$$p(w) = \frac{1}{(2\pi)^{N/2}(\det W)^{1/2}} \exp\left[-\frac{1}{2}(w - \mathrm{E}[w])^{\mathrm{T}}W^{-1}(w - \mathrm{E}[w])\right]$$

where the $N \times N$ matrix W is the covariance matrix of the observations defined by its (i, j)-the element $\mathrm{cov}(w_i, w_j)$ and $\det W$ and W^{-1} are the determinant and the inverse of W, respectively. This probability density function and many others are discussed in Ref. 3. Equation (3) defines the functional dependence of the normal probability density function on the parameters of the function modelling the expectations. For what follows, the logarithm of $p(w)$ as a function of the parameters θ is needed. After substituting $y(\theta)$ for $\mathrm{E}[w]$, it is described by

$$-\frac{N}{2}\ln(2\pi) - \frac{1}{2}\ln(\det W) - \frac{1}{2}[w - y(\theta)]^{\mathrm{T}}W^{-1}[w - y(\theta)] \quad (4)$$

Notice that both first terms of this expression are independent of the parameter vector θ, while the last term is a quadratic form in the elements $w_n - y_n(\theta)$ of $w - y(\theta)$. Observations in practice are often, but not always, normally distributed. One of the reasons is that if the nonsystematic errors are the sum of a number of nonsystematic errors from independent sources, their distribution tends to normal as described by the central limit theorem discussed in Ref. 2.

Example 3. The Poisson probability density function.
This probability density function concerns counting statistics. It is described in Ref. 3. Examples of Poisson distributed stochastic variables are radioactive particle counts and pixel values in electron microscopes. The number of counts is Poisson distributed if the probability that it is equal to w_n is given by

$$p(w_n) = \exp(-\lambda_n)\frac{\lambda_n^{w_n}}{w_n!} \quad (5)$$

Simple calculations show that $\mathrm{E}[w_n] = \lambda_n$, and that the standard deviation of w_n is equal to $\sqrt{\lambda_n}$. If $w_1, \ldots, w_N$ are independent, as is often assumed in applications, their joint probability density function is equal to the product the probabilities described by Eq. (5)

$$p(w) = \prod_n p(w_n) \quad (6)$$

with $n = 1, \ldots, N$. Since $\mathrm{E}[w] = \lambda$ with $\lambda = (\lambda_1 \ldots \lambda_N)^{\mathrm{T}}$ and $\lambda_n = y_n(\theta)$, the logarithm of the probability density function defined by Eq. (6) may be written

$$\sum_n -y_n(\theta) + w_n \ln[y_n(\theta)] - \ln(w_n!) \quad (7)$$

Notice that the last term in this expression is independent of the parameter vector θ.

From Example 2 and Example 3, the general approach to establishing the dependence of the probability density function of the observations on the parameters, that is, the quantities to be measured, is now clear. First, the expectation of the observations w_n is computed. Then, the result is substituted for the relevant quantities in the probability density

function. The probability density function thus obtained is parametric in the parameters of the expectations, that is, of the hypothetical errorless observations. This is the form of the probability density function that will be used hereafter for two purposes. First, it will be used for the computation of the highest attainable precision with which the parameters can be measured from the available observations. It will also be used to find the most precise method to estimate the parameters from the observations.

ATTAINABLE MEASUREMENT PRECISION IN THE PRESENCE OF MEASUREMENT ERRORS

Suppose that a number of N observations $w_1, \ldots, w_N$ is available and that the expectations of the observations are described by the multiexponential model defined by Eq. (1). If this model is fitted to the observations with respect to its parameters, the amplitudes, and the decay constants, one could choose the sum of the squares of the deviations of the model from the observations as a criterion of goodness of fit. Then, this criterion could be minimized with respect to the parameters, and the parameter values for which the criterion would be minimum would be the solution. This is the well-known ordinary least squares solution. Alternatively, one could have chosen the values of the parameters for which the sum of the absolute values of the deviations would be minimum. This is the least absolute values or least moduli solution. Then, if the experiment could be repeated sufficiently often, the experimenter could compare the results of both methods and could decide which of both would be most precise. Seeing that the one method is more precise than the other, the experimenter might wonder what the highest attainable precision from these observations is with any method. It has been found that under general conditions, this question may be answered using the concept Fisher information. For a discussion of Fisher information, see Ref. 4. For the computation of the Fisher information, the probability density function of the observations $p(w;\theta)$ is used. This is done as follows. First, the logarithm of $p(w;\theta)$ is taken. For the normal probability density function and for the Poisson probability density function, the result of this operation is described by Eqs. (4) and (7), respectively. Next, the gradient vector of $\ln p(w;\theta)$ with respect to the elements of θ is calculated. It is defined as

$$\frac{\partial \ln p(w; \theta)}{\partial \theta}$$

If θ is a $K \times 1$ vector, so is the gradient vector. Its k-th element is $\partial \ln p(w;\theta)/\partial\theta_k$. Next, the $K \times K$ matrix

$$\frac{\partial \ln p}{\partial \theta}\frac{\partial \ln p}{\partial \theta^{\mathrm{T}}} \quad (8)$$

is computed where, for simplicity, the arguments of $p(w;\theta)$ have been left out, and $\partial \ln p/\partial\theta^{\mathrm{T}}$ is the transpose of $\partial \ln p/\partial\theta$. The $K \times K$ Fisher information matrix is defined as the expectation of Eq. (8)

$$M = \mathrm{E}\left[\frac{\partial \ln p}{\partial \theta}\frac{\partial \ln p}{\partial \theta^{\mathrm{T}}}\right] = \int \cdots \int \frac{\partial \ln p}{\partial \theta}\frac{\partial \ln p}{\partial \theta^{\mathrm{T}}}p\,dw$$

It is not difficult to show that M may alternatively be written

$$M = -E\left[\frac{\partial^2 \ln p}{\partial\theta\,\partial\theta^T}\right] \qquad (9)$$

In this expression, $\partial^2\ln p/\partial\theta\,\partial\theta^T$ is the Hessian matrix of $\ln p$ defined by its (q, r)-th element $\partial^2\ln p/\partial\theta_q\,\partial\theta_r$.

Example 4. The Fisher information matrix for normally distributed observations.

If the observations are normally distributed, the logarithm of the probability density function as a function of the parameters is described by Eq. (4). Then, elementary computations making use of the fact that $E[w_n - y_n(\theta)] = 0$ yield

$$M = \frac{\partial y^T}{\partial\theta} W^{-1} \frac{\partial y}{\partial\theta^T} \qquad (10)$$

In this expression, the $N \times K$ matrix $\partial y/\partial\theta^T$ is the Jacobian matrix of $y(\theta)$ with respect to θ. Its (n, k)-th element is equal to $\partial y_n(\theta)/\partial\theta_k$. Therefore, for the multiexponential model, the elements of the Jacobian matrix are of the form $\exp(-\beta_\ell x_n)$ or $-\alpha_\ell x_n \exp(-\beta_\ell x_n)$ with $\ell = 1, \ldots, L$.

Example 5. The Fisher information matrix for independent Poisson distributed observations.

If the observations are independent and Poisson distributed, the logarithm of the probability density function of the observations as a function of the parameters is described by Eq. (7). This expression may be used to show that here, the information matrix is also described by Eq. (10), but with $W = \text{diag}(y_1 \ldots y_N)$, where $y_n = y_n(\theta)$.

The importance of the Fisher information matrix is that from it the Cramér Rao lower bound may be computed. This is a lower bound on the variance of all unbiased estimators of parameters or of functions of parameters. An estimator t is said to be unbiased for the parameter θ if its bias, defined as

$$E[t] - \theta$$

is equal to the null vector. Otherwise, it is biased. In measurement terminology: if the model of the expectations of the observations is correctly specified and the estimator used is unbiased for the parameters, the measurement result has no systematic error.

Next, suppose that $t(w)$ is any unbiased estimator of the vector of parameters θ from the observations w. Then the Cramér Rao inequality states that

$$\text{cov}[t(w), t(w)] \geq M^{-1} \qquad (11)$$

In this expression, $\text{cov}[t(w), t(w)]$ is the covariance matrix of the estimator $t(w)$. That is, the (p,q)-th element of this matrix is defined as the covariance of the p-th element $t_p(w)$ and the q-th element $t_q(w)$. Therefore, the diagonal elements are the variances of $t_1(w), \ldots, t_K(w)$, respectively. Ineq. (11) expresses that the difference of the matrix $\text{cov}[t(w), t(w)]$ and the matrix M^{-1} is positive semidefinite. A property of positive semidefinite matrices is that their diagonal elements cannot be negative. Therefore, the diagonal elements of $\text{cov}(t(w), t(w))$, that is, the variances of the elements of the estimator

$t(w)$, must be larger than or be equal to the corresponding diagonal elements of M^{-1}. Consequently, the latter diagonal elements are a lower bound on the variances of the elements of the estimator $t(w)$. The matrix M^{-1} is called the Cramér Rao lower bound. For normally distributed observations and for Poisson distributed observations, the Cramér Rao lower bound may be computed by inverting the Fisher information matrix defined by Eq. (10) with appropriate matrix W, respectively. Notice that the main ingredients are simply the derivatives of the model $y_n(\theta)$ with respect to the parameters in each measurement point. These are quantities that are usually easy to compute.

The Cramér Rao lower bound would be of theoretical value only if there would not exist estimators attaining it. Later in this article, estimators will be introduced that do so, at least asymptotically. Therefore, the Cramér Rao lower bound may be used as a standard to which the precision of any estimator may be compared. Notice that the Cramér Rao lower bound is not related to a particular estimation method. It depends on the statistical properties of the observations, the measurement points, and in most cases, the hypothetical true values of the parameters. This dependence on the true values looks, at first sight, as a serious impediment to the practical use of the bound. However, the expressions for the bound provide the means to compute numerical values for it using nominal values of the parameters. This provides the experimenter with quantitative insight in what precision may be achieved from the available observations, an insight that without the bound would be absent. Thus, using the bound, the experimenter gets a detailed insight in the sensitivity of the precision to the values of the parameters. The experimenter also gets impression if the experimental design, that is, the values and the number of the measurement points x_n, is adequate for the purposes concerned. This means an impression if the precision is sufficient to make conclusions possible. If not, there is no other choice than to change the experimental design. If this is not possible, it is to be concluded that the observations are not suitable for the purposes of the measurement procedure.

In many applications, some of the quantities to be measured are functions of the parameters and not the individual parameters. A simple example is the following.

Example 6. Measurement of peak area and location.

Suppose that a number of error corrupted observations $w_1, \ldots, w_N$ has been made on a spectral peak described by

$$\alpha \exp\left[-\frac{1}{2}\left(\frac{x-\beta}{\gamma}\right)^2\right] \qquad (12)$$

where the parameters α, β, and γ are the peak height, location, and half-width, respectively. Suppose that only the peak location and the peak area are of interest. Then these are described by β and $(2\pi)^{1/2}\alpha\gamma$, respectively.

Fortunately, the Cramér Rao lower bound of functions of the parameters follows relatively easily from the Cramér Rao lower bound for the parameters. Let $r = [r_1(w) \ldots r_L(w)]^T$ be an unbiased estimator of the vector function $\rho(\theta) = [\rho_1(\theta) \ldots \rho_L(\theta)]^T$, that is, $E[r] = \rho(\theta)$. Furthermore, let M be the information matrix for θ. Then it can be shown that

$$\text{cov}[r(w), r(w)] \geq \frac{\partial \rho}{\partial \theta^T} M^{-1} \frac{\partial \rho^T}{\partial \theta}$$

where $\partial \rho / \partial \theta^T$ is the $L \times K$ Jacobian matrix with (p, q)-th element $\partial \rho_p / \partial \theta_q$. Therefore, the Cramér Rao lower bound for unbiased estimation of ρ is described by

$$\frac{\partial \rho}{\partial \theta^T} M^{-1} \frac{\partial \rho^T}{\partial \theta} \quad (13)$$

with M^{-1} the Cramér Rao lower bound for α, β, and γ.

Example 7. The Cramér Rao lower bound for peak area and location. The vector $\rho(\theta)$ for Example 6 is described by $\rho(\theta) = [\beta \ (2\pi)^{1/2} \alpha \gamma]^T$. Then, the Jacobian matrix of $\rho(\theta)$ with respect to $(\alpha \ \beta \ \gamma)^T$ is defined as

$$\frac{\partial \rho}{\partial \theta^T} = \begin{pmatrix} 0 & 1 & 0 \\ (2\pi)^{1/2} \gamma & 0 & (2\pi)^{1/2} \alpha \end{pmatrix}$$

where $\rho = \rho(\theta)$. The Cramér Rao lower bound for unbiased estimation of ρ is then computed from Eq. (13).

The premultiplication and postmultiplication of M^{-1} in Eq. (13) describe what is conventionally called error propagation. To see how this works, suppose that $\rho = [\rho_1(\theta) \ \rho_2(\theta)]^T$, $\theta = (\theta_1 \ \theta_2)^T$, and let the Cramér Rao lower bound for θ be

$$C = \begin{pmatrix} c_{11} & c_{12} \\ c_{12} & c_{22} \end{pmatrix}$$

Then, the diagonal elements of the Cramér Rao lower bound for ρ_1 and ρ_2 are equal to

$$\left(\frac{\partial \rho_i}{\partial \theta_1}\right)^2 c_{11} + 2\left(\frac{\partial \rho_i}{\partial \theta_1}\right)\left(\frac{\partial \rho_i}{\partial \theta_2}\right) c_{12} + \left(\frac{\partial \rho_i}{\partial \theta_2}\right)^2 c_{22}$$

with $i = 1, 2$, respectively. This expression shows how the variances c_{11} and c_{22} and the covariance c_{12} of a hypothetical estimator that attains the Cramér Rao lower bound for θ propagate to the variances of a hypothetical estimator of ρ_1 and ρ_2 that also attains the Cramér Rao lower bound. Similar error propagation schemes are proposed in the literature for covariance matrices of functions of estimators in general, for example in reference 5. These schemes are approximations using the linear Taylor polynomial instead of the nonlinear functions. Equation (13), on the other hand, is exact.

Next, suppose that M is the information matrix for the estimation of $\theta = (\theta_1 \ \ldots \ \theta_K)^T$, and that an additional parameter θ_{K+1} is to be estimated. For example, θ_{K+1} may be a constant term added to the spectroscopic line model described by Eq. (12) to model a constant background contribution. Then, M has to be augmented with one row and one column corresponding to θ_{K+1}. If the augmented information matrix is inverted, all first K diagonal elements can be shown to be larger than or equal to the corresponding diagonal elements of M^{-1}. Equality occurs only if the nondiagonal elements of the $(K+1)$-th row and $(K+1)$-th column of the augmented information matrix happen to be equal to zero. Generally, it is not difficult to show that, typically, the first K diagonal elements of M^{-1} are monotonously increasing with the number of parameters in excess of K.

PRECISELY MEASURING FROM ERROR CORRUPTED OBSERVATIONS

The a priori knowledge of the experimenter about the observations and the extent to which this a priori knowledge is used may considerably influence the precision and accuracy of the measurement result. This concerns both systematic and nonsystematic errors in the observations. Systematic errors in the observations are deviations of the assumed parametric model of the expectations from the true model of the expectations. Even in the absence of nonsystematic errors, discrepancy between both models produces systematic errors, that is, inaccuracy in the measurement result. Since no model fitted will be perfect, there will always be a certain amount of systematic error. Nonsystematic errors are described by their distribution about the expectations of the observations. This distribution is not always known, but if it is, this knowledge may contribute substantially to the reducing of the nonsystematic error in the measurement result, that is, in the parameters estimates.

Suppose that observations $w_1, \ldots w_N$, are available and that their probability density function $f(\omega_1, \ldots, \omega_N; \theta)$ is known where θ is the vector of unknown parameters and $\omega_1, \ldots, \omega_N$ are the independent variables corresponding to the observations $w_1, \ldots, w_N$, respectively. Assume that $w_1, \ldots, w_N$ are substituted for $\omega_1, \ldots, \omega_N$ in $f(\omega_1, \ldots, \omega_N; \theta)$, respectively, and that the fixed true parameters θ are replaced by the vector of corresponding variables t. Then, the resulting function $f(w_1, \ldots, w_N; t)$ of t is called the likelihood function of the parameters t, given the observations $w_1, \ldots, w_N$. The maximum likelihood estimate of the parameters θ is defined as the value $\tilde{t}$ of t that maximizes the likelihood function.

The maximum likelihood estimator has a number of very favorable properties. In the first place, its definition shows that it is simple to find from the known probability density function of the observations. Furthermore, it can be shown to converge under general conditions in a statistically well-defined way to the true values of the parameters as the number of observations increases. Moreover, under general conditions, the covariance matrix of the maximum likelihood estimator approaches asymptotically the Cramér Rao lower bound. Then, the maximum likelihood estimator is asymptotically most precise. Also, a function of a maximum likelihood estimator is the maximum likelihood estimator of the function. This is called the invariance property of maximum likelihood.

Two of these properties are asymptotic; they apply to an infinite number of observations. If they also apply to a finite or even small number of observations can often only be assessed by estimating from artificial, simulated observations. These simulations may reveal that maximum likelihood estimation applied to small numbers of observations may lead to bias, that is, systematic error in the measurement result. This kind of bias, or the major part of it, is usually inversely proportional to the number of observations and may be removed as follows. Let $\tilde{t}_N$ be the biased maximum likelihood estimate of θ obtained from $w_1, \ldots, w_N$, and let $\tilde{t}_{N-1}$ be the average of the N different maximum likelihood estimates computed from the N different sets of $N - 1$ observations obtained by omitting one observation from the set $w_1, . . ., w_N$. Then, it may be shown that

$$N\tilde{t}_N - (N - 1)\bar{t}_{N-1}$$

is an estimator of θ which may only be biased to order $1/N^2$. This is the so-called Quenouille correction, today called jack-knife. A favorable property of this correction is that it hardly affects the variance of the estimator.

Example 8. Maximum likelihood estimation of peak height, width and location from Poisson distributed observations. Suppose that observations $w_1, . . ., w_N$ are available made on the spectral peak of Example 6, and that these observations are independent and Poisson distributed. Then it follows from Example 3 that the likelihood function of the parameters is described by

$$\sum_n -y_n(t) + w_n \ln[y_n(t)] - \ln(w_n!) \qquad (14)$$

with

$$y_n(t) = a \exp\left[-\frac{1}{2}\left(\frac{x_n - b}{c}\right)^2\right]$$

where $t = (a\, b\, c)^{\mathrm{T}}$. To obtain the maximum likelihood estimate of α, β, and γ, Eq. (14) must be maximized with respect to t. This is a nonlinear optimization problem which has to be solved numerically. If the peak area is computed from the maximum likelihood estimates $\tilde{a}$ and $\tilde{c}$ as $(2\pi)^{1/2}\tilde{a}\tilde{c}$ this is, by the invariance property, a maximum likelihood estimate as well.

Example 9. Maximum likelihood estimation from observations disturbed by normally distributed errors. If the errors and, therefore, the observations are normally distributed, Eq. (4) shows that the likelihood function of the parameters is described by

$$-\frac{N}{2}\ln(2\pi) - \frac{1}{2}\ln(\det W) - \frac{1}{2}[w - y(t)]^T W^{-1}[w - y(t)] \quad (15)$$

Since both first terms of this expression do not depend on the vector of parameters t, maximizing Eq. (15) is equivalent to minimizing

$$[w - y(t)]^{\mathrm{T}} W^{-1}[w - y(t)]$$

with respect to t. This shows that with normally distributed observations, maximum likelihood estimation is equivalent to a weighted least squares measurement with W^{-1} as weighting matrix.

Least squares estimation is also often used if the distribution of the observations is not known or is known to be not normal. Then, the general expression for the least squares criterion is

$$[w - y(t)]^{\mathrm{T}}\Omega[w - y(t)] \qquad (16)$$

where Ω is a positive definite weighting matrix to be chosen by the experimenter.

Linear Least Squares

First, as an important special case, models linear in the unknown parameters θ are considered. Then

$$\mathrm{E}[w] = y(\theta) = X\theta$$

with X a known $N \times K$ matrix, that is,

$$y_n(\theta) = x_{n1}\theta_1 + \ldots + x_{nK}\theta_K$$

Notice that

$$x_n = (x_{n1} \ldots x_{nK})^{\mathrm{T}}$$

is the vector independent variable corresponding to the n-th observation w_n.

Example 10. Straight line fitting. If the observations w_n are made on a straight line $y = \alpha x + \beta$ at the points $x_{11}, . . ., x_{N1}$, then X and θ are described by

$$\begin{bmatrix} x_{11} & 1 \\ . & . \\ . & . \\ x_{N1} & 1 \end{bmatrix} \quad \text{and} \quad \theta = (\alpha\, \beta)^{\mathrm{T}}$$

respectively.

The least squares solution $\hat{t}_\Omega$ for θ is

$$\hat{t}_\Omega = (X^{\mathrm{T}}\Omega X)^{-1}X^{\mathrm{T}}\Omega w \qquad (17)$$

It is observed that this solution is a linear combination of the observations. As a result, the propagation of the errors in the observations to the measurement result is perfectly clear. Furthermore, since $\mathrm{E}[w] = X\theta$, $\mathrm{E}[\hat{t}_\Omega] = \theta$ and, hence, $\hat{t}_\Omega$ is an unbiased estimator of θ. Notice that $\hat{t}_\Omega$ has these properties for any distribution of the observations w. It is easily shown that the covariance matrix $\mathrm{cov}(\hat{t}_\Omega, \hat{t}_\Omega)$ is equal to

$$(X^{\mathrm{T}}\Omega X)^{-1}X^{\mathrm{T}}\Omega W\Omega X(X^{\mathrm{T}}\Omega X)^{-1} \qquad (18)$$

The conclusion from this expression is that this covariance matrix and, therefore, the variances of $\hat{t}_\Omega$ depend on the choice of Ω. The question is then which Ω minimizes the covariance described by Eq. (18). The answer has been found to be $\Omega = W^{-1}$. For this choice,

$$\mathrm{cov}(\hat{t}_\Omega, \hat{t}_\Omega) \geq \mathrm{cov}(\hat{t}_{W^{-1}}, \hat{t}_{W^{-1}})$$

As a consequence, the variances of the elements of $\hat{t}_\Omega$ for any choice of Ω are never smaller than those of the corresponding elements of $\hat{t}_{W^{-1}}$. The estimator $\hat{t}_{W^{-1}}$ is called the best linear unbiased estimator. Equation (17) shows that it is described by

$$\hat{t}_{W^{-1}} = (X^{\mathrm{T}}W^{-1}X)^{-1}X^{\mathrm{T}}W^{-1}w$$

Among all estimators that are both linear in the observations and unbiased, it is called best since it has smallest variance.

Notice that only the expectation and the covariance matrix of the observations are specified, not their probability density function. Also notice that $\hat{t}_{W^{-1}}$ is optimal within the class of estimators that are both linear in the observations and unbiased. Therefore, there may be better, that is, more precise estimators among those that are not linear in the observations or are biased.

The covariance matrix of $\hat{t}_{W^{-1}}$ is equal to

$$(X^{\mathrm{T}}W^{-1}X)^{-1} \tag{19}$$

If for normally distributed observations, the maximum likelihood estimator is computed, it is found to be identical to the best linear unbiased estimator $\hat{t}_{W^{-1}}$ and, consequently, to have a covariance matrix equal to the one given by Eq. (19). If next, the Cramér Rao lower bound is computed for the same observations, it is found to coincide with Eq. (19). The conclusion is that for normally distributed observations, the best linear unbiased estimator is identical with the maximum likelihood estimator and attains the Cramér Rao lower bound for any number of observations.

In measurement practice, the weighting matrix Ω of $\hat{t}_\Omega$ is often taken as the identity matrix. The reason may be that the covariance matrix W is unknown. Another reason may be the amount and the complexity of numerical computation involved since with $\Omega = I$, the estimator simplifies to the ordinary least squares estimator

$$\hat{t}_I = (X^{\mathrm{T}}X)^{-1}X^{\mathrm{T}}w \tag{20}$$

which is clearly easier to compute than $\hat{t}_\Omega$. The corresponding ordinary least squares criterion is described by

$$(w - Xt)^{\mathrm{T}}(w - Xt)$$

which is simply the sum of the squares of the deviations

$$w_n - x_{n1}t_1 - x_{n2}t_2 + \ldots - x_{nK}t_K$$

Notice that $\hat{t}_I$ is only the best linear unbiased estimator if the covariance matrix W is equal to $\sigma^2 I$, that is, if the observations are uncorrelated and have equal variance σ^2. The estimator $\hat{t}_I$ is the maximum likelihood estimator and achieves the Cramér Rao lower bound if, in addition, the observations are normally distributed. Therefore, if these conditions are not met, the use of $\hat{t}_I$ may mean an exchange of precision for simplicity.

Finally, it is emphasized that Eq. (20) is a formal description of the ordinary linear least squares estimator. It is not a recipe for its numerical evaluation. Special numerical methods have been designed taking care of the fact that the set of linear equations described by Eq. (20) may be ill-conditioned. References 6 and 7 provide the details.

Nonlinear Least Squares

Nonlinear least squares is the most frequently used method for estimation of the parameters of nonlinear models. The criterion used is described by

$$[w - y(t)]^{\mathrm{T}}[w - y(t)] = \sum_n [w_n - y(x_n; t)]^2 \tag{21}$$

which is Eq. (16) with weighting matrix $\Omega = I$. Notice that generally, the solution $\hat{t}_I$ for t minimizing the least squares criterion defined by Eq. (21) is only the maximum likelihood estimator if the observations are independent and identically normally distributed. This means normally distributed with covariance matrix $\sigma^2 I$. For other distributions, $\hat{t}_I$ is generally not the maximum likelihood estimate since it does not maximize the pertinent likelihood function. As compared with linear least squares, the amount of theory concerning nonlinear least squares is limited. However, if the observations are independent and identically distributed, then under general conditions, the least squares estimator $\hat{t}_I$ is asymptotically normally distributed with covariance matrix

$$\sigma^2 \left(\frac{\partial y^T}{\partial \theta} \frac{\partial y}{\partial \theta^T} \right)^{-1} \tag{22}$$

where $y = y(\theta)$. This result is due to Jennrich (8). Notice that the computation of this covariance matrix requires the parameters to be known. In practice, this is not the case, and nominal or estimated values are substituted for the exact ones. Also notice that for independent and identically normally distributed observations, Eq. (22) is equal to the Cramér Rao lower bound. The general form of the elements of the matrix $(\partial y^{\mathrm{T}}/\partial \theta)(\partial y/\partial \theta^{\mathrm{T}})$ is

$$\sum_n \frac{\partial y(x_n; \theta)}{\partial \theta_p} \frac{\partial y(x_n; \theta)}{\partial \theta_q}$$

This expression shows the dependence of the elements of this matrix upon the values of the independent variable x. Therefore, if the experimenter has some freedom in the choice of the measurement points, it may be used to manipulate the covariance matrix described by Eq. (22) in a desired way. This usually concerns the diagonal elements, that is, the variances and is an example of experimental design: the manipulation of the variances by selecting free experimental variables.

The gradient of the nonlinear least squares criterion with respect to the parameter vector t is equal to

$$-2 \sum_n [w_n - y_n(t)] \frac{\partial y_n(t)}{\partial t} \tag{23}$$

A necessary condition for a point to be a minimum is that the gradient is equal to the null vector. If Eq. (23) is equated to the null vector, this produces a set of K nonlinear equations in K variables. This set must be solved by an iterative numerical method since, typically, it cannot be solved in closed form. For this problem, specialized numerical methods have been developed. Most frequently used are the Gauss–Newton method and the Levenberg–Marquardt method. These are described in references 8 and 6, respectively. Software for their practical implementation is found in references 6 and 7.

Many nonlinear models in engineering practice are linear in some of their parameters. How this special property may be exploited in nonlinear least squares estimation is illustrated in the following example.

Example 11. Least squares estimation of the parameters of a multiexponential model. Suppose that in a least squares estimation problem, the model fitted is described by

$$y_n(t) = a_1 \exp(-b_1 x_n) + \ldots + a_L \exp(-b_L x_n)$$

where $t = (a^T\ b^T)^T$ with linear parameters $\mathrm{a} = (a_1 \ldots a_L)^T$ and nonlinear parameters $b = (b_1 \ldots b_L)^T$. Then Eq. (23) shows that at the minimum of the least squares criterion, the derivatives with respect to the linear parameters a must satisfy

$$\sum_n [w_n - a_1 \exp(-b_1 x_n) + \ldots + a_L \exp(-b_L x_n)] \exp(-b_\ell x_n) = 0$$

with $\ell = 1, \ldots, L$. This may be considered a set of L linear equations in L unknowns a_ℓ. The solution for these unknowns is a function of the unknown nonlinear parameters b_ℓ and is denoted as $a_\ell(b)$. Substitution of the $a_\ell(b)$ for the a_ℓ in the least squares criterion yields

$$\sum_n [w_n - a_1(b) \exp(-b_1 x_n) + \ldots + a_L(b) \exp(-b_L x_n)]^2$$

Thus, the least squares criterion has become a function of the nonlinear parameters b only. Minimization of it with respect to b yields the solution $\hat{b}$ for β and the solution $\hat{a} = a(\hat{b})$ for α.

Nonlinear least squares problems of the kind described in Example 11 are called separable nonlinear least squares problems since the linear and the nonlinear parameters are estimated separately. Notice that in Example 11, the number of parameters involved in the iterative numerical minimization is reduced by a factor of two. This also means that the number of initial values for the procedure is reduced correspondingly.

HANDLING MEASUREMENT ERRORS IN NONSTANDARD PROBLEMS

Complex Parameter Estimation

Many practical measurement problems concern complex valued parameters or mixtures of real and complex valued parameters. In particular, these problems are found in measurement in the frequency domain. Such complex parameter estimation problems can always be transformed into real parameter estimation problems by splitting a complex parameter into its real and imaginary part and estimating these real quantities separately. This, however, leads to unnecessarily complicated expressions for the estimator and, as a result, to complicated numerical procedures. This is avoided by leaving quantities complex if they are complex by nature.

The most important tool in the formulation of complex parameter measurement from error corrupted observations is the following. Suppose that in a measurement problem there are $K + 2L$ parameters

$$\theta = (\eta_1 \ldots \eta_K\ \alpha_1\ \beta_1 \ldots \alpha_L\ \beta_L)^T$$

of which the η_k are intrinsically real, and the α_ℓ and β_ℓ are the real and imaginary parts of the complex parameters $\gamma_\ell = \alpha_\ell + j\beta_\ell$ with $j = \sqrt{-1}$. Then γ_ℓ and its complex conjugate γ_ℓ^* on the one hand and α_ℓ and β_ℓ on the other are connected by

the linear transformation

$$\begin{bmatrix} \gamma_\ell \\ \gamma_\ell^* \end{bmatrix} = J \begin{bmatrix} \alpha_\ell \\ \beta_\ell \end{bmatrix}$$

where

$$J = \begin{bmatrix} 1 & j \\ 1 & -j \end{bmatrix}$$

Therefore, the mixed real complex parameter vector

$$\zeta = (\eta_1 \ldots \eta_K\ \gamma_1\ \gamma_1^*\ \gamma_L\ \gamma_L^*)^T$$

and θ are connected by

$$\zeta = B_{K+2L}\theta \tag{24}$$

where B_{K+2L} is the $(K + 2L) \times (K + 2L)$ block diagonal matrix

$$B_{K+2L} = \mathrm{diag}(I_K\quad A_{2L})$$

with I_K the identity matrix of order K and A_{2L} the $2L \times 2L$ block diagonal matrix

$$A_{2L} = \mathrm{diag}(J \ldots J)$$

The theory, methods, and techniques presented up to now concerned the estimation of real parameters from error corrupted observations. Using the linear transformation described by Eq. (24), transformation of the pertinent expressions into those for estimating a mixed real complex parameter vector is relatively easy. All that is required is observing the mathematical rules governing linear transformation of coordinates in general. For what follows, it is important to notice that Eq. (24) implies that α_ℓ and β_ℓ are transformed into both γ_ℓ and γ_ℓ^*. Also, the definition of the covariance matrix of a vector of complex stochastic variables is needed. Let z be a vector of stochastic variables. Then, the covariance matrix of z is defined as

$$E[(z - E[z])(z - E[z])^H]$$

where the superscript H denotes complex conjugate transposition. The Fisher information matrix defined by Eq. (9) after the transformation of parameters described by Eq. (24) is given by

$$M = -E\left[\frac{\partial^2 \ln f}{\partial \zeta^* \partial \zeta^T}\right]$$

and the corresponding Cramér Rao lower bound on the variance of unbiased estimators of ζ is equal to M^{-1}. Again using Eq. (24), the Cramér Rao lower bound for a vector of real and complex functions $\phi(\zeta)$ of the mixed real complex parameter vector is found to be

$$\frac{\partial \phi}{\partial \zeta^T} M^{-1} \frac{\partial \phi^H}{\partial \zeta^*}$$

where for brevity, the argument of $\phi(\zeta)$ has been omitted. A further example is the weighted least squares estimator defined by Eq. (17). After transformation of θ into ζ and $\hat{t}_\Omega$ into $\hat{z}_\Omega$, respectively, it becomes

$$\hat{z}_\Omega = (R^H \Omega R)^{-1} R^H \Omega w$$

where the complex $N \times (K + 2L)$ matrix R is equal to XB_{K+2L}^{-1}. If Ω is equal to W^{-1}, this is the best linear unbiased estimator. Finally, suppose that the real complex $(P + 2Q) \times 1$ vector of observations u is composed of the elements of the real $(P + 2Q) \times 1$ vector of observations w as follows

$$u = B_{P+2Q} w$$

with $P + 2Q = N$. Then, u is a vector of real and complex observations described by

$$(w_1 \ldots w_P \, w_{P+1} + jw_{P+2} \, w_{P+1} - jw_{P+2} \ldots w_{P+2Q-1} + jw_{P+2Q}$$
$$w_{P+2Q-1} - jw_{P+2Q})^T$$

and

$$\hat{z}_\Psi = (S^H \Psi S)^{-1} S^H \Psi u$$

where S and Ψ are equal to $B_{P+2Q}R$ and $B_{P+2Q}^{-H} \Omega B_{P+2Q}^{-1}$, respectively. The covariance matrix of the mixed real complex observations is defined as $E[(u - E[u])(u - E[u])^H]$ and is, therefore, equal to $B_{P+2Q}WB_{P+2Q}^H$. Hence, the estimator $\hat{z}_\Psi$ with $\Psi = (B_{P+2Q}WB_{P+2Q}^H)^{-1}$ is the best linear unbiased estimator.

The iterative numerical optimization of likelihood functions and nonlinear least squares criteria of mixed real complex parameters may be carried out directly with respect to the vector of mixed real complex parameters. This is discussed in reference 9. In particular, use may be made of the complex gradient. Specifically, the complex gradient of the logarithm of the likelihood function $\ln f$ with respect to the complex parameter vector z is defined as $\partial \ln f / \partial z$. An important property of this complex gradient is that the real gradient $\partial \ln f / \partial t$ is equal to the null vector if and only if the complex gradient is equal to the null vector. Therefore, the complex gradient may be used to find maxima of the likelihood function and minima of the nonlinear least squares criterion in the same way as the real gradient.

Nonstandard Fourier Analysis

Estimation of Fourier coefficients from error disturbed observations made on periodic functions is an important problem in dynamic system identification in general and in specialized applications as crystal structure reconstruction. Suppose that the problem is to estimate the Fourier coefficients γ_k, $k = 0$, $\pm 1, \ldots, \pm K$ and, possibly, the period δ of the real periodic function

$$y_n(\zeta) = \sum_k \gamma_k \exp(-j2\pi kx_n/\delta) \quad (25)$$

from error corrupted observations $w = (w_1 \ldots w_N)^T$ where the measurement points x_n are known, and the vector of unknown parameters ζ is either equal to $\gamma = (\gamma_0 \, \gamma_1 \, \gamma_{-1} \ldots \gamma_K \, \gamma_{-K})^T$ or to $(\gamma^T \, \delta)^T$, where γ_0 and δ are real, while the remaining γ_k are complex and satisfy $\gamma_{-k} = \gamma_k^*$ since the $y_n(\zeta)$ are real. It will not be supposed that the measurement points

are equidistant, nor if they are, that the period is a known integer multiple of the sampling interval, and an integer number of periods is observed. The purpose of this section is to formulate the estimation of the parameters ζ as a complex statistical parameter estimation problem and to describe the special conditions under which this problem simplifies to the standard Discrete Fourier Transform, the DFT.

Example 12. Estimation of Fourier coefficients from Poisson distributed observations. Suppose that observations $w_n \geq 0$, $n = 1, \ldots, N$ are available with expectations described by Eq. (25) and that these observations have a Poisson distribution. Then, by Eq. (14), the likelihood function of the parameters is

$$\sum_n -y_n(z) + w_n \ln[y_n(z)] - \ln(w_n!) \quad (26)$$

with

$$y_n(z) = \sum_k c_k \exp(j2\pi kx_n/d)$$

where the elements of $z = (c^T \, d)^T$ correspond to those of ζ, and those of $c = (c_0 \, c_1 \, c_{-1} \ldots c_K \, c_{-K})^T$ correspond to those of γ. Then, the complex gradient of Eq. (26) with respect to z is

$$-\sum_n \left(1 - \frac{w_n}{y_n(z)}\right) \frac{\partial y_n(z)}{\partial z} \quad (27)$$

Since the maximum of the likelihood function is, by definition, a stationary point, the $2K + 2$ elements of the maximum likelihood estimate $\tilde{z}$ of ζ must satisfy the $2K + 2$ nonlinear equations in z obtained by equating Eq. (27) to the null vector. The numerical solution for z representing the absolute maximum of the likelihood function described by Eq. (26) is the maximum likelihood estimate of the Fourier coefficients.

It follows from Eq. (25) that the expectations $y_n(\zeta)$ of the observations w_n are described by

$$y_n(\zeta) = X\gamma$$

where the n-th row of X is defined as

$$(1 \, \exp(j2\pi x_n/\delta) \, \exp(-j2\pi x_n/\delta) \ldots \exp(j2\pi Kx_n/\delta)$$
$$\exp(-j2\pi Kx_n/\delta)$$

If the observations are normally distributed with covariance matrix W, then the maximum likelihood estimator of ζ has to minimize

$$[w - y(z)]^T W^{-1} [w - y(z)] \quad (28)$$

Hence, if only the Fourier coefficients are unknown, their maximum likelihood estimator is

$$\hat{c}_{W^{-1}} = (X^H W^{-1} X)^{-1} X^H W^{-1} w \quad (29)$$

For observations with a distribution different from normal, this estimator is no longer maximum likelihood but is still best linear unbiased. If, in addition, the period is unknown, the estimation problem is recognized as a separable nonlinear least squares problem. The model is linear in the $2K + 1$ Fourier coefficients and nonlinear in the period d. This means that, in addition to Eq. (29), one further equation must be satisfied. This is the equation resulting from equating the derivative of the least squares criterion with respect to d to zero. If, in this equation, Eq. (29) is substituted for the Fourier coefficients c, a scalar nonlinear equation is obtained in the period d only. Hence, all that needs to be done is real, scalar root finding to estimate the period δ and substitute the estimate in the closed form of Eq. (29) for the Fourier coefficients. The estimates thus obtained are maximum likelihood if the observations are normal with covariance matrix W. They are weighted complex nonlinear least squares estimates with other error distributions.

If the covariance matrix W is unknown, the ordinary least squares estimator

$$\hat{c}_I = (X^H X)^{-1} X^H w \qquad (30)$$

may be chosen, possibly combined with root finding for the period. This is a maximum likelihood estimator only if the w_n are independent and identically normally distributed about their expectations. For other distributions, it is a best linear unbiased estimator if only the Fourier coefficients are to be estimated, and the observations are uncorrelated with equal variance. In other cases, it is simply the ordinary least squares estimator.

A special case occurs if only the Fourier coefficients are to be estimated, the measurement points x_n are equidistant with interval Δ, the period is a known integer multiple of Δ, and an integer number of periods is observed. Under these conditions, the elements of $\hat{c}_I$ described by Eq. (30) may be shown to be equal to the DFT

$$\frac{1}{N} \sum_{n=1}^{N} w_n \exp[-j2\pi k (n-1)/M]$$

where $M\Delta$ is the period. Under the restrictive conditions mentioned, the DFT is, therefore, the maximum likelihood estimator if the observations are independent and identically normally distributed about their expectations. For other distributions, it is best linear unbiased if the observations are uncorrelated and have equal variance.

Measurement Errors and Resolution

Like precision and accuracy, resolution is a key notion in applied science and engineering. It is used in fields as diverse as radar, sonar, optics, electron optics, seismology and various forms of spectroscopy. An extensive review of resolution is presented in reference 10.

The most important form of resolution is two-component resolution.

Example 13. Rayleigh two-component resolution. As discussed in reference 10, Rayleigh considers observations described by

$$\alpha\{\mathrm{sinc}^2[2\pi(x - \beta_1)] + \mathrm{sinc}^2[2\pi(x - \beta_2)]\}$$

with $\mathrm{sinc}(x) = \sin(x)/x$. This is a pair of sinc-square components of equal height and located at β_1 and β_2, respectively.

As the difference in location decreases, the components increasingly overlap and become increasingly difficult to distinguish visually. According to Rayleigh, the components are resolvable if the absolute difference of β_1 and β_2 exceeds 0.5. At this distance, the maximum of the one component coincides with the first zero of the other, and the component sum has two maxima and a relative minimum in between. Then, the ratio of the value at the relative minimum to that at the maxima may be shown to be 0.81. Later, this ratio has been generalized to other component functions, and the distance corresponding to this ratio has been called generalized Rayleigh resolution limit.

From this example, it is clear that this classical resolution limit and comparable ones proposed later are, in fact, measures of component width. Since in definitions such as Rayleigh's, the component functions are known, and the observations are exact, today, the model could be exactly fitted numerically to the observations with respect to the locations, the result would be exact, and there would in fact be no obvious limit to resolution. The reason why in practice unlimited resolution cannot be achieved is that observations exactly describable by two component functions do not occur. Therefore, it is not the distance of the components, but it is the errors in the observations, systematic and nonsystematic, that ultimately limit resolution. During the last decades, a number of measurement error based resolution limits have been proposed in the literature reviewed in Ref. 10. One of the most recent ones will now be described.

Suppose that a number of two-component observations $w = (w_1 \ldots w_N)^T$ has been made, and that the two-component model

$$a[h(x; b_1) + h(x; b_2)] \qquad (31)$$

is fitted to these observations with respect to a, b_1, and b_2. Then, depending on the set of observations available, two essentially different types of solutions for a, b_1, and b_2 may occur. In the first type, the solutions for b_1 and b_2 are distinct. This implies that the two components in Eq. (31) are resolved from the observations. In the second type of solution, the solutions for b_1 and b_2 exactly coincide. Then, the model corresponding to this solution is $2a\, h(x;b)$ with $b_1 = b_2 = b$. Thus, it is concluded that a one-component model is found as solution. This one-component solution is, of course, not found from exact two-component observations of the same functional family as the model fitted. However, it may result from error corrupted, two-component observations if the components seriously overlap.

At first sight, exactly coinciding solutions may look highly improbable. However, their coincidence is not caused by mere chance but by a structural change of the criterion of goodness of fit under the influence of the set of observations. In the N-dimensional Euclidean space of the observations, where the n-th coordinate axis corresponds to the n-th observation w_n, a set of observations is represented by a single point. If two-component models are fitted, this space may be divided into two parts. For observations in the one part, the criterion has an absolute minimum with $b_1 \neq b_2$. For observations in the other part, only a minimum can be shown to exist with $b_1 = b_2$. The boundary of both parts separates the sets of observations from which the components can be resolved from those from which they cannot. Therefore, it is this boundary that constitutes the limit to resolution in terms of the observations. Of course, hypothetical, errorless two-component observations to which a two-component model of the same family is fitted are on the side of the boundary corresponding to resolution. However, nonsystematic and systematic measurement errors may move this point to the other side of the boundary, where resolution is impossible since the solutions coincide. Systematic measurement errors influence the location of the point around which sets of observations are distributed. This point represents the expectations of the observations. The systematic errors may move this point close to the boundary. The kind of distribution of the nonsystematic errors defines how the sets of observations are distributed around this point. Therefore, the probability of resolution is determined by both types of errors combined.

BIBLIOGRAPHY

1. C. Chatfield, *Statistics for Technology,* 3rd ed., London: Chapman and Hall, 1995.

2. A. M. Mood, F. A. Graybill, and D. C. Boes, *Introduction to the Theory of Statistics,* 3rd ed., Auckland: McGraw-Hill, 1987.

3. A. Stuart and J. K. Ord, *Kendall's Advanced Theory of Statistics—Vol.1 Distribution Theory,* London: Arnold, 1994.

4. A. Stuart and J. K. Ord, *Kendall's Advanced Theory of Statistics—Vol. 2 Classical Inference and Relationship,* London: Arnold, 1991.

5. Anonymous, *Guide to the Expression of Uncertainty in Measurement,* 1st ed., Geneva: International Organization for Standardization, 1993.

6. W. H. Press et al., *Numerical Recipes in Fortran; the Art of Scientific Computing,* 2nd ed., New York: Cambridge University Press, 1992.

7. A. Grace, *Optimization Toolbox for Use With MATLAB™, USCL's Guide,* South Natick, MA: The Math Works, 1990.

8. R. I. Jennrich, *An Introduction to Computational Statistics,* Englewood Cliffs, NJ: Prentice-Hall, 1995.

9. A. van den Bos, Complex gradient and Hessian, *IEE Proceedings Vision and Image Signal Processing,* **141** (6): 380–382, 1994.

10. A. J. den Dekker and A. van den Bos, Resolution—A survey, *J. Opt. Soc. Am.,* **14** (3): 547–557, 1997.

ADRIAAN VAN DEN BOS
Delft University of Technology

MEASUREMENT OF FREQUENCY, PHASE NOISE AND AMPLITUDE NOISE

Frequency metrology has the highest resolution of all the measurement sciences. Simple systems readily achieve a fractional frequency resolution of 1 ppm (part per million) and some elaborate systems achieve 1 part in 10^{17} or less. Because of the readily achieved resolution, the growing trend is to convert the measurement of many different parameters to the measurement of frequency or frequency difference.

In the following we describe the basic ideas and definitions associated with various aspects of the specification and measurement of frequency, phase (or time), and the two components of spectral purity—phase modulation (PM) noise and amplitude modulation (AM) noise. The three topics are funda-

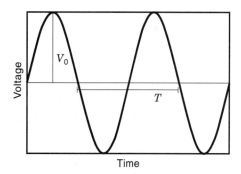

Figure 1. Output voltage of an ideal sinusoidal oscillator.

mentally linked, since as we will see below, frequency is proportional to the rate of change of phase with time, and the degree to which we can specify or measure signal phase, amplitude, or frequency is limited by the spectral purity of the signals.

BASIC CONCEPTS AND DEFINITIONS

Figure 1 shows the output voltage signal of an ideal sinusoid oscillator as a function of time. The maximum value V_0 is the nominal amplitude of the signal. The time required for the signal to repeat itself is the period T of the signal. The nominal frequency ν_0 of the signal is the reciprocal of the period, $1/T$. This voltage signal can be represented mathematically by a sine function,

$$v(t) = V_0 \sin\theta = V_0 \sin(2\pi\nu_0 t) \tag{1}$$

where the argument $\theta = 2\pi\nu_0 t$ of the sine function is the nominal phase of the signal. The time derivative of the phase θ is $2\pi\nu_0$ and is called the nominal angular frequency ω_0. In the frequency domain, this ideal signal is represented by a δ function located at the frequency of oscillation.

In real situations, the output signal from an oscillator has noise. Such a noisy signal is illustrated in Fig. 2. In this example we have depicted a case in which the noise power is much less than the signal power. Fluctuations in the peak values of the voltage result in AM noise. Fluctuations in the zero crossings result in PM noise. Fractional frequency modulation (FM) noise refers to fluctuations in the period of the signal. Since the period (and thus the frequency) of the signal is related to the phase of the signal, FM noise and PM noise are directly related.

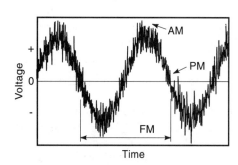

Figure 2. Output voltage of a noisy sinusoidal oscillator.

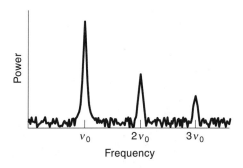

Figure 3. Power spectrum of a noisy signal.

Figure 3 shows the power spectrum of a noisy signal (power as a function of frequency) as measured by a spectrum analyzer. Although the maximum power occurs at the frequency of oscillation, other peaks are observed at frequencies of $2\nu_0$, $3\nu_0$, . . ., $n\nu_0$. These frequencies are called *harmonics* of the fundamental frequency ν_0; $2\nu_0$ is the second harmonic, $3\nu_0$ is the third harmonic, and so on. The power at these harmonic frequencies will depend on the design of the source. The spectrum around the fundamental frequency displays power sidebands at frequencies above the carrier (upper sideband) and at frequencies below the carrier (lower sideband). These power sidebands are the result of PM and AM noise in the signal. While the power spectrum gives an idea of the total noise of a signal, it does not give information about the relative magnitude of the PM and AM noise. Furthermore, at frequencies close to ν_0, it is difficult to separate noise power from the power of the fundamental frequency. Therefore, special measurement techniques are needed to measure PM and AM noise in oscillators.

Characterization of Frequency Stability and Amplitude Stability of a Signal

A noisy signal can be mathematically represented by

$$v(t) = [V_0 + \epsilon(t)]\sin[2\pi\nu_0 t + \phi(t)] \tag{2}$$

where $\epsilon(t)$ represents amplitude fluctuations (amplitude deviation from the nominal amplitude V_0) and $\phi(t)$ represents phase fluctuations (phase deviation from the nominal phase $2\pi\nu_0 t$) (1). The instantaneous frequency of this signal is defined as

$$v(t) = \frac{1}{2\pi}\frac{d}{dt}(\text{phase}) = \nu_0 + \frac{1}{2\pi}\frac{d}{dt}\phi(t) \tag{3}$$

Frequency fluctuations refer to the deviation of the instantaneous frequency from the nominal frequency: $\nu(t) - \nu_0$. Fractional frequency fluctuations, denoted as $y(t)$, refer to frequency fluctuations normalized to ν_0, that is,

$$y(t) = \frac{\nu(t) - \nu_0}{\nu_0} = \frac{1}{2\pi\nu_0}\frac{d}{dt}\phi(t) \tag{4}$$

Equation (4) indicates that there is a direct relation between phase fluctuations and fractional frequency fluctuations. Therefore if the PM noise of a signal is measured, the FM noise can be easily obtained and vice versa. The time deviation or fluctuation $x(t)$ of a signal is equal to the integral of

$y(t)$ from 0 to t. This relation can be expressed as

$$y(t) = \frac{d}{dt}x(t) \tag{5}$$

Units of Measure for PM Noise, FM Noise, and AM Noise

Phase fluctuations in the frequency domain are characterized by the spectral density of the phase fluctuations $S_\phi(f)$, given by

$$S_\phi(f) = \text{PSD}[\phi(t)] = [\phi(f)]^2\frac{1}{\text{BW}} \tag{6}$$

where PSD refers to power spectral density, $[\phi(f)]^2$ is the mean-squared phase deviation at an offset frequency f from the frequency ν_0 (called the carrier in this context), and BW is the bandwidth of the measurement system (1–4). The offset frequency f is also called Fourier frequency. The units for $S_\phi(f)$ are rad^2/Hz. Equation (6) is defined for $0 < f < \infty$; nevertheless it includes fluctuations from the upper and lower sidebands and thus is a double sideband unit of measure.

The PM noise unit of measure recommended by the IEEE (1,2,4) is $\mathscr{L}(f)$, defined as

$$\mathscr{L}(f) = \frac{S_\phi(f)}{2} \tag{7}$$

At Fourier frequencies far from the carrier frequency, where the integrated PM noise from ∞ to f (the Fourier frequency) is less than 0.1 rad^2, $\mathscr{L}(f)$ can be viewed as the ratio of phase noise power in a single sideband to power in the carrier (single-sideband phase noise). When $\mathscr{L}(f)$ is expressed in the form $10\log[\mathscr{L}(f)]$ its units are dB below the carrier in a 1 Hz bandwidth ($d\beta c$/Hz).

Frequency fluctuations in the frequency domain are characterized by the spectral density of the fractional frequency fluctuations $S_y(f)$, given by

$$S_y(f) = \text{PSD}[y(t)] = [y(f)]^2\frac{1}{\text{BW}} \tag{8}$$

where $y(f)^2$ represents the mean-squared fractional frequency deviation at an offset (Fourier) frequency f from the carrier. $S_y(f)$ is defined for Fourier frequencies $0 < f < \infty$, and its units are 1/Hz.

The conversion between $S_y(f)$ and $S_\phi(f)$ can be obtained from Eq. (4). Applying the Fourier transform to both sides of Eq. (4), squaring, and dividing by the measurement bandwidth result in

$$S_y(f) = \left(\frac{1}{2\pi\nu_0}\right)^2(2\pi f)^2 S_\phi(f) = \left(\frac{f}{\nu_0}\right)^2 S_\phi(f) \tag{9}$$

Amplitude fluctuations in the frequency domain are characterized by the spectral density of the fractional amplitude fluctuations $S_a(f)$, given by

$$S_a(f) = \text{PSD}\left(\frac{\epsilon(t)}{V_0}\right) = \left(\frac{\epsilon(f)}{V_0}\right)^2\frac{1}{\text{BW}} \tag{10}$$

where $\epsilon(f)^2$ represents the mean-squared amplitude deviation at an offset frequency f from the carrier (1). $S_a(f)$ is defined for Fourier frequencies $0 < f < \infty$, and its units are 1/Hz.

Figure 4(a) and 4(b) shows the common noise types characteristic of the PM noise and the AM noise of an oscillator (1,2,4–6).

Effects of Frequency Multiplication and Heterodyning on PM, FM, and AM Noise

When the frequency of a signal is multiplied by N, the phase fluctuations are also multiplied by N, as shown in Fig. 5(a).

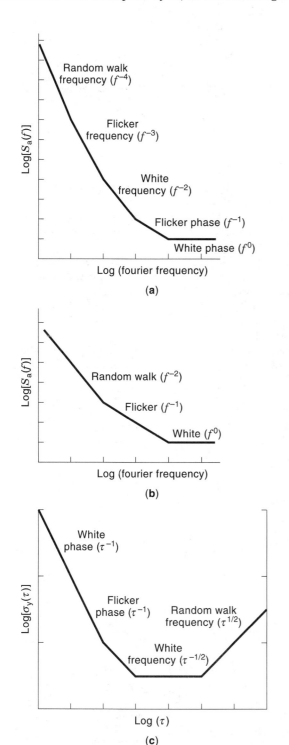

Figure 4. Common types of noise (fluctuations) in oscillators: (a) PM noise; (b) AM noise; (c) $\tau_y(\tau)$ (1,2,4–6).

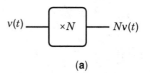

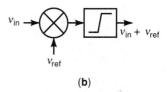

Figure 5. (a) Block diagram of a frequency multiplication system. (b) Block diagram of a frequency heterodyne or translation system.

The PM noise of the multiplied signal is given by

$$S_{\phi_2}(f) = \frac{[\phi_2(f)]^2}{\mathrm{BW}} + S_{\phi,M}(f) = \frac{N^2[\phi_1(f)]^2}{\mathrm{BW}} + S_{\phi,M}(f)$$
$$= N^2 S_{\phi_1}(f) + S_{\phi,M}(f) \tag{11}$$

where $S_{\phi,M}(f)$ is the PM noise added by the frequency multiplier. Similarly, when the frequency of a signal is divided by N, the PM noise $S_\phi(f)$ of the divided signal is divided by N^2. Frequency multiplication and frequency division do not alter the fractional FM noise $S_y(f)$ of a signal since both the frequency fluctuations and the nominal frequency are multiplied by N, and the ratio remains constant (7). Ideally, frequency multiplication or division should not have an effect on AM noise either. Nevertheless, the AM noise of the multiplied or divided signal can be affected and determined by the multiplication or division scheme.

A system that translates or shifts the frequency of an input signal by a fixed frequency is shown in Fig. 5(b). In this system, a mixer is used to multiply the input and reference signals. The output signal after the high-pass filter has a frequency of $\nu_{in} + \nu_{ref}$. (Alternately the lower sideband $\nu_{in} - \nu_{ref}$ could just as well have been chosen.) The input frequency has been shifted by the frequency of the reference. The PM noise of the output signal of a frequency translation or frequency heterodyne system is given by

$$S_{\phi,o}(f) = S_{\phi,in}(f) + S_{\phi,ref}(f) + S_{\phi,T}(f) \tag{12}$$

where $S_{\phi,in}(f)$ is the PM noise of the input signal, $S_{\phi,ref}(f)$ is the PM noise of the reference, and $S_{\phi,T}(f)$ is the PM noise of the translator (in this case the mixer and the high-pass filter). The AM noise of the output signal will depend on the details of the translation scheme.

Time-Domain Fractional Frequency Stability of a Signal

In the time domain, the fractional frequency stability of a signal is usually characterized by the Allan variance, a type of two-sample frequency variance given by

$$\sigma_y^2(\tau) = \frac{1}{2(N-2)\tau^2} \sum_{i=1}^{N-2} (x_{i+2} - 2x_{i+1} + x_i)^2 \qquad (13)$$

$$\sigma_y^2(\tau) = \frac{1}{2(M-1)} \sum_{i=1}^{M-1} (\bar{y}_{i+1} - \bar{y}_i)^2 \qquad (14)$$

$$\sigma_y^2(\tau) = \frac{2}{(\pi \nu_0 \tau)^2} \int_0^\infty S_\phi(f) \sin^4(\pi f \tau) \, df \qquad (15)$$

where N is the number of time deviation samples, x_i is the time deviation over the interval τ, $M = N - 1$ is the number of frequency samples, and $\bar{y}_i$ is the fractional frequency deviation for interval i (1–4). Equation (13) is used when time data are available, Eq. (14) is used when frequency data are available, and Eq. (15) is used to convert frequency domain data (PM noise) to the time domain. The squared root of the Allan variance, $\sigma_y(\tau)$, is generally used to specify the frequency stability of a source. Figure 4(c) shows the slopes of the common noise types characteristic of the $\sigma_y(\tau)$ of oscillators. If the dominant noise type in short-term is flicker PM or white PM, the modified Allan variance given by Eq. (16), (17), or (18) can be used to improve the estimate of the underlying frequency stability of the sources (1,2,8):

$$\mathrm{mod}\,\sigma_y^2(\tau) = \frac{1}{2\tau^2 m^2 (N-3m+1)} \sum_{j=1}^{N-3m+1} \\ \left(\sum_{i=j}^{m+j-1} (x_{i+2m} - 2x_{i+m} + x_i) \right)^2 \qquad (16)$$

$$\mathrm{mod}\,\sigma_y^2(\tau) = \frac{1}{2(N-3m+1)} \sum_{j=1}^{N-3m+1} (\bar{y}'_{j+m} - \bar{y}'_j)^2 \qquad (17)$$

$$\mathrm{mod}\,\sigma_y^2(\tau) = \frac{2}{m^4 (\pi \nu_0 \tau_0)^2} \int_0^\infty S_\phi(f) \frac{\sin^6(\pi \tau f)}{\sin^2(\pi \tau_0 f)} \, df \qquad (18)$$

where

$$\bar{y}'_j = \frac{\bar{x}_{j+m} - \bar{x}_j}{\tau} \qquad (19)$$

and

$$\bar{x}_j = \frac{\sum_{k=0}^{m-1} x_{j+k}}{m} \qquad (20)$$

Here $\bar{x}_j$ is the phase (time) averaged over n adjacent measurements of duration τ_0. Thus mod $\sigma_y(\tau)$ is proportional to the second difference of the phase averaged over a time $m\tau_0$. Viewed from the frequency domain, mod $\sigma_y(\tau)$ is proportional to the first difference of the frequency averaged over m adjacent samples.

The confidence intervals for $\sigma_y(\tau)$ and mod $\sigma_y(\tau)$ as a function of noise type and the number of samples averaged are discussed in Refs. 1 and 4.

MEASUREMENT SYSTEMS

Direct Measurements

Direct Measurements of Frequency and Frequency Stability Using a Counter. Figure 6 shows the timing diagram for the direct measurement of signal frequency relative to a reference frequency using a counter. The normal convention is to start with the signal under test and stop with the reference. The user typically chooses the nominal measurement period, which is an integral number of cycles of the reference. For example $\tau = N_{\mathrm{ref}} / \nu_{\mathrm{ref}} \cong 1$ s. The instrument counts N_{sig}, the nominal number of cycles of the signal under test that occur before the reference signal stops the count. The frequency of the signal averaged over a measurement interval τ is given by

$$\nu_{\mathrm{sig}}(\tau) = N_{\mathrm{sig}}(\tau) \frac{\nu_{\mathrm{ref}}}{N_{\mathrm{ref}}} \qquad (21)$$

where ν_{ref} is the frequency of the reference. Since the measured signal frequency is proportional to the time base or reference frequency, an error in the time-base frequency leads to a proportional error in the determination of the signal frequency.

The intrinsic fractional frequency resolution of a simple counter is $\pm 1/N_{\mathrm{sig}}$. Some sophisticated counters have interpolation algorithms that allow them to improve the intrinsic resolution by a factor of β, which can be 100 or more. If the frequency of the signal under test is less than the frequency of the reference, the resolution can often be improved by reversing the roles of the signal and reference. The counter reading can then be inverted to find the frequency of the signal.

The uncertainty in the frequency from a particular measurement taken over a measurement time τ is the intrinsic resolution plus the combined fractional instability of signal $\sigma_{y,\mathrm{sig}}^2(\tau)$ and the reference $\sigma_{y,\mathrm{ref}}^2(\tau)$. When these factors are inde-

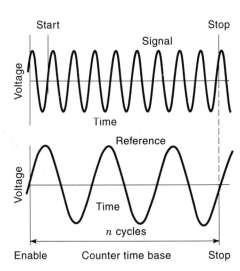

Figure 6. Timing diagram for a direct frequency measurement system.

pendent, they are added in quadrature:

frequency uncertainty

$$= \frac{\Delta \nu(\tau)}{\nu} = \left[\left(\frac{1}{\beta N_{sig}} \right)^2 + \sigma_{y,ref}^2(\tau) + \tau_{y,sig}^2(\tau) \right]^{1/2} \quad (22)$$

The time-domain fractional frequency stability of the signal $\sigma_{y,sig}^2(\tau)$ relative to the frequency stability of the reference can be estimated from a series of consecutive frequency measurements using Eq. (14) (1,2). There is often dead time between frequency measurements in the direct method, which leads to biases in the estimation of fractional frequency stability; these biases depend on noise type (9). More elaborate techniques that eliminate this bias and offer better intrinsic frequency resolution are described in later sections on heterodyne measurements.

Direct Measurements of Phase or Time Using a Counter. For sinusoidal signals, phase or time is usually referenced to the positive-going zero crossing of the signal. For digital signals, time is usually referenced to the mean of the 0 and the 1 states at the positive-going transition. Although the counter can be started with the signal or the reference, we usually start with the signal. Then advancing phase (time) corresponds to a signal frequency that is higher than the reference. The instrument counts N_{ref}, the nominal number of cycles of the counter time base frequency ν_{ref} that occur before the reference signal stops the count. The phase of the signal relative to the reference is

$$\theta_{sig} = 2\pi N_{ref} \frac{\nu_{sig}}{\nu_{ref}} \quad (23)$$

where ν_{sig} is the frequency of the signal. The time of the signal relative to the reference is

$$t_{sig} = \frac{N_{ref}}{\nu_{ref}} \quad (24)$$

Since the measured phase (time) is proportional $1/(\nu_{ref})$, an error in the time base (reference) frequency leads to an error in the determination of the signal phase. There may also be phase errors or time errors in the measurement due to the voltage standing wave ratio (VSWR) on the transmission lines to the counter from the signal and the reference (10).

For a simple counter the intrinsic phase resolution is $2\pi \nu_{sig}/\nu_{ref}$, while the time resolution is $1/\nu_{ref}$. Some sophisticated counters have interpolation algorithms that allow them to improve the intrinsic resolution by a factor of β, which can be 100 or more.

The uncertainty in the measurement of phase $\Delta\phi$ or time ΔT using this approach is given by the intrinsic resolution plus the combined fractional instability of signal $\sigma_{ysig}^2(\tau)$ and the reference $\sigma_{yref}^2(\tau)$. When these factors are independent, they are added in quadrature:

$$\Delta\phi = \theta_{sig} \left[\left(\frac{1}{N_{ref}\beta} \right)^2 + \sigma_{yref}^2(\tau) + \sigma_{ysig}^2(\tau) \right]^{1/2} \quad (25)$$

$$\Delta T = t_{sig} \left[\left(\frac{1}{N_{ref}\beta} \right)^2 + \sigma_{yref}^2(\tau) + \sigma_{ysig}^2(\tau) \right]^{1/2} \quad (26)$$

The time-domain fractional frequency stability of the signal

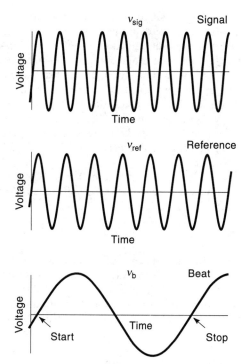

Figure 7. Timing diagram of a heterodyne time measurement system.

$\sigma_{ysig}^2(\tau)$ relative to the frequency stability of the reference can be estimated from a series of consecutive phase measurements separated by a time τ, using Eq. (13) (1,2). More elaborate techniques that offer better resolution are described in later sections on heterodyne measurements.

Heterodyne Measurements

Heterodyne techniques offer greatly improved short-term resolution over direct measurement techniques of frequency, phase, and time (see Fig. 7). In this technique the signal under test is heterodyned against the reference signal ν_{ref} and the difference frequency or beat frequency (lower curve) ν_b measured. The frequency resolution is improved by a factor ν_{ref}/ν_b over direct measurements.

Heterodyne Measurements of Frequency. Using the heterodyne method, the frequency of the signal source is

$$\nu_{sig} = \nu_{ref} \pm \nu_b \quad (27)$$

Additional measurements are required to determine the sign of the frequency difference. The usual method is to change the frequency of the reference by a known amount and determine whether the beat becomes smaller or larger. The resolution for frequency measurements is given by

$$\frac{\Delta\nu}{\nu_{ref}} = \Delta t \frac{\nu_b^2}{\nu_{ref}} \quad (28)$$

where Δt is the timing resolution. The uncertainty is limited by the frequency stability of the reference and the phase variations of the phase detector. The minimum time between data samples is $1/\nu_b$ for clocks that are nearly at the same frequency (a limitation in many situations).

Fifty percent of the time this approach is insensitive to the phase fluctuation between the signal and the reference. This down time, called "dead time" (5,9), biases the calculation of $\sigma_y(\tau)$ and mod $\sigma_y(\tau)$ by an amount that depends on the noise type and the duration of the dead time. Bias tables as a function of noise type and percent dead time are given in Ref. 9. The dead-time limitation can be circumvented by using two counters triggered on alternate cycles of ν_b.

Heterodyne Measurements of Phase (Time). The resolution for heterodyne measurements of time or phase is increased to

$$\Delta\tau = \frac{\nu_b}{\nu_{\text{ref}}}\Delta t \tag{29}$$

where Δt is the timing resolution of the counter. To avoid ambiguity ν_b should be larger than the frequency fluctuations of both the reference and the source under test. Additional measurements are required to determine if the frequency of the source is higher or lower than the reference. The phase of the beat signal goes through zero when the phase difference between the two signals is $\pm(2n + 1) \times 90°$ where $n = 0, 1, 2, 3, \ldots$. The time of the zero crossing is biased or in error by $\Delta\phi$ due to imperfections in the symmetry of the phase detector and/or the VSWR in the reference and signal paths (10). Timing errors due to VSWR effects and typical temperature coefficients for mixer biases are given in Ref. 10 for frequencies of 5 and 100 MHz. These errors generally scale as $1/\nu_{\text{sig}}$.

The time of the source under test is

$$t_{\text{sig}} = T_{\text{ref}} \pm \frac{n}{\nu_{\text{ref}}} \pm \Delta\phi \tag{30}$$

where n is the number of beat cycles that have occurred since the original synchronization. The minimum time between data samples is $1/\nu_b$. For clocks that are at nearly the same frequency, this limitation can be very restrictive.

The time-difference data can be used to characterize the fractional frequency stability of the sources using Eq. (13). The resolution for short-term time domain frequency stability (τ less than 0.1 s) is typically much worse than that obtained from integrating the phase noise using Eq. (15) (8,11)

Basic Configuration of PM and AM Noise Measurement Systems

Figure 8 shows the basic building block used in PM and AM noise measurement systems. It consists of a phase shifter, a mixer, and a low-pass filter. The two input signals to the mixer can be represented as

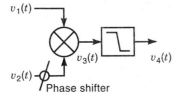

Figure 8. Basic building block for AM and PM noise measurement systems.

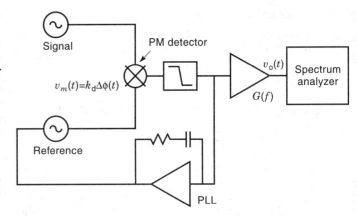

Figure 9. Block diagram for a heterodyne (two-oscillator) PM noise measurement system.

$$\begin{aligned}
v_1(t) &= A_1(t)\cos[2\pi\nu_0 t + \Phi_1(t)] \\
&= [V_1 + \epsilon_1(t)]\cos[2\pi\nu_0 t + \phi_1 + \phi_1(t)]
\end{aligned} \tag{31}$$

$$\begin{aligned}
v_2(t) &= A_2(t)\cos[2\pi\nu_0 t + \Phi_2(t)] \\
&= [V_2 + \epsilon_2(t)]\cos[2\pi\nu_0 t + \phi_1 + \phi_2(t)]
\end{aligned} \tag{32}$$

where $\epsilon_1(t)$ and $\epsilon_2(t)$ represent the amplitude fluctuations of the signals and $\phi_1(t)$ and $\phi_2(t)$ represent the phase fluctuations of the signals. For an ideal mixer, the signal $v_3(t)$ is equal to the product of signals $v_1(t)$ and $v_2(t)$. After some algebraic manipulation, we have

$$\begin{aligned}
v_3(t) = \frac{A_1(t)A_2(t)}{2}\{&\cos[4\pi\nu_0 t + \Phi_1(t) + \Phi_2(t)] \\
&+ \cos[\Phi_1(t) - \Phi_2(t)]\}
\end{aligned} \tag{33}$$

At the output of the low-pass filter, the signal reduces to

$$v_4(t) = \frac{A_1(t)A_2(t)}{2}\{\cos[\Delta\phi + \Delta\phi(t)]\} \tag{34}$$

where $\Delta\phi$ is the difference between the phase angles of the two input signals ($\phi_1 - \phi_2$) and $\Delta\phi(t) = \phi_1(t) - \phi_2(t)$. When $\Delta\phi$ is approximately equal to an odd multiple of $\pi/2$ and the integrated phase noise $[\int \Delta\phi(f)^2 \, df]$ does not exceed 0.1 rad², the two signals are in quadrature and the output voltage of the mixer is proportional to the difference of the phase fluctuations in the signals $v_1(t)$ and $v_2(t)$. When a double-balanced mixer is used, the amplitude fluctuations are suppressed by 25 to 40 dB (11,12). This is also called suppression of AM to PM conversion in a double-balanced mixer.

When the two input signals to the mixer are in phase or $\Delta\phi \cong 0$, the output voltage of the mixer is proportional to the amplitude fluctuations of the signals. The suppression of phase fluctuations when $\Delta\phi \cong 0$ is higher than 90 dB (11,12). The setup in Fig. 8 can be used in either AM noise or PM noise measurement systems by adjusting the phase shifter.

PM Noise Measurement Systems For Oscillators and Amplifiers

Heterodyne Measurements of PM Noise in Signal Sources (Two-Oscillator Method). Figure 9 shows a heterodyne PM noise measurement system for an oscillator. In this system, the signals from the test oscillator and a reference oscillator of similar frequency are fed into a double-balanced mixer. A phase-locked loop (PLL) is used to lock the reference frequency to the test oscillator frequency and to maintain quadrature between the two input signals to the mixer (13). The output voltage of the mixer is proportional to the difference between the phase fluctuations of the two sources. This voltage is amplified and its PSD is measured with a spectrum analyzer. Often this spectrum analyzer is of the fast Fourier transform (FFT) type. The voltage at the output of the amplifier is

$$v_o(t) = k_d G \Delta\phi(t) \tag{35}$$

where k_d is the mixer's phase-to-voltage conversion factor (or mixer sensitivity), G is the gain of the amplifying stage, and $\Delta\phi(t)$ is the difference between the phase fluctuations of the test oscillator and the reference $[\phi_A(t) - \phi_B(t)]$. The PM noise can be obtained from

$$S_\phi(f) = \frac{\text{PSD}[v_o(t)]}{(k_d G)^2} \tag{36}$$

The calibration factor or mixer sensitivity k_d can be found by turning off the PLL to obtain a beat frequency signal at the mixer output. The slope at the zero crossing and the period of this signal can be measured with an oscilloscope or another recording device. The calibration factor k_d in V/rad is

$$k_d = (\text{slope}) \times \frac{T}{2\pi} \tag{37}$$

Ideally this measurement should be made at the output of the amplifier because the measurement then yields $k_d G$, which includes the effect of amplifier input impedance on the performance of the mixer. The calibration of this PM noise measurement system using the beat-frequency method can introduce errors in the measurement if the mixer and the amplifier gains are frequency dependent, as is often the case. Figure 10

shows the variation of k_d with Fourier frequency for different mixer terminations. Capacitive terminations improve the mixer sensitivity, thereby improving the noise floor. However, the frequency response is not nearly as constant as the response obtained with resistive terminations, thus increasing the measurement error (11).

A calibrated Gaussian noise source centered about the carrier frequency can be used to calibrate the frequency-dependent errors (14,15). In this technique Gaussian noise is added to the reference signal by means of a low-noise power summer. Since the Gaussian noise is independent of the reference noise, equal amounts of PM and AM noise are added to the reference signal. When the Gaussian noise is "on," the PSD of the output noise voltage $v_o(t)$ is equal to

$$\text{PSD}[v_{o,\text{on}}(t)] = (k_d G)^2 S_{\text{PMcal}} \tag{38}$$

where S_{PMcal} is the PM noise added by the Gaussian noise source. The calibration factor as a function of frequency is obtained dividing $\text{PSD}[v_{o,\text{on}}(t)]$ by S_{PMcal}. A PSD measurement is then made with the Gaussian noise "off." The calibrated PM noise is obtained from

$$S_\phi(f) = \frac{\text{PSD}[v_{o,\text{off}}(t)]}{(k_d G)^2} = S_{\text{PMcal}} \frac{\text{PSD}[v_{o,\text{off}}(t)]}{\text{PSD}[v_{o,\text{on}}(t)]} \tag{39}$$

This approach greatly reduces the uncertainty of the measurement because it automatically takes into account the frequency-dependent errors. This approach also reduces the time necessary to make routine PM noise measurements as compared to traditional methods since the measurement is now reduced to a simple ratio measurement between noise on and noise off (14,15). The use of this calibration technique is illustrated in the cross-correlation measurements discussed in the following and shown in Fig. 11.

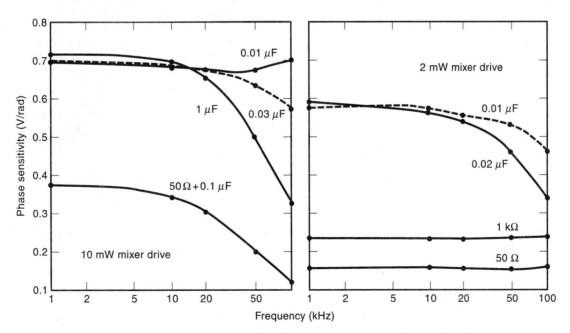

Figure 10. Sensitivity of a low-level double-balanced mixer at 5 MHz as a function of the intermediate frequency (IF) port termination for radio frequency (RF) and local oscillator (LO) inputs of +2 and +10 dBm (11).

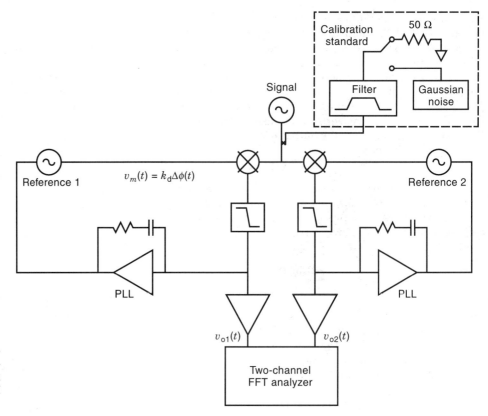

Figure 11. Block diagram of a two-channel cross-correlation system for measuring PM noise in an oscillator. Calibration of the system is accomplished using a PM and AM noise standard (14,15).

The confidence intervals for spectrum analyzers (swept and FFT) as a function of the number of measurements can be estimated from Table 1 (4,16–18). Biases in commonly used FFT window functions are discussed in Ref. 4.

One of the shortcomings of the single-channel, two-oscillator measurement system is that the measured noise includes noise contribution from the test source as well as that from the reference source. The noise terms included in $S_\phi(f)$ are

$$S_\phi(f) = S_{\phi,\mathrm{A}}(f) + S_{\phi,\mathrm{B}}(f) + \frac{v_\mathrm{n}^2(f)}{k_\mathrm{d}^2\mathrm{BW}} + S_{\mathrm{a,A}}(f)\beta_\mathrm{A}^2 + S_{\mathrm{a,B}}(f)\beta_\mathrm{B}^2$$

$$(40)$$

Table 1. Approximate Statistical Confidence Interval for FFT and Swept Spectrum Analyzers (16–18). β = N (the Number of Averages) for FFT Analyzers and β = N(RBW/VBW) for Swept Spectrum Analyzers. (RBW Refers to the Resolution Bandwidth of the Spectrum Analyzer and VBW refers to the video bandwidth).

β	69% Confidence Interval (dB)		95% Confidence Interval (dB)	
4	−2	+3	−3	+6
6	−1.5	+2.3	−2.5	+5
10	−1.2	+1.7	−2	+4
30	−0.72	+0.88	−1.3	+1.8
100	−0.41	+0.46	−0.76	+0.92
200	−0.3	+0.32	−0.54	−0.63
1,000	−0.13	+0.13	−0.25	−0.27
3,000	−0.08	+0.08	−0.15	+0.15
10,000	−0.04	+0.04	−0.08	+0.08

where $S_{\phi,\mathrm{A}}(f)$ is the PM noise of the test source, $S_{\phi,\mathrm{B}}(f)$ is the PM noise of the reference, $v_\mathrm{n}(f)$ is the noise added by the mixer, the amplifier, and the spectrum analyzer, $S_{\mathrm{a,A}}(f)$ and $S_{\mathrm{a,B}}(f)$ refer to the AM noise of the test source and the reference, and β_A and β_B are the factors by which the AM noise is suppressed. If the PM noise of the reference is higher than the PM noise of the test source, then the PM noise of the source cannot be measured accurately.

The last three terms of Eq. (40) constitute the noise floor of the measurement system. The noise floor can be estimated by using a single source (test source or reference) to feed the two inputs of the mixer. A phase shifter placed in one of the channels is used to adjust the phase difference to an odd multiple of 90°. The PM noise of the driving source is mostly canceled and the measured noise at the output is

$$S_\phi(f) = \frac{v_\mathrm{n}^2(f)}{k_\mathrm{d}^2\mathrm{BW}} + 2S_{\mathrm{a,A}}(f)\beta_\mathrm{A}^2 + \eta(f)S_{\phi,\mathrm{A}}(f) \qquad (41)$$

where the factor $\eta(f)$ is due to decorrelation of the source noise and is much smaller than 1 for small Fourier frequencies. [See the section on delay line measurements, especially Eq. (52), for a discussion of this effect.] Equation (41) is approximately equal to the noise-floor components in Eq. (40). Equation (40) indicates that the AM noise of the source and the reference can affect PM noise measurements; thus sources with low AM noise should be used.

Cross-Correlation Heterodyne Measurements of PM Noise in Signal Sources. One of the limitations of the single-channel, two-oscillator method is that the PM noise of the reference contributes to the measured noise. If three different oscillators are available (A, test source; B, reference 1; C, reference 2), PM noise measurements of three different pairs of oscillators can be made and the PM noise of the source can be approximated by

$$S_{\phi,\mathrm{A}}(f) \cong \tfrac{1}{2}[S_{\phi,\mathrm{AB}}(f) + S_{\phi,\mathrm{AC}}(f)$$
$$- S_{\phi,\mathrm{BC}}(f)] - 2S_{\mathrm{a,A}}(f)\beta_\mathrm{A}^2 - \frac{v_\mathrm{n}^2(f)}{k_\mathrm{d}^2\mathrm{BW}} \qquad (42)$$

where $S_{\phi,\mathrm{AB}}(f)$ includes the PM noise of sources A and B, $S_{\phi,\mathrm{AC}}(f)$ includes the PM noise of sources A and C, and $S_{\phi,\mathrm{BC}}(f)$ includes the PM noise of sources B and C. One problem with this approach is that small errors in any of the three measurements taken separately can result in large overall errors. Another problem is that the noise of the measurement system still contributes to the noise floor.

A more effective way of eliminating the PM noise from the reference is to use cross-correlation PM noise measurements. Figure 11 shows a two-channel, cross-correlation PM noise measurement system that uses two reference oscillators and two PM noise detectors operating simultaneously. Each individual channel is a simple heterodyne measurement system. Therefore the noise terms in $\mathrm{PSD}[v_{\mathrm{o}1}(t)]$ and $\mathrm{PSD}[v_{\mathrm{o}2}(t)]$ divided by the respective calibration factors are described by Eq. (40). The PSD of the cross-correlation of the noise voltages $v_{\mathrm{o}1}(t)$ and $v_{\mathrm{o}2}(t)$ divided by the calibration factor is

$$S_\phi(f) = S_{\phi,\mathrm{A}}(f) + S_{\mathrm{a,A}}(f)\beta_\mathrm{A}^2 + \frac{1}{\sqrt{N}}\left(S_{\phi,\mathrm{B}}(f) + S_{\phi,\mathrm{C}}(f) + \frac{v_\mathrm{n}^2(f)}{k_\mathrm{d}^2\mathrm{BW}}\right) \qquad (43)$$
$$+ S_{\mathrm{a,B}}(f)\beta_\mathrm{B}^2 + S_{\mathrm{a,C}}(f)\beta_\mathrm{C}^2$$

where $S_{\phi,\mathrm{A}}(f)$ is the PM noise of the test source, $S_{\phi,\mathrm{B}}(f)$ and $S_{\phi,\mathrm{C}}(f)$ are the PM noise of the references, and N is the number of averages in the measurement (11,14,15,19,20). The contribution of the PM noise of the references and the noise in the detectors and amplifiers are reduced by $\sqrt{N}$, because they are independent. This results in a reduction of the unwanted PM noise in the references and the measurement system of 10 dB for 100 averages and 20 dB for 10,000 averages. This powerful measurement technique makes it possible to obtain an accurate measurement of the PM noise of a source that has lower noise than the references if the AM noise of the source can be neglected.

Heterodyne Measurements of PM Noise in Amplifiers.

Figure 12 shows the block diagram of a PM noise measurement system for a pair of amplifiers. In this system an oscillator signal is split using a reactive power splitter. The outputs of the splitter drive two test amplifiers, and their outputs feed a double-balanced mixer. The mixer output is then amplified and measured by a spectrum analyzer. A phase shifter in one of the channels is used to maintain quadrature. The PM noise of the amplifier pair is obtained dividing the PSD of the noise voltage $v_\mathrm{o}(t)$ by the calibration factor $(k_\mathrm{d}G)^2$,

$$S_\phi(f) = \frac{\mathrm{PSD}[v_\mathrm{o}(t)]}{(k_\mathrm{d}G)^2} = S_{\phi,\mathrm{amp}}(f) + \frac{v_\mathrm{n}^2(f)}{k_\mathrm{d}^2\mathrm{BW}} + S_\mathrm{a}(f)\beta^2 \qquad (44)$$

where $S_{\phi,\mathrm{amp}}(f)$ is the PM noise of the amplifier pair and $S_\mathrm{a}(f)$ is the AM noise of the source. This system assumes that the noise of the amplifiers is higher than the noise floor of the measurement system. If this is not the case, cross-correlation measurement systems should be used. The calibration factor can be easily obtained by adding a Gaussian noise source to one of the channels and making measurements with the noise on and the noise off, as discussed previously. Calibration can also be achieved by using a second source to drive one of the amplifiers to obtain a beat signal at the output of the mixer. An oscilloscope can then be used to measure the zero-crossing slope and the period of the beat signal, and the calibration factor can be computed using Eq. (37).

A similar measurement system can be used to measure the PM noise of a single amplifier if the delay across the amplifier is so small that decorrelation of the source noise is not important to the measurement. See Eq. (52) and associated text for a discussion of decorrelation effects.

Cross-Correlation Heterodyne Measurements of PM Noise in Amplifiers.

Figure 13 shows a cross-correlation PM noise measurement system for a pair of amplifiers. It consists of two channels, each a separate heterodyne measurement system. The PSD of the cross-correlation of the noise voltages $v_{\mathrm{o}1}(t)$ and $v_{\mathrm{o}2}(t)$ divided by the calibration factor is

$$S_\phi(f) = \frac{\mathrm{PSD}[v_{\mathrm{o}1}(t)v_{\mathrm{o}2}(t)]}{(k_\mathrm{d}G)^2}$$
$$= S_{\phi,\mathrm{amp}}(f) + \frac{1}{\sqrt{N}}\frac{v_\mathrm{n}^2(f)}{(k_\mathrm{d}G)^2\mathrm{BW}} + S_{\mathrm{a,A}}(f)\beta_\mathrm{A}^2 \qquad (45)$$

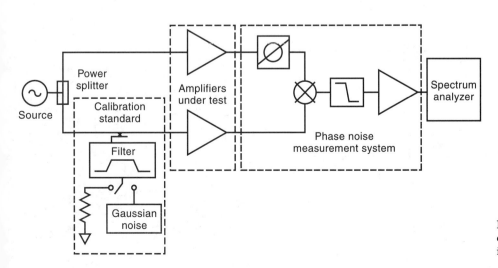

Figure 12. Block diagram of a single-channel system for measuring PM noise in a pair of amplifiers.

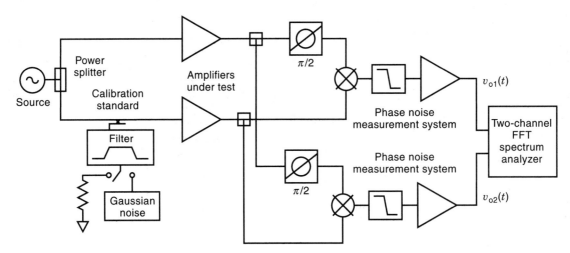

Figure 13. Block diagram of a two-channel cross-correlation system for measuring PM noise in a pair of amplifiers.

The noise added by the phase detectors is reduced by $\sqrt{N}$. The calibration factor k_dG can be obtained by adding a calibrated Gaussian noise source about the carrier frequency as shown in Fig. 13. In this setup the noise floor is generally limited by the AM noise of the source. It is therefore important to select a source with low AM noise and to operate the mixer at the maximum point of AM rejection.

Delay-Line Measurements of PM Noise in Signal Sources. A PM noise measurement system that does not need a second source is the delay-line system shown in Fig. 14 (11,21). In this setup the oscillator signal is split, a delay line of time delay τ_d is placed in one channel, and a phase shifter is placed in the other channel. The two channels are fed into a double-balanced mixer. The phase fluctuations of the combined signal at the mixer output are given by

$$\phi_m(f) = [2 - 2\cos(2\pi f\tau_d)]^{1/2}\phi(f) \tag{46}$$

where $\phi(f)$ are the rms phase fluctuations of the source at an offset frequency f (21,22). If the phase shifter is adjusted so that the phase difference between the two input signals is an odd multiple of 90°, then the output voltage of the mixer is proportional to the phase fluctuations of the source. For $2\pi f\tau_d \ll 1$,

$$v_m(f) \cong k_d\left[2 - 2\left(1 - \frac{(2\pi f\tau_d)^2}{2}\right)\right]^{1/2}\phi(f) = k_d 2\pi f\tau_d\phi(f) \tag{47}$$

where $v_m(f)$ is the output voltage of the mixer at an offset frequency f. Equation (47) can also be expressed in terms of $y(f)$ by multiplying the right side by ν_0/f:

$$v_m(f) \cong k_d 2\pi \nu_0\tau_d y(f) \tag{48}$$

Equation (48) can be equivalently expressed in the time domain as

$$v_m(t) \cong k_d 2\pi \nu_0\tau_d y(t) \tag{49}$$

Equation (49) indicates that the output voltage of the mixer is proportional to the frequency fluctuations in the source, and thus this system measures the FM noise of the test source. The voltage at the output of the amplifier is given by

$$v_o(t) \cong k_d(2\pi \nu_0\tau_d)Gy(t) = \nu_0 k_\nu Gy(t) \tag{50}$$

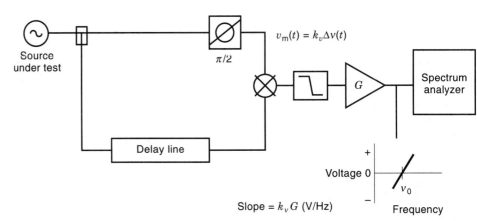

Figure 14. Block diagram of a PM noise measurement system that uses a delay line to measure PM noise in an oscillator.

where k_ν is the mixer sensitivity to frequency fluctuations and G is the voltage gain of the amplifier. The FM and PM noise of the source can then be obtained:

$$S_y(f) \cong \frac{\text{PSD}[v_0(t)]}{(2\pi v_0 \tau_d k_d G)^2} = \frac{\text{PSD}[v_0(t)]}{(v_0 k_\nu G)^2} \qquad (51)$$

$$S_\phi(f) \cong \frac{\text{PSD}[v_0(t)]}{(2\pi f \tau_d k_d G)^2} = \left(\frac{1}{f}\right)^2 \frac{\text{PSD}[v_0(t)]}{(k_\nu G)^2} \qquad (52)$$

Equations (51) and (52) are valid only for $f \ll 1/(2\pi\tau_d)$, where the approximation in Eq. (47) is valid. The calibration factor k_ν can be found by stepping the source frequency up and down and measuring the corresponding voltage change at the output of the amplifier. The voltage change divided by the frequency change is equal to $k_\nu G$ in V/Hz. At $f > 1/(2\tau_d)$, the output of the mixer is approximately sinusoidal with maximums occurring at odd multiples of $1/(2\tau_d)$, and minimums occurring at even multiples of $1/(2\tau_d)$. The first minimum occurs at $f = 1/\tau_d$, and thus a measurement of this occurrence can be used to determine τ_d (21). A calibrated Gaussian noise source can be used to calibrate the system at Fourier frequencies higher than $1/(2\tau_d)$, thus extending the frequency range of this system (14,15).

The close-in noise floor of this measurement system is much larger than that of the two-oscillator system. From Eq. (47), the effective phase sensitivity of the mixer is multiplied by a factor of $2\pi f \tau_d$ and thus is less than the phase sensitivity of the two-oscillator method (21). As discussed previously, the noise floor of the two-oscillator measurement system is given by

$$S_{\phi,\text{floor}}(f) = \frac{1}{k_d^2} \frac{v_n^2(f)}{\text{BW}} + S_{a,A}(f)\beta_A^2 + S_{a,B}(f)\beta_B^2 \qquad (53)$$

For $f \ll 1/(2\pi\tau_d)$, the noise contributions of the mixer, the amplifier, and the spectrum analyzer to the noise floor increase by a factor of $(2\pi f \tau_d)^{-2}$. The noise floor for the delay line measurement system is given by

$$S_{\phi,\text{floor}}(f) = \frac{1}{(2\pi f \tau_d)^2 k_d^2} \left(\frac{v_n^2(f)}{\text{BW}}\right) + S_{a,A}(f)\beta^2 \qquad (54)$$

The first term usually limits the noise floor of this system. The term in large parentheses usually follows a $1/f$ power law at frequencies close to the carrier; thus the overall noise floor at these frequencies follows a f^{-3} dependence on Fourier frequency. In addition, the noise floor is inversely proportional to τ_d^2; therefore longer delays will result in lower noise floors, but also a smaller Fourier frequency span in which Eqs. (51) and (52) are valid. At very long delays, the attenuation of the signal is so large that the noise floor is adversely affected. The noise floor of this measurement system cannot be measured directly since a source is needed and the noise from the source cannot be easily separated from the noise of the measurement system. If the noise contribution of the measurement system (mixer, amplifier, and spectrum analyzer) are known, then the noise floor can be approximated using Eq. (54).

Cavity Discriminator Measurements of PM Noise in Signal Sources. Figure 15 shows a cavity discriminator measurement system (11). This system is similar to the delay-line system, but uses a high-Q cavity in place of the delay line. If the signal frequency and the cavity resonance frequency are close, the cavity causes a phase delay proportional to the signal frequency. In the linear region around v_0, the fractional frequency fluctuations of the source are converted to phase fluctuations according to

$$\phi(t) \cong 2Qy(t) \qquad (55)$$

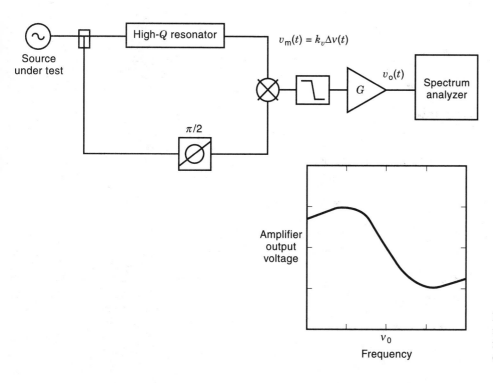

Figure 15. Block diagram of a PM noise measurement system that uses a high-Q-factor cavity to measure PM noise in an oscillator.

If the phase shifter is adjusted so that the two input signals to the mixer are in quadrature, then the output voltage of the mixer is proportional to phase fluctuations of the source. After amplification, the output voltage is given by

$$v_o(t) \cong k_d(2Q)Gy(t) = v_0 k_\nu Gy(t) \tag{56}$$

where the mixer sensitivity to frequency fluctuations k_ν is equal to $2Qk_d/\nu_0$. (Figure 15 shows the output voltage of the amplifier as a function of input frequency.) The FM and PM noise of the source are therefore

$$S_y(f) \cong \frac{\text{PSD}[v_o(t)]}{(2Qk_dG)^2} = \frac{\text{PSD}[v_0(t)]}{(v_0 k_\nu G)^2} \tag{57}$$

$$S_\phi(f) \cong \left(\frac{v_0}{f}\right)^2 \frac{\text{PSD}[v_o(t)]}{(2Qk_dG)^2} = \left(\frac{1}{f}\right)^2 \frac{\text{PSD}[v_o(t)]}{(k_\nu G)^2} \tag{58}$$

Equations (57) and (58) are valid only for $f \ll \nu_0/(2Q)$, where the phase-to-frequency relation of the cavity is linear and k_ν is a constant. The value of k_ν can be found by stepping the frequency of the source up and down and measuring the corresponding voltage change at the output of the amplifier. The voltage change divided by the frequency change is equal to $k_\nu G$ in volts per hertz. At higher Fourier frequencies k_ν changes according to (11)

$$k_\nu \propto \frac{1}{1 + \left(\dfrac{2Qf}{\nu_0}\right)^2} \tag{59}$$

A calibrated Gaussian noise source can be added to the source to calibrate the system at Fourier frequencies higher than $\nu_0/(2Q)$, thus extending the range of this measurement system.

As in the delay-line measurement system, the effective phase sensitivity of the mixer is less than the phase sensitivity of the two-oscillator method. The noise floor is thus given by

$$S_{\phi,\text{floor}}(f) = \frac{1}{(2Qf)^2 k_d^2} \frac{v_n^2(f)}{\text{BW}} + S_{a,A}(f)\beta^2 \tag{60}$$

Close to the carrier the noise floor follows a f^{-3} dependence on Fourier frequency and is higher than the noise of the two-oscillator measurement system. In addition, the noise floor is inversely proportional to Q^2. Therefore, higher Q cavities will result in lower noise floors, but also a smaller Fourier frequency span in which Eqs. (57) and (58) are valid. Other resonant circuits, such as multiple-pole filters, can be used in place of the cavity. The resonant circuit used can add noise to the system and thus limit the resolution or noise floor of the measurement system.

AM Noise Measurement Systems For Oscillators and Amplifiers

Simple AM Noise Measurements For Oscillators. Figure 16 shows a single-channel AM noise measurement system for a source. In this system a source drives an AM detector. The output voltage of the detector is then amplified and fed into a swept or FFT spectrum analyzer. AM detectors commonly used are the mixer detector discussed previously or a diode detector.

The voltage at the output of the mixer is given by

$$v_m(t) \cong k_a \frac{\epsilon(t)}{V_0} \tag{61}$$

where k_a is the detector's sensitivity to fractional amplitude fluctuations. The AM noise of the source is then

$$S_a(f) = \frac{\text{PSD}[v_o(t)]}{(k_aG)^2} \tag{62}$$

where $v_o(t)$ is the voltage into the spectrum analyzer and G is the gain of the amplifier. The sensitivity k_a can be measured by replacing the test source with a source with amplitude modulation capability (12,20). The power of the (calibration) source should be the same as the power of the test source, and it should be adjusted so that the dc voltage at the output of the mixer is the same as when the test source is used. The amplitude modulated input signal is given by

$$v(t) = V_0(1 + \text{AM}_{\text{in}} \cos \Omega t) \cos(2\pi\nu_0 t) \tag{63}$$

where AM_{in} is the peak fractional amplitude modulation and Ω is the modulation frequency. The magnitude of the amplitude modulation $S_a(f)$ at the input signal is $\frac{1}{2}(\text{AM})^2$, where AM is the modulation selected. The amplitude modulation at the output (AM_{out}) is then measured with the spectrum analyzer and the calibration factor is given by

$$(k_aG)^2 = \frac{2\text{AM}_{\text{out}}}{\text{AM}_{\text{in}}} \tag{64}$$

This measurement assumes that the calibration factor is a constant independent of the Fourier frequency. Many times k_a and G vary with frequency, and thus errors are introduced in the calibration. This is especially important for f higher than 100 kHz. To avoid this problem and speed the measurement of k_aG, a Gaussian noise source added to the test source can be used to calibrate the system (12,14,15). To calibrate the system, the PSD of $v_o(t)$ with the Gaussian noise on is measured with the spectrum analyzer. This curve will show any variation of k_aG with Fourier frequency if it exists. The values of $(k_aG)^2$ as a function of Fourier frequency are obtained by dividing the measured $v_o(f)$ by the known calibrated noise power.

The confidence intervals for spectrum analyzer (swept and FFT) measurements, as a function of the number of measurements, can be estimated from Table 1 (4,16–18). Biases in commonly used FFT window functions are discussed in Ref. 4.

One problem of this simple measurement system is that the measured noise, given by Eq. (62), includes the AM noise of the source in addition to noise added by the detector, the amplifier, and the analyzer (noise floor of the system). The noise components included in $S_a(f)$ are

$$S_a(f) = S_{a,\text{src}}(f) + \frac{1}{(k_a)^2} \frac{v_n^2(f)}{\text{BW}} \tag{65}$$

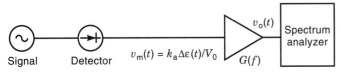

Figure 16. Block diagram of a single-channel AM noise measurement system for measuring AM noise in an oscillator.

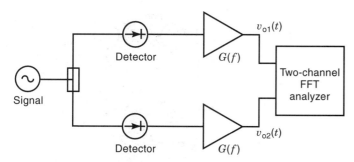

Figure 17. Block diagram of a two-channel cross-correlation system for measuring AM noise in an oscillator.

where $S_{a,src}(f)$ is the AM noise of the source, and the second term represents the noise floor of the system. It is therefore difficult to separate the AM noise of the source from the noise floor.

Cross-Correlation AM Noise Measurements For Oscillators. The noise floor of the AM noise measurement system discussed previously can be considerably reduced by using two-channel cross-correlation techniques as shown in Fig. 17. In this system PSD$[v_{o1}(t)]$ includes AM noise of the source plus noise in channel 1 (detector 1, amplifier 1). PSD$[v_{o2}(t)]$ includes AM noise of the source plus noise in channel 2 (detector 2, amplifier 2). The PSD of the cross spectrum divided by the calibration factor mainly includes the AM noise of the source since the noise that is not common between the two channels is reduced proportionally to $1/\sqrt{N}$, so

$$\frac{\text{PSD}[v_{o1}(t)v_{o2}(t)]}{(k_aG)^2}$$
$$= S_{a,src}(f) + \frac{1}{\sqrt{N}}\left(\frac{v_{n1}^2(t)}{(k_{a1})^2\text{BW}} + \frac{v_{n2}^2(t)}{(k_{a2})^2\text{BW}}\right) \quad (66)$$

This measurement technique is very useful for separating the AM noise of the source from the system noise. This system can be calibrated with a source with AM capability or by using a calibrated Gaussian noise source (12,14).

Simple AM Noise Measurements For Amplifiers. Figure 18 shows a single-channel AM noise measurement system for an amplifier. A source is used to drive the amplifier under test, and the output signal of the amplifier is fed into an AM detector. The signal is amplified, and the output voltage is mea-

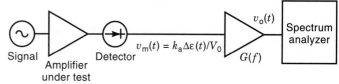

Figure 18. Block diagram of a single-channel system for measuring AM noise in an amplifier.

sured with an spectrum analyzer. The PSD of the noise voltage $v_o(t)$ divided by the calibration factor $(k_aG)^2$ is

$$\frac{\text{PSD}[v_o(t)]}{(k_aG)^2} = S_{a,src}(f) + S_{a,amp}(f) + \frac{1}{(k_a)^2}\frac{v_n^2(t)}{\text{BW}} \quad (67)$$

The first term in Eq. (67) represents the AM noise of the source, the second term is the noise of the test amplifier, and the third term is the noise floor of the measurement system. The calibration factor k_aG can be obtained by using a source with AM capability or by using a calibrated Gaussian noise source to calibrate the system (12,14,20). If the AM noise of the source or the system noise floor is comparable to the amplifier noise, accurate measurements of the amplifier AM noise cannot be made using this system.

Cross-Correlation AM Noise Measurements For Amplifiers. Figure 19 shows a cross-correlation AM noise measurement system for an amplifier that reduces the noise floor of the measurement system. A reference source drives the test amplifier, the amplifier output is split, and each channel is fed into an AM detector. The output signals of the detectors are then amplified and measured with a two-channel FFT spectrum analyzer. The PSD of the cross-correlation $[v_{o1}(t)xv_{o2}(t)]$ divided by the calibration factor is

$$\frac{\text{PSD}[v_{o1}(t)v_{o2}(t)]}{(k_aG)^2} = S_{a,src}(f) + S_{a,amp}(f)$$
$$+ \frac{1}{\sqrt{N}}\left(\frac{v_{n1}^2(t)}{(k_a)^2\text{BW}} + \frac{v_{n2}^2(t)}{(k_a)^2\text{BW}}\right) \quad (68)$$

Equation (68) indicates that part of the noise floor in the system is reduced by a factor of $\sqrt{N}$. If the AM noise of the source dominates the measured noise, a limiter can be placed after the source to reduce its AM noise (6).

CARRIER SUPPRESSION MEASUREMENT SYSTEMS

The concept of carrier suppression was first introduced by Sann for measuring noise in amplifiers (23). Carrier suppres-

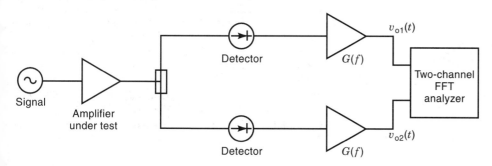

Figure 19. Block diagram of a two-channel cross-correlation system for measuring AM noise in an amplifier.

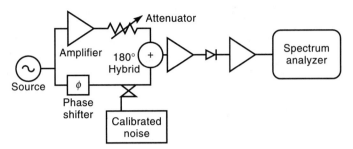

Figure 20. Carrier suppression AM noise measurement system for amplifiers.

sion measurement systems use a bridge network to cancel the power at the carrier frequency, effectively enhancing the noise of the device under test (23–26). This results in a reduction of the noise floor of the measurement system, or in other words, in an increase of the detector sensitivity.

Carrier Suppression AM Noise Measurement Systems for Amplifiers

Figure 20 shows an AM noise measurement system for amplifiers that uses carrier suppression (25). The source signal is split, and the amplifier under test is placed in one of the channels. The variable attenuator is used to match the magnitudes of the two channels. The phase shifter is used to adjust the phase difference between the two channels to 0. When the two signals are combined in the hybrid, the carrier power is mostly cancelled. The degree of cancellation depends on how well the two channels are matched. An additional amplifier is placed at the output of the power summer to increase the power input to the AM detector. The effective AM noise of the device (as seen by the AM detector) is increased by the amount of carrier suppression. An AM noise standard is used to calibrate the gain of the system.

Carrier Suppression PM Noise Measurement System for Amplifiers

Figure 21 shows a carrier suppression PM noise measurement system for amplifiers (25,26). A bridge circuit is used to effectively increase the PM noise of the amplifier with respect to the noise floor of the system. A PM noise standard is used to calibrate the gain of the system.

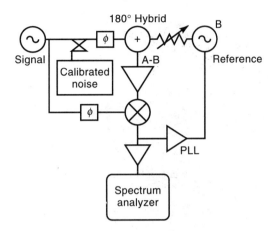

Figure 22. Two-oscillator PM noise measurement system with carrier suppression.

Carrier Suppression PM Noise Measurement Systems for Oscillators

Carrier suppression can also be used in PM measurement systems for oscillators. Figure 22 shows a two-oscillator PM noise measurement system with carrier suppression (25). In this system a bridge is used to raise the magnitude of the PM noise of the two oscillators with respect to the noise in the phase noise detector. A PM noise standard is used to calibrate the gain of the system. Carrier suppression can also be applied to delay line measurement systems (27).

The advantage of carrier suppression measurement systems over cross-correlation systems is that similar noise floors can be achieved at smaller measurement times. The disadvantage is that a very good amplitude and phase match of the bridge channels is required to suppress the carrier power. Furthermore, the match changes with temperature, and thus careful calibrations should be performed before and after a measurement to ensure that the amount of suppression has not changed. The use of a PM/AM noise standard will ease the calibration process. In some specific systems, like AM measurement systems for amplifiers, carrier suppression systems are superior. The reason is that in these systems, the AM noise of an amplifier with lower or comparable noise to the source can be accurately measured. In cross-correlation measurement systems for amplifiers it is difficult to cancel the AM noise of the source, which is often comparable or larger than the noise in the amplifier under test.

BIBLIOGRAPHY

1. E. S. Ferre-Pikal et al., Draft revision of IEEE Std 1139-1988 standard definitions of physical quantities for fundamental frequency and time metrology—random instabilities, *Proc. 1997 IEEE Frequency Control Symp.* 1997, pp. 338–357.

2. D. W. Allan et al., Standard terminology for fundamental frequency and time metrology, *Proc. 42nd Ann. Symp. Frequency Control*, 1988, pp. 419–425.

3. J. A. Barnes et al., Characterization of frequency stability, *IEEE Trans. Instrum. Meas.*, **IM-20**: 105–120, 1971.

4. D. B. Sullivan et al., Characterization of clocks and oscillators, *National Institute of Standards and Technology, Technical Note No. 1337*, 1990.

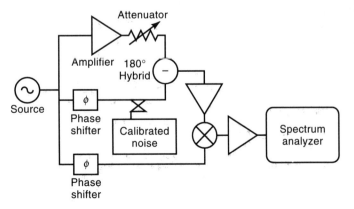

Figure 21. Carrier suppression PM noise measurement system for amplifiers.

5. D. A. Howe, D. W. Allan, and J. A. Barnes, Properties of signal sources and measurement methods, *Proc. 35th Annu. Symp. Frequency Control,* 1981, pp. A1–A47; also in Ref. 4.

6. T. E. Parker, Characteristics and sources of phase noise in stable oscillators, *Proc. 41st Annu. Symp. Frequency Control,* 1987, pp. 99–110.

7. F. L. Walls and A. DeMarchi, RF spectrum of a signal after frequency multiplication; measurement and comparison with a simple calculation, *IEEE Trans. Instrum. Meas.,* **IM-24**: 210–217, 1975.

8. F. L. Walls et al., Time-domain frequency stability calculated from the frequency domain: An update, *Proc. 4th Eur. Frequency Time Forum,* 1990, pp. 197–204.

9. J. A. Barnes and D. W. Allan, Variances based on data with dead time between the measurements, *NIST Tech. Note No. 1318,* 1990; also in Ref. 4.

10. L. M. Nelson and F. L. Walls, Environmental effects in mixers and frequency distribution systems, *Proc. IEEE Frequency Control Symp.,* 1993, pp. 831–837.

11. F. L. Walls et al., Extending the range and accuracy of phase noise measurements, *Proc. 42nd Ann. Symp. Frequency Control,* 1988, pp. 432–441; also in Ref. 4.

12. L. M. Nelson, C. W. Nelson, and F. L. Walls, Relationship of AM noise to PM noise in selected RF oscillators, *IEEE Trans. Ultrason. Ferroelectr. Freq. Control,* **41**: 680–684, 1994.

13. F. L. Walls and S. R. Stein, Servo techniques in oscillators and measurement systems, *National Bureau of Standards (US) Technical Note No. 692,* pp. 1–20, 1976.

14. F. L. Walls, Secondary standard for PM and AM noise at 5, 10 and 100 MHz, *IEEE Trans. Instrum. Meas.,* **IM-42**: 136–143, 1993.

15. F. L. Walls, Reducing errors, complexity, and measurement time for PM noise measurements, *Proc. 1993 Frequency Control Symp.,* 1993, pp. 81–86.

16. D. B. Percival and A. T. Walden, *Spectral Analysis for Physical Applications,* Cambridge, UK: Cambridge University Press, 1993.

17. B. N. Taylor and C. E. Kuyatt, *National Institute of Standards and Technology,* Technical Note No. 1297, 1993.

18. F. L. Walls, D. B. Percival, and W. R. Irelan, Biases and variances of several FFT spectral estimators as a function of noise type and number of samples, *Proc. 43rd Annu. Symp. Frequency Control,* 1989, pp. 336–341; also in Ref. 4.

19. W. F. Walls, Cross-correlation phase noise measurements, *Proc. 1992 IEEE Frequency Control Symp.,* 1992, pp. 257–261.

20. F. L. Walls et al., Precision phase noise metrology, *Proc. National Conf. Standards Laboratories (NCSL),* 1991, pp. 257–275.

21. A. L. Lance, W. D. Seal, and F. Labaar, Phase noise and AM noise measurements in the frequency domain, *Infrared Millimeter Waves,* **11**: 239–289, 1984; also in Ref. 4.

22. Stanley J. Goldman, *Phase Noise Analysis in Radar Systems Using Personal Computers,* New York: Wiley, 1989, chap. 2, pp. 31–40.

23. K. H. Sann, Measurement of near-carrier noise in microwave amplifiers, *IEEE Trans. Microwave Theory and Techniques,* Vol. MTT-16, pp. 761–766, 1968.

24. G. J. Dick and J. Saunders, Method and apparatus for reducing microwave oscillator output noise, US Patent #5,036,299, July 1991. See also D. G. Santiago and G. J. Dick, Microwave frequency discriminator with a cooled sapphire resonator for ultra-low phase noise, *Proc. 1992 IEEE Freq. Control Symp.,* pp. 176–182, 1992.

25. F. L. Walls, Suppressed carrier based PM and AM noise measurement techniques, *Proc. 1997 Intl. Freq. Control Symp.,* pp. 485–492, 1997.

26. C. McNeilage et al., Advanced phase detection technique for the real time measurement and reduction of noise in components and oscillators, *Proc. 1997 IEEE Intl. Freq. Control Symp.,* pp. 509–514, 1997.

27. F. Labaar, New discriminator boosts phase noise testing, *Microwaves,* Hayden Publishing Co., Inc., Rochelle Park, NJ, **21** (3): 65–69, 1982.

FRED L. WALLS
EVA S. FERRE-PIKAL
National Institute of Standards and Technology

MEASUREMENT OF POWER. See POWER METERS.

MEASUREMENT OF WAVELENGTH. See WAVELENGTH METER.

MEASUREMENT, POWER SYSTEM. See POWER SYSTEM MEASUREMENT.

MEASUREMENTS. See CAPACITANCE MEASUREMENT; INSTRUMENTS.

MEASUREMENTS, DIELECTRICS. See DIELECTRIC MEASUREMENT.

MEASUREMENTS FOR SEMICONDUCTOR MANUFACTURING. See SEMICONDUCTOR MANUFACTURING TEST STRUCTURES.

MEASUREMENTS, GRAVITY. See GRAVIMETERS.

MEASUREMENT, SOFTWARE. See SOFTWARE METRICS.

MEASURES OF RELIABILITY. See RELIABILITY INDICES.

MECHANICAL BALANCES. See BALANCES.

MECHANIZED INFERENCE. See THEOREM PROVING.

MECHATRONICS

Mechatronics is a process for developing and manufacturing electronically controlled mechanical devices. Many of today's automated equipment and appliances are complex and smart mechatronics systems, composed of integrated mechanical and electronic components that are controlled by computers or embedded microcomputer chips. As a matter of fact, mechatronic systems are extensively employed in military applications and remote exploratory expeditions (1,2). Industrial mechatronic systems are used extensively in factory automation and robotic applications, while commercial mechatronics products are widely found in office and home appliances as well as in modern transportation. Successful systems and products are the ones that are well designed, well built, and affordable.

The term *mechatronics* was coined in 1969 to signify the integration of two engineering disciplines—*mecha*nics and elec*tronics*. In the early 1970s, Japan was the largest ship and tanker builder in the world and its economy depended heavily on oil-driven heavy machinery and steel industries. The 1973 oil crisis saw the crude oil prices skyrocket from $3.50 per barrel to over $30.00 per barrel. The consequent disastrous impact on its oil-dependent shipping industry prompted Japan to rethink about its national economic survival and strategies. Microelectronics and mechatronics were two emerging technologies embraced by Japan as major industrial priorities after the crisis.

Definition. Several definitions for Mechatronics can be found in the literature (3–17). For example, the *International Journal on Mechatronics: Mechanics–Electronics–Control Mechatronics* (5) defines it as the synergetic combination of precision mechanical engineering, electronic control, and systems thinking in the design of products and manufacturing processes. Others stated it to be a synergetic integration of mechanical engineering with electronics and intelligent computer control in the design and manufacturing of industrial product and processes. Mechatronics requires systems engineering thinking aided by computer simulation technology that enhances complete understanding of how its design decision affects decisions of the other discipline counterparts. It describes ways of designing subsystems of electromechanical products to ensure optimum systems performance.

To be more competitive and innovative, new mechatronics requirements often call for "smart" performance in dealing with operational and environmental variations. So to include for product competitiveness and added value of today's microchip and microprocessor technology, one may generalize the definition of mechatronics as follows:

Mechatronics is a systems engineering process for developing and integrating of computer-based electronically controlled mechanical components in a *timely* and *cost-effective* manner into smart affordable *quality* products that ensure optimum, flexible, reliable, and robust performance under various operating and environmental conditions. We refer to such a well-designed and well-integrated automation system as a *mechatronic system* or *product.*

The three key words in the definition are quality, time, and cost. The product must be safe, reliable, and affordable to consumers. To the manufacturers, the product must be produced quickly and efficiently and must be profitable; profit is indeed the name (purpose) of the game. Readers should not be surprised by the encounter of other contexts of mechatronics. In exploratory research, for instance, mechatronics thinking is used to develop customized systems where cost may not yet be a significant issue.

EXAMPLES OF MECHATRONICS SYSTEMS

There are (almost) endless examples of mechatronic systems and products. It would not be meaningful to attempt a compilation of the available products and automated systems (see AUTOMATION and ROBOTS). However, to emphasize a huge trend in research, development, and engineering effort that adds up to billions of dollars each year, we will take a look at the automotive mechatronics in some detail.

Today's fully loaded modern automobile easily carries over 30 automotive mechatronic systems to provide a high level of ride comfort and road handling, along with devices for safety, fuel economy, and luxury (18,19). Modern cars are controlled by several onboard embedded microcontrollers. A list of automotive mechatronic systems is provided here to emphasize the point: electronic ignition, electronic fuel injection, electronic controlled throttle, emission control, computer-controlled transmission and transaxles, cruise control, anti-lock brakes, traction control, computer-controlled suspension, steering control, body control functions such as power lock, windows, automatic wipers, sunroof and climate control, safety functions such as airbags, security systems, keyless en-

try system, instrument panel display, stereo system, and so on. Some examples of the latest automotive innovations about to hit the market are an anti-squeeze power window and sunroof, vehicle yaw stability control systems, collision warning/avoidance systems, noise and vibration cancellation, anti-roll suspension, hybrid electric vehicles, navigation aids, a built-in automotive personal computer, and others. The automobile industry invests heavily in research to develop these products. It is not surprising to find that several auto companies and suppliers are investigating similar mechatronic products at the same time. Thousands of engineers are employed to work in the area of automotive mechatronics.

THE MECHATRONICS CHALLENGE

Demand for mechatronics products prevail in today's market as consumers become more affluent and opt for gadgets that enhance performance in the products. In addition, as science and technology advance, the requirements for mechatronics systems often become more and more sophisticated and demanding. Many manufacturers realize the trend and the potential highly profitable market. They also realize that the challenge is in getting quality mechatronics product to market in a timely and cost effective manner.

An original equipment manufacturer (OEM) must have well-balanced and well-planned business and engineering strategies to compete in the market. In this article, we will set aside the essential business strategies and address only the relevant engineering aspects of developing a mechatronic product.

Figure 1 shows a simple model for conceptualizing a mechatronic system, emphasizing the control, sensor, actuator, and process entities that make up the system. It presents a general summary for understanding the overall configuration of the system and its connectivity to user command inputs, environmental influence, system states, and commanded outputs. Expected performance requirements for the mechatronic system can be defined at this level.

Figure 2 is basically the same mechatronic system as that shown in Fig. 1, but is conceived as a product made out of an electronic module and a mechanical module. The components in these modules may include the multitechnologies shown in the figure. The final selected components in this design will be used in actual implementation of the mechatronic product.

The engineering challenge to the manufacturers is to transform the concept of the mechatronic system (Fig. 1) into the multitechnology modules that make up the mechatronic product (Fig. 2). The development must be timely and cost effective and must ensure quality in the product.

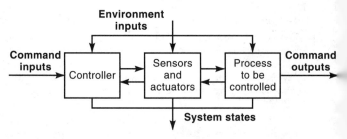

Figure 1. Concept level of a mechatronic system.

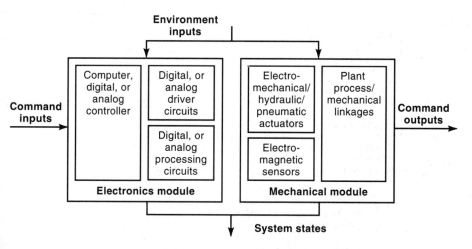

Figure 2. Electronics and mechanical modules of a mechatronic system.

The manufacturers invest in research, development, and manufacturing processes to produce products. A key to successful management of quality, time, and cost lies in a systems engineering perspective and approach (20) to the development process of mechatronic system.

The importance of mechatronics philosophy is quite evident when we reflect on the enormous success of Japan's electronics and automobile industries. The idea has since spread around the world, especially in Europe and the United States. Many industrialists, research councils, and educators have identified *mechatronics* as an emergent core discipline necessary for the successful industry of the next millenium.

SCOPE OF THIS ARTICLE

This article is written with an application engineer in mind. He or she has come up with a viable mechatronics concept or is assigned a mechatronics project. The objective is to build it as well as possible, as inexpensively as possible, and in the shortest amount of time. The idea is to avoid getting bogged down with heavy engineering mathematics and to look for state-of-the-art techniques and tools to expedite the development of the product.

This article emphasizes a systems engineering approach to the development process of mechatronics systems. It stresses the use of computer-aided engineering (CAE) tools for expediting the design and analysis of mechatronic products. It also addresses the foundations needed for dealing effectively with the multitechnology mechatronics. It is written with the assumptions that the reader has some or sufficient background in certain engineering discipline(s) related to mechatronics. It discusses the current trend and practice in *process, techniques, tools,* and *environment* for dealing with mechatronics. Finally, it provides an evaluation of the direction that mechatronics is heading toward in the future. It does not include details of physical system integration and manufacturing processes.

FOUNDATION FOR MECHATRONICS

Science, Technology, Engineering, Business, and Art

In a broad sense, successful mechatronics endeavors often involve one or more of these disciplines:

1. Science—discovery of new materials, methods, and so on
2. Technology—adaptation of technologies for innovation
3. Engineering—development and manufacturing of new products
4. Business—ability to gauge market, opportunity, and profit
5. Art—experience and skills that beat the competition

Coupled with the fact that mechatronics is a multitechnology discipline, the range of the knowledge is usually beyond that of a normal person. An exception may be the case of an exceedingly simple mechatronics endeavor. Mechatronics in general, therefore, is inherently a team effort, rather than a single individual effort.

In the high-technology world of today, however, a paradigm of systems engineering has been applied to improve the efficiency of teamwork. An important point in that paradigm is the use of CAE tools to communicate and explore possibilities of sophisticated ideas among the team members. The CAE tools may include expert knowledge systems, computer simulation, computer graphics, virtual reality immersion, and so on. If we are bold enough to accept it, the CAE tools can effectively be treated as "personal assistants" in the team. When wisely employed, they can assist engineers to ensure quality in the design of the mechatronics system, shorten time for analyses, and reduce cost of development, through countless computer simulations and evaluations of the mechatronics systems.

Multitechnology, Multiengineering, and Systems Engineering

Figure 3 is a pictorial summary of the technologies and engineering that mechatronics can entail. The left half of the figure depicts the mechatronics system as a real-time product that responds to programmed event and user command and reacts to environmental circumstance. It shows the multifunctional interfaces of the mechatronics system including mechanical mechanisms, sensors and actuators, input and output signal conditioning circuits, and computers or embedded microcontrollers (18,19,21).

Shown in the right half of the figure, an integration of such a system would require certain appropriate skills and experience from mechanical, electrical, electronics, and computer engineering. At the implementation level, skills for dealing

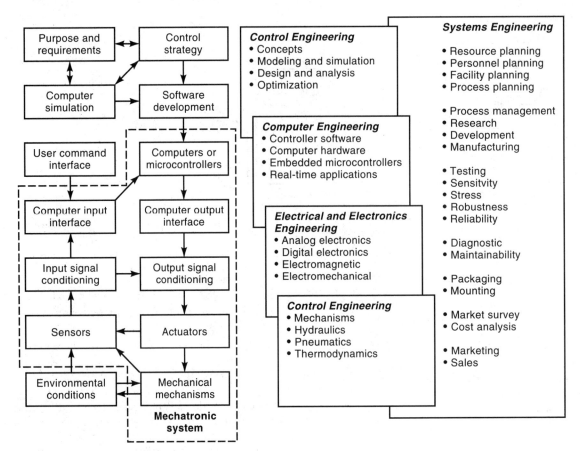

Figure 3. Multitechnology, multiengineering, and systems engineering nature of mechatronics.

with computer, electrical, electronics, electromagnetic, electromechanical, mechanical, hydraulic, pneumatic, and thermal components will be desired (22). At the concept design level, however, background in control theory will be needed to translate the purpose of the product into its technical requirements and define a control strategy with the aid of computer simulation study (23). The software development will then implement the control scheme in the system.

The right half of Fig. 3 also concerns with systems engineering to complete the job—that is, to bring the mechatronic product into being (20). Such an endeavor would entail the following: planning of resource, personnel, facility, and process; management of process; research, development, and manufacturing; product testing; evaluation of sensitivity, stress, robustness, and reliability; packaging and mounting; marketing; maintenance; and cost analysis and management. This reiterates the fact that teamwork is a necessary requirement when dealing with a mechatronic product life cycle (see SYSTEMS ENGINEERING TRENDS).

Computer-Aided Engineering Tools

As mentioned earlier, CAE tools are employed to assist designers and engineers in carrying out the development of mechatronics. Computer-aided design (CAD) packages have been used to render a graphical mock-up of solid models in the design of package, looks, fits, and mounting for mechatronic products. CAD is a widely used technique in mechanical design and analysis in the automobile and aerospace industries.

For the purpose of this article, we are concerned with CAE tools that assist engineers in designing control schemes, conducting performance analysis, and selecting the right components for the mechatronic system. The CAE software therefore must simulate the responses of dynamics system and allow control applications to be evaluated. Examples of such Computer-Aided Control Systems Design (CACSD) packages include Matlab/Simulink®, Matrix$_x$/SystemBuild®, P-Spice®, Electronics Workbench®, Easy-5®, Saber®, and so on. These software packages have a *schematic capture feature* that interprets block diagrams and component schematics for the simulation. This convenient feature lets the engineer concentrate on the engineering problem rather than the mathematical aspects of the simulation.

Saber (24–26). Of the above packages, the one that stands out as the industry standard is the Saber simulator. Saber has been accepted in the automotive industry as the CAE tool for dealing with mechatronics design and analysis. In fact, auto suppliers are now required to use Saber to communicate mechatronics design and analysis problems to General Motors, Ford, and Daimler-Chrysler. Figure 4 explains why Saber is well accepted by the industry. It illustrates a multitechnology nature of mechatronics where interdisciplinary knowledge of engineering and teamwork are key to the endeavor. The Saber simulator can be used to model the cross-disciplinary mechatronic system and provide an interactive platform for experimentation, discussions, and communication among the team of designers, engineers, and managers for the project. It provides a common medium to predict "what if" scenarios for all concerns and leads to "optimal" trade-off decisions.

An easy way to appreciate the Saber simulator is to imagine a virtual "mechatronics superstore" inside the cybernetic space that offers the following products and services:

- The "store" has a large inventory of commercially available electronics and mechanical components for you to choose. It also contains templates with which you can define new specifications for the components. [Saber has libraries of over 10,000 mechatronic parts (represented by component icons).]

- You have unrestricted "shopping" privilege that lets you "buy" and "exchange" any number of parts. (Drag and drop components and templates in a workspace window.)

- The "store" has a "assembly" facility where you can "integrate" the parts together into a working model, according to the schematic of your mechatronic design. (Connect parts to design a schematic.)

- It also has a "testing" facility with signal generators and display scopes for observing, validating, and verifying responses of the newly assembled mechatronics model. (Conduct simulation of system response.)

- It provides a means of conducting performance analysis and component analysis to check how good the selected parts are, and it delivers reports on the results. (Check performance requirements, and investigate components for stress and robustness.)

- You may conduct as many designs, analyses, and experiments in this store as you wish until you satisfy the requirements of the mechatronic product that you plan to build. (Discuss, redesign, and optimize.)

- You may bring your teammates to participate in the above activities. (All players from the start until the end of the mechatronic product life cycle can be included in the discussion using the simulation.)

Saber therefore facilitates *virtual prototyping of mechatronics functions* with realistic models of commercially available parts. Analogy, Inc., the company that produces Saber, collaborates with many OEMs such as Motorola, Texas Instruments, Harris Semiconductors, and Mabuchi Motors to model components and also validate and verify their characteristics as accurately as possible. An application engineer can use the verified models in the schematic without the burden of deriving mathematical formulation, programming, and debugging of codes. He or she can request performance reports from the virtual prototype simulation. As you can imagine, Saber provides support in the form of virtual parts, a facility, and a "personnel assistant."

Driving the Point Home. To illustrate the point further, Fig. 5 illustrates the actual schematic used in Saber simulation to represent the conceptual level design of a servo-positioning system. Figure 6, on the other hand, is the Saber implementation-level schematic for the system with selected components. A major process in mechatronics is to translate Fig. 5 into Fig. 6. The simulator helps the engineers to design the virtual prototype shown in Fig. 6 and analyze the integrity of the selected components. More details of this example will be presented in a later section.

Breadth and Depth of Disciplines in Mechatronics

It is clear that development of a complex mechatronic system will require an experienced engineering specialist with depth of expertise and breadth of experience to lead the team for the project. Of course, this is not necessarily the case if we are dealing with simple mechatronic systems. In either case, learning on the job is often one of the means of getting the job done. Indeed as an added benefit, a multitechnology CAE tool can be a big help in learning and confirming ideas in disciplines other than your own. It complements your knowledge and that of your team.

This article assumes that the reader and his team have certain backgrounds in control, computer, electronics, and

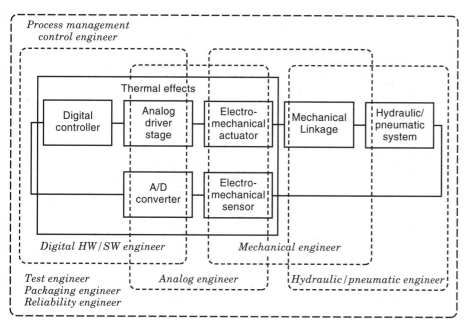

Figure 4. Overlapping disciplines and teamwork in mechatronics.

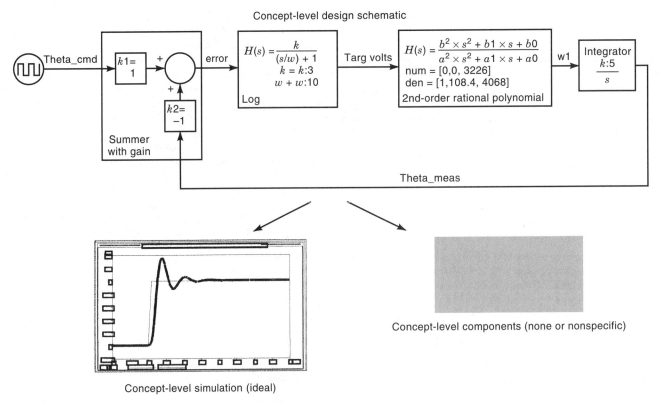

Concept-level design schematic

Theta_cmd

$k1 = 1$

error

$H(s) = \dfrac{k}{(s/w)+1}$
$k = k{:}3$
$w + w{:}10$

Log

Targ volts

$H(s) = \dfrac{b2 \times s^2 + b1 \times s + b0}{a2 \times s^2 + a1 \times s + a0}$
num = [0,0, 3226]
den = [1,108.4, 4068]
2nd-order rational polynomial

w1

Integrator
$\dfrac{k{:}5}{s}$

$k2 = -1$

Summer
with gain

Theta_meas

Concept-level simulation (ideal)

Concept-level components (none or nonspecific)

Figure 5. Concept-level design, analysis, and components.

mechanical engineering. Many of these backgrounds are covered elsewhere in the Encyclopedia. This article chooses to emphasize the systems engineering process (20) for designing and analyzing a mechatronic system. It deals with the problem at the system level, the subsystem level, and the component level with the help of a CAE tool. Although the Saber simulator is the main CAE tool used in developing the illustration, we describe its features and capabilities in a generic way so as to emphasize the concept of the process.

PROCESS AND TECHNIQUES FOR DESIGNING AND ANALYZING MECHATRONIC SYSTEMS

Process in Mechatronics Design and Analysis

The process can be grouped as follows: (1) requirements and specifications process, (2) top-down design process, and (3) bottom-up analysis process.

Requirements and Specifications Process. This is a stage where the engineers use their experience to envision the performance of the mechatronic systems to be built. Technical specifications are derived from nontechnical user requirements.

User requirements are qualitative descriptions of what the users need, want, desire, and expect. They are often stated in nontechnical terms and are not usually adequate for design purposes. However, they provide a subjective qualitative means of characterizing and judging the effectiveness of a system or product.

Technical specifications are derived from the user requirements. They spelled out the required characteristics in clear,

unambiguous, measurable, quantitative technical terms to which the engineering team can refer. The technical specifications define the engineering design problems to be solved and are directly traceable to the user requirements.

The performance design and analysis for a mechatronic system are accountable to two technical specifications: *functional specifications* and *integrity specifications*. A *functional specification* specifies how well the system must perform in normal conditions expected of the system. It seeks a workable scheme for the problem. Functional specifications are a collection of performance measures, which is defined below. An *integrity specification* defines how well the system and its specific components must perform under expected strenuous conditions. It ensures that there are no weak links in the design. Examples of integrity specifications are sensitivity and stress analyses (26) as well as statistical and varying component analyses.

Design and Analysis Process. Mechatronics design and analysis deal with what is achievable through application of engineering technology. They comprise two complementary processes described below.

Top-Down Design Process. This stage is where engineers can become creative in their design to achieve the requirement for the mechatronic system. A *top-down design* is a validation process that ensures that the selected design and components are consistent and complete with respect to the *functional specifications* of the mechatronic system. The validation process is used to ensure that we are working on the *right problem* by guiding the detail design towards the functional requirements (27). The process does the following:

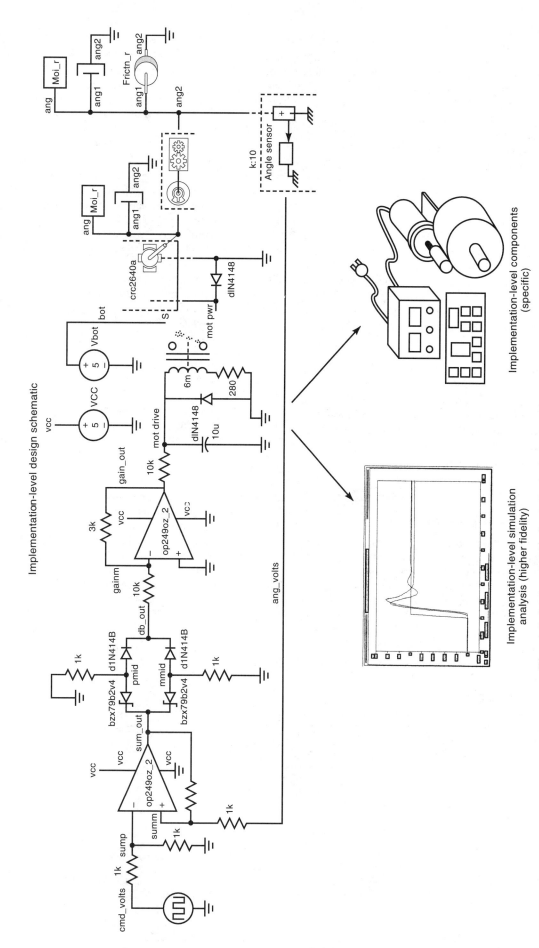

Figure 6. Implementation-level design, analysis, and component selection.

Implementation-level design schematic

Implementation-level simulation analysis (higher fidelity)

Implementation-level components (specific)

- It begins with a schematic of an initial conceptual-level design to establish the operation and technical performance specifications for a mechatronics concept.
- It translates the concept design into a preliminary implementation-level design with specific components, satisfying technical specifications in the presence of the interfacing environments and operating conditions.
- It deals with problems in the intermediate stages of design during the transition through necessary new redesign iterations and requirement variations.

Bottom-Up Analysis Process. This stage is where engineers become critical of the preliminary design and set out to check the soundness or integrity of the design. A *bottom-up analysis* is a verification process that expands on the selected design solution to ensure that it meets the *integrity requirements*. It assures that we have solved the *problem right* by catching potential trouble spots before they become expensive and time-consuming crises (27). The process does the following:

- It carries out sensitivity analysis, stress analysis, and statistical analysis of the selected design under various expected strenuous conditions.
- It checks out feasibility and soundness of the selected design with other engineering groups such as manufacturing, testing, and reliability before commencing to build hardware prototype or "breadboard."
- It deals with problems of component selections and availability through iterations of redesign with the top-down design group.

Techniques for Mechatronics Design and Analysis

There are so many techniques and aspects regarding designing a mechatronics system (6–19,21,22) that it is not possible to mention all of them here. In this section, we have selected to highlights only three basic aspects as examples of design and analysis techniques that engineers should consider in the pursuit of designing a high-performance, robust, and reliable product. The three aspects are attention to (1) technical specifications to ensure that user requirements are met, (2) sensitivity analysis to ensure robustness to parameter variations, and (3) stress analysis to ensure reliability.

Technical Specifications. Derived from user requirements, technical specifications are used to guide the design of the mechatronic system. As explained earlier, we may categorize the technical specifications as functional and integrity specifications. Another useful specification is the term called performance measure.

What a Performance Measure Is. A performance measure, normally denoted by the symbol J, is a scalar numerical index that indicates how well a system accomplishes an objective (23). The index can be measured from the waveform characteristics of signal responses generated by the system in experiments, simulations, or theoretical analyses. A performance measure or index therefore is essentially a score that is used to rank the performance of systems. Simple performance measures that can be directly extracted from an output response of a system are maximums/minimums, rise time/fall time, steady-state value, settling times, initial value, peak-to-peak value, period, duty cycle, and so on. Other indexes require some computational effort—for example, frequency response bandwidth, resonance magnitude and frequency, average, root mean square, sum of weighted squared errors, power and energy, and so on. Figure 7 illustrates the details of some simple performance indexes in a step response. Performance measure J is used to evaluate sensitivity analysis, which is part of the integrity specifications.

Functional Specifications and Performance Measures. Functional specifications are made up of one or more performance measures that can be used to define the desired system performance more rigidly. The selected performance measures should be complementary and not conflict with each other. For example, settling time and percent of maximum overshoot complement one another in defining the specifications,

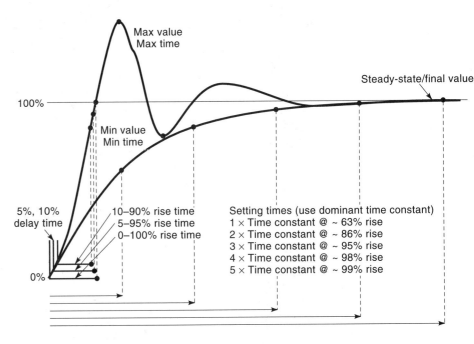

Figure 7. Candidates for performance measures in step responses.

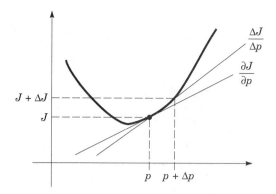

Figure 8. Definitions of sensitivity gradients.

whereas settling time and rise time may conflict in requirements. The functional performance specifications should be validated against "fuzzy" user requirements as well as used to check the performance of the component-level or implementation-level design.

Sensitivity Analysis

What a Sensitivity Analysis Is. A sensitivity analysis is a study that examines how sensitive a specified performance measure is to variation in the values of components or parameters in a system. For example, it can be used to determine how much the speed of a motor is affected by a change in the gain of an amplifier or a drop in the voltage supply.

How a Sensitivity Analysis Can Improve a Design. One can use the information obtained from a sensitivity analysis to identify which part of the system has significant impact on the system performance. Based on the finding, one may redesign the system to reduce the sensitivity and hence improve the robustness with respect to the particular parameter. The analysis can also be used to select appropriate tolerance values for the design to ensure that performance specifications are met.

How Sensitivity Is Defined. Sensitivity analysis of a system can be conducted by examining the gradient of performance measure J with respect to parameter p. This *sensitivity gradient* can be approximated by the ratio of variation ΔJ over perturbation Δp, as shown below:

$$S = \frac{\partial J}{\partial p} \approx \frac{\Delta J}{\Delta p}$$

Figure 8 illustrates the sensitivity gradient for a simple parameter variation. The interpretation of the gradient can be more rigorously observed using the Taylor series expansion of $J(p)$ around $p + \Delta p$:

$$J(p + \Delta p) = J(p) + \frac{\partial J}{\partial p}\Delta p + \frac{1}{2!}\frac{\partial^2 J}{\partial p^2}\Delta p^2 + \frac{1}{3!}\frac{\partial^3 J}{\partial p^3}\Delta p^3 + \cdots$$
$$= J(p) + \Delta J$$

In most cases, it is more meaningful to compute the *normalized sensitivity gradient* as follows:

$$S_{\mathrm{N}} = \frac{\partial J/J}{\partial p/p} = \frac{p}{J}\frac{\partial J}{\partial p} \approx \frac{\Delta J/J}{\Delta p/p} = \frac{p}{J}\frac{\Delta J}{\Delta p}$$

where J and p are the baseline performance measure and parameter, as shown in Fig. 8. However, the normalized sensitivity cannot be evaluated if J or p is 0 or very close to 0; hence the direct sensitivity gradient will be used.

How a Sensitivity Gradient Is Calculated. In certain cases where the performance measure J can be explicitly or implicitly expressed as analytical functions of a parameter p, it is possible to evaluate the sensitivity gradient in closed form. For instance, if

$$J = f(y)$$
$$y = a(u, \, p)$$

where the functions f and a are analytical or differentiable at the points of concern, then the sensitivity gradient can be evaluated as

$$S = \frac{\partial J}{\partial p} = \frac{\partial J}{\partial y}\frac{\partial y}{\partial p}$$

The analytical solution can often shed insights into an analysis. An excellent example of this is the derivation of the well-known backpropagation training algorithm for neural networks, as well as its use in optimization and adaptive control methods. The possible drawbacks of the approach, however, include the need to know the explicit (direct) or implicit (indirect) formula that describes the relationships between J and p, and the necessary condition that the functions be analytical (differentiable) at points of interest.

A less mathematically laborious and yet effective approach for calculating sensitivity gradients is to employ computer simulation. The idea is to simulate and compute the performance measures J and $J + \Delta J$ when the system operates under the parameter p and $p + \Delta p$, respectively, where Δp is a small perturbation. The straightforward calculation $S \sim \Delta J/\Delta p$ approximates the sensitivity gradient. This computational technique can be used in the sensitivity analysis of simple and complex systems.

A Sensitivity Analysis Report. The sample report in Table 1 illustrates how a sensitivity analysis points out the parameters that have high sensitivity impact on the system performance measure. Attention should be given to large sensitivity gradients because they indicate that performance measure is highly sensitive to the parameter variations. Redesign of control scheme or circuit configuration may be required to reduce this effect and improve the robustness of the system. As can be seen, the computations in sensitivity analysis can be very tedious, laborious, and time-consuming. The key to the analysis is to employ a computer program to automatically generate the sensitivity analysis report for selected parameters in a design.

Stress Analysis

What a Stress Analysis Is. A stress analysis checks the conditions of components at operating conditions and compares them against the operating limits of the components. The analysis can pinpoint underrated components that are most likely to fail under expected strenuous operating conditions as well as components that are unnecessarily overrated and costly. It is an important design and analysis step for determining the ratings and rightsizing the components.

Table 1. Sample of a Sensitivity Analysis Report

Parameters	Sensitivity Gradient[a] $S = \Delta J/\Delta p$	Normalized Sensitivity Gradient[a] $S_N = (\Delta J/J)/(\Delta p/p)$	Comments
P1	1.811	1.050	OK. S and S_N are low.
P2	0.010	8.800	S_N is high. Check design
P3	190.0	0.290	S is high. Check design
P4	20.01	5.501	S and S_N are high. Check design

[a] Large values in sensitivity gradients S and S_N signify possible weakness in terms of robustness of the design.

What a Stress Measure Is. A stress measure is the operating level of a component or part that occurs during operation. Examples of stress measures are: power dissipation of a resistor, transistor, or motor; reverse voltage across a capacitor; junction temperature of a bipolar transistor; and maximum temperature and current in the coil of a motor, solenoid, and so on.

What Operating Limits Are. Manufacturers of components test their products and supply ratings of maximum operating limits (MOLs) for the components. The MOL may be a single value, or curve or surface function of the operating variables. Figure 9 shows the maximum power dissipation curve of a resistor alongside with the maximum collector current curve for a transistor. The area below the MOL is the safe operating area (SOA). A component operating within this region will experience no stress, whereas it will be overstressed outside of the SOA. Exceeding the maximum operating limits will lead to malfunction.

What Derating Is. Because the MOL ratings supplied by manufacturers are calibrated at specific test conditions, engineers often adjust the ratings by some derating factors to suit their application. The derating factor depends on the quality standards of the parts such as military, industrial, and com-

mercial standards, and it also depends on the operating condition in which the design will be used. A designer usually reduces the MOL rating of components by a derating factor to decrease the SOA, so that the component will be designed to withstand higher stress. Figure 9 illustrates examples of derated maximum operating limits for the resistor and transistor.

How Stress Is Calculated. Stress ratio is the fundamental quantity for indicating a stress level of a component. It is defined as

$$R = \frac{\text{Measured value} - \text{Reference rating}}{\text{Derated rating} - \text{Reference rating}}$$

where measured value is the worst case (maximum or minimum) or cumulative (average or rms) or other operating values observed during an analysis, and derated rating is the adjusted maximum operating limits as explained. The reference rating is an offset value to which both the measured value and derated rating are referenced, as in the case of temperature calculations; in most cases it is equal to zero. It is obvious that the value of $R \gg 1$ indicates overstress while $R \ll$

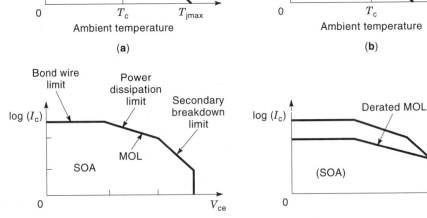

Figure 9. Maximum operating limits and derating of ratings to account for the environment in which the design will be used. (a) Power dissipation rating for a resistor. (b) Sixty percent derating in power dissipation rating implies smaller safe operating area. (c) Maximum current rating for I_c as a function of V_{ce} for a transistor. (d) Maximum current rating derated or reduced so that the stress analysis will select a component that can withstand higher stress.

Table 2. Sample Stress Analysis Report

Components	Derated Rating	Measured Value	Stress Ratio[a], R	Comments
Resistor 1				
Power dissipation	1.44 W	2.00 W	72%	OK
Resistor 2				
Power dissipation	0.12 W	2.00 W	6%	Alert, over designed
Transistor				
Power dissipation	40.0 W	25.0 W	180%	Alert, underdesigned
Junction temperature	250°C	125°C	200%	Alert, underdesigned

[a] The stress ratio points out whether a part is underdesigned, overdesigned, or just right for the application.

1 means understress, and $R \sim 1$ implies that stress is neither overstress nor understress.

A Stress Analysis Report. The sample stress analysis report in Table 2 points out the stress level of components. The stress ratio indicates whether a component has been underdesigned ($R \gg 1$), overdesigned ($R \ll 1$), or correctly designed for the application. The overstressed underdesigned parts can lead to malfunction, whereas the understressed overdesigned parts are unnecessary and can be costly. Stress analysis report checks to see if the selected components are right for the job. As in the sensitivity analysis, the computations in stress analysis can be very tedious, laborious, and time-consuming as well. And as in the preceding case, the key to the analysis is to employ a computer program to automatically generate the stress analysis report for selected components in the design.

ILLUSTRATION OF DESIGN AND ANALYSIS PROCESS

Although a sensible systems engineering approach involves all appropriate engineers (see Fig. 4) at the early and subsequence stages in the development process of a mechatronics system, certain subprocesses, such as design and analysis, may inherently be sequential in nature. The process for dealing with performance requirements, design, and analysis of a mechatronic system is well illustrated by considering a servo-positioning system example shown in Figs. 5 and 6 (24). The servo-system could be part of a product with motion control, such as a robot or vehicle. [See also DC MOTOR DRIVES (BRUSH AND BRUSHLESS).] The example illustrates the following ideas.

Requirements and Specifications Process

This process understands the requirements (needs and expectations) of the users and translates them into specifications that engineers can reference to as guidelines for their design.

1.a. *Functional.* Suppose that the user requirement is to position the output angle of the load accurately and quickly at a reference location specified by the user. A control engineer would translate these nontechnical terms into acceptable technical functional specifications such as settling time, overshoot, and steady-state error, for a step response of the output position. The simulation response diagram in Fig. 5 illustrates the idea. Alternative functional specifications may also be employed.

1.b. *Integrity.* Next also suppose that the user will operate the system at strenuous conditions as in a high-tem-perature environment and at excessively large varying operating levels. An electrical and mechanical engineer must check the integrity of the components used in the system to safeguard against performance deterioration and system failure.

Top-Down Design Process

This process produces a concept-level schematic to validate the requirements and specifications of the mechatronic system. The design process continues to evolve the concept into an implementation-level schematic with selected components for the system.

2.a. *Concept Level.* The top of Fig. 5 shows the concept-level design schematic consisting of a transfer function block diagram representing a simplified ideal model for the system. Here, a control engineer designs and selects a suitable control scheme for servo-positioning system to achieve the functional specifications. Simulation of the ideal model responses validates that the desired servo-positioning requirement for various conditions of operation is achievable. In this example, a step response is used to the specification as shown in the bottom left of the figure. There are no specific hardware components identified at this initial design stage.

2.b. *Implementation Level.* Once the desired functional response from the conceptual block diagram is achieved, the performance specifications and function characteristics are passed on the electrical and mechanical engineers. The top of Fig. 6 shows the implementation-level design schematic for the proposed system where specific electrical and mechanical components for realizing the servo-positioning scheme have been selected. Simulation of the system at this level confirms that the functional specification is met, within acceptable variations, as shown in the bottom left of Fig. 6. The bottom right of the figure shows the *selected components* for the design. This is the *main result* of the top-down design process. At this stage though, we will refer to the result as the *preliminary* implementation-level design since it has yet to pass the component integrity test.

2.c. *Intermediate Level.* The transition between the conceptual and implementation-level designs would require several intermediate stages of design and redesign iterations. For instance, introducing realistic models of mechanical components would introduce undesirable characteristics such as friction, gear backlash, and shaft flexibility, and it can result in the initial control scheme being no longer acceptable. The control, electrical, and mechanical engineers must rework the design to find a solution to the problem. This may involve several iterations of design before arriving at the *implementation-level* schematic.

Bottom-Up Analysis Process

The selected components at the implementation-level schematic are not the final result of the overall design. These selected components must be subjected to rigorous tests to check their integrity or soundness to ensure that they (1) are not the cause of degradation in functional performance under variation, (2) can withstand strenuous operating conditions, and (3) are realistic parts that can be manufactured, tested, and so on.

3.a. *Component Analysis.* The selected components in the implementation-level schematic for the servo-system are subjected to sensitivity and stress analyses. A *sensitivity analysis* report, similar to the one shown in Table 1, will locate the high-sensitivity components in the design, if any. Where necessary, the control scheme, circuits, sensors, and/or actuators will be redesigned or reselected to produce a more robust implementation-level design. Similarly, a *stress analysis* report, similar to the one shown in Table 2, will identify which components in the design are overstressed, understressed, or normal. Resizing of the components will be carried out to improve the reliability of the design in the case of overstress condition, or to possibly reduce cost in the case of understress condition. Alternative solutions to overstress problems may include, as examples, adding a heat sink to cool electronic components, relief valve to limit pressure, damping cushion to reduce stressful impact, and so on.

3.b. *Manufacturability, Test, and Reliability.* The design information and simulation model are shared among manufacturing, test, and reliability engineers for their review. For instance, the manufacturing engineer may question the commercial availability of certain components in the design and may then suggest alternative standard parts and reduced spending. The test engineer may notice that a study may have been overlooked by the design engineers and may then suggest a re-run of simulations to include the new conditions. The reliability engineer may suggest addition of test points in the design for diagnostic purposes.

3.c. *Trade-Off Decisions.* Conducting the top-down design and bottom-up analysis in the virtual prototyping environment let the engineers find potential problems very early in the stage of the development. Modifications are made via rigorous design and analysis development process. At times, trade-off decisions may require modification of the requirements and specifications as well. At the end of arguments, all parties would end up selecting the "optimum" and right component for the job. The decision at this end will produce the recommended implementation-level design, as the main result of the overall design.

Computer-Aided Engineering Tool

As illustrated and described, the development process for designing and analyzing a mechatronic system employs extensive use of CAE software. From the standpoint of the application engineers, this CAE tool is heaven sent, since they must accomplish the design with limited resource and time. Equally important is the fact that it provides a function simulation "blue print" with which the cross-disciplinary team of mechatronics engineers can communicate and verify their multitechnology ideas. The software that has the necessary tools for dealing with the mechatronics development in this case is the Saber simulator. The simulator is a recognized technique that has been adopted as a standard systems engineering practice in the automotive and aerospace industries.

Environment

The last building block to support the above process is the environment. The necessary technical environment includes computing facilities, work group consisting of experts, and organizational, managerial, and technical supports. Another important infrastructure may involve information technology whereby the secure use of *intranet* and *internet* makes it possible to share information rapidly among the mechatronic team. The competitiveness in the high-technology business demands such an enabling environment.

Other Examples

The above example was picked for its simplicity and familiarity to a reader, for the purpose of explaining the development process. There are literally hundreds of other examples that follow similar design and analysis process that is aided by the CAE tools. Readers may refer to Refs. 8 and 24–26 for further reading.

EVALUATION

The domain of mechatronics has expanded from simple electronics and mechanics technologies to complex automation, control, and communication technologies with embedded computer intelligence (20). Mechatronic systems are ubiquitous in military, industrial, and commercial applications. They may exist in the form of unexciting but extremely useful products such as factory robots, household appliances, and so on, or the form of exciting systems such as unmanned vehicles for space and remote exploration, as well as military applications. Consumers have benefited tremendously from mechatronic products such as a video camera with full automatic features, automatic teller machines, and the automobiles. It's what we associate with the term "high tech."

It may be reiterated that successful mechatronics endeavors usually stem from a combined application of science, technology, engineering, business, and art. Evidence of these endeavors can be found in innovative use of materials, parts, and better software techniques. Examples are miniaturization of remote control devices, transponders, micromachines, and so on, and use of more sophisticated methods such as fuzzy logic and neural network to enhance original performance of mechatronic systems. The profitable mechatronic product endeavors are the ones that achieve *quality* products in minimum *time* and *cost*. The systems engineering develop ment process presented here illustrates a means to accomplish this objective.

The path from nurturing a concept to bringing a product into being normally undergoes three stages of development.

1. *Phase 1.* The Basic Research stage, where concept design and analysis are carried out to determine feasibility of the mechatronics concept. This conceptual level stage is the "requirements and specifications" process.

2. *Phase 2.* The Exploratory Research stage, where prototypes are integrated to investigate the feasibility of the mechatronics applications. This stage can be likened to the "top-down design" process to validate that "we are doing the right job."

3. *Phase 3.* The Product Development stage, which deals with manufacturing process, testing, and reliability issues to bring the product to life. This final stage is the "bottom-up analysis" process to verify that "we are getting the job done right."

According to the scale of a US Government research funding agency, the ratio of resource funding for Phase 1 to Phase 2 to Phase 3 is approximately $1:10:30$. This illustrates the relative importance of the processes. Many textbooks and articles in the academic literature describe mainly the *functional* performance design process of building mechatronics systems products. They do not emphasize the importance of the component integrity analysis. On the other hand, the practice in the industry heavily emphasizes integrity analysis verification while maintaining functional design validation. This is necessary to ensure the development of high-quality mechatronic products. This is the key point of this article.

Next, one should review the important role of the CAE tool. The philosophy of computer simulation is simple: It's the ability to predict system performance. With accurate computer models, simulation helps engineers to fully comprehend the problems at hand and enables them to conduct "what if" studies to predict, correct, optimize, and select the right components. The CAE tool used in this mechatronics study was Saber, which is a virtual function prototyping facility. As alluded to in the text, a mechanical CAD tool could be incorporated in the mechatronics study to visualize the motion, the physical layout, the shape, the size, and the color of the mechatronic product. CAD has been adopted in the aerospace and automotive industries. A current trend in the industry is to combine prototypes of virtual functions with virtual mock-ups in a virtual reality environment where a human user can "feel" how the mechatronic product perform, all inside the cyberspace.

Finally, the breadth of disciplines required for a mechatronics project can be quite broad (e.g., electronics, mechanical, hydraulics) and the depth required of a discipline can be quite deep (e.g., details of real-time embedded controller). It is through training and experience that an engineer (from any one of the mechatronics disciplines) will gain sufficient knowledge to lead a mechatronics project and team.

Mechatronic systems and products will keep pace with the progress of technologies and methodologies, and they are here to stay. Mechatronics is the key discipline to the current and future high-tech industries.

BIBLIOGRAPHY

1. *Assoc. of Unmanned Vehicle Syst. Int. Mag., USA,* quarterly issues.

2. *Unmanned Vehicle Syst. Mag., UK,* quarterly issues.

3. *Int. J. Mechatron.*

4. *IEEE / ASME Trans. Mechatron.*

5. *Int. J. Mechatron.: Mech.–Electron.–Control Mechatron.*

6. D. M. Auslander and C. J. Kempf, *Mechatronics: Mechanical Systems Interfacing,* Upper Saddle River, NJ: Prentice-Hall, 1996.

7. W. Bolton, *Mechatronics—Electronic Control Systems in Mechanical Engineering,* Reading, MA: Addison-Wesley, 1995.

8. R. Comerford, Mecha . . . what?, *IEEE Spectrum,* **31** (8): 46–49, 1994.

9. J. R. Hewit (ed.), *Mechatronics,* Berlin: Springer-Verlag, 1993.

10. M. B. Histland and D. G. Alciatore, *Mechatronics and Measurement Systems,* New York: McGraw-Hill, 1997.

11. J. Johnson and P. Picton, *Mechatronics: Designing Intelligent Machines,* Vol. 2: *Concepts in Artificial Intelligence,* London: Butterworth-Heinemann, 1995.

12. L. J. Kamm, *Understanding Electro-Mechanical Engineering: An Introduction to Mechatronics,* New York: IEEE Press, 1996.

13. N. A. Kheir et al., A curriculum in automotive mechatronics system, *Proc. ACE '97, 4th Int. Fed. Autom. Control (IFAC) Symp. Adv. Control Educ.,* Istanbul, Turkey, 1997.

14. D. K. Miu, *Mechatronics: Electromechanical & Contromechanics,* Berlin: Springer-Verlag, 1993.

15. D. Tomkinson and J. Horne, *Mechatronics Engineering,* New York: McGraw-Hill, 1996.

16. G. Rzevski (ed.), *Mechatronics: Designing Intelligent Machines,* Vol. 1: *Perception, Cognition and Execution,* London: Butterworth-Heinemann, 1995.

17. S. Shetty and R. A. Kolk, *Mechatronics Systems Design,* Boston: PWS, 1997.

18. R. Jurgen (ed.), *Automotive Electronics Handbook,* New York: McGraw-Hill, 1995.

19. D. Knowles, *Automotive Computer Systems,* New York: Delmar, 1996.

20. C. J. Harris, *Advances in Intelligent Control,* London: Taylor & Francis, 1994.

21. P. D. Lawrence and K. Mauch, *Real-Time Microcomputer Systems Design: An Introduction,* New York: McGraw-Hill, 1987.

22. C. T. Kilian, *Modern Control Technology: Components and Systems,* St. Paul, MN: West, 1996.

23. B. J. Kuo, *Automatic Control Systems,* Englewood Cliffs, NJ: Prentice-Hall, 1985.

24. Automotive Applications Using the Saber Simulator, Analogy, Inc., 1992.

25. *Proc. Autom. Analogy Saber Simulator Users Resource,* Livernois, MI, 1997.

26. Stress and Sensitivity Option, Release 3.2, Analogy, Inc., 1993.

27. J. N. Martin, *Systems Engineering Guidebook: A Process for Developing Systems and Products,* Boca Raton, FL: CRC Press, 1997.

KA C. CHEOK
Oakland University

MEDICAL COMPUTING

A computer in one form or another is present in almost every instrument used for making measurements or delivering therapeutic or experimental interventions in clinical practice, clinical research, or in any of the fields of the biomedical sciences. In some cases, the computer is a chip that is embedded somewhere in the instrument, and the user has little indication that he or she is actually interacting with a central processing unit (CPU). In other cases, the form of the computer is an engineering workstation or a personal computer with keyboard, mouse, and the other accoutrements that are normally associated with the act of "computing." In either case, the computer allows the accomplishment of tasks that were impossible or overwhelmingly difficult without the use of this ubiquitous technology.

Many of the concepts that were implemented in early computers are still routinely used in modern devices, but with vastly different technological bases. Most digital computers have traditional architectures, with separate memories, central processing units, and input/output and storage devices. A major difference is the miniaturization and increased efficiency of electronics devices, including computers, which have allowed the development of sophisticated medical therapeutic options that can be totally implanted in the body. The increased sophistication of computers has also led to vastly improved user interfaces and quality control in all kinds of medical instruments.

This article reviews some of the historical and contemporary computer applications in medicine and biomedical research, from real-time applications to medical informatics to virtual reality. These applications are as diverse as the physical form of the computer on which they are implemented. They include: real-time signal acquisition and subsequent processing; efficient storage and manipulation of enormous databases for large-scale clinical trials and other clinical and basic research; assistance in clinical decision-making; acquisition, processing, and transmission of diagnostic images from a wide range of technologies; simulation and display of two- and three- dimensional structures and function; animated displays of physiological processes; and access to educational and promotional medical information. Of course, any discussion of the use of computers in biomedicine would be incomplete without reference to the explosion in the use of the Internet for electronic mail, news, and World Wide Web access. Many of these topics will be briefly mentioned; specific applications will be more completely described as examples of how computers can be used to improve medical practice through broadening the scientific underpinnings and making new techniques available to practitioners. Many applications that are implemented on computers will be more fully developed in other articles of this encyclopedia.

MEDICAL INFORMATICS

The study of the use of computers in medicine is often called "medical informatics." Medical informatics has been termed "an emerging discipline that has been defined as the study, invention, and implementation of structures and algorithms to improve communication, understanding, and management of medical information (1)." (Further information can be found on the newsgroup *sci.med.informatics*). This broad area comprises many of the subjects included in medical computing, and there is a vast literature on ways in which computers improve health care delivery through aids in organizing, ac-

cessing, and presenting knowledge. Medical informatics is concerned with the flow of medical information, and how data that can be used in the effective diagnosis and treatment of disease can be made available in a timely fashion and in a format that will provide the most usefulness (2). The integration of medical information systems into routine clinical activities has not been fully achieved (3).

Efficient organization and delivery of biological and medical information have assumed new importance with the explosion in knowledge about the genetic and molecular bases of disease. One of the grand challenges of computing is the deciphering of the human genome, and sophisticated computer algorithms have been developed for identifying genes embedded in long DNA sequences (4). A recent issue of the journal *Science* described some of the opportunities for the application of computers in several areas of medical and biological research. Articles in the aforementioned issue described new approaches to searching the World Wide Web as a digital medical library (5), the application of computing to mathematical ecological studies (6), and the use of a massive database for investigating the relationship between pharmacological agents and cancer (7). In general, the Internet, including news groups and World Wide Web sites, has been a rich resource for all kinds of medical information (8,9); indeed, the challenge is to develop methods for accessing the data in ways that are intuitive and accurate. For example, a system has been written to exploit effectively the digital images of human anatomy that are stored on the Web (10). Figure 1 is an example of a workstation screen from this system which allows flexible interrogation of three-dimensional anatomical databases, providing access to this wealth of information for users without detailed knowledge of computer and database architectures.

The study of artificial intelligence and expert systems is also typically considered a central theme of medical informatics. These systems use learning and advisory strategies to assist in decisions related to diagnosis and therapy of human disease. One of the earliest results in expert systems was the computer program named *MYCIN,* which was designed to assist physicians in the selection of antimicrobial agents for hospital patients with bacterial infections (11). Artificial neural networks have been used to assist in the detection of patterns in clinical data and associate them with clinical conditions and outcomes (12,13). Human beings are quite adept at incorporating uncertainties in their reasoning, but computers have been viewed as unforgiving in the face of ambiguities. The theories of fuzzy sets and fuzzy logic (14) have been developed to address this discrepancy, and they have attracted attention for their wide applicability (15). They have been used in classification of biomedical signals (16) and analysis of biomedical images (17). The study of expert systems is an active area of investigation in the application of computers to medicine.

SIGNAL ACQUISITION AND ANALYSIS

One of the earliest applications of computers in medicine and medical science was their use for acquiring, analyzing, and displaying waveforms that reflect physiological and pathological processes. Prominent examples include the electrocardiogram (ECG) and electroencephalogram (EEG), which record the manifestations at the body surface of the electrical gener-

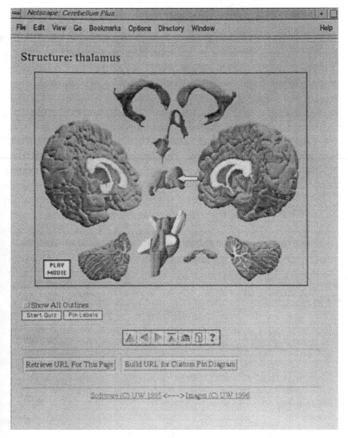

Figure 1. Computer screen from the "Digital Anatomist Information System (10)," an interactive query system for access to three-dimensional anatomic databases using Internet browser principles. This display allows the user to identify particular structures of the brain, in this case the thalamus, by selecting the region of interest with the computer input device. Reprinted from Ref. 10, with permission.

ators in the heart and brain, respectively. For computer analysis, it is necessary to sample the waveform and convert it from analog to digital format (18). In electrocardiology, computers have been used for the analysis of clinical ECGs to identify patterns in the waveform that reflected underlying disorders of electrical activation or structural heart disease (19,20). Similarly, patterns in EEGs have been correlated with normal neurophysiology and with pathologies such as epilepsy (21). Commercial systems are based on similar analyses and are widely used in hospitals and clinics for diagnosing cardiac and neurological abnormalities.

At the same time, the use of computers has expanded the level of investigations that are possible in attempting to understand the scientific basis of normal and abnormal physiological function. Even though much was learned about cardiac electrophysiology by using instruments like string galvanometers (22), improved technology and the application of digital computers have allowed measurements to be made in situations that were not accessible to earlier investigators. For example, it is now possible to record from hundreds of sites on the surface of the heart and within the cardiac tissue to assemble a high-resolution reconstruction of the intrinsic or externally generated bioelectric events in the myocardium (23). Cardiac mapping studies that acquire data from many sites simultaneously have shown that there is considerable order

in heart rhythms formerly thought to be completely disorganized (24,25). Figure 2 is an example of a signal processing result that demonstrates that electrograms recorded from a rectangular array during ventricular fibrillation have a great deal of organization even when they are recorded from sites separated by as much as 5 mm to 11 mm. Thus, computer-based multichannel acquisition and analysis of cardiac arrhythmias have revealed phenomena that might be crucial in the improved prevention and treatment of these often fatal derangements of rhythm.

Multichannel recording from the brain using computer-based systems has resulted in new insights into the way in which the brain's electrophysiological and psychological functions are organized (26). Similar systems have been developed to study the electrical activity associated with the gastrointestinal (27), genitourinary (28), and reproductive (29) systems.

The development and implementation on computers of the fast Fourier transform has allowed the examination of biosignals in the frequency domain, opening the doors for new insight into mechanisms of important clinical entities (30). Spectral methods allow the elucidation of relationships between different physiological systems (31). Wavelet theory has further extended the application of frequency domain techniques by avoiding the limitations imposed by discrete Fourier analysis (32). Wavelets have been applied to electrocardiography, to detect irregularities in heart rhythm; to pho-

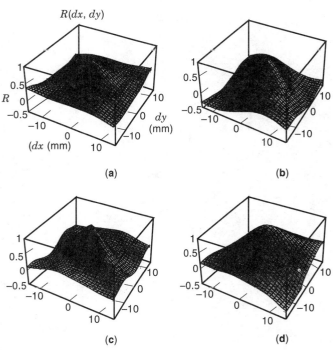

Figure 2. Plots of correlation coefficient between two electrograms recorded from a rectangular plaque array on the ventricular epicardium of a pig during ventricular fibrillation. Panels a to d are data from 1 s to 2 s; 2 s to 3 s; 3 s to 4 s; and 60 s to 61 s after the induction of ventricular fibrillation. The electrical activity in two electrodes appears to be well correlated, out to a distance between recording sites of about 5 mm to 7 mm, but this distance changes as fibrillation continues. Whereas fibrillation has traditionally been considered to be a completely disorganized rhythm, computer-based signal processing techniques have revealed substantial levels of order. Reprinted from Ref. 24, with permission.

nocardiology, to analyze heart sounds in search of turbulence associated with obstructions in coronary vessels; to examine the EEG for evidence of fetal respiratory abnormalities; and in image analysis and compression (32).

COMPUTER-BASED DEVICES

Many medical devices require the acquisition and analysis of analog signals or variables that reflect underlying physiological processes. Typically, a method is developed to transduce the variable of interest to a voltage which can then be converted to digital form and analyzed. The implementation of these systems ranges from integrated chips with analog electronics, analog-to-digital converters, and some signal processing capabilities included to general-purpose computers with specialized or general-purpose electronics designed to carry out these functions. The devices are used in physiological and biomedical research and in instruments used for making clinical decisions. The computers are often used as control systems as well as data acquisition devices, with or without closed-loop feedback.

Embedded Computers

Figure 3 is an example of a highly specialized medical device, the implantable cardioverter-defibrillator, with a fully functional computer embedded in it. The function of this device is to monitor continuously the electrical signal derived from electrodes in contact with the heart, detect the onset of abnormalities in cardiac rhythm that are candidates for treatment, and deliver appropriate therapy. To carry out this role effectively, it is necessary for the instrument to sample and digitize the cardiac electrograms with sufficient accuracy to allow morphological analysis; identify the time at which the electrical wave passes the electrode in order to compute R–R intervals and, thus, heart rate, often with analog processing (33); distinguish between normal rhythms and categories of arrhythmias that require different levels of therapy (34); and control delivery of antitachycardia pacing or a defibrillation shock.

Figure 3(a) is a photograph of such a device. It is intended for implantation beneath the clavicle in patients who are at high risk for sudden cardiac death. In use, leads are directed from the header through the venous system into the right ventricle of the heart. There are electrodes on the catheters for sensing the cardiac activity and delivering energy. Figure 3(b) is a photograph of the die of the microprocessor that is in the defibrillator. The demands on this circuit are extraordinarily stringent, since a malfunction in either software or hardware is likely to be fatal to the patient wearing it. The validation of the operation of the computer is obviously of utmost concern to the manufacturer, the implanting physician, regulatory agencies, and, especially, the patient.

(a)

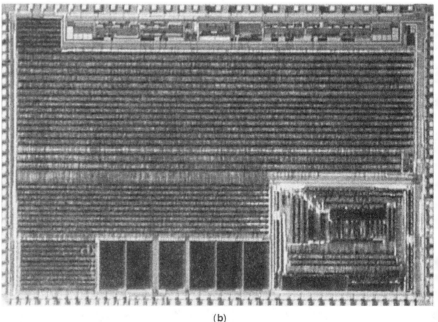

(b)

Figure 3. (a) Photograph of a mockup of an implantable cardioverter-defibrillator. The device is implanted under the clavicular region of a patient who is at high risk of sudden cardiac death. The overall dimensions of the device are 4.5 × 7 × 1.25 cm. (b) A photograph of a die of a digital computer system that is contained in the device shown in panel a. The dimension of the chip that is fabricated from this die is 12.8 by 7.7 mm. The chip, with 2 μ interconnecting lines, contains about 150,000 transistors. In addition to a typical microprocessor, the system contains sense amplifiers, a digital control section for analyzing cardiac electrical activity and initiating therapy when necessary, five banks of 1 Kb read-only memory, and circuits for telemetering data to and from the external world. Photograph courtesy of CPI/Guidant Corporation.

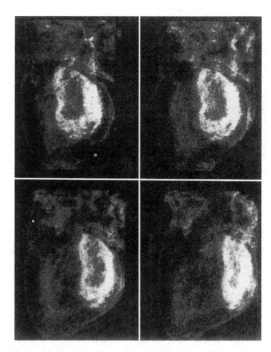

Figure 4. A ray-traced, volume-rendered magnetic resonance image of a canine heart with a myocardial infarction. The image was acquired from a postmortem, formalin-fixed heart after a study to determine whether necrotic tissue could be reliably detected by MRI. Computer-based magnetic resonance microscopy can be used for diagnosis and evaluation of pathology, as well as for investigations into the interplay of structure and function in clinical and experimental studies. Reprinted from Ref. 35, with permission.

Computers are often embedded in other medical instruments that do not play as acutely critical a role as do implanted devices, but are very important in assessing health and disease. For example, a blood gas or electrolyte analyzer might have a microprocessor that controls the user interface, calibration procedures, and data acquisition, analysis, and display. The widespread use of microcomputers in these analyzers provides convenient access to functions such as calibration and standardization that previously were time-consuming and labor-intensive.

Stand-Alone Computers

Other medical instruments are based on general-purpose computers, typically engineering workstations. Examples of these are large imaging systems, such as magnetic resonance imaging (MRI) or computed tomography (CT) systems. In these cases, the computer fulfills a variety of roles. In MRI systems, the computer can provide the user interface to the highly specialized and complex hardware associated with the magnet. The echo sequences that determine the imaging parameters and quality can be controlled through the computer as a front end. These workstations often contain high-performance image processing and graphics hardware that can be used for manipulation of the acquired images and display of the results to the clinician in intuitive, usable formats. Figure 4 is a volume-rendered, ray-traced magnetic resonance image of a canine heart with an experimentally induced myocardial infarction (35). This image was generated on a high-performance workstation, and it demonstrates the kinds of displays

that provide users with scientifically and clinically useful information. Furthermore, the workstation and associated peripheral devices can be used as archival storage systems for managing the overwhelming amounts of data that are generated by modern imaging modalities.

The computers associated with these clinical systems can be networked using industry standard hardware and software to provide convenient access to remote sites, either within a hospital or medical center or outside the center to a referring physician or a specialist who might have more experience in interpreting certain imaging results. This is an example of the interaction between medical imaging and telemedicine (36), often thought of as one of the subdisciplines of medical informatics.

COMPUTER GRAPHICS AND MEDICINE

Another early use of computers in medicine was the application of graphics systems for the acquisition and realistic display of anatomic structures in research and clinical situations. Much of the emphasis has been on two- and three-dimensional reconstructions of data from medical imaging modalities. Early work focused on the use of computers to estimate cardiac function from single- and dual-plane cineangiography (37), as well as estimate the reconstruction of coronary artery anatomy from coronary arteriograms (38,39). As medical imaging technologies have advanced, the computational demands for extracting new information from image analysis and displaying the data in realistic ways have increased. Substantial portions of the techniques developed for nonmedical applications are not immediately applicable to biological and physiological systems because of the inherent variability and irregularity that are not present in, for example, computer-aided design/computer-aided manufacturing (CAD/CAM) structures (40). Another problem that is unique to medical applications is the recent emphasis on reducing health care costs, limiting the unfettered introduction of new technologies (40). At times, there is a problem with the integration of creative, novel algorithms to a community that is sometimes reluctant to modify procedures that have been established as effective, comfortable, and productive (3).

Image Processing

Most of the images from modern techniques, including digital radiography, magnetic resonance imaging, computed axial tomography, and ultrasound imaging, require similar procedures for the production of usable graphics displays. Often, the first step is the segmentation of the images, or the identification of different structures. Much work has been done on computer-based segmentation, and currently most approaches use basically automatic systems with varying amounts of manual editing of the results. There is a wide variety of segmentation algorithms, and many of them have been implemented in readily available software packages. In general, the imaging devices yield two-dimensional images, with $M \times N$ picture elements, or pixels, where M and N are the number of pixels in the horizontal and vertical dimensions, respectively. Typical image sizes are around 256×256. Image resolution, the number of pixels per length, is then determined by the size of the field of view of the device. The computational demands of image processing algorithms in-

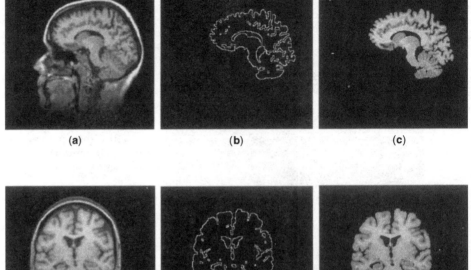

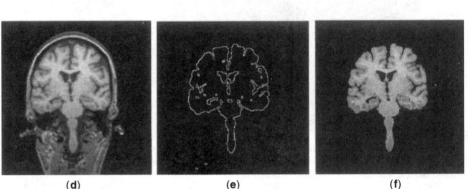

Figure 5. Sagittal (a) and coronal (b) magnetic resonance images of the head. The left two panels are the original images. The middle panels show the outline of the brain as detected by an automatic three dimensional segmentation algorithm. The right panels are the portions of the images that are within the detected brain outline. Reprinted from Ref. 42, with permission.

crease drastically with increasing resolution, since processing generally increases with the square of the resolution for two-dimensional images or as the cube of the resolution for volume analyses.

Each pixel is usually comprised of eight or more bits, each combination of which represents a gray level. The value of the gray level is determined by the method by which the image is formed. For example, in CT scans and radiographs, the intensity of the pixels reflect whether the photons have passed through bone or soft tissue. Magnetic resonance images detect differences in water content in organs and can provide different kinds of information from CT scans. In any case, segmentation algorithms must manually, automatically, or semiautomatically detect the transition from one level of intensity to another (41). Some are based on region growing methods, in which a seed is provided and a region of similar intensities is expanded until the level of the pixel intensities changes beyond a preset limit. Others are based on discontinuities in the pixel intensities, often computing the gradients of the intensities and changing tissue classifications based on the magnitude of the gradient. Figure 5 is a two-dimensional projection of a three-dimensional magnetic resonance image of a head. The brain tissue has been separated from non-brain using a three-dimensional seed growing algorithm (42). The automatic segmentation of images from the various medical imaging technologies remains an active area of research (41).

Modern imaging technologies routinely provide three-dimensional anatomic information (35), and the demands on image processing programs have increased correspondingly. The underlying principles are similar, but the computational constraints become more severe. In addition, even though storage and networking technologies are progressing quite rapidly, image processing software must often incorporate

data compression and decompression capabilities to accommodate extension of images from two to three dimensions, the increased image resolutions achievable with modern devices, and the need to transmit large datasets over networks.

Computer Graphics

Intimately related to the issues of image processing are the techniques by which medical and biological images are displayed with enough realism to achieve the intended results but with enough efficiency to be used in actual clinical situations. Algorithms and programs for accurately portraying anatomy and, to some extent, function have improved steadily, sometimes exceeding the ability of the hardware to meet the demands. Fortunately, the well-known advances in performance and cost of advanced graphics hardware, including general-purpose computers as well as special-purpose graphics processors, have provided the platforms necessary for implementation of state-of-the-art graphics techniques.

The display of two-dimensional images is, in principle, straightforward on a computer output screen with multiple colors or gray levels per pixel. The display programs provide an interface between the user, the image, and the graphics hardware and software of the computer so that one pixel of the image is translated to one pixel of the video screen. Complications arise when there is a mismatch between the image and the screen, so that image pixels must be removed or display pixels must be interpolated. A further complication for the developer of either two- or three-dimensional graphics software is the plethora of data file formats that exist (43). Fortunately, many public domain or proprietary software packages provide excellent format conversion tools, but some experimentation is frequently required to use them properly.

The development of methods for efficient and realistic rendering of three-dimensional images continues to be an area of ongoing research. Early work reduced anatomic structures to wire frame models (44), and that technique is still sometimes used for previewing and rapid manipulation on hardware that is not sufficiently powerful for handling full images in real time or near real time. Several methods require the identification of surfaces through image segmentation, as described above. The surfaces can be triangulated and displayed as essentially two-dimensional structures in three dimensions (45). After initial processing, this is a rather efficient display method, but much of the three-dimensional information is lost. Alternately, the image can be reduced to a series of volumetric structures that can be rendered by hardware specialized for their reproduction (46). One of the most realistic, but computationally expensive, three-dimensional rendering methods is ray tracing, in which an imaginary ray of light is sent through the structures and is attenuated by the opacity of the anatomic structures that it encounters along the way (47). Different effects can be emphasized by modifying the dynamic range of the pixels in the image—that is, by changing the relationship between the opacity of the image and the pixel value to be displayed on the screen.

Medical computer graphics are at their most useful when it is possible to superimpose images from more than one modality into a single display or to superimpose functional information acquired from biochemical, electrical, thermal, or other devices onto anatomical renderings. As an example of the former, images from positron emission tomography (PET) scans, which reflect metabolic activity, can be displayed on anatomy acquired by magnetic resonance imaging. The combination provides a powerful correlation between structure and function, but the technical challenges of registering images from two different devices or taken at different times are significant (48). An example of the combination of functional and anatomic data is the superposition of electrical activity, either intrinsic or externally applied, of the heart onto realistic cardiac anatomy. This kind of technique can provide new insights into the mechanisms and therapy of cardiac arrhythmias (49). Figure 6 is a sequence of still frames from a video showing the progression of a wavefront of electrical activation across a three-dimensional cardiac left ventricle after an unsuccessful defibrillation shock.

Computer graphics and image processing, along with advanced imaging technologies, are making a significant impact in medical knowledge and practice and have the potential for many more applications. A combination of traditional CAD/CAM visualization and advanced imaging can be used for effective assessment of quality of fit of orthopedic prostheses (50). Capabilities and functionality have increased dramatically with the advent of advanced graphics hardware and commercial software packages aimed at scientists and clinicians who are not graphics experts. Full realization of the benefits of these systems will require further advances in these areas, along with adaptation to the needs of clinicians and the constraints of the changing health care climate (51).

COMPUTER SIMULATIONS

Numerical and analytical simulations of physiological processes have intrigued investigators for many decades. The solution of inverse and forward problems in neurophysiology and electrocardiology was considered to be an important exer-

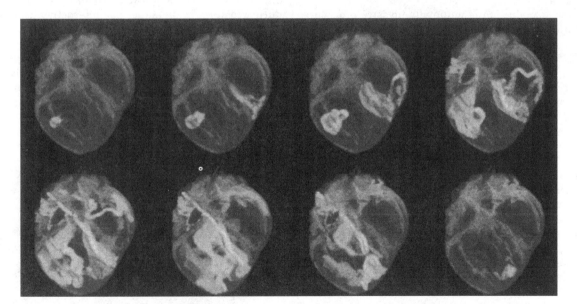

Figure 6. A composite of eight magnetic resonance and isochronal surface images from the second activation wavefront after an unsuccessful defibrillation shock. The electrical data were acquired from about 60 plunge needles with endocardial and epicardial electrodes inserted through the left and right ventricles of the heart of an experimental animal. Successive isochrones (left to right, top to bottom) are shown at 6 ms intervals. Visualization techniques that allow the superposition of function and anatomy are very helpful in understanding the relationships between variables and how they affect physiological mechanisms, and they can potentially lead to improved diagnosis and therapy. Reprinted from Ref. 49, with permission. Copyright CRC Press, Boca Raton, FL.

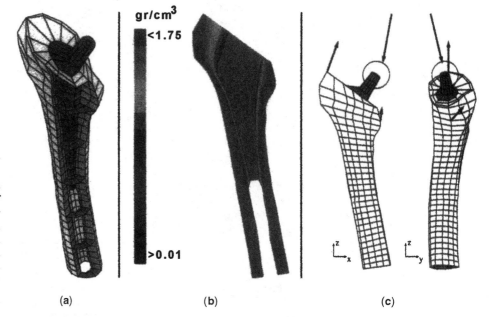

Figure 7. Display of a model of a femur used to study adaptation of bone in response to a stimulus. (a) A finite element mesh of proximal femur with stem of prosthesis numerically implemented for optimal fit. (b) Bone density distribution in the femur at initial implant. (c) Applied and muscle forces in the femoral head with prosthesis in place. Reprinted from Ref. 67, with permission. Copyright Gordon and Breach Publishers, Lausanne, Switzerland.

cise for basic scientific reasons as well as for possible clinical applications. The interpretation of the surface signals recorded in the context of approximate generators in the tissue provided a basis for relating physiology to pathologic and clinical abnormalities in the electrocardiogram (52). The advances in the use of the vectorcardiogram as a diagnostic tool is based on the approximation of the cardiac electrical generator as a current dipole (53). Similarly, patterns in the EEG have been modeled as surface reflections of underlying electrical sources (54).

The recent introduction of minimally invasive procedures for the treatment of diseases, including laparoscopic and thoracoscopic surgical procedures and radio-frequency ablation for cardiac arrhythmias, has intensified the interest of forward and inverse problems as a research area. For example, for radio-frequency ablation of ventricular tachycardia, it would be most helpful to localize, at least approximately, the origins of abnormal cardiac electrical activity from sensors either on the body surface or mounted on a catheter in the blood pool (55–57).

Computers have been used to model tissue at cellular and fiber levels of resolution. It is possible to simulate the propagation of electrical activity either with finite-state automata models (58) or by solving the differential equations that govern the current flow through the cell membrane (59–62). The electrophysiology of other organ systems has also been modeled effectively (63,64).

In addition to electrophysiological simulations, mechanical models have been applied to increase our understanding of the mechanical properties of soft tissue (65) and bone (66). Such models (Fig. 7) are important in simulating surgical procedures and implants (67) for training, planning, and evaluation of surgery.

VIRTUAL MEDICINE

Many of the computing techniques applied to medicine and medical sciences come together in the development of applications of virtual reality to medical practice (68,69). Virtual reality has been applied to surgery planning (70), physical medicine and rehabilitation (71), parkinsonism (72), and psychological disorders (73,74).

Computers have been used in a great many ways to assist in surgical procedures (70). Surgeons can be trained in surgical techniques by using advanced computer graphics and virtual reality methods (75,76); similar techniques can be used for surgical planning (77–79) and for improving the safety and efficacy of the surgical procedure. Computers are used during complex brain surgery as interactive tools for guiding and measuring the progress of the procedure, with the hope that resection of lesions could be performed with less damage to surrounding tissue (80). It is possible to use high-resolution graphics to traverse internal organs virtually, yielding much of the same information that is available from standard endoscopic techniques, as shown in the image in Fig. 8 acquired at the Mayo Clinic (68).

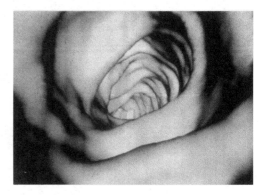

Figure 8. Virtual colonoscopy, with an internal view of the transverse colon. The image was acquired by a helical CT scan, segmented, and reconstructed. Virtual procedures can replace or augment actual endoscopic examinations, reducing or eliminating the attendant risk and discomfort. The image was acquired at the Mayo Clinic. Reprinted from Ref. 68, with permission. Copyright 1998, IEEE.

Computer technology has also made it possible for medical experts to use their knowledge at remote locations. This allows state-of-the-art medical diagnosis and treatment at sites where advanced technology would not normally be available (36). At another level, similar technologies have provided the necessary tools for minimally invasive surgery and microsurgery, using computer graphics and simulated tactile responses to extend the capabilities of the surgeon and to provide simulations for realistic rehearsals of the surgical experience (81,82).

The most sophisticated applications of computer graphics to remote and minimally invasive surgery have depended on the availability of high-resolution, high-quality graphical representations of the anatomy under treatment, either on an individualized basis (83) or as a global atlas of human anatomy (84).

REGULATION, RELIABILITY, AND PRIVACY

Computer software that controls devices used in the diagnosis and treatment of disease is of great interest to a regulatory agency of the US government, namely, the Center for Devices and Radiological Health (CDRH) of the Food and Drug Administration (FDA). This agency maintains a World Wide Web site (http://www.fda.gov/cdrh/swpolpg.html) which provides guidance to device manufacturers in understanding regulatory policy enacted by the Federal Food, Drug and Cosmetic Act through a review of the FDA Software Policy Workshop. The CDRH also sponsors biannual workshops on software policy, the proceedings of which are included at this site. The Food, Drug and Cosmetic Act defines a medical device as any "instrument, apparatus, implement, machine, contrivance, implant, in vitro reagent, or any other similar or related article, including any component, part, or accessory, which is . . . intended for use in the diagnosis of disease or other conditions, or in the cure, mitigation, treatment, or prevention of disease . . . or intended to affect the structure or any function of the body . . ." Software that is excepted from this includes general-purpose programs, such as spreadsheets, which are not intended solely for medical use, and a few other categories.

Independent of the regulatory issues, the design of medical software and systems requires the highest level of reliability and accessibility. The difficulty of developing robust computer programs and documenting their correctness are well known (85,86). The field of software engineering has been instrumental in providing tools for developing medical software that has the highest possible level of accuracy and reliability (87). Device manufacturers have recognized the importance and usefulness of software engineering principles in the design and implementation of control programs for their products.

Finally, ethical use of computer databases and the Internet for the transfer of medical information imposes a need for methods for strict privacy and security in data transmission. The rapid introduction of new technologies requires that the issues related to the sharing of patient information over openly accessible channels be continually evaluated and improved (88).

CARDIAC MAPPING

An area of medical practice and research that has benefited greatly by the use of computers and many computer-based

techniques is cardiac mapping—that is, the use of technology to determine the path of electrical activity in the heart during normal and abnormal cardiac rhythms and to measure the effects of external stimuli and shocks that are applied therapeutically or as test perturbations. The activation sequence and response to interventions can be measured electrically (89,90) or optically (91,92). Electrical mapping is widely applied clinically to guide catheter-based or surgical interventions, and both technologies are important in investigating the mechanisms of electrophysiological phenomena. Both approaches are computer-intensive, requiring real time data acquisition, signal analysis, image acquisition, statistical analysis, data storage and manipulation, and visualization.

In electrical mapping, the first step is the transduction of the ionic currents in the cardiac tissue to electrical signals that can be input to the data acquisition system. The construction of electrodes is crucial to ensure that the appropriate variables are measured (93). Signals in optical mapping are generated by the application of fluorescent dyes that are sensitive to the electrical potential across the membranes of the cardiac cells. In either case, the waveforms are input to analog-to-digital converters.

In one implementation of an electrical cardiac mapping system (23), all of the parameters of the analog front end, including gains, frequency settings, and input range, are controlled by a series of microprocessors (94). The front-end processors assemble the data from 528 independent input channels into a data stream and send it to a data bus for recording on long- or short-term recording devices. Because of the complexity of the analog processing, the user interface has been designed to provide a great deal of direction and intuitive interaction with the investigators. Other microprocessors control stimulators (95) and associated investigational tools, such as defibrillators and waveform analyzers.

After the raw waveforms are acquired, digital signal processing algorithms are applied to them for several purposes. In some cases, the electrograms are analyzed to detect and locate in time local electrical activations, those events that represent the passage of an electrical wavefront in proximity to the electrode (96,97). The local activation times form the basis for other analysis programs. In other cases, the potential generated in the cardiac tissue from an external shock is measured in all electrodes (98) in order to understand the relationship between the electrical potential and gradient (99) distributions in the myocardium and the efficacy of the therapy. The results of a clinical mapping study are shown in Fig. 9. Figure 9(a) is a sequence of cardiac electrograms recorded using a commercial mapping system in a patient undergoing an electrophysiology study. The electrograms were recorded by a "halo" catheter placed in the right atrium, and they demonstrate the progression of an atrial flutter wavefront in a circular pattern. The activation is reentrant—that is, self-sustaining—and continually traverses the same anatomic pathway. Figure 9(b) is a plot of local activation times that were defined by the intrinsic deflection of the electrogram. Computer displays such as these guide the application of radio-frequency energy applied to the heart for ablation of arrhythmogenic tissue. Other useful parameters that emphasize other aspects of the electrophysiology can be derived from data such as these (100–103).

It is often necessary or, at least, helpful to know (1) the three-dimensional anatomy of the heart in which measure-

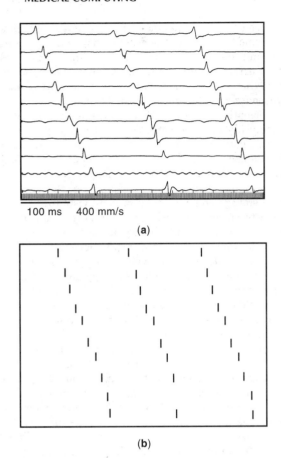

100 ms   400 mm/s

(a)

(b)

Figure 9. Results of a clinical electrical cardiac mapping study in a patient undergoing ablation of atrial flutter. (a) Electrograms recorded during the arrhythmia from a catheter inserted into the right atrium in a loop configuration. (b) Activations derived from the intrinsic deflections in the electrograms shown in panel a. The continuous nature of the activity demonstrates the reentrant mechanism around anatomical obstacles in the right atrium. Ablation can eliminate conductivity in part of the reentrant pathway, curing the atrial flutter.

ments are made and (2) the location of the electrodes used to make the measurements. These variables can then be used for further computations or to make the visualization of the results more compelling and useful (104). Imaging techniques as described above can be used for this purpose, allowing the application of standard image processing packages for better understanding of the electrophysiology (105). Image processing algorithms can also be used to improve our knowledge of the underlying pathology and its relation to abnormalities in electrical phenomena (35).

A traditional way of viewing activation sequences or other variables in the heart is through contour maps—that is, lines of equal values of activation time, potential, or other measured variable. The approach depends on whether the array of recording electrodes is two- or three-dimensional and whether the array is in a regular pattern or is irregularly spaced over the tissue. The variable of interest is typically interpolated over the region in which the measurements were made for more pleasing visual effects (106). Figure 10 is a simple isochronal map of the activation sequence beneath a rectangular array of electrodes on the outer, or epicardial, surface of the right ventricle of an experimental animal. Even

though custom software is often used to produce contour maps (107), most commercial visualization packages have contour generation routines that are efficient and easy to use, especially for regular geometries, as indicated in Fig. 10.

With adequate processing, more complex displays can be produced. As discussed above, it can be most helpful to combine functional electrical information with anatomic data (49). Another approach is the animation of the activation sequence (108). By animating the color-coded values of time since last activation or potential or derivative of potential in each electrode as a function of time, the activation sequence can be effectively followed without explicitly defining times of local activation (Fig. 11) (109). If the electrodes are sufficiently close together, the displays can be produced without interpolation. Thus, animation has the potential of removing two important sources of ambiguity in cardiac mapping (110).

Traditionally, the interpretation of isochronal maps and activation sequences has been subjective and descriptive, with little basis for statistical comparisons between episodes of cardiac arrhythmias. Computer programs and algorithms have recently been developed which describe in quantitative terms the characteristics of supraventricular and ventricular flutter, tachycardia, and fibrillation. Some of them are aimed at inferring the level of organization of the arrhythmia, especially atrial and ventricular fibrillation (25,111). Others are designed to extract and identify wavefronts objectively as they course across the myocardial tissue and to quantitate their characteristics so that the effect of different experimental conditions and interventions can be compared in a rigorous and reproducible manner (112–114).

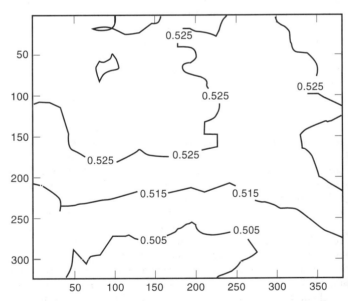

Figure 10. Isochronal map from activation times measured with a 21 × 24 rectangular array of electrodes placed on the right ventricular epicardium of a dog. The contours were interpolated to a 84 × 96 array, with linear interpolation. The lines represent times of equal activation, and they are spaced at 10 ms intervals. The labels of the isochrones are in seconds from an arbitrary reference time. The map was taken from a sinus beat and shows activation spreading from the apical region, at the bottom of the plot, to the base of the right ventricle. The loop in one of the isochrones demonstrates one of the problems with interpolation and isochrone production in data with noise.

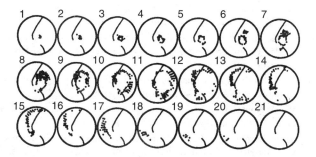

Figure 11. A sequence of frames from an animation of the activation sequence of the several cardiac cycles after an successful defibrillation shock. The circle represents an apical view of the heart, with the apex at the center and the base around the periphery. The data were recorded from a sock containing 510 electrodes which was pulled over the ventricular epicardium. The left anterior descending coronary artery is the line from the top of the circle to near the center and the posterior descending coronary artery is the line at the bottom. Each dot represents a time at which the electrogram recorded by the corresponding electrode is "active," meaning that the absolute value of the derivative of the electrogram exceeded a preselected threshold. The earliest postshock activation is in the apical region of the left ventricle. A secondary early site appears in panel 7. The two wavefronts merge, and then they spread to the right ventricle. Even though the data were sampled at 2 kHz, the sequential displays are at 5 ms intervals. Animation of electrograms in this manner can provide information about the activation sequence without explicitly defining times of local activation in each electrogram. Reprinted from 109, with permission.

CONCLUSION

Computers are used in almost every aspect of clinical medicine and biomedical research. They are indispensable in advanced devices and instrumentation. They are widely used for the collection and analysis of demographic and clinical data which provide a basis for the improved understanding of the causes and epidemiologies of disease. They can be very effectively used for the training and accreditation of physicians and other health care providers. While obviously no substitute for human clinical and scientific judgement, computers have assumed a critical role as facilitators of diagnostic and therapeutic procedures. As the inevitable progress in computer software and hardware occurs, medical professionals will become more dependent on them. There will be continuing improvement in our understanding of methods for increased reliability and safety of computer-based devices and equipment, and regulatory agencies will develop procedures for evaluating these resources routinely and objectively. Advanced imaging technologies, higher performance graphics hardware and software, and new surgical techniques will expand the use of microsurgery and remotely applied surgical and invasive procedures. These prospects will depend on investigators in computer science, biomedical and software engineering, clinical practice, and physiology, but the computer has the potential to be a positive force in improving health care delivery while decreasing the financial burden of the health care system on society.

ACKNOWLEDGMENTS

This work was supported in part by research grant HL-33637 from the National Heart, Lung, and Blood Institute, National Institutes of Health, Bethesda, Maryland, and National Science Foundation Engineering Research Center Grant CDR-8622201.

BIBLIOGRAPHY

1. P. Zickler, Informatics: Transforming raw data into real medical information, *Biomed. Instrum. Technol.*, **32**: 111–115, 1998.

2. J. Gosbee and E. Ritchie, Human-computer interaction and medical software development, *Interactions*, **4**: 13–18, 1997.

3. J. G. Anderson, Clearing the way for physicians's use of clinical information systems, *Commun. ACM*, **40**: 83–90, 1997.

4. A. Kamb et al., Software trapping: A strategy for finding genes in large genomic regions, *Comput. Biomed. Res.*, **28**: 140–153, 1995.

5. B. R. Schatz, Information retrieval in digital libraries: Bringing search to the net, *Science*, **275**: 327–334, 1997.

6. S. A. Levin et al., Mathematical and computational challenges in population biology and ecosystems science, *Science*, **257**: 334–343, 1997.

7. J. N. Weinstein et al., An information-intensive approach to the molecular pharmacology of cancer, *Science*, **275**: 343–349, 1997.

8. W. H. Detmer and E. H. Shortliffe, Using the Internet to improve knowledge diffusion in medicine, *Commun. ACM*, **40**: 101–108, 1997.

9. D. G. Kilman and D. W. Forslund, An international collaboratory based on virtual patient records, *Commun. ACM*, **40**: 111–117, 1997.

10. J. F. Brinkley et al., The digital anatomist information system and its use in the generation and delivery of web-based anatomy atlases, *Comput. Biomed. Res.*, **30**: 472–503, 1997.

11. E. H. Shortliffe et al., Computer-based consultations in clinical therapeutics: Explanation and rule acquisition capabilities of the MYCIN system, *Comput. Biomed. Res.*, **8**: 303–320, 1975.

12. W. G. Baxt, Use of an artificial neural network for the diagnosis of myocardial infarction, *Ann. Intern. Med.*, **115**: 843–848, 1991.

13. M. A. Leon and F. L. Lorini, Ventilation mode recognition using artificial neural networks, *Comput. Biomed. Res.*, **30**: 373–378, 1997.

14. C. A. Kulikowski, History and Development of Artificial Intelligence Methods for Medical Decision Making, in J. D. Bronzino (ed.), *The Biomedical Engineering Handbook*, Boca Raton, FL: CRC Press, 1995, pp. 2681–2698.

15. D. Kalmanson and H. F. Stegall, Cardiovascular investigations and fuzzy sets theory, *Amer. J. Cardiol.*, **35**: 80–84, 1975.

16. J. Piater et al., Fuzzy sets for feature identification in biomedical signals with self-assessment of reliability: An adaptable algorithm modeling human procedure in BAEP analysis, *Comput. Biomed. Res.*, **28**: 335–353, 1995.

17. J.-S. Lin, K.-S. Cheng, and C.-W. Mao, Segmentation of multispectral magnetic resonance image using penalized fuzzy competitive learning network, *Comput. Biomed. Res.*, **29**: 314–326, 1996.

18. J. Macy, Jr., Analog–Digital Conversion Systems, in R. W. Stacy and B. D. Waxman (eds.), *Computers in Biomedical Research*, New York: Academic Press, 1965, pp. 3–34.

19. R. E. Ideker et al., Evaluation of a QRS scoring system for estimating myocardial infarct size: II. Correlation with quantitative anatomic findings for anterior infarcts, *Amer. J. Cardiol.*, **49**: 1604–1614, 1982.

20. G. S. Wagner et al., Evaluation of a QRS scoring system for estimating myocardial infarct size: I. Specificity and observer agreement, *Circulation*, **65**: 342–347, 1982.

21. R. Restak, *The Brain,* New York: Bantam Books, 1984.

22. T. Lewis and M. A. Rothschild, The excitatory process in the dog's heart. Part II. The ventricles, *Philos. Trans. R. Soc. London,* **206**: 181, 1915.

23. P. D. Wolf et al., Design for a 512 Channel Cardiac Mapping System, in D. C. Mikulecky and A. M. Clarke (eds.), *Biomedical Engineering: Opening New Doors,* New York: New York Univ. Press, 1990, pp. 5–13.

24. P. V. Bayly et al., A quantitative measurement of spatial order in ventricular fibrillation, *J. Cardiovasc. Electrophysiol.,* **4**: 533–546, 1993.

25. K. M. Ropella et al., The coherence spectrum. A quantitative discriminator of fibrillatory and nonfibrillatory cardiac rhythms, *Circulation,* **80**: 112–119, 1989.

26. A. Riehle et al., Spike synchronization and rate modulation differentially involved in motor cortical function, *Science,* **278**: 1950, 1997.

27. J. D. Z. Chen, B. D. Schirmer, and R. W. McCallum, Measurement of electrical activity of the human small intestine using surface electrodes, *IEEE Trans. Biomed. Eng.,* **40**: 598–602, 1993.

28. C. R. R. Gallegos and C. H. Fry, Alterations to the electrophysiology of isolated human detrusor smooth muscle cells in bladder disease, *J. Urol.,* **151**: 754–758, 1994.

29. D. Devedeux et al., Uterine electromyography: A critical review, *Amer. J. Obstet. Gynecol.,* **169**: 1636–1653, 1993.

30. M. E. Cain et al., Fast-Fourier transform analysis of signal-averaged electrocardiograms for identification of patients prone to sustained ventricular tachycardia, *Circulation,* **69**: 711–720, 1984.

31. A. Nakata et al., Spectral analysis of heart rate, arterial pressure, and muscle sympathetic nerve activity in normal humans, *Amer. J. Physiol.,* **274** (*Heart Circ. Physiol.,* **43**): H1211–H1217, 1998.

32. M. Akay, Wavelet applications in medicine, *IEEE Spectrum,* **34** (5): 50–56, 1997.

33. D. A. Brumwell, K. Kroll, and M. H. Lehmann, The Amplifier: Sensing the Depolarization, in M. W. Kroll and M. H. Lehmann (eds.), *Implantable Cardioverter Defibrillator Therapy: The Engineering–Clinical Interface,* Norwell, MA: Kluwer, 1996, pp. 275–302.

34. S. M. Bach, Jr., M. H. Lehmann, and M. W. Kroll, Tachyarrhythmia Detection, in M. W. Kroll and M. H. Lehmann (eds.), *Implantable Cardioverter Defibrillator Therapy: The Engineering–Clinical Interface,* Norwell, MA: Kluwer, 1996, pp. 303–323.

35. J. C. M. Hsu et al., Magnetic resonance imaging of chronic myocardial infarcts in formalin-fixed human autopsy hearts, *Circulation,* **89**: 2133–2140, 1994.

36. J. W. Hill and J. F. Jensen, Telepresence technology in medicine: Principles and applications, *Proc. IEEE,* **86**: 569–580, 1998.

37. P. Virot et al., Comparison of 14 methods for analysing left ventricle segmental kinetics by cineangiography, *Arch. Mal Coeur Vaiss.,* **77**: 433–441, 1984.

38. B. G. Brown et al., Quantitative coronary arteriography: Estimation of dimensions, hemodynamic resistance, and atheroma mass of coronary artery lesions using the arteriogram and digital computation, *Circulation,* **55**: 329–337, 1977.

39. C. F. Starmer and W. M. Smith, Problems in acquisition and representation of coronary arterial trees, *Computer,* **8** (7): 36–41, 1975.

40. M. L. Rhodes, Computer graphics and medicine: A complex partnership, *IEEE Comput. Graphics Appl.,* **17**: 22–28, 1997.

41. R. C. Gonzalez and R. E. Woods, *Digital Image Processing,* Reading, MA: Addison-Wesley, 1992.

42. R. K. Justice et al., Medical image segmentation using 3-D seeded region growing, *Proc. SPIE Med. Imag.,* Newport Beach, CA, 1997, pp. 900–910.

43. D. C. Kay and J. R. Levine, *Graphics File Formats.* Blue Ridge Summit, PA: Windcrest/McGraw-Hill, 1992.

44. W. M. Newman and R. F. Sproull, *Principles of Interactive Computer Graphics,* New York: McGraw-Hill, 1979.

45. W. E. Lorensen and H. E. Cline, Marching cubes: A high resolution 3D surface construction algorithm, *Comput. Graphics,* **21**: 163–169, 1987.

46. E. V. Simpson et al., Three-dimensional visualization of electrical variables in the ventricular wall of the heart, *Proc. 1st Conf. Vis. Biomed. Comput.,* Atlanta, GA, 1990, pp. 190–194.

47. K. S. Klimaszewski and T. W. Sederberg, Faster ray tracing using adaptive grids, *IEEE Comput. Graphics Appl.,* **17**: 42–51, 1997.

48. K. R. Castleman, *Digital Image Processing,* Englewood Cliffs, NJ: Prentice-Hall, 1979.

49. T. C. Palmer et al., Visualization of bioelectric phenomena, *CRC Crit. Rev. Biomed. Eng.,* **20**: 355–372, 1992.

50. M. W. Vannier et al., Visualization of prosthesis fit in lower-limb amputees, *IEEE Comput. Graphics Appl.,* **17**: 16–29, 1997.

51. D. P. Mahoney, The art and science of medical visualization, *Comput. Graphics World,* **19**: 25–32, 1996.

52. L. G. Horan et al., On the possibility of directly relating the pattern of ventricular surface activation to the pattern of body surface potential distribution, *IEEE Trans. Biomed. Eng.,* **34**: 173–179, 1987.

53. L. G. Horan and N. C. Flowers, The Relationship Between the Vectorcardiogram and Actual Dipole Moment, in C. V. Nelson and D. B. Geselowitz (eds.), *The Theoretical Basis of Electrocardiology,* London: Oxford Univ. Press, 1976, pp. 397–412.

54. R. J. MacGregor and E. R. Lewis, *Neural Modeling: Electrical Signal Processing in the Nervous System,* New York: Plenum, 1977.

55. B. He and R. J. Cohen, Body surface Laplacian ECG mapping, *IEEE Trans. Biomed. Eng.,* **39**: 1179–1191, 1992.

56. G. Huiskamp and F. Greensite, A new method for myocardial activation imaging, *IEEE Trans. Biomed. Eng.,* **44**: 433–446, 1997.

57. D. S. Khoury and Y. Rudy, A model study of volume conductor effects on endocardial and intracavity potentials, *Circ. Res.,* **71**: 511–525, 1992.

58. K. D. Bollacker et al., A cellular automata three-dimensional model of ventricular cardiac activation, *Proc. Annu. Int. Conf. IEEE Eng. Med. Biol. Soc.,* Piscataway, NJ, 1991, pp. 627–628.

59. G. W. Beeler, Jr. and H. Reuter, Reconstruction of the action potential of ventricular myocardial fibers, *J. Physiol. (London),* **268**: 177–210, 1977.

60. A. L. Hodgkin and A. F. Huxley, A quantitative description of membrane current and its application to conduction and excitation in nerve, *J. Physiol. (London),* **117**: 500–544, 1952.

61. C.-H. Luo and Y. Rudy, A dynamic model of the cardiac ventricular action potential. II. After depolarizations, triggered activity, and potentiation, *Circ. Res.,* **74**: 1097–1113, 1994.

62. C.-H. Luo and Y. Rudy, A dynamic model of the cardiac ventricular action potential. I. Simulations of ionic currents and concentration changes, *Circ. Res.,* **74**: 1071–1096, 1994.

63. E. E. Daniel et al., Relaxation oscillator and core conductor models are needed for understanding of GI electrical activities, *Amer. J. Physiol.,* **266** (Gastrointest. Liver Physiol. 29): G339–G349, 1994.

64. B. O. Familoni, T. Abell, and K. L. Bowes, A model of gastric electrical activity in health and disease, *IEEE Trans. Biomed. Eng.,* **42**: 647–657, 1995.

65. E. S. Almeida and R. L. Spilker, Mixed and penalty finite element models for the nonlinear behavior of biphasic soft tissues in finite deformation: Part I. Alternate formulations, *Comput. Methods Biomech. Biomed. Eng.,* **1**: 25–46, 1997.

66. B. R. McCreadie and S. J. Hollister, Strain concentrations surrounding an ellipsoid model of lacunae and osteocytes, *Comput. Methods Biomech. Biomed. Eng.,* **1**: 61–68, 1997.

67. A. Terrier et al., Adaptation models of anisotropic bone, *Comput. Methods Biomech. Biomed. Eng.,* **1**: 47–49, 1997.

68. R. M. Satava and S. B. Jones, Current and future applications of virtual reality for medicine, *Proc. IEEE,* **86**: 484–489, 1998.

69. W. M. Smith, Scanning the technology: Engineering and medical science chart fantastic voyage, *Proc. IEEE,* **86**: 474–478, 1998.

70. M. L. Rhodes and D. D. Robertson, Computers in surgery and therapeutic procedures, *Computer,* **29** (1): 23, 1996.

71. W. J. Greenleaf, Applying VR to physical medicine and rehabilitation, *Commun. ACM,* **40**: 43–46, 1997.

72. S. Weghorst, Augmented reality and Parkinson's disease, *Commun. ACM,* **40**: 47–48, 1997.

73. G. Riva, L. Melis, and M. Bolzoni, Treating body-image disturbances, *Commun. ACM,* **40**: 69–71, 1997.

74. D. Strickland et al., Overcoming phobias by virtual exposure, *Commun. ACM,* **40**: 35–39, 1997.

75. S. L. Dawson and J. A. Kaufman, The imperative for medical simulation, *Proc. IEEE,* **86**: 479–483, 1998.

76. K. H. Höhne et al., A 'virtual body' model for surgical education and rehearsal, *Computer,* **29** (1): 25–31, 1996.

77. M. Bro-Nielsen, Finite element modeling in surgery simulation, *Proc. IEEE,* **86**: 503, 1998.

78. E. K. Fishman et al., Surgical planning for liver resection, *Computer,* **29** (1): 64–72, 1996.

79. R. A. Robb, D. P. Hanon, and J. J. Camp, Computer-aided surgery planning and rehearsal at Mayo Clinic, *Computer,* **29** (1): 39–47, 1996.

80. L. Adams et al., An optical navigator for brain surgery, *Computer,* **29** (1): 48–54, 1996.

81. E. Chen and B. Marcus, Force feedback for surgical simulation, *Proc. IEEE,* **86**: 524–530, 1998.

82. H. Delingette, Toward realistic soft-tissue modeling in medical simulation, *Proc. IEEE,* **86**: 512–523, 1998.

83. G. E. Christensen et al., Individualizing neuroanatomical atlases using a massively parallel computer, *Computer,* **29** (1): 32–38, 1996.

84. M. J. Ackerman, The visible human project, *Proc. IEEE,* **86**: 504–511, 1998.

85. W. W. Gibbs, Software's chronic crisis, *Sci. Amer.,* **271** (3): 86–95, 1994.

86. B. Littlewood and L. Strigini, The risks of software, *Sci. Amer.,* **267** (5): 62–75, 1992.

87. W.-T. Tsai, R. Mojdehbakhsh, and S. Rayadurgam, Capturing safety-critical medical requirements, *Computer,* **31**: 40–42, 1998.

88. T. C. Rindfleisch, Privacy, information technology, and health care, *Commun. ACM,* **40**: 93–100, 1997.

89. R. E. Ideker et al., Simultaneous multichannel cardiac mapping systems, *Pacing Clin. Electrophysiol.,* **10**: 281–292, 1987.

90. F. X. Witkowski and P. B. Corr, An automated simultaneous transmural cardiac mapping system, *Amer. J. Physiol.,* **247**: H661–H668, 1984.

91. S. B. Knisley et al., Optical measurements of transmembrane potential changes during electric field stimulation of ventricular cells, *Circ. Res.,* **72**: 255–270, 1993.

92. G. Salama, R. Lombardi, and J. Elson, Maps of optical action potentials and NADH fluorescence in intact working hearts, *Amer. J. Physiol.,* **252**: H384–H394, 1987.

93. M. R. Neuman, Biopotential electrodes, in J. G. Webster (ed.), *Medical Instrumentation Application and Design,* 3rd ed., New York: Wiley, 1998, pp. 183–232.

94. P. D. Wolf et al., A 528 channel system for the acquisition and display of defibrillation and electrocardiographic potentials, *Proc. Comput. Cardiol.,* Los Alamitos, CA, 1993, pp. 125–128.

95. D. L. Rollins et al., A programmable cardiac stimulator, *Proc Comput. Cardiol.,* Los Alamitos, CA, 1992, pp. 507–510.

96. K. P. Anderson et al., Determination of local myocardial electrical activation for activation sequence mapping: A statistical approach, *Circ. Res.,* **69**: 898–917, 1991.

97. E. V. Simpson et al., Evaluation of an automatic cardiac activation detector for bipolar electrograms, *Med. Biol. Eng. Comput.,* **31**: 118–128, 1993.

98. A. S. L. Tang et al., Measurement of defibrillation shock potential distributions and activation sequences of the heart in three-dimensions, *Proc. IEEE,* **76**: 1176–1186, 1988.

99. W. Krassowska et al., Finite element approximation of potential gradient in cardiac muscle undergoing stimulation, *Math. Comput. Modelling,* **11**: 801–806, 1988.

100. P. V. Bayly et al., Estimation of conduction velocity vector fields from 504-channel epicardial mapping data, *Proc. Comput. Cardiol.,* Indianapolis, IN, 1996, pp. 133–140.

101. A. H. Kadish et al., Vector mapping of myocardial activation, *Circulation,* **74**: 603–615, 1986.

102. D. S. Rosenbaum, B. He, and R. J. Cohen, New approaches for evaluating cardiac electrical activity: Repolarization alternans and body surface laplacian imaging, in D. P. Zipes and J. Jalife (eds.), *Cardiac Electrophysiology: From Cell to Bedside,* Philadelphia: Saunders, 1995, pp. 1187–1198.

103. F. X. Witkowski et al., Significance of inwardly directed transmembrane current in determination of local myocardial electrical activation during ventricular fibrillation, *Circ. Res.,* **74**: 507–524, 1994.

104. E. V. Simpson, T. C. Palmer, and W. M. Smith, Visualization in cardiac mapping of ventricular fibrillation and defibrillation, *Proc. Comput. Cardiol.,* Los Alamitos, CA, 1992, pp. 339–342.

105. C. Laxer et al., An Interactive Graphics System for Locating Plunge Electrodes in Cardiac MRI Images, in Y. Kim (ed.), *Image Capture, Formatting and Display,* Soc. Photo-Optical Instrum. Eng., 1991, pp. 190–195.

106. E. V. Simpson et al., Discrete smooth interpolation as an aid to visualizing electrical variables in the heart wall, *Proc. Comput. Cardiol.,* Venice, Italy, 1991, pp. 409–412.

107. F. R. Bartram, R. E. Ideker, and W. M. Smith, A system for the parametric description of the ventricular surface of the heart, *Comput. Biomed. Res.,* **14**: 533–541, 1981.

108. C. Laxer et al., The use of computer animation of mapped cardiac potentials in studying electrical conduction properties of arrhythmias, *Proc. Comput. Cardiol.,* Los Alamitos, CA, 1991, pp. 23–26.

109. M. Usui et al., Epicardial shock mapping following monophasic and biphaic shocks of equal voltage with an endocardial lead system, *J. Cardiovasc. Electrophysiol.,* **7**: 322–334, 1996.

110. E. J. Berbari et al., Ambiguities of epicardial mapping, *J. Electrocardiol.,* **24** (Suppl.): 16–20, 1991.

111. P. V. Bayly et al., Spatial organization, predictability, and determinism in ventricular fibrillation, *Chaos,* **8**: 103–115, 1998.

112. J. Rogers et al., Recurrent wavefront morphologies: A method for quantifying the complexity of epicardial activation patterns, *Ann. Biomed. Eng.,* **25**: 761–768, 1997.

113. J. M. Rogers et al., Quantitative characteristics of reentrant pathways during ventricular fibrillation, *Ann. Biomed. Eng.,* **25**: S-62, 1997.

114. J. M. Rogers et al., A quantitative framework for analyzing epicardial activation patterns during ventricular fibrillation, *Ann. Biomed. Eng.,* **25**: 749–760, 1997.

WILLIAM M. SMITH
The University of Alabama at
Birmingham

MEDICAL EXPERT SYSTEMS

The field of medical informatics (also termed health informatics) concerns application of information science and information technology to health care, clinical care, education and biomedical research. Most countries have national societies in this area, and some 40 of them are organized in the International Medical Informatics Association (IMIA). Artificial intelligence (AI) methods specifically refer to the application of computer-based programs simulating human experts. Recent developments in medical informatics benefit from the availability of powerful personal computers (workstations), advanced information processing techniques such as the artificial neural network (ANN), and increased acceptance by the clinical community. The latter seemingly trivial factor should not be underestimated by engineers. In fact, wider acceptance is only partly due to improved user interfaces, but largely by the gradual recognition that computers form a useful tool in the doctor's office. In 1995, still fewer than 1% of the family practitioners in the United States use a computerized patient record, but enthusiasm is increasing. It has been shown that patient satisfaction does not decrease when the computer is employed in the physician's examination room (1).

Information, in general, requires that locally available knowledge can be communicated. Indeed, facts are only meaningful if they can be uniquely described and successfully transmitted from one location or person to another. Trivial examples from everyday life concern the combination of coding of messages by writing news reports and the distribution of newspapers, and the formulation of integrated weather reports and subsequent radio broadcasting. Similar lines of communication apply to medical informatics, although the implementation of advanced techniques started not earlier than around 1975. To understand this delay that surprisingly impeded an important issue such as medical care, and also to appreciate the potential progress that can be realized, it is essential to indicate the circumstances that make health care differ from other areas in the natural sciences. First, the primary information stems from humans (or animals for the sake of veterinary informatics) inflicted with shortcomings regarding their functioning. Second, in medicine it is difficult to define what is "normal." Normality does not refer to a single numerical value, but rather to a certain range defined by reference values. Therefore it seems almost impossible to define a deviating process to begin with, then to assess the severity of any abnormality, to judge whether the defect is dangerous for your health, next to evaluate the impact of therapeutic intervention, and finally to determine the prognosis for each individual. Clearly, a vast number of communication steps are to be taken, thus limiting the efficiency of the process. Moreover, existing knowledge on a particular disease may not be immediately available to any physician, because it is an impossible task to scrutinize weekly or monthly all medical journals published anywhere in the world. Further limitations regarding communication of medical knowledge refer to clinical terminology and classification of health data. With the knowledge that medical informatics deals with enormous amounts of data, often located at widely distributed locations, it is not surprising that computer support in this area will be of great impact on efficiency, accuracy, and advancement of health care. Many projects, often concerted international efforts, address the issue of how to handle an ever-increasing amount of medical information. Universal classifications have been designed and regularly refined, while other approaches aim not only to collect, but also to structure and disclose this exponentially growing body of medical information. The following sections will be devoted to a more general description of various topics of relevance to the field of medical informatics and may be of interest for the average reader.

PATIENT DESCRIPTION AND THE ELECTRONIC PATIENT FILE

The basic goals of the use of computers in medicine concern communication and clinically relevant combination of data. This electronic medium is expected to enhance and facilitate such interaction and data interpretation. Ideally, every citizen should carry a patient data card, which in an emergency case presents valuable information to the physician. The Medical Records Institute is an instrumental force in the movement toward such an electronic patient record. Locally, most hospitals have developed an information system [hospital information system (HIS)]. A patient card may include information on medical history, familial traits, use of prescription drugs, allergies, lifestyle (including sports activities and use of alcohol and/or tobacco), availability of x-ray pictures, electrocardiogram recording, and blood chemistry (2). Obviously, these initiatives involve delicate ethical issues, as well.

MEDICAL TERMINOLOGY AND EPONYMS

Knowledge obviously can be represented by symbols, words, definitions, and their interrelations. Knowledge may be expressed by spoken or written words, flow charts, (mathematical) equations, tables, or figures. Aspects of language and text interpretation are central issues in AI. A powerful abstraction of language also provides a powerful representation of knowledge. Various strategies have been explored: semantic networks offer a versatile tool for representing knowledge of virtually any type that can be captured in words by employing nodes (representing things) and links (referring to meaningful relationships), thus expressing causal, temporal, taxonomic, and associational connections. Other approaches (such as frame systems and production rule systems) have also been investigated. *Conceptual graphs* (3) are an emerging standard for knowledge representation, and the method is particularly suited to the representation of natural language semantics. Free-text data have limitations due to spelling errors, ambiguity, and incompleteness. However, formalisms that collect data in a structured and coded format are more likely to increase the usefulness regarding biomedical research, decision support, quality assessment, and clinical care (4). However, the lack of standardized medical language limits the optimal use of computers in medicine. Incorporation of knowledge bases containing equivalent expressions may be required for the practical use of medical information systems (5). The Generalized Architecture for Language Encyclopaedias and Nomenclature in medicine (GALEN) project, funded by the European Union, develops the architecture and prototypes for a terminology server (6). Indeed, medical language forms one of the greatest obstacles for the practical use of AI in the field of medicine (5,7). Natural language often has remote roots, for example, adrenaline and epinephrine are the same chemical substances. Similarly, (pontine) angle tumor, acoustic (nerve) neurinoma, and acoustic neurilemmoma all have the same meaning. The Latin word *os* means both mouth and bone. The terms heterogenous, heterogeneous, and heterogenic look similar but all have a different meaning. Also, several English words have a dual meaning, for example, apprehension, aromatic, attitude, auricle, bladder, capsule, cast, cervical, cream, and cystectomy. Eponyms (8) further complicate descriptions. These examples illustrate the problem of translating medical phrases into concise "computer-storable" language. In addition to problems inherent to the understanding of natural language, additional difficulties pertaining to medical terminology can be indicated, as follows.

American Versus British Spelling. Two standard differences are evident, namely the use of the digraph in British spelling (e.g., anaemia versus anemia) and preference for using c (e.g., in leucocyte) rather than k (as in the American word leukocyte). Interestingly, the British equivalent of the American spelling of the word leukemia is spelled as leukaemia.

Preferred Terminology. In radiology "air" means gas within the body, regardless of its composition or site, but the term should be reserved for inspired atmospheric gas. Otherwise, the preferred term is "gas". Sometimes the preferred terminology refers to simplicity; the expression "lower extremity" must be replaced by "leg", for example. On other occasions the preferred terminology pertains to technical vocabulary that permits high precision if the available information is exact; the word "clumsiness" describes defective coordination of movement in general, whereas "dysdiadokokinesis" refers to a defect in the ability to perform rapid movements of both hands in unison (9).

Meaning Within a Certain Context. The quality "blue" primarily refers to a particular color. The actual meaning in medical language may, however, relate to a specific noun, for example, blue asphyxia, blue baby, blue bloater, blue diaper syndrome, blue dome, blue line, blue nevus, blue pus, blue sclera, blue stone, and blue toe syndrome (7).

Implicit Information. A particular statement may imply many relevant components, for example, if urinalysis is normal, then this result implies the absence of proteinuria, hematuria, glucosuria, and casts. Also antonyms may apply: leukopenia in particular implies "no leukocytosis". This mutual exclusion principle applies to all terms beginning with hypo- or hyper-.

Imprecise Terminology. (10) Some terms may carry a vague meaning, for example, tumor, swelling, mass, and lump. To a large extent, however, the use of such terms reflects the uncertainty related to an observation. In that respect it is jus-

tifiable: indeed it would be incorrect to specify an observation in greater detail than the facts permit. This notion has consequences for the selection of equivalent expressions.

Certainty Versus Uncertainty. Decision analysis itself does not reduce our uncertainty about the true state of nature, but as long as we must make some choice it does enable us to make rational decisions in the light of our uncertainty (11). Another aspect concerns subjective interpretation of percentage figures about prognosis (12). Outcomes perceived with certainty are overweighted relative to uncertain outcomes. Thus, the formulation of information affects its interpretation by humans.

Limited Scope of a Thesaurus. Thus far, no agreement exists regarding directives for coding diseases. Major sources are organized in different ways, for example, in the International Classification of Diseases[1] (ICD) one finds "Bladder, see condition (e.g., Leukoplakia)," whereas the book *Current Medical Information and Technology* (CMIT) (13) reads "leukoplakia (of bladder), see bladder." Notably, "leukoplakia of the bladder" as such is not listed in the book *Medical Subject Headings* (MeSH) (14).

Knowledge Engineering. This type of engineering implies various levels of translation. Thoughts by the human expert are formulated as precisely as possible, the engineer provides feedback using his or her own phrases to ensure an exact match between both minds, and subsequently the resulting expression is translated into a format usable for the computer program. These steps involve transformations of language while yet assuming that the ultimate user of the program fully appreciates the scope of the original thoughts of the expert.

Information Source Versus Actual Patient. (15) Current medical information sources tend to adhere to preference terminology to promote the use of uniform medical language. However, such standard vocabulary is not used by the average patient to describe individual health problems (16). Then it is left to the clinician to transpose, for example, "puffy face" and "moon face" if appropriate. Indeed, better health care can be realized by educating the patient about the value of structured communication with the physician (17).

Synonyms. For example, "icterus" is identical to "jaundice." Thrombocytosis and thrombocythemia are two words to indicate that the number of platelets in the peripheral circulation is in excess of 350,000 per microliter.

Subspecialty Interpretation. When naming a "hollow space" you may choose anything out of the following set: cavity, crypt, pouch, gap, indentation, dell, burrow, crater, concavity, excavation, gorge, pocket, cave, cavern, cistern, or lacuna. However, every expression may exhibit a nuance within a certain context. Then there is jargon: the terms "show," "engagement," and "station," for example, have a particular meaning within the field of obstetrics (7). The term "streaking" has a different meaning for the microbiologist and the radiologist.

Eponyms. Many disease names refer to the first author (e.g., Boeck's disease for sarcoidosis) who described that particular disorder, to the first patient analyzed in detail (e.g.,

---

[1] The ICD is a widely accepted system that organizes all possible medical diagnoses. The tenth version has been translated for worldwide application. Developed by the Commission on Professional and Hospital Activities, Ann Arbor, Michigan.

Mortimer's disease, again for sarcoidosis), or to the geographical area (e.g., Lyme disease) where the illness was first detected. But variations may occur: The Plummer–Vinson syndrome (sideropenic dysphagia), as it is known in the United States and Australia, is termed Paterson–Kelly syndrome in the United Kingdom, but Waldenström–Kjellberg syndrome in Scandinavia (8).

Multilingual Approaches. The relation between a concept and the various corresponding terms in different languages is in general not unique. This implies that a multitude of different words from different syntactical categories may represent a single concept. Particularly the European countries are confronted with additional natural language problems. The Commission of the European Communities supports research activities in this area through the Advanced Informatics in Medicine (AIM) project, such as EPILEX (a multilingual lexicon of epidemiological terms in Catalan, Dutch, English, etc.) (18), and the development of a multilingual natural language system (19).

Frequency of Occurrence. The meaning of semiquantitative indicators such as "always" and "often" is not transparent when screening a medical text. The intuitive interpretation of some quasinumerical determinants is summarized elsewhere (20).

Noise Terms in Patient Description. When analyzing 104 patient cases, we found (21) that the input consisted on average of 75 terms; the required number of terms for establishing the primary diagnosis was only 15. This implies that 80% of the input data consisted of "noise terms" that may blur the process of hypothesis formation for humans (22).

Illogical Terminology. Certain terms contain paradoxical details, for example, hayfever is usually not accompanied by fever, while acute rheumatic fever typically has a chronic course.

CLASSIFICATION AND CODING SYSTEMS

With the exception of one British project, all classification or coding systems have been developed in the United States. The following survey lists all projects along with some of their characteristics.

- The ICD system just entered its tenth version, although the ninth edition is still used. It is applied worldwide for classifying diagnoses and also permits diagnosis-related group (DRG) assignment employed for billing and reimbursement purposes.

- Another widely accepted system is called Systematized Nomenclature of Medicine (SNOMED) (23), which offers a structured nomenclature and classification for use in human as well as in veterinary medicine. It covers about 132,600 records, with a printed and a CD-ROM version.

- The Current Procedural Terminology volume (CPT) (24) provides a uniform language for diagnostic as well as surgical and other interventional services. The system is distributed by the American Medical Association (AMA) and has been incorporated in the Medicare program.

- MeSH (14) is a systematic terminology hierarchy that is used to index the MEDLINE medical publications system, with annual updates.

- The National Library of Medicine[2] (NLM) in 1986 started a project called the Unified Medical Language System[3] (UMLS) (25), aiming to address the fundamental information access problem caused by the variety of independently constructed vocabularies and classifications used in different sources of machine-readable biomedical information.

- Gabrieli (26) constructed a computer-oriented medical nomenclature based on taxonomic principles. His system covers 150,000 preferred terms and a similar number of synonyms. The partitioning method employed for medical classification readily permits replacement of English names with terms of any other language, thus creating the perspective of a worldwide standard.

- Read from the United Kingdom designed a classification for various computer applications (27). Its design adheres to the following criteria: comprehensive, hierarchical, coded, computerized, cross-referenced, and dynamic. The system is closely connected to the British National Health Service (NHS). Version 2 includes 100,000 preferred terms, 250,000 codes, and 150,000 synonyms.

MEDICAL KNOWLEDGE BASES

Ideally a medical knowledge base (KB) integrates text, graphics, video, and sound. Furthermore, it should be accurate, verifiable, and easily accessible where doctors see patients, and the system should be adaptable to doctors' own preferred terms or abbreviations (28). Future developments will certainly include the use of ANNs (29), and examples realized thus far include myocardial infarction, diabetes mellitus, epilepsy, bone fracture healing, appendicitis, dermatology diagnosis, and electroencephalogram (EEG) topography recognition. An overview will be given of current KB systems. With the exception of the Oxford System of Medicine (OSM) and Medwise, all projects originate in the United States. One approach (CONSULTANT) addresses the field of veterinary medicine. A survey referring to the year 1987 has been published before (15).

Obviously, the Internet, a rapidly expanding network for computer-to-computer communication, nowadays offers a convenient window to medical resources. A useful guide, called *Medical Matrix* is the result of a project devoted to posting, annotating, and continuously updating full content and unrestricted access to this medium. The system can be reached at http://www.medmatrix.org and features a ranking system based on the utility for point-of-care clinical application.

- CMIT developed by the AMA (13) forms a reference for the selection of preferred medical terms including certain synonyms and generic terms with builtin arrangements to provide maximum convenience in usage, currency, and timely publication.

---

[2] 8600 Rockville Pike, Bethesda, MD 20894. The NLM releases news bulletins and provides information on UMLS and contracts for cooperation.

[3] The UMLS is a project initiated by the NLM and distributed on CD-ROM. Cooperation with parties to implement the system within their own environment is encouraged but requires a contract (25).

- Blois was the first to apply CMIT as a diagnostic tool in his RECONSIDER project (30). The application was released in 1981 and covered 3,262 disease entities, while 21,415 search terms were listed in a directory along with their frequency of occurrence. The program is extensively described in his book (30).

- DXplain (31) is also based on CMIT (13). The project had close connections with the AMA, and information is distributed using the World Wide Web. The KB contains information on 2,000 diseases and understands over 4,700 terms, with 65,000 disease-term relationships.

- QMR patient diagnostic software (32) covers some 600 disease profiles, and is the personal computer version of the INTERNIST-I prototype. Unfortunately the size of its disease KB remained remarkably constant over the last few years.

- MEDITEL (33) addresses the issue of diagnosis in adults. Over the last few years not much news was reported in the literature, apart from a comparative study (34).

- ILIAD (35) is a software package designed to aid students and residents in their clinical decision logic. The project stems from the health evaluation through logical processes (HELP) system developed at a major hospital in Salt Lake City. Its KB covers 1,300 diseases and 5,600 manifestations, mainly subspecialties of internal medicine.

- The Oxford System of Medicine (OSM) project, initiated by the Imperial Cancer Research Fund for use in primary care, helps general practitioners during routine work to support decision-making tasks such as diagnosing, planning investigations and patient treatment schedules, prescribing drugs, screening for disease, assessing the risk of a particular disease, and determining referral to a specialist (36).

- Medwise was founded in 1983 and now covers some 3,900 disease entities, with 29,000 different keywords (21). It includes a separate KB with almost 500 equivalent terms that each refer, on average, to three related terms. Equivalent terms are automatically generated to assist the user during the process of data entry. The matrix structure of the Medwise KB permits semantic differentiation, with corresponding weight factors for disease profile matching.

- The Framemed system (37) divides medical information into 26 domains and arranges the items in a hierarchical sequence, thus yielding a logical framework for a standardized terminology. The objective is to achieve a standard coded terminology (including synonyms) to which all existing systems can relate, with obvious use as an electronic encyclopedia and for differential diagnosis.

- STAT!-Ref (38) offers the contents of a first-choice medical library (including several standard textbooks, e.g., on primary care) as well as Medline on CD-ROM.

- MD-Challenger (39) offers a clinical reference and educational software for acute care and emergency medicine (everything from abdominal pain to zygoapophyseal joint arthritis), with nearly 4,000 annotated questions and literature references. MD-Challenger also includes continuing medical education (CME) credits.

- Labsearch/286 is a differential diagnosis program allowing input of up to two abnormal laboratory findings plus information on symptoms and signs. Laboratory data concentrate on body fluids (blood, urine, cerebrospinal, ascitic, synovial, and pleural fluid) entered as high or low (40). The system includes 6,500 diseases and 9,800 different findings.

- CONSULTANT (41), a KB for veterinary medicine, was developed at Cornell University. This database for computer-assisted diagnosis and information management is available on a fee-for-service basis in North America.

- Griffith and Dambro (42) compiled an annually updated book. The information is compiled by a group of contributing authors whose names are listed in conjunction with each disease profile. The first edition appeared in 1993 and contains chartlike presented information on 1,000 topics, along with their ICD code. The printed version shows similarity with CMIT [13]. Publication using electronic media recently became available.

- The *Birth Defects Encyclopedia* (43) is a comprehensive, systematic, illustrative reference source for the diagnosis, delineation, etiology, biodynamics, occurrence, prevention, and treatment of human anomalies of clinical relevance. A unique feature of the printed edition is the Fax service for requesting a current, daily updated version of any article in the KB. The related birth defects information system (BDIS) is a sophisticated computer-based profile-matching system that helps research and diagnostic tasks associated with complex syndromes.

- DiagnosisPro by MedTech, U.S. (44), is a differential diagnosis system including 8,500 diseases, designed by the internist C. Meader and the clinical pathologist H. C. Pribor. It covers all major specialties and is considered a useful tool for primary care professionals.

EXPERT SYSTEMS AND COMPUTER ASSISTED DIAGNOSIS

The term *expert system* refers to a computer program that simulates the professional capabilities of a human expert (45–49). In the field of medicine, the expression *computer-assisted decision system* is also used for an expert system. KB systems are used for interpretation of actual data about a specific problem, considering the knowledge represented in the domain of the KB, to develop a problem-specific model and then to construct plans for problem solution. KBs usually include facts about the problem domain, and procedural knowledge to manipulate facts. In production systems this procedural knowledge adopts the form of IF-THEN or IF-THEN-ELSE rules, where the IF part is the antecedent, the THEN part is the conclusion, and the ELSE part, if exists, is the alternative conclusion. Candidate hypotheses are derived through some pattern-matching system. A reasoning "engine" (also termed *inference machine*) carries out the manipulation specified to obtain an answer. An inference engine is no more than a program, the function of which is to decide what to do at any given moment, that is, it recognizes and activates the appropriate rules. Generally, an inference engine should include an interpreter, which activates the relevant rules at any given moment, taking into account the current state of the active memory; a search strategy, which includes exploration heuristics; a self-knowledge mechanism, which permits the identification of the structures being utilized, the state of the problem, and changes in the active memory; and a termination mechanism for the inferential processes.

The overall functioning of the inference engine occurs in cycles called *basic production system cycles*. The nature of these basic cycles is very different depending on whether the search process is directed by the data or by the objectives. Given that the production systems are essentially based on rules, it will be necessary to define how the propagation of knowledge within the system can be affected. Let us look at two basic propagation methods.

1. Forward chaining of rules, which is implemented as a search process directed by the data, and where the search initiates with the antecedents and leads to the conclusions of the rule
2. Backward chaining, which is implemented as a search process directed by the objectives, and where the search, by means of an evocative process, initiates with the conclusions of the rules, established as hypotheses, and extends to the antecedents

Verification and validation are the two most important stages in the evaluation of an expert system's behavior and functioning (48, 50–53). Verification endeavors to ensure that the system has been constructed correctly. This means that it ensures that the software contains no errors and that the final product satisfies the initial design specifications and requirements. Validation, on the other hand, refers more precisely to a detailed analysis of the quality of the expert system in the context of its work environment, which permits determining whether or not the developed product adequately meets the expectations deposited therein. Verification of an expert system necessarily involves a detailed analysis of the knowledge contained within the system. Particularly if we refer to production systems, the rules may be the origin of many errors, among which the following can be identified:

Conflictive knowledge

$$p(x) \rightarrow q(x)$$
$$p(x) \rightarrow \text{not } q(x)$$

Circular knowledge

$$p(x) \rightarrow q(x)$$
$$q(x) \rightarrow r(x)$$
$$r(x) \rightarrow p(x)$$

Redundant knowledge

$$p(x) \text{ and } q(x) \rightarrow r(x)$$
$$q(x) \text{ and } p(x) \rightarrow r(x)$$

Unnecessary knowledge

$$p(x) \text{ and } r(x) \rightarrow q(x)$$
$$p(x) \text{ and } \text{not } r(x) \rightarrow q(x)$$

Rules included in or contained within others

$$p(x) \text{ and } r(x) \rightarrow q(x)$$
$$p(x) \rightarrow q(x)$$

Rules never executed

$$p(x) \text{ and } r(x) \rightarrow q(x)$$
$$r(x) \text{ cannot be obtained}$$

The preceding erroneous situations must be detected and resolved, and in order to do this, the verification of the expert system may be approached from two different perspectives: first, verification dependent on the domain and second, verification independent of the domain. In the preceding section we have described several systems with KBs that are used for establishing a differential diagnosis. Recently, the diagnostic performance of four such commercially available programs (Dxplain, Iliad, Meditel, and QMR) was evaluated (34). The fraction of correct diagnoses by the computer ranged from 0.52 to 0.72, while half of the candidate diagnoses proposed by the human experts were not generated by the computer. However, on average each program suggested two additional diagnoses per case that the human expert did find relevant but which the researchers failed to include in their original differential diagnosis list. An obscure limitation of the study design is that the researchers themselves created a set of cases to be analyzed by the computer programs. Also, they employed the vocabulary provided by the program's developer. Our written requests to receive a copy of the patient cases used in this study for further testing of other programs were not honored thus far. In his editorial, Kassirer (45) wrote that the results of the study indicate substantial progress, but he found them disappointing from a physician-skeptic point of view. He concluded that the structure of the KBs, the computational *inference engine* that integrates clinical data into a diagnosis, the methods of capturing clinical data from patient's records, and the human-computer interface are still in their infancy. However, several field prototype expert systems have been successfully validated (55–59), and some methodologies for expert systems validation have been proposed (60–62).

ARTIFICIAL NEURAL NETWORKS (ANN)

In 1956, a group of researchers met in Dartmouth college in order to discuss the possibility of constructing genuinely intelligent, technologically advanced machines. This meeting laid the foundations for the science of AI. Principal among the participating researchers of note were Samuel, McCarthy, Minsky, Newell, Shaw, and Simon. Following this meeting two major breakaway groups were formed, both of which continued working more or less independently of each other. Thus, Newell and Simon formed a team with the idea of developing human behavior models, whereas McCarthy and Minsky formed another team dedicated to the development of intelligent machines, not being particularly concerned with human behavior as such. The first approach entailed an emulation of cerebral activity and, wherever possible, the copying of its structure. The second approach implied the construction of systems in which the problem-solving procedures applied are such that, were human beings to apply them, they would be considered intelligent.

However, in practice a combination of both approaches is necessary in order to obtain results that may be considered useful. Both approaches comply with the fundamental objectives of modern AI, namely, the understanding of human intelligence and the use of intelligent machines to acquire knowledge and to resolve complicated problems satisfactorily. Both approaches lead to AI programs, KB systems, expert systems and, finally, to ANNs.

AI programs can be said to exhibit a certain intelligence as a result of the skillful application or use of heuristics in the broadest sense. Heuristic knowledge is considered the fruit of experience, which is difficult to formulate, and which is established implicitly in order to find answers to a specific problem, answers that may be more or less accurate but that are nevertheless always valid. Although the use of heuristic knowledge does not guarantee the finding of optimal solutions, it does offer acceptable solutions, if they exist, by means of so-called inferential processes.

The next epistemological level is that of knowledge-based and expert systems, for which knowledge of the specific domain and knowledge of the control structures used to manipulate said knowledge are physically separate. Therefore, it is necessary to define and implement architectures different from the ones we are accustomed to, in which knowledge and control structures can be developed independently of one another in such a way that one specific control structure can be applied to knowledge from different domains. Expert systems can be considered as specialized knowledge-based systems, in that they resolve real-life problems, which, although limited in terms of size, are complex. The construction of an expert system requires the employment of techniques developed to construct AI programs, in addition to architectures defined for the development of knowledge-based systems. However, it is absolutely essential to place more emphasis on differential aspects such as the acquisition of knowledge and learning.

ANNs can be defined as massively parallel distributed processors with a natural capacity not only for storing experience-based, that is, heuristic, knowledge, but also as a facility for making such knowledge available for use. ANNs allow limitation of the brain in two ways. First, knowledge is acquired by means of a learning process, and second, the synaptic weights are used for storing the knowledge.

It is obvious that, in order to obtain acceptable results at any of the levels of AI described above, we need to draw from other fields such as mathematics (its language and procedures), medicine (especially the neurophysiological models), computer science (particularly software engineering and systems architecture), linguistics (especially syntax and semantics), psychology (which allows us to analyze intelligent behavior models), and finally, even philosophy.

Advanced Aspects of AI

Dealing with Uncertainty. AI is not only concerned with general mechanisms related to the search for solutions within a given space or with how to represent and utilize the knowledge of a specific discourse domain. Another aspect, up to now just mentioned in passing, is that concerning inferential mechanisms and/or processes, which are considered as the starting point for the so-called *reasoning models*.

In any domain, the propagation of knowledge by means of AI programs is always carried out by following a well-defined reasoning model. These reasoning models contribute in a decisive way to the correct organization of the search for solutions. Normally, the domain characteristics and the charac-

teristics of the problems to be solved determine the type of reasoning model to be employed. Thus, there are domains of a markedly *symbolic* nature, in which solutions can be established with absolute confidence. In these cases the use of categorical reasoning models is indicated (63). There are, on the other hand, domains that are of a *statistical* nature, where unique solutions cannot be obtained and where in addition, a decision must be made as to which of the possible solutions arrived at is the most probable. In these cases it is preferable to reason with statistical models of which, given the peculiarities of the inferential processes that AI deals with, the *Bayesian scheme* is the most widely used (64,65).

There are other domains in which the concept of uncertainty appears and which may be inherent to the data of the problem and the facts of the domain, or to the inferential mechanisms themselves. In such cases reasoning models are chosen that are capable of correctly manipulating such uncertainty (66–68).

Finally, there are domains in which the inferential elements include nuances of a linguistic nature where hierarchies and classifications can be established. Indicated in these cases are reasoning models based on *fuzzy sets* (69,70).

Obviously there are domains that manifest more than one of the characteristics just mentioned, in which case the reasoning model most appropriate to the characteristics of the domain or a combination of different models can be used.

The Dempster-Shafer Theory of Evidence

This reasoning scheme has a solid foundation in theory to the extent, in fact, that the original reasoning model proposed by Dempster was subsequently formalized and converted into a genuine theory by Shafer (71). This scheme is attractive, principally for the following reasons: (1) it permits the modeling of uncertainty associated with pieces of evidence and hypotheses in a simplistic manner; (2) it permits the consideration of sets of hypotheses without the confidence in each set having to be distributed in any way between each of the individual hypotheses of the set; (3) it permits an elegant but precise representation of the lack of knowledge so frequently associated with reasoning processes; (4) it deals with the probability theory as a special case; (5) it contains some of the combinatory functions of the certainty factors model.

But how is it possible to deal with both the inexact knowledge and the lack of knowledge in the Dempster-Shafer model? In the first place, and given any discourse universe whatsoever, Dempster and Shafer introduce the concept of a *discernment frame* that can be defined as the finite set of all hypotheses that can be established in the problem domain. The discernment frame should form a complete and thus exhaustive set of hypotheses that are mutually exclusive. On the other hand, the effect of a specific piece of evidence on the overall set of hypotheses is not determined by the contribution of the confidence deposited in the individual hypotheses. On the contrary, each piece of evidence affects, generally speaking, a subset of hypotheses within the discernment frame. This approach is consistent with the reality of almost all real-life routine problems. In real-life problems, the reality is that the evidence e permits discrimination between groups or sets of alternative hypotheses. However, at the same time, within a set, uncertainty with respect to the alternative hypotheses is maintained. According to this argument:

Z is the discernment frame

A is any subset of the frame

$h_1, . . ., h_n$ are the hypotheses of the discernment frame

In this context, the appearance of specific evidence e will favor a determined subset A within Z, in such a way that the degree to which A is favored by e is represented by $m(A)$, where m is indicative of the confidence that the evidence e permits in A. The values of m are represented by the closed interval $[0, 1]$. We will use the following notation:

$$e:\ A = [h_a, h_b, h_c] \to m(A) = x, \ x \in [0, 1]$$

The fact that the evidence e supports the subset A does not imply, as already pointed out, that the individual hypotheses divide, explicitly, the confidence deposited in A itself. This fact diverges considerably from classical theories of probability. Each subset of the discernment frame, for which given evidence e it is established or verified that $m(A) \neq 0$, is called a *focal element*. Returning briefly to the basic probability theory, Dempster and Shafer define the following conditions for m:

$$\sum_{A \subset Z} m(A) = 1$$
$$m(\varnothing) = 0$$

Both conditions are a direct consequence of the restrictions imposed on the discernment frame. Evidence theory provides us with a neat way to deal with the lack of knowledge associated with reasoning processes. Let us take a discernment frame Z and evidence such that

$$e:\ A \subset Z \to m(A) = s, \qquad 0 \leq s \leq 1$$

The first condition required for m establishes that $\Sigma A \subset Z$ $m(A) = 1$.

What happens to the rest of the confidence that has not been assigned to the focal element A? To answer to this question, Dempster and Shafer postulate that if

$$e:\ A \subset Z \to m(A) = s, \qquad 0 \leq s \leq 1$$

then

$$m(Z) = 1 - m(A) = 1 - s$$

This formula should be read as "given that the evidence e supposes the assignation of a given confidence to a specific focal element A within the discernment frame, then the rest of the unassigned confidence represents a 'lack of knowledge' and therefore should be assigned to the discernment frame itself." This situation leads us to reflect as follows: Unassigned confidence is ignorance or lack of knowledge with respect to the importance of the evidence in relation to the focal element under consideration. In other words, it is known that the evidence supports the focal element to the extent of s. However, referring to the unassigned confidence $(1 - s)$, we do not know if it contributes or not to A (or to any other subset within the frame). The unassigned confidence $(1 - s)$ should be assigned to the frame since we constructed the frame; what we do know as a fact is that the solution is within

the frame. The complete formulation for the approach is as follows:

The discernment frame $Z = [h_1, . . ., h_n]$

The focal element $A \subset Z$

Evidence e referring to A

A measure of the "basic probability" of A given e: $m(A)$

$e: A \to m(A) = s$

$m(Z) = 1 - s$

$m(B) = 0 \ \forall \ B \subset Z, B \neq 0, B \neq A$

If the approach were probabilistic, the same evidence would support the focal element A as well as the complement of the focal element. Thus,

$$p(A) = s \to p(\text{not } A) = 1 - s$$

This was precisely one of the major drawbacks of the probabilistic models. With this new theory, if

$$Z = [h_1, h_2, h_3, h_4]$$

and

$$A = [h_1, h_2]$$

with

$$e:\ A \to m([h_1, h_2]) = s$$

then

$$m([h_1, h_2, h_3, h_4]) = 1 - s$$

Generalizing broadly, it could be said that the way in which the lack of knowledge in the evidential theory is managed more than compensates for the defects in the probabilistic models.

Fuzzy Systems. Uncertainty does not only occur as a consequence of an absence of information or of other circumstances that may be, to a greater or lesser extent, formalized. On the contrary, uncertainty could well be associated with the very way in which humans express themselves. In practice, most human statements are ambiguous and this ambiguity is an essential characteristic not only of language but also of the processes of classification, of the establishment of taxonomies and hierarchies, and of the reasoning process in itself. Hence, if we define *living things* as organized molecular structures that are born, grow, reproduce, and die, it is clear that even the humble beetroot does precisely the same and therefore should also be considered a living thing. On the other hand, a sliver of a stone does not behave in the same way, and thus should not be considered a living thing. But what about a virus?

The difficulty in defining a virus so as to include it in the set of living things lies with the selfsame definition of the concept *living things*. In other cases, however, the difficulty arises due to questions of a subjective nature. For example, characterizing the set of *beautiful people* is not easy, since each person has a very personal idea of exactly what attributes an ideal should have in order to be beautiful. But it is particularly difficult to say if one specific object is beautiful or not. In this particular situation, subjective nuances appear that render impossible the very idea of classification.

Furthermore, it is not only problems of definition or of subjective nuances that may complicate categorical classification. In other cases, the context may modify the criteria. For example, the concept of *a tall man*—which in itself is intrinsically ambiguous—differs notably depending on whether one refers to a Scandinavian or to a Pygmy, and to make matters worse, probably both are right!

So it is possible to conclude rapidly that ordinary sets, in which an element of a determinate universe may or may not belong, are not sufficiently complete so as to represent the knowledge normally utilized within the command of a human being, not to speak of reasoning with that knowledge. The fields of mathematics and AI, at their outset concerned with interesting problems of a cognitive nature, were from an early stage intrigued by this concept. Finally, in 1965 Lofti Zadeh published in his famous article "Fuzzy Sets" the results of his investigations in this area (72).

Let us take any universe whatsoever, for example, the universe formed by the set N of natural numbers. Let us define a subset of N called A, characterized by the following description: A is the subset formed by natural even numbers of a value less than 10. Thus A, a subset of N, is perfectly defined as follows: $A = [2, 4, 6, 8]$, and obviously 2 belongs to A, but 3 does not belong to A, and 10 does not either. In this particular example, we have no difficulty in establishing the *degree of belonging* of an element of the discourse universe to a particular subset.

Now let us look at the universe C characterized by the following description: C is the set formed by all human beings; let B be a subset of C characterized by the description "B is a subset of C that includes tall, dark men." In this situation it is difficult to establish the degree of belonging of an element of the universe C to the subset B. Clearly, a common set may be defined as a collection of elements and if an element of the universe is represented in the collection, the element in question belongs to the said set. In these cases, it can be said that the degree of belonging of any particular element of the referential universe has a Boolean value. Thus if the element belongs to the set, the Boolean value is 1. If the element does not belong to the set, the Boolean value is 0. In this way we can construct a function f (which for common sets is a Boolean function) such that, given an element x from the universe U, and given (also) a subset of U, A, then

$$f_A(x) = \begin{cases} 1 \text{ if } x \in A \\ 0 \text{ if } x \notin A \end{cases}$$

We will now extend the discussion to those special kinds of sets that we have called *fuzzy sets*. In reference to these, we said that linguistic, subjective, and other nuances impeded a precise establishment of the degree of belonging to the fuzzy set in question, of particular elements of the universe. Thus there will be elements from the universe that clearly belong to the set, others that clearly do not belong, and yet others that belong to a certain extent, but not totally. Following the approach previously described, the problem is easy to resolve if we consider the function f to have the following values,

given an element x from the universe U, and (from U) a fuzzy subset A (from U):

$$f_A(x) = \begin{cases} 1 \text{ if } x \in A \\ 0 \text{ if } x \notin A \end{cases}$$
$$0 < f_A(x) < 1 \text{ if } x \text{ partially belongs to } A$$

The function f quantifies the degree of belonging of an element from the universe to the fuzzy set in question. Thus, a fuzzy set is one for which there does not exist a clear dividing line between belonging and not belonging of determinate elements from the universe. In order to establish the *fuzzy limits* of the corresponding set, we shall require a criterion, which naturally will be arbitrary. Let us examine the universe of *living persons* along with the fuzzy set A of U answering the description "A is the set of young living persons." a property we consider appropriate for the characterization of the fuzzy subset A is the "age" of the universe elements, but how should we define age? We are faced with the not insignificant problem of the definition of criteria for the "fuzzification" of sets. In our example, we will consider as "young" all those elements from the universe whose age permits them to legally obtain Youth Travel Cards, Inter-Rail tickets, etc. (i.e., elements of age ≤ 25 years old), and "not young" all those elements that can legally obtain Pensioner Travel Cards (i.e., elements of age ≥ 65 years old). Thus,

$$f_A(x) = \begin{cases} 1 \ \forall \ x/\text{age}(x) \leq 25 \\ 0 \ \forall \ x/\text{age}(x) \geq 65 \end{cases}$$

But what happens to all those elements from the universe aged 26 to 64 years old? What exactly is their degree of youth with respect to the criterion of age? We are faced with yet another problem, which is the characterization of the diffuse area or zone. In order to get out of the conundrum, we will construct a linear function as follows:

$$f_A(x) = [65 - \text{age}(x)]/(65 - 25)$$
$$= [65 - \text{age}(x)]/40 \ \forall \ x/\text{age}(x) \in [25, 65]$$

In this way we can segment our numeric space $[0 - 1]$ in three zones, two of which are not fuzzy and which refer to those elements of the universe that clearly belong or that clearly do not belong, to the fuzzy subset in question, and a third zone that is fuzzy and that corresponds to those elements of the universe that belong, to a certain extent, to the fuzzy subset in question.

We will now examine a situation in which Tom is 17 years old, Dick 31, and Harry 73. Conscious of the fact that each one is a "living person," what can we say about their "youth"? In accordance with the established criteria, and carrying out the appropriate substitutions, we can establish the values of their respective functions of belonging to the fuzzy subset of "young living persons" as follows:

$$f_{\text{young}}(\text{Tom}) = 1.00$$
$$f_{\text{young}}(\text{Dick}) = 0.85$$
$$f_{\text{young}}(\text{Harry}) = 0.00$$

It is obvious that as the ages of our friends increase, their degree of belonging to the fuzzy subset decreases. The approach is coherent but is not very natural in linguistic terms. Just to illustrate our meaning, have a look at the following dialogue: "By the way, Sally, how old is Dick?" "I think he is 31." "Oh, so he is 0.85 young!" Absurd! Nobody talks in this way! We are faced again with a new problem, that of *linguistic classification* of fuzzy sets. The basic idea is that once we have managed to segment the numeric space (indicative of the degree of belonging of each element to the fuzzy subset in question), we need to segment the linguistic space by means of labels containing information of a semantic nature and then to match each linguistic label to a specific numeric interval, on the basis of a (minimally) reasonable criterion. Returning to our example, we define a linguistic scale to which we assign concrete values from our function degree of belonging to the fuzzy subset in question. Thus

$$f_A(x) = 0.00 \rightarrow \text{not at all young}$$
$$0.00 < f_A(x) < 0.20 \rightarrow \text{very slightly young}$$
$$0.20 \leq f_A(x) \leq 0.40 \rightarrow \text{a little young}$$
$$0.40 < f_A(x) < 0.60 \rightarrow \text{to some extent young}$$
$$0.60 \leq f_A(x) \leq 0.80 \rightarrow \text{moderately young}$$
$$0.80 < f_A(x) < 1.00 \rightarrow \text{fairly young}$$
$$f_A(x) = 1.00 \rightarrow \text{absolutely young}$$

According to this scale, and using the facts from the example, we may now say that Tom is "totally young" (or simply "young"), Dick is "fairly young," and Harry is "not at all young" (or simply "not young"). These expressions represent a natural, human way of expressing judgments with respect to the ages of our friends.

Although an in-depth discussion of the problems deriving from knowledge representation and from fuzzy reasoning is way beyond the scope of this text, it is, however, necessary to include a reference to both. Remember that, from the perspective of AI, the fuzzy model permits us to represent and manipulate expressions appropriate to the language of human beings. In such expressions we come across fuzzy predicates, fuzzy quantifiers, and fuzzy probabilities. Other, more conventional approaches to the representation of knowledge lack the means for efficiently representing the meaning of fuzzy concepts. Models based on first-order logic, or those based on classical probability theories, do not allow us to manipulate the inappropriately named common-sense knowledge. The reasons for this are as follows.

Knowledge derived from common sense is lexically imprecise.

Knowledge derived from common sense is of a noncategorical nature.

The characteristics of the fuzzy sets examined in the previous paragraphs give us clues as to the procedure to follow if what we require is the application of knowledge representation models and reasoning models, based on fuzzy logic(s). Thus (73)

1. In fuzzy logic, categorical reasoning is a special case of approximate reasoning.

2. In fuzzy logic, it is all a question of degree.

3. Any fuzzy system can be "fuzzified."

4. In fuzzy logic, knowledge should be interpreted as a collection of "fuzzy restrictions" placed on a collection of variables.

5. In fuzzy logic, reasoning problems and therefore inferential processes should be interpreted as "propagations" of the fuzzy restrictions mentioned previously.

Although the theoretical bases for fuzzy logic are quite clear, applications of the latter to systems of an inferential nature is problematic. Even at the time of writing, these difficulties have not been entirely overcome. Nevertheless, it appears that fuzzy-system theories applied to control problems, in place of more conventional approaches, are coming up with solutions that are both brilliant and elegant.

FURTHER READING

The fields of medical KBs and terminology are rapidly developing. There is no single comprehensive source of information available, but the professional reader is advised to scrutinize the following journals and organizations for updated information.

M.D. Computing, published by Springer Verlag (New York and Berlin), reports on research in the field of medical informatics. Of special interest to the clinician is also the journal *Experts Systems with Applications,* published by Pergamon Press (New York).

IEEE Expert appears four times a year and presents the latest on AI and expert systems. Contact P.O. Box 3014, Los Alamitos, CA 90720. This journal is for those interested in technical details on intelligent systems and their applications. IEEE (P.O. Box 1331, Piscataway NJ 08855-1331) also publishes a number of related journals, for examples, on knowledge engineering, fuzzy logic, ANNs, and multimedia.

The NLM releases news bulletins and provides information on UMLS and contracts for cooperation.

Obviously, annual meetings form the forum for presentation of the latest developments:

IMIA conferences are well known, besides the world congress organized every four years. IMIA publishes a *Yearbook of Medical Informatics.* It offers the pearls of medical informatics since it covers influential papers from 100 journals in the field. Contact Schattauer Publishers, P.O. Box 104545, 70040 Stuttgart, Germany.

BIBLIOGRAPHY

1. G. L. Solomon and M. Dechter, Are patients pleased with computer use in the examination room? *J. Fam. Pract.,* **41**: 241–244, 1995.

2. P. L. M. Kerkhof and M. P. van Dieijen-Visser (eds.), *Laboratory Data and Patient Care.* New York: Plenum Publishing, 1988.

3. J. F. Sowa, *Conceptual Graphs, Information Processing in Mind and Machine.* Reading, MA: Addison-Wesley, 1984.

4. P. W. Moorman et al., A model for structured data entry based on explicit descriptional knowledge. *Meth. Inf. Med.,* **33**: 454–463, 1994.

5. P. L. M. Kerkhof, Knowledge base systems and medical terminology. *Automedica,* **14**: 47–54, 1992.

6. A. L. Rector et al., Medical-concept models andl medical records: An approach based on GALEN and PEN&PAD *J. Am. Med. Informatics Assoc.,* **2**: 19–35, 1995.

7. P. L. M. Kerkhof, *Woordenboek der geneeskunde N/E & E/N* (Dictionary of Medicine) 2nd printing. Houten, The Netherlands: Bohn Stafleu Van Loghum, 1996.

8. B. G. Firkin and J. A. Whitworth, *Eponyms.* Park Ridge, NJ: Parthenon Publication Group, 1987.

9. E. A. Murphy, *The Logic of Medicine.* Baltimore: Johns Hopkins University Press, 1976.

10. V. L. Yu, Conceptual obstacles in computerized medical diagnosis. *J. Med. Philos.,* **8**: 67–83, 1983.

11. M. C. Weinstein and H. V. Fineberg, *Clinical Decision Analysis.* Philadelphia: W. B. Saunders, 1980.

12. S. A. Eraker and P. E. Politser, How decisions are reached; physician and patient. *Ann. Intern. Med.,* **97**: 262–268, 1982.

13. A. J. Finkel (ed.), *Current Medical Information and Terminology,* 5th ed. Chicago: American Medical Association, 1981.

14. MeSH, *Medical Subject Headings.* Bethesda: National Library of Medicine, 1993.

15. P. L. M. Kerkhof, Dreams and realities of computer-assisted diagnosis systems in medicine (Editorial). *Automedica,* 8: 123–134, 1987.

16. J. W. Hurst, The art and science of presenting a patient's problems. *Arch. Intern. Med.,* **128**: 463–465, 1971.

17. J. Verby and J. Verby, *How to Talk to Doctors.* New York: Arco Publ., 1977.

18. C. Du V Florey, European Commission *EPILEX,* ISBN 92-826-6211-X, Luxembourg, 1993.

19. R. Baud et al., Modelling for natural language understanding. Symposium on Computer Applications in Medical Care, Washington DC, in *SCAMC,* **93**: 289–293, 1994.

20. A. Kong et al., How medical professionals evaluate expressions of probability. *N. Engl. J. Med.,* **315**: 740–744, 1986.

21. P. L. M. Kerkhof et al., The Medwise diagnostic module as a consultant. *Expert Syst. Applic.,* **6**: 433–440, 1993.

22. F. T. de Dombal, *Diagnosis of Acute Abdominal Pain.* Edinburgh: Churchill Livingstone, 1980.

23. SNOMED International, College of American Pathologists, Northfield, IL, 1993.

24. *Physicians' Current Procedural Terminology* (CPT), 4th ed. Chicago: American Medical Association, 1992.

25. D. A. Lindberg, B. L. Humphreys, and A. T. McCray, The unified medical language system. *Methods Inf. Med.,* **32**: 281–291, 1993.

26. E. R. Gabrieli, A new American electronic medical nomenclature. *Automedica,* **14**: 23–30, 1992.

27. J. Chisholm, Read clinical classification. *Brit. Med. J.* **300**: 1092, 1990.

28. J. Wyatt, Computer-based knowledge systems. *Lancet,* **338**: 1431–1436, 1991.

29. W. G. Baxt, Use of an artificial neural network for the diagnosis of myocardial infarction. *Ann. Intern. Med.,* **115**: 483–488, 1991.

30. M. S. Blois, *Information and Medicine,* Berkeley: University of California Press, 1984.

31. G. O. Barnett et al., DXplain, an evolving diagnostic decision-support *J. Am. Med. Assoc.,* **258**: 67–74, 1987.

32. B. Middleton et al., Probabilistic diagnosis using a reformulation of the INTERNIST-1/QMR knowledge base, part II—Evaluation of diagnostic performance. *Meth. Inform. Med.,* **30**: 15–22, 1991.

33. H. S. Waxman and W. E. Worley, Computer-assisted adult medical diagnosis, subject review and evaluation of a new microcomputer-based system. *Medicine (Baltimore),* **69**: 125–136, 1990.

34. E. S. Berner et al., Performance of four computer-based diagnostic systems. *N. Engl. J. Med.,* **330**: 1792–1796, 1994.

35. B. Bergeron, ILIAD, a diagnostic consultant and patient simulator. *M.D. Comput.,* **8**: 46–53, 1991.

36. P. Krause et al., Can we formally specify a medical decision support system: *IEEE Expert,* **8**: 56–61, 1993.

37. C. W. Bishop and P. D. Ewing, Framemed, a prototypical medical knowledge base of unusual design. *M.D. Comput.,* **10**: 184–192, 1993.

38. STAT!-Ref. Available from Teton Data Systems, P.O. Box 3082, Jackson, WY.

39. MD-Challenger, 70 Belle Meade Cove, Eads, TN 38028.

40. Labsearch/286, software available from REMIND/1 Corp., P.O. Box 752, Nashua, NH 03061.

41. M. E. White and J. Lewkowicz, The CONSULTANT database for computer-assisted diagnosis and information management in veterinary medicine. *Automedica,* **8**: 135–140, 1987.

42. H. W. Griffith and M. R. Dambro, *The 5 Minute Clinical Consult.* Philadelphia: Lea & Febiger, 1997.

43. M. L. Buyse (ed.), *Birth Defects Encyclopedia.* Center for Birth Defects Information Services, Inc., 30 Springdale Ave., Dover, MA 02030. BD Fax service (508) 785-2347.

44. DiagnosisPro by MedTech USA, P.O. Box 67543, Los Angeles, CA 90067.

45. D. A. Waterman, *A Guide to Expert Systems.* Reading, MA: Addison-Wesley, 1986.

46. A. J. Gonzalez and D. Dankel, *The Engineering of Knowledge-Based Systems: Theory and Practice.* NJ: Prentice-Hall, 1993.

47. J. Giarratano and G. Riley, *Expert Systems: Principles and Programming.* PWS Publishing Co., 1994.

48. P. Jackson, *Introduction to Expert Systems.* Reading, MA: Addison-Wesley, 1993.

49. S. Russell and P. Norvig, *Artificial Intelligence: A Modern Approach,* Englewood Cliffs, NJ: Prentice-Hall, 1995.

50. U. Gupta (Ed.), *Validating and Verifying Knowledge-Based Systems.* NJ: IEEE Computer Society Press, 1991.

51. R. M. O'Keefe, O. Balci, and E. P. Smith, Validating Expert System Performance. *IEEE Expert,* **2**: 81–89, 1987.

52. R. M. O'Keefe and D. E. O'Leary, Expert system verification and validation: a survey and tutorial. *Artif. Intell. Rev.* **7**: 3–42, 1993.

53. D. E. O'Leary, Verifying and validating expert systems: A survey, in *Expert Systems in Business and Finance: Issues and Applications,* P. R. Watkins and L. B. Eliot (eds.), Chichester, UK: John Wiley & Sons Ltd., 1993, pp. 181–208.

54. J. P. Kassirer, A report card on computer-assisted diagnosis—the grade: C (editorial). *N. Engl. J. Med.,* **330**: 1824–1825, 1994.

55. A. Alonso-Betanzos et al., FOETOS in clinical practice: A retrospective analysis of its performance, *Artif. Intell. Med.* **1**: 93–99, 1989.

56. V. Moret-Bonillo, et al., The PATRICIA project: A semantic-based methodology for intelligent monitoring in the ICU, *IEEE Eng. Med. Biol.* **12**: 59–68, 1993.

57. A. Alonso-Betanzos, et al., Computerized antenatal assessment: The NST-EXPERT project, *Automedica,* **14**: 3–22, 1992.

58. A. Alonso-Betanzos, et al., The NST-EXPERT project: the need to evolve, *Artif. Intell. Med.,* **7**: 297–314, 1995.

59. C. Hernandez, et al., Validation of the medical expert system RE-NOIR, *Comput. Biomed. Res.* **27**: 456–471, 1994.

60. V. Moret-Bonillo, E. Mosqueira-Rey, and A. Alonso-Betanzos, Information analysis and validation of intelligent monitoring systems in intensive care units, *IEEE Trans. Inf. Technol. Biomed.* **1**: 87–99, 1997.

61. G. Guida and G. Mauri, Evaluating performance and quality of knowledge-based systems: Foundation and methodology, *IEEE Trans. Knowl. Data Eng.* **5**: 204–224, 1993.

62. K. Clarke, et al., A methodology for evaluation of knowledge-based systems in medicine, *Artif. Intell. Med.,* **6**: 107–121, 1994.

63. R. S. Ledley and L. B. Lusted, Reasoning foundations of medical diagnosis, *Science,* **130**: 9–21, 1959.

64. A. R. Feinstein, The haze of Bayes: The aerial palaces of decision analysis and the computerized ouija board. *Clin. Pharma. Therap.* **21**: 482–496, 1980.

65. R. O. Duda, P. E. Hart, and N. J. Nilsson, Subjective Bayesian methods for rule-based inference systems, *Proc. Nat. Comput. Conf.,* pp. 1075–1082, 1976.

66. E. H. Shortliffe and B. G. Buchanan, A model of inexact reasoning in medicine, *Math. Biosci.,* **23**: 351–379, 1975.

67. J. F. Lemmer, Confidence Factors, Empiricism and the Dempster–Shafer Theory of Evidence, in *Uncertainty in Artificial Intelligence,* L. N. Kanal and J. F. Lemmer (eds.), Elsevier, 1986, pp. 117–126.

68. K. Ng and B. Abramson, Uncertainty management in expert systems, *IEEE Expert,* **5**: 29–48, 1990.

69. L. A. Zadeh, The role of fuzzy logic in the management of uncertainty in expert systems, in *Fuzzy Sets and Systems,* 1983, pp. 199–227.

70. H. J. Zimmerman, *Fuzzy Set Theory and Its Applications,* Dordrecht, The Netherlands: Nijhoff, 1987.

71. G. Shafer, *A Mathematical Theory of Evidence,* Princeton University Press, 1976.

72. L. A. Zadeh, Fuzzy sets, *Inf. Contr.* **81**: 338–353, 1965.

73. L. A. Zadeh, Knowledge representation in fuzzy logic, *IEEE Trans. Knowl. Data Eng.* **1**: 89–100, 1989.

Reading List

M. S. Blois, *Information and Medicine.* Berkeley: Univ. California Press, 1984.

P. L. M. Kerkhof and M. P. van Dieijen-Visser (eds.), *Laboratory Data and Patient Care.* New York: Plenum Press, 1988.

E. A. Murphy, *The Logic of Medicine.* Baltimore: Johns Hopkins Univ. Press, 1976.

J. A. Reggia and S. Tuhrim (eds.), *Computer-Assisted Medical Decision Making.* New York: Springer, 1985.

E. H. Shortliffe, Computer programs to support clinical decision making. *J. Am. Med. Assoc.,* **258**: 61–66, 1987.

E. II. Shortliffe and L. E. Perrault (eds.), *Medical Informatics: Computer Applications in Healthcare.* Reading, MA: Addison-Wesley, 1990.

R. H. Taylor, S. Lavallee, G. C. Burdea, and R. Mosges (eds.), Computer-Intergrated Surgery. Cambridge, MA: MIT Press, 1995.

B. T. Williams, *Computer Aids to Clinical Decisions.* Boca Raton, FL: CRC Press, 1982.

PETER L. M. KERKHOF
Medwise Working Group

AMPARO ALONSO-BETANZOS
VICENTE MORET-BONILLO
University Coruna

MEDICAL INFORMATION SYSTEMS

Medical information systems (MIS) support and enable the diagnosis, therapy, monitoring, and management of patients and also the management of health care organizations and their resources delivering care to patients. Medical information systems are used in the same way as information systems in business and manufacturing. They support and automate certain tasks. They provide new ways to carry out patient care. In business, information technology is a strategic tool in improving competitiveness. Similarly, in health care it could empower the actors (clients/patients, care providers, and those paying for and organizing care delivery) to achieve "more health with less cost."

In the United States the application of information technology (IT) to health care started with the need to manage patient admissions, discharges, and transfers (ADT) in order to bill the financiers for the care provided to each patient. In European countries, health systems generally have a different incentive scheme with mostly publicly funded health systems. Consequently, costing and billing were not initially the focus. Instead IT migrated into health care through the different departments that benefited from IT. Microelectronics and digitization revolutionized medical devices. Devices today are embedded computers supporting such applications as computed tomography and magnetic resonance imaging. Picture archiving and communication systems (PACS) grew as a solution to the need to integrate the process of imaging and image interpretation. The same was seen in all technology-intensive domains of health care, such as intensive care, operating rooms, and clinical laboratories. Laboratory medicine was the first to utilize IT, both to automate its processes and to create new improved services to the departments requesting its services.

These opposite approaches have been merging for many years, and currently these interests are highly integrated. At the same time, the role of IT, and of information systems, in particular, has changed. The importance of information (and knowledge) to health care organizations is recognized. Today many health care organizations employ a chief information officer (CIO) and have created and maintained an information management (IM) strategy. The IM concept has evolved to cover the management of data, information, and processes across the enterprise, including the management of the IT and IS infrastructure. Although this trend is clear, its implementations in the United States and Europe are vastly different (at least for the time being). In the United States, health care organizations' investments in IT are on the order of 3% to 4% of operating costs, whereas in the European countries the figure is only about half that (between 1% and 2%).

Also, the medical information systems industry has undergone major changes. The current thinking of IT as a business enabler, through integration of relevant information and knowledge combined with new, improved services, is supported by the availability of distributed heterogeneous computing environments. These make it possible to network departments and health care organizations onto platforms sharing patient data across a continuum of care. These platforms integrate care institutions and extend to the homes of individuals, and through mobile communication anyone can be connected. This trend requires standards of interoperability and common languages, encyclopedias, and nomenclatures, to share patient and other data in context.

A myriad of medical information systems exist today, from home-grown local applications to off-the-shelf software products. The user has a choice in selecting which product to use. The range of choice is also a drawback in that the market is often nationally or regionally organized, with the result that there are too many competing solutions with a limited installation base. Vendors therefore do not generate enough revenue to maintain and update their products, much less invest in R&D for a new generation of solutions. In such highly fragmented markets there is little incentive to push for standard solutions. Consequently, the user has to be well informed to understand all the options and consequences in selecting products.

FRAME OF REFERENCE FOR MIS

Medical Domains and Architectures

A greatly simplified model describing the role of medical information systems in health care service delivery is given in Fig. 1. It divides a health care unit into three parts: one containing the mission, vision, strategies, and goals of the organization; one comprising the operational system of processes, resources, and knowledge; and one with the information systems supporting and enabling operations and achievement of goals. The model illustrates two important concepts. First, it emphasizes that any health care organization must have an explicit strategy on how it intends to meet its goals and fulfill its mission. Second, the strategy must also cover the ways that IT and information systems can be used to further the purposes of the organization.

An organization can also be represented as an architecture. Architectural views differ depending on the way activities are organized and managed. One way to organize them is according to the hierarchical view of departments and clinical units. Another way to represent a health care organization is to present it as service lines supported by service units (e.g., clinical laboratories, operating theaters, imaging services, etc.) and ancillary support services. This latter process view also fits well with several current paradigms, like evidence-based medicine and clinical protocols (guidelines). In either case, an organizational architecture implies an IS architecture that complies with the needs of the operational system.

The health care domain can best be characterized as *federated*. Federation means that parties have negotiated the extent to which they wish to share common resources and thus surrender their exclusive authority over those resources. The alternatives to federation are, at the one extreme, complete

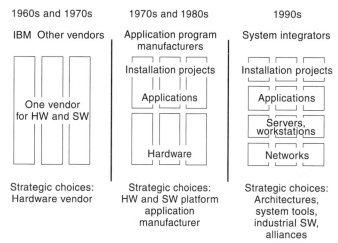

Figure 2. The strategic choices that users have to make have increased in number and complexity.

independence with no common agreements, and, at the other, *unification*, where the parties are governed by one supreme authority. In health care unification is, if not impossible, at least a very difficult and tedious task to achieve. Similarly, isolated IS islands are not any more practical. The concept of federation is an accurate description of current health care systems. Units act independently, but cooperate in patient care by sharing patient data according to mutual interests and agreements.

Consequently, the information systems supporting and enabling operations need not be fully integrated. It is enough if their domains overlap on those areas where data need to be shared. Overlapping of domains means that the organization in question jointly agrees on this. It also means that they agree to use common terminology and classifications in the overlapping domain and that they furthermore agree on a common communications protocol to exchange data. Federation is actually the solution to the dilemma that resulted from the database-centered approach of the 1980s.

At that time, integration was achieved by a common database. As the number of users and needs grew it became gradually impossible to maintain and upgrade such large monolithic systems. Hence industry was forced to find new solutions (Fig. 2). The notion of heterogeneous, federated domains emerged as a way to manage the complexity and to migrate toward meeting the needs of the enterprises. IT vendors have established consortia to develop standards to cope with this, for example, International Standardization Organization/Open Distributed Processing (ISO/ODP), Advanced Networked Systems Architecture (ANSA), and Object Management Group (OMG) (1–3). Distributed client-servers and middleware describe the current approaches.

Current mainstream IT architectures subscribe to the federation principle. Consequently, ISs are otherwise independent, except that they have interfaces through which they communicate with other IS applications. This also means that health care organizations can use different information systems to support different operations, with the provision that these can communicate in their mutual federated domain. This gives users the freedom to select "best-in-class" applications, but at the same time requires that attention is paid to

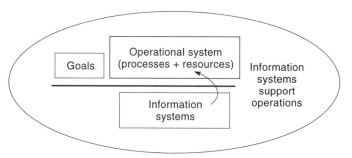

Figure 1. Simplified model of a health care organization.

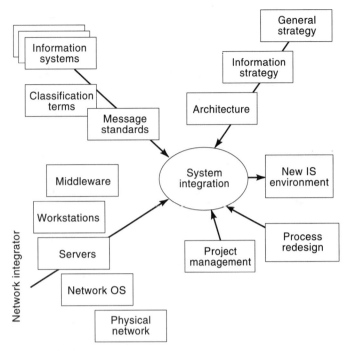

Figure 3. System integration comprises process and IT integration and requires strong abilities in people and project management.

the interconnection of these systems. *Interoperability* means that ISs can communicate among themselves using standard messages and protocols.

The drawback to this is that users today face a much more complex environment than in the past. Whereas in the early days it was enough to select a vendor for the hardware, today the user needs to make an educated choice on a number of issues, such as what IS architecture to have, what software tools to use, what applications to select, and finally with whom to manage and develop this infrastructure (Fig. 2).

Systems integration is the function where information systems are made to work together in their federated domain (Fig. 3). Systems integration comprises, in addition to the technical integration of information systems, the redesign or streamlining of the care processes supported and enabled by these ISs. It also means that a systems integrator must be highly competent in people and project management.

IT—An Agent of Change

As in business and manufacturing industries, health care uses information technology as a strategic change agent to improve efficiency, quality, and effectiveness, and to contain cost. This idea is based on the unique enabling capabilities of IT to provide new ways to diagnose, treat, and monitor patients, and to allow patients to take a more active role in this process (Fig. 4). IT is a vehicle that empowers all actors (citizens, patients, clinical staff, management, third parties, health policy makers, etc.). However, there are no "cookbook" recipes on how to do this. If there were, the health systems of different nations would be similar. As it is, today most countries have recognized the need to reform their respective health care systems. Although certain similarities exist overall, countries have selected different methods and tools to do it. This also highlights the highly heterogeneous nature of health care and underlines the need to model it as a federated domain. Although there are no recipes on how to do this, the first benefit that medical information systems have provided is the ability to make better decisions: Physicians have an integrated view of all data relating to a patient and can combine this with current medical knowledge to arrive at diagnostic and therapeutic decisions or to reevaluate current therapy and diagnosis. Similarly, aggregation of patient resource use and cost data allows managers at all levels to evaluate current practices and to design new ones.

Information Management

Information management (IM) facilitates the business mission of the enterprise by managing its information, processes, and technology (4). An illustration of what IM contains in the case of health care is given in Table 1. IM is not concerned with the management of data and technology only—processes are also included. Information management has a broad focus in seeking ways to improve the functioning of the enterprise in question. Consequently, the role of information technology

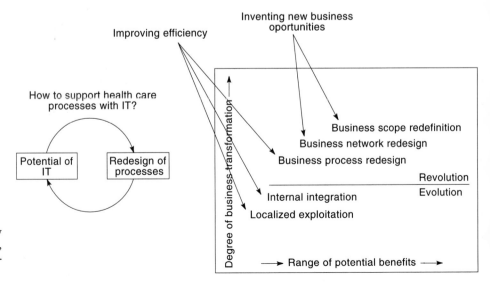

Figure 4. IT enables the creation of new information system products that, in turn, support and enable new ways for providing care to patients.

Table 1. What Information Management Is and Is Not

Is . . .	Is Not . . .
Application of management principles to information	Automation of existing processes
Whole enterprise	Department
Data, process, and technology	Just systems or computers
Resource approach	Technically driven
Information stewardship	Just data management
Management of a critical resource	Isolated islands of data
Business enabler and driver for systems and technology deployment	
All forms of information	

Table 3. IT Leader's Duties

Tactical planner	IT plans must align with the organization's strategies and objectives
System integrator	Initiate and develop an IT plan that facilitates the organization's objectives and has the flexibility to adjust as "marketplace" changes. Optimizing the ground between "best-of-breed" and "one system" extremes
Collaborator	Integration of people and cultures. Using IT to break down interdepartmental, interdisciplinary and interorganizational barriers
Change agent	Facilitator and "perpetrator" for collaboration. Success of quality improvement initiatives depends on the capture of data.
Customer advocate	Meeting both internal and external customer requirements. Being responsive, timely, flexible, and open to improvement

is changing. Whereas in the past it was seen as an expense, it is now an investment (Table 2). This entails the notion that IT costs can be recovered as benefits in improved efficiency, quality, or effectiveness, or in reduced operational costs. A further corollary of this is that the IT leader, or chief information officer, is charged with making this all happen (Table 3). To succeed, the CIO must be a tactical planner, systems integrator, collaborator, change agent, and customer advocate.

MIS Standards

Standardization in MIS is composed of two issues: The first deals with medical practice and medical data; the second, the exchange of medical data among MIS applications. In the heterogeneous changing environment of health care, standards are greatly needed. However, due to its complexity, standards are difficult to establish. This is seen on all levels of activity. The medical profession is concerned with effective and efficient care. In the standardization context this has resulted in a number of classifications and nomenclatures spanning the whole medical domain [e.g., International Classification of Diseases (ICD-10) and Systematized Nomenclature of Human and Veterinary Medicine (SNOMED) (5,6)] and specialized domains (e.g., primary care). In recent years the need for evidence-based medicine has been increasingly recognized (7). This is partly due to the efforts in formalizing care through care protocols, also called clinical pathways and clinical protocols. The results of these activities have been disappointingly slow. This has led to the realization that local care practices need to differ mostly for two reasons. First, the local circumstances need to be respected, as well as local demographics and epidemiology. Second, as medicine is partly an art and

Table 2. Role of Information Technology Has Changed from an Expense to an Investment Approach

From an Expense . . .	To Investment . . .
Small sums of money	Large sums of money
Operational view of benefits	Strategic view of benefits
Must we spend this?	We must spend this!
Discontinuous and discrete spending	Continuous investing
Cost analysis	Investment appraisal
Self-financing by cost savings in other areas	Capital planning
Cost management	Benefits management
Expense accounting	Asset accounting

partly hard science, there is not enough evidence to warrant standardizing care practices across all medical problems.

Countries have initiated health reforms in order to deliver more and better quality health care with fewer resources, that is, cost. Thus care management has become one of the priorities. Means and indicators are needed to measure what resources are needed and what outcomes are produced. Diagnostic related groups (DRGs) have, since their introduction, been applied in many forms across health care systems (8). Managed care and disease management are recent concepts with the same goal.

The heterogeneity of the clinical domain affects medical information systems and ways to integrate them into an interoperable infrastructure. The basic notion is federation, that is, do not try to agree on everything. It is enough to agree on what data and context need to be shared. Communication takes place with messages. Open systems is also a buzzword in health care. Although quite a lot of energy has been invested in international standardization efforts in this area, there is not much to show for it. Instead de facto standards developed by user groups and industries dominate. Health Level 7 (HL7) is the most notable example of data exchange (9), together with Digital Imaging and Communications in Medicine (DICOM) (10). Both have also gained wide acceptance outside United States and are well supported by vendors of medical information systems and imaging modalities and networks. The success of these comes from two factors. First, both allow a certain degree of freedom in implementation (which also means that implementations are not necessarily compatible). Second, both rely on integration engines (message brokers) as the hub receiving and sending messages, thus reducing the number of connections needed between medical information systems in a complex environment. Additionally, some domain specific standards exist, for instance, for exchanging laboratory orders and results (11).

Edifact is the third category of message standards that is being used widely. Whereas HL7 and DICOM are meant mainly for messaging between component systems of a health care organization, Edifact is intended to be used in asynchronous messaging between organizations (12). In Europe several Edifact-based prestandards have been produced by the Health Informatics committee of CEN, for example, for laboratories and drug prescriptions (13).

Middleware is a product of recent years. It is the result of a migration from both the applications and physical layers of services that are common across different environments. Its need has emerged with the three-tier architectures of distributed client-servers and object orientation. A number of international consortia are competing in developing these common services for health care. The main contenders are CORBAmed and Microsoft, and in Europe project consortia STAR and HANSA (14–17). However, these are ongoing activities whose outcomes are still uncertain. For users who need to decide today, the only practical solution are the message brokers based on HL7 and DICOM.

Today the network infrastructure can combine a number of technologies from the TCP/IP based Local Area Network (LAN) and Ethernet to Integrated Services Digital Network (ISDN) and Asynchronous Transfer Mode (ATM) switches. In addition to copper and fiber-optic cabling, connections can also be provided by mobile networks. Consequently bandwidth, as such, is no longer a limiting factor.

As shown above, standardization efforts in the health care domain can be divided into two broad categories: de jure and de facto (industrial) standards. In the former category the work has only begun. The slow progress is due to the nature of medicine. It is evolving at a rapid pace and at the same time a substantial part of it is based on experience rather than hard evidence. This is also the reason why industry standards such as HL7 and DICOM have been so successful. It is also the reason why federation is today the only feasible approach in the creation of an integrated information system infrastructure for a health care enterprise. For middleware and physical infrastructure there are no good reasons why health care needs its own special solutions. Instead health care should use mainstream technologies. This way, health care can leverage the progress and price/performance ratios in mainstream products rather than having to rely on niche vendors specializing in health care.

Frame of Reference

The following are a few suggestions for users to stay competitive and survive in this changing environment:

- Create an information management strategy supporting and aligned with the goals of the health care organization.
- Create an open architecture for the MIS, based on mainstream IT products, system integration, integration engines, HL7, and DICOM.
- Create a plan to migrate from the current ad hoc (possibly monolithic) environment to this open architecture.
- Buy best-of-breed medical information systems and use the architecture and open interfaces to create the necessary interoperability between system components.
- Use IT projects to trigger change in health care processes and in the organization.

STATE-OF-THE-ART IN MIS

A complete description of available medical information systems is beyond the scope of this article. Also, such lists are outdated as soon as they are done. Instead, a generic approach is attempted, where the question of MIS is approached from the medical domain dimension of the health care information framework, that is, what is the medical domain supported by the application, and what capability is included?

The basic task in health care is patient management. This comprises three activities: diagnosis, therapy, and monitoring. The complexity of the task depends on the individual case. The patient may have a problem that can be treated with a single visit (or contact). Or the problem may require referrals to other specialists and care units leading to multiple visits. These extremes illustrate the problem of providing effective, high-quality care efficiently at an acceptable cost to citizens. A wide range of medical information systems has been developed over the years to support patient care and management. As the emphasis in health care has shifted, the MIS have evolved.

The current care environment can be captured in the following phrases: patient-centered, process oriented, clinical guidelines, evidence-based medicine, electronic patient record, management information systems (including case management), structured data (including nomenclatures and encyclopedias), general practitioner/gate keeper, regional seamless care delivery, citizen-centered care and health promotion, telemedicine and the Internet, data confidentiality and data security, and solutions for independent living and security. The following provides a description and discussion of these phrases.

Patient-Centered and Process Oriented

There are at least two ways to look at a health care enterprise: organizational and process oriented. Until recently, hospitals were managed by dividing them into organizational units, like admission/discharge/billing, policlinics, wards, care service units (especially radiology and laboratory), operating rooms, and so forth. As medical information systems are intended to support the organization, the kind of MIS applications that resulted were in accordance with the above organizational structure. Management of the hospital was done through these units, leading to a situation where each unit is optimizing its performance at the cost of others. The job of the clinicians treating patients is to coordinate these in the best interest of the patients. At the same time hospital management is faced with the need to be cost-effective and to contain costs without compromising care.

As these demands are conflicting and cannot easily be reconciled in such an organizational structure, process orientation is winning ground. It has been pioneered with great success in business and industry. Total quality management (TQM), continuous quality improvement (CQI), just-in-time (JIT) delivery, flexible manufacturing, logistics, and others tell of the different facets of the process approach.

A note of warning: Although care processes should be preplanned, they cannot be completely rigid. Each care process is unique. The needs of the patient and the means of the health care unit determine what can and will be done. Furthermore, plans may need to be changed as new evidence emerges.

In health care, solving the problem of a patient can be seen as a process: a chain of partly sequential and partly parallel diagnostic and therapeutic actions. The challenge is to manage these events in order to optimize the outcome and the use

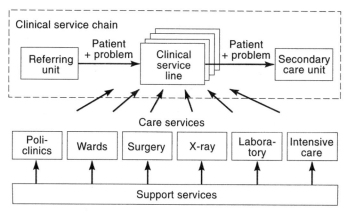

Figure 5. Process-oriented organization of health care delivery.

of resources needed during that process. The process paradigm leads to a new organizational structure for care delivery (Fig. 5). Resources are reorganized to serve the main activity of problem-solving. The core activity is the clinical service line, which uses the skills of service units such as laboratory, radiology, surgery, and wards, according to need.

In the organization unit structure MISs support different units, including laboratories, radiology, picture archiving and communication, pharmacy, intensive care, anesthesia and operating rooms, administration, blood-banking, kitchen, maintenance, cleaning, clinical engineering, and so forth.

This creates a need for a "glue" that integrates the patient data created in the different units and makes it available where and when needed. This is supplied by the application program interface (API) and the message brokering technology utilizing message standards, such as HL7 and DICOM. The patient data store is the electronic patient record. Systems integration is the function where these MIS applications are "glued" together into an interoperable environment.

The resulting MIS architecture has no "order." All applications are equal. This equality has created a further need. The clinicians need to have an overview of what is happening with the patient, that is, how the care plan that they devised is being implemented and with what success. Therefore applications like "clinician's workstation" have been created to provide an integrated picture of the care process. In fact, although the organization is a top-down structure, care is administered in processes.

In the process paradigm, the MIS architecture centers on service. The clinician responsible for a patient designs a care plan. The care plan may contain orders for tests and procedures that are delivered by the care service units. The care plan is reviewed at regular intervals and when new data become available it is adjusted/redesigned according to need. The MIS paradigm for this approach is called "order/entry" (O/E). O/E systems have lately been highly successful, as they provide a way for the clinician to be in control of the procedures performed on the patient and/or on samples taken from the patient.

Common Services

As the understanding of care delivery and of ways to support it with information technology has matured, the need for an infrastructure architecture has emerged. An architecture de-

fines what the total system is, what function it provides, and how the pieces that make up the whole system interact. Integration with message brokering is an architecture. Some MIS applications are used by all, whereas some serve only one function. In other words, there are common and specific services provided by MIS applications. Additionally, there are services that are even more common and that are needed in all IT environments, independent of whether or not that environment has a medical purpose.

Identifying these common and health-care specific common services is attempted by a number of consortia made up of industry and user organizations. The Object Management Group (OMG) is defining a common object request broker architecture (CORBA) and is in the process of identifying these common services (3). As a part of that activity, a health care specific task force has been established with the name CORBAmed (14). Microsoft's OLE/COM and its Healthcare Users Group (HUG) compete with OMG, although some agreements exist on how these can coexist (15). A third approach, known as the Andover Group, combines the strengths of both, building on HL7 (18). HL7 itself may be also a contender as it is moved toward full object orientation (9). In the European context, a prestandard on health-care specific common services exists. It has been produced by the medical informatics committee (TC 251) of the European Standardization Committee (13). The health care common components (HCC) identified are: patient, health datum, activity, resource, authorization and concept (19). The roots of this activity are in a stream of European Union-funded projects that started nearly 10 years ago (16).

Another prestandard by TC 251 presents a health care information framework (HIF) that can be used to view any health care organization and the MIS it uses (20). The HIF comprises three views: (1) health care domain, (2) technology, and (3) performance requirements (Fig. 6). All MIS environments must have the required functions—be dependable and controllable. These requirements are met by technology in three layers. Where they are located depends on the solution. For instance, data privacy and protection can be an integrated feature of an MIS application, or there can be a common middleware service for this across all MIS applications. The management of data privacy and protection in a MIS environment is certainly going to be easier if that function is located in the middleware layer than if changes and updates require manipulation of all MIS applications.

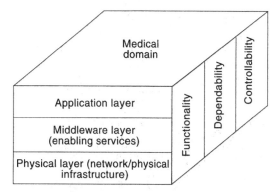

Figure 6. Health care information framework of three views.

Clinical Guidelines and Structuring of Data

The process approach of organizing care into procedures ordered to be performed leads to protocol-based care. The activities needed in handling the problem of a patient can be seen as an instantiation of a template containing that "care package." Clinical protocols are used routinely in clinical research to determine the efficacy and effectiveness of, for example, new drugs and medical procedures. They are also commonly used in cancer therapy and in other medical specialities.

As each patient is an individual, protocols cannot be applied in every detail. Instead, the user must have the freedom to apply a template according to the needs of that particular case. Consequently, what started as protocols and decision trees in the knowledge-based system era of the 1980s has gradually changed so that today they are called clinical guidelines and clinical pathways (21). Strict protocols have been challenged as "cookbook medicine," demonstrating that care cannot be prescribed from a distance.

Guidelines are compiled from medical knowledge. While medical knowledge is universal, clinical practice is local. Therefore, protocols need to take into account the local circumstances. Unfortunately, a large part of medical knowledge is based on experience rather than on hard facts. It has even been claimed that only 40% of medical procedures are based on evidence. "Evidence-based medicine" is one attempt to improve that ratio (7).

The challenge of a physician is to be up-to-date on what are current "best practices" in medicine. Medical textbooks, journals, and other reference materials are the stores of that knowledge. The question is how to access the right source at the right time. Today a lot of this is available on the Internet and on CD-ROMs. Additional efforts are made to improve accessibility through indexing and by setting up knowledge servers. The fame of MYCIN (22) and other favorable conditions led to a boom in medical knowledge-based systems research in the 1980s. As results usable for clinical practice were slow to materialize, interest faded. In hindsight, the reasons for failure were twofold. First, patients are whole human beings and can only in rare cases be treated within the limits dictated by the knowledge-based system. Building systems based on all medical knowledge was, and is still, impossible. Second, clinicians are responsible for the diagnosis and treatment of patients. They cannot be replaced by a machine. A number of approaches were developed to address this, like critiquing and case-based reasoning. However, they did not meet the expectations or the acceptance of clinicians. Today, knowledge-based decision support is used only in embedded systems like diagnostic ECG machines. Similarly, neural networks and fuzzy systems are finding applications into which they can be embedded, thus hiding their existence from the user.

In addition to making medical knowledge readily available in a clinical setting, there is another challenge related to communication between clinicians. A major part of patient data collected during a clinical episode is unstructured. Referrals and discharge letters are summaries of the case accompanied by structured elements like laboratory findings. Each clinician has an individual way of writing these and must include the whole picture with context for them to be useful to other clinicians. Several ways have been developed to summarize this information. These include, among others, ICD-10, SNOMED, Read codes, and diagnosis related groups (DRGs) (5,6,8,23). Some years ago two major projects were launched on both sides of the Atlantic to develop "translators" for communication between clinicians—Unified Medical Language System (UMLS) and Generalised Architecture for Languages, Encyclopaedias and Nomenclatures in Medicine (GALEN), respectively (24,25). The problem they are facing is the same that all those working in medical informatics face. They cannot change medicine; they can only support its advancement. Consequently the path to such translators is a long one and requires among other things progress in the area of evidence-based medicine.

Electronic Patient Record

An integrated electronic patient record that is available anywhere and at any time to those authorized has been the (almost) "holy grail" in MIS research and development for years. Having the relevant parts of patient data available at the point of care is, of course, necessary. The questions, however, are first, which comes first, the electronic patient record or the interoperable MIS environment producing that integration and making data accessible, and second, what is needed to make the integrated patient record clinically useful. Integration can, of course, always be accomplished by a data repository technique, where all data are stored in one database. This is indeed the approach taken by several MIS vendors. However, the usefulness of such a data repository presupposes that the clinical domains involved will have the same interpretation of the data stored. Their domains need to overlap for those data to be useful. This, in turn, implies that such agreements exist and are implemented in clinical practice. The federation of the domain needs therefore to extend across the whole data store. This has large implications for the whole organization. In other words, the issue is not to procure an electronic patient record system, but to achieve the necessary degree of federation among and between the clinical activities to make full use of that system.

The architecture of an electronic patient record system is an issue in itself. There the goal is to provide enough structure and flexibility to allow intelligent storage and retrieval of patient data. In Europe CEN TC 251 has produced a prestandard for the architecture (26) and in the United States the Medical Record Institute is working for the same goal (27).

Resource and Management Information Systems

Care processes also have to be managed in order to use the available resources effectively. This requires assessment of what resources are needed and used in a certain clinical service chain and attaching costs to these. The goal is to gather enough data on how clinical service lines operate, in terms of outcomes and resource utilization, in order to optimize their use of resources and to optimize outcomes and costs. Activity-based costing, DRG, case-mix, and local variations thereof are forms of resource management.

Continuity of Care

In current thinking the process model of care delivery extends from the home of the patient/client to the care facilities and back to the home covering the whole care cycle. Continuity of care (seamless care) is necessary for high-quality care with optimal resources. This means that several service providers have agreed to collaborate in solving the problem of the patient/client. This implies the need for an information network integrating the service providers, the individual care plans, and patient data [community health information networks (CHIN) and regional information networks].

The concepts of and solutions for common services, clinical guidelines, electronic patient records, and resource management apply equally in this environment. The only difference is that, instead of one service provider, there are several who have agreed to collaborate according to an agreement. As the number of actors increases the development and implementation of an information-management strategy is more demanding.

Minimally the chain includes a primary care provider (general practitioner) and specialist unit (hospital). Increasingly nations are using general practitioners as gate keepers and as case managers for specialist services. The GP must then have an information system that assists in keeping track on how the care plan of the patient is being executed/modified, even when the patient has been referred to another service provider. Similarly, telemedicine applications support consultations between clinicians, thus reducing the need for patients to transfer from one site to another. Such applications naturally need to comply with national legislation on data privacy and secrecy.

Personal computers, the Internet, telephone networks, and wireless communication offer additional extensions to continuity of care. Certain medical procedures can be done in the home setting (home care). Patients themselves can perform certain procedures (self-care). The process paradigm of ordering resources for problem-solving is being turned around in order to recognize that the patient has a dual role as both a subject and an object of care. In the 5th Framework Program of the European Union, the health telematics activity revolves around the concept "citizen-centered care" (28). This also aligns with health-promotion goals of making citizens more responsible for their well-being and health (wellness). With mobile communication these services are available everywhere. Finally, for elderly people information systems mean solutions for independent living and security, thus extending their ability to remain in their normal environment when their physical and cognitive abilities are deteriorating.

Telemedicine

Telemedicine uses telecommunications technologies to deliver health care over distances. It provides diagnostic, therapeutic, monitoring, and follow-up activities as well as management, education, and training. While the explosion of interest in telemedicine over the past few years makes it appear new, telemedicine has existed for more than 30 years. Currently it overlaps with what is considered to be covered by the term *medical informatics* (medical information systems). Examples of issues that overlap are regional systems, the integrated electronic patient record, and applications using the Internet. A partial explanation of this overlap is that telemedicine is promoted by teleoperators who are seeking means to add value to their basic services. The medical information systems industry is more fragmented for reasons explained elsewhere in this article. Teleoperators approach health care from the

bottom up [from the telecommunications infrastructure toward the application layer, Fig. 6)]. The MIS industry delivers applications and systems integration services. As health care delivery becomes more integrated and extends to homes and individuals the borders between telemedicine and MIS are disappearing.

The original meaning of telemedicine was "medicine at a distance." Developments in telecommunications, telematics, computers, and multimedia have amended this definition. Now the emphasis is on the access to shared and remote expertise independent of where the patient or the expertise is located, the multimedia nature of such contacts, and the transfer of electronic medical data (e.g., high resolution images, sounds, live video, patient records) from one location to another. Distances and geography are no longer obstacles to delivering timely and quality health care.

Teleradiology is the most often cited telemedicine application. Other applications include dermatology, ophthalmology, pathology, psychiatry, transmission of images and signals generated by ultrasound and endoscopy and by physiological transducers for diagnostic and monitoring purposes (29).

Taylor (30,31) separates telemedicine into systems and services. The first deals with the technology needed to deliver the second. Telemedicine is still mostly in the technology phase. Numerous experiments and pilots have been conducted (and some are still running) that have established that the technology works. However, because they have been mostly closed environments with special funding, the pilots have not survived in real life. Once the pilot is over it has proved to be extremely difficult to build a convincing case to continue with the service on a real cost basis. Teleradiology, however, is an exception. There is evidence that it is cost-effective at case loads that are realistic in typical clinical practice. As teleradiology has been around the longest it is reasonable to assume that as other telemedicine services mature in the coming years they will diffuse into clinical practice. The fact that health care services are becoming integrated on community and regional basis and that the telecommunications infrastructure necessary to support this change is growing also strengthens the case for telemedicine.

A telemedicine system consists of input/output stations and a communications channel. The performance requirements depend on the application. In the case of teleradiology these include:

- Image capture, either directly from digital imaging modalities or indirectly from films scanned with digitizers.
- Transmission of image and associated patient data through a data channel. Depending on the speed requirements, the channel can be an ordinary phone line, an ISDN line, or even ATM. Satellite communications is used.
- Because the size of a digitized X-ray image file is large (an image of 1000 pixels and a 12-bit gray scale means that the file size is 12 Mbit) and the bandwidth of the data channel is limited, the files are usually compressed at the sending side. Efficient compression algorithms are lossy—that is, all image detail cannot be re-created during decompression. Much effort has been invested into researching what compression ratios are acceptable in various radiology applications.

- Workstations to display X-ray images and associated patient data. The features needed are different at the sending and receiving sides.
- User interface to operate the system.
- Teleradiology systems are either standalone or integrated with other ISs at both ends. In such cases, image and patient data communications is usually based on DICOM and HL7 standards, respectively.

Diagnostic and therapeutic telemedicine services include teleconsulting, teleconferencing, telereporting, and telemonitoring (31). Although experiments and pilot projects for numerous telemedicine applications are being conducted, the development of telemedicine services used in routine clinical work has been slow. This is explained by a number of factors, the most important of which is that a telemedicine service is an add-on to existing services. Therefore it must either offer benefits that cannot be disputed or replace a less cost-effective service. The argument that it provides a means to deliver care over a distance is not enough. It must be supplemented with facts about quality, acceptance, and cost in comparison with the services it is replacing or augmenting. So far there is not much data available on the utility of telemedicine with the exception of teleradiology (32). Reimbursement polices are another barrier for telemedicine. However, with community- and regionwide integration of service providers, this barrier will probably disappear.

A further problem is reliability and liability. Can users trust what they access? Who is responsible if something goes wrong? When the expert consulted is a human being, the usual rules of practicing medicine apply. For servers available through the Internet, however, the situation is quite different. Consequently as this concern has been vioced mechanisms have been created to provide guidelines and certification of these servers. Health on the Net (HON) is one such service (33).

Data Confidentiality and Data Security

Confidentiality and security of patient data are issues that cannot be compromised. National legislation defines how the privacy of a person (even a patient) must be protected. Other legislation provides the framework in which health care is practiced. All MISs, MIS environments, and information management strategies must minimally provide what is required by the relevant laws (34). Some countries require that all software used in health care be certified that it meets the national regulations (35).

Organizations should establish an information risk management plan for the implementation of these requirements into operational processes and MIS. Elements to be included in such a plan are:

1. How authorization to access patient data is obtained from the patient (e.g., using individual health cards)
2. How access rights of health professionals are controlled, maintained, and verified (e.g., audit trails and strong authentication with electronic signature)
3. How patient data are grouped with different access rights
4. How patient data are secured (e.g., by encryption)

5. How the training and education of health care professionals and patients in these issues is organized

Data confidentiality however, is not a black-and-white issue. In real life and especially in health care, every situation that will arise cannot be legislated nor can the normative requirements be applied in all situations. Common sense must prevail in such situations.

THE FUTURE

The utilization of information technology applications in health care is influenced by progress in IT, medicine, clinical practice, and health care delivery. These elements are highly intertwined with one feeding the others. In IT the major trends are the Internet, Web technology, and mobile communication. Web browsers are an easy way to provide uniform user interfaces within an organization. Similarly, Extranets are a way for the organization to be in contact with its clients (citizens, patients) without compromising data confidentiality and security (although there are still doubts about the security features of Web implementations). Mobile communication, fueled by the explosive growth in cellular phones and value-added services, seems to offer a limitless range of applications.

However, these are just technologies. They need to be applied in a way that results in benefits for the clients/patients, users, and organizations. User organizations should be careful not to be too enthusiastic about the possibilities offered by new technologies. New technologies "obey" the life cycle of early adaptation by technology enthusiasts and then early adapters. These provide the testing ground to perfect the technology and to make it available at an affordable price to all. If the technology does not survive the tests of the early adapters it dies (36).

The process approach and the need to manage care jointly are pushing service providers toward collaboration in order to meet the needs of their customers and solve the problems of their patients effectively and efficiently. The scenario of Fig. 7 and Table 4 rests with the idea that IT can integrate data

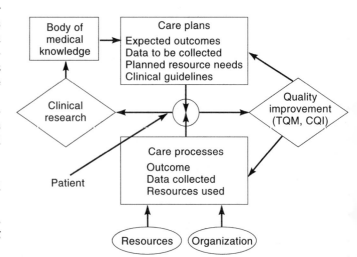

Figure 7. A care scenario combining care processes, plans, and clinical guidelines, and with quality improvement both at the organizational and medical research levels.

Table 4. Characteristics of a Health Care Environment of Today and the Future

Today	Future
Disease and illness management	Health promotion and wellness, independence and security
Hospital-based care	Virtual care (front lines and centers of excellence)
Authoritarian and profession centered	Client-centered care
Patient record centered	Seamless service chains, logistics, community health information networks (CHIN)

and make it and medical knowledge available in the right format anywhere and at any time. From the IT viewpoint, health care will become virtual and transparent.

The development of MIS applications that are transportable and integratable naturally starts with identifying user needs. User involvement in the development, testing, and evaluation phases is equally important. The concept of a user, however, needs to be viewed as widely as possible. This means that one should include all categories of users from daily end users to management. It also means that efforts should be made to involve more than one health care organization. It also means that when the resulting product is taken into use, its costs are offset by benefits and/or savings in other areas, thus justifying the investment in that specific product. According to Gremy and Sessler the key elements in this are the respect of professional identity and a mutual effort for mutual understanding (37). The medical professions should be empowered by the MIS applications instead of being forced into one working pattern.

BIBLIOGRAPHY

1. ISO/IEC DIS 13235-1 Information technology—Open Distributed Processing—Open trading function—Part 1: Specification, ISO Standard.

2. ANSA, Advanced Networked Systems Architecture consortium [Online]. Available http://www.ansa.co.uk/Research/

3. OMG, Object Management Group [Online]. Available http://www.omg.org

4. W. Robson, *Strategic Management and Information Systems: An Integrated Approach,* London: Pitman, 1994.

5. ICD-10, *International Statistical Classification of Diseases and Related Health Problems.* 10th rev., Geneva: WHO, 1992. Also [Online]. Available http://www.who.ch/hst/icd-10/icd-10.htm

6. SNOMED, The Systematized Nomenclature of Human and Veterinary Medicine [Online]. Available http://snomed.org

7. Cochrane, The Cochrane Collaboration [Online]. Available http://hiru.mcmaster.ca/cochrane/default.htm

8. DRG, Health Care Financing Administration (HCFA) DRG version 3 [Online]. Available http://www.hcfa.gov/stats/pufiles.htm

9. HL7, Health Level 7 [Online]. Available http://www.mcis.duke.edu/standards/HL7/hl7.htm

10. DICOM, DICOM—Digital Imaging and Communications in Medicine, ACR-NEMA Digital Imaging and Communications in Medicine (DICOM) Standard version 3.0 [Online]. Available http://www.xray.hmc.psu.edu/dicom/dicom home.html

11. ASTM, The American Society for Testing and Materials [Online]. Available http://www.astm.org

12. Edifact, United Nations directories for Electronic Data Interchange for Administration, Commerce and Transport [Online]. Available http://unece.org/trade/untdid and Electronic Data Interchange [Online]. Available http://www.premenos.com/standards

13. CEN TC 251, European Committee for Standardization, Technical Committee for Health Informatics [Online]. Available http://www.centc251.org

14. CORBAmed, Object Management Group Activity Focusing on Healthcare Services [Online]. Available http://www.omg.org/corbamed/corbamed.htm

15. Microsoft HUG, Microsoft Healthcare Users Group [Online]. Available http://www.mshug.org

16. STAR, Seamless Telematics Across Regions, a European Union Supported Project in the Telematics Applications Program, Sector Health [Online]. Available http://www.mira.demon.ac.uk/star

17. HANSA, Healthcare Advanced Networked System Architecture, a European Union Supported Project in the Telematics Applications Program, Sector Health [Online]. Available http://www.effedue.com/hansa/

18. Andover Group, Andover Working Group for Open Healthcare Interoperability [Online]. Available http://www.dmo.hp.com/mpginf/andover.html

19. HISA, Medical Informatics—Healthcare Information Systems Architecture—Part 1: Healthcare Middleware Layer,European preStandard ENV 12967-1 [Online]. Available http://www.centc251.org/ENV/12967-1/12967-1.htm

20. HIF, Medical informatics, Helathcare Information Framework, European preStandard prENV 12443 [Online]. Available http://www.centc251.org/ENV/12443/12443.htm

21. Prestige, Patient Record Supporting Telematics and Guidelines, A European Union Supported Project in the Telematics Applications Program, Sector Health [Online]. Available http://www.rbh.thames.nhs.uk/rbh/itdept/r&d/projects/prestige.htm

22. MYCIN, E. H. Shortliffe, *Computer-based Medical Consultations: MYCIN,* New York: Elsevier, 1976, and B. G. Buchanan and E. H. Shortliffe, *Rule-Based Expert Systems. The MYCIN Experiments of the Stanford Heuristic Programming Project,* Reading, MA: Addison-Wesley, 1984.

23. M. J. O'Neil, C. Payne, and J. D. Read, Read Codes Version 3: A user led terminology, *Meth. Inform. Med.,* **34**: 187–192, 1995; also [Online]. Available http://www.mcis.duke.edu/standards/termcode/read.htm

24. UMLS, Unified Medical Language System [Online]. Available http://www.nlm.nih.gov/research/umls/UMLSDOC.HTML

25. Galen, Generalised Architecture for Languages, Encyclopaedias and Nomenclatures in medicine, a European Union Supported Project in the Telematics Applications Program, Sector Health [Online]. Available http://www.cs.man.ac.uk/mig/giu

26. EHCRA, Medical Informatics, Electronic Healthcare Record Architecture, European preStandard, prENV 12265 [Online]. Available http://www.centc251.org/ENV/12265/12265.htm

27. MRI, Medical Records Institute [Online]. Available http://www.medrecinst.com

28. Telematics Applications Report of the strategic requirements board, Program, Sector Health Care, 1998 [Online]. Available http://www.ehto.be/ht projects/report board/index.html

29. Telemedicine Information Exchange, TIE [Online]. Available at http://208.129.211.51/

30. P. Taylor, A survey of research in telemedicine. 1: Telemedicine systems. *J. Telemedicine and Telecare,* **4**: 1–17, 1998.

31. P. Taylor, A survey of research in telemedicine. 1: Telemedicine services. *J. Telemedicine and Telecare,* **4**: 63–71, 1998.

32. D. Lobley, The economics of telemedicine. *J. Telemedicine and Telecare,* **3**: 117–125, 1997.

33. Health On the Net Foundation, HON [Online]. Available http://www.hon.ch/

34. R. C. Barrows and P. D. Clayton, Privacy, confidentiality and electronic medical records, *JAMIA,* **3**: 139–148, 1996.

35. CNIL, Commission Nationale pour l'Informatique et les Libertés (CNIL) [Online]. Available http://www.cnil.fr

36. G. A. Moore, *Inside the Tornado: Marketing Strategies from Silicon Valley's Cutting Edge,* New York: HarperCollins, 1995.

37. F. Gremy and J.-M. Sessler, Medical Informatics in Health Organizations and Assessment Issues, Joint Working Conf. of IMIA Working Groups 13 and 15, Helsinki, 1998.

NIILO SARANUMMI
VTT Information Technology

MEDICAL SIGNAL PROCESSING

Medical signal and image processing entails the acquisition and processing of information-bearing signals recorded from the living body. The extraction, enhancement, and interpretation of the clinically important information buried in these physiological signals and images have significant diagnostic value for clinicians and researchers. This is due to the fact that they help probe the state of the underlying physiological structures and dynamics. Because the signals are in general acquired noninvasively (i.e., from the surface of the skin, using sensors such as electrodes, microphones, ultrasound transducers, optical imagers, mechanical transducers, and chemical sensors etc.), medical signal and image processing offers attractive clinical possibilities.

Particularly in noninvasive (and therefore indirect) measurements, the information is not readily accessible in the raw recorded signal. To yield useful results, the signal must be processed to compensate for distortions and eliminate interference. The medical environment is commonly noisy, and the recorded signals are noise corrupted. Signals and systems engineering knowledge, particularly signal processing expertise, is therefore critical in all phases of signal collection and analysis.

Engineers are called upon to conceive and implement processing schemes suitable for medical signals. They also play a key role in the design and development of medical monitoring devices and systems that match advances in signal processing and instrumentation technologies with medical needs and requirements.

This article provides an overview of contemporary methods in medical signal processing that have significantly enhanced our ability to extract information from vital signals.

MEDICAL SIGNALS

The need for noninvasive measurements of signals generated by the body presents unique challenges that demand a clear understanding of medical signal characteristics. Therefore, the success of signal processing applications strongly depends on a solid understanding of the origin and the nature of the signal.

Medical signals are generally classified according to their origin. Bioelectrical, bioimpedance, biomagnetic, bioacoustical, biooptical, biomechanical, and biochemical signals refer to the physiological sources that they originate from or to the induction process imposed by the measurement. The dynamic range and the frequency content of these signals have been tabulated (1).

The source of bioelectric signals detected from single excitable nerve and muscle cells, for example, is the membrane potential. When bioelectric signals are sensed noninvasively by skin electrodes as in the case of surface electrocardiogram (ECG), which emanates from the beating heart (see also ELECTROCARDIOGRAPHY), a multitude of excitable cardiac cells collectively create an electric field that propagates through the biological medium before giving rise to the measured signal. Other clinically used spontaneous bioelectric signals recorded from surface electrodes include the electroencephalogram (EEG), which records the electrical activity of the brain (see also ELECTROENCEPHALOGRAPHY), surface electromyogram (EMG) detected from the muscles (see also ELECTROMYOGRAPHY), electrooculogram (EOG) from the eye, and electrogastrogram (EGG) from the gastrointestinal track.

Evoked potentials (EP) which measure brain response to specific visual, auditory, and somatosensory stimulus are also bioelectric signals. Bioelectric signals measured invasively (i.e., transcutaneously) with needle electrodes include the electroneurogram (ENG) that represents the electrical activity of a nerve bundle, electroretinogram (ERG) of the retina, electrocorticogram of the cortex, and single fiber EMG.

In bioimpedance measurements, by contrast, high-frequency (>50 kHz), low-amplitude (10 μA to 10 mA range) currents are applied to the tissue, organ, or whole body, and the impedance is measured. The impedance characteristics provide information including tissue composition, blood volume, and blood distribution.

The brain, the heart, and the lungs also emit extremely weak magnetic fields that often complement and augment information obtained from bioelectric signals. The main impediment to wide clinical use of biomagnetic signals is the very low signal-to-noise ratio, which presents engineering challenges in detection.

Many physiological phenomena create sound signals. The bioacoustic signals originate from blood flow in vessels, air flow in the lungs, the joints, and the gastrointestinal tract and are noninvasively recorded by microphones. The development of fiber optics and photonics has given rise to a multitude of clinical applications based on biooptical signals. Blood oxygenation, for instance, can be monitored by detecting the backscattered light from tissue at a variety of wavelengths.

The mechanical functions of physiological systems give rise to motion, displacement, flow, pressure, and tension signals. Biomechanical signals are often detected by local transducers and hence require invasive measurements because mechanical signals are rapidly attenuated. Monitoring ion concentration or partial pressure of gases in living tissue or in extracted samples yield biochemical signals. Recent developments in biochemical signal detection underlie the design of new-generation biosensors intended for medical as well as environmental and industrial applications.

Processing of Medical Signals

The rapid pace of development in discrete-time signal processing (2) (see also DISCRETE TIME FILTERS) coupled with the advent of digital computing and, in particular, the rise of virtual instrumentation have given rise to efficient, flexible, and cost-effective methods to acquire and analyze biomedical data in digital form (3). Nonelectrical medical signals are generally transduced to electrical form to readily digitize and process them by digital means. Continuous-time signals are preprocessed by an analog antialiasing filter prior to analog–digital conversion (see also ANALOG-TO-DIGITAL CONVERSION). This presampling amplifier stage is introduced in order to satisfy the sampling theorem, which demands that the sampled signal be band-limited and that the sampling frequency exceed twice the highest frequency contained in the signal. Standards for a minimum sampling rate in clinical recordings have been introduced (4). In medical signal acquisition, it is critically important to preserve the time morphology of the waveforms; thus, phase distortions must be avoided by selecting a presampling amplifier with linear phase within the passband (5).

Although generic signal acquisition, time-domain analysis (see also TIME-DOMAIN NETWORK ANALYSIS), digital filtering [see also DIGITAL FILTERS; FIR FILTERS, DESIGN; IIR FILTERS], and frequency-domain and spectral analysis (see also SPECTRAL ANALYSIS) methods are often used, the low amplitudes and strong interference from other bioelectric signals and noise contamination typical in medical signals require special treatment (6–14). Fourier decomposition implemented using the fast Fourier transform (FFT) is widely used to study the frequency or harmonic content of medical signals. When the frequency content of the interfering signal (e.g., high-frequency muscle contraction or EMG noise) does not overlap with that of the signal under study (e.g., ECG), digital filters can be used to separate the two signals (e.g., ECG can be low-pass filtered to reject the EMG). Similarly, slow baseline drift resulting from low-frequency motion artifacts and effects of breathing in bioelectric recordings can be eliminated by high-pass filtering. The 50 Hz or 60 Hz power-line noise can be drastically reduced by notch filtering. As mentioned earlier, the need to prevent phase distortion in medical signal processing suggests the use of linear-phase finite impulse response (FIR) filters with center symmetric impulse response or, equivalently, filter parameters. Several powerful design methods have been developed for linear-phase FIR filters and are available in commercially available digital signal processing libraries.

Digital filters can not be used, however, when the frequency content of the signal and the interference are not distinguishable. A case in point is the extraction of evoked potentials from the background EEG. An understanding of the deterministic or stochastic nature of the underlying physiological process is necessary for the classification of a medical signal as such, hence the selection of the appropriate processing methods and algorithms. Signals described by exact mathematical formulation, graphical form, rule, or look-up table are recognized as deterministic signals. Nondeterministic or stochastic signals defy mathematical formulation and evolve in an unpredictable random manner (see also STOCHASTIC SYSTEMS). In the detection of EP from the EEG, it is natural to assume that EP is a deterministic signal embedded in EEG, which is assumed to be a zero-mean stochastic signal. Signal averaging with proper time alignment of the EPs is a commonly used technique whenever the signal (EP) and the noise (background EEG) are uncorrelated and the noise is characterized by a white (i.e., constant) spectral distribution.

The nonparametric spectral analysis of medical signals, which can be assumed to be stationary stochastic processes,

can be accomplished using algorithms based on direct Fourier transform of the time series or the inverse Fourier transform of the autocorrelation sequence. To taper the edges of data records to reduce leakage, time-windows are applied to time sequences and lag-windows to autocorrelation sequences. Parametric spectral analysis casts the estimation problem into a framework in which the medical signal is modeled as the response of a rational system function to a white Gaussian noise signal. Powerful methods to identify the model parameters have been introduced, and test criteria to select the appropriate order and type of the model have been developed. Depending on the problem at hand, these models can have all-pole or auto-regressive (AR), all-zero or moving average (MA), or pole-zero or ARMA form (15).

Conventional linear signals and systems theories have been founded on assumptions of stationarity (or, equivalently, time-invariance in deterministic systems) and Gaussianity (or, equivalently, minimum-phaseness). Also, the assumption of a characteristic scale in time constitutes an implicit assumption. Investigators engaged in the study of medical signals have long known that these constraints do not hold under the test of practice. There is evidence that new methods are better suited to capture the characteristics of medical signals.

The investigation of higher-order statistics and polyspectra in signal analysis and system identification fields has provided processing techniques able to deal with signal nonlinearities and non-Gaussian statistics as well as allowing for phase reconstruction. In cases where the power spectrum is of little help, higher-order spectra (HOS) calculated from higher-order cumulants of the data can be used. The aim behind using HOS rather than any other system identification method is threefold:

1. Suppress Gaussian noise of unknown mean and variance in detection and parameter estimation problems;
2. Reconstruct phase as well as the magnitude response of signals or systems;
3. Detect and characterize the nonlinearities in the signal (16,17).

The power of neural-network based methods in medical signal processing is widely acknowledged. In essence, neural networks mimic many of the functions of the nervous system. Simple networks can filter, recall, switch, amplify, and recognize patterns and hence serve well many signal processing purposes. Neural networks have the ability to be trained to identify nonlinear patterns between input and output values of a system. The advantage of neural networks is the ability to solve algorithmically unsolvable problems. Neural networks are trained by examples instead of rules. The most widely used architecture of neural networks is the multilayer perceptron trained by an algorithm called backpropagation (18,19). (See also NEURAL NET ARCHITECTURE.)

Time-frequency representation (TFR) of a one-dimensional signal maps the time-varying (nonstationary) signal into a two-dimensional set where the axes are time and frequency. There are certain properties that a TFR should satisfy: covariance, signal analysis, localization, and inner products. Many innovative TFRs have been proposed to overcome the problems inherent in others, including the spectogram, Wigner distribution, Cohen's class of TFRs, affine class of TFRs, and scalograms (20,21).

Many physiological dynamics are broadband and hence evolve on a multiplicity of scales. This explains the recent success and popularity of medical signal processing applications based on time-scale analysis and wavelet decomposition (22,23). The wavelet transform responds to the need for a tool well suited for the multiscale analysis of nonstationary signals by using short windows for high frequencies and long windows for low frequencies (24). Continuous wavelet transform (CWT) decomposes a random signal onto a series of wide bandpass filters that are time-scaled versions of a parent wavelet. The parent wavelet is in fact a bandpass filter with a given center frequency and parametrized by a scale factor. As this factor is changed, the time domain representation of the parent wavelet is either compressed or expanded. As the scale factor is increased in value, the bandwidth of the signal in the frequency domain is compressed and vice versa. This duality between the time-frequency domains is the advantageous side of the wavelet transform that results in an increased time-frequency localization. However, the signal is expressed in the wavelet-domain as a function of time and scale rather than frequency components as is the case for the Fourier transform. This property is of utmost advantage when the signal in interest has multiscale characteristics.

The discrete wavelet transform (DWT) uses orthonormal parent wavelets for its basis in handling multiscale characteristics of signals. When a dyadic orthonormal basis is chosen, a statistically self-similar, wide sense stationary random signal can be decomposed by DWT to yield wavelet coefficients parametrized by the scaling (dilation) factor and the translation factor. It has been demonstrated that orthonormal bases are well suited for the analysis of $1/f^a$ processes. $1/f^a$ or power-law processes are characterized by a power spectrum that obeys the power-law attenuation (25).

A large class of naturally occurring phenomena exhibit power-law or $1/f^a$ type spectral attenuation over many decades of frequency and, hence, "scale" or reveal new details upon time magnification. Medical examples include biological time series such as voltages across nerve and synthetic membranes and physiological signals such as the EEG and the heart rate variability (HRV) (26,27). HRV is analyzed to diagnose various disorders of the cardiovascular and the autonomous systems. HRV is derived from the heart beat signal by detecting the peaks of the QRS waveforms in the ECG. If the duration between consecutive beats is presented as a function of the beat-number, the resulting HRV is referred to as a tachogram.

The application of scaling concepts and tools such as wavelets to medical signals has uncovered a remarkable scaling order or scale-invariance that persists over a significant number of temporal scales. Methods to capture scaling information in the form of simple rules that relate features on different scales and to relate them to physiological parameters are actively investigated. As in the case of spatial fractals that lack a characteristic length scale and consist of a hierarchy of spatial structures in cascade, $1/f^a$ processes cannot be described adequately within the confines of a characteristic time scale. This property is beginning to be recognized as the dynamical signature of the underlying "complex" physiological phenomena (28).

Compression of Biomedical Signals

Medical signal processing applications generate vast amounts of data, which strain transmission and storage resources. The need for long-term monitoring as well the desire to create and use multipatient reference signal bases places severe demands on storage resources. Future ambulatory recording systems and remote diagnosis will largely depend on efficient and preferably lossless biomedical data compression. Data compression methods tailored to biomedical signals aim to eliminate redundancies while retaining clinically significant information (29).

Signal compression methods can be classified into lossless and lossy coding. In lossless data compression, the signal samples are considered to be realizations of a random variable, and they are stored by assigning shorter or longer codewords to sample values with higher or lower probabilities, respectively. In this way, savings in the memory volume are obtained. Entropy of the source signal determines the lowest compression ratio that can be achieved. In lossless coding, the source signal can be perfectly reconstructed by reversing the codebook assignment. For typical biomedical signals, lossless (reversible) compression methods can achieve compression ratios on the order of 2:1, whereas lossy (irreversible) techniques may produce ratios of 10:1 without introducing any clinically significant degradation. [See also DATA COMPRESSION, LOSSY.]

The promise of medical signal and image processing will be fulfilled when its techniques and tools are seamlessly merged with the rising technologies in multimedia computing, telecommunications, and eventually with virtual reality. Remote monitoring, diagnosis, and intervention will no doubt usher the new era of telemedicine. The impact of this development on the delivery of health care and the quality of life will be profound.

MEDICAL IMAGING

Medical images are of many forms: X ray, ultrasound, and microscopic, to name a few. Increasingly, the formation, storage, and display of these images is based on extensive and intensive utilization of signal and image processing. Some *imaging modalities* such as X-ray imaging onto film are mature and stable, whereas other modalities, such as magnetic resonance imaging, are recent and undergoing rapid evolution. There is a vast range of procedures: projection X-ray images of the body, dynamic studies of the beating heart with contrast medium, and radioisotope images that depict glucose utilization, a biochemical activity.

This section consists of the following subsections:

1. X-ray imaging,
2. Computer tomography,
3. Radionuclide imaging,
4. Magnetic resonance imaging,
5. Ultrasound, and
6. The ecology of medical imaging.

The first five sections deal with the medical imaging modalities. For each modality, we will describe the physical and imaging principles, discuss the role of image processing, and describe recent developments and future trends. The last section will present the clinical and technical setting of medical imaging and image processing and indicate the relative importance of the various image modalities.

An article of this length cannot fully describe all medical image processing areas. We have selected the more important methods that are currently used in modern hospitals. We do not review microscope imaging, endoscopy, or picture archival and communications systems (PACS). Microscopy is the essential tool of pathology, and much image pattern recognition work has been done in this area. Endoscopic imaging is optical visualization through a body orifice (e.g. brachioscopy, viewing lung passages; sigmoscopy, viewing the intestine) or during minimally invasive surgery. PACS, comprising electronic storage, retrieval, transmission and display of medical images, is a very active field, which includes teleradiology (the remote interpretation of radiological images) is described in a separate article on TELEMEDICINE.

X-Ray Imaging

X-ray imaging is the most widely used medical imaging modality. A radiograph is a projection image, a shadow formed by radiation to which the body is partially transparent. In Fig. 1 we show radiographs of a hand. The quantitative relation between the body and the radiograph is quite complex. Some radiographs have extremely high resolution and contrast range. Both static and dynamic studies are performed, and contrast agents are used to enhance the visibility of internal structures. Recent developments in instrumentation are computed radiology and solid-state image enhancement systems. The most commonly used image processing technique is subtraction. Even though extensive research has been performed on image enhancement and recognition of X-ray images, there has been very little practical acceptance of these techniques. [See the article on X-RAY APPARATUS.]

Physical Principles. X rays are electromagnetic waves whose frequencies are higher than those of ultraviolet light. The frequency of X rays is normally not denoted in hertz but rather in photon energy in (kilo) electron volts, which is re-

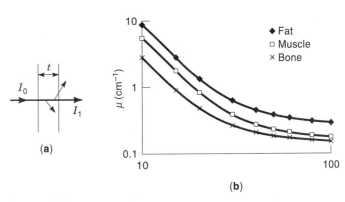

(a)

(b)

Figure 2. (a) Interaction between X-rays and a homogeneous object. An incident X-ray photon can be absorbed or scattered. The X-rays that are neither absorbed nor scattered constitute the attenuated beam, useful for imaging. Scattered radiation impairs image quality. (b) Linear attenuation coefficient for soft tissue (muscle, fat, blood) and for bone as a function of photon energy. Linear attenuation coefficients of different soft tissues are proportional to their densities, which range from about 0.95 g/cm³ for fat to about 1.04 g/cm³ for muscle.

lated to frequency by Planck's formula $E = h\nu$, where E is the photon energy, h is Planck's constant, and ν is the frequency. Photon energies for normal diagnostic use range from 10 keV to 100 keV. X rays are a form of *ionizing radiation;* low exposures to X rays may cause cancer whereas high exposures can be fatal. X-ray imaging systems can be analyzed by using ray optics because diffraction effects are negligible.

Diagnostic X radiation is generated with X-ray tubes, which produce polyenergetic radiation (radiation over a broad range of frequencies). The primary energy (center frequency) is determined by the voltage applied to the tube and the amount of X-ray power by the tube current. The amount of X-ray radiation per unit area is called *exposure,* measured in roentgens, a unit proportional to the energy flux per unit area, where the proportionality factor depends on photon energy. The amount of X-ray energy per unit mass absorbed by a body is called the *dose,* measured in Grays.

If a monoenergetic (single-frequency) X-ray beam falls on a homogeneous object as shown in Fig. 2(a), the relation between the incident and the transmitted flux is

$$I_1 = I_0 e^{-\mu t} \tag{1}$$

where I_0 is the incident flux, I_1 is the transmitted flux, t is the thickness of the object, and μ is the *linear attenuation coefficient,* a property of the object that depends on its atomic composition, density, and the photon energy. Linear attenuation coefficients for body tissues decrease over the diagnostic energy range: plots of attenuation coefficients for soft tissue and bone are given in Fig. 2(b). X rays at different energies combine in an additive way. If a homogeneous body of constant thickness is illuminated with an X-ray beam with spectral intensity $I_0(V)$ in units of energy per unit area and photon energy, the transmitted energy per unit area is given by

$$E = \int I_0(V) e^{-\mu(V)t} \, dV \tag{2}$$

where V is the photon energy, $\mu(V)$ is the attenuation coefficient as a function of photon energy, and t is the thickness of the object.

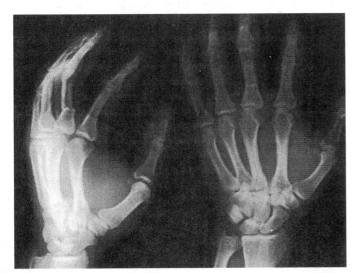

Figure 1. Two radiographic views of a hand. A fracture in the middle finger can be seen more easily in the lateral view on the left. Courtesy of Dr. Harold Kundel, University of Pennsylvania.

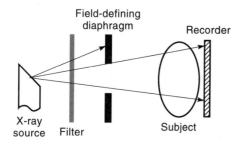

Figure 3. A conventional X-ray imaging system.

X-ray interaction with an object in the diagnostic energy range can occur in two ways: absorption and scattering. In absorption the photon is stopped and gives up all its energy in the vicinity of the interaction point, whereas in scattering the photon gives up part of its energy and continues in a different direction. Equation (1) includes both absorbed and scattered photons. However, scattered photons can leave the body and impinge on the detector. Scattered radiation is undesirable: it contributes to noise in the image and reduces contrast.

X-Ray Imaging. A typical X-ray imaging system is shown in Fig. 3. An X-ray beam from a generator is restricted to the appropriate region with a field-defining diaphragm, passes through the body being imaged, and is detected with a recording system. In *radiography* one or several still images are recorded. In *cineradiography,* a moving image sequence is recorded on film or videotape. In *fluoroscopy,* the image is viewed by a radiologist or a surgeon while the patient is in the X-ray apparatus. Images may be formed from the intrinsic contrast between tissues or by injecting or ingesting a *contrast agent,* a substance that enhances the radiographic difference between tissues (such as between blood and surrounding soft tissues). Contrast studies are *invasive;* they typically cause greater discomfort or danger than *noninvasive* procedures. We will discuss the factors that affect the image contrast and resolution and describe some X-ray image recording systems.

Image Contrast and Resolution. Image contrast is best at low photon energies, and the energy available for image formation increases with photon energy. These considerations lead to the use of lower voltages for imaging thin body sections, such as limbs and the female breast, and high voltages for thick body parts, such as the abdomen. To see how radiological physics affects this choice, consider the configuration shown in Fig. 4, which shows a schematic view of a uniform soft tissue region of thickness t containing a blood vessel of diame-

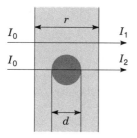

Figure 4. A schematic view of a blood vessel surrounded by soft tissue. The contrast depends on the size of the blood vessel and on the difference between attenuation coefficients (see text).

ter d. The attenuation coefficients of tissue and blood are μ_1 and μ_2, respectively. Applying the law given in Eq. (1), the intensity of the beam passing through the tissue is $I_1 = I_0 \exp(-\mu_1 t)$, whereas the rays passing through the center of the blood vessel have intensity $I_2 = I_0 \exp[-\mu_1 t - (\mu_2 - \mu_1)d]$. For the blood vessel to be visible the *contrast* C

$$C = \frac{|I_1 - I_2|}{I_1} = |1 - \exp[-(\mu_2 - \mu_1)d] \approx |(\mu_2 - \mu_1)d| \quad (3)$$

should be high. To obtain high contrast there should be a large difference between the attenuation coefficients of the blood vessel and surrounding tissue. Also, contrast increases with d, the size of the object being imaged. From Fig. 2(b), we see that it is desirable to use a low photon energy because attenuation coefficient values decrease as energy increases. On the other hand, I_0 is limited by the need to limit the dose to the patient, while the ability to form the image requires a sufficiently large value of transmitted energy, which is proportional to $\exp(-\mu_1 t)$. Because the attenuation coefficient decreases with photon energy, we see that for large values of t (thick body sections) we need to use higher-voltage X rays.

Contrast can be improved by introducing a material that changes the attenuation coefficient. *Contrast angiography* is the visualization of blood vessels by using externally injected agents. Elements with high atomic number, such as iodine, have a much higher attenuation coefficient than normal body tissues. By injecting such a material into the circulatory system, we can increase μ_2 in Eq. (2), making blood vessels visible in thick body sections that require high-voltage X rays. Contrast agents, however, disturb the body so that these procedures may be used only when the risk of not treating the disease outweighs the danger of the diagnostic procedure.

One of the factors affecting resolution (the ability to see small objects) is the focal spot, the size of the area on the generator that emits X rays. The issues in this phenomenon can be seen in Fig. 5(a), which shows a shadow cast by a small object when irradiated with a finite source. The shadow exhibits a penumbra, a region around the edges of the projected image of the objects in which the contrast is reduced. If the object is too small the penumbra may be greater than the image. The contrast of the object is reduced and may be too low to see the object. The extent of degradation depends on the size of the X-ray generator source (focal spot), the size of the structure being imaged, the distance from the source to the object, and the distance from the object to the recording system. For best resolution the focal spot should be small, and the object-recorder distance should be smaller than the source-object distance: if the object is thin and in contact with the recording surface, no blurring is caused by focal spot size. Focal spot size is limited by energy dissipation in the X-ray generator, and X-ray intensity on the detector varies inversely as the square of the source-detector distance, so that an optimal resolution is obtained by an appropriate balance of these factors.

A second limitation on resolution is imposed by the X-ray recording system: currently available technologies require a trade-off between resolution and sensitivity. As an example, consider a screen-film system for radiography, which consists of an *intensifying screen* placed in contact with a photographic film [see Fig. 5(b)]. The intensifying screen is made from a material with a high attenuation coefficient that fluoresces in

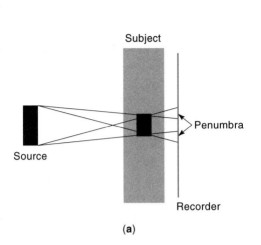

(a)

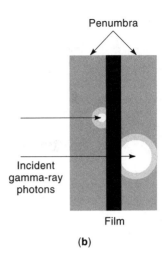

(b)

Figure 5. Source and recorder effect in X-ray resolution. (a) The blurring of an image depends on focal spot size and distances between the object, X-ray source, and the recording medium. (b) A screen-film system for radiography. X-rays are captured by screen materials that emits bursts of light photons which, in turn, expose photographic film. The cassette shown uses two screens to provide higher X-ray detection efficiency.

response to the absorption of X-ray photons: one X-ray photon can produce as many as 1000 light photons. Resolution is limited by the distance between the fluorescing spot and the film: the size of the area illuminated by one X-ray photon is proportional to the distance between absorption point and the film emulsion. An ideal intensifying screen should absorb most incident X-ray photons, should have adequate conversion gain (light photons per X ray), and should be thin enough to produce negligible blurring. Currently available materials and manufacturing techniques require compromises between these factors.

Image Recording Systems. The primary categories we consider are radiographic systems (still images) and fluoroscopic systems (moving images). In radiography, one may use direct film recording, screen-film systems, digital radiography systems, or screens coupled to recording cameras. Direct film recording is used for dental radiography and for imaging thin structures, such as the hand. Even though the sensitivity of direct film recording is not as high as for screen-film systems, these procedures produce a relatively low dose because the structure being imaged is relatively thin. The screen-film recorder was described in the previous section.

A computed radiography system captures (records) X rays on a photostimulable phosphor screen. This forms a latent image that is read by scanning the plate with a laser beam that induces fluorescence proportional to X-ray exposure. The scanner produces a digital image, which may be viewed on a monitor. The plate is cleared by exposing it to visible light and reused. This system has a wider latitude (dynamic range) than film, although the resolution is not as high as the best screen-film combinations. Even though the plates and scanner involve considerable capital investment, computed radiography can lead to savings in operating expense, especially when images are stored and distributed electronically. Some radiology departments are planning for a future when no more film will be used.

In screen-film and computed radiography systems a latent image is formed that requires further processing. In digital radiography the image could be recorded with a sufficiently large electronic camera. Another option is to optically reduce the image produced by an intensifying screen, and to capture the reduced image with a conventional charged-coupled-device (CCD) camera. However, with present technology, this leads to a reduction in image quality.

In fluoroscopy, where a moving image is viewed in real time, best performance is achieved by using image intensifiers. Fluoroscopic exams require prolonged exposure and viewing so that dose reduction is critical. Much of this dose reduction is achieved by using systems that trade resolution for sensitivity. Traditional radiological image intensifiers are vacuum tubes that contain, on the input side, intensifying phosphors coated with photoemissive targets. X rays produce an optical image that falls on the photoemissive material, which in turn produces an electron image emitted into the vacuum on the interior of the intensifying tube, where they are imaged with electron optics onto a small phosphor target (similar to that in a cathode-ray tube display). This small optical image is easily coupled to a television camera with no subsequent loss in image quality. The television signal can be displayed on a monitor, recorded onto videotape, or digitized and stored on computer media. Another option in fluoroscopy is to use digital scan converters: rather than use continuous exposure of the patient, images can be captured occasionally, as required, or at a slow frame rate. These images are digitized and stored in random-access memory (frame buffer) which subsequently generates a conventional television signal that drives a television monitor.

X-ray Image Processing. The most common image processing procedure in X-ray imaging is subtraction angiography, where an image without contrast material is subtracted from a radiogram produced with an injected contrast agent, thereby enhancing the visibility of the structures containing the contrast agent. In Fig. 6 we see a radiogram of the head before and after contrast injection, and the difference image. The arteries are clearly seen. This concept was introduced in the 1930s, when photographic processing was used to produce the difference image, but contemporary digital image processing techniques vastly improve the speed and convenience of these procedures.

Image processing for radiological image enhancement was first proposed in the 1960s. It seems plausible that such processing can improve contrast and correct for image blurring caused by finite focal spot size. Although appropriate contrast manipulation for image enhancement is effective in fluoroscopy systems, where an electronic image is viewed on a monitor, resolution enhancement has met no success, and there are as yet no accepted techniques for enhancing film radiograms. This is probably due to the fact that film radiography

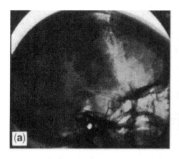
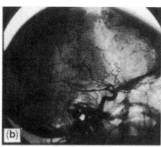
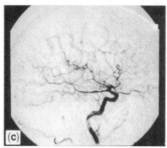

Figure 6. Subtraction angiography. Radiograph of a head before (a) and after (b) contrast agent injection. The difference image (c) shows only the arteries. Courtesy Dr. Barry Goldberg, Thomas Jefferson University.

systems are optimized with the proper choice of X-ray focal spot size and recording system sensitivity and resolution, so that detail enhancement increases noise as well as resolution producing no noticeable improvement in image quality.

One of the hurdles in successful implementation of X-ray image enhancement is the high resolution and contrast range in film images. For example, mammography (imaging of the female breast for cancer screening) produces images 25 cm × 20 cm, which may have significant detail as small as 50 μm in extent. This may require digitization to 6000 × 6000 pixels with 16-bit density resolution. Merely scanning such images in a way that preserves all the visual detail is a substantial task with present technology. Obtaining images that can be enhanced to show detail not apparent on the original film image may require even higher-quality digital image acquisition.

Extensive research has been done on systems for automatic interpretation of radiograms. Some success has been achieved in the measurement of the cardiothoracic ratio (the ratio of the width of the heart to the width of the chest), measurement of arterial stenosis (narrowing) in contrast arteriograms, and detection of degenerative change in lungs. These methods have not yet achieved practical acceptance. Some reasons for the lack of progress is the extreme complexity of X-ray images, which are projections (shadows) of three-dimensional bodies, and many structures may overlap in one area. Radiologists, who undergo extensive training, visualize the three-dimensional structures in the body, compare a specific image with others that they have seen before, and base their diagnosis on an extensive knowledge of anatomy and pathology. It is, at present, not feasible to construct an automatic system that can operate at this level. One recent line of research is in promising computer-assisted diagnosis (CAD) which is being explored for screening mammography interpretation. A processing algorithm examines a scanned image and indicates suspicious areas to a radiologist, who makes the final decision. This type of interactive diagnosis system has the promise of combining a radiologist's extensive knowledge and ability to interpret complex images with a machine's capability to perform a thorough, systematic examination of a large amount of data.

Computer Tomography

A tomogram is an image of a section through the body. Tomographic X-ray images were first formed in the 1930s by moving the source and film in such a way that one plane through the body remained in focus while others were blurred, but they were used for only very specialized investigations. X-ray computer tomography, which is abbreviated as Cat or X-Cat, was introduced in 1973 and gained almost instantaneous acceptance.

In Cat, transmitted X-ray intensity measurements are made for a large number of rays passing through a single plane in the body. These measurements are subsequently processed to compute (reconstruct) the values of attenuation coefficient at each point in the plane. This, in effect, produces a map of the density through the tomographic section plane. The geometric arrangement for performing these measurements is shown in Fig. 7. A ring of X-ray detectors surrounds the body being scanned, and an X-ray generator rotates in an orbit between the body and the detector ring. For each position of the source, intensities of the radiation passing through the body and impinging on the detectors are digitized and captured by a computer. A sectional image is computed from the data collected while the source rotates through a full circle.

Medical CT images typically have a resolution of 500 × 500 pixels, which is about the same as that of television, and much lower than the resolution of radiograms. Radiograms are projection images and superpose many anatomical structures onto one plane, but all structures in a tomogram are shown in their correct geometrical relationship. Figure 8 shows a CT scan through the upper abdomen. The spleen (lower right) contains a hematoma (arrow), a pool of blood caused by an injury. The hematoma density is only about 5% higher than that of surrounding tissues, and it would not be visible in a conventional radiograph.

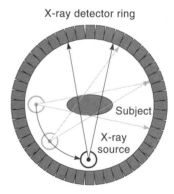

Figure 7. Schematic diagram of an X-ray tomography (CT) scanner. Signals are collected from 500 to 1000 detectors located on a ring. The X-ray source rotates on an orbit between the detectors and the subject.

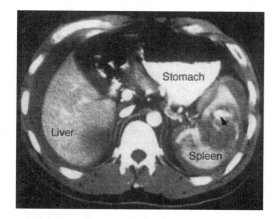

Figure 8. CT scan through the upper abdomen. The arrow points to a hematoma (a pool of blood) in the spleen. The density difference between the blood and surrounding tissues is about 5%. Courtesy of Dr. Richard Wechsler, Thomas Jefferson University.

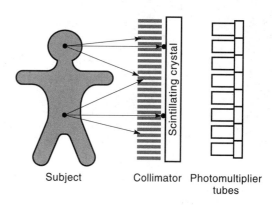

Figure 9. Schematic diagram of a gamma camera. Only gamma rays normal to the camera are admitted by the collimator. An electronic system collects signals from the photomultiplier tubes and computes the coordinates of gamma-ray photons detected by the scintillating crystal.

CT reconstruction algorithms are perhaps the most successful use of image processing in medicine. Unlike X-ray projection images, which are inherently analog, CT images are inherently digital and are impossible to obtain without sophisticated signal and image processing technology.

Radionuclide Imaging

Radionuclides are atoms that are unstable and decay spontaneously. Certain decay reactions produce high-energy photons, which are called *gamma rays.* There is no difference between gamma and X rays; the difference in terminology reflects the process that produces the radiation, and not its physical nature. In nuclear medicine a *radiopharmaceutical agent* (a compound containing a gamma-ray-producing radionuclide) is injected into a body. Images formed from the emitted radiation can be used to visualize the location and distribution of the radioactivity. Radiopharmaceutical agents are designed to migrate to specific structures and to reflect the functioning of various organs or physiological processes. Consequently, radionuclide imaging produces *functional images,* in contrast to the *anatomical images* formed by conventional X rays. Some nuclear medicine procedures require only bulk measurements of radioactivity; for example, the time-course of radioactively labeled hippuric acid in the kidney is an indication of kidney function. In this article, we will concentrate on the procedures that require image formation. We will describe the operation of a *scintillation camera,* the most commonly used device for radionuclide imaging. We will also discuss single-photon emission computer tomography (ELT), an imaging method that produces three-dimensional data. We will not discuss positron emission tomography (PET) scanners, which have unique technical and biological advantages over ECT but are more expensive and are used primarily for research.

Projection images of radionuclide distributions are most commonly formed with a *scintillation camera,* which consists of two major components: a *collimator* and a position-encoding radiation detector, as shown schematically in Fig. 9. The collimator is made form a highly absorbing substance, such as lead, and contains a set of parallel holes that admit only radiation normal to the collimator. These gamma rays fall on

a plate of scintillating material, such as sodium iodide. Each gamma ray photon produces a pulse of light, which is detected by a set of photomultiplier tubes. The current pulses from the tubes differ; the output of the detector closest to the point of scintillation is largest, and those farther from the source are smaller. The pattern of these current pulse intensities is used to compute the location of scintillation at a resolution much finer than that of the tube spacing. The energy of the gamma ray is estimated by computing the total amount of light falling on the protomultiplier tubes. Since scattered gamma rays, which degrade the image, have lower energies and are excluded by the processing system. These gamma rays form a severely defocused image of the object, thus reducing the contrast and increasing the noise level. The image of the radiation distribution is developed by recording individual detected gamma rays and their positions.

Tomographic images of radionuclide distribution can be formed by collecting images while a gamma camera moves along an orbit around the body. A typical system is shown, schematically, in Fig. 10, where three gamma cameras collect projection data while they rotate around a patient. Because each camera collects data for many planes through the body, the data from such an examination facilitates reconstruction of a three-dimensional or volumetric image. The algorithms used for emission computer tomogaphy (ECT) are similar to those for CT, but the reconstruction problem is more complex: emissions from deeper structures are attenuated more than

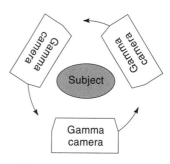

Figure 10. Schematic diagram of an Emission Computed Tomography (ECT) system. Three gamma cameras collect projection data from the subject while rotating about a common center.

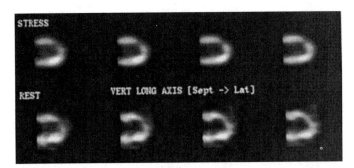

Figure 11. ECAT images of a heart. Shown are a set of slices through the left ventricle obtained after exercise (top) and at rest (bottom). The darker region in the upper right of the stress images indicates an ischemic (reduced blood supply) region of the heart muscle. Courtesy of Dr. David Friedman, Thomas Jefferson University.

those from the surface, and reconstruction algorithms must correct for this effect. There is no elegant solution to the general attenuation correction problem for ECT, although empirically developed methods produce satisfactory results.

Figure 11 shows several sections through the left ventricle of a heart obtained with an ECT scanner. The radiopharmaceutical is absorbed by metabolically active muscle. Two scans are obtained: one after exercise and one after a period of rest. The image has lower intensities in regions that are not receiving blood circulation, such as parts of the heart that may have been injured by a heart attack—a blockage of circulation due to an occlusion in an artery that supplies oxygen and nutrients to the heart muscle. The image is formed over a period of about 10 minutes. This is necessary because the amount of radioactivity is very low, so that photons must be collected for a long period to produce sufficiently high signal-to-noise ratio in the imaging. Blurring caused by heart motion is reduced by *cardiac gating;* the electrocardiogram identifies times at which the heart is in the same position, and data from these time intervals are pooled. The resolution in Fig. 11 is very low, but the technique provides unique information about muscle functioning that is not available from anatomical images.

Signal formation and image processing are an integral part of radionuclide imaging. The formation of the image from a gamma camera requires extensive real-time signal processing. A gamma camera produces a sequence of x–y coordinates of detected gamma rays that are processed to produce a pixel array of gray values. The image so obtained typically contains subtle geometric distortions and nonuniformities, which are corrected with appropriate algorithms. Further processing is required if cardiac gating is used or if tomographic images are produced. Certain studies measure parameters based on dynamic (time-course) data and use sophisticated statistical parameter estimation techniques. The viewing of three-dimensional data from ECT requires rendering methods that produce sectional or projection images.

Magnetic Resonance Imaging

X-ray and radionuclide imaging use electromagnetic radiation of a frequency so high that the human body is partially transparent. Magnetic resonance imaging (MRI) uses electromagnetic frequencies in the range of 40 MHz to 100 MHz, which also penetrate living tissues. At these frequencies, the wavelength is too long to allow focusing as an imaging principle. MRI uses the nuclear magnetic resonance effect to induce signals from the body. Signals from different locations in the body are frequency-encoded by applying spatially varying magnetic fields, and the emitted signals are reconstructed to produce images. The principles used for MRI are unlike those used for any other imaging technique. We will first describe the physical principles of nuclear magnetic resonance, which govern the production of the magnetic resonance signal. We will subsequently describe the methods used to generate the signals that form the images and also describe some of the processes and properties of the body that are viewed in medical MRI.

Nuclear Magnetic Resonance. Nuclear magnetic resonance is a process associated with the magnetic moments possessed by the nuclei of many atoms. There magnetic moments are present, typically, in nuclei whose atomic number (sum of the number of protons and neutrons) is odd. Any of these nuclei can be imaged through MRI. Medical MRI is almost exclusively based on the imaging of hydrogen nuclei or protons, so that in the subsequent discussion we will talk about the magnetic resonance and imaging of hydrogen nuclei. Each nucleus also has an angular momentum, which is aligned with the magnetic moment. Magnetic resonance is based on the joint action of these magnetic and rotational moments.

Thermal motion causes nuclear magnets to take random orientation, but an external magnetic field (the *dc field*) pulls at least some of these nuclei into alignment with the field direction. This alignment process is not instantaneous but follows an approximately exponential time course whose time constant, the so-called the *longitutional relaxation time* (T_1), depends on the physical and chemical state of the material, and ranges from about 0.25 s to 1 s for protons in biological tissues. Should such a nuclear magnet be deflected from the direction of external field, a torque that tends to move it toward alignment is generated. The nuclear angular momentum resists this so that the nucleus behaves like a spinning top: its axis maintains a constant angle in a relation to the dc field and spins (precesses) around it. This phenomenon is called *nuclear magnetic resonance.* The rate of this precession (the Larmour frequency) is proportional to the magnitude of the magnetic field and the proportionality constant, for protons, is 42.6 MHz/T. The magnitude of this oscillating magnetization decays exponentially at the so-called *transverse relaxation time* (T_2), which has a range from 20 ms to 100 ms for protons in biological tissues, depending on tissue type. The resonance of this oscillation is very sharp: the quality factor Q, which is a product of the resonance frequency and T_2, ranges from about 5×10^6 to 25×10^6 for protons in biological tissues.

The precessing magnetic moments set up an external oscillating field that can be detected by placing appropriate conductor loops called *pickup coils* near the body. These detect an induced voltage (due to the Faraday effect), which is proportional to the product of the magnetic moment and the frequency. Both the magnetic moment and the frequency are proportional to the external magnetic field; consequently, the voltage is proportional to the square of the field. It therefore is desirable to have as high a dc field as possible, and magnets with 1.5 T are now fairly common. To deflect (excite) nuclear magnetic moments from their equilibrium direction, an oscillating magnetic field (the *RF field*) is applied at right angles

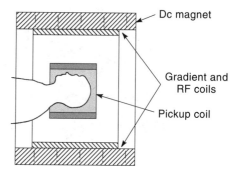

Figure 12. Schematic diagram of a MRI scanner.

to the dc field. If this field is at the Larmour frequency, nuclear magnets tip away from the direction of the dc field. The deflection angle from equilibrium (*flip angle*) is proportional to the product of the RF field strength and duration.

Magnetic Resonance Imaging Principles. A schematic diagram of a magnetic resonance imager is shown in Fig. 12. The bulkiest and most expensive component is the dc magnet. The RF coils excite the MR signal, and the pickup coils detect it. To form an image *gradient coils* impose linearly varying magnetic fields that are used to select the region to be viewed and to modulate the signal during readout. In a typical data collection step, the RF field is applied in the presence of a gradient field (slice selection gradient), so that only one slice through the body is at the Larmour frequency. After the magnetic moments are excited, another gradient field is applied (readout gradient) so that the moments in different portion of the slice radiate at different frequencies. The signal from the pickup coil is amplified and digitized. The Fourier transform of this signal gives a map of the amount of magnetic materials at regions of different gradient field value. A number of such data collection steps, each with a different readout gradient, are required to collect enough data to produce a sectional image. This sequence of data collection is programmed by applying pulses of current to slice selection coils, RF coil, and readout gradient coils. A magnetic resonance imager is a flexible instrument: image characteristics, such as slice thickness and resolution are determined by this *pulse sequence*. The slice orientation, thickness, and resolution can be varied.

Image intensity is basically proportional to proton concentration, or the amount of water in different tissues. Figure 13

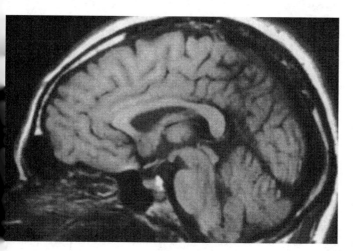

Figure 13. Sectional MRI image of a head.

shows a magnetic resonance image of the human head. This sectional plane, located along the plane of symmetry of the head, cannot be obtained with a single CT scan. Because bone contains less water than the soft tissues, the skull shows a lower image brightness. Further contrast is provided by differences in water content and chemical characteristics of various brain tissues, such as gray matter, white matter, and cerebrospinal fluid. Appropriate pulse sequences can provide images weighted by T_1 and T_2, which show characteristic tissue-specific variations. Other pulse sequences are designed to enhance signals due to differences in chemical composition (chemical shift imaging) which produce small changes in resonance frequency. Pulse sequences can also differentiate between moving and stationary tissues. MRI can produce angiograms (images produced by flowing blood) without using contrast agents. A relatively recent use of MRI is to form *functional images,* which show areas of brain activity that are related to sensory, motor, or cognitive brain activity. Functional imaging can be used to identify areas of the brain affected by stroke before the tissue has undergone a physical change caused by loss of blood circulation. This can guide treatment to restore circulation and avoid brain injury.

From this description, we see that an MRI system functions through electronic and electromagnetic devices such as coils, amplifiers, and digital/analog converters that produce pulse sequences and detect signals. The signals from an MRI scanner are digitally processed to form images and to extract information such as proton density, relaxation times, motion, and chemical composition. Further processing is required to correct for magnetic field nonuniformity and motion-induced artifacts. Postprocessing of magnetic resonance images and image sets is an important area of signal processing and pattern recognition applications. Sets of images obtained with different pulse sequences are combined to enhance signals caused by tumors, neurological plaques, to provide three-dimensional reconstructions of complex structures and to plan surgical procedures.

In contrast to X-ray and radioisotope imaging, MRI uses low-frequency electromagnetic radiation, which is nonionizing. Therefore there is no concern about producing tumors or cancer; the only known source of harm from MRI is heating produced by the RF fields, which can easily be monitored and does not reach harmful levels in clinical equipment. On the other hand, high magnetic and RF fields in MRI may interfere with implanted electronic devices such as cardiac pacemakers, and MRI may be unusable or dangerous in the presence of implanted metal prostheses. MRI exams are long, uncomfortable, and expensive. Because MRI scanners are large and require extensive shielding they are available only in hospitals and other special facilities. Magnetic resonance image resolution is lower than that obtainable from X-rays. Because of these factors, in spite of its versatility and safety, MRI is not likely to replace other modalities in the near future.

ULTRASOUND

Acoustical imaging uses ultrasound (mechanical vibrations at frequencies above the range of human hearing) to form echo images of the interior of the human body. The operation of an ultrasound scanner is illustrated in Fig. 14. A transducer emits acoustic pulses that propagate along narrow beams

Figure 14. Schematic diagram of an ultrasound scanner. In (a), a transducer emits a sequence of acoustic pulses that travel along paths shown in gray. Echoes are produced when these pulses impinge on the front or back surface of the object. In (b) an image is formed by displaying echo signals along monitor scan lines.

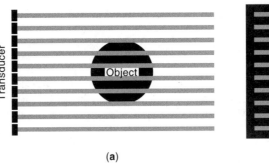

(a)

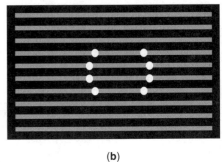

(b)

through the body. Reflections (echoes) from tissues radiate toward the transducer, which converts them to electrical energy. An image is formed by collecting a set of echoes from beams sent out along parallel lines. We will describe the main physical properties that govern propagation of sound in tissues and relate them to factors that affect the quality of images obtained with ultrasound. [See also the article on ULTRASONIC MEDICAL IMAGING.]

Propagation of ultrasound through tissue is governed by the acoustical wave equation, and the wavelength of acoustical waves places critical limits on the resolution of ultrasound scanners. Wavelength, frequency, and acoustic velocity are related by $\lambda f = c$; in soft tissues $c \approx 1500$ m/s, so that, for example, the wavelength at 5 MHz is 0.3 mm. The resolution of the scanner in the depth direction is limited by the duration of the acoustic pulse and in the lateral direction by the width of the sound beam. In Fig. 15 we show the schematic pattern of acoustical energy produced by a concave (focused) transducer excited with a constant frequency. The transducer has diameter D and the focus (center of curvature of the surface) is at a distance F from the transducer. The field pattern is quite complex: approximately, the acoustic beam converges toward the focus where its diameter $w \approx 0.6\lambda F/D$. The width of the beam is relatively constant over a depth of field $d \approx 2wF/D$. To both sides of the region of focus the beam width increases linearly with distance. To obtain high resolution (small w), one should use a short wavelength (high frequency) and or a large transducer of diameter D, which leads to a small depth of field. This physical limitation can be overcome by the principle of dynamic focusing with an array transducer. A flat transducer is decomposed into elements. When an echo is received, signals from various transducer elements are delayed by appropriate amounts, producing the same sort of focusing obtained with a curved transducer surface. These delays are varied with time, allowing focus to be maintained for a large range of depth.

In practice, only soft tissues (muscle, fat, and fluid-filled cavities) can be imaged with acoustical imaging. Bone and

air-filled spaces such as lungs do not propagate ultrasound well enough and are, in effect, opaque. Because resolution is limited by the wavelength, it is desirable to use as high a frequency as possible. Usable frequency is limited by attenuation. Sound power in a plane wave propagating through tissue is given by the formula $P(z) = P_0 \exp(-2\beta z)$, where P_0 is the power at the surface of the body, $P(z)$ is the power at depth z in the body, and β is the attenuation coefficient, which depends on frequency. The attenuation of ultrasound in soft tissue is approximately proportional to frequency, so that higher frequencies are attenuated more strongly and deeper structures can be viewed only with lower frequencies. The frequency used for abdominal ultrasound is about 5 MHz, and a resolution of about 0.3 mm in the depth direction and about 0.4 mm laterally can be obtained to a depth of about 20 cm. Smaller organs, such as the female breast, are scanned with higher frequencies (7.5 MHz), and still higher frequencies are used to examine the eye.

The time to receive echoes from structures at a depth of 20 cm is about 270 microseconds, so that it is possible to emit a sequences of acoustic scanning beams and construct images in real time, at approximately video rates. One of the important uses of ultrasound is to view the heart in real time. Pathologies such as abnormal heart valve motion can be diagnosed from the temporal appearance of the moving image.

The relation between the acoustic echo and tissue characteristics is complex and not well understood. There are two main sources of acoustic echoes. Reflections are produced at tissue boundaries, where there is a change in tissue density or acoustic velocity. These echoes appear as lines at organ boundaries, or at boundaries between tissue and tumor. Such reflections are not seen if the surface is parallel to the beam direction. So-called soft-tissue echoes arise from microscopic inhomogeneities such as blood cells and muscle fibers. An acoustical image typically contains both types of signals. Figure 16 shows an ultrasound image of a pregnant uterus. The head and chest of the fetus are easily seen. The body is outlined with surface echoes, while soft-tissue echoes are produced by the lungs. Much of the image is composed of speckle, which is largely due to soft-tissue echoes, but may be caused by focusing defects or multiple reflections. The interpretation of echosonograms requires extensive training and experience, perhaps more than for other image modalities.

In addition to conventional scanners, there are many other techniques for producing acoustic images. In *Doppler imaging*, frequency differences between incident and reflected pulses are used to measure blood velocity and produce images that allow diagnosis of circulatory problems. *Harmonic imaging* produces signals from nonlinearities of acoustic propa-

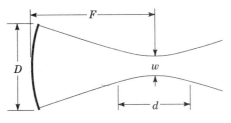

Figure 15. Depth dependent beam width in a focused ultrasound transducer.

gation. *Acoustic contrast agents* allow increased visibility of tissue differences. Very small imaging transducers can be introduced into body cavities or blood vessels. They operate at very high frequencies and produce high resolution images of anatomy of blood flow.

The formation of acoustical images is impossible without signal processing. Originally, acoustical scanners were constructed with analog signal processing circuitry, but modern instruments make extensive use of digital signal processing. Dynamic focusing of the transducer, Doppler signal detection, and harmonic signal analysis must be performed at high rates in real time. There is growth in the use of image processing to enhance images produced by acoustical scanners: for example, promising techniques have been proposed for reducing acoustical speckle.

Medical ultrasound images have a relatively low resolution and noisy appearance. Nonetheless, echosonography offers a number of important advantages. Ultrasound radiation is nonionizing. Even though high levels of ultrasonic energy can produce heating or mechanical injury, there are no known harmful effects from acoustic radiation at the levels used in diagnostic scanners. Consequently, ultrasonography is the preferred—almost exclusive—imaging modality for examining the pregnant uterus, where there is great concern about possible birth defects or tumors in the fetus that may be caused by X radiation. Ultrasound images are formed in real time so that the body can be explored to locate the region to be imaged and moving structures such as the beating heart can be viewed.

ECOLOGY OF MEDICAL IMAGING

What is an ideal medical imaging instrument? Cost, convenience, and safety are important considerations. However, the most important issue is whether the imager can provide definitive information about the medical condition of the patient. To have a better appreciation of these issues, it is necessary to review the conditions under which medical imaging is performed.

Medical imaging may be used as part of therapeutic procedures, to detect disease where no symptoms are present (screening), or to monitor the process of healing. The most common use of medical imaging is for differential diagnosis, where a patient exhibits symptoms that can be caused by one

Table 1. Modalities, Subdivisions, and Annual Case Load in a Large Teaching Hospital

Modalities	Subspecialities	Studies
X ray	Chest, bone, gastrointestinal, genitourinary, neurological, neurosurgical, angiography, mammography	157,958
Computer tomography	Body, neurological (head)	14,343
Nuclear medicine		10,772
Magnetic resonance imaging		12,380
Ultrasound		15,844

or more diseases, and there is a need to ascertain the specific condition so that proper treatment can be applied. The physician orders a specific *imaging procedure,* which may require patient preparation (for example, ingestion of a contrast material), the use of specific imaging instruments and settings (X rays of a certain voltage, intensifying screens, and films), and proper positioning of the patient. Most procedures are performed by properly trained technicians, but some very dangerous or invasive procedures, such as injecting contrast agents directly into the heart, are performed with direct participation of a medical doctor. The majority of imaging procedures are administered by a radiology department of a hospital, and images are interpreted (read) by radiologists, medical doctors who undergo post-MD training (residency) in this specialty. The choice of procedure for a given medical problem is based on medical knowledge and on guidelines promulgated by medical and/or health insurance organizations. We see that most medical imaging tasks are highly specialized and are performed in a very structured setting.

Table 1 shows the divisions of a radiology department of a large teaching hospital. The divisions have been grouped by modality. It also shows the number of studies performed in one year in each modality. It is clear that X-ray imaging is still the predominant technique. Other modalities are growing more rapidly, but none are likely to displace X rays in the near future.

Medical imaging procedures and modalities exist in an extremely competitive environment. For example, heart disease may be diagnosed either with angiography or nuclear medicine. Each procedure, to survive, must fit an ecological niche: it must satisfy a need where it has advantages over other methods. The same issue arises when new image processing techniques are introduced: images produced with these procedures must have an advantage over other imaging procedures. The advantage will probably be application dependent. An image processing technique that is effective for mammography may not be of much value for cardiology, and vice versa.

IMAGE PROCESSING AND INTERPRETATION[1]

Signal and image processing play an essential and growing role in medical imaging. To date, most of the applications of

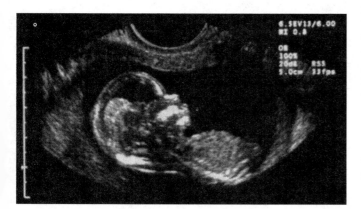

Figure 16. Ultrasound image of a pregnant uterus. The head and chest of a festus are shown. Courtesy of Dr. Barry Goldberg, Thomas Jefferson University.

[1] The work reported in this section is supported by the National Cancer Institute and the National Institutes of Health under grant numbers CA52823 and P41 RR01638.

signal processing have been in design and construction of imaging devices: indeed, novel modalities such as CT, radioisotope imaging, MRI, and acoustic scanning are impossible without signal processing. There is a modest but growing application of *post-processing,* the use of image processing techniques to produce novel images from the output of conventional imaging devices. The most common medical images, radiograms, are largely formed with analog methods. However, there is great promise for digital storage and transmission of these images. This is likely to be a large area for development and application of image compression techniques. There is also the promise, to date largely unrealized, of using pattern recognition techniques to improve on human interpretation of medical images.

As in other biomedical areas, medical image processing is driven by the needs and constraints of the health care system. Innovations must compete with existing methods and require extensive clinical testing before they are accepted.

In conclusion, the fusion of several indices and local decisions leads to a more reliable global decision mechanism to improve on diagnostic imaging.

In this part of the article, we gave a few examples of medical applications that make direct use of various aspects of machine vision technology. We also showed how to formalize many medical applications within the paradigm of image understanding involving low-level as well as high-level vision. The low-level vision paradigm was illustrated by considering the problems of extracting a soft tissue organ's structure from an ultrasound image of the organ, and that of registering a set of histological 2-D images of a rat brain sectional material with a 3-D brain. For the high-level vision paradigm, we considered the problem of combining local decisions based on different aspects or features extracted from an ultrasound image of a liver to arrive at a more accurate global decision.

BIBLIOGRAPHY

1. A. Cohen, Biomedical signals: Origin and dynamic characteristics; Frequency–domain analysis, in J. Bronzino (ed.), *Biomedical Engineering Handbook,* Boca Raton, FL: CRC Press, 1995.

2. A. V. Oppenheim and R. W. Schafer, *Discrete-Time Signal Processing,* Englewood Cliffs, NJ: Prentice-Hall, 1989.

3. W. J. Tompkins (ed.), *Biomedical Digital Signal Processing,* Englewood Cliffs, NJ: Prentice-Hall, 1993.

4. M. D. Menz, Minimum sampling rate in electrocardiography, *J. Clin. Eng.,* **19** (5): 386–394, 1994.

5. L. T. Mainardi, A. M. Bianchi, and S. Cerutti, Digital biomedical signal acquisition and processing, in J. Bronzino (ed.), *Biomedical Engineering Handbook,* Boca Raton, FL: CRC Press, 1995.

6. R. E. Challis and R. I. Kitney, The design of digital filters for biomedical signal processing. I. Basic concepts, *J. Biomed. Eng.,* **4** (4): 267–278, 1982.

7. R. E. Challis and R. I. Kitney, The design of digital filters for biomedical signal processing. II. Design techniques using the z-plane, *J. Biomed. Eng.,* **5** (1): 19–30, 1983.

8. R. E. Challis and R. I. Kitney, The design of digital filters for biomedical signal processing. III. The design of Butterworth and Chebychev filters, *J. Biomed. Eng.,* **5** (2): 91–102, 1983.

9. N. V. Thakor and D. Moreau, Design and analysis of quantised coefficient digital filters: Application to biomedical signal processing with microprocessors, *Med. Biol. Eng. Comput.,* **25** (1): 18–25, 1987.

10. R. E. Challis and R. I. Kitney, Biomedical signal procesing. I. Time-domain methods, *Med. Biol. Eng. Comput.,* **28** (6): 509–524, 1990.

11. R. E. Challis and R. I. Kitney, Biomedical signal processing. II. The frequency transforms and their inter-relationships, *Med. Biol. Eng. Comput.,* **29** (1): 1–17, 1991.

12. R. I. Kitney, Biomedical signal processing. III. The power spectrum and coherence function. *Med. Biol. Eng. Comput.,* **29** (3): 225, 1991.

13. C. L. Levkov, Fast integer coefficient FIR filters to remove the AC interference and the high-frequency noise components in biological signals, *Med. Biol. Eng. Comput.,* **27** (3): 330–332, 1989.

14. M. Akay, *Biomedical Signal Processing,* New York: Academic Press, 1994.

15. M. Akay, *Detection and Estimation of Biomedical Signals,* New York: Academic Press, 1996.

16. C. L. Nikias and A. P. Petropulu, *Higher-order Spectra Analysis: A Nonlinear Signal Processing Framework,* Englewood Cliffs, NJ: Prentice-Hall, 1993.

17. A. P. Petropulu, Higher-order spectra in biomedical signal processing, in J. Bronzino (ed.), *Biomedical Engineering Handbook,* Boca Raton, FL: CRC Press, 1995.

18. A. S. Miller, B. H. Blott, and T. K. Hames, Review of neural network applications in medical imaging and signal processing, *Med. Biol. Eng. Comput.,* **30** (5): 499, 1992.

19. E. M. Tzanakou, Neural networks in biomedical signal processing, in J. Bronzino (ed.), *Biomedical Engineering Handbook,* Boca Raton, FL: CRC Press, 1995.

20. X. Wang, H. H. Sun, and J. M. Van De Water, Time-frequency distribution technique in biological signal processing, *Biomed. Instrum. Technol.,* **29** (3): 203, 1995.

21. G. F. Boudreaux-Bartels and R. Murray, Time-frequency signal representations for biomedical signals, in J. Bronzino (ed.), *Biomedical Engineering Handbook,* Boca Raton, FL: CRC Press, 1995.

22. M. Akay, *Time-Frequency and Wavelets in Biomedical, Signal Processing,* Piscataway, NJ: IEEE Press, 1997.

23. N. V. Thakor and D. Sherman, Wavelet (time-scale) analysis in biomedical signal processing, in J. Bronzino (ed.), *Biomedical Engineering Handbook,* Boca Raton, FL: CRC Press, 1995.

24. M. Vetterli and J. Kovacevic, *Wavelets and Sibband Coding,* Upper Saddle River, NJ: Prentice-Hall, 1995.

25. G. Wornell, *Signal Processing with Fractals,* Upper Saddle River, NJ: Prentice-Hall, 1996.

26. R. W. DeBoer et al., Comparing spectra of a series of point events particularly for heart rate variability data, *IEEE Trans. Biomed. Eng.,* **31**: 384–387, 1984.

27. H. G. Steenis et al., Heart rate variability spectra based on non-equidistant sampling: Spectrum of counts and the instantaneous heart rate spectrum, *Med. Eng. Phys.,* **16**: 355–362, 1994.

28. B. Onaral and J. P. Cammarota, Complexity, scaling, and fractals in biomedical signals, in J. Bronzino (ed.), *Biomedical Engineering Handbook,* Boca Raton, FL: CRC Press, 1995.

29. E. Getin and H. Köymen, Compression of Digital Biomedical Signals, in J. Bronzino (ed), *Biomedical Engineering Handbook,* Boca Raton, FL: CRC Press, 1995.

BANU ONARAL
OLEH TRETIAK
FERNAND COHEN
Drexel University

MEDICAL TELEMETRY. See BIOMEDICAL TELEMETRY.

MEDICAL ULTRASOUND. See BIOLOGICAL EFFECTS OF ULTRASOUND.

MEISSNER EFFECT AND VORTICES IN SUPERCONDUCTORS. See SUPERCONDUCTORS, TYPE I AND II.

MEMORIES, SEMICONDUCTOR. See SRAM CHIPS.

MEMORY ARCHITECTURE

Besides using memory to retain states, a digital system uses memory to store instructions and data. Today the most commonly known digital system is a digital computer. All digital computers being sold commercially are based on the same model: the von Neumann architecture. In this model a computer has three main parts: the central processing unit (CPU), the memory, and the input/output (I/O) unit. There are many ways to design and organize these parts in a computer. We use the term *computer architecture* to describe the art and science of building a computer. We view the *memory architecture* from four different perspectives: (1) memory access interface, (2) memory hierarchy, (3) memory organization, and (4) memory device technology.

First let us examine memory access interface. Logically, computer memory is a collection of sequential entries, each with a unique address as its label. Supplying the address of the desired entry to the memory results in accessing of data and programs. If the operation is to read, after a certain time delay, the data residing in the entry corresponding to the address is obtained. If the operation is to write, data are supplied after the address and are entered into the memory replacing the original content of that entry. Reading and writing can be done asynchronously and synchronously with a reference clock. Other control signals supply the necessary information to direct the transfer of memory contents. Some special memory structures do not follow this general accessing method of using an address. Two of the most frequently used are content addressable memory (CAM) and first-in first-out (FIFO) memory. Another type of memory device, which accepts multiple addresses and produces several results at different ports, is called multiported memory. One of the most common multiported memories, which is written in parallel but is read serially, is called video random access memory (VRAM or VDRAM). It gets its name because it is used primarily in computer graphic display applications.

The second perspective of the memory architecture is memory hierarchy. The speed of memory devices has been lagging behind the speed of processing units. As technology advances, processors become faster and more capable and larger memory spaces are required to keep up with the every increasing program complexity. Due to the nature of increasing memory size, more time is needed to decode wider and wider addresses and to sense the information stored in the ever-shrinking physical storage element. The speed gap between CPU and memory devices will continue to grow wider. The traditional strategy used to remedy this problem is called memory hierarchy. Memory hierarchy works because of the locality property of memory references. Program instructions are usually fetched sequentially, and data used in a program are related and tend to conjugate. Thus, a smaller but fast memory is allocated and brought right next to the processor to bridge the speed gap of the CPU and memory. There can be many levels in the hierarchy. As the distance grows greater between the CPU and memory levels, the performance requirement for the memory is relaxed. At the same time, the size of the memory grows larger to accommodate the overall memory size requirement.

Third we look at memory organization. Most of the time, a memory device is internally organized as a two-dimensional array of cells internally. Usually a cell can store one bit of information. A cell in this array is identified and accessed with row and column numbers. A memory device accepts an address and breaks it down into row and column numbers and uses them to identify the location of the cell being accessed. Sometimes, more than one cell can be accessed at a given time. The size of content that a memory transfers is called the width of the memory device. There are many ways to organize the array in a memory device. By organizing it differently, we can have different widths.

The last aspect of memory architecture is memory technology. Physically, memory can be implemented with different technology. Memory devices can be categorized according to their functionality and fall into two major categories: read-only memory (ROM) and write-and-read memory, more commonly known as random access memory (RAM). There is also another subcategory of ROM, mostly-read-but-sometimes-write memory or flash ROM memory. Within the RAM category there are two types of memory devices differentiated by storage characteristics, static and dynamic RAM or SRAM and DRAM, respectively. DRAM devices represent the stored information with charge. Therefore it needs to be refreshed periodically to prevent the corruption of its contents due to charge leakage. On the other hand, SRAM uses a bistable element to represent the stored information, and thus it does not need to be refreshed. Both of SRAM and DRAM are volatile memory devices, which means that their contents are lost if the power supply is removed from these devices. Nonvolatile memory retains its contents even when the power supply is turned off. All current ROM devices, including mostly-read-sometimes-write devices, are nonvolatile memories.

MEMORY ACCESS INTERFACE

Technology is not the only factor that contributes to the performance of a memory device. Architectural methods also affect the speed of memory. Some of the architectural features are time multiplexing, pipelining, burst mode, clocking methodology, and separated input and output ports. Many times we need to trade off cost with performance when deciding what method to use. We will first discuss several common features used in memory devices.

Asynchronous Versus Synchronous Access

Memory can be accessed asynchronously or synchronously. It is more natural to follow the asynchronous interface. In this mode an address is presented to the memory by a processor. After a certain delay, data are made available at the pin for access. We call the delay between address made available to data ready the *memory access time*. Sometimes the access time is measured from a particular control signal. For example, the time between read control line ready and

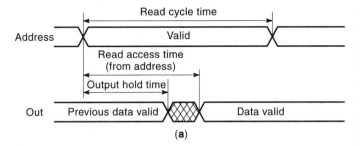

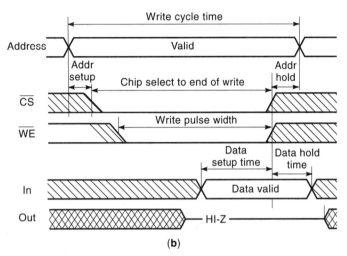

Figure 1. Asynchronous memory access. (a) Asynchronous read cycle. (b) Asynchronous write cycle.

data ready is called *read access time*. Figure 1 shows the timing diagrams of asynchronous memory access schemes. In the first diagram we assume that both the chip select and read enable signals are enabled. The write cycle diagram shown is a write cycle controlled by the write enable control signal. It is important to note that memory access time is different from memory cycle time. The *memory cycle time* is the minimum time between two consecutive memory accesses. The *memory writes command time* is measured from the write control ready to data stored in the memory. The *memory latency time* is the interval between CPU issuing an address and data available for processing. The *memory bandwidth* is the maximum amount of memory capacity being transferred in a given time.

Synchronous access implies a clock signal. Both address and control signals are latched into registers upon the arrival of the clock signal freeing the processor from holding the input to the memory for the entire access time. Instead the processor can initiate the access and continue to perform other important tasks. Figure 2 illustrates generic synchronous access cycles. In this figure we say that the read access has a two-cycle latency, since the data are made available after two clock cycles. Similarly we say that the write operation has zero-cycle latency.

Time Multiplexing

In order to reduce the cost of packaging, many different memory devices use time multiplexing to communicate information to and from other devices. One of the most common time-multiplexing examples is shared input/output (I/O). A memory chip can be configured with either separated or shared

data inputs and outputs. The advantage of having a smaller package when shared inputs and outputs are used is more evident when the width of the data is large. However, the drawback is the possibility of having a slower interface due to contention. For a shared I/O device, either the write enable or the chip select control signal must be off during address transition when writing. Setting one of the control signals off disables the read operation. When the device is not being read, the I/O bus is set to high impedance, thus allowing the data input to be loaded onto the I/O pins. Other common examples of time multiplexing are most of the dynamic random access memory (DRAM) devices. DRAM differs from a static random access memory (SRAM) in that its row and column addresses are time-multiplexed. Again the main advantage is to reduce the pins of the chip package. Due to time multiplexing there are two address strobe lines for the DRAM address: row address strobe (RAS) line and column address strobe (CAS) line. These control signals are used to latch the row and column addresses, respectively. There are many ways to access the DRAM.

When reading, a row address is given first, followed by the row address strobe signal RAS. RAS is used to latch the row address on chip. After RAS, a column address is given followed by the column address strobe CAS. After a certain delay (read access time), valid data appear on the data lines. Memory write is done similarly to memory read, with only the read/write control signal reversed. There are three cycles available to write a DRAM. They are early write, read-modify-write, and late write cycles. Figure 3 shows only the early write cycle of a DRAM chip. Other write cycles can be found in most of the DRAM data books. We list a few of them here: (1) page mode, (2) extended data output (EDO) mode or hyper page mode, (3) nibble mode, and (4) static column mode.

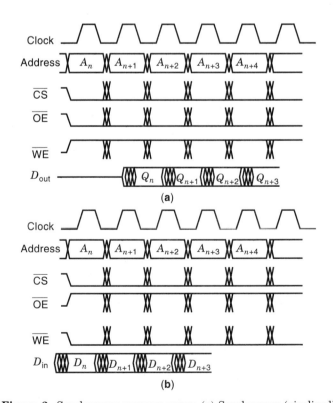

Figure 2. Synchronous memory access. (a) Synchronous (pipelined) read cycle. (b) Synchronous (pipelined) write cycle.

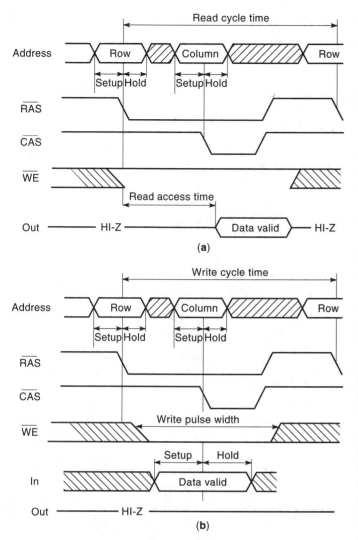

Figure 3. DRAM read and write cycles. (a) DRAM read cycle. (b) DRAM (Early) write cycle.

In page mode (or fast page mode), a read is done by lowering the RAS when the row address is ready. Then, repeatedly give the column address and CAS whenever a new one is ready without cycling the RAS line. In this way a whole row of the two-dimensional array (matrix) can be accessed with only one RAS and the same row address. This is called page mode, since we can arrange the memory device so that the upper part of the memory address specifies a page and the lower portion of the address is used as a column address to specify the offsets within a page. Due to locality, access local to the page does not need to change the row address, allowing faster access. Figure 4 illustrates the read timing cycle of a page mode DRAM chip. Static column is almost the same as page mode except the CAS signal is not cycled when a new column address is given—thus the static column name. In page mode, CAS must stay low until valid data reach the output. Once the CAS assertion is removed, data are disabled and the output pin goes to the open circuit. With EDO DRAM, an extra latch following the sense amplifier allows the CAS line to return to high much sooner, permitting the memory to start precharging earlier to prepare for the next access. Moreover, data are not disabled after CAS goes high. With burst EDO DRAM, not only does the CAS line return to high,

it can also be toggled to step though the sequence in burst counter mode, providing even faster data transfer between memory and the host. IBM originated the EDO mode and called it the hyper page mode (HPM). In the nibble mode after one CAS with a given column, three more accesses are performed automatically without giving another column address (the address is assumed to be increased from the given address).

Special Memory Structures

The current trend in memory devices is toward larger, faster, better-performance products. There is a complementary trend toward the development of special purpose memory devices. Several types of special-purpose memory are offered for particular applications such as content addressable memory for cache memory, line buffers (FIFO or queue) for office automation machines, frame buffers for TV and broadcast equipment or queue, and graphics buffers for computers.

A special type of memory called content addressable memory (CAM) or associative memory is used in many applications such as cache memory and associative processor. CAM is also used in many structures within the processor such as scheduling circuitry and branch prediction circuitry. A CAM stores a data item consisting of a tag and a value. Instead of giving an address, a data pattern is given to the tag section of the CAM. This data pattern is matched with the content of the tag section. If an item in the tag section of the CAM matches the supplied data pattern, the CAM will output the value associated with the matched tag. CAM cells must be both readable and writable just like the RAM cell. Most of the time the matching circuit is built within the memory cell to reduce the circuit complexity. Figure 5 shows a circuit diagram for a basic CAM cell with a "match" output signal. This output signal may be used as input for other logic such as scheduling or used as an enable signal to retrieve the information contained in the other portion of the matched entry.

A FIFO/queue is used to hold data while waiting. It is often called a "buffer" because it serves as the buffering region for two systems, which may have different rates of consuming and producing data. A very popular application of FIFO is in office automation equipment. These machines require high-performance serial access of large amounts of data in each horizontal line such as digital facsimile machines, copiers and image scanners. FIFO can be implemented using shift registers or RAM with pointers.

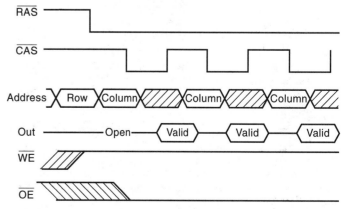

Figure 4. Page mode read cycle.

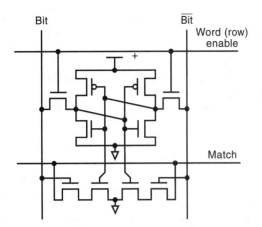

Figure 5. Static CMOS CAM cell.

There is rapid growth in computer graphic applications. The technology which is most successful is termed raster scanning. In a raster scanning display system, an image is constructed with a series of horizontal lines. Each of these lines is connected to pixels of the picture image. Each pixel is represented with bits controlling the intensity. Usually there are three planes corresponding to each primary color: red, green, and blue. These three planes of bit maps are called frame buffer or image memory. Frame buffer architecture affects the performance of a raster scanning graphic system greatly. Since these frame buffers need to be read out serially to display the image line by line, a special type of DRAM memory called video memory or VDRAM is used. Usually this memory is dual ported with a parallel random access port for writing and a serial port for reading. Although synchronous DRAMs are still popular for current PCs, VDRAM is used commonly in high-end graphic systems because of the memory access bandwidth required. We can calculate the memory bus speed as follows. Assume we have a screen size of x by y pixels. Each pixel is made of three colors of z bytes. We further assume that the refresh cycle of the screen is r Hz. Then the total data rate required is the product of all four terms $xyzr$. Now depending on the memory we use, only a certain percentage of the memory access time can be allocated for refresh. Other times we need the interface channel to store new image information. That is, only a portion of the bandwidth is available for reading, since we need to write and refresh the memory. Let's assume that the portion used for refresh (refresh efficiency) is e. We further assume that the width of the memory system is w, and then the memory bus speed required to provide the refresh rate for this graphic screen is $xyzr/we$. For example, in order to refresh a screen size of 1280×1024 pixels with 3 bytes (1 byte for each primary color) at 75 Hz and a 30% refresh efficiency, we need a bus speed of 245 MHz if the bus width is 32 bits. Figure 6 illustrates two designs of a multiple-ported SRAM cell.

New Memory Interface Technique

Until recently, memory interface has progressed with evolution instead of revolution. However, since the memory bandwidth requirement continues to grow, revolutionary techniques are necessary. A new general method uses a packet-type of memory interface. One such interface is proposed by Rambus called Direct RDRAM. Another is termed SLDRM. Both technologies use a narrow bus topology with matched termination operating at high clock frequency to provide the needed bandwidth. In addition, they utilize heavily banked memory blocks to allow parallel access to the memory arrays providing the needed average access time (see paragraph on memory interleaving in the "Memory Organization" section to learn more about memory banks).

MEMORY HIERARCHY

Modern computer systems have ever growing applications. As a result, the application programs running on these computer systems grow in size and require large memories with quick access time. However, the speed of memory devices has been lagging behind the speed of processors. As CPU's speed continues to grow with the advancement of technology and design technique (in particular pipelining), due to the nature of increasing memory size, more time is needed to decode wider and wider addresses and to sense the information stored in the ever-shrinking storage element. The speed gap between processor and memory will continue to grow wider in the future. Cost is another important reason why memory hierarchy is important. Memory hierarchy works because of the locality property of memory references due to the sequentially fetched program instructions and the conjugation of related data. It works also because we perform memory reads much more than memory writes. In a hierarchical memory system there are many levels of memory. A small amount of very fast memory is usually allocated and brought right next to the central processing unit to help match up the speed of the CPU and memory. As the distance becomes greater between the CPU and memory, the performance requirement for the memory is relaxed. At the same time, the size of the memory grows larger to accommodate the overall memory size requirement. Some of the memory hierarchies are registers, cache, main memory, and secondary memory (or disk). When a memory reference is made, the processor accesses the memory at the top of the hierarchy. If the desired

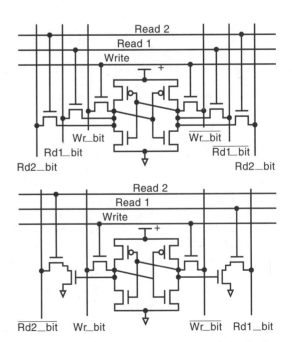

Figure 6. Two designs of Multiported CMOS SRAM cell (shown with 2-read and 1-write ports).

data are in the higher hierarchy, it wins because information is obtained quickly. Otherwise a *miss* is encountered. The requested information must be brought up from a lower level in the hierarchy. We will discuss cache memory and virtual memory in more detail.

Cache

Cache memory provides a fast and effective access time to main memory. A memory reference *hits* if the data are found in the cache. It *misses* if the data are not in the cache and had to be brought in. The amount of misses over the total reference is called the *miss rate*. We may categorize the cache misses in three ways—*compulsory miss, capacity miss,* and *conflict miss.* Compulsory miss rate is independent of the cache organization. It is incurred when a new memory is referenced or after a cache flush. Capacity miss occurs mainly due to the fact that caches are smaller in size compared with main memory. Depending on the cache mapping strategy, there also may be conflict miss even when the cache is not filled. Conflict miss happens because two memory references are mapped into the same cache location. When a miss occurs, a whole block of memory containing the requested missing information is brought in from the lower hierarchy. This block of memory is called a *cache line* or simply a *cache block.* Cache line is the basic unit used in cache. Access to only a part of the line brings the entire line into the cache. Since data and instructions process spatial locality, an entire line acts like pre-fetching, since the nearby addresses are likely to be used soon. Large lines pre-fetch more. However, too large a line may bring unused memory into the cache and pollute the cache unnecessarily and cause the cache to have greater capacity miss. It also wastes memory bandwidth.

Each cache line coexists with a tag that identifies the data held in the line by the data's address. The line hits if the tag matches the requested address. *Sets* comprise lines and do not distinguish among these lines. That is, any lines within a set can be mapped into the same cache location. A cache access takes two steps. The first step is a selection step where the set is indexed. The second step is the tag check step where the tags from the lines are checked and compared against the address. The size of the set gives the *associativity* of the cache. A cache with set size of one is called a *direct mapped* cache. A set size of two is called a *two-way set-associative* cache. A cache with all lines in one set is called *fully associative.* There are several ways to map the cache line into the cache from the main memory. We illustrate these mapping methods with an example. Assume that there are 8 blocks in a cache. An address 11 will map to location 3 in a direct mapped cache. The same address will be mapped to either location 6 or 7 if the cache is two-way set associative. If the cache is a four-way associative cache, then the address 11 may be mapped to locations 4 to 7 of the cache. In a fully associative cache, the address 11 may be mapped into any location of the cache. Figure 7 shows this example in detail. With higher associativity, conflict misses can be reduced. However, such cashes are more complex to build too. In general, associativity trades latency for miss rate. A fully associative cache is a CAM; since each address may be mapped to any location of the cache, a reference to see if an entry is in the cache needs to check every tag of the entire cache. When a memory location needs to be updated with a new result, we must update both the cache and the main memory. The *write-through* cache updates both the cache and the memory simultaneously at the time a write is issued. The *copy-back* (or *write back*) cache does not update immediately the main memory at writing until a block is replaced from the cache. This technique requires an extra bit for each cache block signaling whether the block is *dirty* (has changed the content since reading into the cache) or not. With the dirty bit, we don't have to write the memory every time a cache block is replaced. Only the block with the dirty bit set needs to be written into the main memory while others are simply thrown away. However, in a multi-processor system we need to prevent a processor from reading a stalled cache line, when that cache line has been written by another processor with the copy-back write policy. That is, we need to enforce the coherency of the cache. A popular method is called snooping cache. In this method all caches monitor the memory bus activity. When a cache write occurs, it updates the cache and also issues a memory write cycle for the first word of the cache line. All other caches snooping on the memory bus cycle will detect this write and invalidate the cache line in their cache. Write-through cache requires a larger memory bandwidth and has a longer average write access time.

If the current memory hierarchy level is full when a miss occurs, some existing blocks must be removed and sometimes written back to a lower level to allow the new one(s) to be brought in. There are several different replacement algorithms. One of the commonly used methods is the *least recently used* (LRU) replacement algorithm. Other algorithms are first-in first-out (FIFO) and random. In modern computing systems, there may be several sublevels of cache within the hierarchy of cache. For example, the Intel Pentium PRO system has on-chip cache (on the CPU chip) which is called Level 1 (L1) cache. There is another level of cache which resides in the same package (multichip module) with the CPU chip which is called Level 2 (L2) cache. There could also be a Level 3 (L3) cache on the motherboard (system board) between the CPU chip(s) and main memory chips (DRAMs). Moreover, there are also newer memory devices such as synchronous RAM, which provides enough bandwidth and speed to be interfaced with a processor directly through pipelining. We can express the average memory access time with the following equation:

$$T_{\text{avg}} = \sum_{i=j}^{p_i} \left[p_i \prod_{j=j}^{i-j} (1 - p + j) t_{t_i} \right] + \prod_{i=j}^{p_i} (1 - p_i) t_{\text{m}}$$

For example, a particular computer system has two levels of cache between the processor and the main memory. L1 cache has the same access time as the processor (t). L2 cache an access time 5 times the processor cycle time. Main memory has an access time 50 times the processor cycle time. If we assume that a particular program running on this system has an L1 cache hit rate of 95% and an L2 hit rate of 70%, the average memory access time will be $1.875t$. If we use some kind of cleaver design and increase the hit rate of L2 by 5%, the average access time will reduce to $1.7625t$. On the other hand, if we introduce another level of hierarchy between the main memory and L2 cache, which has a hit rate of 60% and an access time of $20t$, the average access time will reduce further to $1.605t$ instead. By making the cache smarter and having more levels of cache, we can reduce the average memory access time, assuming that the memory

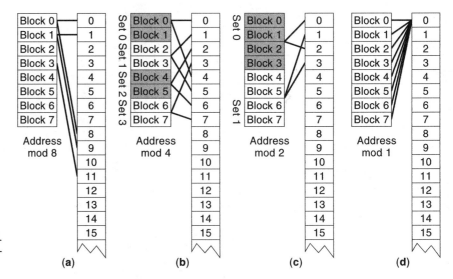

Figure 7. Mapping methods. (a) Direct mapped. (b) Two-way set associative. (c) Four-way set associative. (d) Fully associative.

access time keeps up with the processor cycle time. Unfortunately the trend says otherwise. The speed gap between DRAM and CPU continues to grow. The following scenario explains the effect of this gap. In most programs, 20% to 40% of the instructions reference memory; a particular program that references the memory with 25% of its instruction means that, on average, during execution every fourth instruction references memory. The previous memory system, with three levels of cache, will reach this barrier when the average memory cycle time (in multiples of processor cycle time) reaches $450t$. That is, at the speed ratio the computer system performance running this program is totally determined by memory speed. Making the processor faster will not affect the wall clock to complete the program. We call this the "memory wall."

Virtual Memory

A virtual memory system provides a memory space that is larger than the actual physical memory size of application program being executed. In a computer system the size of the total memory space is usually defined by the instruction set and memory management architecture. The size of the total memory space is typically governed by the width of the computer data path, since a computer uses the arithmetic unit of the CPU to calculate addresses. For example, a 32-bit processor usually has a memory space of size 4 GB (2 to the power of 32). We refer to this type of memory space as linear address space. A clear exception to this rule is the Intel Architecture (or ×86 architecture). The 32-bit Intel Architecture (IA-32) uses segmentation to manage its memory and gives a larger space than 4 GB. Nevertheless, all modern processors divide the entire memory space into chunks which are called *pages*. The size of a memory chunk is called page size. A typical page size is about a few kilobytes. A special program called operation system (OS) manages the pages by setting up a page table. A page table keeps track of pages that are actually in the physical memory. When a process makes a memory reference by issuing a virtual address, this is translated into (1) an index in the page table in order to locate the page this address is in and (2) an offset within the located page. If the page that it looks up is not in the physical memory, a page fault occurs. *Demand paging* brings that page in from the sec-

ondary memory (usually a disk). Since the physical memory is smaller than the total memory space, eventually all space in the physical memory will be filled. After the physical memory is filled and a new page needs to be brought in, we must replace the existing page with a new page. This process of replacing an existing page is called *swapping*. If the total memory space required by a program is much larger than the physical memory space, we may thrash the computer by swapping back and forth pages which have been used recently. There also might be another level of indirection when the number of pages is too many. We call it the *directory table*. In this case a virtual address must be translated and used to look up the directory table to find the page table fist. Then a page entry is located within the page table where it has been located. Then an offset into the page table is used to locate the entry of the physical memory, which is being accessed. The looking up of tables required for every memory access can consume a significant amount of time since each is a memory reference too, not to mention the addition operation it sometimes needs. To speed up the translation time, a *translation lookaside buffer* (TLB) stores frequently used completed translations for reuse.

MEMORY ORGANIZATION

System Level Organization

So far, we have not specified the exact size of a memory entry. A commonly used memory entry size is one byte. For historical reasons, memory is organized in bytes. A byte is usually the smallest unit of information transferred with each memory access. Wider memory entry is becoming more popular as the CPU continues to grow in speed and complexity. There are many modern systems which have a data width wider than a byte. A common size is a double word (32-bit), for example, in current desktop computers. As a result, memory in bytes is organized in sections of multibytes. However, due to need for backward compatibility, these wide datapath systems are also organized to be byte addressable. The maximum width of the memory transfer is usually called *memory word length,* and the size of the memory in bytes is called *memory capacity.* Since there are different memory device sizes, the

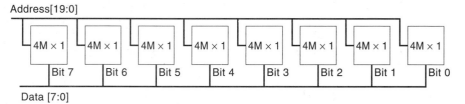

Address[19:0]

| Bit 7 | Bit 6 | Bit 5 | Bit 4 | Bit 3 | Bit 2 | Bit 1 | Bit 0 |

Data [7:0]

Figure 8. Eight 4 M × 1 chips used to construct a 4 Mbyte memory.

memory system can be populated with different-sized memory devices. For example, a 4 Mbyte of main memory (physical memory) can be put together with eight 4 Mbit × 1 chips as depicted in Fig. 8. It can also be designed with eight 512 Kbyte × 8-memory devices. Moreover, it can also be organized with a mixture of different-sized devices. These memory chips are grouped together to form memory modules. SIMM is a commonly used memory module which is widely used in current desktop computers. Similarly, a memory space can also be populated by different types of memory devices. For example, out of the 4MB space, some may be SRAM, some may be PROM, and some may be DRAM. They are used in the system for different purposes. We will discuss the differences of these different types of memory devices later.

There are two performance parameters in a memory system, namely, *memory bandwidth* and *memory latency*. In many cases the important factor in a high-performance computer system is the bandwidth because if we can access more data per access, then the average access time per data is shorter. However, a wider memory system is less flexible. It must increase by a larger chunk when upgraded.

Memory Device Organization

Physically, within a memory device, cells are arranged in a two-dimensional array, with each of the cells capable of storing one bit of information. Specifying the desired row and column addresses will access this matrix of cells. The individual row enable line is generated using an address decoder while the column is selected through a multiplexer. There is usually a sense amplifier between the column bit line and the multiplexer input to detect the content of the memory cell it is accessing. Figure 9 illustrates this general memory cell array

described by an r-bit of row address and a c-bit of column address. With the total number of $r + c$ address bits, this memory structure contains a 2^{r+c} number of bits. As the size of the memory array increases, the row enable lines as well as the column bit lines become longer. In order to reduce the capacitive load of a long row enable line, the row decoders, sense amplifiers, and column multiplexers are often placed in the middle of divided matrices of cells as illustrated in Fig. 10. By designing the multiplexer differently we are able to construct memory with different output width—for example, ×1, ×8, ×16, and so on. In fact, memory designers make great effort to design the column multiplexers so that most of the fabrication masks may be shared for memory devices which have the same capacity but with different configurations. In large memory systems, with tens or hundreds of integrated circuit (IC) chips, it is more efficient to use 1-bit-wide (×1) memory IC chips. This tends to minimize the number of data pins for each chip, thereby reducing the total board area. One-bit-wide memory chips are a disadvantage in small systems, since a minimum of eight chips is needed to implement the desired memory for a memory system with one byte width. Due to the limit of board size, often several memory chips are connected to form a memory module on a specialized package. We called these memory modules. Some examples are SIMM, ZIF, and so on.

Memory Interleaving

Interleaving is a technique for organizing memory into leaves (memory banks) that increases the sustainable memory bandwidth. Each leaf can process a memory request for a processor independently. The latency of DRAM access, which is long

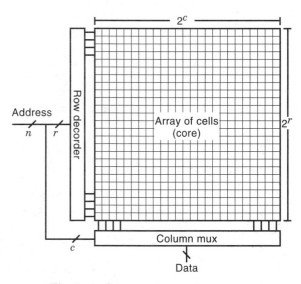

Figure 9. Generic 2-D memory structure.

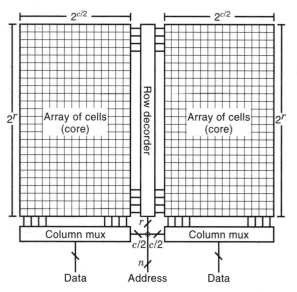

Figure 10. Divided memory structure.

compared with the CPU clock rate, is hidden from the processor when overlapped memory access is initiated in multiple memory leaves.

MEMORY DEVICE TYPES

As mentioned before, according to the functionality and characteristics of memory, we may divide memory devices into two major categories: ROM and RAM. We will describe these different types of devices in the following sections.

Read-Only Memory

In many systems, it is desirable to have the system level software (e.g., BIOS) stored in a read-only format, because these type of programs are seldom changed. Many embedded systems also use ROM to store their software routines because these programs are also never changed during their lifetime in general. Information stored in this ROM is permanent. It is retained even if the power supply is turned off. This memory can be read out reliably by a simple current-sensing circuit without worrying about destroying the stored data. The effective switch position at the intersection of word-line/bit-line determines the stored value. This switch could be implemented using many different technologies resulting in different types of ROM. The most basic type of this ROM is called masked ROM or simply ROM. It is programmed at the manufacturing time using fabrication processing masks. ROM can be produced using many different technologies: bipolar, CMOS, nMOS, pMOS, and so on. Once they are programmed, there is no means to change their contents. Moreover, the programming process is performed at the factory. Some ROM is also one time programmable, but it is programmable by the user at the user's own site. These are called programmable read-only memory (PROM). It is also often referred to as write-once memory (WOM). PROMs are based mostly on bipolar technology, since this technology supports it very nicely. Each of the single transistors in a cell has a fuse connected to its emitter. This transistor and fuse make up the memory cell. When a fuse is blown, no connection can be established when the cell is selected using the ROW line, and thus a zero is stored. Otherwise, with the fuse intact, logic one is represented. The programming is done through a programmer called PROM programmer or PROM burner. It is sometimes inconvenient to program the ROM only once. Thus the erasable PROM is designed. This type of erasable PROM is called EPROM. The programming of a cell is achieved by avalanche injection of high-energy electrons from the substrate through the oxide. This is accomplished by applying a high drain voltage, causing the electrons to gain enough energy to jump over the 3.2 eV barrier between the substrate and silicon dioxide, thus collecting charge at the floating gate. Once the applied voltage is removed, this charge is trapped on the floating gate. Erasing is done using an ultraviolet (UV) light eraser. Incoming UV light increases the energy of electrons trapped on the floating gate. Once the energy is increased above the 3.2 eV barrier, it leaves the floating gate and moves toward the substrate and the selected gate. Therefore these EPROM chips all have a window on their package where erasing UV light can reach inside the package to erase the content of cells. The erase time is usually in minutes. The presence of a charge on the floating gate will cause the metal oxide semiconductor (MOS) transistor to have a high thresh-

old voltage. Thus even with a positive select gate voltage applied at the second level of poly-silicon, the MOS remains to be turned off. The absence of a charge on the floating gate causes the MOS to have a lower threshold voltage. When the gate is selected, the transistor will turn on and give the opposite data bit. EPROM technologies that migrate toward smaller geometry make floating-gate discharge (erase) via UV light exposure increasingly difficult. One problem is that the width of metal bit-lines cannot reduce proportionally with advancing process technologies. EPROM metal width requirements limit bit-lines spacing, thus reducing the amount of high-energy photons that reach charged cells. Therefore, EPROM products built on submicron technologies will face longer and longer UV exposure time.

Reprogrammability is a very desirable property. However, it is very inconvenient to use a separate light-source eraser for altering the contents of the memory. Furthermore, even a few minutes of erase time is intolerable. For this reason, a new type of erasable PROM is then designed, called EEPROM. EEPROM stands for electrical erasable PROM. EEPROM provides new applications where erase is done without removing the device from the system in which it resides. There are a few basic technologies used in the processing of EEPROMs or electrical reprogrammable ROMs. All of them use the Fowler–Nordheim tunneling effect to some extent. In this tunneling effect, cold electrons jump through the energy barrier at a silicon–silicon dioxide interface and into the oxide conduction band through the application of high field. This can only happen when the oxide thickness is of 100 Å or less depending on the technology. This tunneling effect is reversible, allowing the reprogrammable ROMs to be used over and over again.

A new alternative has been introduced recently, namely, flash EEPROM. This type of erasable PROMs lacks the circuitry to erase individual locations. When you erase them, they are erased completely. By doing so, many transistors may be saved, and larger memory capacities are possible. One needs to note that sometimes one does not need to erase before writing. One can also write to an erased, yet unwritten, location, which results in an average write time comparable to an EEPROM. Another important thing to know is that writing zeros into a location charges each of the flash EEPROM's memory cells to the same electric potential so that subsequent erasure will drain an equal amount of free charge (electrons) from each cell. Failure to equalize the charge in each cell prior to erasure can result in the overerasure of some cells by dislodging bound electrons in the floating gate and driving them out. When a floating gate is depleted in this way, the corresponding transistor can never be turned off again, thus destroying the flash EEPROM.

Random Access Memory

RAM stands for random access memory. It is really read-and-write memory because ROM is also random access in the sense that given an address randomly, the corresponding entry is read. RAM can be categorized by the duration its content can last. Static RAM's contents will always be retained as long as power is applied. On the other hand, a DRAM needs to the refreshed every few milliseconds. However, most RAMs by themselves are volatile, which means that without the power supply their content will be lost. All of the ROMs

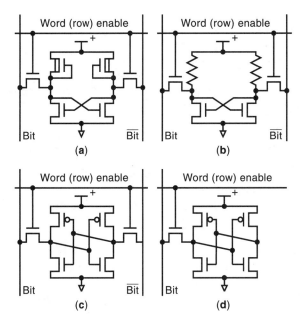

Figure 11. Different SRAM cell circuits. (a) Six-transistor SRAM cell with depletion transistor load. (b) Four-transistor SRAM cell with Poly-resistor load. (c) CMOS Six-transistor SRAM cell. (d) Five-transistor SRAM cell.

mentioned in the previous section are nonvolatile. RAM can be made nonvolatile by using a backup battery.

Figure 11 shows various SRAM memory cells (6T, 5T, and 4T). The six-transistor (6T) SRAM cell is the most commonly used SRAM. The crossed-coupled inverters in a SRAM cell retain the information indefinitely as long as the power supply is on, since one of the pull-up transistors supplies current to compensate for the leakage current. During a read, the bit and bitbar lines are pre-charged while the word enable line is held low. Depending on the content of the cell, one of the lines is discharged a little bit, causing the precharged voltage to drop, when the word enable line is strobed. This difference in voltage between the bit and bitbar lines is sensed by the sense amplifier, which produces the read result. During a write process, one of the bit/bitbar lines is discharged, and by strobing the word enable line the desired data are forced into the cell before the word line goes away.

The main disadvantage of SRAM is in its size since it takes six transistors (or at least four transistors and two resistors) to construct a single memory cell. Thus the DRAM is used to improve the capacity. Figure 12 shows the corresponding circuits for different DRAM cells. There is the four-transistor DRAM cell, the three-transistor DRAM cell, and the one-transistor DRAM cell. In a three-transistor cell DRAM, writing to the cell is accomplished by keeping the Read line low [refer to Fig. 12(b)] while strobing the Write line, and the desired data to be written are kept on the bus. If a one is desired to be stored. The gate of T2 is charged turning on T2. This charge will remain on the gate of T2 for a while before the leakage current discharge it to a point where it cannot be used to turn on T2. When the charge is still there, precharging the bus and strobing the Read line can perform a read. If a one is stored, then both T2 and T3 are on during a read, causing the charge on bus to be discharged. The sense amplifier can pick up the lowering of voltage. If a zero is stored, then there is no direct path from bus to GND; thus the charge

on bus will remain. To further reduce the area of a memory cell, a single transistor cell is often used and is most common in today's commercial DRAM cell. Figure 12(c) shows the one-transistor cell with a capacitor. Usually two columns of cells are the mirror image of each other to reduce the layout area. The sense amplifier is shared. In this one-transistor DRAM cell, there is a capacitor used to store the charge, which determines the content of the memory. The amount of the charge in the capacitor also determines the overall performance of the memory. Putting either a 0 or 1 (the desired data to store) does the writing on the read/writing line. Then the row select line is strobed. A zero or one is stored in the capacitor as charge. A read is performed by precharging the read/write line and then strobing the row select. If a zero is stored due to charge sharing, the voltage on the read/write line will decrease. Otherwise the voltage will remain. A sense amplifier is placed at the end to pick up if there is a voltage change or not. DRAM differs from SRAM in another aspect. As the density of DRAM increases, the amount of charge stored in a cell also reduces. It becomes more subjective to noise. One type of noise is caused by radiation called alpha particles. These particles are helium nuclei, which are present in the environment naturally or are emitted from the package that houses the DRAM die. If an alpha particle hits a storage cell, it may change the state of the memory. Since alpha particles can be reduced but not eliminated, some DRAMs institute error detection and correction techniques to increase their reliability.

Since DRAM loses the charge with time, it needs to be refreshed periodically. Reading the information stored and writing it back does refresh. There are several methods to perform refresh. The first is *RAS-only refresh*. This type of refresh is done row by row. As a row is selected by providing the row address and strobing RAS, all memory cells in the row are refreshed in parallel. It will take as many cycles as the number of rows in the memory to refresh the entire device. For example, a 1M×1 DRAM which is built with 1024 rows and columns will take 1024 cycles to refresh the device. In order to reduce the number of refresh cycles, memory arrays are sometimes arranged to have fewer rows and more columns.

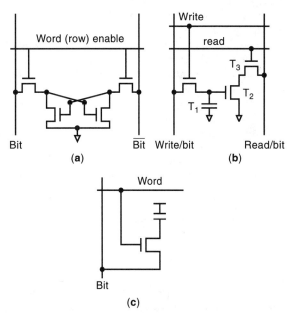

Figure 12. Different DRAM cells.

The address, however, is nevertheless multiplexed as two evenly divided words (in the case of 1M×1 DRAM the address word width is 10 bits each for rows and columns). The higher-order bits of address lines are used internally as column address lines, and they are ignored during the refresh cycle. No CAS signal is necessary to perform the RAS-only refresh. Since the DRAM output buffer is enabled only when CAS is asserted, the data bus is not affected during the RAS-only refresh cycles. Another method is called *hidden refresh*. During a normal read cycle, RAS and CAS are strobed after the respective row and column addresses are supplied. Instead of restoring the CAS signal to high after the read, several RAS may be asserted with the corresponding refresh row address. This refresh style is called the hidden refresh cycles. Again since the CAS is strobed and not restored, the output data are not affected by the refresh cycles. The number of refresh cycles performed is limited by the maximum time that CAS signal may be held asserted. One more method is named *CAS-before-RAS refresh* (self-refresh). In order to simplify and speed up the refresh process, an on-chip refresh counter may be used to generate the refresh address to the array. In such a case, a separate control pin is needed to signal to the DRAM to initiate the refresh cycles. However, since in normal operating RAS is always asserted before CAS for read and write, the opposite condition can be used to signal the start of a refresh cycles. Thus, in modern self-refresh DRAMs, if the control signal CAS is asserted before the RAS, it signals the start of refresh cycles. We called this CAS-before-RAS refresh, and it is the most commonly used refresh mode in 1 Mbit DRAMs. One discrepancy needs to be noted. In this refresh cycle the WE~ pin is a "don't care" for the 1 Mbit chips. However, the 4 Mbit specifies the CAS-before-RAS refresh mode with WE~ pin held at high voltage. A CAS-before-RAS cycle with WE~ low will put the 4 Meg into the JEDEC-specified test mode (WCBR). In contrast, applying a high to the test pin enters the 1 Meg test mode. All of the above-mentioned three refresh cycles can be implemented on the device in two ways. One method utilizes a distributed method, and the second method uses a wait-and-burst method. Devices using the first method refresh the row at a regular rate utilizing the CBR refresh counter to turn on rows one at a time. In this type of system, when it is not being refreshed, the DRAM can be accessed and the access can begin as soon as the self-refresh is done. The first CBR pulse should occur within the time of the external refresh rate prior to active use of the DRAM to ensure maximum data integrity and must be executed within three external refresh rate periods. Since CBR refresh is commonly implemented as the standard refresh, this ability to access the DRAM right after exiting the self-refresh is a desirable advantage over the second method. The second method is to use an internal burst refresh scheme. Instead of turning on rows at a regular interval, a sensing circuit is used to detect the voltage of the storage cells to see if they need to be refreshed. The refresh is done with a serial of refresh cycles one after another until all rows are completed. During the refresh, other access to the DRAM is not allowed.

CONCLUSION

Memory is becoming the determining factor in the performance of a computer. In this section we discussed four aspects of the memory architecture. These four aspects are (1) memory interface access, (2) memory hierarchy, (3) memory organization, and (4) memory devices. As projected, memory device size will continue to shrink and its capacity will continue to increase. Two newly merged memory architecture techniques to speed up computing systems are: (1) synchronous linked high-speed point-to-point connection and; (2) merged DRAM/logic.

GLOSSARY

Cache. A smaller and faster memory that is used to speed up the average memory access time.

CAM. Content addressable memory. This special memory is accessed not by an address but by a key, which matches the content of the memory.

DRAM. Acronym for dynamic random access memory. This memory is dynamic because it needs to be refreshed periodically. It is random access because it can be read and written randomly.

Interleaved memory. Dividing a memory into multiple banks so that access to different banks can be in parallel.

Memory access time. The time between a valid address supplied to a memory device and data becoming ready at output of the device.

Memory bandwidth. Amount of memory access per unit time.

Memory cycle time. The time between subsequent address issues to a memory device.

Memory hierarchy. Organize memory in levels to make the speed of memory comparable to the processor.

Memory latency. The delay between address issue and data valid.

Memory read. The process of retrieving information from memory.

Memory write. The process of storing information into memory.

ROM. Acronym for read-only memory.

SRAM. Acronym for static random access memory. This memory is static because it does not need to be refreshed. It is random access because it can be read and written.

Virtual memory. A method to use a smaller physical memory to support a larger logical memory space.

SHIH-LIEN L. LU
Oregon State University

MESSAGE PASSING. See Distributed memory parallel systems.

METACOMPUTING. See Heterogeneous distributed computing.

METAL-INSULATOR-SEMICONDUCTOR (MIS) TRANSMISSION LINES. See Slow wave structures.

METALLURGY OF BETA TUNGSTEN SUPERCON-DUCTORS. See Superconductors, metallurgy of beta tungsten.

METAL-METAL INTERFACES. See Bimetals.

METAL-SEMICONDUCTOR BOUNDARIES. See Ohmic contacts.

METAL SEMICONDUCTOR FIELD EFFECT TRANSISTORS

The metal-semiconductor field-effect transistor (MESFET) is one of the field-effect transistors in which the conduction process involves predominantly one kind of carrier, and the current transport between the source and drain electrodes is modulated by a voltage applied to the gate electrode. In the MESFET, a metal-semiconductor rectifying contact is used for the gate. There are a few other field-effect transistors: the junction field–effect transistor (JFET) and the metal-oxide-semiconductor field-effect transistor (MOSFET), where the gates are formed by a $p–n$ junction and a metal-oxide-semiconductor structure, respectively. In the Si device, the MOSFET is usually used because a high-quality insulating oxide (SiO$_2$) with a low density of interface states can be fabricated. The MOSFET having an insulated gate allows a higher input-voltage swing and higher input impedance than the other field-effect transistors. Compound semiconductors such as GaAs, InP, and InGaAs have higher electron mobilities and maximum drift velocities than Si, so field-effect transistors fabricated from GaAs etc. show higher operating speed and higher frequency performance. In the compound semiconductors such as GaAs, however, there are no good oxides or insulators to make the MOSFET or the insulated-gate field-effect transistor available now, although some good attempts have been reported recently (1). There exist high densities of interface states between the oxide (or insulator) and the compound semiconductor. Therefore, the MESFET structure is usually adopted for field-effect transistors fabricated from compound semiconductors like GaAs.

Historically, the GaAs MESFET was proposed by Mead (2) in 1966 and subsequently fabricated by Hooper and Lehrer (3) using a GaAs epitaxial layer on the semi-insulating GaAs substrate. In 1971, Turner et al. (4) got useful gain up to 18 GHz. In 1973, a first power GaAs MESFET was fabricated with 1.6 W at 2 GHz (5). Around 1980, the GaAs MESFET technology progressed greatly due to the availability of high-quality semi-insulating substrate and ion-implantation processing techniques. In another development, Mimura et al. (6) demonstrated a new type of field-effect transistor called the high electron mobility transistor (HEMT), where an AlGaAs/GaAs heterojunction with doped AlGaAs and non-doped GaAs layers is utilized. In the AlGaAs/GaAs HEMT or heterojunction field-effect transistor (HFET), the Schottky contact is formed on the AlGaAs layer, so this can be regarded as a kind of MESFET. However, we will only describe the normal MESFET (particularly GaAs MESFET) in this article.

Today, the GaAs MESFET is widely used in both high-speed and high-frequency applications. Particularly, it has been the workhorse of the microwave industry for many years (7). The GaAs MESFET is used as the active device for low noise and power amplifiers as well as for oscillators, mixers, and attenuators. Its microwave performance challenges that of HEMT (8). On the other hand, the integration scale of GaAs MESFET-ICs approaches 10^6 transistors on a chip (9), and GaAs-based 32-bit microprocessors are developed (10).

The superior performance of GaAs MESFET is due to the higher electron mobility and the higher electron velocity of GaAs. However, there are several unfavorable phenomena in GaAs MESFET, such as short-channel effects, sidegating effects, frequency-dependent output conductance and transconductance, slow-current transients, and kink phenomena. The short-channel effect is a phenomenon that the threshold voltage of a MESFET shifts with shortening the gate length, and the sidegating effect is a phenomenon that the drain current of the MESFET is modulated when a negative voltage is applied to an adjacent device in ICs. The kink is a phenomenon that the drain conductance shows an abnormal increase at relatively high drain bias. Almost all phenomena listed above are originated from the fact that the semi-insulating substrate (on which the MESFET is fabricated) is achieved by impurity compensation by deep levels, and that high densities of surface states exist on the active layer of GaAs MESFET. However, the detailed mechanisms are not necessarily made clear.

In this article, we first describe the basic operation principle of the MESFET and its current-voltage characteristics that are derived physically. Next, typical device structures of GaAs MESFETs are described, and their high-speed and high-frequency performances are reviewed. Then we describe parasitic effects in GaAs MESFETs, such as substrate conduction, sidegating effects, slow-current transients, low-frequency anomalies, and kink phenomena. Finally, some modeling methods for GaAs MESFETs are presented which are important for circuit design and for understanding physical phenomena in GaAs MESFETs.

BASIC PRINCIPLES

Operation Principle

Figure 1 shows a schematic diagram of a GaAs MESFET. A conductive n-layer is formed on the semi-insulating GaAs substrate which has a high resistivity of $\sim 10^8$ Ωcm. So usually, current does not flow in the substrate region. On the n-layer, two ohmic contacts are provided. One acts as the source and the other as the drain. When a positive voltage V_D is applied to the drain with respect to the source, electrons flow from source to drain. Hence, the source supplies carriers, and the drain acts as the sink.

The third electrode, the gate, forms a rectifying Schottky contact with the n-layer, and so the depletion region exists around the gate. Because the positive voltage is applied to the drain, the depletion layer extends deeper at the drain side. The width of depletion layer can change by applying the gate voltage, so the thickness of conductive channel is varied. Therefore, current from source to drain can be modulated by the gate voltage, leading to the three terminal device.

For a given gate voltage V_G, the channel current increases as the drain voltage increases. Eventually, for sufficient large

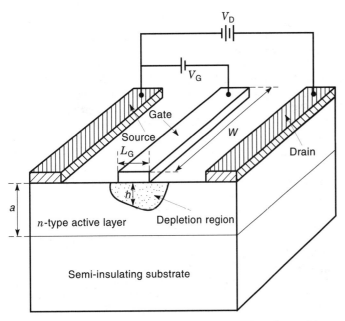

Figure 1. Schematic diagram of a (GaAs) MESFET on the semi-insulating substrate. The depletion region is formed under the gate.

V_D, the current saturates due to the pinching of the channel at the drain side or the electron velocity saturation there. The pinching of the channel means that the n-layer is fully depleted (at the drain side) due to the reverse gate-to-drain voltage. In Fig. 1, the basic device dimensions are the gate length L_G, the gate width W, the channel depth a, and the depletion-layer width h.

The operation of a MESFET is identical to that of a JFET, which was first analyzed by Shockley (11) in 1952. We will next describe current-voltage characteristics of a MESFET that are derived physically.

Current-Voltage Characteristics

Contact Mobility Model. The simplest but most essential method for deriving I–V characteristics of a MESFET is based on that by Shockley (11,12). A long-channel MESFET is considered ($L_G \gg a$), and the following assumptions are adopted: (1) gradual channel approximation, (2) abrupt depletion layer, and (3) constant mobility. As shown in Fig. 2, we consider a region under the gate and assume that the semi-insulating layer is perfectly insulating. Now, we treat a case with uniform doping N_D. Under the gradual channel approximation, the depletion layer width h varies only gradually along the x direction, and it can be obtained by solving the one-dimensional Poisson's equation in the y direction:

$$\frac{d^2\psi}{dy^2} = -\frac{qN_D}{\epsilon} \tag{1}$$

Using the boundary condition that $\psi = V_G - V_b$ at $y = a$ and $\psi = V(x)$ at $y = a - h$, we obtain

$$h(x) = \sqrt{\frac{2\epsilon\{V(x) + V_b - V_G\}}{qN_D}} \tag{2}$$

where V_b is built-in potential at the Schottky contact, and $V(x)$ is the potential at x in the channel region. The depletion widths at the source and drain ends of the gate are

$$h_1 = \sqrt{\frac{2\epsilon(V_b - V_G)}{qN_D}} \qquad (x = 0) \tag{3}$$

$$h_2 = \sqrt{\frac{2\epsilon(V_D + V_b - V_G)}{qN_D}} \qquad (x = L_G) \tag{4}$$

The maximum value of h_2 is equal to a, and in such a case, the drain end of the gate pinches off and is depleted of carriers. The corresponding voltage is called the pinch-off voltage and defined as

$$V_P \equiv \frac{qN_D a^2}{2\epsilon} = V_{DSS} + V_b - V_G \tag{5}$$

where V_{DSS} is the drain voltage at which the pinch-off occurs.

The current density in the x direction along the channel is given by

$$J_x = q\mu N_D E_x = -q\mu N_D \frac{dV}{dx} \tag{6}$$

where the diffusion current is neglected. E_x is the electric field along the x direction, and μ is the electron mobility which is assumed constant. The channel current at x (or the drain current I_D) is then given by

$$I_D = q\mu N_D \frac{dV}{dx}(a - h)W \tag{7}$$

From Eq. (2), we obtain

$$dV = \frac{qN_D}{\epsilon} h\, dh \tag{8}$$

and hence,

$$I_D = \frac{q^2 N_D^2}{\epsilon}\mu W(a - h)h\frac{dh}{dx} \tag{9}$$

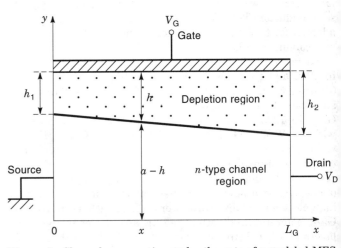

Figure 2. Channel cross-section under the gate of a modeled MESFET. The depletion region extends deeper at the drain side, and hence, the channel becomes thinner there.

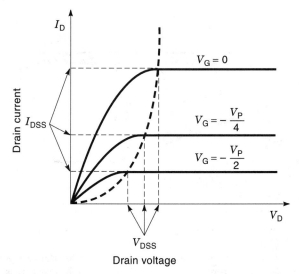

Figure 3. Basic I–V characteristics of a MESFET. The drain current saturates at $V_{DSS} = V_P + V_G - V_b$, and the saturation current I_{DSS} decreases as the gate voltage V_G becomes negative. V_P is the pinch-off voltage, and V_b is built-in potential at the Schottky contact.

Integrating from $x = 0$ $(h = h_1)$ to $x = L_G$ $(h = h_2)$ yields

$$\int_0^{L_G} I_D \, dx = \frac{q^2 N_D^2 \mu W}{\epsilon} \int_{h_1}^{h_2} (a - h) h \, dh \qquad (10)$$

Therefore,

$$I_D = \frac{W \mu q^2 N_D^2 a^3}{6 \epsilon L_G} \left\{ \frac{3}{a^2} (h_2^2 - h_1^2) - \frac{2}{a^3} (h_2^3 - h_1^3) \right\} \qquad (11)$$

or

$$I_D = I_P \left[3\frac{V_D}{V_P} - 2 \left\{ \left(\frac{V_D + V_b - V_G}{V_P} \right)^{3/2} - \left(\frac{V_b - V_G}{V_P} \right)^{3/2} \right\} \right] \qquad (12)$$

where

$$I_P = \frac{W \mu q^2 N_D^2 a^3}{6 \epsilon L_G} \qquad (13)$$

These expressions relate the current up to the point of pinch-off of the channel. At this bias, which occurs when $h_2 = a$, the drain current saturates and remains constant. This current I_{DSS} is given by

$$I_{DSS} = I_P \left[1 - 3\left(\frac{V_b - V_G}{V_P} \right) + 2 \left(\frac{V_b - V_G}{V_P} \right)^{3/2} \right] \qquad (14)$$

The current-voltage characteristics calculated from Eq. (12) are schematically shown in Fig. 3, where the saturation voltage is given by $V_{DSS} = V_P + V_G - V_b$.

From the current-voltage characteristics, we can obtain important device parameters such as transconductance g_m and drain conductance g_D. In the region before saturation,

from Eq. (12), we obtain

$$g_m \equiv \frac{\partial I_D}{\partial V_G} = \frac{3I_P}{V_P} \left\{ \left(\frac{V_D + V_b - V_G}{V_P} \right)^{1/2} - \left(\frac{V_b - V_G}{V_P} \right)^{1/2} \right\} \qquad (15)$$

$$g_D \equiv \frac{\partial I_D}{\partial V_D} = \frac{3I_P}{V_P} \left\{ 1 - \left(\frac{V_D + V_b - V_G}{V_P} \right)^{1/2} \right\} \qquad (16)$$

In the saturation region, $g_D = 0$ and from Eq. (14), g_m becomes

$$g_m = \frac{\partial I_{DSS}}{\partial V_G} = \frac{3I_P}{V_P} \left\{ 1 - \left(\frac{V_b - V_G}{V_P} \right)^{1/2} \right\} \qquad (17)$$

So g_m decreases when V_G becomes more negative.

The model presented here is useful when understanding the basic principle of the MESFET. However, in itself, this model cannot treat the characteristics beyond the pinch-off. Also, usually, the estimated drain current is rather higher than the experimental one. This is attributed to the fact that the electric-field dependence of electron mobility is neglected here.

Field-Dependent Mobility Model. Lehovec and Zuleeg (13) extend the previous model by considering electric-field dependence of electron mobility. They use the function:

$$\mu_n = \frac{\mu}{1 + \mu |E_x|/v_s} \qquad (18)$$

where v_s is the saturation velocity and takes a value of about 10^7 cm/s for GaAs at $T = 300$ K. As shown in Fig. 4, the drift velocity $v = \mu_n E_x$ saturates at high E_x. From Eq. (7), the drain current in this case is given by

$$I_D = qN_D \frac{\mu(dV/dx)}{1 + (\mu/v_s)(dV/dx)} (a - h)W \qquad (19)$$

Substituting Eq. (8) into Eq. (19), we obtain

$$I_D \left(1 + \frac{qN_D}{\epsilon} \frac{\mu}{v_s} h \frac{dh}{dx} \right) = \frac{q^2 N_D^2}{\epsilon} \mu W(a - h) h \frac{dh}{dx} \qquad (20)$$

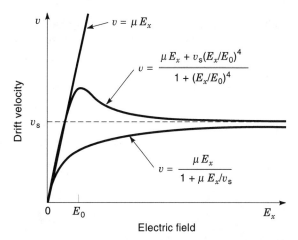

Figure 4. Three kinds of velocity versus electric field characteristics. One $(v = \mu E_x)$ is a case of constant mobility, and the other two are cases of field-dependent mobilities where the velocities saturate.

Integrating from $x = 0$ $(h = h_1)$ to $x = L_G$ $(h = h_2)$ yields

$$I_D = \frac{I_P}{1 + \mu V_D / v_s L_G}$$
$$\left[3\frac{V_D}{V_P} - 2\left\{ \left(\frac{V_D + V_b - V_G}{V_P} \right)^{3/2} - \left(\frac{V_b - V_G}{V_P} \right)^{3/2} \right\} \right] \quad (21)$$

Comparing Eq. (21) and Eq. (12) shows that the drain current is reduced by a factor of $(1 + \mu V_D / v_s L_G)$ due to the field-dependent mobility.

The above model successfully explains the reduction of the drain current. However, the used mobility model does not include the negative differential mobility observed in GaAs. Often, the electron drift velocity of GaAs is expressed analytically by the following-type function (14), as is also shown in Fig. 4.

$$v = \frac{\mu |E_x| + v_s (E_x / E_0)^4}{1 + (E_x / E_0)^4} \quad (22)$$

where E_0 is a parameter. Please also note that this model is only effective before the current saturates, as is the same in the previous model.

Two-Region Model. Statz et al. (15) developed a model that is effective also beyond current saturation. They used a velocity-field curve as shown in Fig. 5, where the mobility is assumed constant up to a critical field E_C, and the velocity is assumed constant beyond E_C. The MESFET is divided into two regions as shown in Fig. 6. Region I near the source is the constant-mobility region, and the gradual channel approximation described previously is applicable. Region II near the drain is the velocity saturation region, where a conductive channel of finite width is postulated to account for current continuity. The point $x = L_1$, which corresponds to the onset of velocity saturation, is allowed to move depending on the drain voltage V_D. Its position is determined by the location at which the longitudinal electric field E_x equals the critical field E_C. So the two-region model is applicable to operation conditions for all I-V characteristics including the saturation region.

In Region I, an expression of the current is essentially the same as that of Eq. (11) or Eq. (12). Integrating Eq. (9) from

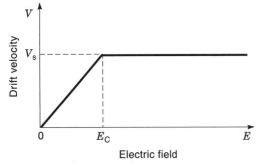

Figure 5. Velocity-field curve used for the two-region model. Below the critical field E_C, the mobility is constant, and the velocity is constant beyond E_C.

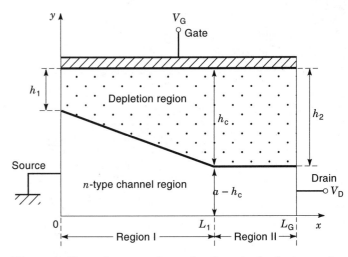

Figure 6. Channel cross-section under the gate for the two-region model. Region I is the constant mobility region, and Region II is the velocity saturation region where the channel thickness is constant.

$x = 0$ $(h = h_1)$ to $x = L_1$ $(h = h_c)$, we obtain

$$I_D = I_P \frac{L_G}{L_1} \left\{ \frac{3}{a^2} (h_c^2 - h_1^2) - \frac{2}{a^3} (h_c^3 - h_1^3) \right\} \quad (23)$$

where

$$h_c = \sqrt{\frac{2\epsilon (V_{DL1} + V_b - V_G)}{qN_D}} \quad (24)$$

Here, V_{DL1} is the potential at $x = L_1$ in the channel. We can determine L_1 by utilizing current continuity between Regions I and II. In Region II, electrons are assumed to travel at the saturation velocity v_s, and hence,

$$I_D = qN_D v_s (a - h_c) W \quad (25)$$

From Eq. (23) and Eq. (25), we obtain

$$L_1 = \frac{qN_D \mu \{3a (h_c^2 - h_1^2) - 2(h_c^3 - h_1^3)\}}{6\epsilon v_s (a - h_c)} \quad (26)$$

Once the h_c is known, the length L_1 is specified, and the current I_D is determined.

For a given I_D, the potential drop from the source to drain can be obtained by integrating the longitudinal electric field from $x = 0$ to $x = L_G$. In Region I, the potential drop V_{DL1} is, from Eq. (24) and Eq. (3),

$$V_{DL1} = \frac{qN_D}{2\epsilon} (h_c^2 - h_1^2) \quad (27)$$

In Region II, the potential drop V_{DL2} is determined by solving Laplace's equation. By taking the lowest space harmonic, we obtain (15)

$$V_{DL2} \simeq \frac{2a}{\pi} \frac{v_s}{\mu} \sinh \left\{ \frac{\pi}{2a} (L_G - L_1) \right\} \quad (28)$$

The drain voltage V_D is the sum of Eq. (27) and Eq. (28), and hence,

$$V_D = \frac{qN_D}{2\epsilon}(h_c^2 - h_1^2) + \frac{2a}{\pi}\frac{v_s}{\mu}\sinh\left\{\frac{\mu}{2a}(L_G - L_1)\right\} \qquad (29)$$

Eq. (29) and Eq. (26) allow one to determine L_1 and h_c for given V_D and V_G, yielding the current I_D. So $I_D - V_D$ curves as a parameter of V_G are obtained. The two-region model described here has been the basic of several physics-based analytical GaAs MESFET models later developed (16).

In the discussions above, we derived the current-voltage characteristics of a MESFET in closed or analytical forms based on various assumptions. Particularly, we used the gradual channel approximation and assumed the one-dimensional current flow. As the gate length becomes shorter and the drain voltage becomes larger, two-dimensional effects will dominate the device characteristics, and current flow in the depletion layer and in the substrate should be considered. In such cases, two-dimensional numerical simulation is required, where the Poisson's equation and the transport equations are solved simultaneously. Many works on this subject have been done to predict the I–V curves or to understand physical phenomena in GaAs MESFETs (17).

GaAs MESFET STRUCTURES

In the analysis done in the previous section, only the intrinsic region (under the gate) is considered. In real devices, there exist parasitic source resistance and drain resistance which originate from bulk regions between source and gate electrodes and between gate and drain electrodes, respectively. If the source resistance is high and the potential drop there becomes significant, the effective potential drop along the gate junction becomes smaller. So the degree of current modulation by the gate voltage is reduced, leading to a lower transconductance g_m. It should be also noted that high densities of surface states exist on the active layer, and so a surface depletion region is formed between source (drain) and gate electrodes. This contributes to increasing the source resistance.

Therefore, in real GaAs MESFETs, some methods to reduce the source resistance are adopted. There are many kinds of GaAs MESFET structures depending on their desired application. But we may classify GaAs MESFET structures in two main categories—the recessed gate structure and the self-aligned structure, as shown schematically in Fig. 7.

Recessed-Gate Structure

Recessing is a technique for adjusting the pinch-off (or threshold) voltage by reducing the active-layer thickness under the gate while maintaining a relatively low resistance between gate and source (drain) electrodes. This is achieved by using a rather thick n-type active layer in which the actual channel thickness is defined by controlled etching of a trench. The position and the shape of the recess are important design issues.

In power devices, the recess and the position of gate electrode are often asymmetrically located with a shorter distance to the source electrode than to the drain electrode. This has two advantages. It reduces the source resistance to maintain high transconductance g_m. And it increases the drain-to-source breakdown voltage and the gate breakdown (Schottky diode breakdown) voltage by allowing additional expansion

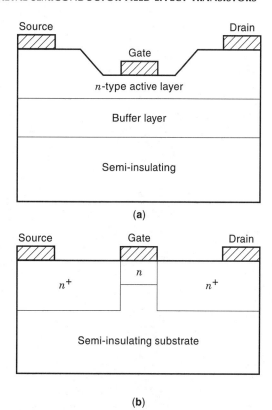

Figure 7. Schematic cross-section of the two main categories of GaAs MESFET structures: (a) recessed-gate structure, (b) self-aligned structure. Both of them have low source resistances.

space for the high-field region at the drain side of the gate. It is recognized that a graded recess is more effective than an abrupt recess in reducing electric field at the recess edge, and it provides higher breakdown voltage (18). As for the details on the breakdown phenomena, please refer to Ref. 18.

For microwave application, the semiconductor layer is typically grown by a molecular beam epitaxy (MBE) method to obtain desired doping profiles. A variety of doping profiles in the active layer may be used—from uniform doping to delta doping. Often, the gate is given a T-shape which combines a short gate length with a large gate metal cross section. The latter leads to a reduced parasitic gate resistance, which is particularly important in microwave devices.

A drawback of the recessed gate technique, especially in the context of ICs, is the limited alignment accuracy of the recess and of the gate electrode. The accuracy of recess etching is also a problem. These inaccuracies lead to non-uniformities in the source resistance, the transconductance and the threshold voltage over a wafer (and from wafer to wafer).

Self-Aligned Structures

The self-aligned gate technique is a method to self-align the source and drain n^+-layers to the gate as shown in Fig. 7(b). This structure is usually realized by first forming the gate region and then utilizing n^+-implantation. This is a planar structure and has a low source resistance and a high transconductance because of the n^+ source region which also reduces the surface-state effects. The self-aligned structure is particularly used for digital FETs, where the active layer is also fabricated by direct ion-implantation into the semi-insulating substrate, and relatively uniform threshold voltage over a wafer is realized. A drawback of this structure is the low breakdown voltage of the Schottky diode. To overcome this problem, the lightly doped drain (LDD) structure is adopted (19).

The basic feature of the self-aligned gate process is shown in Fig. 8 (20). First, an n-type active layer is formed in the semi-insulating substrate by Si ion-implantation and subsequent annealing. Next, TiW (a refractory metal) is deposited by sputtering and etched to form the gate electrode. The gate

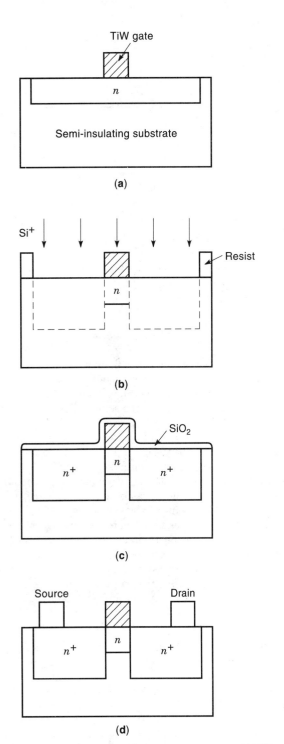

Figure 8. Fabrication process of self-aligned GaAs MESFET using TiW refractory gate, developed by Yokoyama et al. (20).

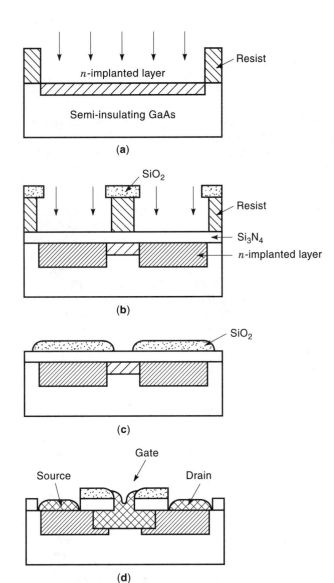

Figure 9. Fabrication process of SAINT MESFET. This process allows any choice of gate metal because the gate metalization is done after the high temperature anneal. (Reproduced with permission from K. Yamasaki, K. Asai, and K. Kurumada, GaAs LSI-directed MES-FET's with self-aligned implantation for n+-layer technology (SAINT), *IEEE Trans. Electron Devices,* **ED-29**(11): 1772–1777 (© 1982 IEEE).)

then served as a mask for the subsequent n^+ source and drain implant, which is followed by another annealing stage using a SiO$_2$ cap. The device is completed using AuGe/Au ohmic contacts formed by liftoff. The gate metal must be capable of surviving the high temperature anneal (about 850 °C) without damaging the Schottky barrier properties. Various alloy metals such as TiW-based alloys and WSi-based alloys have proven suitable for this purpose. These compositions are not very conductive, but this is usually not a severe problem for digital FETs in which the gate width may be only 10 to 20 μm. It would be a severe problem for analog FETs having much wider gate fingers.

Another class of self-aligned process known as SAINT (self-aligned implantation for n^+-layer technology) (21) involves the use of a complex mask structure acting as a "dummy gate" for the n^+ implantation, as shown in Fig. 9.

The process starts by the selective implantation of the active layer and the deposition of Si₃N₄ cap layer. Then, the dummy gate is fabricated from layers of resist and SiO₂, patterned in a T-shape by undercutting the lower resist using plasma etching. The n^+-implantation is followed by the sputter deposition of a layer of SiO₂, a lift-off step, and the annealing of the implanted dopants. Then the ohmic contacts are fabricated. Finally, the remaining Si₃N₄ in the gate area is removed, and the gate metal is deposited. The process allows any choice of gate metalization because the gate metal is placed on the wafer after the high temperature anneal.

In the self-aligned MESFETs with n^+ source and drain regions, current-flow via the semi-insulating substrate between the n^+-layers becomes remarkable when the gate length becomes shorter. So the threshold voltage shifts with shortening the gate length, showing a remarkable short-channel effect. To overcome this problem, so-called BP (Buried p-layer)—SAINT (22) was proposed as shown in Fig. 10. Here, the p-implanted layer is formed under the n and n^+ regions and acts as a barrier for electrons injected into the substrate. In

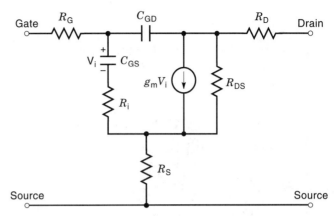

Figure 11. Small-signal equivalent circuit of a MESFET. C_{GS} and C_{GD} are the gate-to-source and gate-to-drain capacitances, respectively. R_i is the charging resistance, and R_{DS} is the output resistance. R_G, R_S, and R_D are the parasitic gate, source, and drain resistances, respectively.

fact, it is shown experimentally and theoretically (23) that the short-channel effect is greatly reduced by introducing the p-layer. However, the high dose buried p-layer may lead to the degradation of device performance due to its parasitic capacitance (24).

HIGH-SPEED AND HIGH-FREQUENCY PERFORMANCE

How fast the GaAs MESFET operates or switches is an interesting point for a practical viewpoint. In a logic circuit, however, the switching time depends on the load capacitance, and it is not a unique measure of high-speed performance. As to standard high-speed and high-frequency figures of merit for FETs, there are the cutoff frequency f_T and the maximum frequency of oscillation f_{max}. f_T is defined as the frequency at which the short-circuit current gain falls to unify, and f_{max} is the highest frequency at which power gain can be obtained from the FET. These are correlated to the small-signal equivalent circuit of the FET, and they are also easily estimated by microwave measurements.

A typical small-signal equivalent circuit of a GaAs MESFET is shown in Fig. 11 (7,12). From this, the cutoff frequency f_T is derived as

$$f_T \simeq \frac{g_m}{2\pi C_{GS}} \tag{30}$$

where C_{GS} is the gate-source capacitance. The approximate expression for f_{max} is given by

$$f_{max} \simeq \frac{f_T}{2}\sqrt{\frac{R_{DS}}{R_G + R_i + R_S}} \tag{31}$$

where R_{DS} is the output resistance, and R_i is the charging resistance. R_G and R_S are the parasitic gate and source resistances, respectively. f_T is also expressed, by using the transit time through the channel τ, as (12)

$$f_T = \frac{1}{2\pi\tau} \tag{32}$$

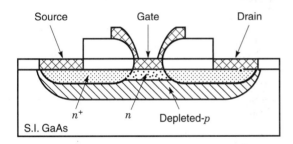

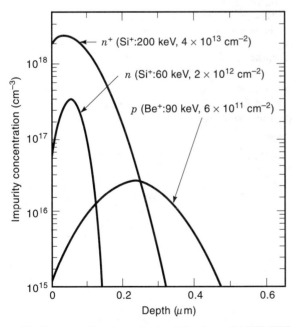

Figure 10. Cross-section view of a buried p-layer SAINT FET and calculated impurity concentration profiles. The p-layer acts as a barrier against electron injection from the channel into the semi-insulating substrate. (Reproduced with permission from K. Yamasaki, N. Kato, and M. Hirayama, Buried p-layer SAINT for very high-speed GAs LSI's with submicrometer gate length, *IEEE Trans. Electron Devices*, ED-32(11): 2420–2425 (© 1985 IEEE).)

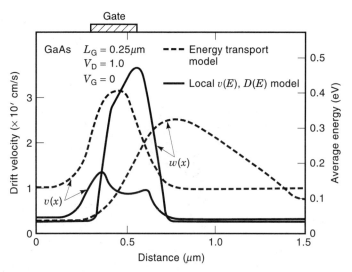

Figure 12. Drift velocity v and average electron energy w as functions of distance along the channel for a 0.25 μm gate-length GaAs MESFET, calculated by using energy-transport model (dashed lines) and quasi-equilibrium model (solid lines). In the case of energy transport model, the velocity becomes much higher than the saturation velocity (10^7 cm/s) under the gate, showing a remarkable velocity-overshoot effect. (Reproduced with permission from R. K. Cook and J. Frey, Two-dimensional numerical simulation of energy transport effect in Si and GaAs MESFET's, *IEEE Trans. Electron Devices*, **ED-29**(6): 970–977 (© 1982 IEEE).)

If we assume that electrons travel under the gate with the saturation velocity v_s, f_T becomes

$$f_T = \frac{v_s}{2\pi L_G} \qquad (33)$$

From this, we can say that f_T should become higher as the gate length L_G becomes shorter.

As is understood from Eq. (32) and Eq. (33), f_T depends on how the electrons travel through the channel. There are some theoretical calculations on electron velocity profiles in short-channel GaAs MESFETs. Figure 12 shows such an example (25). According to the model that includes energy transport effects, the electron velocity becomes much higher than the saturation velocity ($\sim 10^7$ cm/s). This so-called velocity overshoot effect is more pronounced in GaAs-based devices than in Si-based devices. Therefore, higher f_T than that estimated by assuming the velocity saturation is expected in short-channel GaAs MESFETs.

Das (26) theoretically estimated f_T, f_{max}, and g_m of a GaAs MESFET with a short gate (0.1 ~ 0.25 μm) by using the concept of charge control (27). Table 1 shows some of the results.

Table 1. Physical Parameters and Estimated Performance of GaAs MESFETs (Ref. (26)).

L_G	L_{GD}	a	N_D	v_S	g_m	f_T	f_{max}
(μm)	(μm)	(nm)	(cm^{-3})	(cm/s)	(mS/mm)	(GHz)	(GHz)
0.25	0.10	48.5	5×10^{17}	1.4×10^7	241	54	128
0.20	0.10	41	7×10^{17}	1.7×10^7	331	77	181
0.15	0.08	36	9×10^{17}	2.1×10^7	461	125	266
0.10	0.07	30	1.3×10^{18}	2.6×10^7	648	213	424

Here, L_{GD} is the gate-drain distance, a is the active-layer thickness, N_D is the donor density in the active layer, and v_s is the saturation velocity. The velocity overshoot was not taken into account explicitly in the calculations, but it was included in the value of v_s as an effective saturation velocity. It was also assumed that the gate was located as far from the drain and as close to the source as possible so that the gate-drain capacitance and the source resistance would be minimized. High values of $f_T = 213$ GHz, $f_{max} = 424$ GHz, and $g_m = 648$ mS/mm are predicted for the gate length L_G of 0.1 μm.

Golio et al. (28) collected and examined the experimental data for f_T, f_{max}, and g_m of GaAs MESFETs published in the literatures between 1966 and 1988. They have projected limits to the ultimate frequency performance which can be realized with GaAs MESFETs. The data projected at $L_G = 0.1$ μm are $f_T = 80 \sim 200$ GHz, $f_{max} = 300 \sim 1000$ GHz, and $g_m = 300 \sim 1000$ mS/mm. Recently, Feng et al. (8) obtained f_T values of 55 GHz for 0.5 μm, 89 GHz for 0.25 μm, and 109 GHz for 0.15 μm gate-length GaAs MESFETs utilizing ion-implantation technology. These are comparable to those for GaAs-base HEMTs. As for f_{max}, a high value of 120 GHz was reported for a 0.25 μm gate-length GaAs MESFET in Ref. 29.

In the 1980s, GaAs MESFETs for digital ICs were studied extensively. The high-speed performance was characterized by the propagation delay time of the ring oscillator. A delay of 9.9 ps/gate for a 0.4 μm gate-length (BP-SAINT) GaAs MESFET was reported in Ref. 22. The performance of power GaAs MESFETs were also improved in the 1980s. The details are found in Ref. 18.

PARASITIC EFFECTS

The high-speed and high-frequency performance of GaAs MESFETs is due to the high electron velocity of GaAs. However, there are several unfavorable phenomena or parasitic effects in GaAs MESFETs such as short-channel effects, side-gating effects, slow-current transients, low-frequency anomalies (frequency-dependent transconductance and output conductance), and kink phenomena. These phenomenon are originated from the fact that the semi-insulating GaAs substrate (on which the MESFET is fabricated) is achieved by

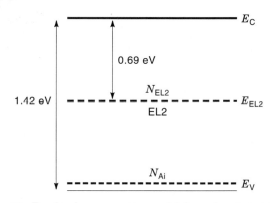

Figure 13. Two-level compensation model for undoped semi-insulating LEC GaAs. N_{EL2} and N_{Ai} are densities of deep donor "EL2" and shallow acceptor, respectively. The deep donors donate electrons to the shallow acceptors, and hence, the ionized EL2 density N_{EL2}^+ becomes nearly equal to the shallow acceptor density N_{Ai} under equilibrium.

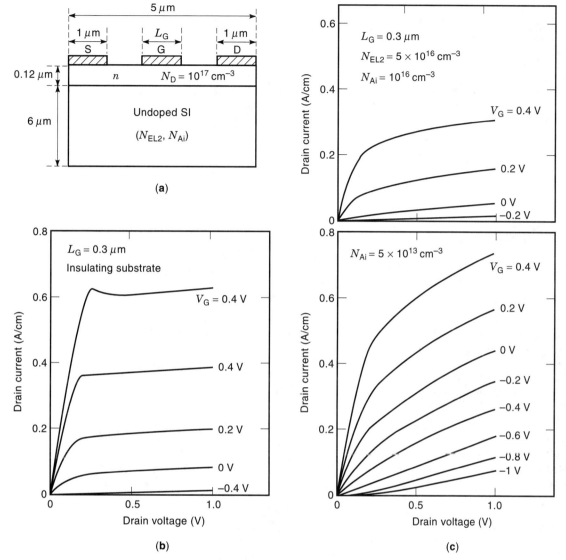

Figure 14. (a) Simulated GaAs MESFET structure. (b) Calculated drain characteristics for a case with perfectly insulating substrate. (c) Calculated drain characteristics for the two cases with different shallow-acceptor densities N_{Ai} in the semi-insulating substrate, where the deep-donor density N_{EL2} is 5×10^{16} cm^{-3}. With the semi-insulating substrate, the drain current does not saturate, particularly for lower N_{Ai}, because the substrate current becomes large (33).

impurity compensation at deep levels (30) and that high densities of surface states exist on the active layer of GaAs MES-FETs. We will discuss these phenomena below.

Substrate Conduction

The analysis done before for deriving I–V characteristics of the MESFET is based on the assumption that the semi-insulating substrate is a perfect insulator, and current does not flow through it. But, in fact, the substrate is "semi-insulating" and not a perfect insulator. The semi-insulating nature is achieved by impurity compensation by deep levels. For an example, in the undoped semi-insulating LEC (liquid-encapsulated Czochralski) GaAs, which has been widely used since early 1980s, it is thought that deep donors "EL2" (N_{EL2}) compensate shallow acceptors due to residual carbon (N_{Ai}) (31), as shown in Fig. 13. In this case, the deep donors donate electrons to the shallow acceptors, and hence, semi-insulating

properties are realized. In equilibrium, the ionized deep-donor density N_{EL2}^+ becomes nearly equal to the shallow acceptor density N_{Ai}, and the ionized deep donors act as trap centers. If the n-layer is attached to the semi-insulating substrate, electrons are injected into the substrate and are captured by the traps. So if the ionized deep-donor density N_{EL2}^+ (or N_{Ai}) is low, the trap-filled region (where all traps are filled with electrons) extends deeper into the substrate (32). The resistance in this region is low, and hence, the current can flow through the semi-insulating substrate.

Figure 14 shows examples of I–V characteristics of GaAs MESFETs on the undoped semi-insulating substrate, calculated by two-dimensional (2-D) numerical simulation in which Poisson's equation and continuity equations are solved self-consistently (33). Figure 14(a) is the simulated structure, and Fig. 14(b) corresponds to a case with perfectly insulating substrate. In Fig. 14(c), two cases with different shallow-acceptor

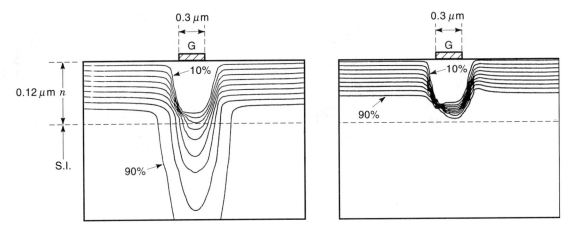

Figure 15. Comparison of current distributions of 0.3 μm gate-length GaAs MESFETs with different N_{Ai} in the semi-insulating substrate, corresponding to Fig. 14(c). $V_D = 1$ V and $V_G = 0$ V. $N_{EL2} = 5 \times 10^{16}$ cm^{-3}. (a) $N_{Ai} = 5 \times 10^{13}$ cm^{-3}, and (b) $N_{Ai} = 10^{16}$ cm^{-3}. For lower N_{Ai}, the substrate current component becomes larger, because the barrier for electrons at the channel–substrate interface is less steep.

density in the substrate ($N_{Ai} = 10^{16}$ cm^{-3} and 5×10^{13} cm^{-3}) are shown. The gate length L_G is 0.3 μm, and the field-dependent mobility expressed in Eq. (22) is used. The surface states are not considered in this calculation. In the case with perfectly insulating substrate, the drain current almost saturates with the drain voltage. In the cases with semi-insulating substrates, however, the drain currents do not saturate in general, and increase with the drain voltage, particularly for lower acceptor density N_{Ai} in the substrate. This is because, as shown in Fig. 15, the substrate current component becomes larger for lower N_{Ai}. This increase in substrate current leads to lower transconductance at a given drain current. It should be also noted that in the case with high N_{Ai}, the drain currents become lower than those for the case with perfectly insulating substrate. This is because, as schematically shown in Fig. 16, a space-charge layer is formed at the active layer–substrate interface, and the effective channel thickness becomes thinner for higher N_{Ai}. From the above considerations, we can say that to consider impurity compensation by deep levels in the semi-insulating substrate is important for evaluating I-V characteristics of GaAs MESFETs.

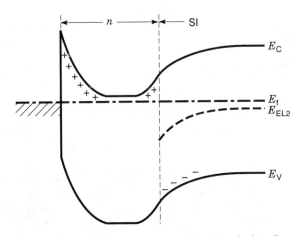

Figure 16. Schematic energy band diagram along the line from gate electrode to the substrate. The space-charge region is formed at the channel–substrate interface, because the semi-insulating substrate is achieved by impurity compensation by deep levels.

As seen in Fig. 14(c), when N_{Ai} is low and the substrate current becomes large, the threshold voltage of GaAs MESFETs shifts toward deeply negative. The shift of threshold voltage becomes more remarkable when the gate length becomes shorter. This phenomenon is one of the so-called short-channel effects. To reduce this, the substrate current must be reduced. For this purpose, the shallow acceptor density N_{Ai} in the semi-insulating substrate should be made relatively high. It is also effective to introduce a buried p-layer or a p-buffer layer because the acceptors in the p-layer should have the same electrical role as acceptors in the semi-insulating substrate. In fact, it is shown experimentally and theoretically that to introduce a buried p-layer (p-buffer layer) is effective to reduce the shore-channel effects in GaAs MESFETs (22,23).

Sidegating Effects

The sidegating effect is a phenomenon that the drain current of a GaAs MESFET is modulated when a negative voltage is applied to an adjacent device in ICs. This was also called the backdating effect because initially the current modulation was studied by attaching an electrode to the backside of the substrate. This effect is detrimental in GaAs digital, analog and microwave ICs because of unintentional electrical interactions between closely spaced devices. Numerous studies have suggested that this effect is caused by modulation of the space-charge region at the interface between the MESFET active layer and the buffer layer or the semi-insulating substrate which is achieved by impurity compensation by deep levels.

Two representative experimental data about sidegating (backgating) effects in the early 1980s are shown in Fig. 17 (34) and in Fig. 18 (35). In Fig. 17, a Cr-doped HB (horizontal Bridgman) semi-insulating substrate was used, and the electrode was attached to the bottom of the substrate. The group X' corresponded to a case without a buffer layer, and the group A' and B' corresponded to cases with different buffer layers. In all three cases, the drain currents decreased without threshold as the substrate bias voltage became negative. The authors detected hole drops due to Cr both in the buffer layer and in the semi-insulating substrate. In Fig. 18, a LEC

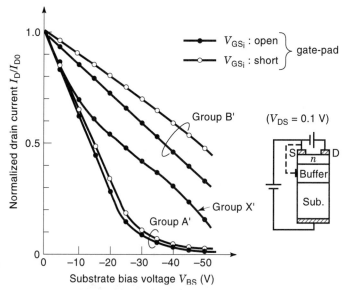

Figure 17. Experimental results of drain-current reduction due to substrate bias. The Cr-doped HB semi-insulating substrate is used. The drain currents decrease without threshold when the substrate bias voltage becomes negative. (Reproduced with permission from T. Itoh and H. Yanai, Stability of performance and interfacial problem in GaAs MESFET's, *IEEE Trans. Electron Devices*, **ED-27**(6): 1037–1045 (© 1980 IEEE).)

Cr-doped semi-insulating substrate was used, and the side-gate (backgate) electrode was attached to the same surface as the MESFET. The substrate current showed ohmic behavior at low voltages and showed a sudden rise at a certain threshold voltage. Just at this voltage, the drain current began to decrease. Thus, the threshold voltage for the sudden increase in the substrate current was exactly the same as the threshold voltage for the sidegating effect. This threshold behavior was typical also for cases of using undoped semi-insulating LEC substrates extensively studied later. The threshold behavior was qualitatively explained by Lampert's carrier injection model (32,36).

The above difference in sidegating behavior for different types of substrates can be explained as follows. In the undoped semi-insulating LEC GaAs, as described before, deep donors "EL2" compensate shallow acceptors, and the deep donor acts as an electron trap because its capture cross section for electrons is much larger than that for holes (30). In the Cr-doped semi-insulating substrate, deep acceptors "Cr" compensate shallow donors, and the deep acceptor acts as a hole trap because its capture cross section for holes is much larger than that for electrons (30). Fig. 19 shows a comparison of calculated energy band diagrams of n–i–n structures with different i-layers (substrates) (37). The left n-layer corresponds to the MESFET active layer. Part (a) is for a case with EL2, and part (b) is for a case with Cr. In (b), the voltage is entirely applied along the reverse-biased n–i junction because electrons as well as holes are depleted there, and hence, the drain current of the MESFET decreases without threshold when negative voltage is applied to an adjacent n-layer. In (a), the voltage is applied along the bulk i-region because electrons are not depleted at the reverse-biased n–i junction because of the electron-trap nature of EL2 (32). Then, the substrate current shows ohmic behavior at low voltages, and the drain current of the MESFET changes little. When the applied voltage

becomes so high that the traps are all filled with electrons, the substrate current increases suddenly, and the voltage becomes applied along the reverse-based n–i junction. Therefore, the drain current begins to decrease with threshold in this case. The corresponding voltage is called the trap-fill-limited voltage (36) and given by

$$V_{\text{TFL}} = \frac{qN_{\text{Ai}}}{2\epsilon}d^2 \tag{34}$$

where d is the i-layer thickness. However, the voltage given by Eq. (34) was usually too high as compared to the experimental voltage for threshold, and this model did not necessarily give quantitative explanations.

Recently, it was pointed out that the Schottky pad (of GaAs MESFET) directly attached to the semi-insulating substrate should play an important role in the sidegating effect. Liu et al. (38) studied effects of Schottky contact on the semi-insulating substrate by using the test structure shown in Fig. 20, where the results of sidegating effects are also shown when the Schottky contact (SC) is floating, or its voltage V_{SC} is set to 0 V. It is seen that the current in the n–i–n structure, formed by the MESFET and the sidegate, is very low within the voltage range of the measurements. On the other hand, the current in the Schottky-i–n structure, formed by the Schottky contact and the sidegate, show a sudden increase at a relatively low voltage, indicating a remarkable sidegating effect. The most likely cause of this high leakage current might be the hole injection from the Schottky contact, as has been suggested by computer simulation (39). The existence of the Schottky pad on the semi-insulating substrate could explain, to some extent, the reduction of the threshold voltage for sidegating. However, there are no quantitative models to predict the onset voltage adequately.

To improve the sidegating threshold, two approaches were adopted (40). One was to use isolation implantation techniques such as oxygen, boron, and protons. This technique increased the threshold voltage for sidegating by about a factor of three. The other approach was to shield the MESFET channel from the offending sidegating electrodes. A Schottky metal shield between the sidegate electrode and the FET and a Be-implanted p-type shield tied to the source contact were shown to improve sidegate immunity. Recently, Chen and Smith et al. (41) showed that a new buffer layer grown by molecular-beam epitaxy (MBE) at a substantially low temperature (~200 °C) could greatly reduce the sidegating effect. However, the basic mechanism responsible for its unique semi-insulating property is not yet clarified. Many studies on this so-called LT (low temperature)–GaAs buffer are being made.

Slow Current Transients

GaAs MESFETs are essentially high-speed and high-frequency devices. However, slow current transients are often observed experimentally even if the drain voltage or the gate voltage is changed abruptly. These are called "drain-lag" or "gate-lag" and could seriously limit the performance of power MESFETs as well as pulse operating integrated circuits. For example, gate-lag affects digital circuits such as inverter chains by causing pulse narrowing, which finally leads to function error. It is suggested that these phenomena occur due to the slow responses of deep traps in the semi-insulating substrate or surface states on the active layer.

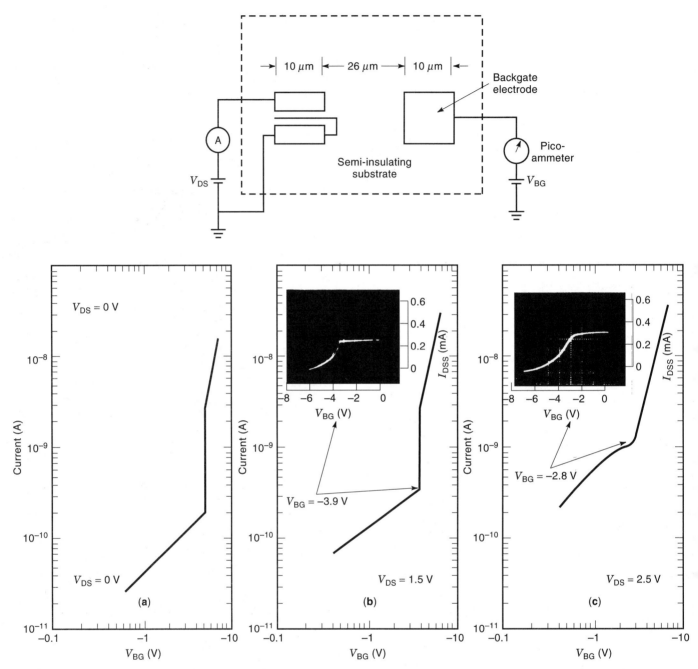

Figure 18. The test structure used for sidegating measurements, and the current-voltage relation of the substrate conduction between the MESFET and the backgate (sidegate) electrode when (a) $V_{DS} = 0$ V, (b) $V_{DS} = 1.5$ V, and (c) $V_{DS} = 2.5$ V. The LEC Cr-doped semi-insulating substrate is used. The sidegating characteristics, I_{DSS} vs. V_{BG} at $V_{DS} = 1.5$ V and $V_{DS} = 2.5$ V, are shown by the photographs in (b) and (c), respectively. The sudden increase in the substrate conduction corresponds to the sudden decrease in the drain current. (Reproduced with permission from C. P. Lee, S. J. Lee, and B. M. Welch, Carrier injection and backgating effect in GaAs MESFET's, *IEEE Electron Device Lett.*, **EDL-3**(4): 97–98 (© 1982 IEEE).)

Figure 21 shows an example of drain-lag phenomenon experimentally reported by Mickanin et al. (42). They used a 300 μm wide enhancement-mode GaAs MESFET on undoped semi-insulating LEC substrate. The gate voltage was set to the pinch-off voltage, defined as $I_D = 1$ μA/μm at $V_D = 1$ V. As shown in Fig. 21, 10 Hz and 100 kHz drain-voltage pulses from 0 to 5 V were applied, and the drain-current transients were traced. It is seen that the current with a 10 Hz pulse rate shows overshoot behavior. This type of overshoot has been commonly observed by other researchers. The absence of overshoot with a 100 kHz pulse rate indicates that this phenomenon is frequency-dependent. This phenomenon was usually explained by trapping dynamics at the channel-substrate interface or in the semi-insulating substrate. An example of drain-lag phenomena calculated by 2-D simulation is shown in Fig. 22 (43). The device structure is the same as that

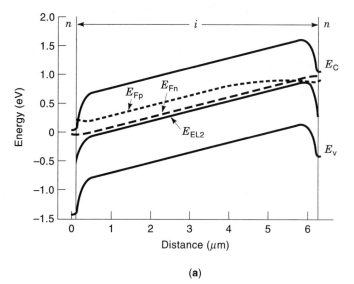

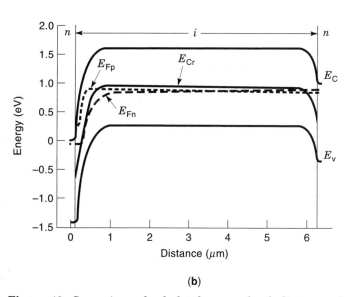

Figure 19. Comparison of calculated energy band diagrams of $n–i–n$ structures with different deep levels in the i-layer. (a) Case with deep donor "EL2" ($N_{\text{EL2}} = 5 \times 10^{16}$ cm^{-3}) and shallow acceptor ($N_{\text{Ai}} = 10^{16}$ cm^{-3}), and (b) case with deep acceptor Cr ($N_{\text{Cr}} = 10^{16}$ cm^{-3}) and shallow donor ($N_{\text{Di}} = 10^{15}$ cm^{-3}). "EL2" acts as an electron trap, and "Cr" acts as a hole trap. This difference leads to the difference in the energy-band diagrams (32,37).

in Fig. 14(a). The gate voltage is 0 V, and the drain voltage is changed abruptly from 0 to 1 V or from 1 to 0.5 V. The drain currents become constant temporarily (a "quasi-steady state") around $t = 10^{-11}$ sec, and after some periods, they begin to decrease or increase, reaching real steady-state values. In fact, the current overshoot is observed when the drain voltage is raised. It is interpreted that the quasi-steady state is a state where the deep donors "EL2" in the substrate do not respond to the voltage change, and electrons move under the same ionized-impurity densities as those for $V_{\text{D}} = 0$ V or 1 V. When the deep donors begin to capture or emit electrons, the drain currents begin to decrease or increase, reaching steady-state values. Therefore, the deep donors in the substrate can be also a cause of hysteresis in I–V curves.

Figure 23 shows schematic diagram of gate-lag measurement and $V_{\text{gs(off)}}$ dependence of drain current transients re-

cently reported by Kohno et al. (44). They used a single recessed-gate MESFET with the pinch-off voltage of -2.4 V. When the gate voltage was changed from $V_{\text{gs(off)}}$ to $V_{\text{gs(on)}} = 0$ V, the drain currents remained low for some periods and began to increase gradually between 10^{-4} and 10^{-1} s, reaching steady-state values. As $V_{\text{gs(off)}}$ was lower, the gate-lag became more pronounced. The gate-lag phenomenon was usually correlated to the surface states on the active layer. Lo and Lee (45) simulated the gate-lag phenomenon by considering surface states, which were assumed to consist of a pair of deep donors and deep acceptors. The results are shown in Fig. 24. They used a planer MESFET with the threshold (pinch-off) voltage of -2.5 V. Here, the Gate lag rate was defined as $(I_{\text{D}}(t = 250 \text{ ms}) - I_{\text{D}}(t = 1 \text{ ns}))/I_{\text{D}}(t = 250 \text{ ms})$. In fact, as the initial gate voltage was lower, the gate-lag percentage increased. It was understood that the gate-lag arose because the response of surface deep levels were slow. It was also shown that when the initial gate voltage was lower, the negative surface charge density was higher to enhance the gate-lag phenomenon.

To reduce the above lag phenomena, effects of deep traps in the substrate and surface states should be minimized. Canfield et al. (46) used the p-well technology where the p-well potential was constrained by connecting it to the source and showed that the drain-lag was eliminated. To reduce the gate-lag, several methods to minimize the surface-state effects have been proposed, but no conclusive way has been realized.

Low-Frequency Anomalies

Many of the electrical characteristics of GaAs MESFETs shift dramatically in values at relatively low frequencies (<1 MHz) (7). Device parameters which have been observed to shift include output conductance (drain conductance) g_{D}, transconductance g_{m}, and device capacitances. As the frequency is increased, the measured output conductance is seen to increase by as much as a factor of two (47). The characteristic frequencies at which this increase occurs ranges less than 10 Hz to about 100 kHz. The transconductance usually decreases with the frequency, and the decrease rate is typically 5 to 30 percent. The frequency dependences of g_{D} and g_{m} can be correlated to the drain-lag and the gate-lag, respectively. Therefore, these frequency dependences are attributed to the existence of deep traps in the semi-insulating substrate and surface states on the active layer.

A typical example of measured frequency dependence of g_{D} in a GaAs MESFET is shown in Fig. 25 by symbols (47). The device was a standard, recessed-gate depletion-made MESFET fabricated on undoped semi-insulating LEC GaAs. The gate length was 1 μm, and the n-channel was formed by ion implantation to a peak concentration of approximately 2×10^{17} cm^{-3}. As seen, the output conductance also indicated temperature dependence. These frequency and temperature dependences of g_{D} were attributed to deep donor "EL2" in the semi-insulating substrate. By assuming a temperature-dependent time constant of electron emission form EL2:

$$\tau_{\text{e}} \simeq \frac{3.5 \times 10^{-8}}{T^2} \exp\left(\frac{9450}{T}\right) \qquad (35)$$

theoretical curves of frequency-dependent g_{D} were derived. They are shown in Fig. 25 by solid curves. These fit well with the experimental results. It was shown experimentally that

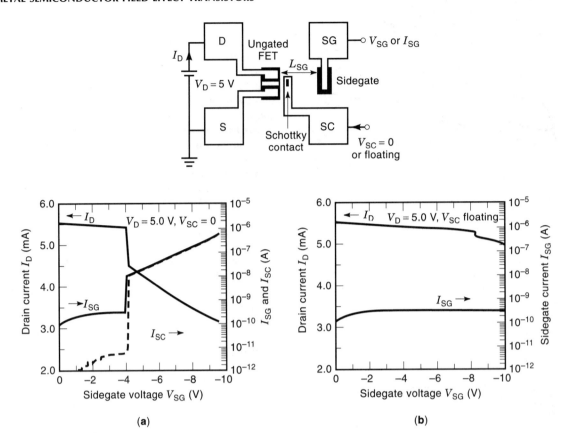

Figure 20. Schematic top view of the test structure used in the measurements, and results of the voltage-controlled sidegating effect measurements. (a) Results measured with $V_{SC} = 0$, and (b) results measured with V_{SC} floating. With $V_{SC} = 0$, the sidegating effect is observed at a low sidegate voltage V_{SG}. (Reproduced with permission from Y. Liu, R. W. Dutton, and M. D. Deal, Schottky contact effects in the sidegating effect of GaAs devices, *IEEE Electron Device Lett.*, **13**(3):149–151 (© 1992 IEEE).)

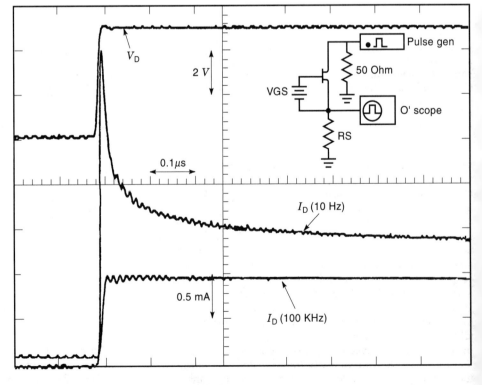

Figure 21. Schematic measurement system of drain-lag and measured waveforms. The upper is the drain voltage pulse. The lower curves show drain current transients at 10 Hz and 100 KHz pulses, respectively. The current overshoot and the subsequent slow transient are observed at the 10 Hz pulse. (Reproduced with permission from W. Mickanin, P. Canfield, E. Finchem, and B. Odekirk, Frequency-dependent transients in GaAs MESFETs: process, geometry and material effects, *IEEE GaAs IC Symposium Technical Digest*, 211–214 (© 1989 IEEE).)

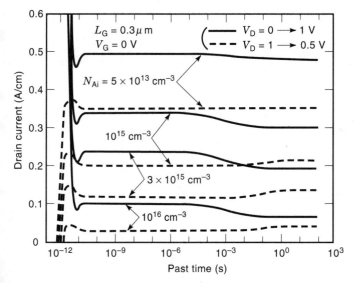

Figure 22. Calculated responses of drain currents for 0.3 μm gate-length GaAs MESFETs on undoped semi-insulating substrate [Fig. 14(a): $N_{EL2} = 5 \times 10^{16}$ cm^{-3}] when V_D steps from 0 to 1 V and when V_D steps from 1 to 0.5 V (43). The current overshoot and the subsequent slow transient are qualitatively reproduced for the former case (solid lines).

by using the p-well GaAs MESFET technology, the frequency dependence of output conductance was not observed up to 1 MHz (46).

An example of measured frequency dispersion of transconductance for a GaAs MESFET is shown in Fig. 26 (48). The device was fabricated by Si ion implantation into undoped semi-insulating LEC GaAs. The gate length was 1 μm. Both the drain bias and the superimposed gate modulation signal δv_{gs} were kept small ($\sim$50 mV). (The drain bias was kept low in order to make the assumption that the channel depth is constant, and so the channel region may be treated as a resistance.) As seen in the figure, the transconductance decreased with the frequency and showed dispersion over a limited range of temperature. This behavior was explained by the surface-state dynamics. At temperatures below 150 K, the surface states responded so slowly that they could not follow even the lowest modulation frequency used (10 Hz). As the temperature was raised, their response time decreased, falling within the window of used measurement frequencies. At still higher temperature ($\sim$400 K), the response of the surface states was so fast that the characteristic frequency became much higher than the highest used frequency here (20 kHz), so there was no dispersion again. Zhao et al. (49) developed an analytical model for frequency dependence of transconductance in GaAs MESFETs. Assuming a single surface state ES1 ($E_C - E_{S1} = 0.4$ eV, $N_{S1} = 10^{12}$ cm^{-2}, $\sigma_{S1} = 10^{-11}$ cm^2), he obtained the temperature dependence of transconductance frequency dispersion as shown in Fig. 27. A general agreement is seen between the modeling results and experimental results in Fig. 26.

Kink Phenomena

High-voltage behavior of GaAs MESFETs has always been of interest for microwave applications where the maximum power is limited in part by the breakdown voltage of the de-

vice. Recently, it was reported that the GaAs MESFETs showed an abnormal increase in the output conductance (kink phenomenon) at relatively low voltages (3 $\sim$ 4 V) (50,51). This phenomenon may limit the operation voltage of GaAs MES-FETs. It was recognized that the kink was associated with impact ionization of carriers in the channel. It was also suggested that the kink was not due to direct gate breakdown but could be regarded as a phenomenon related to the substrate.

An example of measured drain characteristics of GaAs MESFETs reported by Harrison (51) is shown in Fig. 28. The device was a 0.8 μm gate-length self-aligned GaAs MESFET fabricated on undoped LEC semi-insulating substrate. The output conductance in the saturated region showed an increase for $V_D > 3.5$ V. It was shown that in this kink region, the sidegating effect became remarkable. This strongly suggests that the semi-insulating substrate should play an important role in the kink phenomenon. It was proposed that holes which were generated by impact ionization and injected into the semi-insulating substrate were the origin of these phenomena.

Drain characteristics of a 0.3 μm gate-length GaAs MES-FET, calculated by 2D simulation considering impact ionization of carriers, are shown in Fig. 29(a) (52). The simulated structure is the same as that in Fig. 14(a), where the shallow acceptor density in the substrate N_{Ai} is 10^{16} cm^{-3}. The characteristic show kink behavior at $V_D = 3 \sim 5$ V and at $V_D = 10 \sim 15$ V. However, these are not due to direct gate breakdown because the gate current is much lower than the drain current as shown in Fig. 29(b). Figure 30 shows calculated hole density profiles at $V_D = 4$ V and 12 V for $V_G = 0$ V. It is understood that holes generated by impact ionization flow into the substrate and are captured by deep donors "EL2," and hence, the ionized deep-donor density N_{EL2}^+ increases. This increase in positive charges in the substrate increases the channel thickness, resulting in the first kink. At $V_D = 12$ V, hole densities in the substrate become very high and comparable to N_{EL2}^+, but the hole current is much lower than the electron current. In this case, we can interpret that the increase in the positive hole charges in the substrate widens the channel thickness, leading to the steep increase in the drain current. It was also ascertained by 2D simulation (53) that the sidegating effects should become remarkable in the kink region. In another work (54), it was suggested that the kink could be reduced by decreasing the acceptor density in the substrate.

MESFET MODELING

Modeling for Circuit Simulation

For circuit design applications, accurate but simple device models are required. Purely physical models such as those described when deriving the I–V characteristics are usually not accurate as required for most applications. The inaccuracies arise from the assumptions and approximations required to perform the device analysis. In contrast, empirical models can be accurate enough to fit the experimental data, though large amount of tedious characterization data are often required to obtain the accuracy.

An early and basic empirical model for GaAs MESFETs was proposed by Curtice (55). The current-voltage character-

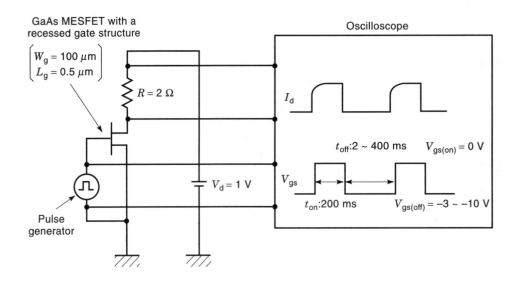

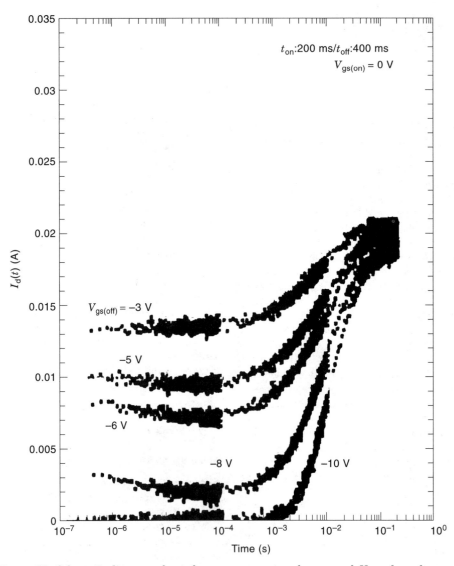

Figure 23. Schematic diagram of gate-lag measurement and measured $V_{gs(off)}$ dependence of drain-current transients. The gate-lag is remarkable when the off-state gate voltage $V_{gs(off)}$ is deeply negative. (Reproduced with permission from Y. Kohno et al., Modeling and suppression of the surface trap effect on drain current frequency dispersions in GaAs MESFETs, *IEEE GaAs IC Symposium Technical Digest*, 263–266 (© 1994 IEEE).)

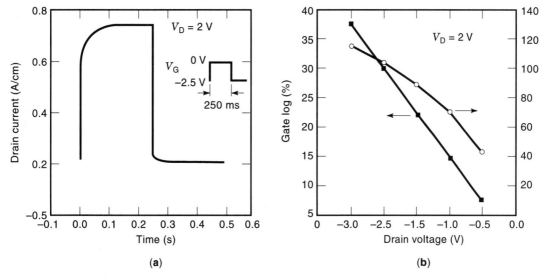

Figure 24. (a) Calculated example of gate-lag phenomenon. (b) Calculated lag percentage and lag time when the gate voltage is changed from -2.5, -2, -1.5, -1, and -0.5 V to 0 V. The gate-lag is enhanced when the off-state gate voltage is more negative. (Reproduced with permission from S. H. Lo and C. P. Lee, Analysis of surface trap effect on gate lag phenomena in GaAs MESFET's, *IEEE Trans. Electron Devices*, **41**(9):1504–1512 (© 1994 IEEE).)

istics were expressed as

$$I_D = \beta(V_G - V_T)^2(1 + \lambda V_D)\tanh(\alpha V_D) \qquad (36)$$

where V_T is the threshold voltage or the pinch-off voltage. β, λ, and α are the parameters. Eq. (36) can be separated into three components. The first component $\beta(V_G - V_T)^2$ is used to model the approximately square-law behavior of the I_D–V_G relationship. In the second component $1 + \lambda V_D$, the parameter λ is used to model the drain conductance. The third component $\tanh(\alpha V_D)$ is used because the hyperbolic tangent ap-

proximates the I_D–V_D characteristics observed in GaAs MESFETs. α determines the voltage at which the drain current saturates. Curtice and Ettenberg (56) altered the original Curtice model to get a closer fit to the relationship between I_D and V_G. The new equation is

$$I_D = (A_0 + A_1 v_1 + A_2 v_1^2 + A_3 v_1^3)\tanh(\gamma V_D) \qquad (37)$$

$$v_1 = V_G\{1 + \beta(V_{D0} - V_D)\}$$

where β and γ are the parameters. V_{D0} is the drain voltage at which the A_i coefficients are evaluated.

Statz et al. (57) developed a model based on Eq. (36). They thought that the square-law approximation of the $I_D - V_G$ relationship was only valid for small values of $V_G - V_T$, and that I_D became almost linear for larger values of $V_G - V_T$. To model this behavior, in place of $\beta(V_G - V_T)^2$, they adopted the

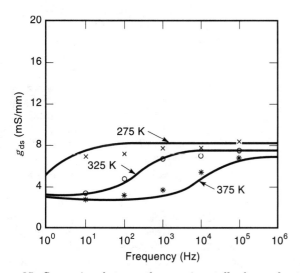

Figure 25. Comparison between the experimentally observed output conductance ($\times$, $\circ$, $*$) and the theoretical curves (solid lines) at three different temperatures for $V_D = 3$ V and $V_G = 0.2$ V. The increase in output conductance with frequency is often seen experimentally. (Reproduced with permission from P. C. Canfield, S. C. F. Lam, and D. J. Allstot, Modeling of frequency and temperature effects in GaAs MESFETs, *IEEE J. Solid-State Circuits*, **25**(1):299–306 (© 1990 IEEE).)

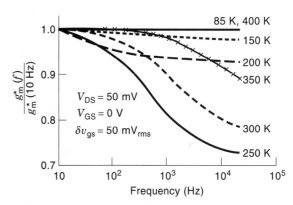

Figure 26. Measured frequency dispersion of transconductance as a function of temperature for a 1 μm gate-length GaAs MESFET. The decrease in transconductance with frequency is often seen experimentally. (Reproduced with permission from S. R. Blight, R. H. Wallis, and H. Thomas, Surface influence on the conductance DLTS spectra of GaAs MESFET's, *IEEE Trans. Electron Devices*, **ED-33**(10): 1447–1453 (© 1986 IEEE).)

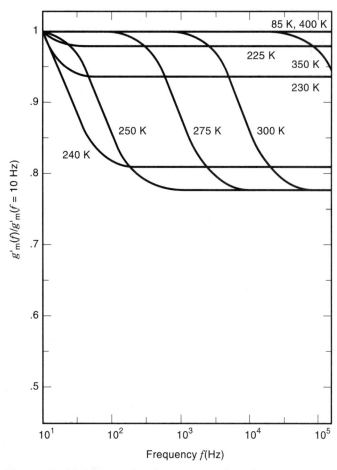

Figure 27. Modeling results of temperature dependence of transconductance dispersion for a single surface state case. A general agreement is seen between these modeling results and the experimental results in Fig. 26. (Reproduced with permission from J. H. Zhao, R. Hwang, and S. Chang, On the characterization of surface states and deep traps in GaAs MESFETs, *Solid-State Electron.*, **36**(12):1665–1672 (© 1993 Elsevier Science Ltd.).)

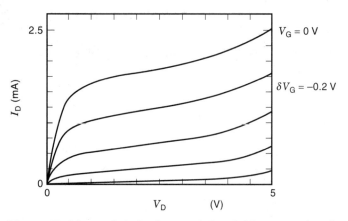

Figure 28. Measured drain characteristics of 0.8 μm gate-length GaAs MESFET, showing the onset of the kink phenomenon at $V_D = 3.5$ V. (Reproduced with permission from A. Harrison, Backgating in submicrometer GaAs MESFET's operated at high drain bias, *IEEE Electron Device Lett.*, **13** (7): 381–383 (© 1992 IEEE).)

empirical expression

$$\frac{\beta(V_G - V_T)^2}{1 + b(V_G - V_T)}$$

In addition, they found that the tanh function in Eq. (36) consumed considerable computer time. The tanh function below saturation was modified using a polynominal of the form

$$1 - (1 - (\alpha V_D/n))^n$$

with $n = 3$. In the saturated region ($V_D > n/\alpha$), the tanh function was replaced by unity. These modifications led to a new form for the drain current.

For $0 < V_D < 3/\alpha$,

$$I_D = \frac{\beta(V_G - V_T)^2}{1 + b(V_G - V_T)}\left[1 - \left(1 - \frac{\alpha V_D}{3}\right)^3\right](1 + \lambda V_D) \qquad (38)$$

For $V_D \geq 3/\alpha$

$$I_D = \frac{\beta(V_G - V_T)^2}{1 + b(V_G - V_T)}(1 + \lambda V_D) \qquad (39)$$

The model was compared with the experimental data. Figure 31 shows such an example (57). The agreement between the model and the experiment was satisfactory.

Besides the models mentioned above, some other models for circuit simulation such as SPICE were proposed. About these models and the applicability of them, please refer to Ref. 7, where the modeling of device capacitance not mentioned here is also found.

Device Simulation

When the gate length of GaAs MESFETs became short, the analytical one-dimensional approach with several assumptions became inadequate for estimating the *I–V* characteristics or other device performance. Then, the 2-D numerical simulation that solved Poisson's equation and transport equations self-consistently became used for predicting the device performance and for understanding physical phenomena observed in GaAs MESFETs. Historically, 2-D simulation of GaAs MESFETs was already made in the middle 1970s, in particular for understanding effects of negative differential mobility in GaAs on the device performance.

The so-called drift-diffusion type simulation method is now a mature and standard tool for evaluating the performance of GaAs devices as well as Si devices (17,58). As to recent topics regarding this method, there are numerical simulations of trapping effects on GaAs MESFET performance. If we now treat a GaAs MESFET on undoped semi-insulating LEC substrate including deep donors "EL2," the basic equations for device analysis can be written as follows (33,43).

(a) Poisson's equation

$$\nabla^2\psi = -\frac{q}{\epsilon}(p - n + N_{EL2}^+) \qquad (40)$$

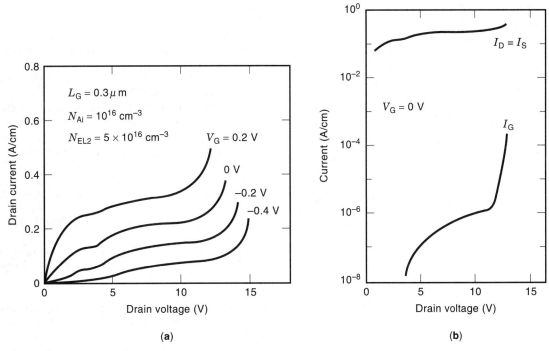

(a)

(b)

Figure 29. Simulated (a) drain characteristics and (b) terminal currents versus drain voltage curves of 0.3 μm gate-length GaAs MESFET on undoped semi-insulating substrate shown in Fig. 14(a). Two kinks are seen at $V_D = 3 \sim 5$ V and at $V_D = 10 \sim 15$ V (52).

(b) Continuity equations for electrons and holes

$$\frac{\partial n}{\partial t} = \frac{1}{q}\nabla \cdot J_n - \{C_n N_{EL2}^+ n - e_n(N_{EL2} - N_{EL2}^+)\} \quad (41)$$

$$\frac{\partial p}{\partial t} = -\frac{1}{q}\nabla \cdot J_p - \{C_p(N_{EL2} - N_{EL2}^+)p - e_p N_{EL2}^+\} \quad (42)$$

(c) Rate equation for deep levels

$$\frac{\partial}{\partial t}(N_{EL2} - N_{EL2}^+) = \{C_n N_{EL2}^+ n - e_n(N_{EL2} - N_{EL2}^+)\} \\ - \{C_p(N_{EL2} - N_{EL2}^+)p - e_p N_{EL2}^+\} \quad (43)$$

(d) Current equations for electrons and holes

$$J_n = -q\mu_n n\nabla\psi + qD_n\nabla n \quad (44)$$

$$J_p = -q\mu_p p\nabla\psi - qD_p\nabla p \quad (45)$$

where N_{EL2}^+ represents the ionized EL2 density, C_n and C_p are electron and hole capture coefficients of EL2, respectively, e_n and e_p are electron and hole emission rates of EL2, respectively, and other symbol have their normal meanings. By solving these equations, the deep-trap effects on the substrate conduction, the sidegrating effects, the slow-current transients, and the frequency-dependent small-signal parameters

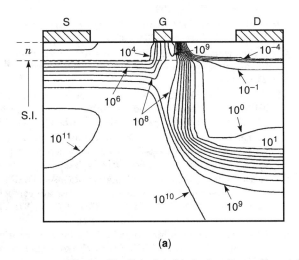

(a)

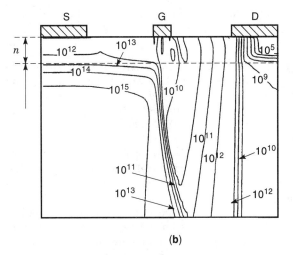

(b)

Figure 30. Calculated hole density profiles at (a) $V_D = 4$ V and (b) $V_D = 12$ V, corresponding to Fig. 29. $V_G = 0$ V. Holes generated by impact ionization flow into the substrate and are captured by deep donors "EL2." The increase in N_{EL2}^+ is the origin of the first kink, and the increase in hole charges themselves (b) is the cause of the second kink.

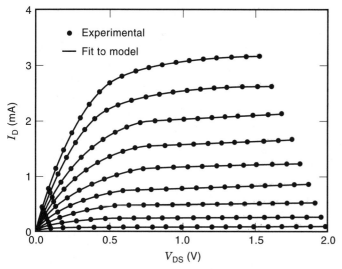

Figure 31. Comparison of modeled and measured drain characteristics of a GaAs MESFET. The very good agreement is observed. (Reproduced with permission from H. Statz et al., GaAs FET device and circuit simulation in SPICE, *IEEE Trans. Electron Devices,* **ED-34**(2): 160–169 (© 1987 IEEE).)

can be analyzed. By modeling the surface states as deep levels, surface-state effects on these phenomena can also be analyzed.

It is well recognized that the drift-diffusion model is inadequate for treating the nonequilibrium carrier transport that becomes important in shorter gate-length GaAs MESFETs. To treat this problem, a hydrodynamic model that uses three conservation equations derived from the Boltzmann transport equation have been adopted (17,25,59). These equations for electrons can be written as

$$\frac{\partial n}{\partial t} + \nabla \cdot (nv) = 0 \qquad (46)$$

$$\frac{\partial v}{\partial t} + v \cdot \nabla v = -\frac{qE}{m^*} - \frac{2}{3m^*n} \nabla \left[n \left(w - \frac{m^*}{2}v^2 \right) \right] - \frac{v}{\tau_p} \qquad (47)$$

$$\frac{\partial \omega}{\partial t} + v \cdot \nabla w = -qv \cdot E - \frac{2}{3n} \nabla \cdot \left[nv \left(w - \frac{m^*}{2}v^2 \right) \right] - \frac{w - w_0}{\tau_w} \qquad (48)$$

where v is the average electron velocity, w is the average electron energy, and w_0 is the equilibrium value of w. m^*, τ_p, and τ_w are the effective mass, the momentum relaxation time, and the energy relaxation time, respectively, and these are usually given as a function of w. As a more fundamental method to treat the nonequilibrium carrier transport, there is a Monte Carlo simulation method (60). This method is suitable for studying fundamental carrier transports in the device, but may not be suited for device design use because it requires large computer resources.

Finally, it should be pointed out that in the compound semiconductor device fields such as GaAs MESFETs, the simulation results have not necessarily been compared with experimental data. This situation is quite contrary to that in Si devices. To compare the simulation results with experiments will contribute much to refining the models and optimizing the device structures.

BIBLIOGRAPHY

1. F. Ren et al., III–V compound semiconductor MOSFETs using Ga₂O₃(Gd₂O₃)As gate dielectric, *IEEE GaAs IC Symp. Tech. Dig.,* 18–21, 1997.

2. C. A. Mead, Schottky barrier gate field-effect transistor, *Proc. IEEE,* **54**: 307–308, 1966.

3. W. W. Hooper and W. I. Lehrer, An epitaxial GaAs field-effect transistor, *Proc. IEEE,* **55**: 1237–1238, 1967.

4. J. Turner et al., *Inst. Phys. Conf. Ser.,* **9**: 234–239, 1971.

5. M. Fukuta et al., Mesh source type microwave power FET, *IEEE Int. Solid-State Circuit Conf. Tech. Dig.,* 84–85, 1973.

6. T. Mimura et al., A new field-effect transistor with selectively doped GaAs/*n*-Al$_x$Ga$_{1-x}$As heterojunctions. *Jpn. J. Appl. Phys.,* **19**: L225–L227, 1980.

7. J. M. Golio, *Microwave MESFETs & HEMTs,* Norwood, MA: Artech House, 1991.

8. M. Feng and J. Laskar, On the speed and noise performance of direct ion-implanted GaAs MESFET's, *IEEE Trans. Electron Devices,* **40**: 9–17, 1993.

9. *Vitesse Data Book.*

10. V. Milutinovic (ed.), *Microprocessor Design for GaAs Technology,* New Jersey: Prentice Hall Advanced Reference Series, Engineering, 1990.

11. W. Shockley, Unipolar field-effect transistor, *Proc. IRE,* **40**: 1365–1376, 1952.

12. S. M. Sze, *Physics of Semiconductor Devices,* 2nd ed., New York: Wiley, 1981.

13. K. Lehovec and R. Zuleeg, Voltage-current characteristics of GaAs JFET in the hot electron range, *Solid-State Electron.,* **13**: 1415–1426, 1970.

14. H. W. Thim, Computer study of bulk GaAs devices with random and dimensional doping fluctuations, *J. Appl. Phys.,* **39**: 3897–3904, 1968.

15. H. Statz, H. A. Haus, and R. A. Pucel, Noise characteristics of gallium arsenide field-effect transistor, *IEEE Trans. Electron Devices,* **ED-21**: 549–562, 1974.

16. M. Shur, *GaAs Devices and Circuits,* New York: Plenum Press, 1987.

17. C. M. Snowden and R. E. Miles (eds.), *Compound Semiconductor Device Modelling,* London: Springer-Verlag, 1993.

18. J. L. B. Walker (ed.), *High Power GaAs FET Amplifiers,* Norwood, MA: Artech House, 1993.

19. J. Mikkelson, GaAs digital VLSI device and circuit technology, *IEDM Tech. Dig.,* 231–234, 1991.

20. N. Yokoyama et al., A self-aligned source/drain planar device for ultra-high-speed GaAs VLSIs, *ISSCC Dig. Tech. Papers,* 218–219, 1981.

21. K. Yamasaki, K. Asai, and K. Kurumada, GaAs LSI-directed MESFET's with self-aligned implantation for *n*⁺-layer technology (SAINT), *IEEE Trans. Electron Devices,* **ED-29**: 1772–1777, 1982.

22. K. Yamasaki, N. Kato, and M. Hirayama, Buried *p*-layer SAINT for very high-speed GaAs LSI's with submicrometer gate length, *IEEE Trans. Electron Devices,* **ED-32**: 2420–2425, 1985.

23. K. Horio et al., Numerical simulation of GaAs MESFET's with a *p*-buffer layer on the semi-insulating substrate compensated by deep traps, *IEEE Trans. Microw. Theory Tech.,* **37**: 1371–1379, 1989.

24. R. A. Sadler et al., A high yield buried *p*-layer fabrication process for GaAs LSI circuits, *IEEE Trans. Electron Devices,* **38**: 1271–1279, 1991.

25. R. K. Cook and J. Frey, Two-dimensional numerical simulation of energy transport effect in Si and GaAs MESFET's, *IEEE Trans. Electron Devices,* **ED-29**: 970–977, 1982.

26. M. B. Das, Millimeter-wave performance of ultrasubmicrometer-gate field-effect transistors: a comparison of MODFET, MESFET and PBT structures, *IEEE Trans. Electron Devices, ED-34*: 1429–1440, 1987.

27. M. B. Das, Charge-control analysis of MOS and junction-gate field-effect transistors, *Proc. IEE, 113*: 1240–1248, 1966.

28. J. M. Golio and J. R. J. Golio, Projected frequency limits of GaAs MESFET's, *IEEE Trans. Microw. Theory Tech., 39*: 142–146, 1991.

29. R. M. Nagarajan et al., Design and fabrication of 0.25-μm MESFET's with parallel and π-gate structures, *IEEE Trans. Electron Devices, 36*: 142–145, 1989.

30. G. M. Martin et al., Compensation mechanism in GaAs, *J. Appl. Phys., 51*: 2840–2852, 1980.

31. D. E. Holmes et al., Compensation mechanism in liquid encapsulated Czochralski GaAs: Importance of melt stoichiometry, *IEEE Trans. Electron Devices, ED-29*: 1045–1051, 1982.

32. K. Horio, T. Ikoma, and H. Yanai, Computer-aided analysis of GaAs n-i-n structures with a heavily compensated i-layer, *IEEE Trans. Electron Devices, ED-33*: 1242–1250, 1986.

33. K. Horio, H. Yanai, and T. Ikoma, Numerical simulation of GaAs MESFET's on the semi-insulating substrate compensated by deep traps, *IEEE Trans. Electron Devices, 35*: 1778–1785, 1988.

34. T. Itoh and H. Yanai, Stability of performance and interfacial problem in GaAs MESFET's, *IEEE Trans. Electron Devices, ED-27*: 1037–1045, 1980.

35. C. P. Lee, S. J. Lee, and B. M. Welch, Carrier injection and backgating effect in GaAs MESFET's, *IEEE Electron Device Lett., EDL-3*: 97–98, 1982.

36. M. A. Lampert and P. Mark, *Current Injection in Solids*, New York: Academic Press, 1970.

37. K. Horio, K. Asada, and H. Yanai, Two-dimensional simulation of GaAs MESFETs with deep acceptors in the semi-insulating substrate, *Solid-State Electron., 34*: 335–343, 1991.

38. Y. Liu, R. W. Dutton, and M. D. Deal, Schottky contact effects in the sidegating effect of GaAs devices, *IEEE Electron Device Lett., 13*: 149–151, 1992.

39. S. J. Chang and C. P. Lee, Numerical simulation of sidegating effect in GaAs MESFET's, *IEEE Trans. Electron Devices, 40*: 698–704, 1993.

40. R. Y. Koyama et al., Parasitic effects and their impact on gallium arsenide integrated circuits, *Proc. 5th Conf. Semi-insulating III-V Materials*, 203–212, 1988.

41. C. L. Chen et al., Reduction of sidegating in GaAs analog and digital circuits using a new buffer layer, *IEEE Trans. Electron Devices, 36*: 1546–1556, 1989.

42. W. Mickanin et al., Frequency-dependent transients in GaAs MESFETs: process, geometry and material effects, *IEEE GaAs IC Symposium Tech. Dig.*, 211–214, 1989.

43. K. Horio and F. Fuseya, Two-dimensional simulations of drain-current transients in GaAs MESFET's with semi-insulating substrates compensated by deep levels, *IEEE Trans. Electron Devices, 41*: 1340–1346, 1994.

44. Y. Kohno et al., Modeling and suppression of the surface trap effect on drain current frequency dispersions in GaAs MESFETs, *IEEE GaAs IC Symp. Tech. Dig.*, 263–266, 1994.

45. S. H. Lo and C. P. Lee, Analysis of surface trap effect on gate lag phenomena in GaAs MESFET's, *IEEE Trans. Electron Devices, 41*: 1504–1512, 1994.

46. P. C. Canfield and D. J. Allstot, A p-well GaAs MESFET technology for mixed-mode applications, *IEEE J. Solid-State Circuits, 25*: 1544–1549, 1990.

47. P. C. Canfield, S. C. F. Lam, and D. J. Allstot, Modeling of frequency and temperature effects in GaAs MESFETs, *IEEE J. Solid-State Circuits, 25*: 299–306, 1990.

48. S. R. Blight, R. H. Wallis, and H. Thomas, Surface influence on the conductance DLTS spectra of GaAs MESFET's, *IEEE Trans. Electron Devices, ED-33*: 1447–1453, 1986.

49. J. H. Zhao, R. Hwang, and S. Chang, On the characterization of surface states and deep traps in GaAs MESFETs, *Solid-State Electron., 36*: 1665–1672, 1993.

50. H. I. Fujishiro et al., Modulation of drain current by holes generated by impact ionization in GaAs MESFET, *Jpn. J. Appl. Phys., 28*: L1734–L1736, 1989.

51. A. Harrison, Backgating in submicrometer GaAs MESFET's operated at high drain bias, *IEEE Electron Device Lett., 13*: 381–383, 1992.

52. K. Horio and K. Satoh, Two-dimensional analysis of substrate-related kink phenomena in GaAs MESFET's, *IEEE Trans. Electron Devices, 41*: 2256–2261, 1994.

53. K. Usami and K. Horio, 2-D simulation of kink-related sidegating effects in GaAs MESFETs, *Solid-State Electron., 39*: 1737–1745, 1996.

54. W. Wilson et al., Understanding the cause of IV kink in GaAs MESFET's with two-dimensional numerical simulation, *IEEE GaAs IC Symp. Tech. Dig.*, 109–112, 1995.

55. W. R. Curtice, A MESFET model for use in the design of GaAs integrated circuits, *IEEE Trans. Microw. Theory Tech., MTT-28*: 448–456, 1980.

56. W. R. Curtice and M. Ettenberg, A nonlinear GaAs FET model for use in the design of output circuits for power amplifiers, *IEEE Trans. Microw. Theory Tech., MTT-33*: 1383–1394, 1985.

57. H. Statz et al., GaAs FET device and circuit simulation in SPICE, *IEEE Trans. Electron Devices, ED-34*: 160–169, 1987.

58. S. Selberherr, *Analysis and Simulation of Semiconductor Devices*, Wien, Germany: Springer-Verlag, 1984.

59. K. Blotekjaer, Transport equations for electrons in two-valley semiconductors, *IEEE Trans. Electron Devices, ED-17*: 38 47, 1970.

60. C. Jacoboni and P. Lugli, *The Monte Carlo Method for Semiconductor Device Simulation*, Wien: Springer-Verlag, 1989.

KAZUSHIGE HORIO
Shibaura Institute of Technology

METAL-SEMICONDUCTOR FIELD EFFECT TRANSISTOR.

See METAL SEMICONDUCTOR FIELD EFFECT TRANSISTORS.

METAL SEMICONDUCTOR METAL PHOTODETECTORS

The two-terminal device discussed here converts optical power into an electric current. This basic photodetector, which is mainly based on an interdigitated electrode structure, benefits from a simple fabrication process. It can offer a large photosensitive surface area and leads to high speed performance. This article describes the basic physical mechanisms of this device, the material used in its fabrication and technology, and its main performance.

The metal–semiconductor–metal (MSM) photodetector is one of the solid-state photodetectors fabricated on semiconductor slices. The general principle of photodetection in semiconductor material is to collect electron-hole pairs that have been generated by light absorption. The main condition for

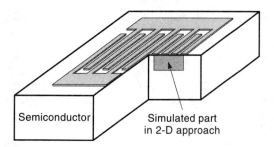

Figure 1. Typical structure of metal–semiconductor–metal photodiode.

this absorption is

$$h\nu \geq E_g$$

where ν is the optical wave frequency, h the Planck constant, and E_g the semiconductor energy bandgap. Once photogenerated, the carriers move to the photodetector electrodes by diffusion or conduction. The PIN, Avalanche, and MSM photodiodes use a depleted zone where a high electric field separates photocarriers and drives them as quickly as possible to the electrodes. The MSM photodiode consists of two metallic Schottky contacts deposited on a low-doped absorbing semiconductor material. Usually, the structure is planar and the two contacts take the form of interdigitated fingers, as presented in Figure 1 (it can be top, back, or side illuminated). This type of design allows for a large photosensitive surface area, while keeping a short distance between the fingers. Indeed, this type of photodetector is generally intended to work under very short light pulses (some picoseconds or even smaller) or with high-frequency–modulated light signals (up to several tens of gigahertz). Moreover, the number of technological steps needed to make such a photodetector is small. In some cases, only the Schottky contact deposition by electron beam evaporation is needed to make the photodetector. These features make this photodetector particularly attractive for monolithic integration with microelectronic devices, for example, amplifiers, to constitute receivers. This device is interesting for a large range of applications, such as high data bit rate fiber-optic receivers in the field of optical telecommunications (short or long distance), optical interconnects, or high-speed optical sampling.

MSM photodetector performance depends on the depleted zone extension, the electric field strength in the different regions of the semiconductor, and the dark current. Moreover, as for all other photodetectors, the different device properties to be considered are responsivity, bandwidth, and noise. We will first consider the essentials to understanding the total behavior of this photodetector. So, we look at the current-voltage characteristic in darkness in relation to the depletion region extension; then we present an overview of the properties due to the planar interdigitated structure, particularly the carrier paths between the electrodes and the electric field distribution. In this way, we examine the carrier transit time distribution inside the photodetector. Then we present an overview of the different fundamental sources of the dark current. This first part concludes with a short paragraph on the optimum bias voltage of the MSM.

The second part of this article is devoted to semiconductor epitaxial structure and device features. After the description of the whole epitaxial structure in the different material sys-

tems used to fabricate MSM photodetectors, we analyze their responsivity for various illumination conditions, the unintentional gain observed in this device (like in photoconductors), the different phenomena that limit the bandwidth, the source of the dominant noise, and some scaling rules. For each parameter, experimental results reflect the typical characteristics obtained in each material system.

Finally, we briefly introduce some types of integrated devices in which the MSM photodiode is the key element, that is, electrooptic sampling cells and monolithic integrated receivers.

PRINCIPLES

Metal–Semiconductor–Metal Photodiode Without Illumination Versus Bias Voltage

Let us consider two metallic Schottky contacts on a uniformly low or unintentionally doped n-type semiconductor. The energy-band diagram of this structure at thermodynamic equilibrium, presented in Fig. 2, shows the two depletion regions (width W_1 and W_2, respectively) inherent to the metal-semiconductor junctions. Each depletion region space charge is due to the ionized donor atoms whose electrons now act as surface charge at the metal-semiconductor interfaces. ϕ_{m1} and ϕ_{m2} are the work functions of metal 1 and 2, respectively, χ is the semiconductor electron affinity, and E_f is the Fermi level. The potentials across these two depleted zones are V_{b1} and V_{b2}, built-in potentials of the contacts. This whole structure is equivalent to two Schottky barriers connected back to back. Their barrier height is ϕ_{n1}, ϕ_{n2} for electrons, and $(\phi_{p1} + V_{b1})$, $(\phi_{p2} + V_{b2})$ for holes. We have (1):

$$\phi_{n1,2} = \phi_{m1,2} - \chi \tag{1}$$

and

$$\phi_{p1,2} = E_g - (\phi_{m1,2} - \chi) \tag{2}$$

At thermodynamic equilibrium, the summation of all currents due to carrier transport through barriers is equal to zero. When a voltage is applied, the dark current of the structure

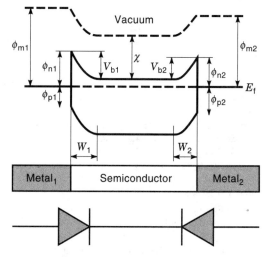

Figure 2. Energy-band diagram of an MSM structure at thermodynamic equilibrium; the semiconductor is n-type, unintentionally or low doped.

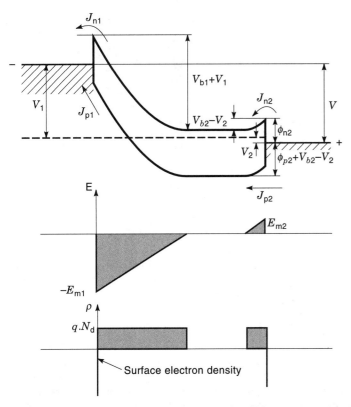

Figure 3. Energy-band diagram of a biased MSM structure with electric field and charge density distributions; the applied voltage is $V = V_1 + V_2$, $V < V_{RT}$.

is due to carriers passing through the metal-semiconductor interfaces. Considering the energy-band diagram of the biased structure presented in Fig. 3, for equal contact surface areas, and neglecting the recombination processes inside the neutral zone between the two depletion regions, we have:

$$J_{n1} = J_{n2} \tag{3}$$

$$J_{p1} = J_{p2} \tag{4}$$

and

$$J_{dark} = J_n + J_p \tag{5}$$

As a consequence, contact 1 is reverse biased (potential V_1) while contact 2 is forward biased (potential V_2), with

$$V = V_1 + V_2 \tag{6}$$

The electron current is limited by the first contact while the hole current is limited by the second. These limitations depend on the transport process through the barriers. Following the analysis presented in Ref. 2, whether this transport process is due to thermionic emission or tunneling effect (assisted or not), there are two noteworthy bias voltages for this structure (Fig. 4).

The read-through voltage V_{RT}, when the semiconductor (width L) is just fully depleted,

$$W_1 + W_2 = L$$

and the flat-band voltage V_{FB} (Fig. 4), corresponding to the relations

$$W_1 = L \text{ and } W_2 = 0$$

Knowing that:

$$W_1 = \sqrt{\frac{2\epsilon_s(V_{b1} + V_1)}{qN_d}} \tag{7}$$

and

$$W_2 = \sqrt{\frac{2\epsilon_s(V_{b2} - V_2)}{qN_d}} \tag{8}$$

where ϵ_s is the semiconductor permittivity and N_d its doping level, we obtain

$$V_{FB} = \frac{qN_dL^2}{2\epsilon_s} - V_{b1} + V_{b2} \tag{9}$$

V_{RT} is close to V_{FB} (2). For bias voltage lower than V_{RT}, the holes injected through contact 2 diffuse in the neutral region between the two depleted zones and some of them recombine. Increasing bias voltage results in a decrease of the neutral region width and so of the recombinations. Consequently, the dark current increases quickly. Above V_{RT}, almost all injected carriers at one electrode are collected at the other. The increase of dark current with bias voltage is slower. It depends on the nature of the transport process through Schottky barriers and on the electric field at each contact. Generally, for $V > V_{FB}$, the electric field increases in the device and particularly near electrode 1 (the cathode). If it is sufficiently high, tunneling or impact ionization probability increases sharply at this place; the breakdown process occurs and the dark current increases very quickly with bias voltage. The typical I-V characteristic of a GaAs MSM photodiode with low dark current is reported in Fig. 5. We will complete this brief analysis by a review of the different transport processes reported in MSM photodiodes. But any analysis of the MSM-I-V characteristic cannot be carried out without considering the specific problems due to its planar structure.

Specific Properties Due to Planar Structure

For a bias voltage higher than the flat-band voltage, the electric field distribution in the semiconductor between two fin-

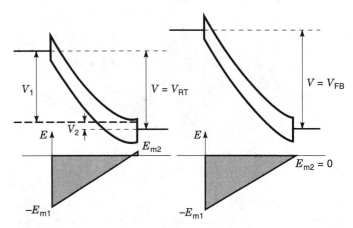

Figure 4. Energy-band diagrams of a biased MSM structure with electric field for applied voltages of V_{RT} and V_{FB}.

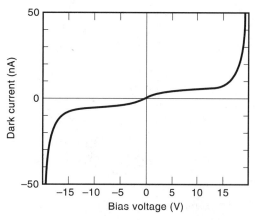

Figure 5. Measured dark current voltage characteristic of a GaAs MSM photodetector.

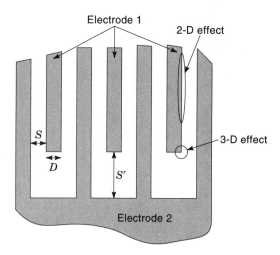

Figure 7. Two-dimensional and 3d corner effects in interdigitated structure. S is the electrode spacing, D the finger width, and S' the electrode distance.

gers, calculated by bidimensional Poisson equation solution, is similar to the one presented in Fig. 6. This calculation is made for a GaAs doping level of $10^{14}\mathrm{cm}^{-3}$. There are two electric field maxima at the contact edges. This bidimensional effect becomes three dimensional at the tip of each finger (Fig. 7). As a result of this particular electric field distribution, breakdown will occur first

- At the finger tips
- Near the semiconductor surface between two finger edges.

In order to avoid parasitic breakdown located at the finger tips, the common method is to set a distance S' between the two metallic electrodes larger than the electrode spacing S, as shown in Fig. 7. This solution appears clearly in Refs. 3 to 5. It is also possible to round the finger tips, as reported, for example, in Ref. 6, or to design a specific finger form without corners (7). Anyway, the effect due to finger edges cannot be avoided, and the main part of the dark current will flow near the semiconductor surface between two different fingers. The electric field is high (approximately 20 kV/cm in our example)

in the vicinity of the semiconductor surface and the deeper we go in the semiconductor, the weaker the electric field is. The field lines (perpendicular to the equipotential lines) of the modeled structure presented in Fig. 8 illustrate the carrier paths between fingers. Obviously, the deeper the carrier's location, the longer the distance it has to cover to join the electrode, and since its velocity is electric field dependent, the longer the time to be collected by the electrode. For example, supposing electron–hole pair photogeneration along the O–y line of Fig. 8, the transit time needed for an electron or hole to join the electrodes is plotted in Fig. 9 versus the depth at which it has been photocreated. The electron and hole field-dependent velocity in GaAs used for this calculation is from (8). Compared to carriers generated near the semiconductor surface, those generated at 2 μm depth have a transit time to electrodes 5 times and 3 times greater for, holes and electrons, respectively.

Analysis of Dark Current

The MSM photodetector dark current has to be as small as possible in order to reduce excess noise in the optoelectronic receiver and, consequently, to lower the minimum detectable

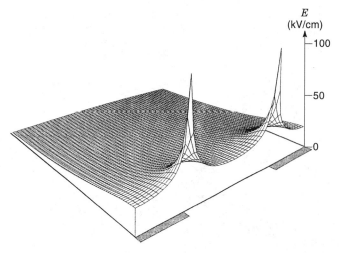

Figure 6. Electric field distribution in a GaAs MSM photodetector between two fingers. In this example, electrode spacing and finger width are 2 μm, absorbing layer is 3.5 μm thick and the bias voltage 8 V.

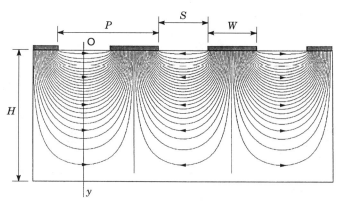

Figure 8. Example of field lines in a GaAs MSM photodetector. The simulated zone is that of Fig. 1. The arrows indicate the direction of electron transport on each line. O–y line is used in the following figure. P is the Period, S the electrode Spacing and D the finger Width.

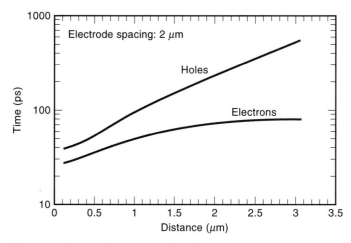

Figure 9. Required transit time for electrons and holes photogenerated at a certain depth on the O–y line of Fig. 8 to join the electrode.

power (9). More generally, this current depends strongly on the shape of the potential barrier at electrodes under bias. The different mechanisms that can occur in these conditions are (1,10):

1. Emission over the Schottky barrier (thermionic emission process)
2. Tunneling and thermionic field emission
3. Thermal generation of carriers in the depleted zone or in the neutral region and, in case of defects inducing states in the forbidden bandgap, mainly:
   1. Carrier generation assisted by states within the forbidden bandgap
   2. Tunneling through the metal–semiconductor barrier assisted by states in the forbidden bandgap

Assuming at first that the material is lattice matched and without defects inducing states in the forbidden bandgap, we are limited to the three first mechanisms. Assuming also that there is no unintentional thin insulating layer between the semiconductor and the electrode and that the material doping level is low, the thermionic emission process dominates over tunneling processes due to the strong dependence of the tunneling probability on doping level; moreover, it dominates over thermal generation processes because the material resistivity is high. This predominant process is the dark current limiting factor. If V_B is the breakdown voltage, for $V_{FB} < V < V_B$, following the analysis of Sze et al. (2), we have:

$$J_{dark} = \left[A_n^* T^2 \cdot \exp\left[-\frac{q(\phi_{n1} - \Delta\phi_{n1})}{kT}\right] + A_p^* T^2 \right.$$
$$\left. \cdot \exp\left[-\frac{q(\phi_{p2} - \Delta\phi_{p2})}{kT}\right]\right] \cdot \left[1 - \exp\left(-\frac{qV}{kT}\right)\right] \quad (10)$$

where $A_{n,p}^*$ are the effective Richardson constants for electrons and holes, T is the absolute temperature, and $\Delta\phi_{n1,p2}$ are the barrier lowering values due to the image force effect. This barrier lowering depends on the electric field $E_{m1,2}$ at each contact (1)

$$\Delta\phi_{n1,p2} = \sqrt{\frac{qE_{m1,2}}{4\pi\epsilon_s}} \quad (11)$$

which explains the slow increase of the dark current with bias voltage. The dark current is due to electrons emitted at the cathode (J_{n1}) and to holes emitted at the anode (J_{p2}) (Fig. 3). It depends essentially on the barrier heights at the two contacts. For example, in the case of MSM photodiodes fabricated on a low-doped GaAs layer, if the electrodes 1 and 2 are made of the same metal, we have:

$$\phi_{n1} + \phi_{p2} = E_g = 1.42 \text{ eV} \quad (12)$$

So, a Schottky barrier lower than $E_g/2 = 0.71$ eV makes the electron current predominant by increasing the probability of electron transport over the barrier of the contact 1, while a barrier higher than 0.71 eV makes the hole current predominant by increasing the hole emission at the contact 2. As a result, as shown by Wada & al and as presented in Fig. 10, the tungsten Silicide WSi$_x$ is the most suitable contact to decrease the GaAs MSM photodiode dark current because its barrier height is near 0.71 eV. Such a study has been confirmed by Koscielniak et al. (11). By the same way of analysis, an interesting solution to decrease the dark current consists in the deposition of an electrode metal with a high electron barrier at the cathode and another one, with a high hole barrier, at the anode. Such an approach has been successfully applied by Wohlmuth et al. in the case of InGaAs/AlInAs MSM photodiodes with Pt/Ti/Pt/Au and Ti/Au for the cathode and anode metalizations, respectively. The dark current of this assymetric structure was 0.2 nA at 5 V (10).

These results demonstrate the existence of a dominant thermionic process, but the other mechanisms can become dominant, depending on the perturbation of the metal-semiconductor interface during the technological process. All phenomena leading to hole accumulation in the vicinity of the

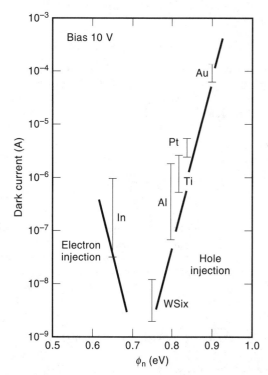

Figure 10. Dark current versus metal-semiconductor energy barrier for a biased GaAs MSM structure. This curve is reprinted from Wada et al. (Ref. 6) with permission of *IEEE Journal of Quantum Electronics*.

cathode, or electron accumulation near the anode, will modify the energy band structure at the interfaces and increase the tunneling probability at these electrodes. Within this context, electron tunneling assisted by hole accumulation in a thin native insulating layer between metal and semiconductor (GaAs) has been reported by Sugeta et al. (12), and hole tunneling due to electron accumulation in a hollow conduction band in the bidimensional potential distribution has been suggested by Wada et al. (6).

Tunneling assisted by deep levels inside the forbidden bandgap has also been reported by Wehmann et al. (13) in the specific case of heteroepitaxy of InGaAs on silicon.

Finally, a review of the dark current problems must mention the gain phenomena (see section on gain). Moreover, like most micro- and optoelectronic devices, an MSM photodetector has to be carefully passivated. Usually, the thickness of the dielectric layer used (SiO_2 or Si_3N_4) is chosen to act also as an antireflection coating (6,14). This insulating layer can be deposited

1. Above the fingers and the pads
2. Above the fingers and under the pads
3. Under the fingers and the pads

The two last solutions need comment. The dielectric insulating layer under pads as well as fingers is used as a barrier enhancement layer, to decrease the dark current (15), and to avoid photocurrent gain due to surface states (see section on gain). But in most cases, the barrier enhancement layer is a large bandgap lattice matched or mismatched semiconductor epilayer. Insulating only under the pads leads to a decrease of the dark current (16,17) and also to a decrease of the photodetector parasitic capacitance (18). This solution has been adopted mainly when the photodiode is integrated in a coplanar microwave line (9).

Optimum Bias Conditions of MSM Photodetector

From the above descriptions, we can conclude that the electric field has to be sufficiently high to allow the quick collection of photogenerated carriers. Because of the specific bidimensional distribution, this requires a high bias voltage, particularly for large interelectrode spacing. Moreover, because the dark current has to be kept as low as possible, we must avoid breakdown conditions. This leads to

$$V_{FB} < V < V_B$$

But an optimum bias voltage can be difficult to find since the electric field distribution is not homogeneous and the edge effect can lead to breakdown while the electric field is still small in the electrode gap. The larger the interelectrode spacing, the higher the difference between the electric field in the gap and those at finger edge. This makes the small interelectrode spacing interesting.

EPITAXIAL STRUCTURE AND DEVICE PROPERTIES

Material Systems

Obviously, the MSM photodetector epitaxial structure is designed starting from the absorbing layer whose gap is small

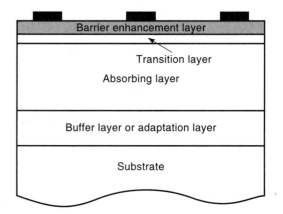

Figure 11. Whole MSM photodetector epitaxial structure.

enough to absorb at the wavelength of operation. The successive structure evolutions have led to the whole epitaxial structure presented in Fig. 11. This structure is the most that can be encountered; usually all epilayers are not required or used.

The different absorbing materials already used are presented in the Table 1 with the corresponding enhancement barrier and transition epilayers. If the Schottky barrier obtained by direct metal deposition on the absorbing epilayer is sufficiently high to induce low dark current (some tens of nanoamperes at 5 or 10 V bias voltage), the MSM photodetector structure is reduced to its simplest form (i.e., two Schottky contacts on an absorbing layer). This is, for example, the case in GaAs MSM photodetectors, as already explained in the previous section. But in the (1.0 to 1.6 μm) wavelength domain, the Schottky barrier obtained on N-$Ga_{0.53}In_{0.47}As$ is low, near 0.2 eV (40), which compels introduction of a large-gap enhancement-barrier layer. This layer has been grown in lattice-mismatched GaAs, AlGaAs, InP-GaInP, or in lattice-matched Fe:InP, P+ GaInAs, $Al_{0.48}In_{0.52}As$. Its thickness is a few hundred angstroms. The enhancement-barrier epilayer has also been introduced on GaAs (often in lattice-matched AlGaAs) in order to reduce the dark current and to improve the photodetector reliability (32). Furthermore, in the case of the GaAs-based photodiode, a large-gap bottom insulating layer can be used to limit the absorbing layer thickness and, consequently, to decrease the transit time of photogenerated carriers (Fig. 9). In the case of silicon, this insulating layer is silicon dioxide and the photodetector is then directly fabricated on commercially available SIMOX (separation by implanted oxygen) wafer (35,37). It is also possible to use sapphire substrate (34). The heterointerface between the absorbing layer and the enhancement-barrier layer leads to electron and hole pile-up with recombinations, a phenomenon which decreases the responsivity as well as the bandwidth. This is why a transition layer, a compositionally graded layer or a graded superlattice, is sometimes added. The impact of this layer on the photodetector dynamic behavior will be considered later. In order to demonstrate the state of the art, we sum up in Table 2 some typical realizations with their metallization and dark current. This table demonstrates the possibility obtaining dark current of a few nanoamperes or even less than 1 nA in each wavelength domain for an MSM photodetector active area, allowing the absorption of the whole light beam issued from a fiber. The active area is indi-

Table 1. MSM Photodetector Epitaxial Structure (The Heteroepitaxial Structures on Silicon Substrate Are Not in This Table)

Wavelength	Absorbing Material	Substrate	Enhancement-barrier Layer	Transition Layer
Ultraviolet $\lambda < 0.3~\mu m$	GaN	Saphire	—	
	SiC	SiC	-(33)	
$\lambda < 1.1~\mu m$	Si	Si	-(34–37)	
$\lambda < 0.87~\mu m$	GaAs	GaAs	-(6, 9)	
			AlGaAs (4, 18, 19, 20)	—
			GaInP (21)	
Long wavelength $1.0~\mu m < \lambda < 1.6~\mu m$	$Ga_{0.53}In_{0.47}As$	InP	$Al_{0.48}In_{0.52}As$ (22, 23)	In(GaAl)As graded (16, 23, 28) AlInAs-GaInAs GSL (28, 29)
			P+ $Ga_{0.53}In_{0.47}As$ (31)	
			Fe:InP (7)	
			GaAs * (24)	
			InP-GaInP * (25)	
			AlGaAs * (26)	
Far infrared $\lambda > 1.3~\mu m$	HgCdTe	GaAs*	CdTe (27)	
	GaSb	GaSb	AlGaSb (30)	

*Lattice mismatched GSL: graded superlattice

cated even if we have to keep in mind that the dark current is correlated with the contact area value. In most commonly reported designs, this later value is close to the half of the active area value. MSM photodiodes with transparent electrodes have been included. We did not introduce the GaInAs-based devices heteroepitaxially grown on Si or on GaAs substrate. By using these growth techniques, high responsivity (1.0 to 1.6 μm) wavelength photodetectors whose fabrication is compatible with GaAs or Si foundries were obtained (43,44).

Illumination Conditions and Responsivity

The photodetector can be top-, back-, or side-illuminated through an integrated optical waveguide. For top-illuminated structures, the light has to propagate through the interelectrode region into the semiconductor. Neglecting the recombi-

nations, the quantum efficiency is then given by

$$\eta = (1 - R) \cdot \frac{S}{(S + D)} \cdot [1 - \exp(-\alpha \cdot W_a)] \quad (13)$$

where R is the reflection coefficient of light at the air-semiconductor interface. R is reduced to a few percent in case of the use of an antireflection coating. The ratio $S/(S + D)$ introduces the finger shadowing effect, which is the main drawback of top-illuminated interdigitated structures. The absorption coefficient is α and the absorbing layer thickness is W_a. The responsivity is given by

$$\Re = \frac{q}{h\nu} \cdot \eta \quad (14)$$

where q is the electron charge (absolute value), and $h\nu$ the photon energy of the incident light. These expressions show

Table 2. Typical MSM Photodetector Material Structures With the Metallization and the Corresponding Dark Current

Ref.	Wavelength Domain	Epitaxial Structure	Growth	Metallization	Active Surface (μm^2)	Dark Current
(35)	$<0.9~\mu m$	crist.-Si/SiO$_2$/Si (SIMOX wafer)	—	Ti-Au	5×5	0.05nA (1V)
(38)	$<0.9~\mu m$	crist.-Si/amorph.-Si-H	(PECVD)	Cr-Au	100×100	0.4 nA(10V)
(6)	$<0.85~\mu m$	GaAs(S.I.)/GaAs	VPE	Wsi$_x$	100×100	1nA (10V)
(18)	$<0.85~\mu m$	GaAs(S.I.)/GaAs/AlGaAs	MOCVD	Ti-Pt-Au	75×75	0.04nA (10V)
(32)	$<0.85~\mu m$	GaAs(S.I.)/GaAs/AlGaAs/GaAs/ AlGaAs/GaAs	MOCVD	Ti-Pt-Au	20×50	0.3nA (10V)
(22)	$1.3–1.55~\mu m$	InP(S.I.)/GaInAs/AlInAs	LPMOCVD	Ti-Au	50×50	20nA (10V)
(41)	$1.3–1.55~\mu m$	InP(S.I.)/InP/GaInAs/InP	GSMBE	Pt-Au	50×50	200nA (10V)
(43)	$1.3–1.55~\mu m$	InP(Fe)/GaInAs(Fe)/InP(Fe)	LPMOCVD	Ti-Au	100×100	200nA (10V)
(25)	$1.3–1.55~\mu m$	InP(Fe)/InP/GaInAs/InP/GaInP	LPMOCVD	Au**	100×100	30nA (10V)
(23)	$1.3–1.55~\mu m$	InP(S.I.)/AlInAs/GSL/GaInAs/ GSL/AlInAs	GSMBE	Ti-Pt-Au	30×30	5nA (10V)
(16)	$1.3–1.55~\mu m$	InP(S.I.)/AlInAs/GaInAs/ In(GaAl)As/AlInAs	MBE	Ti-Au	50×50	109nA (10V)
(42)	$1.3–1.55~\mu m$	InP(Fe)/AlInAs/GaInAs/ In(GaAl)As/AlInAs	MOVPE	ITO **	50×50	4nA (5V)
(27)	$>1.5~\mu m$	GaAs*/HgCdTe/CdTe	MOCVD	Pt-Au	48×40	70 nA (5V)
(33)	$0.33–0.39~\mu m$	6H-SiC/SiC	—	Cr-Au	100×100	1nA (80V)

*Lattice mismatched; **Transparent fingers; GSL: graded superlattice; S.I.: semi-insulating that the MSM photodetector responsivity strongly depends on the absorbing layer thickness, the material absorption coefficient, and the interdigitated structure.

In the long wavelength region, GaInAs absorbing coefficient is 1.16 μm^{-1} at 1.3 μm and 0.68 μm^{-1} at 1.55 μm wavelength (45). The absorbing coefficient of GaAs is 1 μm^{-1} around 0.8 μm wavelength and it is lower than 0.2 μm^{-1} for $\lambda > 0.7$ μm (close to 0.2 μm^{-1} at 0.8 μm) and higher than 1 μm^{-1} for $\lambda < 0.5$ μm (34) for crystalline silicon. All these material properties mean a penetration length close to 1 μm for III–V materials. Because of indirect bandgap transition, the silicon penetration length is higher than 10 μm near 0.8 μm wavelength and lower than 1 μm only for $\lambda < 0.5$ μm. The calculated external quantum efficiency and responsivity of MSM photodetectors with different absorbing materials is presented in Table 3. Two absorbing layer thicknesses of 1 μm and 1000 Å have been used for calculation allowing to appreciate the photodetector sensitivity in typical cases. For these calculations, we supposed a perfect antireflection coating ($R = 0$) and equal finger width and spacing ($S = D$).

As for the finger shadowing effect, it seems that the finger width must be decreased as far as possible, but the finger resistance then increases, which reduces the photodetector bandwidth (Fig. 16). The finger resistance depends obviously on the finger dimensions, but, as noted by Chou et al. (5) in the case of fingers made of 150 Å/350 Å thick Ti/Au deposited on SiO$_2$ substrate, the finger resistance per unit length is higher than the one calculated using bulk resistivities. This difference may be due to the electron scattering with the metallic boundaries. Indeed, for fingers wider than 0.1 μm, only the top and bottom boundaries induce an increase of the resistance, while for fingers narrower than 0.1 μm, the side boundaries produce an effect. For example, the measured resistance for 0.06 μm finger width is 80 $\Omega/\mu m$ (11 $\Omega/\mu m$ with bulk resistivity) while it is 4.3 $\Omega/\mu m$ for 0.5 μm (1.2$\Omega/\mu m$ with bulk resistivity) (5). All these results permit us to estimate finger resistance in a great number of cases.

All these reasons generally lead to a finger width not far from the interelectrode spacing ($D \geq S \geq D/4$). However, a way to avoid the finger shadowing effect consists of the use of transparent metallizations in indium tin oxide (ITO), cadmium tin oxide (CTO), tungsten or gold. High–carried-concentration ITO is used with success in the short wavelength domain ($\lambda < 0.85$ μm) where its optical transmittance is greater than 87% (46) and its resistivity is poor. Its transmittance around 1.3 μm and 1.5 μm wavelength is respectively near 70% and 50%, but at 1.3 μm, it increases strongly while decreasing carrier concentration. Nevertheless, a good trade-off between transmittance and resistivity can be obtained by improving the technological deposition process (42), it is then possible to get a 99.5% transmittance, with a resistivity which is only an order of magnitude higher (0.015 $\Omega \cdot$ cm at room temperature). In the same way, the CTO has a transmittance over 85% in the long wavelength domain, and minimum resistivity can be obtained under specific deposition conditions (46). However, we must keep in mind that sputtering often produces defects at the metal-semiconductor interface, which increase the leakage current (47) and also induce photocurrent gain (this is documented in references 42 and 46). For all these reasons, Chu et al. (47) proposed to deposit on an AlInAs enhancement-barrier layer a thin tungsten silicide (WSi$_x$:200 Å) layer under the ITO (550 Å); this dual layer structure allows a 57% optical transmittance at 1.55 μm wavelength. Avoiding sputtering, Matin et al. (21) deposited 300 Å thick tungsten fingers by electron beam evaporation on an InGaP cap layer and obtained a 95% GaAs photodetector quantum efficiency. Yuang et al. (25) improved, the responsivity of a GaInAs photodetector from 0.4 to 0.7 A/W, with an InGaP enhancement barrier layer by depositing 100 Å thick gold fingers. To sum up, the use of transparent fingers allows the improvement of the responsivity, but the holes photogenerated under the anode (and the electrons photogenerated under the cathode) take a very long time to join the other electrode (see Fig. 8); this reduces the photodetector bandwidth by increasing the average transit time of the photogenerated carriers. As a consequence, in case of top illumination, transparent metallization is generally used when the high-speed operation of the device is not of prime importance.

On the other hand, Eq. (13) is acceptable for a period ($P = S + D$) larger than the wavelength. In case of smaller finger period, the electrodes constitute a metallic grating through which light propagation must be studied. This complicated problem, already suggested by Sano (48), has been experimentally studied by Kuta et al. (49). The transmission coefficient through an interdigitated structure depends on:

- The ratios λ/P and S/D
- The optical polarisation: electric field of the optical wave parallel ($\|$) or perpendicular ($\perp$) to the fingers
- The finger thickness
- The semiconductor refraction index n_s

For $S \approx D$, the dominant factor is the ratio λ/P with two particular values: $\lambda/P = 1$ and $\lambda/P = n_s$, where the transmission coefficient reaches a minimum (49). Moreover, the transmission coefficient is higher for perpendicular polarisation than for a parallel one. For $\lambda/P > 1.5$, the external quantum effi-

Table 3. Calculated Quantum Efficiency (η) and Responsivity ($\Re$) of MSM Photodetectors with Different Absorbing Materials for Absorption Layer Thicknesses of 1 μm and 1000 Å. We Supposed $R = 0$ and $S = W$.

| Wavelength | Material | $W_a = 1$ μm | | | |
		η	$\Re$ (A/W)	η	$\Re$ (A/W)
0.8 μm	GaAs	31.5%	0.195	4.7%	0.03
1.3 μm	GaInAs	34.3%	0.343	5.5%	0.055
1.55 μm	GaInAs	24.7%	0.29	3.3%	0.04
0.8 μm	crist.-Si	4.7%	0.029	0.5%	0.003
0.6 μm	crist.-Si	16.5%	0.076	2%	0.012
0.4 μm	crist.-Si	50%	0.154	31.6%	0.097

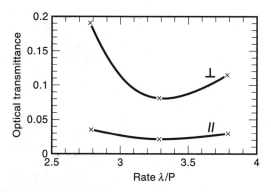

Figure 12. Measured transmittance of a GaAs MSM photodiode with $P = 4000$ Å and $D/P = 0.57$ versus the rate λ/P for linearly polarised light oriented parallel ($\parallel$;) or perpendicular ($\perp$) to the fingers. The lines are guides to the eye. Reprinted with permission from J. J. Kuta et al., Polarization and wavelength dependence of Metal-semiconductor-metal photodetector response, *Appl. Phys. Lett.* **64** (2):140–472, 1994. Copyright 1994 American Institute of Physics.

ciency of linearly parallel polarized light is lower than a few percents. A typical behaviour, presented in Fig. 12, shows the measured optical transmittance of such a grating on GaAs material ($P = 4000$ Å, $D/P = 0.57$) versus λ/P for wavelength higher than 0.85 μm. The interdigitated structure in 550 Å thick Ti–Au has been patterned using electron beam lithography. Because the GaAs is then transparent, the grating influence appears, this means a very low efficiency obtained for parallel linearly polarized light, while a higher one is measured for perpendicular polarization. A minimum exists at $\lambda/P = n_s = 3.4$ because of a resonance effect with the substrate. The studies that can be carried out with modeling tools such as finite difference time domain beam propagation method (FDTD-BPM) also show the great influence of the finger thickness because it is close to the finger width in these conditions. We cannot here describe all the consequences of this complicated problem but we must point out that the modification of photodetector responsivity due to small finger period is sufficiently high to allow the fabrication of a wavelength discriminator using several MSM photodiodes with different periods. Chen et al. (50) did this using two simultaneously illuminated parallel photodetectors with periods of 4000 Å and 6000 Å, respectively ($S = D$). Under these conditions, the photocurrent ratio (I_{4000}/I_{6000}) depends strongly on the wavelength, while it remains unchanged when the optical power varies.

Because of the photodetector planar structure and its specific transit characteristic (Figs. 8 & 9), top-illuminated high-speed devices need a thin absorption layer, which is not compatible with high quantum efficiency. In order to overcome this trade-off, a back reflector can be introduced. Instead of being placed on the back side of the wafer, it is positioned on the bottom of the absorption layer. In the case of III–V materials, this is of a Bragg reflector, while in case of silicon in the 0.8 μm wavelength region, it can be made from metal and needs wafer bonding technique (51). In order to increase the quantum efficiency, another method is roughening the front or back surface of the absorbing layer (37,52), but controlling the absorbing layer thickness is difficult, especially when this thickness has to be of a few thousand angstroms.

Let us consider now the back and side illumination. For back illumination, the substrate must be made thin and polished in order to deposit an antireflection layer. Obviously, the shadowing effect disappears; moreover, the finger presence increases the quantum efficiency by reflecting the light into the absorbing region. Nevertheless, compared to top illumination, the photodetector speed is reduced because of two different phenomena. First, because the generation rate decreases exponentially starting from the absorbing layer bottom, the greater part of the electron-hole pairs is photogenerated near the absorbing layer bottom, which increases the average transit time. Second, as already explained, the carriers photogenerated just under the electrodes have a long transit time. All these considerations explain that back-illuminated MSM photodiodes with very high responsivity (0.96 A/W at 1.3 μm wavelength) have already been made (see, eg., Ref 28) with an MSM photodetector directly integrated with the optical fiber, and also that this illumination condition is chosen if the high speed is not of prime importance.

Finally, the side illumination through an optical waveguide is interesting because it allows decreasing the absorbing layer thickness while absorbing nearly 100% of the light that propagates inside the waveguide (53). However, because a high coupling efficiency between the fiber and the waveguide is difficult to achieve, the responsivity of the whole structure constituted by the photodetector with its waveguide is generally not high. In spite of this problem, the waveguide MSM photodetector is a good candidate in the field of integrated photonic circuits (53,54).

Gain and Recombinations

Among the different electrical mechanisms that influence MSM photodetector responsivity, the recombinations tend to decrease responsivity, whereas trapping effects increase it. The recombinations occur in bulk material, particularly in cases of high impurity or high defect density. Such material as amorphous silicon, Cr-doped GaAs, low-temperature GaAs, Fe-doped InP, or Fe-doped GaInAs holds a short recombination time, which can be used to improve the photodetector dynamic behavior, as will be seen in the next section. But if the recombination time is shorter than the average interelectrode transit time, only a part of the photocarriers will be collected. Anyway, the recombinations abate the responsivity in all cases if the light penetration depth exceeds the depleted zone thickness; the carriers photogenerated outside the depletion region slowly diffuse and some of them recombine. This situation is encountered particularly in crystalline silicon MSM photodetectors in the (0.6 μm to 0.9 μm) wavelength domain, where the absorption coefficient is small. In this case, the quantum efficiency is only a few percent and increases with bias voltage because the depletion region depth increases with bias.

Recombinations also occur at the heterointerfaces, that induce carrier pileup. This is notably often the case between the small-gap absorbing layer and the large-gap Schottky enhancement layer. In this situation, the recombination rate at the heterointerface increases with carrier density and decreases with the electric field, which allows the carriers to surmount the barrier. An example of such behavior, already reported by Yang et al. (55), is given by Burroughes et al.

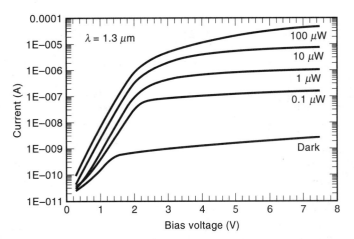

Figure 13. Dark and photocurrent of a top-illuminated InP/GaInAs/AlInAs MSM photodetector versus dc bias voltage. The GaInAs-AlInAs interface is abrupt and with a low defect density. The finger spacing is 3 μm and there is no AR-coating © 1991 IEEE. This curve is reprinted from Ref. 56 with permission.

(56) in Fig. 13 where, for a bias voltage higher than 2 V, the responsivity of a Fe:InP/GaInAs/AlInAs MSM photodiode depends on the optical power as well as on the dc bias voltage. Such a mechanism incites to introduce a transition layer in order to avoid the carrier pile-up; in this case, the abrupt interface with low defect density makes the photodetector compatible with a GaInAs/AlInAs H-MESFET process (56). Moreover, these curves show a responsivity of 0.7 A/W at 7 V bias, which corresponds, according to the authors, with an external quantum efficiency of 160%. Therefore, a gain phenomenon is combined with the effect of abrupt heterointerface. The gain often observed at low bias voltage in MSM photodetectors is due to trapping rather than to impact ionization in the high electric field region near the electrode edges. As reported by Klingenstein et al. (57) in the case of GaAs based photodetectors, two mechanisms can occur with different behavior versus frequency. The first has a specific cut-off frequency lower than a few hundred megahertz and can be compared to the low-frequency gain observed in GaAs photoconductors (58). It varies weakly versus temperature and disappears if a large-gap cap layer is grown on the absorbing layer. This is shown in Fig. 14, which compares the photocurrent-voltage characteristic of an (S.I.) GaAs/GaAs/Al$_{0.6}$Ga$_{0.4}$As MBE–grown MSM photodetector to one of an (S.I.) GaAs/GaAs. The traps located at the semiconductor surface between the fingers are responsible for this mechanism. The photogenerated carriers trapped at this surface create a dissymmetric electric charge distribution, the electrons being trapped rather near the anode. This charge modifies the potential distribution at the metal-semiconductor contact, which induces additional carrier injection from the electrode. The injection mechanism is tunneling rather than thermionic emission process, because the latter would make the gain strongly temperature dependent, which is not the case. For every trapped hole, an electron is injected, drifts in the semiconductor and is collected, and so on, until the trapped hole is re-emitted. Therefore, this phenomenon is limited by the trap density at the semiconductor surface, which leads to a gain decrease when the optical power and thus the photogenerated carrier density increase. Gain values of some hundreds then can be observed for a frequency below a few hundred megahertz or even lower.

The second gain mechanism existing at higher frequencies is due to hole pile-up or trapping in the vicinity of the cathode. These holes induce electron injection by tunneling from the electrode. Furthermore, gain due to impact ionization in the semiconductor bulk can also be observed when the electric field strength is sufficiently high. This occurs generally at high bias voltage (12).

Usually, the trapping effect of the surface is cancelled by the Schottky enhancement layer (18,57) and the quality of the metal-semiconductor interface. By doing this, it is possible to reduce the trap density strongly and to suppress the formation of a native thin insulating layer, so to avoid the gain phenomena that increase the photodetector pulse response time by adding a long tail to the short pulse due to collected photocarriers.

Dynamic Behavior

As with all photodetectors, the different phenomena leading to the MSM photodiode dynamic behavior are photodetector capacitance and resistance and carrier transit time. The main advantage of the MSM photodetector is its low capacitance C_{PD}, which can be calculated taking into account the planar structure by using conformal mapping technique (59). For two electrodes with width D and spacing S, (the period is $P = S + D$) above a semi-infinite semiconductor with relative permittivity ϵ_r, the capacitance per unit length is

$$C_0 = \epsilon_0 \cdot (1 + \epsilon_r) \cdot \frac{K(k)}{K(k')} \qquad (15)$$

with

$$K(k) = \int_0^{\pi/2} \frac{d\phi}{\sqrt{1 - k^2 \sin^2 \phi}} \qquad k = \tan^2 \left[\frac{\pi}{4} \cdot \frac{D}{P} \right] \text{ and }$$
$$k' = \sqrt{1 - k^2}$$

and the capacitance of the MSM photodetector active zone is given by

$$C_{PD} = C_0 \cdot \frac{A}{P} \approx C_0 \cdot (N - 1) \cdot L \qquad (16)$$

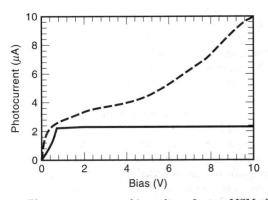

Figure 14. Photocurrent versus bias voltage for two MSM photodetectors with and without cap layer. Dashed line: (SI) GaAs/GaAs-Ti-Pt-Au structure, straight line: (SI) GaAs/GaAs/Al$_{0.6}$Ga$_{0.4}$As-Ti-Pt-Au structure. These MBE grown photodetectors with 1.5 μm finger spacing are illuminated at 632.8 nm with 20 μW optical power. These experimental results are reprinted from M. Klingenstein et al., Photocurrent gain mechanisms in metal-semiconductor-metal photodetectors, *Solid State Electron*, **37** (2): 333–340. Copyright 1993 with kind permission from Elsevier Science Ltd, The Boulevard, Langford Lane, Kidlington OX5 1GB, UK.

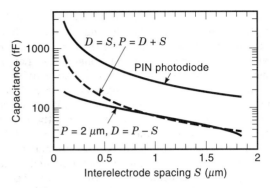

Figure 15. MSM photodetector capacitance versus finger spacing S. The active area is 2500 μm^2. Straight line: constant period so $D = P - S$, dashed line: $D = S$ so $P = D + S$. For comparison, the capacitance of a *pin* photodiode with an active layer thickness equal to S and the same area is plotted.

where A is the photodetector active area, P the period, N the number of fingers, and L the finger length. The comparison of the capacitance of an MSM photodiode with a spacing between 0.2 μm and 2 μm to the capacitance of a *pin* photodiode with an active layer thickness equal to S is shown in Fig. 15. Under these conditions, in spite of their different structures, these two photodetectors have a similar carrier transit time and the same active area—2500 μm^2. As is clear, the MSM photodiode capacitance is more than three times lower than that of the corresponding *pin* photodiode. The third curve of this figure, corresponding to a 2 μm period, shows the decrease of the capacitance when decreasing the finger width while keeping the period constant. This demonstrates the interest in decreasing the finger width compared to the finger spacing, since it also allows enhancement of the responsivity by diminishing the shadowing effect. However, the finger resistance can become important. Such a problem necessitates taking into account the whole MSM photodetector equivalent circuit represented in Fig. 16. I_{PH} represents the photocurrent, R_{PD} the leakage resistance, R_F the finger resistance, C_{PA} the parasitic capacitance including the ground-finger capacitance, that between pads, and the ground-pads capacitance. L_{BW} and C_{BW} are the parasitic elements due to the bond wire and R_L the load. The finger resistance is given by

$$R_F = 2R_0 L/N \qquad (17)$$

where R_0 is the finger resistance per unit length, L the finger length, and N the number of fingers on each electrode. The typical values of these elements are presented in Table 4.

In most cases, we can neglect the finger resistance and the bond-wire parasitic elements. Thus, only the classical elements R_L and $C_{MSM} = C_{PD} + C_{PA}$ remain. The characteristic

Table 4. Typical Values of the Elements to Be Introduced in the Equivalent Circuit

Element	Typical Value
C_{PD}	10–500 fF (see Fig. 15)
R_{PD}	$10^7 - 10^9$ Ω
R_F	1–100 Ω (41, 60, 61)
C_{PA}	10–100 fF (60, 61) ($<C_{PD}$)
L_{BW}	some tens of pH
C_{BW}	some fF

time constant corresponding to the photodetector impedance is then the well-known RC time constant.

For example, the RC time constant of a ($D = S = 1$ μm) MSM photodetector with a 50 $\times$ 50 μm^2 active area, loaded on 50 Ω resistance, is 4 ps, leading to an RC cut-off frequency as high as 40 GHz. On the other hand, assuming a saturation velocity of 5 $\times$ 10^6 cm/s, the corresponding transit time delay is 20 ps. All these values explain that the RC time constant is often not the dominant limiting factor of the MSM photodetectors' dynamic behavior. This is the reason we will now consider the carrier transit phenomenon.

Starting from the electric field distribution and the carrier paths described in Figs. 6 and 8, the transit time delay depends first on the interelectrode spacing, the absorbing layer thickness, and the bias voltage. It increases with the spacing and the absorbing layer thickness. On the other hand, in the regions where the carriers are photogenerated, the electric field must be sufficiently high to induce a saturation drift velocity for electrons and holes. Assuming at first that the absorbing layer (thickness W_a) is fully depleted, theoretical studies (62) predict that for a given finger spacing, the transit-time limited cut-off frequency is the highest when

$$\frac{S}{W_a} \approx 0.5$$

An example of such a characteristic is given in Fig. 17, which presents the cut-off frequency of GaInAs/AlInAs MSM photo-

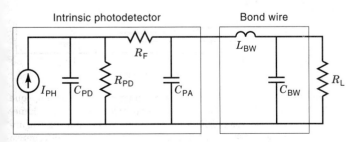

Figure 16. MSM photodetector equivalent circuit.

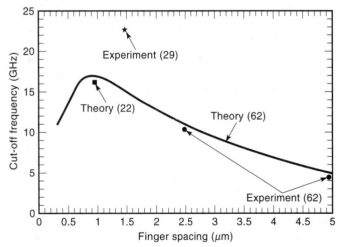

Figure 17. Cut-off frequency versus finger spacing for GaInAs/AlInAs MSM photodetector. The straight line represents theoretical result after (62), $W_a = 2$ μ, 500 Å AlInAs; the dots represent the experimental results after (62), $W_a = 2$ μm, 500 Å AlInAs; the square represents the theoretical result after (22), $W_a = 2$ μm, 500 Å AlInAs; and the star represents the experimental result after (29), $W_a = 1.5$ μm, 700 Å AlInAs, graded superlattice.

detectors versus finger spacing. We have joined theoretical (8,22,62) and experimental (29,62) results. The comparison between these results is difficult to carry out because generally experiments are performed with relatively high optical power (some milliwatts). This induces high carrier densities in the absorbing layer and as consequence an electric field screening (63) that modifies the carrier velocity distribution and thus increases the transit time delay. This phenomenon also occurs for high-input modulated light power (64) and depends strongly on the optical spot width. The specific electric field distribution due to the planar structure makes the MSM photodetector particularly sensitive to this factor. Moreover, for experiments under pulse operation, the cut-off frequency is derived from Fourier transform of the time response, which introduces additional errors. These reasons explain the differences between the results even if the overall behavior is the same.

Obviously, for thinner absorption layers, the obtained transit-time limited cut-off frequency is higher (22), particularly in the submicronic electrode spacing domain. More generally, in case of low penetration depth, because the majority of electron-hole pairs are photogenerated near the semiconductor surface (for III-V materials, the penetration depth is around 1 μm near cut-off wavelength) a semi-infinite absorption layer allows short transit time leading to very high cut-off frequencies. For example, a bandwidth of 105 GHz has been obtained on an MSM photodetector made by aluminum deposition on bulk GaAs with 0.5 μm electrode spacing (65); 300 GHz has been recorded by Chou et al. (5) on bulk GaAs with 0.1 μm electrode spacing by using high-resolution electron beam lithography. But for this electrode spacing domain, Monte Carlo simulations predict that the electron and hole pulse currents are separated because of the lower saturation drift velocity of holes. Moreover, the influence of parasitic elements becomes predominant, modifying the shape and the width of the output pulse (66). In the extreme case, for electrode spacing lower than the mean free path, the electronic transport is nonstationary and the transit time strongly decreases due to the electron velocity overshoot. This phenomenon permits reaching the terahertz frequency domain.

Anyway, the decrease of the absorbing layer thickness permits to reduction of the transit time. For example, Chou et al. (67) introduced an insulating AlGaAs-GaAs superlattice between the semi-insulating GaAs substrate and the 0.4 μm thick absorption layer in order to avoid the collection of carriers photogenerated in the substrate. In silicon, around 0.8 μm wavelength, a cut-off frequency higher than 100 GHz cannot be obtained because of the very low absorption coefficient. Particularly, the carriers photogenerated below the depletion region, which are collected after a long diffusion process toward the depleted zone, introduce a long tail to the photodetector pulse response. All these problems can be overcome by a local etching of the wafer back so as to get a very thin absorbing layer (52), or by using a specific insulating wafer, such as SIMOX (35) or sapphire (34). Obviously, these solutions lead to bandwidth increase at the expense of responsivity. If this latter is not of prime importance, it is possible to use an absorbing material with a very short carrier lifetime. The cut-off frequency is then limited by the recombination time and not by the carrier transit time. With this objective, low-temperature GaAs. Cr-doped GaAs, Fe-doped InGaAs, and amorphous silicon have been employed. The responsivity of such MSM photodiodes is lower than that obtained on pure

bulk materials, but the bandwidth is enhanced. For these recombination time-limited MSM photodetectors, a submicronic electrode spacing is not required. All the above considerations concern very high-speed MSM photodetectors. Let us now consider the photodetectors were a high responsivity, as well as a bandwidth of a few tens of gigaherz, is needed.

Especially in the (1.3 μm to 1.55 μm) wavelength domain, where the absorption layer bandgap is small, the heterointerface between the enhancement barrier layer and the absorbing layer leads to carrier pileup, increasing the transit time. The carrier pileup also influences the photodetector photocurrent-voltage characteristic because of the recombinations that occur if the heterointerface electric field is not high enough for carriers to pass through (Fig. 13). As demonstrated theoretically and experimentally (23,68), the introduction of a transition layer smoothing the conduction and valence band discontinuities allows us to avoid this problem. As a typical example, the fall time of the pulse response measured on a GaInAs/AlInAs MSM photodetector ($D = 2.5$ μm, $S = 2.5$ μm, $A = 30 \times 30$ μm^2, $W_a = 0.8$ μm) decreases from 19.3 ps to 15.7 ps by introducing a graded superlattice GaInAs-AlInAs (23). This results in a bandwidth enhancement from 18.1 GHz to 22.3 GHz (more than 20%).

Figure 18 presents the frequency and time responses with opaque and transparent fingers for a top-illuminated GaInAs-AlInAs MSM photodiode ($D = 3$ μm, $S = 3$ μm, $A = 50 \times 50$ μm^2, $W_a = 1$ μm) grown on InP:Fe substrate (42). In this case, the carriers photogenerated under the electrodes lead to increasing the responsivity from 0.3 to 0.6 A/W (without antireflection coating) but also to reducing the bandwidth from 13 GHz to 6 GHz. In case of back illumination, the bandwidth reduction will be of the same order of magnitude (obviously, whatever the transparency of the electrodes is).

In order to review the various typical characteristics of the MSM photodetectors, we have gathered in Table 5 typical reported structures with their measured responsivity and cut-off frequency. We included silicon, GaAs, and InP based devices, with transparent or opaque fingers, top or back illuminated. Most of these photodetectors are transit-time limited and the one on low-temperature GaAs is recombination-time

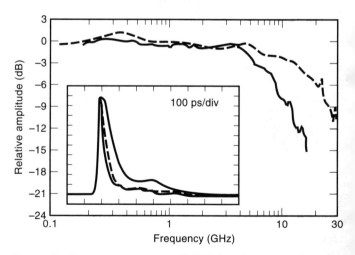

Figure 18. Frequency responses of MSM photodetectors with opaque Ti-Au fingers (dashed line) and transparent ITO(N2/H2) (dashed line) fingers. The active area is 2500 μm^2, the finger width and spacing are 3 μm, and the wavelength is 1.3 μm. This curve is reprinted from Seo et al., Ref. 42 with permission. ©1993 IEEE.

Table 5. Typical MSM Photodetector Material Structures with the Corresponding Measured Responsivity and Cut-off Frequency. The Noted Wavelength Is That of the Dynamic Measurement System. Some Structures of This Table Are also in Table 2 with Their Dark Current.

Ref.	Wavelength (μm)	Epitaxial Structure	Active Area (μm^2)	$D \times S$ (μm^2)	W_a (μm)	$\Re$ (A/W)	Cut-off Frequency
(5)	0.633	crist-Si substrate (P type)	10×10	0.1×0.1	semi-inf.	0.4	41 GHz
(35)	0.78	crist.-Si/SiO$_2$/Si (SIMOX wafer)	5×5	0.1×0.1	0.1	0.0057	140 GHz
(5)	0.633	GaAs(S.I.) substrate	10×10	0.1×0.1	semi-inf.	0.2	300 GHz
(5)	0.633	GaAs(S.I.)/LT-GaAs (1 μm)	10×10	0.3×0.3	semi-inf.	0.1	510 GHz
(18)	0.82	GaAs(S.I.)/GaAs/AlGaAs	75×75	2×4	semi-inf.	0.32	3.5 GHz
(22)	1.55	InP(S.I.)/GaInAs/AlInAs	50×50	1×1	1	0.35	17 GHz
(39)	1.3	InP(S.I.)/InP/GaInAs/InP	50×50	2×2	1	0.73	10 GHz
(41)	1.3	InP(Fe)/GaInAs(Fe)/InP(Fe)	100×100	1.5×1.5	2.5	0.35	4.8 GHz
(25)	1.54	InP(Fe)/InP/GaInAs/InP/GaInP	100×100	3×3	1	0.7**	2 GHz
(23)	1.3	InP(S.I.)/AlInAs/GSL/GaInAs/ GSL/AlInAs	30×30	2.5×1.5	0.8	0.36	20 GHz
(42)	1.3	InP(Fe)/AlInAs/GaInAs/ In(GaAl)As/AlInAs	50×50	3×3	1	0.8**	10 GHz
(28)	1.3	InP(S.I.)/AlInAs/GaInAs/ In(GaAl)As/AlInAs	17700	1×2	1	0.85*	4 GHz

*Back illumination; **Transparent fingers; GSL: graded superlattice; S.I.: semi-insulating

limited. This table gives also an idea of the trade-off between responsivity and cut-off frequency existing for this type of photodetector. In the (1.3 μm to 1.55 μm) wavelength domain, this trade-off leads to a low responsivity for bandwidths exceeding 20 GHz. For example, numerical modelling results show a responsivity lower than 0.15 A/W for a cut-off frequency higher than 45 GHz in the case of a InP/GaInAs/AlInAs photodetector with 0.3 μm electrode spacing (62). In fact, as for *pin* photodetectors, this trade-off can be overcome by using side-illuminated structures grown on an optical waveguide. Indeed, the absorbing layer grown on top of the waveguide can be very thin without giving the internal quantum efficiency up since the guided light is then progressively absorbed during its propagation. This is the main advantage of this type of structure.

Noise

Like all other semiconductor photodetectors, the MSM photodetector noise is related to different sources:

1. Thermal noise
2. $1/f^\alpha$ noise
3. Shot noise

Thermal noise intensity is theoretically frequency independent (white noise) and depends on the resistances existing in the whole equivalent circuit. In fact, the load resistance is generally the main part of the resistance that has to be taken into account, but finger resistance could have an influence, especially in the case of small finger width or if the electrode material has a relatively high resistivity. The thermal noise spectral density is given by

$$\langle I_{\mathrm{TH}2} \rangle = 4kT/R \quad \text{(in A}^2\text{/Hz)} \quad (18)$$

where T is the temperature and R the resistance of interest.

The $1/f^\alpha$ noise has a complex origin. Its spectral distribution is in $1/f^\alpha$ where α is close to 1, which makes it important for low frequencies. It has been notably related to material defects inducing various local lifetimes, these defects being

localized at the semiconductor surface or interfaces. It is often higher in devices with horizontal current flow (such as MESFET e.g.) rather than in devices with vertical current flow. As for the shot noise, it is related to the randomness of photogeneration and transport processes. Its spectral density is given by

$$\langle I_{\mathrm{SH}2} \rangle = 2q(I_d + I_{\mathrm{PH}}) \quad \text{(in A}^2\text{/Hz)} \quad (19)$$

where I_d and I_{PH} are, respectively, the dark and photocurrent. Its spectral distribution is flat (white noise) up to the photodetector cut-off frequency, where it behaves like the photocurrent (69).

Consequently, the MSM photodetector noise depends, for a given device, on the frequency, the bias voltage, and obviously, the dark and photocurrents.

Assuming an illuminated photodiode with small dark current (some tens of nanoamperes) at the optimal bias voltage, the $1/f$ noise has been reported to be higher than the shot noise only for frequencies lower than 1 MHz or even lower (70). For higher frequencies, the shot noise is dominant and can be considered the main noise source of this type of photodetector (70,71). The reduction of this noise current leads thus to better sensitivity receivers. For high bias voltage, the shot noise is strongly enhanced if a gain phenomenon occurs. We know, as encountered for avalanche photodiodes, that the noise factor F_N depends particularly on the nature and on the distribution of the gain phenomenon in the material. Assuming that $I_{\mathrm{PH}} \gg I_d$, the noise density is given by

$$\langle I_{\mathrm{SH}2} \rangle = 2q I_{\mathrm{PH}0} G^2 F_N = 2q I_{\mathrm{PH}0} G^{2+x} \quad (20)$$

where $I_{\mathrm{PH}0}$ is the primary photocurrent (without gain), and G the gain. The different reported measurements demonstrate values of x much higher than those observed in the InP/GaInAs avalanche photodiodes ($x \le 1$). For example, Vinchant et al. (53) measured $x = 2.4$ and Wada et al. (71), $x = 1.6$. Moveover, close to breakdown, especially in GaInAs-based devices with an enhancement barrier layer, the $1/f$ noise increases with the internal electric field strength, exceeding the shot noise at higher frequencies. This behavior related to traps at the heterointerfaces makes the $1/f$ noise significant at frequencies in order of 100 MHz (22).

These results explain that the noise behavior of the MSM photodetector under optimum bias voltage is similar to those of the *pin* photodiode; moreover, the use of MSM photodetectors near breakdown is not of interest.

Scaling Rules

In spite of the great number of parameters influencing the MSM photodetector behavior, we will give some simple rules necessary to design such a structure. The responsivity being related to the absorbing layer thickness and the shadowing effect [Eq. (14)], the major problem is the bandwidth. Indeed, the transit phenomenon, generally simulated by using complicated bidimensional models (48,62,63), is difficult to predict realistically with simple calculations. However, it is possible to give a simple formula in order to have an idea of the structure suitable to a given application. In this way, the photodetector transfer function can be written

$$H(\omega) = \frac{1}{(1 + j\omega R_L C_{\mathrm{MSM}})} \frac{1}{(1 + j\omega\tau)} \qquad (21)$$

where the first term introduces the RC time constant, C_{MSM} being calculated with Eq. (16), and the second term represents the transit influence. The transit-time delay is then given by (72)

$$\tau = \frac{S}{2v_{\mathrm{sat}}}\delta \qquad (22)$$

where v_{sat} is the carrier saturation velocity (the same for electrons and holes) and δ a number generally between 1 and 2, which introduces a correction due to the curved nature of the field lines (see Fig. 8). Under these assumptions, the cut-off frequency is

$$f_c = \frac{1}{2\pi \sqrt{(R_L C_{\mathrm{MSM}})^2 + \tau^2}} \qquad (23)$$

For example, Fig. 19 presents the cut-off frequency versus the photodiode area and the period in different cases: $S = W$ and

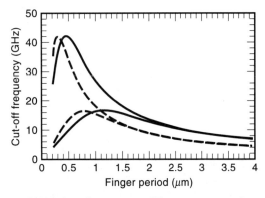

Figure 19. MSM photodetector cut-off frequency versus finger period for various surfaces and ratios S/W. The higher curves correspond to a 20×20 μm^2 surface while the lower ones to 50×50 μm^2. For all curves, we have taken: $\delta = 1.4$. Straight lines: $S = D$, dashed lines: $S = 3D$ and the load is 50 Ω.

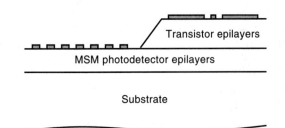

Figure 20. Integration scheme of MSM detector and transistors. The ohmic contacts of transistors are more often made before the common Schottky contact of MSM and FET's gate so as not to destroy these during the high temperature anneal of ohmic contacts.

$S = 3D$, for $R_L = 50$ Ω. The shape of each curve is due to the influence of capacitance and transit, the RC time delay being dominant for small periods (and small interelectrode spacing) and the transit-time delay becoming dominant for large periods. Obviously a trade-off leads to a maximum cut-off frequency, which increases strongly when decreasing area. Furthermore, a comparison of the cases $S = D$ and $S = 3D$ demonstrates that for a given period, the decrease of the finger width can be interesting since the resulting reduction of the shadowing effect is not at the expense of the bandwidth. Obviously, this result is acceptable as long as the finger resistance is not too high.

INTEGRATION

As can be seen in the above descriptions, several aspects make the MSM photodetector a favored candidate for the fabrication of high bandwidth monolithic integrated circuits. These are:

1. Planar configuration of electrodes
2. Use of undoped layers
3. Low capacitance
4. Low dark current

All these points make this photodiode particularly suitable for integration in a microwave strip or coplanar line. Indeed, this integration is needed for high-speed operation in order to reduce parasitics due to interconnections. On this topic, coplanar designs have been reported, such as those of Nakajima et al. (9), proposing a photodetector in a coplanar line directly integrated in a coaxial cable, or those of Kim et al. (28). In the first example, the photodetector is in the middle of the microwave line and it is connected by two accesses in quadripole configuration. On the contrary, in the second example, the photodiode has one electrode to the ground and needs only one microwave connector. It can be tested by using microwave measurement probes. The first configuration is used especially in cases of optical sampling [see, e.g., the optoelectronic AND gate and the inhibitor fabricated by Sugeta et al. (12)].

Optoelectronic integrated circuits (OEIC) combining both optoelectronic and electronic functions always constitute attractive technical and commercial subjects since they allow simultaneously the increase in performance of components by the decrease of parasitics and the decrease of manufacturing costs by reducing the handling and assembling of separate devices. The counterpart of this is sometimes and more often

Table 6. Advances in OEICs Associating MSM Detector and FETs. The Upper Part of the List Is Dedicated to 0.8 μm Wavelength Devices, the Lower One to 1.3 μm and 155 μm.

Ref.	MSM Characteristics: Active Area, $D \times S$ (μm)	Transistor (Type, Gate Length, Amp. Type)	Bandwidth	MSM Responsivity (A/W)	IC Responsivity (V/W)
(74)	100 × 100, 3, 3	MESFET, 2 μm, TZ	2 Gbits/s	0.2	400
(75)	75 × 75, 1, 3.25	MESFET, 1 μm, TZ diff. pair	3.2 GHz	0.45	
(76)	10 × 10, 1, 1	MESFET, 0.35 μm, TZ	5.2 GHz	0.2	60
(77)	25 × 25, 1, 1.5	c-HEMT, 0.3 μm, TZ	14 GHz	0.25	170
(78)	InP, 1.5, 2.5	FET, 1.1 μm	200 Mbits/s	0.9	
(79)	30 × 30, InAlAs, 1,1	InGaAs HEMT, 1 μm (GaAs subst.)		0.45	
(20)	25 × 25, GaAs, 0.75, 1	GaAs HEMT, 0.3 μm, TZ (GaAs subst.)	430 MHz	0.08	2100
(80)	50 × 50, InP, 0.5, 0.5	InGaAs HEMT, 0.3 μm to 1 μm	16 GHz	0.26	

a larger complexity in technological process. The four previously mentioned characteristics of the MSM photodetector are significant assets compared to its main competitor for integration objectives, the *pin* photodiode. Nevertheless, the *pin* advances a better responsivity, which is not a trivial detail. But from a technological point of view, the integration scheme proves easy: the MSM detector is built on the buffer layer (or even more simply on the substrate) and transistor—mainly field effect transistors (FET)—epilayers are grown above (Fig. 20). This often leads to a very weak difference in height between MSM and transistor planes, which facilitates the interconnections. Moreover, since transistor gate and MSM finger metallization is often the same, they are usually deposited simultaneously, thus reducing the number of technological steps.

Owing to a very simple manufacturing process on GaAs material system and a total compatibility with FET technology, the first integrated photoreceivers combining MSM and metal-semiconductor field effect transistors (MESFET) have been produced in this material system. After the first attempt—association with a high-impedance (HZ)–type amplifier (73)—which mainly demonstrated the feasibility of such integration, all subsequent ICs used a transimpedance (TZ) design. The MSM detector is made either on the substrate or on the buffer layer, which, in any case, is needed for the transistor fabrication. Various forms of FETs can be used, namely more frequently MESFET or high electron mobility transistor (HEMT), different ICs are listed in Table 6 (upper part). More recent devices exhibit a bandwidth greater than 10 GHz and very good performance. Unfortunately, the 0.8 μm wavelength is not in favor with optical communications either in long-haul (high bit rate) systems or even in short-haul (distribution) ones. It is then preferable to move to longer wavelengths: 1.3 μm and 1.55 μm. Use of GaInAs(P)/InP MSM detectors is then required. For these materials (see section on material systems), MSM loses its technological simplicity since a much more complicated epitaxy is needed. Different enhancement layers can be used, which lead to many integrable devices. Different kinds of transistors can also be made, either in the InP or GaAs material system. Depending on the available device technology, expertise of heteroepitaxy, and knowledge of the research teams involved, several attempts (some of them are listed in the above table) tend to demonstrate that integrated photoreceivers can be made mixing various transistor and MSM types. In all cases, the frequency response of the long-wavelength MSM detector (see Section

on dynamic behavior) governs the response of the integrated photoreceiver.

Because of the very large range of possible design rules and the related trade-offs between the performance of the optoelectronic (the MSM detector) and electronic (the transistors) components, no real conception scheme has emerged up to now, and so no integrated photoreceiver of this type is currently commercially available.

BIBLIOGRAPHY

1. S. M. Sze, *Physics of Semiconductor Devices,* 2nd ed., New York: Wiley, 1981.

2. S. M. Sze, D. J. Coleman Jr., and A. Loya, *Current transport in Metal-Semiconductor-Metal structure,* Solid State Electronics, New York: Pergamon Press, 1971, vol 14, pp 1209–1218.

3. J. B. D. Soole et al., High speed performance of OMCVD grown InAlAs/InGaAs MSM photodetectors at 1.5 μm and 1.3 μm wavelengths, *IEEE Photon. Tech. Lett.* 1: 250–252, 1989.

4. A. Aboudou et al., GaAlAs/GaAs planar photoconductors and MSM photodetectors monolithically integrated with HIGFETs: application for optical dock distribution, *IEE Proc. J* 139 (1): 83–87, 1992.

5. S. Y. Chou and M. Y. Liu, Nanoscale tera-hertz Metal-Semiconductor-Metal photodetectors, *IEEE J. Quantum Electr.,* 28: 2358–2368, 1992.

6. M. Ito and O. Wada, Low dark current GaAs Metal-Semiconductor-Metal photodiodes using Wsi$_x$ contacts. *IEEE J. Quantum Electr.,* 22: 1073–1077, 1986.

7. L. Yang et al., High performance of Fe:InP/InGaAs Metal-Semiconductor-Metal photodetector grown by metalorganic vapor phase epitaxy. *IEEE Photon. Tech. Lett.* 2: 56–58, 1990.

8. I. S. Ashour et al., Comparison between GaAs and AlInAs/GaInAs MSM PD for microwave and millimeter-wave applications using a two dimensional bipolar physical model. *Micr. Opt. Tech. Lett.* 9 (1): 52–57, 1995.

9. K. Nakajima et al., Properties and design theory of ultrafast GaAs Metal-Semiconductor-Metal photodetector with symmetrical Schottky contacts, *IEEE Trans. Electron Devices,* 37: 31–35, 1990.

10. W. Wohlmuth et al., Engineering the Schottky barrier heights in InGaAs Metal-Semiconductor-Metal photodetectors, *Proc. SPIE,* 3006: 52–60, 1997.

11. W. C. Koscielniak et al., Dark current characteristics of GaAs Metal-Semiconductor-Metal photodetectors, *IEEE Trans. Electron. Devices,* 37: 1623–1629, 1990.

12. T. Sugeta et al., Metal-Semiconductor-Metal photodetector for high speed optoelectronic circuits. *Proc. 11th Conf. Solid State Devices Jpn. J. Appl. Phys.* 19 Suppl 19-1: 459–464, 1980.

13. H. H. Wehmann et al., Dark current analysis of InGaAs MSM photodetectors on Silicon substrate. *IEEE Trans. Electron Devices,* **43**: 1505–1509, 1996.

14. S. Kollakowski et al., Fully passivated AR coated InP/InGaAs MSM photodetectors, *IEEE Photon. Tech. Lett.* **6**: 1324–1326, 1994.

15. W. Wohlmuth, P. Fay, C. Caneau, and I. Abesida, Low dark current, long wavelength Metal-Semiconductor-Metal photodetectors, *Electron. Lett.* **32** (3): 249–250, 1996.

16. H. T. Griem et al., Long wavelength (1.0-1.6 μm) InAlAs/In(GaAl)As/InGaAs Metal-Semiconductor-Metal photodetector. *Appl. Phys. Lett.* **56** (11): 1067–1068, 1990.

17. J. H. Burroughes, H-Mesfet compatible GaAs/AlGaAs MSM photodetector, *IEEE Photon. Techn. Lett.* **3**: 660–662, 1991.

18. C. X. Shi et al., High performance undoped InP/N InGaAs MSM photodetectors grown by LP-MOVPE, *IEEE Trans. Electron Devices,* **39**: 1028–1031, 1992.

19. O. Vendier, N. M. Jokerst, and R. P. Leavitt, High efficiency thin-film GaAs MSM Photodetectors, *Electron. Lett.* **32** (4): 394–395, 1996.

20. V. Hurm et al., 1.3 μm monolithic integrated optoelectronic receiver using an InGaAs MSM photodiode and AlGaAs/GaAs HEMTs grown on GaAs, *Electron. Lett.* **31** (1): 67–68, 1995.

21. M. A. Matin et al., Very low dark current InGaP/GaAs MSM Photodetector using semi transparent and opaque contacts, *Electron. Lett.* **32** (8): 766–767, 1996.

22. J. B. D. Soole and H. Schumacher, InGaAs Metal-Semiconductor-Metal photodetectors for long wavelength optical communications. *IEEE J Quantum Electr.* **27**: 737–752, 1991.

23. Y. G. Zhang, A. Z. Li, and J. X. Chen, Improved performance of InAlAs-InGaAs-InP MSM photodetectors with graded superlattice structure grown by gas source MBE. *IEEE Photon. Tech. Lett.* **8**: 830–832, 1996.

24. H. Schumacher et al., An investigation of the optoelectronic response of GaAs/InGaAs MSM photodetectors. *IEEE Electr. Device Lett.* **9**: 607–609, 1988.

25. R. H. Yuang et al., High responsivity InGaAs MSM photodetectors with semi transparent Schottky contacts. *IEEE Photon. Tech. Lett.* **7**: 1333–1335, 1995.

26. W. P. Hong, G. K. Chang, and R. Bhat, High performance AlGaAs/InGaAs MSM photodetectors grown by OMCVD. *IEEE Trans. Electron. Devices* **36**: 659–662, 1989.

27. P. W. Leech et al., HgCdTe Metal-Semiconductor-Metal photodetectors, *IEEE Trans. Electron. Devices,* **40**: 1365–1370, 1993.

28. J. H. Kim et al., High performance back-illuminated InGaAs/InAlAs MSM photodetector with a record responsivity of 0.96A/W, *IEEE Photon. Tech. Lett.* **4**: 1241–1245, 1992.

29. O. Wada et al., Very high speed GaInAs Metal-Semiconductor-Metal photodiode incorporating an AlInAs/GaInAs graded superlattice, *Appl. Phys. Lett.* **54** (1): 16–17, 1989.

30. S. Tiwari et al., 1.3 μm GaSb Metal Semiconductor Metal photodetectors, *IEEE Photon. Tech. Lett.* **4**: 256–25, 1992.

31. S. V. Averin et al., Low dark current quasi-Schottky barrier MSM photodiodes structures on N-GaInAs with P+ GaInAs cap layer, *Electron. Lett.* **28** (11): 992–995, 1992.

32. A. Aboudou et al., Ultralow dark current GaAlAs/GaAs MSM photodetector, *Electron. Lett.* **27** (10): 793–794, 1991.

33. Y. G. Zhang, A. Z. Li, and A. G. Milnes, Metal-Semiconductor-Metal ultraviolet photodetectors using 6H-SiC, *IEEE Photon. Tech. Lett.* **9**: 363–364, 1997.

34. C. C. Wang et al., Comparison of the picosecond characteristics of silicon and silicon-on-sapphire Metal-Semiconductor-Metal photodiodes, *Appl. Phys. Lett.* **64** (26): 3578–3580, 1994.

35. M. Y. Liu, E. Chen, and S. Y. Chou, 140GHz Metal-Semiconductor-Metal photodetectors on silicon-on-insulator substrate with a scaled active layer, *Appl. Phys. Lett.* **65** (7): 887–888, 1994.

36. S. Y. Chou, Y. Liu, and T. F. Carruthers, 32GHz Metal-Semiconductor-Metal photodetectors on crystalline silicon, *Appl. Phys. Lett.* **61** (15): 1760–1762, 1992.

37. B. F. Levine et al., 1 Gb/s Si high quantum efficiency monolithically integrable λ = 0.88 μm detector, *Appl. Phys. Lett.* **66** (22): 2984–2986, 1995.

38. L. H. Laih et al., High performance Metal-Semiconductor-Metal photodetector with a thin hydrogenated amorphous silicon layer on crystalline silicon, *Electron. Lett.* **31** (24): 2123–2124, 1995.

39. M. A. Matin et al., High responsivity InGaAs/InP based MSM Photodector operating at 1.3 μm wavelength, *Microw. Opt. Technol. Lett.* **12** (6): 310–313, 1996.

40. K. Kajiyama, Y. Mizushima, and S. Sakata, Schottky barrier height of N-InGaAs diodes, *Appl. Phys. Lett,* **23** (8): 458–459, 1973.

41. E. H. Böttcher et al., Ultrafast semiinsulating InP:Fe-InGaAs:Fe-InP MSM photodetectors: Modeling and performance, *IEEE J. Quantum Electr.* **28**: 2343–2357, 1992.

42. J. W. Seo et al., Application of Indium-Tin-Oxide with improved transmittance at 1.3 μm for MSM photodetectors, *IEEE Photon. Tech. Lett.* **5**: 1313–1315, 1993.

43. A. Bartels et al., Performance of InGaAs Metal-Semiconductor-Metal photodetectors on Si, *IEEE Photon. Techn. Lett.* **8**: 670–672, 1996.

44. D. L. Rogers et al., High speed 1.3 μm GaInAs Detectors fabricated on GaAs substrates, *IEEE Electr. Device Lett.* **9**: 515–517, 1988.

45. D. A. Humphreys et al., Measurement of absorption coefficient of GaInAs over the wavelength range 1.0–1.7 μm, *Electron. Lett.* **21** (25): 1187–1189, 1985.

46. W. Gao et al., InGaAs Metal-Semiconductor-Metal photodiodes with transparents Cadmium Tin Oxide Schottky contacts, *Appl. Phys. Lett.* **65** (15): 1930–1932, 1994.

47. C. C. Chu et al., Performance enhancement using Wsix/ITO electrodes in InGaAs/InAlAs MSM photodetectors, *Electron. Lett.* **31** (19): 1692–1694, 1995.

48. E. Sano, Two-dimensional ensemble Monte Carlo calculation of pulse responses of submicrometer GaAs Metal-Semiconductor-Metal photodetectors, *IEEE Trans. Electron. Devices* **38**: 2075–2081, 1991.

49. J. J. Kuta et al., Polarization and wavelength dependence of Metal-Semiconductor-Metal photodetector response, *Appl. Phys. Lett.* **64** (2): 140–142, 1994.

50. E. Chen and S. Y. Chou, A wavelength detector using monolithically integrated subwavelength Metal-Semiconductor-Metal photodetectors, *Proc SPIE* **3006**: 61–67, 1997.

51. E. Chen and S. Y. Chou, High efficiency and high speed Metal-Semiconductor-Metal photo-detectors on Si-on-insulator substrates with buried backside reflectors, *Proc SPIE* **3006**: 74–82, 1997.

52. H. C. Lee and B. V. Zeghbroeck, A novel high speed silicon MSM photodetector operating at 830nm wavelength, *IEEE Electr. Device Lett.* **16**: 175–177, 1995.

53. J. F. Vinchant et al., Monolithic integration of a thin and short Metal-Semiconductor-Metal photodetector with a GaAlAs optical inverted rib waveguide on a GaAs semi insulating substrate, *Appl. Phys. Lett.* **55** (19): 1966–1968, 1989.

54. J. B. D. Soole et al., Waveguide integrated MSM photodetector on InP, *Electron. Lett.* **24** (24): 1478–1480, 1988.

55. L. Yang, A. S. Sudbo, and W. T. Tsang, GaInAs Metal-Semiconductor-Metal photodetectors with Fe : InP barrier layers grown by chemical beam epitaxy, *Electron. Lett.* **25** (22): 1479–1481, 1989.

56. J. H. Burroughes and M. Hargis, 1.3 μm InGaAs MSM photodetector with abrupt InGaAs/AlInAs interface, *IEEE Photon. Tech. Lett.* **3**: 532–534, 1991.

57. M. Klingenstein et al., Photocurrent gain mechanisms in Metal-Semiconductor-Metal photodetectors, *Solid State Electron* **37** (2): 333–340, 1994.

58. J. P. Vilcot, J. L. Vaterkowski, and D. Decoster, Temperature effects on high gain photoconductive detectors, *Electron. Lett.* **20** (2): 86–87, 1984.

59. Y. C. Lim and R. A. Moore, Properties of alternately charged coplanar parallel strips by conformal mapping, *IEEE Trans. Electron. Devices* **15**: 173–180, 1968.

60. W. C. Koscielnak, J. L. Pelouard, and M. A. Littlejohn, Intrinsic and extrinsic response of GaAs Metal-Semiconductor-Metal photodetector, *IEEE Photon. Tech. Lett.* **2**: 125–127, 1990.

61. J. W. Chen, D. K. Kim, and M. B. Das, Transit time limited high frequency response characteristics of MSM photodetectors. *IEEE Trans. Electron. Devices* **43**: 1839–1843, 1996.

62. I. S. Ashour et al., Cutoff frequency and responsivity limitation of AlInAs/GaInAs MSM PD using a two dimensional bipolar physical model, *IEEE Trans. Electron. Devices* **42**: 231–237, 1995.

63. C. Moglestue et al., Picosecond pulse response characteristics of GaAs Metal-Semiconductor-Metal photodetectors, *J Appl. Phys.* **70** (4): 2435–2448, 1991.

64. I. S. Ashour et al., High optical power nonlinear dynamic response of AlInAs/GaInAs MSM Photodiode, *IEEE Trans. Electron. Devices* **42**: 828–834, 1995.

65. B. J. Van Zeghbroeck et al., 105GHz bandwidth Metal-Semiconductor-Metal photodiode. *IEEE Electr. Device Lett.* **9**: 527–529, 1988.

66. W. C. Koscielnak, J. L. Pelouard, and M. A. Littlejohn, Dynamic behavior of photocarriers in a GaAs Metal-Semiconductor-Metal photodetector with sub-half-micron electrode pattern, *Appl. Phys. Lett.* **54** (6): 567–569, 1989.

67. S. Y. Chou, Y. Liu, and P. B. Fischer, Tera-hertz GaAs Metal-Semiconductor-Metal photodetectors with 25 nm finger spacing and width, *Appl. Phys. Lett.* **61** (4): 477–479, 1992.

68. E. Sano et al., Performance dependence of InGaAs MSM photodetectors on barrier enhancement layer structure, *Electron. Lett.* **28** (13): 1220–1221, 1992.

69. J. P. Gouy et al., Microwave noise performance and frequency response of PIN GaInAs photodiodes, *Microw. Opt. Technol. Lett.* **3** (2): 47–49, 1990.

70. H. Schumacher et al., Noise behavior of InAlAs/GaInAs MSM photodetectors, *Electron. Lett.* **26** (9): 612–613, 1990.

71. O. Wada et al., Noise characteristics of GaAs Metal-Semiconductor-Metal Photodiodes, *Electron. Lett.* **24** (25): 1574–1575, 1988.

72. J. Burm et al., Optimization of high speed Metal-Semiconductor-Metal photodetectors, *IEEE Photon. Tech. Lett.* **6**: 722–724, 1994.

73. M. Ito et al., Monolithic integration of a Metal-Semiconductor-Metal photodiode and a GaAs preamplifier, *IEEE Electr. Device Lett.* **5**: 531–532, 1984.

74. D. L. Rogers, Monolithic integration of a 3Ghz detector/preamplifier using a refractory gate ion implanted MESFET process, *IEEE Electr. Device Lett.* **7**: 600–601, 1986.

75. H. Hamaguchi et al., GaAs optoelectronic integrated receiver with high output fast response characteristics, *IEEE Electr. Device Lett.* **8**: 39–41, 1987.

76. C. S. Harder et al., 5.2 GHz bandwidth monolithic GaAs optoelectronic receiver, *IEEE Electr. Device Lett.* **9**: 171–173, 1988.

77. V. Hurm et al., 14 GHz bandwidth MSM photodiode AlGaAs/GaAs HEMT monolithic integrated optoelectronic receiver, *Electron. Lett.* **29** (1): 9–10, 1993.

78. L. Yang et al., Monolithically integrated InGaAs/InP MSM-FET photoreceiver prepared by chemical beam epitaxy, *IEEE Photon. Tech. Lett.* **2**: 59–61, 1990.

79. W. P. Hong et al., InAlAs/InGaAs MSM photodetectors and HEMT's grown by MOCVD on GaAs substrates, *IEEE Trans. Electron. Devices* **39**: 2817–2818, 1992.

80. M. Horstmann et al., 16Ghz bandwidth MSM photodetector and 45/85GHz ft/fmax HEMT prepared on an identical InGaAs/InP layer structure, *Electron. Lett.* **32** (8): 763–764, 1996.

J. A. HARARI
J. P. VILCOT
D. J. DECOSTER
Institut d'Electronique et de
Microélectronique du Nord

METALS INDUSTRY

Metals, and especially iron, were some of the oldest materials processed by our ancestors. The commercial exploitation of iron ores began in central Europe about 1000 B.C. In the Middle Ages, iron was produced by directly heating a mixture of iron ore and charcoal in a small shallow pit dug in the ground and lined with clay. In these so-called bowl furnaces, the ore was reduced to a mushy mass of relatively pure iron including cinder and ash, and then hammered and forged into a finished product. Blast furnaces were introduced in Europe around the end of the fifteenth century. These furnaces were capable of continuously producing metal in much larger quantities. In the sixteenth and seventeenth centuries the output of blast furnaces was mostly in the form of pig iron for subsequent working at forges.

The operation of the forges remained essentially the same until 1784 when Henry Cort introduced the process of puddling and rolling iron. This process involved decarburizing pig iron to an appropriate carbon content by melting it directly in a puddling furnace and then passing it through rollers of special design. The top roll in Cort's process contained a properly shaped collar that was made to fit into a specially designed grove in the bottom roll. Using this mill, it was possible to produce iron bars at a rate ten to fifteen times faster than with tilt hammer technology, which predominated then.

The Industrial Revolution of the nineteenth century provided considerable impetus for the development of improved hot rolling processes for iron and also for other metals, such as aluminum. The development of the continuous rod mill did not begin until about 1870, and the length of the rod it produced was limited to less than 100 m. Yet, around the same time, the rapid expansion of the new electric telegraph required wire in the greatest possible length. The first reversing plate rolling mill and the first universal mill, which combined horizontal and vertical pairs of rolls in one stand, were placed in operation in the mid 1850s. During the second half of the century, considerable progress was made in mills for rolling structural shapes. Universal reversing mills for rolling Z-bars, H-beams, and flanged beams came into existence around that time.

Simultaneously with the progress in steel rolling, advancements were also made in steel conversion. The Bessemer converter was invented in 1857. This steelmaking process consisted essentially of blowing hot air under pressure through a bath of molten iron in a vessel. Initially, the Bessemer converter was lined with acid refractories and was used to refine iron that was high in manganese and low in phosphorus. In later years, the acid-based converter was continuously improved until the 1950s when oxygen, instead of hot air, was introduced to remove carbon from the hot metal. This new oxygen-based process was readily adapted for blast-furnace iron of medium to high phosphorus content. It was extensively used in the 1960s and 1970s, especially with the introduction of computers for automatically controlling the steelmaking process.

The rapid progress of the metals industry in the early part of the twentieth century included the development of the open-hearth and electric furnaces and major advancements in rolling and shaping of metals. This required an in-depth understanding of the fundamentals of metallurgy to improve metals properties and expand their usefulness. Open-hearth furnaces were very popular in the first half of the twentieth century and were the main producers of steel in the United States. However, since around 1970, production from open-hearth furnaces has been decreasing, and production from electric furnaces has been steadily increasing. In 1977, approximately 22% of the steel produced in the United States was melted in electric furnaces. The 1970s also witnessed considerable improvements in metal rolling technology. Larger, faster, and more powerful mills were being produced capable of rolling metals of larger sizes to closer dimensional tolerances and improved surface finishes. Factors that contributed to these improvements include changes in the mill housing, the availability of better rolls, more powerful drive motors, and enhancements in the instrumentation and control systems. Today's mills are more productive and are generally completely computer-controlled. For example, in a modern hot-strip rolling mill of the 1990s, the computer decides when a slab may be charged or discharged from the reheat furnaces. The computer also decides when the cooling system on the run-out table and coilers should be turned on or off to provide the desired finishing and coiling temperatures.

In the next several sections, we review some basic concepts associated with the metals industry, in general, with particular emphasis on the steel industry. The topics discussed include an overview of the steelmaking process, iron ores, furnaces, continuous casting machines, rolling mills, automatic gauge control, run-out table temperature control, ac/dc drives, and product quality.

OVERVIEW OF THE STEELMAKING PROCESS

The manufacture of steel products, or of metal products in general, is a complicated procedure that involves a series of sophisticated interacting operations (1,2). Figure 1 is a diagram of an oversimplified steelmaking process. This figure does not include the part that corresponds to the product processing lines, which is a continuation to the displayed process. The process line operation is illustrated in a different figure shown at the end of this section. Nearly all products made of steel today fall into the sequence of operations shown in Fig. 1.

Iron oxides, such as iron ore, are contained in iron-bearing materials. The blast furnace reduces the iron ore to molten iron, known as pig iron, by using charcoal or the carbon of coke as the reducing agent. During the process, approximately 3 to 4.5% of carbon is absorbed by the iron, which is subsequently processed to make cast iron. Modern day demands require steel to be produced at even further reduced carbon content on the order of less than 1%. To meet these requirements, today's steelmaking furnaces remove the excess carbon by employing controlled oxidation of mixtures of molten pig iron, molten iron, and scrap steel. Furthermore, while the carbon removal process takes place, the addition of certain chemical elements, such as, chromium, nickel, and manganese to the molten pig iron produces the so-called alloy steels. After the molten steel has attained the desired chemical composition, it is in liquid form and is poured into a ladle from where it is teemed down into a large mold and eventually forms into a solid structure, termed an ingot. Ingots are removed from the mold and taken to reheating furnaces, where they are uniformly reheated, before they are rolled into semifinished steel structures, such as slabs, blooms, and billets. Then semifinished steel is further processed in rolling mills to produce finished steel products.

Nowadays, most liquid steel is taken to continuous casting machines, where it is poured into the top of open-bottomed molds and then withdrawn continuously from the bottom of the mold in solid forms of various shapes and dimensions (1,2,4,5). After heat treating to attain certain mechanical properties, the slabs, blooms, and billets are subsequently mechanically processed in hot rolling mills, resulting in finished steel products (i.e., bars, plates, sheets, wires, rods, structural beams, and tubes). Moreover, these products can be used in their present form or further processed in cold mills, temper mills, forging, extruding, and so on, for additional surface and hardness improvement. If coating, pickling, tinning, annealing, or cleaning is necessary, the finished steel products are taken to the corresponding process line (1,2,19). Figure 2 is a schematic arrangement of a typical horizontal electrolytic cleaning line.

IRON ORES

It is well known that a significant percentage of the earth's crust (approximately 4%) is composed of iron. However, it needs to be processed to become suitable for use. The part of the iron-bearing material that could be processed and sold is called "ore." Iron ore is classified according to the processing requirements that transform it into marketable form. The ore's quality is characterized by the following chemical and physical properties:

Chemical: high iron content, low acid gangue content (i.e., combined silica and alumina), low phosphorus, low in deleterious elements, mainly sulfur, titanium, arsenic and the base metals, and low or free combined moisture

Physical: uniform size with particle diameter varying from 0.5 to 3 cm which contributes to higher blast furnace productivity

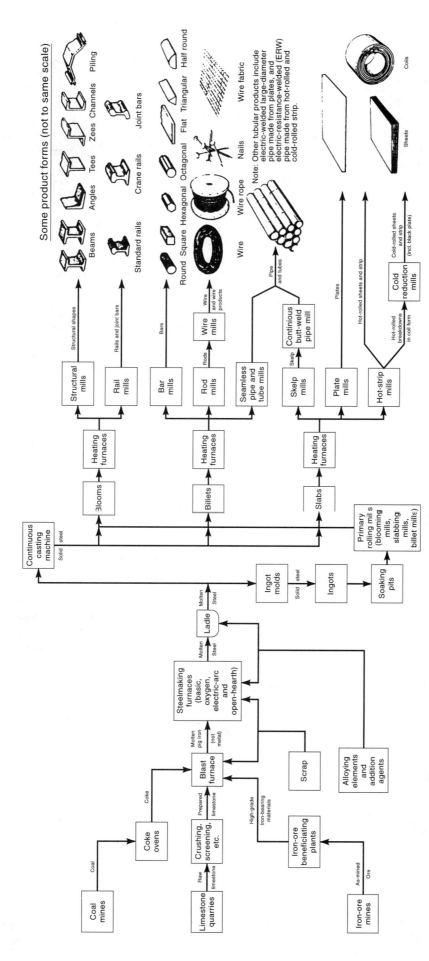

Figure 1. A flow diagram of an oversimplified steelmaking process (1).

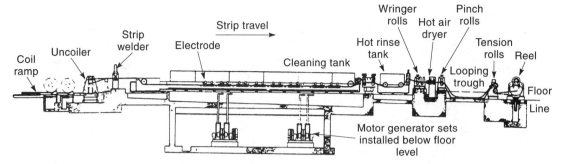

Figure 2. Schematic arrangement of a cleaning process line (1).

Before iron ore is fed to a blast furnace, it undergoes treatment to produce a more desirable blast furnace feed. This treatment, called beneficiation, may include crushing, grinding, screening, classifying, sintering, pelletizing, nodulizing, and concentrating.

FURNACES

Blast Furnace

A very simple description of a blast furnace operation is as follows: Ore is charged at the top of the furnace, and molten iron is tapped close to the bottom by incompletely burning fuel in combination with iron ore. In modern days coke is used as a fuel instead of charcoal. Heated pressurized air is supplied, promoting partial combustion of the fuel and evolution of carbon monoxide, generating heat in excess of 1900°F. Figure 3 shows a plant and an idealized cross section of a typical American blast furnace, respectively. The blast furnace is a tall shaft-type structure with a vertical stack superimposed over a crucible-like hearth. Iron-bearing materials are charged into the top of the shaft. At the bottom crucible of the shaft, directly above the hearth, there are openings, which introduce blasts of heated air and fuel. The injected fuel and the majority of the charged coke are burned by the heated air to generate the required temperature and reducing gas that remove oxygen from the ore. The reduced iron melts and runs down to the bottom end of the furnace's hearth. Flux and impurities in the ore combine to create a slag which in turn melts and accumulates on top of the liquid iron in the hearth. Both slag and iron are drained out of the blast furnace through tapping holes (1,2).

Open-Hearth Furnace

The open-hearth furnace employs a regenerative technique to generate the high temperatures necessary to melt the charged raw material, that is, the hot combustion products leave the furnace chamber through passages guiding them to checker chambers containing firebrick. The large brick area arrangement contributes to heat generation when it contacts hot gases. Part of the generated heat is transferred back to the brick. In the open-hearth furnace the charge is also melted on a shallow refractory hearth of molten metal by a flame passing over the charge so that both the charge and the roof are heated by the flame. Despite the unique features offered by the open-hearth furnace, it also has some major drawbacks, such as low productivity and high installation and maintenance costs. As a result, the electric-arc furnace and the basic

oxygen process have largely replaced it. Figure 4 shows a typical vertical section across an open-hearth furnace (1,2,9).

Electric-Arc Furnace

Raw materials and scrap steel are charged into the electric-arc furnace and then melted via an electric-arc generated onto these materials in the furnace, thus, generating heat and high temperature which are important elements in the steelmaking process. The electric furnaces are designed to remove impurities as gases or liquid slags and to tap the molten steel into a ladle for further processing, as described previously in Overview of the Steelmaking Process. Although, there is a plethora of development in the electric furnace area, electric furnaces can be broadly categorized into two major types, arc furnace and induction furnace. However, the most practical and readily applicable types are (1,9)

Alternating current (ac) direct-arc electric furnaces

Direct current (dc) direct-arc electric furnaces

Induction electric furnaces

Electric heating is basically achieved in two ways, first, by current circulation through a medium and second by bombardment of a surface with a high-intensity electron beam. The latter method is not widely developed and has been applied only to low production capacities. Circulating current through an iodized gaseous medium or through a solid conductor generates enormous heat, which can be utilized to heat the steel. For completeness, we briefly mention some additional commonly known types of electric furnaces. These are typically named for the method of arc heating used, either indirect-arc heating or direct-arc heating. The direct-arc heating method is applied to both nonconducting and conducting bottom furnaces. Similarly, based on the method of resistance heating, furnaces are of the indirect, direct, and induction types. Figure 5 illustrates the cross section of an electric-arc furnace. In direct-arc furnaces, passing electric arcs from the electrodes to a metal charge circulates electric current through the metal charge, resulting in heat generation due to the inherent electrical resistance of the metal. The generated heat, along with the heat radiated from the arcs, constitutes the required furnace heat.

A fundamental difference between the ac and dc direct-arc furnace is that the former is designed with nonconducting bottoms whereas the latter has conducting bottoms. Nonconducting bottoms imply that the current passes from one electrode down through an arc to the metal charge and through an arc to another electrode. Conducting bottoms, on the other hand, means that current passes from an electrode through an arc to the metal charge to an electrode located in the bottom of the electric-arc furnace.

In induction electric-arc furnaces, electric current is induced in the metal charge via an oscillating magnetic field. The primary winding of a transformer is formed by inductors attached to the vessel, and the secondary winding is formed by a loop of liquid metal confined in a closed refractory channel.

CONTINUOUS CASTING MACHINES

Continuous casting caused the world steel production to skyrocket because of tremendous improvements in the efficiency

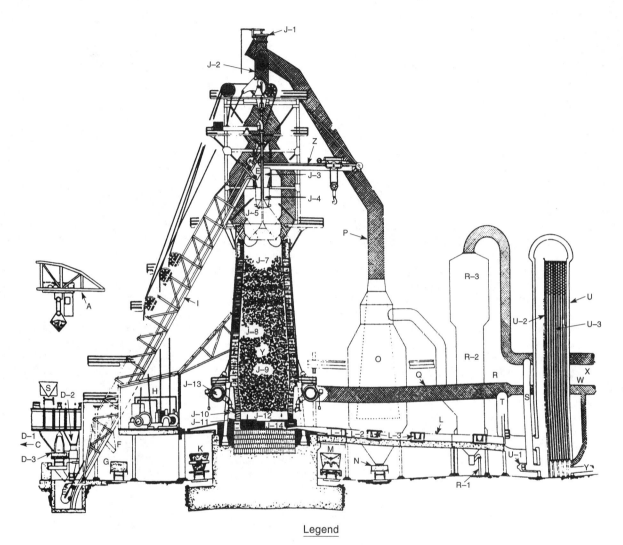

Legend

A. Ore bridge
B. Ore transfer car
C. Ore storage yard
D. Stockhouse
  D-1 Ore and limestone bins
  D-2 Coke bin
  D-3 Scale car
E. Skip
F. Coke dust recovery chute
G. Freight car
H. Skip and bell hoist
I. Skip bridge
J. Blast furnace
  J-1 Bleeder valve
  J-2 Gas uptake
  J-3 Receiving hopper
  J-4 Distributor

J-5 Small bell
J-6 Large bell
J-7 Stock line
J-8 Stack
J-9 Bosh
J-10 Tuyeres
J-11 Slag notch
J-12 Hearth
J-13 Bustle pipe
J-14 Iron notch
K. Slag ladle
L. Cast house
  L-1 Iron
  L-2 Slag skimmer
  L-3 Iron runner
M. Hot-metal ladle
N. Flue dust car
O. Dust catcher

P. Downcomer
Q. Hot blast line to furnace
R. Gas washer
  R-1 Sludge line to thickener
  R-2 Spray washer
  R-3 Electrical precipitator
S. Gas offtake to stove burner
T. Hot blast connection from stove
U. Stove
  U-1 Gas burner
  U-2 Combustion chamber
  U-3 Checker chamber
V. Exhaust gas line to stack
W. Cold blast line from blower
X. Surplus gas line
Y. Stock—Iron ore, coke, limestone
Z. Jib boom crane

Figure 3. A diagram of a plant and an idealized cross section of a typical blast furnace (1).

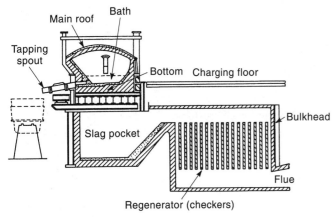

Figure 4. A vertical section of a typical open-hearth furnace (1).

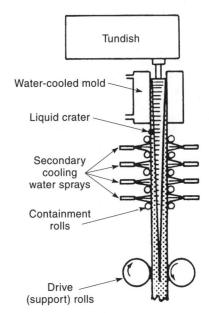

Figure 6. Major components of a continuous casting machine (1).

of material utilization. A dramatic increase of more than 15% in process yield and significant improvements in product quality are attributed to the continuous casting process. For example, the process yield is better than 95% for continuous casting as compared with about 80% in ingot, slabbing, or bloom casting. In addition, other major benefits, such as significant energy savings, less environmental pollution and substantially reduced operating costs, are derived from the continuous casting process (1,2,4,5,9).

Casting machines are classified according to the product shapes they produce. It is important to mention that because of the advantages of continuous casting, the modern minimill concept (i.e., combinations of continuous casters with powerful electric-arc furnaces) has been extensively applied throughout the metals industry in recent years. There are four major categories to which casting machines belong, billet, bloom, round, and slab continuous casters. The major components of a continuous caster are illustrated in Fig. 6. The tundish is a reservoir for delivering liquid steel. The principal function of the water-cooled mold is to contain the liquid steel and initialize the solidification process. The secondary cooling system controls the cooling rate through a series of cooling zones associated with a containment section as the strand progresses through the machine. The function of the drive

rolls is to support, bend, and guide the strand through a prescribed arc and the straightener, that is, from the vertical to the horizontal plane, as shown in Fig. 7. Note that there can be casting machines with multiple casting strands operating in parallel, each one having its own mold, secondary cooling water sprays drive rolls, and straightener.

The continuous casting process can be described as follows. Before starting the casting process, a long mechanical withdrawal system shaped like a slab, and known as the dummy bar, is inserted from the straightener (i.e., the horizontal plane) in the bottom of the mold to facilitate the initial extraction of the strand. Liquid steel is delivered in a ladle and poured at a controlled rate into the tundish. Nozzles guide the liquid metal flow in the bottom of the tundish so that the mold can be filled. When a certain liquid level limit in the mold is reached, the dummy bar, which is attached to the solidified metal, is withdrawn, pulling along the solidified cast. When the dummy bar exits the curved rack section, the

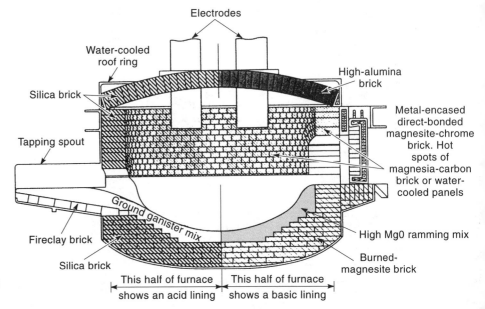

Figure 5. A cross section of a typical electric-arc furnace (1).

solidified metal is mechanically disconnected, removed, and cut to desired lengths.

ROLLING MILLS

Rolling mills are the workhorse of the entire metals industry. They are responsible for producing the largest percentage of finished metals (steel, aluminum, etc.) in many forms and shapes. A considerable number of various types of rolling mills exist today, but it is beyond the scope of this article to cover them all. Instead, we mention the most general classes of rolling mills and focus on those most widely used in the steel industry. The list of references at the end of this article provides a wealth of valuable and detailed information for the interested reader (1–3,6–12).

Many of the components, accessories, and systems of rolling mills are common to all types. They differ in design, performance, and operation to conform to the special conditions and specifications of a particular mill. Because most technical challenges, innovative solutions, and high performance specifications have been in the domain of flat product rolling, we refer to concepts and systems associated with this type of rolling mill with special emphasis on hot and cold strip rolling, such as automatic gauge control (AGC), run-out table (ROT) strip cooling, ac/dc mill drives, and product quality. In particular, we consider hot and cold strip mills because this area has experienced explosive research and development in recent years.

The following are basic components common to most types of rolling mills:

work and backup rolls with their bearings
mill foundation and housing
roll balance system
roll-gap adjustment system
roll change system
mill protection devices

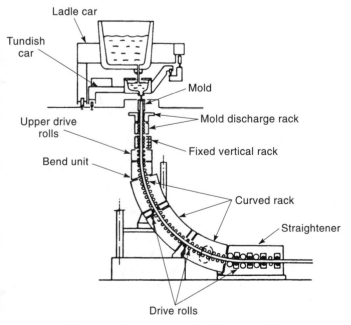

Figure 7. Cross section of a slab caster (1).

roll cooling and lubrication systems
drive spindles and coupling
pinions and gearing
drive motors and motor couplings
electrical power supply and control systems
idle and bridle rolls
uncoiler and coiler systems
coil handling equipment
mill instrumentation, monitoring, and operating control devices

Figure 8 shows a small experimental rolling mill with most of the previously listed components identified. The most important component of a rolling mill is the mill stand and its associated auxiliary equipment. Figure 9 shows a typical mill stand arrangement. Mill stands are mainly categorized with respect to the rolling temperature, direction of roll axes, direction of rolling, main motor type, mechanical drive arrangement, and special design.

Roll arrangement types are based on the way the rolls are arranged. There are two, three, four, five, and six-high, cluster mill stands and mill stands with off-set rolls (see Fig. 10).

Direction of roll axes types include horizontal, vertical crossed-roll and mill stands with parallel tilted rolls.

Direction of rolling types include reversing, nonreversing, and back-pass mill stands.

Main motor type: The work and backup rolls of a mill stand are driven either by ac or by dc motors.

Mechanical drive train arrangement types include direct drive, gear drive, pinion stand drive, and independent drive. In addition, they can also be identified with respect to the type of driven rolls, that is, drive train with driven work rolls, backup rolls, and intermediate rolls.

Special design types: specially designed mill stands have been developed, such as planetary, rolling-drawing, contact-bend-stretch, and reciprocating.

In general, rolling mills can be classified in terms of rolling temperature, type of rolled product, and mill stand arrangement. For example, if the range of material temperature during the rolling process is between approximately 900° and 1300°C, then it is rolled in a hot rolling mill. Conversely, if the range is from 120° to 150°C, then it is rolled in a cold rolling mill. In warm rolling mills (for low-carbon steel), the material temperature is around 700°C. Finally, depending on the way mill stands are arranged a few more rolling mill types can be mentioned. For example, an open mill stand arrangement implies that the rolled piece is in one rolling stand at a time, whereas a close-coupled arrangement implies that the material is rolled simultaneously by more than one stand. In this latter arrangement, the stands must be appropriately speed-synchronized. Further types of mill stand arrangements include two very well known types, the universal and tandem rolling mills.

Hot Strip Mills

Figure 11 shows a multistand continuous hot strip mill. Slabs are heated in two or more continuous reheating furnaces. A

SCREWS
DRIVE MOTOR
MILL HOUSING
BACKUP ROLLS
PINION STAND
UNCOILER
ROLLING LUBRICANT RESERVOIR
WORK ROLLS
BACKUP ROLL CHOCKS
SPINDLES
WORK ROLL CHOCKS
STRIP
COILER
MANDRIL
CONTROL PANEL

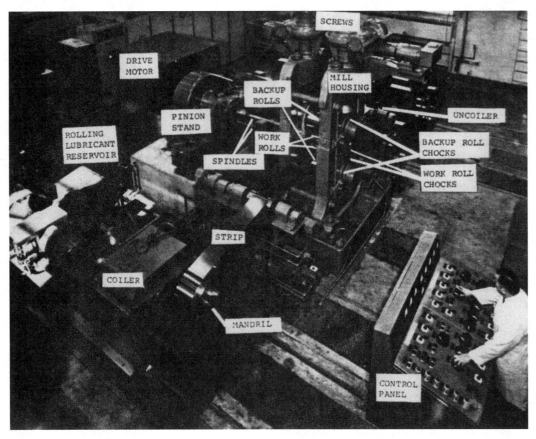

Figure 8. A picture of a small experimental rolling mill (7).

typical rolling train consists of a roughing scale breaker, one or more roughing stands, a finishing scale breaker, five to seven finishing stands, and one to three coilers. Driven table rolls convey the slab from furnace to the roughing mill and through the stands. Separating the roughing and finishing stands is a finishing table. High-pressure hydraulic sprays are located after the two scale breakers and after each roughing stand. As the steel exits the finishing mill, it crosses over a long table, consisting of many driver rolls, called a run-out table. On the run-out table, laminar jets or water sprays apply water to both top and bottom strip surfaces to reduce the strip temperature to a desired level before the strip is coiled. After cooling, the strip is carried to one of the coilers where it is wrapped into coils.

Housing
Roll bending cylinder
Roll balance and bending cylinders
Work roll assembly
Roll bending cylinder
Roll gap adjustment mechanism
Roll gap adjustment cylinder
Backup roll assembly
Work roll assembly
Backup roll balance cylinder
Mae West block
Roll bending cylinder
Backup roll assembly

Figure 9. Arrangement drawing of a typical four-high mill stand (9).

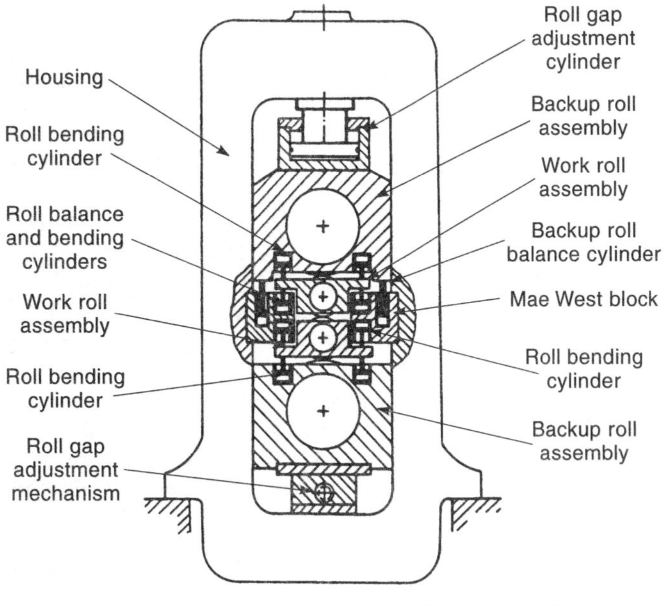

Mathematically, the operation of hot flat rolling can be described as follows (6,8–10): Given the distribution of normal pressure P_x in the deformation zone, we can determine the roll separating force P as (see Figs. 12 and 13):

$$P = \int_0^{l_d} P_x d_x = \int_0^{\alpha} RP_\theta d_\theta$$

where P_x is normal pressure at distance x from the exit plane, R is the work roll diameter, P_θ is the normal pressure at roll angle θ, l_d is the projected contact arc between the work roll and the material, and α is the roll bite angle.

If the entry and exit strip tensions are taken into account, an approximation of the roll separating force F is given by the expression

$$F = F_d(K_w - \beta_1 S_1 - \beta_2 S_2)$$

where F_d is the projected area of contact between the roll and the material, K_w is the material's resistance to deformation, S_1 and S_2 are the entry and exit strip tensions, and β_1 and β_2 are the entry and exit strip tension coefficients, respectively. The projected area of contact between the roll and the rolled

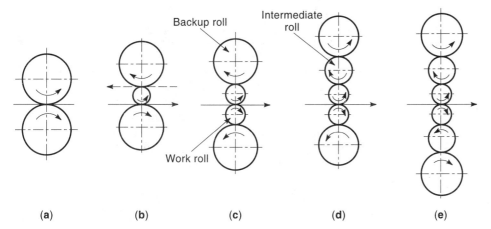

Figure 10. Types of roll arrangements in mill stands (9).

material is given by the expression

$$F_d = W l_d$$

where W is the mean width and l_d is the projected contact arc defined by the expression

$$l_d = \sqrt{R(h_1 - h_2) - \frac{(h_1 - h_2)^2}{4}} \approx \sqrt{R(h_1 - h_2)}$$

where h_1 and h_2 are the entry and exit thicknesses, respectively. The following relationship describes the mass flow through the roll bite:

$$V_1 h_1 = V_2 h_2 = V_\alpha h_\alpha = V h$$

where V is the workpiece velocity, h is the workpiece thickness at any point, and the subscripts 1, 2, and α denote entry, exit, and neutral points, respectively. Approximating the curve of spread with a parabola gives what is known as the parabolic mean width

$$w_p = \frac{w_1 + w_2}{3}$$

where w_1 and w_2 are the entry and exit workpiece widths, respectively. The strain rate differential equation can be written as

$$\frac{d\epsilon}{dt} = \frac{\Delta L}{L_0 \, \Delta t}, \quad L_0 = L(0)$$

where ΔL is the change in length of a deformed body, Δt is the time needed to deform the body, and L_0 is the initial length. There are several proposed solutions to this differential equation, but we mention only Sims' solution without deriving it. For further details, the interested reader should con-

Figure 11. A multistand continuous hot strip mill (1).

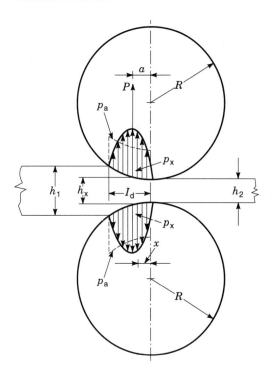

Figure 12. Distribution of normal pressure and rolling force (9).

sult Ref. 9. Sims' solution for mean strain rate is

$$\lambda = \frac{\pi N}{30} \sqrt{\left(\frac{R}{h_1}\right)} \frac{1}{\sqrt{r}} \ln\left(\frac{1}{1-r}\right)$$

where N is the rotational speed of the roll in rpm and r is the reduction which can be expressed as

$$r = \frac{h_1 - h_2}{h_1}$$

Cold Strip Mills

After hot rolling, the hot strip is wound into hot coils and the coils are further processed into what is called cold rolling or

cold reduction. The hot-rolled coils are uncoiled, passed through a continuous pickle line, dried out, oiled, and finally recoiled. Oiling facilitates the cold reduction process and provides protection against rust. Cold rolling is done for one or a combination of the following three requirements: (1) to reduce thickness, (2) to harden and smooth the surface, and (3) to develop certain mechanical properties. After cold reduction, the cold strip goes through a cleaning and annealing process, and then it may or may not go through another cold rolling stage, known as temper rolling. The principal objective of temper rolling is to impart certain mechanical properties and surface characteristics to the final product. It is not intended to drastically reduce its thickness. Thickness reduction during temper rolling is most often less than 1% but may not exceed 10%.

It is interesting to note that the original purpose of cold rolling was to let the rolled material attain certain desired surface and mechanical properties. Reduction of thickness was of incidental importance. However, because thickness reduction in the hot strip mill is limited to no less than 1.2 mm, it is through cold rolling that ultrathin thicknesses of flat-rolled products can be achieved. A typical batch cold mill consists of an entry reel (payoff reel), multiple stand mill train, and an exit reel (tension reel). As the name implies, the coil is "paid off" (i.e., uncoiled) by the payoff reel, fed into the mill train, and recoiled on a mandrel at the exit side. A continuous cold mill, however, is directly coupled to a continuous pickle line, where the coils are continuously fed, so the payoff reel is not necessary.

Mathematically, the operation of cold flat rolling can be described as follows (7–9,11): Unlike hot rolling, where temperature is the germane variable affecting resistance to deformation, in cold rolling temperature is a function of work-hardening and roll contact friction. Furthermore, strip tension is a significant factor because it is of higher magnitude compared with hot rolling tension. As a result, the roll force and torque calculations differ from the corresponding calculations in hot rolling. Figure 14 illustrates elastic flattening of a cylinder on a plate. The length L of the region of contact is

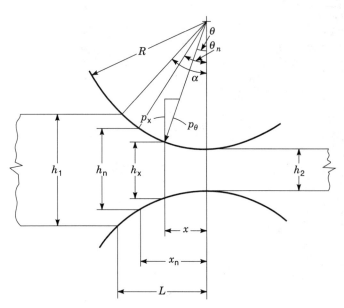

Figure 13. Deformation zone parameters (9).

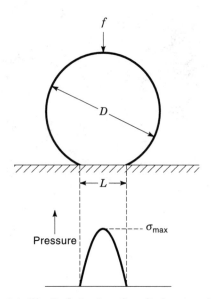

Figure 14. Elastic flattening of a cylinder on a plate (7).

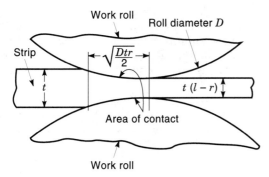

Figure 15. Length of contact arc for a rigid roll (7).

expressed as

$$L = 1.6 \left[fD \left(\frac{1-v_1^2}{E_1} + \frac{1-v_2^2}{E_2} \right) \right]^{1/2}$$

where f is the force per unit length of the cylinder, D is the cylinder diameter, v_1 and v_2 are Poisson's ratio for the cylinder and plate, respectively, and E_1 and E_2 are the corresponding Young's modulus constants. The maximum stress at the center of the contact region is given by

$$\sigma_{max} = 0.798 \left[\frac{f}{D \left(\frac{1-v_1^2}{E_1} + \frac{1-v_2^2}{E_2} \right)} \right]^{1/2}$$

Figure 15 shows the length L of the contact arc for rigid rolls, which is given by the expression

$$L = \left(\frac{Dtr}{2} \right)$$

If the work rolls are not rigid, D should be replaced by the deformed roll diameter D_d, that is,

$$L = \left(\frac{D_d tr}{2} \right)$$

In the two previous expressions, t and r are entry thickness and reduction, respectively. The two diameters of iron work rolls are related by the following expression:

$$D_d = D \left(1 + 68 \times 10^{-4} \frac{f}{\Delta h} \right)$$

where Δh is the draft, that is the reduction in strip thickness resulting from passage through the roll bite. There are many complicated mathematical models for calculating roll force and torque. However, here we consider only one without referring to details. SKF Industries (18) proposed the following formula for calculating roll force with strip tension:

$$F_T = F \left(1 - \frac{2T_{s1} + T_{s2}}{3S_m} \right)$$

where F is the roll separating force without strip tension, T_{s1} and T_{s2} are the entry and exit specific strip tensions, respectively, and S_m is the average yield stress of the compressed

material. Assuming no entry and exit tensions and unit width, the specific total spindle torque (i.e., top and bottom) may be approximated by the following equation:

$$G_T = 0.5Dtr\sigma_c$$

where σ_c is the dynamic constrained yield strength of the strip in the roll bite. With this determined, the total torque that must be supplied to the mill stand is given by

$$G_{TW} = 0.5WDtr\sigma_c$$

Now, if entry and exit stresses due to tension are introduced, the forces acting on the strip result in the following expression for total torque (see Fig. 16)

$$G_{TW} = WDtr\sigma_c \left[r \left(1 + \frac{\sigma_2}{\sigma_c} \right) + \frac{\sigma_1 - \sigma_2}{\sigma_c} \right]$$

where σ_1 and σ_2 are entry and exit stresses due to tension, respectively. A very important parameter of flat processing of strip as it leaves the roll bite is what is known as forward strip differential v, defined by

$$v = \frac{(V_s - V_R)}{V_R}$$

where V_s and V_R are the strip and roll speeds, respectively. Assuming equal entry and exit stresses due to tension, then the relationship denoting the forward slip as a function of actual coefficient of friction μ and the minimum coefficient of friction μ_m can be written as

$$v = \frac{r}{r(1-r)} \left(1 - \frac{\mu_m}{\mu} \right)^2$$

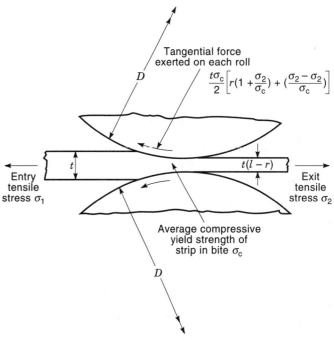

Figure 16. Tangential force on a roll surface (7).

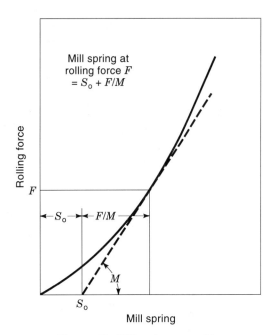

Figure 17. Mill-spring curve (8).

AUTOMATIC GAUGE CONTROL (AGC)

Measuring the thickness of a rolled strip accurately involves special, highly expensive, computer-controlled equipment, which applies various sophisticated measurement principles. Modern rolling mills for flat products are expected to deliver extremely high gauge performance under stringent tolerances. Factors that disturb the gauge of a rolled strip include strip tension, strip temperature, rolling speed, mill vibration, roll eccentricity, oil film in the roll bearings, thermal expansion and wear, and most importantly mill stand stretching. A well-known and widely used equation to analyze and synthesize the control systems to minimize gauge variations is referred to as the gaugemeter equation. It utilizes knowledge of the mill spring, obtained during mill calibration, rolling force, and the nominal position of the work rolls with no strip present. Assuming a no-load gap, then the roll gap h is expressed as (8,9)

$$h = s_0 + s + \frac{F}{M}$$

where s_0 is the intercept of the extrapolated linear position of the mill-spring curve, F is the total rolling force, and M is the mill modulus (see Fig. 17). The control system that regulates the material thickness to the desired value is called gaugemeter automatic gauge control (AGC). It is also sometimes refered to as BISRA compensation named after its developers. Figure 18 depicts an oversimplified block diagram of such system, and its equation is

$$s = h - \frac{F}{M} - \{other\ higher\ order\ compensations\}$$

In this "gaugemeter mode," the mill is considered to be a spring that stretches according to the rolling load. Then the stretch of the mill is added to the unloaded roll opening to provide a measure of material thickness in the roll bite. The relationship between load and stretch is defined by the mill stretch curve, which is automatically measured during roll gap calibration. Because this curve is measured with full-face roll contact, bending of the rolls is not included in the basic curve. Therefore this curve is further adjusted for the roll diameters and strip width to ensure the most accurate estimate of centerline thickness.

During rolling in gaugemeter mode, the required on-gauge cylinder position reference is continuously calculated by using the gaugemeter equation and is issued to the position control loops. This reduces gauge variations due to mill stretch variation with changes in force. In principle, this makes the mill appear stiff in maintaining a constant loaded roll gap. The delivered gauge closely follows the loaded roll gap and therefore is also held nearly constant.

RUN-OUT TABLE TEMPERATURE CONTROL (ROT)

Temperature control of hot rolled strip has always been of great interest in the steel/metals industry. The cooling process is directly correlated to the grain structure formation of the product to be cooled. The grain structure defines the mechanical properties of the strip which in turn dictate the temperature requirements (1,2,6,12). It has been shown (13) possible to obtain algorithms that achieve uniform temperature throughout the material at a specified target temperature within tight tolerances. Such algorithms based on two-point boundary problem and dynamic programming theory minimize the temperature error with respect to a target temperature.

Two linearized models describe the system. Temperature loss due to radiation is given by

$$\Delta T_r = \frac{CA_r}{V_r}[(T + 460)^4 - (T_a + 460)^4]t_r$$

where

$$C = \frac{s\xi}{\rho(T)c(T)}$$

s is the Stefan-Boltzmann constant, ξ is the emissivity, A_r is the surface area of a body subjected to radiation, T is the material temperature in °F at time t, T_a is the ambient temperature in °F, ρ is the specific gravity of the rolled material, V_r is the body volume, c is the specific heat of the rolled material, and t_r is the time interval of radiation.

The linearized model describing temperature change due to cooling water is

$$\Delta T_d = \frac{2k}{\rho(T)c(T)h}(T - T_w)\left[\frac{\Delta l}{\pi a v}\right]^{1/2} t_w$$

where k is the thermal conductivity of the surface layer, h is the material thickness, Δl is the water contact length, a is the thermal diffusivity, v is the material velocity, T is the material temperature, T_w is the water temperature, and t_w is the water contact time.

Using these models, both dynamic programming and feed-forward algorithms can be used to calculate the coiling temperature on the ROT, before the strip is delivered to the coiler. The difference is that the former performs better than the latter.

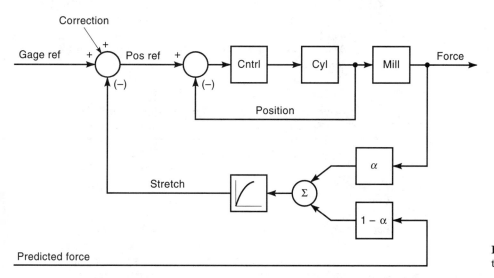

Figure 18. Block diagram of a gaugemeter AGC system.

The resulting boundary temperature T_2 can be written as

$$T_2 = T_1 - \sum_{i=1}^{N} \Delta T_{r_i} - \sum_{i=0}^{N-1} u_{i+1} \Delta T_{d_{i+1}}$$

where

$$u_{i+1} = \begin{cases} 1 & \text{if} \quad \Delta T_{d_{i+1}} \leq \Delta T_{w_i} \\ 0 & \text{Otherwise} \end{cases}$$

and ΔT_{w_i} is the temperature that needs to be lost by turning the water sprays ON. The dynamic programming algorithm solves a set of recursive difference equations:

$$T_{k+1} = a_k T_k - b_k \Delta T_{d_k}, \quad k = 0, 1, \ldots, N - 1$$

with the following corresponding cost function:

$$J = \alpha T_N^2 - \sum_{k=0}^{N-1} \beta \Delta T_{d_k}^2$$

where a_k, b_k, α, and β are appropriate coefficients.

AC/DC DRIVES

DC Drives

Mill stands and other major associated moving components of a rolling mill require variable speed drives to control the dc motor speed. The dc drives are connected to an ac power line and apply controlled voltage to a dc motor at the load. As a result, some means of ac to dc conversion is necessary. The theory of converters/inverters is a broad and exceedingly complex subject (14–17).

In general, the basic characteristic of an all ac to dc converter is that its voltage source comes from an ac line, and that causes what is known as natural commutation. If the direction of power flow is from a dc source to an ac load, then the power converter is an inverter. In contrast to naturally commutated converters, the inverters used in dc drives are force-commutated unless the load has a leading power factor. Conceptually, the inverter can be thought of as a group of

switches that connect the load to the dc bus and then alternate the polarity of the connections in a regular cycle. In large inverters, various types of power electronic semiconductor devices are used as switching components. Most such devices suffer from a serious limitation called forced-commutation, which occurs during the on-off switching process. What is known as a four-quadrant converter, or dual converter, can operate with both positive and negative polarities of both voltage and current at the dc bus. As a result, the dc motor can be driven and braked regeneratively in both forward and reverse directions. Figure 19 shows a simple six-pulse bridge converter with its waveforms. The speed of a dc motor can be adjusted by adjusting the armature voltage or by controlling the motor field. The drive is inherently of a constant horse-

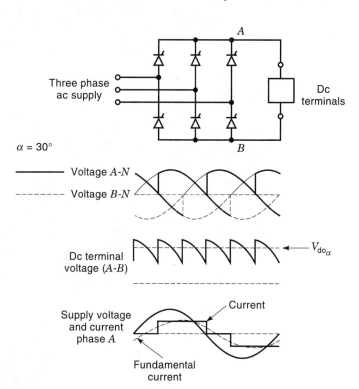

Figure 19. A simple six-pulse bridge converter with its waveforms (16).

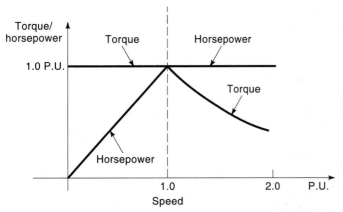

Figure 20. Torque/horsepower vs. speed for an adjustable-speed dc motor.

power type if the speed range of a dc motor is covered by its field control, that is, the torque varies inversely with the speed. Alternatively, if the dc motor speed range is covered by armature voltage control, the drive is inherently of a constant horsepower type, that is, the horsepower varies proportionally with speed. The steady-state voltage characteristic V_t of a dc motor can be expressed as follows:

$$V_t = R_a I_a + E + V_B$$

where R_a and I_a are the armature resistance and current, respectively, E is the counter emf voltage and V_B is the voltage drop at the commutator brushes. The steady-state Torque T can be expressed as

$$T = K\Phi I_a$$

where K is the motor torque constant and Φ is the air gap flux (i.e., per unit of full field value). The previous equation it shows that the torque delivered by the motor is directly proportional to the armature and field current.

The speed of the motor N can be expressed as

$$N = \frac{E}{K_E \Phi}$$

where K_E is the counter emf constant, lumping together certain design parameters, such as number of turns and connection type. The previous equation states that the speed of a dc motor is directly proportional to the counter emf and inversely proportional to the motor field. Considering the horsepower equation

$$HP = \frac{(rpm)T}{5250}$$

and the steady-state characteristics, we can obtain the torque/horsepower versus speed characteristic of an adjustable speed dc motor, as depicted in Fig. 20.

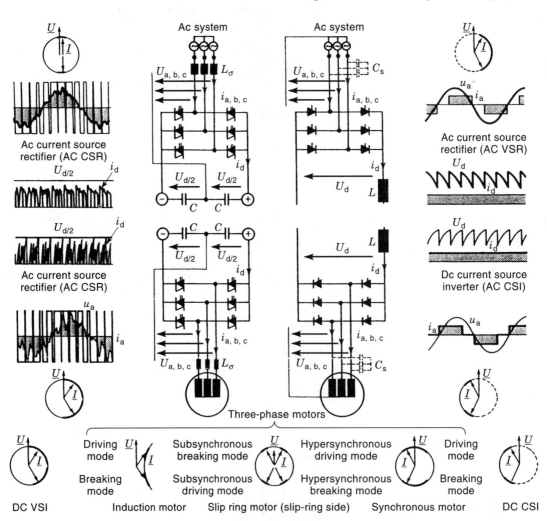

Figure 21. Ac voltage source rectifier and dc voltage source inverter circuits (14).

Ac Drives

For technical and economic reasons, the industry trend in recent years has been to replace dc drives with their ac counterparts. First, ac motors are more robust, reliable, and require less maintenance than dc motors. Secondly, ac drives save energy and provide increased motor output. Finally, ac motors are smaller in size for a given rating compared to dc motors, hence, are more cost effective. However, it should be mentioned that the high upfront cost of ac drives is offset by much lower lifecycle costs due to low maintenance and lower energy consumption. For these reasons, ac drives are the preferred choice.

The main types of variable speed ac drives used in rolling mills today, although rapidly changing, are cyclo converter current and voltage source drives. There are five main types of power semiconductor devices used to implement the bridges in the drives. These are (1) normal thyristors using natural commutation, (2) fast thyristors with forced commutation, (3) gate turn-off thyristors (GTO), (4) gain transistors (BJT) and (4) insulated-gate bipolar transistors (IGBTs) (14,15)

Converters are used to connect the ac power system to the motors and to convert the voltages and currents of the ac system to meet the motor requirements. In addition, they control the power flow from the ac system to the motors in the driving phase, and conversely they control the flow from the motor to the ac power system in the regenerative or braking phase. Figure 21 shows the ac voltage source rectifier and dc voltage source inverter circuits of converters used in large drives. The ac voltage source rectifier rectifies an impressed ac voltage and generates a switched dc voltage. However, a dc current source inverter, as the name implies, also inverts an impressed dc current and generates a switched ac current. The dc voltage source inverter, inverts an impressed dc voltage and generates a switched ac voltage, and at the same time it is also an ac current source rectifier, that rectifies an impressed ac current and generates a switched dc current, which, however, is not so common.

PRODUCT QUALITY

Product quality is of prime importance in every process of any metals industry. For the final product to be of high quality, each particular process in the industry must meet its specific quality requirements. Otherwise, the succeeding process might not be able to achieve maximum performance and/or product quality, resulting in economic loss. For example, a consistently well-proportioned coal blend is essential to produce the highest quality, ultimately uniform coke from the supplied coals, which in turn contributes to maximized blast furnace performance.

A major objective in the production of high quality steel in the secondary steel process is appropriate degassing of the gases (mainly, oxygen and hydrogen) that the liquid steel absorbs from the steelmaking material and the atmosphere. Continuous casting should be able to deliver good surface quality of the cast with minimum shape variations. Of equal importance are metallurgical qualities, such as minimized variability of chemical composition and solidification characteristics.

Performance objectives, and in particular how to achieve them in rolling mills, is a very interesting and challenging problem of enormous importance. In addition to the references mentioned (4,11,12), nearly all yearly proceedings of the Association of Iron and Steel Engineers (AISE) provide a wealth of information. The primary performance definitions in rolling mills, particularly flat rolling, are gauge, shape, cross-sectional profile, and width tolerances. One of the main reasons why a rolled product may not conform to the required tolerances of these three parameters is temperature variation in both transverse and longitudinal directions. The causes of temperature variations are many and include reheat furnace problems, poor surface quality of the slabs delivered from the caster, improper operating practices, and excessive edge radiation. It is an industry standard to have different sets of performance requirements for steady-state and nonsteady-state (i.e., accelerating, decelerating) conditions. Naturally, tolerances during transient conditions are usually relaxed. In addition, quite often, the tolerances differ for the head-end, the body, and the tail-end of the rolled piece. There are innovative solutions for optimizing various conditions that enhance performance, for example, optimizing mill configuration, operating parameters, and practices.

In the 1990s many sophisticated computer-based tools and technologies have been developed to monitor, identify, analyze, and eventually correct one or more special problems in any metals production process. Such software tools include automatic recognition of "out-of-control" features in critical process variables, rule-based diagnosis of special causes, a model-based search for symptoms where a diagnosis is not possible, and automated reporting of special problems (20).

Statistical process control methods (SPC) have been used to limit process variability and thus to produce higher quality products (21). SPC methods are intended to identify a variation in a process signal that differs significantly from the usual variability of the process. The statistical process control model assigns such a variation to special causes (i.e., a collection of charts), events that are not part of the normal operation of the process. Such events might include material changes, equipment failures, operator error, or environmental changes. Figure 22 illustrates a special cause management

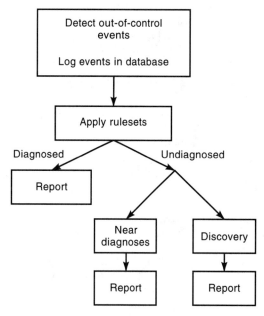

Figure 22. Special cause management block diagram.

block diagram of a statistical process control system. Process instabilities and abnormalities can also be diagnosed and their causes identified with knowledge-based procedures that automatically recognize primitive variations or changes by observing process signals (22).

BIBLIOGRAPHY

1. U.S. Steel, *The Making, Shaping and Treating of Steel,* Pittsburgh: Assoc. Iron and Steel Engineers, 10th ed., 1985.

2. *Watkins cyclopedia of the Steel Industry, 13th ed.,* Pittsburgh: Steel Publications, 1971.

3. J. Lysaght, *An Introduction to Hot Strip Mill,* Port Kembla, N.S.W. Australia: Limited, 1976.

4. *METEC Congr. 94, Proc. 2nd Eur. Continuous Casting Conf., 6th Int. Rolling Conf.,* Dusseldorf, 1994, Vol. 2.

5. Continuous casting of steel, *Proc. Int. Conf. Organized by the Metals Soc.,* Biarritz, France, 1976.

6. W. L. Roberts, *Hot Rolling of Steel,* New York: Dekker, 1983.

7. W. L. Roberts, *Cold Rolling of Steel,* New York: Dekker, 1978.

8. W. L. Roberts, *Flat Rolling of Steel,* New York: Dekker, 1988.

9. V. B. Ginzburg, *Steel-Rolling Technology,* New York: Dekker, 1989.

10. Z. Wusatowski, *Fundamentals of Rolling,* Oxford, UK: Pergamon, 1969.

11. Cold rolling fundamentals, *Specialty Conf., Assoc. of Iron and Steel Engineers,* Pittsburgh, PA: 1998.

12. *4th Int. Steel Rolling Conf.,* Deauville, France, 1987, Vols. 1 and 2.

13. N. S. Samaras and M. A. Simaan, Two-point boundary temperature control of hot strip via water cooling, *ISA Trans.,* **36** (1): 11–20, 1997.

14. B. K. Bose, *Power Electronics and Variable Frequency Drives Technology and Applications,* Piscataway, NJ: IEEE Press, 1997.

15. R. W. Lye, *Power Converter Handbook Theory, Design and Application,* Peterborough, Ontario, Canada, Canadian General Electric Co., 1976.

16. D. G. Fink and W. H. Beaty, *Standard Handbook for Electrical Engineers,* New York: McGraw-Hill, 12th ed., 1987.

17. R. Stein and W. T. Hunt, Jr., *Electric Power System Components,* New York: Van Nostrand-Reinhold, 1979.

18. *SKF Calculation of Rolling Mill Loads,* King of Prussia, PA: SKF Industries, Inc., 1982.

19. D. A. McArthur, *Strip Finishing,* Warren, OH: Wean Engineering Company, 1962.

20. K. R. Anderson et al., Special cause management: A knowledge-based approach to statistical process control, *Ann. Math. Artif. Intell.,* **2**: 21–38, 1990.

21. H. M. Wadsworth, K. S. Stephens, and A. B. Godfrey, *Modern Methods for Quality Control and Improvement,* New York: Wiley, 1986.

22. P. L. Love and M. A. Simaan, Automatic recognition of primitive changes in manufacturing process signals, *Pattern Recognition,* **21** (4): 333–342, 1988.

NICHOLAS S. SAMARAS
Danielli Automation Inc.

MARWAN A. SIMAAN
University of Pittsburgh

METASTABILITY. See CIRCUIT STABILITY OF DC OPERATING POINTS.

METEOR BURST COMMUNICATION

Billions of small meteors enter the earth's atmosphere daily. Upon entering the atmosphere, these meteors quickly vaporize, leaving behind a trail of ionized particles. Ionized particles reflect, or actually reradiate, radio signals; so if properly oriented, a meteor trail can be used to establish a long-range communication link between a pair of radios. Meteor trails diffuse quickly, however, resulting in a rapid decay in signal strength. In many systems, usable trail lifetimes are on the order of 1 s. Communication that takes place via meteor trails is known as *meteor-burst communications.*

The utility of meteor-burst communications is evidenced by the snowpack telemetry system (SNOTEL) (1), which has been in operation for over 15 years, as well as by the commercial success of companies such as Meteor Communications Corporation and StarCom, which design meteor-burst communication systems. SNOTEL is a system that collects snowfall and weather data for 12 states in the western United States. In addition to remote telemetry, applications of meteor-burst communications include vehicle tracking and two-way messaging (1). The U.S. military was one of the first proponents of meteor-burst systems because it was determined that such systems would be among the first beyond-line-of-sight media to resume operability after a nuclear war.

The two primary alternatives for wireless communication at ranges beyond 100 km are satellite communications and terrestrial communication in the high-frequency (HF) band of 3 MHz to 30 MHz. For certain commercial applications, however, satellite communication is prohibitively expensive and HF communication is too unreliable. With regard to military applications, satellite and HF systems have additional disadvantages. For example, satellite and HF communication signals can be received over a large area and are therefore easy to intercept. In addition, satellite and HF communication links can be relatively easy to disrupt. These problems can be overcome by meteor-burst systems, which provide relatively low-cost, reliable, survivable, long-range communication.

PROPERTIES OF METEOR-BURST COMMUNICATION SYSTEMS

A typical meteor-burst protocol and network topology is as follows. The meteor-burst terminals are arranged in a network with one master station and many remote stations. The master station transmits a probe signal continuously using 1 kW to 5 kW of power. Once a meteor of appropriate size and trajectory enters the atmosphere, the probe signal is reflected from the meteor trail down to a remote station, which remains idle until the probe signal is detected. Once the probe signal is detected, the remote station transmits a burst of digital data to the master station via the meteor trail. Once the trail has diffused, the master station begins transmitting the probe signal again, and the process repeats. This basic protocol is illustrated in Fig. 1.

Meteor trails are generally categorized as either underdense or overdense based on their electron line density (1). For underdense trails, the received signal power decreases

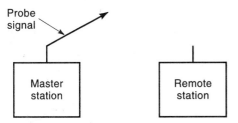

Step 1: Probe signal is transmitted until
response from remote station is received.

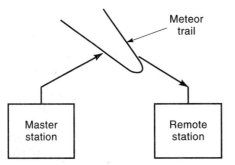

Step 2: Probe signal is detected at remote station.

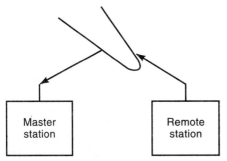

Step 3: Confirming message is sent from remote
station to master station.

Figure 1. Basic meteor-burst communication protocol.

with time and is generally modeled as an exponential decay. Underdense trails occur much more frequently than overdense trails, and the underdense trail model is used almost exclusively in the analysis of system performance.

Underdense trails are formed by meteors whose mass is between 10^{-5} g and 10^{-3} g. The number of meteors of a given size that enter the atmosphere per day is inversely proportional to mass. For example, the number of earth-bound meteors per day that have a mass of 10^{-5} g is 10^{10}, and the number of earth-bound meteors per day that have a mass of 10^{-3} g is 10^8 (2), where all numbers are approximate.

The average data rate of a meteor-burst system is a function of the frequency of occurrence of useful meteor trails, which has a daily and a seasonal variation. The maximum to minimum daily variation is approximately 4 : 1, and the maximum to minimum seasonal variation is between 2 : 1 and 4 : 1 (1). A meteor-burst system is usually designed based on pessimistic assumptions about the frequency of useful meteor trails.

The selection of the operating frequency for a meteor-burst system is a compromise between opposing criteria. As the operating frequency increases, the strength of the received sig-

nal decreases and the trail decay rate increases, so the operating frequency should be as small as possible according to this argument. Important practical considerations favor higher frequencies, however. For example, the desired signal may be reflected by the ionosphere at frequencies below approximately 30 MHz, and these lower-frequency signals are subject to higher atmospheric noise and are unreliable. Also, antenna size and cost increase as the operating frequency is decreased. As a result of the opposing criteria, the operating frequency of most meteor-burst systems lies between 40 MHz and 60 MHz. Several key characteristics of meteor-burst systems and channels are presented in Table 1.

THE METEOR-BURST CHANNEL

As a meteor enters and descends into the earth's atmosphere, it collides with increasing numbers of atmospheric particles. The collisions convert kinetic energy into a combination of heat, light, and ionization. As collisions continue, the outer layers evaporate and the meteor loses velocity and breaks up. Although portions of the very largest meteors retain solid form all the way to the earth's surface, the vast majority of meteors completely evaporate long before this point. Because the outer layers of the meteor are the first to lose velocity, the result is a "tail" of ionized particles that forms behind the meteor. It is the ionized particles that provide a communication medium, because free electrons in the meteor tail are capable of re-radiating a radio-frequency signal.

The geometry between the meteor tail and the transmit and receive antennas plays a crucial role in the degree to which the transmitted signal is received, if at all. Maximum signal power is received if two things occur: (a) The tail forms a tangent line to an ellipsoid for which the foci coincide with the two antennas, and (b) substantial ionization occurs at and around the point of tangency (3). Ionized particles all along the tail contribute to the signal to some extent, but most important are the particles close to the point of tangency. These particles add constructively to the received signal provided that the distance from transmitter to meteor particle to receiver does not vary more than a half-wavelength from the corresponding distance for the point of tangency.

The length of the region on the tail for which this constructive contribution occurs is proportional to the square root of the wavelength of the signal. Because wavelength is equal to the speed of light divided by the carrier frequency, it follows that the length of the region decreases as the carrier frequency increases. A reduction in the length of the region corresponds to a reduction in the number of ionized particles that contribute to the signal strength.

Table 1. Typical Meteor-Burst System and Channel Parameters

Carrier frequency:	40–60 MHz
Transmitter power:	100–5000 W
Communication range:	Up to 2000 km
Trail height:	80–120 km
Typical application:	Vehicle tracking, environmental monitoring, two-way messaging
Trail lifetime:	0.2–1.0 s
Throughput:	1000 bits/min

For those ionized particles for which the distance from transmitter to particle to receiver is between a half-wavelength and a full wavelength different from the corresponding distance for the point of tangency, *destructive* interference results. Differences in distance of additional multiples of a half-wavelength alternatively add constructively and destructively to the signal.

When the meteor first enters the atmosphere, very little ionization is present, and the received signal level is zero. Fairly quickly, ionization begins, and while the tail is relatively short, most of the ionized particles add constructively. The optimal time for a meteor's usefulness typically occurs when the meteor is at an altitude between 120 km and 80 km (4). As the meteor continues to descend, the tail elongates and a higher percentage of particles add destructively until the numbers tend to cancel out and the received signal level again drops to zero. The period of time for which a usable signal is present is called a *meteor burst*.

The length of a meteor burst ranges from milliseconds to seconds. The time duration is a function of many factors including the ionization density along the trail. If the density is sufficiently low, the trail is termed *underdense*. Theoretical analysis and experimental data show that underdense trails provide a received signal level that very rapidly rises to a peak value and then decays exponentially. Most trails are underdense. If the density is not low enough to result in the exponentially shaped signal level, the trail is termed *overdense*. The received signal level of overdense trails typically rises and decays more slowly, although eventually the trail usually exhibits an exponential decay.

For longer bursts, multipath fading may occur. In terms of the received signal, the result is multiple short dropouts of the signal occurring before the burst ends. Multipath can be caused by upper-atmosphere winds that bend the meteor trail in such a way that multiple communication paths between transmitter and receiver result. Because of their shorter duration, meteor bursts due to underdense trails typically are less affected by multipath than overdense trails.

Between meteor bursts, communication is not possible, unless other beyond-line-of-sight propagation mechanisms occur. Additional propagation mechanisms seen to occur on meteor links include sporadic-E propagation, ionospheric scatter, auroral scatter, diffraction, and troposcatter (3). Some of these effects, when present, can provide much higher data rates than can be obtained when propagation occurs solely by the meteor-burst mechanism.

PERFORMANCE MODELING AND CHARACTERIZATION

One of the factors that affect the performance of a meteor-burst communication system is the time waveform for the received signal power. For underdense trails, the received signal power is given by

$$P(t) = P_0 \exp\left(-\frac{t}{\tau}\right), \qquad t \geq 0$$

where P_0 is the peak received signal power and τ is a time constant related to the decay rate. Both P_0 and τ depend on the trail characteristics and vary from trail to trail. Since most trails are underdense, this model is often used in the design and analysis of meteor-burst communication systems. For overdense trails, the received signal power is more diffi-

cult to model. One model that does reasonably well for overdense trails has the form (5)

$$P(t) = P_0 \exp\left(\frac{1}{2}\right) \sqrt{\frac{at+b}{c} \ln\left(\frac{c}{at+b}\right)}, \qquad t \geq 0$$

where a, b, and c are constants that vary from trail to trail. Unfortunately, a significant number of overdense trails are not well-fit to this model. The general shape of $P(t)$ for an overdense trail is an initial rapid rise followed by a slower rise, which is in turn followed by a slow decay and, finally, an exponential decay. Although meteor bursts due to overdense trails usually have a longer duration than those due to underdense trails, they may or may not have larger peak power.

With regard to underdense trails, because P_0 varies from trail to trail, it can be modeled as a random variable (6). Experimental studies indicate that P_0 is reasonably well-modeled as a Gaussian random variable (3). The parameter τ can also be modeled as a random variable, and experimental studies indicate that either a Rayleigh or a log-normal distribution provides a reasonably good fit (7). An apparently untested assumption in the literature is that P_0 and τ are statistically independent.

The number of meteor bursts in a given time window is well-modeled as a Poisson random variable; that is, the probability that i bursts occur in T seconds is $(aT)^i \exp(-aT)/i!$, where a is a parameter that depends on the time of day, the time of year, and the region of the sky to which the antennas point. This model corresponds to a Poisson process (6), and it follows that the time between meteor bursts is exponentially distributed; that is, the probability that the time between bursts is at least T seconds is $\exp(-aT)$.

An additional important characterization of the performance of meteor-burst systems relates to the geographical region for which communication is possible using a particular meteor trail. The *footprint* of a meteor trail can be defined as the region on earth for which, at a particular time and a particular location of the transmitter, the received signal power exceeds some specified threshold. This footprint might more properly be termed the instantaneous footprint, because the size and location of the footprint changes over the course of the burst. As might be expected, the size grows and then shrinks, and the location drifts as a result of the fact that the meteor trail drifts with the winds in the upper atmosphere.

The *trail footprint* can be defined as the total accumulated region in which the instantaneous footprint ever extends over the course of the burst. The size of the trail footprint gives an indication of the degree that the communication is secure. A small footprint implies that the communication link has low probability of detection (LPD).

The size of a footprint is significantly affected by the location of the trail relative to the locations of the transmitter and receiver. Meteor trails close to the transmitter tend to have larger footprints than trails close to the receiver (3). This fact implies that the footprints for two radios transmitting back and forth to each other may be of different sizes.

CODING AND MODULATION FOR METEOR-BURST SYSTEMS

The nature of the meteor-burst channel implies that for all trails, the received signal power is time-varying and the trails are short-lived. Therefore, the capability to adapt transmission parameters, such as the ratio of data to redundancy in an error-control coding system or the rate at which channel symbols are transmitted, is desirable provided that the following two conditions are satisfied: (1) The channel conditions can be estimated quickly and accurately, and (2) the cost of adaptivity is not prohibitive.

Error-control coding, also referred to as forward error correction coding, must be used in many communication systems to provide acceptable levels of performance. It is not clear at the outset, however, that error-control coding will improve the performance of a meteor-burst system. This is because the use of error-control coding requires the inclusion of redundant symbols in the encoded packet, and this in turn requires a larger transmission time than an uncoded packet if the symbol transmission rate is fixed. The received signal strength may be large during early portions of a trail, but the power decays rapidly so there is a penalty for longer transmission times. The following question therefore arises: Do the benefits of error-control coding outweigh the penalty that results from a longer transmission time? As shown in several research articles, the answer is "yes." Several approaches to the use of error-control coding in meteor-burst systems along with selected results are discussed below.

There are at least three approaches to the use of error-control coding for meteor-burst communication systems. In the first approach we consider, referred to as standard coding, a fixed-rate code is used. In the second approach, which we refer to as singly adaptive-rate coding, a fixed-rate code is used for each trail but the rate of the code is allowed to vary from trail to trail. In the third approach, which we refer to as doubly adaptive-rate coding, the ratio of data to redundancy is allowed to vary throughout a trail lifetime as well as from trail to trail. In the singly and doubly adaptive-rate approaches, the channel characteristics must be measured in order to adapt the code rate correctly, but these approaches have the potential to offer much higher throughput than the standard coding approach.

In Refs. 8 and 9 the authors study the performance of systems that use standard coding with block codes and confirm that systems that use error-control coding outperform systems that do not use coding. In Ref. 10 a study of doubly adaptive-rate coding is presented for systems that use Reed–Solomon codes, and in Ref. 11 a doubly adaptive-rate system that uses adaptive trellis-coded modulation is presented and analyzed. The latter two articles demonstrate that a doubly adaptive-rate coding can have a much larger throughput than uncoded systems and standard coding systems. In Ref. 12 an implementation of an adaptive-TCM system is presented, and a standard coding system has been fielded (13) that uses a rate $\frac{1}{2}$ convolutional code with Viterbi decoding.

Some form of automatic-repeat-request (ARQ) may be required in some applications, and ARQ may be used with any of the three error-control-coding approaches described above. Additional ways of using ARQ with error-control coding are presented in Refs. 14 and 15. In Ref. 14 a system is investigated in which only part of the redundancy of the error-control code is transmitted initially along with the data. If errors are detected, the transmitter is alerted and additional redundancy is transmitted thereby increasing the error-correcting capability of the code that was sent previously. In Ref. 15 two practical methods for implementing doubly adaptive-rate coding are investigated which use feedback from ARQ to determine when to change the code rate. Other work in ARQ (with-

out error-control coding) for meteor-burst communications includes Ref. 16, in which three different ARQ protocols are compared. The protocols are designed to study the relative advantages of a simple stop-and-wait ARQ scheme as well as the ability to detect the presence of a meteor channel. Symbol interleaving can also be used effectively with error-control coding (17). The idea is that interleaving can provide additional burst error correction capability. This is especially important near the end of the packet where the rapidly decreasing received signal power results in a large number of errors.

Adaptive-rate coding is one approach to adapting the transmission parameters to the channel conditions. Another approach is to adapt the rate at which channel symbols are transmitted, and this approach is known as variable-rate signaling. The goal is to vary the signaling rate in direct proportion to the received signal power so that the received symbol energy is maintained within a desired range. In Ref. 18 an analysis of a system that uses variable-rate signaling is presented. Furthermore, at least two systems (19,20) that use variable-rate signaling have been implemented. In the approach used for these systems, a feedback channel is maintained for each meteor trail that allows the receiver to communicate information about the new signaling rate to the transmitter. (Note that the systems are designed to change the signaling rate during lifetimes of each usable trail.) Demodulator outputs are used to estimate the current signal-to-noise ratio, and the signal-to-noise ratio estimate is used in turn to determine the new signaling rate.

Finally, note that it is generally believed that channel disturbances, such as multipath propagation, are not severe enough to make tracking of the carrier phase impossible. The existence of several commercial systems (1) that use a carrier-phase tracking system (e.g., a phase-locked loop) for the purposes of coherent demodulation supports this point. Therefore, the evidence suggests that coherent demodulation should be used if the additional cost of a phase tracking device is not prohibitive and if the phase can be acquired in a small period of time.

BIBLIOGRAPHY

1. J. Z. Schanker, *Meteor Burst Communications,* Norwood, MA: Artech House, 1990.

2. G. R. Sugar, Radio propagation by reflection from meteor trails, *Proc. IEEE,* 1964, pp. 116–136.

3. J. A. Weitzen, Meteor Scatter Communication: A New Understanding, in D. L. Schilling (ed.), *Meteor Burst Communications: Theory and Practice,* New York: Wiley, 1993, pp. 9–58.

4. R. A. Desourdis, Jr., Modeling and Analysis of Meteor Burst Communications, in D. L. Schilling (ed.), *Meteor Burst Communications: Theory and Practice,* New York: Wiley, 1993, pp. 59–342.

5. C. O. Hines and P. A. Forsythe, The forward scattering of radio waves from overdense meteor trails, *Can. J. Phys.,* **35**: 1033–1041, 1957.

6. A. Papoulis, *Probability, Random Variables, and Stochastic Processes,* 3rd ed. New York: McGraw-Hill, 1991.

7. J. A. Weitzen and W. T. Ralston, Meteor scatter: An overview, *IEEE Trans. Antennas Propag.,* **AP-36**: 1813–1819, 1988.

8. K. Brayer and S. Natarajan, An investigation of ARQ and hybrid FEC-ARQ on an experimental high latitude meteor burst channel, *IEEE Trans Commun.,* **COM-37**: 1239–1242, 1989.

9. S. L. Miller and L. B. Milstein, Error correction coding for a meteor burst channel, *IEEE Trans. Commun.*, **COM-38**: 1520–1529, 1990.

10. M. B. Pursley and S. D. Sandberg, Variable-rate coding for meteor-burst communications, *IEEE Trans. Commun.*, **COM-37**: 1105–1112, 1989.

11. J. M. Jacobsmeyer, Adaptive trellis-coded modulation for bandlimited meteor-burst channels, *IEEE J. Select. Areas Commun.*, **10**: 550–561, 1992.

12. J. M. Jacobsmeyer, Adaptive data rate modem, U.S. Patent No. 5,541,955, 1996.

13. E. J. Morgan, Meteor burst communications: An update, *Signal,* **42**: 55–61, 1988.

14. M. B. Pursley and S. D. Sandberg, Incremental-redundancy transmission for meteor-burst communications, *IEEE Trans. Commun.*, **COM-39**: 689–702, 1991.

15. M. B. Pursley and S. D. Sandberg, Variable-rate hybrid ARQ for meteor-burst communications, *IEEE Trans. Commun.*, **COM-40**: 60–73, 1992.

16. S. L. Miller and L. B. Milstein, A comparison of protocols for a meteor-burst channel based on a time-varying channel model, *IEEE Trans. Commun.*, **COM-37**: 18–30, 1989.

17. C. W. Baum and C. S. Wilkins, Erasure generation and interleaving for meteor-burst communications with fixed-rate and variable-rate coding, *IEEE Trans. Commun.*, **COM-45**: 625–628, 1997.

18. S. Davidovici and E. G. Kanterakis, Performance of Meteor Burst Communication Using Variable Data Rates, in D. L. Schilling (ed.), *Meteor Burst Communications: Theory and Practice*, New York: Wiley, 1993, pp. 383–410.

19. D. L. Schilling et al., The FAVR Meteor Burst Communication Experiment, in D. L. Schilling (ed.), *Meteor Burst Communications: Theory and Practice*, New York: Wiley, 1993, pp. 367–381.

20. D. K. Smith and T. G. Donich, Maximizing throughput under changing channel conditions, *Signal,* **43** (10): 173–178, 1989.

CARL W. BAUM
CLINT S. WILKINS
Clemson University

METEOROLOGICAL RADAR

The origins of meteorological radar (RAdio Detection And Ranging) can be traced to the 1920s when the first echoes from the ionosphere were observed with dekametric (tens of meters) wavelength radars. However, the greatest advances in radar technology were driven by the military's need to detect aircraft and occurred in the years leading to and during WWII. (A brief review of the earliest meteorological radar development is given in Ref. 1; a detailed review of the development during and after WWII is given in Ref. 2.) Precipitation echoes were observed almost immediately with the deployment of the first decimeter ($\approx$0.1 m) wavelength military radars in the early 1940s. Thus, the earliest meteorological radars used to observe weather (i.e., weather radars) were manufactured for military purposes. The military radar's primary mission is to detect, resolve, and track discrete targets such as airplanes coming from a particular direction, and to direct weapons for interception.

Although weather radars also detect aircraft, their primary objective is to map the intensity of precipitation (e.g., rain, or hail), which can be distributed over the entire hemisphere above the radar. Each hydrometeor's echo is very weak; nevertheless, the extremely large number of hydrometeors within the radar's beam returns a continuum of strong echoes as the transmitted pulse propagates through the field of precipitation. Thus, the weather radar's objective is to estimate and map the fields of reflectivity and radial velocities of hydrometeors; from these two fields the meteorologists need to derive the fall rate and accumulation of precipitation and warnings of storm hazards. The weather radar owes its success to the fact that centimetric waves penetrate extensive regions of precipitation (e.g., hurricanes) and reveal, like an X-ray photograph, the morphology of weather system.

The first U.S. national network of weather radars, designed to map the reflectivity fields of storms and to track them, were built in the mid 1950s and operated at 10 cm wavelengths. 1988 marked the deployment of the first network of Doppler weather radars (i.e., the WSR-88D), which in addition to mapping reflectivity, have the capability to map radial (Doppler) velocity fields. This latter capability proved to be very helpful in identifying those severe storm cells that harbor tornadoes and damaging winds.

If a hydrometeor's diameter is smaller than a tenth of the radar's wavelength, its echo strength is inversely proportional to the fourth power of the wavelength. Thus, shorter wavelength (i.e., millimeter) radars are usually the choice to detect clouds. Cloud particle diameters are less than 100 μ, and attenuation due to cloud particles is not overwhelming. However, if clouds bear rain of moderate intensity, precipitation attenuation can be severe (e.g., at a wavelength of 6.2 mm, and rainrate of 10 mm h^{-1}, attenuation can be as much as 6 dB km^{-1} (3). Spaceborne meteorological radars also operate in the millimetric band of wavelengths in order to obtain acceptable angular resolution with reasonable antenna diameters required to resolve clouds at long ranges (4). Airborne weather radars operate at short wavelengths of approximately 3 and 5 cm; these are used to avoid severe storms that produce hazardous wind shear and extreme amounts of rainwater (which extinguish jet engines), and to study weather phenomena (5). The short wavelength waves are used to obtain acceptable angular resolution with small antennas on aircraft; however, short waves are strongly attenuated as they propagate into heavy precipitation. At longer wavelengths (e.g., >10 cm), only hail and heavy rain significantly attenuate the radiation.

Weather radars are commonly associated with the mapping of precipitation intensity. Nevertheless, the earliest, of what we now call, meteorological radars detected echoes from the nonprecipitating troposphere in the late 1930s (1,2). Scientists determined that these echoes are reflected from the dielectric boundaries of different air masses (1,6). The refractive index n of air is a function of temperature and humidity, and spatial irregularities in these parameters, caused by turbulence, were found to be sufficiently strong to cause detectable echoes from refractive index irregularities.

FUNDAMENTALS OF METEOROLOGICAL DOPPLER RADAR

The basic principles for meteorological radars are the same as for any radar that transmits a periodic train of short duration pulses (i.e., with period T_s called the pulse repetition time (PRT), and duration τ) of microwaves, and measures the delay between the time of emission of the transmitted pulse and the time of reception of any of its echoes. The PRT (i.e., T_s) is typically of the order of milliseconds, and pulsewidths τ are of the order of microseconds. The radar has Doppler capability if it can measure the change in frequency or wavelength between the backscattered and transmitted signals.

The Doppler radar's microwave oscillator (Fig. 1) generates a continuous wave sinusoidal signal, which is converted to a sequence of microwave pulses by the pulse modulator. Therefore, the sinusoids in each microwave pulse are coherent with those generated by the microwave oscillator; that is, the crests and valleys of the waves in the pulse bear a fixed or known relation between the crests and valleys of the waves emitted by the microwave oscillator. The microwave pulses are then amplified by a high-power amplifier (a klystron is used in the WSR-88D) to produce about a megawatt of peak power. The propagating pulse has a spatial extent of $c\tau$, and travels at the speed of light c along the beam (beamwidth θ_1 is the one-way, 3 dB width of the beam, and is of the order of 1 degree). The transmit/receive (T/R) switch connects the transmitter to the antenna (an 8.53 m diameter parabolic reflector is used for the WSR-88D) during τ, and the receiver to the antenna during the interval $T_s - \tau$. The echoes are mixed in the synchronous detectors with a pair of phase quadrature signals (i.e., sine, 90°, and cosine, 0°, outputs from the oscillator). The pair of synchronous detectors and filter amplifiers shift the carrier frequency from the microwave band to zero frequency in one step for the homodyne radar and allow measurement of both positive and negative Doppler shifts (most practical radars use a two step process involving an intermediate frequency).

A hydrometeor intercepts the transmitted pulse and scatters a portion of its energy back to the antenna, and the echo voltage

$$V(r,t) = Ae^{j[2\pi f(t - 2r/c) + \psi]}U(t - 2r/c) \qquad (1)$$

at the input to the synchronous detectors is a replica of the signal transmitted; A is the echo amplitude that depends on the hydrometeor's range r and its backscattering cross section σ_b, and $2\pi f(t - 2r/c) + \psi$ is the echo phase. The microwave carrier frequency is f, t is time after emission of the transmitted pulse, and ψ is the sum of phase shifts introduced by the radar system and by the scatterer; these shifts are

usually independent of time. The function U locates the echo; it is one when its argument is between zero and τ, and zero otherwise. The output of one synchronous detector is called the in-phase (I) voltage and the other is called the quadrature-phase (Q) voltage (Fig. 1); these are the imaginary and real parts of the echo's complex voltage [Eq. (1)] after its carrier frequency f is shifted to zero. Thus

$$I(t,r) = A\cos\psi_e U(t - 2r/c), \quad Q(r,t) = A\sin\psi_e U(t - 2r/c) \qquad (2)$$

where

$$\psi_e = -\frac{4\pi r}{\lambda} + \psi \qquad (3)$$

is the echo phase, and $\lambda = c/f$ is the wavelength of the transmitted microwave pulse. The time rate of the echo phase change is related to the scatterer's radial (Doppler) velocity v,

$$\frac{d\psi_e}{dt} = -\frac{4\pi}{\lambda}\frac{dr}{dt} = -\frac{4\pi}{\lambda}v = \omega_d \qquad (4)$$

where ω_d is the Doppler shift (in radians per second). For typical transmitted pulse widths (i.e., $\tau \approx 10^{-6}$ s) and hydrometeor velocities (tens of m s^{-1}), the changes in phase are extremely small during the time that $U(t - 2r/c)$ is nonzero. Therefore, the echo phase change is measured over the longer PRT period ($T_s \approx 10^{-3}$ s) and, consequently, the pulse Doppler weather radar is both an amplitude– and phase–sampling system. Samples are at $\tau_s + mT_s$, where τ_s is the time delay between a transmitted pulse and an echo, and m is an integer; τ_s is a continuous time scale and always lies in the interval $0 \le \tau_s \le T_s$, and mT_s is called sample time, which increments in T_s steps.

Because the transmissions are periodic, echoes repeat, and thus, there is no way to determine which transmitted pulse produced which echo (Fig. 2). That is, because τ_s is measured with respect to the most recent transmitted pulse and has values $<T_s$, the apparent range $c\tau_s/2$ is always less than the unambiguous range $r_a = cT_s/2$. However, the true range r can be $c\tau_s/2 + (N_t - 1)r_a$, where N_t is the trip number, and $N_t - 1$ designates the number of $cT_s/2$ intervals that need to be

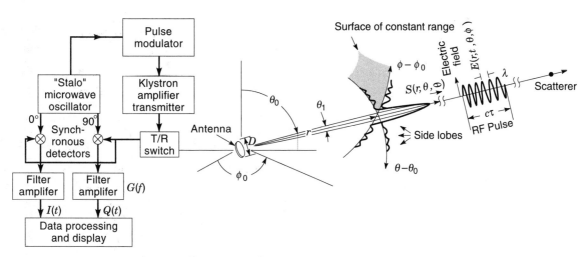

Figure 1. Simplified block diagram of a homodyne radar (no intermediate-frequency circuits are used to improve performance) showing the essential components needed to illustrate the basic principles of a meteorological Doppler radar.

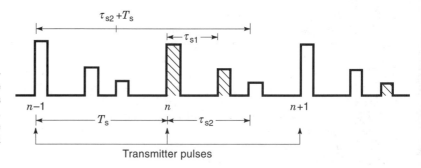

Figure 2. Range-ambiguous echoes. The nth transmitted pulse and its echoes are crosshatched. This example assumes that the larger echo at delay τ_{s1} is unambiguous in range, but the smaller echo, at delay τ_{s2}, is ambiguous. This smaller second trip echo, which has a true range time delay $T_s + \tau_{s2}$, is due to the $(n-1)$th transmitted pulse.

added to the apparent range to obtain r. There is range ambiguity if $r \geq r_a$.

The I and Q components of echoes from stationary and moving scatterers are shown in Fig. 3 for three successive transmitted pulses. The echoes from the moving scatterer clearly exhibit a systematic change, from one mT_s period to the next, caused by the scatterers' Doppler velocity, whereas there is no change in echoes from stationary scatterers. Echo phase $\psi_e = \tan^{-1}(Q/I)$ is measured, and its change over T_s is proportional to the Doppler shift given by Eq. (4).

The periodic transmitted pulse sequence also introduces velocity ambiguities. A set of ψ_e samples cannot be related to one unique Doppler frequency. As Fig. 4 shows, it is not possible to determine whether $V(t)$ rotated clockwise or counterclockwise and how many times it circled the origin during the interval T_s. Therefore, any of the frequencies $\Delta\psi_e/T_s + 2\pi p/T_s$ (where p is a $\pm$ integer, and $-\pi < \Delta\psi_e \leq \pi$) could be correct. All such Doppler frequencies are called aliases, and $f_N = \omega_N/2\pi = 1/(2T_s)$ is the Nyquist frequency (in units of Hertz). All Doppler frequencies between $\pm f_N$ are the principal aliases, and frequencies higher or lower than $\pm f_N$ are ambiguous with those between $\pm f_N$. Thus, hydrometeor radial velocities must lie within the unambiguous velocity limits, $v_a = \pm\lambda/4T_s$, to avoid ambiguity. Signal design and processing methods have been advanced to deal with range-velocity ambiguities (1).

REFLECTIVITY AND VELOCITY FIELDS OF PRECIPITATION

A weather signal is a composite of echoes from a continuous distribution of hydrometeors. After a delay (the roundtrip time of propagation between the radar and the near boundary of the volume of precipitation), echoes are continuously received (Fig. 5) during a time interval equal to twice the time it takes the microwave pulse to propagate across the volume containing the hydrometeors. Because one cannot resolve each of the hydrometeor's echoes, meteorological radar circuits sample the I and Q signals at uniformly spaced intervals along τ_s, and convert the analog values of the I, Q voltages to digital numbers. For each sample, there is a resolution volume V_6 (i.e., the volume enclosed by the surface on which angular and range-weighting functions (1) are smaller than 6 dB below their peak value) along the beam within which hydrometeors contribute significantly to the sample. Each scatterer within V_6 returns an echo and, depending on its precise position to within a wavelength, its corresponding I or Q can have any value between maximum positive and negative excursions. Echoes from the myriad of hydrometeors constructively or destructively (depending on their phases) interfere with each other to produce the composite weather signal voltage $V(mT_s, \tau_s) = I(mT_s, \tau_s) + jQ(mT_s, \tau_s)$ for the mth T_s interval. The random size and location of hydrometeors cause the I and Q weather signals to be a random function of τ_s. How-

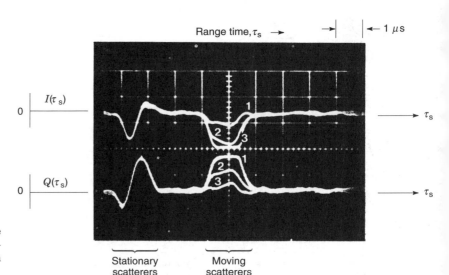

Figure 3. $I(\tau_s)$ and $Q(\tau_s)$ signal traces vs τ_s for three successive sampling intervals T_s have been superimposed to show the relative change of I, Q for both stationary and moving scatterers.

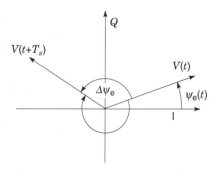

Figure 4. A phasor diagram used to depict frequency aliasing. The phase of the signal sample $V(t)$ could have changed by $\Delta\psi_e$ over a period T_s.

ever, these random signals have a correlation time τ_c (Fig. 5), dependent on the pulsewidth τ and the receiver's bandwidth (1). Thus, $V(mT_s, \tau_s$ has noise-like fluctuations along τ_s even if the scatterer's time averaged density is spatially uniform.

The sequences of $M(m = 1 \rightarrow M)$ samples at any τ_s are analyzed to determine the motion and reflectivity of hydrometeors in the corresponding V_6. The dashed line in Fig. 5 depicts a possible sample time mT_s, dependence of $I(mT_s, \tau_{s1})$ for hydrometeors having a mean motion that produces a slowly changing sample amplitude along mT_s. The rate of change of I and Q vs mT_s is determined by the radial motion of the scatterers. Because of turbulence, scatterers also move relative to one another and, therefore, the I, Q samples at any τ_s change randomly with a correlation time along mT_s dependent on the relative motion of the scatterers. For example, if turbulence displaces the relative position of scatterers a significant fraction of a wavelength during the T_s interval, the weather signal at τ_s will be uncorrelated from sample to sample, and Doppler velocity measurements will not be possible; Doppler measurements require a relatively short T_s.

The random fluctuations in the I, Q samples have a Gaussian probability distribution with zero mean; the probability of the signal power $I^2 + Q^2$ is exponentially distributed (e.g., the weakest power is most likely to occur). Using an analysis of the $V(mT_s, \tau_s)$ sample sequence along mT_s, the meteorological radar's signal processor to estimate both the average sample power and the power-weighted velocity of the

scatterers accurately. The samples' average power $\overline{P}$ is

$$\overline{P}(r_o) = \iiint \eta(\mathbf{r})I(\mathbf{r}_o, \mathbf{r})\,dV \tag{5}$$

in which the reflectivity η, the sum of the hydrometeor backscattering cross sections σ_b per unit volume, is

$$\eta(\mathbf{r}) = \int_0^\infty \sigma_b(D)N(D, \mathbf{r})\,dD \tag{6}$$

The factor $I(\mathbf{r}_o, \mathbf{r})$ in Eq. (5) is a composite angular and range-weighting function; its center at $\mathbf{r}_o$ depends on the beam direction as well as τ_s. The values of $I(\mathbf{r}_o, \mathbf{r})$ at $\mathbf{r}$ depend on the antenna pattern, transmitted pulse shape and width, and the receiver's frequency or impulse transfer function (1). In general, $I(\mathbf{r}_o, \mathbf{r})$ has significant values only within V_6; $N(D)$, the particle size distribution, determines the expected number density of hydrometeors with equivolume diameters between D and $D + dD$.

The meteorological radar equation,

$$\overline{P}(r_o) = \frac{P_t g^2 g_s \lambda^2 \eta c \tau \pi \theta_1}{(4\pi)^3 r_o^2 l^2 l_r 16 \ln 2} \tag{7}$$

is used to determine η from measurements of $\overline{P}$, wherein P_t is the peak transmitted power, g is the gain of the antenna (a larger antenna directs more power density along the beam and hence has larger gain), and g_s is the gain of the receiver (e.g., the net sum of losses and gains in the T/R switch, the synchronous detectors, and the filter/amplifiers in Fig. 1). Here $r_o \approx c\tau_s/2$, l is one-way atmospheric transmission loss, and l_r is the loss due to the receiver's finite bandwidth (1). Equation (7) is valid for transmitted radiation having a Gaussian function dependence on distance from the beam axis, and for uniform reflectivity fields.

Radar meteorologists have related reflectivity η, which is general radar terminology for the backscattering cross section per unit volume, to a reflectivity factor Z which has meteorological significance. If hydrometeors are spherical and have diameters much smaller than λ (i.e., the Rayleigh approximation), the reflectivity factor,

$$Z = \int_0^\infty N(D, r)D^6\,dD \tag{8}$$

is related to η by

$$\eta = \frac{\pi^5}{\lambda^4}|K_m|^2 Z \tag{9}$$

where $K_m = (m^2 - 1)/(m^2 + 2)$, and $m = n(1 - j\kappa)$ is the hydrometeor's complex refractive index, and κ is the attenuation index (1).

The relation between radial velocity $\nu(\mathbf{r})$ at a point $\mathbf{r}$ and the power–weighted Doppler velocity $\nu(\mathbf{r}_o)$ is

$$\overline{\nu}(\mathbf{r}_o)\frac{\iiint \nu(\mathbf{r})\eta(\mathbf{r})I(\mathbf{r}_o, \mathbf{r})\,dV}{\overline{P}(\mathbf{r}_o)} \tag{10}$$

It can be shown (1) that $\overline{\nu}(\mathbf{r}_o)$ is the first moment of the Doppler spectrum. An example of a Doppler spectrum for echoes

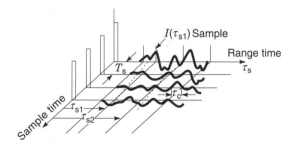

Figure 5. Idealized traces for $I(\tau_s)$ of weather signals from a dense distribution of scatterers. A trace represents $V(mT_s, \tau_s)$ vs τ_s for the mth T_s interval. Instantaneous samples are taken at sample times τ_{s1}, τ_{s2}, etc. The signal correlation time along τ_s is τ_c. Samples at fixed τ_s are acquired at T_s intervals and are used to compute the Doppler spectrum for scatterers located about the range $c\tau_s/2$.

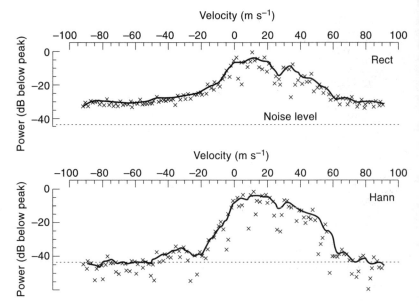

Figure 6. The spectral estimates (denoted by ×) of the Doppler spectrum of a small tornado that touched down on 20 May 1977 in Del City, Oklahoma. V_6 is located at azimuth: 6.1°; elevation: 3.1°; altitude: 1.9 km. Rect signifies the spectrum for weather signal samples weighted by a rectangular window function (i.e., uniform weight), whereas Hann signifies samples weighted by a von Hann window function (1).

from a tornado is plotted in Fig. 6. This power spectrum is the magnitude squared of the spectral coefficients obtained from the discrete Fourier transform for M = 128 $V(mT_s, \tau_s)$ samples at a τ_s corresponding to a range of 35 km. The obscuration of the maximum velocity (i.e., ≈60 m s⁻¹) of the scatterers in this tornado, and the power of stronger spectral coefficients leaked through the spectral sidelobes of the rectangular window (i.e., uniform weighting function) are evident. The von Hann weighting function reduces this leakage and better defines both true signal spectrum and maximum velocity. Where spectral leakage is not significant (e.g., samples weighted with the von Hann window function), the spectral coefficients have an exponential probability distribution and hence there is a large scatter in their power. Thus, a 5-point running average is plotted to show the spectrum more clearly. The spectral density of the receiver's noise is also obscured by the leakage of power from the larger spectral components if voltage samples are uniformly weighted.

RAIN, WIND, AND OBSERVATIONS OF SEVERE WEATHER

Fields of reflectivity factor Z and the power-weighted mean velocities $\bar{\nu}(\boldsymbol{r}_o)$ are displayed on color TV monitors to depict the morphology of storms. The Z at low elevation angles is used to estimate rain rates because hydrometeors there are usually rain drops, and vertical air motion can be ignored so that the drops are falling at their terminal velocity w_t, a known function of D. The rainfall rate R is usually measured as depth of water per unit time and is given by

$$R = \frac{\pi}{6} \int_0^\infty D^3 N(D) \omega_t(D) \, dD \, ms^{-1} \qquad (11)$$

where mks units are used. To convert to the more commonly used units of millimeters per hour, multiply Eq. (11) by the factor 3.6×10^6. The simple and often observed $N(D)$ is an exponential one, and even in this case, we need to measure or specify two parameters of $N(D)$ to use Eq. (11). A real drop-size distribution requires an indefinite number of parameters to characterize it and thus, the radar-determined value of Z

alone cannot provide a unique measurement of R. Although radar meteorologists have attempted for many years to find a useful formula that relates R to Z, there is unfortunately no universal relation connecting these parameters. Nonetheless, it is common experience that larger rainfall rates are associated with larger Z. For stratiform rain, the relation

$$Z = 200R^{1.6} \qquad (12)$$

has often proved quite useful.

Although the Doppler radar measures only the hydrometeor motion toward and away from the radar, the spatial distribution of Doppler velocities can reveal meteorological features such as tornadoes, microbursts (i.e., the divergent flow of strong thunderstorm outflow near the ground), buoyancy waves, etc. For example, the observed Doppler velocity pattern for a tornado in a mesocyclone (a larger-scale rotating air mass) is shown in Fig. 7. The strong gradient of Doppler velocities associated with a couplet of closed ± isodops (i.e., contours of constant Doppler velocity) is due to the tornado having a diameter of about 700 m, and the larger-scale closed isodop pattern (i.e., the −30 and +20 contour) is due to the larger scale (i.e., 3.8 km diameter) mesocyclone. In practice, the data are shown on color TV displays wherein the regions between ± isodops are often colored with red and green hues of varying brightness to signify values of $\pm\bar{\nu}(\boldsymbol{r}_o)$.

A front is a relatively narrow zone of strong temperature gradients separating air masses. A dry line is simply the boundary between dry and moist air masses. Turbulent mixing along these boundaries creates relatively intense irregularities of refractive index, which return echoes through the Bragg scatter mechanism (1,2,6; also described in the following section). Figure 8 shows the reflectivity fields associated with a cold front and a dry line, as well as the storms that initiated at the intersection of these boundaries. From Doppler velocity fields and observations of the cold front position at subsequent times, it was established that the cold air mass to the northwest of the front is colliding with the ambient air flowing from the SSW. The convergence along the boundary creates a line of persistent vertical motion that can lift scat-

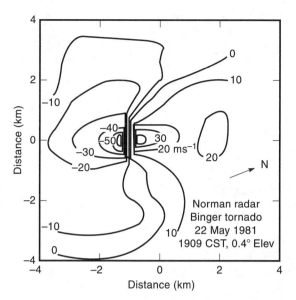

Figure 7. The isodops for the Binger, Oklahoma tornadic storm on 22 May 1981. The center of the mesocyclone is 70.8 km from the Norman Doppler radar at azimuth 284.4°; the data field has been rotated so that the radar is actually below the bottom of the figure.

tering particles, normally confined to the layers closer to the ground, making them visible as a reflectivity thin line. Thus, the reflectivity along the two boundaries could be due to these particles as well as to Bragg scatter. The intersection of cold fronts and dry lines is a favored location for the initiation of storms (seen to the northeast of the intersection). As the cold front propagates to the southeast, the intersection of it and the relatively stationary dry line progresses south-southwest-

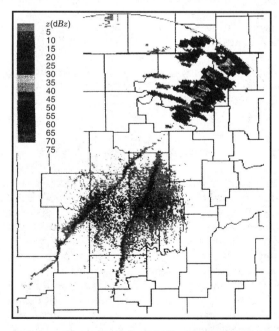

Figure 8. Intersecting reflectivity thin lines in central Oklahoma on 30 April 1991 at 2249 U. T. The thin line farthest west is along a NE-SW oriented cold front; the thin line immediately east is along a NNE-SSW oriented dry line. The reflectivity factor (dBZ) categories are indicated by the brightness bar. (Courtesy of Steve Smith, OSF/NWS.)

ward, and storms are initiated at this moving intersection point.

WIND AND TEMPERATURE PROFILES IN CLEAR AIR

In addition to particles acting as tracers of wind, irregularities of the atmosphere's refractive index can cause sufficient reflectivity to be detected by meteorological radars. Although irregularities have a spectrum of sizes, only those with scales of the order of half the radar wavelength provide echoes that coherently sum to produce a detectable signal (1). This scattering mechanism is called stochastic Bragg scatter because the half-wavelength irregularities are in a constant state of random motion due to turbulence, and thus the echo signal intensity fluctuates exactly like signals scattered from hydrometeors. The reflectivity η is related to the refractive index structure parameter C_n^2 that characterizes the itensity of the irregularities (1,2,7)

$$\eta = 0.38 C_n^2 \lambda^{-1/3} \qquad (13)$$

Mean values of C_n^2 range from about 10^{-14} m$^{-2/3}$ near sea level, to 10^{-17} m$^{-2/3}$ at 10 km above sea level.

Meteorological radars that primarily measure the vertical profile of the wind in all weather conditions and, in particular during fair weather, are called profilers. A wind profile is obtained by measuring the Doppler velocity vs range along beams in at least two directions (about 15 ° from the vertical) and along the vertical beam and by assuming that wind is uniform within the area encompassed by the beams. The vertical profile of the three components of wind can be calculated from these three radial velocity measurements along the range and the assumption of wind uniformity (7). A prototype network of these profilers has been constructed across the central United States to determine potential benefits for weather forecasts (8).

Temperature profiles are measured using a Radio-Acoustic Sounding System (RASS; 1,7). This instrument consists of a vertically pointed Doppler radar (for this application the wind profiling radar is usually time-shared) and a sonic transmitter that generates a vertical beam of acoustic vibrations, which produce a backscattering sinusoidal wave of refractive index propagating at the speed of sound. The echoes from the acoustic waves are strongest under Bragg scatter conditions (i.e., when the acoustic wavelength is one-half the radar wavelength). The backscatter intensity at various acoustic frequencies is used to identify those frequencies that produce the strongest signals.

Because the acoustic wave speed (and thus wavelength) is a function of temperature, this identification determines the acoustic velocity and hence the temperature. Allowance must be made for the vertical motion of air, which can be determined by analyzing the backscatter from turbulently mixed irregularities.

TRENDS AND FUTURE TECHNOLOGY

Networks of radars producing Doppler and reflectivity information in digital form have had a major impact on our capability to provide short-term warnings of impending weather hazards. Still, there are additional improvements that could enhance the information derived from meteorological radars significantly. Resolution of velocity and range ambiguities, faster coverage of the surveillance volume and better resolution, estimates of cross beam wind, and better measurements of precipitation type and amounts are some of the outstanding problems. Signal design techniques that encode transmitted pulses or stagger the PRT are candidates to mitigate the effects of ambiguities (1). Faster data acquisition can be achieved with multiple-beam phase-array radars, and better resolution can be obtained at the expense of a larger antenna. The cross beam wind component can be obtained using a bistatic dual-Doppler radar (i.e., combining the radial component of velocity measured by a Doppler weather radar with the Doppler shift measured at the distant receiver), or by incorporating the reflectivity and Doppler velocity data into the equations of motion and conservation. Vector wind fields in developing storms could be used in numerical weather prediction models to improve short-term forecasts. We anticipate an increase in application of millimeter wavelength Doppler radars to the study of nonprecipitating clouds (5). For better precipitation measurements, radar polarimetry offers the greatest promise (1,2). Polarimetry capitalizes on the fact that hydrometeors have shapes that are different from spherical and a preferential orientation. Therefore, differently shaped hydrometeors interact differently with electromagnetic waves of different polarization. To make such measurements, the radar should have the capability to transmit and receive orthogonally polarized waves, e.g., horizontal and vertical polarization. Both backscatter and propagation effects depend on polarization; measurements of these can be used to classify and quantify precipitation. Large drops are oblately shaped and scatter more strongly the horizontally polarized waves; they also cause larger phase shift of these waves along propagation paths. The differential phase method has several advantages for measurement of rainfall compared to the reflectivity method [Eq. (12)]. These include independence from receiver or transmitter calibrations errors, immunity to partial beam blockage and attenuation, lower sensitivity to variations in the distribution of raindrops, less bias from either ground clutter filtering or hail, and possibilities to make measurements in the presence of ground reflections not filtered by ground clutter cancelers. Polarimetric measurements have already proved the hail detection capability, but discrimination between rain, snow (wet, dry), and hail also seems quite possible. These areas of research and development could lead to improved short-term forecasting and warnings.

BIBLIOGRAPHY

1. R. J. Doviak and D. S. Zrnić, *Doppler Radar and Weather Observations,* 2nd ed., San Diego: Academic Press, 1993.

2. D. Atlas, *Radar in Meteorology,* Boston: Amer. Meteorol. Soc., 1990.

3. L. J. Battan, *Radar Observations of the Atmosphere,* Chicago: Univ. of Chicago Press, 1973.

4. R. Meneghini and T. Kozu, *Spaceborne Weather Radar,* Norwood, MA: Artech House, 1990.

5. *Proc. IEEE; Spec. Issue Remote Sens. Environ.,* **82** (12): 1994.

6. E. E. Gossard and R. G. Strauch, *Radar Observations of Clear Air and Clouds,* Amsterdam: Elsevier, 1983.

7. S. F. Clifford et al., Ground-based remote profiling in atmospheric studies: An overview, *Proc. IEEE,* **82** (3): 1994.

8. U.S. Department of Commerce, National Oceanic and Atmospheric Administration, *Wind profiler assessment report and recommendations for future use 1987–1994,* Prepared by the staffs of the National Weather Service and the Office of Oceanic and Atmospheric Research, Silver Spring, MD, 1994.

RICHARD J. DOVIAK
DUŠAN S. ZRNIĆ
National Severe Storms Laboratory
The University of Oklahoma

METER, PHASE.   See PHASE METERS.

METERS.   See OHMMETERS; POWER SYSTEM MEASUREMENT.

METERS, ELECTRICITY.   See WATTHOUR METERS.

METERS, REVENUE.   See WATTHOUR METERS.

METERS, VOLT-AMPERE.   See VOLT-AMPERE METERS.

METHOD OF MOMENTS SOLUTION.   See INTEGRAL EQUATIONS.

METHODS, OBJECT-ORIENTED.   See BUSINESS DATA PROCESSING.

METHODS OF RELIABILITY ENGINEERING.   See RELIABILITY THEORY.

METRIC CONVERSIONS.   See DATA PRESENTATION.

METRICS, SOFTWARE.   See SOFTWARE METRICS.

METROPOLITAN AREA NETWORKS

Metropolitan area networks (MAN) have been studied, standardized, and constructed for less than 20 years. During that time the capabilities of the telecommunications network and the requirements of users have changed rapidly. What a MAN is supposed to do has changed as quickly as MANs are designed. Recent changes in user requirements, resulting from the growing use of the Internet at home, are likely to redefine MANs once again.

To understand the evolution and predict the future of MAN, one must consider the applications and alternative technologies. There are also inherent differences in the capabilities of local, metropolitan, regional, and wide area networks.

HISTORY

The first mention of MANs, that I am aware of, occurred at a workshop on local area networks (LAN) in North Carolina in the late 1970s. One session at the workshop was dedicated to customer experience with LANs. One of the customers, from a New York bank, described a successful application of LANs but complained about the difficulty he had transferring data between branches of the bank in the same city.

The feeling among the workshop participants was that we could do better connecting sites in the same city than using technology that was designed for a national network. In the 1970s telephone modems were expensive, about a buck a bit per second, and the highest rate modem that was generally available was 9.6 kbit/s. High-rate private lines, such as the current T-carrier system, were not widely deployed. Using the available technology was expensive and created a bottleneck between LANs. With the customer and application identified, work on MANs began.

The first MANs were designed to interconnect LANs. They evolved from LANs and looked very much like LANs. Two MAN standards that are clearly related to LANs, fiber distributed data interface (FDDI) and distributed queue, dual bus (DQDB), are described later in this article. A third network, which is based on a mesh structure, the Manhattan street network (MSN), is also described. The MSN is a network of two-by-two switches that operate on fixed-size cells. The MSN straddles the middle ground between a LAN and a centralized asynchronous transfer mode (ATM) switch (1).

Interest in MANs waned as the useful functions performed by MANs were subsumed by wide area networking (WAN) technology. WANs had a much larger customer base than MANs. It became more economical to interconnect LANs in a city with routers and private lines than to deploy new, special-purpose networks.

There is a resurgence in interest in MANs because of the World Wide Web (Web). The time required to download Web pages using WAN technologies is frustrating many users and constraining the growth of this service. As more and more individual homes are connected to the Internet, there is a rapidly growing demand for bursty, high-rate data to a large number of locations in a metropolitan area. Therefore, there is a renewed interest in MANs, although the requirements and customer set are completely different from those of the earlier MANs.

DIFFERENCES AMONG LANS, MANS, AND WANS

The distance spanned by WANs is greater than that by MANs, and the distance spanned by MANs is greater than that by LANs. It is useful to define the maximum distance spanned by the various network technologies as increases of an order of magnitude. LANs span distances up to 3 miles, and include most networks that are installed in a building or on a campus. MANs span distances up to 30 miles (50 km according to the standards committees) and can cover most cities. RANs (regional area networks) span distances up to 300 miles, the area serviced by the telephone operating companies in the United States. And WANs span distances up to 3000 miles, the distance across the United States. The next order of magnitude increase covers international networks. The distances spanned by networks affects the transmission costs, access protocols, ownership of the facilities, and the other users who share the network.

The transmission costs usually increase with distance. This cost affects both the applications that are economically viable and the protocols that are used to transfer data. For instance, access protocols have been designed for LANs that trade efficiency for processing complexity. Carrier sense multiple access/collision detection (CSMA/CD) protocols, in which many users share a channel by continuing to try until the data successfully get through, are used on LANs. WANs use reservation mechanisms that require more processing, but pack the transmission facility as fully as possible. With CSMA/CD protocols the propagation delay across the network must be much less than a packet transmission time, which precludes using these protocols in WANs.

Traditionally, LANs are networks that are owned and installed by a single company or organization. An organization can choose to try new technologies. MANs are less expensive to install than WANs and may not be interconnected. There is more freedom to experiment with new technologies on MANs than on WANs. The expense of installing MANs relative to installing LANs has resulted in far fewer experimental MANs than LANs.

In a LAN the other network users are generally more trusted than the users in a more open environment. Traditionally, MANs, such as CATV networks, and RANs and WANs, such as the telephone network, service an unrelated community of users. The users in these networks do not trust one another as much as the users on a LAN, and greater measures must be taken to protect data.

There are increasing numbers of wide area networks that are owned and controlled by a single organization. Corporate networks and intranets, which use Internet technology within a corporate network, are becoming common. These networks have trust structures and flexibility that is more closely related to LANs than to WANs. The differences between general WANs and intranets are reflected in the applications of the networks and are leading to different implementations.

Many of the economic tradeoffs that are related to the distance spanned by networks change with time. However, the difference in propagation delay can never change. As the size of the network increases, the maximum useful transmission rate that a user can access to transfer a particular size packet decreases. This phenomenon is demonstrated in Fig. 1. The lines in this figure show when the propagation delay and transmission time are equal in LANs, MANs, RANs and WANs. The calculations are performed assuming that the propagation delay in the medium is 80% of the speed of light in free space, which is common for optical fibers. To the right of these lines, the time it takes to get the message from the source to the destination is dominated by the propagation delay rather than the transmission time.

Increasing the transmission rate when operating to the right of the line does not bring a commensurate decrease in the time it takes to deliver a message. For instance, on a 3000 mile WAN the equilibrium point on a 1.5 megabits/s T1 circuit is about 30.3 kbit. For a message of this size, the propagation delay and transmission time are equal. If the user's rate is increased to 45 megabits/s, a T3 circuit, which is 30

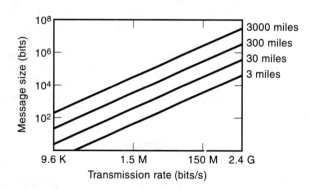

Figure 1. Equal delay lines when the distance between the source and destination is 3, 30, 300, and 3000 miles. Along the line the propagation delay for light and the transmission time of the message are equal.

times faster, the time to transmit the message decreases by a factor of 30, but the propagation time remains the same. The time it takes the message to get to the destination decreases by less than a factor of 2 as the transmission rate increases by a factor of 30. At T3 rates the message delivery time is almost entirely due to the propagation delay, and if the user's rate increases to 155 Mbits/s, an ATM circuit, there is virtually no decrease in the time it takes to receive the message. If the user sends the same message on a MAN and increases the rate from T1 to T3, the time to deliver the message decreases by almost a factor of 30, and increasing the ATM rates decreases the delivery time by another factor of 2. Therefore, ATM rates may be used to obtain faster delivery of this size message in a MAN, but not in a WAN.

THE MAN ANOMALY

Generally, transmission links cost more as the distance increases. At present, high-rate channels are readily available in LANs and WANs, but not in MANs. The use of computers in offices has become ubiquitous and has resulted in most office buildings being wired for high-speed communications. The backbone of the wide area telephone network is shared by a large number of users. Even though only a fraction of the users require high-rate facilities, the number is large enough to warrant providing those facilities between central offices.

Fiber to the curb and other methods to provide high data rates in a MAN exist, but most users do not require these rates. The lines running down a street in a MAN are not shared by as large a number of users as the lines between central offices. As a result, when high-rate channels are installed in an office, the line that spans the final mile or two from the central office to the office building is frequently an expensive, custom installation.

The current networks must be modified to provide high data rates to homes until the demand increases to the point where new facilities are justified. Two possible technologies are digital subscriber loops (DSL) and the CATV network. DSL and ADSL (asymmetric DSL) use adaptive equalizers to transmit between 1.5 and 6.3 Mbits/s over current local loops in the telephone network. ADSL provides higher rates in one direction than in the other.

To date, the only really successful MAN for distributing information to a large number of homes has been the CATV network. CATV networks are mainly used to distribute entertainment video; however, experimental networks are being deployed to deliver data to homes. Standards organizations and working groups are actively considering these networks. The multimedia requirements of the Web may well make CATV technology the correct solution for the next MAN. Several of the early proposals for MANs used CATV networks to deliver point-to-point voice and data services, as well as broadcast TV. Later we describe one of these techniques, which is still one of the most forward-looking CATV solutions.

THE FIBER DISTRIBUTED DATA INTERFACE

FDDI (2) is a token passing loop network that operates at 100 megabytes/s. It is the American National Standards Institute (ANSI) X3T9 standard and was initially proposed as the suc-

cessor to an earlier generation of LANs. FDDI started as a LAN and has been primarily used as a LAN; however, it is capable of transmitting at the rates and spanning the distances required in a MAN. Therefore, it has become common to discuss FDDI in the context of MANs.

Baseband Transmission

FDDI uses a baseband transmission system. Baseband systems transmit symbols, ones and zeros, on the medium rather than modulating the symbols on a carrier, as in a radio network. Baseband systems are simpler to implement than carrier systems; however, the signal does not provide timing and there may be a dc component that is incompatible with some system components. For instance, a natural string of data may have a long sequence of ones or zeros. If the medium stays at the same level for a long period, it is difficult to decide how many ones or zeros were in the string and the dc level of the system will drift toward the value of that symbol. To tailor the signal to have desirable characteristics, the data are mapped into a longer sequence of bits.

A common code for transmitting baseband data on early twisted pair networks is a Manchester code. Each data bit is mapped into a 2 bit sequence, a one is mapped into $+1, -1$ and a zero into $-1, +1$. There is at least one transition per bit, which provides a strong timing signal. There is no dc component in this code. Twisted pairs are connected to receivers and transmitters by transformers to protect the electronics from energy picked up by the wires during lightning storms, and transformers do not pass dc. A framing signal is needed to identify bit boundaries and the beginning of a sequence of bits. Framing signals occur infrequently and are sequences that do not occur in the data. Framing can be obtained by alternately transmitting $-1, -1$ and $+1, +1$ every n bits. With a Manchester code the bit rate on the medium is twice the bit rate from the source; however, this is a very simple coding system to implement.

The FDDI standard uses a rate 4/5 code that maps 4 data bits into 5 transmitted bits. The constraints on codes used for fiber optics are different from those on twisted pairs. Fibers do not act as antenna during lightning storms, are not coupled to electronics through transformers, and can tolerate a dc component. In addition, logic costs have decreased since the early twisted pair networks were designed, so that it is now reasonable to implement more complex bit mappings and to design signal extraction circuits with fewer transitions. In the FDDI code there are 16 possible data patterns per symbol and 32 possible transmitted patterns. The 16 patterns are selected to guarantee at least one transition every 3 bits. Some of the remaining patterns serve control and framing functions. The use of the transmitted patterns is listed in Ref. 3.

Token Passing Protocol

A token is a unique sequence of bits following a framing sequence. When a station on the loop receives the token, it may transmit data after changing the token to a different pattern of bits. When the station has completed its data transmission, it transmits a framing sequence and the token so that the next station has a chance to transmit. When a station does not have the token, it forwards the data it receives on the loops. A station may remove the data destined for itself. When a station has the token and is transmitting, it discards any data it receives on the loop. The discarded data either passed this station prior to the token, and has circulated around the loop or was transmitted by this station after accepting the token. In either case, the data have circulated around the loop at least once and every station has had a chance to receive the data.

In FDDI there is a time for the token to circulate around the loop, a target token rotation time (TTRT). In a simple token passing protocol one station can hold the token for a very long period of time. That station can obtain a disproportionate fraction of the bandwidth and delay other stations for long periods. In the FDDI protocol, station i is entitled to send $S(i)$ bits each time it receives the token. The TTRT is set so that every station can send at least the bits it is entitled to send in a single rotation.

$$\sum_i S(i) + \Delta < TTRT$$

where Δ is the propagation delay around the loop plus the maximum time that is added by the transmission format. The TTRT provides guaranteed bandwidth and delay for each station.

After a station i transmits $S(I)$, it can continue to transmit data if the time since it last forwarded the token is less than TTRT, which indicates that the token is circulating more quickly than required. How long a station holds the token depends on the priority of the data it is transmitting. If the data are high priority, the station can hold the token until all of the surplus time in the token rotation has been used. If the data are lower priority, the station may leave surplus time on the token to give stations with higher-priority data a chance to acquire the surplus time.

The TTRT is set individually for each FDDI system depending on the requirements of the stations on that system. The maximum delay that can occur is 2*TTRT, and the average token rotation time is less than TTRT (4). Therefore, TTRT can be set to provide the guaranteed bits, $S(i)$, and an upper bound on delay for synchronous applications. When best effort traffic, rather than traffic that requires service guarantees, is the dominant traffic type, then TTRT is set to trade access delay and efficiency. As TTRT is made smaller, the time delay until a token arrives decreases. As TTRT is made larger, the amount of data transmitted before passing the token increases, less time is spent passing the token, and the efficiency of the system increases. The increase in efficiency is greatest when there is only one active station.

The operation of the token protocol is depicted in Fig. 2. When a station transfers the token, it sets the local counter TRT(i) to TTRT. The station decrements TRT(i) every unit of time, whether or not it has the token. If TRT(i) reaches $-$TTRT before the token is received, then the token has not visited this station in 2*TTRT and is presumed to be lost.

When station i receives the token it has $W_j(i) \geq 0$ units of data waiting to be transmitted at priority level j, where j_{max} is the highest level and $W_{j_{max}}(i) \leq S(i)$. A station transmits $W_{j_{max}}$ no matter how much time is left on TRT(i). For the priority levels $j < j_{max}$, a station transmits a data unit as long as TRT(i) is greater than the threshold T_j. $T_{j-1} < T_j$ so that we never transmit data at level $j - 1$ if we cannot transmit data at level j. We process priority j_{max} data in the same structure

as the other levels by setting $T_{j_{max}} < -(2*TTRT + S(i))$, so that data at that level are never inhibited by the threshold.

Tokens that circulate the loop and can be used by any station are called unrestricted tokens. The FDDI standard also supports a mode of operation in which all of the capacity that is not being used by synchronous traffic is assigned to a single station, possibly for a large file transfer. To support this mode of operation, a restricted token is defined. A station that enters this mode forwards the restricted token rather than the standard token. Another station that receives a restricted token transmits its synchronous traffic, but does not transmit traffic at level $j < j_{max}$. Therefore, all of the capacity available for asynchronous traffic is given to a single station until that station forwards an unrestricted token.

Isochronous Traffic

FDDI-II adds the ability to send isochronous traffic to FDDI. An isochronous channel provides a regularly occurring slot. The channel is assigned to a specific station on a circuit switched basis.

FDDI-II is implemented by periodically switching the network between a circuit switched and packet switched mode. There is a central station that sends out a framing signal every 125 μs. A portion of the interval following a frame signal is assigned to circuit switched traffic, the isochronous mode, and the remaining time in the frame is assigned to the token passing protocol. An isochronous station that is assigned one byte per frame has a 64 kb/s channel. This channel is adequate for telephone quality voice.

In FDDI-II the stations that implement the token passing protocol must switch between the two modes of operation when they receive framing signals. A station that enters the circuit switched mode must forward whatever bits it receives. When the circuit switched mode ends, the station must resume the token passing protocol where it left off.

Architecture

A single failure of a node or a link disconnects the stations on the loop. In a LAN, loops are made more reliable by using normally closed relays to bypass individual stations that lose power or fail in an obvious manner. It is also common to arrange stations in subloops that are chained together at a central location. Subloops with failures are removed from the network (5), so that the stations on the other subloops can continue to communicate.

Poor reliability prevents loop networks from spanning the distances and connecting the number of users associated with MANs. In FDDI the reliability is improved with a second loop. The second loop does not carry data during normal operation, but is available when failures occur.

Figure 3 shows three components that are used in FDDI networks:

A. Units that implement the token rotation protocol
B. Units that manage the reliability
C. A unit that is responsible for signal timing and framing

Type A units connect user devices to the primary loop. There can be more than one type A unit attached to a type B unit. The type A units do not have to be collocated with the type B

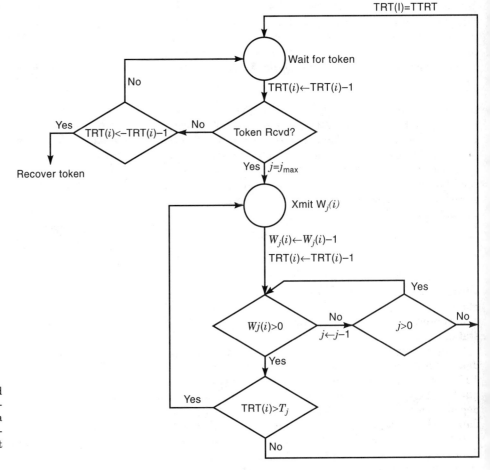

Figure 2. The flow diagram for the timed token rotation protocol in FDDI. The diagonal boxes are decision points where a terminal decides whether or not to transmit when a token is received dependent upon a local timer.

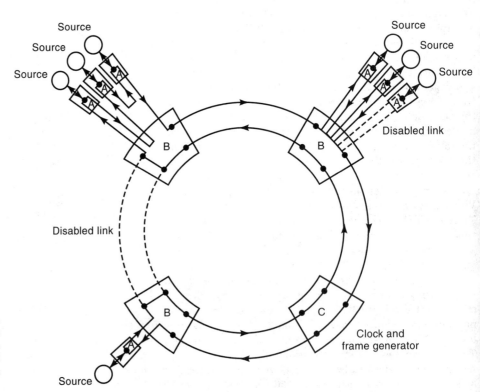

Figure 3. The topology of an FDDI loop. This shows the application of the three types of units that are defined for an FDDI in unidirectional and bidirectional loops. The dashed lines show how the loops are reconfigured after failures.

unit, and they can be daisy-chained together to form a subloop.

Type B units are connected to both loops. These units monitor the signal returning from type A units and bypass type A units that have stopped operating. Type B units also monitor the signal on the two loops and bypass failed loop components.

The secondary loop is used to bypass failed links, or failed type B units. The signal on the secondary loop is transmitted in the opposite direction from the primary loop. Normally type B units patch through the signal that they receive on the secondary loop. When a primary loop failure is detected, by a loss of received signal on that loop, a type B unit replaces the lost signal with the signal it receives from the secondary loop and stops transmitting on the secondary loop. When a type B unit stops receiving the signal on the secondary loop, it replaces that signal with the signal it would have transmitted on the primary loop and stops transmitting on the primary loop. As an example, in Fig. 3 the X signifies a link failure and the — (horizontal bar) indicates the link on which the type B unit has stopped transmitting. The entire secondary loop replaces the single failed link on the primary loop.

The configuration of type A and type B units reflects the way loop networks are installed. Loop networks are installed by running wires or fibers from an office to a wiring cabinet. The type A units are located in offices and the type B units are located in the wiring cabinets. Physically, the topology is a star, but the connection of wires in the wiring cabinet form a logical loop. The star topology makes it possible for stations to be added to a loop, or moved from one loop to another, without rewiring a building.

THE DISTRIBUTED QUEUE, DUAL BUS PROTOCOL

DQDB (6,7) is the IEEE 802.6 standard for MANs. It uses two buses that pass each station. The stations use directional taps to read and write data on a bus without breaking the bus. Directional taps transmit or receive data from one, rather than both, directions at the point of connection, and are common components in both CATV and fiber optic networks.

DQDB transmits information in fixed-size slots and uses the distributed queue protocol to provide fair access to all of the stations. Signals on the two buses propagate in opposite directions. A station selects one bus to communicate with another specific station and uses the other to place reservations for that bus.

The DQDB standard was preceded by two earlier protocols, Express-Net (8) and Fasnet (9), that used directional taps. In the two earlier systems there was a single bus that passed each station twice. On the first pass each station could insert signals and on the second pass a station could receive signals from all other stations. Both of the earlier protocols provided fair access by guaranteeing that every station had the opportunity to transmit one slot before any station could transmit a second slot.

Passive Taps

Passive taps distinguish directional buses from loop networks, which use signal regenerators. In loop networks, there is a point-to-point transmission link between each station.

Each station receives the signal on one link and transmits on the next link. A station can add or remove the signal on the loop. A failure in the electronics in a station breaks the communications path. By contrast, the stations on a directional bus network do not interrupt the signal flow. A passive tap reads the signal as it passes the station, and another tap adds signal to the bus. If a station with passive taps stops working, the rest of the network is not affected.

The protocols that can be used on bus networks are a subset of the protocols that can be used on loop networks, since the stations can add but not remove signals. The inability to remove signals makes it necessary for the the bus to have a break in the communications path, where signals can leave the system.

The protocols for directional buses can be implemented using regenerators rather than passive taps when it is advantageous. Passive taps remove energy from the signal path, and the signal must be restored to its full strength after passing several stations. In addition, by removing information from the transmission medium after it is received, the medium can be reused and to support more communications.

The DQDB standard provides for erasure nodes (10,11), which remove information that has already been received. Erasure nodes are regenerators that read slots and, depending on the location of the destination, regenerate the slot or leave the bus empty.

Architecture

The dual bus in a DQDB network is configured as a bidirectional loop, as shown in Fig. 4. The signal on the outer bus propagates clockwise around the loop, and the signal on the inner bus propagates counterclockwise. The signal does not circulate around the entire loop, but starts at a head-end on each bus and is dropped off the loop before reaching the head-end.

To communicate, a station must know the location of the destination and the head-ends and transmit on the proper bus. For instance, station A transmits on the outer bus to communicate with station B, and station B transmits on the inner bus to communicate with station A.

The dual bus is configured as a loop so that the head-end can be repositioned to form a contiguous bus after a failure occurs. The head-end for each bus is moved so that the signal is inserted immediately after the failure and drops off at the failure. This system continues to operate after any single failure.

The ability to heal failures increases the complexity of stations on the DQDB network. To heal failures, the station that assumes the responsibility of the head-end must be able to generate clock and framing signals. In addition, after a failure each station must determine the new direction of every other station. For instance, after the failure in Fig. 4 is repaired, station A must use the inner bus, rather than the outer bus, to transmit to station B.

The Access Protocol

In a DQDB network transmission time is divided into fixed-size slots that stations acquire to transmit data. A station at the beginning of the bus periodically transmits a sync signal that each station uses to determine slot boundaries. The first bit in the slot is a "busy" bit. It is one when the slot is being

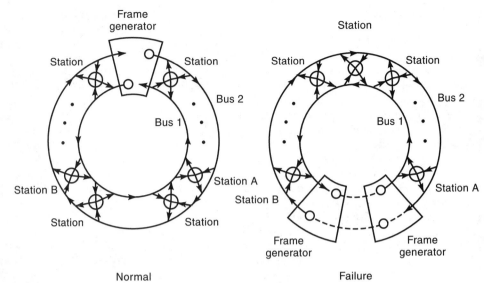

Figure 4. The topology of the DQDB network. This shows dual bus configured as a dual loop in order to survive single failures.

Normal   Failure

used, and zero when it is empty. When a station has data to send, it writes a one into the busy bit. A read tap precedes the write tap at each station. When a station writes a one into the busy bit, it also reads the bit from upstream to determine if it was already set. If the busy bit was zero, the station transmits data. If the busy bit was one, then the slot is full, but there is no harm in over writing the busy bit.

The problem with this type of a protocol is that stations that are closer to the head-end see more empty slots than stations that are farther away. To prevent unfair access to the medium, stations send reservations to the stations that can acquire slots before they have a chance. Reservation requests are transmitted to the upstream stations on the opposite bus as the data. There are two separate reservation systems, one for transmitting data on each bus. In each system, the bus that is used to transmit data is referred to as the data bus, and the other bus is the reservation bus.

Reservations are used to prevent upstream stations from acquiring all of the empty slots when downstream stations are waiting. A station places reservations in a queue with its own requests and services the entrys in this queue whenever an empty slot passes. If the entry is a reservation, the busy bit in the slot is left unchanged so that the slot is available for a downstream station. If entry is from the local source, the busy bit is set and data are inserted in the slot.

A station with a very large message may still dominate the bus by requesting a large number of slots. To prevent this, a station is only allowed to have one outstanding request at a time. When a message requires several slots, one request is transmitted on the reservation bus and one data slot is placed in the local queue. When the data slot is removed from the queue, a second request is transmitted on the upstream bus and the second data slot is placed in the queue. The process continues until the entire message is transmitted. When there is more than one active user, the active users are given one slot at a time and serviced in a round-robin order.

The queue in each reservation system is implemented with two counters that track the number of reservations. Counter C_B counts the reservations that precede this station's data slot, and counter C_A counts the reservations that follow this station's slot. When a station is not actively transmitting, it

increments C_B whenever a reservation is received and decrements it whenever an empty slot passes. When a station has a data slot to transmit, it increments C_A whenever a reservation is received. When C_B is zero and an empty slot appears on the data bus, the station transmits its data slot and then transfers the count from C_A to C_B. If the station has another data slot, transferring the count from C_A to C_B gives downstream stations, which placed reservations while this data slot was queued, a chance to acquire the data bus before this station transmits a second data slot.

In a DQDB network with multiple priority levels for data, there is one reservation bit and two counters for each priority level. When empty slots are received, the counters for the higher priority levels are emptied first.

Protocol Example

Figure 5 shows the operation of five stations in the middle of a DQDB bus as data arrive and are then transmitted. The data bus propagates from left to right and the reservation bus from right to left. The figure shows the value of the busy bit on the data bus and the reservation bit on the reservation bus for each time slot, as the bus passes each station. The values in counters C_A and C_B, and whether or not a station is waiting to transmit data, are shown in between the data and request bus at each station. The operation is simplified and ignores questions of relative timing and propagation delay.

Starting at slot time 1, a single data slot arrives at station 5, followed by two data slots at station 2, and another single data slot at station 3. When the data slots arrive, the network is busy transmitting other data slots, so the data is queued rather than being transmitted immediately. The service order for the messages is station 5, station 2 slot 1, station 3, and station 2 slot 2. The protocol operates as a first in first out (FIFO) queue with a round-robin strategy for multiple slot messages.

In the first three slots, while the reservations arrive, the data bus is busy carrying slots from stations that are closer to the head-end, which were previously queued. Since the stations are upstream, the reservations are not in the queues at these five stations. In the first slot, station 5 inserts a one on the reservation bus that is entered in the queues at stations 1 to 4. In the second slot station 2 places a reservation that is only entered in the queue at station 1. In the third slot station 3 places a reservation that is entered in the queues at stations 1 and 2. After three time slots, station 1 has 3 reservations and station 2 has two reservations, one that will be serviced before its own request and one after. At station 3 there is only one reservation, since station 3 did not receive the reservation from station 2. The fact that station 3 cannot contribute to seeing that station 2 is serviced in its fair turn will not matter, since station 2 has earlier access to the data bus. Stations 4 and 5 have 1 and 0 reservations, respectively.

The reservations are serviced starting in slot 4. Since C_B is greater than zero in stations 1 to 4, these stations let the empty slot pass by, and each removes one reservation by decrementing C_B. Station 5 transmits in slot 4. In slot 5 both stations 2 and 3 are poised to transmit, with C_B at zero and a data slot waiting. Station 2 has the first access at the slot and acquires it, demonstrating that it is unimportant that station 3 did not have an entry for station 2. Station 2 has a second slot that it must transmit, but C_A indicates that one other downstream station has not been serviced. Station 2 moves the count from C_A to C_B before entering its next request in the queue. Slot 6 is acquired by station 3, since C_B is greater than

zero at station 2, and slot 7 is acquired by the second slot from station 2.

Isochronous Traffic

There is no mechanism in the distributed queueing protocol to provide service guarantees on delay or bit rate. This problem has been sidestepped in the standard by creating a separate protocol that shares the bus with the DQDB protocol, as in FDDI. The slots leaving the head-end are grouped into 125 μs frames. In some slots the busy bit is zero and the slots are available for the data transfer protocol. In other slots the busy bit is one so that stations practicing the data transfer protocol do not try to access these slots.

The busy slots that are generated by the head-end occur at regular intervals and contain a unique header so that they are recognized by stations that require guaranteed rates. These slots are partioned into octets (bytes) that can be reserved. A station that reserves a single octet in a frame acquires a guaranteed 64 kbits/s channel with at most a 125 μs delay. This is the same guarantee provided by the telephone system. As in FDDI, the channels are circuit switched and referred to as isochronous channels.

Protocol Unfairness

Soon after the IEEE 802.6 standard was passed, it was noted that because of the distance-bandwidth product of the network, there was a potential for gross unfairness (12,13). In this work we will explain the source of the problem, and a particularly simple solution, called bandwidth balancing (BWB) (14), which eliminated most of the problem and was added to the standard.

The IEEE 802.6 standard is designed to operate at 155 megabits/s, with 53 byte slots, and is compatible with ATM. At these rates, a cell is only about 0.4 miles long. The standard spans up to 30 miles. Therefore, there may be 75 cells simultaneously on the bus. Assume that a station near the head-end of the bus has a long file transfer in progress when a station 50 cells away requests a slot. In the time it takes the request to propagate to the upstream station, that station transmits 50 slots. When the request arrives, the upstream station lets an empty cell pass and then resumes transmission. An additional 50 slots are transmitted before the empty cell arrives at the downstream station. When the empty slot arrives, the downstream station transmits one slot and submits a request for another. The round trip for this request to get to the upstream station and return an empty slot is another 100 slots. As a result, the upstream station obtains 100 times the throughput of the downstream station.

Although the reason is less obvious, a similar imbalance can occur in favor of the downstream station when that station starts transmitting first. While the downstream station is the only source, it transmits in every cell, while placing a reservation in every slot. When the upstream station begins transmitting, there are no reservations in its counter, but there are 50 reservations on the bus. While the upstream source transmits a slot, a reservation is received. Therefore, the upstream station must allow one slot to pass before transmitting its second slot. During the time it takes to service the reservation and the upstream station's next slot, two more reservations arrive. Therefore, the upstream station lets two empty slots pass before transmitting its third slot. The reser-

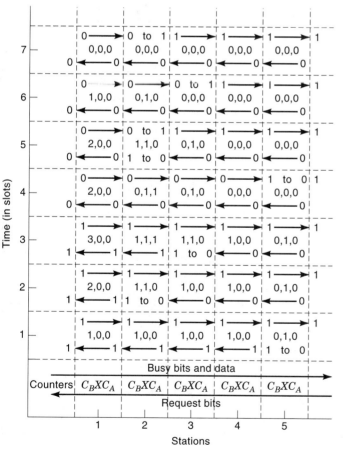

Figure 5. The simplified operation of the DQDB protocol. This shows the operation of the counters at five stations on the bus as data packets arrive at the stations and are transmitted on the bus.

C_B, C_A = Counters at station
X = 1 When station has data to transmit

vation queue at the upstream station continues to build up each time it transmits a slot, and the upstream station takes fewer of the available slots. The imbalance between the upstream and downstream station is sustained after the 50 reservations pass the upstream station because, at the other end of the bus, the downstream station places a reservation on the bus for each of the empty slot that the upstream station releases. The imbalance is not as pronounced as when the upstream station starts first, but it is considerable. The exact imbalance depends on the distance between two stations and the time that they start transmitting relative to one another (14).

Bandwidth Balancing

The bandwidth balancing (BWB) mechanism, which was added to the standard to overcome the unfairness, is based on two observations:

1. Each station can calculate the exact number of slots that are used, whether or not the data physically pass the station.
2. It is possible to exchange information between stations by controlling the fraction of the slots that are not used.

A station sees a busy bit for every slot transmitted by an upstream station and a reservation for every slot transmitted by a downstream station. By summing the busy bits and reservation bits that apply to a data bus and adding the number of slots that the station transmits, the station calculates the total number of slots transmitted on the bus.

Table 1 shows an example of how stations can communicate and achieve a fair utilization by using the average number of unused slots. Two stations, station A and station B, each try to acquire 90% of the unused bandwidth on a channel. Station A starts first and acquires 90% of the total slots. When station B arrives, only 10% of the slots are available, but station B does not know if the slots are being used by a single station taking its allowed maximum share, or many stations transmitting sporadically. Station B uses 90% of the available slots, or 9% of the slots in the system. Station A now has 91% of the slots available. When station A adjusts its rate to 90% of 91% of the slots, it uses 82% of the slots, making 18% of the slots available to station B. Station B adjusts its rate up to 90% of 18%, which causes station A to adjust its rate down, and so on until both stations arrive at a rate of 47.4%. Note that this mode of communications can

only be used when stations try to acquire less than 100% of the slots.

The implementation of BWB in the standard is particularly simple. A station acquires a fraction of the slots available by counting the slots it transmits and placing extraneous reservations in the local reservation queue when the count reaches certain values. In this way, a station lets slots pass that it would have acquired. For instance, if stations agree to take 90% of the slots that are available, they count the slots that they transmit and insert an extra reservation in C_A after every ninth slot that they transmit. As a result, every tenth slot that the station would have taken passes unused.

With BWB, the fraction of the throughput that station i acquires, T_i, is a fraction α of the throughput left behind by the other stations:

$$T_i = \alpha \left\{ 1 - \sum_{j \neq i} T_j \right\}$$

When N stations contend for the channel, they each acquire a throughput:

$$T = \frac{\alpha}{1 + \alpha(N - 1)}$$

The total throughput of the system increases as α approaches one or the number of users sharing the facility becomes large. The disadvantage with letting α approach one is that it takes the network longer to stabilize. We can see from the example in Table 1 that the network converges exponentially toward the stable state. However, as $\alpha \to 1$ the time for convergence goes to infinity. When $\alpha = 1$, BWB is removed from the network and the original DQDB protocol is implemented.

THE MANHATTAN STREET NETWORK

The Manhattan Street Network (MSN) (15) is a network of two-by-two switches. A source and destination may be attached to each switching node. The logical topology of the network resembles the grid of one-way streets and avenues in Manhattan. Fixed size cells are switched between the two inputs and outputs using a strategy called deflection routing.

The MSN resembles a distributed ATM switch. The two-by-two switching elements may be in a large number of wiring centers rather than a central location. However, the same interconnection structure is used for switching elements in different wiring centers as for elements in the same location. Routing is simpler in the structured MSN than in a general network of small switches.

In deflection routing, packets can be forced to take an available path rather than waiting for a specific path. Each packet between a source and destination is routed individually and may take different paths. The packets may arrive at the destination out of order and have to be resequenced.

Deflection routing has several advantages over virtual circuit routing. The overhead associated with establishing and maintaining circuits is eliminated, and the capacity is shared between bursty sources without large buffers and without losing packets because of buffer overflows. Deflection routing is also being used for routing inside some ATM switches (16,17)

Table 1. Convergence of Rates When Two Stations Use 90% of the Slots Available to Them

Station A		Station B	
Measure Bsy + Rqst	Take	Measure Bsy + Rqst	Take
0	0.9 * 1 = 0.9	—	—
0	0.9 * 1 = 0.9	0.9	0.9 * 0.1 = 0.09
0.19	0.9 * 0.91 = 0.82	0.82	0.9 * 0.18 = 0.16
0.16	0.9 * 0.84 = 0.76	0.76	0.9 * 0.24 = 0.22
0.22	0.9 * 0.78 = 0.7	0.7	0.9 * 0.3 = 0.27
0.27	0.9 * 0.73 = 0.66	0.66	0.9 * 0.34 = 0.31
0.31	0.9 * 0.69 = 0.62	0.62	0.9 * 0.38 = 0.34
. . .		. . .	. . .
0.474	0.9 * 0.526 = 0.474	0.474	0.9 * 0.526 = 0.474

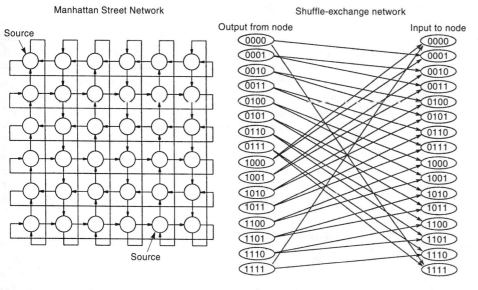

Manhattan Street Network

Shuffle-exchange network

Figure 6. Regular Mesh Topologies. This figure shows the connectivity of the nodes in the Manhattan Street Network and the Shuffle-Exchange Network. The same nodes appear in the left and right columns of the Shuffle–Exchange Network in order to make the regular structure easier to see.

Topology

The MSN is a two-connected topology. In two-connected networks, other than linear structures like the dual bus or dual loop, a path must be chosen at every intermediate node. Two two-connected topologies that have simple routing rules are the MSN and the shuffle-exchange network (15), shown in Fig. 6.

The MSN is a grid of one-way streets and avenues. The directions of the streets alternate. By numbering the streets and avenues properly, it is possible to get to any destination without asking directions or having a complete map.

The grid is logically constructed on the surface of a torus instead of a flat plane. The wraparound links on the torus decrease the distance between the nodes on the edges of the flat plane and eliminate congestion in the corners.

In the shuffle exchange network node i is connected to nodes $2*i$ and $2*i + 1$, modulo the number of nodes in the network. In the figure for the shuffle exchange network each node appears in both the left- and right-hand column in order to make it easier to draw the connections. The two links leaving each 2×2 switching element are shown in the left-hand column and the two links arriving at each switching element are shown in the right-hand column. The shuffle exchange network has a simple routing rule, based on the address of the destination. The simple routing rule is one of the reasons why this structure is used in most ATM switches.

Deflection Routing

Deflection routing is a rule for selecting paths for fixed-size cells at network nodes with the same number of inputs and outputs, as shown in Fig. 7. The cells are aligned at a switching point. If both cells select the same output, and the output buffer is full, one cell is selected at random and forced to take the other link. The cell that takes the alternate path is deflected. Deflection routing can operate without any buffers in the nodes.

Deflection routing gives priority to cells passing through the node. Cells are only accepted from the local source when there are empty cells arriving at the switch. Cells are never dropped due to insufficient buffering because the number of

cells arriving at the node never exceeds the number of cells that can be stored or transmitted. There is an implicit assumption that the source can be controlled. This assumption is common in data networks with variable throughputs, such as the Ethernet.

The MSN is well suited for deflection routing for three reasons:

1. At any node many of the destinations are equidistant on both output links. Cells headed for these destinations have no preference for an output link and do not force other cells to be deflected.

2. When a cell is deflected, only four links are added to the path length. The worse that happens is that the cell must travel around the block.

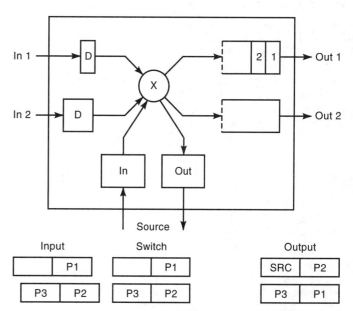

Figure 7. A block diagram of a node in a deflection routing network. The arriving packets are delayed so that they aligned. The packets can be exchanged and placed in their respective output buffers.

3. Deflection routing can guarantee that cells are never lost, but cannot guarantee that they will not be deflected indefinitely and never reach their destination. It has been proven that this type of livelock will never occur in the MSN (18).

Deflection routing is similar to the earlier hot potato routing (19), which operated with variable-size packets, on a general topology, with no buffers. Fixed-size cells, the MSN topology, and two or three cells of buffering converted the earlier routing strategy, which had very low throughputs, to a strategy that can operate at levels exceeding 90% of the throughput that is achieved with infinite buffering.

Reliability

The MSN topology has several paths between each source and destination. The alternate paths can be used to communicate after nodes or links have failed.

There are two simple mechanisms to survive failures in the MSN. Both mechanisms, shown in Fig. 8, are adopted from loop networks. Node failures are bypassed by two normally closed relays that connect the rows and columns through. The missing node in the grid in Fig. 8 has failed. Link failures are detected by a loss of signal, as in loop networks. Nodes respond to the loss of signal by not transmitting on the link at right angles to the link that has stopped. When one link fails, three other links are removed from service and the node at the input to the failed link stops transmitting on it. The dotted link in Fig. 8 has failed and nodes stop transmitting on the dashed links. This link removal procedure works with any number of link failures.

Since the number of input and output links are equal at all of the operating nodes, deflection routing continues to operate without losing cells. In addition, it has been found that the simple routing rules that are designed for complete MSNs continue to work on networks with failures.

COMPARISON OF FDDI, DQDB, AND THE MSN

The FDDI, DQDB, and MSN networks are all two-connected topologies. There are two links entering and leaving each node. Some units in the FDDI network may not be two-connected, but the main part of the network is a dual loop. Two-connectivity distinguishes these networks from the earlier loop and bus networks, used in LANs, which were one-connected. The increased connectivity makes these networks better suited for the increased number of users and longer distances spanned in a MAN.

Like the earlier LANs, DQDB and FDDI are linear topologies. Logically, the nodes are arranged in a one-dimensional line. In linear topologies the throughput per user and reliability are not as high as they are in other two-connected topologies, which have shorter distances and more paths between nodes. There were early proposals (20) to connect the second paths in loop networks to achieve higher throughputs and reliability than the bidirectional loop, but these more complicated topologies lost some of the more desirable characteristics of loops and buses. The MSN and shuffle-exchange topology achieve both goals while approaching the ease of routing and growth that is associated with linear topologies.

Throughput

A disadvantage with linear topologies is that the average throughput that each user can obtain decreases linearly with the number of users. In linear topologies, the average distance between nodes increases linearly with the number of nodes in the network. With uniform traffic, the number of users sharing a link increases linearly with the number of users in the network and the average rate per user decreases accordingly. The protocols used in FDDI and some DQDB networks do not reuse slots. The throughput in these networks is reduced for all traffic distributions.

By contrast, in the MSN the distance between nodes increases as the square root of the number of nodes in the network, and in the shuffle exchange the distance increases as the log of the number of nodes. As a result, the reduction in the throughput per user, which occurs as networks become large, is much less in the MSN and the shuffle-exchange networks than in the FDDI or DQDB network.

In the DQDB network, the penalty for large networks can be reduced by breaking the network into segments and erasing data that have already been received when they reach the end of a segment. This strategy works particularly well when the traffic requirements are nonuniform. When there are communities of users that communicate frequently, if those users are placed on the same segment of the bus, the traffic between them does not propagate outside the segment and interfere with users in other segments. A similar strategy for reusing capacity does not exist for FDDI. In addition, the concept of communities is not as meaningful in FDDI. FDDI operates as a single loop. When user A has a short path to user B, user B

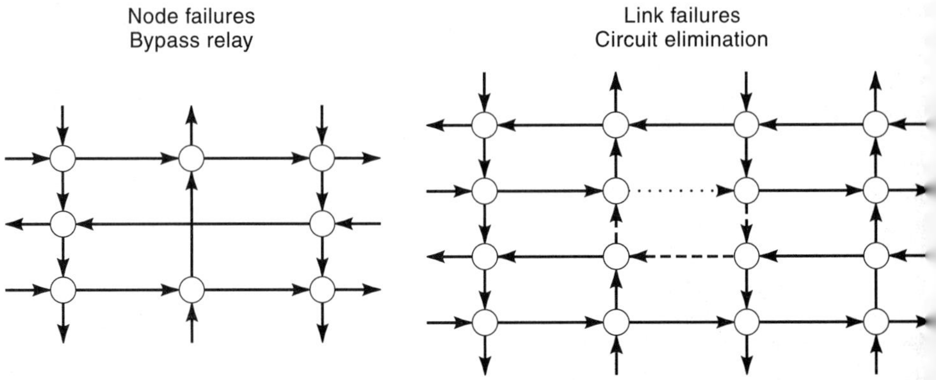

Figure 8. Failure recovery mechanisms in the MSN. Node failures are survived with bypass relays at the failed node. Link failures are survived by eliminating a circuit that includes the failed link, thus preserving the criterion necessary to perform deflection routing at every node.

Node failures
Bypass relay

Link failures
Circuit elimination

must traverse the rest of the loop to get to user A, and therefore has a long path.

In the MSN and shuffle-exchange network special erasure nodes are not needed because the protocol removes cells that reach their destination. In the MSN, communities of users are supported in a very natural way. If nodes that communicate frequently are located within a few blocks, they only traverse the paths in those few blocks and do not affect the rest of the network. The concept of communities does not exist in shuffle-exchange networks.

There is less interference between neighborhoods in the MSN than in DQDB networks. In the MSN, when there is heavy traffic within a neighborhood, communications between other neighborhoods can continue without passing through that neighborhood. In deflection routing, because of the random selection process when there are equal length paths or oversubscribed nodes, cells naturally avoid passing through congested neighborhoods. By contrast, when a community in the middle of the bus becomes congested in a DQDB network, communications between nodes at opposite edges of the bus must still pass through that community.

Reliability

Both DQDB and FDDI have a structure that enables them to survive single failures without losing service. FDDI has a redundant loop that is pressed into service to bypass single failures, and DQDB repositions the head-end of the bus. In both of these networks if two or more failures occur in adjacent components, the mechanism bypasses those components without affecting the operating components. However, in the more likely event that the multiple failures occur in components that are separated from one another, the network is partitioned into islands of nodes that cannot communicate with one another.

Nodes in the MSN are not cut off from one another until at least four failures occur. When four failures have occurred, the likelihood of nodes being disconnected, and the number that are actually disconnected, is small. A quantitative comparison of the MSN, DQDB, and FDDI networks is presented in Ref. 21.

Routing Complexity

An advantage of linear topologies is that routing is relatively simple. All of the data that enter an FDDI system are transmitted on a single path, and there is only one path to select at any intermediate node. In a DQDB network the source must decide which of the two paths leads to the destination, but once the data are in the network there are no choices to make. The MSN and shuffle-exchange networks have simple rules to select a path, but a choice must be made at each node.

In a linear topology everything that is destined for an output link either originates at the source at the node or is received from a single input link. When the data that arrive on the input link have precedence over the data from the source, there is no need to delay or buffer the data from the input link. Deflection routing has the same result in any network where the in-degree equals the out-degree at each node.

Growth

An important consideration in any large network is how easily it can be modified to add or delete users. In the early days of LANs the correspondence between the topology of the network and the physical distribution of users was considered important. A loop network was supposed to be a daisy chain between adjacent offices and a bus network was supposed to pass down a hallway. With experience we realized that it is more difficult to change the wiring between offices as users move than to have all offices connected to a wiring cabinet and change the interconnection in that cabinet. As a result, most LANs are physically a star network, or a cluster of stars, of connections between users and a wiring cabinet. The users are connected together inside the wiring cabinet to form a logical loop or bus or mesh network. The number of wires that must be changed in the wiring cabinet determines how difficult it is to add or delete users from a network.

In a bidirectional loop or bus network, adding or deleting a user is a relatively simple operation. To add a user, the connection between two users is broken and the new user is inserted between them. In the wiring cabinet, two wires are deleted and four are added. In the FDDI loop, there is only one path to the new user. In the DQDB network, every user must determine which bus to use to transmit to the new user.

In the shuffle-exchange network, adding or deleting nodes is very complex. Complete networks are only defined for certain numbers of users. The shuffle–exchange network shown in Fig. 6 is only defined when the number of users is 2^n. When a network is replaced by the next larger network, virtually every wire must be removed and the network reconfigured from scratch.

In a complete MSN, two complete rows or columns must be added to retain the grid structure. There are, however, partial MSNs in which rows or columns do not span the entire grid, and a technique is known for adding one node at a time to a partial MSN to construct eventually a complete MSN (22). With this technique the number of links that must be changed in the wiring cabinet is the same as in the loop or bus network. In addition, an addressing scheme is known that allows new partial rows or columns to be added without changing the addresses of existing nodes or the decisions made in the routing strategy.

Multimedia Traffic

Multimedia traffic is playing an increasingly more important role in networks. Non-real-time video or audio, as is currently delivered by the Web sites on the Internet, is adequately handled by the data modes on any of the MANs that have been described. However, the data mode of operation cannot provide the guarantees on throughput or delay that are required for real-time voice or video applications; nor can it guarantee the sustained throughput that is needed to view movies while they are being received. DQDB and FDDI have an isochronous mode of operation that provides dedicated circuits to support real-time traffic.

The isochronous mode is well integrated into the DQDB protocol. Nodes that only require the data mode do not have to change any protocols or hardware when isochronous traffic is added to the network. The only change that these nodes notice is that some slots seem to enter the network in a busy state. By contrast, when isochronous traffic is added to an FDDI network, every node must be able to perform context switches to move between the data and circuit modes. The MSN does not have an isochronous mode of operation. It is unlikely that dedicated circuits can be added to the mesh structure without constructing a separate circuit switch at every node in the network.

CATV

The current CATV network is an existing MAN that delivers TV programs to a large number of homes. The network is designed for unidirectional delivery of the same signal to a large number of receivers. The channels have a very wide bandwidth relative to telephone channels. In bidirectional CATV systems there are many more channels from the head-end to the home than in the opposite direction.

The growth of the use of the Web in homes has created the need for wide-band channels to homes. The traffic on the Web is predominantly from servers to clients, and most homes are clients rather than servers. Increasingly, the Internet is being used for multicast communications (23) of video or audio programming, in which the same signal is received by more than one receiver. CATV networks are naturally suited for the Web and multicast communications.

The simplest CATV systems are hybrid networks that use the CATV network to send addressed packets to the home, and the telephone network to receive data from the home. The CATV network provides a means of quickly getting a large amount of data to the home. The home terminals share a CATV channel. They receive all of the packets and filter out the packets that are intended others. If the data are considered sensitive, encryption is used to keep sensitive data from being intercepted. Shared CATV channels are particularly useful when high data rates are needed for short periods, as when receiving new pages from Web servers. The telephone channel is a relatively low bandwidth channel that adequately handles the traffic to Web servers. The IEEE 802.14 working group is currently considering standards for hybrid networks.

The application of the Internet is not stationary. Packet telephony or an increase in the number of servers in homes for publishing or small businesses can quickly change the current unidirectional traffic demands. Data MANs that are implemented on CATV networks should be flexible so that they can track changes in the applications. In the early MANs, experimental CATV systems were used for two-way voice and data (24). There was not sufficient demand for data to homes at that time, and work on these networks stopped. The renewed interest in data applications of CATV networks makes it reasonable to reconsider the earlier work. In this section, we describe one such network, the Homenet (25).

Homenet

The Homenet transmission strategy partitions the CATV network into smaller areas, called Homenets. Stations only contend with stations in their own Homenet to gain access to the network. Because of the size of the Homenets, the contending stations satisfy the distance constraints imposed on the CSMA/CD protocol used in Ethernet LANs. Any station on the CATV network can receive the signal on any Homenet. Two stations communicate by transmitting on their own Homenet and receiving on the other stations Homenet.

The Homenet strategy is readily tailored to load imbalances, such as the directional imbalance in Web traffic. There

is no relationship between the bandwidth that is available on a station's transmit and receive channel. Bandwidth is assigned to transmitters by adjusting the number of stations that they contend with for a Homenet. For instance, an Internet service provider (ISP) can be given one or more Homenets, so that traffic from the servers can access the network without contention. Many clients, in homes, may share the same Homenet, since the traffic level from a client to the servers is much lower.

A convenient characteristic of the Homenet strategy, in comparison with hybrid networks, is that the network can be modified easily to match the changes in load imbalances. If the traffic distribution changes and more traffic originates in some homes, those homes can be placed on less heavily populated Homenets to increase the amount that they can transmit.

Access Strategy. The stations in a Homenet access the channel using Movable Slot Time Division Multiplexing (MSTDM) (26), which is a variation on the CSMA/CD (Carrier Sense Multiple Access/Collision Detection) protocol used in Ethernet. MSTDM is implemented by a very small change in the standard Ethernet access unit, and sets up telephone quality voice connections on an Ethernet.

Figure 9 shows the transmission strategy that is used within a Homenet. The CATV taps are directional so that a station only receives signals from downstream and can only transmit upstream. The stations in a Homenet transmit upstream in an assigned frequency band. At a reflection point the upstream frequency band is received and transmitted downstream in a different frequency band. Before transmitting, a station listens to the downstream channel to determine if it is busy; CSMA then transmits on the upstream channel. When a station receives the signal that it transmitted, it knows that it has not collided with any other station, CD, so that any station that receives this channel can receive its data.

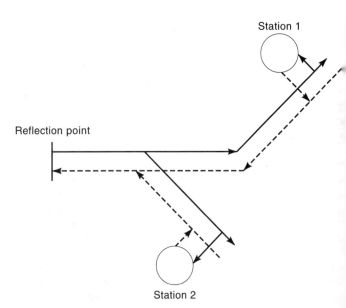

Figure 9. Transmission strategy within a Homenet. Stations transmit upstream on the CATV tree using one channel. At the root of the homenet the signal that is received on this channel is retransmitted downstream on a different channel.

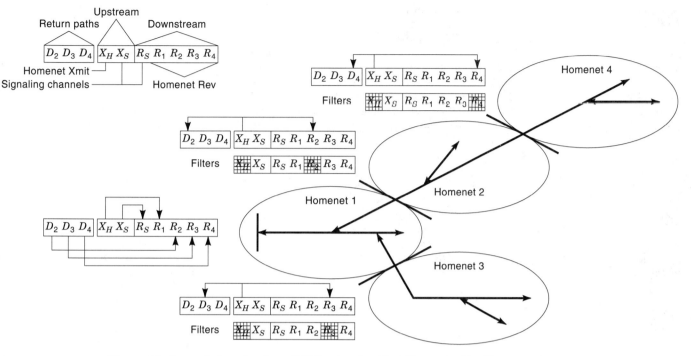

Figure 10. Transmission plan for a CATV network with 4 Homenets. The X's are upstream, transmit channel and the R's are downstream, receive channels. The cross-hatched channels are filtered out at the edge of the homenets.

Interconnection Strategy. Figure 10 shows the boundaries of Homenets, the assignment and filtering of frequencys, and the interconnections for a four-Homenet system. There are many fewer upstream channels in a CATV network than downstream channels. The same upstream data channel is reused in every Homenet by filtering that channel at the boundaries between Homenets. The stations in every Homenet contend for the same upstream channel, but the transmissions from stations in different Homenets do not interfere with one another.

Each Homenet is assigned a unique receive channel. At the Homenet's reflection point the upstream signal on the transmit channel is received and retransmitted on the Homenet's receive channel. The signal is also carried by a point-to-point link to the head-end of the CATV network, where it is again transmitted on the Homenet's receive channel. A Homenet's receive channel is filtered at the reflection point for the Homenet so that the signal from the head-end does not interfere with the signal that is inserted. The reflection point of each Homenet appears to be the root of the entire CATV tree for that Homenet's receive channel, and any station on the CATV network can receive the signal that a station transmits in its Homenet.

To see how the transmission strategy works, consider a station in Homenet 2. The station transmits on the common upstream channel. This transmission may collide with transmissions from other stations in Homenet 2. The transmission will not collide with transmissions from stations in Homenet 4 because the transmit channel is filtered at the boundary between Homenets 2 and 4. At the reflection point for Homenet 2, the signal on the transmit channel is placed in the downstream channel 2. A station that has transmitted detects collisions by listening to the downstream signal on channel 2.

A successfully transmitted packet on Homenet 2 can be received by any station in Homenets 2 and 4. The signal on the transmit channel is also transferred to the head-end of the CATV network, where it is placed in channel 2, so that stations in Homenets 1 and 3 can also receive the signal from the station in Homenet 2. The signal from the head-end does not interfere with the signal inserted in channel 2 at the reflection point because the signal from the head-end is removed at the entry to Homenet 2.

In Fig. 10 there is a second upstream and downstream channel that is used by all of the stations on the CATV network. This is a signaling channel and is used for communications between stations during call setup. A station places a call by transmitting on the upstream signaling channel. At the head-end of the CATV network this signal is placed on the downstream signaling channel. The CSMA/CD contention rule cannot be used on this channel because of the larger distances involved, but the low utilization of the channel makes it reasonable to use a less efficient Aloha (27) contention rule. A station does not listen to the channel before transmitting, but does listen to the receive channel to determine if its signal collided with the signal from another station. If a station can receive its own signal, so can all other stations. When a station is not busy, it listens to the downstream signaling channel and responds to a connect request.

CONCLUSION

The principal reasons for using different technologies in different area networks are economic. As MANs have developed, the economics have changed, and continue to change. The initial application of MANs was interconnecting LANs in a city.

When the economics changed, WAN technology was used for this function.

At present there is a growing demand for wider bandwidth channels to individual homes in a MAN. The bandwidth is not available. In the short term we should expect the bandwidth to be made available using existing networks, either the CATV infrastructure or ADSL technology over the telephone local loop. As the demand continues to grow, additional capacity will be installed. Until the demand reaches a level where individual fibers to a home are justified, channel sharing techniques, such as FDDI, DQDB, and the MSN, are likely to be used to share the added capacity.

Just as the capabilities of WANs have affected MANs, the capabilities of MANs can affect LANs. An interesting question is whether or not the future growth of MANs will lead to the end of LANs. When inexpensive, wide-bandwidth channels are available from every desk to the telephone central office or an ISP, will it still be economical for companies to install and maintain private LANs?

BIBLIOGRAPHY

1. E. W. Zegura, Architectures for ATM switching systems, *IEEE Comm. Mag.*, **31** (2): 28–37, 1993.

2. F. E. Ross, An overview of FDDI: The fiber distributed data interface, *IEEE JSAC*, **7** (7): 1043–1051, 1989.

3. W. Stallings, *Local and Metropolitan Area Networks*, Upper Saddle River, NJ: Prentice Hall, 1997.

4. M. J. Johnson, Proof that timing requirements of the FDDI token ring protocol are satisfied, *IEEE Trans. Comm.*, **COM-35**: 620–625, 1987.

5. H. E. White and N. F. Maxemchuk, An experimental TDM data loop exchange, *Proc. ICC '74*, June 17–19, 1974, Minneapolis, MN, pp. 7A-1–7A-4.

6. R. M. Newman, Z. L. Budrikis, and J. L. Hullett, The QPSX man, *IEEE Comm. Mag.*, **26** (4): 20–28, 1988.

7. R. M. Newman and J. L. Hullett, Distributed queueing: A fast and efficient packet access protocol for QPSX, *Proc. 8th Int. Conf. on Comp. Comm.*, Munich, F.R.G., Sept. 15–19, 1986, published by North-Holland, pp. 294–299.

8. L. Fratta, F. Borgonovo, and F. A. Tobagi, The Express-Net: A local area communication network integrating voice and data, *Proc. Int. Conf. Perf. Data Commun. Syst.*, Paris, Sept. 1981, pp. 77–88.

9. J. O. Limb and C. Flores, Description of Fasnet—A unidirectional local area communications network, *BSTJ*, **61** (7): 1413–1440, 1982.

10. M. Zukerman and P. G. Potter, A protocol for Eraser Node Implementation within the DQDB framework, *Proc. IEEE GLOBECOM '90*, San Diego, CA, Dec. 1990, pp. 1400–1404.

11. M. W. Garrett and S.-Q. Li, A study of slot reuse in dual bus multiple access networks, *IEEE JSAC*, **9** (2): 248–256, 1991.

12. J. W. Wong, Throughput of DQDB networks under heavy load, *EFOC/LAN-89*, Amsterdam, The Netherlands, June 14–16, 1989, pp. 146–151.

13. J. Filipiak, Access protection for fairness in a distributed queue dual bus metropolitan area network, *ICC '89*, Boston, June 1989, pp. 635–639.

14. E. L. Hahne, A. K. Choudhury, and N. F. Maxemchuk, Improving the fairness of distributed-queue dual-bus networks, *INFOCOM '90*, San Francisco, June 5–7, 1990, pp. 175–184.

15. N. F. Maxemchuk, Regular mesh topologies in local and metropolitan area networks, *AT&T Tech. J.*, **64** (7): 1659–1686, 1985.

16. S. Bassi et al., Multistage shuffle networks with shortest path and deflection routing for high performance ATM switching: The open loop shuffleout, *IEEE Trans. Commun.*, **42**: 2881–2889, 1994.

17. A. Krishna and B. Hajek, Performance of shuffle-like switching networks with deflection, *Proceedings INFOCOM '90*, June 1990, pp. 473–480.

18. N. F. Maxemchuk, Problems arising from deflection routing: Live-lock, lockout, congestion and message reassembly, In G. Pujolle (ed.), *High Capacity Local and Metropolitan Area Networks*, Springer-Verlag, 1991, pp. 209–233.

19. P. Baran, On distributed communications networks, *IEEE Trans. Comm. Sys.*, **cs-12**, 1–9, 1964.

20. C. S. Raghavendra and M. Gerla, Optimal loop topologies for distributed systems, *Proc. Data Commun. Symp.*, 1981, pp. 218–223.

21. J. T. Brassil, A. K. Choudhury, and N. F. Maxemchuk, The Manhattan Street Network: A high performance, highly reliable metropolitan area network, *Computer Networks and ISDN Systems*, 1994.

22. N. F. Maxemchuk, Routing in the Manhattan Street Network, *IEEE Trans. Commun.*, **COM-35**: 503–512, 1987.

23. S. Deering, Multicast routing in internetworks and extended LANs, *Proceedings of ACM SIGCOMM '88*, Aug. 1988, Stanford, CA, pp. 55–64.

24. A. I. Karchmer and J. N. Thomas, Computer networking on CATV plants, *IEEE Network Mag.*, pp. 32–40, 1992.

25. N. F. Maxemchuk and A. N. Netravali, Voice and data on a CATV network, *IEEE J. Sel. Areas Commun.*, **SAC-3** (2): 300–311, 1985.

26. N. F. Maxemchuk, A variation on CSMA/CD that yields movable TDM slots in integrated voice/data local networks, *BSTJ*, **61** (7): 1527–1550, 1982.

27. N. Abramson, The Aloha system—Another alternative for computer communications, *Fall Joint Computer Conference, AFIPS Conference Proceedings*, **37**, pp. 281–285, 1970.

N. F. MAXEMCHUK
AT&T Labs–Research

MHDCT.   See HADAMARD TRANSFORMS.

MHD POWER PLANTS.   See MAGNETOHYDRODYNAMIC POWER PLANTS.

MICROBALANCES.   See BALANCES.

MICROCOMPUTER.   See MICROPROCESSORS.

MICROCOMPUTER APPLICATIONS

This article reviews the field of microcomputer applications. We will discuss basic concepts and provide examples of microcomputers used in the design of embedded systems. We begin with an overall discussion of the topic and introduce relevant terminology. Next, we present the fundamental hardware and software building blocks required to construct a microcomputer system. Then, we customize our computer system by interfacing specific devices to create the desired functionality. We conclude with a systems-level approach to microcomputer applications by presenting a few case studies that illustrate the spectrum of applications which employ microcomputers.

OVERVIEW OF MICROCOMPUTER APPLICATIONS

The term *embedded microcomputer system* refers to a device that contains one or more microcomputers inside. To get a better understanding, we break the expression "embedded microcomputer system" into pieces. In this context, the word *embedded* means "hidden inside so we can't see it." A computer is an electronic device with a processor, memory, and input/output ports, as shown in Fig. 1. The processor performs operations (executes software). The processor includes registers (which are high-speed memory), an arithmetic logic unit (ALU) (to execute math functions), a bus interface unit (which communicates with memory and I/O), and a control unit (for making decisions.) Memory is a relatively high-speed storage medium for software and data. Software consists of a sequence of commands (functions) which are usually executed in order. In an embedded system, we use read only memory (ROM) (for storing the software and fixed constant data,) and random access memory (RAM) (for storing temporary information.) The information in the ROM is nonvolatile, meaning the contents are not lost when power is removed. I/O ports allow information to enter via the input ports and exit via the output ports. The software, together with the I/O ports and associated interface circuits, give an embedded computer system its distinctive characteristics.

The term *microcomputer* means a small computer. Small in this context describes its size not its computing power, so a microcomputer can refer to a very wide range of products from the very simple (e.g., the PIC12C08 is an 8-pin DIP microcomputer with 512 by 12 bit ROM, 25 bytes RAM, and 5 I/O pins) to the most powerful Pentium. We typically restrict the term embedded to systems which do not look and behave like a typical computer. Most embedded systems do not have a keyboard, a graphics display, or secondary storage (disk). In the context of this article we will focus on the microcomputers available as single chips, because these devices are more suitable for the embedded microcomputer system.

We can appreciate the wide range of embedded computer applications by observing existing implementations. Examples of embedded microcomputer systems can be divided into categories:

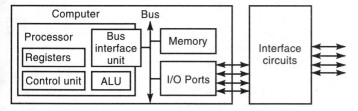

Figure 1. An embedded computer system performs dedicated functions.

1. Consumer
   - Washing machines (computer controls the water and spin cycles)
   - Exercise bikes (computer monitors the workout)
   - TV remotes (computer accepts key touches and sends IR pulses)
   - Clocks and watches (computer maintains the time, alarm, and display)
   - Games and toys (computer entertains the child)
   - Audio/video (computer interacts with the operator and enhances performance)

2. Communication
   - Telephone answering machines (record and play back messages)
   - Cellular phones and pagers (provide a wide range of features)
   - Cordless phones (combine functionality and security)
   - ATM machines (provide both security and banking convenience)

3. Automotive
   - Automatic braking (optimizes stopping on slippery surfaces)
   - Noise cancellation (improves sound quality by removing background noise)
   - Theft-deterrent devices (keyless entry, alarm systems)
   - Electronic ignition (controls spark plugs and fuel injectors)
   - Power windows and seats (remember preferred settings for each driver)
   - Instrumentation (collects and provides the driver with necessary information)

4. Military
   - Smart weapons (don't fire at friendly targets)
   - Missile-guidance systems (direct ordnance at the desired target)
   - Global positioning systems (can tell you where you are on the planet)

5. Industrial
   - Set-back thermostats (adjust day/night thresholds, saving energy)
   - Traffic-control systems (sense car positions and control traffic lights)
   - Robot systems used in industrial applications (computer controls the motors)
   - Bar code readers and writers for inventory control
   - Automatic sprinklers for farming (control the wetness of the soil)

6. Medical
   - Monitors (measure important signals and generate alarms if patient needs help)
     - Apnea (monitor breathing and alarms if baby stops breathing)
     - Cardiac (monitor heart functions)
     - Renal (study kidney functions)
   - Therapeutic devices (deliver treatments and monitor patient response)
     - Drug delivery
     - Cancer treatments (radiation, drugs, heat)
   - Control devices (take over failing body systems providing life-saving functions)

- Pacemakers (help the heart beat regularly)
- Prosthetic devices (increase mobility for the handicapped)
- Dialysis machines (perform functions normally done by the kidney)

MICROCOMPUTER COMPONENTS

Hardware Components

Digital Logic. There are many logic families available to design digital circuits. Each family provides the basic logic functions (and, or, not), but differ in the technology used to implement these functions. This results in a wide range of parameter specifications. Some of the basic parameters of digital devices are listed in Table 1. Because many microcomputers are high-speed CMOS, typical values for this family are given. In general, it is desirable to design digital systems using all components from the same family.

Speed. There are three basic considerations when using digital logic. The first consideration is speed. For simple combinational logic, speed is measured in propagation delay or the time between changes in the input to resulting changes in the output. Another speed parameter to consider is the rise time of the output (time it takes an output signal to go from high to low or from low to high). A related parameter is slew rate (dV/dt on outputs during transitions). For memory devices, speed is measured in read access time, which is how long it takes to retrieve information. For communication devices, we measure speed in bandwidth, which is the rate at which data are transferred.

Power. The second consideration is power. Many embedded systems run under battery power or otherwise have limited power. High-speed CMOS is often used in embedded applications because of its flexible range of power supply voltages and low power supply current specifications. It is important to remember that CMOS devices require additional current during signal transitions (e.g., changes from low to high or from high to low). Therefore, the power supply current requirements will increase with the frequency of the digital signals. A dynamic digital logic system with many signal transitions per second requires more current than a static system with few signal transitions.

Loading. The third consideration is signal loading. In a digital system, where one output is connected to multiple inputs, the sum of the I_{IL} of the inputs must be less than the available I_{OL} of the output which is driving those inputs. Similarly, the sum of the I_{IH}'s must be less than the I_{OH}. Using the above data, we might be tempted to calculate the fanout (I_{OL}/I_{IL}) and claim that one 74HC04 output can drive 4000 74HC04 inputs. In actuality, the input capacitance's of the inputs will combine to reduce the slew rate (dV/dt during transitions). This capacitance load will limit the number of inputs one CMOS output gate can drive. On the other hand, when interfacing digital logic with external devices, these currents (I_{OL}, I_{OH}) are very important. Often in embedded applications we wish to use digital outputs to control non-CMOS devices like relays, solenoids, motors, lights, and analog circuits.

Application-Specific Integrated Circuits. One of the pressures which exist in the microcomputer embedded systems field is the need to implement higher and higher levels of functionality into smaller and smaller amounts of space using less and less power. There are many examples of technology developed according to these principles. Examples include portable computers, satellite communications, aviation devices, military hardware, and cellular phones. Simply using a microcomputer in itself provides significant advantages in this faster-smaller race. Since the embedded system is not just a computer, there must also be mechanical and electrical devices external to the computer. To shrink the size and power required of these external electronics, we can integrate them into a custom IC called an application-specific integrated circuit (ASIC). An ASIC provides a high level of functionality squeezed into a small package. Advances in integrated circuit design allow more and more of these custom circuits (both analog and digital) to be manufactured in the same IC chip as the computer itself. In this way, systems with fewer chips are possible.

Microprocessor. In the last 20 years, the microprocessor has made significant technological advances. The term microprocessor refers to products ranging from the oldest Intel 8080 to the newest Pentium. The processor, or CPU, controls the system by executing instructions. It contains a bus interface unit (BIU), which provides the address, direction (read data from memory into the processor or write data from processor to memory), and timing signals for the computer bus. The registers are very high-speed storage devices for the computer. The program counter (PC) is a register which contains the address of the current instruction which the computer is executing. The stack is a very important data structure used by computers to store temporary information. It is very easy to allocate temporary storage on the stack and deallocate it when done. The stack pointer (SP) is a register which points into RAM specifying the top entry of the stack. The condition code (CC) is a register which contains status flags describing the result of the previous operation and operating mode of the computer. Most computers have data registers which contain information and address registers which contain pointers. The arithmetic logic unit (ALU) performs arithmetic (add, subtract, multiply, divide) and logical (and, or, not, exclusive or, shift) operations. The inputs to the ALU come from registers and/or memory, and the outputs go to registers or memory. The CC register contains status information from the previous ALU operation. Typical CC bits include:

Table 1. Some Typical Parameters of a High-Speed CMOS 74HC04 Not Gate

Parameter	Meaning	Typical 74HC04 Value
V_{cc}	Power supply voltage	2 V to 6 V
I_{cc}	Power supply current	20 μA max (with V_{cc} = 6 V)
t_{pd}	Propagation delay	24 ns max (with V_{cc} = 4.5 V)
V_{IH}	Input high voltage	3.15 V min (with V_{cc} = 4.5 V)
I_{IH}	Input high current	1 μA max (with V_{cc} = 6 V)
V_{IL}	Input low voltage	0.9 V max (with V_{cc} = 4.5 V)
I_{IL}	Input low current	1 μA max (with V_{cc} = 6 V)
V_{OH}	Output high voltage	4.4 V min (with V_{cc} = 4.5 V)
I_{OH}	Output high current	4 mA max (with V_{cc} = 4.5 V)
V_{OL}	Output low voltage	0.33 V max (with V_{cc} = 4.5 V)
I_{OL}	Output low current	4 mA max (with V_{cc} = 4.5 V)
C_I	Input capacitance	10 pF

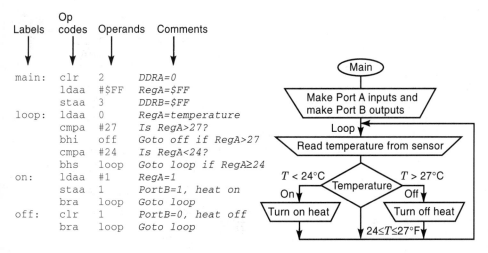

```
         Op
Labels   codes   Operands   Comments

main:    clr     2          DDRA=0
         ldaa    #$FF       RegA=$FF
         staa    3          DDRB=$FF
loop:    ldaa    0          RegA=temperature
         cmpa    #27        Is RegA>27?
         bhi     off        Goto off if RegA>27
         cmpa    #24        Is RegA<24?
         bhs     loop       Goto loop if RegA≥24
on:      ldaa    #1         RegA=1
         staa    1          PortB=1, heat on
         bra     loop       Goto loop
off:     clr     1          PortB=0, heat off
         bra     loop       Goto loop
```

Figure 2. This program implements a bang-bang temperature controller by continuously reading temperature sensor on port A (location 0), comparing the temperature to two thresholds, then writing to the heater connected to port B (location 1) if the temperature is too hot or too cold.

- Z result was zero
- N result was negative (i.e., most significant bit set)
- C carry/borrow or unsigned overflow
- V signed overflow (some computers do not have this bit)

Software is a sequence of commands stored in memory. The control unit (CU) manipulates the hardware modules according to the software that it is executing. The CU contains an instruction register (IR), which holds the current instruction. The BIU contains an effective address register (EAR) which holds the effective address of the current instruction. The computer must fetch both instructions (op codes) and information (data). Both types of access are controlled by the bus interface unit.

When an instruction is executed, the microprocessor often must refer to memory to read and/or write information. Often the I/O ports are implemented as memory locations. For example, on the Motorola 6812, I/O ports A and B exist as locations 0 and 1. Like most microcomputers, the I/O ports can be configured as inputs or outputs. The 6812 Port A and B have direction registers at locations 2 (DDRA) and 3 (DDRB), respectively. The software writes 0's to the direction register to specify the pins as inputs, and 1's to specify them as outputs. When the 6812 software reads from location 0 it gets information from Port A, and when the software writes to location 1, it sends information out Port B. For example, the Motorola 6812 assembly language program, shown in Fig. 2, reads from a sensor which is connected to Port A, if the temperature is above 27 °C, it turns off the heat (by writing 0 to Port B). If the temperature is below 24 °C, then it turns on the heat by writing 1 to Port B.

Microcomputer. The single-chip microcomputer is often used in embedded applications because it requires minimal external components to make the computer run, as shown in Fig. 3. The reset line (MCLR on the PIC or RESET on the 6805) can be controlled by a button, or a power-on-reset circuit.

During the development phases of a project, we often would like the flexibility of accessing components inside the single-chip computer. In addition, during development, we are often unsure of the memory size and I/O capabilities that will be required to complete the design. Both of these factors point to the need for a single-board computer like the one shown in Fig. 4. This board has all of the features of the single-chip computer but laid out in an accessible and expandable manner. For some microcomputer systems, the final product is delivered using a single-board computer. For example, if the production volume is small and the project does not have severe space constraints, then a single-board solution may be cost-effective. Another example of a final product delivered with a single-board occurs when the computer requirements (memory size, number of ports, etc.) exceed the capabilities of any single-chip computer.

Choosing a Microcomputer

The computer engineer is often faced with the task of selecting a microcomputer for the project. Figure 5 presents the relative market share for the top twelve manufacturers of 8 bit microcontrollers. Often the choice is focused only on those devices for which the engineers have hardware and software experience. Because many of the computers overlap in their cost and performance, this is many times the most appropriate approach to product selection. In other words, if a microcomputer that we are familiar with can implement the desired functions for the project, then it is often efficient to bypass that more perfect piece of hardware in favor of a faster development time. On the other hand, sometimes we wish to evaluate all potential candidates. It may be cost-effective to hire or train the engineering personnel so that they are proficient in a wide spectrum of potential computer devices. There are many factors to consider when selecting an embedded microcomputer:

- Labor costs include training, development, and testing
- Material costs include parts and supplies
- Manufacturing costs depend on the number and complexity of the components
- Maintenance costs involve revisions to fix bugs and perform upgrades
- ROM size must be big enough to hold instructions and fixed data for the software
- RAM size must be big enough to hold locals, parameters, and global variables

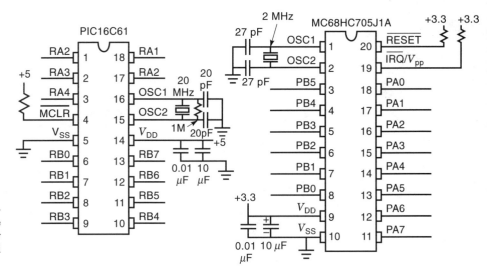

Figure 3. These PIC and 6805 single-chip microcomputer circuits demonstrate that to make the computer run, usually all we need to add is an external crystal for the clock.

- EEPROM to hold nonvolatile fixed constants which are field configurable
- Speed must be fast enough to execute the software in real time
- I/O bandwidth affects how fast the computer can input/output data
- 8, 16, or 32 bit data size should match most of the data to be processed
- Numerical operations, like multiply, divide, signed, floating point
- Special functions, like multiply&accumulate, fuzzy logic, complex numbers
- Enough parallel ports for all the input/output digital signals
- Enough serial ports to interface with other computers or I/O devices

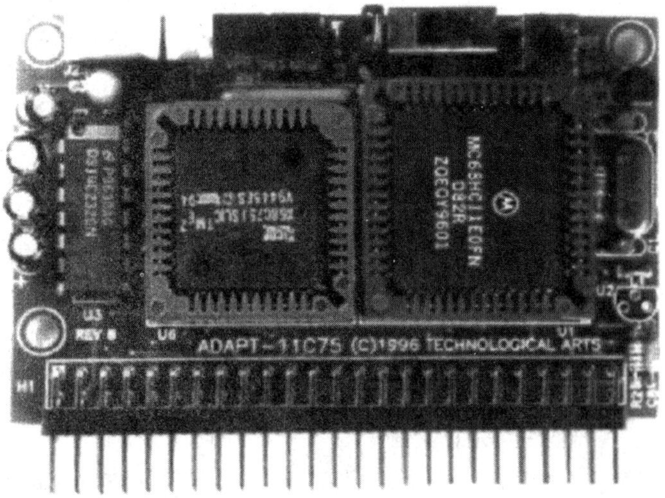

- Timer functions generate signals, measure frequency, measure period
- Pulse width modulation for the output signals in many control applications
- ADC is used to convert analog inputs to digital numbers
- Package size and environmental issues affect many embedded systems
- Second source availability
- Availability of high-level language cross-compilers, simulators, emulators
- Power requirements, because many systems will be battery operated

When considering speed it is best to compare time to execute a benchmark program similar to your specific application, rather than just comparing bus frequency. One of the difficulties is that the microcomputer selection depends on the speed and size of the software, but the software cannot be written without the computer. Given this uncertainty, it is best to select a family of devices with a range of execution speeds and memory configurations. In this way a prototype system with large amounts of memory and peripherals can be purchased for software and hardware development and, once the design is in its final stages, the specific version of the

Figure 4. The Adapt-11C75 board from Technological Arts is a typical example of a single-board microcomputer used to develop embedded applications. It is based on the Motorola MC68HC11 computer, and has 8 K of external EEPROM. Additional I/O ports and memory can be easily added to the 50-pin connector.

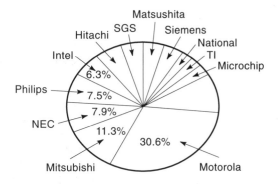

Figure 5. 1995 worldwide market share in dollars for 8 bit microcontrollers (from 1997 Motorola University Symposium, Austin, TX).

computer can be selected, knowing the memory and speed requirements for the project.

Software

Assembly Language. An assembly language program, like the one shown in Fig. 2, has a 1 to 1 mapping with the machine code of the computer. In other words, one line of assembly code maps into a single machine instruction. The label field associates the absolute memory address with a symbolic label. The op code represents the machine instruction to be executed. The operand field identifies the data itself or the memory location for the data needed by the instruction. The comment field is added by the programmer to explain what, how, and why. The comments are not used by the computer during execution, but rather provide a means for one programmer to communicate with another, including oneself at a later time. This style of programming offers the best static efficiency (smallest program size), and best dynamic efficiency (fastest program execution). Another advantage of assembly language programming is the complete freedom to implement any arbitrary decision function or data structure. One is not limited to a finite list of predefined structures as is the case with higher level languages. For example one can write assembly code with multiple entry points (places to begin the function).

High-Level Languages. Although assembly language enforces no restrictions on the programmer, many software developers argue that the limits placed on the programmer by a structured language, in fact, are a good idea. Building program and data structures by combining predefined components makes it easy to implement modular software, which is easier to debug, verify correctness, and modify in the future. Software maintenance is the debug, verify, and modify cycle, and it represents a significant fraction of the effort required to develop products using embedded computers. Therefore, if the use of a high-level language sacrifices some speed and memory performance, but gains in the maintenance costs, most computer engineers will choose reliability and ease of modification over speed and memory efficiency.

Cross-compilers for C, C++, BASIC, and FORTH are available for many single-chip microcomputers, with C being the most popular. The same bang-bang controller presented in Fig. 2 is shown in Fig. 6 implemented this time in C and FORTH.

One of the best approaches to this assembly versus high-level language choice is to implement the prototype in a high-level language, and see if the solution meets the product specifications. If it does, then leave the software in the high-level language because it will be easier to upgrade in the future. If the software is not quite fast enough (or small enough to fit into the available memory), then one might try a better compiler. Another approach is to profile the software execution, which involves collecting timing information on the percentage of time the computer takes executing each module. The profile allows you to identify a small number of modules which, if rewritten in assembly language, will have a big impact on the system performance.

Software Development

Simple Approach. Recent software and hardware technological developments have made significant impacts on the software development for embedded microcomputers. The simplest approach is to use a cross-assembler or cross-compiler to convert source code into the machine code for the target system. The machine code can then be loaded into the target machine. Debugging embedded systems with this simple approach is very difficult for two reasons. First, the embedded system lacks the usual keyboard and display that assist us when we debug regular software. Second, the nature of embedded systems involves the complex and real-time interaction between the hardware and software. These real-time interactions make it impossible to test software with the usual single-stepping and print statements.

Logic Analyzer. A logic analyzer is a multiple channel digital oscilloscope. For the single-board computer which has external memory, the logic analyzer can be placed on the address and data bus to observe program behavior. The logic analyzer records a cycle-by-cycle dump of the address and data bus. With the appropriate personality module, the logic analyzer can convert the address&data information into the corresponding stream of executed assembly language instructions. Unfortunately, the address and data bus singles are not available on most single-chip microcomputers. For these computers, the logic analyzer can still be used to record the digital signals at the microcomputer I/O ports. The advantages of the logic analyzer are:

- Very high bandwidth recording (100 MHz to 1 GHz)
- Many channels (16 to 132 inputs)
- Flexible triggering and clocking mechanisms
- Personality modules, which assist in interpreting the data

```
// bang-bang controller in C
void main(void) { unsigned char T;
   DDRA=0;    // Port A is sensor
   DDRB=0xFF; // Port B is heater
   while(1){
     T=PORTA; // read temperature
     if(T>27)
       PORTB=0;     // too hot
     else
       if(T<24)
         PORTB=1; // too cold
   }
}
```

```
\ bang-bang controller in FORTH
: main ( - )
   DDRA 0 !    \ Port A is senso
   DDRB 0xFF ! \ Port B is heate
   begin
     PORTA @  \ read temperature
     dup 27 > if
       PORTB 0 !      \ too hot
     else
       dup 24 < if
         PORTB 1 !    \ too col
     then
   then drop 0 until ;
```

Figure 6. Bang-bang controllers implemented in C and FORTH, showing that both languages have well-defined modular control structures and make use of local variables on the stack.

Simulation. The next technological advancement which has greatly affected the manner in which embedded systems are developed is simulation. Because of the high cost and long times required to create hardware prototypes, many preliminary feasibility designs are now performed using hardware/software simulations. A simulator is a software application which models the behavior of the hardware/software system. If both the external hardware and software program are simulated together, even although the simulated time is slower than the actual time, the real-time hardware software interactions can be studied.

In-Circuit Emulator. Once the design is committed to hardware, the debugging tasks become more difficult. One simple approach, mentioned earlier, is to use a single-board computer which behaves similarly to the single-chip. Another approach is to use an in-circuit emulator. An in-circuit emulator (ICE) is a complex digital hardware device which emulates (behaves in a similar manner to) the I/O pins of the microcomputer in real time. The emulator is usually connected to a personal computer, so that emulated memory, I/O ports, and registers can be loaded and observed. Figure 7 shows that to use an emulator we first remove the microcomputer chip from the circuit, then attach the emulator pod into the socket where the microcomputer chip used to be.

Background Debug Module. The only disadvantage of the in-circuit emulator is its cost. To provide some of the benefits of this high-priced debugging equipment, some microcomputers have a background debug module (BDM). The BDM hardware exists on the microcomputer chip itself and communicates with the debugging personal computer via a dedicated 2- or 3-wire serial interface. Although not as flexible as an ICE, the BDM can provide the ability to observe software execution in real time, the ability to set breakpoints, the ability to stop the computer, and the ability to read and write registers, I/O ports, and memory.

Segmentation. Segmentation is when you group together in physical memory information which has similar logical properties. Because the embedded system does not load programs off disk when started, segmentation is an extremely important issue for these systems. Typical software segments include global variables, local variables, fixed constants, and machine instructions. For single-chip implementations, we store different types of information into the three types of memory:

1. RAM is volatile and has random and fast access
2. EEPROM is nonvolatile and can be easily erased and reprogrammed
3. ROM is nonvolatile but can be programmed only once

In an embedded application, we usually put structures which must be changed during execution in RAM. Examples include recorded data, parameters passed to subroutines, global and local variables. We place fixed constants in EEPROM because the information remains when the power is removed, but can be reprogrammed at a later time. Examples of fixed constants include translation tables, security codes, calibration data, and configuration parameters. We place machine instructions, interrupt vectors, and the reset vector in ROM because this information is stored once and will not need to be reprogrammed in the future.

Real-Time Systems. The microcomputer typically responds to external events with an appropriate software action. The time between the external event and the software action is defined as the latency. If we can guarantee an upper bound on the latency, we characterize the system as real time, or hard real time. If the system allows one software task to have priority over the others, then we describe it as soft real time. Since most real-time systems utilize interrupts to handle critical events, we can calculate the upper bound on the latency as the sum of three components: (1) maximum time the software executes with interrupts disabled (e.g., other interrupt handlers, critical code); (2) the time for the processor to service the interrupt (saving registers on stack, fetching the interrupt vector); and (3) software delays in the interrupt handler before the appropriate software action is performed. Examples of events which sometimes require real-time processing include:

- New input data ready to when the software reads the new input
- Output device is idle to when the software gives it more data
- An alarm condition occurs until the time the alarm is processed

Sometimes the software must respond to internal events. A large class of real-time systems involve performing software tasks on a fixed and regular rate. For these systems, we employ a periodic interrupt which will generate requests at fixed intervals. The microcomputer clock guarantees that the interrupt request is made exactly on time, but the software response (latency) may occur later. Examples of real-time systems which utilize periodic interrupts include:

- Data acquisition systems, where the software executes at the sampling rate
- Control systems, where the software executes at the controller rate

Figure 7. To use an in-circuit emulator, remove the microcomputer chip from the embedded system, and place the emulator connector into the socket.

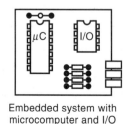

Embedded system with microcomputer and I/O

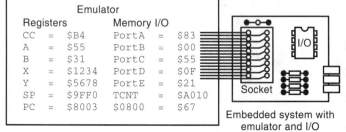

Emulator			
Registers		Memory I/O	
CC	= $B4	PortA	= $83
A	= $55	PortB	= $00
B	= $31	PortC	= $55
X	= $1234	PortD	= $0F
Y	= $5678	PortE	= $21
SP	= $9FF0	TCNT	= $A010
PC	= $8003	$0800	= $67

Embedded system with emulator and I/O

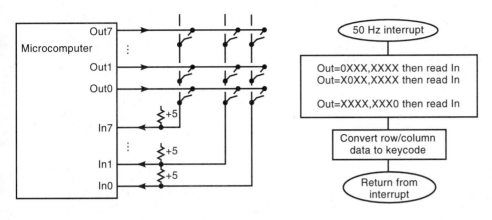

Figure 8. The matrix-scanned keyboard allows many keys to be interfaced using a small number of I/O pins.

- Time-of-day clocks, where the software maintains the date and time

MICROCOMPUTER INTERFACING AND APPLICATIONS

Keyboard Inputs

Individual buttons and switches can be interfaced to a microcomputer input port simply by converting the on/off resistance to a digital logic signal with a pull-up resistor. When many keys are to be interfaced, it is efficient to combine them in a matrix configuration. As shown in Fig. 8, 64 keys can be constructed as an 8 by 8 matrix. To interface the keyboard, we connect the rows to open collector (or open drain) microcomputer outputs, and the columns to microcomputer inputs. Open collector means the output will be low if the software writes a zero to the output port, but will float (high impedance) if the software writes a one. Pull-up resistors on the inputs will guarantee the column signals will be high if no key is touched in the selected row. The software scans the key matrix by driving one row at a time to zero, while the other rows are floating. If there is a key touched in the selected row, then the corresponding column signal will be zero. Most switches will bounce on/off for about 10 ms to 20 ms when touched or released. The software must read the switch position multiple times over a 20 ms time period to guarantee a reliable reading. One simple software method to use a periodic interrupt (with a rate slower than the bounce time) to scan the keyboard. In this way, the software will properly detect single key touches. One disadvantage of the matrix-scanned keyboard is the fact that three keys simultaneously pressed sometimes "looks" like four keys are pressed.

Finite State Machine Controller

To illustrate the concepts of programmable logic and software segmentation, consider the simple traffic light controller illustrated in Fig. 9. The finite state machine (FSM) has two inputs from sensors in the road which identify the presence of cars. There are six outputs, red/yellow/green for the north/south road and red/yellow/green for the east/west road. In this FSM, each state has a 6 bit output value, a time to wait in that state, and four next states, depending on if the input is 00 (no cars), 01 (car on the north/south road), 10 (car on the east/west road), or 11 (cars on both roads).

In the software implementation, presented in Fig. 10, the following three functions are called but not defined: *InitializeHardware();* is called once at the beginning to initialize the hardware. The function *Lights()* outputs a 6 bit value to the lights. The function *Sensor()* returns a 2 bit value from the car sensors. The software implementation for this system exhibits the three classic segments. Since the global variable *Pt* and the local variable *Input* have values which change during execution, they must be defined in RAM. The finite state machine data structure, *fsm[4],* will be defined in EEPROM, and the program *main()* and its subroutines *InitializeHardware(); Lights()* and *Sensor()* will be stored in ROM. You should be able to make minor modifications to the finite state machine (e.g., add/delete states, change input/output values) by changing the linked list data structure in EEPROM without modifying the assembly language controller in ROM.

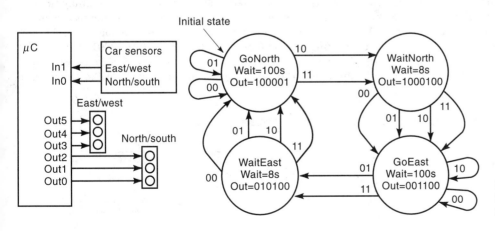

Figure 9. A simple traffic controller has two inputs and six outputs, and is implemented with a finite state machine.

```
struct State
{   unsigned char Out;       /* 6 bit Output                      */
    unsigned char Time;      /* Time to wait in seconds           */
    struct State *Next[4];} /* Next state if input=00,01,10,11 */
typedef struct State StateType;
typedef StateType * StatePtr;
StatePtr Pt;  /* Current State                                RAM */
#define GoNorth &fsm[0]
#define WaitNorth &fsm[1]
#define GoEast &fsm[2]
#define WaitEast &fsm[3]
StateType fsm[4]={                                            /*
{0x21,100,{GoNorth, GoNorth,WaitNorth,WaitNorth}},  /* GoNorth    EEPROM*/
{0x22,  8,{ GoEast,  GoEast,   GoEast,   GoEast}},  /* WaitNorth  EEPROM*/
{0x0C,100,{ GoEast,WaitEast,   GoEast, WaitEast}},  /* GoEast     EEPROM*/
{0x0C,100,{GoNorth, GoNorth,  GoNorth,  GoNorth}}}; /* WaitEast   EEPROM*/
void Main(void){                                             /*          ROM*/
    unsigned char Input;                                     /*          RAM*/
    Pt=GoNorth;              /* Initial State                          ROM*/
    InitializeHardware(); /* Set direction registers, clock           ROM*/
    while(1){                                                /*          ROM*/
        Lights(Pt->Out);     /* Perform output for this state          ROM*/
        Wait(Pt->Time);      /* Time to wait in this state             ROM*/
        Input=Sensor();      /* Input=00 01 10 or 11                   ROM*/
        Pt=Pt->Next[Input];}};                              /*          ROM*/
```

Figure 10. C implementation of the finite state machine and controller.

Two advantages of segmentation are illustrated in this example. First, by placing the machine instructions in ROM, the software will begin execution when power is applied. Second, small modifications/upgrades/options to the finite state machine can be made by reprogramming the EEPROM without throwing the chip away. The RAM contains temporary information which is lost when the power is shut off.

Current-Activated Output Devices

Many external devices used in embedded systems activate with a current, and deactivate when no current is supplied. Examples of such devices are listed in Table 2. The control element describes the effective component through which the activating current is passed. dc motors which are controlled with a pulse width modulated (PWM) signal also fall into this category and are interfaced using circuits identical to the EM relay or solenoid. Figure 11 illustrates the similarities between the interface electronics for these devices.

The diode-based devices (LED, optosensor, optical isolation, solid-state relay) require a current-limiting resistor. The value of the resistor determines the voltage (V_d), current (I_d) operating point. The coil-based devices (EM relay, solenoid, motor) require a snubber diode to eliminate the large back EMF (over 200 V) that develops when the current is turned off. The back EMF is generated when the large dI/dt occurs across the inductance of the coil. The microcomputer output pins do not usually have a large enough I_{OL} to drive these devices directly, so we can use an open collector gate (like the 7405, 7406, 75492, 75451, or NPN transistors) to sink current to ground or use an open emitter gate (like the 75491 or PNP transistors) to source current from the power supply. Darlington switches like the ULN-2061 through ULN-2077 can be configured as either current sinks (open collector) or sources (open emitter). Table 3 provides the output low currents for some typical open collector devices. We need to select a device with an I_{OL} larger than the current required by the control element.

Stepper Motors

The unipolar stepper motor is controlled by passing current through four coils (labeled as B′ B A′ A in Fig. 12) exactly two at a time. There are five or six wires on a unipolar stepper motor. If we connect four open collector drivers to the four coils, the computer outputs the sequence 1010, 1001, 0101, 0110 to spin the motor. The software makes one change (e.g., change from 1001 to 0101) to affect one step. The software repeats the entire sequence over and over at regular time intervals between changes to make the motor spin at a constant rate. Some stepper motors will move on half-steps by outputting the sequence 1010, 1000, 1001, 0001, 0101, 0100, 0110, 0010. Assuming the motor torque is large enough to overcome the mechanical resistance (load on the shaft), each

Table 2. Output Devices Which Can Be Controlled by an Open Collector Driver

Device	Control Element	Definition	Applications
LED	Diode	Emits light	Indicator light, displays
EM relay	Resistor + inductor coil	μC-controlled switch	Lights, heaters, motors, fans
Solid-state relay	Diode	μC-controlled switch	Lights, heaters, motors, fans
Solenoid	Resistor + inductor coil	Short binary movements	Locks, industrial machines

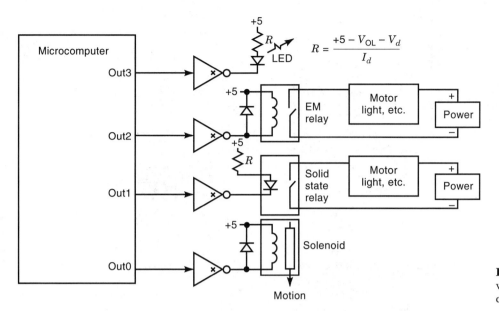

$$R = \frac{+5 - V_{OL} - V_d}{I_d}$$

Figure 11. Many output devices are activated by passing a current through their control elements.

output change causes the motor to step a predefined angle. One of the key parameters which determine whether the motor will slip (a computer change without the shaft moving) is the jerk, which is the derivative of the acceleration (i.e., third derivative of the shaft position). Software algorithms which minimize jerk are less likely to cause a motor slip. If the computer outputs the sequence in the opposite order, the motor spins in the other direction. A bipolar stepper motor has only two coils (and four wires.) Current always passes through both coils, and the computer controls a bipolar stepper by reversing the direction of the currents. If the computer generates the sequence (positive, positive) (negative, positive) (negative, negative) (positive, negative), the motor will spin. A circular linked list data structure is a convenient software implementation which guarantees the proper motor sequence is maintained.

Microcomputer-Based Control System

Basic Principles. A control system, shown in Fig. 13, is a collection of mechanical and electrical devices connected for the purpose of commanding, directing, or regulating a physical plant. The real state variables are the actual properties of the physical plant that are to be controlled. The goal of the sensor and data-acquisition system is to estimate the state variables. Any differences between the estimated state variables and the real state variables will translate directly into controller errors. A closed-loop control system uses the output of the state estimator in a feedback loop to drive the errors to zero. The control system compares these estimated state variables, $X'(t)$, to the desired state variables, $X^*(t)$, in order to decide appropriate action, $U(t)$. The actuator is a transducer which converts the control system commands, $U(t)$, into driving forces, $V(t)$, which are applied the physical plant. The goal of the control system is to drive $X(t)$ to equal $X^*(t)$. If we define the error as the difference between the desired and estimated state variable:

$$E(t) = X^*(t) - X'(t) \qquad (1)$$

then the control system will attempt to drive $E(t)$ to zero. In general control theory, $\boldsymbol{X}(t)$, $\boldsymbol{X}'(t)$, $\boldsymbol{X}^*(t)$, $\boldsymbol{U}(t)$, $\boldsymbol{V}(t)$, and $\boldsymbol{E}(t)$ refer to vectors (multiple parameters), but the example in this article controls only a single parameter. We usually evaluate the effectiveness of a control system by determining three properties: (1) steady-state controller error, (2) transient response, and (3) stability. The steady-state controller error is the average value of $E(t)$. The transient response is how long does the system take to reach 99% of the final output after X^* is changed. A system is stable if steady-state (smooth constant output) is achieved. An unstable system may oscillate.

Pulse Width Modulation. Many embedded systems must generate output pulses with specific pulse widths. The internal microcomputer clock is used to guarantee the timing accuracy of these outputs. Many microcomputers have built-in hardware which facilitate the generation of pulses. One classic example is the pulse-width modulated motor controller. The motor is turned on and off at a fixed frequency (see the *Out* signal in Fig. 14). The value of this frequency is chosen to be too fast for the motor to respond to the individual on/off signals. Rather, the motor responds to the average. The computer controls the power to the motor by varying the pulse width or duty cycle of the wave. The IRF540 MOSFET can sink up to 28 A. To implement Pulse Width Modulation

Table 3. Output Low Voltages and Output Low Currents Illustrate the Spectrum of Interface Devices Capable of Sinking Current

Family	Example	V_{OL}	I_{OL}
Standard TTL	7405	0.4 V	16 mA
Schottky TTL	74S05	0.5 V	20 mA
Low-power Schottky TTL	74LS05	0.5 V	8 mA
High-speed CMOS	74HC05	0.33 V	4 mA
High-voltage output TTL	7406	0.7 V	40 mA
Silicon monolithic IC	75492	0.9 V	250 mA
Silicon monolithic IC	75451 to 75454	0.5 V	300 mA
Darlington switch	ULN-2074	1.4 V	1.25 A
MOSFET	IRF-540	Varies	28 A

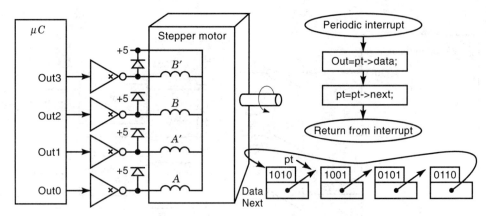

Figure 12. A unipolar stepper motor has four coils, which are activated using open collector drivers.

(PWM), the computer (either with the built-in hardware or the software) uses a clock. The clock is a simple integer counter which is incremented at a regular rate. The Out signal is set high for time T_h then set low for time T_l. Since the frequency of *Out* is to be fixed, $(T_h + T_l)$ remains constant, but the duty cycle $[T_h/(T_h + T_l)]$ is varied. The precision of this PWM system is defined to be the number of distinguishable duty cycles that can be generated. Let n and m be integer numbers representing the number of clock counts the *Out* signal is high and low, respectively. We can express the duty cycle as $n/(n + m)$. Theoretically, the precision should be $n + m$, but practically the value may be limited by the speed of the interface electronics.

Period Measurement. In order to sense the motor speed, a tachometer can be used. The ac amplitude and frequency of the tachometer output both depend on the shaft speed. It is usually more convenient to convert the ac signal into a digital signal (*In* shown in the Fig. 14) and measure the period. Again, many microcomputers have built-in hardware which facilitate the period measurement. To implement period measurement the computer (either with the built-in hardware or the software) uses a clock. Period measurement simply records the time (value of the clock) of two successive rising edges on the input and calculates the time difference. The

period measurement resolution is defined to be the smallest difference in period which can be reliably measured. Theoretically, the period measurement resolution should be the clock period, but practically the value may be limited by noise in the interface electronics.

Control Algorithms
Incremental Control. There are three common approaches to designing the software for the control system. The simplest approach to the closed-loop control system uses incremental control, as shown in Fig. 15. In this motor control example, the actuator command, U, is the duty cycle of the pulse-width modulated system. An incremental control algorithm simply adds or subtracts a constant from U, depending on the sign of the error. To add hysteresis to the incremental controller, we define two thresholds, X_H X_L, at values just above and below the desired speed, X^*. In other words, if $X' < X_L$ (motor is spinning too slow) then U is incremented and if $X' > X_H$ (motor is spinning too fast), then U is decremented. It is important to choose the proper rate at which the incremental control software is executed. If it is executed too many times per second, then the actuator will saturate resulting in a bang-bang system like Fig. 6. If it is not executed often enough, then the system will not respond quickly to changes in the physical plant or changes in X^*.

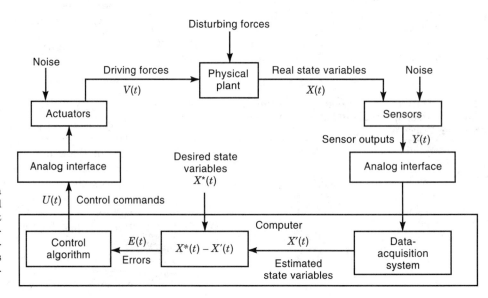

Figure 13. The block diagram of a closed-loop control system implemented with an embedded computer shows that the computer: (1) estimates the state variable, (2) compares it with the desired values, then (3) generates control commands which drive the physical plant to the desired state.

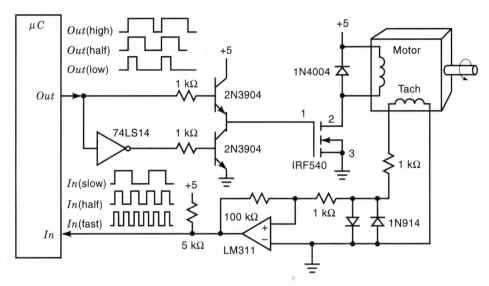

Figure 14. A dc motor can be controlled by varying the duty cycle, and the computer can sense the shaft speed by measuring the frequency or period from the tachometer.

Proportional Integral Derivative (PID) Control. The second approach, called proportional integral derivative, uses linear differential equations. We can write a linear differential equation showing the three components of a PID controller.

$$U(t) = K_P E(t) + K_I \int_0^t E(\tau)\,d\tau + K_D \frac{dE(t)}{dt} \qquad (2)$$

To simplify the PID controller, we break the controller equation into separate proportion, integral and derivative terms, where $P(t)$, $I(t)$ and $D(t)$ are the proportional, integral, and derivative components, respectively. In order to implement the control system with the microcomputer, it is imperative that the digital equations be executed on a regular and peri-

odic rate (every Δt). The relationship between the real time, t, and the discrete time, n, is simply $t = n\,\Delta t$. If the sampling rate varies, then controller errors will occur. The software algorithm begins with $E(n) = X'(n) - X^*$. The proportional term makes the actuator output linearly related to the error. Using a proportional term creates a control system which applies more energy to the plant when the error is large. To implement the proportional term we simply convert the above equation into discrete time.

$$P(n) = K_P \cdot E(n) \qquad (3)$$

The integral term makes the actuator output related to the integral of the error. Using an integral term often will improve the steady-state error of the control system. If a small error accumulates for a long time, this term can get large. Some control systems put upper and lower bounds on this term, called anti-reset-windup, to prevent it from dominating the other terms. The implementation of the integral term requires the use of a discrete integral or sum. If $I(n)$ is the current control output, and $I(n-1)$ is the previous calculation, the integral term is simply

$$I(n) = K_I \cdot \sum_1^n [E(n) \cdot \Delta t] = I(n-1) + K_I \cdot E(n) \cdot \Delta t \qquad (4)$$

The derivative term makes the actuator output related to the derivative of the error. This term is usually combined with either the proportional and/or integral term to improve the transient response of the control system. The proper value of K_D will provide for a quick response to changes in either the set point or loads on the physical plant. An incorrect value may create an overdamped (very slow response) or an underdamped (unstable oscillations) response. There are a couple of ways to implement the discrete time derivative. The simple approach is

$$D(n) = K_D \cdot \frac{E(n) - E(n-1)}{\Delta t} \qquad (5)$$

In practice, this first-order equation is quite susceptible to noise. In most practical control systems, the derivative is cal-

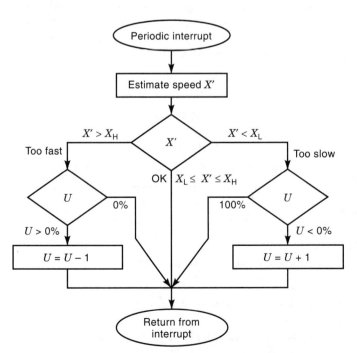

Figure 15. An incremental controller simply adds or subtracts a constant to the actuator control, depending on whether the motor is too fast or too slow.

culated using a higher-order equation like

$$D(n) = K_D \cdot \frac{E(n) + 3E(n-1) - 3E(n-2) - E(n-3)}{6\Delta t} \qquad (6)$$

The PID controller software is also implemented with a periodic interrupt every Δt. The interrupt handler first estimates the state variable, $X'(n)$. Finally, the next actuator output is calculated by combining the three terms.

$$U(n) = P(n) + I(n) + D(n) \qquad (7)$$

Fuzzy Logic Control. The third approach uses fuzzy logic to control the physical plant. Fuzzy logic can be much simpler than PID. It will require less memory and execute faster. When complete knowledge about the physical plant is known, then a good PID controller can be developed. That is, if you can describe the physical plant with a linear system of differential equations, an optimal PID control system can be developed. Since the fuzzy logic control is more robust (still works even if the parameter constants are not optimal), then the fuzzy logic approach can be used when complete knowledge about the plant is not known or can change dynamically. Choosing the proper PID parameters requires knowledge about the plant. The fuzzy logic approach is more intuitive, following more closely to the way a "human" would control the system. If there is no set of differential equations which describe the physical plant, but there exists expert knowledge (human intuition) on how it works, then a fuzzy system can be developed. It is easy to modify an existing fuzzy control system into a new problem. So if the framework exists, rapid prototyping is possible. The approach to fuzzy design can be summarized as

- The physical plant has real state variables (like speed, position, temperature, etc.).
- The data-acquisition system estimates the state variables.
- The preprocessor calculates relevant parameters, called crisp inputs.
- Fuzzification will convert crisp inputs into input fuzzy membership sets.
- The fuzzy rules calculate output fuzzy membership sets.
- Defuzzification will convert output sets into crisp outputs.
- The postprocessor modifies crisp outputs into a more convenient format.
- The actuator system affects the physical plant based on these outputs.

The objective of this example is to design a fuzzy logic microcomputer-based dc motor controller for the above dc motor and tachometer. Our system has two control inputs and one control output. S^* is the desired motor speed, S' is the current estimated motor speed, and U is the duty cycle for the PWM output. In the fuzzy logic approach, we begin by considering how a "human" would control the motor. Assume your hand were on a joystick (or your foot on a gas pedal) and consider how you would adjust the joystick to maintain a constant speed. We select crisp inputs and outputs on which to base our control system. It is logical to look at the error and the

change in speed when developing a control system. Our fuzzy logic system will have two crisp inputs. E is the error in motor speed, and D is the change in motor speed (acceleration).

$$E(n) = S^* - S'(n) \qquad (8)$$
$$D(n) = S'(n) + 3S'(n-1) - 3S'(n-2) - S'(n-3) \qquad (9)$$

Notice that if we perform the calculations of D on periodic intervals, then D will represent the derivative of S', dS'/dt. To control the actuator, we could simply choose a new duty cycle value U as the crisp output. Instead, we will select, ΔU which is the change in U, rather than U itself because it better mimics how a "human" would control it. Again, think about how you control the speed of your car when driving. You do not adjust the gas pedal to a certain position, but rather make small or large changes to its position in order to speed up or slow down. Similarly, when controlling the temperature of the water in the shower, you do not set the hot/cold controls to certain absolute positions. Again you make differential changes to affect the "actuator" in this control system. Our fuzzy logic system will have one crisp output. ΔU is the change in output:

$$U = U + \Delta U \qquad (10)$$

Next we introduce fuzzy membership sets which define the current state of the crisp inputs and outputs. Fuzzy membership sets are variables which have true/false values. The value of a fuzzy membership set ranges from definitely true (255) to definitely false (0). For example, if a fuzzy membership set has a value of 128, you are stating the condition is half way between true and false. For each membership set, it is important to assign a meaning or significance to it. The calculation of the input membership sets is called fuzzification. For this simple fuzzy controller, we will define six membership sets for the crisp inputs:

1. *Slow* will be true if the motor is spinning too slow.
2. *OK* will be true if the motor is spinning at the proper speed.
3. *Fast* will be true if the motor is spinning too fast.
4. *Up* will be true if the motor speed is getting larger.
5. *Constant* will be true if the motor speed is remaining the same.
6. *Down* will be true if the motor speed is getting smaller.

We will define three membership sets for the crisp output:

1. *Decrease* will be true if the motor speed should be decreased.
2. *Same* will be true if the motor speed should remain the same.
3. *Increase* will be true if the motor speed should be increased.

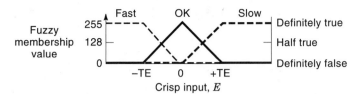

Figure 16. These three fuzzy membership functions convert the speed error into the fuzzy membership variables *Fast, OK,* and *Slow.*

The fuzzy membership sets are usually defined graphically (see Fig. 16), but software must be written to actually calculate each. In this implementation, we will define three adjustable thresholds, TE, TD, and TN. These are software constants and provide some fine-tuning to the control system. If TE is 20 and the error, *E*, is −5, the fuzzy logic will say that *Fast* is 64 (25% true), *OK* is 192 (75% true), and Slow is 0 (definitely false.) If TE is 20 and the error, *E*, is +21, the fuzzy logic will say that *Fast* is 0 (definitely false), *OK* is 0 (definitely false), and *Slow* is 255 (definitely true.) TE is defined to be the error above which we will definitely consider the speed to be too fast. Similarly, if the error is less than −TE, then the speed is definitely too slow.

In this fuzzy system, the input membership sets are continuous piecewise linear functions. Also, for each crisp input value, *Fast, OK, Slow* sum to 255. In general, it is possible for the fuzzy membership sets to be nonlinear or discontinuous, and the membership values do not have to sum to 255. The other three input fuzzy membership sets depend on the crisp input, *D*, as shown in Fig. 17. TD is defined to be the change in speed above which we will definitely consider the speed to be going up. Similarly, if the change in speed is less than −TD, then the speed is definitely going down.

The fuzzy rules specify the relationship between the input fuzzy membership sets and the output fuzzy membership values. It is in these rules that one builds the intuition of the controller. For example, if the error is within reasonable limits and the speed is constant, then the output should not be changed, [see Eq. (11)]. If the error is within reasonable limits and the speed is going up, then the output should be reduced to compensate for the increase in speed. If the motor is spinning too fast and the speed is constant, then the output should be reduced to compensate for the error. If the motor is spinning too fast and the speed is going up, then the output should be reduced to compensate for both the error and the increase in speed. When more than one rule applies to an output membership set, then we can combine the rules using the *or* function.

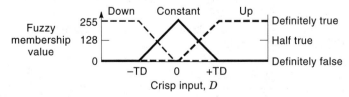

Figure 17. These three fuzzy membership functions convert the acceleration into the fuzzy membership variables *Down, Constant,* and *Up.*

$$Same = OK \text{ and } Constant \quad (11)$$

$$Decrease = (OK \text{ and } Up) \text{ or } (Fast \text{ and } Constant)$$
$$\text{or } (Fast \text{ and } Up) \quad (12)$$

$$Increase = (OK \text{ and } Down) \text{ or } (Slow \text{ and } Constant)$$
$$\text{or } (Slow \text{ and } Down) \quad (13)$$

In fuzzy logic, the *and* operation is performed by taking the minimum and the *or* operation is the maximum. The calculation of the crisp outputs is called defuzzification. The fuzzy membership sets for the output specifies the crisp output, ΔU, as a function of the membership value. For example, if the membership set *Decrease* were true (255) and the other two were false (0), then the change in output should be −TU (where TU is another software constant). If the membership set *Same* were true (255) and the other two were false (0), then the change in output should be 0. If the membership set *Increase* were true (255) and the other two were false (0), then the change in output should be +TU. In general, we calculate the crisp output as the weighted average of the fuzzy membership sets:

$$\Delta U = [Decrease \cdot (-\text{TU}) + Same \cdot 0 + Increase \cdot \text{TU}]/$$
$$(Decrease + Same + Increase) \quad (14)$$

A good C compiler will promote the calculations to 16 bits, and perform the calculation using 16 bit signed math, which will eliminate overflow on intermediate terms. The output, ΔU, will be bounded in between −TU and +TU. The Motorola 6812 has assembly language instructions which greatly enhance the static and dynamic efficiency of a fuzzy logic implementation.

Remote or Distributed Communication

Many embedded systems require the communication of command or data information to other modules at either a near or a remote location. We will begin our discussion with communication with devices within the same room, as presented in Fig. 18. The simplest approach here is to use three or two wires and implement a full duplex (data in both directions at the same time) or half duplex (data in both directions but only in one direction at a time) asynchronous serial channel. Half-duplex is popular because it is less expensive (two wires) and allows the addition of more devices on the channel without change to the existing nodes. If the distances are short, half-duplex can be implemented with simple open collector TTL-level logic. Many microcomputers have open collector modes on their serial ports, which allow a half-duplex network to be created without any external logic (although pull-up resistors are often used). Three factors will limit the implementation of this simple half-duplex network: (1) the number nodes on the network, (2) the distance between nodes; and (3) presence of corrupting noise. In these situations a half-duplex RS485 driver chip like the SP483 made by Sipex or Maxim can be used.

To transmit a byte to the other computers, the software activates the SP483 driver and outputs the frame. Since it is half-duplex the frame is also sent to the receiver of the computer which sent it. This echo can be checked to see if a collision occurred (two devices simultaneously outputting.) If more than two computers exist on the network, we usually

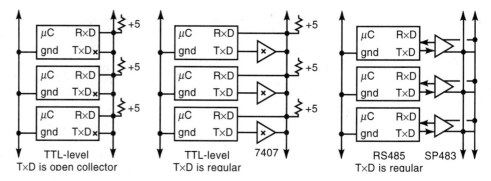

Figure 18. Three possibilities to implement a half-duplex network. The first network requires that the serial transmit output be open collector.

send address information first, so that the proper device receives the data.

Within the same room, infrared (IR) light pulses can be used to send and receive information. This is the technology used in the TV remote control. In order to eliminate background EM radiation from triggering a false communication, the signals are encoded as a series of long and short pulses which resemble bar codes.

There are a number of techniques available for communicating across longer distances. Within the same building the X-10 protocol can be used. The basic idea is to encode the binary stream of data as 120 kHz pulses and mix them onto the standard 120 V 60 Hz ac power line. For each binary one, a 120 kHz pulse is added at the zero crossing of the first half of the 60 Hz wave. A zero is encoded as a 120 kHz pulse in the second half of the 60 Hz wave. Because there are three phases within the ac power system, each pulse is repeated also 2.778 ms, and 5.556 ms after the zero crossing. It is decoded on the receiver end. X-10 has the flexibility of adding or expanding communication capabilities in a building without rewiring. The disadvantage of X-10 is that the bandwidth is fairly low (about 60 bits/s) when compared to other techniques. A typical X-10 message includes a 2 bit start code, a 4 bit house code, and a 5 bit number code requiring 11 power line cycles to transmit. A second technique for longer distances is RF modulation. The information is modulated on the transmitted RF, and demodulated at the receiver. Standard telephone modems and the internet can also be used to establish long-distance networks.

There are two approaches to synchronizing the multiple computers. In a master/slave system, one device is the master, which controls all the other slaves. The master defines the overall parameters which govern the functions of each slave and arbitrates requests for data and resources. This is the simplest approach but may require a high-bandwidth channel and a fast computer for the master. Collisions are unlikely in a master/slave system if the master can control access to the network.

The other approach is distributed communication. In this approach each computer is given certain local responsibilities and certain local resources. Communication across the network is required when data collected in one node must be shared with other nodes. A distributed approach will be successful on large problems which can be divided into multiple tasks that can run almost independently. As the interdependence of the tasks increase, so will the traffic on the network. Collision detection and recovery are required due to the asynchronous nature of the individual nodes.

Data-Acquisition Systems

Before designing a data-acquisition system (DAS) we must have a clear understanding of the system goals. We can classify system as a quantitative DAS, if the specifications can be defined explicitly in terms of desired range, resolution, precision, and frequencies of interest. If the specifications are more loosely defined, we classify it as a qualitative DAS. Examples of qualitative DAS include systems which mimic the human senses where the specifications are defined, using terms like "sounds good," "looks pretty," and "feels right." Other qualitative DAS involve the detection of events. In these systems, the specifications are expressed in terms of specificity and sensitivity. For binary detection systems like the presence/absence of a burglar or the presence/absence of cancer, we define a true positive (TP) when the condition exists (there is a burglar) and the system properly detects it (alarm rings). We define a false positive (FP) when the condition does not exist (there is no burglar) but the system thinks there is (alarm rings). A false negative (FN) occurs when the condition exists (there is a burglar) but the system does not think there is (alarm is silent). Sensitivity, TP/(TP + FN), is the fraction of properly detected events (burglar comes and alarm rings) over the total number of events (number of burglars). It is a measure of how well our system can detect an event. A sensitivity of 1 means you will not be robbed. Specificity, TP/(TP + FP) is the fraction of properly detected events (burglar comes and alarm rings) over the total number of detections (number of alarms.) It is a measure of how much we believe the system is correct when it says it has detected an event. A specificity of 1 means when the alarm rings, the police will arrest a burglar when they get there.

Figure 19 illustrates the basic components of a data-acquisition system. The transducer converts the physical signal into an electrical signal. The amplifier converts the weak transducer electrical signal into the range of the ADC (e.g., −10 V to +10 V). The analog filter removes unwanted frequency components within the signal. The analog filter is required to remove aliasing error caused by the ADC sampling. The analog multiplexer is used to select one signal from many sources. The sample and hold (S/H) is an analog latch used to keep the ADC input voltage constant during the ADC conversion. The clock is used to control the sampling process. Inherent in digital signal processing is the requirement that the ADC be sampled on a fixed time basis. The computer is used to save and process the digital data. A digital filter may be used to amplify or reject certain frequency components of the digitized signal.

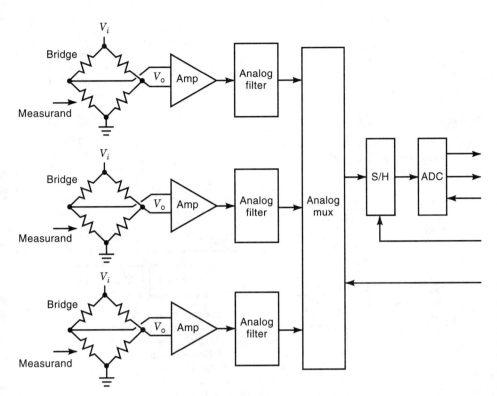

Figure 19. Block diagram of a multiple-channel data-acquisition system, where the transducer and bridge convert the measurands into electrical signals (V_o), the analog circuits amplify and filter the signals, and the multiplexer-ADC system converts the analog signals into digital numbers.

The first decision to make is the ADC precision. Whether we have a qualitative or quantitative DAS, we choose the number of bits in the ADC so as to achieve the desired system specification. For a quantitative DAS this is a simple task because the relationship between the ADC precision and the system measurement precision is obvious. For a qualitative DAS, we often employ experimental trials to evaluate the relationship between ADC bits and system performance.

The next decision is the sampling rate, f_s. The Nyquist Theorem states we can reliably represent, in digital form, a band-limited analog signal if we sample faster than twice the largest frequency that exists in the analog signal. For example, if an analog signal only has frequency components in the 0 Hz to 100 Hz range, then if we sample at a rate above 200 Hz, the entire signal can be reconstructed from the digital samples. One of the reasons for using an analog filter is to guarantee that the signal at the ADC input is band-limited. Violation of the Nyquist Theorem results in aliasing. Aliasing is the distortion of the digital signal which occurs when frequency components above $0.5 f_s$ exist at the ADC input. These high-frequency components are frequency shifted or folded into the 0 to $0.5 f_s$ range.

The purpose of the sample and hold module is to keep the analog input at the ADC fixed during conversion. We can evaluate the need for the S/H by multiplying the maximum slew rate (dV/dt) of the input signal by the time required by the ADC to convert. This product is the change in voltage which occurs during a conversion. If this change is larger than the ADC resolution, then a S/H should be used.

BIBLIOGRAPHY

1. H. M. Dietel and P. J. Dietel, *C++ How to Program,* Englewood Cliffs, NJ: Prentice-Hall, 1994.

2. R. H. Barnett, *The 8951 Family of Microcomputers,* Englewood Cliffs, NJ: Prentice-Hall, 1995.

3. Brodie, *Starting FORTH,* Englewood Cliffs, NJ: Prentice-Hall, 1987.

4. G. J. Lipovski, *Single- and Multiple-Chip Microcomputer Interfacing,* Englewood Cliffs, NJ: Prentice-Hall, 1988.

5. J. B. Peatman, *Design with Microcontrollers,* New York: McGraw-Hill, 1988.

6. J. B. Peatman, *Design with PIC Microcontrollers,* New York: McGraw-Hill, 1998.

7. C. H. Roth, *Fundamentals of Logic Design,* Boston, MA: West, 1992.

8. J. C. Skroder, *Using the M68HC11 Microcontroller,* Upper Saddle River, NJ: Prentice-Hall, 1997.

9. K. L. Short, *Embedded Microprocessor Systems Design,* Upper Saddle River, NJ: Prentice-Hall, 1998.

10. P. Spasov, *Microcontroller Technology The 68HC11,* Upper Saddle River, NJ: Prentice-Hall, 1996.

11. H. S. Stone, *Microcomputer Interfacing,* Reading, MA: Addison-Wesley, 1982.

12. R. J. Tocci, F. J. Abrosio, and L. P. Laskowski, *Microprocessors and Microcomputers,* Upper Saddle River, NJ: Prentice-Hall, 1997.

13. J. W. Valvano, *Real Time Embedded Systems,* Pacific Grove, CA: Brooks/Cole, 1999.

14. J. G. Webster (ed.), *Medical Instrumentation,* Application and Design, 3rd ed., New York: Wiley, 1998.

15. W. C. Wray and J. D. Greenfield, *Using Microprocessors and Microcomputers,* Englewood Cliffs, NJ: Prentice-Hall, 1994.

Reading List

L. Steckler (ed.), *Electronics Now,* Boulder, CO: Gernsback, 1993–current.

S. Ciarcia (ed.), *Circuit Cellar INK—The Computer Applications J.,* Vernon, CT: Circuit Cellar Inc., 1991–current.

JONATHAN W. VALVANO
University of Texas at Austin

MICROCOMPUTERS

A microcomputer is a small, inexpensive computer that contains a single-chip processing unit called a microprocessor. Another name for a microcomputer is personal computer (PC), reflecting the fact that microcomputers are designed to be used by one person at a time. A microcomputer is a general-purpose computer, meaning it can be programmed to perform a wide range of computational tasks, and has low to moderate processing power. Laptop and notebook computers are two types of portable microcomputer.

In contrast to microcomputers, workstations and servers (formerly called minicomputers) are more powerful and more expensive. These systems use more circuitry to implement the central processing unit (CPU) and other subsystems, and have higher capacities for moving and storing information. These midrange computers are designed to support one or two users that have high computational requirements, or several users with moderate requirements. Two still more powerful classes of computers are supercomputers and main-frames. Supercomputers are designed to support the very highest requirements for computational power, while main-frames are designed to support many users simultaneously.

At the other end of the computational spectrum are computing devices with less power than microcomputers. These also use microprocessors to perform computation, but may have limited or no general-purpose programmability and have fewer peripheral devices with which to access and store data. Graphics terminals, network computers, and palmtop computers are examples of such devices.

TYPICAL MICROCOMPUTER SYSTEM

Like all computers, a microcomputer consists of electronic circuitry along with a variety of physical devices used to store, display, and move information from one place to another. Collectively, these components comprise the hardware. Microcomputer hardware consists of three main subsystems: (1) the processor and (2) memory, which comprise the central electronics, and (3) the input/output (I/O) subsystem composed of the peripheral electronics (adapters) and devices (see Fig. 1).

The memory stores information, both programs (code) and data. Programs are sequences of instructions that specify some desired behavior for the computer. In general, that behavior involves moving data into the computer, manipulating it in some fashion, and moving the results back out of the computer. The processor comprises a single integrated circuit (IC), or chip—the microprocessor. It is responsible for fetching instructions out of memory and executing them. The processor instructions specify particular operations to be performed on data held in the processor or in memory. The I/O subsystem provides the means for moving data into and out of the computer, under control of the processor. The processor, memory, and I/O are connected together by busses that pro-

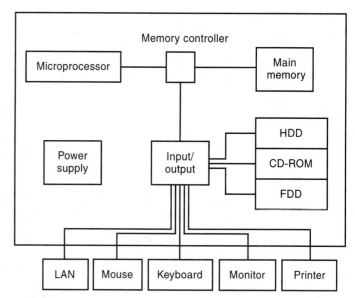

Figure 1. The hardware inside a typical microcomputer system includes the central electronics, the peripheral electronics, some peripheral devices, and the power supply. The central electronics consist of the microprocessor and main memory. The peripheral electronics control the I/O devices. The memory controller is responsible for communications among the subsystems. Devices commonly built into the enclosure include a hard disk drive (HDD), a floppy disk drive (FDD), and a compact disk read-only memory (CD-ROM) player. Other devices connected through external cables include a local area network (LAN), a mouse, a keyboard, a monitor, and a printer.

vide pathways for the movement of data among the subsystems.

Stored on peripheral devices and in electronic memory is information, in the form of instructions and data, which control the behavior of the physical components. This stored information is called software. When it is being moved from one place to another or stored, the term data refers to any kind of information, including instructions. When being contrasted with instructions, the term data refers to the information that is manipulated by the instructions.

Inside the Box

Most of the electronics that implement the various subsystems are contained in a single enclosure. These consist of various components, such as transistors, capacitors, resistors, and integrated circuits, mounted on printed circuit boards (PCB) that are attached to one another by connectors and cables. The core electronics—processor, memory controller, standard peripheral adapters—are typically mounted on a single large PCB called the motherboard. Also mounted on the motherboard are several different kinds of connectors, allowing other components to be installed in the system as needed.

For example, memory chips are mounted on one or both sides of small PCBs called single inline memory modules (SIMM) or dual inline memory modules (DIMM), respectively. These memory modules fit into the memory connectors on the motherboard. DIMMs provide a data width that is twice that of SIMMs. By choosing to install cheaper memory modules with low storage capacity or more expensive memory modules with higher storage capacity, the microcomputer can be configured to fit the needs of the user.

Similarly, while the core electronics on the motherboard provides support for basic input/output devices, peripheral adapter cards can be installed in corresponding connectors on the motherboard. These support additional functionality, including graphics displays, local area networks (LAN), hi-fi sound, and external storage devices.

Also packaged in the enclosure is a power supply that develops the required voltages for the different components, fans to dissipate heat from the ICs, and built-in peripheral devices, such as disk drives.

Instruction Execution

Computers use sequential electronic circuits to perform operations as specified by the software. In sequential circuits, a clock signal defines the beginning of each processing cycle. The state of processing, the information associated with the progress of a computation, can only change from one cycle to the next, not within a given cycle. A faster clock allows more computation to be performed in a given amount of time. Clock speed is measured in hertz (Hz), a unit of measure equal to one cycle per second. A microcomputer driven by a 400 MHz (megahertz) clock can change computational states 400 million times each second—once every 2.5 ns.

The basic unit of computation in any particular microprocessor is an instruction. A given microprocessor has a defined set of instructions that it can execute. The overall behavior of the microcomputer is defined by the sequence of instructions—the program—that it is executing. When a program is being executed, its instructions and data are stored in memory. The microprocessor contains circuitry to fetch each instruction from memory, fetch any data needed by the instruction from memory, execute the instruction, and put the results of executing that instruction back in memory. Different instructions take varying amounts of time (numbers of cycles) to execute. An indicator of the relative processing power of two microprocessors within a family (executing the same instruction set) is how many million instructions per second (MIPS) they can execute. To compare microprocessors from different families, execution time for certain standard applications, called *benchmarks,* can be used.

Data Storage

Computers manipulate digital information. A digital representation of a value is discrete, meaning it can take on only a fixed number of possible values. The basic unit of digital representation is the bit, which can have a value of 0 or 1. Combinations of bits can be used to represent larger values. For example, eight bits can be used to represent a value from 0 to 255. Eight bits is a standard unit for representing information in computers, and so has its own name, the byte. Storage capacities for memories and disk drives are usually expressed in megabytes (Mbyte—millions of bytes) or gigabytes (Gbyte—thousand millions of bytes). Transfer speeds for data are usually expressed in Mbyte/s, or in the case where data is transfered serially, a single bit at a time, the unit bits per second (bit/s—often referred to as the baud rate)—or kilobits per second (kbit/s—thousand bits per second) is used.

As it is being manipulated, the information in the computer, both code (instructions) and data, is stored in a variety of ways until needed. The processor itself stores the information for which it has an immediate need in registers. Main memory stores the code and data for the currently active program(s) so that the processor can access it. Main memory also contains the operating system (see below) along with a variety of data structures (organized collections of data) maintained by the operating system to keep track of the overall state of the microcomputer. Programs and data that are not currently active are stored on various peripheral devices, such as disk drives, CD-ROM, and tapes. When needed, these data are copied from the peripheral device to main memory, and if new data is generated, it may be copied from main memory back to a (writeable) peripheral device.

Different storage devices exhibit different combinations of several characteristics that are important to the proper functioning of the microcomputer. First, a storage system may allow only sequential access or it may be a random-access system. In the first case, the individual storage elements can be read or stored only in a particular order, while in the second case any order is allowed. Second, a storage system may be read-only or it may be writeable (read-write). In the first case, the information that is stored can never be changed, while in the second case, new information can replace the current data. Third, a storage system may be volatile or nonvolatile. Volatile memory loses its information when power is turned off, while nonvolatile memory maintains its information in the absence of power.

The memory subsystem is organized hierarchically, using fast, expensive, low capacity devices that are directly accessible to the processor, and successively slower, less expensive, higher capacity devices as that access becomes more remote. Main memory is composed of several different types of IC memory, including two kinds of random-access memory (RAM)—static (SRAM) and dynamic (DRAM)—as well as read-only memory (ROM).

Flow of Information

To be useful, a computer must manipulate data that come from outside the system itself. Similarly, it must be able to make the results of its computations known to the external world. The various systems that provide the data input and output functions to the central system (processor and main memory) are called peripherals. Each peripheral consists of the device itself, which is generally an electromechanical system that originates input, accepts output or stores data, and an adapter, which is an electronic component that allows the processor to control the device.

A basic user interface to the microcomputer is provided by the keyboard and monitor. The keyboard is an input device that allows the user to type information—commands, programs, text, numeric data—into the microcomputer. The monitor is an output device that displays information generated by the microprocessor in a user-readable form. A basic monitor might display only alphanumeric characters in fixed rows and columns on its screen; more typically information is displayed in a graphical form. The monitor itself may be either a cathode ray tube (CRT), like that in a television set, or, particularly for portable computers, it may be a liquid crystal display (LCD) flat panel. Another input device, the mouse, provides a means of pointing to graphical objects displayed on the monitor screen. In addition to the user interface, a hard disk drive (HDD), floppy disk drive (FDD) and compact disk read-only memory (CD-ROM) player are commonly used to load programs into memory.

Microcomputers can be configured with a variety of other peripherals to provide better functionality or performance. For example, alternative pointing devices include joysticks, trackballs, and tablets. Output devices for producing hardcopies (images on paper) of text and figures include printers and plotters. Input devices for capturing image data include scanners and digital cameras. Input/output devices for connecting to other computers include modems and network controllers. Input/output devices for processing sounds include microphones and speakers as well as musical instrument digital interface (MIDI) and other digital audio devices.

Software

The microprocessor gets work done by following sequences of instructions that specify how to access and manipulate particular sources of data to accomplish desired tasks. The term program is used to describe the set of instructions that performs a particular task. The term code is also often used to distinguish instructions from the data they manipulate.

Two main classes of software are system software and application programs. System software includes the base operating system (OS), device driver code that provides an interface between the OS and each peripheral component, library code that serves as an interface between the OS and an application, and the boot code that is responsible for initializing the computer when it is first turned on.

Application programs are designed to perform some particular task for a user. Applications commonly found on microcomputers include programs for word processing and spreadsheets, publishing and presentation, web browsing and e-mail access, bookkeeping and games, as well as accessory and utility programs. Accessories—applications that remain in memory for ongoing use—include clock, calendar, and calculator programs. Utilities—applications that perform maintenance functions—include antivirus and file-compression tools.

To execute an application program, or any other software, it must first be copied from a peripheral device into main memory. The processor is then given the memory address where the first instruction of the application is stored, and program execution begins. The operating system has the task of loading applications, as directed by the user, and then supporting the execution of each application in a number of ways. The OS manages the allocation and security of microcomputer resources such as processor time, memory space, and access to peripherals. It also provides a set of services that allow applications programs to access these resources through simple procedure calls which hide the complexity of the hardware details from the application. In this way, the OS mediates the execution of the application on the particular microcomputer hardware.

MICROCOMPUTER HARDWARE

The microprocessor is the principal component in a microcomputer. All other components are designed to support the efficient operation of the microprocessor. The peripheral subsystem transfers data to and from outside sources to be used by the processor, while the memory subsystem provides a staging area for those data on their way to and from the processor.

Memory Subsystem

The memory subsystem is used to store programs, and the data that are manipulated by the programs, so that the processor can have direct access to them. At any given time, main memory may hold the operating system, including device drivers, dynamic libraries, and tables of configuration and status data, and one or more application programs, including instructions and several areas used to store program data. Whenever the need for main memory space exceeds the available capacity, some contents are copied to backing store (hard disk) temporarily. This costly operation can be minimized by having a large-capacity memory.

The majority of main memory is implemented as random-access memory (RAM), using a technology called dynamic RAM (DRAM). The advantage of DRAM memory is that each unit of storage, or bit cell, is small, and so a high capacity can be achieved with a few ICs. One disadvantage of the small cell is that the stored information must be periodically (dynamically) rewritten into the cell in order to persist. The other disadvantage of DRAM is that it has a slow access time, meaning that there is a significant delay from the time data are requested to the time they are available.

A faster but less dense RAM technology is static RAM (SRAM). This type of RAM is used to implement a smaller-capacity memory called cache memory. Cache memory is placed between the processor and main memory, and holds a copy of some of the information stored in main memory. Since not all of main memory can be cached, some means is needed to decide what should be stored in the cache at any given time. While there are many answers to this question of how to manage the cache, they are all based on the fact that memory access patterns exhibit locality rather than randomness.

For example, if a particular piece of data has recently been accessed, there is a high probability that it will soon be accessed again. This behavior is referred to as temporal locality. Similarly, if a particular piece of data has recently been accessed, there is a high probability that another piece of data stored at a nearby address will be accessed soon. Thus, memory access patterns are said to exhibit spatial locality. Based on locality, the guiding principle for cache management is to retain in cache a copy of any block of data containing an element that has recently been accessed.

Most microprocessors today have a relatively small cache memory on the chip itself. On-chip caches, called level one (L1) caches, range from 8 kbyte to 64 kbyte while main memories are roughly 1000 times larger. In many cases, an additional level of memory is placed between the on-chip cache and main memory. This level two (L2) cache has characteristics somewhere between those of L1 and main (L3 in this case) memory. L2 is slower to access than L1, but faster than L3, and its size may be 10 to 100 times larger than the L1 cache.

Processor Subsystem

The microprocessor chip contains the electronics for the processor and the L1 cache. For the processor itself, there are two main tasks: fetching instructions and data into (and writing data out of) the processor, and executing those instruc-

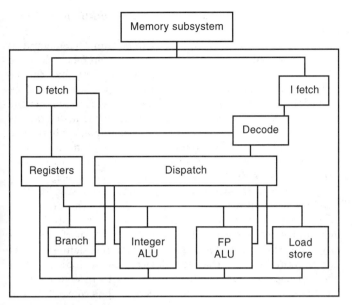

Figure 2. Two main tasks performed by the microprocessor are fetching of instructions and data, and executing instructions. The execution sequence starts with fetching the next instruction from memory (I fetch) then decoding the instruction (Decode) and fetching operand data (D fetch). Once operands are available, the instruction is dispatched (Dispatch) to one of the execution units (Branch, Int ALU, FP ALU, or Load Store). The result is stored back in the registers.

tions. Figure 2 shows the major hardware units in the processor that support these activities.

Registers are storage elements that hold operands and temporary results of computations. These storage elements are referenced in the instructions, and accessed directly by the execution units, providing fast and predictable access times compared to the slower and more variable times required to access memory. In some microprocessor designs operands are required to go through the registers prior to execution, while in other designs operands can be retrieved directly from memory.

Computer memory is organized as an array of storage elements, each of which is identified by its location in the array, referred to as its address. Instructions to be executed in sequence are stored at successive locations in memory. A branch instruction at the end of such a sequence indicates the starting address of the next sequence of instructions to be executed. To execute a given instruction, the following sequence of operations must be performed by the processor: instruction fetch, instruction decode, operand fetch, execution, operand store.

Instruction fetch involves the determination of the next instruction address, followed by a request to memory for that instruction. Once the instruction is in the processor, it can be decoded. Instruction decode involves the determination of the instruction type, and identification of operands (data) that the instruction operates on. The instruction type determines which of the execution units will be used to process the instruction. Prior to executing the instruction, its operands must be made available.

Once all operands are available, the instruction is executed. The execution portion of the processor is generally partitioned into separate computational units corresponding to the different instruction types. For example, fixed-point or integer arithmetic and logical operations would be performed in one unit; floating-point arithmetic, used to manipulate noninteger operands, in another. A separate unit might be used for data movement operations, and another for instructions that change the flow of instructions to another sequence. After the instruction has been executed, the result of any computation is stored back to a register or to memory.

To perform useful work, data from outside the microcomputer must be manipulated. There are two ways for the processor to access peripheral devices. Some microprocessors have instructions specifically for I/O operations. The instruction specifies which I/O device is being accessed and what type of operation is to be performed. If the operation involves a transfer of data, the data are then moved between a register and the I/O device. A second way to perform I/O operations is to allocate a block of memory addresses for use by I/O devices. In this memory-mapped I/O method, each device has one or more control and data registers accessed using an address in the block. A normal instruction that reads or writes memory can then be used to access the I/O device using the appropriate address.

I/O Subsystem

The I/O, or peripheral, subsystem is a heterogeneous system of busses, controllers and devices, whose characteristics vary according to the access times and bandwidth (rate at which data are transferred) requirements associated with different types of input and output devices. A peripheral adapter for each device is attached to the system by a bus, providing a data and command path back to the processor. The adapter controls the operation of the device, and enforces the bus protocol (the rules that define correct use of bus control signals) for transferring data between the device and the central system.

The user interface, consisting of the monitor, keyboard, and mouse, exhibits different bandwidth requirements for input and output. On the high end of the spectrum is the graphics video display. The amount of information displayed at one time on this output device depends on the number of picture elements (pixels) used to fill the screen, and the number of bytes used to represent the color or intensity of each pixel. A 640×480 pixel display that uses three bytes of data per pixel to specify its color requires over 900 kbyte of data per screen image. To support video playback at 30 frames per second, the bandwidth requirement is over 27 Mbyte/s. At the low end of the spectrum are the mouse and keyboard. A user typing on the keyboard at a relatively high rate of 80 words per minute will require a bandwidth of less than 10 byte/s to keep pace with this input device.

Another key I/O device is the hard disk drive. The hard drive is both an input and an output device that stores information on a spinning magnetized disk. It stores programs and data that can be copied into memory for execution and manipulation, and it stores data that have been generated by programs and then copied from memory. Hard drives can also be used to temporarily store data from memory when the memory capacity would otherwise be exceeded. The hard drive is then an extension of the memory hierarchy, and referred to as backing store.

Other I/O devices used to store programs and data include the floppy disk drive, the compact disk read-only memory (CD-ROM) player, and magnetic tape drives. Floppy disks use nearly the same technology as hard disks, except that the magnetized disk is nonrigid. Floppy disks are slower and have less storage capacity than hard disks, but are less expensive and are removable, providing a portable medium for data storage. Removable hard disks having higher capacity and higher cost are also available. CD-ROMs store data optically, rather than magnetically, but otherwise are a form of spinning disk storage. The optical technology prevents the disk from being rewritten with new data, so the CD-ROM player is strictly an input device. Magnetic tape was once used in microcomputers to store programs and data that could then be copied to memory for use. It has a low cost per unit of storage, but is slow and requires sequential access. It is now used for archiving infrequently used data and for hard drive backup—storing a copy of hard drive data in case the hard drive experiences problems.

Some other common peripheral devices found on microcomputers are modems, LAN controllers, sound cards, and printers. A modem uses a serial—one bit at a time—data path to transfer data over phone lines, providing a connection to other computers (and to FAX machines). Data compression is used to achieve a higher bandwidth than phone lines would otherwise support. A LAN controller is used to transfer data over a local area network, such as ethernet or a token ring, also providing a connection to other computers. These network connections allow one microcomputer to share data or to receive services from other computers.

Printers are attached to the microcomputer via a standard interface called a parallel port. Dot-matrix printers represent a low-quality technology that has now been replaced by laser printers and ink-jet printers. Laser printers produce high-quality images at a relatively rapid rate, but are not economical for color printing. Ink-jet printers are slower but support affordable color printing. A scanner is an input device that can be attached to the parallel port to provide a means of capturing image data for display and manipulation.

Various other peripheral adapters are now available to support computationally intensive multimedia processing. Multimedia capabilities include display of 2-D images, 3-D graphics and video clips, along with playback and synthesis of multiple channel music, voice, and other sounds, and two-way audiovisual communication (teleconferencing). Adapters to support these capabilities comprise processing subsystems that may include several megabytes of memory and special purpose processors, such as digital signal processors (DSP) or even an additional microprocessor. The computation performed on-board these adapters is tuned to the requirements of the peripheral task, and reduces the computational load on the microcomputer's CPU.

Busses

A bus provides a pathway for the movement of data from one component to another in the microcomputer. Different types of bus, exhibiting different characteristics, are used to connect the various components, depending on the communication requirements among the components. In general, the choice of what type of bus to use for a particular purpose involves a trade-off between the cost of implementing the bus and its

controller, and the amount of data that can be moved in a given period of time.

The number of bits of data that can be transferred simultaneously is called the bus width. Some common bus widths are 1, 2, 4, and 8 bytes. The number of transfers that can be achieved in a specified period of time depends on the clock rate of the bus. If the minimum time between transfers is one clock cycle, then the maximum bandwidth, or transfer rate, is the bus width times the clock rate. For example, a 2 byte wide bus running at 33 MHz would have a maximum bandwidth of 66 Mbyte/s. Bus overhead due to arbitration or collision resolution will reduce the actual bandwidth of a bus.

Associated with each bus is a protocol, or set of rules, that is followed by all devices that share the bus. The protocol is used to determine which device is currently in control of the bus, and what particular function the bus is performing at any given time. A protocol may, for example, define a set of handshaking signals that the devices use to indicate their need to use the bus, or their readiness to receive data. In most cases, there is one device, called the bus controller, that provides a central mechanism for arbitrating requests from the devices sharing the bus.

Figure 3 shows a typical arrangement of busses connecting the components in a microcomputer. Included in the figure are a processor bus, a memory bus, and several I/O busses. Components that connect one type of bus to another are called bridges. The memory controller controls the processor (or system) bus, the memory bus, and the PCI bus. The ISA bridge controls the ISA bus. The SCSI bus has no central controller. The SCSI protocol defines a fixed priority scheme that devices use to arbitrate bus conflicts among themselves. The memory controller in Fig. 3 is also a PCI bridge, providing a path from the processor and memory busses to the PCI bus. Attached to the PCI bus is a SCSI bridge and an ISA bridge.

If an L2 cache is present in the system, it is attached in one of several ways directly to the microprocessor. The microprocessor with or without L2 is connected to the rest of the

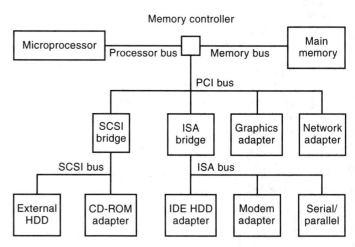

Figure 3. Busses provide pathways for data movement among microcomputer components. The processor bus and memory bus are high-bandwidth busses used within the central system. The PCI bus carries I/O data at moderate rates for devices such as graphics and network adapters. The ISA, SCSI, and IDE busses carry I/O data at a lower rate for slower devices such as the keyboard, modem, printer, and disk drives.

system through the processor bus. This bus carries both the instructions and data needed by the processor to execute applications. To keep the processor busy, the bus must be able to maintain a high rate of data movement. A typical processor bus has a bus width of 8 bytes and a clock speed of 66 MHz (528 Mbyte/s bandwidth), while more recent designs use a 100 MHz bus clock (800 Mbyte/s maximum bandwidth).

Attached to the other side of the processor bus is the memory controller. This component, usually comprising a pair of ICs, is the central arbiter for all data movement within the computer. In addition to the processor, the memory controller connects to the main memory and to the I/O devices. The memory bus connects the memory controller to the system memory. It initiates transfers to read and write memory at a rate that is compatible with the memory type and access time of the particular memory chips being used.

While the processor bus protocol is specific to a particular microprocessor family, it is desirable to define standard I/O busses so that peripheral adapters can be designed to work with any microprocessor. Different I/O device characteristics call for different bus protocols, and so several different bus standards have become generally accepted.

The peripheral component interconnect (PCI) bus is used to connect the central components (processor and memory) to peripherals that have relatively high bandwidth requirements. For example, a graphics adapter would be attached to the PCI bus, as might an adapter for a LAN connection. Connectors on the motherboard allow PCI-compliant adapters to be attached to the PCI bus, to improve the functionality or performance of the microcomputer. Bridges to slower busses are often connected to the PCI bus as well. The standard PCI bus width is 4 bytes, and the clock speed is 33 MHz, so the maximum bandwidth is 132 Mbyte/s.

The Industry Standard Architecture (ISA) bus protocol is older than PCI and supports a lower bandwidth. However, it is easier and cheaper to build an ISA-based adapter, so the ISA bus remains popular for use with peripherals that have only moderate bandwidth requirements. For example, adapters used for the keyboard and mouse, modems, and printers would all be attached to the ISA bus. The ISA bus width is 2 bytes, and the clock speed is 8.33 MHz, but ISA can only transfer data once every 2 clock cycles, yielding a maximum bandwidth of 8.33 Mbyte/s.

Two busses commonly used to connect the various disk drive peripherals to the system are the integrated device electronics (IDE) bus and the Small Computer System Interface (SCSI) bus. IDE provides a relatively cheap interface to hard drives, CD-ROMs, and floppy drives that are contained within the system enclosure. IDE has a maximum bandwidth of 5.5 Mbyte/s. SCSI is more expensive to implement, but it is faster and allows external as well as internal disk drives and other peripheral devices to be attached to the bus. Maximum bandwidth for a SCSI bus is typically 10 Mbyte/s or 20 Mbyte/s, though a wider range of protocols exist.

MICROCOMPUTER SOFTWARE

The information that controls the behavior of a computer is called software. It consists of both instructions and the data used by those instructions for decision-making. Software is often categorized as either an application program or system software. Applications are designed to be run by users to accomplish some task. System software, in particular the operating system (OS), is designed to supervise the execution of applications, and to provide services for those applications. Some programs, such as programming language translators—compilers, assemblers, interpreters—share characteristics of both application and system code.

Application Programs

Microcomputers are most often used by a single user in an interactive mode. Many applications have been developed for microcomputers specifically aimed at this interactive style of computation. For example, what-you-see-is-what-you-get (WYSIWYG) word processors format text as it is input rather than through a postprocessing step. Spreadsheet programs calculate tabular data on-the-fly, providing immediate feedback for testing alternative hypotheses or investigating how a change in one parameter affects the values of other parameters. Image-processing programs allow interactive analysis and enhancement of image data. Media-editing applications support unlimited experimentation with cuts, joins, and special effects to obtain suitable sequences of audio and video streams. Games, educational applications, and desktop publishing programs are also designed around the interactive aspect of microcomputer use. Even applications development itself is well supported through integrated development environments in which program editors, compilers, and debuggers are combined to streamline program development. Of course, noninteractive applications, such as scientific (numeric) programs and data-processing programs—bookkeeping, inventory, database—are also available for microcomputers.

To run an application, it must have space allocated for it in memory for both the instructions and the data that it will use. The program and data are then loaded into memory and linked to (supplied with the actual memory location of) any dynamic libraries that it calls. The processor then branches to the first instruction in the application, and it begins to execute. While executing, if data or instructions are referenced that are not currently in memory, they must be moved into memory. If an application needs data from a hard disk, or prints a message to the screen, or checks to see if a key on the keyboard has been pressed, the corresponding I/O operation must be performed. All of these functions—managing memory space, loading applications, controlling I/O operations, among others—are performed by the processor executing instruction sequences that are part of the operating system (OS).

Operating System

There are several major subsystems in an operating system, including the process scheduler, various resource managers (file system, I/O, memory), and the program loader. An application gains access to resources managed by the OS through calls to dynamic libraries. The OS, in turn, uses device drivers to provide control functions for specific I/O devices. In addition to supporting applications by providing common functions that would otherwise have to be replicated in every application, these OS modules provide security to the system and its users. This is accomplished through the use of certain instructions and certain data areas in memory that can only be accessed by the operating system.

Process Scheduler. The process scheduler determines which of perhaps several available instruction streams, called runnable processes, should be executed next. Early microcomputer operating systems were single-tasking, meaning that there was only one runnable process at any given time. This process was either a command shell in the operating system waiting for user input, or an application that the user chose to run.

More recent operating systems allow nonpreemptive, or cooperative, multitasking. This means that multiple processes may be runnable at any given time, but once the scheduler chooses one to execute, that process executes until it completes. The operating system has no mechanism to stop it.

As with microcomputer hardware, operating systems for microcomputers have evolved and grown more complex, and have inherited functionality from main frames and minicomputers. The most recently developed microcomputer operating systems support preemptive multitasking. This means that multiple processes may be runnable, and that once a process starts to run, it may be suspended by the operating system at any time, to allow another process to run. This capability is particularly important for multiuser systems, where it provides time-sharing of the processor in such a way that each user has the impression that their application is progressing at a steady rate. However, it is also important in a single-user microcomputer, both to support particular styles of programming (multithreading), and to allow efficient and convenient background execution (e.g., spooling), at the same time that one or more interactive applications are running.

Memory Manager. Main memory is physically organized as a one-dimensional array of storage elements, each identified by its order in the array, called its address. All of the information used by the processor to do work, including both instructions and data, must be stored in main memory in order to be accessible to the processor. The memory manager must partition main memory so that each of the different software components that require this resource at any given time have the needed space. Among the different partitions required are those for base OS code and data, for applications code and data, and for dynamic libraries and device drivers.

Today's microprocessors provide hardware support for memory managers to implement a virtual memory. The idea is that the memory manager can behave as if it had a very large memory to work with, and each application has its own memory distinct from that of other applications. This simplifies the memory-management task. However, more space may be allocated in this very large virtual memory than is actually available in the physical memory. The virtual memory system, a combination of microprocessor hardware and OS code, solves this problem by moving information as needed between main memory and backing store. This gives the appearance of having a very large main memory.

Dynamic Libraries. Application programs request operating system services by calling library routines. Each of the services has associated with it an application programming interface (API), which defines the format the application must use to interact with the service. The API provides a level of abstraction between the application and the library. This allows the details of the library software or the hardware involved in the service to change, while the application software remains unchanged.

A library is simply a collection of software functions commonly used by applications. Dynamic libraries, also called shared libraries, are loaded into memory once and retained, so that any application that needs them can access them. Such a library function is dynamically linked to an application that references it when the application is loaded. This dynamic linking reduces the size of the application, and allows the library routine to change without a corresponding change to the application.

Device Drivers. Among the services that an operating system provides to an application program is I/O processing. When an application specifies that a particular data stream is to be written to the display, or that a new file should be created on the hard disk, or the next keystroke should be read in, operating system code is executed to perform the requested function.

The request from the application is abstract, in the sense that it is made independent of which particular device or even class of device will be involved in satisfying the request. The I/O manager has knowledge of different classes of devices, but does not have specific information on how to control every possible I/O device that might be attached to the microcomputer.

The device driver is the piece of code that does have device specific information. When a particular device is installed, the corresponding device driver software is installed as well. When the I/O manager gets a request to perform a particular function on a particular type of device, it passes the request to the appropriate device driver, which turns the request into the correct control sequence for that device.

Booting the Computer. RAM memory, used for most of main memory and caches, is volatile. That is, it loses its information whenever the power is turned off. When a microcomputer is first turned on, main memory has no information in it. In order for the operating system to load a program, the OS must already be in memory. But how does the operating system itself get loaded? In a reference to the expression "picking oneself up by the bootstraps," the process of getting the computer to bring itself to a state where it can run programs is called bootstrapping, or just booting.

The set of instructions for booting the computer, the boot code, is stored in ROM memory, a nonvolatile, nonwriteable form of IC memory. Since instructions in ROM cannot be changed, programs in ROM are often referred to as hardwired, or hard-coded. Since boot code has this property of being hard-coded software, it is also referred to as firmware.

The boot code performs two functions. First, it checks the hardware to determine that enough of it is functional to begin loading the operating system. In particular, it exercises the basic functionality of the microprocessor, writes and reads the RAM memory to check for data errors, and tests the display adapter, disk drives, and keyboard to verify that they are operational. Second, the boot code loads the operating system. Although loading the operating system can be involved, the boot code itself need only get the process started. Once it locates the device from which the operating system is to be loaded (usually a hard disk, sometimes a CD-ROM, or even the LAN), the boot code loads a program from that device containing information needed to load other pieces of software. These, in turn, may take part in the loading of the rest of the operating system. In this way, the computer "picks itself up by its bootstraps."

EVOLUTION OF THE MICROCOMPUTER

Early electronic computers used vacuum tubes as the switches that implement the calculation and storage circuitry. In the next generation, computers used transistors. Given the sizes of the components, these computers had to be quite large to be capable of doing useful work. Third-generation computers used integrated circuits (IC), consisting of many transistors on a single piece of silicon. At this point, more powerful large computers could be built, but a smaller computer could be built and still do useful work. Fourth-generation computers used higher levels of integration of transistors on single IC chips, referred to as large-scale integration (LSI) and very large scale integration (VLSI). At this point, the entire central processing unit (CPU) of the computer could be implemented on a single chip. Such a chip is called a microprocessor, and the computer that contains it is a microcomputer.

The first microprocessor, the Intel 4004, was introduced in the early 1970s. It had a 4 bit-wide data bus, a 740 kHz clock that required eight clock cycles to execute each instruction, and could address 4 kbyte of memory. Combined with several other chips for memory and I/O, the 4004 was part of the first microprocessor-based computer kit, the MCS-4.

For the next decade, microcomputers evolved from 4 bit and 8 bit hobby kits consisting of a motherboard with chips, some switches and 7-segment displays, to several complete 8 bit microcomputer systems. The Altair 8800, Apple II, and TRS 80 are examples of early microcomputers. These systems generally included a keyboard and a monitor, and could have a floppy disk drive or a printer attached as well. In addition, operating systems, such as CP/M, and programming language translators, such as BASIC, were available for these systems, allowing users to develop applications more quickly and easily. While Intel continued to develop more complex 8-bit and then 16-bit microprocessors, other manufacturers developed their own designs, including TI's TMS1000 and TMS9900, MOS Technology's 6502, and Motorola's 6800 and 68000.

In the early 1980s, IBM announced their PC, a microcomputer based on the Intel 8088 microprocessor (16 bit processing inside the chip, 8 bit bus externally), running the Microsoft disk operating system MS-DOS. To encourage third-party hardware and software vendors to develop products for the PC, IBM published details of its design. This encouraged not only the development of add-on hardware and software, but of PC clones—copies of the entire microcomputer built by other manufacturers.

Over the next decade the market for microcomputers grew rapidly. Dozens of companies introduced complete microcomputer systems and many more developed hardware and software to be used on these systems. During this time there was little standardization, so a hardware adapter or a piece of software had to be developed for one particular microcomputer system. Both functionality and performance improved steadily. Systems based on 16 bit processors replaced 8 bit systems, and 32 bit microprocessors were in development. Hard disk drives were uncommon on early systems, but became more common with capacities growing from 5 Mbyte to 40 Mbyte and higher. Dot matrix printers were replaced by laser and ink jet printers. Modem speeds increased from 300 bit/s to 9600 bit/s. CD-ROM drives became available.

Other developments included networking hardware and software, allowing data and other resource sharing among clusters of microcomputers. Large portable computers and then laptop computers also appeared during this period. In addition, the SCSI, ISA, and EISA bus standards became established, allowing peripherals to be more easily added to a system. Also, user interfaces evolved from primarily text-based to graphics-based. The graphical user interface (GUI) first appeared on microprocessor systems on the Apple Lisa and Macintosh systems, and then later in the decade in Microsoft's Windows operating system.

By the early 1990s, the IBM PC family of microcomputers, based on the 8088 microprocessor and its successors, and the MS-DOS operating system and its successors, had become the dominant microcomputer platform in the industry. As this decade has progressed, further improvements in functionality and performance have been achieved. These include faster modems and CD-ROM drives, higher capacity main memories and hard disks, hardware adapters to speed up display of 2-D and 3-D graphics, and playback and synthesis of sounds, and the availability of scanners and digital cameras for image capture. Another important development has been the emergence of the World Wide Web (WWW), and the availability of browser programs for microcomputers. These allow access to a wide range of information sources, many taking advantage of the multimedia capabilities of today's microcomputers.

The original IBM PC, introduced in 1981, contained a microprocessor running at 4.88 MHz, with 64 kbyte of DRAM for main memory, along with a keyboard, a monitor that displayed text only, and a 160 kbyte floppy disk drive. A 300 bit/s modem and a low-quality (dot-matrix) printer could be added. As performance and functionality increased over the years, the price for a typical system has dropped to about 50% the price of the original PC. In mid-1998, that typical PC would have a microprocessor running at 266 MHz with 32 Mbyte of RAM, along with a keyboard, a graphics monitor, a mouse, a 1.4 Mbyte floppy disk drive, a 4 Gbyte hard disk drive, a 56 kbit/s modem, a CD-ROM drive, and a color printer.

CURRENT TRENDS IN MICROCOMPUTER DEVELOPMENT

The trend toward higher performance—faster cycle times, higher capacities, higher bandwidths—is expected to continue for some time to come. At the same time, there is renewed interest in low-cost computing devices having lower capabilities and capacities for users that do not require the power of current microcomputers.

Performance-Driven Developments

Each of the major computer subsystems—processor, memory, and I/O—are being developed for high performance. The factors driving these high-performance developments are the desire to run current applications more quickly, and to run new applications that have higher computational requirements than could previously be satisfied. For example, such applications as video playback, 3-D graphics, voice input, and teleconferencing, could not have been run on the microcomputers of several years ago. In addition, microcomputers are now being used as servers—systems that manage a particular resource so that other computers (clients) can access them—for more computationally intensive tasks, such as database and transaction processing.

Microprocessor Performance. Since microcomputers use a single-chip microprocessor as the central processing unit, the processing power available to the system is always limited by the size of chips that can be fabricated, and the density of the devices on the chip. As both chip size and density have increased over the years, the larger numbers of available semiconductor devices on a chip have led to increases in both performance and functionality. Often, the increase in circuit count has allowed mechanisms previously used in minicomputers or even main frames to be used in microprocessors. Among the mechanisms used to achieve high performance are pipelining, superscalar processing, out-of-order instruction execution, prefetching, branch prediction, and speculative execution.

One measure of raw microprocessor performance is the number of instructions it can execute in a given period of time, usually expressed in millions of instructions per second (MIPS). This measure is a function of the clock speed in cycles per second, and the number of instructions per cycle (IPC) that can be executed. Improving performance requires that the clock speed or IPC rating (or both) be increased.

Clock speeds have been increasing at a steady rate due to the decreasing sizes of semiconductor devices on silicon chips. Clock speed can be further increased by reducing the amount of computation done on each cycle. This reduction is achieved by using an instruction pipeline. The pipeline consists of a series of processing stages, each stage responsible for only one of the operations needed to execute an instruction. For example, a typical breakdown of instruction execution into stages would include fetching the next instruction from memory, decoding the instruction to determine its type, fetching any operands used by the instruction from memory, performing the specified computation, and writing the results of the computation back to memory.

A given instruction will go from one stage to the next on each cycle, completing the process in five cycles. This is about the same amount of time it would take the instruction to complete if execution were not pipelined. However, after the first instruction completes the first stage, the next instruction can enter that stage. Thus there are five instructions in the pipeline at a time, with one finishing every cycle. Using a pipeline with more stages allows a faster clock, since less work is done at each stage.

IPC can be increased by using superscalar execution. A superscalar processor can execute more than one instruction in each cycle. This is done by fetching and decoding the next two or more sequential instructions, and providing multiple execution units to perform the specified computations in parallel. Each execution unit contains the circuitry for performing one particular class of computation, such as integer arithmetic, floating-point arithmetic, shifting and rotating bit patterns, loading data into registers, and so on. By allowing them to operate independently, and providing more than one of such highly used units as the integer arithmetic unit, two or more instructions that are adjacent in the instruction stream can be executed at the same time.

For a superscalar processor that can execute two instructions per cycle running at 400 MHz, the maximum performance rating is 800 MIPS. There are several reasons why microprocessors do not achieve such maximum performance. One is that there are not enough of the right kind of execution units to process the next two (or more) adjacent instructions. For example, if there is a single floating-point unit, and the next two instructions are both floating-point instructions, they will have to execute sequentially. This problem can be reduced by allowing out-of-order execution of instructions. That is, if an instruction appearing later in the sequence is of a type for which an execution unit is available, and if that later instruction does not depend on any intervening instructions for operands, then the later one can be executed early to avoid IPC degradation.

IPC also degrades because of data dependencies. Two adjacent instructions cannot be executed in parallel if they depend on each other in one of several ways. For example, if an instruction that uses the value in a particular register is followed by an instruction that stores a new value in that same register, the second instruction must not write to the register before the first one reads it. This apparent data dependency, apparent because it is due to a shared register resource, not due to sharing the data value itself, can be solved by reassigning the registers accessed by the two instructions, making use of some additional registers called *rename registers*. The processor must still detect real data dependencies and sequentialize processing to resolve them.

A third reason that maximum IPC is not achieved is that data are not always available when needed. Data that are in the processor registers are immediately available for use. If the data are in memory, there will be a delay to retrieve them. This delay might be one cycle if the data are in L1, several cycles if in L2, and tens of cycles if in main memory. Hardware prefetching of instructions reduces this problem as does software prefetching of data.

Finally, a major source of performance degradation is associated with branch instructions. A branch instruction corresponds to a jump from one sequence of instructions to another. For conditional branch instructions, the branch is taken only if a particular condition is met. If the condition is not met, execution of the current instruction sequence continues. A branch is said to be resolved when it is known whether it will be taken or not.

Many of the performance enhancements described above take advantage of the fact that instructions are executed sequentially. When a branch occurs, this assumption is defeated, and the performance enhancements break down. For example, instructions that have entered the pipeline after the branch instruction must be flushed out and their partial execution discarded if the branch is taken. The pipeline then starts to fill with the first instruction of the new sequence, but nothing comes out of the pipeline for several cycles. This is referred to as a bubble in the pipeline, corresponding to lost instruction throughout.

There are several mechanisms used in today's microprocessors to reduce the degradation caused by branches. First, there is branch prediction. If it is known that a branch will be taken, the target address of the branch can be used to begin fetching the new sequence of instructions. Several sources of information are used for predicting branches, including target addresses of previously taken branches and a history of whether conditional branches have been taken before. This information can be maintained in a table indexed by the branch instruction address. Second, there is speculative execution, involving the execution of one or more instruction streams that may or may not be on the correct execution path, depending on the outcome of upcoming branch instructions. The complexity of such a mechanism comes from the need to undo the effects of any instruction that was speculatively executed and later found to be on a wrong path. The performance gain comes from the fact that as long as one of the paths executed was the correct one, there is no delay due to the branch.

Currently, developments aimed at increasing the computational power of microcomputers are focused on increasing clock speed, increasing IPC using approaches just described, and combining multiple processors in a single system (multiprocessing). In the future, alternative approaches to keeping the processor busy, such as multithreading and the use of a very long instruction word (VLIW) may become popular. Multithreading involves maintaining several independent streams of instructions in the processor so that data dependencies can be reduced and pipeline bubbles from one stream can be filled in by instructions from another stream. VLIW processors use wide instructions to specify multiple operations per instruction that have been determined prior to execution not to be interdependent. These operations can be executed simultaneously to achieve a high level of parallel computation.

Memory Performance. The raw computational power of the processor is not the only factor that determines the overall performance of a microcomputer system, as is clear from the discussion above. If the processor cannot be supplied with enough instructions and data to keep it busy, it will waste many cycles doing nothing. The various components of the memory subsystem are characterized by their capacity, bandwidth, and latency. Because no device exists that optimizes all three of these attributes, the memory system is composed of a variety of components that are combined in such a way that the advantageous characteristics of each component are emphasized.

For example, small and fast caches are used close to the processor to provide low-latency responses to most memory requests, while larger main memory modules are used to provide high capacity. If the cache can be managed so that the requested data are almost always in the cache, the overall memory subsystem appears to the processor as a low-latency, high-capacity storage system. The use of an even smaller and faster cache on the microprocessor chip, and of hard disk backing store at the other end of the memory hierarchy, provide even better latency and capacity, respectively.

While multilevel caches help to alleviate the memory latency problem, main memory latencies have become an ever growing problem in recent years, due to the rapid increases in processor clock speeds. DRAM latencies of 60 ns represent a nearly tenfold improvement over the 500 ns latencies of two decades ago. However, in that time, microprocessor clock speeds have increased from about 5 MHz to over 200 MHz,

and will continue to rise quickly for at least the next few years. A processor clock speed of 266 MHz corresponds to a clock period of 3.8 ns. A 60 ns memory latency then corresponds to a 16 cycle delay.

A number of recent developments in DRAM design have been aimed at improving the memory bandwidth that is otherwise degraded by poor memory access time. DRAM chips are organized as two-dimensional arrays of memory cells, each cell storing one bit of information. A memory access consists of first reading the entire row of cells containing the bit of interest, and then choosing the column containing that bit as the data to be transferred. The overall latency in accessing the data is due to the row access time followed by the column access time.

Fast page mode (FPM) DRAM accesses multiple columns once a row has been accessed, reducing the average access time per bit. Extended data out (EDO) DRAM allows overlapping of data transfer with the next memory request to reduce the effective latency. Burst EDO (BEDO) memory allows multiple data transfers per request, reducing the amount of time spent sending addresses to memory. The current trend in DRAM system development is toward the use of synchronous DRAM (SDRAM) memory.

For SDRAM memory, the memory subsystem is clocked at the same frequency as the rest of the system (the microprocessor itself has its own clock that may be some multiple of the system clock frequency). The memory controller puts an address on the DRAM address bus and receives the corresponding data a fixed number of cycles later, with no additional protocol overhead. While today's asynchronous busses typically run at 66 MHz, the first SDRAM busses run at 100 MHz, with higher frequencies expected. What is not yet clear is which of several proposals for synchronous DRAM architectures will become prominent in the coming years.

Peripheral Performance. Early microcomputers connected peripheral adapters directly to the memory or I/O bus on the processor. To support development of peripherals by third parties, standard bus protocols were later defined. The ISA bus, which was introduced in the mid 1980s and standardized in the late 1980s, has a maximum bandwidth of 8.33 Mbyte/s. This is sufficient for connecting keyboard, text-based and low-resolution graphics monitors, modems, printers, and other devices with moderate bandwidth requirements. The PCI bus was developed to support higher bandwidth requirements, such as those of high-resolution and 3-D graphics adapters and high-speed networks. It has a maximum bandwidth of 132 Mbyte/s.

Current trends in peripheral developments are toward both additional functionality and increasing performance. Image rendering for multimedia and 3-D graphics is supported by graphics adapters with on-board microprocessors that can process several hundred million bytes of data per second. Sound cards have DSP chips for real-time processing of multichannel audio signals or synthesis of stereo sounds in virtual worlds. CD-ROM, modem, and network data rates continue to increase. In some cases, current peripheral busses are sufficient to handle the higher bandwidth requirements, but in other cases faster busses are needed. One way to increase bandwidth is to enhance current bus capabilities. For example, a 64-bit wide PCI standard has been defined to double its previous bandwidth. However, as with the memory subsystem, new I/O bus protocols are being developed to significantly increase the data transfer rate.

Cost-Driven Developments

The design of low-cost microcomputers generally involves innovative uses of technology, rather than innovations in the technology itself. The way to achieve low cost is to leave out functionality, reduce capacities, and use parts (such as the microprocessor) that are no longer at the leading edge of technology. The challenge in designing such devices is to find a combination of components that has a significant cost advantage, but also provides a sufficient and balanced functionality and performance to support some useful class of computations.

For example, the network computer (NC) is a low-cost microcomputer designed for users who need only a subset of the functionality available in a PC. For instance, an NC could connect over phone lines or a LAN to the World Wide Web (WWW) or other network resources, allowing the user to browse information, fill in electronic forms, and execute programs that can be downloaded off the network. An NC would not be used, on the other hand, for most applications development, for running computationally intensive applications, or for running applications with large memory requirements. Among the current developments in microcomputer software, the Java programming language is aimed at supporting the NC model of computation.

Reducing functionality further yields a system that can no longer be called a microcomputer. There are a growing number of uses for microprocessor-based systems that contain some of the other components of microcomputers as well. These systems are referred to as embedded, meaning that the computation is done in the service of some fixed control mechanism, rather than being used for general-purpose processing. Examples of systems that incorporate embedded processors include automobiles, microwave ovens, digital cameras, video games, and telephone switches.

The cost constraints of embedded applications encourages higher levels of integration of functions on chips. For example, the integration of the processor, the memory controller, and the L2 cache on one chip has been proposed, as has the integration of the processor and DRAM. Any successful low cost integration of these functions is likely to find its way into future microcomputer designs, particularly in the portable computer segment, where physical space, rather than cost, is at a premium.

BIBLIOGRAPHY

1. A. S. Tanenbaum, *Modern Operating Systems,* Chap. 8, Englewood Cliffs, NJ: Prentice-Hall, 1992.

2. K. Polsson, Chronology of events in the history of microcomputers [Online], 1998. Available http://www.islandnet.com/kpolsson/comphist.htm.

3. J. L. Hennessy and D. A. Patterson, *Computer Organization and Design,* 2nd ed., San Francisco: Morgan Kaufmann, 1998.

4. M. Pietrek, *Windows Internals,* Chap. 6, Reading, MA: Addison-Wesley, 1993.

5. T. Shanley and D. Anderson, *PCI System Architecture,* 3rd ed., Reading, MA: Addison-Wesley, 1995.

6. PowerPC 603/604 Reference Design, Order No. MPRH01TSU-02, IBM Corporation, 1995, available through IBM branch offices.

7. R. White, *How Computers Work,* Emeryville, CA: Ziff-Davis Press, 1997.

8. N. Randall, A RAM primer, *PC Magazine,* **16** (18): 1997.

9. M. J. Zulich, DRAM: The next generation, *Computer Shopper,* June 1997.

10. R. Jain, *The Art of Computer Systems Performance Analysis,* Part I, New York: Wiley, 1991.

11. M. Johnson, *Superscalar Microprocessor Design,* Englewood Cliffs, NJ: Prentice-Hall, 1991.

12. J. Walrand and P. Varaiya, *High-Performance Communication Networks,* San Francisco: Morgan Kaufmann, 1996, Chap. 3.

Reading List

D. Burger and J. R. Goodman, eds., Special issue on Billion-Transistor Architectures, *Computer,* **30** (9): 1997.

Y. Patt, ed., Special issue on Trends in Architecture, *Computer,* **30** (12): 1997.

PETER A. SANDON
IBM Microelectronics Division

MICROCONTROLLER.   See MICROPROCESSORS.

MICROELECTRODES

Microelectrodes have traditionally been developed and used to measure voltage inside and outside biological cells, and much of our understanding of the nervous system has come from these recordings. Microelectrodes have been adapted with some clever arrangements to measure membrane voltage and ion concentrations simultaneously. More recently, microelectrodes capable of measuring partial pressures of various gases and concentrations of physiologically relevant chemical substances, such as neurotransmitters, have also been designed. The advantages of these microelectrodes is that they allow direct measurements within biological tissues to give information about the local microscopic milieu. Therefore, these electrodes must be small to minimize interference with physiological function and the damage generated by their insertion. Although these small structures are more fragile than macrosensors, they usually have better time constants.

MEMBRANE POTENTIALS

All cells in the body have a nucleus and a cytoplasm surrounded by a lipid membrane. The cytoplasm is a good conductor with a resistivity varying between $50 \ \Omega \cdot cm$ and $300 \ \Omega \cdot cm$. The cytoplasm is separated from the outside of the cell by a thin (7.5 nm to 10 nm), resistive and capacitive membrane composed almost entirely of proteins and lipids. Electrical potentials exist across this membrane in practically all cells of the body. The resting potential difference is generally around -70 mV, and the inside of the membrane is negative with respect to the outside (See BIOELECTRIC POTENTIALS). This resting potential is generated by the equilibrium of two

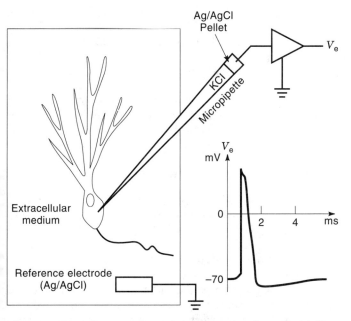

Figure 1. Recording neural activity with a microelectrode. (a) Neuron and intracellular electrode. (b) Intracellular voltage obtained with electrode showing the resting potential and the action potential.

forces, the diffusion of ions across the membrane through protein channels and the electrical force generated by accumulation of ions at the membrane. The resting potential is sustained by pumps that maintain diffusion gradients across the membrane. Some cells, such as muscle fibers and neurons, are excitable and generate large voltage signals (about 100 mV) either spontaneously or when stimulated. Figure 1 shows the resting potential and action potential in a neuron. The action potential is generated by nonlinear voltage-sensitive ion channels. Sodium (Na) channels are normally at rest and open when the membrane voltage reaches a threshold value. A large influx of Na current depolarizes the membrane to positive potentials. The Na channels turn themselves off (inactivation), and the potassium channels then open bringing the membrane voltage down to it resting value (1). This action potential is an all-or-none phenomenon and carries information from sensory inputs to the brain, from the brain to the muscles, and within various parts of the brain. Therefore, the measurement of the membrane potential is crucial to our understanding of the activity of excitable and nonexcitable cells. To measure the electrical activity of a cell, the transmembranous voltage or transmembranous current must be evaluated. An electrode must be located directly inside or make electrical contact with the inside of the cell. Because the cells recorded from can be as small as 1 μm, electrodes with submicron dimensions must be used to collect information without damaging the cell. This can be achieved with micropipettes. A micropipette as a microelectrode to record membranous voltage in muscle cells was first used by Graham and Gerard in 1946 (2).

GLASS MICROELECTRODES (MICROPIPETTES)

Micropipettes are made of very thin glass tubes filled with a conducting electrolytic solution. A 1 mm diameter borosilicate or aluminosilicate glass tube is placed in a pipette puller. The

two ends of the glass tube are clamped to a pulling device, and a heating filament is placed in the middle of the glass [Fig. 2(a)]. As the heating filament melts the glass, the ends of the tube are pulled apart until separation takes place, forming two electrodes with very small diameter (<0.2 μm) [see Fig. 2(b)]. The length of the electrode's shank is controlled by cooling the glass with an air puff following heating (3). The tip diameter and the taper of the electrode are also controlled to make various types of microelectrodes. Intracellular potential is recorded with submicron diameter electrodes filled with an electrolyte similar to that found within the cell. KCl is often used because the intracellular potassium concentration is high with respect to other ions. To facilitate filling the electrode, a small glass capillary is inserted in the tube by the manufacturer (Fig. 2). The other end of the electrode is inserted into an electrode holder which contains a reference electrode made of sintered Ag/AgCl. Then the electrode is connected to the input of a high input impedance amplifier. The potential inside the cell is measured with respect to the voltage outside the cell using a reference electrode located in the extracellular space, also made of Ag/AgCl (see Fig. 1). Microelectrodes are also used to measure the voltage outside the cell (extracellular recording). The diameter of extracellular electrodes is larger than intracellular electrodes (around 1 μm or 2 μm), and the electrodes are filled with NaCl because the sodium chloride concentration is significantly higher outside than inside the cell. A third type of microelectrode, the patch clamp electrode, is made to measure the current through the membrane and through ion channels (4). These electrodes are also pulled to a diameter of 1 μm to 2 μm but have a steep taper generated by pulling the electrode in a two-step procedure. Then the pipette is filled and placed on the surface of the cell. The glass forms a high impedance seal (GΩ) with the membrane. Then the patch of membrane inside the electrode is removed by applying a small amount of suction, and direct access to the cell is pro-

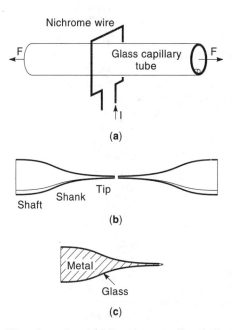

Figure 2. Microelectrodes. (a) Micropipette puller. A glass capillary tubing is heated and pulled. (b) Two small diameter electrodes are produced (c) Supported metal microelectrode with glass insulation.

vided for voltage or current measurement (whole-cell patch clamp). By pulling the electrode and patch from the cell, an inside–outside patch is formed (cytoplasmic surface toward the bath). By pulling the electrode slowly away from the cell, an outside-out patch can sometimes form (extracellular surface toward the bath). The membranous current from these various combinations is obtained by clamping the voltage at various values and measuring the current.

The impedance of these micropipettes significantly influences the measurement of both voltage and current. Intracellular electrodes, in particular, have high impedance and capacitance and require special circuits for signal processing (see later).

METAL MICROELECTRODES

Although micropipettes can be made very small, they have large impedance and are also very fragile. Microelectrodes for recording the activity of small neurons can also be made with sharpened and insulated metal wires. These electrodes have very small tips and can be placed within a few micrometers of the cell to be studied. Moreover, they can be stiff enough to penetrate tough tissue, such as the protective membrane around the brain or nerves. These electrodes can record neural activity but also provide very localized stimulation by passing current large enough to excite the neurons in the vicinity of the electrode (5).

The most common type of electrode uses tungsten etched to a sharp tip and insulated to within a few micrometers of the tip. These electrodes are shaped with a shallow taper to ease penetration. A more recent design uses iridium because it is stiffer than tungsten, is extremely resistant to corrosion, and can be electrochemically "activated (6)." In the recording mode, these metal electrodes operate at small voltages, and the impedance between the metal and the tissue is dominated by the double-layer capacitance. Therefore, the electrical model of the electrode consists of a small capacitance in series with the access resistance and the tissue resistance (see next section). The impedance is determined mainly by the size of the tip and can be in the $M\Omega$ range. The noise generated by the electrode is typical thermal noise but does not come entirely from the resistance of the electrode. The electrochemical process, which generates the double-layer capacitance, involves movement of charge carriers and contributes to the noise (7). Therefore, low-noise recording amplifiers must be used to maximize the signal-to-noise ratio.

Metal microelectrodes can also be used to stimulate neuronal activity. Electrons are charge carriers in the metal, but in the body's aqueous solution the current is carried by ions. A conversion must be made at the interface, which necessarily involves electrochemical reactions. Some of these reactions are reversible, but others are not. The irreversible reactions cause pH changes, gas formation, metal dissolution, and corrosion (8). All of these by-products are potentially damaging to the tissue but can be minimized by using biphasic current pulses which keep the voltage across the interface below threshold values for these reactions and at least partially reverse some of the reactions. For microelectrodes designed to stimulate neural tissue, the charge-carrying capacity is the most important characteristic. This capacity is proportional to the surface area of the electrode. To increase the amount

of charge which can be delivered by a small electrode, a layer of metal oxide (iridium oxide for an iridium electrode) is deposited on the surface. The oxide layer formation or activation is achieved by cycling the electrode between the anodic and cathodic voltages which generate electrolysis. Iridium oxide is a conductive layer which exchanges an electron for an hydroxyl ion across the interface. The charge-carrying capacity of the electrode is effectively increased by adding an iridium oxide layer. The impedance of the electrodes is also reduced by this activation process by an order of magnitude (6). The impedance decreases as a function of frequency and is best modeled by a capacitor in series with a resistor which has a 300 Hz cutoff.

Metal electrodes are insulated with various types of polymeric materials, such as Teflon or Parylene-C. Insulation at the tip of the electrode is removed with electric fields or lasers to burn the material away. By combining the properties of glass and metals, supported metal electrodes are fabricated [Fig. 2(c)]. A micropipette is filled with a metal which has a melting point below that of glass and is pulled to a fine tip. Thin metal films can also be deposited onto a solid glass rod pulled to a sharp point. Then the electrode is insulated with a polymer, except at the tip, to form a sharp microelectrode (9).

ELECTRICAL PROPERTIES OF MICROELECTRODES

Micropipettes and metals microelectrodes consist of a conductive cylindrical core surrounded by an insulating layer made of polymer or glass. The electrode is inserted within a tissue which is a relatively good conductor (about $100 \ \Omega \cdot cm$). The electrode is pulled to a very small diameter, and therefore the narrow shank region of the electrode generates large resistance. This region is usually tapered, and assuming a small amount of taper, the resistance is given by

$$R_E = \frac{4\rho L}{\pi d^2} \tag{1}$$

where ρ is the specific resistivity of the electrode's conducting material, L is the length of the electrode shank, and d is the internal diameter of the electrode. A distributed capacitance is also formed along the immersed part of the shank between the interior and the extracellular (or intracellular) space. This capacitance, again assuming a small amount of taper, is given by

$$C_E = \frac{2l\pi\epsilon}{\ln\left(\frac{D}{d}\right)} \tag{2}$$

where D is the outside diameter of the electrode, l is the length of the shank inserted into the tissue, and ϵ is the dielectric constant of the insulating material. There are several impedances which affect the signal, and detailed equivalent circuits of these microelectrodes have been presented (10).

For micropipettes, the impedance characteristic is dominated by the resistance R_E (about $100 \ M\Omega$ for an intracellular microelectrode) and the small capacitance C_E (a few pF) of the shank. The cell is modeled by an equivalent membranous resistance R_m and capacitance C_m in series with a voltage source V_m. A simplified equivalent circuit is shown in Fig.

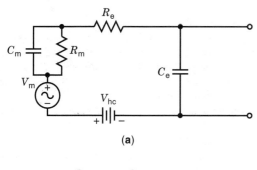

(a)

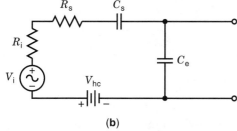

(b)

Figure 3. Microelectrode impedance. (a) Equivalent circuit for micropipette. (b) Equivalent circuit for metal microelectrode.

3(a). The resistance of the cells R_m is often as large as the source resistance of the electrode. This can cause saturation of the amplifier when current is injected within the cell. Special circuits have been designed to remove this effect (see bridge circuit in next section).

For metal microelectrodes, the impedance is dominated by the metal–liquid capacitance C_s and the series resistance R_s. For extracellular recordings with a metal microelectrode, the source is modeled by a voltage generator V_i with a source resistance R_i. A simplified equivalent circuit is shown in Fig. 3(b). In both cases, there is a dc battery V_{hc} added to the equivalent circuit which takes into account all of the half-cell potentials throughout the circuit.

ELECTRONIC CIRCUITS FOR MICROPIPETTES

The high resistance and capacitance of micropipettes can severely affect the signals recorded and require special circuits for processing (11). In particular, most researchers inject current inside through the cell to measure its input resistance or to depolarize the membrane. To inject current and record voltages with the same electrode, a circuit is used similar to that shown in Fig. 4(a). The membranous voltage is measured with the high impedance buffer amplifier (A). The injected current I_i is given by the following equation assuming that the input impedance of A is infinite,

$$I_i = \frac{BV_s + V_m(AB - 1)}{R_s} \tag{3}$$

where B is a finite gain summing amplifier. If AB is set equal to 1, the circuit becomes a current source because the injected current equals BV_s/R_s and depends only on V_s and R_s. A is implemented as a high-impedance buffer and B as a summing amplifier with a gain of 1. The input resistance R_{in} of the circuit is obtained by considering that the resistance R_s is bootstrapped between the input and the output. Then the Miller

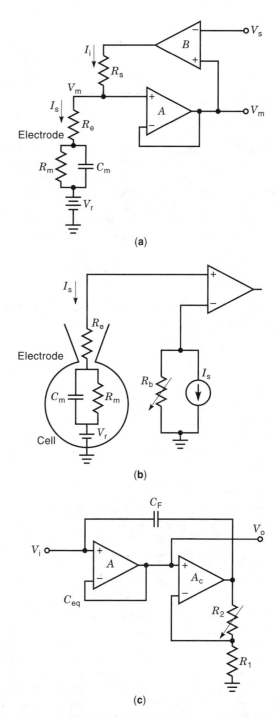

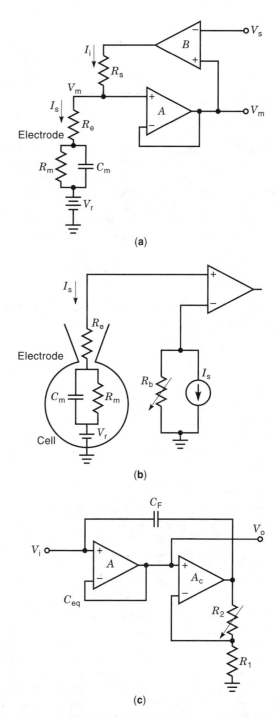

theorem gives the resistance R_m as

$$R_{in} = \frac{R_s}{1 - AB} \tag{4}$$

The product AB can be designed to be less than but very close to 1 thereby generating a very high input resistance. Then this circuit allows voltage recording with high input impedance and current injection with a grounded voltage source.

(a)

(b)

(c)

Figure 4. Circuits for microelectrodes. (a) Current injection and voltage measurement. (b) Bridge circuit for electrode resistance measurement and compensation. (c) Negative capacitance compensation.

The high resistance of the electrode generates a large voltage in response to an applied current which is added to the cell's response. To remove the effect of the electrode resistance from the response of the cell, a bridge circuit shown in Fig. 4(b) is implemented. The circuit takes advantage of the fact that the electrode's time constant is significantly faster (in most cases) than the response of the cell to a step increase in current. A current generator injects the same current into a variable resistor that is injected in the cell. The voltage recorded by the electrode is subtracted from the voltage across the variable resistance. By varying the resistance, the effect of the electrode resistance can be removed from the recorded voltage. When the recorded signal is properly compensated for, then the resistance of the electrode is equal to the value of the variable resistance. Therefore, this simple circuit can compensate for the electrode resistance and allows direct measurement of the electrode resistance.

The bridge circuit depends on the relative speed of the rise time of the amplifier and the membranous voltage of the cell. The capacitance of the micropipette can be large enough to slow down the response of the amplifier until it is no longer possible to distinguish between the cell response and the electrode response. Therefore, a circuit compensating for the capacitance of the electrode has been developed. The circuit [Fig. 4(c)] introduces a negative capacitance at the amplifer input. The input impedance of the circuit, assuming an amplifier with an infinite input resistance is given by

$$Z_{in} = \frac{V_{in}}{I_f} = \frac{1}{(1 - AA_c)jC_F\omega} \qquad (5)$$

where C_F is the feedback capacitor. Therefore, the input impedance of the circuit is the impedance of capacitor C_{eq} given by

$$C_{eq} = (1 - AA_c)C_F \qquad (6)$$

If the product AA_c is adjusted to a value greater than 1, the input capacitance becomes negative and is placed in parallel with the capacitance of the electrode. Then the net capacitance is decreased, improving the frequency response and the rise time of the electrode-amplifier recording system (12). To implement the circuit, the amplifier (A) is a buffer and is the same as in Fig. 4(a). The gain A_c amplifier is made variable and greater than 1.

Microelectrodes have also been used for voltage and patch clamping of cells. These techniques allow researchers to fix the voltage and measure the membraneous current of a whole cell, a patch of membrane, or individual channels. The methods required specialized circuits which measure a current while maintaining the voltage of the membrane (4).

ION-SELECTIVE GLASS MICROELECTRODES

By adding a membrane material to a glass micropipette, the activity of ions in sample solutions can be measured. The selectivity depends mainly on the type of membrane used. The pH of solutions can be measured with a glass membrane, and recent developments in polymer membranes have made it possible to measure the activities of Na^+, K^+, Ca^{2+}, Cl^-, and other ions. These ion-selective microelectrodes are inexpensive and are small enough to measure the activity of ions in both extracellular and intracellular spaces. Abnormal cellular states often result from an imbalance of ionic concentration. Therefore, the fact the activity of ions (not just the concentration) can be measured with very small electrodes compatible with glass micropipette technology has provided researchers with a powerful method to analyze both the normal and diseased physiological states.

A membrane is chosen to allow the diffusion of certain ions selectively and then is inserted into the glass micropipette. The potential difference generated between an external reference electrode (Ag/AgCl, for example) and the ion-selective microelectrode immersed in the sample solution is given by the Nicolsky equation (13):

$$E_{measured} = V_r + \frac{2.3RT}{zF} \log(a_i + k_{ij}a_j^{2/x}) \qquad (7)$$

where V_r is a constant and takes into account the voltages within the electrochemical cell, a_i is the activity of the ion in the sample, a_j is the activity of the interferention j, k_{ij} is the selectivity constant of the electrode for ion j relative to ion i, x the charge of ion i, R is the gas constant, T is the absolute temperature, z is the charge of the ion, and F is the Faraday constant. Typically, electrochemical cells have very large impedances, and the voltage difference must be measured with high input resistance amplifiers ($\sim 10^{15}\ \Omega$), such as electrometers. The selective membranes can be grouped into three classes: glass, liquid, and solid state. Glass membranes are very sensitive to the concentration of H^+ ions. Liquid membranes are commonly used in biological preparations for other ions. For example, potassium ion activity can be measured using a liquid membrane containing the antibiotic Valinomycin combined with a resin. When applied to the extracellular space of brain tissue undergoing epileptiform activity, the potassium concentration observed rises from its resting value of 7 mM to 11 mM. The rise in potassium activity is accompanied by increased neuronal activity and is similar to neural activity observed during epilepsy. The last class of membranes is made of solid-state material, such as crystals or insoluble salts. For example, a pellet of silver sulfide can be used to detect Ag^+ with very high sensitivity.

MICROELECTRODES FOR CHEMICAL MEASUREMENTS

The small size of microelectrodes makes them ideal for measuring local concentrations of various chemical species, such as neurotransmitters, or the partial pressure of various gases (see BIOSENSORS). The carbon fiber microelectrode is very popular. These fibers have diameters as small as a few microns and can be inserted into glass micropipettes for insulation by exposing a short length of the electrode. Contact between the fiber and the electronic circuit is made by filling the electrode with mercury or silver paint and sealing it with epoxy. Then the assembly is inserted within physiological tissue and selective measurement of various compounds is carried out using a combination of selective membranes, enzyme coatings, surface modifications, and various electrochemical techniques. For example chronoamperometry has been used to detect the release of easily oxidized neurotransmitters, such as dopamine, serotonin, and epinephrine. In chronoampersmetry, a constant voltage is applied between the carbon fiber and a ground electrode. The current is measured and is directly related to the level of oxidation. Carbon fiber microdisks are typically more useful than microcylinders since they provide a better geometry and excellent resolution. They are also more difficult to fabricate and have lower current amplitudes. A three-electrode system is used to generate a constant voltage between the working electrode and the medium around it. A reference electrode is located close to the working electrode to estimate the medium's voltage. Using feedback, a current is applied between the working electrode and a ground electrode to maintain a constant voltage. For the very low currents of microelectrodes and microdisks, a two electrode system is sufficient since the ohmic drop and polarization of the reference electrode are negligeable. The amplitude of the current is modulated by oxidation of the compound to be measured and is proportional to its activity. To improve selectivity, a differential method is used whereby the current amplitudes obtained at two voltages are subtracted and the contribution of the interfering species, such as ascorbic acid, can be reduced or eliminated. The electrode can also be coated with selective membranes. Nafion is a commonly used membrane because it prevents various charged molecules from interfering with measurement. The surface of the carbon fiber can also be modified chemically or with a laser to improve the selectivity and sensitivity of the electrode (14). By depositing appropriate enzymes on the surface of the electrode, very selective electrodes can be made to measure glucose, nitric oxide, acetylcholine, etc.

Field effect transistors (FET) have also been adapted to allow measuring various ions or chemicals. Ion-sensitive field-effect transistors (ISFET), for example, are enhanced MOS-FET transistors but use an ion-sensitive membrane instead of gate metallization (14). Then the transistor is immersed in a solution containing the ions to be measured. An electrochemical potential is established at the interface between the solution and the gate dielectric. This potential is established with respect to a reference electrode located in the solution and can modulate the conductance of the channel under the gate. The electrodes can be made very small but have drift and selectivity problems.

MICROELECTRODE ARRAYS

The silicon technology used to make integrated circuits can be adapted to manufacture arrays of microelectrodes. The activity of single cells can be recorded with the micropipette technology discussed previously. Neuroscientists are now increasingly interested in recording simultaneously from a large number of cells. Moreover, by stimulating a large number of cells in the spinal cord or in the brain selectively, it should be possible, in principle, to restore motor function in paralyzed patients or vision in blind patients, for example. Therefore, multiple arrays of electrodes capable of recording or stimulating the nervous system are clearly important to understanding nervous system function and to designing neural prostheses. Three silicon-based types of microelectrode arrays have been developed: (1) A 1-D beam electrode where a thin-film platinum-iridium is deposited on a thin layer of silicon substrate (15). This thin substrate provides a surprising amount of flexibility and can be utilized for the leads and the electrode pads. (2) A 2-D array for recording the activity of neurons grown in cultures and axons in nerves. A thin film

microelectrode array is made of gold electrodes covered with platinum black on a silicon substrate. The assembly is built into the bottom of a neuron culture dish. Neurons grow over these electrodes and make direct contact with them (16). In another design, micromachining of a silicon wafer generates a matrix of 64 square holes with a side dimension of 90 μm. Gold pads and leads are deposited near each hole. Then the thin wafer is inserted between the two sides of a severed nerve. As the axons grow inside the holes, it is possible to record from a selectively small groups of axons (17). (3) A 3-D array for cortical recording and stimulation. The 1-D beam electrode discussed previously can be assembled to form three-dimensional arrays. The longitudinal probes are inserted perpendicularly into a silicon platform. The leads from each probe are transferred to the silicon probe and are routed to a digital processing unit (18). Current work also involves including low noise amplification directly on the platform. Another implementation involving micromachining and etching techniques was used to fabricate a 10×10 electrode array. One hundred conductive needlelike electrodes (80 μm at the base and 1.5 mm long) are micromachined on a 4.2 mm $\times$ 4.2 mm substrate (19). Aluminum pads are deposited on the other side of the substrate and make contact with each needle electrode. The tips of the electrodes are coated with gold or platinum. A high-speed pneumatic device is used to place the array into cortical tissue because the high density of the electrodes makes insertion difficult. Then the microelectrode arrays are available for recording from a large number of cortical sites.

ACKNOWLEDGMENTS

This chapter was prepared with the financial support of NSF grants IBN 93-19591 and INT 94-17206. I would also like to thank Zhenxing Jin for his help with drawing the circuit diagrams.

BIBLIOGRAPHY

1. E. R. Kandel and J. H. Schwartz, *Principles of Neural Science,* 3rd ed., Norwalk, CT: Appleton and Lance, 1991.

2. J. Graham and R. W. Gerard, Membrane potentials and excitation of impaled muscle cells, *J. Cell. Comp. Physiol.,* **28**: 99–117, 1946.

3. K. T. Brown and D. G. Fleming, New microelectrode techniques for intracellular work in small cells, *Neuroscience,* **2**: 813–827, 1977.

4. T. G. Smith, Jr., et al. (eds.), *Voltage and Patch Clamping with Microelectrodes,* Bethesda, MD: Americal Physiological Society, 1985.

5. I. A. Silver, Microelectrodes in medicine, *Philos. Trans. R. Soc. Lond. Series B. Biol. Sci.* **316**: 161–167, 1987.

6. G. E. Loeb, R. A. Peck, and J. Martyniuk, Toward the ultimate metal microelectrode, *J. Neurosci. Meth.,* **63**: 175–183, 1995.

7. J. Millar and G. V. Williams, Ultra-low noise silver plated carbon microelectrodes, *J. Neurosci. Meth.,* **25**: 59–62, 1988.

8. D. M. Durand, Electrical stimulation of excitable tissue, in J. D. Bronzino (ed.), *Handbook of Biomedical Engineering,* Boca Raton: CRC Press, 1995, pp. 229–251.

9. J. G. Webster (ed.), *Medical Instrumentation, Application and Design,* 3rd. ed., New York: Wiley, 1998.

10. L. A. Geddes (ed.), *Electrodes and the Measurement of Bioelectric Events,* New York: Wiley-Interscience, 1972.

11. A. D. McClellan, Theoretical and practical considerations concerning an active current source for intracellular recording and stimulation, *Med. Biol. Eng. Comput.,* **19**: 659–661, 1981.

12. R. S. C. Cobbold, *Transducers for Biomedical Measurements, Principles and Applications,* New York: Wiley, 1974.

13. M. E. Meyerhoff and W. N. Opdycke, Ion-selective electrodes, *Adv. Anal. Chem.,* **25**: 1–47, 1986.

14. D. G. Buerk, *Biosensors, Theory and Applications,* Lancaster, PA: Technomic Publishing, 1993.

15. K. D. Wise, J. B. Angell, and A. Starr, An integrated circuit approach to extracellular microelectrodes, *IEEE Trans. Biomed. Eng.,* **17**: 238–247, 1970.

16. W. Nish et al., *Biosensors and Bioelectronics,* **9**: 734–741, 1994.

17. G. T. A. Kovacs et al., Silicon-substrate microelectrode arrays for parallel recording of neural activity in peripheral and cranial nerves, *IEEE Trans. Biomed. Eng.,* **41**: 567–577, 1994.

18. A. C. Hoogerwerf and K. D. Wise, A three-dimensional microelectrode array for chronic neural recording, *IEEE Trans. Biomed. Eng.,* **41**: 1136–1146, 1994.

19. K. E. Jones, P. K. Campbell, and R. A. Normann, A glass/silicon composite intracortical electrode array, *Ann. Biomed. Eng.,* **20**: 423–437, 1992.

DOMINIQUE M. DURAND
Case Western Reserve University

MICRO-ELECTRO-MECHANICAL SYSTEMS

(MEMS).   See MICROMACHINED DEVICES AND FABRICATION
TECHNOLOGIES.

MICROELECTRONICS RELIABILITY DESIGN.   See

DESIGN FOR MICROELECTRONICS RELIABILITY.

MICROMACHINED DEVICES AND FABRICATION TECHNOLOGIES

Micromachined (or micromechanical) devices are miniature structures designed for sensing, actuation, and packaging. They are created by micromachining, which constitutes a broad class of fabrication techniques that grew out of the technology and materials of the microelectronics industry. Photolithography, oxidation, diffusion, chemical vapor deposition, evaporation, and wet and dry chemical etching (1) are used not only to create transistors, resistors, and capacitors, but to also create sensors, actuators and packaging structures. By using integrated circuit (IC) processing to create micromachined devices, the integration of electronics with sensors and actuators on a single substrate is possible. Single crystal silicon is the substrate material of choice both because of its dominant use in ICs and because of its superb mechanical and etching characteristics. Applications of micromachining have required the development of augmented and new process technologies to accommodate the unique critical parameters (e.g., stress control) as well as nonstandard materials. The combination of small size and batch fabrication characteristic of IC processing makes micromachined devices particularly applicable for low-cost, high-volume applications. Expanded use of these devices allows more sensors to be in-

corporated into single or distributed systems and thus allows more widespread use of data collection and electronic control.

Micromachined devices have a minimum feature size in the range of 1 μm to 1 mm. The term "micromachined" seems to imply that the devices are incredibly precise, but this is not the case. While micromachined devices are quite small, they are not created to the same relative tolerance as macroscopic objects made with conventional machining. Objects on the order of a centimeter to a meter can have a relative tolerance of 10^{-4} to 10^{-5}, whereas micromachined devices have a relative tolerance of 10^{-2} to 10^{-3}. So although micromachined devices are small, they are not as accurately fabricated as macroscopic devices.

One of the early commercial applications of a micromachined device was a pressure sensor (2) that appeared in 1962. Diaphragms of silicon were formed by combining wet chemical etching, plasma etching, and oxidation. The piezoresistive effect in silicon was used as the transduction mechanism from the mechanical to electrical domain. Through the 1970s, many companies commercialized other micromachined products such as the Texas Instruments thermal print head (1977) (3), IBM ink jet nozzle arrays (1977) (4) and Hewlett Packard thermally isolated diode detectors (1980) (5). The first commercial application of surface micromachining was the ADXL50 accelerometer from Analog Devices, as shown in Fig. 1 (6). Other materials have also been utilized in micromachined devices such as quartz wristwatch tuning fork resonators (7).

In this article, the fabrication technology of micromachining is presented with the two important techniques, bulk and surface micromachining, being described in detail. In the next section, the fundamental building blocks of micromachined devices are outlined followed by a discussion of the main sensing and actuating physical mechanisms. Finally, some applications of micromachined devices are discussed.

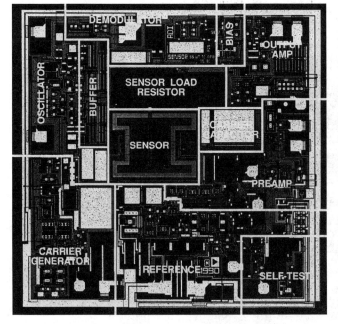

Figure 1. A photomicrograph of the Analog Devices ADXL50, the first surface micromachined accelerometer. (Courtesy of Analog Devices).

BULK MICROMACHINING

Bulk micromachining, which typically refers to any etching of the substrate (isotropically or anisotropically) or bonding of substrates, has been used for over 20 years and remains the most prevalent fabrication process for micromachined devices.

In bulk micromachining, the silicon substrate itself is machined to form a functional component in a sensor, actuator, or package. Bulk micromachined structures include diaphragms, membranes, cantilever beams, V-grooves, and through-wafer holes that are created primarily with anisotropic etching of the substrate. The anisotropic etch produces tight dimensional control even in structures that are as thick as the substrate. By selecting different crystal orientations for the initial substrate, different etch profiles can be produced (see Wolf (1) for a description of crystals and crystal notation). For example, a ⟨110⟩ wafer will produce vertical sidewalls when etched in an anisotropic etchant. Thin films can be deposited on the substrate surface to create transducing layers (i.e., piezoresistive or piezoelectric films), etch stop layers, or isolation layers. Two-sided processing is common and has the benefit of separating the electronics from the sensing environment, thus allowing the sensors to be used in hazardous environments. More complicated structures, such as accelerometers, gyros, valves, and pumps, can be created by incorporating wafer bonding techniques.

Materials

The most commonly used substrate for bulk micromachining is single-crystal silicon. Silicon has a diamond-cubic atomic structure (1). The crystalline structure lends itself well to anisotropic etching along the crystal planes. By changing the orientation of the substrate to the crystal planes, different resulting etch profiles can be created. Silicon wafers are also abundant and inexpensive. Because of the single crystal nature and purity of silicon wafers, the mechanical properties are well controlled (8). Silicon is stronger than steel, but is also very brittle. The elastic behavior of silicon makes the fabrication of reliable and repeatable structures possible. Many sensors can be created from silicon because it is sensitive to stress, temperature, radiation, and magnetic fields as discussed in the sections below.

Silicon dioxide (glass) substrates are also commonly used, especially in bonding with other substrates. The mechanical characteristics and melting points of glass can be modified by changing the level of doping impurities in the glass. The addition of sodium creates substrates that work well with anodic bonding. Corning #7750 glass is best suited for bonding to ⟨100⟩ silicon substrates because the coefficient of thermal expansion is matched to that of silicon. The melting point of doped glass puts an upper limit on the temperature that can be used in subsequent processing and on the operating temperature of the device.

Quartz is the single-crystal form of silicon dioxide. Like silicon, quartz can be anisotropically etched (9). The most important characteristic of quartz is that it has a large piezoelectric effect (see below). Quartz, however, is quite fragile and must be handled with care. As with silicon, the crystal orientations of the quartz produce different mechanical and electrical characteristics. Table 1 shows some of the mechanical parameters of both single-crystal silicon and quartz. Other

Table 1. Mechanical Properties of Bulk Silicon and Quartz

Property	Silicon	Quartz
Young's modulus (GPa)	129.5×10^9 [100]	
	168.0×10^9 [110]	
	186.5×10^9 [111]	
Yield strength	7 GPa	
Coefficient of thermal expansion (°C⁻¹)	2.33×10^{-6}	$7.1 \times 10^{-6} \parallel Z$
		$13.2 \times 10^{-6} \perp Z$
Thermal conductivity (W/mK) at 300°C	15.7	$29 \parallel Z$
		$16 \perp Z$
Density (kg/m³)	2300	2660
Melting point	1350°C	1710°C

substrates such as gallium arsenide have been utilized for micromachining, but their high cost has limited their applications.

Wet Anisotropic Silicon Etching

Wet anisotropic etchants create three-dimensional structures in a crystalline silicon substrate because the etch rate depends on the crystal orientation. Typically, the etch rate is fastest along the ⟨100⟩ and ⟨110⟩ directions and slowest along the ⟨111⟩ directions. Since silicon dioxide (SiO_2) and silicon nitride (Si_3N_4) are etched more slowly than silicon, they can be used as masking films or etch stops. Some silicon etchants also have a reduced etch rate in the presence of heavy boron doping, thus allowing a boron-doped layer to act as an etch stop in silicon.

A tremendous variety of structures can be created using combinations of silicon dioxide, silicon nitride, and boron diffused etch stops, front- and back-side lithography, and silicon substrates with different crystal orientations. For a ⟨100⟩ silicon substrate the mask is aligned to the [110] direction. This will create an etched feature that terminates at the mask edge. Figure 2 shows the progress of the etch of three different types of simple structures in ⟨100⟩ silicon: a V-groove, a diaphragm, and a through-wafer hole. As the etch proceeds {111} planes form along the edges of the silicon nitride etch masks. These planes form a 54.7° angle with the (100) surface of the wafer. If allowed, the etch will continue until only {111} planes are exposed. The angle between the {100} and {111} planes causes the size of the etch pit for the diaphragm and through hole to be much larger than the whole opening. This drawback can be compensated for by using a bonded wafer for the membrane. If the substrate is not aligned well to the [110] direction, the etched feature will underetch the mask until only the {111} planes are visible. Figure 3 illustrates the resulting structure due to a mask misalignment. Some applications require vertical sidewalls in the silicon substrate. This is accomplished with ⟨110⟩ substrates. The alignment of the mask is to the [111] direction. The {111} planes are perpendicular to the wafer surface. The etch proceeds as before, etching all planes except the {111} planes. It is thus easy to create through wafer slits with ⟨110⟩ substrates. For a more detailed review of bulk micromachining, see Ref. (10).

High etch selectivities between different crystal orientations and masking materials allows lateral as well as vertical dimensions to be controlled to within 0.5 μm or better. However, while the dimensions of the structures are accurate, there are some problems with wet anisotropic etching of sili-

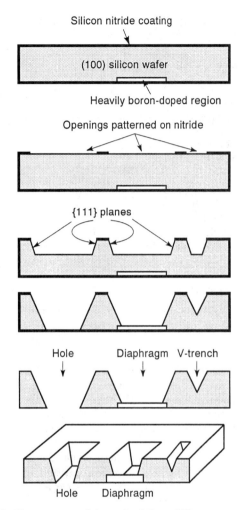

Figure 2. The progress of the etch of three different types of simple structures in ⟨100⟩ silicon: a V-groove, a diaphragm, and a through-wafer hole. Note the use of the boron-diffused silicon membrane. (After Ref. 10.)

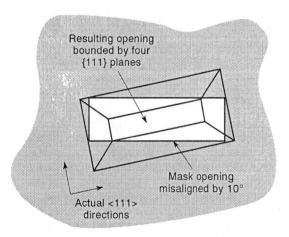

Figure 3. Effect of misalignment between the mask and the actual ⟨111⟩ directions on the final etch profile. Note that the etched feature always increases in size due to misalignment. Also, that the etch feature is bounded by {111} planes.

con. All of the chemistries etch silicon at ≈1 μm/min in the fast etching orientation, ⟨100⟩. Thus, to etch through an entire 500 μm wafer over 8 hours are required. In addition, the etch rates and performance depends on temperature and concentration of the solutions. The three main wet anisotropic silicon etchants and their characteristics are listed in Table 2. Finally, although the anisotropically etched structures are three-dimensional, there are only a few types of vertical features that can be created (i.e., vertical and angled at 54.7°).

The most common anisotropic etchant is a mixture of potassium hydroxide (KOH), water, and isopropyl alcohol at 80°C (11,12). At higher temperatures the uniformity of the etch decreases. The concentration of KOH can be varied from 10 wt% to 50 wt% (2 M to 12 M). High KOH concentrations result in smooth structures. Hydrogen gas is generated as a byproduct of the etch and forms bubbles which are thought to cause the surface roughness seen at low KOH concentrations. The main advantage of KOH is that it has the highest etch rate ratio between the {100} and {111} planes, 400 to 600. On the other hand, KOH is not compatible with IC fabrication because of the presence of the mobile ion, K^+. KOH also etches silicon dioxide at ≈28 Å/min, which makes oxide unusable as a masking material for through-wafer etches. Silicon nitride, which is not attacked, must be used as the masking material in long KOH etches.

Another common anisotropic etchant is a mixture of ethylenediamine/pyrocatechol (EDP) and water at 115°C (13). This etchant is a thick, opaque liquid that ages quickly when exposed to oxygen. A reflux system is required to keep the composition of the solution constant as well as provide a nitrogen atmosphere to prevent aging. When mixing this etchant the water is added last because the water triggers the oxygen sensitivity. The main advantage of EDP is that it etches oxide much more slowly than KOH. The oxide to (100) silicon etch rate ratio is 5000 for EDP and 400 for KOH. Also, EDP is more selective to boron doped masking than KOH. Care must be taken when mixing the solution because EDP is toxic.

Tetramethyl ammonium hydroxide (TMAH) (14) is of interest because it is more stable than EDP, has a high selectivity of oxide to silicon, and is compatible with complementary metal-oxide-semiconductor (CMOS) circuit fabrication. This makes TMAH useful for high-volume production applications. However, the etch rate ratio of (100)/(111) silicon is only 12.5 to 50. Many other alkaline solutions have been studied: hydrazine, sodium hydroxide, ammonium hydroxide, cesium hydroxide, and tetraethyl ammonium hydroxide.

Dry Anisotropic Silicon Etching

An area of active research is high aspect ratio microstructures (HARM). Either bulk silicon, silicon on insulator, or thick (i.e., >10 μm) polysilicon is etched anisotropically to form structures that have a thickness-to-width ratio as high as 100. Whereas anisotropic etching of ⟨110⟩ silicon can form tall, narrow structures, plasma etching is easier to use. The plasma etch is performed by alternating between a silicon etch (e.g., Cl_2 or SF_6) and a plasma polymerization step based on a fluorocarbon. One drawback of this etch is the 1 μm/min to 2 μm/min etch rate. Thus for a 650 μm thick silicon wafer it would take approximately 5 h to etch through the wafer. Much research is being carried out to increase the etch rate,

Table 2. Common Wet Anisotropic Silicon Etchants and Characteristics (10)

Etchant	Etch Rate (100) μm/min	Etch Rate Ratio (100)/(111)	Masking Materials	Boron Etch Stops
KOH/water, isopropyl alcohol, 85°C	1.25	400	Si$_3$N$_4$ (not etched), SiO$_2$ (28 A/min)	B > 10^{20} cm^{-3} reduces e/r by 20
EDP/wafer., pyrazine, 115°C	1.25	35	Si$_3$N$_4$ (2–5 A/min), SiO$_2$ (1 A/min), many metals	B > 5×10^{19} cm^{-3} reduces e/r by 50
TMAH/wafer, 90°C	1.0	1.25–50	Si$_3$N$_4$, SiO$_2$ (1 A/min), many metals	B > 4×10^{20} cm^{-3} reduces e/r by 40

but a value of 10 μm/min is probably the limit for the near future.

An interesting application of this technology uses the anisotropic plasma etch to form a mold in which a thin sacrificial oxide layer is deposited followed by a polysilicon thin film (termed Hexil) (15). The oxide is then removed with hydrofluoric acid (HF) as in surface micromachining, and the resulting polysilicon structure is detached from the substrate. This allows the creation of structures with milliscale dimensions (50 μm to 100 μm tall) using 2 μm films.

Wafer Bonding

In wafer bonding processes, two substrates are bonded together either with an intermediate layer to promote adhesion or directly together without an intermediate layer. Pressure and heat are applied to the two or more substrates during the bonding process. Higher-temperature processing typically results in bonds with higher levels of hermiticity and strength. Anodic bonding, fusion bonding, low-temperature glass bonding, and reactive metal sealing are common techniques for wafer bonding. For all bonding techniques, alignment of the substrates is accomplished either by two-sided alignment or by infrared alignment. The quality of the bond can be ascertained by looking at the wafers in visible wavelengths if transparent substrates are used, or in infrared (IR) wavelengths if silicon substrates are used.

Anodic bonding, also known as electrostatic bonding (16) or field-assisted thermal bonding, is used to bond silicon and sodium-rich glass substrates. A commonly used glass is Corning #7740 (Pyrex) because it has a thermal expansion coefficient that is matched to silicon. This helps to prevent failure of the bond due to residual thermal stresses. The anodic bonding process can take place in air or vacuum (Fig. 4). The silicon and glass wafer are brought in contact and placed on a hot plate at a relatively low temperature, 350° to 450°C. An electrostatic potential of 400 V to 700 V is then applied between the substrates using the glass substrate as the anode. The mobile sodium ions are thus depleted from the interface. This creates a depletion region of about 1 μm with high electric fields of 4–7 $\times$ 10^6 V/m. An electrostatic pressure is thus induced between the substrates and is the driving force for the bond (17).

Fusion bonding (18) is the fusing of two wafers thermally without the need for a intermediate adhesion layer. The wafers must be cleaned and placed in contact. The wafers will initially stick together because of van der Waals forces. A force is then applied to the stack of wafers and the assembly is annealed at high temperatures. The resulting bond is very strong and hermetic, but the high temperatures preclude the

use of integrated electronics on the wafers. Also, the flatness and cleanliness of the wafers are critical in producing a quality bond.

Both anodic and fusion bonding are not compatible with electronics. To bond in the presence of electronic circuits,

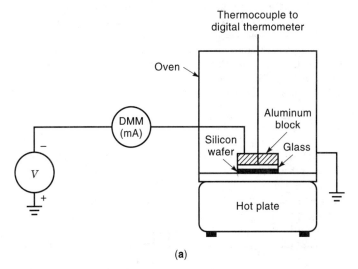

(a)

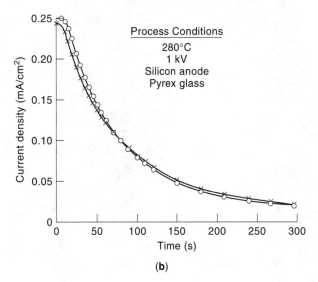

(b)

Figure 4. (a) Diagram of the anodic bonding apparatus. When the electric field is applied, the wafer assembly is pressed together and the bond is formed. (b) Plot of typical current through the wafer assembly versus time. When the current falls to 10% of the initial value, the bond is complete. (After Ref. 87.)

techniques which use an intermediate bonding layer are required. This adhesive layer can be low-temperature glass, reactive metal, or organic films (19). All these bonds occur at low temperatures, but each have different thermal characteristics and can thus generate thermal stresses. Also, none of these bonds are hermetic and all require flat wafers to ensure quality bonds.

Isotropic Silicon Etches

Xenon difluoride (XeF_2), when exposed to silicon at a reduced pressure (2 torr), will selectively etch silicon at several hundred micrometers per minute. The selectivities of XeF_2 to oxides, nitrides, and metals are 100:1, 100:1, and 200:1, respectively. XeF_2 has been used to integrate standard circuit processes with microstructures such as accelerometers (20). Typical structures are created as composites of metal and polysilicon encased in silicon dioxide or silicon nitride. The substrate under these structures is isotropically removed to create freestanding structures. Although this approach is very inexpensive, it suffers because the structure is made from a composite of oxide, polysilicon, and metals and is less reliable than silicon or polysilicon alone. Also, the isotropic nature of the etch creates a large undercut around the perimeter of the etch hole. Silicon can also be etched in mixtures of nitric acid (HNO_3) and hydrofluoric acid (HF) (21), but the selectivities are much worse than those for XeF_2.

SURFACE MICROMACHINING

In surface micromachining, devices are created from thin films that are deposited and patterned on the surface of much thicker substrates. The underlying layers are then selectively removed, thus creating a free-standing structure which is attached to the substrate. Many layers can be used to sequentially build up complicated structures such as gears and motors (22) (see Fig. 5 for an example). The thin films are typi-

Figure 5. A scanning electron micrograph of a complex surface micromachined mechanism. The mechanism in the center converts the linear motion of the long beam on the right to a rotary motion of the small gear. Note that two different structural polysilicon layers are used. (Courtesy of Sandia National Labs.)

Figure 6. A scanning electron micrograph of the Analog Devices ADXL76, an integrated, micromachined accelerometer. The sensor is fabricated from a 2 μm thick polysilicon structure and is ~500 μm along the long axis. (Courtesy of Analog Devices.)

cally 0.1 μm to 10 μm thick, as compared to the substrate, which is 500 μm to 700 μm thick. The resulting devices are comprised of many stacked thin films, but are still essentially two-dimensional (2-D) planar structures. This is in contrast to bulk micromachining, which shapes the substrate to create a more truly three-dimensional (3-D) structure.

Surface micromachining was first demonstrated in 1965. An electromechanical filter was created using a gold cantilever as the free-standing gate of a field-effect transistor (23). Only input signals near one-half the resonant frequency of the cantilever would create any output of the field effect transistor. In 1984, the first polycrystalline silicon surface micromachined devices were fabricated (24). By the early 1990s, the first integrated surface micromachined process was developed by Analog Devices (25). Figure 6 shows the ADXL76 structure, a ±50 g full-scale integrated lateral accelerometer created from a 2 μm thick polysilicon layer. Also, in the 1980s, the digital micromirror device (DMD) was created using aluminum as the mechanical structure (26).

The most basic surface micromachining process produces a single structural layer suspended over the substrate. Figure 7 outlines the process for polysilicon surface micromachining. The process begins with the deposition and patterning of a silicon dioxide to form anchor points for the micromechanical films. The oxide acts as a sacrificial film that will be removed later in the process. Next, polysilicon is deposited and patterned. Polysilicon is the structural micromechanical film out of which the micromachined device will be created. The lateral dimensions are defined by photolithography and etching while the vertical dimensions are defined by film thickness. The final step is the selective removal of the underlying film using a selective chemical etch (i.e., hydrofluoric acid, HF) and the drying of the devices. Preventing the structures from sticking to the substrate or to each other during drying has been the subject of much research. Drying techniques such as supercritical point drying (27) or sublimation (28) have been used to eliminate the surface tension forces. In addition, special anti-stick coatings like Teflon and self-assembling molecules (SAMs) have been deposited after the removal of the sacrificial oxide (29).

One of the major advantages of surface micromachining is the possible economical integration of electronics with the sensor. Surface micromachining produces structures that have smaller dimensions than those fabricated by bulk techniques and thus have smaller signals to external stimulus. The integration of the electronics can make up for this small signal because it reduces the parasitic losses between the sensor and the electronics.

Materials

Any thin film can be used in surface micromachining, although chemical vapor deposition (CVD) films offer the best repeatability. The key is finding suitable pairs of materials where an etchant exists that will selectively remove the lower layer and thus create a free-standing micromachined structure. The most prevalent CVD films are silicon dioxide, silicon nitride, and polycrystalline or amorphous silicon. These films are most commonly deposited using low-pressure chemical vapor deposition (LPCVD). Other thin films that have been used include polyimides, electroplated or deposited metals (aluminum, nickel, tungsten, and nickel-iron), and polymers. The mechanical characteristics of the thin films differ from the corresponding bulk materials because surface effects begin to dominate. Also, the assumption that the grain size is small relative to the dimensions of the devices is no longer valid, and statistical variations in the mechanical parameters like Young's modulus become significant. Finally, in thin films both the axial stress and the stress gradient through the thickness of the film are critical. Compressive stresses can

cause buckling of constrained mechanical structures. Excessive stress gradients can cause the films to curl up off of the substrate.

Polysilicon has similar material characteristics as single-crystal silicon and is therefore a desirable mechanical material. However, the material parameters of polysilicon change with different deposition, doping, and annealing conditions. For example, Young's modulus and yield stress vary somewhat due to processing, but the amount of variation is small compared with that of residual stress and stress gradient. Extensive studies have explored polysilicon including original work by Guckel (30) and Howe (31) and subsequent investigations of in situ doping (32), fine-grained polysilicon (33), and correlations between texture and stress and stress gradient (34).

Integration

Integration of electronics with surface micromachined devices is intrinsically simpler because only one side of the wafer is utilized. In addition, the wafers are not as fragile, at least not until the surface micromachined structures are released from the substrate. There are three approaches to integration of microstructures: precircuit, mixed, and postcircuit processing. Almost all integrated technologies involve some level of mixing between the circuit and micromechanical processing steps. It is thus by the division of the majority of the processing steps that the techniques are characterized. By fabricating the micromachined devices prior to the circuit processing, the high-temperature anneals which are required to relieve the stresses in the thin films will not detrimentally affect the circuit processing. Similarly, the circuit processing must also not effect the microstructures, which is typical as the circuit processes use lower temperatures. The topography of the surface micromachined devices requires planarization. Sandia National Labs has demonstrated a "structures-first" process using CMOS electronics (35). Unless the structures are released from the substrate and encapsulated, the structures will need to be released from the substrate after the circuit processing is complete.

A mixed integration strategy, like that used by Analog Devices (25), further interweaves the circuit and microstructure processing. For example, the high-temperature circuit processing would be completed first, followed by the microstructure fabrication, then the rest of the circuit processing. Polysilicon/silicon dioxide surface micromachining is an example of a technique that works well with preintegration or mixed integration. The high-temperature anneals and topology issues are minimized by gradually building up the structures and choosing the timing of the anneals to minimize their impacts on the already created devices.

Integrating the microstructures after the circuit processing is the closest to a modular technique. After the metal is etched and passivated, the microstructures are fabricated. The presence of the metal precludes the use of high-temperature anneals unless special metalization like tungsten is used (36). Electroplating techniques or deposited metal/polymer surface micromachining allows the use of conventional circuits, but the mechanical characteristics of the metal structures are not as desirable as polysilicon. As described above, XeF$_2$ can be used to selectively remove the silicon substrate to release a composite microstructure of oxide, polysilicon,

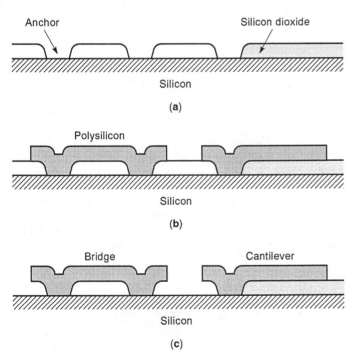

Figure 7. Illustration of surface micromachining—a cantilever–bridge example. (a) Deposit and pattern the sacrificial oxide film; (b) deposit and pattern the structural film (e.g., polysilicon); (c) selectively remove the underlying sacrificial film to create a free-standing micromechanical element. Note that the thickness of the structure is defined by the deposition thickness and that the in-plane dimensions are defined by photolithography and etching.

and metal. The mechanical characteristics of this composite structure are also not as desirable as those of polysilicon, but the simplicity of the process makes some low-cost applications very attractive.

Other Micromachining Technologies

Although surface and bulk micromachining can be used to create a large variety of structures, there are applications and materials that require other micromachining technologies. For example, many applications require truly 3-D structures such as the miniaturization of mechanical systems (e.g., clocks) and microrobotics. In general, surface micromachining is limited to structures in the plane of the substrate. The in-plane dimensions are controlled by lithography and thus have great design freedom, but the thickness is controlled by a deposition thickness and is thus fixed. Bulk micromachining, on the other hand, does create 3-D structures, but the shapes are constrained by the crystallographic planes.

Surface micromachining has created hinged polysilicon structures (37) that can be erected above the substrate, but the thickness of each element is still limited to the thickness of the deposition. An active area of research is high-aspect ratio structures. These thick planar structures can be created with electroplating or highly anisotropic silicon etches. Although these structures are still planar, the additional thickness produces structures with more mass and additional robustness. Many other technologies exist including electron discharge machining (38), focused ion-beam milling (39), ultrasonic machining (40), laser-assisted etching and deposition, 3-D photolithography, and ultrahigh-precision mechanical machining.

LIGA. The German Lithographie Galvanoformung Abformtechnik (LIGA) is a technique that consists of lithography, electroplating, and molding (41). The lithography process uses a 100 μm to 500 μm thick layer of photoresist [e.g., polymethylmethacrylate (PMMA)] on a conductive substrate. High-energy synchrotron X-ray radiation is used to expose the photoresist. A special X-ray mask is required that uses gold to absorb the X-ray radiation. After development, the desired thick, high-aspect-ratio resist structure is obtained. Metal is subsequently electroplated on the conductive substrate. After resist removal, a free-standing metal structure is obtained. This structure can be used as the final product or as an injection mold for plastic parts. The injection mold can be reused and is thus an inexpensive way of creating precision plastic parts, although the fabrication of the mold itself is expensive.

Although LIGA produces relatively thick microstructures, they are 2-D. To create more complex, multilevel devices, structures produces by the basic process will require assembly (42). There have been many modifications to LIGA that include placing a patterned sacrificial film such as silicon dioxide, photoresist, or polysilicon under the metal LIGA structures to create a free-standing structure (43). In addition, multiple layers of LIGA are now possible by repeating the basic process with a planarization process between the two LIGA processes. Applications such as electromagnetic micromotors (44) and an electromagnetic dynamometer (45) have been produced using LIGA.

MICROMACHINED DEVICES AND APPLICATIONS

The purpose of this section is twofold: (1) to examine quantitatively some of the fundamental building blocks (or models) used in typical micromachined device design and (2) to examine several broad classes of micromachined devices to give the reader a feeling for the possible application areas to which micromachined devices can be applied. We will start with the building blocks because they form the physical underpinning of the functional mechanisms of actual micromachined devices.

The Building Blocks of Micromachined Devices

Since almost any structure built using the broad class of MEMS fabrication techniques can be considered a micromachined device, it is rather presumptuous to make a list of building blocks for MEMS. However, as micromachined devices have evolved, a group of canonical structures have become ubiquitous. These structures, and the simple models used for their first-cut design, form the basis for the design of many micromachined systems.

The Diaphragm. Arguably, the most basic micromachined structure is the thin-film diaphragm. The fundamental function of a diaphragm in a micromachined device is to deform under a load. This deformation is then sensed, typically with either (1) piezoresistors, which sense changes in stress, or (2) a capacitance measurement that is sensitive to changes in the deflection itself. Figure 8 is a sketch of a diaphragm cross section with an embedded (diffused) piezoresistor. The fundamental design relationship for a thin diaphragm is the deflection versus applied load relation (9):

$$\frac{\partial^4 w}{\partial x^4} + \frac{\partial^4 w}{\partial x^2 \, dy^2} + \frac{\partial^4 w}{\partial y^4} = \frac{12(1 - \nu^2)p}{Et^3} \tag{1}$$

where w is the z-axis deflection as a function of the x and y position, p is a uniform applied pressure load, t is the diaphragm thickness, ν is Poisson's ratio, and E is Young's modulus (or the modulus of elasticity). Equation (1) must be solved in conjunction with the appropriate boundary conditions for the case under study (for example, the diaphragm is fixed along its entire edge). In general, such solutions are best performed by numerical methods such as finite element analysis (FEA) (46), especially in cases where residual stress effects must be added to Eq. (1). However, approximate design relations have been developed (47). For a square membrane, the maximum absolute value of the stress occurs at the center

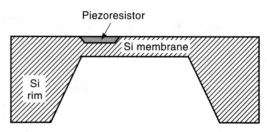

Figure 8. Cross-sectional sketch of a micromachined diaphragm with a diffused piezoresistor. Note that the thickness of the diaphragm is not to scale. (After Ref. 10.)

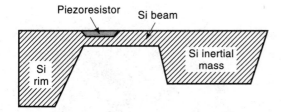

Figure 9. Cross-sectional sketch of a micromachined accelerometer with piezoresistive sensing. The device is made up of a bulk micromachined silicon beam with an inertial ("proof") mass attached to its free end. Note that the beam width (into the page) is typically much thinner than the proof mass. Often there are multiple beams supporting one proof mass. (After Ref. 10.)

of the sides of the membrane and decreases toward the corners and the center of the membrane. The surface stress in the middle of a side can be approximated as

$$\sigma_{\max} \approx 0.31p\,\frac{a^2}{t^2} \tag{2}$$

where a is the membrane side length. Note that the maximum stress magnitude is proportional to the square of the side length-to-thickness ratio. This ratio is constrained by sensor area constraints and the strength and manufacturability of the thin membrane.

Beams and the Spring-Mass System. We saw that the fundamental relationship for the diaphragm was the force displacement relation. Beams behave in a similar manner. The deflection characteristics of beams under various loading conditions and various boundary constraints have been tabulated (48).

Beams have an even more powerful building-block function in the context of lumped tether-plate (spring-mass) systems. The typical arrangement contains a plate, which constitutes the mass, and a set of beams (tethers) which form the spring. Figure 9 shows a cross-sectional sketch of a cantilever beam which is fixed to the silicon substrate (rim) on one side and has an attached inertial ("proof") mass on the other. Modeling this type of structure as a lumped spring-mass system assumes that the plate mass moves as a rigid body and all of the bending occurs in the beams (tethers). In addition to a simple force versus displacement relation, such systems are resonant, with the resonant frequency given by

$$f_{\mathrm{r}} = \frac{1}{2\pi}\sqrt{\frac{k}{M}} \tag{3}$$

where k is the spring constant and M is the mass of the system. In the case where the beams can be treated as idealized linear lumped elements, the spring constant can be found:

$$k = \frac{3EI}{L^3} \tag{4}$$

where I is the area moment of inertia around the centroidal axis of the beam cross-section, and L is the beam length. More complex spring structures can be created by combining these simple cantilever elements in series or parallel. This relationship ignores the effect of residual stress in the beam and is

valid for small displacements. Residual stresses will exist for any structure with more than one anchor point. To build structures which do not suffer from the complication of significant residual stress effects, micromachined tethers are often built as folded tethers, as shown in Fig. 10. Such structures have the benefit of being compact; but even more important, any residual or built-in stresses in the beams are allowed to relax. The spring constant of a folded tether (such as one of the four tethers in Fig. 10) can be approximated as two long beams in series. Clearly, a parallel set of relations can be developed for a beam in torsion yielding a torsional spring, moment of inertia system.

Capacitive Elements. A fundamental element for both sensing and actuation is the air-gap capacitor. Such a capacitor consists of a pair of conductors separated by a gap that allows one or both of the conductors to move. If the conductor is moved due to an applied force, the capacitance between the conductors changes. This change in capacitance can be measured electrically. Given an applied voltage, the same capacitance structure yields a force, which can be exploited for actuation.

The canonical capacitance structure is the parallel-plate capacitor. Such structures are often made up of a conductive surface on a bulk micromachined structure (e.g., a diaphragm) and a counterelectrode on an adjacent wafer-bonded surface. In surface micromachining, one capacitor plate is often made from the suspended structural film and the other plate is a conducting electrode on the substrate surface. For cases where the lateral extent of the capacitor is large compared to the gap between the electrodes (i.e., fringing fields can be neglected), the relation for the capacitance of an infinite parallel plate capacitor is an appropriate model:

$$C_{\mathrm{pp}} = \frac{\epsilon A}{g} \tag{5}$$

where A is the area of the capacitor, g is the gap between the capacitor plates, and ϵ is the permittivity of the material in the capacitor gap.

The other canonical microfabricated capacitance structure occurs in cases where the motion to be sensed or actuated is in the plane of the thin film. This leads to the interdigitated

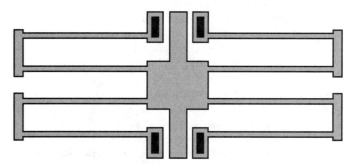

Figure 10. Sketch of a spring-mass system where the four springs are beam structures known as a folded flexures. These compact spring structures have the additional advantage that they allow residual stresses to relax. Thus their stiffness is not a function of residual stress.

Figure 11. A scanning electron micrograph of a typical surface micromachined interdigitated finger array. (Courtesy of MCNC.)

Figure 12. A scanning electron micrograph of a surface micromachined hinge joint. The movable portion (U-shaped part from center to right) is made from the first polysilicon layer. The fixed hinge (jumps up and over the central beam) is made from the second polysilicon layer. The depressed areas to the left and right of the fixed hinge are the anchor areas that fix the hinge to the substrate. (Courtesy of MCNC.)

finger structure as shown in Fig. 11. Calculation of the capacitance in this case is more difficult because the aspect ratio of the gap to the characteristic electrode dimension is close to 1. Consequently, the electric field has a large fringing field component and the parallel plate capacitance approximation is no longer valid. In general, 3-D numerical computations are required to obtain accurate values in this case (46). It is possible to use a correction factor in Eq. (5) to get a reasonably accurate closed-form solution for small perturbations to the geometry. However, these correction factors must, in general, be computed numerically, and their accuracy is generally limited to small displacements.

Pivots and Bearings. In addition to constrained, spring-mass type motions, structures have been fabricated that allow untethered motions. Constructions which allow free rotations are typically planar pin joints (22) or out-of-plane hinges (37). Hinges can be fabricated by using two polysilicon layers, with one layer as the stationary hinge and one layer as the rotating part. Figure 12 shows a close-up scanning electron micrograph of an out-of-plane hinge joint. Figure 13 shows an example of a pin bearing in the form of a variable capacitance micromachined motor. In addition, linear sliding constraints have been constructed (22). Further increasing the number of polysilicon layers allows the creation of ever more complicated joints and bearings. One application of hinges is in microptical systems such as bar code readers (49).

Isolated Thermal Mass. The most basic structure in micromachined thermal sensors is the thermally isolated (or "floating") thermal mass. Micromachining allows masses to be built with mechanical supports which have very small cross sections and thus very small thermal conductivities. Such isolation allows physical transduction to occur without significant thermal influence from the surrounding support wafer. The use of a metal film or a diffused conductive "wire" allows heaters and thermal sensors to be placed on these floating thermal masses.

Sensing and Actuation Mechanisms

A sensing mechanism is any physical effect that converts energy from one form to another. In most practical sensors (or actuators), conversion into (or from) an electrical signal is the desired goal. There are many, many physical transduction mechanisms, which might be exploited for sensors or actuators. Here we will examine several that are widely used in

Figure 13. A scanning electron micrograph of a typical surface micromachined rotary pin bearing. The mushroom-shaped structure in the center is the pin, which is fixed to the substrate by the anchor in the very center. The central section of the rotor is under the cap of the pin and thus is constrained to rotary motions around the pin. (Courtesy of MCNC.)

MEMS. Other mechanisms include floating gate transistor devices (23), electron tunneling from atomic tips (50), acoustic wave interactions (51), and so on.

Variable Capacitance. As a sensing mechanism, variable capacitance is very straightforward. As implied by Eq. (5), a change in the gap spacing or electrode area causes a change in capacitance. There is a multitude of mechanisms for sensing this change in capacitance, such as using a bridge circuit or using switched capacitance techniques (52,53).

Changes in capacitance also form a basic actuation mechanism. The electromechanical energy transduction associated with such systems can be examined by accounting for energy conservation in the electromechanical system. In the case of a simple capacitive system, the electrostatic co-energy can be found as the integral of (54):

$$dW'_e = q \cdot dv + f^e \cdot dx \tag{6}$$

where q and v are the charge and voltage on the capacitor, f^e is the force of electrical origin, and x is the direction of motion. In the electrically linear case where voltage is the independent variable, integration of Eq. (6) yields

$$f^e = \frac{\partial W'_e(v, x)}{\partial x} = \frac{1}{2} v^2 \frac{dC}{dx} \tag{7}$$

This relationship forms the basis for the electrostatic actuation in micromachined devices (55). Clearly this relation can be generalized to any number of electromechanical transduction elements or degrees of freedom (54). If we substitute the relation for the infinite parallel plate capacitor, Eq. (5), into Eq. (7), we find

$$f^e = -\frac{v^2 \epsilon A}{2g^2} \tag{8}$$

What is important to note here is that the force is a function of the square of the capacitive gap spacing. In cases where the mechanical restoring force is a linear spring (as would be the case for devices with beam tethers, for example), the system will become unstable at some value of the gap spacing (56).

For small motions about an operating position, the electrostatic force can be linearized:

$$f^e = -\frac{v^2 \epsilon A}{2(g + \Delta x)^2} \approx -\frac{v^2 \epsilon A}{2g^2} \left(1 - 2\frac{\Delta x}{g}\right) \tag{9}$$

where Δx is an incremental motion in one of the capacitor plates. This allows the electrostatic force to be cast in the form of an electrostatic spring constant:

$$k_e = \frac{\Delta f^e}{\Delta x} = \frac{v^2 \epsilon A}{g^3} \tag{10}$$

Note that this electrostatic spring acts in opposition to the mechanical spring.

Piezoresistance. Piezoresistance is the change in electrical resistivity of a material due to mechanical stresses. Semiconductors, such as silicon and germanium, show a particularly large piezoresistive effect which is more than an order of magnitude higher than that of metals. Hence, piezoresistance is frequently used as a sensing mechanism in micromachined devices. The geometry and placement of the piezoresistors are easily defined by selectively doping the semiconductor surface or by depositing a piezoresistive thin film on a nonsilicon surface. The ability to diffuse the dopant into the surface or deposit a thin film (i.e., polysilicon) allows the fabrication of thin piezoresistive layers (0.5 μm to 3.0 μm, for example). This allows the current to be restricted to the volume of maximum stress and thus yields the maximum signal.

The piezoresistive effect is an anisotropic relation between the stress tensor and the relationship between the electric field and the current density at a point. A full mathematical description of piezoresistance is beyond the scope of this article. The reader is referred to Sze (10) or Middelhoek (57) for a mathematical treatment. The relationship between electric field and current density can be written as

$$\frac{E_i}{\rho_0} = J_i + \pi_{ijkl} \sigma_{kl} J_j \tag{11}$$

where E is the electric field, ρ_0 is the unstrained resistivity, J is the current density, σ is the stress tensor, and π is the fundamental piezoresistive coefficient. Due to the cubic symmetry of silicon, π can be reduced to a 6-by-6 matrix with only three independent coefficients. The values of these coefficients for silicon are shown in Table 3 (58). Of all the orientations of stress and current that can be described by Eq. (11), two are particularly useful, the longitudinal and transverse components. To examine these, we can write the change in resistance in a piezoresistor as

$$\frac{\Delta R}{R} = \sigma_1 \pi_1 + \sigma_t \pi_t \tag{12}$$

where σ_1 is stress component parallel to the direction of the current, σ_t is stress component perpendicular to the direction of the current, π_1 is the longitudinal piezoresistance coefficient, and π_t is the transverse piezoresistance coefficient. Table 4 shows these piezoresistance coefficients as a function of various crystal orientations and the fundamental material parameters of Table 3.

Two canonical piezoresistive sensing structures are the diaphragm, typically used for pressure sensing, and the cantilever beam-proof mass, typically used for inertial sensing. Schematic cross-sectional drawings of these structures are shown in Figs. 8 and 9. In the diaphragm structure, piezoresistors are typically placed near the center of the diaphragm edge where the stress is highest [see Eq. (2)]. In the cantilever structure, the maximum stress caused by the deflection of the proof mass is at the beam surface.

A typical resistor configuration is the Wheatstone Bridge. In this configuration, two resistors are oriented so that they are most sensitive to stresses along their current carrying axis. Two more resistors are oriented to be most sensitive to stresses at right angles to their current flow. Thus, the resistance change of each pair is opposite. When electrically connected in a Wheatstone Bridge circuit, a large differential output voltage, which is independent of the absolute value of the piezoresistor's resistance, is obtained. One difficulty with piezoresistive sensors is their large temperature sensitivity.

Table 3. Piezoresistive Coefficients for *n*- and *p*-Type Silicon at Room Temperature (56)

Silicon	ρ_0 (Ω-cm)	π_{11} (10^{-11} Pa^{-1})	π_{12} (10^{-11} Pa^{-1})	π_{44} (10^{-11} Pa^{-1})
n-Type	11.7	-102.2	53.4	-13.6
p-Type	7.8	6.6	-1.1	138.1

These effects can be reduced by use of Wheatstone bridge circuits and careful resistor matching (10).

Piezoelectricity. Piezoelectricity relates to the crystalographic strains and polarization charge of an ionic crystal. When external force is applied to a piczoclcctric crystal, a polarization charge is induced on the surface. If the force is time-varying, this polarization charge can be sensed as a time-varying voltage or current. Similarly, if an electric potential is applied, the crystal is deformed. The mathematical description of piezoelectricity is a description of the coupling terms that enter the stress–strain and polarization–electric-field relations. The reader is referred to Madou (10) for an introduction to these relations or to Auld (59) for a more in-depth treatment.

The most common piezoelectric materials used in micromachined devices are crystalline quartz, ceramics such as zinc oxide, lead zirconate titanate (PZT), barium titanate, and lead niobate, and polymers such as polyvinylidene fluoride (PVDF). Piezoelectric materials are most often used as actuators because they are capable of producing high stresses (but low strains) and can achieve large forces with a small amount of input power. Piezoelectric sensors often use a bimorph beam structure whose output is proportional to the bending of the bimorph.

Thermal Mechanisms. Thermal sensing can be divided into methods that directly produce an electrical signal and those that are mediated by another, typically mechanical, mechanism. Direct electrical transduction occurs with thermocouples, temperature-dependent resistors, and the various transistor-based mechanisms (60). Mechanically mediated mechanisms typically take advantage of the thermal expansion of a material or the difference in expansion of two bonded materials (bimorph) to achieve a mechanical stress or deformation. This mechanical signal is then measured with one of the mechanisms above such as a piezoresistor. Thermal actuation is also mechanically mediated as in the case where a

bimorph is used to generate forces. Other actuation mechanisms take advantage of the large forces associated with a liquid–gas phase change.

Resonant Sensing. Another general sensing mechanism is resonant sensing. Here an energy storage element is driven to resonance by a feedback mechanism. The system is designed so that variations in the quantity of interest alter the resonant frequency of this feedback system. This change in frequency is then measured and converted to an output signal. Since frequencies can be measured with high accuracy, very high precision sensors can often be realized using this technique (61). Actuators can also make use of the large stored energy in a resonant system to obtain large nonvibratory motions from small vibratory motions (62).

Viscous Damping. Although not a sensing mechanism, damping is a very important physical phenomenon in micromachined devices. Because of the small dimensions of micromachined devices, surface forces, such as viscous fluid damping, tend to dominate over momentum-based forces. From a fluid mechanics point of view, micromachined devices operate in low Reynolds number regimes, even for large velocities. There are approximations to the full Navier–Stokes equations which often apply to micromachined devices. For example, for plate-like structures that move laterally over each other, the approximations of Stokes and Couette flow shear damping are appropriate (63). For parallel plates that move toward (or away from) each other, a squeeze-flim damping model is appropriate (64). Since micromachined devices operate in a low Reynolds number regime, the damping can be quite large when the device operates in liquids or in atmospheric pressure gas. For low-bandwidth devices, this damping can be helpful. However, high-bandwidth and resonant devices typically require evacuated packaging to achieve the required high values of the quality factor.

MICROMACHINED DEVICES

The breadth of micromachined devices that have been proposed or demonstrated is large and continues to grow. A general characteristic of these devices is that they interact or facilitate interaction across physical domains. Cataloging these devices can be quite difficult since they span a broad range of physical domains. However, these devices can, in general, be characterized as either sensors or actuators. Sensors typically translate energy from the energy domain being sensed into a signal (typically electrical), possibly through intermediate domains. Actuators, on the other hand, typically convert an electrical signal into an energy or action in the physical domain to be actuated. As can be seen from the similarity of these definitions, both sensors and actuators are fundamentally the same thing: transducers of energy from one physical

Table 4. Longitudinal and Transverse Piezoresistance Coefficients for Various Directions in Cubic Crystals (after Ref. 9)

Longitudinal Direction	π_l	Transverse Direction	π_t
[1 0 0]	π_{11}	[0 1 0]	π_{12}
[0 0 1]	π_{11}	[1 1 0]	π_{12}
[1 1 1]	$\frac{1}{3}(\pi_{11} + 2\pi_{12} + 2\pi_{44})$	[1 $-$1 0]	$\frac{1}{3}(\pi_{11} + 2\pi_{12} - \pi_{44})$
[1 1 0]	$\frac{1}{2}(\pi_{11} + \pi_{12} + \pi_{44})$	[1 1 1]	$\frac{1}{3}(\pi_{11} + 2\pi_{12} - \pi_{44})$
[1 1 0]	$\frac{1}{2}(\pi_{11} + \pi_{12} + \pi_{44})$	[0 0 1]	π_{12}
[1 1 0]	$\frac{1}{2}(\pi_{11} + \pi_{12} + \pi_{44})$	[1 $-$1 0]	$\frac{1}{2}(\pi_{11} + \pi_{12} - \pi_{44})$

After Ref. 10, p. 166.

domain to another. The real distinction between a sensor and an actuator is one of intent. And of course, there are devices which contain elements of both, such as the force re-balance accelerometer which uses actuation driven by a feedback loop to null balance the sensor.

The commercial successes of micromachined devices remain few, mostly due to manufacturing reproducibility problems and packing complexity. The successes mostly fall into the category of sensors. The following sections overview many of the broad classes of micromachined sensors and actuators.

Pressure Sensors

The silicon micromachined pressure sensor was one of the first commercially successful micromachined devices. As late as 1989, bulk micromachined pressure sensors accounted for most of the revenue in silicon micromachined sensors. Most of the silicon pressure sensors manufactured today are for automobile applications, most notably the manifold absolute pressure (MAP) sensor. These pressure sensors can sense relative to a vacuum or relative to a fixed pressure such as atmospheric, or they can be differential.

The general functional characteristic of the pressure sensor is quite simple. Applied pressure causes a thin-film diaphragm to deflect. The deflection is then sensed, typically by one of two methods: capacitive or piezoresistive. Figure 8 shows a schematic cross section of a bulk micromachined, silicon diaphragm pressure sensor. Piezoresistively sensed devices place a piezoresistor at one or more locations which encounter maximum stress when the diaphragm deflects. The piezoresistors are often placed in a bridge circuit to increase sensitivity. In the case of a capacitive sensor, the deflection of the diaphragm changes the gap between it and a counterelectrode (not shown). The change in capacitance is sensed electronically.

Piezoresistive sensing is simple and requires little, if any, external circuitry. Capacitive sensing requires an external circuit; and in the case where this circuit is not integrated on the same piece of silicon as the diaphragm, the capacitance must be large enough to be sensed in the presence of the large parasitic capacitance associated with the interconnection to the circuit. These issues often cause capacitive sensors to be more expensive to produce. However, capacitive sensing is inherently more sensitive and significantly less temperature-dependent than piezoresistive sensing. Thus, for many applications, where the sensor will see significant temperature variations in its environment, capacitive sensing is chosen.

Inertial Sensing

Inertial sensing implies the sensing of inertial forces such as acceleration, gravitation, or angular acceleration. Current commercial examples are accelerometers and gyroscopes, or angular rate sensors, used in automobile and military applications. Other related applications are shock sensors, vibration sensors, and gravitometers.

Accelerometers. Accelerometers have seen their largest market in automobile airbag sensing. Micromachined accelerometers have found a niche here because of their ability to be cheaply produced in large volumes. This has given them a significant advantage over their larger electromechanical predecessors.

The majority of micromachined accelerometers work in a similar manner. They are constructed as a proof mass and spring system. When the device is subject to an acceleration load, the proof mass moves. This motion is typically sensed by capacitive or piezoresistive means. A good figure of merit for accelerometers is the resonant frequency of the sensitive mode [see Eq. (3)]. In most cases, the accelerometer output is proportional to the displacement of the proof mass. Thus, the scale factor is proportional to one over the square of the resonant frequency. Hence, a lower resonant frequency yields a more sensitive device for a given proof mass deflection to output gain.

Bulk micromachined accelerometers have the advantage of a large proof mass which can be as thick as the wafer from which it is constructed. In fact, wafer-bonded structures allow the proof mass to be multiple wafer thicknesses. As with bulk pressure sensors, the deflection of the proof mass can be measured capacitively with an adjacent counterelectrode. In the case of piezoresistive sensing, piezoresistors are typically placed on the tether springs which support the proof mass (see Fig. 9). Since the proof mass is thick and thus rigid, the tethers are the only part of the structure where significant strains are obtained. The tethers in this type of system are typically thin beams which are left unetched during the silicon etch, or are made from a thin film which was patterned and etched prior to the proof mass etch.

Surface micromachined accelerometers are thin-film planar devices. They can move perpendicular to the plane of the proof mass, as bulk micromachined accelerometers do, or they can move in the plane. In the perpendicular case, they can use capacitive sensing in an analogous fashion as the bulk structures. The mass of a surface micromachined structure is typically much smaller than a bulk micromachined structure. Thus, to keep the resonant frequency of a surface micromachined accelerometer at an acceptably low value, the tethers must be significantly more compliant. As a consequence, the strain level in the thethers will be small and usually insufficient for piezoresistive sensing.

In the case of lateral, or in-plane, accelerometers, the capacitive counter electrodes must be in the same plane as the proof mass motion. This implies that the "plates" are formed by the edge of the structural layer's thin-film material. In order to obtain sufficient capacitance for sensing, interdigitated fingers are used to increase the capacitor electrode area and thus the variable capacitance. Figure 6 shows a SEM of the Analog Devices ADXL76 accelerometer. This is an example of a surface micromachined accelerometer showing the interdigitated finger electrodes.

Given the basic accelerometer structure and sensing mechanism, there are many ways to construct the sensing system. A fundamental difference is the closed-loop versus the open-loop system. In an open-loop system, the proof mass is free to deflect according to the applied acceleration and the tether's compliance. In a closed-loop system, a feedback loop applies a force to the proof mass which balances the inertial force and keeps the proof mass close to its zero acceleration position. The output of the sensor is then proportional to the magnitude of the feedback signal. The benefits of such a scheme are that the sensor will avoid any nonlinearities associated with large proof mass deflections. However, this comes at the expense of a more complicated, larger, and more expensive to produce circuit.

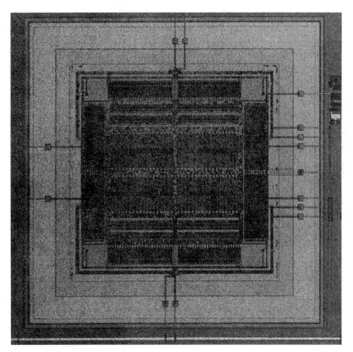

Figure 14. A photomicrograph of a surface micromachined vibratory gyro. The device is driven into resonance in the up/down axis. An angular rate about the out-of-plane axis produce a coriolis acceleration in the left/right axis which is sensed as a left/right motion. (From Ref. 66.)

Other types of accelerometers have been devised. In particular, the resonant accelerometer is of interest because of its potential for high accuracy. In a typical resonant accelerometer, the motion of a proof mass caused by the acceleration causes the stiffness of a resonant beam to change. This, in turn, causes its resonant frequency to change. Another type of accelerometer is the tunneling accelerometer (65). Here displacements of a proof mass are measured with high precision by measuring the tunneling current between a sharp tip and an adjacent ground plane. Micromachining techniques are well suited to fabricating tips with appropriately small tip radii.

Gyroscopes. Another important class of inertial sensor is the gyroscope or angular rate sensor. The most familiar type of gyroscope is the rotating disk gyro. However, this type of free rotational motion remains impractical for micromachined devices. More practical for micromachined devices is the vibrating gyro. The fundamental operating principle is the sensing of Coriolis acceleration:

$$a_c = 2\Omega \times v \qquad (13)$$

where a_c is the Coriolis acceleration, Ω is the angular rate, and v is velocity of the proof mass. This velocity is usually obtained as the vibration of a spring-mass system in an oscillatory feedback loop. The Coriolis acceleration then acts on the inertial mass to generate a motion which is sensed.

Figure 14 shows a vibratory rate gyroscope which is sensitive to rotation rates about the direction perpendicular to the plane of the structure (z axis) (66). The moving mass is driven into resonance in the left–right, in-plane direction (x axis).

Rotations in the z axis cause y-axis-directed forces. The y-axis-directed motions are sensed capacitively in the same manner as the lateral accelerometer described above.

Although the vibratory gyro avoids many of the problems associated with rotating gyros, its concept is based on the assumption of an ideal lossless vibrating member with perfect symmetry. In a practical case, the most obvious loss mechanism is viscous damping. This can be eliminated or reduced by operation in a vacuum. This requires a stable vacuum package which is difficult and expensive to produce (67). Also of significance are acoustic losses into the body to which the gyro is fixed (i.e., the substrate). Such coupling can lead to linear accelerations or vibrations appearing as rate signals. Constructing a balanced oscillator in which reactions to the driving force are not felt by the device's mountings can reduce these losses (68).

In addition to vibrating mass structures as described above, there is a class of structures which use rings or thin axisymmetric shells as the vibrating structures. Figure 15 shows a vibrating ring gyro made with electroplated nickel (69). Vibrating shell gyros pick off the rotation of a vibrational mode's antinode position, which is caused by the input rotation rate. Note that the rotation of the antinodes corresponds to a rotation-induced transfer of energy between two identical vibrational modes (70). In vibrating mass gyros, the output signal depends on the rotation-induced transfer of energy between two different vibration modes. This difference allows the vibrating shell gyro to avoid the temperature sensitivities caused by different temperature variations in the vibration modes of the vibrating mass gyro.

Thermal Sensors

There are a large number of thermal sensors which are silicon-based, including the temperature-sensing capabilities of the transistor itself. Here we will focus on those thermal devices that rely on micromachining for their function.

Thermal flow sensors are often based on the idea that a fluid, which flows past a hot surface, will carry heat away from the surface. A typical micromachined device using this type of mechanism is the flow anemometer. Here, a temperature-dependent resistor is used to detect the heat lost to the flow from a resistive heat source. The rate of heat loss is proportional to the flow velocity. A related mechanism uses two thermally isolated elements. The first is a heater, which provides a pulsed heat flux. The second is a downstream temperature sensor which detects the heat pulse generated by the upstream heater. The fluid velocity is given by the time of flight (10). Figure 16 shows a characteristic implementation (71). The heater and the sensor are suspended in the center of the flow channel by thin, thermally isolating supports. The thermal sensing element is often a temperature-dependent resistor in a bridge configuration. The IR bolometer is another device that uses an isolated thermal resistor (72).

Another basic construction is the stress-based thermal sensor. Here two material layers with different thermal expansion coefficients are used to form a bimorph structure that exhibits a bending moment that is a function of the temperature of the device. Examples include IR sensors where the heating caused by the IR energy flux causes a deflection. Thermal actuation devices are also based on the bimorph effect. Thermally actuated valves have been built that take ad-

Figure 15. A scanning electron micrograph of a ring gyro device. The ring is driven into resonance by the surrounding electrodes (T-shaped structures). An angular rate about the out-of-plane axis causes a shift in the position of the vibration nodes which is sensed (capacitively) by the surrounding electrodes. (From Ref. 69.)

vantage of the large forces associated with a liquid–gas phase change (73). Other examples include the actuation of optical mirror devices (74) and thermally driven resonators.

Chemical and Biological Sensors

A full discussion of chemical and biological sensors is an intricate one that is far beyond the scope of this article. However, it is worth pointing out a few common threads. These sensors can be categorized into devices which detect gases, devices which detect liquids, and devices which detect complicated biological molecules such as DNA or proteins. One thing to note is that most of these sensors do not rely fundamentally on micromachining for their construction. The fundamental functional mechanisms are typically surface chemical in nature (often requiring catalysts, etc.). To date, most innovation in these systems has come from exploration of chemical reac-

tions and materials. Semiconductor and micromachined device fabrication techniques simply provide a way of making the devices smaller, cheaper, faster, etc. These chemical issues are beyond the scope of this article. The reader is referred to Ref. (10) for a discussion of these issues.

The detection of combustible gases such as CO, H_2, alcohols, hydrocarbons, and so on, is most frequently performed with either semiconducting metal oxides or field-effect transistors (FETs). In the case of the metal oxide semiconductors, a surface reaction occurs between oxygen and the gas, which changes the resistance of the semiconductor. In the case of the FET, a reaction occurs on the surface over the channel, which alters the potential at the "gate" and modulates current flow in the channel. This same FET mechanism can be used to detect ionic species. Biosensors are an extension of chemical sensors, which rely on biological materials and biochemical reactions. These reactions have the benefit of high selectivity and sensitivity. Thus a biosensor can be described as the addition of a biological sensing mechanism to a chemical (or physical) transducer.

One fundamental difficulty with both chemical and biological sensors is that they must come into contact with the chemical or biological environment they wish to sense. It is difficult to make them reversibly reactive to the desired species while, at the same time, unreactive to other species present in the environment. Selectively permeable membranes and arrays of sensors have been used to overcome this problem.

Micromachined Actuators

Actuators have long been envisioned as allowing the true revolution in micromachined devices. To date, however, their use has been limited. The most widespread use of actuation is in force feedback inertial sensors (6). One of the first devices that constituted an actuator in its own right was the so-called micromotor. These devices were typically rotary, variable-capacitance motors as shown in Fig. 17 (75). Variable capacitance was chosen as a drive mechanism because of its compatibility with micromachining processes and materials.

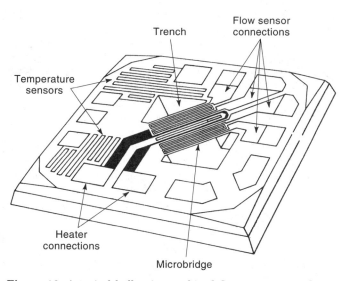

Figure 16. A typical bulk micromachined flow anemometer device. Heat is generated by a resistive element (black) and sensed by the upstream and downstream flow sensor elements. (After Ref. 71.)

Figure 17. A scanning electron micrograph of a salient-pole variable-capacitance micromotor. Attractive electrostatic forces between the rotor poles and the fixed stator poles (around the periphery) cause rotary motion. (Courtesy of MCNC.)

However, electrostatic systems have some scaling advantages over magnetic systems as the characteristic length decreases to the micron range (55). The most significant difficulty in the construction of such devices is the formation of a bearing structure that can restrain the rotor to rotational motion without significant friction or wear. Because of this difficulty, and in an effort to overcome such friction (as well as viscous air damping), various motor configurations have been constructed (76). In spite of these design variations, micromachined motors continue to suffer from short lifetimes due to wear. Also, they suffer from restricted usefulness due to the relatively high applied voltages required to overcome friction and supply useful output torque.

Electrostatic linear actuators have also been developed. Devices with relatively unconstrained linear motion suffer from many of the same problems as their rotary predecessors. Constrained motion devices—that is, actuators whose moving member is supported and whose limited motion is typically due to a balance between the electrostatic drive force and an elastic restraint—have been more successful. Such devices form the basis for resonant sensors, where the spring support and the moving mass form the resonant system (77). Hybrid systems have also been developed such as impact systems which use a resonator to strike a second moving mass (for example, a rotor) and cause it to move (78). Repeated, high-frequency impacts yield quasi-continuous motion. Another re-

lated concept is the use of electrostatic forces to modulate the effective stiffness and hence resonant frequency of a spring-mass resonant system (79).

One class of constrained motion actuating devices is the optical mirror and optical grating devices. The simplest type is the torsional mirror device (80). Here a mirror, supported on a torsional tether, is actuated to control the angle of reflection of a laser beam or other light source. Both continuous and bi-stable control have been used. Various devices which use thin cantilever beams or bridges to form controllable defraction gratings have been demonstrated (81).

The use of piezoelectric materials in actuator devices is very common due to their ability to provide large forces and small displacements with low applied voltages. Typical applications are fluid pumps (82), linear and rotary motors (83), and surface acoustic wave (SAW) sensors (10).

Another device, which has garnered considerable interest, is the micromachined switch, or relay. The ability to achieve a true open circuit switch, integrated with electronics could be very useful for integrated circuit testing devices, communications systems, and RF systems. Simple electrostatically actuated cantilever beams or bridges have been designed for this purpose (84). The difficulty with these structures, as with macroscopic switches, are the material properties of the contacting surfaces. Their sticking properties and their ability to withstand millions of open/close cycles is critical.

Software Design Tools for Micromachined Devices

The design of micromachined devices involves examining devices that operate in a broad set of physical domains. The 3-D nature of micromachined devices typically requires full 3-D numerical simulation based on the Finite Element Method (FEM) or the Boundary Element Method (BEM). FEM and BEM tools for specific physical domains, such as mechanics or magnetics, were developed for macroscopic systems. Although these tools can provide useful simulation results for some types of uncoupled behavior, micromachined devices usually perform a transduction of energy between one physical domain and another which often requires the solution of coupled physics problems (85). In this case, coupled physical domain solvers are required. A classic example for micromachined devices is the electrostatic instability of a bending beam electrode above a ground plane. As a voltage is applied to the electrode, a force develops between the electrode and the ground plane. This causes the beam to bend, closing the gap between the electrode and the ground plane. This, in turn, causes the force to increase, which bends the beam some more. At some "pull-in" voltage, this interaction is unstable and the beam collapses down to the ground plane. Knowledge of this pull-in voltage is often an important design parameter. Simulation of this value requires that the mechanics and the electrostatics of this system be solved together in a coupled, self-consistent way. There are many other coupled domains that are relevant to micromachined devices, such as magnetomechanics and fluid–structure interaction. Work on micromachined devices has accelerated efforts to develop coupled solvers for these domains.

From the perspective of a design tool environment, MEMS involve making 3-D mechanical devices with the manufacturing techniques of integrated circuit fabrication. This creates a need for design tools that join the functionality of 3-D mechanical design tools (MDA) with integrated circuit design tools (EDA). For example, 3-D micromachined devices are typically generated photolithographically from a series of 2-D layout masks. The masks are made in an EDA layout editor. However, FEM-based simulations require a 3-D solid model as would be generated in a MDA solid model editor. Micromachined device design software environments such as MEM-CAD (46) bridge this gap by generating the 3-D solid model directly from the 2-D layout masks and a description of the fabrication process. In addition, micromachined devices are most often used within a larger system which typically contains one or more micromechanical elements and circuit elements. In order to simulate such complex systems in a reasonable amount of time, the FEM/BEM-based device model must, in general, be reduced to a lower-order lumped model (86). Software tools which automatically produce such low-order models are beginning to appear (46).

Another important area for simulation is the interaction of the MEMS system with the package. Integrated circuits interact with their packages primarily through thermal dissipation and the mechanical stresses caused by thermal expansion coefficient mismatches. Due to their mechanical nature, micromachined devices are typically even more susceptible to these issues. In addition, many micromachined devices, such as pressure or flow sensors, must interact directly with their environment. Thus, the MEMS designer often has to design application-specific packaging. Optimizing such coupled MEMS/package systems requires considerable simulation. These application-specific packages are often more expensive than the underlying MEMs device. Therefore, an optimized package/device design can be critical to overall price and performance.

Further Reading

The reader is directed to several books and chapters discussing micromachining technology (10). Other very useful books are listed in the Reading List. The two major journals of the field are the IEEE *Journal of Micro Electromechanical Systems* (JMEMS) and *Sensors & Actuators*. Although there are at least 10 conferences that discuss micromachined devices, the three major conferences are the *International Conference on Solid-State Sensors and Actuators* (Transducers), *The Solid-State Sensor and Actuator Workshop,* and the *International Workshop on Micro Electro Mechanical Systems* (MEMS).

BIBLIOGRAPHY

1. S. Wolf and R. N. Tauber, *Silicon processing for the VLSI era,* Vol. 1, *Process technology,* Sunset Beach, CA: Lattice Press, 1986.

2. O. N. Tufte, P. W. Chapman, and D. Long, Silicon diffused element piezoresistive diaphragms, *J. Appl. Phys.,* **33**: 3322, 1962.

3. Editorial, *Thermal Character Print Head,* Austin: Texas Instruments, 1977.

4. E. Bassous, H. H. Taub, and L. Kuhn, Ink jet printing nozzle arrays etched in silicon, *Appl. Phys. Lett.,* **31**: 135–137, 1977.

5. P. O'Neill, A monolithic thermal converter, *Hewlett-Packard J.,* **31**: 12–13, 1980.

6. S. J. Sherman et al., A low cost monolithic accelerometer: Product/technology update, *IEEE Int. Electron Devices Meet.,* San Francisco, 1992, pp. 501–504.

7. B. Studer and W. Zingg, Technology and characteristics of chemically milled miniature quartz crystals, *4th Eur. Freq. Time Forum,* Neuchatel, Switzerland, 1990, pp. 653–658.

8. K. E. Peterson, Silicon as a mechanical material, *Proc. IEEE,* **70**: 420, 1982.

9. J. S. Danel, F. Michel, and G. Delapierre, Micromachining of quartz and its applications to an acceleration sensor, *Sensors Actuators,* **A21-A23**: 971, 1990.

10. S. M. Sze (ed.), *Semiconductor Sensors,* New York: Wiley, 1994; M. Madou, *Fundamentals of Microfabrication,* Boca Raton, FL: CRC Press, 1997.

11. H. L. Seidel et al., Anisotropic etching of crystalline silicon in alkaline solutions: I. Orientation dependence and behavior of passivation layers, *J. Electrochem. Soc.,* **137**: 3612, 1990.

12. H. L. Seidel et al., Anisotropioc etching of crystalline silicon in alkaline solutions: II. Influence of dopants, *J. Electrochem. Soc.,* **137**: 3626, 1990.

13. A. Reisman et al., The controlled etching of silicon in catalyzed ethylene-diamine-pyrocathechol-water solutions, *J. Electrochem. Soc.,* **126**: 1406–1414, 1979.

14. O. Tabata et al., Anisotropic etching of silicon in TMAH solutions, *Sensors Actuators,* **A34**: 51–57, 1990.

15. C. Keller and M. Ferrari, Milli-scale polysilicon structures, *Tech. Dig., 1994 Solid State Sensor Actuator Workshop,* Hilton Head Island, SC, 1994, pp. 132–137.

16. G. Wallis and D. L. Pomerantz, Field assisted glass-metal sealing, *J. Appl. Phys.,* **40**: 3946, 1969.

17. T. R. Anthony, Anodic bonding of imperfect surfaces, *J. Appl. Phys.,* **54**: 2419–2428, 1983.

18. M. Shimbo et al., Silicon-to-silicon direct bonding method, *J. Appl. Phys.,* **60**: 2987, 1986.

19. W. H. Ko, J. T. Suminto, and G. J. Yeh, Bonding techniques for microsensors, in C. D. Fung et al. (eds.), *Micromachining and Micropackaging of Transducers,* Amsterdam: Elsevier, 1985, pp. 41–61.

20. E. J. J. Kruglick, B. A. Warneke, and K. S. J. Pister, CMOS 3-axis accelerometers with integrated amplifier, *Proc. IEEE Micro Electro Mechn. Syst. Workshop (MEMS '98),* Heidelberg, 1998, pp. 631–636.

21. D. L. Klein and D. J. D'Stefan, Controlled etching of silicon in the HF–HNO₃ system, *J. Electrochem. Soc.,* **109**: 37–42, 1962.

22. L. S. Fan, Y.-C. Tai, and R. S. Muller, Integrated movable micromechanical structures for sensors and actuators, *IEEE Trans. Electron Devices,* **35**: 724–730, 1988.

23. H. C. Nathanson et al., The resonant gate transistor, *IEEE Trans. Electron Devices,* **ED-14**: 117–133, 1967.

24. R. T. Howe and R. S. Muller, Resonant polysilicon microbridge integrated with NMOS detection circuitry, *IEEE Int. Electron Devices Meet.,* San Francisco, 1984, pp. 213–216.

25. T. A. Core, W. K. Tsang, and S. J. Sherman, Fabrication technology for an integrated surface-micromachined sensor, *Solid State Technol.,* **October**: 39–44, 1993.

26. L. J. Hornbeck, Deformable-mirror spatial light modulators, *SPIE Crit. Rev.,* **1150**: 86, 1989.

27. G. T. Mulhern, S. Soane, and R. T. Howe, Supercritical carbon dioxide drying for microstructures, in *1993 Int. Conf. Solid-State Sensors Acuators (Transducers '93),* Yokohama, Japan: p. 296.

28. H. Guckel, J. J. Sniegowski, and T. R. Christenson, Advances in processing techniques for silicon micromechanical devices with smooth surfaces, in *1989 Int. Workshop on Micro Electromechanical Systems (MEMS '89),* Salt Lake City, UT: p. 71.

29. U. Srinivasan, M. R. Houston, R. T. Howe, and R. Maboudian, Self-assembled fluorocarbon films for enhanced silicon reduction,

1997 Int. Conf. Solid-State Sensors Actuators (Transducers '97), Chicago, IL: pp. 1399–1402.

30. H. Guckel, T. Randazzo, and D. W. Burns, A simple technique for the determination of mechanical strain in thin films with application to polysilicon, *J. Appl. Phys.,* **57**: 1671–1675, 1985.

31. R. T. Howe and R. S. Muller, Polycrystalling silicon and amorphous silicon micromechanical beams: Annealing and mechanical properties, *Sensors Actuators,* **4**: 447–454, 1983.

32. M. Biebl, G. T. Hulhern, and R. T. Howe, Low in-situ phosphorous doped polysilicon for integrated MEMS, *8th Int. Conf. Solid-State Sensors Actuators (Transducers 95),* Stockholm, Vol. 1, 1995, pp. 198–201.

33. H. Guckel et al., The application of fine-grained tensile polysilicon to mechanically resonant transducers, *Sensors Actuators* **A21-A23**: 346–350, 1990.

34. P. Krulevitch and G. C. Johnson, Stress gradients in thin films used in micro-electro-mechanical systems, *ASME Winter Annu. Meet.,* New Orleans, LA, 1993, DSC-46: pp. 89–95.

35. J. H. Smith et al., Embedded micromechanical devices for the monolithic integration of MEMS with CMOS, *Int. Electron Devices Meet.,* Washington, DC, 1995, pp. 609–612.

36. J. M. Bustillo et al., Process technology for the modular integration of CMOS and polysilicon microstructures, *Microsyst. Technol.,* **1**: 30–41, 1994.

37. K. S. J. Pister et al., Microfabricated hinges: 1 mm vertical features with surface micromachining, *Tech. Dig., 6th Inter. Conf. Solid-State Sensors Actuators (Transducers '91),* 1991.

38. T. Masaki, K. Kawata, and T. Masuzawa, Micro electro-discharge machining and its applications, *Proc. IEEE Micro Electro Mech. Syst. Workshop (MEMS '90),* Napa, CA, 1990, pp. 21–26.

39. M. J. Vasile, C. Biddick, and A. S. Schwalm, Microfabrication by ion milling: The lathe technique, *J. Vac. Sci. Technol.,* **B12**: 2388–2393, 1994.

40. M. A. Moreland, Ultrasonic machining, in S. J. Schneider (ed.), *Engineered Materials Handbook,* Metals Park, OH: ASM International, 1992, pp. 359–362.

41. E. W. Becker et al., Production of separation nozzle systems for uranium enrichment by a combination of x-ray lithography and galvanoplastics, *Naturwissenschaften,* **69**: 520–523, 1982.

42. H. Guckel et al., Fabrication of assembled micromechanical components via deep x-ray lithography, *Proc., IEEE Micro Electro Mech. Syst. Workshop (MEMS '91),* Nara, Japan, 1991, pp. 70–74.

43. C. Burbaum et al., Fabrication of capacitive acceleration sensors by the LIGA technique, *Sensors Actuators,* **A25**: 559–563, 1991.

44. H. Guckel et al., A first functional current excited planar rotational magnetic micromoter, *Proc., IEEE Micro Electro Mech. Syst. Workshop (MEMS '93),* Fort Lauderdale, FL, 1993, pp. 7–11.

45. T. R. Christenson, J. Klein, and H. Guckel, An electromagnetic micro dynamometer, *Proc., IEEE Micro Electro Mech. Syst. Workshop (MEMS '94),* Amsterdam, 1994, pp. 386–391.

46. *Memcad 5.0 Users Manual,* Cambridge, MA: Microcosm Technologies, Inc., 1999.

47. S. P. Timoshenko and S. Woinowsky-Krieger, *Theory of Plates and Shells,* New York: McGraw-Hill, 1970, 2nd ed.

48. W. C. Young, *Roark's Formulas for Stress & Strain,* New York: McGraw-Hill, 1989, 6th ed.

49. M.-H. Kiang et al., Micromachined polysilicon microscanner for barcode readers, *IEEE Photon. Technol. Lett.,* **8** (12): 1707–1709, 1996.

50. J. Wang et al., Study of tunneling noise using surface micromachined tunneling tip devices, *Proc. 1997 Int. Conf. Solid-State Sensors Actuators (Transducers '97),* Chicago, 1997, Vol. 1, pp. 467–470.

51. B. A. Martin, S. W. Wenzel, and R. M. White, Viscosity and density sensing with ultrasonic plate waves, *Sensors Actuators,* **A21-A23**: 704–708, 1989.

52. J. T. Kung, H.-S. Lee, and R. T. Howe, A digital readout technique for capacitive sensor applications, *IEEE J. Solid-State Circuits,* **SC-23**: 972–977, 1988.

53. M. Lemkin, B. E. Boser, and D. Auslander, Fully differential lateral ΣΔ accelerometer with digital output, *Tech. Dig., 1996 Solid State Sensor Actuator Workshop,* Hilton Head Island, SC, 1996.

54. H. H. Woodson and J. R. Melcher, *Electromechanical Dynamics,* New York: Wiley, 1968, Part 1, Chap. 3.

55. S. F. Bart et al., Design considerations for microfabricated electric actuators, *Sensors Actuators,* **14**: 269–292, 1988.

56. P. M. Osterberg and S. D. Senturia, M-TEST: A test chip for MEMS material property measurement using electrostatically actuated test structures, *J. Microelectromech. Syst.,* **6**: 107–118, 1997.

57. S. Middelhoek and S. A. Audet, *Silicon Sensors,* London: Academic Press, 1989, Chap. 3.

58. C. S. Smith, Piezoresistance effect in germanium and silicon, *Phys. Rev.,* **94**: 42–49, 1954.

59. B. A. Auld, *Acoustic Fields and Waves in Solids,* Malabar, FL: Krieger Publishing, 1990, 2nd ed., Chap. 8.

60. L. Ristic (ed.), *Sensor Technology and Devices,* Boston: Artech House, 1994, Chap. 8.

61. R. T. Howe and R. S. Muller, Resonant microbridge vapor sensor, *IEEE Trans. Electron Devices,* **ED-33**, 499–506, 1986.

62. M. J. Daneman et al., Linear microvibromotor for positioning optical components, *Proc. IEEE Micro Electro Mech. Syst. Workshop (MEMS '95),* Amsterdam, 1995, pp. 55–60.

63. Y.-H. Cho, A. P. Pisano, and R. T. Howe, Viscous damping model for laterally oscillating microstructures, *J. Microelectromech. Syst.,* **3**: 81–87, 1994.

64. Y.-J. Yang and S. D. Senturia, Numerical simulation of compressible squeezed-film damping, *Tech. Dig., 1996 Solid State Sensor Actuator Workshop,* Hilton Head Island, SC, 1996, pp. 76–79.

65. H. K. Rockstad et al., A miniature high-sensitivity broad-band accelerometer based on electron tunneling transducers, *Sensors Actuators,* **A43**: 107–114, 1994.

66. W. A. Clark, R. T. Howe, and R. Horowitz, Surface micromachined Z-axis vibratory rate gyroscope, *Tech. Dig., 1996 Solid State Sensor Actuator Workshop,* Hilton Head Island, SC, 1996, pp. 283–287.

67. M. B. Cohn et al., Wafer-to-wafer transfer of microstructures for vacuum packaging, *Tech. Dig., 1996 Solid-State Sensor Actuator Workshop,* Hilton Head Island, SC, 1996, pp. 32–35.

68. C. H. J. Fox and D. J. W. Hardie, Vibratory gyroscopic sensors, in H. Sorg (ed.), *Proceedings of the Symposium on Gyro Technology,* Berln: Stuttgart, Germany: 1984, pp. 13.0–13.30.

69. M. W. Putty and K. Najafi, A micromachined vibrating ring gyroscope, *Tech. Dig., 1996 Solid-State Sensor Actuator Workshop,* Hilton Head Island, SC, 1994, pp. 213–220.

70. A. Lawrence, *Modern Inertial Technology,* New York: Springer-Verlag, 1993.

71. R. G. Johnson and R. E. Higashi, A highly sensitive silicon chip microtransducer for air flow and differential pressure sensing applications, *Sensors Actuators,* **11**: 63–72, 1987.

72. W. Lang, K. Kuhl, and E. Obermeier, A thin-film bolometer for radiation thermometry at ambient temperature, *Sensors Actuators,* **A21-A23**: 473–477, 1990.

73. M. J. Zdeblick et al., Thermopneumatically actuated microvalves and integrated electro-fluidic circuits, *Tech. Dig., 1994 Solid State Sensor Actuator Workshop,* Hilton Head Island, SC, 1994, pp. 251–255.

74. W. D. Cowan and V. M. Bright, Thermally actuated piston micromirror arrays, *Proc. SPIE,* **3131**: 1997.

75. S. F. Bart et al., Electric micromotor dynamics, *IEEE Trans. Electron Devices,* **39**: 566–575, 1992.

76. M. Mehregany et al., Principles in design and microfabrication of variable-capacitance side-drive motors, *J. Vac. Sci. Technol.,* **A8**: 3614–3624, 1990.

77. C. T.-C. Nguyen, Microelectromechancal devices for wireless communications, *Proc. IEEE Micro Electro Mech. Syst. Workshop (MEMS '98),* Heidelberg, 1998, pp. 1–7.

78. A. P. Lee, P. B. Ljung, and A. P. Pisano, Polysilicon micro vibromotors, *Proc. IEEE Micro Electro Mech. Syst. Workshop (MEMS '91),* Nara, Japan, 1991, pp. 177–182.

79. S. G. Adams, F. Bertsch, and N. C. MacDonald, Independent tuning of the linear and nonlinear stiffness coefficients of a micromechanical device, *Proc. IEEE Micro Electro Mech. Syst. Workshop (MEMS '96),* San Diego, CA, 1996, pp. 32–37.

80. J. M. Younse, Projection display systems based on the Digital Micromirror Device (DMD), *Proc. SPIE, Microelectron. Struct.,* **2641**: 64–75, 1995.

81. D. M. Bloom, Grating light valves for high resolution displays, *Tech. Dig., 1994 Int. Electron Devices Meet.,* San Francisco, 1994, p. 343.

82. H. T. G. Van Lintel, F. C. M. van der Pol, and S. Bouwstra, A piezoelectric micropump based on micromachining of silicon, *Sensors Actuators,* **15**: 153–167, 1988.

83. K. R. Udayakumar et al., Ferroelectric thin film ultrasonic micromotors, *Proc. IEEE Micro Electro Mech. Syst. Workshop (MEMS '91),* Nara, Japan, 1991, pp. 109–113.

84. S. Majumder et al., Measurement and modeling of surface micromachined, electrostatically actuated microswitches, *Proc. 1997 Int. Conf. Solid-State Sensors Actuators (Transducers '97),* Chicago, 1997, pp. 1145–1148.

85. S. D. Senturia, CAD for microelectromechanical systems, *Proc. Int. Conf. Solid-State Sensors Actuators (Transducers '95),* Stockholm, 1995.

86. N. R. Swart et al., AutoMM: Automatic generation of dynamic macromodels for MEMS devices, *Proc. IEEE Micro Electro Mech. Syst., Workshop (MEMS '98),* Heidelberg, 1998, pp. 178–183.

87. K. B. Albaugh, P. E. Cade, and D. H. Rasmussen, "Mechanisms of anodic bonding of silicon to pyrex glass," in *Int. Workshop on Solid-State Sensors and Actuators* (Hilton Head '88), p. 109.

Reading List

R. S. Muller et al., *Microsensors,* Piscataway, NJ: IEEE Press, 1991.

W. S. Trimmer (ed.), *Micromechanics and MEMS classic and seminal papers to 1990,* Piscataway, NJ: IEEE Press, 1997.

G. T. A. Kovacs, *Micromachined Transducers sourcebook,* New York: McGraw-Hill, 1998.

STEPHEN F. BART
Microcosm Technologies

MICHAEL W. JUDY
Analog Devices

MICROMECHANICAL DEVICES AND FABRICATION.

See MICROMACHINED DEVICES AND FABRICATION TECHNOLOGIES.

MICROMECHANICAL RESONATORS

Micromechanical resonators have been used for several years as a time base in electronic and mechanical systems and in sensors for a wide range of applications. This article details the basic principles of resonant sensing and reviews the fundamentals involved in designing a micromechanical resonator. The various materials previously used to realize miniature resonant transducers are discussed, after which the article concentrates on micromachined silicon resonant transducers. As explained in the article, the fact that single crystal silicon is not piezoelectric means that alternative methods of exciting and detecting the resonator's vibrations must be fabricated as an integral part of the transducer. Various mechanisms can be used, and these are reviewed in detail.

The quality factor of a micromechanical resonator is a figure of merit describing its resonance. The importance of the quality factor and the damping effects that limit it are discussed, enabling the reader to gain an in-depth knowledge of the design principles involved. Other important resonator behavioral characteristics, such as nonlinear behavior, are also included. Finally, a comprehensive literature survey of key resonant silicon sensors developed to date is presented, enabling the reader to access a full background of the technology.

MICROMECHANICAL RESONATORS

A resonator is a mechanical structure designed to vibrate at its resonant frequency. Micromechanical resonators are miniature structures with dimensions typically varying between a few microns and a centimeter. The structure can be as simple as a cantilever beam, a fixed–fixed beam clamped at each end, or a tuning fork. The frequency of the vibrations of the structure at resonance are extremely stable, enabling the resonator to be used as a time base (the quartz tuning fork in watches, for example) or as the sensing element of a resonant sensor. In each application the behavior of the resonator is of fundamental importance to the performance of the device.

A resonant sensor is designed such that the resonator's natural frequency is a function of the measurand (1). The measurand typically alters the stiffness, mass, or shape of the resonator, hence causing a change in its resonant frequency. The other major components of a resonant sensor are the vibration drive and detection mechanisms. The drive mechanism excites the vibrations in the structure, while the detection mechanism "picks up" these vibrations. The frequency of the detected vibration forms the output of the sensor, and this signal is also fed back to the drive mechanism via an amplifier maintaining the structure at resonance over the entire measurand range. A typical resonant sensor is shown diagrammatically in Fig. 1.

The most common coupling mechanism is for the resonator to be stressed in some manner by the action of the measurand. The applied stress effectively increases the stiffness of the structure, which results in an increase in the resonator's natural frequency. This principle is commonly applied in force sensors (2), pressure transducers (3), and accelerometers (4). Coupling the measurand to the mass of the resonator can be achieved by surrounding the structure by a liquid or gas, or by contact with small solid masses. The presence of the sur-

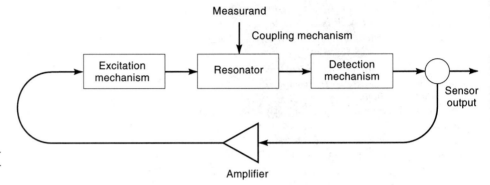

Figure 1. Block diagram of resonant sensor layout showing essential device components.

rounding media increases the effective inertia of the resonator and lowers its resonant frequency. Density sensors and level sensors are examples of mass coupled resonant sensors (1). One example of coupling to the resonator via the shape effect is a micromachined silicon pressure sensor, which uses a hollow, square diaphragm supported at the midpoints of each edge as the resonator (5). The internal cavity within the diaphragm allows each side to be squeezed as applied pressure increases, which alters the curvature of the diaphragm, in turn altering its resonant frequency.

MICROMECHANICAL RESONATOR MATERIALS

Since the design and behavior of the resonator is a major influence on the performance of the sensor, the properties of the material from which it is constructed become a fundamental factor. This section discusses the common materials used to fabricate micromechanical resonators: silicon, polysilicon, quartz, and gallium arsenide. The material properties of these materials are given in Table 1. Full details of these materials can be found in references (6,7,8).

Single Crystal Silicon

Silicon is a single crystal material possessing a face-centered diamond cubic structure. Silicon atoms are covalently bonded, with four atoms bonded together forming a tetrahedron. As these tetrahedra combine, they form a large cube or unit cell, as shown in Fig. 2. It is an anisotropic material with many of its physical properties, such as Young's modulus, varying with the crystalline direction. Directions and planes within crystals are identified using the Miller indices. The indices use the principal axes within the crystal and parentheses to denote each feature. For example, (111) denotes a particular plane that bisects the x, y, and z axes at 1, as shown in Fig. 2.

An important feature of silicon is that it remains elastic up to fracture, exhibiting no plastic behavior and therefore no hysteresis. This is an ideal material characteristic for resonant structures, since any plastic deformation would permanently alter the shape, and therefore resonant frequency, of the structure. Single crystal silicon is also intrinsically very strong. Given the fact that it is elastic to fracture, the practical strength of silicon is, however, extremely dependent on the number and size of crystalline defects and the nature of the surface. Given silicon wafers with a low level of defects, sensible structural design, and careful handling and processing, strengths close to intrinsic and far in excess of alloyed steels can be obtained (6). The mechanical fatigue characteristics of the silicon is also excellent. Fatigue can result in structures fracturing at applied stresses much lower than should otherwise cause such failure. Fatigue failure begins at a microscopic level with a minute defect or crack in the material, which propagates due to fluctuating applied stresses. The varying stresses caused by the flexural vibrations of a resonating structure would be a typical environment for such fatigue to occur.

Other important favorable considerations include the low temperature cross sensitivity of resonant frequency, being measured at -29 ppm/°C (9). The long-term stability of silicon resonators has also been tested in numerous applications, and these results show no significant frequency shift over long periods of time. Silicon wafers are readily available in a variety of sizes, currently from 50 to 200 mm in diameter. Wafers are relatively cheap in their basic form, especially con-

Table 1. Material Properties for Common Resonator Materials (6–8)

	Young's Modulus (N/m²)	Fracture Strength (N/m²)	Density (kg/m³)	Thermal Expansion (10⁻⁶/°C)	Thermal Conductivity (W/cm °C)
Silicon	1.9×10^{11}	7×10^9	2330	2.33	1.57
Quartz[a]	9.7×10^{10}	8.4×10^9	2650	0.55	0.014
Stainless steel	2×10^{11}	2.1×10^9	7900	17.3	0.329
GaAs	8.53×10^{10}	2.7×10^9	5360	6.4	0.44

[a]Values given for Z-cut quartz.

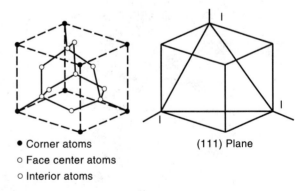

Figure 2. Diamond cubic unit cell and the (111) plane shown within it.

sidering that hundreds of devices can be realized on each wafer. Standard wafers are available in two forms distinguished by the crystal plane from which the wafer is cut, the most common wafer orientation being (100). More expensive wafers are available that include buried layers of silicon dioxide, and these are potentially very useful for the fabrication of many types of micromachined devices, including resonant sensors.

The single drawback associated with silicon with regard to its application as a resonator material is that it is not piezoelectric. When a voltage is applied to a piezoelectric material it deforms, and when it is forced to deform a potential gradient is generated. Piezoelectric behavior is ideal for exciting and detecting vibrations. To excite and detect vibrations in silicon resonators, alternative mechanisms such as electrostatic forces or thermal effects have to be employed. While their use is well established, it does complicate the fabrication of the resonator in comparison to similar structures fabricated from a piezoelectric material such as quartz.

Polysilicon

Polysilicon is an alternative material suitable for resonator fabrication and, as with single crystal silicon, much of the process development for polysilicon has been carried out by the semiconductor industry. It is often used where electronics is being integrated on the sensor chip due to the compatibility of the fabrication processes. In many applications, however, the material properties required for polysilicon micromechanical devices are different from those required in the fabrication of microelectronic devices. This is especially true in the case of resonant sensing, where the mechanical properties of the polysilicon will play a vital role in sensor performance. Also, reproducibility and repeatability of these properties are essential.

Polysilicon is a film of silicon atoms deposited on the top surface of a substrate typically using chemical vapor deposition (CVD) processes. The film either has a granular or amorphous structure depending on the process parameters (10). In the case of resonator applications, the film is typically deposited on a sacrificial layer of silicon dioxide. The polysilicon layer is then patterned and the sacrificial layer removed to leave the free-standing resonant structure. This is called surface micromachining (11).

The deposition process is of prime importance in determining the resulting mechanical properties of the deposited film. Process parameters such as substrate temperature, gas flow rate, deposition rate, deposition pressure, and reactor design all affect the structure and behavior of the film. Another key factor affecting the performance of a resonator fabricated from polysilicon is the amount of residual stress built into in the layer during deposition. Residual stresses result from thermal expansion coefficient mismatches between the film and substrate, and also from the grain growth process, which can trap atoms in positions that induce stress in the lattice. For resonant structures such stresses will naturally alter the performance of the device and will make prior modeling inaccurate. Residual stresses can be controlled and reduced by annealing the deposited films.

The mechanical properties of polysilicon will vary depending on the process parameters, but general values have been reported for Poisson's ratio, Young's modulus, and tensile strength of 0.226, $1.75 \pm 0.21 \times 10^{11}$ Nm^{-2} and 1×10^9

Nm^{-2}, respectively (12). The long term stability of the material is reported to be good, with resonators fabricated in polysilicon being tested over 7000 h of temperature cycling showing no detectable change in resonant frequency. Also, no signs of fatigue failure have been reported. Such polysilicon films are well suited to resonant sensor design as long as residual stresses are controlled (10). The repeatability of the film's properties will also be an important consideration when attempting to mass produce a resonant silicon sensor. Like single crystal silicon, polysilicon is not piezoelectric and vibration excitation and detection mechanisms must be fabricated at wafer level. Again, many possible mechanisms are possible.

Quartz

The application of quartz as a resonant sensor material has evolved from its use as a crystal control for oscillating circuits. Quartz crystals were first used in radio communications equipment and now can be widely found in many applications, most noticeably as the time base in clocks and watches (13). The single most important feature of quartz is its piezoelectric behavior. Piezoelectric materials deform when an electric field is applied and conversely generate a potential field when forced to deform. Quartz is a piezoelectric material because it possesses groups of atoms with an unbalanced charge (these are known as dipoles). These dipoles are permanently orientated in the same direction, and quartz is therefore permanently polarized. When external stress is applied to the material, its crystal structure is deformed. This deformation shifts the dipoles, changing the polarization of the material and inducing a voltage in the process. Conversely, if an external voltage is applied to the crystal, the orientation of the dipoles is changed, deforming the crystal structure. Silicon is not piezoelectric because it possesses a covalently bonded symmetrical structure and therefore has no dipoles.

Quartz resonators employ the piezoelectric effect to excite the vibrations in the resonant structure. An applied alternating electric field will produce strain in alternating directions, and this is a very simple way of exciting vibrations. A quartz resonator consists of a precisely dimensioned resonant structure with electrodes patterned on its surface. The nature of the vibration excited by the applied field depends on the geometry of the resonator, its crystalline orientation, and the electrode pattern (14). Resonators can be designed so that torsional vibrations can also be simply excited (see Fig. 3). Torsional vibrations are advantageous since they suffer less damping effects within the material.

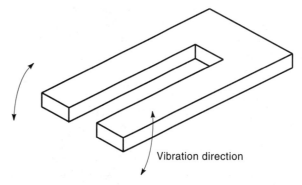

Vibration direction

Figure 3. Quartz tuning fork resonator vibrating in a torsional mode. Such vibrations can be simply excited in piezoelectric materials.

Quartz possesses many other attractive features. It has excellent stability and long-term aging characteristics. It possesses excellent material properties, and resonators can be batch fabricated using photolithographic techniques similar to those used in silicon. It also benefits from very low temperature cross sensitivity in certain crystal orientations.

The main drawbacks of quartz when compared to silicon revolve around its fabrication technology and the fact that integrated circuitry cannot be formed on the sensor chip. Also, quartz wafers are more expensive than silicon wafers. Nevertheless, quartz is an excellent resonator material, and a wide range of quartz resonant sensors exist. Applications include sensing temperature, thin film thickness, force, pressure, and fluid density (13).

Gallium Arsenide

Gallium arsenide (GaAs) is used in the semiconductor industry for high-frequency, high-temperature (>125°C) electronic and optoelectronic applications. As with silicon, the development of the material for these applications has benefited the microengineering community, and GaAs is increasingly being considered as a micromechanical material. It is especially well suited to resonant sensing applications since it is piezoelectric and electronic circuitry can also be integrated onto the sensor chip. It may therefore be considered to combine the benefits of quartz and silicon as a resonator material.

GaAs atoms are arranged in a zincblende crystal structure and are held in place by ionic bonds. The presence of the arsenide atoms in the lattice draws the electrons, and this results in dipoles orientated along the (111) directions. These dipoles result in the piezoelectric nature of GaAs (15). As with quartz, longitudinal, flexural, torsional, and shear vibrations can be excited piezoelectrically, and this makes GaAs well suited for resonant applications.

GaAs is a brittle material and, as with silicon, its mechanical strength will ultimately depend on the presence of stress concentrating defects. GaAs wafers contain more crystalline defects than single crystal silicon wafers, and structures will therefore not be as strong as their silicon counterparts. Tests have been carried out on GaAs cantilever beams that failed at 2.7 GNm^{-2} (16); while this is well below the values obtained for silicon structures, it is nevertheless stronger than steel. Fatigue failure has not been reported in any GaAs structures, but long-term stability studies are yet to be carried out. As with the other single crystal materials, the stability characteristics should be excellent.

There is a wide range of micromachining techniques available, including various wet and dry etches, bonding, and selective etch stops. Structures can also be fabricated on GaAs substrates using thin films of aluminum gallium arsenide (AlGaAs) and sacrificial layers as described in the case of polysilicon structures (17). Internal stresses mainly due to thermal mismatches are, however, a problem with this technique. There are certainly many more micromachining options available for GaAs resonant sensors than for their quartz counterparts.

The main drawbacks associated with GaAs as a microresonator material result from the increased costs involved in its use. GaAs wafers, are more expensive than silicon wafers, and processing costs are also higher than for the equivalent silicon operation. Silicon also possesses a much wider range of processes. Silicon and quartz are also well characterized materials, and their use as resonator materials is longstanding and fully documented.

RESONANT SILICON SENSORS

The mechanical properties of the pure silicon wafers used in the fabrication of integrated circuits and the ability to form microstructures using some of the processes developed by the semiconductor industry have enabled silicon to be used in a variety of microengineered devices. At the forefront of such microengineered devices are silicon sensors, including resonant silicon sensors.

In the case of micromachined silicon resonant sensors, the resonator can be fabricated from silicon and the vibration drive and detection mechanisms fabricated on, or adjacent to, the resonator. The resonator is then located in a silicon structure specifically designed for the sensing application. This is illustrated by the approach used in the design of the majority of resonant silicon pressure sensors, where the resonator is fabricated on the top surface of a silicon diaphragm (see Fig. 4). In such a case the resonator is being used as a strain gauge sensing the strain induced due to the deformation of the diaphragm caused by an applied pressure (18).

The use of silicon overcomes the major disadvantage of traditional resonant sensors—namely, the manual fabrication of individual resonating elements. Fabricating the sensor in silicon enables the resonator to be batch fabricated, and hundreds of devices can be realized on each silicon wafer simultaneously. The wafers themselves can also be processed simultaneously in the majority of the fabrication steps.

Comparison of Silicon Sensing Technologies

Silicon strain sensors monitoring deflections in the sensor structure can employ piezoresistive, capacitive, or resonant sensing techniques (19). Of these, resonant sensing is inherently the more complex approach. This is highlighted by the fact that the vibration excitation and detection mechanisms are commonly based on piezoresistive or capacitive techniques. However, the typical performance figures for the three strain sensing techniques listed in Table 2 clearly show the advantages of resonant sensing.

VIBRATION EXCITATION AND DETECTION MECHANISMS

There are many potential vibration excitation and detection mechanisms possible in silicon. Many of the mechanisms discussed in this section can be used both to excite and detect a

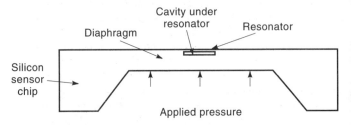

Figure 4. Resonant pressure sensor configuration with the resonator located on the top surface of the diaphragm.

Table 2. Performance Features of Resonant, Piezoresistive, and Capacitive Sensing (19)

Feature	Resonant	Piezoresistive	Capacitive
Output form	Frequency	Voltage	Voltage
Resolution	1 part in 10^8	1 part in 10^5	1 part in 10^4–10^5
Accuracy	100–1000 ppm	500–10,000 ppm	100–10,000 ppm
Power consumption	0.1–10 mW	$\approx$10 mW	<0.1 mW
Temperature cross sensitivity	-30×10^{-6}/°C	-1600×10^{-6}/°C	4×10^{-6}/°C

resonator's vibrations, either simultaneously or in conjunction with another mechanism. Devices where a single element combines the excitation and detection of the vibrations in the structure are termed one-port resonators. Devices that use separate elements are termed two-port resonators (20).

It is impossible to state which mechanisms are the best for driving or detecting a resonator's vibrations since it depends on several important factors. These factors include the magnitude of the drive forces generated, the coupling factor (or drive efficiency), sensitivity of the detection mechanism, the effects of the chosen mechanism on the performance and behavior of the resonator, and practical considerations pertaining to the fabrication of the resonator and the sensor's final environment (21).

Electrostatic Excitation and Detection

The electrostatic excitation of a resonator relies on electrostatic forces between two electrodes. These electrodes may be separated by an air gap, with one plate being located on the resonator and the other located on the surrounding structure. Alternatively, a three-layer sandwich can be formed on resonator's surface consisting of two electrodes separated by a dielectric layer. The associated vibration detection mechanism relies on the change of capacitance between the electrodes as the resonator deflects. One drawback to be considered when using the dielectric sandwich on the resonator surface is the increase in the temperature cross sensitivity of the resonator.

Lateral vibrations in the plane of the wafer have also been excited using the electrostatic mechanism combined with a comb structure (22). The resonator must be designed with suitable comb fingers that align with comb fingers fabricated adjacent to the resonator. The resonator is typically fabricated from polysilicon on the top surface of the wafer.

One-port resonators using the same pair of electrodes both to excite and detect vibrations (23) and two-port configurations using separate electrode pairs (24) have both been achieved. Electrostatic excitation is also commonly used alongside alternative vibration detection techniques, such as piezoresistive (25).

Piezoelectric Excitation and Detection

This approach uses a deposited layer of piezoelectric material both to excite and detect the resonator's vibrations (26). The piezoelectric material, typically zinc oxide (ZnO), is formed in a sandwich between two electrodes, creating a piezoelectric bimorph. Applying an oscillating voltage to the bimorph causes periodic deformation, thereby exciting the resonator. Detection is provided by the corresponding potential generated by the deformation in the bimorph. The piezoelectric bimorph is formed and patterned on the resonator surface in a manner similar to the dielectric electrostatic mechanism.

Magnetic Excitation and Detection

This technique places a current-carrying resonator in a permanent magnetic field, the excitation being due to Lorenz forces acting on the resonator. Applying an alternating current to the resonator results in alternating forces, and hence vibrations are induced in the resonator. The associated detection mechanism utilizes the change in magnetic flux caused by the resonator moving in the magnetic field (27).

Electrothermal Excitation

This approach relies on the heating effect caused by passing a current through a patterned resistor located on the surface of the resonator. The heat energy generated causes a thermal gradient across the resonator's thickness, with the top surface at a higher temperature than the bottom. The induced thermal expansion of the material results in a bending moment on the resonator, thereby deforming the structure. Vibrations can simply be excited by modulating the current through the resistor. Electrothermal excitation is widely used and can be found on many resonant devices (4).

Optothermal Excitation

The thermal drive technique described previously can also be achieved using the heating effect resulting from a light source incident on the resonator. Modulating the light source will result in periodic thermal expansion on the surface of the resonator and hence induce vibrations in the structure (28). The light source is commonly aligned over the resonator using an optical fiber, and this method can be conveniently used in conjunction with optical vibration detection techniques.

Optical methods can also be used to obtain self-oscillation of the resonator using unmodulated light. The unmodulated light source is directed on to the resonator via an optical fiber, the end of which forms a Fabry Perot interferometer with the reflective surface of the silicon resonator. The interferometer output varies as a function of the resonator's displacement, and this has the effect of modulating the incident light as the resonator vibrates. By correctly setting the interferometric spacing and optical power, self-resonance can be achieved (29). It is, however, difficult to sustain these vibrations, and doing so requires critical and stable alignment of the fiber.

Optical excitation techniques are attractive since they enable a passive resonant structure with no on chip excitation mechanisms required. The use of optical fibers is compatible with the miniature nature of micromachined silicon devices. Furthermore, they are immune to electromagnetic interference and can allow resonant silicon sensors to be used in harsh, high-temperature environments. However, the accurate integration and alignment of the optical fiber onto the sensor chip is difficult to achieve, especially in mass-produced sensors.

Optical Detection

There are several optical detection techniques suitable for monitoring the vibrations of a resonator. This approach commonly uses an optical fiber to couple the light to the resonator and can therefore be readily combined with the optical drive mechanisms. Interferometric techniques rely on two interfering reflections, one from a fixed reference and the other from the vibrating resonator. The combination of the reflected light results in interference fringes, the number of which gives a direct measure of the amplitude of displacement (30). Intensity modulation techniques uses changes in the intensity of the reflected light to monitor the amplitude of the resonator vibrations. The reflections can be modulated, for example, by the angular displacement of the resonator as it vibrates.

Piezoresistive Detection

This technique uses the inherent piezoresistive nature of silicon to detect the vibrations of the resonator. Resistors can be fabricated by diffusing or implanting them into the silicon, or by depositing polysilicon resistors on the top surface of the resonator. The resistor can simply be connected in a Wheatstone bridge circuit, the value of the resistor varying sinusoidaly as the resonator vibrates. The frequency of the changing resistance forms the output of the sensor, and hence the actual value of the resistance and the behavior of the resistor becomes largely unimportant. This approach is simple to achieve and is widely used in many resonant sensors (31). It is compatible with integrated circuit fabrication techniques and can be readily used either with electrothermal excitation or an alternative drive excitation mechanism, such as electrostatic.

THEORY OF RESONANCE

To understand resonance, the propagation of mechanical waves within a solid must be understood. A mechanical wave may be defined as the propagation of a physical quantity (e.g., energy or strain) through a medium (solid or fluid) without the net movement of the medium. As the wave travels through the medium, its particles are displaced from their equilibrium position, thereby distorting the medium. The form of wave will depend on the nature of its source and the material through which it travels. The speed of a wave in a solid is dependent on the mechanical properties of the material, and therefore its wavelength is a function of the frequency of the wave source.

As the wave travels through a solid, it can meet boundaries or regions of nonuniformity of material or geometrical form. Upon meeting such discontinuities, the nature of the wave will be changed due to phenomena known as refraction, diffraction, reflection, and scattering. When considering the phenomena of natural frequencies and resonance, reflection of a mechanical wave at a system boundary becomes important. Reflection is the reversal in direction of the wave back along its original path.

If the reflected wave exactly coincides with the incoming wave, then a standing wave is created. The superposition of these waves results in the amplitude of each wave combining to become twice that of the initial wave. Taking a string fixed at each end as an example, a mechanical wave will be reflected back from the fixed boundary at each end. The standing wave phenomena will occur at specific frequencies known as the natural, or modal, frequencies of the fixed-fixed string. For each natural frequency, the string will have a characteristic distorted form, or mode shape, as shown in Fig. 5. These are the modes of vibration of the string and also the mode shapes associated with a fixed-fixed beam.

Points of zero displacement are called nodes, while the points of maximum displacement are known as antinodes.

The amplitude of the standing waves will decline with time due to damping effects in the system. The amplitude will be maintained, however, if the wave source (for example, a harmonic driving force), at that frequency is maintained. If a harmonic driving force is applied to a system and a mode of vibration is excited, then the system is in resonance. Resonance will occur when the frequency of the driving force matches the natural frequency of the system. Resonance is a phenomenon associated with forced vibrations, while natural frequencies are associated with free vibrations. The motion of the oscillating body will not usually be in phase with the driving force. When resonance has occurred, the motion of the oscillating body will lag the driving force by a phase angle of $\pi/2$.

QUALITY FACTOR

As a structure approaches resonance, the amplitude of its vibration will increase (its resonant frequency is defined as the point of maximum amplitude). The magnitude of this amplitude will ultimately be limited by the damping effects acting on the system. The level of damping present in a system can be defined by its quality factor (Q factor). The Q factor is a ratio of the total energy stored in the system to the energy lost per cycle due to the damping effects present, as in Eq. (1):

$$Q = 2\pi \, (\text{maximum stored energy/energy lost per cycle}) \quad (1)$$

A high Q factor indicates a pronounced resonance easily distinguishable from nonresonant vibrations, as illustrated in Fig. 6. Increasing the sharpness of the resonance enables the

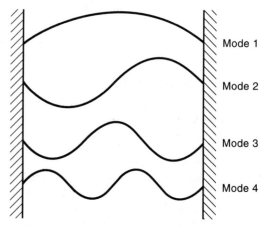

Mode 1

Mode 2

Mode 3

Mode 4

Figure 5. First four modes of vibration of a fixed-fixed string.

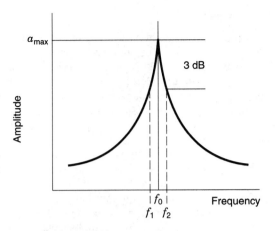

Figure 6. Typical amplitude frequency relationship at a structural resonance.

resonant frequency to be more clearly defined and will improve the performance and resolution of the resonator. It will also simplify the operating electronics since the magnitude of the signal from the vibration detection mechanism will be greater than that of a low Q system. A high Q means little energy is required to maintain the resonance at a constant amplitude, thereby broadening the range of possible drive mechanisms to include weaker techniques. A high Q factor also implies that the resonant structure is well isolated from its surroundings and therefore the influence of external factors (e.g., vibrations) will be minimized.

The Q factor can also be calculated from Fig. 6 using Eq. (2):

$$Q = \frac{f_0}{\Delta f} \qquad (2)$$

where resonant frequency f_0 corresponds to a_{max}, the maximum amplitude, and Δf is the difference between frequencies f_1 and f_2. Frequencies f_1 and f_2 correspond to amplitudes of vibration 3 dB lower than a_{max}.

From Eq. (1) it is clear the Q factor is limited by the various mechanisms by which energy is lost from the resonator. These damping mechanisms arise from three sources (9):

1. The energy lost to a surrounding fluid ($1/Q_a$)
2. The energy coupled through the resonator's supports to a surrounding solid ($1/Q_s$)
3. The energy dissipated internally within the resonator's material ($1/Q_i$)

Minimizing these effects will maximize the Q factor as shown:

$$\frac{1}{Q} = \frac{1}{Q_a} + \frac{1}{Q_s} + \frac{1}{Q_i} \qquad (3)$$

Gas Damping Mechanisms

Energy losses associated with $1/Q_a$ are potentially the largest, and therefore the most important, of the loss mechanisms. These losses occur due to the interactions of the oscillating resonator with the surrounding gas. There are several distinguishable loss mechanisms and associated effects. The magni-

tude of each depends primarily on the nature of the gas, surrounding gas pressure, size and shape of the resonator, direction of its vibrations, and its proximity to adjacent surfaces. Gas damping affects can be negated completely by operating the resonator in a suitable vacuum, and this is used in most micromechanical resonator applications.

Molecular Damping. At low pressures between 1 and 100 Pa the surrounding gas molecules act independently of one another. The damping effect arises from the collisions between the molecules and the resonator's surface as it vibrates. This causes the resonator and molecules to exchange momentum according to their relative velocities (32). The magnitude of the loss is directly proportional to the surrounding fluid pressure and also close proximity of the oscillating structure to adjacent surfaces will exaggerate the damping effects. This is due to molecules reflecting off the surrounding surfaces increasing the frequency of molecular collisions with the resonator's surface above the free space number expected for the given gas pressure.

Viscous Damping. At pressures above 100 Pa the molecules can no longer be assumed to act independently and the surrounding gas must be considered as a viscous fluid. The viscosity of the surrounding gas introduces two separate damping mechanisms (33,34). The first is related purely to the viscosity of the surrounding fluid, viscosity representing the resistance of one fluid layer to movement over another layer. The vibrations of a resonator will therefore be subject to a similar resistance to movement as the fluid travels over its surface. This drag force consists of two components, the first being proportional to the surface velocity of the vibrating structure, and is responsible for part of the energy lost from the resonator to the gas. The second component is proportional to the acceleration of the resonator's surface and has an inertial effect on the resonator, reducing its resonant frequency.

The second source of viscous damping arises from the formation of boundary layer around the structure. The structure's vibrations will set up a transverse wave that travels into the fluid medium, as shown in Fig. 7. The amplitude of this wave decreases with distance as it travels through the medium (35) at a rate depending on viscosity, frequency of vibration, and fluid density.

The magnitude of each viscous damping effect will depend on the geometry and size of the resonator and the pressure of the surrounding fluid. Initially the damping will primarily arise from the drag force exerted by the fluid, the energy lost due to this effect being independent of the pressure. At a cer-

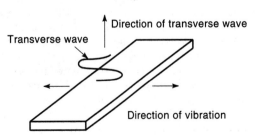

Figure 7. Formation of transverse wave off the surface of the vibrating resonator.

tain pressure, termed P_c, the formation of the boundary layer will become the dominant loss mechanism. The critical pressure P_c is a function of the resonator's geometry.

Acoustic Radiation. The motion of the resonator as it vibrates will cause a displacement of surrounding gas particles in the direction of the resonator's vibrations. The oscillatory motion of the gas particles can form an acoustic wave, depending on the size and geometry of the resonator, its frequency of operation, its wavelength of vibration, and the properties of the fluid surrounding it. This phenomenon is called acoustic radiation. When a dimension, such as the width of a beam, is much smaller than the acoustic wavelength, the fluid has time to slip around the edges of the vibrating structure (36). Hence the fluid avoids being compressed and the acoustic wave is not formed. It has also been shown that when the standing waves forming the resonator's vibrations are of a wavelength less than the acoustic wavelength, acoustic radiation is reduced to almost zero (37). Taking a micromachined resonator vibrating in air at a typical operating frequency of 100 kHz, the acoustic wavelength will be 3.43 mm. Given the inherently small size of micromechanical resonators, acoustic losses can be ignored in the vast majority of cases.

The displacement of the fluid particles will also have an effect on the resonator's natural frequency. This is because the mass of the adjacent fluid particles is effectively added to the mass of the resonator, and hence its resonant frequency will be reduced. The order of magnitude of the added mass effect is dependent on the relative densities of the fluid and resonator material (38), and there is a much smaller dependence on the fluid pressure and the aspect ratio of the structure (39). The result of this added mass effect is the same as the inertial effect due to the gas viscosity, but the mechanisms by which each occurs are unrelated.

Squeezed Film Damping. Squeezed film damping occurs when the resonator is positioned close to an adjacent surface and vibrates in a direction perpendicular to the surface. This form of damping often occurs in electrostatically driven resonators where close positioning of the drive and detection electrodes is essential. In such an arrangement it can become the major damping mechanism (40). The loss mechanism occurs due to the pumping action displacing the fluid from between the two surfaces and then drawing it back as the resonator vibrates (32).

The magnitude of this effect depends on the frequency of operation, the pressure of the surrounding fluid, and the geometry of the resonator. The frequency dependence arises from the time taken to displace the fluid. At low frequencies the fluid has time to move between the surfaces and energy is lost from the resonator's vibrations. As the frequency rises, however, some of the fluid at the center of the resonator does not have enough time to move and becomes trapped between the surface. As the frequency rises further, an increasing proportion of the fluid remains trapped. The trapped volume of fluid acts as a spring, resulting in a slight increase in resonant frequency. As the frequency continues to rise, the spring effect becomes increasingly predominant and the damping effect progressively declines (41). The squeeze film effect can be alleviated to a certain extent by incorporating apertures in the resonator, allowing some fluid transport through the resonator.

Structural Damping

Structural damping, $1/Q_s$, is associated with the energy coupled from the resonator through its supports to the surrounding structure and must be minimized by careful design of the resonant structure. Minimizing the energy lost from the resonator to its surroundings can be achieved by designing a balanced resonant structure, supporting the resonator at its nodes, or employing a decoupling system between the resonator and its support.

The coupling mechanism between the resonator and its support can be illustrated by observing a fixed-fixed beam vibrating in its fundamental mode. Following Newton's second law, that every action has an equal and opposite reaction, the reaction to the beam's vibrations is provided by its supports. The reaction causes the supports to deflect, and as a result energy is lost from the resonator. This is shown in Fig. 8.

The degree of coupling of a fixed-fixed beam can be reduced by operating it in a higher-order mode. For example, the second mode in the plane of vibrations shown in Fig. 8 will possess a node halfway along the length of the beam. The beam will vibrate in antiphase either side of the node, and the reactions from each half of the beam will cancel out at the node. There will inevitably still be a reaction at each support, but the magnitude of each reaction will be less than for mode 1. The use of such higher-order modes is limited by their reduced sensitivity to applied stresses and the fact that there will always be a certain degree of coupling.

Balanced resonator designs operate on the principle of providing the reaction to the structure's vibrations within the resonator. Multiple-beam-style resonators, for example, incorporate this inherent dynamic moment cancellation when operated in a balanced mode of vibration. Examples of such structures are the double-ended tuning fork (DETF), which consists of two beams aligned alongside each other, and the triple-beam tuning fork (TBTF), which consists of three beams aligned alongside each other, the center tine being twice the width of the outer tines. Figure 9 shows these structures and their optimum modes of operation. Resonant frequencies and performance characteristics of these structures can be modeled using finite element analysis techniques (42).

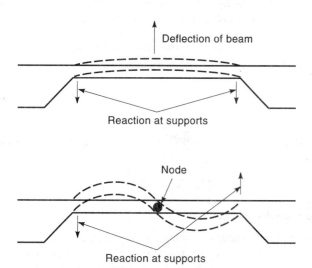

Figure 8. Support reactions: (a) Fundamental mode, (b) second mode.

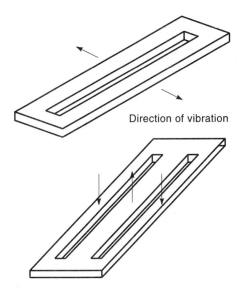

Figure 9. DETF vibrating in the plane of the resonator and TBTF vibrating in a perpendicular direction to the plane of the resonator.

The term $1/Q_s$ is of fundamental importance since it not only affects the Q factor of the resonator, but it provides a key determinant of resonator performance. A dynamically balanced resonator design that minimizes $1/Q_s$ provides many benefits:

High resonator Q factor and therefore good resolution of frequency

A high degree of immunity to environmental vibrations

Immunity to interference from surrounding structural resonances.

Improved long-term performance since the influence of the surrounding structure on the resonator is minimized

Internal Loss Mechanisms

The Q factor of a resonator is ultimately limited by the energy loss mechanisms within the resonator material. This is illustrated by the fact that even if the external damping mechanisms $1/Q_a$ and $1/Q_s$ are removed, the amplitude of its vibrations will still decay with time. There are several internal loss mechanisms by which vibrations can be attenuated.

Movement of Dislocations and Scattering by Impurities. A dislocation is an imperfection within the crystal lattice, and it falls into two categories: edge or screw dislocations. Attenuation of the resonator vibrations can occur due to movement of these dislocations. The magnitude of such losses depends on the number of dislocations, the resonator's frequency, and temperature. Only at temperature above 130°C is there sufficient energy present to move the dislocations. Also, silicon is available in a very pure form and typical wafers contain a very low number of dislocations. Given likely operating temperature ranges and the low dislocation density, this form of internal damping can effectively be ignored in silicon resonators.

Impurities are foreign atoms either inadvertently trapped in the silicon lattice during crystal growth or intentionally added during the fabrication process. The presence of these impurities can result in point defects within the lattice, introducing regions of anelasticity scattering of the structural wave within the solid and increasing the energy lost within the material. The levels of impurities present within silicon are low and they have been shown to have a negligible effect below 200 °C (43). This effect can therefore also be ignored for typical device operating conditions.

Phonon Interaction. Each atom contained within the crystal lattice vibrates about its mean position due to thermal energy. When determining the heat capacity and thermal conductivity of the solid, these atomic vibrations are viewed collectively as a series of traveling lattice waves known as phonons (44). These phonons are subject to the scattering effect of, and they can also be similarly scattered by, variations in the lattice strain field due to the presence of other phonons. This is known as phonon-phonon interaction. The mechanical wave forming the structure's resonance can also interact with variations in the strain field due to the presence of phonons. This results in the acoustic energy being converted to thermal energy (i.e., phonons are created). This mechanism only occurs, however, at frequencies far in excess of that used at present in resonant silicon devices.

Akheiser Loss. At lower frequencies Akheiser theory must be used to analyze the damping effects of phonons on the mechanical wave. The thermal phonons within the material are analyzed as a gas described by a number of macroscopic parameters. The energy loss mechanism arises from the influence of the structural wave of the resonator on the phonon gas.

The structural wave of the resonator will introduce fluctuations in the strain field of the material, causing a modulation of the phonons. The modulated phonons collide with each other and with impurities, and this has the effect of throwing the phonon gas out of equilibrium. This is an irreversible process and therefore energy is lost from the resonator's vibrations.

Thermoelastic Effect. The flexural vibrations of silicon beam resonators lead to cyclic stressing of the top and bottom surfaces. The majority of the energy employed in displacing the beam is stored elasticity but some of the work is also converted into thermal energy. Material on the surface in compression will rise in temperature while the surface in tension falls in temperature. Hence a temperature gradient is formed across the thickness of the beam. If the deformation is maintained for a sufficient length of time, heat energy will flow across the gradient in order to equilibrate the system. Any thermal energy transferred in this manner is lost to entropy and the resonator's vibrations are attenuated.

The magnitude of this effect is dependent on the mechanical properties of the resonator material, temperature and the resonant frequency. The damping fraction ($\delta = \frac{1}{2}Q$) can be expressed as a product of two different functions, $\Gamma(T)$ and $\Omega(F)$ (45):

$$\delta = \Gamma(T)\Omega(F) \qquad (4)$$

$$\Gamma(T) = \frac{a^2 T E}{4\rho C} \qquad (5)$$

$$\Omega(F) = 2\left[\frac{F_0 F}{F_0^2 + F^2}\right] \qquad (6)$$

where a is the thermal expansion coefficient, T is the beam temperature, E is Young's modulus, ρ is the material density, C is the specific heat capacity, F is the resonant frequency, and F_0 is the characteristic damping frequency given by Eq. (7):

$$F_0 = \frac{\pi K}{2\rho C t^2} \tag{7}$$

where t represents the beam thickness and K the thermal conductivity of the resonator material. This form of damping can be reduced or eliminated by cooling the resonator or by designing a structure with a resonant frequency removed from F_0. Shear deformation, associated with torsional modes, also does not produce thermoelastic damping.

NONLINEAR BEHAVIOR

Nonlinear behavior becomes apparent at higher vibration amplitudes when the resonator's restoring force becomes a nonlinear function of its displacement. This effect is present in all resonant structures. In the case of a flexurally vibrating fixed-fixed beam, the transverse deflection results in a stretching of its neutral axis (46). A tensile force is effectively applied and the resonant frequency increases. This is known as the hard spring effect. The magnitude of this effect depends on the boundary conditions of the beam. If the beam is not clamped firmly, the nonlinear relationship can exhibit the soft spring effect whereby the resonant frequency falls with increasing amplitude. The nature of the effect and its magnitude also depends on the geometry of the resonator (47).

The equation of motion for an oscillating force applied to an undamped structure is given by Eq. (8), where m is the mass of the system, F the applied driving force, ω the frequency, y the displacement, and $s(y)$ the nonlinear function.

$$m\ddot{y} + s(y) = F_0 \cos \omega t \tag{8}$$

In many practical cases $s(y)$ can be represented by Eq. (9), with the nonlinear relationship being represented by the cubic term.

$$s(y) = s_1 y + s_3 y^3 \tag{9}$$

Placing Eq. (9) in Eq. (8), dividing through by m, and simplifying gives Eq. (10):

$$\ddot{y} + s_1/m(y + s_3/s_1 y^3) = F_0 \cos \omega t \tag{10}$$

where s_1/m equals ω_{or}^2 (ω_{or} representing the resonant frequency for small amplitudes of vibration) and s_3/s_1 is denoted β. The restoring force acting on the system is therefore represented by Eq. (11):

$$R = -\omega_{\mathrm{or}}^2(y + \beta y^3) \tag{11}$$

If β is equal to zero, the restoring force is a linear function of displacement; if β is positive, the system experiences the hard spring nonlinearity; a negative β corresponds to the soft spring effect. The hard and soft nonlinear effects are shown in Fig. 10. As the amplitude of vibration increases and the nonlinear effect becomes apparent, the resonant frequency exhibits a quadratic dependence on the amplitude, as shown in Eq. (12):

$$\omega_{\mathrm{r}} = \omega_{\mathrm{or}}(1 + \tfrac{3}{8}\beta y_0^2) \tag{12}$$

The variable β can be found by applying Eq. (12) to an experimental analysis of the resonant frequency and maximum amplitude for a range of drive levels.

The amplitude of vibration is dependent on the energy supplied by the resonator's drive mechanism and the Q factor of the resonator. Driving the resonator too hard or a high Q factor that results in excessive amplitudes at minimum practical drive levels can result in undesirable nonlinear behavior. Nonlinearities are undesirable since they can adversely affect the accuracy of a resonant sensor. If a resonator is driven in a nonlinear region, then changes in amplitude (due, for example, to amplifier drift) will cause a shift in the resonant frequency indistinguishable from shifts due to the measurand. The analysis of a resonator's nonlinear characteristics is therefore important when determining a suitable drive mechanism and its associated operating variables.

Hysteresis

A nonlinear system can exhibit hysteresis if the amplitude of vibration increases beyond a critical value. (See Fig. 11.) Hysteresis occurs when the amplitude has three possible values at a given frequency. This critical value can be determined by applying Eq. (13) (46):

$$y_0^2 > \frac{8h}{3\omega_{\mathrm{or}}|\beta|} \tag{13}$$

where h is the damping coefficient and can be found by measuring the Q factor of the resonator at small amplitudes and

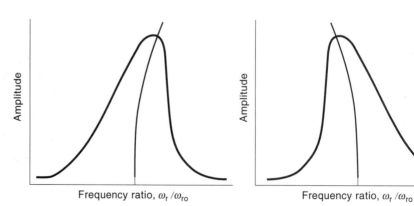

Figure 10. Hard and soft nonlinearities. Hard nonlinearities cause the resonant frequency to rise, soft cause it to fall.

Frequency ratio, $\omega_{\mathrm{r}}/\omega_{\mathrm{ro}}$ Frequency ratio, $\omega_{\mathrm{r}}/\omega_{\mathrm{ro}}$

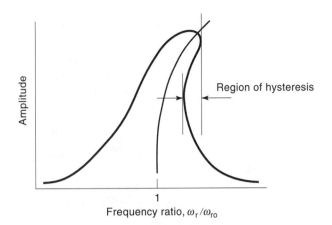

Figure 11. Frequency/amplitude relationship exhibiting hystersis.

applying Eq. (14):

$$Q = \frac{\omega_{\mathrm{or}}}{2h} \qquad (14)$$

SUMMARY

Micromechanical sensors are miniature structures that can use a variety of sensing principles to monitor a wide range of measurands, such as pressure, acceleration, density, and flow. Micromechanical resonator sensors use a resonator excited at resonance to sense the measurand. The micromechanical resonator can be fabricated in batches from silicon, quartz, and gallium arsenide wafers. Such sensors offer high accuracy, high resolution, and high stability.

The frequency of the mechanical resonator typically changes with the measurand, and the output of the sensor is thus a change in frequency. Mechanisms to excite the resonator's vibrations and detect them must be included in the sensor, and a wide variety of techniques are available to the sensor designer. A simple method is the use of the piezoelectric effect in quartz.

BIBLIOGRAPHY

1. R. M. Langdon, Resonator sensors—a review, *J. Phys. E. Sci. Instrum.,* **18**: 103–115, 1985.

2. E. P. Eernisse and J. M. Paros, *Resonator force transducer,* United States Patent No. 4,372,173, October 20, 1980.

3. J. C. Greenwood and T. Wray, High accuracy pressure measurement with a silicon resonant sensor, *Sensors and Actuators A,* **37-38**: 82–85, 1993.

4. D. W. Satchell and J. C. Greenwood, A thermally excited silicon accelerometer, *Sensors and Actuators A,* **17**: 145–150, 1989.

5. E. Stemme and G. Stemme, A balanced dual-diaphragm resonant pressure sensor in silicon, *IEEE Trans. Electron Devices,* **37**: 648–653, 1990.

6. K. E. Petersen, Silicon as a mechanical material, *Proc. IEEE,* **70**: 420–457, 1982.

7. D. Salt, *Hy-Q Handbook of Quartz Crystal Devices,* Wokingham, Berkshire, England: Van Nostrand Reinhold.

8. K. Hjort, J. Söderkvist, and J.-Å. Schweitz, Gallium arsenide as a mechanical material, *J. Micromech. Microeng.,* **4**: 1–13, 1994.

9. M. J. Tudor et al., Silicon resonator sensors: interrogation techniques and characteristics, *IEEE Proc.,* **135**, D: 364–368, 1988.

10. H. Guckel et al., The application of fine grained, tensile polysilicon to mechanically resonant transducers, *Sensors and Actuators,* **A21-A23**: 346–351, 1990.

11. C. Linder et al., Surface micromachining, *J. Micromech. Microeng.,* **2**: 122–132, 1992.

12. H. Guckel et al., Mechanical properties of fine grained polysilicon the repeatability issue, *Technical Digest IEEE Solid State Sensor and Actuator Workshop,* 96–99, June 6–9, 1988.

13. E. P. Eernisse, R. W. Ward, and R. B. Wiggins, Survey of quartz bulk resonator sensor technologies, *IEEE Trans. Ultrason. Ferroelectr. Freq. Control,* **35**: 323–330, 1988.

14. L. D. Clayton et al., Miniature crystalline quartz electromechanical structures, *Sensors and Actuators A,* **20**: 171–177, 1989.

15. J. Söderkvist and K. Hjort, Flexural vibrations in piezoelectric semi-insulating GaAs, *Sensors and Actuators A,* **39**: 133–139, 1993.

16. K. Hjort, F. Ericson, and J-Å. Schweitz, Micromechanical fracture strength of semi-insulating GaAs, *Sensors and Materials,* **6** (6): 359–367, 1994.

17. S. Adachi, GaAs AlAs and $Al_xGa_{1-x}As$: material parameters for use in research and device applications, *J. Appl. Physics,* **58** (3): R1–R12, 1985.

18. K. Petersen et al., Resonant beam pressure sensor fabricated with silicon fusion bonding, *Proc. 6th Solid-State Sensors and Actuators (Transducers '91),* 664–667, June 1991.

19. J. C. Greenwood, Silicon in mechanical sensors, *J. Phys. E. Sci. Instrum.,* **21**: 1114–1128. 1988.

20. H. A. C. Tilmans, D. J. Ijntema, and J. H. J. Fluitman, Single element excitation and detection of (micro-) mechanical resonators, *Tech. Digest of the Int. Conf. on Solid-State Sensors and Actuators (Transducers '91),* 533–537, June 24–27, 1991.

21. A. Prak, T. S. J. Lammerink, and J. H. J. Fluitman, Review of excitation and detection mechanisms for micromechanical resonators, *Sensors and Materials,* **5** (3): 143–181, 1993.

22. W. C. Tang, T-C. H. Nguyen, and R. T. Howe, Laterally driven polysilicon resonant microsensors, *Sensors and Actuators A,* **20**: 25–32, 1990.

23. M. W. Putty et al., One port active polysilicon resonant microstructures, *Proc. Micro Electro Mechanical Systems (MEMS 89),* 60–65, February 1989.

24. J. C. Greenwood, Etched silicon vibrating sensor, *J. Phys. E. Sci. Instrum.,* **17**: 650–652, 1984.

25. J. J. Sniegowski, H. Guckel, and T. R. Christenson, Performance characteristics of second generation polysilicon resonating beam force transducers, *Proc. IEEE Solid-State Sensors and Actuators Workshop,* 9–12, June 1990.

26. Th. Fabula et al., Triple beam resonant silicon force sensor based piezoelectric thin films, *Sensors and Actuators A:* **41-42**: 375–380, 1994.

27. K. Ikeda et al., Silicon pressure sensor integrates resonant strain gauge on diaphragm, *Sensors and Actuators,* **A21-A23**: 146–150, 1990.

28. R. M. A. Fatah, Mechanisms of optical activation of micromechanical resonators, *Sensors and Actuators A,* **33**: 229–236, 1992.

29. R. M. Langdon and D. L. Dowe, Photoacoustic oscillator sensors, *Fibre Optics II Conf.,* The Hague, March 30–April 3, 1987.

30. M. V. Andres, K. H. W. Foulds, and M. J. Tudor, Analysis of an interferometric optical fibre detection technique applied to silicon vibrating sensors, *Electronic Lett.,* **23**: 774–775, 1987.

31. M. B. Othman and A. Brunnschweiler, Electrothermally excited silicon beam mechanical resonators, *Electronics Lett.,* **23**: 728–730, 1987.

32. W. E. Newell, Miniaturisation of tuning forks, *Science,* **161**: 1320–1326, 1968.

33. Y-H. Cho, A. P. Pisano, and R. T. Howe, Viscous damping model for laterally oscillating microstructures, *J. Microelectromechanical Systems,* **3** (2): 81–86, 1994.

34. F. R. Blom et al., Dependance of the quality factor of micromachined silicon beam resonators on pressure and geometry, *J. Vac. Sci. Technol. B,* **10** (1): 19–26, 1992.

35. L. D. Landau and E. M. Liftshitz, *Fluid Mechanics,* New York: Pergamon Press, 1959.

36. F. Fahy, *Sound and Structural Vibration,* Chapter 3, London: Academic Press, 1985.

37. R. A. Johnston and A. D. S. Barr, Acoustic and internal damping in uniform beams, *J. Mech. Eng. Sci.,* **11** (2): 117–127, 1969.

38. A. P. Wenger, Vibrating fluid densimeters: a solution to the viscosity problem, *IEEE Trans. Ind. Electron. Control Instrum.,* **IECI-27**: 247–253, 1980.

39. M. Christen, Air and gas damping of quartz tuning forks, *Sensors and Actuators A,* **4**: 555–564, 1983.

40. H. Hosaka, K. Itao, and S. Kuroda, Damping characteristics of beam shaped micro-oscillators, *Sensors and Actuators A,* **49**: 87–95, 1995.

41. M. Andrews, I. Harris, and G. Turner, A comparison of squeeze film theory damping with measurements on a microstructure, *Sensors and Actuators A,* **36**: 79–87, 1993.

42. S. P. Beeby and M. J. Tudor, Modelling and optimisation of micromachined silicon resonators, *J. Micromech. Microeng.,* **5**: 103–105, 1995.

43. W. P. Mason, *Physical Acoustics,* Vol. III, part B, Chapter 6, New York: Academic Press, 1965.

44. J. M. Ziman, *Electrons and Phonons,* New York: Oxford University Press, 1960.

45. T. V. Roszhart, The effect of thermoelastic internal friction on the Q of micromachined silicon resonators, *Proc. IEEE Solid State Sensor and Actuator Workshop,* 13–16, June 4–7, 1990.

46. J. G. Eisley, Nonlinear vibration of beams and rectangular plates, *J. Appl. Math. Phys.,* **15**: 167–175, 1964.

47. M. V. Andres, K. H. W. Foulds, and M. J. Tudor, Nonlinear vibrations and hysteresis of micromachined silicon resonators designed as frequency out sensors, *Electronics Lett.,* **23** (18): 952–954, 1987.

S. P. Beeby
University of Southampton

M. J. Tudor
ERA Technology Ltd.

MICROPIPETTES.   See Microelectrodes.

MICROPROCESSORS

In 1972, Intel Corporation sparked an industrial revolution with the world's first microprocessor, the 4004. The 4004 replaced the logic of a numeric calculator with a general-purpose computer, implemented in a single silicon chip. The 4004 is shown in Fig. 1. The 4004 integrated 2300 transistors and ran at a clock rate of 108 kHz (108,000 clock cycles per second). In 1997, the 4004's most recent successor is the Pentium II processor, running at 300 MHz (300 million clock cycles per second) and incorporating nearly 8 million transistors. The Pentium II processor is shown in Fig. 2.

Figure 1. The world's first microprocessor, the Intel 4004, ca. 1971. Originally designed to be a less expensive way to implement the digital logic of a calculator, the chip instead spawned a computing revolution that still shows no signs of abating.

From the 4004's humble beginning, the microprocessor has assumed an importance in the world's economy similar to that of the electric motor or the internal combustion engine. Microprocessors now supply more than 90% of the world's computing needs, from small portable and personal desktop computers to large-scale supercomputers such as Intel's Teraflop machine, which contains over 9000 microprocessors. A variant of the microprocessor, the microcontroller, has become the universal controller in machines from automobile engines to audio systems to wristwatches.

MICROPROCESSORS AND COMPUTERS

Microprocessors are the processing units or the "brains" of the computer system. Every action that microprocessors perform is specified by a computer program that has been encoded into "object code" by a software program known as a compiler. Directed by another software program known as the operating system (e.g., Microsoft's Windows 95), the microprocessor locates the desired application code on the hard drive or compact disk and orders the drive to begin transferring the program to the memory subsystem so that the program can be run.

Digital electronic computers have at least three major subsystems:

- A memory to hold the programs and data structures
- An input/output (I/O) subsystem
- A central processor (CPU)

A microprocessor is the central processor subsystem, implemented on a single chip of silicon.

In microprocessor-based computer systems, the I/O subsystem moves information into and out of the computer system. I/O subsystems usually include some form of nonvolatile storage, which is a means of remembering data and programs even when electrical power is not present. Disk drives, floppy drives, and certain types of memory chips fulfill this requirement in microprocessor-based systems. Keyboards, trackballs, and mice are common input devices. Networks, modems, and compact discs are also examples of I/O devices. The memory subsystem, a place to keep and quickly access programs or data, is usually random-access memory (RAM) chips.

Microprocessors and microcontrollers are closely related devices. The differences are in how they are used. Essentially, microcontrollers are microprocessors for embedded control applications. They run programs that are permanently encoded into read-only memories and optimized for low cost so that they can be used in inexpensive appliances (printers, televisions, power tools, and so on). The versatility of a microcontroller is responsible for user-programmable VCRs and microwave ovens, the fuel-savings of an efficiently managed automobile engine, and the convenience of sequenced traffic lights on a highway and of automated bank teller machines.

Microprocessor software is typically created by humans who write their codes in a high-level language such as C or Fortran. A compiler converts that source code into a machine language that is unique to each particular family of microprocessors. For instance, if the program needs to write a character to the screen, it will include an instruction to the microprocessor that specifies the character, when to write it, and where to put it. Exactly how these instructions are encoded into the 1s and 0s (bits) that a computer system can use determines which computers will be able to run the program successfully. In effect, there is a contract between the design of a microprocessor and the compiler that is generating object code for it. The compiler and microprocessor must agree on what every computer instruction does, under all circumstances of execution, if a program is to perform its intended function. This contract is known as the computer's instruction set. The instruction set plus some additional details of implementation such as the number of registers (fast temporary storage) are known as the computer's instruction set architecture (ISA). Programs written or compiled to one ISA will not run on a different ISA. During the 1960s and 1970's, IBM's System/360 and System/370 were the most important ISAs.

With the ascendancy of the microprocessor, Intel's x86 ISA vied with Motorola's MC68000 for control of the personal computer market. By 1997, the Intel architecture was found in approximately 85% of all computer systems sold.

Early microprocessor instruction set architectures, such as the 4004, were designed to operate on 8-bit data values (operands). Later microprocessors migrated to 16-bit operands, including the microprocessor in the original IBM PC (the Intel 8088). Microprocessors settled on 32-bit operands in the 1980s, with the Motorola 68000 family and Intel's 80386. In the late 1980s, the microprocessors being used in the fastest servers and high-end workstations began to run into the intrinsic addressability limit of 4GB (four gigabytes, or four billion bytes, which is 2 raised to the power 32). These microprocessors introduced 64-bit addressing and data widths. It is likely that 64-bit computing will eventually supplant 32-bit microprocessors. It also seems likely that this will be the last increase in addressability that the computing industry will ever need because 2 raised to the power 64 is an enormous number of addresses.

Prior to the availability of microprocessors, computer systems were implemented in discrete logic, which required the assembly of large numbers of fairly simple digital electronic integrated circuits to realize the basic functions of the I/O, memory, and central processor subsystems. Because many (typically thousands) of such circuits were needed, the resulting systems were large, power-hungry, and costly. Manufacturing such systems was also expensive, requiring unique tooling, hand assembly, and a large amount of human debug effort to repair the inevitable flaws that accumulate during the construction of such complex machinery. In contrast, the fabrication process that underlies the microprocessor is much more economical. As with any silicon-integrated circuit, microprocessor fabrication is mainly a series of chemical processes performed by robots. So the risk of introducing human errors that would later require human debugging is eliminated. The overall process can produce many more microprocessors than discrete methods could.

Figure 2. The 1998 successors to the line of microprocessors started by the 4004, Intel's Pentium II processor, mounted within its Single-Edge Cartridge Connector (SECC). This picture shows the cartridge with its black case removed. On the substrate within the cartridge, the large octagonal package in the center is the Pentium II CPU itself. The rectangular packages to the right and left of the CPU are the cache chips. The small components mounted on the substrate are resistors and capacitors needed for power filtering and bus termination.

MOORE'S LAW

In 1964, Gordon Moore made an important observation regarding the rate of improvement of the silicon-integrated circuit industry. He noted that the chip fabrication process permitted the number of transistors on a chip to double every 18 months. This resulted from the constantly improving silicon process that determines the sizes of the transistors and wiring on the integrated circuits. Although he made the initial observation on the basis of experience with memory chips, it has turned out to be remarkably accurate for microprocessors as well. Moore's Law has held for well over 30 years. Figure 3 plots the number of transistors on each Intel microprocessor since the 4004.

These improvements of the underlying process technology have fueled the personal computer industry in many different ways. Each new process generation makes the transistors smaller. Smaller transistors are electrically much faster, allowing higher clock rates. Smaller wires represent less electrical capacitance, which also increases overall clock rates and reduces power dissipation. The combination of both permits far more active circuitry to be included in new design. Constant learning in the silicon fabrication plants have also helped drive up the production efficiency, or yield, of each new process to be higher than its predecessor, which also helps support larger die sizes per silicon chip.

The impact of this progression has been profound for the entire industry. The primary benefit of a new microprocessor is its additional speed over its predecessors, at ever better price points. The effect of Moore's Law has been for each new microprocessor to become obsolete within only a few years after its introduction. The software industry that supplies the applications to run on these new microprocessors expects this performance improvement. The industry tries to design so that its new products will run acceptably on the bulk of the installed base but can also take advantage of the new performance for the initially small number of platforms that have the new processor. The new processor's advantages in price/performance will cause it to begin to supplant the previous generation's volume champion. The fabrication experience gained on the new product allows its price to be driven ever

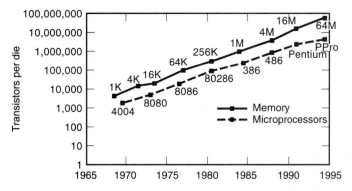

Figure 3. Moore's Law has accurately predicted the number of transistors that can be incorporated in microprocessors for over 25 years. Since this transistor count strongly influences system performance, this remarkable "law" has become one of the central tenets in the field of computers and integrated electronics. It guides the design of software, hardware, manufacturing production capacity, communications, and corporate planning in nearly every major area.

downward until the new design completely takes over. Then an even more advanced processor on an even better process technology is released, and the hardware/software spiral continues.

MICROPROCESSOR ARCHITECTURES

Another factor in the performance improvement of microprocessors is its microarchitecture. Microarchitecture refers to *how* a microprocessor's internal systems are organized. The microarchitecture is not to be confused with its instruction set architecture. The ISA determines what kind of software a given chip can execute. The earliest microprocessors (e.g., Intel 4004, 4040, 8008, 8080, 8086) were simple, direct implementations of the desired ISA. But as the process improvements implied by Moore's Law unfolded, microprocessor designers were able to borrow many microarchitectural techniques from the mainframes that preceded them, such as caching (Intel's 486, MC68010), pipelining (i486 and all subsequent chips), parallel superscalar execution (Pentium processor), superpipelining (Pentium Pro processor), and out-of-order and speculative execution (Pentium Pro processor, MIPS R10000, DEC Alpha 21264).

Microprocessor designers choose their basic microarchitectures very carefully because a chip's microarchitecture has a profound effect on virtually every other aspect of the design. If a microarchitecture is too complicated to fit a certain process technology (e.g., requires many more transistors than the process can economically provide), then the chip designers may encounter irreconcilable problems during the chip's development. The chip development may need to wait for the next process technology to become available. Conversely, if a microarchitecture is not aggressive enough, then it could be very difficult for the final design to have a high enough performance to be competitive.

Microarchitectures are chosen and developed to balance efficiency and clock rate. All popular microprocessors use a synchronous design style in which the microarchitecture's functions are subdivided in a manner similar to the way a factory production line is subdivided into discrete tasks. And like the production line, the functions comprising a microprocessor's microarchitecture are pipelined, such that one function's output becomes the input to the next. The rate at which the functions comprising this pipeline can complete their work is known as the pipeline's clock rate. If the functions do not all take the same amount of time to execute, then the overall clock rate is determined by the slowest function in the pipeline.

One measure of efficiency for a microarchitecture is the average number of clock cycles required per instruction executed (CPI). For a given clock rate, fewer clocks per instruction implies a faster computer. The more efficient a microarchitecture is, the fewer the number of clock cycles it will need to execute the average instruction. Therefore, it will need fewer clock cycles to run an entire program. However, the desire for high microarchitectural efficiency is often in direct conflict with designing for highest clock rate. Generally, the clock rate is determined by the time it takes a signal to traverse the slowest path in the chip, and adding transistors to a microarchitecture to boost its efficiency usually makes those paths slower.

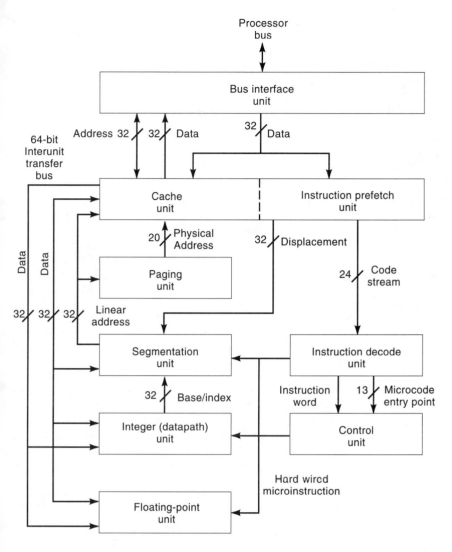

Figure 4. Block diagram of the most popular microprocessor of the early 1990s, the Intel i486. The various blocks shown work together to execute the Intel Architecture instruction set with approximately 1.1M transistors. Newer designs, such as the Pentium processor, or the P6 microarchitecture at the core of the latest Pentium II processor, are much more complicated.

Figure 4 illustrates the functional block diagram of the Intel 486, a very popular microprocessor of the early 1990s. (The microarchitectures of microprocessors that followed the 486, such as the Pentium processor or the Pentium Pro processor, are too complex to be described here.) The prefetch unit of the 486 fetches the next instruction from the instruction cache at a location that is either the next instruction after the last instruction executed or some new fetch address that was calculated by a previous branch instruction. If the instruction requested is not present in the cache, then the bus interface unit generates an access to main memory across the processor bus, and the memory sends the missing instruction back to the cache. The requested instruction is sent to the instruction decode unit, which extracts the various fields of the instruction, such as the opcode (the operation to be performed), the register or registers to be used in the instruction, and any memory addresses needed by the operation. The control unit forwards the various pieces of the instruction to the places in the microarchitecture that need them (register designators to the register file, memory addresses to the memory interface unit, opcode to the appropriate execution unit).

Certain very complex instructions are implemented in an on-chip read-only memory called the microcode. When the instruction decoder encounters one of these, it signals a micro-

code entry point for the microcode unit to use in supplying the sequence of machine operations that correspond to that complex macroinstruction.

Although it is not obvious from the block diagram, the Intel 486 microarchitecture is pipelined, which allows the machine to work on multiple instructions at any given instant. While one instruction is being decoded, another instruction is accessing its registers, a third can be executing, and a fourth is writing the results of an earlier execution to the memory subsystem.

See References 1–4 for sources of more details on designing microarchitectures.

THE EVOLUTION OF ISAS

Although microprocessor ISAs are crucial in determining which software will run on a given computer system, they are not static and unchangeable. There is a constant urge to develop the ISA further, adding new instructions to the instruction set or (much more rarely) removing old obsolete ones. Almost all old ISAs have many instructions, typically hundreds, some of which are quite complicated and difficult for compilers to use. Such architectures are known as Complex

Instruction Set Computers (CISC). In the early 1980s, substantial academic research was aimed at simplifying ISAs [Reduced Instruction Set Computers (RISC)], and designing them with the compiler in mind, in the hopes of yielding much higher system performance. Some important differences remain, such as the number of registers, but with time the differences in implementations between these two design philosophies have diminished. RISC ISAs have adopted some of the complexity of the CISC ISAs, and the CISC designers borrowed liberally from the RISC research. Examples of the CISC design style are the Intel x86, the Motorola MC68000, the IBM System/360 and /370, and the DEC VAX. RISC ISAs include MIPS, PowerPC, Sun's Sparc, Digital Equipment Corp. Alpha, and Hewlett-Packard PA-RISC.

COPROCESSORS AND MULTIPLE PROCESSORS

Some microprocessor systems have included a separate chip known as a coprocessor. This coprocessor was intended to improve the system's performance at some particular task that the main microprocessor was unsuited for. For example, in the Intel 386 systems, the microprocessor did not implement the floating-point instruction set; that was relegated to a separate numerics coprocessor. (In systems that lacked the coprocessor, the microprocessor would emulate the floating-point functions, albeit slowly, in software.) This saved die size and power on the microprocessor in those systems that did not need high floating-point performance, yet it made the high performance available in systems that did need it, via the coprocessor. However, in the next processor generation, the Intel 486, enough transistors were available on the microprocessor, and the perceived need for floating-point performance was large enough, that the floating-point functions were directly implemented on the microprocessor.

Floating-point coprocessors have not reappeared, but less-integrated hardware for providing audio (sound generation cards) and fast graphics are quite common in personal computers of the 1990s, which are similar to the coprocessors of the past. As the CPUs get faster, they can begin to implement some of this functionality in their software, thus potentially saving the cost of the previous hardware. But the audio and graphics hardware also improves, offering substantially faster functionality in these areas, so that buyers are tempted to pay a small amount extra for a new system.

HIGH-END MICROPROCESSOR SYSTEMS

Enough on-chip cache memory and external bus bandwidth is now available that having multiple microprocessors in a single system has become a viable proposition. These microprocessors share a common platform, memory, and I/O subsystem. The operating system attempts to balance the overall computing workload equitably among them. Dedicated circuits on the microprocessor's internal caches monitor the traffic on the system buses, in a procedure known as "snooping" the bus, to keep each microprocessor's internal cache consistent with every other microprocessor's cache. The system buses are designed with enough additional performance so that the extra microprocessors are not starved.

In the late 1990s, systems of 1, 2, and 4 microprocessors became more common. Future high-end systems will probably continue that trend, introducing 8, 16, 32, or more microprocessors organized into clusters. As of the mid 1990s, the fastest computers in the world no longer relied on exotic specialized logic circuits but were composed of thousands of standard microprocessors.

FUTURE PROSPECTS FOR MICROPROCESSORS

From their inception in 1971, microprocessors have been riding an exponential growth curve in the number of transistors per chip, delivered performance, and growth in the installed base. But no physical process can continue exponential growth forever. It is of far more than academic interest to determine when microprocessor development will begin to slow and what form such a slowdown will take.

For example, it is reasonable to surmise that the process technology will eventually hit fundamental limitations in the physics of silicon electronic devices. The insulators most commonly used in an integrated circuit are layers of oxide, and these layers are only a few atoms thick. To keep these insulators from breaking down in the presence of the electric fields on an integrated circuit, designers try to lower the voltage of the chip's power supply. At some point, the voltage may get so low that the transistors no longer work.

Power dissipation is becoming an increasingly important problem. The heat produced by fast microprocessors must be removed so that the silicon continues to work properly. As the devices get faster, they also generate more heat. Providing the well-regulated electrical current for the power supply, and then removing the heat, means higher expense in the system. With the 486 generation, aluminum blocks with large machined surface areas, known as heat sinks, became commonplace. These heat sinks help transfer the heat from the microprocessor to the ambient air inside the computer; a fan mounted on the chassis transfers this ambient air outside the chassis. With the Pentium processor generation, a passive aluminum block was no longer efficient enough, and a fan was mounted directly on the heat sink itself. Future microprocessors must find ways to use less power, transfer the heat more efficiently and inexpensively to the outside, and modulate their operations to their circumstances more adroitly. This may involve slowing down when high performance is temporarily unnecessary, changing their power supply voltages in real time, and managing the program workload based on each program's thermal characteristics.

Microprocessor manufacturers face another serious challenge: complexity, combined with the larger and less technically sophisticated user base. Microprocessors are extremely complicated, and this complexity will continue to rise commensurate with, among other things,

- Higher performance
- Higher transistor counts
- The increasing size of the installed base (which makes achieving compatibility harder)
- New features to handle new workloads
- Larger design teams
- More difficult manufacturing processes

This product complexity also implies a higher risk that intrinsic design or manufacturing flaws may reach the end user undetected. In 1994, such a flaw was found in Intel's Pentium processor, causing some floating-point divides to return slightly wrong answers. A public relations debacle ensued, and Intel took a $475 million charge against earnings, to cover the cost of replacing approximately 5 million microprocessors. In the future, if existing trends continue, microprocessor manufacturers may have tens or even hundreds of millions of units in the field. The cost of replacing that silicon would be prohibitive. Design teams are combating this problem in a number of ways, most notably by employing validation techniques such as random instruction testing, directed tests, protocol checkers, and formal verification.

What really sets microprocessors apart from the other tools that humankind has invented is the chameleonlike ability of a computer to change its behavior completely under the control of software. A computer can be a flight simulator, a business tool for calculating spreadsheets, an Internet connection engine, a household tool to balance the checkbook, and a mechanic to diagnose problems in the car. The faster the microprocessor and its supporting chips within the computer, the wider the range of applicability across the problems and opportunities that people face. As microprocessors continue to improve in performance, there is ample reason to believe that the computing workloads of the future will evolve to take advantage of the new features and higher performance, and applications that are inconceivable today will become commonplace.

Conversely, one challenge to the industry could arise from a saturated market that either no longer needs faster computers or can no longer afford to buy them. Or perhaps the ability of new software to take advantage of newer, faster machines will cease to keep pace with the development of the hardware itself. Either of these prospects could conceivably slow the demand for new computer products enough to threaten the hardware/software spiral. Then the vast amounts of money needed to fund new chip developments and chip manufacturing plants would be unavailable.

However, negative prognostications about computers or microprocessors have been notoriously wrong in the past. Predictions such as "I think there is a world market for maybe five computers" (Thomas Watson, chairman of IBM, 1943) or "photolithography is no longer useful beyond one micron line widths" have become legendary for their wrongheadedness. It is usually far easier to see impending problems than to conceive ways of dealing with them, but computer history is replete with examples of supposedly immovable walls that turned out to be tractable.

In its short life, the microprocessor has already proven itself to be a potent agent of change. It seems a safe bet that the world will continue to demand faster computers and that this incentive will provide the motivation for new generations of designers to continue driving the capabilities and applications of microprocessors into areas as yet unimagined.

ACKNOWLEDGMENTS

Various trademarks are the property of their respective owners.

BIBLIOGRAPHY

1. D. A. Patterson and J. L. Hennessy, *Computer Architecture: A Quantitative Approach, 2nd edition,* San Francisco: Morgan Kaufmann, 1996.

2. G. A. Blaauw and F. P. Brooks, Jr., *Computer Architecture Concepts and Evolution,* Reading, MA: Addison-Wesley, 1997.

3. D. P. Siewiorek, C. G. Bell, and A. Newell, *Computer Structures: Principles and Examples,* New York: McGraw-Hill, 1981

4. M. S. Malone, *The Microprocessor: A Biography,* Santa Clara, CA: Springer-Verlag, 1995.

ROBERT P. COLWELL
Intel Corporation